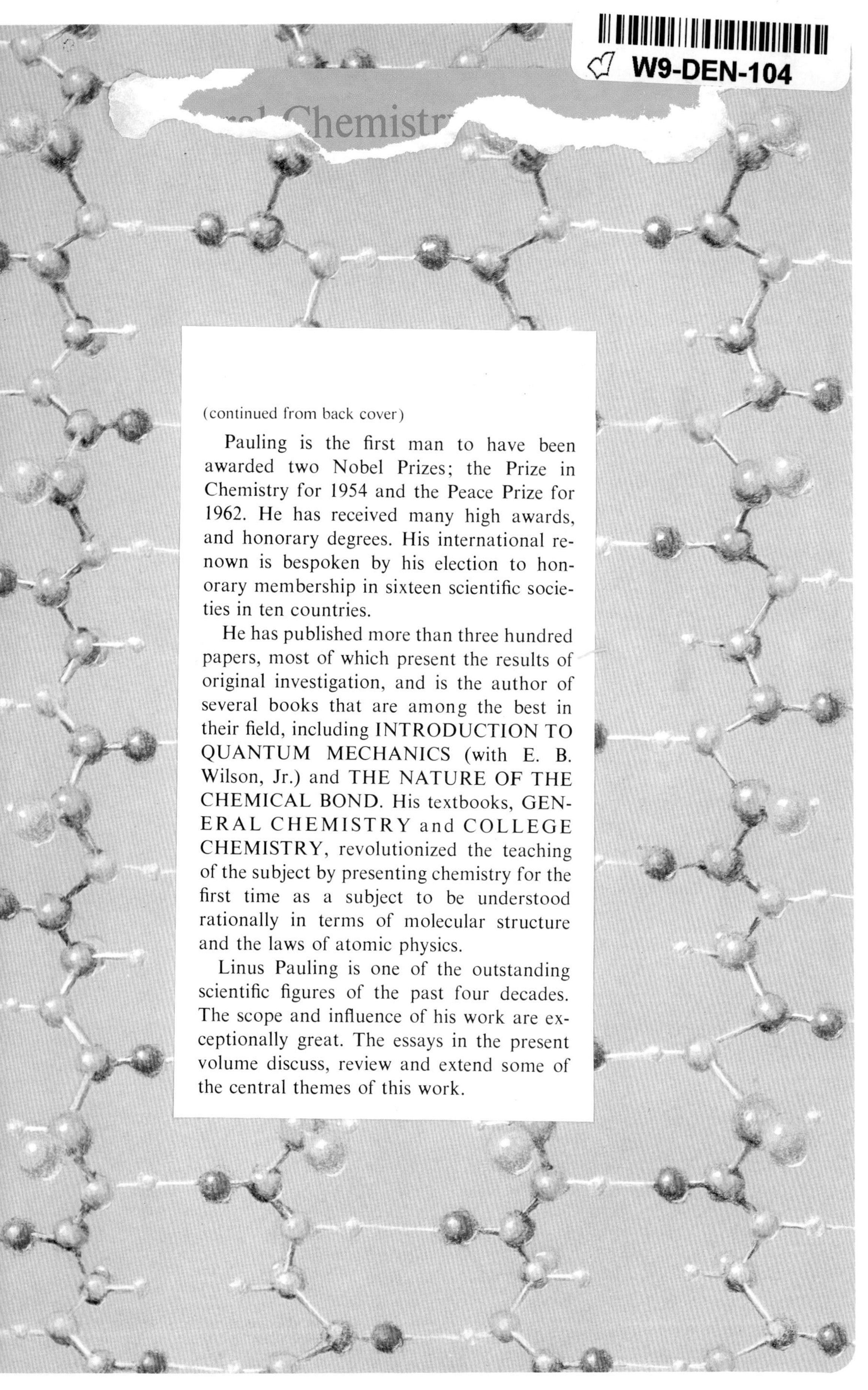

(continued from back cover)

Pauling is the first man to have been awarded two Nobel Prizes; the Prize in Chemistry for 1954 and the Peace Prize for 1962. He has received many high awards, and honorary degrees. His international renown is bespoken by his election to honorary membership in sixteen scientific societies in ten countries.

He has published more than three hundred papers, most of which present the results of original investigation, and is the author of several books that are among the best in their field, including INTRODUCTION TO QUANTUM MECHANICS (with E. B. Wilson, Jr.) and THE NATURE OF THE CHEMICAL BOND. His textbooks, GENERAL CHEMISTRY and COLLEGE CHEMISTRY, revolutionized the teaching of the subject by presenting chemistry for the first time as a subject to be understood rationally in terms of molecular structure and the laws of atomic physics.

Linus Pauling is one of the outstanding scientific figures of the past four decades. The scope and influence of his work are exceptionally great. The essays in the present volume discuss, review and extend some of the central themes of this work.

Structural Chemistry and Molecular Biology

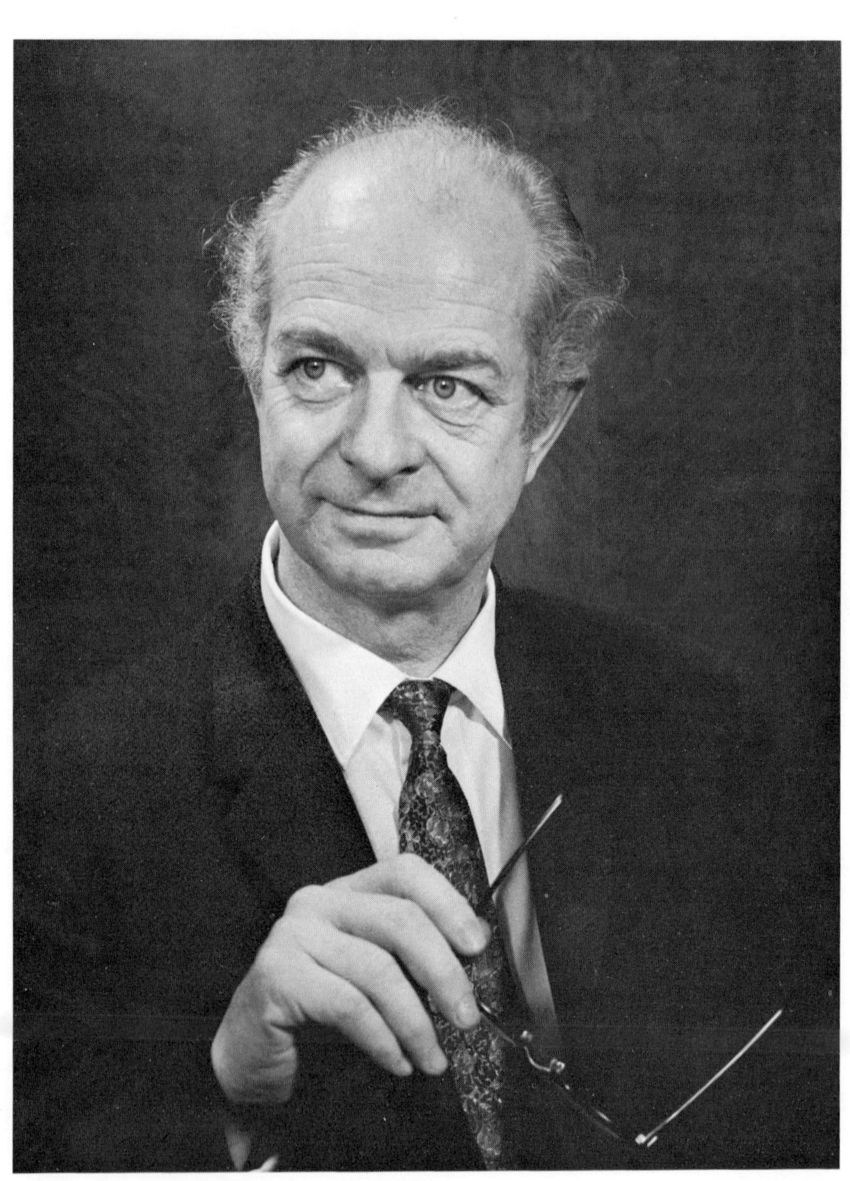

A volume dedicated to LINUS PAULING by his students, colleagues, and friends

Structural Chemistry and Molecular Biology

Edited by Alexander Rich *Massachusetts Institute of Technology*

and Norman Davidson *California Institute of Technology*

W. H. Freeman and Company *San Francisco*

PRINTED IN THE UNITED STATES OF AMERICA. (C2)

LIBRARY OF CONGRESS CATALOG CARD NUMBER 67-21127

Preface

Linus Pauling is one of the great scientific figures of our time. His work, more than that of any other individual, has led to the general realization that molecular structure is the central and most fruitful theme of modern chemistry.

His bibliography, appended at the end of this volume, documents a remarkable intellectual journey—a journey that started with the quantum mechanics of atomic energy levels, progressed into the quantum theory of the nature of the chemical bond and the determination and interpretation of the structures of small molecules in simple systems, then into the elucidation of the structures of biological macromolecules, and finally into the creation of simple structural theories for complex biological phenomena, such as the molecular nature of certain genetic diseases, memory, and anesthesia. His work has helped to transform and rationalize chemistry so that there is in effect today a continuum of understanding that extends from the energy levels in atomic nuclei to the behavior of complex macromolecules in biological systems.

The advent of his sixty-fifth birthday in 1966 has stimulated a number of his students, colleagues, and friends to contribute the essays in this volume in an attempt to express their admiration for the scope and influence of his work. His central scientific interests, molecular structure, the nature of the chemical bond, and molecular biology, are well represented here. Certain other topics such as nuclear structure and antiferromagnetism that have attracted his active, curious mind, unfortunately, are not covered. As the last entry, we have reprinted an important 1931 article of Pauling's on the nature of the chemical bond.

Linus Pauling was born in Portland, Oregon, on February 28, 1901, and grew up there and in a smaller town nearby. His father, a pharmacist, died when Linus was nine. He worked his way through the Oregon Agricultural College in Corvallis where he majored in chemical engineering. As an undergraduate he showed great promise and helped to support himself by teaching chemistry. Because of his financial responsibility for his mother and two younger sisters, he dropped out of college for one year and worked as a full-time teaching assistant in a quantitative analysis course. Indeed, he first met his future wife, Ava Helen Miller, when she was a student in a class in chemistry which he taught. Summers were occupied with a variety of jobs, including one in which he managed to overturn a steamroller. Although he had some uncertainty about his ultimate career, this was resolved when he came to the California Institute of Technology in 1922 to begin graduate work in chemistry in the young department there, which was headed by A. A. Noyes.

Pauling describes himself as having had "the greatest good luck in having gone to Pasadena in 1922." Conditions there were excellent for preparing for a career in physical chemistry. Noyes had recently founded the department and it had a capable staff, including such people as Richard C. Tolman and Roscoe G. Dickinson. The number of graduate students was small, and many of them developed into highly productive scientists. Noyes suggested that Pauling work with Dickinson on the determination of the structure of crystals by X-ray diffraction, a subject of great interest to Pauling as a student since he had always speculated about the relationship between properties of substances and their molecular structure. He was also caught up with the excitement of quantum theory and as a graduate student he felt it was time to attack chemical problems by applying quantum theory. Pauling has written an account of the early years in physical chemistry at Caltech in which he describes "the general feeling of the excitement of discovery in those days" (*Annual Reviews of Physical Chemistry*, 1965, pp. 1-14). After receiving his doctorate, he spent nineteen months studying in Europe; most of this time was spent with Sommerfeld in Munich, but there were brief visits to Bohr's Institute in Copenhagen and to Zurich where Schrödinger and Debye were lecturing. On returning to Pasadena in 1927, he joined the staff of the chemistry department and remained there for thirty-five years carrying out research and teaching.

An account by J. H. Sturdivant of Pauling's scientific work up until 1963 is reproduced in this volume. A few comments may be added here. We doubt that we can fully describe the characteristics of Linus Pauling's scientific work that give it its special style. There is a boldness and audacity in the suggestion of new ideas that are often, at first blush, outrageous in the simplicity with which they explain complicated phenomena. The essence of his approach to chemistry lies in the way in which he relates quantitative physical chemical measurements, especially molecular structural parameters, to theoretical interpretations which, in turn, suggest new experiments. Pauling is aided by his enormous knowledge of chemical properties, his phenomenal memory, and a theoretical understanding that allows him to make correlations, often unexpected, between physical measurements, theoretical interpretations, and chemical properties. This can be seen, for example, in his early work in formulating a simple set of rules governing the structure of silicate minerals. In one step he was able to order and make rational a large field of structural chemistry dealing with complex ionic materials. Likewise, in his work on the nature of the chemical bond he started with such known chemical facts as bond distances, bond angles, and molecular dipole moments—the elementary data of structural chemistry—and developed a theoretical structure, based on the hybridization of bond orbitals, that predicted the directed valence of atoms. With considerable chemical intuition, he showed the effects of resonance on the shape and stability of molecules and made further correlations between the magnetic properties of atoms and their chemical bonding. The primary effect of this work was to aid in predicting chemical properties of molecules. His work also established an outlook that has influenced nearly everyone who has subsequently entered this field of science. This is simply the attitude that one can explain and correlate by theory the many diverse physical and chemical properties of atoms and molecules.

In the mid 1930's, Pauling's interest began to spread beyond small molecules. His interest in magnetism led him to initiate experiments on the magnetic state of the iron atoms in hemoglobin, and from this he began working on antibodies, and then on proteins in general. In the course of this work, he began to recognize the central role of hydrogen bonding in determining and maintaining macromolecular structure. In 1940, he and Max Delbrück developed the idea of the importance of molecular complementarity in macromolecular interactions. They recognized that such interactions were, very probably, an essential part of the molecular basis of genetics.

Perhaps one of the best examples of Pauling's audacious thinking is his proposals enumerating the possible structures of polypeptides. These discoveries were made by the intelligent use of molecular structure data for simpler compounds, thought, intuition, and model building—all this being done, as Pauling has said, "with one's feet higher than one's head, for convenience." The rules for determining possible configurations were simple, and, in essence, they minimized the total energy of the structures. They were quantitative so that it was possible to make a number of predictions. (Pauling's use of an *a priori* approach, which he later called the stochastic method, is alluded to in several contexts in J. H. Sturdivant's article.) His approach to structure solving has served as a model for many workers in the field, including those who have solved the structure of a variety of polymeric molecules ranging from DNA and collagen to polypropylene.

In the late 1940's, Pauling attended a medical lecture in which William Castle described sickle-cell anemia, a disease of the blood that is characterized by a peculiar intracellular aggregation of hemoglobin molecules in nonaerated blood. While listening to the lecture, he imagined that this phenomenon might be explained by a hereditary alteration of the surface of a hemoglobin molecule that would make it complementary to another part of the surface, so that in the absence of oxygen these altered hemoglobin molecules might combine together. This simple thought was the beginning of the idea of a molecular disease, an explanation of malfunction based upon a genetically determined macromolecular modification.

The diversity of the fields in which Pauling has made significant contributions is truly remarkable. In addition to those already alluded to, mention should be made of his contributions in the areas of metallic structure, antibodies, nuclear structure, antiferromagnetism, the nature of anesthesia, memory, and macromolecular evolution. He has received many honors in recognition of his scientific accomplishments, including the Nobel Prize for Chemistry in 1954.

Working near Pauling as a colleague or as a student is frequently very exciting. His productivity in new ideas is very large. A student of his in the mid 1930's asked him how he managed to originate so many; his answer was that it is not difficult to have new ideas, it is just difficult to find out which of them are good ones. Pauling's active imagination is evident even in casual discussion. There have been many occasions when a student has had the impressive experience of starting a discussion about his research with Pauling, only to find that, in the course of the discussion, new ideas that had a substantial effect in modifying the research program were developed.

Pauling is a stimulating and lively lecturer; for many years he taught the first-year general chemistry course at Caltech. He greatly enjoyed direct contact with freshmen. His text, *General Chemistry*, revolutionized the teaching of the subject because it presented chemistry as a subject to be understood rationally in terms of molecular structure and the laws of atomic physics.

Even when he had substantial teaching and administrative duties at Caltech, it was relatively easy for students or postdoctoral fellows to see him. Writing a scientific paper or thesis with Pauling was stimulating both from a scientific and a literary point of view. Pauling's prose style is very simple and direct and when he found it necessary he taught his young associates how to use Fowler's *Modern English Usage* as well as the *Chemical Handbook*. Pauling's command of the English language is impressive. This was especially evident to those of his colleagues who watched him when he dictated his elementary chemistry texts, an entire chapter at one sitting. The dictated version of the text required very little alteration.

Pauling has had another public life outside of science. He has always been deeply moved by social injustice whether it affects a single person or a class of people. In more recent years this has led him to become increasingly active in the political arena, often on issues which were highly controversial. He was deeply moved by the senseless genetic damage which he could see occurring as a consequence of nuclear bomb testing. His activities in publicizing this danger were recognized by the award of the Nobel Peace Prize in 1962. However, it did not seem practicable to us to attempt to include in this volume essays that adequately dealt with his deep, concerned involvement in these problems.

We wish to thank Ava Helen Pauling for making available to us the photographs of her husband reproduced in the book, and we are indebted to Gustav Albrecht for compiling the bibliography.

The editors and authors dedicate this volume to Linus Pauling as an expression of admiration and of appreciation for his scientific contributions and his personal qualities.

ALEXANDER RICH
NORMAN DAVIDSON

November 1967

Contents

MOLECULAR BIOLOGY

NUCLEIC ACIDS

HISTORY

J. H. STURDIVANT
Division of Chemistry and Chemical Engineering
California Institute of Technology
Pasadena, California

The Scientific Work of Linus Pauling

Between a determination by X-ray diffraction in 1923 of the arrangement of atoms in the crystal molybdenite and a chemical paleogenetic study in 1963 on the reconstruction of amino-acid sequences in the polypeptide chains of ancestors of contemporary organisms, some 350 scientific papers by Linus Pauling and many more by his students and collaborators delineate the advance of his knowledge, intuition, understanding, and imagination, and record his elucidation in structural terms of the properties of stable molecules of progressively higher significance to chemistry and biology.

Pauling's scientific work has unfolded with a strong, logical coherence and interdependence of the successive extensions; but for the purposes of this summary an artificial compartmentation into two bodies of work is useful. The older body of work comprises Pauling's development of the theory of molecular structure and the nature of the chemical bond, "in considerable part empirical—based upon the facts of chemistry—but with the interpretation of these facts greatly influenced by quantum mechanical principles and concepts;" this work is summarized in his Nobel Lecture "Modern Structural Chemistry" (1954) and in *The Nature of the Chemical Bond* (1960). The body of work that has dominated in later years is concerned with chemical biology.

The first two sections of this paper describe two principal achievements from the earlier body of work, and the third section outlines the later.

THE COORDINATION THEORY OF THE STRUCTURE OF IONIC CRYSTALS

After the discovery in 1912 of X-ray diffraction by crystals the new technique was applied with great success by W. H. and W. L. Bragg, Wyckoff, Westgren, Goldschmidt, Dickinson, Pauling himself, and many others. By 1926 the crystal structures of most elements, many simple salts, and a small number of more complicated substances, including one or two organic compounds, had been determined; but the difficulty of finding additional substances whose structures could be completely analyzed by the accepted methods was becoming apparent. In 1928 Pauling made a highly successful attack upon

This paper was completed in December, 1963, and was published in Spanish translation in *Folia Humanistica*, **2**, 197–208 (1964) in the second of two issues dedicated "To Linus Pauling and to Peace." Contribution No. 3061 from the Gates and Crellin Laboratories of Chemistry.

this impasse in a paper entitled "The Coordination Theory of the Structure of Ionic Crystals" (Pauling, 1928). He wrote as follows.

In the study of the structure of a crystal with x-rays the effort has been made by many workers, especially in America, to eliminate rigorously all but one of the possible atomic arrangements consistent with the smallest unit of structure permitted by the experimental data, without reference to whether or not the structures were chemically reasonable or were in accord with assumed interatomic distances. The importance of this procedure arises from the certainty with which its results can be accepted. . . . but unfortunately the labor involved in its application to complex crystals is insuperable. . . .

But complex crystals are of great interest, and it is desirable that structure determinations be carried out for them even at the sacrifice of rigor. The method which has been applied in these cases is this: one atomic arrangement among all the possible ones is chosen, and its agreement with the experimental data is examined. If the agreement is complete or extensive, it is assumed that the structure is the correct one. The principal difficulty underlying this treatment is the selection of the structure to be tested.

Simple ionic substances, such as the alkali halides, have very little choice of structure; there exist only a very few relatively stable ionic arrangements corresponding to the formula M^+X^-, and the various factors which influence the stability of the crystal are pitted against one another, with no one factor finding clear expression in the final decision between the sodium chloride and the cesium chloride arrangement. For a complex substance, such as mica, $KAl_3Si_3O_{10}(OH)_2$, or zunyite, $Al_{13}Si_4O_{20}(OH)_{18}Cl$, on the other hand, very many conceivable structures differing only slightly in nature and stability can be suggested, and it might be expected that the most stable of these possible structures, the one actually assumed by the substance, will reflect in its various features the various factors which are of significance in the determining of the structure of ionic crystals. . . . [Pauling found it possible] to formulate a set of five rules relating to the stability of complex ionic crystals. They were obtained in part by induction from the structures known in 1928 and in part by deduction from the equations for the crystal energy. . . . The substances to which the rules apply are those in which the bonds are essentially ionic in character, and in which all or most of the cations are small and multivalent, the anions being large and univalent or bivalent. The anions which are most important are oxygen and fluorine. . . . The rules are based upon the concept of the coordination of anions at the corners of a tetrahedron, octahedron, or other polyhedron about each cation, and they relate to the nature and interrelations of these polyhedra.

The first rule is the following: "A coordinated polyhedron of anions is formed about each cation, the cation-anion distance being determined by the radius sum and the coordination number of the cation by the radius ratio." Thus it is possible, with the aid of a table of ionic radii, and after theoretical or empirical determination of the transition values of the radius ratio, to predict that Si^{++++}, say, will have a coordination number of four with oxygen ions, Al^{+++} of four or six, Ca^{++} of eight, and so on.

The strength of the electrostatic bond from a cation to each coordinated anion is defined as the electric charge of the cation in units of e divided by the coordination number. Now the second and most important rule, the electrostatic valence principle, postulates that the state of maximum stability of an ionic crystal is that in which the valence of each anion, with changed sign, is equal to the sum of the strengths of the electrostatic bonds to it from the adjacent cations. It is seen that this rule controls the number of polyhedra with a common corner.

Sir Lawrence Bragg (1937) has discussed the electrostatic valence rule in the following terms.

> The rule appears simple, but it is surprising what rigorous conditions it imposes upon the geometrical configuration of a structure. In a silicate, for instance, each silicon atom is surrounded by four oxygen atoms. These atoms have half their valency satisfied by the silicon, and so are left with an electrostatic charge which is unity. . . . Aluminum within an octahedral group of six contributes one-half to each oxygen. Magnesium or ferrous iron within an octahedral group contributes one-third. Hence we may link a corner of a silicon tetrahedron to another silicon tetrahedron, to two Al octahedra, or three Mg octahedra. . . . Proceeding to link tetrahedra and octahedra in this way, we find that very few alternative structures which obey Pauling's law remain open to a mineral of given composition, and one of these alternatives always turns out to be the actual structure of the mineral. . . .
>
> The rule may be termed the cardinal principle of mineral chemistry. It often accounts for the non-existence of certain types of compound, although the formulae of these compounds would be quite possible according to the ordinary laws of valency.

The electrostatic valence rule and the other principles of the coordination theory discovered by Pauling have been used in the analysis of numerous crystal structures both by him and his students, and by workers elsewhere.

THE NATURE OF THE CHEMICAL BOND

The Coulomb attractions and kernel repulsions that stabilize the infinite molecules, the ionic crystals, discussed in the first section, represent only a special and extreme case of the chemical bond. In 1928 Pauling published in a review article a unified treatment of the application of the quantum mechanics to two diatomic systems, the hydrogen molecule and the hydrogen molecule ion (1928b). He emphasized the function of resonance in stabilizing a molecule, and stated, "The interchange energy of electrons is in general the energy of the nonpolar or shared-electron chemical bond." This paper signalized the directing of his energies to the solution of the general problem of the chemical bond. It was also the beginning of his effort to make the results of quantum mechanics accessible and familiar to chemists untrained in the new theoretical physics, and to fuse the results into the foundations of chemical theory.

Before the discovery of the electron the valence bond was represented by a line drawn between the symbols of two chemical elements, which expressed in a concise way many qualitative chemical facts. The nature of the bond was completely unknown. The discovery of the electron led to numerous attempts to develop an electronic theory of the chemical bond. Of these the most successful was the work of G. N. Lewis, who discussed in 1916 the formation of ions by the completion of stable shells of electrons, and the formation of a chemical bond by the sharing of electrons between two atoms. His ideas were further developed by many other investigators. All of these early treatments contained, however, many suggestions since discarded. The electronic theory of valence in its present form arises almost entirely from the theory of quantum mechanics, which has provided a method for the calculation of the properties of simple molecules, leading to the complete elucidation of the phenomena involved in the formation of a covalent bond and has also introduced into chemical theory a new concept, that of resonance, which, if not entirely unanticipated in its applications to chemistry, nevertheless had not before been clearly recognized and understood.

In a brief paper published in 1928 Pauling (1928c) discussed for the first time the dominant role of changes in quantization, or hybridization of orbitals, in the production of stable bonds in ordinary chemical compounds. He had discerned that the quantization of electronic levels developed by physicists interested in atomic spectra was unsuitable for the discussion of stable chemical compounds, and that the perturbations incident to the approach of atoms would produce directed atomic orbitals whose overlapping would determine the bonds to be formed.

A paper "The Nature of the Chemical Bond," which was of the greatest importance, followed in 1931.* In it Pauling succeeded in obtaining from the quantum mechanical equations results of broad chemical significance that went beyond the earlier electronic valence theories. This involved the extension of quantum mechanical concepts to systems much too complex for rigorous mathematical discussion, but of correspondingly more essential interest to chemistry. Pauling stated six rules for the electron-pair bond. The first three were derived by inference from quantum-mechanical results on the interaction of helium atoms, hydrogen atoms, and lithium atoms: (1) The electron-pair bond is formed through the interaction of an unpaired electron on each of two atoms. (2) The spins of the electrons are opposed when the bond is formed, so that they cannot contribute to the paramagnetic susceptibility of the substance. (3) Two electrons which form a shared pair cannot take part in forming additional pairs.

The remaining three postulates were essentially new, and are the basis of Pauling's analysis of the electron-pair bonds in terms of atomic orbitals. They were justified at the time by the qualitative consideration of the factors influencing bond energies; their validity may now be appraised by the repeated experimental confirmation of the numerous theorems derivable from them. One of these postulates states that the main resonance terms for a single electron-pair bond are those involving only one wave function from each atom; it follows that a treatment by atomic orbitals is of significance. Another states that only stable atomic orbitals form stable bonds, the 1*s* orbital for hydrogen, the 2*s* and 2*p* orbitals for the first-row atoms, and so on.

The fifth rule deserves the most extended attention. It asserts that of two orbitals in an atom the one that can overlap more with an orbital of another atom will form the stronger bond with that atom, and furthermore that the bond formed by a given orbital will tend to lie in that direction in which the orbital is concentrated. From this rule Pauling derived, with the use of general quantum mechanical principles, the whole body of results concerning directed valance, including not only the tetrahedral arrangement of the four single bonds of the carbon atom, but also square and octahedral configurations (as well as other configurations), together with rules regarding the

*Reproduced in this volume, pp. 849–884

occurrence of these configurations, the strengths of the bonds, and the relation of configuration to magnetic properties. Thus a single reasonable postulate was made the basis of a large number of the rules of stereochemistry, and was found to lead to several new stereochemical results.

In the process of assigning valence-bond structures to molecules, it was early recognized that molecules can be divided roughly into two classes. For a molecule of one class a single obvious valence-bond structure could be formulated that would account for the properties of the substance. For a molecule of the second class several more or less equally stable valence-bond structures would present themselves, and only a linear combination of these would account satisfactorily for the properties of the substance. The ethane molecule belongs to the first class; the benzene molecule to the second. The fundamentals of the quantum mechanical resonance of a molecule among several valence-bond structures were treated by Pauling between 1933 and 1935. In 1933 he discovered that by restricting himself to thermochemical data obtained on molecules of the first class above he was able to derive a satisfactory set of essentially constant bond energies. By applying these bond energies to thermochemical data obtained on molecules of the second class, he determined the contribution of resonance to the energy of formation and established one experimental criterion for the existence of resonance. He established and applied repeatedly other experimental criteria for the assignment of resonating structures to a molecule; these included the use of evidence provided by interatomic distances, bond angles, dipole moments, and force constants.

The sureness of Pauling's progress toward a comprehensive explanation of the nature of the chemical bond arose to a considerable extent from his constant control of theory and postulate with experiment, and particularly with the type of experiment that yields the most unambiguous and direct information on bond angles and interatomic distances. In 1931 he already had at hand a mass of accurate information from crystal structure analysis, to which frequent reference is made in the paper of that year. In the next few years he and his associates extended this very greatly by the preparation and interpretation of electron diffraction patterns of substances in the gaseous state.

CHEMICAL BIOLOGY

Pauling began to turn his basic discoveries about the nature of the chemical bond to account in the field of chemical biology in the middle 1930's. This work therefore overlapped and interacted strongly with the work on the chemical bond for twenty years, adapting to its purposes during that period the methods of attack and the principles that had been and were being developed and refined in the now simpler field of the nature of the chemical bond. The major advances are summarized in the following paragraphs.

The problem of the structure of proteins began in the early 1930's to absorb the energies of many of the ablest workers and the strongest laboratories in X-ray diffraction. It was long suspected that polypeptide chains that were folded to about half of their extended length were the important structural elements in the great majority of natural protein fibers. X-ray evidence suggested similarly folded chains in some globular proteins. Studies of synthetic polypeptides proved that the fold represents the state of lowest free energy of an isolated —CO—CH_2—NH— chain, and that it is held together by hydrogen bonds. The element was named the α-fold. Its structure remained unknown.

From about 1939 Linus Pauling and R. B. Corey carried on accurate studies by X-ray diffraction of the structures of amino acids and small peptides. The results led them to formulate a small set of structural conditions that any model of a polypeptide chain of lowest free energy must satisfy. One condition, for example, is the planarity of all amide groups due to resonance, a characteristic of amides that Pauling had discussed in 1932. Pauling and Corey then adopted the purely stereochemical approach of building a chain by rigorous application of their structural principles. In a note, "Two Hydrogen-Bonded Spiral Configurations of the Polypeptide Chain" (1950), they announced that there was strong evidence that one of these structures, subsequently called the α-helix, a helical structure with about 3.7 residues per turn of the helix, and with each residue hydrogen bonded to the third residue from it in each direction along the chain, is present in α-keratin, contracted myosin, some other fibrous proteins, and hemoglobin and other globular proteins. Detailed discussions of precisely described structures for proteins were presented in a series of papers published in 1951 and 1952 by Pauling and Corey (one with H. R. Branson as coauthor). M. F. Perutz soon predicted X-ray reflections from fibers of the α-type on the basis of the α-helix and observed them (1951). Other confirmation followed. In 1960 Kendrew and coworkers published the electron-density map of myoglobin at 2 Å resolution; it leaves no doubt that the α-helix is the most important single element of the folding.

The theory of the relation between hybrid bond orbitals and magnetic properties was applied to the investigation of the structure of hemoglobin in a paper by Pauling and Coryell (1936), in which the discovery of a change in magnetic properties of hemoglobin on oxygenation was reported. The summary of this paper reads as follows:

> It is shown by magnetic measurements that oxyhemoglobin and carbonmonoxyhemoglobin contain no unpaired electrons; the oxygen molecule, with two unpaired electrons in the free state, accordingly undergoes a profound change in electronic structure on attachment to hemoglobin. The magnetic susceptibility of hemoglobin itself (ferrohemoglobin) corresponds to an effective magnetic moment of 5.46 Bohr magnetons per heme, calculated for independent hemes. This shows the presence of four unpaired electrons per heme, and indicates that the heme-heme interaction tends to stabilize to some extent the parallel configuration of the moments of the four hemes in the molecule. The bonds from iron to surrounding atoms are ionic in hemoglobin, and covalent in oxyhemoglobin and carbonmonoxyhemoglobin.

Work on magnetic properties of hemoglobin and related molecules, in relation to the structures of the molecules, was continued, and the results were described in a series of about a dozen papers. The magnetic method was also applied to cytochrome and related substances by H. Theorell, with great success.

With the stimulus of Pauling's interest, highly productive chemical studies utilizing chromatography, electrophoresis, centrifugation, and so on, have been continued on hemoglobins by associates and colleagues. For example, Rhinesmith, Schroeder, and Pauling (1957) established that normal adult human hemoglobin is built up from four chains, two each of two kinds; Zuckerkandl, Jones, and Pauling (1960) compared animal hemoglobins by tryptic peptide pattern analysis, with particular interest in questions in the realm of genetics and evolution.

Until thirteen years ago it was thought that all human beings manufacture the same kind of adult human hemoglobin. At about that time Pauling heard a description of the disease, sickle-cell anemia (first described in 1910), by William B. Castle, who writes of the sequel: "In 1949 Pauling recognized the scent as that of an abnormal hemoglobin molecule and with his associates took up the trail by means of electrophoretic and other physical measurements with the brilliant results . . . ." This first discovery of a molecular basis of a disease was described by Pauling, Itano, Singer, and Wells (1949). A detailed mechanism for the sickling process was formulated with the hypothesis that

there is a surface region on the globin of the sickle-cell-anemia hemoglobin molecules which is absent in the normal molecule and which has a configuration complementary [see a few paragraphs below] to a different region of the surface of the hemoglobin molecule. This situation would be somewhat analogous to that which very probably exists in antigen-antibody reactions. The fact that sickling occurs only when the partial pressures of oxygen and carbon monoxide are low suggests that one of these sites is very near to the iron atom of one or more of the hemes, and that when the iron atom is combined with either one of these gases, the complementariness of the two structures is considerably diminished. Under the appropriate conditions, then, the sickle-cell-anemia hemoglobin molecules might be capable of interacting with one another at these sites sufficiently to cause at least a partial alignment of the molecules within the cell, resulting in the erythrocytes's becoming birefringement, and the cell membrane's being distorted to accommodate the now relatively rigid structure within its confines. The addition of oxygen or carbon monoxide to the cell might reverse these effects by disrupting some of the weak bonds between the hemoglobin molecules in favor of the bonds formed between gas molecules and iron atoms of the hemes.

Substantiation of this picture was soon obtained through microscopic investigations. . . . It accordingly is probable that sickle-cell anemia can be described as a molecular disease, resulting from the difference in molecular structure of sickle-cell-anemia hemoglobin and normal adult human hemoglobin, the properties of the abnormal hemoglobin being such that when deoxygenated the molecules combine with one another to form long molecular strings, which, through intermolecular attraction, aggregate into tactoids, which have enough mechanical strength to distort the red cell, changing the viscosity of the blood, and causing the clinical and pathological manifestations of the disease.

The discoveries relative to sickle-cell-anemia had an impact upon several disciplines that is described in the following quotation from an article entitled "Abnormal Hemoglobins," by Zuelzer, Neel, and Robinson (1956).

Today the sickling phenomenon has ceased to be the exclusive concern of the hematologist and pathologist. It has come to occupy a position at the crossroads of medicine, biochemistry, genetics, and anthropology. . . . [The] discovery in 1949 of an abnormal variant of hemoglobin in erythrocytes possessing the sickling property by Pauling and his associates immediately proved to be of the greatest importance . . . and at once opened up new and fertile fields of investigation.

The new concept of sickling as the expression of a "molecular" disorder contributed greatly to the understanding of the clinical, hematologic, and pathogenetic aspects of sickle cell disease but its greatest significance emanated from the fact that the production of the abnormal hemoglobin was simultaneously shown to be under genetic control. . . .

The findings of Pauling and his colleagues lent dramatic support to the new genetic theory. In their original publication these workers showed that the (heterozygous) sickle trait carriers actually possessed two kinds of hemoglobin, sickle cell hemoglobin and normal hemoglobin, . . . while the (homozygous) individuals with sickle cell anemia had only one kind of hemoglobin, the abnormal. . . . In the heterozygotes each of the two genes, that for normal hemoglobin and that for sickle hemoglobin, expressed itself in a qualitatively demonstrable and even quantitatively measurable way. . . . Not only did the electrophoretic analysis provide an objective basis for the distinction, formerly somewhat uncertain, between sickle-cell trait and sickle-cell anemia, it also promised to give insight into the mechanisms and quantitative aspects of a human gene occurring with an appreciable frequency, and thus lending itself to the exploration of fundamental problems in human biology.

Several other abnormal varieties of hemoglobin have subsequently been discovered by various investigators. The abnormal hemoglobins, either alone or in synergistic interaction with one another or with thalassemia, have been observed in association with several diseases that had not previously been recognized as clinical entities.

A paper entitled "A Theory of the Structure and Process of Formation of Antibodies," by Pauling (1940), and a second, "The Nature of the Intermolecular Forces Operative in Biological Processes," by Pauling and Delbrück (1940) treated for the first time in precise terms the role of complementarity in serological phenomena. The introduction to the former paper contains these sentences:

The field of immunology is so extensive and the experimental observations are so complex (and occasionally contradictory) that no one has found it possible to induce a theory of the structure of antibodies from the observational material. As an alternative method of attack we may propound and attempt to answer the following questions: What is the simplest structure which can be suggested, on the basis of the extensive information now available about intramolecular and intermolecular forces, for a molecule with the properties observed for antibodies, and what is the simplest reasonable process of formation of such a molecule? Proceeding in this way, I have developed a detailed theory of the structure and process of formation of antibodies and the nature of serological reactions which is more definite and more widely applicable than earlier theories, and which is compatible with our present knowledge of the structure and properties of simple molecules as well as with most of the direct empirical information about antibodies.

The second paper contains the following statement.

It is our opinion that the processes of synthesis and folding of highly complex molecules in the living cell involve, in addition to covalent-bond formation, only the intermolecular interactions of van der Waals attraction and repulsion, electrostatic interactions, hydrogen-bond formation, etc., which are now rather well understood. These interactions are such as to give stability to a system of two molecules with *complementary* structures in juxtaposition, rather than of two molecules with necessarily identical structures; we accordingly feel that complementariness should be given primary consideration in the discussion of the specific attraction between molecules and the enzymatic synthesis of molecules.

A series of approximately thirty papers by Pauling and collaborators, involving a large amount of experimental work, developed the consequences of the 1940 papers.

The concept of complementarity, applied with the precision that modern structural knowledge permits, continues to prove fruitful in studies of the structures of proteins (such as those by Pauling, Corey, and associates) and of other complex biological molecules. The lineage of such an important conception as the Watson-Crick model for desoxyribonucleic acid can be clearly traced back to Pauling's discussion of complementarity.

There is current work on other problems of chemical biology, such as the search for a chemical basis for mental disease, and the development and testing of a molecular theory of general anaesthesia by nonhydrogen-bonding agents. This work is obviously too new for evaluation, but the vigor, originality, and clarity of the ideas can hardly fail to stimulate progress in chemical biology.

REFERENCES

Bragg, L. (1937). *Atomic Structure of Minerals*, p. 35. Ithaca: Cornell Univ. Press.

Kendrew, J. C., R. E. Dickerson, B. E. Strandberg, R. D. Hart, D. R. Davies, D. C. Phillips, and V. C. Shore (1960). *Nature* **185**, 422.

Lewis, G. N. (1928). *J. Am. Chem. Soc.* **38**, 762.

Pauling, L. (1928a). In *Festschrift zum 60. Geburtstage Arnold Sommerfelds*, pp. 11–17. Leipzig: Verlag Hirzel.

Pauling, L. (1928b). *Chem. Rev.* **5**, 173.

Pauling, L. (1928c). *Proc. Nat. Acad. Sci. U.S.* **14**, 259.

Pauling, L. (1931). *J. Am. Chem. Soc.* **53**, 1367.

Pauling, L., and C. D. Coryell (1936). *Proc. Nat. Acad. Sci. U.S.* **22**, 210.

Pauling, L., and M. Delbrück (1940). *Science* **92**, 77.

Pauling, L. (1940). *J. Am. Chem. Soc.* **62**, 2643.

Pauling, L., H. A. Itano, S. J. Singer, and I. C. Wells (1949). *Science* **110**, 543.

Pauling, L., and R. B. Corey (1950). *J. Am. Chem. Soc.* **72**, 5349.

Pauling, L., R. B. Corey, and H. R. Branson (1951). *Proc. Nat. Acad. Sci. U.S.* **37**, 205.

Pauling, L. (1954). Modern Structural Chemistry. Nöbel Lecture, delivered 11 December. In Liljestrand, M. G., ed., *Les Prix Nöbel en 1954*, pp. 91–99. Stockholm: Kungl. Boktryckeriet P. A. Norstedt and Söner, 1955. This lecture also appears in *Angew. Chem.* (1955) **67**, 241 and *Science* (1956) **123**, 255.

Pauling, L. (1960). *The Nature of the Chemical Bond and the Structure of Molecules and Crystals*, 3rd ed. Ithaca: Cornell Univ. Press.

Perutz, M. F. (1951). *Nature* **167**, 1053.

Rhinesmith, H. S., W. A. Schroeder, and L. Pauling (1957). *J. Am. Chem. Soc.* **79**, 609, 4682.

Zuckerkandl, E., R. T. Jones, and L. Pauling (1960). *Proc. Nat. Acad. Sci. U.S.* **46**, 1349.

Zuelzer, W. W., J. V. Neel, and A. R. Robinson (1956). *Prog. Hematol.* **1**, 91.

THE STRUCTURE OF PROTEINS

DOROTHY CROWFOOT HODGKIN
DENNIS PARKER RILEY
Department of Chemical Crystallography
Oxford University
Oxford, England

Some Ancient History of Protein X-Ray Analysis

The history of the X-ray analysis of protein crystals began for many of us when the first X-ray diffraction photographs of single pepsin crystals were taken in 1934. The crystals were hexagonal bipyramids, 2 mm long or more, prepared by John Philpot while he was working for a short time at Uppsala. He had left his preparations in the refrigerator while he was off on a skiiing holiday, and on his return was astonished to find how large his crystals had grown. He showed them to Glen Millikan, a visiting physiologist from California and Cambridge, who said, "I know a man in Cambridge who would give his eyes for those crystals." Philpot naturally offered him some crystals to take back in his coat pocket and so Millikan took them to J. D. Bernal.

It was very lucky for protein crystallography that Millikan took the crystals in the tube in which they were growing in their mother liquor. This enabled Bernal to make his first critical observation: that the crystal lost birefringence when removed from their liquid of crystallization. He observed only a vague blackening of the film when X-rays were passed through the dry crystal. Therefore he mounted some crystals in their mother liquor in Lindemann glass tubes. The wet crystals gave individual X-ray reflections, which were rather blurred owing to the large size of the crystals and the large size of the crystal unit cell, but which extended all over the films to spacings of about 2 Å. That night, Bernal, full of excitement, wandered about the streets of Cambridge, thinking of the future and of how much it might be possible to know about the structure of proteins if the photographs he had just taken could be interpreted in every detail.

During the next few days, when D.C. returned to the laboratory from a brief absence, many more X-ray photographs were taken and calculations were made on the possible weight of the asymmetric unit in the pepsin crystal. There were also consultations with friends among the biochemists to find out what was already known about proteins and protein crystals. The picture

This paper grew out of the suggestion that some of the early unwritten history of protein X-ray crystallography should be described at the protein session at the International Congress of Crystallography in Moscow in 1966. It is compiled largely from old notebooks and old letters to Dorothy Crowfoot Hodgkin (in the text referred to as D.C.), and necessarily gives a very one-sided record of all that occurred.

obtained was complicated; it was clear the crystals, as crystals, had unusual properties. Professor Hopkins described how he had isolated large quantities of lecithin from partly purified crystals of egg albumin; the lecithin molecules presumably fitted in between the protein molecules in water-containing spaces. Bernal found an old paper by Schimper (1881) which recorded the swelling and shrinking of protein crystals under different conditions of humidity and their penetration by large dye molecules. Svedberg had recently begun to measure the molecular weights of protein molecules in the ultracentrifuge. The order of magnitude he found for pepsin, about 40,000, was large; it fitted with the size of unit cell indicated by the X-ray data, given that the *c*-dimension of the unit cell was probably twice the minimum value recorded.

Certain accidents contributed to some of the speculations made in the first note published about pepsin in *Nature* (Bernal and Crowfoot, 1934). The crystals have high symmetry, space group $C6_122$, and long cell dimensions (a = 67 Å, c = 291 Å), and they collapse to very disordered structures on drying; the correct lattice constants and symmetry for both wet and dry crystals were determined many years later by Max Perutz (1949). The high degree of disorder in the dried crystals and the limited X-ray diffraction techniques that were employed combined to give the vague blackening observed when the first X-rays were passed through the dried crystal. Astbury had previously obtained a fiber pattern from a preparation of dried pepsin; no trace of this was observed on the wet crystal photograph. It therefore seemed possible that the pepsin molecule had altered radically on drying. Working on the *Annual Report of the Chemical Society* in 1933, Bernal and D.C. had read of the remarkable transformation of a single crystal of trioxymethylene into oriented fibers of polyoxymethylene by ultraviolet light (Kohlschütter and Sprenger, 1932; Sauter, 1932). It seemed conceivable that something similar had happened in pepsin. The letter to *Nature* comments on the first X-ray photographs:

> From the intensity of the spots near the centre, we can infer that protein molecules are relatively dense bodies, perhaps joined by valency bridges but in any event separated by relatively large spaces which contain water. From the intensity of the more distant spots, it can be inferred that the arrangement of atoms inside the protein molecule is also of a perfectly definite kind, although without the periodicities characterising the fibrous proteins ... we may imagine degeneration to take place by the linking up of amino acid residues in such molecules to form chains as in the ring-chain polymerisation of polyoxymethylenes. Peptide chains in the ordinary sense may exist only in the more highly condensed or fibrous proteins, while the molecules of the primary soluble proteins may have their constituent parts grouped more symmetrically around a prosthetic nucleus.

(One suspects hexagonal symmetry played a part in giving us this idea. There was a time also when we tried very hard to make a protein by shining ultra violet light on crystals of diketopiperazine.)

These ideas, admittedly speculative in their original form, were abandoned very rapidly. A letter from John Philpot (dated June 28, 1934) describes measurements of the specific enzyme activity of our pepsin crystals which showed that X-rays and crystal drying had little, if any, effect on the enzyme activity. Philpot commented, "one can conclude that an exposure to X-rays sufficient to destroy the crystal structure has very little effect on the peptic activity. This result is so surprising that I think it ought to be repeated." It does not seem, from our old correspondence, that such measurements were ever repeated. Soon both Philpot and D.C. moved to Oxford University, and soon X-ray measurements of new crystals made it clear that crystal drying was accompanied only by packing disorder and not by any internal transformation of the protein molecules. These new crystals were insulin and lactoglobulin, followed very soon afterwards by hemoglobin and chymotrypsin (Bernal, Fankuchen, and Perutz, 1938).

The research on insulin began in 1935, when Professor Robert Robinson gave D.C. a sample of finely crystalline insulin he had received from Dr. Pyman. To obtain crystals large enough for X-ray work, D.C. somewhat slavishly followed Scott's method of recrystallization of the rhombohedral zinc containing crystals, including (as would any well-trained organic chemist) washing them and drying them with alcohol. The crystals so treated were birefringent and gave single crystal diffraction effects; it was not immediately realized that the limitation of these effects to a sphere of about 7 Å spacing implied disorder in the crystal. The most important first consequence of the X-ray measurements was the correlation of the size of the rhombohedral unit cell with the molecular weight of insulin as measured by Svedberg, and this correlation also limited the shape of the Svedberg unit. The trigonal symmetry of the crystal implied the presence of 3 n subunits within the Svedberg molecule; since trigonal symmetry is well known as a possible type of molecular symmetry, there was some hesitation about immediately making the obvious deduction that the molecular weight of insulin was 12,000 or less. Professor K. Freudenberg therefore wrote to D.C. as follows (October 4. 1935):

The Svedberg misst keine Molekulargewicht in chemischen Sinne (einzelne durch Hauptvalenzen zusammengehaltene Moleküle) sondern Teilchengrössen . . . Im Falle des Insulins misst er eine Teilchengrösse 35–36,000. Er gibt selbst an, dass dieser Wert nur bei einem p_H dicht am isoelektrischen Punkt konstant ist, und dass bei grösserem und kleinerem p_H die Dissoziation in kleinere Teilchen wahrgenommen wird. Demnacht ist das Molekulargewicht (im strengen, chemischen Sinne) 36,000:n = 18,000 oder 12,000 oder 9,000 etc.

Die chemische Untersuchung hat ergeben (Zeitschr, f. physiol. chem. 233, 165, (1935)), dass eine Blutzucker-wirksame Gruppe in einem Aequivalent von 18,000 vorhanden ist. Wenn dieser Wert genau wäre, so wäre er die unterste Grenze für das Molekulargewicht, ist n = 2, so ist mit 36,000: 2 die oberste Grenze ermittelt aus den Befunden von Svedberg. Es ist jedoch möglich, dass das physiologische Wirkungsäquivalent kleiner als 18,000 ist, jedoch nicht kleiner als 9,000. Fur das Molekulargewicht des Insulins kommen daher die Grössen 9,000, 12,000 oder 18,000 in Betracht, von denen 9,000 weniger wahrscheinlich ist.

The number 12,000 appeared the obvious choice for the molecular weight of insulin from chemical and crystallographic observations combined.

At about this time, we also discussed the problem of the structure of insulin with C. R. Harington at University College Hospital, who pointed out that insulin was doubtfully to be classified as a protein and suggested lactoglobulin as a more typical protein for study by X-rays and one very recently crystallized. R. A. Kekwick, from the same laboratory, produced crystals in two modifications, orthorhombic and tetragonal, that were the next crystals to be X-ray photographed. Actually, the most beautiful lactoglobulin crystals, orthorhombic plates 2–3 mm across, arrived unlabeled through the mail one day and were recognized immediately. They came from Lindeström-Lang by way of Bernal to Oxford and they marked our first contact with Lang and Copenhagen. Yet another preparation was made in Oxford by Ogston. With these large crystals one could easily see with the naked eye the shrinking of the plates in one direction on drying, measure the amount with a traveling microscope, and follow the fall in birefringence. As a consequence, lactoglobulin was the first protein from which single crystal diffraction data were obtained when the crystals were both wet and air-dried (Crowfoot and Riley, 1938).* The relations between the two suggested strongly that the protein molecule was essentially unchanged on drying. Again the crystal asymmetric unit in both examples was related to the Svedberg molecule but not identical with it. In the orthorhombic form it was approximately twice the Svedberg weight, and in the tetragonal form it was equal to it.

Arguments about the kind of structure present in the protein molecule began to develop in this period. These were partly stimulated by the remarks in the first letter to *Nature* about pepsin which stated that the periodicities characteristic of peptide chains in fiber patterns were not observed on the pepsin photographs. On the one hand, attempts were made to devise structures for the protein molecule in which peptide chains did not appear as such; these included the cyclol structure of Dr. Wrinch and, somewhat later, the layer structure of Dervichian. On the other hand, it was realized that only nearly parallel chains would be likely to give a fiber type pattern with the periodicities observed by Astbury or by Meyer and Mark. A very suggestive observation was made by Astbury, Dickinson, and Bailey (1935) that a drying crystal of excelsin gave an X-ray photograph in which a fiber pattern was superimposed on a single crystal pattern. Astbury wrote to D.C. (April 15, 1935):

> I should like to ask you, too, if you can give me a definite assurance that in both the case of pepsin and of insulin it is impossible to generate the usual protein powder photograph (showing principally the backbone ($4\frac{1}{2}$ Å) and side chain (10 Å) spacings) simply by rotating the rotation photographs. You will realize that it is absolutely essential to clear this possibility out of the way once and for all and I should like to receive a definite statement from you that you have considered the point. . . . I am now convinced that the phenomenon of denaturation is not other than the breaking away of peptide chains from the particular configurations specific to any protein. I hardly think now that there can be any other interpretation than that the photographs you and Bernal have obtained do really arise from rather globular molecules packed in the simplest close packing. These molecules must be built up each from some very specific configuration of chains which breaks up as soon as the rather narrow conditions of stability are departed from. On heating the chains they simply coagulate to form parallel bundles analogous in their structure to that of the stable natural fibres. . . .

*By an unfortunate accident the unit cell dimensions were recorded wrongly here. They were corrected later, but better values are now available (see Aschaffenburg et al., 1965).

Very similar ideas to those of Astbury's were put forward by Pauling and Mirsky (1936) in a paper on denaturation phenomena and the structure of proteins:

Our conception of the native protein is the following. The molecule consists of one polypeptide chain which continues without interruption throughout the molecule (or in certain cases, of two or more chains); this chain is folded in a uniquely defined configuration in which it is held by hydrogen bonds between the peptide nitrogen and oxygen atoms and also between the free amino and carboxyl groups of the diamino and dicarboxyl amino acid residues.

An early suggestion that the chain configuration in a globular protein might be related to α-keratin is contained in a letter Astbury wrote to Bernal (July 15, 1935). In it he said,

If you are going to have a go at oxyhaemoglobin, here's a point that looks interesting. On page 188 of Katz's book it says that in the ordinary ring photos of oxyhaemoglobin, the backbone spacing is absent as in α keratin and α myosin. It says also that Abel has shown that air dried insulin gives a similar photo, i.e., without backbone spacing.

We did make a rather feeble effort to take powder photographs of our protein crystals and we detected a certain strengthening of the lines in the 10 Å region, as had been hoped for by Astbury. Better evidence on the relation between the fibrous and globular proteins was obtained soon afterwards directly from the intensity data through the calculations of Patterson distributions.

In 1935, A. L. Patterson published, in the *Zeitschrift für Kristallographie*, "A direct method for the determination of the components of interatomic distances in crystals." This seemed to provide just what was needed—a direct way of treating the X-ray data from single protein crystals to show the interatomic distances in the molecules. Patterson calculations were tried out first on the data from air-dried insulin crystals. The data were very limited but so were the means available for computing. The first sums were made without even Beevers-Lipson strips as an aid, and were made by two alternative routes to check the calculations. From the outset it was clear that the calculations ought to be made in three dimensions; owing to the rhombohedral symmetry of the crystal, this could be effected on a coarse 5 Å grid by the calculation of two sections only (at $z = O$ and $z = \frac{1}{2}$)—sections which Harker (1936) had shown were easy to compute. A little thought showed that there would not necessarily be any simple means for interpreting the Patterson functions for insulin such as those used by Patterson himself for copper sulfate or hexachlorbenzene. The insulin maps did however indicate that there were, within the insulin crystals, objects concentrated at interatomic distances of 10 and 22 Å, distances observed by Astbury in the fibrous proteins. There were other features too of the insulin vector patterns that invited attempts at further interpretation. Patterson himself wrote in his first letter to D.C. (March 13, 1938), "There seems to be so much detail in the F^2 patterns for insulin that it ought to be possible to get a more detailed description of the shape of the molecule, but I must confess that for the moment I do not see how this can be done." In the next letter (July 15, 1938) he wrote, "My great regret at the moment is that our laboratory is now torn into small pieces and that it is therefore impossible for me to start madly calculating and possibly experimenting on proteins." Others looked at the patterns and a number of attempts at interpretation were made. There is a very interesting letter from P. P. Ewald (August 2, 1939), tentatively suggesting and then rejecting an idea for a protein structure.

The idea was simply that a sphere of constant density gives a diffraction effect of a wavy kind and that by the general reciprocity between physical space v. Fourier space, a wavy density distribution in crystal space would give a sudden drop in intensity in Fourier space, i.e., with increasing order of diffraction. If the protein molecule might be conceived to have a structure of shells inside one another, like chinese carved ivory balls or certain radiolariae skeletons, such a wavy density distribution would be approximated.

He goes onto discuss the effects on the intensity distribution of filling in spaces between the "balls" with water.

The calculation of Patterson maps for wet insulin as well as for air-dried insulin (Crowfoot and Riley, 1939) seemed at first of most importance in confirming the view that the protein molecule was a rigid unit; from the comparison of the vector patterns of the wet and air-dried crystals shown in Figure 1 it was clear that the molecules rotated around the three-fold axis as the crystals dried, and an estimate could be made of their extent. In retrospect it is an interesting fact that there is in the first Patterson maps direct information about the exact size of the insulin molecule, which we might have deduced long ago but failed to recognize. The symmetry of both of the first Patterson distributions implies the presence of six, not three, asymmetric units in the crystal. This information was missed in our early thinking although every model any of us proposed for insulin at this period (for example, Wrinch and Langmuir, 1938; Bernal, 1939) had the correct symmetry, 32. It was not until after Sanger's work had established the chemical structure of insulin and shown that it had a weight of about 6,000 that it was realized that the two insulin molecules in the rhombohedral asymmetric unit were related by noncrystallographic two-fold axes of symmetry (Low and Einstein, 1960). The relations which have been established by an application of the rotation function of Rossman and Blow (1962) can in fact be seen very easily by looking at the old Patterson distributions.

Something of the same sort can be said about our early work on lactoglobulin. This was very much disrupted by the onset of the war in 1939. That year, vector maps were calculated from very limited data for both wet and air-dried orthorhombic and tetragonal crystals, and many efforts were made to interpret these. These were abandoned after the war as it become clearer how much more data it was both necessary and possible to collect to solve the structures in detail. However, some of the conclusions about lactoglobulin that were reached with considerable hesitation in that period agree almost surprisingly well with the molecular structure now appearing in the quite different lactoglobulin crystal structures that are being studied in detail by Green and his coworkers. The two original modifications were tetragonal and orthorhombic; the unit cells were similar in volume, and this suggested that the asymmetric unit in the orthorhombic form contained two

FIGURE 1.
Patterson projections, *Pxy*, for (*a*) wet insulin and (*b*) air-dried insulin, showing similarity of the vector distributions within 25 Å of the origin (the peaks are at distances of roughly 10 and 22 Å from the origin); (*c*) section in the Patterson distribution, *Pxyo*, calculated for wet insulin. The strong inner peak is at 5 Å from the origin.

molecules, while that in the tetragonal cell contained one of the order of size, 35,000, required by Svedberg's measurements. Patterson projections were calculated for both forms and also for a partly shrunk modification of the orthorhombic structure. They showed complicated patterns in which it was possible to distinguish peaks that must be due to structure within the molecules from those due to distances between them (compare Figures 2 and 3). In the tetragonal Patterson maps the peaks were much drawn out parallel with the *c*-axis, suggesting long molecules of the general shape of a prolate ellipsoid 15–17 Å in radius and 60 Å long. Molecules of the same general shape could be fitted in four layers in the wet orthorhombic unit cell. Here the elongation of the Patterson peaks suggested that the lengths of the molecules might be nearly parallel with *b*.

The air-dried lactoglobulin crystal structures were very much disordered. Of the two varieties, that of the tabular orthorhombic form looked the most promising. The intensities of all the 142 X-ray reflections observable on

FIGURE 2.
Tetragonal (needle) lactoglobulin: Patterson projections along (*a*) [001], (*b*) [100], (*c*) [110].

2a

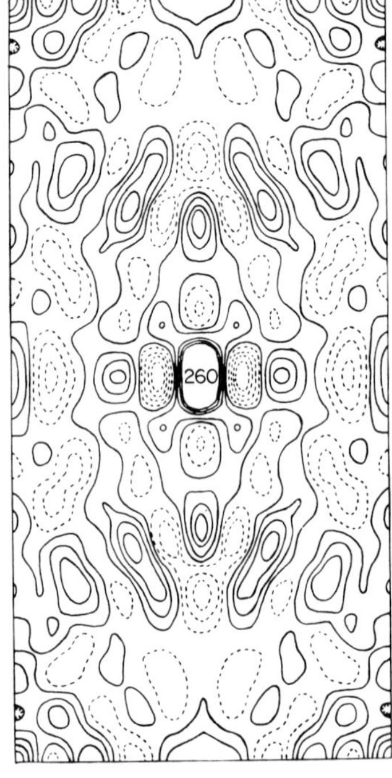

2b

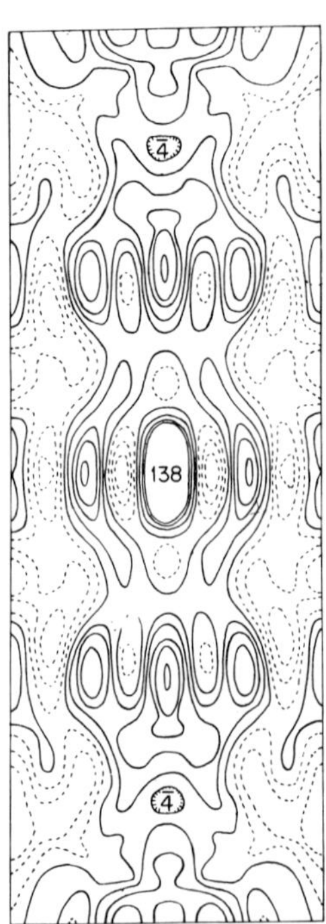

2c

oscillation photographs were measured, and calculations were made of sections of the three-dimensional Patterson distribution. These were quite simple in character as shown in Figure 4. The main peak distribution could be accounted for by arrangements of four layers parallel with the *c*-plane, with four "points" in each layer, that is, sixteen points in all. These points could be taken as representing the scattering centers of protein masses, providing each of these masses had a molecular weight equal to half the Svedberg weight of the lactoglobulin molecule. One of the possible alternative sixteen-point distributions was consistent with the two submolecules making contact with one another to compose the overall ellipsoidal molecule suggested from the wet-crystal data. Although the first interpretation of the air-dried crystal data therefore "suggested an arrangement of half molecules" it did not seem to require the splitting of the molecule as a whole into two on crystal drying. Somehow we did not think that the most natural conclusion of our observations would be that the two "half" molecules were identical and that the true molecule of lactoglobulin "in chemische Sinne" might be the half-molecule. Although we spent a great deal of time thinking of possible nonspace-group ways of arranging the molecules in disordered lattices, we did not recognize the nonspace-group two-fold axes that must be present in all these crystal structures, judging by the work of Aschaffenburg, Green, and Simmons (1965). One can only hope that now that the shape and size of the lactoglobulin molecules are known, these particularly interesting crystal structures may be re-examined.

FIGURE 3.
Orthorhombic (tabular) lactoglobulin: Patterson projections along (*a*) [001], (*b*) [100], (*c*) [010], (*d*) [110]; (*e*) partly wet crystal, projection along [100].

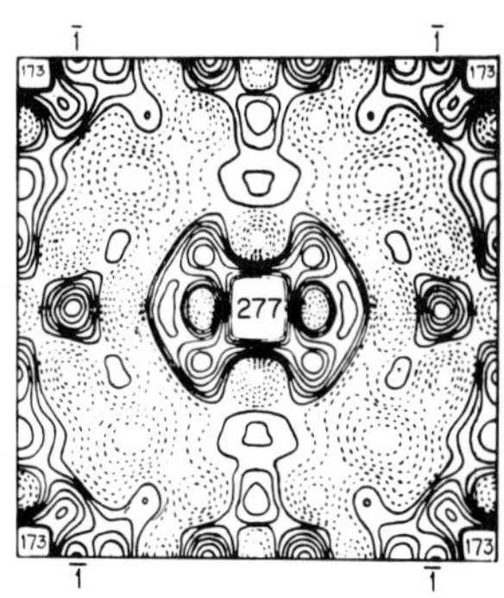

3a

3b

3c

3d

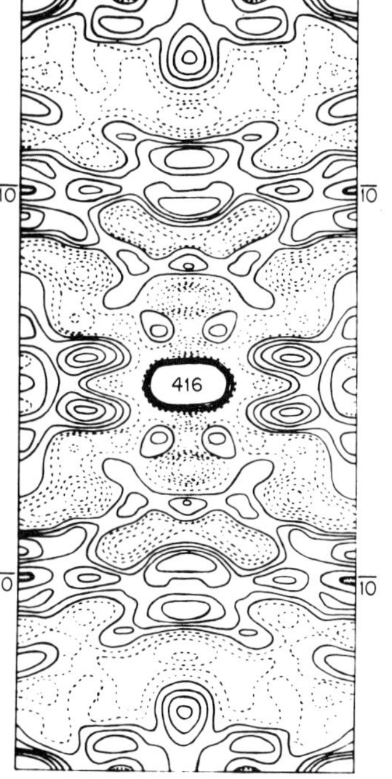

3e

FIGURE 4.
Dry orthorhombic lactoglobulin: Patterson projections along (*a*) [001], (*b*) [100], (*c*) [010]; sections in the three-dimensional distribution at (*d*) $z = 0$, (*e*) $z = \frac{1}{4}$, (*f*) $z = \frac{1}{2}$.

We realized from the first X-ray photographs we took that most of the general problems of protein chemistry, such as size and shape and hydration of the molecules, would be solved, in addition to the internal structure, once the complete solution of any protein crystal structure was achieved. In 1934, however, we were still very close to the beginning of the X-ray analysis of quite simple organic compounds. The crystal structures of naphthalene and anthracene were published by J. M. Robertson in 1933: they followed from an application of trial and error analysis. In 1939, when Pauling and Niemann discussed problems of protein structures in general, they pointed out that the first amino acid crystal structure, glycine, had only just been solved, by Albrecht and Corey (1939), through the application of the Patterson synthesis. It was madness (as Patterson said) to think of solving protein structures in any detail by such means, and indeed we did not really expect to do so. At the backs of our minds, we had very early the idea of introducing heavy atoms into the crystals and using the changes of intensity that we expected to see to determine phases and so to proceed directly to the calculation of electron densities and atomic positions. The general idea, derived from Cork's work on the alums, was current in Bernal's laboratory in 1933; the plan was to test it out on Rochelle salt. After insulin was photographed in 1935, Bernal sent a note which ran, "Zn insulin 0.52 wt. percent., Cd 0.77 wt. percent, Co 0.44 wt. percent. This gives slightly less than 3 in all cases . . . I will send you the cadmium stuff as soon as I get back to Cambridge." Later, in 1939, after J. M. Robertson had solved the structure of phthalocyanine by the isomorphous introduction of nickel into the phthalocyanine molecule, Robertson also suggested that the same method might be used to determine the crystal structure of insulin. However, the intensity changes produced by replacing zinc with cadmium in insulin proved too small to measure on our early X-ray photographs, and we began to discuss with insulin chemists how to get heavier atoms into insulin. Experiments on the iodination of insulin had already been carried out; it seemed that these should be followed up (Harington and Neuburger, 1936). Meanwhile, the report of the preparation of pale yellow crystals of iodobenzene-azo-insulin by Reiner and Lang came into our hands. Dr. Reiner very kindly sent us samples of a few milligrams of very tiny crystals from which it was just possible to grow slightly larger crystals and to record a few spectra. These were enough to indicate that the crystal lattice was essentially unchanged as were the relative intensities of the inner strong X-ray reflections. The experiments were carried out in 1941, and were reported briefly to the Rockefeller Foundation in 1941–1942. It is a little odd to look back on them today; nothing about them was really quite right. According to the records there was only 0.9 mole of iodine per "molecule" insulin (that is, per 36,000 molecular weight). Although we meant to repeat the experiment later with a sample containing a higher iodine content, somehow we never did; we were partly inhibited by Dr. Reiner's opinion that the iodine would be scattered on different tyrosine groups throughout the insulin molecule. Even if the iodine present in the first crystals we looked at had been concentrated at a single residue in the insulin molecules we examined, it would, according to our present knowledge of the structure, have been spread over six crystallographic sites; our crystals were too small and our methods of measurement too weak to have recorded at that time the effects for which we sought. Perhaps it was just as well that our somewhat pessimistic comments on our experiments were buried in notebooks and reports.*

Our pessimism in the early 1940's was itself a consequence of our limited experience in structure analysis. We realized that the X-ray diffraction effects from protein crystals were weak, volume for volume, compared with common salt or even with glycine. Precisely how weak they were only gradually emerged with the consideration of the statistics of the distribution of X-ray intensities by A. J. C. Wilson in 1942, followed closely by E. W. Hughes (1949) and D. Harker (1948). This weakness was to prove a source of strength, since heavy atoms were found in practice by Perutz and Kendrew to be much more effective in phase determination than at one moment we had imagined. By the early 1950's the necessary background experience had been gained in many directions. Great improvements had been made in intensity measuring techniques and in computing. Bijvoet and Peerdeman had demonstrated the

*The reports were those made to the Rockefeller Foundation in 1942.

efficiency of phase determination in isomorphous derivatives of medium complicated asymmetric molecules and had added observations on anomalous dispersion effects. A great number of structure analyses of simple molecules had established stereochemical principles limiting the arrangements of the atoms in space. These were to make it easier than we had previously supposed, to recognize the arrangement of the atoms in the actual first electron density maps obtained for myoglobin, blurred as they were at 2 Å resolution. Particularly important in this connection was the theoretical derivation of the α-helix conformation for the peptide chain. It is nice to record here that this was first thought of by Linus Pauling, rolling paper scrolls on a sick bed in Oxford in 1948, before it was built as a model structure by Branson and Corey in Pasadena in 1950, or proved to exist in synthetic fibers such as α-polyalanine in 1953, or seen in myoglobin in 1960.

There was a period in the early 1940's, before the real breakthrough in protein X-ray analysis, when we all turned to model building. By comparison with other proteins, the molecule was so small that it seemed that, at least for insulin, we might be able to devise a model that would fit the diffraction effects. The first problem we considered was the form of chain folding present. Partly led on by the trigonal symmetry of the crystal, we first built a model of a spiral peptide chain with a three-fold screw axis of symmetry, which we soon discovered had been built before us, both by M. L. Huggins (1943) and by H. S. Taylor (1941). We then built trial structures in which short lengths of such chains were placed parallel with the insulin three-fold axis in positions suggested by Bernal's observations on the insulin Patterson projections and the correlation of parts of the distribution with particles in close packing. Luckily we had with us Dr. Käthe Schiff, who destroyed this model practically at birth. She convinced us that the rather even distribution of observed intensities in insulin excluded any structure based predominantly on chains parallel to the c-axis and also any chain structures having a regular repeat along the chain axis. Later we examined also model structures in which four insulin chains were arranged parallel with one another in a 12,000 molecular weight unit and placed roughly either at 45° or 80° to the c-axis. This type of structure was less easy to exclude out of hand; one very similar, with chains built predominantly as α-helixes, was independently considered by Lindley and Rollett in 1954. The general chain orientation was one favored by observations with polarized infrared light on insulin crystals. Other models were also constructed, based largely on a β-pleated sheet chain configuration for the A chains.

It is curious to look again at the protein structures we know today after looking back at our old observations and speculations. The course of the main peptide chain through lysozyme or even through myoglobin is so complicated and irregular, it is not surprising that for a brief time we thought that it was not there at all. The particular geometries, and the intermolecular attractive forces—polar and nonpolar—of the amino acid residue within a protein molecule are studied today as factors that may determine the detailed configuration of the peptide chain itself. In different regions of lysozyme, in small stretches of the molecule, there are examples of some of the types of peptide chain configuration suggested long ago by Huggins, by Astbury, and by others from model-building experiments, in addition to stretches of the well-defined α-helical and β-pleated sheet units we have since come to expect.

Of the three proteins of which X-ray photographs were taken first, lactoglobulin is well on the way to solution (Camerman et al., 1966), and a promising new start has been made with pepsin (Borisov et al., 1966). There has also been some progress in the determination of the crystal structure of rhombohedral zinc insulin, which bears on our old speculations. It has proved possible first to remove the zinc from the crystals by soaking in EDTA solution and then to replace the zinc with the heavier metal, lead. The changes of intensity, zinc to lead insulin, are marked and can be accurately recorded together with observations on anomalous dispersion effects. The lead atoms can be placed in the crystal by the use of Patterson difference maps. Even more marked are the changes that occur when zinc-free crystals are compared with lead-containing crystals. Measurements have therefore been made on the whole of the isomorphous series of crystals, zinc-free insulin, zinc insulin, cadmium insulin, and lead insulin. The data collection has so far been extended to planes of spacing of about 2.5 Å. Whether the structure can be solved, as long ago proposed, from this series of crystals alone, remains to be seen. Preliminary calculations of the electron density at a resolution of about $4\frac{1}{2}$ Å have been made and show many suggestive features. We must wait for further calculation for this detailed interpretation.

There are many problems concerning protein structures that we discussed in the 1930's but think about no longer. Scientific papers then were full of numerology: Bergmann and Niemann's idea of regular repetition of amino acid residues with relative weights $2n \times 3n$, Svedberg's idea that protein molecules might have weights that were some simple multiple of a basic unit of weight, perhaps about 17,500. These ideas in their original form disappeared when accurate chemical analysis and sequence studies came in. All the same, they were expressions of the need we still have to discover any systems there may be that determine the construction of proetin molecules. Clearly, before we understand these, we need the detailed X-ray analyses of many more varieties of protein molecules; these analyses in turn require a tremendous scientific cooperative effort. Perhaps we also need Linus Pauling to look once again at this problem to see the answers that elude us.

REFERENCES

Adams, M. J., E. Coller, G. Dodson, D. C. Hodgkin, and S. Ramaseshan. *Seventh International Congress of Crystallography.*

Albrecht, G. A., and R. B. Corey (1939). *J. Am. Chem. Soc.* **61**, 1087.

Aschaffenburg, R., D. W. Green, and R. M. Simmons (1965). *J. Mol. Biol.* **13**, 194.

Astbury, W. T., S. Dickinson, and K. Bailey (1935). *Biochem. J.* **29** 2, 2351,

Bernal, J. D. (1939). *Proc. Roy. Soc.* (London) **170**, 50.

Bernal, J. D., and D. Crowfoot (1934). *Nature* **133**, 794.

Bernal, J. D., I. Fankuchen, and M. Perutz (1938). *Nature* **141**, 523.

Borisov, V. V., W. R. Melik-Adamyan, N. E. Shutskever, and N. S. Andreeva (1966). *Seventh International Congress of Crystallography*, Abstracts, 166.

Camerman, A., R. D. Diamond, D. W. Green, and R. M. Simmons (1966). *Seventh International Congress of Crystallography*, Abstracts, A 167.

Crowfoot, D. M. (1935). *Nature* **135**, 591.

Crowfoot, D. M., and D. P. Riley (1938). *Nature* **141**, 521.

Crowfoot, D. M., and D. P. Riley (1939). *Nature* **144**, 1011.
Dervichian, D. G. (1943). *J. Chem. Phys.* **11**, 236.
Harker, D. (1936). *J. Chem. Phys.* **4**, 381.
Harker, D. (1948). *Am. Min.* **33**, 764.
Harington, C. R., and A. Neuberger (1936). *Biochem. J.* **30**, 809.
Huggins, M. L. (1943). *Chem. Rev.*, 32.
Hughes, E. W. (1949). *Acta Cryst.* **2**, 34.
Kohlschütter, H. W., and L. Sprenger (1932). *Z. Phys. Chem.* B **16**, 284.
Lindley, H., and J. S. Rollett (1955). *Biochem. Biophys. Acta* **18**, 183.
Low, B. W., and J. R. Einstein (1960). *Nature* **186**, 470.
Meyer, K. H., and H. Mark (1938). *Ber.* B **61**, 1932.
Patterson, A. L. (1935). *Z. Krist.* **90**, 517.
Pauling, L., and A. Mirsky (1936). *Proc. Nat. Acad. Sci. U.S.* **22**, 439.
Pauling, L., and C. Niemann (1939). *J. Am. Chem. Soc.* **61**, 1860.
Perutz, M. F. (1949). *Research* **2**, 52.
Riley, D. P. (1942). *The Crystal Structure of Some Proteins*, Ph.D. thesis, Oxford.
Riley, D. P. (1947). *Médicine et Biologie* (Liege), no. 5, p. 35.
Roberston, J. M. (1939). *Nature* **143**, 75.
Rossman, M. G., and D. M. Blow (1962). *Acta Cryst.* **15**, 24.
Sauter, E. (1932). *Z. Phys. Chem.* B **18**, 417.
Schimper, A. F. W. (1881). *Z. Krist.* **5**, 131.
Svedberg, T. (1939). *Nature* **127**, 438.
Svedberg, T. (1939). *Proc. Roy. Soc.* (London) A **170**, 40.
Taylor, H. S. (1941). *Proc. Am. Phil. Soc.* **85**, 1.
Wilson, A. J. C. (1942). *Nature* **150**, 152.
Wrinch, D. (1936). *Nature* **137**, 411. *Proc. Roy. Soc.* (197) A **160**, 59; A **161**, 505.
Wrinch, D., and I. Langmuir (1938). *J. Am. Chem. Soc.* **66**, 2247.

G. KARTHA
J. BELLO
D. HARKER
Center for Crystallographic Research
Roswell Park Memorial Institute
Buffalo, New York

Binding Site of Phosphate Ion to Ribonuclease Molecules

A model representing the tertiary structure of bovine pancreatic ribonuclease based on a 2 Å resolution electron density map has recently been put forward (Kartha, Bello, and Harker, 1967), and it is of interest now to correlate the specific properties of this enzyme in terms of its three-dimensional structure. In addition to the description of the amino acid sequence and the position of the disulfide bridges, a wealth of other chemical data have accumulated during the past few years regarding some of the noncovalent interactions in this protein in its native form (Anfinsen, 1962). A physical approach that gives, in effect, a photograph of the chemical interaction between the protein and other small molecules or ions, and that has been successful in some earlier studies (Stryer, Kendrew, and Watson, 1964; Johnson and Phillips, 1965), is the use of the isomorphous replacement technique of X-ray crystallography. In this method, X-ray diffraction data from two crystals are studied, one of the native protein and the second of a complex of the protein with some other small molecule or ion that crystallizes in a form closely isomorphous with that of the native protein. In principle, this is possible even when a small part of the contents of the unit cell of the crystal is replaced by some other material of differing electron density. In such a situation, the comparison of the X-ray diffraction intensities from the two crystals enables us to locate the binding site of the additional atom or groups in the crystal. If, in addition, the details of the three-dimensional arrangement of the enzyme molecule in the unit cell of the crystal are known, we can identify the groups involved in binding the substrate molecule to it.

The aim of these studies is to establish the binding sites of some of the substrates and competitive inhibitors of the enzymic activity of ribonuclease. Even though what we see in these studies is a static picture of the protein and the molecule in the bound state, it is possible that from a detailed study of the groups involved in the binding and their stereochemistry, we could come up with a reasonable mechanism of the enzymic activity. The dynamics of this mechanism could be worked out from the three-dimensional structure of the complex and from the forces called into play in achieving the binding between the enzyme and its substrate molecule and the subsequent catalytic action.

PHOSPHATE BINDING SITE IN RIBONUCLEASE

In this paper, we discuss the result of our studies in establishing the binding site of phosphate ion in the ribonuclease molecule. This ion is one of a large number of polyanions that interact very specifically with ribonuclease (Sela and Anfinsen, 1957; Sela, Anfinsen, and Harrington, 1957). Stein and Barnard (1959) have obtained chemical evidence that points to at least some of the groups involved in this specific interaction. These studies suggest that the binding of these polyanions to riboncleas makes several functional groups inaccessible to some specific chemical reactions. It was also noted that it is, in general, extremely difficult to remove completely the phosphate from ribonuclease, and that in the presence of sufficient phosphate ion, the enzyme is inactive. All these results suggest that the phosphate binding site is likely to be near the region of ribonuclease activity. Hence, the study of the molecular neighborhood of this site is of importance in the correlation of the structure of this enzyme with its specificity.

MATERIALS USED IN THIS STUDY

The protein crystals under study were crystallized in the presence of phosphate, because phosphate-free RNase could not be crystallized unless at least one molecule of phosphate (or sulfate) were added per molecule of RNase. This suggests that the phosphate ions are specifically attached to the molecules in these crystals. As it was known (Sela and Anfinsen, 1957; Sela, Anfinsen, and Harrington, 1957) that the protein shows similar interaction towards the electrostatically very similar arsenate ion, it was hoped that crystals obtained in the presence of these ions would be isomorphous with the phosphate crystals, but with the arsenate ion—of greater electron density—replacing the phosphate ion.

RNase arsenate crystals were made as follows: a stock solution of phosphate-free RNase (Worthington RAF) 50 mg/ml was adjusted to *p*H 5 with dilute HCl, and the solution was distributed into small test tubes, 0.4 ml per tube. On top of the aqueous solution was carefully placed a layer of 0.48 ml 2-methyl-2,4-pentanediol (MPD). After standing several days to allow the two layers to mix by diffusion, the solution was seeded with crushed RNase crystals and 0.03 ml of 0.1 *M*, *p*H 5 sodium arsenate was added. Crystals were grown, and when they reached the appropriate size, were transferred to 75 percent MPD. An additional 0.03 ml solution of arsenate was added to bring the total amount of arsenate used up to 4 molecules per molecule of RNase.

X-RAY DIFFRACTION

The unit cell dimensions of the phosphate and arsenate crystals are given below and these show satisfactory agreement.

	Phosphate RNase	Arsenate RNase
$a =$	30.13 Å	30.11 Å
$b =$	38.11 Å	38.01 Å
$c =$	53.29 Å	52.95 Å
$\beta =$	105.75°	105.70°

The space group of both crystals is $P2_1$ with two molecules in the unit cell.

Diffraction data from the phosphate crystals had already been collected in the course of our studies of the structure of ribonuclease (Kartha, Bello, and Harker, 1967). The data to 4 Å resolution from the arsenated crystal were first collected manually on a General Electric XRD-3 diffractometer equipped with a goniostat with a crystal mounted with the c^* axis vertical. Cu$K\alpha$ radiation was obtained by using a Co-Ni Ross filter pair. The intensities were corrected for Lorentz and polarization factors and an empirical absorption correction, as usual, and were scaled against the squared amplitudes from the phosphate protein. A similar set of data was also measured on an automated GE XRD-6 diffractometer, and this enabled us to have an estimate of the reproducibility of the measurements; both sets of which were collected by the stationary crystal, stationary counter method, and processed in a very similar manner. The scaled amplitudes $|F_{H_1}|$ and $|F_{H_2}|$ of the two sets of measurements were compared with each other as well as with the amplitudes $|F_P|$ from the phosphate protein and gave the following values for the average percentage differences R_1 and R_2.

$$R_1 = \frac{\sum ||F_{H_1}| - |F_{H_2}||}{\frac{1}{2}\sum ||F_{H_1}| + |F_{H_2}||} = .071$$

$$R_2 = \frac{\sum ||F_{H_1}| - |F_P||}{\sum |F_P|} = .102$$

Thus, the agreement between two measurements on the arsenate crystals is seen to be 7 percent, whereas, the average difference in amplitude between the phosphate and arsenate crystals is 10 percent. It may be worth mentioning at this point that the average amplitude differences calculated in a similar manner for heavy atom derivatives used in the original phase evaluation gave values of 20 to 30 percent in the region under study. It is thus seen that the average difference between the phosphate and arsenate protein crystals is only slightly greater than the reproducibility of data from crystals of the same type.

LOCATION OF ARSENATE GROUP

If in the arsenated crystal, the arsenate ion replaces the phosphate ion and the positions of most of the remaining atoms are unchanged, then we can locate the site of the replaceable group by the isomorphous series technique. Two methods are available for achieving this:

1. *The electron density difference (EDD) method:* Since we are assuming that the two crystals are isomorphous, the electron density at every part of the unit cell of the crystal is the same, except at the binding site of the phosphate ion. At this region, however, the arsenate crystal has a density increment corresponding to the electron density difference between the arsenic and phosphorous atoms. Hence if we calculate a map of the electron density difference between the arsenate and phosphate crystals, it will show a peak at the site of the phosphate ion, with the rest of the crystal volume devoid of much detail. If we have a good a priori knowledge of the phase angles α_{hkl} corresponding to each reflection *hkl* of the phosphate crystals, then the electron density difference (EDD) map is given by the three-dimensional Fourier series:

$$D\rho(x, y, z) = \frac{1}{v} \sum_{h} \sum_{k} \sum_{l-\alpha}^{\alpha} \{|F_{\mathrm{H}}| - |F_{\mathrm{P}}|\} \exp\left[-2\pi i(hx + ky + lz - \alpha_{hkl})\right] \quad (1)$$

where $D^{-}\rho\ (x, y, z)$ is the difference between the densities in the two crystals at the point (x, y, z).

Using the amplitudes and phase angles for the phosphated protein and the amplitude of the arsenated protein, electron density difference maps were computed from Equation (1) using data to 4 Å resolution; these maps are shown in Figures 1–5.

These maps very clearly show a single peak 1200 units high on a background that nowhere else reaches a height of even 300 units. This peak has fractional coordinates

$$x = .045 \quad y = 0.46 \quad \text{and } z = 0.39$$

and presumably corresponds to the arsenate group.

2. *The difference Patterson method.* The procedure described earlier assumes that we have a reasonably correct set of phases for the protein reflections to start with, so that we can directly compute the electron density difference map from Equation (1). Even though this information is usually available at this stage of the work, it may be pointed out that reliability of the site found from the EDD map depends entirely on the availability of a reasonably accurate set of phases derived from other work. Hence, it would be highly satisfying if the site could also be established independently of these phases.

It is indeed possible to obtain an independent check of at least two of the coordinates of the binding site by calculating a difference Patterson map, which uses only the measured amplitudes of the two sets of reflections and not their phase angles. Such a map shows peaks at positions corresponding to the vectors between replaceable groups in the crystal. In actual practice these vector maps are difficult to interpret if: (1) the isomorphism between the two crystals is poor; (2) the electron density difference of the replacing groups is not large enough; or (3) the sites of replacement are neither few nor specific. Luckily, in the present study the first and third difficulties do not occur and it was therefore decided to establish the site from a vector map $P(uvw)$ computed using Equation (2)

$$P(uvw) = \frac{1}{V} \sum\sum\sum (|F_{\mathrm{H}}| - |F_{\mathrm{P}}|)^2 \cos 2\pi(hu + kv + 1w) \quad (2)$$

For the space group $P2_1$, the vector between the symmetry related arsenate groups occurs in the section $P(u\frac{1}{2}w)$, and this section computed with data to 4 Å resolutions is shown in Figure 6. This map contains one peak of height 1150 at $u = 11/104$, $w = 28/120$, and two other peaks of height less than 700. All other peaks, except the origin peak $P(0, 0, 0)$, are much lower. Assuming that the (11/104, 28/120) peak is the peak $(-2x, -2z)$ of the arsenate group, the coordinates of this group are $x = .447$, $z = .383$, in close agreement with the values obtained from the EDD map. Note, however, that the y parameter of the group as obtained from the difference Patterson map is arbitrary for this space group, whereas no such arbitrariness exists in the coordinate obtained by the first method, in which the position of the site is found in relation to the position of the protein molecule.

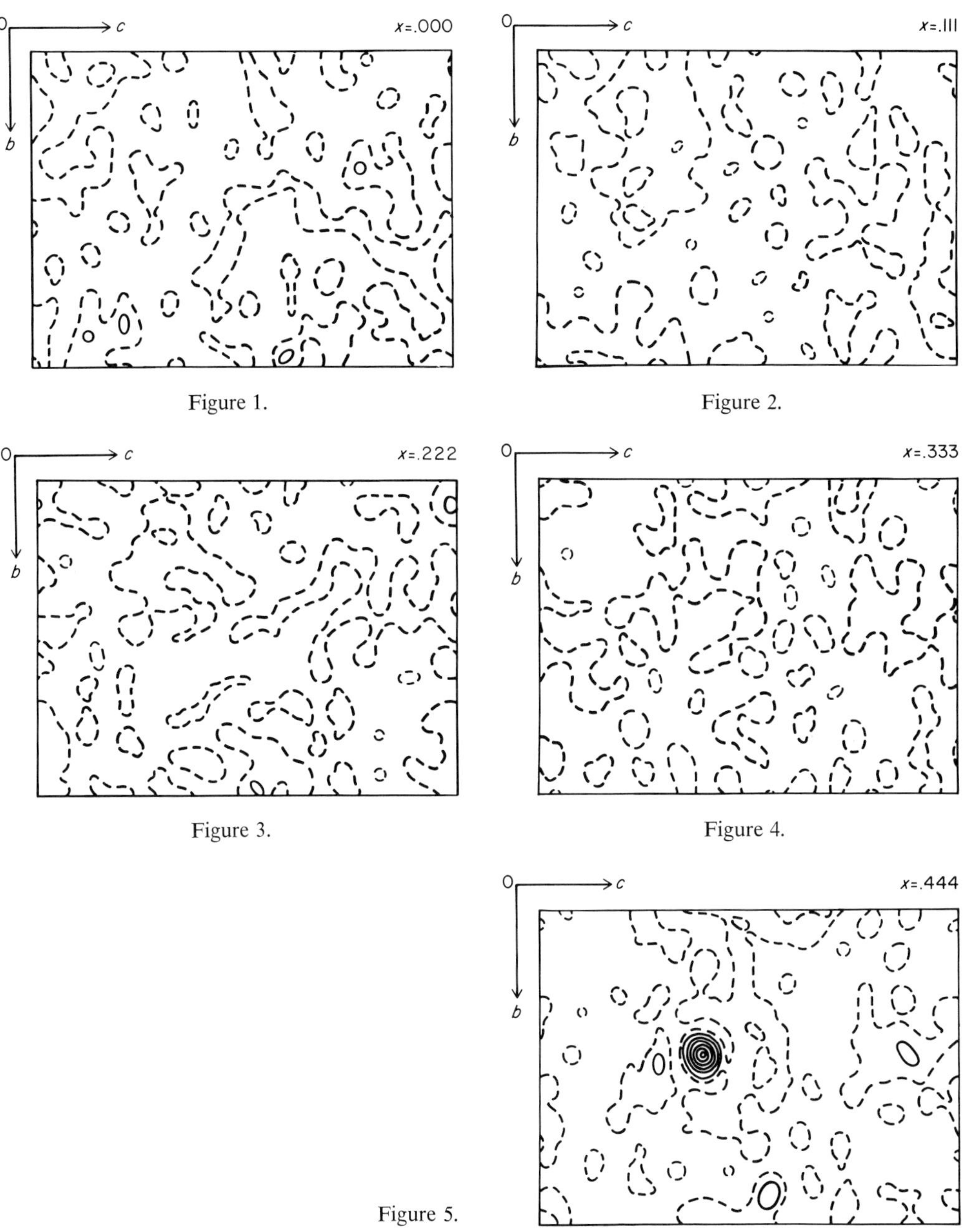

Figure 1.

Figure 2.

Figure 3.

Figure 4.

Figure 5.

FIGURES 1–5.
Map of the electron density difference between arsenate and phosphate ribonuclease crystals. The maps are bounded projections along the *a*-axis of density difference slices about 3 Å thick. Contours are drawn at equal but arbitrary intervals.
Figure 1. $x = .000 \pm .055$
Figure 2. $x = .111 \pm .055$
Figure 3. $x = .222 \pm .055$
Figure 4. $x = .333 \pm .055$
Figure 5. $x = .444 \pm .055$
The peak in Figure 5 has fractional coordinates $x = .45$, $y = .46$, and $z = .39$.

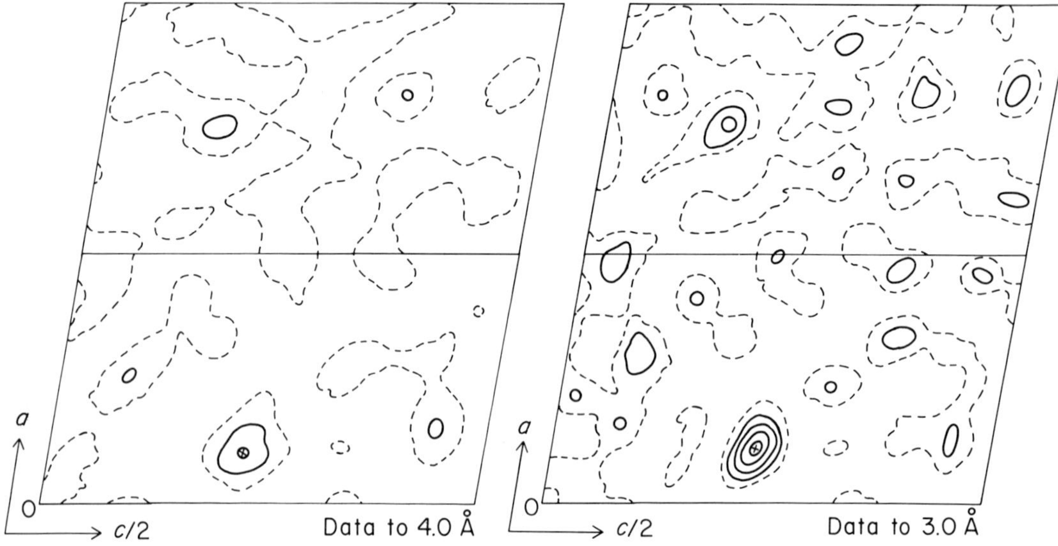

Figure 6. Difference Patterson section with data to 4 Å.

Figure 7. Difference Patterson section with data to 3 Å.

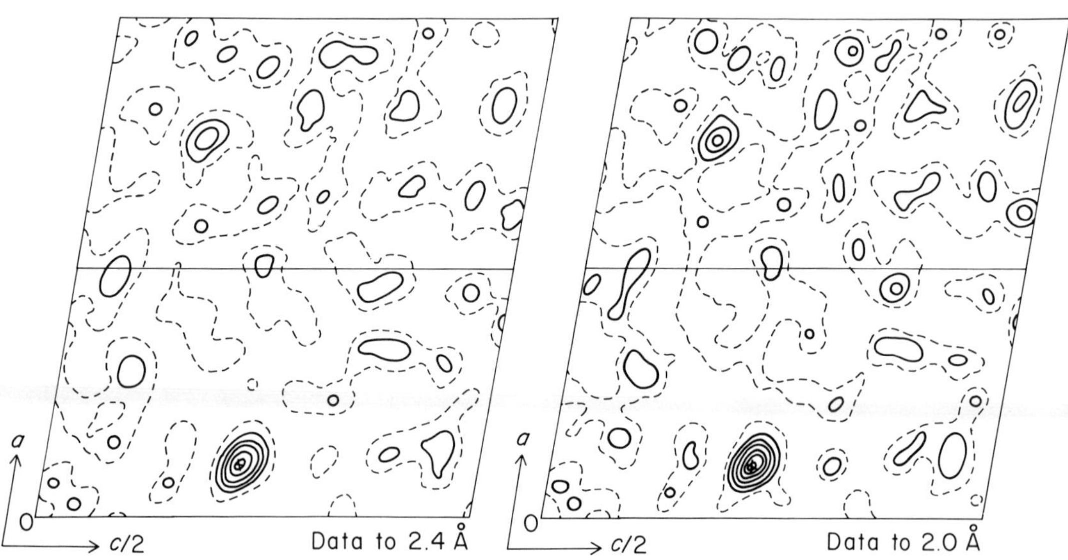

Figure 8. Difference Patterson section with data to 2.4 Å.

Figure 9. Difference Patterson section with data to 2.0 Å.

FIGURES 6–9.
Section $P(u,\frac{1}{2},w)$ of the difference Patterson map computed with $(|F_H|-|F_P|)^2$ as Fourier coefficients. The peak due to the symmetrically related arsenate groups is denoted by +. The fractional x and z coordinates of this peak are: $x = .447, z = .383$.

HIGH-RESOLUTION (2Å) STUDIES

The sections $P(u\frac{1}{2}w)$ of difference Patterson maps using data from phosphate and arsenate crystals at 3, 2.4, and 2 Å resolutions are shown in Figures 7, 8, and 9. These maps show a progressive sharpening of the peak corresponding to the arsenate group, as is seen from Table 1. Perusal of this table suggests that the peak-to-background ratio increases at higher resolutions and that, in spite of the low electron density difference between the phosphate and arsenate groups, the average isomorphism between the two crystals is good enough to make the intensity difference between the two crystals significant even at 2 Å resolution. Hence, it was decided to consider the arsenate crystals also as heavy atom derivatives in establishing the protein phases to 2 Å.

From high-resolution maps it was, however, seen that the phosphate and arsenate groups do not occupy exactly the same positions in the two crystals, but are separated by about 1.4 Å. The respective fractional coordinates of the phosphate and the arsenate groups from the 2 Å resolution maps are given below:

Phosphate group $x = .475$ $y = .460$ $z = .366$

Arsenate group $x = .450$ $y = .462$ $z = .386$

Even though these two groups do not have identical coordinates, the overall isomorphism between the two crystals is excellent and hence, the arsenate crystals are considered as heavy atom derivative crystals for purposes of phase angle evaluation.

CORRELATION WITH CHEMICAL DATA

The binding site of the phosphate group was compared with the electron-density models of the protein at 4 Å resolution and also with a 3 Å model that was available at that time. It was seen that the phosphate ions bind in a cleft in the kidney-shaped molecule. A 2 Å resolution map computed recently (Kartha, Bello, and Harker, 1967) has enabled us to trace the folding of the main chain in the molecule, and we can now identify the groups that are in the vicinity of this phosphate binding site. The position of the phosphate group relative to the main chain is shown in Figure 10.

TABLE 1.
Progressive sharpening of arsenate peak in the difference Patterson map with increasing resolution. Coefficients used are: $(|F \text{ Arsenate}| - |F \text{ Phosphate}|)^2$.

Resolution of data	Number of reflections	Peak height	Average background	Next highest peak in map
4.0	946	115	50	70
3.0	2214	210	60	90
2.4	4450	240	70	110
2.0	7174	290	80	120

From the recent maps it seems that the group nearest the phosphate ion is the histidine, residue 119, near the carboxyl end of the polypeptide chain. The next nearest group appears to be histidine 12 near the amino end; it projects towards the phosphate. These observations agree well with the earlier results (Anfinsen, 1962) that the presence of phosphate ions prevents carboxymethylation of both histidines 12 and 119. Other functional groups that are also within short distances from the phosphate are: lysine 41, lysine 7, and glutamine 11. All these are close enough to the phosphate so that the presence of the ion in this region can block the free access of other molecules to these functional groups. The proximity of some of these sites, even though they are widely separated in the amino acid sequence, was predicted by the dinitrophenylation studies of Hirs (1962) and polyalanination studies of Anfinsen (1962) and co-workers.

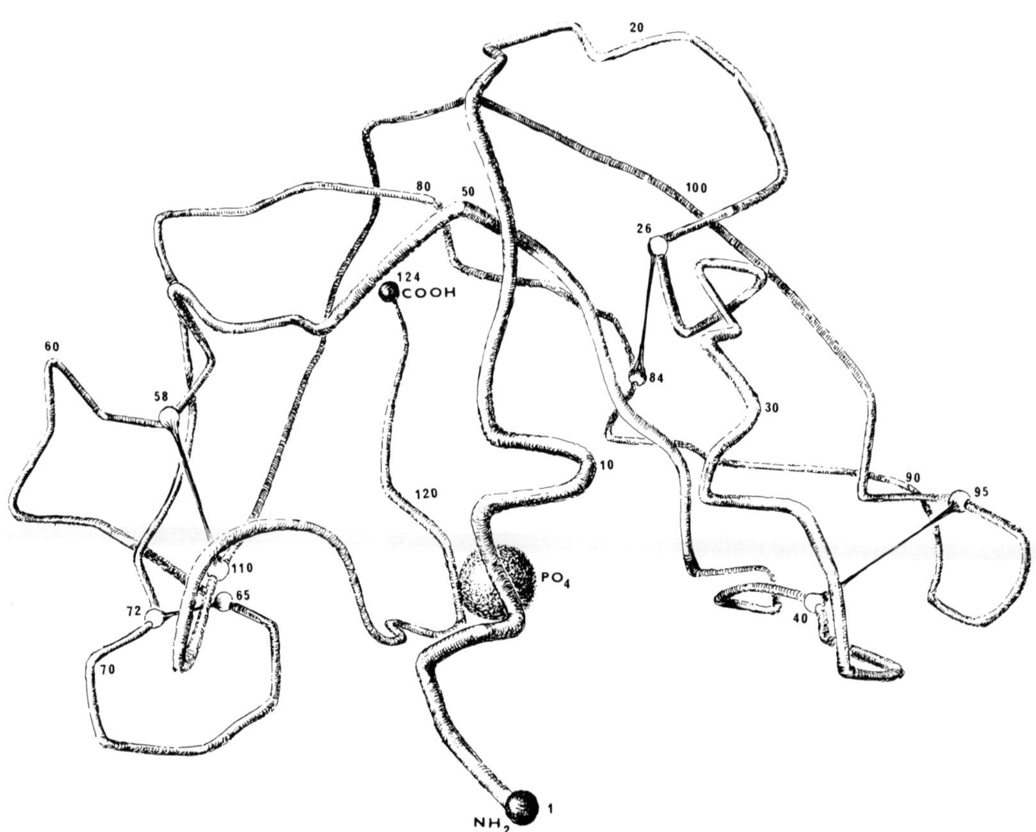

FIGURE 10.
The course of the polypeptide chain of ribonuclease and the position of the phosphate group.

ACKNOWLEDGMENTS

The authors wish to thank the following Institutions for supporting the research described here: The National Science Foundation for a series of research grants; the National Institutes of Health for additional research grants and for Public Health Service Award GM-K3-16737 which paid the salary of one of the authors (G. K.). We are grateful for the excellent asistance we obtained from Mrs. T. Falzone, Miss F. E. DeJarnette, Mrs. C. Vincent, and Miss Go in the preparation and handling of the crystals and the processing of the data used in these investigations.

REFERENCES

Anfinsen, C. B. (1962). *Brookhaven Symposium*, 184.
Barnard, E. A., and W. A. Stein (1960). *Biochim. Biophys. Acta* **37**, 371.
Hirs, C. H. W. (1962). *Brookhaven Symposium*, 154.
Johnson, L. N., and D. C. Phillips (1965). *Nature* **206**, 761.
Kartha, G., J. Bello, and D. Harker (1967). *Nature* **213**, 862.
Sela, M., and C. B. Anfinsen (1957). *Biochim. Biophys. Acta* **24**, 229.
Sela, M., C. B. Anfinsen, and W. F. Harrington (1957). *Biochim. Biophys. Acta* **26**, 502.
Stein, W. D., and E. A. Barnard (1959). *J. Mol. Biol.* **1**, 350.
Stryer, L., J. C. Kendrew, and H. C. Watson (1964). *J. Mol. Biol.* **8**, 96.

WILLIAM N. LIPSCOMB
Department of Chemistry
Harvard University
Cambridge, Massachusetts

Structure of an Enzyme

It is a special pleasure to present this preliminary account of the low-resolution structure in a volume dedicated to Linus Pauling, both because of my association with him when I was a graduate student at the California Institute of Technology during part of the years 1941–1946, and because of his remarkable contributions to protein structures. Indeed, my interest in protein structures stemmed from this association.

During the past five years my colleagues—Martha L. Ludwig (since May 1962), Jean A. Hartsuck (since April 1963), Thomas A. Steitz (since June 1963), Hilary Muirhead (since February 1964), James Searl and James C. Coppola (both since October 1964)—and I have carried out the protein preparations, the finding of suitable heavy atom derivatives, and the X-ray diffraction measurements leading to solution of the three-dimensional structure of bovine pancreatic carboxypeptidase A (CPA) (Lipscomb et al., 1966).

The CPA enzyme, discovered by Anson (1937), cleaves peptide and ester bonds, at a C-terminal residue of L configuration where the carboxyl group is free and is α to the peptide or ester bond. If the side chain of the C-terminal residue is aromatic, the reaction is favored. The molecule has a molecular weight of 34,600, including a replaceable Zn atom, two half cystines (but no S—S bond), and three methionines (Vallee and Neurath, 1954; Bargetzi et al., 1963). Further work on the sequence is presently under way in Neurath's laboratory. Removal of the Zn atom inactivates the enzyme and has been shown to expose a cysteine residue (Williams, 1960; Vallee, Coombs and Hoch, 1960). The Zn atom of the enzyme has also been said to bind to the N-terminal end of the single polypeptide chain (Vallee, Riordan, and Coleman, 1963; Coombs, Omote, and Vallee, 1964), which, in the form called CPA_{α}, has 307 amino acid residues (Bargetzi et al., 1963).

Our study began with confirmation of Kraut's survey (see Rupley and Neurath, 1960), which established that the space group is $P2_1$, and that the monoclinic unit cell has parameters a = 51.4 Å, b = 59.9 Å, c = 47.2 Å, and β = 97.6°, and contains two molecules. Our earlier work was carried out on crystals supplied by Neurath, but nearly all of the results presented here have been obtained from single crystals prepared in our laboratory from samples of pancreatic juice or from samples of procarboxypeptidase A which had been donated to us by Hirs of the Brookhaven National Laboratories. Satisfactory heavy-atom derivatives were obtained only after long and careful chemical and X-ray studies, telescoped here into two observations.

First, it is important not to give up too soon if a heavy-atom derivative does not appear satisfactory. More than once, derivatives in which we first found the difference Patterson analysis to be unsatisfactory, and in which we suspected that the derivative crystals were not isostructural with the native protein crystals, proved later to be excellent at 6 Å resolution, provided only that further search was made for all of the other, sometimes minor, sites of heavy-atom occupancy. Second, high-quality X-ray data are essential, especially on the parent protein.

We have located at least ten sites of heavy-atom substitution in derivatives prepared by treatment of protein crystals with $PbCl_2$, K_2PtCl_6, p-acetoxymercurianiline (HgAn), or silver acetate, or with 8-hydroxyquinoline-5-sulfonic acid in order to remove the Zn atom. Phases for the electron density map were found by developments of Bijvoet's methods that were similar to the methods used by Kendrew and Perutz (see Cullis et al., 1961). Only four derivatives were used for phasing: the Pb derivative (two sites only 4 Å apart), the $PtCl_4^{--}$ derivative (one major and three minor sites), a first HgAn derivative (four sites), and a second HgAn derivative (three sites). Anomalous scattering data, included for three of these derivatives permitted unambiguous choice of the correct enantiomorph. A summary of the various criteria is given in Table 1 for all but the Ag derivative. Occupancies and atomic positions of heavy atoms are listed in Table 2.

The electron density map at 6 Å resolution has been computed, plotted, and drawn up in sections. After addition of the constant F_{000} term of 0.37 electrons/Å^3, a set of contour levels ranging from 0.5 eÅ^{-3} to 0.8 eÅ^{-3} was selected. Balsa sections of all electron densities greater than 0.5 e/Å^3 are shown in Figures 1 and 2. All densities are to be compared with the standard deviation of 0.06 eÅ^{-3}. Helical regions, taken as those greater than 0.76 Å^{-3}, are relatively straight, about 5 or 6 Å in diameter and no closer than 10 Å to another helical region or 8 Å to a nonhelical region. In the figures we also show a clay-covered wire model of a tentative course of the single polypeptide chain. In this model the regions that satisfy the criteria for helix are

TABLE 1.
Criteria for results on derivatives (at 6 Å).

	Pb	Pt	HgAnI	HgAnII	Zn-free
Closure Error[a] (E) in electrons	35	38	42	34	31
RMS f	109	111	100	53	37
$R^b = \dfrac{\Sigma\lvert\lvert F_H\rvert_o - \lvert F_H\rvert_c\rvert}{\Sigma\lvert F_H\rvert_o}$	0.08	0.10	0.11	0.09	0.07

Average figure of merit[c] = 0.88, assuming half weight to anomalous scattering errors of Pb, Pt, and HgAnII derivatives.

[a]Closure errors of phase triangles are estimated from $E_j = [\sum_{hkl}(\lvert F_H\rvert_{obs} - \lvert F_H\rvert_{calc})^2/n_j]^{1/2}$ where n_j is the number of intensities for derivative j.
[b]In the calculation of R the values of $\lvert F_H\rvert_{calc} = \lvert\lvert F_p\rvert e^{i\alpha} + f_c\rvert$ include the estimate of phase angle.
[c]The figure of merit is the mean value of $\cos(\alpha - \alpha_o)$.

TABLE 2.
Occupancies and positions of heavy atoms (at 6 Å).

Atoms	Occupancy, electrons	Fractional Coordinates		
		x	y	z
Pt	130	.085	0	.114
Pt_1	92	.668	−.098	.962
Pt_2	34	.442	−.219	.569
Pt_3	52	.291	−.446	.857
Pt_4	44	.496	−.032	.487
$HgAnI_1$	67	.072	−.055	.108
$HgAnI_2$	58	.517	−.442	.260
$HgAnI_3$	63	.484	−.414	.140
$HgAnI_4$	66	.448	−.007	.575
$HgAnII_1$	46	.087	−.038	.097
$HgAnII_2$	36	.522	−.439	.250
$HgAnII_4$	37	.468	−.018	.574
Zn-free	27	.078	−.079	.122

enlarged. Six or seven helical regions occur (none near the active site), making up a length of about 110 Å and containing about 75 amino-acid residues. The helix content is thus about 25 $\pm$ 5 percent. Positions of the heavy metals are also shown in the figures. A possible, but not necessarily unique, tracing of the polypeptide chain (Figure 2) includes all electron densities above 0.55 eÅ^{-3}, and is at least 0.45 eÅ^{-3} at all points along the chain. The molecule, roughly 50 $\times$ 44 $\times$ 40 Å in dimensions, is thus a bit ellipsoidal. In spite of ambiguities in the trace of the chain, some of which are noted as 1, 2, 3, and 4 in Figure 2, our present interpretation tentatively places the ends E_1 and E_2 at least 25 Å away from the Zn atom.

This Zn atom is 1 to 2 Å from a mercury site, which, if occupied, drastically reduces the peptidase activity. This result is consistent with earlier experiments from Vallee's and Neurath's laboratories on requirement of a metal at the Zn site for activity, failure of binding of the competitive inhibitor (phenylacetate) to the Zn-free enzyme, and the influence of inhibitors on metal exchange. The Zn atom lies in a groove on the surface of the molecule, and it is adjacent to a pocket, 8 or 10 Å in diameter, that extends into the molecular interior. We are inclined to believe that this pocket is hydrophobic, and, occasionally for visitors to our laboratory, we place molecular models of typical substrates or inhibitors in this region so that the aromatic group is inside the pocket and the carbonyl oxygen of the peptide bond is near the Zn atom. Such speculation is surely consistent with the dedication of this chapter.

Our studies are progressing along biochemical lines, both at 6 Å resolution and at 2 Å resolution. Removal of Zn occurs readily when crystals are treated with the chelating agent 8-hydroxyquinoline-5-sulfonic acid. A difference electron density map shows very clearly the Zn atom of the parent protein —and no other observable structural change—at 6 Å resolution, when the Zn is removed in the crystalline phase. At least the Zn is not, therefore,

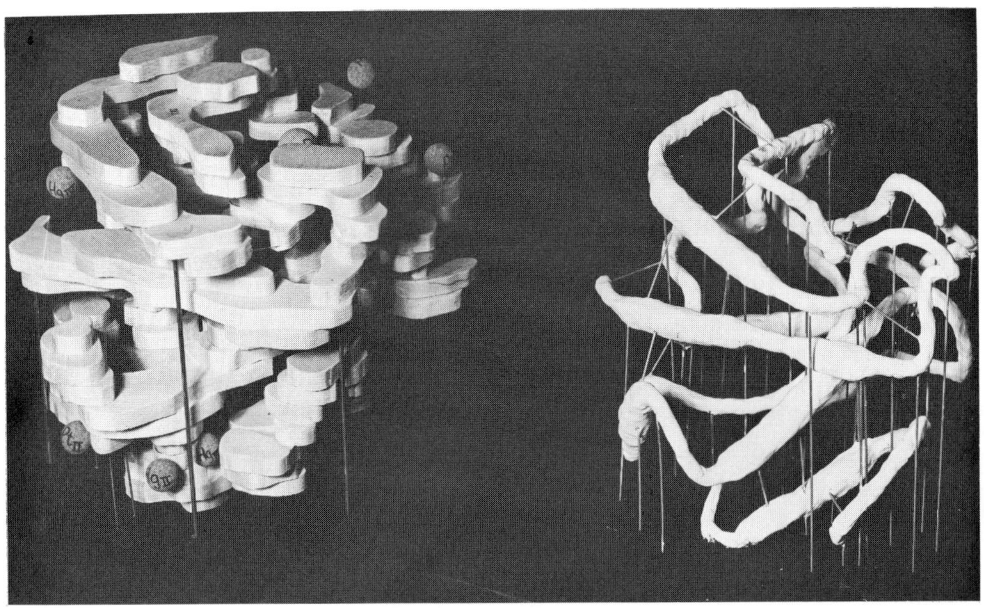

FIGURE 1.
Balsa model and clay-covered wire model of the carboxypeptidase A molecule, emphasizing helical regions at 6 Å resolution.

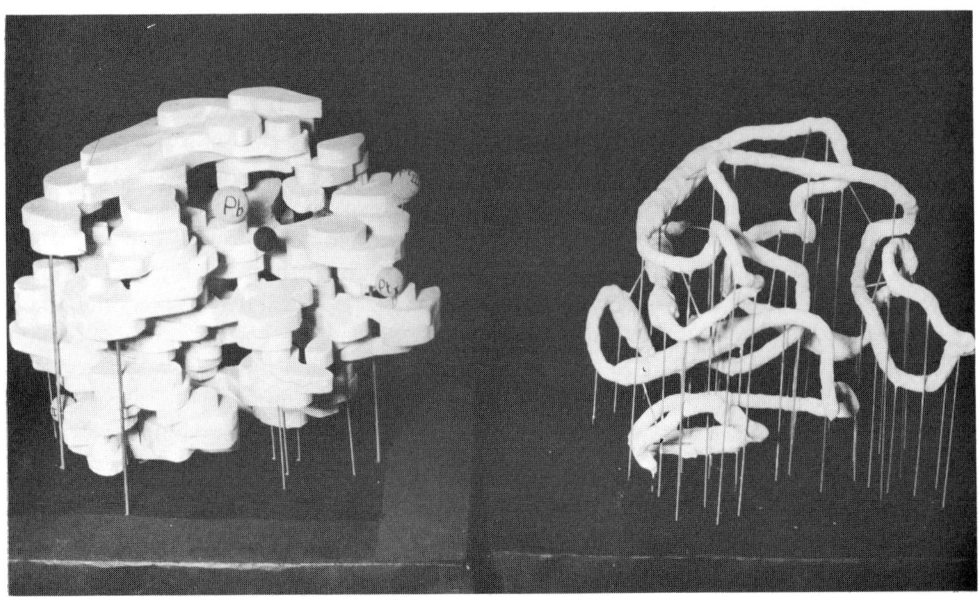

FIGURE 2.
Models of the carboxypeptidase A molecule at 6 Å resolution, emphasizing the active site region around Zn, some possible ambiguities (1, 2, 3, and 4) in the trace of the polypeptide chain, and the probable locations (E_1 and E_2) of the ends of the chain.

required to fix the position of the N-terminal end of the polypeptide chain. Indeed, difference maps of proteins that differ in N-terminal residues show a residual peak in the E_2 region of the structure. Hence, in our present interpretation of this low-resolution structure, we differ from Vallee et al. on the placement of this N-terminus, and suggest that for their results there may be a different explanation, such as a conformational change of the molecule. Further X-ray diffraction studies at higher resolution are being undertaken to elucidate details of the structure and to reinvestigate this three-dimensional trace and the tentative location of the ends of the chain.

Enzymatic activity of the crystals, recently demonstrated by Quiocho and Richards (1964), is reasonably expected, in view of the large region of low electron density around the active site in the crystal structure. Nevertheless, we find that certain inhibitors and substrates produce changes of as much as two percent in cell dimensions, even when the molecules are cross-linked by reaction with glutaraldehyde. In our present attempts to obtain an electron density map of a bound inhibitor or substrate, we hope to establish also whether these changes are due to molecular reorientation or to actual conformational changes in the molecule. Rotatory dispersion studies of these systems in solution are also in progress. So far as we are aware, these are the first data that suggest conformational changes in this molecule.

ACKNOWLEDGMENT

I would like to thank the National Institutes of Health for support of this research, and to acknowledge a stimulating discussion with F. H. Westheimer at a Colloquium at Harvard on March 2, 1966, where these results, interpretations, and speculations were first presented.

APPENDIX

Added January 29, 1967. Our progress following the communication of the above manuscript on last March 30, 1966, has left us a choice of an extensive updating, or of allowing the historical record to remain as it is and adding supplementary material. We choose the latter course.

Three-dimensional electron density maps have been computed at 2.8 Å resolution in June 1966 from the Zn enzyme [7028 reflections], the Pb(2) and Pt(4) derivatives, where the number of sites of the heavy atoms is indicated in parentheses. In August 1966, we added data for the Hg(1) derivative, in which Hg replaces Zn at the active site, and in January 1967 we included also the Hg(3) derivative to 2.8 Å resolution. Both of these Hg derivatives were prepared from action of $HgCl_2$ solution on the enzyme: Hg(1) from the enzyme in solution, and Hg(3) from crystals of the enzyme. The figure of merit is 0.70 (no anomalous scattering data included). We are just beginning to analyze in detail the results of this work.

The Pb(2), Hg(1) and Hg(3) derivatives are quite effective in phasing to 2.8 Å, and, very likely, to considerably higher resolution. The Pt(4) derivative is not effective beyond about 3.5 Å, but it does contribute to the phasing of the inner reflections. Particularly clear features of the new map are the molecular boundaries, the long stretches of α-helix, the direction of the chain, and the largest peak at the Zn position, which is adjacent to a pocket extending into the molecular interior. We are not yet inclined to alter our conclusions at 6 Å concerning the general regions of the chain ends. On the complete tracing of the chain, we feel that the more detailed study that is now under way is essential, but we have emphasized the ease with which the tracing at low resolution can be ambiguous. From our attempts to fit published bits of the sequence to our electron density at 2.8 Å resolution, we believe that if the complete sequence were available we could now construct a three-dimensional model of the detailed molecular structure of carboxypeptidase A.

Our present lines of research on the structure at 2.0 Å resolution are as follows. First, we have already collected all X-ray diffraction data out to 2.0 Å on the parent enzyme. Those 6,000 largest reflections from these data will also be measured on the Pb(2) and the Hg(3) derivative, for phasing by the multiple isomorphous method. Second, we shall attempt to calculate phases in the 2.8 Å to 2.0 Å range, using contributions from the Zn atom, from the five S atoms, from the location of some of the tyrosines in difference maps after iodination (and similar modifications on other residues that can yield derivatives which do not destroy the crystal), from the locations of the published 53 of the 307 residues, and from the fitting of the remainder of the chain as polymeric units of alanine or glycine. In total, these contributions contain a very large fraction of the ordered scattering matter, and should fix the phases to 2.0 Å reasonably well. Third, we shall attempt to extrapolate the phases ϕ, known to 2.8 Å, to the 2.8 Å to 2.0 Å range with the use of the relation

$$\phi_h = \langle \phi_k + \phi_{h-k} \rangle$$

where the average is taken over the larger reflections k (Karle and Hauptman, 1950; Coulter, 1965). Fourth, we are attempting at 2.8 Å resolution least square matches, using translation and rotation functions, in order to fit known density of the amino acid residues to our electron density map in the region of each unidentified residue. Thus, we feel that we can develop the X-ray method to a high degree of reliability in the direct determination of the amino acid sequence in protein structures.

The discovery, after several years of exploration, of just the right conditions of cross-linking and concentrations of materials led us, in December, 1966, to the calculation of a three dimensional map at 6 Å resolution of what we believe to be the first organic substrate (glycyl-tyrosine) complexed with an enzyme. Several studies of complexes of inhibitors and substrates with the enzyme and with the apoenzyme (Zn removed) have now been completed. Aside from comment on the multiple regions of binding, we state here our present decision: we shall attempt to carry the enzyme-(Gly-Tyr) complex to high resolution, in the expectation that the mechanism of action of this enzyme on the peptide bond will become clearly elucidated.

Finally, especially in honor of Linus Pauling on this occasion, it seems most appropriate to add Figure 3, which shows one of the regions of α-helix in the carboxypeptidase A structure at 2.8 Å resolution. Also, in Figure 4 a through g we show composites of sections of electron density normal to the crystallographic b-axis. The individual sections are 0.75 Å apart in these composites. Helical regions can be seen, and the active site cavity is above and to the right of the Zn atom, which is the largest feature in Figure 4c.

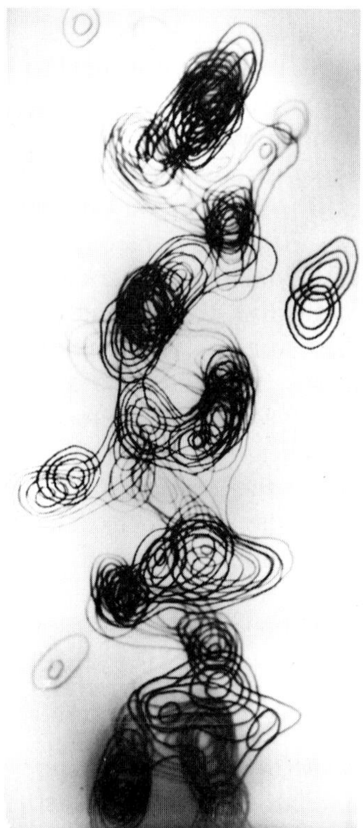

FIGURE 3.
A region of α-helix in carboxypeptidase A at 2.8 Å resolution.

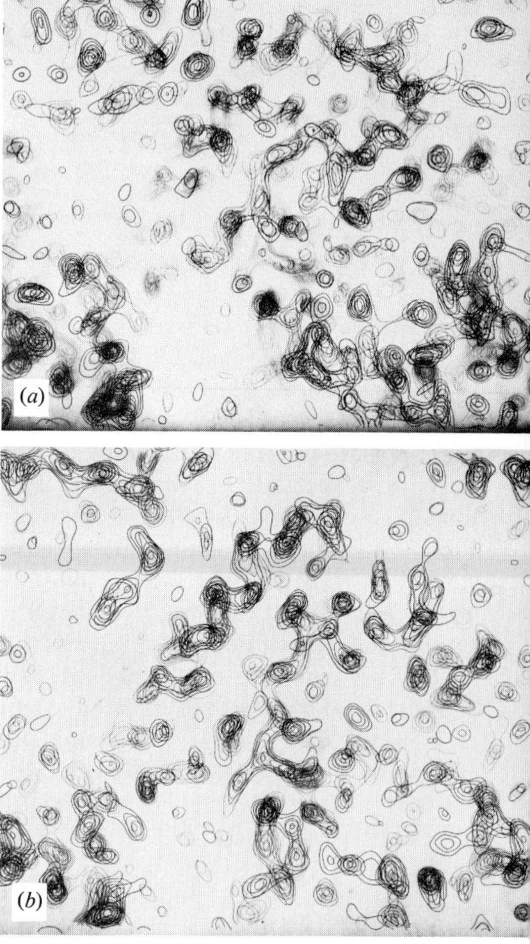

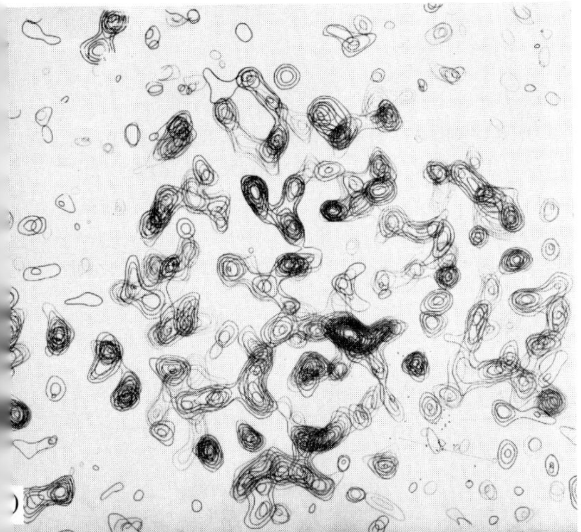

FIGURE 4.
Electron-density map at 2.8 Å resolution of the three-dimensional structure of carboxypeptidase A.
(*a*) Sections 47–61
(*b*) Sections 48–55
(*c*) Sections 40–47
(*d*) Sections 32–39
(*e*) Sections 24–31
(*f*) Sections 16–23
(*g*) Sections 8–15

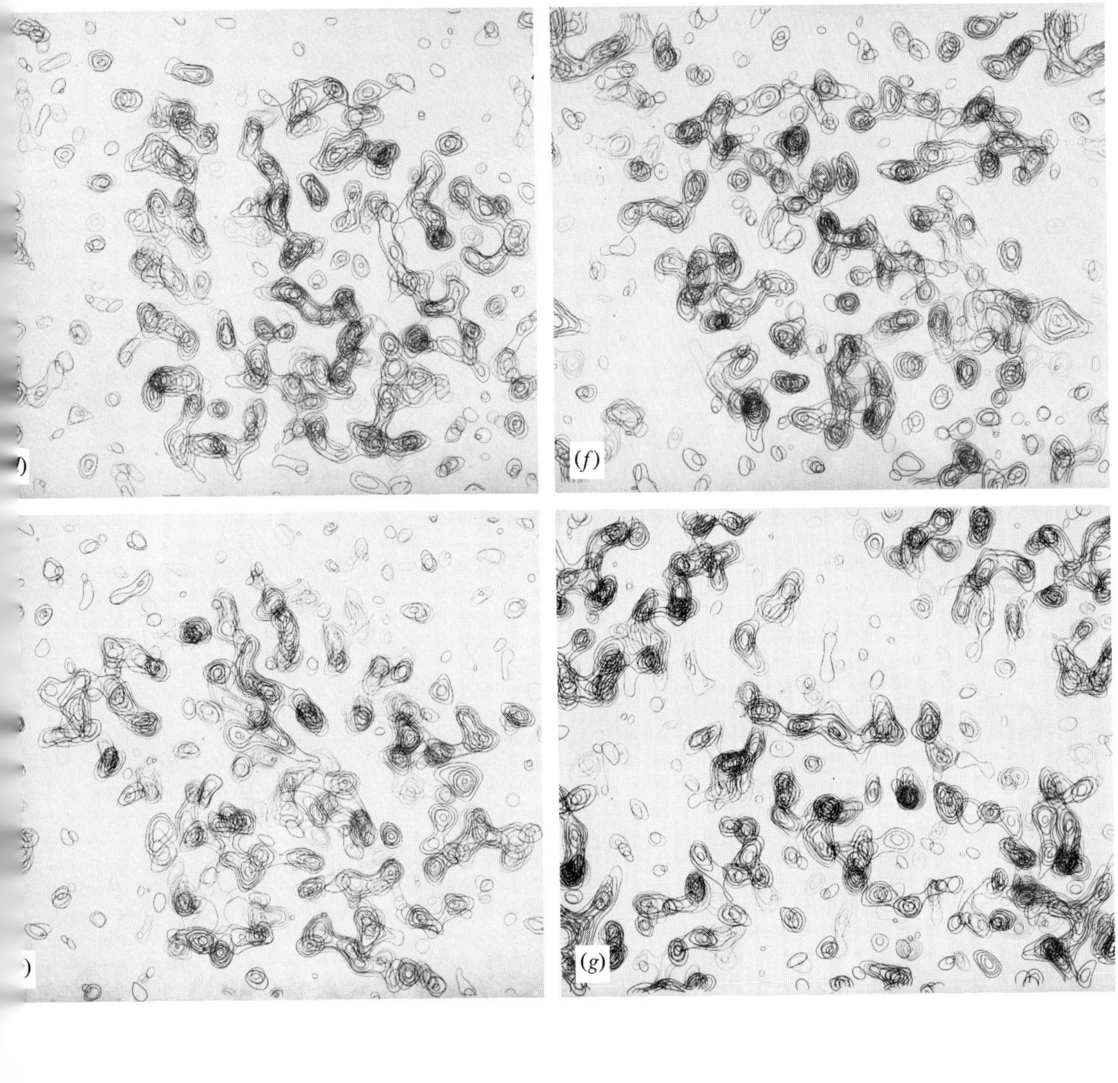

(*f*)

(*g*)

REFERENCES

The major reference on the molecular structure of carboxypeptidase A at 6 Å resolution is Lipscomb et al. (1966). Complete references to the earlier studies will be found therein.

Anson, M. L. (1937). *J. Gen. Physiol.* **20**, 663.

Bargetzi, J.-P., K. S. V. Sampath Kumar, D. J. Cox, K. A. Walsh, and H. Neurath (1963). *Biochemistry* **2**, 1468.

Coombs, T. L., Y. Omote, and B. L. Vallee (1964). *Biochemistry* **3**, 653.

Coulter, C. L. (1965). *J. Mol. Biol.* **12**, 292.

Cullis, A. F., H. Muirhead, M. F. Perutz, M. G. Rossmann, and A. T. C. North (1961). *Proc. Roy. Soc.* (London) A **265**, 15.

Karle, J., and H. Hauptman (1950). *Acta Cryst.* **3**, 181.

Lipscomb, W. N., J. C. Coppola, J. A. Hartsuck, M. L. Ludwig, H. Muirhead, J. Searl, and T. A. Steitz (1966). *J. Mol. Biol.* **19**, 423.

Quiocho, F. A., and F. M. Richards (1964). *Proc. Nat. Acad. Sci. U.S.* **52**, 833.

Rupley, J. A., and H. Neurath (1960). *J. Biol. Chem.* **235**, 609.

Vallee, B. L., H. Neurath (1954). *J. Am. Chem. Soc.* **76**, 5006.

Vallee, B. L., T. L. Coombs, and F. L. Hoch (1960). *J. Biol. Chem.* **235**, PC45.

Vallee, B. L., J. F. Riordan, and J. E. Coleman (1963). *Proc. Nat. Acad. Sci. U.S.* **49**, 109.

Williams, R. J. P. (1960). *Nature* **188**, 322.

RICHARD H. STANFORD, JR.
ROBERT B. COREY
Gates and Crellin Laboratories of Chemistry
California Institute of Technology
Pasadena, California

Determination of the Structure of Proteins by X-Ray Diffraction: Possible Use of Large Heavy Ions in Phase Determination

A knowledge of the structure of proteins has long been recognized as fundamental to an understanding of their chemical and biological functions. The mechanism of the action of enzymes and hormones, the formation of antigen-antibody complexes, and the behavior of many viruses are directly related to the three-dimensional configuration of the proteins of which they are composed. During the past two decades, great progress has been made in the determination of the primary structure of proteins—the number and sequence of the amino acid residues of which the molecule is composed (Canfield, 1963; Edmundson, 1965; Harris et al., 1956; Hartley et al., 1965; Jolles et al., 1963; Matsuda et al., 1963; Walsh and Neurath, 1964). Progress has also been made in the description of certain fundamental elements of the secondary structure—the manner of folding and coiling of polypeptide chains (Pauling, Corey, and Branson, 1951; Pauling and Corey, 1951, 1953). Although this information is extremely important, it does not provide an accurate and detailed description of the mechanisms of enzymes, antibodies, and other proteins mentioned above. These mechanisms all depend upon the spatial interrelationships of various atoms and groups of atoms, and therefore upon the so-called tertiary structure (Linderstrom-Lang and Schellman, 1959)—the details of the three-dimensional configuration of the protein molecules.

Precise determinations of the primary structures of many protein molecules, combined with indications of their secondary structures, have provided a basis for speculation and tentative explanations of their molecular behavior. However, the only direct approach to a determination of the tertiary structure, and hence to a definite understanding of functional behavior, seems at the present time to be the technique of X-ray diffraction analysis.

Current X-ray methods for the successful determination of the structure of protein crystals were initiated by Perutz (Green, Ingram, and Perutz,

Contribution No. 3376 from the Gates and Crellin Laboratories of Chemistry. This work was supported principally by a contract (Nonr-220(38)) between the U.S. Office of Naval Research and the California Institute of Technology and by grants G-9467 and GB-3053 from the National Science Foundation.

1954) in his application of the heavy-atom technique which had been proposed by Robertson (1940) and used by him and numerous other investigators for the analysis of somewhat smaller and less complicated organic molecules. Perutz (Cullis et al., 1962a), Kendrew (1960), and their colleagues showed that heavy atoms or ions could be introduced by chemical combination, co-crystallization, or diffusion into definite positions in the crystal lattice, without significant alteration of the dimensions of the unit cell or its protein contents. They also demonstrated that if the positions of these heavy atoms could be determined, they could be used to calculate the phase angles of the diffracted X-ray beams, and thus lead to the derivation of the detailed structure of the protein crystal.

The outstanding example of the success of this technique is the determination of the structure of sperm whale myoglobin by Kendrew and his colleagues (1960), in which the positions of about 70 percent of the atoms in the molecule were established. A simultaneous study by Perutz and his co-workers of a much larger and more complex molecule, horse hemoglobin, yielded a definite picture of the secondary structure of its two polypeptide chains (Cullis et al., 1962b). Thus far, the only other detailed study of a protein in which definite spatial positions are assigned to the amino acid residues, is that of hen egg lysozyme carried out by Phillips and his co-workers (Blake et al., 1965). Investigations of other proteins are in progress in many laboratories throughout the world. An excellent review of this subject has been published by Dickerson (1964).

SOME PROBLEMS IN THE APPLICATION OF THE ISOMORPHOUS HEAVY-ATOM TECHNIQUE TO PROTEINS

One of the prime requirements of the heavy-atom technique is that the atoms or groups of atoms introduced into the structure for determining the phases of the X-ray reflections shall not significantly distort the crystal lattice or cause any displacement or change in configuration of the molecules. In early applications of this technique to proteins, the heavy atom (generally a metal) was chemically attached to an organic molecule or radical, which in turn was combined with a particular amino acid residue of the protein (Cullis et al., 1962a). Crystals were then grown of the resulting metallo-organic-protein complex. Conspicuous success was attained in certain instances, but the organic molecular link between heavy atom and protein frequently caused significant changes in the spatial geometry of the crystal. This was especially true when the organic compound was a relatively large dye molecule. Later it was found that heavy ions could be introduced into specific positions in the crystal lattice by cocrystallization with the protein or by direct diffusion into the protein crystal. The introduction of comparatively small heavy ions, especially by diffusion, has little if any significant effect on the molecular configuration or packing in the crystal; larger ions, which will be discussed in detail below, may cause spatial changes in the crystal, and give rise to other problems associated with their orientation.

X-ray data from the protein crystal and from a single isomorphous derivative are sometimes sufficient for the calculation of approximate phase angles from which a significant low-resolution picture of the protein molecule may be obtained (Blow and Rossmann, 1961). For even moderate accuracy, phase angles derived from at least two isomorphous derivatives are required. To minimize the effect of the unavoidable errors in even the best methods for data collection, it is very desirable that the phase angles used for the calculation of high-resolution electron density plots be based on data obtained from several isomorphous heavy-atom derivatives. There is a great difference between various proteins in the ease with which such derivatives can be obtained. Some have been especially uncooperative in this respect (King et al., 1956, 1962; Avey et al., 1962).

X-ray data from a heavy-atom derivative, when compared with those from the parent crystal, may indicate that the heavy atom has taken up a specific position in the crystal lattice and that its occupancy of this position, or site, extends to all equivalent positions related by the symmetry, and throughout all unit cells of the crystal. In an ideal situation of this sort, the number of heavy atoms per unit cell may be related to the number of protein molecules by the simple ratio of one to one. Sometimes the data will indicate that more than a single crystallographic site is occupied, and the ratio may be two to one, or slightly higher, but so long as the occupancy of all sites is close to 100 percent the situation shows little ambiguity and the positions of the heavy atoms, together with their high occupancies, can be used with confidence for the determination of phase angles (Bodo et al., 1959). However, the situation sometimes arises in which the data suggest that several different crystallographic sites are occupied by the heavy atom, and that the degree to which the various sites are filled differs widely (Cullis et al., 1962a; Kartha et al., 1963). In such a case a great deal of caution must be exercised in the interpretation of the data. Subsequent attempts to refine simultaneously the parameters and occupancies of the sites could lead to relatively meaningless results, and ultimately to correspondingly unreliable phase angles and electron density plots.

Proteins selected for detailed structural investigation are all of relatively small molecular weight (35,000 or less). In fact, the two proteins that have so far yielded information regarding the configuration of the polypeptide chain and the identity of individual amino acid residues—myoglobin and lysozyme—have molecular weights of only 17,000 and 14,500, respectively. However, many, if not most, of the proteins of outstanding biological interest have molecular weights that are greater by an order of magnitude, 150,000 or more. The higher the atomic number of the heavy atom or group of atoms introduced, the more it will contribute to the X-ray scattering and therefore to the differences in the intensities of corresponding reflections from crystals of the native protein and of the heavy-atom derivative. These differences will in turn determine the ease and definiteness with which the position of the heavy atom or group of atoms can be established. Similarly, if the atoms introduced do not scatter sufficiently, their contribution to the intensities of the reflections may be of almost the same order of magnitude as the errors of measurement, and the certainty with which their positions can be fixed will be greatly reduced. This resulting uncertainty could give rise to corresponding errors in the phase angles, with consequent significant errors and ambiguities in the resulting structure. There is some doubt if the introduction of mercury or uranium ions into a crystal of gamma globulin, for example, would contribute enough to the X-ray spectrum of the resulting isomorphous derivative to permit the accurate determination of the atom's position and the phases necessary for the delineation of the structure. On the other hand, complex ions made up of several heavy atoms would scatter much more strongly and might provide a solution to this particular problem. But these complex ions would doubtless introduce other problems, one of which is mentioned in the following paragraph.

If isomorphous derivatives of protein crystals contain single heavy atoms only, these atoms will scatter as point sources, so that no orientation effect is involved. However, ions made up of several heavy atoms, such as HgI_4^{--} or PtI_6^{--}, scatter as a complex, and X-ray reflections corresponding to spacings of the order of magnitude of the interatomic distances between the constituent atoms will be more or less strongly affected by the orientation of the ion in the crystal. The intensities of reflections corresponding to smaller spacings, and thus to a higher order of resolution, will be even more sensitive to this effect. It is interesting to note that although HgI_4^{--} ions were apparently successfully introduced into crystals of both myoglobin and lysozyme, these ions were abandoned as a means of phase determination in the later stages of both investigations (Blake et al., 1965; Bodo et al., 1959). However, if a practical method could be worked out that would permit the use of heavy, complex ions of this sort, they might be found to be extremely helpful in getting at the structures of certain proteins, especially those having large molecular weights. Thus, if such ions not only assumed specific positions in the crystal lattice but also assumed specific orientations, and if these orientations could be definitely determined, then the X-ray scattering contributed by them could be calculated and used with confidence in the calculation of the phase angles of all observed reflections.

For several years we have been engaged in an investigation of the structure of crystals of lysozyme chloride by the use of large isomorphous complex ions of tantalum and niobium, $Ta_6Cl_{12}^{++}$ and $Nb_6Cl_{12}^{++}$, in which the possibility of solving the problem of ion orientation has become the principal objective (Corey et al., 1962; Stanford et al., 1962). The course of this investigation is outlined briefly in the following section. Although a definite solution must await further study, the results thus far are sufficiently encouraging to justify reasonable optimism.

A STUDY OF LYSOZYME CHLORIDE CRYSTALS CONTAINING THE COMPLEX IONS $Ta_6Cl_{12}^{++}$, $Nb_6Cl_{12}^{++}$, AND PtI_6^{--}

Our investigations of the structure of lysozyme chloride crystals began at a time when several X-ray crystallographic laboratories working on the structure of protein crystals were having particular difficulty in obtaining isomorphous derivatives. Professor Linus Pauling and his collaborators had recently determined the configurations and dimensions of the complex ions $Ta_6Cl_{12}^{++}$ and $Nb_6Cl_{12}^{++}$ by X-ray diffraction studies of concentrated alcoholic solutions of the chlorides (Vaughan et al., 1950). They had found these ions to be so nearly identical in dimensions that crystals which differed only in containing one or the other of them might be expected to be precisely isomorphous throughout their structure. Professor Pauling therefore suggested them to us

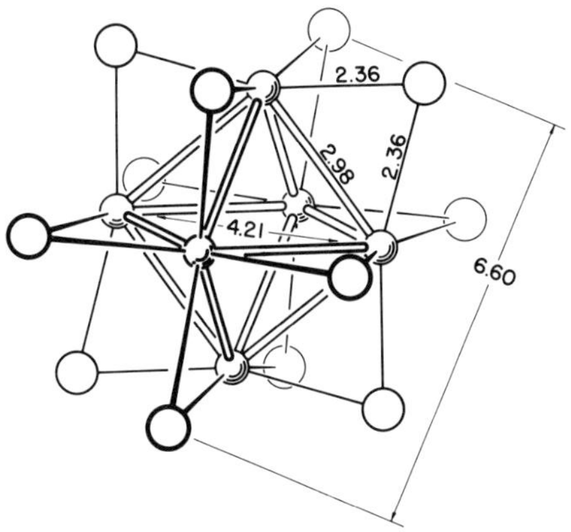

FIGURE 1.
The configuration and dimensions of the complex ions $Ta_6Cl_{12}^{++}$ and $Nb_6Cl_{12}^{++}$. Small shaded spheres represent the metal atoms at corners of an octahedron; unshaded circles represent chlorine atoms. Dimensions are in Ångströms.

for incorporation into lysozyme chloride crystals. The dimensions of these ions are shown in Figure 1. Although the ions are apparently too large to diffuse readily into preformed crystals of lysozyme chloride, they are easily incorporated by cocrystallization. Determinations of the crystal densities showed that the number of each complex ion in the tetragonal unit cell was eight, the same as the number of lysozyme molecules.

The coordinates of the centers of the ions in the crystals were readily derived from vector plots calculated from the intensities of the X-ray reflections. Inspection of the X-ray data also showed that the arrangement of the protein in crystals of lysozyme chloride had been significantly altered by the addition of the large heavy ions; therefore the crystals containing the tantalum and niobium complexes could not be regarded as isomorphous derivatives of those of the parent protein. However, the crystals containing the two different ions were found to be highly isomorphous, and hence those containing $Ta_6Cl_{12}^{++}$ could be regarded as an isomorphous derivative of those containing $Nb_6Cl_{12}^{++}$, the difference being the difference in scattering powers of the tantalum and niobium octahedra (Corey et al., 1962).

It had been shown that for reflections extending to a minimum spacing of 5 Å the contribution of the tantalum and niobium octahedra to the X-ray scattering was practically independent of their orientation (Stanford, 1962). X-ray intensity data extending to 5 Å resolution were therefore used to calculate a three-dimensional electron density plot based on the approximate phase angles obtainable from the single isomorphous replacement. As anticipated, no details of the protein structure were revealed at this resolution (Stanford, et al., 1962).

In order to obtain more definite and accurate phase angles, attempts were made to obtain another isomorphous derivative by the diffusion of other ions into the crystals of lysozyme which contained $Ta_6Cl_{12}^{++}$. The greatest success was attained with the iodoplatinate ion, PtI_6^{--}. Data from crystals containing both the Ta_6 Cl_{12}^{++} and the PtI_6^{--} ions were used to derive specific and more accurate phase angles for the X-ray reflections. A new three-dimensional electron density plot calculated to 5 Å resolution with the improved phase angles was in general similar to the one described above. In both of these calculations the $Ta_6Cl_{12}^{++}$ ion was assumed to be in a particular orientation with respect to the crystal lattice. To test the effect of the orientation of the ions at higher resolution, two additional plots were calculated from data extending to 4 Å. In the first, the tantalum octahedron was assumed to be definitely oriented as before; in the second a random orientation was assumed. These two plots were so strikingly different in the regions of protein density that it was very evident that the orientation of the heavy ion must be established before any further progress toward the details of the protein structure could be made.

An attempt was first made to find the orientation of the tantalum and niobium octahedra by calculating the contributions of tantalum and niobium to the X-ray scattering for 27 nonequivalent orientations generated by successive rotations of the octahedra through 30°. The differences between the scattering of the tantalum octahedra and the niobium octahedra in each orientation were then compared with the observed differences in the scattering derived from the intensities of selected reflections. When the discrepancies between the calculated and the observed values for each orientation were compared, they were found to vary in a regular fashion with the angle of orientation and to approach a minimum value corresponding to a specific orientation of the octahedron. Finally, this position was made more definite by a least-squares refinement based on observed intensity data of nearly 1000 centric reflections extending to 2.0 Å resolution. An orientation for the PtI_6^{--} ion was obtained directly by a least-squares procedure.

The calculated contributions of the heavy metal ions in their newly refined orientations and positions were used to determine the phases of the reflections from the crystals of lysozyme containing the tantalum ions. These phases were used for the calculation of a three-dimensional electron density plot based on data extending to minimum spacings of 1.8 Å. An attempt to interpret this plot in terms of the known composition of the lysozyme molecule (Canfield, 1963; Jolles et al., 1963; Brown, 1964) was unsuccessful. Even though sequences of 3 or 4 amino acid residues could be fitted quite well, it was not possible to proceed with confidence much further along the chain.

The nearly spherical shape of the tantalum and niobium complex ions still raised the question of possible random orientation. An electron-density plot was therefore calculated from the same data as those used in the 1.8 Å plot described above, but the phase angles were those obtained from randomly oriented ions. This new plot showed almost none of the features that made it possible for us to attempt an interpretation of the former one. Indeed, they did not even suggest the features of a polypeptide chain. The assumption of an intermediate structure, one in which the orientation of the ions varied statistically around the specifically preferred orientation, resulted in a plot which was closely related in general appearance to that based on the oriented ions, although the features were less definite and less suggestive of specific amino acid residues.

Finally, an attempt was made to completely circumvent the problem of the orientation of the ions by using a statistical treatment of the X-ray intensity data, a method that has recently been used successfully in the analysis of much simpler crystals (Hauptman and Karle, 1956; Karle and Karle, 1964). The resulting electron density plots bore no resemblance to the previous ones and were not even suggestive of protein material.

The best electron density plot is certainly that based on the selected specific orientations of the ions. Many attempts were made to identify sequences of amino acid residues that might lead to a complete interpretation of this plot. Certain regions strongly suggested disulfide bridges and their adjacent sections of polypeptide chain. Comparisons between molecular models and contours of the plot were used for tentative identification of particular disulfide bridges; additional residues were then added in their proper sequence until no further progress could be made because of lack of agreement between the known chemical composition and the contours of the map. Even though these attempts at interpretation were unsuccessful, they did indicate strongly that this electron density plot was close to an interpretable representation of the structure. The inaccuracies of the contours of the plot are without doubt to be attributed to two principal sources of error: inaccuracies in the quantitative X-ray data on which the entire determination is based, and small, but significant, errors in the determination of the orientation of the ions. Tests indicate that our measurements of X-ray intensities are probably as accurate as one could expect to obtain from visual estimations of reflections on Weissenberg photographs, which have been our only source of data. An X-ray diffractometer has now been installed in our laboratories and is available for the collection of an equivalent set of more accurate data. Hopefully, such data will permit significant refinement of the parameters and the orientations of the large heavy ions, thus contributing added accuracy to the protein portions of the electron density plot.

Plans are already under way for the collection of new data. We hope that use of these data may lead to a satisfactory understanding of the behavior of large heavy ions when incorporated into protein crystals, and may possibly establish a helpful technique for the use of such ions in determining the detailed structure of large protein molecules of biological importance.

ACKNOWLEDGMENTS

We wish to express our appreciation to Professor Linus Pauling for suggesting the use of the tantalum and niobium complexes and for his interest and encouragement, to Dr. R. E. Marsh, Mr. G. E. Haven, and Mrs. B. J. Stroll for providing helpful assistance in computer programming, to Mrs. P. S. Clauser and Mrs. L. Samson for the prepartaion of crystals and the collection of intensity data, to Miss L. I. Casler and Mr. D. D. Dill for the preparation of electron density plots and the construction of molecular models, and to Miss A. M. Kimball for the recording of data and reports and the preparation of manuscripts.

REFERENCES

Avey, H. P., C. H. Carlisle, and P. D. Shukla (1962). *Brookhaven Symp. Biol.*, No. 15 (BNL 738(C-34)), 199.

Blake, C. C. F., D. F. Koenig, G. A. Mair, A. C. T. North, D. C. Phillips, and V. R. Sarma (1965). *Nature* (London) **206**, 756.

Blow, D. M., and M. G. Rossmann (1961). *Acta Cryst.* **14**, 1195.

Bodo, G., H. M. Dintzis, J. C. Kendrew, and H. W. Wyckoff (1959). *Proc. Roy. Soc.* (London) A **253**, 70.

Brown, J. R. (1964). *Biochem. J.* **92**, 13P.

Canfield, R. E. (1963). *J. Biol. Chem.* **238**, 2698.

Corey, R. B., R. H. Stanford, Jr., R. E. Marsh, Y. C. Leung, and L. M. Kay (1962). *Acta Cryst.* **15**, 1157.

Cullis, A. F., H. Muirhead, M. F. Perutz, M. G. Rossmann, and A. C. T. North (1962a). *Proc. Roy. Soc.* (London) A **265**, 15.

Cullis, A. F., H. Muirhead, M. F. Perutz, M. G. Rossmann, and A. C. T. North (1962b). *Proc. Roy. Soc.* (London) A **265**, 161.

Dickerson, R. E., (1964). In Neurath, H., ed., The Proteins, 2nd ed., vol. 2, p. 603. New York: Academic Press.

Edmundson, A. B. (1965). *Nature* (London) **205**, 883.

Green, D. W., V. M. Ingram, and M. F. Perutz (1954). *Proc. Roy. Soc.* (London) A **225**, 287.

Harris, J. I., F. Sanger, and M. A. Naughton (1956). *Arch. Biochem. Biophys.* **65**, 427.

Hartley, B. S., J. R. Brown, D. L. Kauffman, and L. B. Smillie (1965). *Nature* (London) **207**, 1157.

Hauptman, H., and J. Karle (1956). *Acta Cryst.* **9**, 45.

Jolles, J., J. Jauregui-Adell, and P. Jolles (1963). *Biochim. Biophys. Acta* **78**, 68.

Karle, I. L., and J. Karle (1964). *Acta Cryst.* **17**, 835.

Kartha, G., J. Bello, D. Harker, and F. E. DeJarnette (1963). Protein Structure, *Proc. Symp. Madras*, 13.

Kendrew, J. C., R. E. Dickerson, B. E. Strandberg, R. G. Hart, D. R. Davies, D. C. Phillips, and V. C. Shore (1960). *Nature* (London) **185**, 422.

King, M. V., B. S. Magdoff, M. B. Adelman, and D. Harker (1956). *Acta Cryst.* **9**, 460.

King, M. V., J. Bello, E. H. Pignataro, and D. Harker (1962). *Acta Cryst.* **15**, 144.

Linderstrom-Lang, K. U., and J. A. Schellman (1959). In Boyer, P. D., H. Lardy, and K. Myrback, eds., *The Enzymes*, 2nd ed., vol. 1, p. 443. New York: Academic Press.

Matsuda, G., R. G. Mueller, and G. Braunitzer (1963). *Biochem. Z.* **338**, 669.

Pauling, L., R. B. Corey, and H. R. Branson (1951). *Proc. Nat. Acad. Sci. U.S.* **37**, 205.

Pauling, L., and R. B. Corey (1951). *Proc. Nat. Acad. Sci. U.S.* **37**, 729.

Pauling, L., and R. B. Corey (1953). *Proc. Nat. Acad. Sci. U.S.* **39**, 253.

Robertson, J. M., and I. Woodward (1940). *J. Chem. Soc.*, 36.

Stanford, R. H., Jr. (1962). *Acta Cryst.* **15**, 805.

Stanford, R. H., Jr., R. E. Marsh, and R. B. Corey (1962). *Nature* (London) **196**, 1176.

Vaughan, P. A., J. H. Sturdivant, and L. Pauling (1950). *J. Am. Chem. Soc.* **72**, 5477.

Walsh, K. A., and H. Neurath (1964). *Proc. Nat. Acad. Sci. U.S.* **52**, 884.

JOSEPH KRAUT
GERALD STRAHS
STEPHAN T. FREER
Department of Chemistry
University of California
San Diego, California

An X-Ray Anomalous Scattering Study of an Iron Protein from *Chromatium D*

Within the past few years, the entire class of nonheme iron proteins has received steadily increasing attention. This interest has been stimulated by the realization that proteins of this type are widely distributed, and that they serve important functions in a variety of biological oxidation-reduction processes. Nonheme iron proteins have been isolated from such diverse sources as aerobic bacteria (Shethna et al., 1964), anaerobic photosynthetic bacteria (Bartsch, 1963; Yamanaka and Kamen, 1965; Arnon, 1965), non-photosynthetic anaerobic bacteria (Mortenson et al., 1962; Lovenberg et al., 1963; Lovenberg and Sobel, 1965), blue-green algae (Arnon, 1965), chloroplasts of green plants (San Pietro and Lang, 1958), and mammalian tissues (Omura et al., 1965). One or another of these proteins is thought to participate in photosynthesis (Arnon, 1965), nitrogen fixation (Mortenson, 1964; Hardy et al., 1965), CO_2 assimilation (Buchanan et al., 1965; Raeburn and Rabinowitz, 1965), mitochondrial electron transport (Ringler et al., 1963) and oxidative phosphorylation (Butow and Racker, 1965).

Undoubtedly, any attempt to generalize concerning a subject undergoing such rapid development would be fruitless, especially since the proteins of this group are so diverse. Indeed, they seem to share few characteristics, other than the obvious ones that they all contain iron without heme and that they all undergo reversible oxidation and reduction. For example, most of these proteins contain "labile" sulfur—that is, sulfur that can be liberated as H_2S upon acidification, in an amount equivalent to the iron content of the protein; but rubredoxin does not (Lovenberg and Sobel, 1965). Further, their iron content varies over a wide range, even within the ferredoxin group alone; clostridial ferredoxin, with molecular weight 6000, contains 7 iron atoms, but spinach ferredoxin, with molecular weight 13,000, has only 2 (Arnon, 1965). Again, many of these proteins have quite low molecular weights, in the range just cited for the ferredoxins, but others—as, for example, liver aldehyde oxidase (Rajagopalan et al., 1962)—are complex molecules with molecular weight around 300,000, and contain FAD, CoQ, and molyb-

This work was supported by grants from the National Institutes of Health, the National Science Foundation, and the American Cancer Society.

denum in addition to iron and labile sulfur. Finally, the redox potentials of these proteins also vary widely. Those of the ferredoxin type have an E_0' of about -0.40 volts (Arnon, 1965), rubredoxin has $E_0' = -0.057$ volts (Lovenberg and Sobel, 1965), and the iron protein of *Chromatium D*, which is the subject of this study, has $E_0' = +0.35$ volts (Bartsch, 1963). In the face of such diversity, it is not surprising that Palmer and Brintzinger (1966) have concluded, from a survey of the available physicochemical data, that the chemical environments of the iron atoms in the various proteins of this class must all be somewhat different.

Some of the nonheme iron proteins are especially attractive subjects for study by the methods of X-ray crystallography. They can be induced to grow well-formed single crystals, they have low molecular weights, and the iron atoms should be readily identifiable, even without complete solution of the structure at atomic resolution. Sieker and Jensen (1965) have already reported that they can see no evidence, in a three-dimensional Patterson function, for the presence of a linear array of iron atoms in *Micrococcus aerogenes* ferredoxin. Such an arrangement had been proposed for *Clostridium pasteurianum* ferredoxin by Blostrom et al., (1964). However, Sieker and Jensen did not draw any conclusions concerning the correct iron atom arrangement.

The nonheme iron protein that is the subject of this study was first isolated by Bartsch (1963) from the photosynthetic purple sulfur bacterium *Chromatium D*. It has, since then, also been isolated from the photosynthetic purple nonsulfur bacterium *Rhodopseudomonas gelatinosa* (de Klerk and Kamen, 1966). Because of its high redox potential of $+0.35$ volts, it has been christened high-potential-iron-protein (HiPIP) to distinguish it from the ferredoxins. The molecule contains four iron atoms and four equivalents of acid-labile sulfur, and has a molecular weight of 10,000 (Dus et al., 1967). Redox titrations indicate that only one electron is removed upon oxidation (Bartsch, 1963). Although HiPIP can be isolated in high yields from both organisms, at the moment of this writing it is without known function.

The objective of the study reported here was to find out whether the geometrical arrangement of the iron atoms in HiPIP could be determined from Bijvoet-difference measurements (measurements of the intensity differences between reflections hkl and $\bar{h}\bar{k}\bar{l}$) without attempting to solve the rest of the structure. According to theory, such differences will occur whenever the crystal contains atoms with an absorption edge slightly to the long-wavelength side of the X-ray wavelength. This is because scattering from those atoms takes place with a small "anomalous" advance in phase (Buerger, 1960). The phenomenon was first employed by Bijvoet to establish the absolute configuration of optically active molecules.

The magnitude of the anomalous scattering effect is always small relative to normal scattering, and indeed it is quite negligible for carbon, nitrogen, and oxygen atoms at the usual X-ray wavelengths. But it begins to be more readily measurable for some of the elements of the third row and beyond. When anomalous scattering is not negligible, the scattering factor for an atom may be written as

$$f = f_0 + \Delta f' + i\Delta f''$$

where f_0 is the scattering factor appropriate to a wavelength far from an absorption edge, $\Delta f'$ is the real part of the correction due to anomalous scattering, and $\Delta f''$ is the imaginary part of the correction term. It is this imaginary part that gives rise to the Bijvoet differences.

For iron atoms and Cu K_α radiation, $\Delta f' = -1.1$ electrons and $\Delta f'' = 3.4$ electrons, at low scattering angles. One is therefore dealing with a small perturbation, such as might, for example, be caused by isomorphous substitution of four nitrogen atoms into the protein molecule. Based upon the assumption that Wilson statistics will hold for the actual distribution of both iron and the other atoms, a simplified estimate predicts the average relative Bijvoet difference will be in the neighborhood of 10 percent in the intensities. Since this is about the level of experimental error in protein crystallography, there is some reason to question whether we can hope to locate the iron atoms by analysis of their anomalous scattering effects. However, Zachariasen (1965) has shown that reliable measurements of Bijvoet differences can be made for quartz, in which anomalous scattering due to the silicon atoms, with Cu K_α X-rays, gives an average $\Delta I/I$ of only 7 percent. Extreme care had to be exercised, however, including such steps as grinding a quartz crystal to an almost perfect sphere in order to minimize absorption errors. In the realm of protein crystallography, Cullis et al. (1961) have employed anomalous scattering effects due to the iron atoms in hemoglobin to establish its absolute configuration, and similar use has been made of the HgI_4^{--} group in a derivative of chymotrypsinogen (Kraut et al., 1964). In both cases the locations of the anomalous scatterers were already known.

What of the case in which the anomalous scatterers are in unknown locations? Rossmann (1961) has used anomalous scattering to locate the mercury atoms in two hemoglobin derivatives, independently of isomorphous replacement data. And, most recently, Matthews (1966) has combined anomalous scattering with isomorphous replacement to locate iodine and platinum atoms in α-chymotrypsin derivatives. Thus there has been adequate precedent to suggest that similarly small anomalous effects due to the iron atoms in HiPIP may be put to good use.

Finally, a further motivation for this study has been the potential application of anomalous scattering to provide supplementary phasing information in isomorphous replacement work. It has been suggested by North (1965)— and the idea has been further developed by Matthews (1966)— that even such small intensity differences as we are dealing with here may be more accurately measurable, and therefore more useful in phasing, than might at first be supposed. To explore the practical limitations, we have exercised considerable care in the measurement of reflection intensities, including application of empirical absorption corrections and extensive replication. We therefore feel that our results represent the level of precision and accuracy that may reasonably be expected from the present state of experimental technique in protein crystallography.

EXPERIMENTAL

Crystals

Suitably large prismatic crystals ($0.1 \times 0.1 \times 1.0$ mm) of reduced HiPIP were grown in about two to three weeks from a solution of the protein in approximately 40 percent saturated ammonium sulfate. The solution was prepared as follows. Enough reduced protein to give a concentration of 0.5 percent was dissolved in 0.01 M Tris buffer, *p*H 8.0, containing a 10-fold excess of mercaptoethanol. Solid ammonium sulfate was then added slowly until a permanent precipitate of protein was obtained. Crystals grew during subsequent slow evaporation of the solution.

The crystals are dark greenish brown, the larger ones being almost opaque. In contrast to the heme proteins, no dichroism is visible under the polarizing microscope, suggesting that the chromophoric center is not highly anisotropic.

The space group is $P2_12_12_1$, with a = 42.5 Å, b = 42.0 Å, and c = 38.0 Å. The long prism axis of the crystal corresponds to the unit cell *b*-axis. Given a molecular weight of 10,000 (Bartsch, 1963; Dus et al., 1967), and assuming one molecule per asymmetric unit (four per unit cell in this space group), the volume per molecular weight unit is 1.7 $Å^3$—somewhat low, but within the range found for other protein crystals.

Precession photographs show reflections out to a Bragg spacing of 2 Å or less.

Crystals of *oxidized* HiPIP can also be obtained by a similar method. The oxidized crystals are reddish brown, rather than greenish brown. Precession photographs taken with oxidized crystals are almost indistinguishable from those taken with reduced crystals in each of four reciprocal lattice planes, suggesting that only very small structural changes, if any at all, accompany the change in oxidation state of the HiPIP molecule. All intensity data were collected on reduced crystals.

Intensity data

Copper K_α intensities were measured, out to 4 Å, with a General Electric XRD-5 equipped with a quarter χ-circle single-crystal orienter. The θ-2θ scan technique was used, with stationary left and right backgrounds subtracted for each reflection. Crystals were cut to a length of 0.3-0.5 mm and were mounted in thin-walled glass capillaries with the *b*-axis parallel to the ϕ-axis of the single-crystal orienter. Six reference reflections were monitored periodically, and crystals were discarded whenever their intensities decreased by more than 10 percent.

For HiPIP there are a total of 675 unique reflections in an octant of reciprocal space out to 4 Å; 443 of these are general *hkl*'s and may show Bijvoet differences. A total of 6171 intensity measurements (5001 on the general *hkl*'s) were made, using 13 different crystals. Thus each Bijvoet difference was measured, on the average, 5.6 times. The largest differences were redetermined more often than the smaller, some as many as ten or more times. No Bijvoet difference was determined less than 3 times, or with fewer than 2 crystals. Replications were made of a given reflection on an individual crystal, a given reflection on different crystals, and of symmetry-related reflections on an individual crystal. Intensity data for different crystals were scaled together by a least-squares procedure similar to that proposed by Hamilton, Rollett, and Sparks (1965).

Absorption correction

An empirical absorption correction was applied to all data in a way now described. For each crystal, 36 initial intensity measurements were made on the 0,2,0 reflection ($\chi = 90°$), with ϕ set at intervals of 10° from 0° to 350°. The resulting function of ϕ was normalized to a maximum of unity. For convenience, we refer to these normalized functions as τ-curves. Each reflection intensity subsequently measured on this crystal was then corrected for absorption effects by dividing the observed intensity by a factor interpolated from the τ-curve at the appropriate value of ϕ. As a check, the τ-curve was redetermined just before discarding a crystal in several cases, and was always found to be unchanged.

For most HiPIP crystals, the τ-curve has a form that crudely approximates the function

$$\tau = (1 - a)\cos^2(\phi + \alpha) + a$$

where a and α are constants characteristic of the crystal size and orientation. For a typical case, the value of a is about 0.6.

An empirical correction of this type is probably more accurate than calculating absorption corrections from crystal size, shape, and orientation, as it would be difficult to take into account the effects of the capillary wall and of adhering liquid. The procedure used here is expected to be valid so long as all reflections are at small 2θ and so long as the τ-curve truly represents an absorption effect. It is probably worthwhile to consider this last assumption rather carefully, since the τ-curve could arise, at least in part, from Renninger effects, extinction, or instrumental misalignment. Our reasons for believing it is primarily a result of absorption are as follows.

1. The depth of the trough in a typical τ-curve—the constant a is generally in the range 0.5 to 0.7—corresponds well with an effective difference in X-ray path length of 0.3 mm between the directions of maximum and minimum absorption. Even for the largest crystals employed, about half of the absorption is due to X-rays passing tangentially through the 0.01-mm-thick capillary wall.
2. Larger, more anisometric crystals yield τ-curves with deeper troughs.
3. When the capillary containing a crystal is rotated with respect to the ϕ-circle, the τ-curve is translated by the same amount.
4. The τ-curve for a given crystal has a deeper trough when determined with chromium K_α instead of copper K_α X-rays.
5. The 0,2,0 and 0,10,0 reflections yield almost identical τ-curves.

In any case, the ultimate test of any absorption correction is its influence on the results. As will be seen in the following, our procedure appears to have been at least partially effective. A similar empirical absorption correction has been used by Ueki, Zalkin, and Templeton (1965).

Precision of intensities

Various statistics were compiled in the hope of identifying some of the major sources of error and unreproducibility. We will refer to several types of relative mean $\sigma(I)$, defined for our purposes as $\Sigma\,\sigma(I)/\Sigma\,\bar{I}$, where the I have been drawn from some specified portion of the measurements on that intensity, and the summation is over all reflections. Several significant facts are apparent from the statistics.

1. The general precision of an intensity measurement is even poorer than we had thought, the relative mean $\sigma(I)$ for all replicates being 17 percent.

Based on counting statistics alone, a figure of 5 percent would have been expected.

2. A large part of this unreproducibility is due to differences among crystals, even though all were presumably identical. The relative mean $\sigma(I)$ for sets of measurements on different crystals was 18 percent, the relative mean $\sigma(I)$ for sets of measurements on symmetry-related reflections from a given crystal was 9 percent, and for repeated measurements of a given reflection on a single crystal it was 8 percent. Indeed, a few reflections were observed to be remarkably variable; the 10,0,0, for example, had a relative $\sigma(I)$ of 72 percent over 13 crystals. This variability probably results from slightly differing internal salt concentrations caused by the inevitable irregularities of crystal growing and mounting.

3. The empirical absorption correction improves reproducibility in all cases, but it most important for relating intensities measured on different crystals. The absorption correction reduced the relative mean $\sigma(I)$ for symmetry-related replicates from 10.4 to 9.0 percent and the relative mean $\sigma(I)$ for crystal-to-crystal replicates from 24.8 to 17.8 percent.

Bijvoet differences

The observed intensity for each reflection was taken to be the unweighted mean for all replicate measurements. The relative mean Bijvoet difference

$$\frac{\Sigma\,|I(h) - I(\bar{h})|}{\Sigma\,\frac{1}{2}\{I(h) + I(\bar{h})\}}$$

was 9 percent, about as expected.

During the course of collecting the data, the 52 largest and most obvious Bijvoet differences were selected for extra replications. A striking peculiarity was that only 4 of the 52 lay between the limits of 5 Å and 4 Å. Indeed, a plot of the mean Bijvoet differences as a function of sin θ/λ shows a much more rapid decline with increasing Bragg angle than a similar plot for intensities.

The basic analytical tool employed in this study was the Bijvoet-difference Patterson function—that is, the Fourier summation with coefficients $\{|F(h)| - |F(\bar{h})|\}^2$. This function was first discussed by Rossmann (1961), who employed it to locate the mercury atoms in two derivatives of hemoglobin. Roughly speaking, the Bijvoet-difference Patterson can be thought of as having coefficients proportional to $F_i^2 \cos^2 \gamma$, where F_i is the structure factor for the "imaginary" parts of the anomalously scattering atoms, and γ is the difference in phase angle between the latter and the structure factor for the "real" atoms. If $\cos^2 \gamma$ always has a fixed value, then, of course, the Bijvoet-difference Patterson would be just the straightforward Patterson function of the anomalously scattering atoms, and it would not be difficult, in the present case, to solve it by inspection. However, γ is effectively a random variable, and so the Bijvoet-difference Patterson represents a folding of the desired Patterson function with an unknown "noise Patterson" whose coefficients are $\cos^2 \gamma$. Fortunately, the "noise Patterson" will *probably* have an overwhelmingly dominant origin peak, and thus the resulting folded function will *probably* resemble the true Patterson being sought.

Three different Bijvoet-difference Patterson functions were calculated. They contained, respectively, all terms to 5 Å and all terms to 4Å, and only the 52 terms with the largest and most reliable coefficients. In addition, similar calculations were made from a single set of unreplicated data, and from data that had not been corrected for absorption. Finally, conventional Patterson functions were also computed for comparison.

RESULTS AND DISCUSSION

Three very prominent peaks situated at (u,v,w) coordinates of $(\frac{1}{2},0,\frac{1}{2})$, $(0,\frac{1}{2},\frac{1}{3})$, and $(\frac{1}{2},\frac{1}{2},\frac{1}{6})$ appear on all Bijvoet-difference Patterson maps we have calculated. On the 52-term map, which we consider to be the most reliable, they have peak heights of 64, 34, and 29, relative to an origin peak height of 100. Each of the three peaks is almost spherical. The obvious interpretation is that all four iron atoms are clustered more or less symmetrically about a single site, centered at $(x,y,z) = (0,0,\frac{1}{12})$.

If there were no other features of comparable height in the Bijvoet-difference Patterson maps, this interpretation would be unavoidable. Unfortunately, other moderately strong features are apparent. For example, in the 52-term map, the next three largest peaks have heights of 22, 18, and 17, and there are many somewhat lower still. Nevertheless, we believe these features are spurious, probably resulting from the nature of the Bijvoet-difference Patterson function, as described in the previous section. There are three principal reasons for this conclusion.

1. The three highest peaks on the 5-Å conventional Patterson map coincide with the locations of the three characteristic peaks on the Bijvoet-difference Patterson map. The conventional Patterson function was calculated with coefficients $\frac{1}{2}\{I(h) + I(\bar{h})\}$. Presumably the iron-cluster to iron-cluster vectors are its most prominent features. The remaining peaks of the Bijvoet-difference map do not correlate at all with the convential Patterson map. This is illustrated in Figure 1.

2. Bijvoet-difference Patterson maps were computed successively from (a) unreplicated data with no absorption correction; (b) replicated data, but still with no absorption correction; (c) unreplicated data, but *with* absorption corrections; (d) replicated, absorption-corrected data; and (e) the 52 largest terms out of 443 in (d). The three characteristic peaks increased steadily in height relative to the origin, in the order (a) to (e), but other features generally decreased. The order here is worth noting, as it is strong evidence that the empirical absorption correction was at least partially effective.

3. Early in the investigation, before the presently proposed structure was recognized, Patterson superposition methods were applied in a determined effort to arrive at an arrangement for the iron atoms, involving four separate sites. Two somewhat different solutions were achieved, both with pseudo-space-group symmetry $B22_12$, and both with the present cluster site treated as a single iron atom lying on a pseudo-two-fold axis parallel to c. In both solutions, the atoms were widely separated, the closest approach being 8 Å. Neither solution was really satisfactory, however. For one thing, neither could adequately explain all the features of the Bijvoet-difference Patterson. For another, both necessarily involved more than four iron sites—one five and

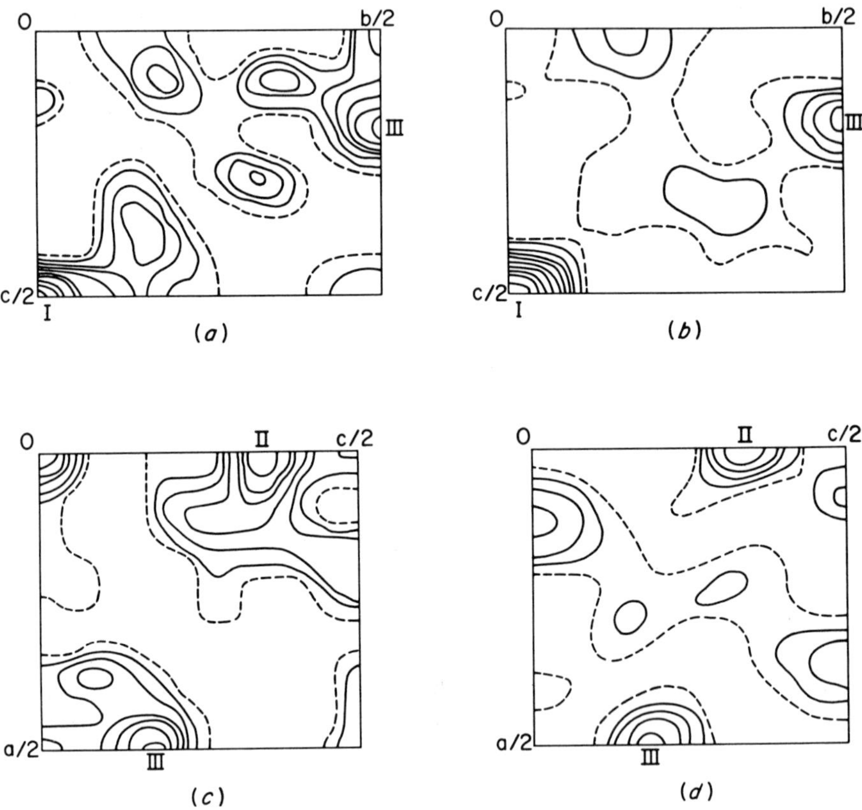

FIGURE 1.
Comparison of conventional 5 Å Patterson function with 52-term Bijvoet-difference Patterson function. The three characteristic peaks are labeled I, II, and III. Contours are drawn at arbitrary intervals, such that peak I has the same number of contours on both maps. The dashed contour is at level zero. (*a*) Patterson, section $u = \frac{1}{2}$. (*b*) Bijvoet-difference Patterson, section $u = \frac{1}{2}$. (*c*) Patterson, section $v = \frac{1}{2}$. (*d*) Bijvoet-difference Patterson, section $v = \frac{1}{2}$.

the other six. Finally, there was no apparent reason for the observed falloff of Bijvoet differences.

For these reasons, we conclude that all four iron atoms of HiPIP lie clustered about a single site, possibly in a tetrahedral arrangement, with the individual atoms too close together (perhaps 3–4 Å) to be resolved. This structure would also explain the rapid decline of the Bijvoet differences with increasing Bragg angle, as resulting from an expected rapid falloff of the structure factor for such a cluster (Stanford, 1964).

If we may be permitted to conclude with a flurry of speculation, it should be mentioned that the black Roussinate ion, $Fe_4S_3(NO)_7^-$, is a suggestive model for the iron cluster of HiPIP. Blomstrom et al. (1964) have focused attention on the Roussin compounds in discussing the iron binding of *C. pasteurianum* ferredoxin. We note that the four iron atoms of the black Roussinate ion are arranged in a trigonal pyramid, with Fe(apex)· · ·Fe(base) = 2.7 Å, and Fe(base)· · ·Fe(base) = 3.6 Å. Each of the three sulfur atoms is, in turn, situated slightly above a pyramid face (Johansson and Lipscomb, 1958). A structure of this kind would, of course, appear to be more or less spherical at 4 Å resolution. If a similar moiety were present in HiPIP, it would account for the strong color and for the fact that HiPIP slowly forms complexes with nitric oxide (Bartsch and Strahs, unpublished data).

To wax even more speculative, consider the disposition of iron clusters in the HiPIP crystal. By virtue of the space-group symmetry $P2_12_12_1$ and of the peculiar coordinates $(x,y,z) = (0,0,\frac{1}{12})$ of the iron cluster in the first asymmetric unit, it follows that the clusters are arranged in a way that approaches cubic close packing. This may, of course, be pure coincidence. A more interesting explanation, however, would be that the iron clusters lie at the approximate centers of roughly spherical protein molecules, which are in turn stacked in an almost cubic-close-packed array. Indeed, this is further suggested by the near equality of the unit-cell edges and by the fact that, for the low-order (to 10 Å) 19 reflections, the $h + l$ evens are, on the average, 2.5 times more intense than the $h + l$ odds.

Again, all this may be coincidence. Fortunately we need not suffer the pangs of uncertainty forever, since the question can be settled, eventually, by isomorphous replacement methods.

ACKNOWLEDGMENTS

We are grateful to Dr. Robert G. Bartsch for supplying us with generous amounts of HiPIP, and for his advice on methods of crystallization. We thank Mr. Jim Lance for help with data collection.

REFERENCES

Arnon, D. I. (1965) *Science* **149**, 1460.

Bartsch, R. G. (1963). In Gest, H., A. San Peitro, and L. P. Vernon, eds., *Bacterial Photosynthesis*, p. 315. Yellow Springs, Ohio: The Antioch Press.

Blomstrom, D. C., E. Knight, W. D. Phillips, and J. F. Weiher (1964). *Proc.·Nat. Acad. Sci.* **51**, 1085.

Buchanan, B. B., M. C. W. Evans, and D. I. Arnon (1965). In San Pietro, A., ed., *Non-Heme Iron Proteins*, p. 175. Yellow Springs, Ohio: The Antioch Press.

Buerger, M. J. (1960). *Crystal Structure Analysis*, p. 542. New York: John Wiley & Sons.

Butow, R., and E. Racker (1965). In San Pietro, A., ed., *Non-Heme Iron Proteins*, p. 383. Yellow Springs, Ohio: The Antioch Press.

Cullis, A. F., H. Muirhead, M. F. Perutz, M. G. Rossman, and A. C. T. North (1961). *Proc. Roy. Soc.* (London) **A 265**, 15.

Dus, K., H. de Klerk, K. Sletten, and R. G. Bartsch (1967). *Biochim. Biophys. Acta.* **140,** 291.

Hamilton, W. C., J. S. Rollett, and R. A. Sparks (1965). *Acta Cryst.* **18**, 129.

Hardy, R. W. F., E. Knight, C. C. McDonald, and A. J. D'Eustachio (1965). In San Pietro, A. ed., *Non-Heme Iron Proteins*, p. 275. Yellow Springs, Ohio: The Antioch Press.

Johansson, G., and W. N. Lipscomb (1958). *Acta Cryst.* **11**, 594.

de Klerk, H., and M. D. Kamen (1966). *Biochim. Biophys. Acta* **112**, 175.

Kraut, J., D. F. High, and L. C. Sieker (1964). *Proc. Nat. Acad. Sci. U.S.* **51**, 839.

Lovenberg, W., and B. E. Sobel (1965). *Proc. Nat. Acad. Sci. U.S.* **54**, 193.

Lovenberg, W., B. B. Buchanan, and J. C. Rabinowitz (1963). *J. Biol. Chem.* **238**, 3899.

Matthews, B. W. (1966). *Acta Cryst.* **20**, 230.

Mortenson, L. E. (1964). *Biochim. Biophys. Acta* **81**, 473.
Mortenson, L. E., R. C. Valentine, and J. E. Carnahan (1962). *Biochem. Biophys. Res. Commun.* **7**, 448.
North, A. C. T. (1965). *Acta Cryst.* **18**, 212.
Omura, T., E. Sanders, D. Y. Cooper, O. Rosenthal, and R. W. Estabrook (1965). In San Pietro, A., ed., *Non-Heme Iron Proteins*, p. 401. Yellow Springs, Ohio: The Antioch Press.
Palmer, G., and H. Brintzinger (1966). *Nature* **211**, 189.
Raeburn, S., and J. C. Rabinowitz (1965). In San Pietro, A., ed., *Non-Heme Iron Proteins*, p. 189. Yellow Springs, Ohio:The Antioch Press.
Rajagopalan, K. V., I. Fridovich, and P. Handler (1962). *J. Biol. Chem.* **237**, 922.
Ringler, R. L., S. Minakami, and T. P. Singer (1963). *J. Biol. Chem.* **238**, 801.
Rossmann, M. G. (1961). *Acta Cryst.* **14**, 383.
San Pietro, A., and H. M. Lang (1958). *J. Biol. Chem.* **231**, 211.
Shethna, Y. I., P. W. Wilson, R. E. Hansen, and H. Beinert (1964). *Proc. Nat. Acad. Sci. U.S.* **52**, 1263.
Sieker, L. C., and L. H. Jensen (1965). *Biochem. Biophys. Res. Commun.* **20**, 33.
Stanford, R. H. (1964). *Acta Cryst.* **17**, 1180.
Ueki, T., A. Zalkin, and D. H. Templeton (1965). *Acta Cryst.* **19**, 157.
Yamanaka, T., and M. D. Kamen (1965). *Biochem. Biophys. Res. Commun.* **18**, 611.
Zachariasen, W. H. (1965). *Acta Cryst.* **18**, 714.

THE CHEMISTRY OF PROTEINS

GIOVANNI GIACOMETTI
Institute of Physical Chemistry
University of Padua
Padua, Italy

Recent Experimental Approaches to the Thermodynamics of Conformational Transitions in Polypeptides

More than fifteen years ago Linus Pauling (1951) suggested the α helix as a most stable secondary structure for protein molecules. Since then, experimental evidence as to the real existence of helical structures in many native protein molecules has accumulated at a very fast rate, and it has become more evident that the disruption of this characteristic structure is one of the important features of the process of protein denaturation. This belief, together with the understanding of the biological necessity of particular conformations in functional proteins, makes it extremely important to study the forces involved in stabilizing these secondary structures.

Decisive steps forward in the investigation of these forces were made with the measurements of physical properties of solutions containing synthetic polypeptides such as the poly-α-aminoacids. Not only are these systems very simple models of proteins but it was soon recognized that they constitute as well ideal substrates for a model process of protein denaturation.

For a number of these polymers there exists sufficient evidence, mainly from X-ray investigations, that their solid state structure is an α helix of the Pauling type (Schellman and Schellman, 1964; Katchalski et al., 1964). The structural situation of these polymer molecules when dispersed in a solvent is known with much less certainty, owing to the fact that such knowledge must rely on the discussion of solution properties such as viscosity, light scattering, flow birefringence, optical rotatory dispersion, ultraviolet absorption, and so on. Nonetheless it is now generally agreed that, in a number of cases, the conformation of the molecules when dispersed in weakly interacting solvents (dimethylformamide, chloroform, ethylene dichloride, and others) or in aqueous media of appropriate acidity, is still a rigid structure of the helical type (Schellman and Schellman, 1964; Katchalski et al., 1964).

By varying gradually the nature of the medium (addition of a strongly interacting solvent such as dichloro- or trifluoroacetic acid, in the case of organic solvents; *p*H changes, in the case of aqueous solutions) or by changing the temperature, we often observe a sudden change in the physical properties of the solutions. These sharp "transitions" are ascribed to a conformational change occurring in the polymer molecules from the rigid helical structure to a nonrigid, random situation, and they are referred to as "helix to coil transitions."

The reader is referred to a number of comprehensive reviews in which both the experimental and the theoretical features of this class of facts are most authoritatively discussed (Schellman and Schellman, 1964; Katchalski et al., 1964; Scheraga, 1963; Singer, 1962).

The purpose of the present article is to discuss the information that can be obtained as to the nature and entity of the forces responsible for the stability of the rigid conformations in these systems and, especially, to point out how much of this information relies on the theoretical model chosen to describe the transition. It will be shown that the crucial test of the theoretical models is the direct measurement of the changes in the thermodynamic functions involved. A review of the very few measurements of this type made up to the present will be given and lines of possible future research will be proposed.

THE MODEL OF THE HELIX-COIL TRANSITION

The numerous statistical approaches to the thermodynamics of the helix to coil transition in polypeptides* are essentially based on the same monodimensional Ising model, differing only in the mathematics used for computing the partition function and in minor details.

The model consists of considering the polypeptide as a directional chain of segments (equal in number to the number of monomers in the polymer molecule) whose possible states are as the following.

1. Segment in the "coil" state.
2. Segment in the "helix" state, preceded (with respect to a given direction of the chain) by one or more segments in the "coil" state.
3. Segment in the "helix" state, preceded (with respect to the same direction of the chain) by one or more segments in the "helix" state.

Each of these segment "states" has its own statistical weight. The characteristic of direction provides a model of the possible helical nature of the real chain. The difference in statistical weight between a "coil" and a "helix" state is a model for describing the interaction which stabilizes the particular conformation of the amino-acid residue giving origin to the helical structure. (In the Pauling α helix this interaction is thought to be a hydrogen bond between the NH group of the amino-acid residue and the CO group of the third preceding residue.) The difference among "helix" states—according to their being or not being at the beginning of a helical section of the chain—gives a model of the configurational constrictions, created by the existence of the particular interaction which stabilizes the helical state, on the preceding residues. (In the Pauling α helix, for example, it is impossible to have a hydrogen bond to the third preceding residue if the two residues in between are not in some particular conformation.)

Beside minor differences in the details of the models employed by the various authors, which bear no influence on the results (at least in the limit of long chains), it is important to note that the model is absolutely unspecific as to the nature of the stabilizing forces. Furthermore, the model is absolutely

*A very comprehensive list of the literature on this subject is contained in Scheraga (1963).

unspecific with respect to the functioning of the solvent in its interaction with the polypeptide.

Perhaps the clearest interpretation of the two basic parameters of the model is that given by Applequist (1963). The first parameter is the equilibrium constant for the transformation of a coil segment into a helix segment *after* (in the direction of the chain) already existing helix segments; it is given by

$$s = \exp\left(\frac{-\Delta F_{\mathrm{I}}}{RT}\right) \tag{1}$$

Here ΔF_{I} is the free-energy change in process (I)

$$\cdots\mathrm{EE}\overrightarrow{\mathrm{EC}}\cdots \rightleftarrows \cdots\mathrm{EE}\overrightarrow{\mathrm{EE}}\cdots \tag{I}$$

where E indicates a helix segment and C a coil segment. The second parameter may be described as the equilibrium constant for the formation of an interruption in a sequence of helix segments by a process that maintains a constant number of helix segments; it is given by

$$\sigma = \exp\left(\frac{\Delta S_{\mathrm{II}}}{R}\right) \tag{2}$$

where ΔS_{II} is the entropy change in process (II)

$$\cdots\mathrm{EEEE}\overrightarrow{\mathrm{EC}}\cdots \rightleftarrows \cdots\mathrm{EEC}\overrightarrow{\mathrm{EEE}}\cdots \tag{II}$$

Equation (2) suggests that this process is one in which enthalpy can be thought of as remaining constant given a constant number of stabilizing bonds, which are thought to be formed when a segment is in the helix state.

The evaluation of the partition function (Applequist, 1963) allows us to obtain the average fraction θ of helix segments in a long chain as a function of s and σ:

$$\theta = \frac{1}{2} + \frac{(s-1)}{2[(1-s)^2 + 4s\sigma]^{1/2}} \tag{3}$$

Figure 1 shows the behavior of θ as a function of s for different values of the parameter σ, which, as may be noted, determines the "sharpness" of the transition.

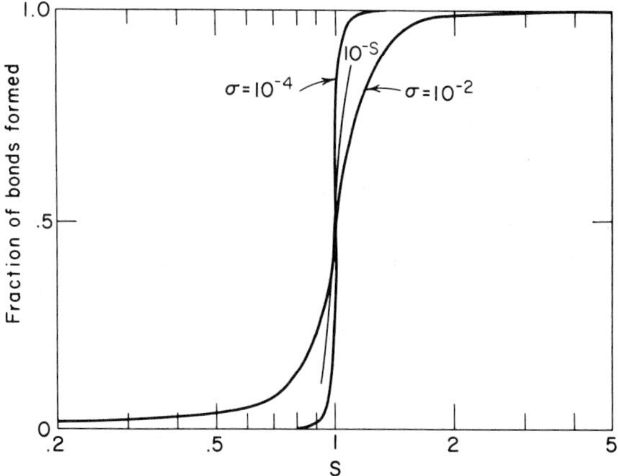

FIGURE 1.

Equation (3) can be expanded in a Taylor series in θ and, for $\sigma \ll 1$, we obtain for the derivative

$$\left(\frac{\partial \theta}{\partial \ln s}\right)_{s=1} \cong \frac{1}{4\sigma^{1/2}} \tag{4}$$

This equation can be used to discuss the thermodynamics of the process.

THE THERMODYNAMICS OF THE HELIX-COIL TRANSITION

There are two possible general ways of utilizing Equation (4) to obtain information on the changes in the thermodynamic functions during the transition. The first is to study the variation of a property of the solution, which is proportional to θ, as a function of the temperature at constant solvent composition. The second way is to hold the temperature constant and to vary the solvent composition.

The emphasis has been up to now on the use of the first approach, which gives directly the clue to the information desired.

By inserting Equation (1) into (4) and writing

$$\Delta F_{\mathrm{I}} = \Delta H_{\mathrm{I}} - T\,\Delta S_{\mathrm{I}} \tag{5}$$

we obtain

$$\left(\frac{\partial \theta}{\partial (1/T)}\right)_{T=T_0} = -\frac{\Delta H_{\mathrm{I}}}{4R\sigma^{1/2}} \tag{6}$$

where T_0 is the transition temperature at which

$$\Delta H_{\mathrm{I}} = T_0\,\Delta S_{\mathrm{I}} \tag{7}$$

Zimm, Doty, and Iso (1959) were able to separate the values of ΔH_{I} and of σ by analyzing values of optical rotatory power of solutions of poly-ϕ-benzyl-L-glutamate (PBG) in a solvent mixture of dichloroethane (DCE) and dichloroacetic acid (DCA) (20:80 weight percent) at various temperatures and for molecular weights running from a few thousands up to 350,000. They had to use also the results of the theory valid for short chains and to assume that the value of σ was independent of molecular weight. They found $\Delta H_{\mathrm{I}} = 890 \pm 120$ calories per mole of monomer unit (cal/m.u.) and $\sigma = 2 \times 10^{-4}$, with a somewhat larger margin of percent error. I shall only observe at this point that these results may be very much dependent on the model, which for short chains is sensitive also to minor details. Another important point to be noted is that a positive value of ΔH_{I} ("inverted transition") is only understandable if the solvent system plays an essential role in the transition in such a way that formation of bonds within the polypeptide chain is compensated by disruption of bonds between the polymer and the solvent.

This observation brings us to the second approach to thermodynamic information—from θ versus solvent composition curves. This approach necessitates a little more detailed discussion of the model with respect to the role of the solvent. The problem has been tackled by Peller (1959), both for changes in the solvent mixture and in the pH (in case of aqueous solutions), and more recently by Snipp, Miller, and Nylund (1965) for the aqueous solution. I shall only discuss the case of changes in the solvent mixture in a very simple way, but the methodological conclusions hold rather generally.

Referring, for example, to the PBG-DCE-DCA system, we may consider the two components of the solvent as having very different roles: one (DCE) is a real inactive diluent; the other (DCA) is an active component of the solution with the ability of forming bonds (perhaps hydrogen bonds) with polypeptide residues not bonded intramolecularly, and hence in the "coil" state. If we symbolize by A one molecule of the active solvent, we may redescribe process (I) as

$$\cdots \mathrm{E{-}E{-}E{-}}\overset{\displaystyle \mathrm{A}}{\overset{|}{\mathrm{C}}}\cdots \rightleftarrows \cdots \mathrm{E{-}E{-}E{-}E}\cdots + \mathrm{A} \tag{I}$$

The parameter s must be reinterpreted as

$$-RT \ln s = \Delta F_{\mathrm{I}} + RT \ln X_{\mathrm{A}} \tag{8}$$

where X_{A} is the mole fraction of active solvent and any departures from ideality of the solution are neglected. Also, the value of ΔH_{I} must be reinterpreted accordingly and is now comprehensive also of the heat of dissociation of the C—A bond, so that a positive value may be understood.

Equations (4) and (8) then give us, at constant temperature,

$$-\left(\frac{\partial \theta}{\partial \ln X_{\mathrm{A}}}\right)_{X_{\mathrm{A}}=X_{\mathrm{A}}{}^{\circ}} = \frac{1}{4\sigma^{1/2}} \tag{9}$$

where $X_{\mathrm{A}}{}^{\circ}$ is the solvent composition at mid-transition for the given temperature.

From a plot of a quantity proportional to θ versus X_{A} at constant temperature, it is then possible to obtain a value of σ. By performing such an experiment on the PBG-DCE-DCA system at 30°C ($X_{\mathrm{A}}{}^{\circ} = 0.8$), we obtain for σ a value of $\sim 6.0 \times 10^{-4}$ (Giacometti and Turolla, unpublished). Recalculating ΔH_{I} from the experiment of Zimm et al. (1959) ($T_0 = 28.7$°C) for the largest molecular weight and from the value of σ now obtained, we get $\Delta H_{\mathrm{I}} = 400$ cal/m.u. It is apparent that a combination of two sets of experiments, one at constant temperature and another at constant solvent composition, are able to give both σ and ΔH_{I}, the two thermodynamic parameters of the statistical model.

Karasz and O'Reilly (1967) have proposed a very similar kind of approach, obtaining for the same system, however, a quite higher ΔH value (around 800 to 1000 cal/m.u.).

Other models of solvent interaction would give different results. The already cited Peller (1959) model, for example, differs from the very simple one given above both for the stoichiometry (each coil site may associate two molecules of solvent), and for allowing solvent unbonded coil sites. In this case process (I) can be described by a series of steps:

$$\cdots\text{—}\overrightarrow{\mathrm{EEEC}}\text{—}\cdots \rightleftarrows \cdots\text{—EEEE—}\cdots \tag{I$_1$}$$

$$\cdots\text{—EEEC—}\cdots + \mathrm{A} \rightleftarrows \cdots \mathrm{EEEC}(\text{—A}) \cdots \tag{I$_2$}$$

$$\cdots\text{—EEEC}(\text{—A})\cdots + \mathrm{A} \rightleftarrows \cdots \mathrm{EEEC}(\text{—A})(\text{—A}) \cdots \tag{I$_3$}$$

Step [I_1] describes the conversion of a coil segment into a helix segment in the absence of solvent interactions (as if only the inert solvent were present); the other two steps describe the attachment of two molecules of solvent to the two hydrogen-bonding sites that are free when the intramolecular hydrogen bond is not formed (in the α helix these are the CO group of a peptide unit and the NH group of the third nearest neighbor peptide unit). If we assume the same association constant K for steps I_2 and I_3 and take into account the fact that there are two sites available for bonding in step I_2, the reinterpretation of the parameter s (the ratio of the number of E segments to the total number of C segments, bonded and unbonded) gives

$$RT \ln s = -\Delta F_{I_1} - 2RT \ln (1 + KX_A) \tag{10}$$

with the usual assumption of ideality of the solvent mixture. From Equations (10) and (4) we now obtain

$$-\left(\frac{\partial \theta}{\partial X_A}\right)_{X_A = X_A^{\circ}} = \frac{K}{2\sigma^{1/2}(1 + KX_A^0)} \tag{11}$$

which can give information about σ if we know the value of K. If the value of K is very large (all coil sites are solvent-bonded) we again obtain Equation (9) but for a stoichiometric factor of 2.

A third approach to ΔH_I and σ, but experimentally much more laborious, may be mentioned. It is represented by the study of the pressure dependence of the transition together with a dilatometric estimation of the volume changes involved. This approach has been also applied to the system PBG-DCE-DCA (Gill and Glogovsky, 1965), but the results are still inconclusive because of the small size of the volume changes to be measured.

In concluding this section we note that all these indirect approaches to the value of the thermodynamic parameters are very much dependent on the models chosen, both for the chain as a whole and for the solvent interaction mechanism, and that it becomes quite desirable to have direct measurements of the more accessible of the thermodynamic parameters—that is, ΔH_I.

CALORIMETRIC MEASUREMENTS OF ΔH_I

Although the phenomenon of the helix-to-coil transition received much attention after its discovery in 1954 (Doty et al.), it was ten years before direct measurement of the enthalpy change during the process was attempted. The first measurements were made almost at the same time on the system PBG-DCE-DCA by Karasz, O'Reilly, and Bair (1964) and by Ackermann and Rüterjans (1964).

These authors measured by a calorimeter the excess heat capacity of a 2 percent weight solution of PBG in a DCE-DCA mixture (80 percent volume DCA) at various temperatures. The kind of results they obtained is illustrated in Figure 2. The measurements are very delicate owing to the small amount of measured heat, which is two or three orders of magnitude smaller than a typical heat of fusion.

Ackermann and Rüterjans discovered a concentration effect on ΔH_I; their value, extrapolated to zero concentration, is 950 cal/m.u. The value found by Karasz and co-workers is much smaller, being 525 cal/m.u. for a concentration at which the value of the former authors is around 800 cal/m.u. Other measurements of this kind were made on the system poly-ϵ-carbobenzoxy-L-lysine in DCA-chloroform (Karasz et al., 1965) and on the system ribonuclease A in water at *p*H 2.8 (Beck et al., 1965), but the results need not be discussed here.

A totally different approach to the calorimetric measurements of ΔH_I was attempted by Giacometti and Turolla (1966), who measured the integral heats of solution of standardized films of PBG in different mixtures of DCA-DCE at constant temperature. The Calvet-Prat (1963) microcalorimeter is an ideal tool for such experiments, in which a small quantity of heat is released over a long time (the time necessary for the solution process to be completed). It allows the very precise measurement of a heat power as small as 0.1 cal/hour, thus allowing the measurements on PBG solutions to be carried out even at 0.2 percent weight concentration—too small for the excess heat capacity method.

The results obtained by Giacometti and Turolla are illustrated in Figure 3. The jump in the heat of solution at 80 percent volume DCA should be equivalent to ΔH_I and is measured to be approximately 650 cal/m.u., a value that is very near to the result of Karasz and co-workers if the difference in polymer concentration is taken into account. The method employing the heat of solution provides a further very important bit of information. It appears from the shape of the curve of Figure 3 that a plateau is reached at very small DCA concentrations, starting from the almost negligible heat of solution in pure DCE. This phenomenon brings into the picture effects arising from the aminoacid side chains, which have been neglected in our discussion so far but to which we must assign an essential role in the process, in consideration of the striking changes in the stability of the helices formed by different polyaminoacids. Nemethy, Leach, and Scheraga (1966) give a very interesting description of the way in which side chains, owing to their different steric requirements, influence both the entropy and the enthalpy of the process. The only direct thermal data available, besides those for PBG, are for PCBL

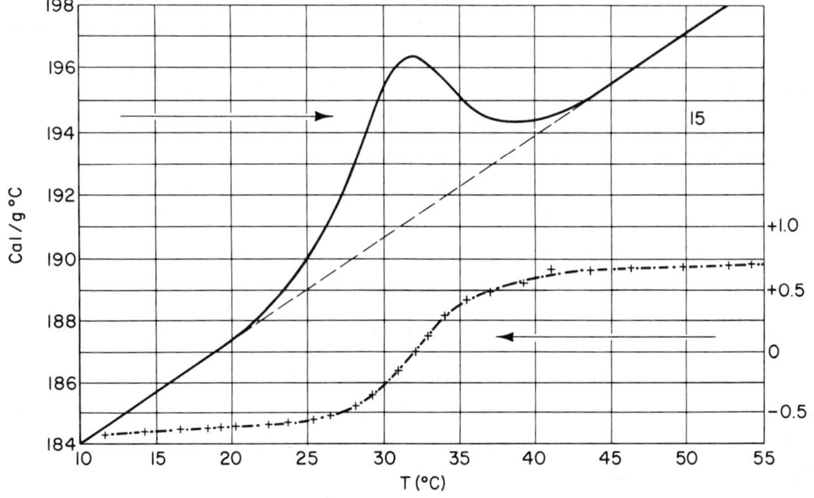

FIGURE 2.

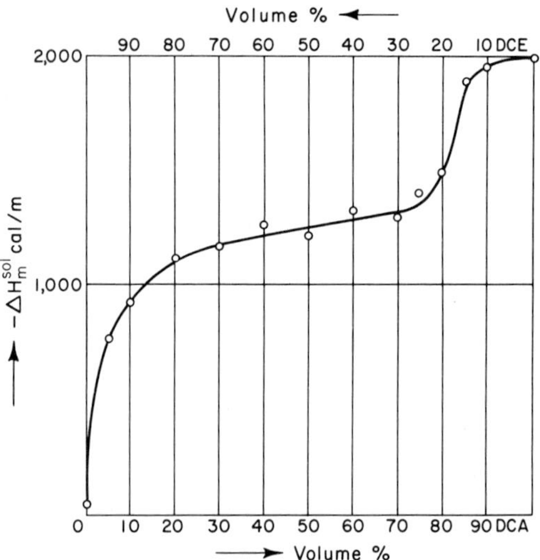

FIGURE 3.

(Karasz et al., 1965), and they indicate a value of ΔH_I around 200 cal/m.u. (to be compared with a lower limit of about 600 cal/m.u. for PBG). This means that interactions between the side chains or between the side chains and the solvent are of importance.

The evidence from the experiment illustrated in Figure 3 demonstrates that important changes occur in PBG solutions at low DCA concentrations, as compared with the solution in pure DCE. These changes obviously have to do with a preferential interaction of DCA with the side chains of the polymer, because they occur in a range of concentrations in which the helix is still a very stable structure.

The picture thus arises of the active solvent competing first with side chains interactions and then—but only after its activity has reached a larger value—competing with the hydrogen-bonding interactions of the backbone chain. This picture fits very nicely with some NMR evidence (Marlborough et al., 1965) showing, for PBG in a solvent mixture of $CDCl_3$ and trifluoroacetic acid, that the intensity of the signals develops gradually at first only for the protons of the side chains and then later for those of the backbone as the active solvent concentration is increased. Since the absence of lines in the NMR spectrum of a polymer is usually ascribed to the complete rigidity of the molecules, it seems clear that the gradual development of the various lines is due to "unfreezing" of the side chains from the periphery of the molecule inward, a process that has certainly to do with solvent interactions, at first with the side chains, and later with the backbone groups.

CONCLUSIONS

After this brief survey of the essential features of the experimental approaches to the problem of the thermodynamics of the helix-coil transition in polypeptides, we are in a position to make a few comments and to outline the needs of further work.

The existing data are far from being complete for quantitative explanation. Even those on the most studied PBG system, as we have recalled above, must be considered only as indicative because of the frequent lack of the necessary uniformity in the experimental conditions in different laboratories.

Furthermore, a thoroughly correct analysis of the θ versus composition curves will not be possible in the absence of the true thermodynamic activities of the active solvent component.

What we can say with some degree of certainty is that in general the monodimensional Ising model is an adequate phenomenological description of the helix-coil transition in polypeptides.

It is not yet proved that the parameter σ for a given system is really independent of the temperature within the margin required for its being purely entropic in character.

What we know for the moment about the σ's is that their value is certainly around 10^{-4} (within perhaps two orders of magnitude) and that it certainly depends (just as does ΔH_{I}) on the nature of the side chain. It is very probable that it depends also on the nature of the active solvent, while, on the contrary both σ and ΔH_{I} are likely to be independent on the nature of the inactive solvent.* This behavior would be in line with the postulated general model of solvent action. The order of magnitude of σ corresponds to a decrease in entropy for the coil segments when followed by helix segments ($-\Delta S_{\mathrm{II}}$) running between 23 and 13 e.u. per mole of residue. This is to be compared with the entropy change of process (I), which is of one order of magnitude smaller, being given according to Equation (5) by $\Delta S_{\mathrm{I}} = \Delta H_{\mathrm{I}}/T_0$, and hence, for the indicative values known up to the present, running around ± 1–3 e.u. per mole of residue (the positive sign corresponds to the case of inverted transitions).

In this respect the necessity of invoking the solvent as an essential component of the picture in order to understand the data is still more evident than through the mere existence of inverted transitions. ΔS_{II} corresponds to the entropy change when a truly "free" coil segment becomes a "constricted" coil segment, because it is followed by an helix segment. It follows that ΔS_{II} should give an upper limit to the change in entropy when a truly "free" coil segment "freezes" into an helix segment, a change that is formally given by ΔS_{I}. The small absolute values of ΔS_{I} are therefore understandable only if during process (I) solvent molecules are liberated, with a corresponding quite noticeable gain in configurational entropy. It is also apparent that considerations of the type invoked by Scheraga and co-workers (Nemethy et al., 1966) to estimate side chain effects on the entropy of the transition cannot neglect to take into account the solvation state of the system.

The road to the quantitative knowledge of the forces stabilizing helical structures in solution is laid out but is still to be paved, and the vehicle for traveling the road must be a detailed knowledge of the solvent action mechanism. It is possible that careful systematic investigations along the indirect approaches, together with very precise measurements of the enthalpy changes of many different systems, may allow a decision to be made on this point.

The idea underlying this suggestion is that through direct measurements of ΔH_{I} we may get a value of σ *without postulating a solvent action mechanism*, and that through the indirect approaches we are able to get both ΔH_{I} and σ, but *after making assumptions as to the solvent action mechanism.*

*Unpublished data both from Karasz's and the author's laboratory seem to substantiate this point.

REFERENCES

Ackermann, T., and M. Ruterjans (1964). *Zeit. fur Phys. Chem.* **41**, 116.
Applequist, T. (1963). *J. Chem. Phys.* **38**, 934.
Beck, K., S. T. Gill, and M. Downing (1965). *J. Am. Chem. Soc.* **87**, 901.
Calvet and Prat (1963). *Recent Progress in Microcalorimetry*. London.
Doty, P., A. M. Holtzer, J. M. Bradbury, and E. R. Blout (1954). *J. Am. Chem. Soc.* **76**, 4493.
Giacometti, G., and A. Turolla (1966). *Zeit. Phys. Chem.* **51**, 108.
Gill, S. J., and R. L. Glogovsky (1965). *J. Phys. Chem.* **69**, 1515.
Karasz, F. E., J. M. O'Reilly, and H. E. Bair (1964). *Nature* **202**, 693.
Karasz, F. E., J. M. O'Reilly, and H. E. Bair (1965). *Biopolymers* **3**, 214.
Karasz, F. E., and J. M. O'Reilly (1967). *Biopolymers* **5**, 27.
Katchalski, E., M. Sela, H. I. Silman, and A. Berger (1964). In Neurath, H., ed., *The Proteins*, vol. 2, p. 405. New York: Academic Press.
Marlborough, D. T., K. G. Orrell, and H. N. Rydon (1965). *Chem. Comm.*, 518.
Nemethy, G., S. J. Leach, and M. A. Scheraga (1966). *J. Phys. Chem.* **70**, 998.
Pauling, L., R. B. Corey, and H. R. Branson (1951). *Proc. Nat. Acad. Sci. U.S.* **37**, 205.
Peller, L. (1959). *J. Phys. Chem.* **63**, 1199.
Schellman, J. A., and C. Schellman (1964). In Neurath, H., ed., *The Proteins*, vol. 2, p. 1. New York: Academic Press.
Scheraga, M. A. (1963). In Neurath, H., ed., *The Proteins*, vol. 1, p. 476. New York: Academic Press.
Singer, S. J. (1962). *Adv. in Protein Chem.* **17**, 1.
Snipp, R. L., W. G. Miller, and R. E. Nylund (1965). *J. Am. Chem. Soc.* **87**, 3547.
Zimm, B. H., P. Doty, and K. Iso (1959). *Proc. Nat. Acad. Sci. U.S.* **45**, 1601.

G. N. RAMACHANDRAN
Centre of Advanced Study in Biophysics
University of Madras
Madras, India

Analysis of Permissible Conformations of Biopolymers

A new chapter was opened in the study of protein structure with the enunciation by Pauling and co-workers (Pauling et al., 1951; Pauling and Corey, 1951) of the existence of nonintegral helices in fibrous proteins and the postulation of the rules that govern the formation of such helical and other stable structures of polypeptides. Corey and Pauling (1953) also worked out the dimensions of the planar peptide group that forms the building block in the constitution of a protein or a polypeptide chain. These studies have been the foundations on which all later work on the stability of polypeptide chain conformations has been built. For instance, Pauling and his co-workers themselves (1951) worked out the now well-known α helix and the extended β structure, which together form the structural basis of the keratin-myosin-epidermin-fibrinogen group of fibrous proteins. Following the same two criteria—that only *trans* peptide groups can occur in protein chains and that as many NH groups as possible in the chain must form hydrogen bonds—I and my colleagues were successful in working out the molecular structure of the collagen group of fibrous proteins in the form of a triple helix (Ramachandran and Kartha, 1954, 1955; see Ramachandran and Sasisekharan, 1965, for the latest version of this). Almost all the fibrous proteins and polypeptides known so far exhibit either the α helix, the extended β structure, or the triple helical structure. The only fibrous protein that has yet defied a full understanding is feather keratin, for which widely different structures have been proposed—by Krimm and Schor (1956) and by Ramachandran and Dweltz (1962), both of which are unlikely to be correct, and recently by Fraser and his co-workers (1959, 1963, 1965). It is to be hoped that a rational application of the stereochemical principles laid down by Pauling will lead to a full interpretation of the beautiful X-ray pattern exhibited by this protein.

CONFORMATION OF POLYPEPTIDES

Although these studies succeeded in discovering some of the stable hydrogen-bonded conformations of polypeptide chains, the problem of working out all the allowed conformations of such a chain was not tackled until recently.

Contribution No. 208 from the Centre of Advanced Study in Biophysics, University of Madras, Madras 25.

The first of such studies—the working out of the allowed ranges of conformation for a pair of peptide units—was undertaken in our laboratory (Ramachandran et al., 1963), making use of certain criteria for the allowed lower limits for interatomic contacts between different types of atoms. The criteria developed in this paper (in terms of the so-called "normally allowed" and "outer limit" contacts; see Table 1) required contacts appreciably shorter than the classical sum of van der Waals radii, as given by Pauling (1945). However, the correctness of the criteria put forward by us has been amply verified by the experimental data on the conformations observed in various peptides and, in particular, in myoglobin and lysozyme.

In this approach, the polypeptide chain is considered to consist of "peptide units" having the Pauling-Corey dimensions, and the variety of conformations arises essentially from the freedom of rotation about the two single bonds N—C^{α} and C^{α}—C′ by which neighboring peptide units are joined at an α carbon atom. For a pair of peptide units, these two angles were denoted by Ramachandran et al. (1963) as ϕ and ϕ'. Recently, the conventions for representing these and other dihedral angles, such as those in the side groups, have been standardized (Edsall et al., 1966). A pair of linked units labeled according to these new conventions is shown in Figure 1, and the angles ϕ and ψ (ψ replaces ϕ' defined earlier) have each in principle a range of 0° to 360°. However, due to the contact criteria, the allowed regions are highly restricted in the (ϕ, ψ)-plane (Ramachandran et al., 1963; see Figure 2). It is found that most of the area of the map is forbidden; only about 20 percent is allowed when there is a side group with a β carbon atom attached to C^{α}. For a glycyl α carbon atom, the allowed range is rather larger (Ramakrishnan and Ramachandran, 1965).

TABLE 1.
List of normally allowed and outer limit contact distances (In Ångstroms).

Contact	Normally allowed	Outer limit
C···C	3.20	3.00
C′···C′	2.95	2.90
C···N	2.90	2.80
C···O	2.80	2.70
C···P	3.40	—
N···N	2.70	2.60
N···O	2.70	2.60
N···P	3.20	—
O···O	2.70	2.60
O···P	3.20	—
P···P	3.50	—
C···H	2.40	2.20
N···H	2.40	2.20
O···H	2.40	2.20
P···H	2.65	—
H···H	2.00	1.90

SOURCE: All contacts involving phosphorus atoms are from Sasisekharan et al. (1967); others are from Ramakrishnan and Ramachandran (1965).

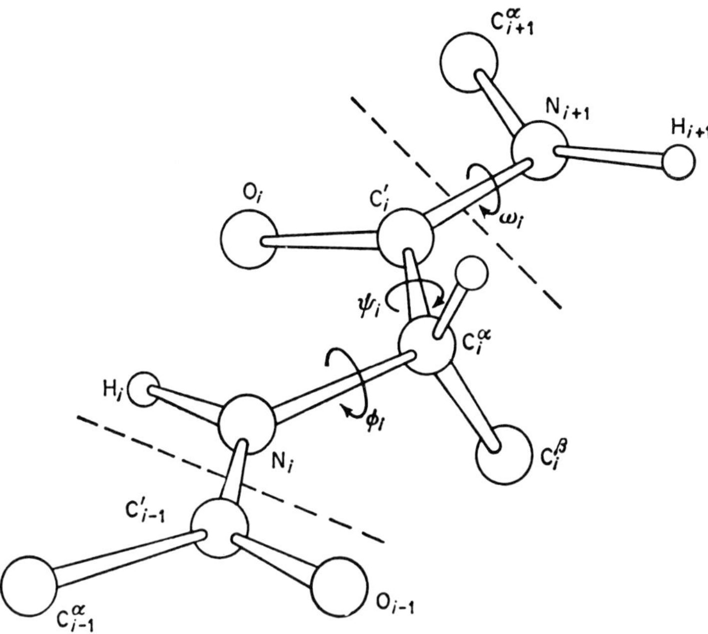

FIGURE 1.
Definition of the rotation angles ϕ and ψ for a pair of peptide units: $\phi = \psi = 0$ corresponds to the conformation shown in the diagram, which is that of a fully extended chain. [From Edsall et al., 1966.]

It is interesting to note that the (ϕ, ψ) map of allowed conformations is fully supported by the available data on the two crystalline proteins whose structure has been fully worked out (Figure 2, a and b). It will be noticed that almost all the points occur within the three allowed regions I, II, and III and the bridge connecting I and III across the line $\psi = 180°$. The few points that occur outside these allowed regions (such as G, G in Figure 2a) are conformations associated with a glycyl α carbon atom and are allowed for that case.

The contact criteria can readily be extended to longer side chains and to helices (Leach et al., 1966; Ramachandran et al., 1966), but it can be improved upon by considering the atoms to be compressible spheres and not "hard spheres," so that the interaction between a pair of atoms may be represented by a suitable potential function. The first attempt at applying the potential method to polypeptide conformation was by Liquori and his co-workers (De Santis et al., 1965), who calculated the potential energy per residue of a helical chain of poly-L-alanine and polyglycine with varying (ϕ, ψ). The interesting result emerged that the right-handed α helix is theoretically more stable than the left-handed α helix. More recently, the potential energy method has been extended by a series of workers (for example, Brant and Flory, 1965; Ramachandran et al., 1966; Scott and Scheraga, 1966), and they all lead to the result that the right-handed α helix is the more stable of the two. This is confirmed by the fact that only this form of the helix has been

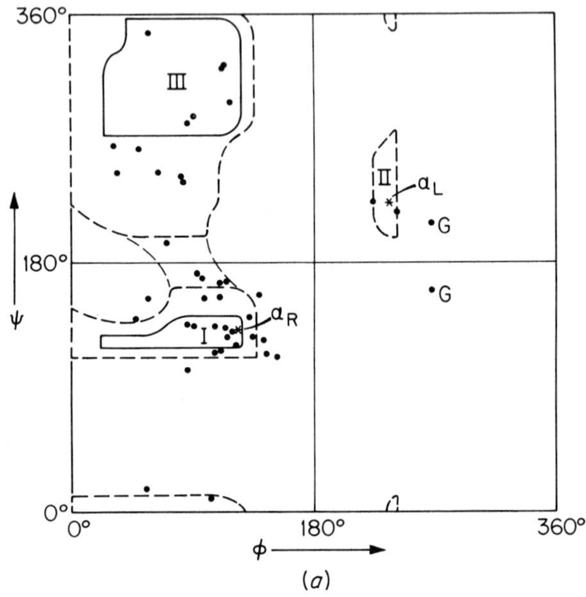

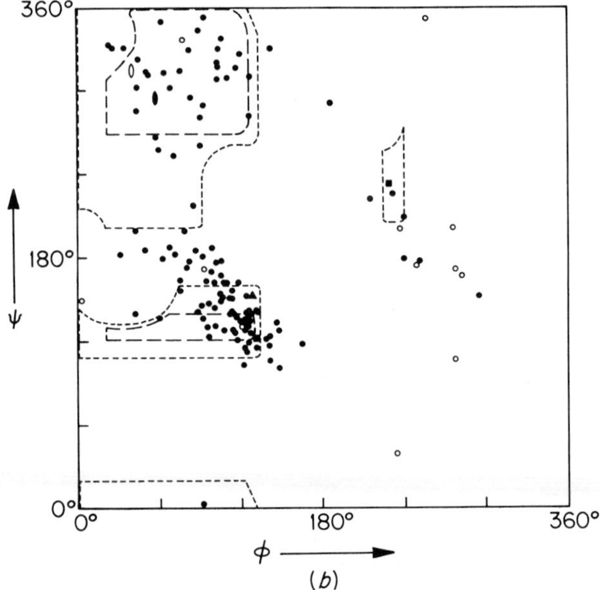

FIGURE 2.
(*a*) The (ϕ, ψ) map for a pair of linked peptide units with a β-carbon atom; —— normally allowed, ---- outer limits. The observed conformations in the nonhelical regions of myoglobin are also plotted as points. The conformation of right-handed and left-handed α helices are denoted by α_R and α_L. [Data by courtesy of Dr. H. C. Watson; from Ramachandran et al., 1966.] (*b*) The (ϕ, ψ) map with the observed conformations in lysozyme plotted on it. [Data by courtesy of Dr. D. C. Phillips; from Venkatachalam and Ramachandran, 1967.]

found in myoglobin and lysozyme and in all polypeptides except poly-β-benzyl aspartate (Bradbury et al., 1962).

The α helix has, however, not been observed for polyglycine, which has always been found to take up either the extended polyglycine I or the triple-helix-like polyglycine II structure. Therefore, an energy calculation was recently made for these three structures for polyglycine, and it indicated the greater stability of the last two (Venkatachalam and Ramachandran, 1967).

Recently a comparative study of the different potential functions that have been proposed has been undertaken in our laboratory (Venkatachalam and

Ramachandran, 1967). It shows that the main conclusions derived from contact criteria are supported by all the potential functions, but that the numerical values of the potential energy differ very much from one set of functions to the other. For example, the difference in energy per residue between right- and left-handed α helices is -2.0 kcal/mole/residue for the Flory functions (Brant and Flory, 1965; Ramachandran et al., 1966) and -2.5 for the Liquori functions (De Santis et al., 1965), but it is only -0.5 and -0.2 for the Scheraga functions (Scott and Scheraga, 1967) and the Kitaigorodsky functions (Kitaigorodsky, 1961; Venkatachalam and Ramachandran, 1967). It appears that reliable conclusions regarding the actual energies of different conformations can only be drawn after more detailed studies are made on the potential functions themselves to obtain the best set of these.

There is no doubt that the contact criteria can be used to eliminate a large number of disallowed conformations of a polypeptide or protein chain. This seems to be the method adopted in a generalized program being developed by Levinthal and his co-workers (C. Levinthal, private communication). However, one should be rather sceptical of the correctness of the conformations of *minimum energy* derived for large peptide chains (including side groups), because of the uncertainties regarding potential energy functions mentioned above. In fact, in a study of the conformation of sugars (see below, Conformations of Polysaccharide Chains) the Flory and Kitaigorodsky potential functions were found to give slightly different conformations for the minimum energy (Rao et al., 1967).

CONFORMATION OF NUCLEOTIDES AND NUCLEIC ACIDS

The success of the application of contact criteria to polypeptide chains has prompted us to apply them to polynucleotides and polysaccharides. In the polynucleotide chain, the repeating unit is not a rigid plane, as it is in a peptide unit. The unit of the backbone of this—the ribose-phosphate unit, which is taken to extend from the phosphorus atom P3 at the 3′-end to the phosphorus atom P5 at the 5′-end of the ribose—is shown in Figure 3, a and b. This unit is capable of three internal rotations about single bonds, which are marked $\theta 1$, $\theta 2$, $\theta 3$ in Figure 3a, following Sasisekharen at al. (1967). Although the ring of the ribose itself may take up more than one conformation, so far as the backbone is concerned, it may be taken to be a rigid unit. The angles $\theta 1$, $\theta 2$, $\theta 3$ may be suitably defined, and a set ($\theta 1$, $\theta 2$, $\theta 3$) defines the conformation of the ribose-phosphate unit. When a base is attached at C1′ to the ribose through the glycosidic bond, a fourth rotation, χ, about this bond is also possible, and the value of χ, together with the values of $\theta 1$, $\theta 2$, and $\theta 3$, defines the conformation of a nucleotide unit.

If now we wish to specify the conformation of a dinucleotide, we would require a pair (analogous to the case of two linked peptide units) of dihedral angles ϕ and ψ, which define the rotations about the single bonds O3′ (of previous unit)—P3 and P3—O5′ (of next unit) at the bridge phosphorus atom. A set of five such angles ($\theta 1$, $\theta 2$, $\theta 3$, ϕ, ψ) per nucleotide unit would completely specify the conformation of the "backbone" of a polynucleotide chain and the specification of the conformation of the ribose moiety and the angle χ in each unit would fix the corresponding "side chain."

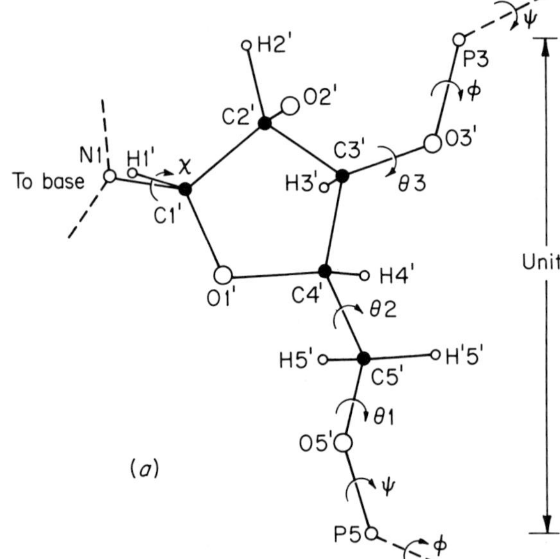

FIGURE 3.
(*a*) Atoms in the ribose-phosphate unit, showing the rotation angles $\theta 1$, $\theta 2$, $\theta 3$, χ, ϕ, and ψ. (*b*) Bond lengths and bond angles of the standard unit.

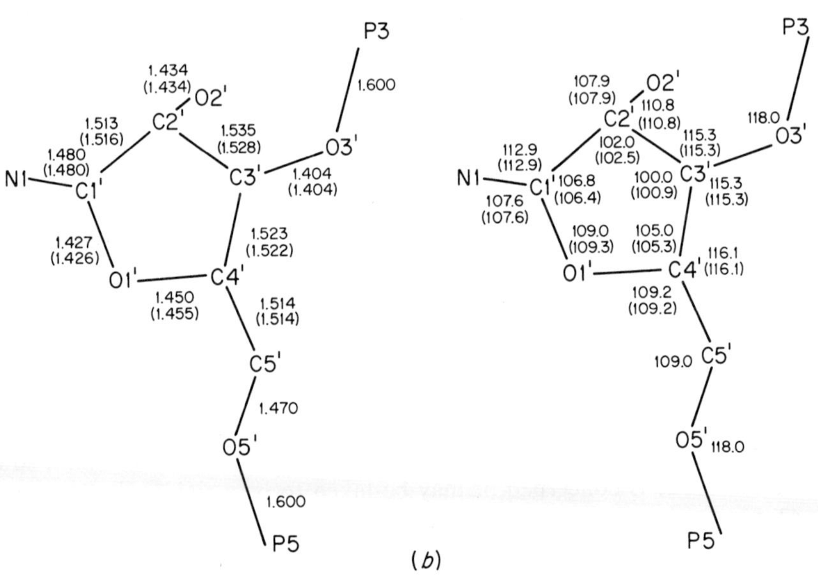

The allowed ranges of the angles $\theta 1$, $\theta 2$, $\theta 3$, and χ have been investigated recently in our laboratory (Sasisekharan et al., 1967), using the contact criteria. The most interesting result is that the allowed ranges are practically independent, and that, in each case, the allowed range or ranges are quite limited. For the "side group" dihedral angle χ, the allowed range is interestingly the same both for purines and pyrimidines, the rotation being about the bond C1′—N9 in the first and the bond C1′—N9 in the second. These are listed in Table 2 along with the observed values in a number of cases, and it will be seen that all the observations closely agree with the predictions of theory.

Because of these severe restrictions on the allowed ranges of $\theta 1$, $\theta 2$, and $\theta 3$, the working out of all the allowed conformations of a polynucleotide chain has been brought within the realm of possible investigation. The allowed conformations of dinucleotides and of helical chains are now under study in this laboratory.

CONFORMATIONS OF POLYSACCHARIDE CHAINS

The units out of which the polysaccharide chains are built up are sugar residues. As is well known, glucose has two cyclic forms, α-glucose and β-glucose. Corresponding to these, we can have an α, 1—4′ or a β, 1—4′ linkage in disaccharides and polysaccharides. The linear polymer of glucose obtained through the α, 1—4′ linkage is amylose, and that obtained through the β, 1—4′ linkage is cellulose. They have very different properties and they probably have entirely different conformations. Recent studies (Rao et al., 1967; Ramakrishnan, unpublished), in fact, show that amylose has a flat open-coil conformation, and the cellulose chain is ribbonlike.

TABLE 2.
Observed values (in degrees) of $\theta 1$, $\theta 2$, $\theta 3$, and χ in nucleosides and nucleotides.

Structure	Ribose conformation	$\theta 1$	$\theta 2$	$\theta 3$	χ	Reference
Theory	C3′—*endo* (assumed)	−30 to 30	−20 to 10; 110 to 140; 220 to 280	270 to 280	130 to 170; 320 to 330[a]	
Deoxy adenosine H_2O	C3′—*exo*	—	353.1	—	175.0	1
Adenosine-5′-phosphate	C3′—*endo*	2.8	140.0	—	154.3	2
Adenylic acid b	C3′—*endo*	—	8.3	302.7	176.2	3
Adenosine (in AU complex)	C3′—*endo*	—	139.1	—	167.6	4
Deoxyguanosine (in GC complex)	C2′—*endo*	—	137.2	—	328.7	5
Cytidine	C3′—*endo*	—	132.9	—	161.6	6
Cytidylic acid b	C2′—*endo*	—	136.2	271.5	138.5	7
5-bromo deoxy cytidine (in GC complex)	C2′—*endo*	—	354.7	—	121.3	5
Calcium thymidylate	C3′—*endo*	335.6	123.0	—	136.5	8
5-iodo deoxyuridine	C2′—*endo*	—	128.0	—	116.8	9
5-fluoro deoxyuridine	C2′—*endo*	—	248.2	—	120.9	10
Barium uridine-5′-phosphate	C2′—*endo*	4.0	124.8	—	136.7	11
5-bromo uridine (in AU complex)	C3′—*endo*	—	143.8	—	160.0	4

REFERENCES: (1) Watson et al. (1965); (2) Kraut and Jensen (1963); (3) Sundaralingam (1966); (4) Haschemeyer and Sobell (1965a); (5) Haschemeyer and Sobell (1965b); (6) Furberg et al. (1965); (7) Sundaralingam and Jensen (1965); (8) Trueblood et al. (1961); (9) Camerman and Trotter (1965); (10) Harris and Macintyre (1964); (11) Shefter and Trueblood (1965).

[a] Only for Purines.

α-Glucose and Amylose

In a careful study, making use of potential functions, Rao et al., (1967) have found that the chair form desiganted as C1 (Reeves, 1950) is the most stable form of α-D-glucopyranoside. Further, using the contact criteria as well as potential energy calculations, they find that two glucose residues in the C1 form may readily be joined through an α, 1—4′ linkage without any steric strain. A set of dihedral angles (ϕ, ϕ') can be defined for this linkage in a manner very similar to the case of the polypeptide; the allowed region in the (ϕ, ϕ') plane is shown in Figure 4a. As will be seen from this diagram, a flat helix with small pitch is possible for amylose. It is known that in the presence of complexing agents such as iodine, amylose has such a helical structure (the so-called "V form") with an outer diameter of the order of 13 Å and pitch of about 1.33 Å (Valletta et al., 1964). The conformation occurs well within the allowed region and, for a left-handed helix, it is very close to the minimum of energy and is further stabilized by an OH···O hydrogen bond between neighboring residues. So also, the stereochemical studies show that six glucose residues in the C1 form can be readily joined by α, 1—4′ linkages to form a cyclic chain. In fact, cyclohexaamylose is known and the values of (ϕ, ϕ') in its crystal structure [average (349°, 189°)] agree closely with the values deduced in this study of (336°, 198°). Thus, not only is the behavior of amylose explained by these theoretical studies, but the basic result—that two glucose residues can be joined by an α, 1—4′ linkage—will have interesting repercussions in the understanding of the stereochemistry of sugars.

Cellobiose and Cellulose

When a similar study is made of two glucose residues in the chair form joined together by a β, 1—4′ linkage, it is again found that only a small region of the (ϕ, ϕ') map is permitted around (30°, 180°). A preliminary study was made by Ramachandran et al. (1963), and the results of a more detailed investigation made in our laboratory by Ramakrishnan (unpublished) are shown in Figure 4b.* This figure shows, as for the α linkage, both the limits of the allowed regions, as well as the helical parameters of a regular structure with a sequence of residues having the same conformation at every bridge oxygen atom. It is interesting that the crystal structure of cellobiose agrees with this diagram, for the values of ϕ and ϕ' in this disaccharide (24°, 174°) are within the allowed region. It is in fact further stabilized by an OH···O hydrogen bond between the two glucose units.

The most interesting result is that—unlike the α linkage, which leads to flat helices with small pitch—in the β linkage the chain is nearly straight, with the pitch never coming below 4.9 Å from a maximum of 5.15 Å. The twist can, however, vary somewhat, with the number of residues per turn

*Because of an error in sign, the data of Ramachandran et al. (1963) refer to L-glucose residues. The data in the present paper refer to D-glucose residues.

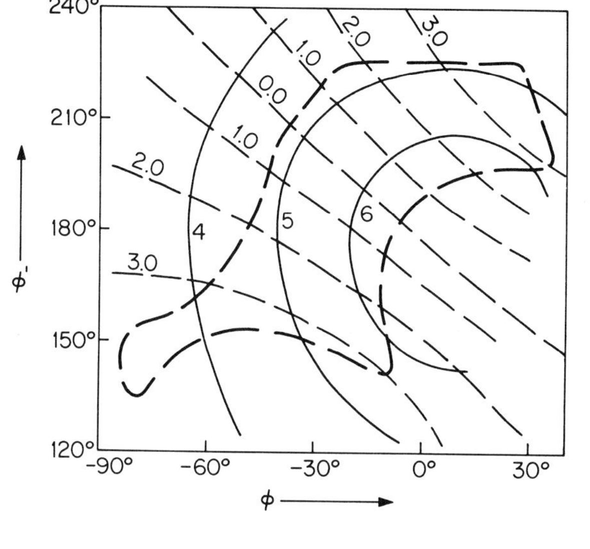

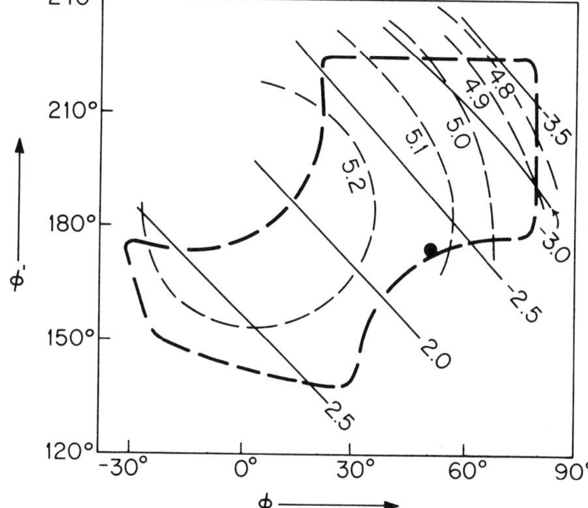

FIGURE 4.
(*a*) The allowed region (outer limits) in the (ϕ, ϕ')-plane for the α, 1—4′ linkage of two glucose residues. The values of n (the number of residues per turn) and h (the residue height) in the allowed helical structures are also marked (—— constant n, ---- constant h). The conformation of cyclohexaamylose corresponds to $n = 6$, $h = 0$. (*b*) The allowed region in the (ϕ, ϕ') plane for the β, 1—4′ linkage. The conformation observed in the disaccharide cellobiose is indicated by a black circle.

being from about -3.5 to -2.0 ($+2.0$) and going up to about $+2.5$. In other words, it would be impossible to have a cyclic saccharide having throughout the β, 1—4′ linkage.

As is well known, cellulose is a linear polymer of glucose with the β, 1—4′ linkage. The most commonly observed form of cellulose is a chain with a twofold axis (Meyer, 1950) and this has $n = 2$, $h = 5.15$ Å, with the conformation (39°, 158°). Two other structures are known (Meyer, 1950) with $n = 3$ and $n = 2.5$ and h approximately 5 Å; they also lie within the allowed range of conformations in Figure 4. A still further structure with an identity period of approximately 7×5 Å (that is, probably $n = 3.5$) has been reported, but only one example of this is known (Meyer, 1950). This form would occur very near the boundary of the allowed region and calls for re-examination.

The β linkage occurs also in chitin, a polysaccharide whose monomer is the amino sugar glucosamine. The structures of two modifications of the α and β chitin have been worked out in our laboratory (Dweltz, 1961; Carlstrom, 1962; Ramakrishnan, 1965).

CONCLUSION

The set of contact distances worked out by Ramachandran et al. (1963) for the allowed ranges of interatomic contacts in biopolymers has proved to be very fruitful in working out the conformations of a variety of such materials. The predictions of these contact criteria are in general borne out by more detailed potential energy calculations. There is no doubt that the contact method will turn out to be a valuable first step in the solution of newer problems in this field.

REFERENCES

Bradbury, E. M., L. Brown, A. R. Downie, A. Elliot, R. D. B. Fraser, and W. E. Hanby (1962). *J. Mol. Biol.* **5**, 230.

Brant, D. A., and P. J. Flory (1965). *J. Am. Chem. Soc.* **87**, 2791.

Camerman, N., and J. Trotter (1965). *Acta Cryst.* **18**, 203.

Carlström, D. (1962). *Biochim. Biophys. Acta* **59**, 361.

Corey, R. B., and L. Pauling (1953). *Proc. Roy. Soc.* (London) B **141**, 10.

De Santis, P., E. Giglio, A. M. Liquori, and A. Ripamonti (1965). *Nature* (London) **206**, 456.

Dweltz, N. E. (1961). *Biochim. Biophys. Acta* **51**, 283.

Edsall, J. T., P. J. Flory, J. C. Kendrew, A. M. Liquori, G. Némethy, G. N. Ramachandran, and H. A. Scheraga (1966). *J. Mol. Biol.* **15**, 339.

Fraser, R. D. B., and T. P. MacRae (1959). *J. Mol. Biol.* **1**, 387.

Fraser, R. D. B., and T. P. MacRae (1963). *J. Mol. Biol.* **7**, 272.

Fraser, R. D. B., and E. Suzuki (1965). *J. Mol. Biol.* **14**, 279.

Furberg, S., C. S. Peterson, and Chr. Romming (1965). *Acta Cryst.* **18**, 313.

Harris, R. D., and W. M. Macintyre (1964). *Biophys. J.* **4**, 203.

Haschemeyer, A. E. V., and H. M. Sobell (1965a). *Acta Cryst.* **18**, 525.

Haschemeyer, A. E. V., and H. M. Sobell (1965b). *Acta Cryst.* **19**, 125.

Kitaigorodsky, A. I. (1961). *Tetrahedron* **14**, 230.

Kraut, J., and L. H. Jensen (1963). *Acta Cryst.* **16**, 79.

Krimm, S., and R. Schor (1956). *J. Chem. Phys.* **24**, 922.

Leach, S. J., G. Némethy, and H. A. Scheraga (1966). *Biopolymers* **4**, 369.

Meyer, K. (1950). In *Natural and Synthetic High Polymers*, vol. 4, pp. 300, 340.

Pauling, L. (1945). *Nature of the Chemical Bond.* New York: Cornell Univ. Press.

Pauling, L., and R. B. Corey (1951). *Proc. Nat. Acad. Sci. U.S.* **37**, 235, 241, 251, 256, 261, 272, 282, 729.

Pauling, L., R. B. Corey, and H. R. Branson (1951). *Proc. Nat. Acad. Sci. U.S.* **37**, 205.

Ramachandran, G. N., and N. E. Dweltz (1962). In Ramanthan, N., ed., *Collagen*, p. 147. New York: John Wiley & Sons.

Ramachandran, G. N., and G. Kartha (1954). *Nature* (London) **174**, 269.

Ramachandran, G. N., and G. Kartha (1955). *Nature* (London) **176**, 593.

Ramachandran, G. N., and V. Sasisekharan (1965). *Biochim. Biophys. Acta* **109**, 314.

Ramachandran, G. N., C. Ramakrishnan, and V. Sasisekharan (1963). *J. Mol. Biol.* **7**, 95.

Ramachandran, G. N., C. M. Venkatachalam, and S. Krimm (1966). *Biophys. J.* **6**, 849.

Ramakrishnan, C. (1965). Ph.D. Thesis. Univ. of Madras.

Ramakrishnan, C., and G. N. Ramachandran (1965). *Biophys. J.* **5**, 909.

Rao, V. S. R., P. R. Sundararajan, C. Ramakrishnan, and G. N. Ramachandran (1967). In Ramachandran, G. N., ed., *Conformation of Biopolymers*, p. 721. London: Academic Press.

Reeves, R. E. (1950). *J. Am. Chem. Soc.* **72**, 1499.

Sasisekharan, V., A. V. Lakshminarayanan, and G. N. Ramachandran (1967). In Ramachandran, G. N., ed., *Conformation of Biopolymers*, p. 61. London: Academic Press.

Scott, R. A., and H. A. Scheraga (1966). *J. Chem. Phys.* **45**, 2091.

Shefter, E., and K. N. Trueblood (1965). *Acta Cryst.* **18**, 1067.

Sundaralingam, M. (1966). *Acta Cryst.* **21**, 495.

Sundaralingam, M., and L. H. Jensen (1965). *J. Mol. Biol.* **13**, 914.

Trueblood, K. N., P. Horn, and V. Luzzati (1961). *Acta Cryst.* **14**, 965.

Valletta, R. M., F. J. Gernino, R. E. Lang, and J. Mosphy (1964). *J. Polymer Sci.* **2**, 1085.

Venkatachalam, C. M., and G. N. Ramachandran (1967). In Ramachandran, G. N., ed., *Conformation of Biopolymers*, p. 83. London: Academic Press.

Watson, D. G., D. J. Sutor, and P. Tollin (1965). *Acta Cryst.* **19**, 111.

JOHN T. EDSALL
Biological Laboratories
Harvard University
Cambridge, Massachusetts

Thoughts on the Conformation of Proteins in Solution

We owe to Pauling, Corey and Branson (1951) and to Pauling and Corey (1951, 1953) the first clear vision of certain major kinds of three-dimensional order in proteins. The great pioneer work of Astbury had indeed revealed the existence, and many of the characteristics, of the α-structure in the fibrous proteins, and of the α-β transformation; but it was Linus Pauling above all who formulated the criteria for acceptable conformations of polypeptide chains and described the α-helix and the pleated sheets with such precision that the proposed structures could be subjected to rigorous experimental test, both in fibrous and in globular proteins. I will say nothing of the great success of these proposed structures in deciphering the nature of fibrous proteins; my concern here is with the globular proteins, and with these the vindication of Pauling's ideas materialized somewhat later. With the solution of the myoglobin structure (Kendrew, 1962) came the clear demonstration that a globular protein could be mostly composed of segments of α-helix; with the solution of the lysozyme structure (Phillips, 1966, 1967) came further evidence of some rather distorted α-helix in a very different kind of globular protein. The work on lysozyme also revealed a section of the molecule that doubles back on itself to form an antiparallel pleated sheet, in essentially the pattern described by Pauling and Corey (1951).

Thus there is no doubt that the ordered patterns described by Pauling and Corey are fundamental to our understanding of both fibrous and globular protein molecules. Yet the X-ray study of the known globular proteins reveals strange forms, with unanticipated twists and convolutions separating the ordered segments, yielding patterns unlike anything previously known, in nature or in art. Naturally, the endeavor has begun to discern the principles that govern the organization of these structures, and to relate form to function.

The conformation of a single peptide chain of a protein molecule in a crystal can be reasonably well described by the orientations of the successive peptide units. The orientation of each such unit is specified by the two angles ϕ (N—C$^{\alpha}$) and ψ (C$^{\alpha}$—C′), which are defined and illustrated by Ramachandran in the preceding paper in this volume. By specifying these two angles for each such unit, proceeding from the amino terminal to the carboxyl terminal end of the peptide chain, we specify the entire course of the chain in space, relative to coordinate axes fixed in the molecule, which we choose at our convenience.

This statement is subject, of course, to certain restrictions. It assumes that the peptide units

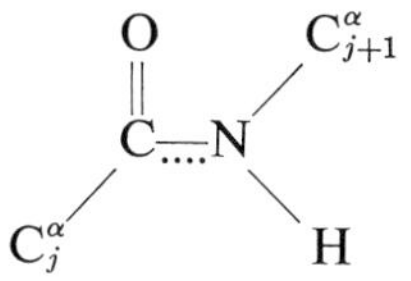

$$\begin{array}{ccc} & \mathrm{O} & & \mathrm{C}^{\alpha}_{j+1} \\ & \| & & / \\ & \mathrm{C} & \overset{\cdots\cdot}{=} \mathrm{N} & \\ / & & & \backslash \\ \mathrm{C}^{\alpha}_{j} & & & \mathrm{H} \end{array}$$

are planar, that the two α-carbon atoms of residue j and residue j + 1 in the chain, are *trans* to each other, and that the interatomic distances and bond angles are essentially those originally proposed by Pauling and Corey, with small refinements resulting from later work. Even so, this description specifies only the course of the main peptide chain, not the orientations of the side chains relative to the main chain. If a region of the chain forms an α-helix, the values of ϕ and ψ for the residues in that region will all be nearly identical; similarly for a region of pleated sheet. In "disordered" regions of the chain* there is no simple repeating relation between the successive ϕ and ψ values. However, even in such regions steric factors impose drastic restrictions on the possible values that these pairs of angles can assume. These restrictions have been analyzed in detail, especially by Ramachandran and by Scheraga, and there is no need to discuss them further here. Ramachandran's article shows maps of the distribution of the ϕ and ψ values for the residues in sperm whale myoglobin and egg white lysozyme.

Of course protein molecules are not rigid structures. There is some freedom of rotation of the side chains, especially of polar side chains that project outward from the protein molecule into the surrounding liquid. Beyond this, however, there must be local fluctuations of conformation in the main peptide chain itself, with temporary loosening of the structure, now in one region, now in another. Such small fluctuations in the conformation of the native protein molecule appear to be required in order to explain many phenomena—notably, the slow exchange rates of many hydrogen atoms (Hvidt and Nielsen, 1966). As the temperature rises, such fluctuations become more frequent and intense, until they lead eventually to disruption of the native structure. As yet, however, we can say little of the detailed nature of these fluctuations, and as a first approximation we can work with a rigid model of the native protein, in which the ϕ and ψ values of the amino acid residues are regarded as fixed.

In a protein composed of several associated peptide chains, such as hemoglobin, there is of course a higher order of organization. The distances and relative orientations of the various subunits, as well as the successive ϕ and ψ values within each subunit, must be known in order to specify the conformation of the whole molecule.

Structures of this sort remain uniquely characteristic for the globular proteins; no synthetic polymer yet made is like this. There are of course synthetic polymers that exist in ordered forms: poly-L-glutamic acid, for instance, is

*A "disordered" region is defined as such by the statement here given. Such a region is of course still a part of the ordered arrangement of the entire peptide chain—an order fixed by the fact that ϕ and ψ values can be assigned to each peptide unit. The adjective "random" is clearly a misnomer, if applied to such a precisely specified sequence.

an α-helix at *p*H 4, and poly-L-lysine at *p*H 11. It is the existence of "disordered" sequences of residues (which are nevertheless fixed and specified as components of a well-defined three-dimensional pattern) that is characteristic of proteins and—as yet—of no other known molecules.

The existence of these well-defined structures in protein crystals inevitably raises questions. Does the specified conformation in the crystal correspond in detail to the conformation of the protein molecule in solution, in a medium similar to its "natural" environment—that is, for most proteins, an aqueous buffered solution, not far from *p*H 7 and at a temperature not higher than that of its natural surroundings? Is this conformation subject to minor reversible modifications by binding of substrates, prosthetic groups, or other ligands to the protein, as proposed for example by Koshland (1960) in his hypothesis of "induced fit" in enzyme-substrate interactions? Is the native conformation truly a state of minimum free energy for the protein in its "native" medium, to which it will spontaneously return, even after being subjected to the action of a denaturing agent that has so disorganized it as to convert it into a random coil? I will consider, briefly and tentatively, possible answers to each of these questions.

CONFORMATION IN THE CRYSTAL AND IN NONDENATURING SOLVENTS. IS IT THE SAME IN BOTH?*

Protein crystals, unlike most crystals, are much like highly concentrated solutions, except that the protein molecules are arranged in an ordered lattice. Usually about half the weight of the crystal is water; if the mother liquor contains salt, the crystal also is salty, though the ratio of salt ions to water is commonly lower in the crystal than in the mother liquor—it may in some cases be higher, if certain ions are bound selectively to the protein. Moreover, the X-ray data as yet have given no indication of extensive order in the arrangement of water molecules in the interstices between the protein molecules. There may be water molecules in a monolayer, around the protein surface or portions of it; such a layer would presumably persist, essentially unchanged, when the protein molecule leaves the crystal and passes into solution. The major change, when the crystal dissolves, is the breaking of a limited number of protein-protein contacts and their replacement by protein-solvent contacts. The former usually affect only a few corners of the surface of each protein molecule, where it touches its nearest neighbors. Since this outer surface is commonly composed predominantly of charged polar side chains, the change on dissolution of the protein will commonly involve breaking the interactions between two polar side chains, and replacing them with interactions between the individual side chains and the polar solvent water. This qualitative argument suggests that changes in protein conformation, on dissolution of the crystal, should be small.

Urnes (1965) has discussed in detail the conformation of myoglobin in solution, relative to that in the crystal. His discussion centers largely on optical rotatory dispersion, and leads to the conclusion that the helix content (75–80 percent) is, within the limits of experimental error, the same in crystal

*Any investigator concerned with the problems discussed here should consider carefully the very thoughtful discussion by Richards (1963, pp. 278–282), which remains highly relevant.

and solution. We can of course readily imagine detailed conformational changes that leave the net helix content unaltered, increasing it in some regions and decreasing it in others, but such occurrences seem improbable. Urnes also presents arguments based on the solubility and salting out of proteins, somewhat akin to those I have presented above.

A significant test of these ideas is provided by the work of Muirhead and Perutz (1963) and Perutz et al. (1964) on the conformational changes accompanying the oxygenation of hemoglobin. The four subunits of the hemoglobin molecule are in considerably closer contact with each other than are separate protein molecules in a crystal. On oxygenation the two β-chains move closer together by about 7 Å and both α and β chains undergo changes in relative orientation. Yet, as far as can be observed at 5.5 Å resolution, the conformation of the individual α and β subunits remains unaltered during the process. If this is true here, there are strong reasons for believing that the removal of protein molecules from the framework of a crystal lattice can occur with very slight conformational changes, indeed.*

Detailed evidence of conformation in solution is provided most clearly by studies of the residues at the active sites of enzymes. For ribonuclease and chymotrypsin we are fortunate in now having direct evidence from the crystal structure that is entirely concordant with the data obtained by chemical modification of those enzymes in solution.

The structure of tosyl-α-chymotrypsin at 2 Å resolution (Matthews, Sigler, Henderson, and Blow, 1967) shows a picture of the active site region that is quite in harmony with expectations from previous chemical and enzymatic studies, with open regions in the surface near the active serine, residue 195, to which the tosyl group is bound, with serine 195 and histidine 57 close together, as would be expected if they act in concert in the catalytic mechanism (Bender and Kezdy, 1965; Cunningham, 1965), and with the negative charge of the aspartic acid residue 194 forming an ion pair with the positive charge on isoleucine 16, in harmony with the findings of Oppenheimer, Labouesse, and Hess (1966). Although there may well be some modification, associated with the binding of substrate, in the local conformation at the active site, the available data do not suggest a change associated with the transition from the crystal to solution.

In the crystals of ribonuclease A (Harker, Kartha, and Bello, 1967) and of ribonuclease S (Wyckoff et al., 1967) one can see clearly, along with many other features, the close relation between the two reactive histidine residues at positions 12 and 119, in accord with the chemical evidence of Crestfield, Stein, and Moore (1963), the spatial relation between the lysine residues 7 and 41, which had been inferred from the linking of these residues through chemical cross linking agents (Marfey, Uziel, and Little, 1965), and the distinctive arrangement of each of the 6 tyrosine residues, which from studies in solution are known to belong to two classes—3 in one class being freely ionizable whereas the other 3 are shielded from the solvent in the molecular interior.

*We must add, however, that on oxygenation there *must* be changes in the subunits of hemoglobin that will surely be detected in a high-resolution X-ray study. The conformational changes within the individual subunits may be very small, but they must exist or the rearrangement in the relations between the subunits would not occur.

The development of knowledge concerning these tyrosine residues of ribonuclease is instructive to follow. Shugar (1952) and Tanford et al. (1955) established by spectrophotometric titrations the existence of the two classes of tyrosine residues. Bigelow and Geschwind (1960) and Bigelow (1960, 1961) studied the difference spectra of ribonuclease produced by denaturation in acid and in urea solutions, and showed that the three "buried" residues could be selectively exposed to solvent by proper choice of denaturing conditions. They labeled the residues in question A, B, and C, but could not specify their position in the amino acid sequence. Scheraga and his collaborators (for their recent papers, see Woody et al., 1966; Friedman et al., 1966), by iodination experiments, showed that tyrosine residues 73, 76, and 115 could be fairly readily iodinated, whereas residues 25, 92, and 97 could not. The latter were presumably the "buried" residues. The crystal structure studies confirmed these findings, and defined the environment of each of the six residues. Woody et al. further presented evidence that they interpreted as involving hydrogen bonds between the three "buried" tyrosyl residues and three specific aspartyl residues. How far this hypothesis is verified by the crystal structure data is still not entirely clear as I write this, although the necessary information will probably appear before this article is published. We would expect, on general grounds, that tyrosyl-carboxylate hydrogen bonds would be rather weak if these groups are on the surface of the protein, because of competition from the hydrogen bonds formed by each group separately with surrounding water molecules. On the other hand, a tyrosine-carboxylate pair in the molecular interior, a region of low effective dielectric constant, would have a high electrostatic energy because of its negative charge and thus tend to destabilize the protein structure (Tanford, 1961). Hydrogen-bonded structures with zero net charge, such as that in myoglobin (Kendrew, 1962) between the tyrosyl phenolic group at H22 and the peptide carbonyl group at FG 5, would be expected to be more stable.

Detailed correlation of structure can be carried further by the study of the reactivity of specific groups, in the crystal and in solution. The most systematic study of this sort is that of Gurd and his collaborators on the chemical modification of the histidine residues in the sperm whale myoglobin molecule (for a survey of the results, see Gurd et al., 1966). In the crystalline state they could regularly carboxymethylate 6 histidine residues per molecule with bromoacetate, and in the dissolved state 7, without apparently altering the native conformation. The researches are still in progress, but as far as they go—and they have gone a considerable way—they suggest no significant change of conformation when myoglobin passes into solution.

CONFORMATIONAL CHANGES ASSOCIATED WITH BINDING OF SUBSTRATES, COENZYMES, OR SUBSTRATE ANALOGUES

Interactions of this sort are inherently more likely to involve conformational changes in the protein—local changes, at least—than are the changes involved when a protein crystal dissolves. A substrate or substrate analogue, or a coenzyme, associates very closely with a particular region of a protein; even if no covalent bonds are formed, there are numerous points of close interaction, involving both polar and nonpolar interactions. As yet, however, there are few data to indicate how large the resulting conformational changes may be. For lysozyme, Phillips (1967) notes that the binding of N-acetylglucosamine shifts the tryptophan residue 62 by about 0.75 Å, in such a way as to narrow the cleft in which the bound molecule lies, and that there are related small shifts in other residues, especially on one side of the cleft. It appears highly probable that the binding of substrates or related compounds to ribonuclease induces a conformational change in the active site region of the enzyme, but the full picture is not yet available.

Rosenberg, Theorell, and Yonetani (1965) have reported a striking change in the optical rotatory dispersion of liver alcohol dehydrogenase on the binding of nicotinamide adenine dinucleotide (NAD^+) and pyrazole, or of the reduced coenzyme (NADH) and isobutyramide. The value of the Moffitt parameter b_0 was close to -185 degree cm^2 per decimole for these complexes, whereas it was only about -100 for the enzyme alone, and -119 for the enzyme plus NADH. The results strongly suggest a large change in helix content of the enzyme when *both* the coenzyme and a suitable small molecule are bound at the active site, although the authors are careful to point out the uncertainties involved in such an inference. Brändén (1965) reports that the crystal structure of such ternary complexes of liver alcohol dehydrogenase is markedly different from that of crystals of the free enzyme or of its complex with NADH.

Certainly, within two or three years, we shall know far more about such induced conformational changes for many proteins.

CONCERNING THE EVOLUTION OF COOPERATIVE INTERACTIONS

Ingram (1961, 1962) has portrayed a possible course for the evolution of the hemoglobins, with their four peptide chains and four heme groups, from a myoglobin-like molecule with only one of each. It is clear from the X-ray studies of Kendrew and Perutz that the tertiary structure of the individual chains is very similar in myoglobin and hemoglobin. What must alter, if these chains are to associate to form stable tetramers and not remain as isolated monomers, is the character of certain regions of the surface of the myoglobin-like subunits. Certain groups of side chains, which in the original monomer prefer water, must be changed in the hemoglobins to others that prefer to be in contact with a neighboring protein subunit. Presumably this means that such side chains become more hydrophobic on the average (see, for instance, Fisher, 1964) but other more subtle changes may be involved as well. Whatever the requirements may be for the formation of a "good" hemoglobin molecule, with cooperative interactions between the subunits in oxygen binding and with the heterotropic interactions between oxygen and proton binding that give rise to the Bohr effect (Wyman, 1964, 1967; Guidotti, 1967), it is certain that multiple mutations, giving rise to different amino acid residues in a number of different positions in the peptide chain sequences, must be required. Such changes cannot happen all at once; they must occur in many stages, over a long period of time. The intermediate mutational forms, we would suppose, must be rather unsatisfactory proteins, with imperfect tendencies to form loose aggregates of monomers, but without the advantageous cooperative interactions of hemoglobin as we know it. They would be neither good myoglobin nor good hemoglobin, and the organisms that possessed such molecules would scarcely be expected to do well in the ordeal of natural selection. Such molecules as the hemoglobin of the lamprey, studied by Briehl (1963), may represent rather satisfactory intermediate way stations between myoglobin and mammalian hemoglobin. However, even lamprey hemoglobin is a long way from myoglobin. Such problems are of course familiar to the evolutionary biologists (see, for instance, Huxley, 1942; Mayr, 1960). The transition from one type of adaptation to another at a more advanced level often appears to lead through intermediate stages that are less satisfactory than either the initial or the final adaptation.

In bacteria, with their brief generation time, mutation and selection can occur rapidly. Thus the remarkable enzyme aspartate transcarbamylase of *Escherichia coli*, is a highly organized cooperative system of catalytic and regulatory subunits. Gerhart (1964) has observed a variety of mutants in which the cooperative interactions are profoundly modified; yet all of them, on destroying the interactions by dissociating the enzyme into subunits with mercurials, give typical Michaelis-Menten kinetics and become indistinguishable in kinetic behavior.

FROM ORDER TO DISORDER AND BACK AGAIN

The native state of globular proteins in neutral aqueous solution, at reasonably low temperatures, clearly represents some sort of free-energy minimum—at least, a relative minimum—to which the molecule tends to return spontaneously from a more disordered condition. The work of Anfinsen and his colleagues (Anfinsen, 1965–1966) has shown clearly that ribonuclease, lysozyme, and other proteins can be reversibly unfolded into a state at least closely resembling that of a random coil, after breaking all disulfide bonds by reduction. When brought back into neutral aqueous buffer, in the presence of oxygen, they will then refold spontaneously into the native conformation, with reformation of the correct disulfide bonds. The work of Harrison and Blout (1965) and of Breslow et al. (1965) has provided similar evidence for sperm whale myoglobin, which contains no disulfide bonds. In this protein, of course, the presence of the heme group is essential if the protein is to regain its native conformation.

These and many other experiments have led to the hypothesis that primary sequence determines three-dimensional conformation. Is the hypothesis valid, and if so with what restrictions?

One important restriction is immediately apparent. It is only in certain media that it holds at all. In 8–10 M urea, or in 5–6 M guanidinium chloride, there is no single preferred conformation. Conceivably we might have found a new conformation for the protein in such media, with a set of well-defined ϕ and ψ values for the successive peptide groups, quite different from those found for the native protein in water. In fact, of course, we find no such thing, but instead encounter a disorganized unfolded chain, capable of assuming a vast number of different conformations, and undergoing incessant transitions from one such conformation to another.

Are we really justified in calling such a peptide chain a random coil? In 6 M guanidinium chloride, at least, the work of Tanford et al. (1967) indicates that the answer is "yes." In this medium the intrinsic viscosities and sedimentation coefficients of a considerable number of proteins were found to vary, with the length of the peptide chain, in the manner expected for random coils. Moreover, the hydrogen ion titration curve of ribonuclease in this solvent (Nozaki and Tanford, 1967) is in full accord with the picture of the protein as a random coil. In 8 M urea some residual secondary and tertiary structure may remain, for some proteins; in 6 M guanidinium chloride all this has apparently vanished.

We must remember that the formation of such a random coil involves no relaxation of the rigorous steric restrictions on the local conformations of the peptide groups. Most combinations of ϕ and ψ values, for any given residue, are still forbidden as sterically impossible; the peptide units are still planar and presumably still *trans*. Within the allowed domain of possible conformations, however, there is no longer a unique value of ϕ or ψ for any given residue that is maintained within narrow limits, as in the native protein. Instead, the links in the chain undergo constant transitions from one set of values to another, with no correlation of orientations between residues that are some distance away from one another along the chain. Yet when such a random chain is removed from the denaturing solvent and returned to neutral aqueous buffer, it finds its way back to the native conformation within minutes, or a few hours at most.

Is the native conformation therefore a state of minimum free energy? Or is it only a secondary minimum, which the molecule reaches during the refolding process, and then finds itself unable to leave because it cannot cross the energy barriers that separate it from the true minimum? As yet we can give no assured answer to this question. However, it is an impressive fact that a molecule of lysozyme or ribonuclease, after its initial synthesis and release from the ribosome, assumes the same conformation as a molecule that has been denatured in guanidinium chloride and then transferred to a more physiological medium. From such diverse starting points the molecules attain the same final state. Unless there are drastic restrictions on the possible paths that the random coils must follow, when refolding into the native state, we would suppose that the latter state must truly be one of minimum free energy. If, on the other hand, there are such drastic restrictions, then presumably we must conclude that the "random coils" are not truly random.

The tendency to assume a unique stable conformation in a "physiological" medium is characteristic of the globular proteins found in nature. I would suggest that it is probably not a property of polypeptide chains in general. The natural proteins, numerous as they are, represent only a tiny fraction of the vast number of possible combinations of amino acid residues. The great majority of these possible sequences may have no tendency, in any solvent medium whatever, to assume a unique conformation. Natural selection would in general eliminate such sequences, if they arose in an organism by mutation, as biologically useless. There might then be three categories of peptide chains: (1) those forming functional globular proteins, such as enzymes; (2) those forming nonfunctional globular proteins, that also tend to assume a unique conformation but are biologically ineffective because of local alterations by mutation at specific sites (Helinski and Yanofsky, 1966); (3) nonfunctional sequences that are incapable of assuming a unique conformation at all, although certain regions of the chain may adopt preferred conformations in aqueous solution.

We must note, of course, that many polypeptide hormones, such as adrenocorticotropic hormone, appear to exist as random coils in neutral aqueous solution. Presumably they assume some more specific conformation during their biological action, when the active portion of the hormone would fit itself specifically to some structure within the cell.

Some native proteins may perhaps exist, not in a single conformation but in a set of closely related but distinct conformations. This, for instance, may be true of serum albumin, with its well-known "configurational adaptibility" that permits it to bind reversibly so many diverse ions and molecules. A certain looseness of the molecular framework may be desirable here, rather than the more close-knit structure generally characteristic of the enzymes with their specific sites. Conceivably this may have something to do with the microheterogeneity of the molecules in serum albumin preparations, so convincingly demonstrated by Foster and his colleagues (Foster et al., 1965; Petersen and Foster, 1965). They have separated albumin preparations into subfractions that differ greatly in their solubility and other physical properties, and in the exact location of the critical *p*H region within which the molecules undergo the characteristic reversible conformational change in acid solution (the "N-F transition") that Foster has so thoroughly studied. The molecules in the subfractions are not all alike; presumably they too are capable of still further subfractionation. The individual molecules, however, apparently are not simply conformational isomers; each subfraction retains its individual properties, and they do not interconvert when left to stand, even for a long time. Whether the various subclasses of albumin molecules differ in primary sequence, in the arrangement of disulfide bonds, or in some other respect, is still not clear.

In general it does seem clear that, within the limits sketched above, primary sequence does determine three-dimensional conformation. To predict the latter, given the former, is now the aspiration of an increasing number of protein chemists. I will say nothing here of the current studies of Ramachandran, Scheraga, Liquori, Levinthal, and others that are directed to this end. The problems involved are most formidable; even to make the attempt is bold in the extreme. However, the protein molecule itself knows how to solve the problem; in a few minutes or hours it finds its way from disorder to order. If the molecule knows how to do it, we can learn how it does it. That is the faith that sustains the attackers of this problem, and we may believe that in due time their faith will be justified.

REFERENCES

Anfinsen, C. B. (1965–1966). Harvey Lectures, Series 61, p. 95.

Bender, M. L., and F. J. Kézdy (1965). *Ann. Rev. Biochem.* **34**, 48.

Bigelow, C. C. (1960). *Compt. rend. trav. lab. Carlsberg* **31**, 305.

Bigelow, C. C. (1961). *J. Biol. Chem.* **236**, 1706.

Bigelow, C. C., and I. Geschwind (1960). *Compt. rend. trav. lab. Carlsberg* **31**, 283.

Brandén, C.-I. (1965). *Arch. Biochem. Biophys.* **112**, 215.

Breslow, E., S. Beychok, K. D. Hardman, and F. R. N. Gurd (1965). *J. Biol. Chem.* **240**, 304.

Briehl, R. W. (1963). *J. Biol. Chem.* **238**, 2361.

Crestfield, A. M., W. H. Stein, and S. Moore (1963). *J. Biol. Chem.* **238**, 2413, 2421.

Cunningham, L. (1965). In *Comprehensive Biochemistry*, vol. 16. M. Florkin and E. H. Stotz, eds., Amsterdam: Elsevier, p. 85.
Fisher, H. F. (1964). *Proc. Nat. Acad. Sci. U.S.* **51**, 1285.
Foster, J. F., M. Sogami, H. A. Petersen, and W. J. Leonard (1965). *J. Biol. Chem.* **240**, 2495.
Friedman, M. E., H. A. Scheraga, and R. F. Goldberger (1966). *Biochemistry* **5**, 3770.
Gerhart, J. C. (1964). *Brookhaven Symp. in Biology* **17**, 222.
Guidotti, G. (1967). *J. Biol. Chem.* **242**, 3704.
Gurd, F. R. N., L. J. Banaszak, A. J. Veros, and J. F. Clark (1966). In B. Chance, R. W. Estabrook, and T. Yonetani, eds., *Hemes and Hemoproteins*. New York: Academic Press, p. 221.
Harrison, S. C., and E. R. Blout (1965). *J. Biol. Chem.* **240**, 299.
Helinski, D. R., and C. Yanofsky (1966). In H. Neurath, ed., *The Proteins*, 2nd ed., vol. 4. New York: Academic Press.
Huxley, Julian. (1942). *Evolution: The Modern Synthesis*. New York: Harper & Sons.
Hvidt, A., and S. O. Nielsen (1966). *Adv. in Protein Chem.* **21**, 287.
Ingram, V. M. (1961). *Nature* **189**, 704.
Ingram, V. M. (1962). *Fed. Proc.* **21**, 1053.
Kartha, G., J. Bello, and D. Harker (1967). *Nature* **213**, 862.
Koshland, D. E., Jr. (1960). *Adv. in Enzymology* **22**, 46.
Li, Lu-Ku, J. P. Riehm, and H. A. Scheraga (1966). *Biochemistry* **5**, 2043.
Marfey, P. S., M. Uziel, and J. Little (1965). *J. Biol. Chem.* **240**, 3270.
Matthews, B. W., P. B. Sigler, R. Henderson, and D. M. Blow (1967). *Nature* **214**, 652.
Mayr, E. (1960). "The Emergence of Evolutionary Novelties" in Sol Tax, ed., *Evolution after Darwin*, vol. I. University of Chicago Press, p. 349.
Muirhead, H., and M. F. Perutz (1963). *Nature* **199**, 633.
Nozaki, Y., and C. Tanford (1967). *J. Am. Chem. Soc.* **89**, 742.
Oppenheimer, H. L., B. Labouesse, and G. P. Hess (1966). *J. Biol. Chem.* **241**, 2720.
Pauling, L., and R. B. Corey (1951). *Proc. Nat. Acad. Sci. U.S.* **37**, 729.
Pauling, L., and R. B. Corey (1953). *Proc. Nat. Acad. Sci. U.S.* **39**, 253.
Pauling, L., R. B. Corey, and H. R. Branson (1951). *Proc. Nat. Acad. Sci. U.S.* **37**, 205.
Perutz, M. F., W. Bolton, R. Diamond, H. Muirhead, and H. C. Watson (1964). *Nature* **203**, 687.
Petersen, H. A., and J. F. Foster (1965). *J. Biol. Chem.* **240**, 3858.
Phillips, D. C. (1966). *Scientific American*, **215**, 78.
Phillips, D. C. (1967). *Proc. Nat. Acad. Sci. U.S.* **57**, 484.
Richards, F. M. (1963). *Ann. Rev. Biochem.* **32**, 269.
Riehm, J. P., and H. A. Scheraga (1966). *Biochemistry* **5**, 99.
Shugar, D. (1952). *Biochem. J.* **52**, 142.
Tanford, C. (1961). *Physical Chemistry of Macromolecules*, Chapters 7 and 8. New York: Wiley.
Tanford, C., J. D. Hauenstein, and D. G. Rands (1955). *J. Am. Chem. Soc.* **77**, 6409.
Tanford, C., K. Kawahara, and S. Lapanje (1967). *J. Am. Chem. Soc.* **89**, 729.
Urnes, P. (1965). *J. Gen. Physiol.* **49**, no. 1, part 2, p. 75.
Woody, R. W., M. E. Friedman, and H. A. Scheraga (1966). *Biochemistry* **5**, 2034.
Wyckoff, H. W., K. D. Hardman, N. M. Allewell, T. Inagami, L. N. Johnson, and F. M. Richards (1967). *J. Biol. Chem.* **242**, In press.
Wyman, J. (1964). *Adv. in Protein Chem.* **19**, 223.
Wyman, J. (1967). *J. Am. Chem. Soc.* **89**, 2202.

SIDNEY A. BERNHARD
GIAN LUIGI ROSSI
Institute of Molecular Biology and Department of Chemistry
University of Oregon
Eugene, Oregon

On the Substrate-Induced Stabilization of Native Enzyme Protein Conformation

The phenomenon of stabilization of enzyme protein structure by substrate has been recognized for as long as biochemists have been undertaking the isolation of enzyme proteins away from their native physiological environment. In this context, "stabilization of enzyme protein" is usually taken to mean stabilization of the catalytic activity. In most (but by no means all) instances, the substrate stabilization mechanism involves an inhibition of the *denaturation* of protein by the in vitro solvent environment.

Intensive interest in the effects of enzyme-substrate interactions on the *enzyme* protein structure has been periodically revived in light of various hypotheses concerning the role of polypeptide conformations in enzyme catalysis. Recent research in regard to the mechanism of regulatory enzyme action has prompted a renewed interest in the effect of substrates on the conformation of enzyme proteins. Consider an enzyme molecule, which in its native conformation consists of a discrete number of similar or identical polypeptide subunits, each subunit containing the requisite amino acid sequence for catalysis: a substrate-induced change in the conformation of *one* subunit may induce complementary conformational changes within the discrete aggregate via *inter*subunit polypeptide interactions. Such *inter*subunit, *intra*molecular interactions may result in a change in the catalytic activity of all the sites within a single enzyme molecule (via changes in either catalytic rates or binding affinities). Such interactions can lead to a *cooperative effect* of substrate concentration on catalytic activity—that is, a dependence of reaction velocity on a greater than first power of the substrate concentration. *Substrate cooperativity* of this type has often been noted. Recently, it has been extensively analyzed in terms of interacting polypeptide subunits (Gerhart and Pardee, 1963; Gerhart and Schachman, 1965; Changeux, 1963; Monod et al., 1965; Atkinson et al., 1965).

A change in the conformation of a polypeptide upon interaction with substrate does not necessarily depend on the existence of *multiple* polypeptide subunits in an enzyme molecule. Similar conformational changes may occur in an enzyme composed of a single polypeptide chain and/or a single (non-interacting) catalytic site. However, in the case of these latter structurally simpler enzymes, the usual procedures of enzyme kinetics (steady state;

$S_o \gg E_o$) would *in principle* yield no result bearing on this conformational change. Hence, investigations directed toward an understanding of the *effect of substrate-induced changes in the polypeptide conformation* on catalytic activity have been largely confined to the multiunit enzymes (by virtue of the observable substrate cooperativity, as measured by conventional enzyme kinetic procedures). Such investigations are now in progress in our laboratory but, as a preliminary, we have felt it advisable to undertake a study of the effect of substrate on the structure of single-sited enzymes. For such studies we have had necessarily to abandon the conventional "steady-state" kinetic procedures and examine, in detail, single (transient) enzyme-substrate turnovers. The results we shall consider relate primarily to this latter type of experiment.

THE ROLE OF CHEMICAL INTERMEDIATES IN THE STABILIZATION OF NATIVE CONFORMATION

Examples of the enhanced stability of enzyme-substrate complexes over the free enzymes toward denaturants are well documented. These enzyme-substrate complexes can be of two distinct types, complexes of substrate *adsorbed* to the specific (catalytic) site of the enzyme, and *chemical* complexes of substrate covalently attached to the catalytic site (Racker and Krimsky, 1952; Snell and Jenkins, 1959; Bender et al., 1962*a*; Schwartz, 1963; Laursen and Westheimer, 1966).

In the following discussion, we will demonstrate that the formation of particular covalent enzyme-substrate intermediates is relevant to the phenomenon of "substrate stabilization" of the native enzyme conformation.

The catalyzed reaction pathway can be generalized by Equation (1):

$$\mathrm{E+S} \underset{}{\overset{FAST}{\rightleftharpoons}} \underset{(\mathrm{I})}{\mathrm{ES}} \overset{SLOW}{\rightleftharpoons} \underset{(\mathrm{II})}{\mathrm{ES'}}\overset{\mathrm{P'}}{\underset{}{+}} \overset{SLOW}{\rightleftharpoons} \underset{(\mathrm{III})}{\mathrm{EP}} \overset{FAST}{\rightleftharpoons} \mathrm{E+P} \qquad (1)$$

In some (perhaps many) of these systems, the chemical intermediate, II, rather than I or III, has been demonstrated to be the *principal* enzyme-substrate species during most of the course of catalysis. No straightforward distinction as to the relative stoichiometric significance of the various enzyme-substrate intermediates (I, II, III) can be inferred from conventional steady-state kinetic analysis (Hearon et al., 1959) in which the initial substrate concentration, S_o, is very much greater than the total concentration of catalytic sites, E_o. This would likewise be true in the case of multisited enzymes. A priori, we cannot ascribe a substrate-induced stabilization (or change) in the polypeptide conformation to any particular transformation in the reaction pathway [$S \rightleftharpoons$ I, I $\rightleftharpoons$ II, or II $\rightleftharpoons P$ in Equation (1)].

The actual chemical transformations associated with the formal scheme of Equation (1) have been plausibly identified in a few specific enzyme-substrate reactions from a combination of transient and steady-state kinetic data. For example, in the reactions of proteolytic and esterolytic enzymes with substrate, the formation of an acyl enzyme intermediate has been demonstrated

(Bender et al., 1962a; Caplow and Jenks, 1962; Bernhard, Lau, and Noller, 1965). The reaction is

$$E + RC\overset{O}{\underset{X}{\lessgtr}} \overset{K_s}{\rightleftharpoons} E\left(RC\overset{O}{\underset{X}{\lessgtr}}\right) \overset{k_a}{\rightleftharpoons} E-C\overset{O}{\underset{R}{\lessgtr}} + X^- \overset{k_w}{\rightleftharpoons} E\left(RCO_2H\right) \rightleftharpoons E + RCO_2^- + H^+ \quad (2)$$

A good electron-withdrawing group, X, will facilitate the formation of the covalently linked acyl-enzyme, whereas the leaving group (X) will have no effect on the conversion of acyl enzyme to product in the virtually irreversible pathway. Hence, the reactions with enzyme of a set of substrates (RCOX) with "X" groups of variable electron-withdrawing potential should result in a variety of "X-dependent" rates of acyl enzyme formation, but to a unique rate of conversion of the intermediate to product (Nelson et al., 1962; Bender and Zerner, 1964; Epand and Wilson, 1964, 1965). For a single (non-interacting) catalytic site, the dependence of reaction velocity on substrate concentration will follow the substrate dependence described by

$$v_0 = \frac{k_p E_0 S_0}{K_M + S_0} \quad (3)$$

where S_0 = initial substrate concentration and E_0 = total concentration of enzyme sites.

According to the model of Equation 2, the apparent binding constant, K_M, is composed of two sets of constants, one reflecting the formation of the noncovalent intermediate (I) and the other reflecting the extent of formation of the covalent intermediate, as summarized by

$$K_M(\mathit{Apparent}) = \frac{k_w}{k_a + k_w}\left(\frac{k_{-s} + k_a}{k_s}\right) \simeq \frac{k_w}{k_a + k_w} K_s$$

and, at substrate saturation, by

$$V_{Max}^{S\to\infty} = \frac{k_a k_w}{k_a + k_w} \quad (4)$$

The plausibility of such arguments is demonstrated by the fact that specific ester substrates (RCO_2R') of α-chymotrypsin (and other proteolytic and esterolytic enzymes) are hydrolyzed under conditions of substrate saturation at a rate that is independent of the specific nature of the alkoxide (OR′) derivative, when all of the esters contain the same specific configuration amino acyl substituent (R) (Nelson et al., 1962; Epand and Wilson, 1964). The parameter K_M, on the other hand, varies with the particular alkoxide in these instances. These results are readily explainable on the basis of a common acyl enzyme intermediate, the hydrolysis of which is rate-determining. In other situations, where the leaving group (X) is very much poorer than alkoxide (as for example in amides and peptides), the rate of hydrolysis under conditions of substrate saturation is very much slower (Bender and Zerner, 1964; Foster and Niemann, 1955; Hein and Niemann, 1961). Moreover, acyl derivatives containing amide leaving groups of different electron withdrawing capabilities exhibit different rates of enzyme catalyzed hydrolysis at substrate saturation

(Inagami et al., 1965). Once again, this is plausibly explained on the basis of Equations (3) and (4) and on the assumption that, with a poorer leaving group, the formation rate (k_a) rather than the hydrolysis rate (k_w) of the acyl enzyme intermediate will become rate-determining.

Identifiable acyl enzyme intermediates have been reported to be markedly more stable toward a variety of denaturants than is the enzyme chymotrypsin in the absence of substrate (Wooten and Hess, 1960; Bender et al., 1962b; Martin and Bhatnagar, 1966). This suggests that covalent intermediates between enzyme and substrate, rather than or in addition to adsorbed substrate complexes [such as I of Equation (1)], contribute to the stabilization of enzyme protein structure.

As noted above, a steady-state kinetic analysis would not distinguish among the various types of enzyme-substrate intermediates. Moreover, with the usual specific substrates of α-chymotrypsin the lifetimes of intermediates ($\approx 10^{-2}$ sec) are much shorter than the lifetime of the native conformation of the unprotected enzyme in typical denaturing solvents (see below).

We have therefore chosen a simpler enzyme-substrate system—namely a metastable covalently bound acyl enzyme intermediate, for which a relatively slow single (*transient*) turnover to product can be measured under a variety of conditions of solvent environment. Such acyl enzyme derivatives can be prepared by utilizing *nonspecific* acyl (R) groups (Bender et al., 1962a; Caplow and Jencks, 1962; Bernhard, Lau, and Noller, 1965). The observed rates of hydrolysis of *nonspecific* acyl enzymes are far slower than the turnover numbers for specific substrates. These slow reactions are, however, chemically analogous to the rapid enzyme-catalyzed hydrolysis of specific substrates, except in the absolute magnitude of the time constants for chemical transformation (Bender, 1962).

SPECIFIC PROPERTIES OF ACYL ENZYMES

In the experiments to be described, we compare the chromophoric acyl enzyme, β-(3-indole)acryloyl α-chymotrypsin (IV) (Bernhard and Tashjian, 1965),

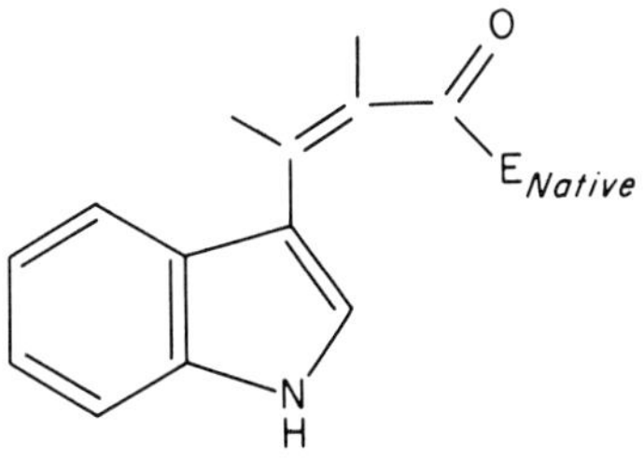

O

E_{Native}

N
H

(IV)

with free (nonacylated) enzyme, in regard to their relative stabilities toward a variety of denaturants. The intramolecular rate of (enzyme-catalyzed) hydrolysis of the acyl enzyme can, in addition, be measured in the presence of these denaturants. The resultant acyl product compositions (denatured acyl enzyme and acylate anion) can be readily analyzed. Our choice of particular acyl derivative is on the following basis:

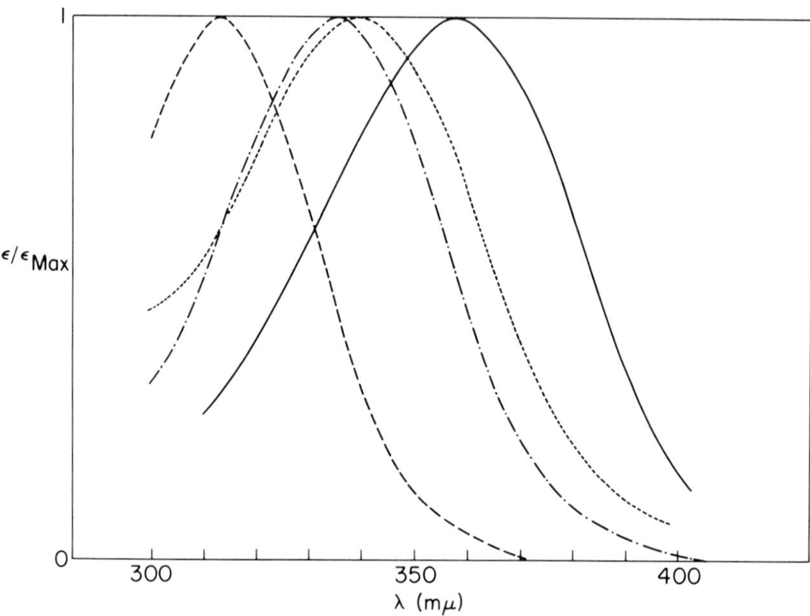

FIGURE 1.
Electronic spectra of indoleacryloyl derivatives. ——— Acylate anion in water; — · — O-acyl, N–acetyl serineamide in water or SDS; - - - - - - Denatured acyl enzyme in SDS (0.017 *M*); ———— Native acyl enzyme in water.

1. The intense ultraviolet absorption spectrum of the chromophoric acyl enzyme is nearly in the visible region ($\lambda_{max} = 360$ mμ; $\epsilon_{max} = 2.1 \times 10^4$ OD/cm *M*), far removed from the electronic absorption of the inherent chromophores of the protein. The two processes—chemical transformation of the acyl enzyme and protein denaturation—can hence be measured independently.

2. Two readily identifiable transformations in the spectrum of the native acyl enzyme are possible a priori. (*a*) The conversion of native acyl enzyme to denatured acyl enzyme, and (*b*) the catalyzed intramolecular hydrolysis of the native acyl enzyme to the corresponding acylate (indoleacrylate) anion. The native acyl enzyme is peculiar in that its spectrum differs from the observed spectra (Figure 1) of indoleacryloyl-O-seryl peptides (Noller and Bernhard, 1965), although, upon denaturation and degradation of the acyl enzyme, a specific O-acyl-seryl peptide has been identified. The spectrum of *denatured* chromophoric *acyl* enzyme is the same as that observed with the corresponding O-*acyl*-seryl peptides. The transformation of native to denatured acyl enzyme can thus be measured. The spectrum of indoleacrylate anion (the product of hydrolysis of the native acyl enzyme) is distinct from *both* the denatured and the native acyl enzyme, as also illustrated in Figure 1. It is interesting to note that neither the denatured acyl enzyme nor the acylate product are particularly sensitive in spectrum to the aqueous denaturing solvent.

3. The first-order rate of hydrolysis of the native acyl enzyme is convenient to follow, varying in half-life from hours to minutes over the *p*H range 6–9, the *p*H region of catalytic interest.

The large changes in absorption spectrum concomitant with either of the potential transformations of native acyl enzyme allow for measurement of the rate of the reaction regardless of the chemical nature of the products. The chemical nature of these products, however, can always be assayed from the absorption spectrum of the final products.

That acylation of a specific serine residue of the enzyme confers a stabilization toward denaturation has already been well documented (Bernhard, Lau, and Noller, 1965; Wooten and Hess, 1960; Bender et al., 1962b); the present experiments are designed to quantitate and thereby to draw mechanistic inferences regarding the nature of this stabilization.

The *p*H dependence of the intramolecular catalyzed hydrolysis rate, in the absence of denaturants, is illustrated in Figure 2. In the absence of substrate, the rate of denaturation of the enzyme can be measured by standard procedures involving either changes in optical rotation, or in the ultraviolet absorption at 280–290 mμ (the region of tyrosine and tryptophan absorption), or at 235 mμ (a region in which absorption is due to hydrogen-bonded peptides and to the second electronic transitions of the aromatic amino acids).

At all *p*H's, the rate of catalyzed hydrolysis of the acyl enzyme is significantly slower than the rate of denaturation of the native enzyme in the denaturants we have selected, as is illustrated in Figures 2 and 3.

We may therefore proceed immediately to inquire as to the extent of sub-

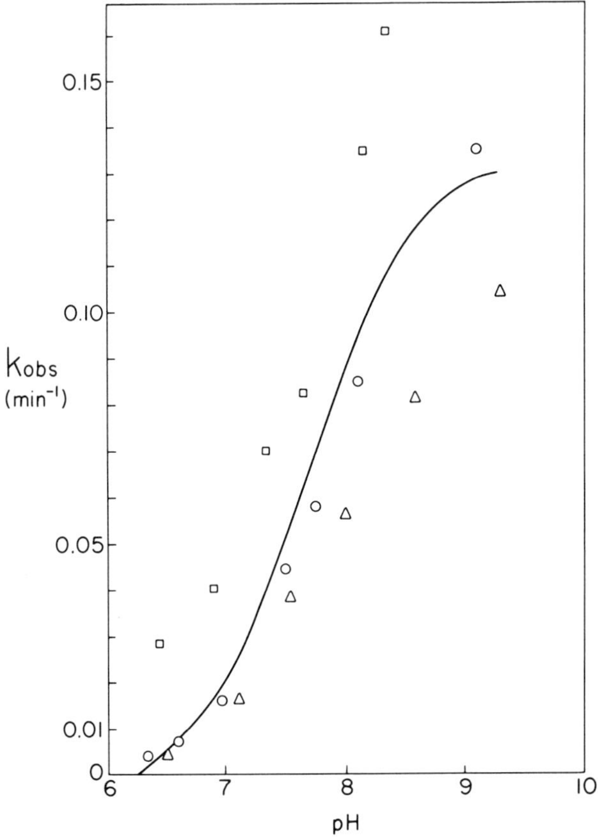

FIGURE 2.
First-order rates of change in the optical density of the native acyl enzyme in various solvents over the *p*H range of catalytic significance. Solid line, water; □, 5.6 *M* guanidine HCl; ○, 0.017 *M* SDS; △, 8 *M* urea.

strate stabilization of the native enzyme conformation by examining the chemical composition of the chromophoric acyl products of reaction in a "denaturing solvent." The product composition is qualitatively described by the observed λ_{max} of the reaction products. Such a comparison (Table 1) yields rough quantitative information on the composition of products, since the absorption spectra of the two potential products are of comparable intensity and symmetry (the intensities are each essentially Gaussian-distributed about the particular λ_{max}). A cursory examination of Table 1 indicates that solvent conditions are available to us for obtaining complete hydrolysis to the exclusion of acyl enzyme denaturation, and vice versa. It should be noted that under these conditions of *p*H and over these total reaction times, the denatured acyl enzyme (ester) and also the acylate anion are totally inert (Noller and Bernhard, 1965; Bender et al., 1962b), and the enzyme protein is totally denatured at the end of the reaction whether or not it contains a covalently linked acyl group.

The dramatic stabilization of native enzyme configuration is well illustrated by the fact that no denaturation of the acyl enzyme occurs in *some* of the "denaturing solvents." Note that at higher *p*H, where the rate of catalytic hydrolysis would be optimal in the absence of denaturation, substantial yields of denatured *acyl enzyme* can be obtained only with the most drastic denaturants (Table 1).

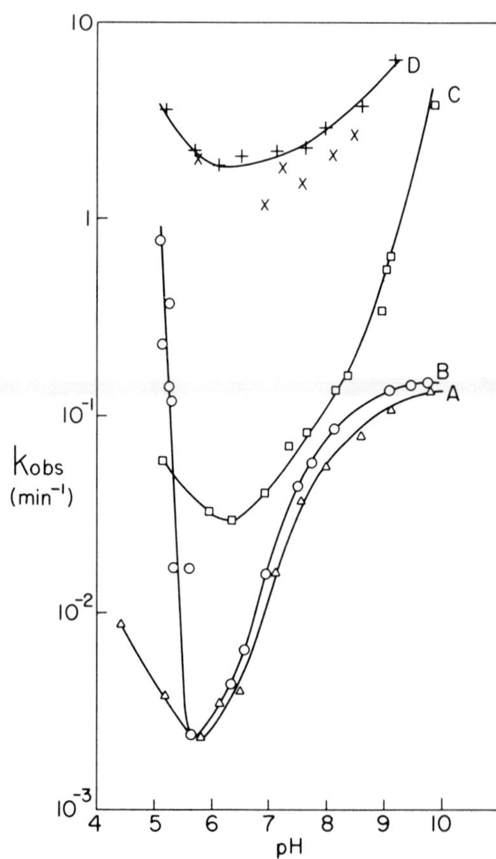

FIGURE 3.
Disappearance of native acyl enzyme (A–C) and native enzyme (D) as measured by changes in optical density. (A) In 8 *M* urea, (B) in 0.017 *M* SDS, and (C) in 5.6 *M* guanidine HCl, at 370 mμ. (D) In 8 *M* urea (+) and 0.017 *M* SDS (×) at 235 mμ.

TABLE 1.
λ_{max} (*in* mμ) *of products of reaction of indoleacryloyl chymotrypsin in various solvents.*

*p*H	Solvent		
	0.017 *M* SDS[a]	8 *M* Urea[a]	5.6 *M* guanidine hydrochloride[a, b]
4.5–5.5	338	342, 342[c]	342
5.63	338	—	—
5.74	324	335	—
5.79	321	—	—
6.32	317	317	342
6.92	317	—	333
7.35	—	—	333
7.51	315	315	335
8.00	315	315	323[d], 337, 342[e]
8.34	—	—	337
9.11	320	—	342
9.46	321	319, 328[c]	—
9.75	323	326	342
9.97	—	—, 336[c]	—

[a]In acetate and pyrophosphate buffers at ionic strengths of 0.1–0.3 *M*.
[b]In "tris" and glycineamide buffers at *p*H $\geq$ 8.
[c]In 10 *M* urea.
[d]In 4.2 *M* guanidine hydrochloride. $k_{obs} = 0.1$ min^{-1}.
[e]In 6 *M* guanidine hydrochloride. $k_{obs} = 0.35$ min^{-1}.

The rate of catalyzed hydrolysis of the native acyl enzyme at any specified *p*H is independent of the nature of the solvent under all conditions that lead to 100 percent transformation to acylate ion. Thus, for example, the ionic environment can be varied over as much as a hundredfold (Bender et al., 1962a), the polarity of the solvent can be changed appreciably (Clement and Bender, 1963), and a variety of denaturants can be added to the solvent without in any significant way affecting the absolute rate of hydrolysis of the acyl enzyme at a particular *p*H. This is somewhat surprising in light of the fact that the binding affinities for substrates, the rates of formation of acyl enzymes (Clement and Bender, 1963), the rates of denaturation of the free enzyme (Martin, 1964), and the rates of hydrolysis of specific amide substrates (Shine and Niemann, 1955) are all highly sensitive to the particular nature of the solvent environment.

A COMPARISON OF THE RELATIVE STABILITIES OF ENZYME AND ACYL ENZYME

The rate of denaturation of native α-chymotrypsin has been extensively investigated (Martin, 1964; Martin and Bhatnagar, 1966). It is abundantly clear that a variety of environmental factors (polarity of the solvent, ionic strength, specific ions, and others) affect the stability of the native enzyme protein. Moreover, the *p*H-dependent influence of each of these environmental factors on the stability of the native protein is distinctive. That such complexity exists is not particularly surprising, in light of both the variety of molecular interactions involved in defining the *native* polypeptide conformation and the variety of mechanisms operative in different denaturation procedures. For example, the detergent sodium dodecyl sulfate (SDS), is generally assumed to function by solvation of hydrophobic amino acid side chains by the hydrocarbon, and interaction of the charged sulfate anion with the aqueous solvent. This type of denaturation would be expected to be different from the denaturing action of guanidine hydrochloride, in which the principal interaction between denaturant and native polypeptide results in a disruption of the hydrogen-bonding patterns in the native enzyme conformation. Moreover, the positively charged guanidinium ions would be expected to interact with the charged protein to a greater extent, as the net positive charge of the protein is reduced (that is, at alkaline *p*H's), whereas the anionic SDS would be expected to interact most favorably with the protein as the positive charge on the protein increases. Because of these Coulombic interactions, the effectiveness of either denaturant should be sensitive to the ionic strength of the aqueous environment, and to the extent to which the constituent charged groups of the protein are neutralized by complex formation with environmentally supplied ligands (for example, Ca^{2+}) (Martin, 1964; Martin and Bhatnagar, 1966). The effectiveness of *any* denaturant will depend on the stability of the native conformation of the protein. To the extent that this stability is dependent on the charge distribution within the protein, the denaturation rate will be *p*H-dependent. We might therefore anticipate a host of complicating factors in a study of the rates of denaturation of the acyl enzyme as a function of the solvent environment.

The actual rates of disappearance of the native acyl enzyme as a function of the *p*H of the solvent medium, in the presence of a variety of denaturants, are shown in Figure 2. The region of *p*H over which the enzyme activity is optimal, or nearly optimal, is of special interest. It so happens that over this region of *p*H electrostatic factors are minimized. Note that over a portion of this *p*H region (*p*H 7–9) the rates of disappearance of native acyl enzyme in a variety of denaturants are nearly independent of the particular denaturant. A straightforward explanation for these results would be evident if either hydrolysis of the acyl enzyme *preceded* denaturation, or if both the rates of denaturation and of hydrolysis were common at any specified *p*H in all of the various solvents. Were either of these situations the case, we would anticipate similar or identical acyl product compositions independent of the solvent environment. This explanation is clearly negated by the data in Table 1; common rates of reaction result in highly disparate product compositions. Note, for example, that at *p*H 8 the rate of reaction is essentially the same in the absence of denaturant (where the product is 100 percent acylate anion), in the presence of sodium dodecyl sulfate (where the product is again nearly completely acylate anion), and in varying concentrations of urea and guanidine hydrochloride, where with increasing concentration and "potency" of the denaturant the percentage composition of denatured acyl enzyme product rises (to greater than 90 percent in the most potent denaturant). It is also noteworthy that over a limited *p*H range the rate of denaturation of the acyl enzyme (as indicated by the rate of formation of the high wavelength acyl product) follows the same *p*H dependence as does the catalytic hydrolysis reaction. In the region of catalytically optimal *p*H the composite rate of denaturation *and* hydrolysis of the acyl enzyme can *not* be described by what might, a priori, have appeared to be the most plausible scheme:

$$\mathrm{RC}(=\mathrm{O})\text{—}E_{Native} \xrightarrow[\text{Denaturant}]{k_d} \mathrm{RC}(=\mathrm{O})\text{—}E_{Denatured}$$
$$\mathrm{RC}(=\mathrm{O})\text{—}E_{Native} \xrightarrow[k_w]{H_2O} E_{Native} + RCO_2^{\ominus} + H^{\oplus}$$
$$E_{Native} \xrightarrow[\text{Denaturant}]{\text{FAST}} E_{Denatured} \quad (5)$$

The model of Equation (5) would predict that the *total* rate of disappearance of acyl enzyme is given by

$$-\frac{d\,[\textit{Native Acyl Enzyme}]}{dt} = (k_d + k_w)\,[\textit{Native Acyl Enzyme}] \quad (6)$$

In 5.6 *M* guanidine hydrochloride at *p*H 8.0, the final *acyl* spectrum has a λ_{max} at 337 mμ (Table 1), indicating a product mixture containing about 16 percent acylate anion and 84 percent denatured acyl enzyme. On the basis of Equation (5), the total rate of disappearance of native acyl enzyme in this solvent should exceed that in the absence of denaturant by at least a factor of six.

ON THE MECHANISM OF ACYL ENZYME DENATURATION IN THE CATALYTICALLY OPTIMAL *p*H REGION

As the *p*H is lowered into regions far from optimal for catalytic hydrolysis of the acyl enzyme, denaturation of the acyl enzyme occurs at rates progressively more rapid than the rate of catalyzed hydrolysis, in all of the denaturants we have studied. Increasingly, the product of reaction becomes the denatured acyl enzyme (λ_{max} = 338–342 mμ). At sufficiently low *p*H, the rates of denaturation of the enzyme and of the acyl enzyme are comparable. These results at lower *p*H are indicative of independent phenomena—*hydrolysis* and *denaturation.* We have not investigated the properties of the acyl enzyme at low *p*H in detail. Of far greater interest to us is the behavior of the acyl enzyme in various denaturants at *p*H values that lie in the optimal catalytic region, since, as will be discussed below, *hydrolysis* and *denaturation* are not independent processes in this latter range of *p*H.

The bulk of the data in the optimal *p*H range *is* reconcilable with a reaction scheme in which acyl enzyme denaturation and acyl enzyme hydrolysis are dependent on a common structural transformation:

$$\textit{Native Acyl Enzyme} \underset{}{\overset{\text{SLOW}}{\rightleftharpoons}} Y \xrightarrow[\text{Denaturants}]{\text{FAST}} \textit{Denatured Acyl Enzyme}$$
$$Y \xrightarrow[\text{FAST}]{H_2O} \textit{Acylate}^{\ominus} + \textit{Denatured Enzyme} \quad (7)$$

The chemical and conformational nature of the common intermediate is of considerable concern. Basically, two types of intermediates may be envisaged: (1) another enzyme-substrate chemical compound—for example, the tetrahedral adduct (V) (Bernhard and Gutfreund, 1957; Bernhard, Coles, and Nowell, 1960; Moon, Sturtevant, and Hess, 1965; Bernhard, Feinman, Straub, and Swanson, 1967), or (2) a conformationally altered acyl enzyme (VI).

$$\underset{\text{(V)}}{\mathrm{E{-}C(OH)_2{-}R}} \qquad \underset{\text{(VI)}}{\mathrm{E'{-}C({=}O){-}R}}$$

That an intermediate of type V exists is plausible in light of the reaction pathway. That it precedes, or in fact is, the common intermediate Y of Equation (7) cannot be established from the data presented above.

To investigate further the possible relevance of a tetrahedral intermediate to the conformational stability, we took advantage of the fact that *nucleophiles* other than water can successfully compete for the activated acyl group:

$$\mathrm{RC({=}O){-}E} + \begin{cases} \mathrm{N:} \longrightarrow \mathrm{RC({=}O){-}N} \\ \mathrm{HOH} \longrightarrow \mathrm{RC({=}O){-}OH} \end{cases} \tag{8}$$

In this regard hydroxylamine is a particularly effective nucleophile. The initial deacylation reaction, leading to *O*-acyl hydroxylamine, follows first-order kinetics and is rapid relative to hydrolysis at concentrations above 0.1 *M* in hydroxylamine. If, as we have postulated, the deacylation and denaturation processes involve a common intermediate *whose rate of formation is*

TABLE 2.
Effect of hydroxylamine on the rates of deacylation and denaturation of the native indoleacryloyl chymotrypsin.

*p*H	$[NH_2OH]$ (*M*)	[Guanidine hydrochloride] (*M*)	$k_{obs}(\min^{-1})$	% of acyl enzyme denatured without deacylation[a]
6.4	—	—	0.003	—
6.4	0.5	—	0.08	—
6.4	—	5.6	0.033	85 ± 5%
6.4	0.5	5.6	0.23	85 ± 5%
8.0	—	—	0.09	—
8.0	0.5	—	0.96	—
8.0	—	5.6	0.13	50 ± 10%
8.0	0.5	5.6	0.37	50 ± 10%

[a]In the *absence* of NH_2OH, the denatured acyl enzyme and acylate anion could both be estimated directly by spectral analysis. In the presence of NH_2OH, the *O*-acyl hydroxylamine (λ_{max} = 332 mμ) and the secondary product, N-acyl hydroxamate (λ_{max} = 323 mμ), complicate the estimation of denatured acyl enzyme. By extraction with ethyl acetate, however, all small molecule products could be removed, permitting an estimation of the denatured acyl enzyme.

rate-limiting, "hydroxylaminolysis" of the acyl enzyme should lead to a higher transient concentration of such an intermediate, since it is a faster reaction, and hence to an increase in *both* the rate of deacylation and the rate of denaturation of the acyl enzyme. On the other hand, if a tetrahedral intermediate is not relevant to the denaturation process, the fraction of the acyl enzyme denatured prior to deacylation should drop sharply (in a denaturing solvent) in the presence of an effective competitive nucleophile. Experimental results are summarized in Table 2. These results establish that the competitive reactant, hydroxylamine, increases the rate of denaturation of the acyl enzyme to an extent concomitant with its effect on the rate of deacylation. For example, at *p*H 6.4, and at 0.5 *M* NH_2OH, the deacylation rate is nearly an order of magnitude faster than the rate of hydrolysis of the acyl enzyme in the absence of NH_2OH. Likewise, the rate of denaturation of the acyl enzyme in 5.6 *M* guanidine hydrochloride is increased (by nearly an order of magnitude) in 0.5 *M* NH_2OH. The rate of denaturation of the native (nonacylated) enzyme is insignificantly affected by the presence of 0.5 *M* NH_2OH. On the basis of these results, it is reasonable to draw the parallels in time sequence summarized by Equation 9.

$$
\begin{array}{ccccc}
 & & RC(OH)(ONH_2){-}E_{Native} & \xrightarrow{FAST} & RC(=O)ONH_2 + E_{Native} \\
 & \overset{SLOW}{\rightleftharpoons}\nearrow & \downarrow \text{Denaturant, FAST} & & \downarrow \text{Denaturant, FAST} \\
RC(=O){-}E_{Native} \;\; \begin{matrix} + NH_2OH \\ + H_2O \end{matrix} & & RC(=O){-}E_{Denatured} & & E_{Denatured} \\
 & \underset{SLOW}{\rightleftharpoons}\searrow & \uparrow \text{Denaturant, FAST} & & \uparrow \text{FAST} \\
 & & RC(OH)_2{-}E_{Native} & \xrightarrow{FAST} & RCO_2H + E_{Native}
\end{array}
\qquad (9)
$$

A similar tetrahedral intermediate has been postulated, on kinetic grounds, in the *formation* of the acyl enzyme (Moon et al., 1965):

$$
E + RC(=O)X \rightleftharpoons E\left(RC(=O)X\right) \rightleftharpoons \underset{\text{(VII)}}{E{-}\overset{OH}{\underset{R}{C}}{-}X} \rightleftharpoons E{-}C(=O)R + X \qquad (10)
$$

Previously arguments have been presented that the acylation and deacylation reaction mechanisms should be "symmetrical" (Bender and Kézdy, 1964). The composite of results thus far obtained tends to substantiate such symmetry to the extent that we can confidently set forth the *minimal* catalytic mechanism in the symmetrical form given by Equation (11). Evidence has been presented in support of the view that a change in protein conformation accompanies the formation of an intermediate which we have designated as VII in Equation (10) (Moon et al., 1965). The present results on the effect of denaturants tend to substantiate a corresponding return toward the original

Chemical

$$E+S \rightleftharpoons ES \rightleftharpoons E-\overset{\displaystyle OH}{\underset{\displaystyle R}{\overset{|}{\underset{|}{C}}}}-X \rightleftharpoons E-C\genfrac{}{}{0pt}{}{\nearrow O}{\searrow R} \rightleftharpoons E-\overset{\displaystyle OH}{\underset{\displaystyle R}{\overset{|}{\underset{|}{C}}}}-OH \rightleftharpoons EP \rightleftharpoons E+P \qquad (11)$$

Conformational

E ⇌ E ⇌ E

native enzyme conformation in the step following acyl enzyme formation. Assuming the over-all symmetrical model of Equation (11), it is tempting to relate the observed changes in conformational stability with a conformational change leading to a stabilization of the transition state in the formation of the presumed tetrahedral intermediates. If conformational changes are indeed involved in the intermediate chemical transformations, and if such conformational transitions are *essential* to enzymatic catalysis, a *dynamic* role for the enzyme protein in the catalytic process becomes obvious. Over the past ten years much effort has been devoted to the attempted synthesis of "enzyme models"—that is, small rigid organic molecules with catalytic properties analogous to enzymes. The singular lack of success in catalyzing the making and breaking of *stable* bonds (the usual situation in enzyme catalysis) may reflect the impossibility of coupling an energetically favored conformational transition in the catalyst to an energetically unfavorable reaction pathway. The uniformly large dimensions of enzyme proteins are notable. A conformational transition providing significant energy may not be possible even with smaller polypeptides.

ON THE INSTABILITY OF NATIVE ENZYMES

To function effectively an enzyme must not only contain a site for its substrate, but the site must be accessible to the substrate from the solvent environment. The rates of substrate adsorbtion have now been measured in a number of instances (Eigen, 1964). In every instance, this rate is approximately that predicted on the basis of free diffusion of substrate to the active site. This implies that there are no serious obstacles to enzyme-substrate complex formation. The native enzyme protein will *tend* to assume that conformation which leads to a maximal close-packed structure. The region of the enzyme site must represent a deviation from this general tendency toward close-packed structures, since it must be accessible to substrate. This simple intuitive argument is borne out by the recent structural studies on a number of enzymes, most notably the three-dimensional structure (to atomic resolution) of egg-white lysozyme recently completed by Phillips and his collaborators (1966). Electron density maps (to a lower resolution) have been carried out on a variety of other crystalline enzymes—α-chymotrypsin Kraut et al., 1967; Matthews et al., 1967), carbonic anhydrase (Tilander et al., 1965; Fridborg et al., 1967), carboxypeptidase (Lipscomb et al., 1966), ribonuclease (Kartha et al., 1967; Wycoff et al., 1967). These results reveal a common geometrical pattern: a close-packed folded polypeptide containing a single cavity (the active site region). This "empty" cavity, in addition to providing accessibility to substrate, can give rise to an instability owing to the general accessibility to denaturants. The enzyme site may thus serve as the nucleus for the cooperative denaturation process. Indeed, very many enzymes are "unstable" proteins, very readily susceptible to the action of denaturants. Presumably, the crystallographically observed cavities contain a *loose* collection of water molecules. These water molecules must necessarily be poorly organized within the cavity; otherwise they would show up as part of the enzyme crystal structure (and would prevent free accessibility of substrate). Since protein denaturation is known to be a highly cooperative process, it is perhaps not surprising that specific chemical modifications of the protein that tend to alter the conformation of the site, such as acylation reactions, exert a strong influence on the stability of the entire molecule toward denaturation. In α-chymotrypsin, a specific and highly polar serine hydroxyl residue is involved in the catalytic process. Acylation of this polar hydroxyl group might have a profound influence on the loosely organized water molecules within the site, as schematically depicted in Figure 4. In this regard it is interesting that the stabilization

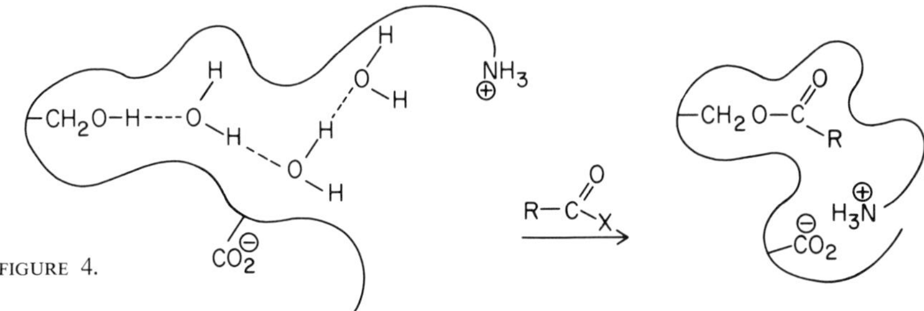

FIGURE 4.

against denaturation afforded by acylation of this serine residue is not sensibly dependent on the geometry of the particular acyl group. [Rather, it is dependent on the nucleophilicity of the carbonyl carbon atom (Bernhard, Hershberger, and Keizer, 1966)]. It is not implausible to assume that acylation of the active site serine hydroxyl residue results in the expulsion of water molecules from the site, forcing the closure of the cavity about the acyl group.* In this way, the "nucleus" for denaturation (the site) would disappear upon acylation, in accord with the observed increased stability toward denaturants. Inert chromophoric *competitive inhibitors* of α-chymotrypsin combine with enzyme to form complexes that are sensitive to the external solvent environment in both stability (dissociability) properties and spectra (Bernhard and Lau, unpublished). In contrast, the spectral properties of the covalently bound acyl group of the acyl enzyme are not perturbable by the external environment. For example, the second-order rate of acylation of α-chymotrypsin by the acylating agent N-β-(3-indole)acryloyl imidazole, a process which is presumably dependent on the pre-equilibrium binding of substrate to enzyme, is highly sensitive to the presence of low concentrations of organic (nonpolar) solvent molecules, whereas the spectrum of the resultant native acyl enzyme is unperturbable by solvents that significantly perturb the spectra of small molecule indoleacryloyl derivatives (Table 3).

*One of us (S.A.B.) is indebted to Professor Bruno Zimm for an enlightening discussion on this subject.

TABLE 3.
Spectral characteristics of indoleacryloyl derivatives in various solvents.

Compound	$\epsilon_{max} \times 10^{-4}$ (λ_{max} in mμ)		
	H_2O	4 *M* KCl	20% aqueous CH_3CN
Indoleacryloyl—			
Methyl ester	2.7 (329)	1.4 (330)	2.7 (325)
Imidazole	3.0 (380)	1.0 (381)	3.0 (374)
Native chymotrypsin	2.0 (359)	2.0 (359)	2.0 (359)

ON THE POTENTIAL SIGNIFICANCE OF ENZYME SUBSTRATE COMPOUND FORMATION TO PROTEIN-PROTEIN INTERACTIONS

We have seen that in the system α-chymotrypsin-substrate, covalent compound formation has a profound effect on polypeptide conformation. In an enzyme containing more than one polypeptide chain, the extent and specific manner of aggregation must depend on the conformation of the polypeptide subunits. Two currently popular views of the structural effects of interactions of substrate with enzyme are (1) the "induced-fit" mechanism (Koshland, 1963), in which enzyme site configuration adapts to the molecular geometry of the substrate, and (2) the "allosteric mechanism" (Monod et al., 1965), in which substrate adsorption alters the pre-existent equilibrium between two different polypeptide conformations. Both these views may be subject to modification owing to the induction of conformational transitions upon formation of a covalent enzyme-substrate intermediate. Conventional steady-state kinetic techniques afford no useful distinction among these proposed mechanisms.

Recent crystallographic studies of hemoglobin have revealed changes in intersubunit interactions concomitant with the formation of the specific heme-oxygen covalent bonds (Perutz et al., 1964). The specific chemical reaction of wet chymotrypsin crystals with organic phosphate esters and sufonyl halides causes noticeable changes in the unit cell dimensions (Sigler et al., 1966). The binding of both a specific competitive inhibitor and a (catalytically) poor substrate to the enzyme egg white lysozyme results in only a minute change in the configuration of amino acid side chains (most notably, tryptophan) which define the active site (Phillips, 1966). It should be noted that the lysozyme-substrate complex examined crystallographically is almost certainly a physically adsorbed complex—that is, covalent interactions do not occur to any significant extent. On the basis of extremely fragmentary information, it would appear to us that minute changes in polypeptide conformation can result from physical adsorbtion of substrate, but that functionally more profound changes in conformation can result from the formation of specific covalent bonds.

It is tempting to speculate on the possible significance of rapid chemical bond formation to physiological response mechanisms, for example, in the contraction of muscle and in the induction of neural responses. In muscular contraction, chemical reaction of the phosphorylating agent ATP with the protein myosin may plausibily involve the formation of an enzyme phosphate ester or phosphoamide. In the induction of neural responses, the enzyme acetylcholinesterase may itself be the chemoreceptor. This enzyme is known to be of very high molecular weight ($\sim 10^6$) and is presumably an aggregate of smaller polypeptide subunits. Cholinesterase very probably reacts with substrate to form a metastable acyl enzyme intermediate (Wilson et al., 1950). Both these enzyme-substrate reactions are analogous to the reaction for which we have proposed the model of Equation (11). Conformational changes in the enzyme proteins, produced as a consequence of chemical reaction, may alter the orderly arrangement of polypeptide subunits in the physiological systems so as to "trigger" chemomechanical and chemoelectric responses.

REFERENCES

Atkinson, D. A., J. A. Hathaway, and E. C. Smith (1965). *J. Biol. Chem.* **240**, 2682.
Bender, M. L. (1962). *J. Am. Chem. Soc.* **84**, 2582.
Bender, M. L., and F. J. Kézdy (1964). *J. Am. Chem. Soc.* **86**, 3704.
Bender, M. L., and B. Zerner (1964). *J. Am. Chem. Soc.* **86**, 3669.
Bender, M. L., G. R. Schonbaum, and B. Zerner (1962a). *J. Am. Chem. Soc.* **84**, 2540.
Bender, M. L., G. R. Schonbaum, and B. Zerner (1962b). *J. Am. Chem. Soc.* **84**, 2562.
Bernhard, S. A., and H. Gutfreund (1957). *Proc. Int. Symp. Enzyme Chemistry*, p. 124, Tokyo.
Bernhard, S. A., and Z. H. Tashjian (1965). *J. Am. Chem. Soc.* **78**, 1807.
Bernhard, S. A., W. C. Coles, and J. F. Nowell (1960). *J. Am. Chem. Soc.* **82**, 3043.
Bernhard, S. A., E. Hershberger, and J. Keizer (1966). *Biochemistry* **5**, 4120.
Bernhard, S. A., S. J. Lau, and H. Noller (1965). *Biochemistry* **4**, 1108.
Bernhard, S. A., R. D. Feinman, B. A. Straub, and E. Swanson (1967). *Science*. In press.
Caplow, M., and W. P. Jencks (1962). *Biochemistry* **1**, 883.
Changeux, J.-P. (1963). *Cold Spring Harbor Symposia Quantitative Biology* **28**, 497.
Clement, G. E., and M. L. Bender (1963). *Biochemistry* **2**, 836.
Eigen, M. (1964). *Proc. 6th Int. Congress Biochemistry*, vol. 4, S14. New York.
Epand, R. M., and I. B. Wilson (1964). *J. Biol. Chem.* **239**, 4138.
Epand, R. M., and I. B. Wilson (1965). *J. Biol. Chem.* **240**, 1104.
Foster, R. J., and C. Niemann (1955). *J. Am. Chem. Soc.* **77**, 1886.
Fridborg, K., K. K. Kannan, A. Liljas, J. Lundin, B. Strandberg, R. Strandberg, B. Tilander, and G. Wirén (1967). *J. Mol. Biol.* **25**, 505.
Gerhart, J. C., and A. B. Pardee (1963). *Cold Spring Harbor Symp. Quant. Biol.* **28**, 491.
Gerhart, J. C., and H. K. Schachman (1965). *Biochemistry* **4**, 1054.
Hearon, J. Z., S. A. Bernhard, S. L. Friess, D. J. Botts and M. F. Morales (1959). In Boyer, P. D., H. Lardy, and K. Myrbäck, eds., *The Enzymes*, 2nd ed., vol. 1, p. 49. New York: Academic Press.
Hein, G., and C. Niemann (1961). *Proc. Nat. Acad. Sci. U.S.* **47**, 1341.
Inagami, T., S. S. York, and A. Patchornik (1965). *J. Am. Chem. Soc.* **87**, 126.
Kartha, G., J. Bellow, and D. Harker (1967). *Nature* **213**, 862.
Koshland, D. E., Jr. (1963). *Cold Spring Harbor Symp. Quant. Biol.* **28**, 473.
Kraut, J., H. T. Wright, M. Kellerman, and S. T. Freer (1967). *Proc. Nat. Acad. Sci. U.S.* **58**, 304.
Laursen, R. A., and F. H. Westheimer (1966). *J. Am. Chem. Soc.* **88**, 3426.
Lipscomb, W. N., J. C. Coppola, J. A. Hartsuck, M. L. Ludwig, H. Muirhead, J. Searl and T. A. Steitz (1966). *J. Mol. Biol.* **19**, 423.
Martin, C. J. (1964). *Biochemistry* **3**, 1635.

Martin, C. J., and G. M. Bhatnagar (1966). *Biochemistry* **5**, 1230.

Matthews, B. W., P. B. Sigler, R. Henderson, and D. M. Blow (1967). *Nature* **214**, 652.

Monod, J., J. Wyman, and J. P. Changeux (1965). *J. Mol. Biol.* **12**, 88.

Moon, A. Y., J. M. Sturtevant, and G. P. Hess (1965). *J. Biol. Chem.* **240**, 4204.

Nelson, G. H., J. L. Miles, and W. J. Canady (1962). *Arch. Biochem. Biophys.* **96**, 545.

Noller, H. F., and S. A. Bernhard (1965). *Biochemistry* **4**, 1118.

Perutz, M. F., W. Bolton, R. Diamond, H. Muirhead, and H. C. Watson. (1964). *Nature*, **203**, 687.

Phillips, D. C. (1966). *Scientific American* **215**, p. 78.

Racker, E., and I. Krimsky (1952). *J. Biol. Chem.* **198**, 731.

Schwartz, J. H. (1963). *Proc. Nat. Acad. Sci. U.S.* **49**, 871.

Shine, H. J., and C. Niemann (1955). *J. Am. Chem. Soc.* **77**, 4275.

Sigler, P. B., B. A. Jeffery, B. W. Matthes, and D. M. Blow (1966). *J. Mol. Biol.* **15**, 175.

Snell, E. E., and W. T. Jenkins (1959). *J. Cell. Comp. Physiol.* **54**, suppl. 1, 161.

Tilander, B., B. Strandberg, and K. Fridborg (1965). *J. Mol. Biol.* **12**, 740.

Wilson, I. B., F. Bergmann, and D. Nachmansohn (1950). *J. Biol. Chem.* **186**, 781.

Wooten, J. F., and G. P. Hess (1960). *Nature* **188**, 726.

Wycoff, H. W., K. D. Hardman, N. M. Allewell, T. Inagami, L. N. Johnson, and F. M. Richards (1967). *J. Biol. Chem.* **242**:3749.

CAROLE L. HAMILTON
HARDEN M. McCONNELL
Stauffer Laboratory for Physical Chemistry
Stanford University
Stanford, California

Spin Labels

The paramagnetic resonance spectrum of a free radical is sensitive to its environment. Motion relative to the laboratory fixed axis system can affect the line shape if there are anisotropic nuclear hyperfine and/or spin-orbit interactions. The electronic structure can be modified by electrostatic interactions with neighboring molecules, producing changes in the spin distribution and thus in the nuclear hyperfine spectrum. Exchange and magnetic dipolar interactions with the unpaired electron spins of other paramagnetic species can also modify the spectrum, as can magnetic dipolar interactions with surrounding nuclei. All of these phenomena have been the subject of intensive quantitative study by chemists and physicists during the past decade, and truly spectacular achievements have been made in this area and in the associated theories of molecular electronic structure, spin distributions, and electron and nuclear magnetic relaxation, as well as in the area of magnetic resonance instrumentation.

Since an exceedingly small proportion of the components of biological systems is paramagnetic, the resonance spectra of free radicals introduced into such systems are essentially free from interference and accurately reflect the properties of the paramagnetic molecules and their environment. In this laboratory, a technique has been developed for using stable organic free radicals, "spin labels," for probing the structure, function, and chemistry of biological systems.

At present, because the spin label method is only a little more than a year old, a full-fledged review would be premature. On the other hand, the applications of the method have proved to be so numerous and so diverse that complete coherent coverage of the field will soon be impossible. The present contribution is therefore designed to summarize current and past work in our laboratory, as well as to indicate likely areas of future research.

For our purposes, it is convenient to define a "spin label" as a synthetic organic free radical that can be incorporated within or attached to a molecule or system of biological interest to provide information concerning structure, conformation change, or chemical reactions. For convenience we exclude paramagnetic metal ions as well as certain "natural" free radicals, such as those present in electron transport pathways. The great practical difficulty in devising useful spin labels arises from the fact that nearly all organic free

Supported by the National Science Foundation under Grant No. GB-4911 and by the Office of Naval Research under Contract No. NONR-225(88).

radicals are extremely reactive, especially in aqueous solutions under physiological conditions. The first spin label employed in our work was the positive ion radical of chlorpromazine, which was used to study the problem of intercalation of aromatic molecules in native (double-helical) DNA. The hyperfine

S

N Cl

$(CH_2)_3$

N

CH_3 CH_3

$^+$

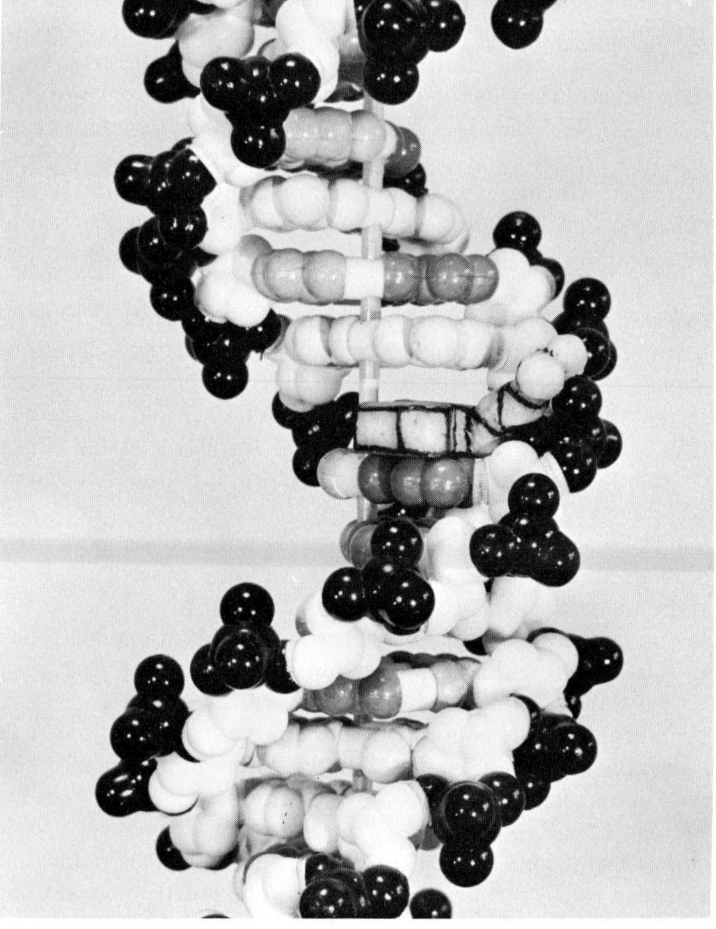

FIGURE 1.
Molecular model of a portion of a DNA double helix, showing the intercalation of a chlorpromazine ion radical.

interaction and *g*-factor anisotropy of this radical are sufficiently large that its binding to DNA has a pronounced effect on the paramagnetic resonance spectrum; by observing the paramagnetic resonance in a flowing system to produce orientation of the helix axis it was concluded that the aromatic plane of the ion radical is in fact nearly perpendicular to the helix axis, as indicated in the model in Figure 1 (Ohnishi and McConnell, 1965). The ion radical of chlorpromazine has many drawbacks for use as a general purpose spin label, however. It is only moderately stable in aqueous solution, and only in a limited *p*H range; the hyperfine structure is so complicated that quantitative analyses of line shape are very difficult if not impossible.

Much of the rapid progress with the spin label method has been due to the use of remarkably stable and unreactive nitroxide radicals that have an extremely simple nuclear hyperfine interaction. The properties of these radicals are so important for the application of the method that we give below a rather extensive survey of the chemistry relevant to the preparation and use of spin labels.

CHEMISTRY OF NITROXIDE RADICALS

The di-*t*-alkyl nitroxides have characteristics that make them ideal for use as spin labels. The first such radical that was synthesized, di-*t*-butyl nitroxide (I)

$$(CH_3)_3C-N(O)-C(CH_3)_3$$

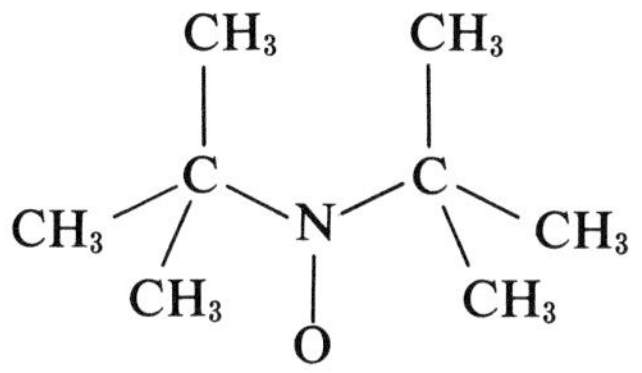

I

(Hoffmann and Henderson, 1961), may be fractionally distilled, for example, or subjected to vapor phase chromatography at 118° without decomposition. Its solutions in organic solvents, water, and aqueous alkali are stable, and it does not react with oxygen. Shortly after this first synthesis, it was shown that a functional group incorporated into a nitroxide-containing molecule can act independently of the free electron (Neiman, Rozantzev, and Mamedova, 1962). Extensive research on the synthesis and chemistry of a variety of hindered nitroxides has followed.

Di-*t*-butyl nitroxide (I) has been prepared by the reaction of *t*-nitrobutane with sodium metal (Hoffmann and Henderson, 1961; Hoffmann, Feldman, Gelblum, and Hodgson, 1964), with *t*-butyllithium (Hoffmann, Feldman, and Gelblum, 1964), and with excess *t*-butylmagnesium chloride in the presence of magnesium metal (Briere and Rassat, 1965). Several functionally substituted analogues of di-*t*-butyl nitroxide have been synthesized by thermal decomposition of appropriate azo compounds in the presence of 2-nitroso-2-methylpropane (Hoffmann, 1964).

A more generally utilized method of synthesis involves oxidation of the appropriate cyclic secondary amine. The nitroxide of triacetonamine, 2,2,6,6-tetramethyl-4-piperidone-1-oxyl (II), has been prepared by treatment of the

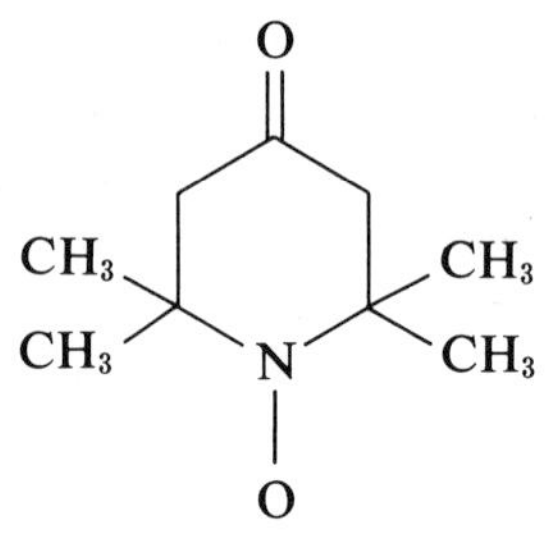

II

amine with silver oxide (Rozantzev and Neiman, 1964), but oxidation with hydrogen peroxide in the presence of a catalyst has become the method of choice. Sodium tungstate with EDTA (Neiman et al., 1962; Rozantzev and Neiman, 1964; Krinitskaya, Rozantzev, and Neiman, 1965; Rozantzev and Krinitskaya, 1965) and phosphotungstic acid (Briere, Lemaire, and Rassat, 1965) have been employed to catalyze the formation of a large number of functionally substituted nitroxides. The reaction rate with phosphotungstic acid seems to be significantly higher than that produced by sodium tungstate (Briere et al., 1965), but in some cases the yield of product is poorer. For example, the oxidation of 2,2,6,6-tetramethylpiperidine is reported to give 70 percent of the theoretical nitroxide with sodium tungstate (Rozantzev and Neiman, 1964), while phosphotungstic acid results in a yield of 54 percent (Briere et al., 1965). 2,2,5,5-Tetramethyl-3-carboxamidopyrroline-1-oxyl is formed in 92 percent yield by sodium tungstate (Rozantzev and Krinitskaya, 1965) and in 61 percent yield by phosphotungstic acid (Briere et al., 1965).

Some limitations are imposed upon nitroxide formation by the accompanying functional groups, of course. One substituent sensitive to the conditions of nitroxide formation is the cyanide group, which is converted to an amide (Krinitskaya et al., 1965; Rozantzev and Krinitskaya, 1965). Also, oxidation of 2,2,5,5-tetramethyl-3-aminopyrrolidine leads to the corresponding oxime nitroxide III (Krinitskaya et al., 1965; Rozantzev and Krinitskaya, 1965).

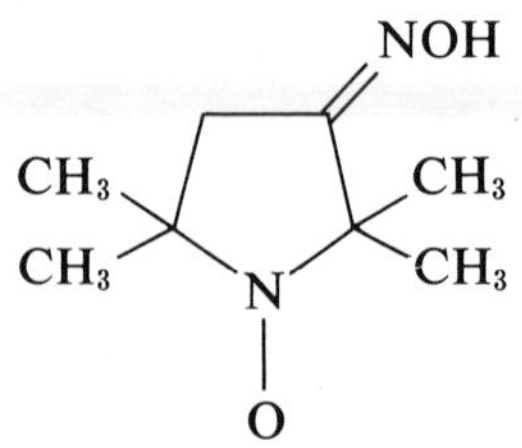

III 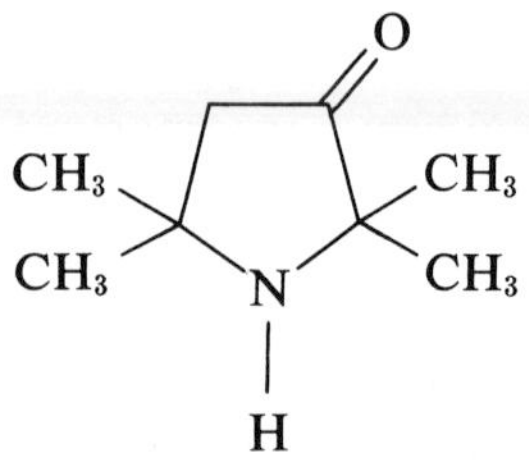 IV

2,2,5,5-Tetramethyl-3-pyrrolidinone (IV) did not survive attempts at sodium tungstate/EDTA-catalyzed oxidation; no paramagnetic products were formed (Krinitskaya et al., 1965). It has been reported that when phosphotungstic acid was used as a catalyst (Dupeyre, Rassat, and Rey, 1965) some free radical was formed but could not be isolated from the crude product mixture. The nitroxide of IV could be successfully prepared, however, by oxidation of its semicarbazone and subsequent hydrolysis (Dupeyre et al., 1965) or by

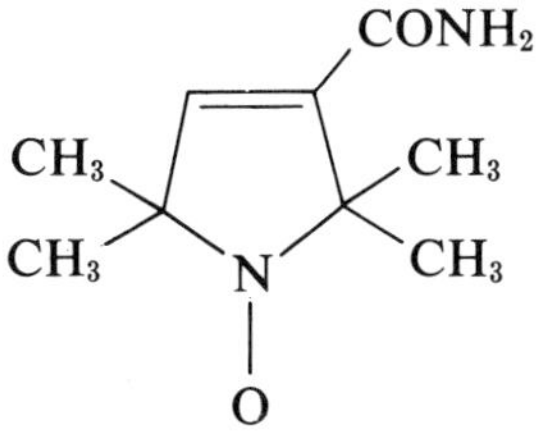

V

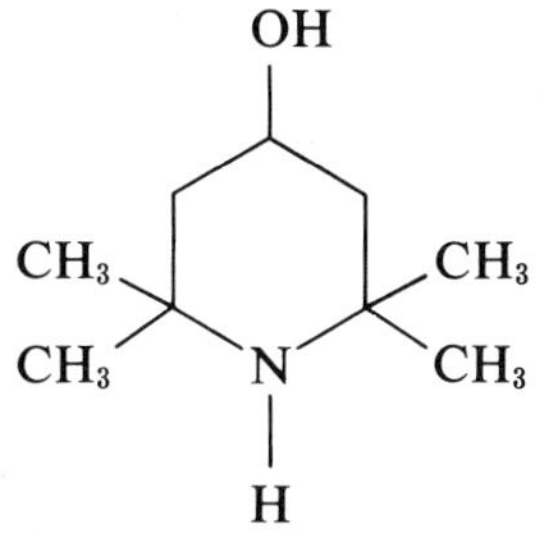

VI

Hoffmann degradation of 2,2,5,5-tetramethyl-3-carboxamidopyrroline-1-oxyl, V (Krinitskaya et al., 1965). It has also been reported (Rozantzev and Krinitskaya, 1965) that attempted oxidation of 2,2,5,5-tetramethyl-3-hydroxypyrrolidine causes destruction of the heterocycle. The six-membered alcohol, 2,2,6,6-tetramethyl-4-hydroxypiperidine (VI), however, is converted smoothly to its nitroxide in the presence of either catalyst (Briere et al., 1965; Rozantzev, 1964), as is the corresponding ketone, triacetonamine (unpublished result from these laboratories).

Once synthesized, the functionally substituted stable nitroxides undergo a wide variety of reactions without involvement of the odd electron. Derivatives can be made of the ketones (Neiman et al., 1962; Rozantzev and Neiman, 1964) and alcohols (Briere et al., 1965; Rozantzev, Golubev, and Neiman, 1965), and an amide substituent may be hydrolyzed to a carboxyl group (Rozantzev and Krinitskaya, 1965) or dehydrated to produce a nitrile (Krinitskaya et al., 1965; Rozantzev and Krinitskaya, 1965). Esterification of a nitroxide-carboxylic acid with diazomethane goes quantitatively with no loss of spin (Rozantzev and Krinitskaya, 1965). *Para*-nitrophenol (Berliner and McConnell, 1966) and imidazole (C. L. Hamilton, unpublished) have been acylated with nitroxide acids by using dicyclohexylcarbodiimide as a coupling agent. Likewise, nitroxide amines have been acylated (C. L. Hamilton, unpublished; S. Ogawa, unpublished; Griffith and McConnell, 1966; Smith and McConnell, 1967), alkylated (W. J. Deal, unpublished), and incorporated into N-alkyl maleimides (C. L. Hamilton, unpublished; Griffith and McConnell, 1966), as well as converted to isocyanate (Stone, Buckman, Nordio, and McConnell, 1965). The nitroxide group survived reaction of 2,2,6,6-tetramethyl-4-piperidinone-l-oxyl (II) with potassium-*t*-butoxide and dialkyl succinates in the Stobbe condensation (Golubev and Rozantzev, 1965). Grignard reagents have also been made to react with the carbonyl of II without alteration of the nitroxide moiety (Rozantzev and Nieman, 1964; Briere et al., 1965).

It must be remembered, though, that for all their model behavior, nitroxides are still free radicals and not immune to attack in many situations. Oxidation and reduction are facile; for di-*t*-butyl nitroxide the respective one-electron potentials are $+0.55$ v and -1.63 v, measured polarographically in acetonitrile (Hoffmann and Henderson, 1961). Catalytic hydrogenation takes a nitroxide group to the corresponding hydroxylamine (Hoffmann and Henderson, 1961; Neiman et al., 1962; Rozantzev and Shapiro, 1964) or amine (Rozantzev and Shapiro, 1964), depending on conditions. The hydroxylamine, which is readily reconverted to nitroxide in air (Hoffmann and Henderson,

1961; Rozantzev et al., 1965), may also be formed by treatment of the radical with phenylhydrazine (Rozantzev et al., 1965; Rozantzev and Shapiro, 1964). It is interesting, however, that bulky agents such as aluminum isopropoxide may be used to reduce ketone nitroxides to secondary alcohols (Rozantzev and Krinitskaya, 1965; C. L. Hamilton, unpublished) without extensive destruction of the paramagnetic center.

Nitroxide-containing molecules react reversibly with sodium dithionite to form a diamagnetic species, which then undergoes a slow irreversible conversion to products as yet unidentified (C. L. Hamilton, unpublished). Although bis(trifluoromethyl) nitroxide forms an adduct with nitric oxide in the gas phase (Blackley and Reinhard, 1965), the tetramethylated cyclic radicals do not seem to react with it. The paramagnetic resonance spectrum of N-(2,2,6,6-tetramethyl-1-oxyl-4-piperidinyl)maleamic acid in aqueous solution was not affected by treatment with oxygen-free nitric oxide (J. Chien, unpublished). Little has been published about the stability of the nitroxide group to acids except that anhydrous hydrogen chloride converts di-*t*-butyl nitroxide (I) to N, N-di-*t*-butylhydroxylammonium chloride (VII). Upon neutralization with

$$(CH_3)_3C-\overset{+}{N}H(OH)-C(CH_3)_3 \quad Cl^-$$

VII

aqueous alkali, VII reverts to the hydroxylamine, from which I can be recovered (Hoffmann and Henderson, 1961). Several nitroxide-containing molecules have survived solution in aqueous media of *p*H 1 for moderate lengths of time, but this stability may not be a general phenomenon (unpublished result from these laboratories).

The overall lack of reactivity of the di-*t*-alkyl nitroxides merits comment. It has been suggested that the intrinsic stability of the three-electron N—O bond (Linnett, 1961) and the steric hindrance provided by the adjacent *t*-alkyl groups are responsible for the molecules' resistance to dimerization (Hoffmann and Henderson, 1961). Evidence has also been presented that decomposition of nitroxides involves intermediate formation of nitrones,

$$(CH_3)_3C-\overset{+}{N}(O^-)=C(CH_3)_2$$

a process inhibited when none of the α-substituents is an easily eliminated group. In norpseudopelletierine-N-oxyl,

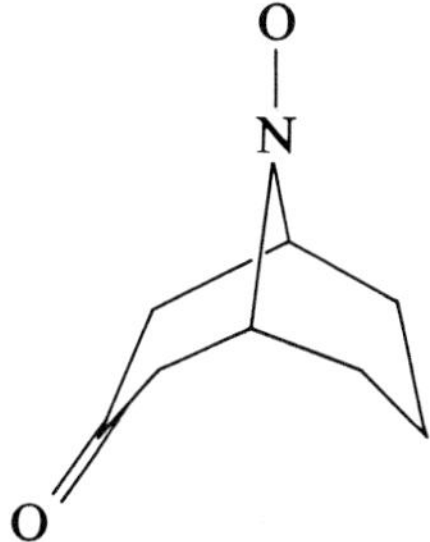

the nitroxide is flanked by secondary bridgehead carbons; this radical is stable in the solid state and in solution in benzene or neutral water (Dupeyre and Rassat, 1966).

RESONANCE SPECTRA OF NITROXIDE RADICALS

The electron spin resonance spectrum of a typical hindered nitroxide

$$CH_3-\underset{CH_3}{\overset{R}{C}}-\underset{O}{\dot{N}}-\underset{CH_3}{\overset{R}{C}}-CH_3$$

in solution at room temperature is simple, consisting of three sharp lines produced by nitrogen hyperfine interaction, as shown in Figure 4, *a*. The isotropic splitting constant a_N varies slightly with the structure of the rest of the molecule and with the solvent (Briere et al., 1965; Briere, Lemaire, and Rassat, 1964; Dupeyre, Lemaire, and Rassat, 1964). Table 1 lists values of a_N that have been measured for various methyl-protected nitroxides. Observed effects of solvent on a_N are somewhat larger than those associated with structure, and are illustrated in Figure 2 (Briere et al., 1965). The observed increase of the hyperfine interaction with polarity of the solvent is proportional to Kosower's (1958) Z factor (an empirical constant based on the charge-transfer band of a pyridinium iodide), and probably reflects increased stabilization of the negative charge on oxygen, hence a greater spin density on nitrogen (Briere et al., 1965; Stone et al., 1965).

In some cases, further fine structure has been observed in the spectra of nitroxides. A high-gain spectrum of triacetonamine nitroxide II shows doublet satellites with a splitting of 21 gauss produced by N^{15} in natural abundance (Briere et al., 1965; Briere et al., 1964). If the line width is narrow enough, the three main peaks are each accompanied by a doublet (splitting of approximately 4.5 to 7 gauss) due to C^{13} in the methyl and ring methylene groups (Briere et al., 1965; Briere et al., 1964).

The spectroscopic *g* factors are independent of structure (Briere et al., 1965; Griffith, Cornell, and McConnell, 1965), but they decrease with increasing polarity of solvent (Briere et al., 1965).

As pointed out at the beginning of this paper, the overall shapes of paramagnetic resonance spectra of radicals in general depend on several environ-

TABLE 1.
Isotropic nuclear hyperfine splitting constants for various methyl-protected nitroxides.

Compound	a_N (gauss)[a]	Reference
$(CH_3)_3CNC(CH_3)_3$ with N–O	15.1[b]	Griffith et al., 1965
O; CH_3, CH_3, CH_3, CH_3; N; O	15.1	Briere et al., 1964; 1965
	15.3[c]	Rozantzev and Neiman, 1964
	14.3[b]	Griffith et al., 1965
	16.1[e]	Ogawa, unpublished
N–OH; CH_3, CH_3, CH_3, CH_3; N; O	15.5	Briere et al., 1964; 1965
OH; CH_3, CH_3, CH_3, CH_3; N; O	16.1	Briere et al., 1964; 1965
	15.1[d]	Rozantzev, 1964
	17.1[e]	Ogawa, unpublished
$OCOCH_3$; CH_3, CH_3, CH_3, CH_3; N; O	15.9	Briere et al., 1965

Compound	a_N (gauss)[a]	Reference
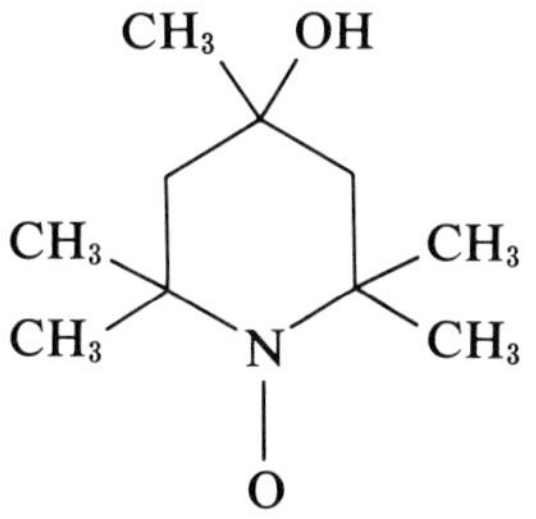 CH_3 OH CH_3 CH_3 CH_3 N CH_3 O	15.96	Briere et al., 1964; 1965
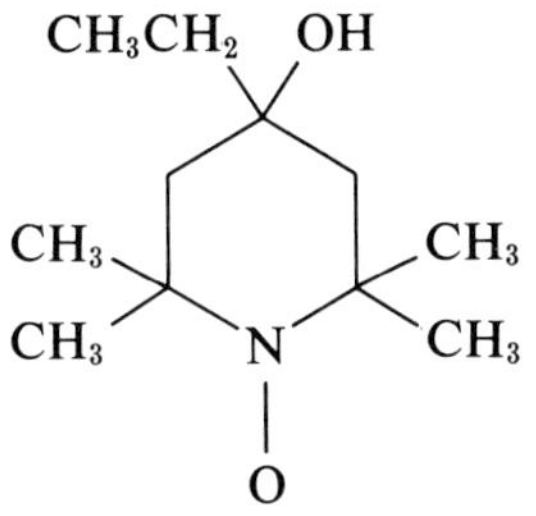 CH_3CH_2 OH CH_3 CH_3 CH_3 N CH_3 O	15.3[d]	Rozantzev, 1964
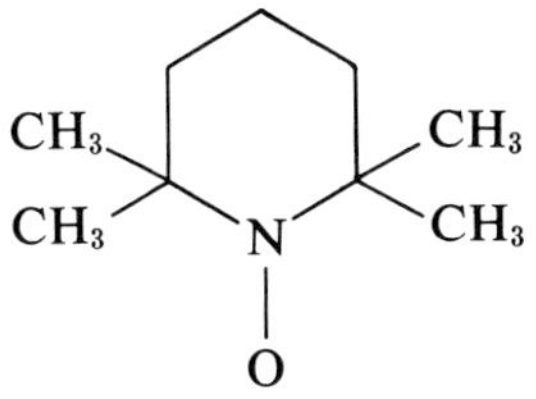 CH_3 CH_3 CH_3 N CH_3 O	16.3	Briere et al., 1964; 1965
$CONH_2$ CH_3 CH_3 CH_3 N CH_3 O	14.9 14.0[b] 16.0[c]	Briere et al., 1965 Griffith et al., 1965 Briere et al., 1965
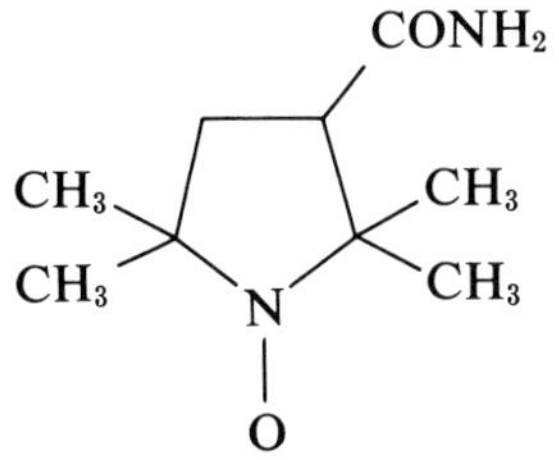 $CONH_2$ CH_3 CH_3 CH_3 N CH_3 O	16.0[c]	Ogawa, unpublished
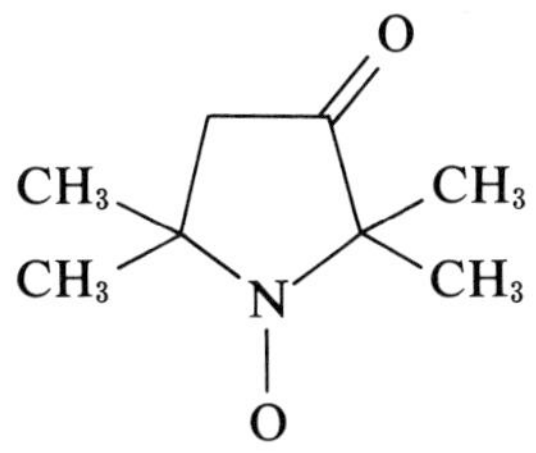 O CH_3 CH_3 CH_3 N CH_3 O	14.76	Dupeyre et al., 1964

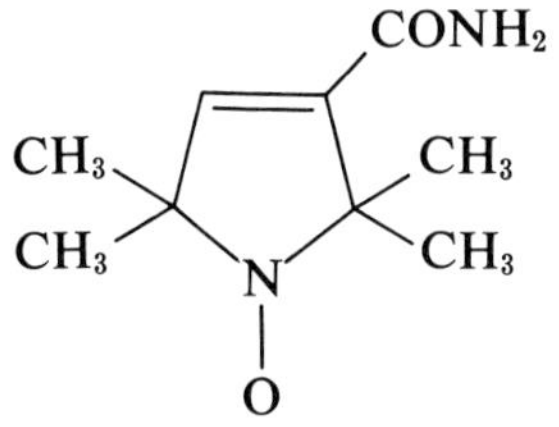

TABLE 1. (*continued*)
Isotropic nuclear hyperfine splitting constants for various methyl-protected nitroxides.

Compound	a_N (grauss)[a]	Reference
NOH; CH_3, CH_3, CH_3, CH_3; N; O	15.01	Dupeyre et al., 1964
OH; CH_3, CH_3, CH_3, CH_3; N; O	15.2	Dupeyre et al., 1964
NH_2; CH_3, CH_3, CH_3, CH_3; N; O	15.6[e] 16.2[f]	Ogawa, unpublished Ogawa, unpublished
CH_3, CH_3, CH_3, CH_3; N; O	15.3	Dupeyre et al., 1964
COOH; CH_3, CH_3, CH_3, CH_3; N; O	15.8[g] 15.5[h]	Hamilton, unpublished Hamilton, unpublished

[a]Splittings measured in diethylene glycol unless otherwise noted.
[b]Measured in di-*t*-butyl ketone.
[c]Measured in water.
[d]Measured in benzene.
[e]Measured in water at *p*H 1.9.
[f]Measured in water at *p*H 11.2.
[g]Measured in 5% sodium bicarbonate.
[h]Measured in 0.1 *N* HCl.

mental factors. The conditions of biological experiments are such that the analysis of nitroxide resonance line shapes is particularly simple. The great magnetic dilution of biological systems, the modest temperature ranges involved, and the relative invariance of hyperfine and *g*-factor anisotropies to solvent permit line shapes to be analyzed in terms of the effect of local environment on radical motion. This analysis requires knowledge of the anisotropies of the hyperfine and *g*-factor tensors; single crystal experiments that have been used to obtain this information are described below.

The paramagnetic resonance of oriented single crystals (hosts) containing stable alkyl nitroxides (Griffith et al., 1965) as substitutional impurities (guests) can be accounted for by the spin Hamiltonian

$$\mathcal{H} = |\beta|\mathbf{S}\cdot\mathbf{g}\cdot\mathbf{H}_0 + hAS_zI_z + hBS_xI_x + hCS_yI_y \tag{1}$$

In equation (1), x, y, z are the principal axes of the hyperfine interaction, $\mathbf{g}$ is the spectroscopic splitting factor tensor, $|\beta|$ is the absolute value of the Bohr magneton, $\mathbf{H}_0$ is the applied magnetic field vector, and $\mathbf{S}$ and $\mathbf{I}$ are the electron and nuclear (N^{14}) spin operators, in units of $\hbar$. The elements of the *g*-tensor (for compound I) are $g_{xx} = 2.0089$, $g_{yy} = 2.0061$, and $g_{zz} = 2.0027$. The measured N^{14} nuclear hyperfine splittings are $|A| = 87$ Mc, $|B| \simeq |C| \simeq 14$–17 Mc. These hyperfine interactions yield splittings in a high-field, field-swept spectrum of 32 gauss, 6 ± 1.5 gauss, and 6 ± 1.5 gauss when the applied magnetic field is along the z-, x-, and y-axes, respectively. The experimental absolute value of 43 Mc (or 15.5 gauss) for the isotropic hyperfine splitting constant, a, is consistent with equation (2) and establishes that A, B, and C

$$|a| = \tfrac{1}{3}|A + B + C| \tag{2}$$

have the same sign. These results, together with additional data given below, permit one to deduce a great deal about the molecular and electronic structure of these radicals.

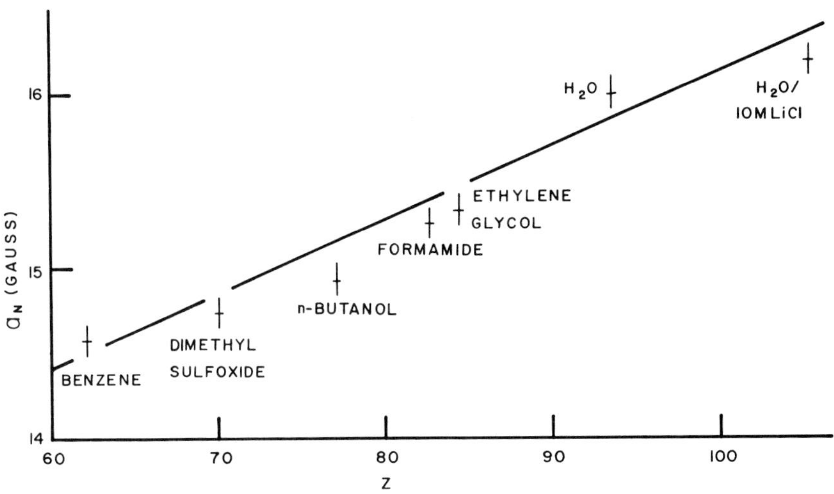

FIGURE 2.
Variation with solvent of the nuclear hyperfine splitting constants for 2,2,6,6-tetramethyl-4-piperidone-1-oxyl. The abscissa represents values of the Kosower Z factor for each solvent (Briere et al., 1965).

In the experiments of Griffith et al. (1965) radicals I and II as well as the nitroxide derived from IV were incorporated into crystals of tetramethyl-1, 3-cyclobutanedione, a host of known crystal structure (Friedlander and Robertson, 1956). The structure is monoclinic with two magnetically equivalent molecules per unit cell. The cyclobutane ring and the two oxygen atoms all lie in a plane (Friedlander and Robertson, 1956). Since the spin resonance spectra observed for nitroxide I in tetramethylcyclobutanedione show the symmetry expected from the host crystal structure, it is reasonable to assume that nitroxides enter the crystals as oriented substitutional impurities. The structure of I is certainly compatible with oriented substitution for the dione.

As a second assumption we take the nitrogen, the oxygen, and the two tertiary carbon atoms of I to be coplanar. The two assumptions then require one symmetry axis of the hyperfine interaction to be perpendicular to the cyclobutanedione ring. This is observed; the principal axis z is found to be normal to the plane of the ring. Furthermore, the first and second assumptions require that a second principal axis of the hyperfine interaction be found in the cyclobutanedione oxygen-oxygen direction, as observed (the x-axis). Since the spin resonance spectra of II and the nitroxide of IV in tetramethylcyclobutanedione are very similar to those observed for I in the same host crystal, similar assumptions can be applied to these nitroxide radicals as well.

If the nitrogen, oxygen, and two equivalent tertiary carbon atoms are all coplanar, then the odd electron must certainly be in a molecular orbital of π symmetry (antisymmetric to reflection in the xy plane). In fact, the N^{14} nuclear hyperfine interaction indicates that the odd electron is largely confined to a $2p\pi$ atomic orbital on the nitrogen atom (Stone et al., 1965). The approximate axial symmetry of the hyperfine interaction ($B \simeq C$) is consistent with this picture. Also, the isotropic hyperfine interaction observed for aliphatic nitroxides (15.5 gauss) is close to the value of 19.5 gauss measured for NH_3^+ (Cole, 1961); the isotropic splitting calculated from theory for planar NH_3^+ is 20.9 gauss (Giacometti and Nordio, 1963). By arguments analogous to those originally made for the C^{13} splitting in CH_3 (Cole, Pritchard, Davidson, and McConnell, 1958), one may conclude that NH_3^+, and therefore the nitroxide radicals, are essentially planar. Otherwise, any significant first-order s-orbital character for the odd electron would produce a much larger isotropic splitting.

Carrington and Longuet-Higgins (1962) have used a Slater $2p\pi$ orbital for calculations on anisotropic nitrogen hyperfine interactions, and have obtained a value of 76 Mc for the purely anisotropic electron-nucleus dipole interaction for spins aligned along the z-direction. Taking the spin density in the $2p\pi$ orbital of an aliphatic nitroxide as $15.5/19.5 = 0.80$, we can estimate that the splitting parallel to the π-orbital axis should be

$$A = 43 \text{ Mc} + 0.80 \times 76 \text{ Mc} \simeq 100 \text{ Mc}$$

The agreement between the calculated ($\sim$100 Mc) and observed (87 Mc) values of A is quite reasonable, considering the known crudeness of the Slater orbitals for this particular calculation. Similarly, using -38 Mc as the anisotropic interaction in the x- and y-directions (Carrington and Longuet-Higgins, 1962), we can estimate the splittings perpendicular to the π-orbital axis to be

$$B \simeq C = 43 \text{ Mc} + 0.80\,(-38 \text{ Mc}) = 13 \text{ Mc}$$

again in good agreement with the values of 14–17 Mc measured for nitroxide radicals. Finally, the ratio $a(N^{14})/A(N^{14}) = 0.5$ observed for nitroxide radicals is very close to the ratio of the isotropic C^{13} splitting in $C^{13}H_3$(115 Mc) to the value for $A(C^{13})$(210 Mc) in the extensively studied malonic acid radical, $a(C^{13})/A(C^{13}) = 0.55$ (McConnell and Fessenden, 1959). Thus, in view of this overwhelming body of remarkably self-consistent evidence, we may safely conclude that the odd electron in the nitroxide radicals under discussion is largely confined (80–90 percent) to a $2p\pi$ atomic orbital on nitrogen, corresponding to a polar valence bond structure,

A sketch of the $2p\pi$ orbital and the local molecular geometry for nitroxide radicals is given in Figure 3.

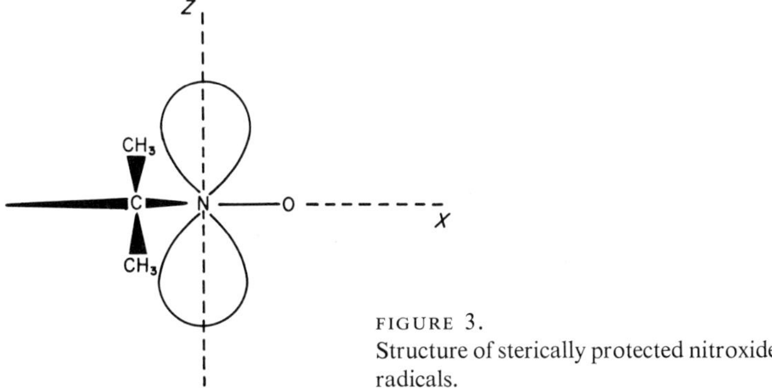

FIGURE 3.
Structure of sterically protected nitroxide radicals.

As is evident from the foregoing discussion, the paramagnetic resonance of a nitroxide radical depends on how the radical is oriented relative to an applied magnetic field. If one has a random distribution of orientations of such radicals in space, one obtains a "powder" spectrum, illustrated in Figure 4,*f*. Such spectra can be obtained experimentally by grinding up the tetramethylcyclobutanedione crystals containing nitroxide I as a substitutional impurity, or by observing the resonance of the radicals dissolved in a rigid glass. Spectra like that in Figure 4,*f* are obtained when the nitroxide motion is slow—when the correlation time τ for radical tumbling is related to the anisotropy of the nuclear hyperfine interaction by the inequality

$$\tau^{-1} \ll |A - B| \tag{3}$$

(Strictly speaking, certain field-dependent terms involving the g-factor anisotropy must also be considered, as is evident from the following discussion.)

The resonance spectrum in Figure 4,*f* can be reproduced almost exactly by a computing machine calculation that takes sums of spectra for an isotropic statistical distribution of orientations (Itzkowitz, 1966). Although one cannot give a simple interpretation of all the details of the line shape, certain features of these powder spectra are easily understood. The two outer peaks (separated

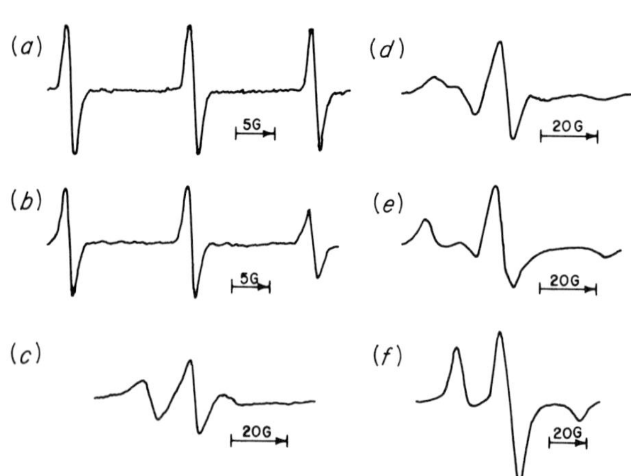

FIGURE 4.
Effect of viscosity on the spectra of nitroxide radicals: (*a*) 2,2,5,5-tetramethyl-3-carboxamidopyrrolidine-1-oxyl (XXII) in water at room temperature (Stone et al., 1965); (*b*) XXII in 60 percent sucrose solution at room temperature (Stone et al., 1965); (*c*) dansyl nitroxide (XXIII) in 76 percent glycerol-19 percent water-5 percent ethanol at 23°C (Stryer and Griffith, 1965); (*d*) XXIII in 90 percent glycerol-5 percent water-5 percent ethanol at 35°C (Stryer and Griffith, 1965); (*e*) XXIII in 90 percent glycerol-5 percent water-5 percent ethanol at −15°C (Stryer and Griffith, 1965); (*f*) XXII in glycerol at 77°K (Stone et al., 1965).

by ~64 gauss), as well as a portion of the central peak, arise from radicals whose π-orbitals (z axes) are parallel to the applied magnetic field (parallel, say, to within 15° of the applied field). The central peak of the spectrum in Figure 4,*f* also receives a large contribution from radicals whose π-orbitals are nearly perpendicular to the applied magnetic field. In this case the N^{14} splitting, equal to $B \simeq C$, is not resolved because of the large effective line width in this region of the spectrum. Other qualitative features of the spectrum, such as the upward and downward deviations of the outer peaks, are characteristic of anisotropic terms in a spin Hamiltonian.

Figure 4 also shows the variation of the paramagnetic resonance spectra of nitroxide radicals in solutions of varying viscosity. In the region intermediate between high viscosity (characterized by the spectrum in Figure 4,*f*) and low viscosity (spectra in Figure 4, *a*, *b*) it is exceedingly difficult to make *a priori*

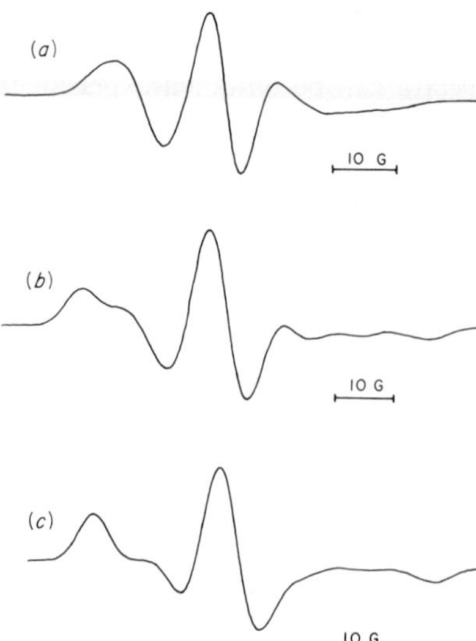

FIGURE 5.
Spectra calculated for nitroxide radicals with intermediate correlation times: (*a*) $\tau \sim 1.4 \times 10^{-8}$ sec; (*b*) $\tau \sim 3.6 \times {}^{-8}$ sec; (*c*) $\tau \sim 7.2 \times 10^{-8}$ sec. Values of τ were estimated by the method of Itzkowitz (1966).

theoretical calculations of line shapes. An approximate semiempirical, semirigorous method has been developed by Itzkowitz (1966), and comparisons between observed and calculated spectra are given in Figures 4, *c–e* and 5. It deserves special mention that in this region spectra are extremely sensitive to changes in the motion of the radical.

In the region of very fast tumbling (see Figure 4, *a, b*) one may use the theory of paramagnetic resonance line shapes developed by McConnell (1956), Freed and Fraenkel (1963), and Kivelson (1960). In this case each resonance line of the hyperfine triplet is characterized by a line width parameter $T_2(M)$, where $M = 1, 0, -1$ corresponds to the three N^{14} nuclear spin orientations. For nitroxide radicals, $T_2(M)$ is given by the expression

$$(T_2(M))^{-1} = \tau\left\{[3I(I+1) + 5M^2]\frac{b^2}{40} + \frac{4}{45}(\Delta\gamma H_0)^2 - \frac{4}{15}b\Delta\gamma H_0 M\right\} + X \quad (4)$$

Here τ is the correlation time for isotropic molecular tumbling, H_0 is the applied field in gauss, $I = 1$ (the nuclear spin quantum number for N^{14}), $b = 4\pi/3(A - B)$, and

$$\Delta\gamma = \frac{-|\beta|}{\hbar}\left[g_{zz} - \tfrac{1}{2}(g_{xx} + g_{yy})\right] \quad (5)$$

The quantity X denotes contributions from broadening mechanisms that do not involve M. Equation (4) should represent line shapes satisfactorily if the anisotropic hyperfine interaction possesses axial symmetry ($B = C$), if the tumbling motion is isotropic and slow enough so that $\omega^2\tau^2 \gg 1$ ($\omega = g|\beta|H_0\hbar^{-1}$), and if $(\pi a)^2\tau^2 \ll 1$ and $b^2\tau^2 \ll 1$. A more convenient form of equation (4) is

$$\frac{T_2(0)}{T_2(M)} = 1 - \frac{4\tau}{15}b\Delta\gamma H_0 T_2(0)M + \frac{\tau}{8}b^2 T_2(0)M^2 \quad (6)$$

since the ratio $T_2(0)/T_2(\pm 1)$ can be obtained experimentally. The correlation time τ calculated from this relationship for the spectrum in Figure 4, *b* is 3×10^{-10} seconds; τ for the conditions producing the spectrum in Figure 4, *a* is 2.5×10^{-11} seconds (Stone et al., 1965).

SPIN-LABELING REAGENTS

Compounds of two types have been used as labeling reagents in the study of biological macromolecules. The first and main class comprises cyclic hindered nitroxides that possess a ring substituent capable of forming a covalent bond with active groups on the macromolecule. Many of the members of this set of labels have been tailored to be analogous to known protein reagents of well characterized specificity. There is no guarantee, of course, that the radicals will exhibit the same specificity as their "pattern" compounds, but the behavior of the traditional reagents should provide a starting point for consideration of the habits of the spin-labeling compounds.

The first nitroxide spin label was 2,2,5,5-tetramethyl-3-isocyanatopyrrolidine-1-oxyl, VIII (Stone et al., 1965). The isocyanate is primarily of historical interest only because, although it did label, it showed little specificity, could not be purified, and was rapidly destroyed by the aqueous solutions in which the experiments were done.

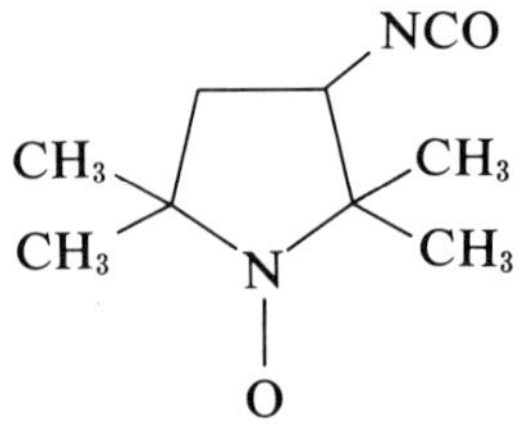

VIII

More successful reagents are the N-nitroxyl maleimides IX (Griffith and McConnell, 1966) and X (Hamilton and McConnell, 1967).

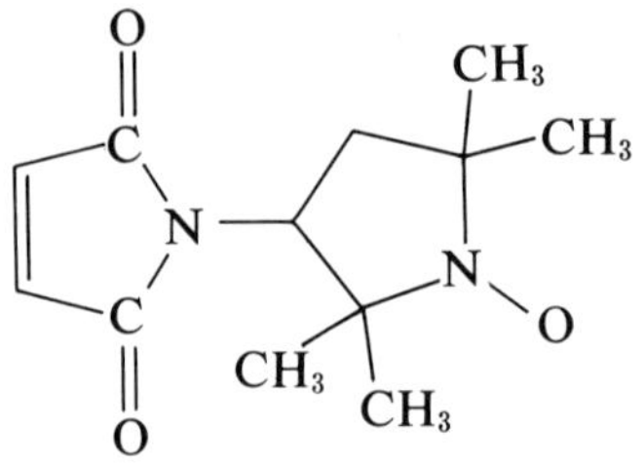

IX

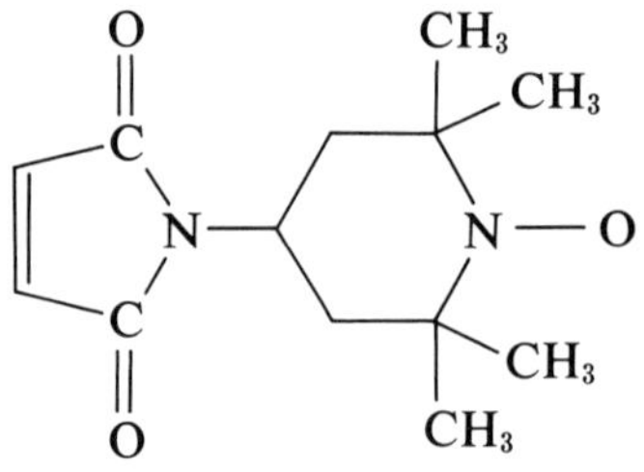

X

Intended as sulfhydryl reagents, these labels show somewhat more diverse behavior than their model compound, N-ethylmaleimide. The maleimide radicals do seem to react preferentially with "exposed" SH groups, but are more susceptible than N-ethylmaleimide to ring opening (Griffith and McConnell, 1966; Boeyens and McConnell, 1966; Ohnishi, Boeyens, and McConnell, 1966). Compound IX is converted to the corresponding maleamic acid after a few minutes in water (Griffith and McConnell, 1966); X is hydrolyzed within an hour (C. L. Hamilton, unpublished). This sensitivity to attack at the carbonyl is also reflected in the facile reaction of nitroxyl maleimides with amino groups (Griffith and McConnell, 1966).

The five- and six-membered haloacetamides XI, X = Br, I (S. Ogawa, unpublished) and XII, X = I (C. L. Hamilton, unpublished) are not fully characterized as labeling reagents, but seem to act as general alkylating agents. The point of attachment depends on the protein and reaction conditions (S. Ogawa, unpublished; Smith and McConnell, 1967), much as with iodoacetic acid (Cecil, 1963). A dinitrofluorobenzene derivative, XIII, containing the nitroxide moiety has been prepared (W. J. Deal, unpublished) for tagging terminal and ϵ-amino groups. In analogy with N-acetylimidazole (Riordan,

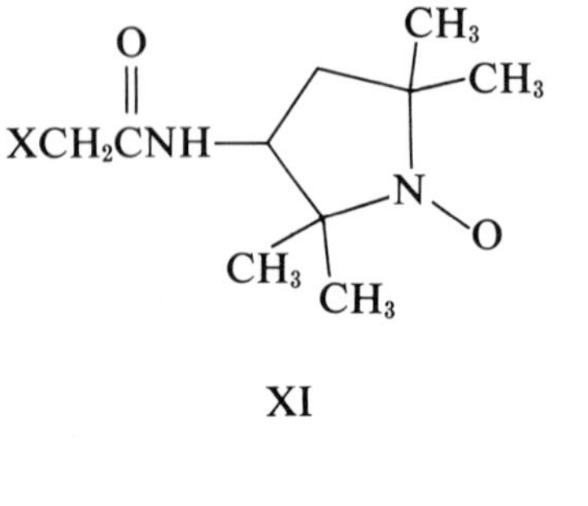

XI

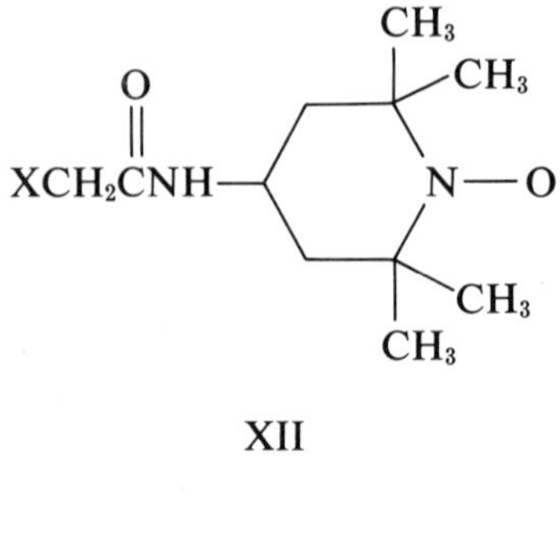

XII

Wacker, and Vallee, 1965), N-(2,2,5,5-tetramethyl-l-oxyl-3-pyrrolinoyl)imidazole (XIV) has been made to react with tyrosine residues (C. L. Hamilton, unpublished).

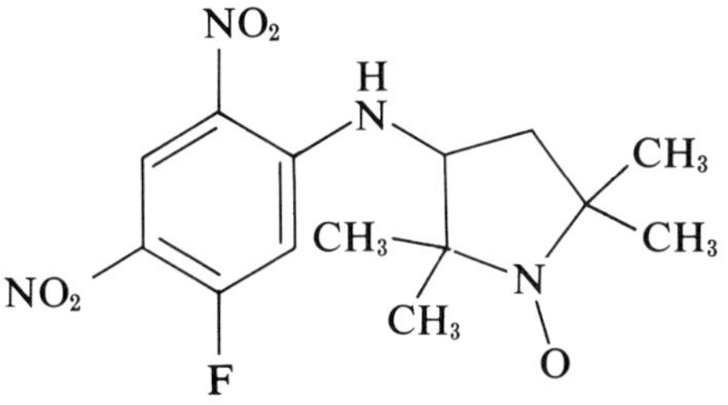

XIII

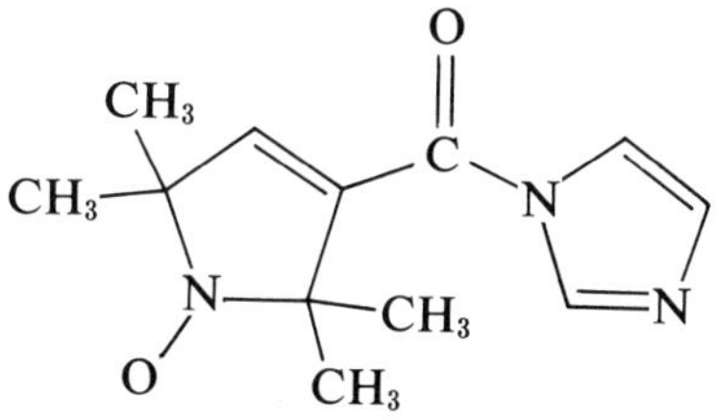

XIV

Two reagents have been made to take advantage of special enzyme specificities. The *p*-nitrophenyl ester of 2,2,5,5-tetramethyl-3-carboxypyrrolidine-1-oxyl (XV) acylates the active serine of α-chymotrypsin (Berliner and McConnell, 1966), following the precedent set by *p*-nitrophenyl acetate (Hartley and Kilby, 1952; Kézdy and Bender, 1962). An apparently similar reaction has been observed with the active site SH group in triosephosphate dehydrogenase (W. J. Deal, unpublished). A mixture of the mononitroxylamides of α- and β-bromosuccinic acid, XVI (Smith and McConnell, 1967) has been used to mimic the α-bromo acids that react with known amino acids at the active site of ribonuclease (Heinrickson, 1966).

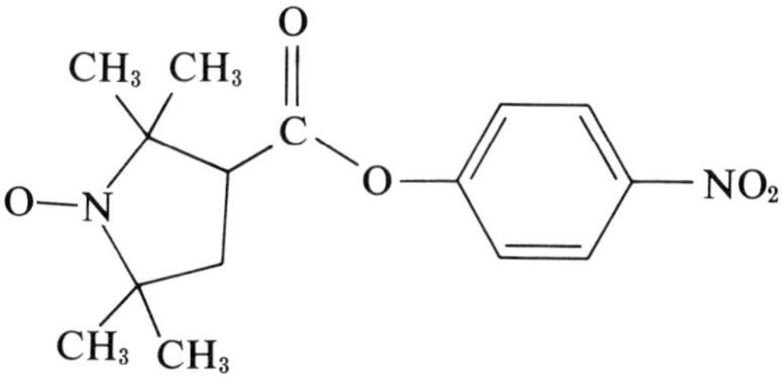

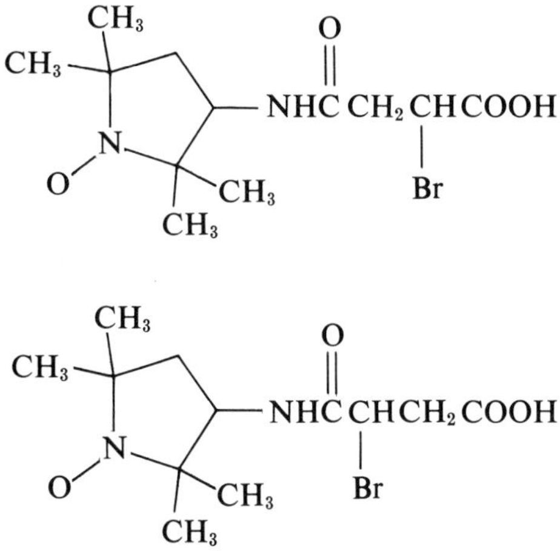

XV

XVI

The second class of spin-labeling compounds is made up of those that form complexes not involving covalent bonds with biological molecules. Two dye-like species, N-(2,2,5,5-tetramethyl-1-oxyl-3-pyrrolidinyl)-2,4-dinitroaniline XVII (W. J. Deal, unpublished) and 2,2,4,4-tetramethyl-1-2,3,4-tetrahydro-γ-carboline-3-oxyl XVIII (Rozantzev and Shapiro, 1964; S. Ogawa, unpublished) have been found to bind to proteins, and a paramagnetic hapten, XIX (Stryer and Griffith, 1965) has been prepared. Studies on a specific fully competitive inhibitor, XX, of α-chymotrypsin are in progress (C. L. Hamilton, unpublished). The cholesterol derivative XXI has been synthesized to be incorporated into the lipid portions of membranes (W. Hubbell, unpublished).

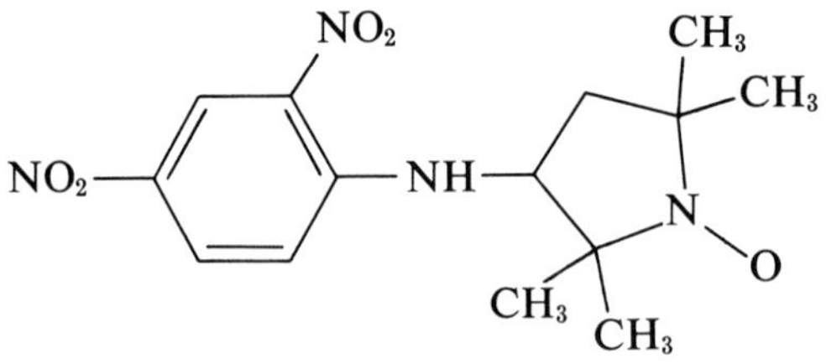

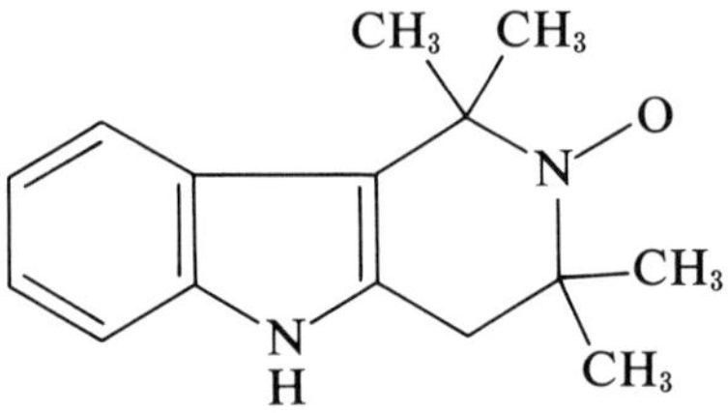

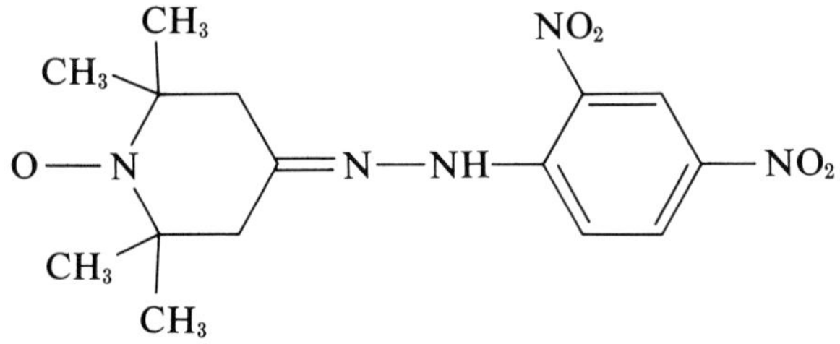

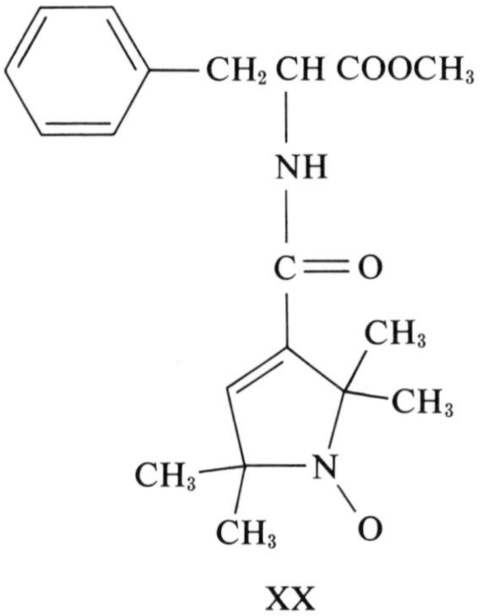

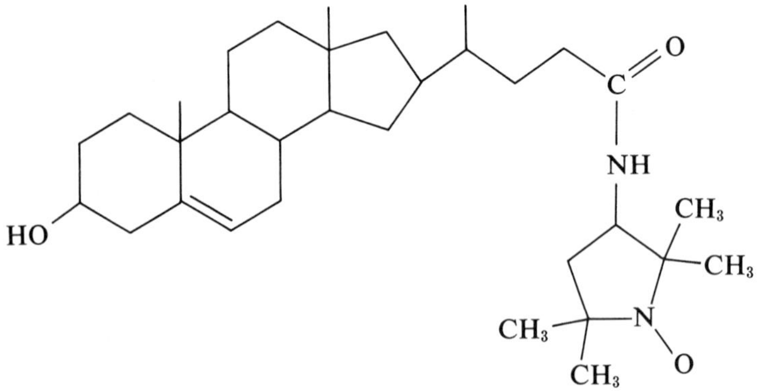

XVII

XVIII

XIX

XX

XXI

RESONANCE SPECTRA OF SPIN LABELS ATTACHED TO MACROMOLECULES

The paramagnetic hapten XIX is strongly complexed by antidinitrophenyl antibody, with the same stoichiometry and very nearly the same binding constant as the standard hapten, ε-N-dinitrophenyl-L-lysine (Stryer and Griffith, 1965). Whereas the unbound hapten has a resonance spectrum characteristic of a freely rotating species (see Figure 4, *a*) the complexed radical gives a spectrum closely resembling Figure 4, *d*. The marked increase in correlation time (from $\sim 10^{-11}$ to $\sim 10^{-8}$ seconds) reflects the reduction in tumbling rate that is to be expected when a small molecule becomes tacked to a large, slow-moving protein. The antibody has a rotational relaxation time of the order of 10^{-7} seconds (Krause and O'Konski, 1963).

The same sort of effect is observed when nitroxide isocyanate (VIII) is allowed to react with poly-L-lysine (Stone et al., 1965). The esr spectrum of the labeled polyamino acid in solution at *p*H 8, shown in Figure 6, *a*, is indicative of a radical with motion that is somewhat restricted but not nearly as slow as that of the polymer itself.

The term "weakly immobilized" has been coined to describe spectra like that of labeled poly-L-lysine, where the lines are only slightly broadened compared with those from a completely freely tumbling radical. The appearance of a weakly immobilized spectrum in this case reflects the flexibility of the lysine side chains on which the label probably resides. Even a motion so loosely coupled with that of the main body of the polymer can yield information on the status of the macromolecule, however. For example, at *p*H 11, where poly-L-lysine has a helical structure (Applequist and Doty, 1962), the esr spectrum is significantly *less* weakly immobilized (see Figure 6, *b*) than at *p*H 8, where the polyamino acid exists as a random coil. In fact, plots of correlation time τ versus *p*H for labeled poly-L-lysine (Stone et al., 1965) bear a marked resemblance to titration curves for the specific rotation of the same polymer (Applequist and Doty, 1962).

The reaction of bovine serum albumin (BSA) with spin labels results in a paramagnetic resonance spectrum more complex that that of poly-L-lysine (Stone et al., 1965; Griffith and McConnell, 1966). The albumin was treated with the nitroxide maleimide IX and the unattached radicals were removed by dialysis. The spectrum of the labeled BSA, recorded in Figure 7, is obviously a superposition of two spectra. In addition to a weakly immobilized component,

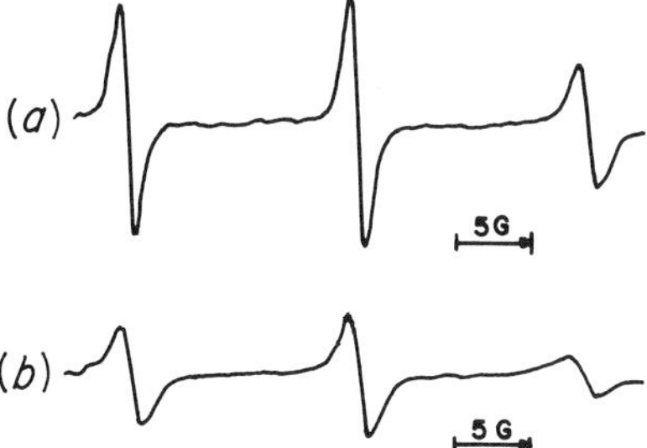

FIGURE 6.
Resonance spectra of spin-labeled poly-L-lysine (*a*) at *p*H 8 and (*b*) at *p*H 11 (Stone et al., 1965).

FIGURE 7.
Spectrum of bovine serum albumin labeled with nitroxide maleimide IX (Griffith and McConnell, 1966).

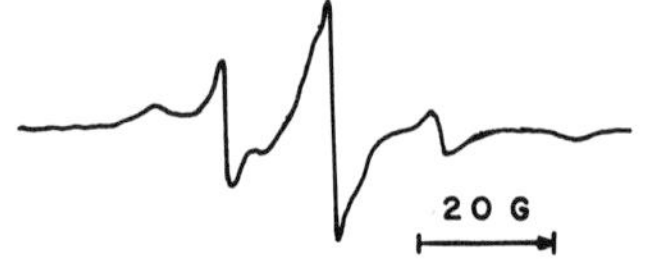

there appear broad lines closely resembling the spectrum of a nitroxide in a rigid glass (Figure 4,*f*). This broad portion arises from label that is "strongly immobilized" in our terminology.

Resonance spectra of the strongly immobilized type are quite commonly observed in spin-labeled proteins. The occurrence of such spectra depends on two important physical conditions. First, the protein as a whole must be rigid and must tumble so slowly in solution that the condition of equation (3) holds. That is, the correlation time for tumbling of the rigid protein must be more than $\sim 10^{-7}$ seconds. Secondly, the spin label must not wiggle relative to the protein, but must be held rigidly to it. That is, the spin label must be held "rigidly" to the "rigid" protein for at least 10^{-7} seconds.

The rotational relaxation time of BSA is approximately 10^{-6} seconds (Krause and O'Konski, 1959). The existence of a strongly immobilized component in the spectrum of labeled BSA clearly implies that part of the paramagnetic tag is severely constrained in its motion relative to the protein. Treatment of the protein with N-ethylmaleimide prior to its reaction with the labeling reagent prevents the appearance of the strongly immobilized portion of the spectrum, so this constrained label is probably attached to a sulfhydryl group (Griffith and McConnell, 1966).

As with poly-L-lysine, a known structural change in BSA may be followed with the paramagnetic resonance spectrum. As the *p*H of its environment is lowered from 4 to 2 this protein undergoes a dramatic expansion that is evidenced by changes in several physical parameters (Weber, 1953; Yang and Foster, 1954; Harrington, Johnson, and Ottewill, 1956; Krause and O'Konski, 1959). As *p*H is varied in this range, the strongly immobilized portion of the paramagnetic resonance spectrum of labeled BSA is gradually converted to weakly immobilized signal (Stone et al., 1965), nearly in parallel with the change in viscosity and optical rotation (Yang and Foster, 1954).

It has also been found possible to observe an enzyme in action, using the resonance of a paramagnetic substrate (Berliner and McConnell, 1966). The esterase activity of α-chymotrypsin involves the reaction sequence

$$E + S \underset{}{\overset{K}{\rightleftarrows}} ES \overset{k_2}{\underset{+P_1}{\longrightarrow}} ES' \overset{k_3}{\longrightarrow} E + P_2$$

For many substrates, particularly active esters such as those of *p*-nitrophenol, k_3 is rate-determining and the intermediate ES', which is the enzyme acylated at a serine in the active site, has an appreciable lifetime (Bender and Kézdy, 1959). When the *p*-nitrophenyl "nitroxylate" XV is acted upon by α-chymotrypsin at *p*H 4.5, where k_3 is very small, its esr spectrum becomes strongly immobilized, corresponding to formation of the acyl enzyme. At higher *p*H, k_3 can be measured by the rate of decay of the broad resonance lines into free signal (Figure 8); the value obtained agrees well with that measured spectrophotometrically for rate of release of *p*-nitrophenolate under steady-state conditions.

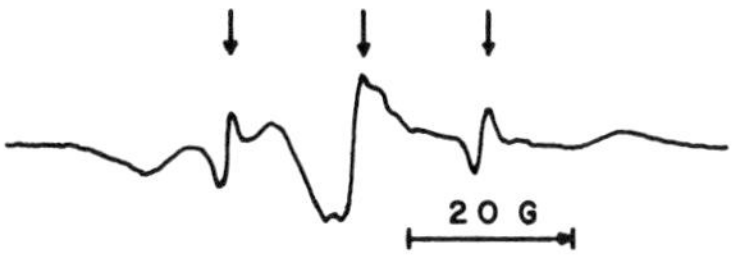

FIGURE 8.
Paramagnetic resonance spectrum of deacylating "nitroxyl chymotrypsin." Arrows indicate the appearance of free signal due to the "nitroxylate" ion being released (Berliner and McConnell, 1966).

The foregoing experiments have shown that the spin-label spectra can accurately reflect previously known conformational and biochemical changes. Now extensive studies have been undertaken on other systems. Preliminary observations of interest have been made with several proteins in the course of work in these laboratories. Yeast alcohol dehydrogenase, when labeled with the N-nitroxyl iodoacetamide XI, exhibits a strongly immobilized esr spectrum that is altered upon addition of the coenzyme NAD. Also, the rate of the labeling reaction is lowered in the presence of NAD, suggesting that the locus of reaction is the active site (S. Ogawa, unpublished). The paramagnetic resonance spectrum of ribonuclease labeled at the active site with α-haloacids or -amides (compounds XI, X = Br, and XVI) is broadened when RNA is introduced into the solution. Detailed studies of the variation of this effect with experimental conditions have been carried out (Smith and McConnell, 1967). Bovine serum albumin has been found to complex several aromatic nitroxide compounds so that the spins become strongly immobilized. Binding is reversible with the carboline nitroxide XVIII (S. Ogawa, unpublished) and the N-nitroxyldinitroaniline XVII (W. J. Deal, unpublished). The hydrophobic binding site in BSA carries in it a reactive group of some sort, since interaction with the dinitrofluorobenzene derivative XIII is irreversible (W. J. Deal, unpublished). Preliminary experiments indicate that RNA can be labeled with nitroxide alkylating agents (I. C. P. Smith, unpublished), suggesting that the spin-label technique may be applicable to the study of nucleic acids as well as proteins.

Horse hemoglobin reacts with the nitroxide maleimide IX to produce a paramagnetic protein that exhibits quite complex behavior (Boeyens and McConnell, 1966). The major site of attachment is to the cysteine β93*, the residue that has been specifically labeled previously with N-ethylmaleimide (NEM) (Benesch and Benesch, 1961). Nitroxide IX reacts more rapidly with oxyhemoglobin (HbO_2) than with deoxyhemoglobin (Hb), as does NEM (Riggs, 1961), giving a product with a weakly immobilized signal in both cases. Within a short time after the initial combination with HbO_2, a secondary reaction converts the SH-bound label to a species with a strongly immobilized spin that does not become freed by deoxygenation. Methemoglobin gives a similar spectrum. Oxygenation of Hb immediately after labeling produces the same effect, but if the conversion to HbO_2 is postponed for 30 minutes no immobilization will occur upon oxygen uptake.

The spin-label results indicate that there is a definite difference in conformation near β93 between horse Hb and HbO_2 and complement the observation that NEM also undergoes a secondary reaction after attachment to human HbO_2 (Benesch and Benesch, 1961), accompanied by modification of the Bohr

*Some reaction with other groups, probably ϵ-NH_2, to give weakly immobilized spins always occurs. This portion of the protein spectrum remains invariant throughout the experiments discussed here.

effect. It has been postulated that, in HbO_2, the imide ring is opened through catalytic action by a histidine and that the label becomes immobilized by hydrogen bonding of the resulting carboxyl group and by hydrophobic interaction between the nitroxide ring and portions of the protein (Boeyens and McConnell, 1966; Ohnishi, Boeyens, and McConnell, 1966). The secondary reaction taking place with maleimide-labeled horse Hb is probably an uncatalyzed hydrolysis by water of the imide ring; the sensitivity of the maleimide nitroxides to ring opening is known (see above), and the reaction does not seem to occur as rapidly with Hb that is labeled with the more stable NEM (Benesch and Benesch, 1961).

FIGURE 9.
The β-chain of horse oxyhemoglobin.

The molecular models in Figure 9, constructed from the X-ray data of Perutz (1965), give a rough idea of how the spin label is situated at β93. For reference, the paper model at the left represents an entire β-chain. Helical regions are denoted by cylinders and the heme (made of gray cardboard) can be seen behind the F-helix and the ends of helices G and H. Cysteine β93 is at the bottom end of helix F; a rod slightly in front of it marks the position of the mercury used as a heavy atom in the X-ray analysis. The portion of the β-chain containing the helices F, G, and H was reproduced with Corey-Pauling-Koltun models, with the nitroxide label attached. Both models are built to the same scale and are seen at the same orientation in Figure 9, except that the partial atomic model is mounted a few centimeters lower than the paper one. The sulfur of cysteine β93, striped for identification, is visible near the lower left of the partial atomic model, and immediately to the right of it the nitroxide oxygen (marked with an X) and the four flanking methyl groups of the spin label can be seen. The positioning of the label is consistent with the resonance spectra of labeled HbO_2 single crystals (Ohnishi et al., 1966).

The β-chain is repeated by rotation around the vertical two-fold axis marked on the base of the paper model (Figure 9). Thus it appears that the spin label is not only tucked against its own chain but also sandwiched between it and the other β-chain. If all facets of this picture are to be accepted, the free spin of the nitroxide label may be close enough to the heme iron for interaction between the two paramagnetic species to be observable under the proper conditions. Work in progress is directed toward detecting this interaction (J. Chien and H. M. McConnell, unpublished).

Hemoglobin has also been labeled at β93 with the N-nitroxyliodoacetamides XI and XII (X = I) (S. Ogawa, experiments in progress). These reagents provide a situation analogous to that in the maleimide-labeled protein without the constraint of an extra carboxyl group, and produce spectra that are somewhat immobilized in Hb and very strongly immobilized in HbO_2. At intermediate stages of oxygen saturation, both components are present, and the paramagnetic resonance spectra show an isosbestic point much as the optical absorption spectra do. The existence of two hemoglobin conformations is clearly indicated. These results provide an opportunity to investigate directly the relationship between oxygenation and conformational change and to help establish which of the theories regarding allosteric transitions (Monod, Wyman, and Changeux, 1965; Koshland, Némethy, and Filmer, 1966) most closely represents the real situation.

SPIN-LABELED ALLOSTERIC PROTEINS

The concept of allosteric proteins was introduced (Monod, Changeux, and Jacob, 1963) to explain why the activities of a large number of metabolic enzymes are profoundly altered by small molecules (effectors) that are often totally unrelated to the enzymes' substrates, products, or coenzymes. In essence, an allosteric enzyme is thought to possess distinct stereospecific binding sites for substrate and effector that are physically well removed from each other. The binding of effector to its site is imagined to produce a specific conformational change in the protein (an allosteric transition) that affects the affinity of the other site for substrate.

Such allosteric transitions may not only account for feedback inhibition in metabolic pathways, but could also be involved in genetic control of protein synthesis as well as in more familiar phenomena such as the cooperative binding of oxygen by hemoglobin (heme-heme interaction). Since allosteric conformational changes are thought to be widespread and important in biological systems they are obvious targets for spin-label studies.

From a qualitative physical point of view, interactions between sites removed from one another by, say, 10–100 Å in a compact molecule are not too surprising, at least in retrospect. That is, if we regard a protein molecule as merely a three-dimensional collection of balls and springs, the forced displacement of any one set of balls (produced by the effector molecule) must in general produce displacements at all other points in the protein, including the active site region that acts on the substrate. Thus, we presume that in an allosteric protein (Figure 10), a conformational change brought about by combination of effector E with its binding site e propagates not only to the substrate site s but also to any general position l (Figure 10, *b*). If a spin label L

is attached at *l* (or at *s*, for that matter) we may anticipate that the alteration in configuration induced by *E* will be reflected in the resonance spectrum of the bound label. A reciprocal effect associated with the binding of substrate *S* should also be observed (Figure 10, *c*).

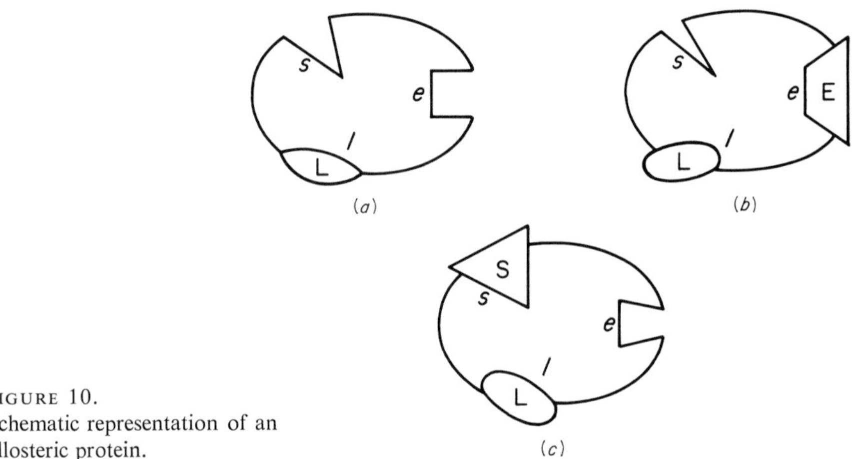

FIGURE 10.
Schematic representation of an allosteric protein.

The detailed mechanism of how distant sites are coupled on a molecular level in proteins has been a subject of intensive study. Allosteric effects fall into two classes, heterotropic (interaction between different ligands, illustrated schematically in Figure 10) and homotropic (interaction between identical ligands). Two general theories have been advanced to explain homotropic allosteric effects, one by Monod, Wyman, and Changeux (1965) and the other by Koshland, Némethy, and Filmer (1966). Since systems exhibiting heterotropic effects almost invariably show homotropic effects too, it may be assumed that similar transitions are involved in both (Monod, Wyman, and Changeux, 1965) and therefore that the models are probably applicable also to heterotropic interactions. The two theories of homotropic interactions are discussed briefly below, with particular regard to how they can be tested with the use of spin-label techniques.

Monod and his coworkers (Monod, Wyman, and Changeux, 1965) have pointed out that allosteric proteins are usually oligomers, consisting of small numbers of chemically equivalent subunits or protomers. They conclude that the protomers in these oligomers are arranged in a symmetrical fashion, and postulate that each protomer possesses only one stereospecific binding site for any given ligand. Their model assumes that at least two states differing in the distribution and/or energy of interprotomer bonds are reversibly accessible to the oligomers. It is also assumed that these states differ in their affinity for ligands and that the transitions between them take place with conservation of molecular symmetry. A further assumption is that the binding of a ligand to a protein in any one state is independent of the binding of any others.

The situation postulated by Monod, Wyman, and Changeux may be illustrated by the following example. Consider a tetrameric protein existing in tautomeric equilibrium, characterized by the constant $L = [T]/[R]$, between two forms, "round" *R* and "square" *T*, each having a four-fold symmetry axis (Figure 11). The affinity of each subunit for a ligand *S* is expressed by one

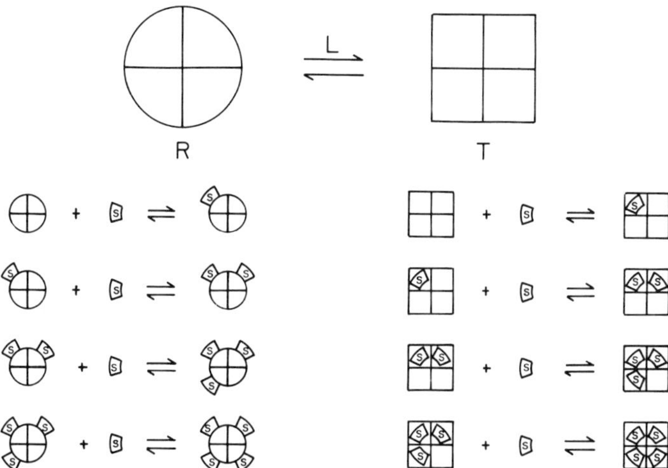

FIGURE 11.
Schematic illustration of a tetrameric allosteric protein according to the model of Monod, Wyman, and Changeux (1965).

of the microscopic dissociation constants K_R or K_T (depending on whether the subunit is in the R or T form). A system containing the protein and ligand would then be described in terms of the following equilibria (see also Figure 11):

$$R \overset{L}{\rightleftarrows} T$$

$$\begin{array}{ll} R + S \rightleftarrows RS_1 & T + S \rightleftarrows TS_1 \\ RS_1 + S \rightleftarrows RS_2 & TS_1 + S \rightleftarrows TS_2 \\ RS_2 + S \rightleftarrows RS_3 & TS_2 + S \rightleftarrows TS_3 \\ RS_3 + S \rightleftarrows RS_4 & TS_3 + S \rightleftarrows TS_4 \end{array}$$

The microscopic dissociation constants may be seen to be in this case

$$K_R = \frac{4-(n-1)}{n}\frac{[RS_{n-1}][S]}{[RS_n]} \qquad K_T = \frac{4-(n-1)}{n}\frac{[TS_{n-1}][S]}{[TS_n]}$$

The allosteric behavior of this system is characterized by the tautomeric equilibrium constant L and the ratio $c = K_R/K_T$ (reflecting the relative affinities of the tautomers for S). When L is large and c is small, the allosteric effect exerted by ligand binding is large; if $\bar{Y}$, the fraction of sites (both forms) occupied by ligand, is plotted against a measure of ligand concentration, $\alpha = [S]/K_R$, the sigmoid curve characteristic of cooperative substrate binding (Figure 12) results.

Monod et al. (1965) have shown that their model can account for experimental data on oxygen uptake by hemoglobin, as well as for the kinetics of a number of enzymatic reactions.

An alternative theory of allosteric effects, that of Koshland, Némethy, and Filmer (1966), is sufficiently general that it can represent the Monod-Wyman-Changeux model as one special case. However, the Koshland theory can also be applied to other limiting situations in which the experimental data can be

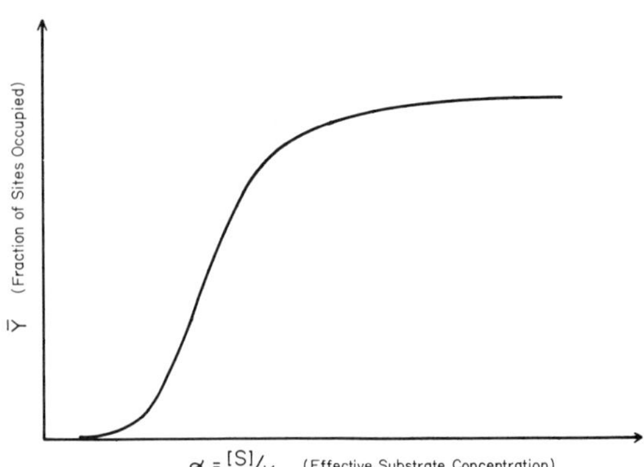

FIGURE 12.
Typical cooperative substrate saturation curve.

represented just as well and which present a quite different physical picture for allosteric interaction.

In all the limiting models treated by Koshland et al. (1966) it is assumed that individual protomers exist in two tautomeric conformations, only one of which combines significantly with ligand. In each case the equilibrium constants for the conformational transition in each subunit and for the uptake of ligand by the binding tautomer are considered. In addition, terms representing the strength of interaction between subunits are included in the treatment. The geometrical relationships between subunits (governing which pairs of subunits interact) are varied from model to model.

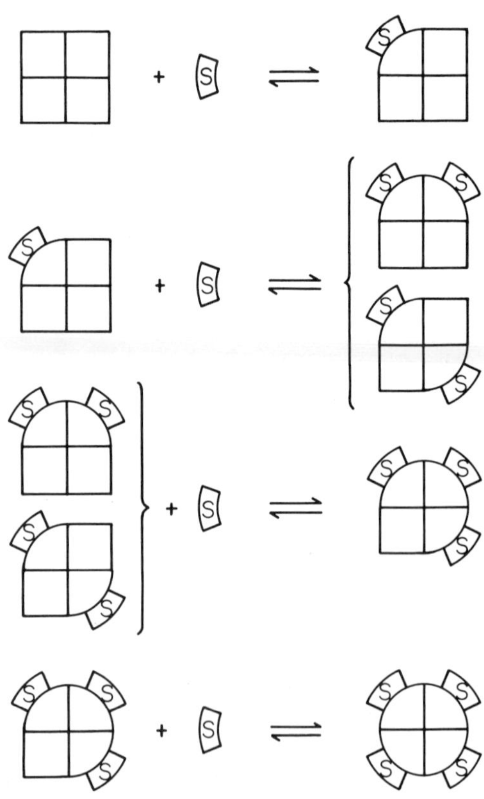

FIGURE 13.
Schematic illustration of a tetrameric allosteric protein for one of Koshland's (1966) limiting models.

The limiting model giving the best fit in the treatment of Koshland, Némethy, and Filmer (1966) to experimental oxygen saturation curves for hemoglobin is represented in Figure 13. This case assumes that the "round" form, which is the one capable of binding the ligand S, is present in significant amounts only when combined with S; interactions are allowed between adjacent subunits but not between diagonal pairs. If a "square" subunit is designated as t (so that t_4 is equivalent to T from the preceding schematic illustration) and a "round" subunit, which by definition here is combined with S, is denoted by r, we may write the following equilibria to describe the system:

$$T + S \rightleftarrows (rS)t_3$$

$$(rS)t_3 + S \rightleftarrows \begin{cases} (rS)_2t_2 \text{ (adjacent)} \\ (rS)_2t_2 \text{ (diagonal)} \end{cases}$$

$$(rS)_2t_2 + S \rightleftarrows (rS)_3t$$

$$(rS)_3t + S \rightleftarrows RS_4$$

Analysis of the system is carried out in terms of the microscopic equilibrium constants,

$$K_S = \frac{[rS]}{[r][S]}$$

$$K_t = \frac{[r]}{[t]}$$

and interaction constants

$$K_{rt} = \frac{[rt][t]}{[tt][r]}$$

$$K_{rr} = \frac{[rr][t][t]}{[tt][r][r]}$$

where rt refers to interacting subunits and r and t to noninteracting ones.

There is one crucial difference between the two theories for the allosteric binding of substrate to protein described above. In the Koshland scheme there is one subunit conformational change for every molecule of ligand bound, whereas with the Monod theory the number of subunit transitions in general does *not* equal the number of molecules of substrate taken up. The spin-label technique appears to be an ideal method for distinguishing between alternatives of this type, since the spin label resonance spectra are sensitive functions of local conformation.

We take as an example the hypothetical protein used above to illustrate the two theories, this time with a spin label attached to each protomer. In a solution containing ligand, this protein would exist, according to the Monod-Wyman-Changeux scheme, as species of the type illustrated in Figure 14, *a*. Here the arrows indicate the two types of resonance spectra that can arise from each label. By considering the distribution of all possible complexes of this type, it becomes evident that the intensities of the two different spin label spectra will not be simply proportional to the number of occupied and empty binding sites. On the other hand, according to the Koshland-Némethy-Filmer treatment the solution would be populated with complexes of the sort shown

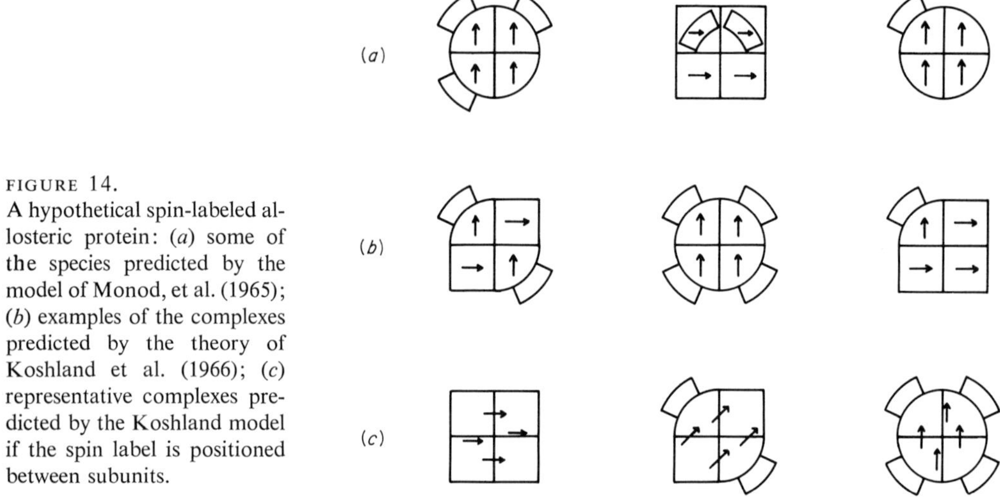

FIGURE 14.
A hypothetical spin-labeled allosteric protein: (*a*) some of the species predicted by the model of Monod, et al. (1965); (*b*) examples of the complexes predicted by the theory of Koshland et al. (1966); (*c*) representative complexes predicted by the Koshland model if the spin label is positioned between subunits.

in Figure 14, *b*. Each spin label spectrum of the type → would be associated with a subunit with no conformation change and no substrate, and each spin-label spectrum of the type ↑ would come from a subunit that is complexed with a substrate molecule. Thus, by comparing the effects of substrate on spin-label resonance spectra with the number of bound substrate molecules (determined by conventional biochemical or physical chemical procedures), one can distinguish between these two simple models of allosteric interactions.

A possible complication in the above analysis could arise for systems following the Koshland scheme if the spin label were located along a junction between subunits in such a position that it would be influenced by more than one protomer. In this case one could conceive of there being at least three spin-label resonance spectra (↑, ↗, →) as illustrated in Figure 14, *c*. The absence of resonance spectral isosbestic points as well as other spectral features can be used to detect such complications if they are present.

In current work in these laboratories (Ogawa and McConnell, 1967) experiments of the sort outlined above have been carried out in which the β-chains of horse hemoglobin have been labeled at the reactive SH groups with the iodoamide label XI. In these experiments the series of spin-label resonance spectra obtained with increasing oxygenation shows a well-defined isosbestic point. This in itself is consistent with the theories of both Monod et al. (1965) and Koshland et al. (1966). However, there is precisely one spin resonance spectral change (for example, → to ↑) for each oxygen molecule bound, and this is consistent only with the Koshland scheme. Even more direct evidence for the Koshland scheme could be obtained if it *were* possible to put labels in the "*cracks*" between subunits, because then one could anticipate that there should be *three* types of spin-label spectra, and no isosbestic points.

In the experiments of Ogawa and McConnell it was found that the "cooperativity" of uptake of oxygen by hemoglobin was not markedly affected by the spin labels attached to the reactive SH groups at β93. However, the absolute affinity of the labeled molecule for oxygen was increased by a factor of approximately 10. It is extremely interesting that this increase in oxygen affinity seems to involve all the chains equally, even though only half of them are modified; there is no evidence that the α-hemes take up oxygen with a binding coefficient significantly different from that of the β-hemes.

In trying to get a feeling for these diverse phenomena, we have found it convenient to consider a not entirely preposterous physical analogue to protein changes linked with ligand binding, the "oil drop effect." This is not proposed seriously as an explanation of how proteins work. However, it does show homotropic and heterotropic allosteric behavior and is consistent with the above observations on spin-labeled hemoglobin. Its function is to suggest directions of research that might lead to a more realistic physically based theory of allosteric effects.

Consider a drop of oil in water, with a radius R_0 of the same order of magnitude as the average radius of an allosteric protein. Let the oil-water interfacial tension be γ. Then the surface free energy of the oil drop-water system is

$$\begin{aligned} F &= \gamma A_0 \\ &= 4\pi R_0^2 \gamma \\ &= 4\pi \left(\frac{3}{4\pi}\right)^{2/3} V_0^{2/3}\gamma \end{aligned} \tag{7}$$

where A_0 is the surface area of the drop, and where V_0 is the volume of the drop. Now, consider the change of surface free energy when the drop of oil takes up a molecule ("substrate") so that the volume changes from V_0 to $V_0(1 + \delta)$, where δ is small (for example, 0.1). The corresponding increase of surface free energy is

$$\begin{aligned} \Delta F &= \gamma 4\pi \left(\frac{3}{4\pi}\right)^{2/3} V_0^{2/3}\left(1 + \tfrac{2}{3}\,\delta - \tfrac{1}{9}\,\delta^2 + \cdots\right) \\ &= \gamma A_0\left(1 + \tfrac{2}{3}\,\delta - \tfrac{1}{9}\,\delta^2 + \cdots\right) \end{aligned} \tag{8}$$

To consider "allosteric" interactions in this system, we introduce two molecules with volumes $\delta_{\mathrm{I}} V_0$ and $\delta_{\mathrm{II}} V_0$ into the oil drop and look at the contributions to the free energy change, $\Delta F^{(2)}$, that are quadratic in δ [equation (9)].

$$\Delta F^{(2)} = -\tfrac{1}{9}\,\gamma A_0(\delta_{\mathrm{I}}^2 + \delta_{\mathrm{II}}^2 + 2\delta_{\mathrm{I}}\delta_{\mathrm{II}}) \tag{9}$$

The interaction between "substrate" molecules in the oil drop comes from the term involving $\delta_{\mathrm{I}}\delta_{\mathrm{II}}$; note that this is always attractive for identical molecules ($\delta_{\mathrm{I}} = \delta_{\mathrm{II}}$), regardless of the sign of δ_{I}. (A negative sign for δ_{I} has no physical significance for a molecule introduced into an oil drop, but an authentic substrate introduced into a protein could quite conceivably produce a volume contraction). Thus, in this oil drop model, "homotropic" interactions are always attractive, whereas "heterotropic" interactions can be attractive or repulsive, depending on the relative signs of δ_{I} and δ_{II}. In proteins, only attractive cooperative homotropic effects have been observed, while both cooperative and antagonistic heterotropic interactions have been seen (Monod, Wyman, and Changeux, 1965).

It is also of interest to consider the introduction of three molecules into the oil drop. In this case the quadratic term takes the form,

$$\Delta F^{(2)} = -\tfrac{1}{9}\,\gamma A_0(\delta_{\mathrm{I}}^2 + \delta_{\mathrm{II}}^2 + \delta_{\mathrm{III}}^2 + 2\delta_{\mathrm{I}}\delta_{\mathrm{II}} + 2\delta_{\mathrm{I}}\delta_{\mathrm{III}} + 2\delta_{\mathrm{II}}\delta_{\mathrm{III}}) \tag{10}$$

In equation (10) if $\delta_{I} = \delta_{II}$ (corresponding to an authentic substrate) and δ_{III} comes from a chemical perturbation, we see that in this second-order term the chemical perturbation affects the affinity of the oil drop for substrate (through the $\delta_{I}\delta_{III}$ and $\delta_{II}\delta_{III}$ terms) but does not alter the substrate-substrate interaction ($\delta_{I}\delta_{II}$). The reader may recall that just this sort of effect occurs in modified hemoglobin (where $\delta_{I} = \delta_{II}$ would arise from O_2 and δ_{III} from an attached label).

The oil drop analogy can be carried a step farther, to yield semiquantitative information about an allosteric interaction. For an order of magnitude estimate of interfacial tension we may take $\gamma \simeq 35$ dynes/cm (8.4×10^{-7} cal/cm^2), the value for benzene in water, to be representative of a general hydrophobic substance. Deoxyhemoglobin is spheroidal, with dimensions of 50 Å by 55 Å by 69 Å (Muirhead and Perutz, 1963), giving an "average radius" of ~29 Å; it contains four heme groups. Oxyhemoglobin has an "average radius" of ~26 Å, estimated as above from the dimensions given by Muirhead and Perutz (1963), suggesting that $\delta_{I} = \delta_{II} = \ldots$ should fall in the range -0.075 to -0.15. This corresponds to a heme-heme interaction energy of 1100 to 4000 cal/mole; the observed homotropic heme-heme interaction energy is ~3000 cal/mole per heme group (Wyman, 1964).

The lesson to be learned from the above calculation is that except for the active site structure, allosteric interactions could well be more a general property of all protein structures than a consequence of sophisticated, "well-planned," detailed interactions between protein subunits, at least within the range of substrate or chemical interactions that can be regarded as relatively weak perturbations of the overall protein. More generally, if the free energy of a protein in solution can be represented for these purposes by an expansion in a single variable δ,

$$F = F_0 + \frac{\partial F}{\partial \delta}(\Sigma_i \delta_i) + \frac{1}{2}\frac{\partial^2 F}{\partial \delta^2}(\Sigma_i \delta_i)^2 + \cdots \tag{11}$$

where δ_i is a stress parameter associated with the introduction of molecule i (substrate or otherwise) into the protein, then homotropic and heterotropic interactions between substrates, and between substrates and chemical perturbations, can be understood from a qualitative if not quantitative point of view. Although it would certainly be improper to push the analogy between an allosteric protein and our oil drop too far, two points of similarity should be recognized. At least some proteins, especially the so called globular proteins, must be quite tightly packed structures. Under these circumstances the protein, like the oil drop, may be expected to undergo a volume change when a molecule is introduced into the structure. A change in volume of a protein may, as in the oil drop, bring about an increase or decrease in the hydrophobic area exposed to water. Such effects may be detectable with suitably designed labels whose spectra are sensitive to solvent polarity.

SPIN-LABELED PROTEIN CRYSTALS

When a spin label attached to a protein in solution gives a strongly immobilized spectrum, it may be expected that the label will have a well-defined orientation when the protein is crystallized (McConnell, 1966; McConnell and Boeyens, 1967). This has been found to be the case for horse hemoglobin labeled with the nitroxide maleimides IX and X (Ohnishi, Boeyens, and McConnell, 1966) and for α-chymotrypsin acylated at the active site with the spin-labeled substrate XV (Berliner and McConnell, 1967). In the case of horse hemoglobin, the analysis of the single crystal resonance spectra gives the following spin-Hamiltonian parameters: $A = 91$, $B = 22$, $C = 28$ Mc; $g_{xx} = 2.0078$, $g_{yy} = 2.0060$, $g_{zz} = 2.0027$. The radical is oriented so that the hyperfine z-axis is parallel to the crystallographic **b**-axis, which is also the twofold molecular axis (Cullis, Muirhead, Perutz, and Rossmann, 1962), and so that the x component of the hyperfine tensor is parallel to the crystallographic **a**-axis. Also observable is another set of oriented labels related by the two-fold molecular axis **b** with hyperfine axes x' and x'' displaced from $x = \mathbf{a}$ by approximately $+15°$ and $-15°$. The two distinct orientations of the radical must arise from some nonequivalence in labeled β-subunits; this nonequivalence may be due to intrinsic heterogeneity in the protein (Perutz, Steinrauf, Stockell, and Baugham, 1959), to isomeric forms of the label itself, to effects exerted on the β-chain by the presence of the label, or, most probably, to crystal imperfections (M. F. Perutz, private communication, 1966).

The paramagnetic resonance spectra of labels attached to protein molecules in single crystals must show the appropriate crystallographic and molecular symmetry properties. Crystallographic symmetry of protein crystals can be determined relatively easily using X-ray diffraction. When molecular symmetry axes coincide with crystallographic axes, the presence of such molecular axes in a protein can also be determined relatively easily using X-ray diffraction (Green and Aschaffenburg, 1959; Perutz et al., 1960; Pickles, Jeffery, and Rossmann, 1964; Watson and Banaszak, 1964). When some or all of the molecular symmetry axes do not coincide with the crystallographic axes, the detection of the molecular symmetry axes with the use of X-ray diffraction is a formidable problem. Since the paramagnetic resonance of labeled crystals must show molecular as well as crystallographic symmetry, the resonance spectra do provide information on molecular protein symmetry. Experiments are in progress in this laboratory on α-chymotrypsin crystals that have been labeled at the active site with a paramagnetic acyl group by soaking the crystals in a solution of XV (Berliner and McConnell, 1967) following the procedures introduced by Sigler, Jeffery, Matthews, and Blow (1966), in connection with X-ray studies of derivatives of this enzyme. These studies are directed toward the spin-label determination of the previously known two-fold symmetry axis that relates the two halves of the α-chymotrypsin dimer in this crystal (Sigler et al., 1966).

One special trick is involved in the use of nitroxides to determine molecular symmetry in crystals. Because the odd electron is contained primarily in a cylindrically symmetric $2p\pi$ atomic orbital, the hyperfine Hamiltonian [see Equation (1)] has nearly axial symmetry. Since in addition the spin Hamiltonian has inversion symmetry, any two electrons in $2p\pi$ orbitals can be converted into one another by *three* two-fold axes (illustrated in Figure 15). If a real molecular symmetry axis relates the orbitals to each other, it must coincide with one of the three. Two topologically distinct spin-labeled derivatives are required to identify which is the unique axis.

This particular interest in molecular symmetry axes stems from their importance in theoretical molecular biology, such as the theory of allosteric transitions (Monod et al., 1965) and the theory of interallelic complementa-

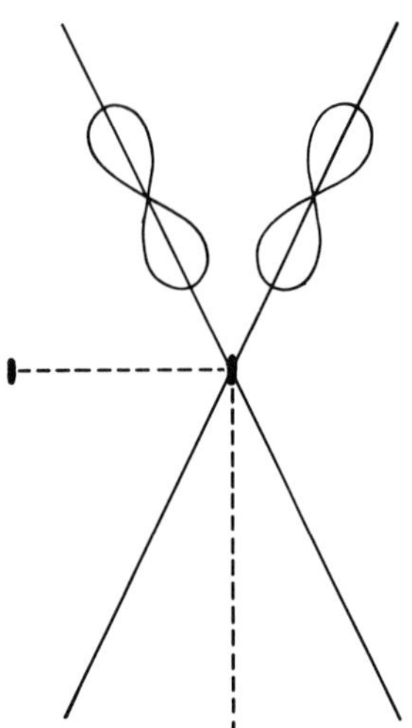

FIGURE 15.
Two arbitrarily placed $2p\pi$ orbitals and the three twofold axes relating them. Solid lines represent the unique axis of each orbital. Dashed lines show the twofold axes in the plane of the paper; the third is perpendicular.

tation (Crick and Orgel, 1964). Single crystal spin-label studies obviously offer many other opportunities for investigation of molecular structure. Of particular interest is the possibility of determining distances between labels from their dipole-dipole interaction.

CONCLUSION

We have surveyed here the developing technique of spin labeling. The paramagnetic resonance of spin labels has been shown to reflect changes in molecular conformation and chemical reactions previously revealed by other methods in molecular biology and biochemistry. Now well established, the technique is being applied to the solution of new problems, such as the measurement of allosteric conformational changes.

It may be appropriate in conclusion to summarize some overall features of spin labeling as a probe of biological systems. The criticism has been leveled that many of our studies involve chemically modified, and therefore somewhat perturbed, macromolecules. The technique must of course be used with some judgment. However, such perturbations as may occur will not often defeat the purpose of an experiment. The symmetry of a symmetrical enzyme, for example, will in all probability not be destroyed if all the subunits are labeled. The allosteric proteins that have been spin-labeled have remained allosteric, and it is doubtful that the basic interaction mechanisms are changed qualitatively by the reaction. Synthetic spin-labeled substrates certainly show spectral effects that must be closely related to the native activity of the enzymes that act upon them. The spin-labeled single crystals of at least two proteins (hemoglobin and α-chymotrypsin) are isomorphous with unmodified crystals. Finally, helix-coil transitions and other conformational changes that have been observed by optical rotation and other techniques show no significant modification of their behavior when they are followed by means of spin-label resonance spectra.

One of the potentially most significant aspects of spin-label spectra are that they can be observed in many situations—in protein single crystals, in solution, and in more complex structures such as membranes. This means that molecular structural information obtained from the combined sources of X-ray diffraction and spin-label resonance spectra of protein crystals can at least in certain cases be correlated with spin-label spectra in solutions (and possibly membranes). Our current work on protein symmetry in crystals and in solution, and on the several conformations of hemoglobin in solution, are examples of this approach.

ACKNOWLEDGMENTS

We are indebted to the Advanced Research Project Agency through the Center for Materials Research at Stanford University for facilities made available for this work. We are greatly indebted to Dr. M. F. Perutz and Dr. J. C. A. Boeyens for help leading to the construction of the hemoglobin models shown here. We are also indebted to Seiji Ogawa for many helpful discussions.

In concluding our contribution to this Festschrift, it is pertinent to remark that there are few topics in modern structural chemistry that have not been profoundly influenced by Linus Pauling. The present work is no exception. As one evident example, Pauling (1935) was the first to interpret cooperative oxygen binding to hemoglobin in terms of an interaction between the oxygen binding sites, and Coryell, Pauling, and Dodson (1939) were the first to distinguish between certain symmetry-breaking and symmetry-conserving models for this interaction (using magnetic susceptibility measurements). A less evident example is the important role that has been played by Pauling's pioneering studies of orbital hybridization and spin correlation in atomic valence states. These studies have provided a foundation for the subsequent development of theories of nuclear hyperfine interactions in molecules, which in turn are important in a number of applications of the spin-label method. It is thus with pleasure and gratitude that this contribution is dedicated to Linus Pauling.

REFERENCES

Applequist, J., and P. Doty (1962). In Stahmann, M. A., ed., *International Symposium on Polyamino Acids, Polypeptides and Proteins*. Madison, Wisconsin: University of Wisconsin Press.

Bender, M. L., and F. J. Kézdy (1959). *Ann. Rev. Biochem.* **34**, 49.

Benesch, R., and R. E. Benesch (1961). *J. Biol. Chem.* **236**, 405.

Benesch, R., and R. E. Benesch (1963). *J. Mol. Biol.* **6**, 498.

Berliner, L. J., and H. M. McConnell (1966). *Proc. Nat. Acad. Sci. U.S.* **55**, 708.

Berliner, L. J., and H. M. McConnell (1967). In preparation.

Blackley, W. D., and R. R. Reinhard (1965). *J. Am. Chem. Soc.* **87**, 802.

Boeyens, J. C. A., and H. M. McConnell (1966). *Proc. Nat. Acad. Sci. U.S.* **56**, 22.

Briere, R., H. Lemaire, and A. Rassat (1964). *Tetrahedron Letters* **27**, 1775.

Briere, R., H. Lemaire, and A. Rassat (1965). *Bull. Soc. Chim. France* **32**, 3273.

Briere, R., and A. Rassat (1965). *Bull. Soc. Chim. France* **32**, 378.

Carrington, A., and H. C. Longuet-Higgins (1962). *Molecular Physics* **5**, 447.

Cecil, R. (1963). In Neurath, H., ed., *The Proteins*, vol. 1, p. 380. New York: Academic Press.

Cole, T. (1961). *J. Chem. Phys.* **35**, 1169.

Cole, T., H. O. Pritchard, N. R. Davidson, and H. M. McConnell (1958). *Mol. Phys.* **1**, 406.

Coryell, C. D., L. Pauling, and R. W. Dodson (1939). *J. Phys. Chem.* **43**, 825.

Crick, F. H. C., and L. E. Orgel (1964). *J. Mol. Biol.* **8**, 161.

Cullis, A. F., H. Muirhead, M. F. Perutz, and M. G. Rossmann (1962). *Proc. Roy. Soc.* (London) A **265**, 161.

Dupeyre, R.-M., H. Lemaire, and A. Rassat (1964). *Tetrahedron Letters* **27**, 1781.
Dupeyre, R.-M., and A. Rassat (1966). *J. Am. Chem. Soc.* **88**, 3180.
Dupeyre, R.-M., A. Rassat, and P. Rey (1965). *Bull. Soc. Chim. France* **32**, 3643.
Freed, J. H., and G. K. Fraenkel (1963). *J. Chem. Phys.* **39**, 326.
Friedlander, P. H., and J. M. Robertson (1956). *J. Chem. Soc.*, 3083.
Giacometti, G., and P. L. Nordio (1963). *Mol. Phys.* **6**, 301.
Golubev, V. A., and E. G. Rozantzev (1965). *Izv. Akad. Nauk SSSR, Ser. Khim.*, 716.
Green, D. W., and R. Aschaffenburg (1959). *J. Mol. Biol.* **1**, 54.
Griffith, O. H., D. W. Cornell, and H. M. McConnell (1965). *J. Chem. Phys.* **43**, 2909.
Griffith, O. H., and H. M. McConnell (1966). *Proc. Nat. Acad. Sci. U.S.* **55**, 8.
Hamilton, C. L., and H. M. McConnell (1967). In preparation.
Harrington, W. F., P. Johnson, and R. H. Ottewill (1956). *Biochem. J.* **62**, 569.
Hartley, B. S., and B. A. Kilby (1952). *Biochem. J.* **50**, 672.
Heinrickson, R. L. (1966). *J. Biol. Chem.* **241**, 1393.
Hoffmann, A. K. (1964). *C. A.* **61**, 8191 (P).
Hoffmann, A. K., A. M. Feldman, and E. Gelblum (1964). *J. Am. Chem. Soc.* **86**, 646.
Hoffmann, A. K., A. M. Feldman, E. Gelblum, and W. G. Hodgson (1964). *J. Am. Chem. Soc.* **86**, 639.
Hoffmann, A. K., and A. T. Henderson (1961). *J. Am. Chem. Soc.* **83**, 4671.
Itzkowitz, M. S. (1966). Ph. D. Thesis, California Institute of Technology.
Kézdy, F. J., and M. L. Bender (1962). *Biochemistry* **1**, 1097.
Kivelson, D. (1960). *J. Chem. Phys.* **33**, 1094.
Koshland, D. E., Jr., G. Némethy, and D. Filmer (1966). *Biochemistry* **5**, 365.
Kosower, E. M. (1958). *J. Am. Chem. Soc.* **80**, 3253.
Krause, S., and C. T. O'Konski (1959). *J. Am. Chem. Soc.* **81**, 5082.
Krause, S., and C. T. O'Konski (1963). *Biopolymers* **1**, 503.
Krinitskaya, L. A., E. G. Rozantzev, and M. B. Neiman (1965). *Izv. Akad. Nauk SSSR, Ser. Khim.*, 115.
Linnett, J. W. (1961). *J. Am. Chem. Soc.* **83**, 2643.
McConnell, H. M. (1952). *J. Chem. Phys.* **20**, 701.
McConnell, H. M. (1956). *J. Chem. Phys.* **25**, 709.
McConnell, H. M. (1967). Talk to be published in Wenner-Gren Center International Symposium Series.
McConnell, H. M., and J. C. A. Boeyens (1967). *J. Phys. Chem.* **71**, 12.
McConnell, H. M., and R. W. Fessenden (1959). *J. Chem. Phys.* **31**, 1688.
Monod, J., J.-P. Changeux, and F. Jacob (1963). *J. Mol. Biol.* **6**, 306.
Monod, J., J. Wyman, and J.-P. Changeux (1965). *J. Mol. Biol.* **12**, 88.
Muirhead, H., and M. F. Perutz (1963). *Nature* **199**, 633.
Neiman, M. B., E. G. Rozantzev, and Yu. G. Mamedova (1962). *Nature* **196**, 472.
Ogawa, S., and H. M. McConnell (1967). *Proc. Nat. Acad. Sci. U.S.* **58**, 19.
Ohnishi, S., J. C. A. Boeyens, and H. M. McConnell (1966). *Proc. Nat. Acad. Sci. U.S.* **56**, 809.
Ohnishi, S., and H. M. McConnell (1965). *J. Am. Chem. Soc.* **87**, 2293.
Pauling, L. (1935). *Proc. Nat. Acad. Sci. U.S.* **21**, 186.
Perutz, M. F. (1965). *J. Mol. Biol.* **13**, 646.
Perutz, M. F., M. G. Rossmann, A. F. Cullis, H. Muirhead, G. Will, and A. C. T. North (1960). *Nature* **185**, 416.
Perutz, M. F., L. K. Steinrauf, A. Stockell, and A. D. Baugham (1959). *J. Mol. Biol.* **1**, 402.
Pickles, B., B. A. Jeffery, and M. G. Rossmann (1964). *J. Mol. Biol.* **9**, 598.
Riggs, A. (1961). *J. Biol. Chem.* **236**, 1948.
Riordan, J. F., W. E. C. Wacker, and B. L. Vallee (1965). *Biochemistry* **4**, 1758.
Rozantzev, E. G. (1964). *Izv. Akad. Nauk SSSR, Ser. Khim.* 2187.
Rozantzev, E. G., V. A. Golubev, and M. B. Neiman (1965). *Izv. Akad. Nauk SSSR, Ser. Khim.* 379.

Rozantzev, E. G., and L. A. Krinitskaya (1965). *Tetrahedron* **21**, 491.
Rozantzev, E. G., and M. B. Neiman (1964). *Tetrahedron* **20**, 131.
Rozantzev, E. G., and A. B. Shapiro (1964). *Izv. Akad. Nauk SSSR, Ser. Khim.* 1123.
Sigler, P. B., B. A. Jeffery, B. W. Matthews, and D. M. Blow (1966). *J. Mol. Biol.* **15**, 175.
Smith, I. C. P., and H. M. McConnell (1967). In preparation.
Stone, T. J., T. Buckman, P. L. Nordio, and H. M. McConnell (1965). *Proc. Nat. Acad. Sci. U.S.* **54**, 1010.
Stryer, L., and O. H. Griffith (1965). *Proc. Nat. Acad. Sci. U.S.* **54**, 1785.
Watson, H. C., and L. J. Banaszak (1964). *Nature* **204**, 918.
Weber, G. (1953). *Discussions Faraday Soc.* **13**, 33.
Wyman, J. (1964). *Adv. in Protein Chem.* **19**, 268.
Yang, J. T., and J. F. Foster (1954). *J. Am. Chem. Soc.* **76**, 1588.

ANTIBODIES

DAVID PRESSMAN
ALLAN GROSSBERG
Department of Biochemistry Research
Roswell Park Memorial Institute
Buffalo, New York

The Chemical Nature of the Combining Regions of Antibody Molecules

The combining sites of antibody molecules are formed by portions of the polypeptide chain(s) of a γ-globulin molecule, and there results a region that is complementary in configuration to the antigenic region against which it is directed. This configuration makes a close fit with the antigenic region, so that the various weak forces between molecules can be active in holding the antigen and antibody molecules together. In this combining region of the antibody, certain amino acid residues are present and are so oriented that they contribute to the charge or dipolar interaction, to hydrogen bond formation, to the formation of nonpolar bonds by close approximation of nonpolar regions, and to van der Waal's interaction (Pauling, Campbell, and Pressman, 1943; Pressman, 1957). This close and complementary fit results in the specificity observed.

In recent years, information has been forthcoming about the chemical nature of the combining regions of antibody molecules and the amino acid residues in this region.

Most of this information has been obtained from studies of the effect of chemical alteration of antibody on its activity. Such studies are carried out by reacting antibody with a reagent that is known to react with a particular chemical group. Loss of activity is indicative of an effect in the binding site. However, even if the antibody is inactivated by reaction with a particular reagent, it must be proven that the loss of activity is caused by a reaction in the binding site and not by chemical alteration of the antibody molecule elsewhere, followed by a change in the general structure of the protein, which in turn disrupts the site. Moreover, although reagents are available that react predominantly with certain groups or amino acid residues, at least some reaction usually takes place with other residues. Therefore it is important to demonstrate just which residue is attacked and to relate the attack of the particular residue to the loss of antibody activity.

Nearly all the information about the chemical composition of antibody sites has been obtained with four different antibodies: those directed against the *p*-azobenzenearsenate group (anti-R_p antibody), the *p*-azobenzoate group (anti-X_p antibody), the 3-azopyridine group (anti-P_3 antibody), and the *p*-azophenyltrimethylammonium group (anti-A_p antibody). Information is thus available about antibodies against two negative groups—a neutral group and a positively charged group.

Previous experiments (Pressman, Grossberg, Pence, and Pauling, 1946) not involving chemical alteration, had indicated the presence of a negative charge in the antibody site of the anti-positive ion antibody (anti-A_p), since the antibody combined ten times as strongly with a compound containing a trimethylphenylammonium group as with the corresponding compound in which the trimethylammonium group was replaced by a tertiary butyl group. The tertiary butyl group has the same size and shape but lacks the positive charge. The ratio of the combining constants is a measure of the differences in free energy of combination of the haptens with antibody. Taking this difference to be due to charge interaction of the positively charged hapten with a charge in the site, the distance between the charges was calculated and the result showed that there appeared to be a negative charge on the antibody within the distance of closest approach to the positive charge of the hapten.

Similarly, the antibody to the negative azobenzoate group (anti-X_p antibody) appeared to have a positive charge in the combining site. Further studies have revealed additional information, which allows interpretation of experimental data in terms of structural characteristics of the antibody site.

THE NATURE OF THE NEGATIVE CHARGE IN THE COMBINING SITE OF THE ANTI-A_p ANTIBODY

The negative charge appears to be due to a carboxylate group, whose presence has been shown by chemical means (Grossberg and Pressman, 1960). Anti-*p*-azophenyltrimethylammonium antibodies were treated with the reagent diazoacetamide, which is known to esterify carboxyl groups according to the reaction

$$\text{Protein—COOH} + N_2CHCONH_2 \longrightarrow \text{Protein—COOCH}_2CONH_2 + N_2$$

which removes the negative charge on the carboxylate. When this was done, the anti-A_p antibody lost its ability to bind specific hapten, as measured by the method of equilibrium dialysis. The loss of activity parallels the degree of reaction: the greater the number of carboxylate groups esterified, the greater the loss of activity. By the time that 25 percent of the carboxyl groups were esterified, the antibody lost 65 percent of its binding activity.

However, this does not necessarily mean that there is a carboxylate group in the active site. That a reaction in the active site is actually involved in the loss of activity must be proved. This was done by a "protection" experiment in which the antibody is combined with a simple hapten such as the phenyltrimethylammonium ion, and the mixture is esterified with diazoacetamide. In the experiment, esterification is carried out to an extent that is known to destroy antibody activity in the absence of hapten. When the solution is freed of hapten by dialysis it is found that antibody activity is retained; the hapten has blocked the chemical alteration of the site. The reactive group of the antibody site is protected by hapten against esterification. When the hapten is removed, the antibody site is recovered unaltered, although many carboxyl groups elsewhere on the antibody have been esterified.

Experiments of this type are useful in proving the presence of a particular group in the antibody site. If treatment with a reagent destroys antibody activity in the absence of hapten but does not do so in the presence of hapten, then the reagent affects a group present in the binding site.

In a typical experiment involving protection of anti-A_p antibody against esterification, the antibody with no protection lost 58 percent of its binding activity when 18 percent of its carboxyls were esterified. When the esterification was carried out to the same extent in the presence of .005 *M* *p*-iodophenyltrimethylammonium ion, only 19 percent of antibody activity was lost.

The loss of activity due to esterification was found only for the anti-A_p antibody. None of the other antibodies tested, that is, the anti-R_p, anti-X_p or anti-P_3, suffered loss of activity on esterification. This result indicates that no carboxyl group, important for combination of site with hapten, exits in the combining regions of these latter antibodies.

The inactivation of anti-A_p antibody by esterification can be reversed. Activity can be recovered from the esterified antibody following hydrolysis of the ester according to the reaction

$$\text{Protein—COOCH}_2\text{CONH}_2 + \text{OH}^- \longrightarrow \text{Protein—COO}^- + \text{HOCH}_2\text{CONH}_2$$

Hydrolysis, which liberates the carboxylate and is accomplished (though not completely) by bringing the esterified antibody to *p*H 11, results in recovery of some binding activity, about 20 percent of the binding capacity lost by the original esterification. This recovery of activity shows further that esterification merely blocks the site rather than irreversibly denaturing the antibody.

Esterified anti-A_p antibody was analyzed to determine whether groups other than carboxyl had been modified. At the level of 30 percent esterification of carboxyls, there appeared to be no decrease in amino content of the protein by van Slyke analysis, indicating that less than 3 to 4 percent of amino groups had reacted. It was estimated that less than 5 percent of the tyrosine or histidine groups had been alkylated. One or two sulfhydryls were alkylated.

From the rate of decrease of binding activity with esterification—70 percent loss of activity when 30 percent of the carboxyls are esterified—it appears that a carboxyl in the site is more easily esterified than are some of the carboxyls elsewhere on the antibody.

When binding activity is affected by a chemical reaction, such as by esterification in hapten absence, the attack may be directed to an important part of the site as designated by position A in the hypothetical site shown in Figure 1. Such an attack will drop the binding constant to essentially zero so that the site becomes undetectable (that is, is "lost"). If esterification is at B or C, on the periphery of the binding site, the constant for the site is decreased, depending on the degree of interference with binding exerted by the ester group. However, the number of sites present will not be altered, because binding still takes place, although a higher concentration of hapten is required to show binding. Esterification farther from the site, such as at D, will produce little effect on the binding constant and of course no effect on the number of sites detectable.

When the chemical alteration is carried out in the presence of hapten,

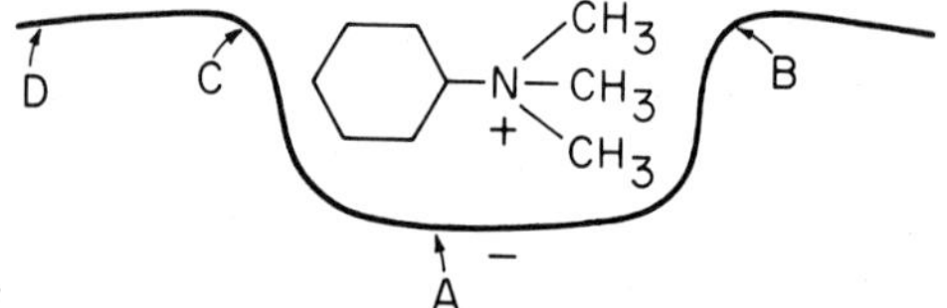

FIGURE 1.
Effect of chemical alterations on hypothetical combining site of anti-A_p antibody. Alteration at A would have a major effect; at B and C, a lesser effect; and at D, a minor effect.

there is protection against loss of sites by interference with alteration at point A. Interference by hapten at points B and C results in protection against the decrease in binding constant caused by alteration at these points. When the point of alteration is further removed from the site, as that at point D, the effects of alteration and of protection are less.

Chemical alteration, particularly when extensive, can cause changes in the shape of a protein molecule (Habeeb, Cassidy, and Singer, 1958). However these effects would be expected to be the same for protected or unprotected antibody molecules, since all of the alteration (except for the single group that may be in the site) must take place elsewhere than in the site. Therefore the demonstration of a protection effect gives evidence that the loss of activity is not due to nonspecific general alteration of the antibody molecule. Moreover, the anti-X_p, anti-R_p, and anti-P_3 antibodies showed no loss of activity on esterification, and if distortion of the site were nonspecific, we would expect these antibodies as well as anti-A_p to be affected.

Another specific type of conformational change could explain the protection of antibody activity by hapten during a chemical modification. The apparent properties of a protein are the average of the properties of the various conformations of the molecule existing in the equilibrium mixture. It is possible that when a particular carboxylate elsewhere than in the site is esterified, conformations of the molecule, in which the binding site is so altered that it no longer binds hapten, are stabilized. If, in addition, the presence of hapten in the binding site of unaltered antibody causes stabilization of conformations in which the reactivity toward esterification of the particularly important carboxylate mentioned above is decreased, the protection effect would be observed, without any reaction having taken place in the site.

This mechanism does not seem to be applicable to the situation under consideration, however, since esterification affects only anti-A_p antibodies. If esterification exerted its effect at a distance by the above allosteric mechanism, it is difficult to see why antibodies to other haptens would not be similarly affected. Although it is reasonable that anti-A_p might differ from the other antibodies by having a carboxylate in the site, it is not likely that it should differ from other antibodies by having an important carboxylate elsewhere than in the site.

Esterification of anti-A_p antibody causes some decrease in the binding constant of remaining antibodies. In one experiment, when 42 percent of the sites were lost (20 percent of the carboxyls esterified) there was a reduction in average binding constant from 8.7×10^5 l/mole for the unaltered antibody to 5.6×10^5 l/mole for the esterified antibody. This reduction in constant is probably partially of an electrostatic nature, owing to the increased positive charge of the antibody as a whole, which results in a nonspecific decreased attraction (or increased repulsion) for the positive hapten. Steric effects due to esterification of carboxyl group close to the antibody site but not in it, or alteration of the whole molecule due to increased positive charge, may also be contributing factors to the decrease in binding constant. Another possibility is that the antibody sites having higher binding constants are those which are more effectively inactivated. Then the antibody sites remaining after esterification would be those with a lower average binding constant.

The second possibility seems most plausible because when anti-R_p antibody is esterified to essentially the same extent (20 percent of the carboxyls esterified), there is no loss of sites and there is no difference in binding constant for esterified and unesterified anti-R_p antibody.

THE EFFECT OF IODINATION ON ANTIBODY ACTIVITY

When the four antibodies studied are iodinated there is a loss of binding activity in each case. The presence of the homologous haptens protects against this loss, showing that the loss of activity in each antibody is caused by chemical reaction of the iodine with a group in the binding site (Pressman and Sternberger, 1951) and (Grossberg, Radzimski, and Pressman, 1962). The reaction of iodine with protein results in the incorporation of iodine as a substituent on tyrosine or, to a minor extent, on histidine.

During iodination there may also be some oxidation of tyrosine, histidine, tryptophan, cysteine, cystine, and methionine. The degree to which tryptophan, histidine, and methionine are affected when up to 50 iodine atoms per mole protein are incorporated has been shown to be negligible by amino acid analyses of the hydrolysates of the iodinated proteins (Koshland, Englberger, and Gaddone, 1963).

When a globulin molecule is altered by chemical reaction, such as incorporation of iodine, chemical analysis yields a figure for the average number of atoms incorporated per protein molecule. The number of iodine atoms per single molecule will vary from this figure—some having more, some having less, and some having none. The shape of the distribution curve for any particular degree of iodination will depend on the relative iodination rates of the various residues. The particular residues iodinated will vary even for molecules that incorporate identical numbers of iodine atoms.

At very low levels of iodination where the average number of iodine atoms is much less than one per molecule and most molecules are uniodinated, those molecules that are iodinated have only one iodine atom each. This iodine appears on different residues of the individual molecules, depending on the relative iodination rates of the residues. The effect of iodination on the binding site of a particular antibody molecule will depend, of course, on which residues are iodinated.

The iodination reaction is further complicated because tyrosine first forms monoiodotyrosine and then diiodotyrosine. Thus at a particular level of iodination—for example at an average of ten iodine atoms per molecule—certain tyrosines of each antibody molecule are completely converted to diiodotyrosines, other tyrosines are partially converted to diiodotyrosine so that they are present as a mixture of uniodinated, monoiodinated, and di-iodinated forms, and still others may be essentially uniodinated or only partially converted to the monoiodotyrosine form.

The effect of iodination on antibody activity is illustrated by the effect on the activity of anti-X_p antibodies. As the degree of iodination is increased, the activity decreases.

The presence of a simple hapten protects against the loss of activity and shows that the loss of activity upon iodination is caused by reaction in the anti-X_p site.

The other antibodies studied—anti-R_p, anti-A_p, and anti-P_3—also show loss of activity following iodination. A very close comparison of degree of inactivation of the different antibodies can be obtained by iodinating a mixture of two antibodies in the same solution and determining the effect on the binding activity of each individual antibody. When the iodination of the two antibodies is done in the same solution, they necessarily obtain the same exposure to the reagent. For example, it has been found that anti-R_p activity is more sensitive to loss by iodination than is anti-A_p activity.

Iodination in the absence of hapten has produced preparations in which the remaining sites of anti-R_p, anti-X_p, and anti-A_p antibodies show decreased binding constants and the remaining anti-P_3 sites show increased binding constants. It should be noted that all the binding curves indicate that the preparations were heterogeneous with respect to binding constants. It appears that, for each system, the various antibodies composing a heterogeneous population have somewhat different binding sites. The different sites in each system may also have different susceptibilities to alteration as well as different binding constants. If iodination causes a preferential loss of binding sites of high binding constant, there results a population with a lower average binding constant after iodination. On the other hand, if iodination causes a preferential loss of binding sites with a low constant, then the average constant observed after iodination is larger.

A hapten protects only those sites it occupies. The degree of protection depends on the strength of binding and on the importance for binding of the iodinatable group in the site. From these considerations, it follows that with the preparations of anti-R_p, anti-X_p, and anti-A_p antibodies studied, iodination in the presence of hapten results in preparations that bind more strongly than do those obtained after iodination in the absence of hapten, because the stronger binding sites, although more sensitive to iodination than are the weaker binding sites, are better protected.

Loss of activity by iodination in the presence of hapten does take place. The most plausible explanation for this loss is that because of the equilibrium state, iodination takes place in the sites of antibodies that are uncombined with hapten. Other possible reasons for change of activity include iodination of the antibody, or antibody-hapten complex elsewhere than in the site. Iodination or oxidation elsewhere in the molecule may cause loss of some sites by unfolding the protein in the region composing the site. This destruction would not be prevented by the presence of hapten. Thus, specifically purified anti-X_p, which was exposed to 285 moles of HOI per mole of protein, incorporated about 70 iodine atoms—only a few more than the iodine atoms incorporated by samples that were exposed to only 94 moles of HOI. The former samples lost much more activity than the latter, and the loss could not be prevented by the presence of hapten. It seems a reasonable assumption that the lack of increased incorporation in the more highly exposed samples can be correlated with the occurrence of increased oxidative reactions leading to destruction of the site.

Another possibility is that a tyrosine contributing to a site occupied by hapten may still be iodinated even though it is blocked on one side by hapten. This could cause the binding energy of hapten-site interaction to change, owing to the increased ionization of the tyrosine hydroxyl to yield a negative charge in the site. Such ionization would decrease binding of an anionic hapten. Any reduction of binding would increase the ability of a susceptible site to be iodinated, owing to increased dissociation of the hapten.

THE ISOLATION OF PEPTIDES FROM THE BINDING SITES OF ANTIBODY MOLECULES

The iodination reaction has made it possible to isolate, from a digest of iodinated antibody, peptides that appear to have come from the combining region of the antibody molecules. These peptides contain iodine-labeled tyrosine, whose formation is affected by the presence of hapten during the iodination. The technique used is the "Paired Label" iodination technique (Pressman and Roholt, 1961).

One portion of a preparation is iodinated directly with ^{125}I. The other portion is similarly iodinated with ^{131}I, but in the presence of hapten. Iodine incorporation is carried out to the same level for each portion. If there is a tyrosine in the site, it is iodinated with ^{125}I in proportion to its reactivity along with other tyrosines on the molecule. However, in the mixture containing the hapten, the rate of iodination by ^{131}I of the tyrosine protected by the hapten is reduced. The two preparations are then mixed and digested with pepsin or another proteolytic enzyme. A portion is placed on a sheet of filter paper, and the peptides are separated by chromatography in one direction and then subjected to high voltage electrophoresis at right angles to the chromatography. The radioactive peptides are located by making a radioautograph of the paper. The portions of the filter paper corresponding to the location of radioactivity are then cut out and the amounts of ^{125}I and ^{131}I present are determined for each. The amounts of the two isotopes are in the same ratio for most of the spots and for the original digest, but some spots have a high or low ratio of ^{125}I to ^{131}I. These then are peptides containing iodinatable residues which are iodinated at different rates in the presence and absence of hapten.

When the peptides containing these aberrant ratios are hydrolyzed to amino acids with pronase and then subjected to high voltage electrophoresis, it is found that the radioactive iodine is present either as mono- or diiodotyrosine. It is interesting that the number of binding sites lost for these preparations, as determined from ability of the iodinated antibody to bind hapten, is essentially equivalent to the amount of iodinated tyrosine residues recovered from the "deviant ratio" peptides. This indicates that each lost site yields an iodinated tyrosine.

The reason that both high and low ratio peptides are associated with the protected tyrosine (of decreased reactivity) in the site is that some peptides contain diiodotyrosine and others contain monoiodotyrosine. All diiodotyrosine-containing peptides that come from the site will have a high ratio of ^{125}I to ^{131}I because the protected site is of reduced reactivity. Some of the monoiodotyrosine peptides will have a low ratio because there is more monoiodotyrosine peptide from the protected (less reactive) tyrosine than there is from the unprotected tyrosine. The reason for this is that, in the protected tyrosine, not as much of the monoiodotyrosine has been converted to the diiodotyrosine as has been converted in the more rapidly iodinating unprotected tyrosine.

The peptides of aberrant ratio are of particular interest because they contain the tyrosine present in the antibody site. By determining the amino acid sequence of such peptides, it is possible to obtain information about the composition of the chain contributing to the binding site.

There remains, however, the possibility that when hapten combines with an antibody, there is some change in the conformation of the antibody molecule so that a form is stabilized in which a tyrosine elsewhere than in the site is reduced in reactivity. If this were to take place, deviant ratio peptides would be obtained, but they would be caused by the conformation change rather than by the protection effect. However we do know that the number of sites disappearing is roughly equivalent to the amount of mono- and diiodotyrosine isolated from deviant ratio peptides. This is evidence that the tyrosine isolated came from the antibody site.

THE EFFECT OF CYANATE ON ANTIBODY ACTIVITY

Cyanate converts amino (ammonium) groups to urea groups that do not carry the positive charge at physiological *p*H values. A modification reaction of this sort is of particular interest in connection with antibodies, such as anti-X_p and anti-R_p, which are directed against negatively charged groups and for which there appears to be a positive charge in the specific binding site. Reaction with cyanate can thus give information about the presence of an amino (ammonium) group in the combining site of such antibodies.

Treatment of anti-R_p antibodies with cyanate until 70 to 80 percent of the free amino groups were carbamylated was found to cause the loss of 20 to 30 percent of the antibody sites. That this loss of activity was due to an attack on a group in a site was shown by the fact that the loss could be partially prevented when hapten was present during the carbamylation reaction (Chen, Grossberg, and Pressman, 1962).

Anti-azobenzoate antibody loses some activity when reacted with cyanate but this appears to be due to an attack in the antibody elsewhere than in the specific site, since the loss of activity could not be prevented by the presence of benzoate.

Anti-A_p antibody sites are not attacked at all by cyanate under the conditions in which anti-R_p sites are attacked, as shown by experiments in which mixtures of anti-A_p and anti-R_p antibodies were carbamylated.

Cyanate is relatively specific in its reaction with amino groups. It does not seem to react with tyrosine residues or with guanidinium groups. Although cyanate reacts with —SH groups to form a thiourethane, the inactivation of anti-R_p observed is probably not due to such a reaction because the SH group does not appear to be present in this antibody site.

THE EFFECT OF IODOACETAMIDE ON ANTIBODY ACTIVITY

The presence of a sulfhydryl group in binding sites was investigated by treating anti-R_p, anti-X_p, and anti-A_p antibodies with iodoacetamide under conditions known to alkylate one SH group in γ-globulins.

The reaction with these antibodies does not cause appreciable loss of binding activity (Grossberg, Stelos, and Pressman, 1962) and indicates that a sulfhydryl does not appear to be involved in the binding sites of any of these antibodies.

THE EFFECT OF IODOACETATE ON ANTIBODY ACTIVITY

Because of the possible presence of an amino group in some of the sites of certain antibodies, anti-X_p, anti-R_p, and anti-A_p antibodies have been treated at *p*H 8 with iodoacetate at concentrations greater than those required to react with SH groups. This reaction then involves mainly amino groups.

There is a loss of activity of the anti-benzenearsonate antibodies that can be prevented by the presence of hapten. It would appear that an amino group in some of the anti-R_p sites is alkylated. The presence of an amino group in the site of antibenzoate antibodies is uncertain because some loss of activity occurs but this is not prevented by the presence of hapten. There was no effect on the anti-A_p sites indicating absence of an amino group in those sites.

It is interesting to note that the carboxymethylation of amino groups leads to a large increase in net negative charge on the protein molecule. Although an increase in net charge as effected by succinylation leads to expansion of γ-globulin molecules, such expansion in carboxymethylation, if it takes place, apparently does not affect the binding sites as shown by the lack of effect on the anti-A_p site.

THE EFFECT OF AMIDINATION ON ANTIBODY ACTIVITY

One of the most specific reactions involving modification of amino groups on proteins under relatively mild conditions is that of amidination.

Wofsy and Singer (1963) extensively amidinated the amino groups in anti-R_p antibody and found a loss of up to 27 percent in binding activity. This loss might have been due to the amidination of an important amino group in the site of some anti-R_p antibodies, but since protection experiments were not carried out, this point was not clear.

We have investigated this problem (results unpublished) and have confirmed that amidination of specifically purified anti-R_p antibody preparation causes a loss of from 15 to 40 percent of binding activity in different preparations. The presence of 0.1 *M* phenylarsonate during the amidination of these preparations prevents some if not all of this loss, thus providing evidence that a certain proportion of antibody sites contain an amino group that is important for the combination of hapten with site.

THE EFFECT OF ACETYLATION ON ANTIBODY ACTIVITY

The effect of reaction with acetic anhydride on anti-hapten antibody activity has been studied with the four antibodies, anti-X_p, anti-R_p, anti-A_p, and anti-P_3. Acetic anhydride reacts with amino and hydroxy groups of proteins to give acetamino groups, and acetyl esters, respectively. The reaction with the amino groups is much more rapid than that with hydroxy groups, so that nearly all amino groups are acetylated before many hydroxyls react.

It was found that when anti-X_p antibody is acetylated, the precipitating activity is lost much more rapidly than the ability to bind hapten (Nisonoff and Pressman, 1958). Thus precipitating activity was reduced 60 percent and binding activity only 5 percent upon acetylation of 20 percent of the amino groups. Precipitating activity was lost completely, whereas binding activity was reduced only 12 percent on acetylation of 85 percent of the amino groups. This indicates that a reactive amino is apparently not in the anti-X_p site. At higher levels of acetylation, however, when hydroxy groups are acetylated, binding activity is lost and the loss can be prevented by the presence of hapten (Grossberg and Pressman, 1963). Moreover, activity can be recovered by exposure of the acetylated protein to alkali (*p*H 11) or to hydroxylamine—substances that do not affect acetylated amines but do split acetylated hydroxyl groups.

The loss of activity only at high levels of acetylation, its recovery following hydrolysis, and the protection of the site by hapten indicate, therefore, that hydroxy groups are in the anti-X_p sites.

Low levels of acetylation of anti-R_p antibodies have demonstrated some inactivation, which appears to be due to acetylation of an amino group in some of the sites. Only some of the anti-R_p antibodies appear to have an amino group in the binding site in accord with the observations of carbamylation and amidination.

Low levels of acetylation have practically no effect on the binding activity of anti-A_p and anti-P_3 antibodies, but high levels do cause loss of activity in these antibodies as well as in anti-X_p and anti-R_p antibodies. In each antibody the loss can be partially prevented by protecting the site with hapten during acetylation. Moreover, the effect on each can be largely reversed by mild hydrolysis.

There is thus considerable evidence for the participation of a hydroxyamino acid residue in the site of each of the four antibodies studied. The fact that the four can be protected by their specific haptens from loss of activity or sites caused by extensive acetylation indicates that a residue in each antibody site is attacked by acetic anhydride. In addition, the reversal by hydroxylamine of the loss of activity caused by acetylation indicates that a hydroxyl group in the sites is being attacked. In view of the evidence from iodination studies that tyrosines are present in these sites, it would appear that the hydroxyl groups involved in acetylation are from these tyrosines.

THE EFFECT OF DIAZONIUM COUPLING ON ANTIBODY ACTIVITY

Because of the apparent presence of tyrosine in their sites, two of the antibodies, anti-X_p and anti-R_p, were coupled with *p*-sulfobenzenediazonium chloride. This compound couples with proteins on the tyrosine, histidine, and lysine groups. For anti-R_p antibody, it was found that antibody activity was lost with coupling and that this loss could be prevented by the presence of hapten. This indicates that a residue is present in the binding site that reacts with diazonium ion. For anti-X_p, there was loss of activity that was not prevented by the presence of hapten (Grossberg and Pressman, unpublished). Thus the effect of diazonium coupling on anti-X_p is more complicated than simple coupling with the tyrosine known to be present in its site.

REDUCTION OF EXPOSED DISULFIDE BONDS

Mild reduction of several disulfide bonds and subsequent alkylation does not affect the activity of anti-X_p, anti-R_p, and anti-A_p antibodies, indicating that an easily reduced disulfide bond is not important for maintaining the configuration of the sites of these antibodies. Since the alkylation of the cysteine residues thus liberated does not affect antibody activity either, it is also apparent that the cysteine residues concerned cannot be present in the antibody sites (Grossberg, Stelos, and Pressman, 1962).

AFFINITY LABELING

A novel method of modifying residues of antibody molecules is that of affinity labeling, introduced by Wofsy, Metzger, and Singer (1962). The objective of this method is to carry out a reaction specifically in the combining site by taking advantage of the binding properties of anti-hapten antibodies. A similar approach has been applied to the modification of enzyme molecules by Baker in 1959, Schoellman and Shaw in 1962, and Lawson and Shramm in 1962 (see Baker, 1964). A reagent that specifically binds to the active site and that contains a chemically reactive group is employed. First the chemically reactive hapten binds to the antibody site, and then the reactive group forms a covalent bond with a particular residue of the protein. The residue must be of the kind, and be so situated, that it can react with the hapten fixed in the antibody site.

The studies have employed reagents with a diazonium group on benzenearsonate, nitrobenzene, or phenyltrimethylammonium. These substances are reacted with the specific antibodies anti-R_p, anti-dinitrophenyl, or anti-A_p, respectively. The reaction is much more rapid than the action of the diazonium compounds with normal γ-globulin or with the antibody molecule whose binding sites are reversibly blocked by hapten. Although the rapid reaction appears to take place with tyrosine, there is a question whether a tyrosine within the site or close to the site is being attacked. Thus the diazonium group on the hapten may possibly react with a tyrosine close to the site and in the proper orientation to react while the hapten is combined with antibody. Whether this is the same tyrosine found in the site by iodination studies remains to be determined.

It has been difficult to establish that antibody sites are irreversibly blocked by the affinity-labeling reaction. Although reduction of binding activity of anti-dinitrophenyl antibodies by this reaction has been demonstrated, it cannot be said that this is actually due to the modification of a group in the antibody site, since a dinitrophenyl group on one antibody molecule could reversibly bind with the antibody sites of a second molecule to give the observed loss of binding activity. Nonetheless, inasmuch as this method of labeling an antibody molecule apparently can give relatively restricted reaction with certain residues of the molecule, it should prove useful in mapping out antibody structure.

EFFECTS OF CHEMICAL ALTERATION ON THE PRECIPITIN REACTIONS

Chemical alteration can have marked effects on the specific precipitation of antibody without affecting the binding site.

Marrack and Orlans (1954) first pointed out that acetylation of an antiprotein antibody reduced its ability to precipitate but that the antibody so acetylated could be precipitated when mixed with intact antibody. What apparently happens is that the increased negative charge of the antibody, due to acetylation, prevents precipitation. Normally the antigen and antibody are both negatively charged, and there is just enough energy of interaction to hold these large charged particles together. However, a single antigen molecule has to hold several antibodies around it. Acetylation of antibody increases its net negative charge so that the energy of interaction is no longer large enough to hold the antigen and the mutually repelling antibodies together. Although complexes of antigen and acetylated antibody form, they are not large enough to precipitate. Increasing the net negative charge on an antibody decreases the binding of a simple negative hapten to an extent proportional to this negative charge, whereas antibody-antibody repulsion in antigen-antibody complexes is proportional to the square (or more) of the net charge.

Iodination of antibody produces an effect opposite to that of acetylation, in that iodinated antibody is more precipitable. Ordinarily, anti-X_p antibody does not completely precipitate with antigen; some molecules form soluble complexes and precipitate only with difficulty. When this antibody is iodinated it becomes more precipitable, presumably because of the increased hydrophobic attraction induced by the additional iodine atoms (Pressman and Radzimski, 1962). This means that in experiments in which iodination of an antibody has been shown to decrease the amount of specific precipitate formed, the effect cannot be used to determine quantitatively the number of antibody sites affected. Some antibody molecules that would not have precipitated in any event are now brought down, owing to the nonspecific increased precipitability of iodinated antibody.

CONCLUSIONS

Studies on the chemical nature of the antibody site have revealed the presence of certain groups in antibodies to *p*-azobenzenearsonate, *p*-azobenzoate, *p*-azophenyltrimethylammonium ion, and 3-azopyridine.

A negatively charged group in the combining site of anti-A_p antibodies is a carboxylate. The other antibodies studied do not have a carboxylate in the site.

A positively charged group in the combining site of anti-X_p antibody seems to be guanidinium since it does not appear to be an ammonium or a histidinium group. The charged groups in the site of some anti-R_p antibodies appear to be ammonium, and the charged groups in the sites of other anti-R_p antibodies would have to be guanidinium.

A tyrosine group appears to be present in the combining sites of all the antibodies inevstigated by the authors. It is implicated to differing extents. Koshland, Englberger, and Gaddone (1965) have found it absent in one pool of anti-A_p antibody.

There does not seem to be a disulfide in any of the combining sites.

REFERENCES

Baker, B. R. (1964). *J. Pharm. Sci.* **53**, 347.
Chen, C. C., A. L. Grossberg, and D. Pressman (1962). *Biochem.* **1**, 1025.
Grossberg, A. L., and D. Pressman (1960). *J. Am. Chem. Soc.* **82**, 5478.
Grossberg, A. L., and D. Pressman (1963). *Biochem.* **2**, 90.
Grossberg, A. L., G. Radzimski, and D. Pressman (1962). *Biochem.* **1**, 391.
Grossberg, A. L., P. Stelos, and D. Pressman (1962). *Proc. Nat. Acad. Sci. U.S.* **48**, 1203.
Habeeb, A. F. S. A., H. G. Cassidy, and S. J. Singer (1958). *Biochim. Biophys. Acta* **29**, 587.
Koshland, M. E., F. M. Englberger, and S. M. Gaddone (1963). *J. Biol. Chem.* **238**, 1349.
Koshland, M. E., F. M. Englberger, and S. M. Gaddone (1965). *Immunochem.* **2**, 115.
Marrack, J. R., and E. S. Orlans (1954). *Brit. J. Exptl. Pathol.* **35**, 389.
Nisonoff, A., and D. Pressman (1958). *Science* **128**, 659.
Pauling, L., D. H. Campbell, and D. Pressman (1943). *Physiol. Rev.* **23**, 203.
Pressman, D. (1957). *Mol. Struc. Biol. Spec.* **2**, 1.
Pressman, D., and G. Radzimski (1962). *J. Immunol.* **89**, 367.
Pressman, D., and O. A. Roholt (1961). *Proc. Nat. Acad. Sci. U.S.* **47**, 1606.
Pressman, D., and J. A. Sternberger (1951). *J. Immunol.* **66**, 609.
Pressman, D., A. L. Grossberg, L. H. Pence, and L. Pauling (1946). *J. Am. Chem. Soc.* **68**, 250.
Wofsy, L., and S. J. Singer (1963). *Biochem.* **2**, 104.
Wofsy, L., H. Metzger, and S. J. Singer (1962). *Biochem.* **1**, 1031.

DAN H. CAMPBELL
Division of Chemistry and Chemical Engineering
California Institute of Technology
Pasadena, California

Antibody Formation: From Ehrlich to Pauling and Return

At the turn of the century, enough studies had been made to establish quite conclusively that the serum of both normal and immunized animals contained materials that were involved in immune mechanisms such as hemolysis and neutralization of certain toxins. Paul Ehrlich, who was one of the leading immunologists during this period, speculated on the origin of the neutralizing substance which arose as a result of immunization and was referred to as "antibody" (Ehrlich, 1897). Ehrlich presented his ideas before the Royal Society of London in 1900. A simplified scheme of his proposal is presented in Figure 1. Since he was basing his ideas on the neutralization of toxin, although antibodies could be induced with the nontoxic toxoid form, the scheme is somewhat complicated. Also, he attempted to explain the mechanism of alexin (complement), but for the present discussion the concept of receptor sites is the only point of interest. It will be seen in Ehrlich's scheme that cells have "receptors," which combine with the toxin (antigen). This combination causes a release of the receptor, which is replaced by a new, but

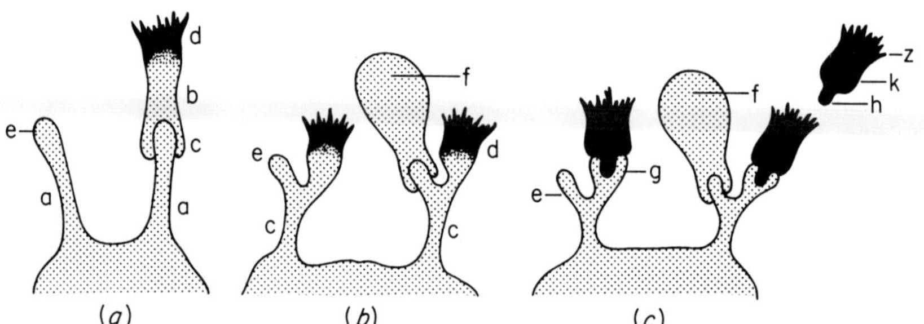

FIGURE 1.
The various types of receptors according to Ehrlich. (I) Receptors of the first order. Here *a* is the receptor, with *e* its haptophore group; *b* is a toxin or enzyme, with *c* its haptophore group and *d* its toxophore group. (II) Receptors of the second order. Here *c* is the receptor, with *e* its haptophore group and *d* its ergophore group; *f* is a food molecule or antigen. (III) Receptors of the third order. Here *e* is the antigen-combining haptophore group and *g* is the complement-combining (complementophilic) haptophore group of the receptor (amboceptor); *k* is complement, with its haptophore group *h* and its ergophore group *z*. [After P. Ehrlich, *Erkrankungen des Blutes*. Nothnagel's Specielle Pathologie und Therapie, Vol. VIII. Vienna, 1901.]

identical, receptor. If the stimulation is continued, receptors are made in excess and subsequently released into the circulation to become circulating antibodies.

During the same period Hans Buchner proposed another theory of antibody formation, which in some respects might be considered the forerunner of what was to become the template theory. Buchner postulated that foreign material combined with serum proteins and that the resulting combination produced the specific antibody. This theory was quite popular with many chemists.

The ideas of both Ehrlich and Buchner had a profound influence in stimulating speculation and experimentation on the nature of antibody formation. However, two lines of chemical investigation developed that caused both theories to lose popularity by 1920. One line of research was developed by Karl Landsteiner, who firmly established the concept of antigenic specificity by using relatively simple chemical substances conjugated to proteins for immunization. The resulting antiserum contained antibody to the simple chemical group (hapten) that had been attached to the protein carrier for immunization. Since antibodies could be produced against practically any simple native or artificial group such as phenyl arsonic acid, it seemed illogical to expect cells normally to contain an almost unlimited number of different receptors (antibodies). The second important development was the general advance of the field of protein chemistry, especially with regard to analytical methods and concepts of protein structure. As a result, Buchner's theory was also discarded, because antigen or antigenic components were not detected in serum proteins. The significant information that was discovered about proteins was the concept of amino acids being linked together by peptide bonds to form long chains of polypeptides. These chains would then fold into various configurations, depending to some extent on the amino acid sequence, and form the final characteristic protein molecule. About 1930 several laboratories published theories of antibody formation based on the chain-like structure of proteins and the steric (three-dimensional) configuration of the antibody combining site. These theories were grouped under the general term of the "template theory." The use of template arose from the idea that the foreign antigen in some way influenced the steric configuration of the normal protein molecule at some stage during synthesis in such a manner that the final antibody molecules contained areas that were complementary and specific for the foreign antigen. Although several laboratories agreed in principle with the template theory, they disagreed about whether the antibody specificity was the result of differences in amino acid sequence or merely a difference in final folding of the polypeptide chain.

In 1940 Linus Pauling postulated, and schematically presented, a possible simple mechanism of antibody formation based on the template theory (Pauling, 1940). The scheme is presented in Figure 2. In most respects Pauling's ideas were basically the same as those previously proposed by Breinl and Haurowitz. However, certain aspects of Pauling's ideas were unique: first, they were presented diagrammatically with characteristic simple clarity that could be understood by students as well as investigators; second, it was assumed that the terminal ends of a polypeptide chain could assume an infinite number of different steric configurations that were specific and complementary

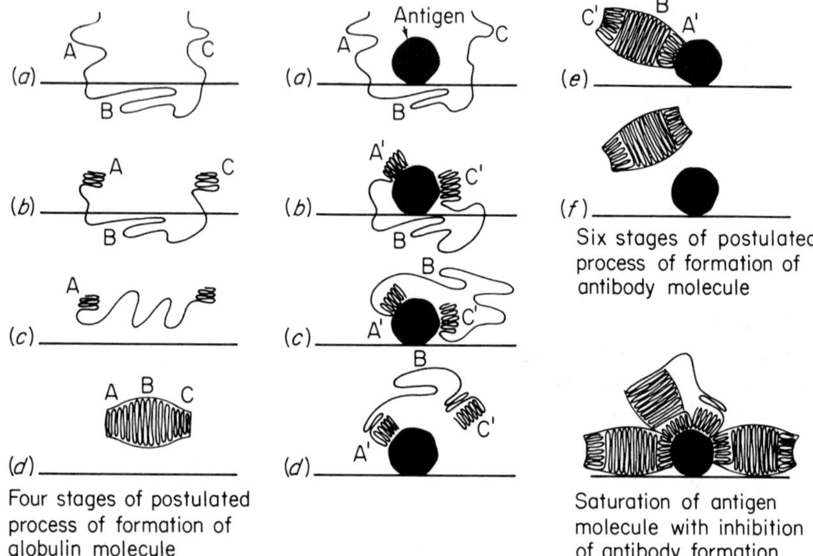

FIGURE 2.
Diagrams representing four stages in the process of formation of a molecule of normal serum globulin (left side of figure) and six stages in the process of formation of an antibody molecule as the result of interaction of the globulin polypeptide chain with an antigen molecule. There is also shown (lower right) an antigen molecule surrounded by attached antibody molecules or parts of molecules and thus inhibited from further antibody formation.

for any given antigenic determinant. Thus, the general assumption was that the antibody molecule had two combining sites and the specificity of combination with antigen was the result of the antigen's influence on the final folding of the polypeptide chain, and not upon a difference in the amino acid sequence.

Pauling's scheme had a tremendous impact on many aspects of immunochemistry aside from antibody formation. For example, later investigations established that precipitating antibodies were predominantly bivalent as predicted, and until very recently there was little or no evidence that the antibody combining site with one specificity differed from any other in amino acid sequence. The template theory is extremely attractive but, like most theories, is subject to certain logical objections that can be raised until more experimental facts have accumulated. One of the major objections was the failure to find antigen after a few days following injection. However, it has now been firmly established that antigens are rapidly broken down and that the molecular fragments, which are about the size of antigenic determinants, persist in tissue for many months or years. Until more is known about the biosynthesis of proteins and the natural replication of protein molecules, one can only speculate on the possible role of antigen fragments in antibody formation. It seems significant that most studies indicate that the antigen fragments are combined with ribonucleic acid. This suggests the possibility that antigen acts as a template by modifying the function of RNA. Although current studies indicate that 7S gamma globulin consists of four major chains (two "light" and two "heavy") instead of one, this in no way negates Pauling's idea of bivalence. It would appear that a single antibody combining site occurs on each of the heavy chains and that the structure of both heavy chains is identical. Since the manufacture of both heavy chains is controlled by the same mechanism, the specificity of each would be expected to be the same and the final molecule thus to have two combining sites resulting from the same RNA-antigen template. This would also explain the failure of antigen to form heteroligating antibody molecules.

Regardless of much of the current information that was developed in Pauling's laboratory as well as in others in support of the template theory, speculation about the role of genetic mechanisms in antibody formation started to gain impetus about 1950. The first hint of a possible return to Ehrlich's ideas, but in the light of more knowledge and modern concepts of the genetic control of biosynthesis of proteins, was expressed by Burnet and Fenner (1949). Their idea was based on the adaptive enzyme concept that an organism could respond specifically *de novo* to the addition of a foreign substrate.

The adaptive enzyme theory was soon discarded when Monod proved that the organism had the genetic constitution to synthesize a given enzyme and probably did so at a low level until stimulated by the addition of the specific substrate. This finding led to an increased interest in the idea that specific genetic components were responsible for the biosynthesis of any given specific antibody molecule as well as any other specific protein. Subsequently, this led Burnet to formulate his well-known "clonal" theory of antibody formation, which is shown diagrammatically in the following figure taken from a publication by Mackay and Burnet (1963). It will be seen that each individual immunological competent cell has a different potential for antibody production, which is dependent upon its genetic composition. Thus, only cells labeled E will produce anti E. This is not necessarily true because there are two alleles, each with a different potentiality so that each cell, if heterozygous, could produce two different antibodies. However, in general the idea is that information for antibody formation originates from the gene and not directly from the antigen. Although many investigators question the validity of the clonal concept, the idea of one allele, one antibody combining site, is currently being given more attention than the template idea of one antigen template for each antibody molecule.

At the present time there are two general classes of theories about the mechanism of antibody formation. One class consists of the selection theories, which assume: (1) the complete preexistence of the genetic information at a cellular level, as proposed by Ehrlich and later by Burnet; (2) a selection at a molecular level involving the successive synthesis of antibodies of increasing fitness, as proposed by Jerne. The second general class consists of the induction or instruction theories, which assume: (1) that information or modification of protein structure is the result of the presence of a foreign antigen or a fragment of the antigen which functions as a "template," an idea originated by earlier investigators and currently modified by Campbell and Garvey on the basis of persistence of antigen fragments in combination with ribonucleic acid; (2) that information is transferred to autocatalytic intermediary units, as proposed by Burnet and by Schweet and Owen.

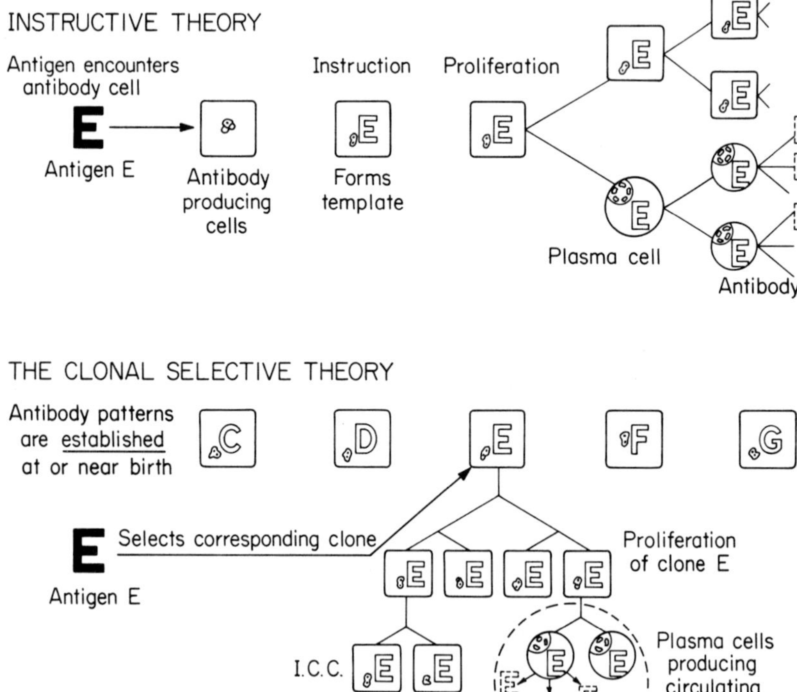

FIGURE 3.
The difference between instructive and selective theories of antibody production. In instructive theories, the antigen E forms a template within a potential antibody-producing cell: this template determines the subsequent pattern of antibody produced by that cell. In selective theories, as illustrated by the clonal selective theory, the antigen E stimulates the proliferation of a clone of cells with a preordained capacity to produce the corresponding antibody. This and similar figures are simplified and formalized diagrams to indicate only the basic principles of the two theories. [From J. R. Mackay and F. M. Burnet, *Autoimmune Diseases, Pathogenesis, Chemistry, and Therapy*, 1963. Courtesy of Charles C. Thomas, Publisher, Springfield, Illinois.]

Although the present literature is dominated by the immunogenetic concept of the preexisting genetic control of antibody formation such as Ehrlich visualized in an uninformed manner, I cannot help but look forward to a more favorable acceptance of the template theory when more is known about the fate of antigen material and the role of nucleic acids, and about intracellular enzymes in the biosynthesis of antibody molecules. The possible influence of antigenic fragments in the structure of otherwise genetically controlled synthesis of native gamma globulin cannot be ignored. The ultimate answer lies in the understanding of the role of enzymes in handling foreign material and the mechanism of the biosynthesis of proteins. Is antibody a natural or a foreign protein?

One wonders how Ehrlich would have modified his so-called "side chain" theory in the light of current knowledge. On the basis of the genetic control of the biosynthesis of proteins, perhaps he would have defended his concept that specific receptors are present and dependent upon the genetic constitution of the cell. Whether Ehrlich would have accepted the clonal concept is questionable, since he may have considered that the genetic constitution of any one antibody-forming cell would be the same as every other cell.

The problem of the mechanism of antibody formation remains usnolved, but valuable information from biologists and chemists is being rapidly accumulated. Perhaps we must await another Paul Ehrlich or Linus Pauling before we can put the pieces of data into place and finally decide whether antibody molecules arise *de novo* as a modified foreign molecule or as a natural protein molecule under genetic control.

REFERENCES

Breinl, F., and F. Z. Haurowitz (1930). *Physiol. Chem.* **192**, 45.

Burnet, F. M., and F. Fenner (1949). *The Production of Antibodies*, 2nd ed. London: MacMillan and Co.

Ehrlich, P. (1897). *Klin. Jaohr.* **6**, 299.

Mackay, J. R., and F. M. Burnet (1963). *Autoimmune Diseases: Pathogenesis, Chemistry, and Therapy*. Springfield: Charles C. Thomas.

Pauling, L. (1940). *J. Am. Chem. Soc.* **62**, 2643.

S. J. SINGER
Department of Biology
RUSSELL F. DOOLITTLE
Department of Chemistry
University of California, San Diego
La Jolla, California

Evolution and the Immunoglobulins

Information about the structure of immunoglobulins has been accumulating at a remarkable rate during the past few years. In particular, amino acid sequence studies, especially on Bence-Jones proteins, have given tremendous impetus to the field and offered the tantalizing possibility that the fundamental nature of the immune response may be revealed. The discovery of Hilschmann and Craig (1965) that the light chains of human Bence-Jones proteins are divided into amino-terminal halves, which are variable in their amino acid sequences, and carboxy-terminal halves, which are virtually invariant, has been the basis for a number of hypotheses that attempt to explain the genesis of the great multiplicity of an individual's immunoglobulins. The most popular hypotheses may be divided into three categories:

1. The "*genetic load*" or "zygote complete information" hypothesis. According to this view, all the DNA segments (cistrons) coding for all the myriads of immunoglobulin chains are present in each zygote. This hypothesis, with certain additional features, has been particularly developed by Dreyer and Bennett (1965), and Hood, Gray, and Dreyer (1966a).

2. The "*translation*" hypothesis. In this hypothesis, it is proposed that the variable portions of immunoglobulins arise as the result of perturbations of the translation phase of protein biosynthesis. Precedent exists in at least one other system (von Ehrenstein, 1966) that transfer RNA molecules may vary from individual to individual and thereby perhaps from tissue to tissue, with the result that different amino acid sequences may arise from the same DNA base sequence. It may seem a logical extension, then, that the many different immunoglobulins are the result not of changes in the immunoglobulin DNA cistrons but of changes at some other stage in immunoglobulin synthesis. A proposal of this type has been advanced by Potter, Appella, and Geisser (1965).

3. The "*somatic hypermutability*" hypothesis. According to this hypothesis, a small number of DNA segments—presumably one for each major class of immunoglobulin chain—exists in the genome of the zygote. During development, however, the replicative features of these cistrons are impaired in some way so that the result is a vast population of cellular types, each with a slightly different DNA base sequence in that region of the genome. Smithies (1965) suggested that any such differences might be the result of intrachromosomal rearrangements; and Brenner and Milstein (1966) have resurrected a

scheme suggested and discarded by Dreyer and Bennett (1965), in which various phenomena observed in the disruption and repair of DNA in micro-organisms are employed.

These hypotheses have been largely oriented to the early experimental findings with the Bence-Jones proteins. Recently, however, we have proposed that the light and heavy chains of immunoglobulins have homologous variable regions, and, in particular that both kinds of chains have arisen by evolution from a common ancestral gene (Singer and Doolittle, 1966). This proposal has significant implications for the problem of immunoglobulin biosynthesis. It is our object in this article to explore these evolutionary ideas and their implications in some detail. In this process, we take full advantage of the invitation of the Editors of this Festschrift to speculate even more freely than usual.

PROTEIN EVOLUTION IN HIGHER ANIMALS

It is necessary first to review briefly some of the prevailing ideas about the general problem of protein evolution. Numerous attempts have been made and are being made to reconstruct the evolutionary history of certain proteins on the basis of their present day structures. In general, the data fall into two categories: (*a*) comparisons of amino acid sequences of the "same" polypeptide chain from different species, or (*b*) comparisons of "different" but related polypeptides (which can usually be isolated from a single individual), such as the α and β chains of hemoglobin. In fact, the hemoglobin molecule is the protein whose evolution has been most studied; it provides most of our examples about what kinds of changes have probably occurred during evolution. For example, estimates have been made concerning the average frequency of base substitutions leading to single amino acid replacements (Zuckerkandl and Pauling, 1965). From these data, we also get some notion of the frequency of insertions and deletions, and perhaps of how often a segment of DNA is likely to duplicate itself (gene duplication) independently of the rest of the genome. Generally, however, because of the unknown influence of natural selection on the survival of all these mutational events, it has been difficult to draw firm conclusions. Furthermore, the complications of dealing with this type of microevolution in diploid organisms have only recently begun to be appreciated (for example, Boyer et al., 1966).

We can, however, safely draw a few general conclusions at this time. One is that closely related organisms generally have proteins whose amino acid sequences are more similar than those of distantly related organisms. An apparently logical assumption stemming from this finding is that the more closely the two different polypeptide chains resemble each other in an individual, the more recent has been the gene duplication which resulted in their divergence. Secondly, the evidence is that the duplication of sizable portions of gametic DNA (gene duplications) has been a successful device for introducing new proteins as organisms have evolved. The series of duplications leading to the various types of hemoglobin chain is the most common instance cited, but there is little doubt that trypsinogen and chymotrypsinogen (Walsh and Neurath, 1964; Hartley, 1964) or glucagon and secretin (Bromer et al., 1957; Mutt et al., 1965) are examples of the same phenomena.

Cistronic duplications seem to fall into two general categories. In one category, the genetic material coding for a certain polypeptide chain is duplicated in its entirety and results in the production of *two* identical gene products of the original type. Polypeptide biosynthesis is evidently initiated independently in each gene. In this article we will refer to these events as *discrete duplications*. Subsequently, base substitutions and other minor changes in the DNA lead to phenotypic differences between the two polypeptide chains. In the other category, the segment of DNA is duplicated and comes to lie adjacent to the original cistron in such a way that chain initiation occurs only at a single point. The result is a single polypeptide chain, which is larger —in theory, as much as twice as large—as the original gene product. An excellent example is afforded by a scrutiny of the present day amino acid sequence of *Clostridium* ferredoxin, a polypeptide chain that clearly must have resulted from such a duplication (Figure 1). We will refer to these genetic events as *contiguous duplications*. Again, subsequent base substitutions will lead to amino acid replacements which will be different in the two regions of the molecule. Recently, computer methods have been developed that can rapidly scan amino acid sequences in search of any internal repeat units that have not been completely blurred out by other changes during the course of evolution (Fitch, 1966).

EVOLUTION OF THE FUNDAMENTAL UNITS OF IMMUNOGLOBULINS

Each of the major classes of immunoglobulins (IgG, IgA, IgM) is made up of heavy and light chains. In humans and mice, there exist two immunologically distinct classes of light chains (κ and λ), which are common to all immunoglobulin molecules. Bence-Jones proteins are homogeneous light chains of either the κ or λ types. The heavy chains of the different immunoglobulins determine the major class of the molecule. The IgG molecule has a molecular weight of approximately 160,000 and is composed of two heavy (γ) chains and two light chains (either both κ or both λ). On the other hand, the IgM molecule—sometimes referred to as macroglobulin—has a molecular weight of the order of 900,000 and is apparently a disulfide-linked pentameric aggregate of a 7S subunit, which is homologous to the IgG molecule (Miller and Metzger, 1965).

Our first clue to the evolutionary history of mammalian immunoglobulin chain types arose from a comparison of the amino acid sequence data provided by Hilschmann and Craig (1965) on human κ light chains and preliminary data on the carboxy-terminal sequence of human and rabbit γ heavy

```
  1  2  3  4   5   6   7   8   9  10  11  12  13  14  15  16  17  18  19  20  21  22  23  24  25  26  27  28  29
N-Ala-tyr-lys-ilu-ALA-ASP-ser-CYS-val-ser-CYS-GLY-ala-CYS-ALA-ser-glu-CYS-PRO-VAL-asn-ALA-ilu-ser-GLN-gly-asp-ser-ilu-

-phe-val-ilu-asp-ALA-ASP-thr-CYS-ilu-asp-CYS-GLY-asn-CYS-ALA-asn-val-CYS-PRO-VAL-gly-ALA-pro-val-GLN-glu-C
  30  31  32  33  34  35  36  37  38  39  40  41  42  43  44  45  46  47  48  49  50  51  52  53  54  55
```

FIGURE 1.
The amino acid sequence of ferrodoxin of *Cl. pasteurianum* (Tanaka et al., 1964), showing the homology between its two halves (Eck and Dayhoff, 1966) and suggesting that a contiguous gene duplication occurred in the evolution of the ferrodoxin structural gene.

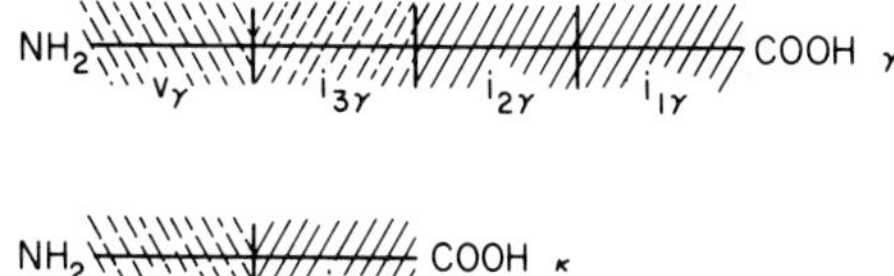

FIGURE 2.
The homologous units within the γ heavy and κ light chains of immunoglobulins, i designating regions of essentially invariant, and v of variable amino acid sequences.

chains (Press et al., 1966). At the time we were convinced there was a relationship between light and heavy chains on the basis of the similar physicochemical properties of peptide fragments isolated from both chains after affinity labeling of antibodies (Doolittle and Singer, 1965). These fragments were presumably from the "variable" regions of the two chains, however, and the homology between the amino acid sequences of light and heavy chains at their carboxy-terminal ends, although initially surprising to us, led us to postulate that heavy and light chains of immunoglobulins are evolutionary related and are the result of a series of gene duplications. Furthermore, it gave rise to a prediction about the nature of amino acids at the carboxy-terminal of μ chains (the heavy chain type of the macroglobulin class) and its structural implications. The forecast was made that a cysteine residue might exist at or near the carboxy-terminal of μ chains just as it does in both classes of light chains, where it links light to heavy chains. This evolutionary remnant of a past gene duplication might be the functional residue that links the 7S subunits into the pentameric IgM aggregate. We have since indeed found cysteine at or next to the carboxy-terminal of a human μ chain (Doolittle et al., 1966).

Recently a very large portion of the Fc segment of rabbit heavy (γ) chains has been sequenced (Hill et al., 1966). These studies confirm our earlier inference about the homology between the carboxy-terminal halves of light chains and the carboxy-terminal regions of heavy chains (the regions i_κ and $i_1\gamma$ in Figure 2). Furthermore, the data strongly suggest that the Fc segment itself is the product of a duplication of the type which we have termed "contiguous." Visual inspection clearly indicates a repeat sequence of about 100 amino acids corresponding to the regions $i_{1\gamma}$ and $i_{2\gamma}$ in Figure 2. In addition, the data of Hill et al. suggest that another such region of homology, labeled $i_3\gamma$ in Figure 2, exists within the heavy chain, although here the results are still fragmentary.

It is a striking fact, of course, that the variable and invariant halves of the κ light chains are also about 100 amino acids in length. However, visual inspection of the amino acid sequences published thus far does not indicate any close homology between the variable and invariant regions. When the sequences are put through a modified computer program of the type designed by Fitch (1966), no significant relationship between the v_κ and i_κ (Figure 2) regions is evident (Tideman and Doolittle, unpublished). Nevertheless, closer examination of the sequence data does support the notion that the two regions are related. Note the invariance in position of the two cysteine residues in the variable regions of all human and mouse κ chains (Milstein, 1966); note also the essential invariance in position of the disulfide bridge in the v_κ and i_κ halves (Figure 3), and the closely similar positions of the two tryptophan residues in the human v_κ and i_κ regions (Figure 3) and their counter-

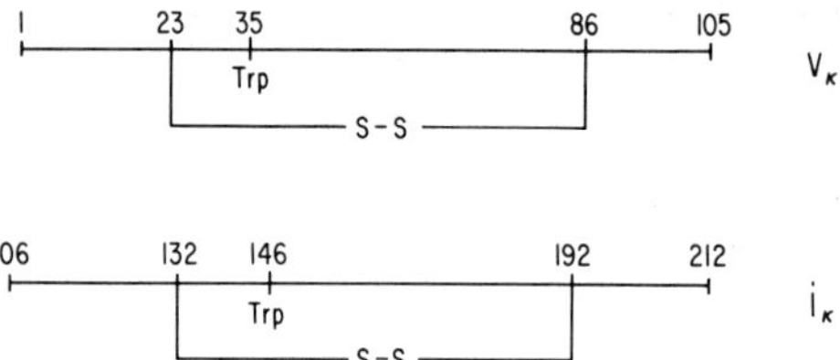

FIGURE 3.
A diagram showing the structural similarities between the v_κ and i_κ regions of κ Bence-Jones proteins. The numbers are the positions of the amino acid residues in question relative to the amino-terminal of the chain.

parts in mouse κ chains (W. R. Gray, personal communication). Furthermore, it is interesting that the tripeptide sequence –Pro–Pro–Ser– appears near the amino-terminal of two of the human λ chains that have been studied (Hood et al., 1966b) (Figure 4). We have previously noted (Singer and Doolittle, 1966) that this tripeptide, which is found in a roughly similar portion of the invariant region of human κ chains (Hilschmann and Craig, 1965) and in the Fc region of human γ chains (Thorpe and Deutsch, 1966), is sufficiently uncommon to suggest some homology between related peptide chains containing it.

Taking these structural features together, we conclude there is only a very low probability that the relationship between the v and i regions is random. It is therefore highly likely that the v and i regions had a common evolutionary origin, but that only a relatively small degree of homology has been retained between them.

Although the data are fragmentary at present, the amino acid sequences near the amino-terminals of several human λ-type, and of mouse and human κ-type, Bence-Jones proteins (Hood et al., 1966b) are very striking. These data are reproduced in Figure 4. They permit several important conclusions to be drawn. (*a*) At least half of the amino acid positions in the v region are likely to be variable, that is, they can be occupied by more than one amino acid residue in different chains of a given class (see also Hood et al., 1966a; Putnam et al., 1966; Milstein, 1966). (*b*) The different residues appearing at any given variable position are few in number (for example, see Milstein, 1966), and in the majority of cases, codons can be assigned to these alternative residues so that only single nucleotide base changes may be involved. (*c*) The different λ variable sequences are closely similar to one another, as are the different κ variable sequences to one another. "Prototype" v_κ and v_λ sequences can be assigned, from which the multitude of different v_κ and v_λ sequences, respectively, can be derived as "perturbations." Of particular interest, for example, is the fact that the v_λ regions all appear to have a deletion at the ninth position in from the amino-terminal of the λ-chain, as compared to the different κ chains. (*d*) In spite of the distinctive differences between prototype v_κ and v_λ sequences, these sequences are clearly related, suggesting that they arose from a common ancestral gene (Hood et al., 1966b). (*e*) It is remarkable that the prototype v_κ sequences of mouse and human κ chains are hardly distinguishable.

Little detailed information is available at this time about the amino-terminal region of heavy chains. There is good evidence, however, that this is also a variable region (Frangione and Franklin, 1965; Doolittle and Singer, 1965; Porter and Weir, 1966). We have therefore designated the amino-terminal

COMPARISON OF N-TERMINAL REGIONS OF λ AND K CHAINS

	Protein	Sequence
λ	HBJ2	PCA.Ser.Ala.Leu.Thr(5).Gln.Pro.Pro.Ser.[ ](10).Ala.Ser.Gly.Ser.Pro(15).Gly.Gln.Ser.Val.Thr
	HBJ7	PCA.Ser.Val.Leu.Thr(5).Gln.Pro.Pro.Ser.[ ](10).Ala.Ser.Gly.Thr.Pro(15).Gly.Gln.Gly.Val.Thr
	HBJ8	PCA.Ser.Ala.Leu.Ala(5).Gln.Pro.Ala.Ser.[ ](10).Val.Ser.Gly.Ser.Pro(15).Gly.Gln.Ser.Ilu.Thr
	HBJ11	PCA.Ser.Val.Leu
K	HS-4	Glu.Ilu.Val.Leu.Thr(5).Gln
	Ag	Asp.Ilu.Gln.Met.Thr(5).Gln.Pro.Ser.Ser.Ser(10).Leu.Ser.Ala.Ser.Val(15).Gly.Asp.Arg.Val.Thr(20)
	MBJ41	Asp.Ilu.Gln.Met.Thr(5).Gln.Ser.Pro.Ser.Ser(10).Leu.Ser.Ala.Ser.Leu(15).Gly.Glu.Arg.Val.Ser(20)
	MBJ70	Asp.Ilu.Val.Leu.Thr(5).Gln.Ser.Pro.Ala[Ser(10).Leu][Ala.Val.Ser.Leu(15)]Gly.Glu.Arg.Ala.Thr(20)

FIGURE 4.
A comparison of the amino-terminal sequences of λ and κ Bence-Jones proteins. The last two proteins of the κ series are mouse proteins, the others are of human origin. Data are from Hood et al. (1966b), except for Ag, which is from Titani, et al. (1966).

region of the γ heavy chain v_γ in Figure 2. When specific labels are attached to the active sites of various antibodies by the method of affinity labeling, it is found that both heavy and light chains acquire label (Metzger, Wofsy, and Singer, 1964). Labeled peptide fragments produced by tryptic digestion of the two kinds of chains are very heterogeneous and appear to originate from the variable regions of the chains. Furthermore, the fragments from the two chains have very similar physicochemical properties (Doolittle and Singer, 1965). We have therefore suggested that the variable regions (presumably v_γ and v_κ, Figure 2) on heavy and light chains are chemically similar and evolutionary related (Singer and Doolittle, 1966). It seems likely that v_γ regions have properties similar to v_κ and v_λ regions; that is, a distinctive prototype sequence probably exists for that region, and different γ chains in a population exhibit variations on the prototype. Nothing is known at this time about the presumed v-type regions of α and μ heavy chains.

To summarize up to this point, our present structural information is consistent with the following conclusions. (*a*) A polypeptide chain of about 100 amino acids appears to be a fundamental unit of all immunoglobulin chains. Four nonidentical but related units are connected linearly to make up a heavy chain; two units, a light chain. (*b*) The amino-terminal units of both heavy and light chains are of the variable, or v-, type, whereas all the others are essentially invariant, or i-type. (*c*) The v-type units of κ and λ light chains, and probably of γ heavy chains, although similar, are distinctive; that is, a prototype v sequence can be assigned to each class of chain; (*b*) The i-type units are chemically much more similar to one another than they are to v-type units.

To account for these facts, we suggest that at an early stage in the evolution of the immunoglobulins, there existed a primitive gene corresponding to a polypeptide chain of about 100 amino acids, and that by a series of discrete and contiguous duplications originating from this primitive gene, accompanied by other mutational events, present day immunoglobulins evolved. DNA cistrons corresponding to v-type polypeptide units, namely *v*-type genes,* also originated from this primitive gene, but then must in some manner have become differentiated from the primitive *i*-type gene. This differentiation involved acquiring a set of properties unique within the entire genome. It is clear that the nature of this differentiation process is intimately connected with the basic mechanism of antibody biosynthesis, and we will return to this point later.

At this juncture, it is appropriate to introduce several reasonable postulates.

1. It is assumed that *prototype* v_κ and *prototype* v_λ *genes* exist or existed, to account for the fact that prototype v_κ and v_λ amino acid sequences are observed.

2. By the principle of parsimony, it is assumed that the process which differentiated a *v*-type gene from its *i*-type ancestor occurred only once in the evolution of the immunoglobulins; that is, that the prototype v_κ, prototype v_λ, and other prototype genes arose from a *primitive v-type gene.*

3. The differentiation process that produced the primitive *v*-type gene occurred much earlier in evolution than the divergence of the different prototype *v* genes or the divergence of the different *i*-type genes from one another. This postulate is required to account for the only slight amino acid sequence homologies between the v- and i-regions, and for the strong homologies among the different i-regions and between the prototype v_λ and v_κ sequences.

The third postulate follows if the mutational events by which the primitive *v*- and *i*-type genes diverged in evolution were of the same kinds, and appeared with roughly the same frequency, as those by which the different *i*-type genes diverged (discussed earlier in this paper). On the other hand, it is possible that a radically different mutational event, such as a reading frame shift, occurred to a primitive *v*-type gene. Such a single event could produce an amino acid sequence for v-type polypeptide regions that would be grossly different from that for i-type regions, and it could have occurred late in evolution. However, analysis of v_κ and i_κ sequences for possible reading frame shifts (Tideman and Doolittle, unpublished) has provided no evidence for such an event. Alternatively, a primitive *v*-type gene may have arisen late in evolution, but the selection pressures operating on the primitive *i*-type gene may have been so relaxed that they allowed the accumulation of a large number of mutations within the *v*-type gene in a relatively short evolutionary period. If such a hypermutability characterizes *v*-type genes, however, it is difficult to understand how stable prototype v_κ and v_λ structures continue to exist, or the remarkable fact that the prototype v_κ structures in mice and in humans are so similar.

For these reasons, we feel that the third postulate is a reasonable one.

*Regions of DNA are designated by italics to distinguish them from regions of polypeptide chains.

EVOLUTION AND THE GENESIS OF IMMUNOGLOBULIN MULTIPLICITY

These evolutionary considerations have a significant bearing on the nature of immunoglobulin multiplicity, and its essential features must be accommodated satisfactorily into any hypothesis of immunoglobulin biosynthesis. To illustrate this point, we will consider briefly the consequences of these evolutionary ideas for each of the three hypotheses mentioned at the beginning of this article.

In the version of the genetic load hypothesis of Dreyer and Bennett (1965), as we interpret it, there exists in each somatic cell of an individual all the myriad genes corresponding to the complete set of v_κ, v_λ, v_γ, and other similar units elaborated by that individual. These many genes are normally in a state corresponding to the prophage state of lysogenic phages in bacteria; that is, they are incorporated in the genomic structure and replicated with the rest of the genome, but their expression is repressed. Elsewhere in the genome are the discrete genes for each of the i-type units. At some particular stage in the development of an immunocompetent cell, a single *v*-type gene of, let us say, the v_μ kind, is specifically excised from its resting position in the genome and becomes an episomal element. Following this the episome then attaches to a specific region of DNA immediately adjacent to its appropriate *i*-type gene, and it is incorporated into the genome at that site. Messenger RNA corresponding to the now stable *v-i* contiguous gene is then synthesized. In this manner, although each zygotic cell starts with the same genetic information, each cell that becomes immunocompetent acquires a small number of distinctive genes, which are expressed as heavy and light immunoglobulin polypeptide chains.

For this hypothesis, we would infer that the property acquired by the primitive *v*-type gene, which early in evolution differentiated it from its *i*-type ancestor, was the potential to undergo very extensive discrete duplications (hyperduplicability). If, however, this potential was acquired only once in evolution, it would have had to have been repressed until all of the prototype genes v_κ, v_λ, v_γ, and so on, had evolved from the primitive *v*-type gene. After this divergence of the different prototype *v* genes, the repression would have had to be released in each case, and each prototype *v* gene would then have begun a long chain of discrete duplications during subsequent evolution to generate the complete sets of genes v_κ, v_λ, and v_γ. The recognition by different *v*-type episomes of their appropriate *i*-type genes presents another problem. This recognition is difficult to rationalize if the divergence of the different prototype *v* genes took place independently of the different *i*-type genes. Therefore, one is led to suggest that a primitive *v-i* contiguous gene might have first undergone discrete duplications to prototype contiguous genes v_κ-i_κ, v_λ-i_λ, and so forth, following which each of the cistrons v_κ, v_λ, and so on, of the genes would have had to have been translocated and lysogenized elsewhere in the genome, and *only then* would they have undergone hyperduplication.

We emphasize once again that it is not our object to belabor these speculations, but rather to confront the reader with the problems raised by the evolutionary ideas presented in this article. The main conclusion to be drawn from the foregoing remarks is that any genetic-load hypothesis of immunoglobulin genesis must provide a mechanism whereby (*a*) *v* and *i* genes can have undergone such radically different and independent evolutionary development and yet lead to the production of a contiguous v-i polypeptide chain, and (*b*) the very large numbers of each class of v polypeptide chains become contiguous with their appropriate i-type polypeptide chain and no other.

The translation hypothesis of Potter et al. (1965) proposes that a *v*-type gene contains a cluster of specific and unusual codons. In this hypothesis, a specific unusual codon is one which, in the process of its translation to form the polypeptide chain, can be read alternatively by a group (two or more) of different transfer RNA molecules. Transfer RNA molecules of a given group are characterized by having the same codon recognition site but different amino acid acceptor specificities. In this manner, a single cistronic message can be read in different ways to produce a large set of related polypeptide chains.

A particular unusual codon may occur relatively infrequently and at random throughout the genome. What is therefore unique about a *v*-type gene in this hypothesis is that a large set of such unusual codons is clustered within that gene. Only one structural gene is required for each class of immunoglobulin chain in the gamete, the *v*-cistron of which presumably corresponds to the prototype v-sequence of that class of chain. Thus, the differentiation of a primitive *v*-type gene from its *i*-type ancestor during evolution must have involved the acquisition of the potential to accumulate a cluster of specific unusual codons.

Several problems arise with this hypothesis. In the first place, sequence information for the κ-type chains already indicates that at least half of the residues in the v_κ region are variable (Figure 4). Almost all the 20 naturally occurring amino acids are involved in the variability. Secondly, the variability occurring in the λ-type chains is different at any given position when it is compared with the κ type. This fact would suggest that this highly improbable clustering of ambiguous codons occurred to some extent independently with the two different classes of light chain and presumably also with the v_γ regions of heavy chains. The cumulative improbability of this multiple, but at least partly independent, clustering of special codons would seem to preclude a translation mechanism as being responsible for immunoglobulin variability. A number of other difficulties—unrelated to our primary evolutionary argument—can also be raised against translation hypotheses, but we will not concern ourselves with these here.

Because of its relatively simple mechanism for generating immunoglobulin multiplicity, the somatic hypermutability hypothesis of Brenner and Milstein (1966) provides a much simpler basis for our evolutionary postulates. In this hypothesis, the key to the differentiation of a *v*-type cistron is a sequence of bases in its *immediately adjacent i-type cistron*. This base sequence is unique to certain portions of the genes coding for immunoglobulin chains. This sequence is recognized at the appropriate stage in the somatic development of an immunocompetent cell by a specific deoxyribonuclease, which cleaves one strand of the DNA, always at the same phosphodiester bond.

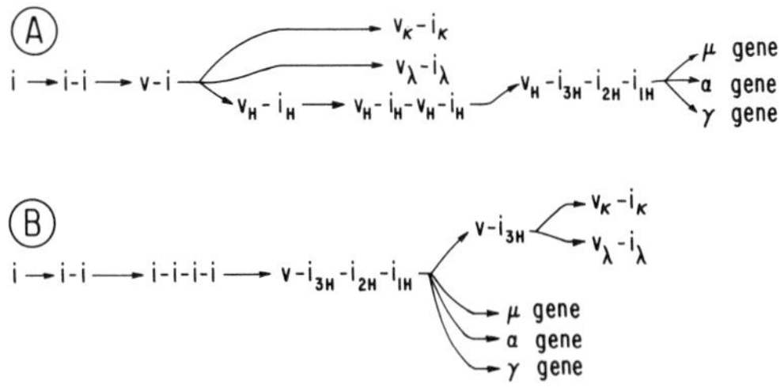

(A) $i \rightarrow i\text{-}i \rightarrow v\text{-}i \rightarrow v_\kappa\text{-}i_\kappa$; $v_\lambda\text{-}i_\lambda$; $v_H\text{-}i_H \rightarrow v_H\text{-}i_H\text{-}v_H\text{-}i_H \rightarrow v_H\text{-}i_{3H}\text{-}i_{2H}\text{-}i_{1H} \rightarrow$ μ gene, α gene, γ gene

(B) $i \rightarrow i\text{-}i \rightarrow i\text{-}i\text{-}i\text{-}i \rightarrow v\text{-}i_{3H}\text{-}i_{2H}\text{-}i_{1H} \rightarrow v\text{-}i_{3H} \rightarrow v_\kappa\text{-}i_\kappa$, $v_\lambda\text{-}i_\lambda$; μ gene, α gene, γ gene

FIGURE 5.
Hypothetical pathways for the evolution of mammalian immunoglobulin genes starting from the ancestral gene *i*.

The broken strand is then degraded by an exonuclease in the direction corresponding to the *v*-cistron. (The polarity of this digestion is insured by the fact that one of the exposed strand ends has a terminal phosphate group and the other does not.) Finally, the degraded strand is restored by a "repair" enzyme, a polymerase which, using the unbroken strand as a template, is presumably somewhat less efficient than ordinary polymerases, and allows a number of errors to be incorporated onto the repaired strand. After several rounds of cell divisions there could thus result a population of cells, each of which might have unique but similar base sequences in their *v*-type cistrons.

It follows from this hypothesis and our evolutionary considerations that each of the prototype amino acid sequences v_κ, v_λ, v_γ, and so forth, corresponds to a prototype structural gene *v* in the zygote. If there is only a single kind of cleavage deoxyribonuclease, then those base sequences that are recognized by the deoxyribonuclease of *all* the *i*-cistrons that are immediately juxtaposed to *v*-cistrons must be essentially the same. The prediction is that the corresponding amino acid sequences in the appropriate i-type polypeptide regions (such as i_κ, i_λ, $i_{3\gamma}$, and so on, Figure 2) must be essentially the same. On the other hand, it is equally important that all the i-type polypeptide regions (such as $i_{2\gamma}$, $i_{1\gamma}$, and so on) not immediately adjacent to v-type regions (and presumably the v-regions themselves) must have significantly altered amino acid sequences corresponding to the deoxyribonuclease-recognition region, so that, for example, v-type regions are not generated in the Fc fragment of γ chains. At the present time, there are not enough amino acid sequence data available to test these predictions.

To consolidate the considerations of this article, it is useful at this point to present somewhat more detailed hypothetical schemes for the evolution of the immunoglobulins. These schemes should be thought of as possible, rather than unique, evolutionary pathways. Although they are specifically related to the Brenner-Milstein hypothesis of immunoglobulin biosynthesis, they can be readily modified to fit any other hypothesis of somatic hypermutability. The schemes are illustrated in Figure 5.

In scheme A (Figure 5), starting with the primitive *i*-type gene, we assume that a contiguous duplication produced an *i-i* gene. We may suppose next that by mutational events extending over a long stretch of evolutionary time, one of the *i*-cistrons of the *i-i* gene acquired a base sequence quite different from the other, and also different from any other cistron in the genome, and that this unique sequence was capable of being recognized by a specific deoxyribonuclease. At this stage, the *i-i* gene had differentiated to a primitive *v-i* gene. Subsequently, a series of discrete duplications occurred that eventually led to the prototype genes v_κ-i_κ, v_λ-i_λ, and to a precursor gene for heavy chains v_H-i_H. At a later stage in evolution, this precursor gene might have undergone a contiguous duplication to yield a gene v_H-i_H-v_H-i_H. It would be at this juncture in historical time that it would be appropriate for the i_H cistron corresponding to the carboxy-terminal of the heavy chain to have lost through mutation the base sequence which was the recognition site for the deoxyribonuclease, eventually resulting in a gene, v_H-i_{3H}-i_{2H}-i_{1H}. The genes for the production of present-day heavy chains would then presumably have arisen as the result of a series of discrete duplications of this last gene in a fashion analogous to the evolution of the various hemoglobin chains.

Alternatively, as indicated in scheme B (Figure 5), one could imagine that at the stage corresponding to the *i-i* gene, a contiguous duplication resulted in the gene we have designated *i-i-i-i*, *before* the differentiation of a *v*-cistron. Upon the accumulation of enough appropriate mutations in an internal *i*-cistron, the generation of a primitive gene, v-i_{3H}-i_{2H}-i_{1H}, could then have occurred. At this juncture, a *partial* and discrete duplication of this chain might have generated a precursor gene, v-i_{3H}, for the present-day light-chain genes.

An interesting difference between the two schemes is that they predict a different type of chain (a light chain in scheme A, a heavy chain in scheme B) as the most primitive form of immunoglobin containing a variable region.

We have attempted in this article to develop the thesis that the polypeptide chains of immunoglobulins have arisen from a common ancestral gene, and that these evolutionary considerations carry profound implications for the still unknown mechanisms of immunoglobulin biosynthesis. It is evident that much more amino acid sequence data will be required to flesh out the skeleton of evolutionary ideas which has been presented. The acquisition of such data will be of great interest, not only for our concepts of immunoglobulin genesis, but also for the general problem of protein evolution, of which the immunoglobulins constitute such an extraordinary example.

ACKNOWLEDGMENT

It is a pleasure to acknowledge our indebtedness to Dr. L. E. Hood, Dr. W. R. Gray, Dr. W. J. Dreyer, and Dr. R. L. Hill, for allowing us to see their important amino acid sequence data before their publication.

REFERENCES

Boyer, S. H., P. Hathaway, F. Pascasio, C. Orton, J. Bordley, and M. A. Naughton (1966). *Science* **153**, 1539.
Brenner, S., and C. Milstein (1966). *Nature* (London) **211**, 242.
Bromer, W. W., L. G. Sinn, and O. K. Behrens (1957). *J. Am. Chem. Soc.* **79**, 2807.
Doolittle, R. F., and S. J. Singer (1965). *Proc. Nat. Acad. Sci. U.S.* **54**, 1773.
Doolittle, R. F., S. J. Singer, and H. Metzger (1966). *Science* **154**, 1561.
Dreyer, W. J., and J. C. Bennett (1965). *Proc. Nat. Acad. Sci. U.S.* **54**, 864.
Eck, R. V., and M. O. Dayhoff (1966). *Science* **152**, 363.
Fitch, W. M. (1966). *J. Mol. Biol.* **16**, 9.
Frangione, B., and E. C. Franklin (1965). *J. Exptl. Med.* **122**, 1.
Hartley, B. S. (1964). *Nature* (London) **201**, 1284.
Hill, R. L., R. Delaney, H. E. Lebovitz, and R. E. Fellows, Jr. (1966). *Proc. Roy. Soc.* (London) B **166**, 159.
Hilschmann, N., and L. C. Craig (1965). *Proc. Natl. Acad. Sci. U.S.* **53**, 1403.

Hood, L. E., W. R. Gray, and W. J. Dreyer (1966a). *Proc. Nat. Acad. Sci. U.S.* **55**, 826.
Hood, L. E., W. R. Gray, and W. J. Dreyer (1966b). *J. Mol. Biol.* **23**, 179.
Metzger, H., L. Wofsy, and S. J. Singer (1964). *Proc. Nat. Acad. Sci. U.S.* **51**, 612.
Miller, F., and H. Metzger (1965). *J. Biol. Chem.* **240**, 4740.
Milstein, C. (1966). *Nature* (London) **209**, 370.
Mutt, V., S. Magnusson, J. E. Jorpes, and E. Dahl (1965). *Biochem.* **4**, 2358.
Porter, R. R., and R. C. Weir (1966). *J. Cell. Comp. Physiol.* **67**, suppl. 1, 51.
Potter, M., E. Appella, and S. Geisser (1965). *J. Mol. Biol.* **14**, 361.
Press, E. M., P. J. Piggot, and R. R. Porter (1966). *Biochem. J.* **99**, 356.
Putnam, E. W., K. Titani, and E. Whitley, Jr. (1966). *Proc. Roy. Soc.* (London) B **166**, 124.
Singer, S. J., and R. F. Doolittle (1966). *Science* **153**, 13.
Smithies, O. (1965). *Science* **149**, 151.
Tanaka, M., T. Nakashima, A. Benson, H. F. Mower, and T. Yasunobu (1964). *Biochem. Biophys. Res. Commun.* **16**, 422.
Thorpe, N. O., and H. F. Deutsch (1966). *Immunochem.* **3**, 329.
Titani, K., E. Whitley, Jr., and F. W. Putnam (1966). *Science* **152**, 1513.
Von Ehrenstein, G. (1966). *J. Cell. Comp. Physiol.* **67**, suppl. 1, 46.
Walsh, K. A., and H. Neurath (1964). *Proc. Nat. Acad. Sci. U.S.* **52**, 884.
Zuckerkandl, E., and L. Pauling (1936). In V. Bryson, and H. J. Vogel, eds., *Evolving Genes and Proteins*, p. 97. New York: Academic Press.

MOLECULAR BIOLOGY

JOHN C. KENDREW
Medical Research Council
Laboratory of Molecular Biology
Cambridge, England

Information and Conformation in Biology

Scientific advance goes by way of revolution and counterrevolution, by way of a dialectical process. Advances in biology have been no exception. In recent years there has come a new revolution in biological thinking that has brought with it conflicts both external and internal. It is with this new revolution I wish to concern myself—a revolution not yet completed. The smoke of battle still conceals some of the dispositions of the troops, and the exact nature of the final outcome cannot yet be discerned. Indeed, as one of the participants I have my own bias, my own point of view which, for want of a better phrase, might be described as that of a molecular biologist.

It has often been pointed out that this really a very unsatisfactory term: it appears to claim too much, for certainly biochemists concerned themselves with molecules long before anyone had heard of a molecular biologist; and, on the other hand, some classical biologists have been known to suggest—though perhaps significantly this criticism has been heard less and less often in the last few years—that anyone who concerns himself with objects so static, so unlifelike as crystals cannot be properly described as a biologist at all. But unsatisfactory as it is the term has stuck, *faute de mieux*, and even if its boundaries are hard to define molecular biology does represent a distinctive, recognizably new, and powerful approach—all the more powerful just because its frontiers are obscure and have continually been extended. I chose not to use "molecular biology" in my title, partly because it does indeed fail to reveal the real nature of this approach, partly because I wanted at the start to emphasize its dialectical nature. I had contemplated other formulations—geometry and biology, form and information, three dimensions and one dimension—but in the end it seemed to me that the study at the molecular level of the interaction interplay of information and conformation best conveyed the essence of what I hope to say. In the life cycle of every living organism we can recognize this interaction between conformation, or form (in the sense that D'Arcy Thompson used the term) and information, the genetic message passed on from one generation to the next that determines conformation in the progeny. Both these aspects of living organisms are principal preoccupations of molecular biologists; indeed, progress has come from the interplay between those mainly interested in conformation and those concerned with information—another

Based on the Herbert Spencer Lecture delivered at Oxford University, 1965.

example at a different and human level of the dialectical process at work. In a wider context it is interesting to ponder on the relationships, not always too amicable, between molecular biologists and their colleagues who approach biology from other angles. In each of these three contexts we find a conjunction of opposites, an interplay between thesis and antithesis, which is the very essence of change and of progress, whether in the adaptation of a species to its environment or in man's understanding of living organisms. Sometimes the interplay may seem like a conflict. I have even found it natural to use phrases like "a new revolution" or "the smoke of battle." But whatever name we give to this dialectical process, it is in my view a cause and a condition of progress rather than a frustration and an obstacle.

Let us try to follow the notions of information and conformation in the growth and reproduction of living organisms. Plants and animals are complicated; a human being contains more than a million million cells, each in itself an intricate and in some ways a self-sufficient chemical factory conducting numerous chains of reactions that produce chemical or mechanical energy and other specialized outputs, and yet all are subservient to the whole, under a common system of chemical and nervous control. Each individual, however, stems from only two single tiny cells, the spermatozoon and the ovum, which between them contain all the information needed to direct and specify the formation of the adult organism. Biologists have always been struck by this paradox—the contrast between the adult organism and the germ cells, between phenotype and genotype. They have long realized that the complexities of the phenotype must be specified by an enormous store of information in the genotype, that the minute germ cells must contain some kind of gigantic blueprint. Microscopical observations of the processes accompanying cell division have shown that in fact each and every living cell contains the whole blueprint, passed on in duplicate, one copy to each of the daughter cells. The key to an understanding of the nature of this process was the correlation between, on the one hand, Mendel's mathematical formulation of the statistics governing the appearance of particular characteristics in successive generations of crossbred strains, and on the other hand, the microscopical observations of the replication and segregation of the chromosomes during cell division. It became clear that the chromosomes themselves must be the repositories of the hereditary information store, and from this realization there stemmed the whole of modern genetics. It was shown that microscopically identifiable regions of each chromosome, named genes, carried the information specifying visible characteristics of the organism, and the study of the relationship between these two made up of what one might call the observational phase of genetics. Meanwhile, the development of biochemistry as a new discipline in biology carried with it the idea that observable characteristics of organisms were the consequence of chemically identifiable features of the metabolism of each cell, that every chain of chemical reactions was mediated and controlled by enzymes, specific catalysts each responsible for its own link in the chain, so that alterations in the reaction chains would be due to the presence or absence of particular enzymes or to changes in their properties. Thus it became possible to formulate the "one gene–one enzyme" hypothesis: each chromosomal gene determines the synthesis of a particular enzyme, and alterations in the gene result in alterations of its corresponding enzyme and hence, at second remove, in alterations of the external appearance, the phenotype of the organism. So there came into being a new phase, the biochemical phase of genetics. The subject had now reached a point at which its main concern was to correlate observable events during cell division, and observable features of the chromosomes, with changes in the metabolic pattern.

Like observational genetics before it, biochemical genetics has developed into a most powerful discipline, which established a firm conceptual framework, embracing a great range of observations and having the most important practical consequences in all branches of applied biology. It did, however, suffer from two serious limitations. The first of these concerned the metabolism of cells. The reaction chains themselves were generally understood; the molecules whose transformations they encompassed were for the most part of only moderate complexity, such as sugars and lipids and amino acids—molecules consisting of a few tens of atoms with chemical structures that had been established by classical methods. Besides these there were a few very important and rather more complicated molecules such as adenosine triphosphate, ATP, which is a universal reservoir of chemical energy in all types of cell, and these too had chemical structures which could be fully determined. But the controlling catalysts, the enzymes, were not so well understood. All known enzymes are proteins; all proteins are very complex molecules containing thousands of atoms.

In many ways proteins are the molecules most characteristic of living systems. The organization of proteins is at the same time simple and complex: simple in that they are made out of only twenty different types of amino acids, and complex in that these are organized into very long strings, which may contain hundreds of amino acids strung into a long chain. All living organisms are found to use the same set of twenty amino acids. However, since they can be arranged in these chains in an infinite variety of sequences, this in turn leads to an infinite variety of functions that can be found in the proteins. Thus one might consider the amino acids like the standard parts in a Meccano set, which can be arranged to make an endless variety of structures.

However, the long strings of amino acids in proteins can be rolled together to form a ball-like structure. The active site of an enzyme is produced by a particular constellation of amino acids on its surface and, although the particular amino acids involved in the active site are all organized around the same area, they may of course be found on many different parts of the polypeptide chain because of the complex nature of the folding.

Chemists have worked on proteins for over a hundred years, but it is only as recently as the early 1950's that Fred Sanger first determined the sequence of amino acids in the protein insulin. Since that time a large number of sequence determinations have been carried out on a variety of proteins. It is important, however, to distinguish between the structure of the protein in the chemist's sense, the sequence of amino acids, and its conformation. The chemical structure is basically a topological description, defined by the connectiveness between atoms. In contrast, the conformation of the protein is determined by the three-dimensional positions of all of the atoms. Thus the conformation is a description of all of the interactions between the amino acids relative to each other.

For simple molecules containing twenty or thirty atoms, the chemical structure may be an adequate guide to the function of the molecule, especially if the chemist's experience and intuition are used to interpret the function. This is so because the degrees of freedom are limited and the number of conformations the molecule can assume is small. However, as the complexity of the molecule increases, the topological description of it becomes more and more inadequate, and the number of degrees of freedom becomes enormous. The inadequacy of a topological description for proteins is clearly evident, since the number of ways in which a protein chain can be folded up is very large, and a description of the "structure" in the chemical sense is useless for predicting function. This is the reason why classical chemical methods were inadequate to explain the function of enzymes. This represented a definite limitation to the development of biochemistry.

Another limitation of the classical approach was the fact that the chemistry of the genetic material was unknown even at the structural level. Chromosomes were known to contain protein and nucleic acid, DNA, but until the middle 1940's it was not known which of these contained the information. While working on the transformation of pneumococci, Avery found that an extract of DNA was capable of permanently changing the inheritance of these organisms. His work led to the idea that DNA, a long-chain molecule containing four different kinds of bases, is able to act as an information carrier. However, it was not known how this information was carried or reproduced.

The development of molecular biology has served to open up these two major blocks in the understanding of biological systems: the relationship between protein conformation and protein function, and the elucidation of the mechanism whereby DNA acts as an information carrier in biological systems. Molecular biology had its beginnings in the application of X-rays to determine the conformations of biologically important large molecules. Starting in the early 1930's and continuing into the 1950's, a number of enthusiastic attempts were made by several workers, especially Astbury, to interpret molecular conformation through the application of X-ray diffraction methods. Even though their solutions were usually wrong, they were successful in establishing a new outlook, which emphasized the importance of molecular geometry in understanding biological function.

Beginning in the early 1950's, a new phase of molecular biology developed, characterized by the correct determination of a number of structures. One of the first and most dramatic successes in this phase was that of Pauling, who, with his collaborators, was working on the conformations of the polypeptide chain. He recognized that the number of the stable conformations was limited and in particular pointed out that certain helical configurations having a nonintegral number of amino acids per turn should be stable. The most important of these is the α helix, which has approximately 3.6 amino acids per turn of the helix and in which the helical turns are held together by hydrogen bonds. The next most important step was taken in 1953 by Watson and Crick, who developed for the structure of DNA a molecular model that had obvious biological implications. The molecule had the form of a two-stranded helix with specific pairing or hydrogen bonding between the bases. Finally, toward the end of the 1950's, the X-ray diffraction work on the conformation of proteins began to reach completion. The structure of myoglobin was determined and was followed shortly by the structure of hemoglobin. For the first time it became possible to understand the conformation of protein molecules in concrete terms; the relation between the chemical structure or sequence and the actual conformation of the molecule became apparent.

During this rapid development of molecular biology in the 1950's the informational side of the subject benefited first from the application of X-ray diffraction techniques. There are three characteristics necessary for the DNA molecule. First of all, it must be capable of acting as an information carrier. For DNA this is brought about through the use of the four different kinds of bases in the structure of the molecule. Although the molecules are organized with a regular phosphate-sugar backbone, the sequence of the four different bases clearly allows the molecules to contain information based on the varying order of these. The second important characteristic is that the molecule must be able to replicate in order to reproduce the information. This is brought about by utilizing the complementary hydrogen bonding between the purine and pyrimidine bases. This geometric complementarity is utilized in the organization of the double helical array, since only the particular relationships involving the two base pairs used in DNA are capable of giving rise to the ordered structure revealed by the X-ray diffraction studies. A third characteristic of this molecule is that it must have the capacity to direct the translation of the sequence of bases in the molecule into the sequence of amino acids in a polypeptide chain. This is carried out by using one of the two strands of DNA as a template for assembling the closely related nucleotides of the messenger RNA molecule. This contains a sequence of bases complementary to that found in one strand of the DNA, and it is able to act as a template for organizing the amino acids into specific sequences. This process is carried out on ribosomes, the complex cellular organelles that act to assemble the amino acids after they have been activated by attachment to a transfer RNA molecule. In the ribosomal structure, the transfer RNA molecules with their attached amino acids are able to assemble on the messenger RNA by binding on to triplets of nucleotides to bring about the translation of information. The amino acids are then polymerized together. Thus the information in polynucleotides, containing four different kinds of bases, becomes transformed into a polypeptide chain involving twenty different amino acids, the bases being read out in groups of three.

After the polypeptide is formed, the chain then begins spontaneously to roll up into a globular form, which may form an enzyme. Not only can a single polypeptide chain fold into one nearly spherical object, but several of these can then associate to make even more complex organized structures, involving several polypeptide chains that may be identical or different. In these larger aggregates, there is still another degree of conformational freedom: many different surfaces of the individual subunits may be used in the association process. This makes possible another level of conformational complexity, which does not depend upon the covalent chemical bonds between individual amino acids but rather upon the secondary or weaker chemical bonds that are used to hold the polypeptide chains together in the larger aggregate. This principle of the association of completed polypeptide chains is used to build even larger aggregates as, for example, those making up the viruses. In these the individual protein subunits associate in a regular array, either in helical fashion to form elongated rodlike particles or in the form of spherical shells to build up the spherical viruses.

As this knowledge was uncovered by molecular biologists it became apparent that some new concepts were being introduced. The information is stored in the DNA molecules in the form of a one-dimensional array. This is, of course, a familiar mechanism of information encoding, since it is the basis for writing in systems that use an alphabet, and it is familiar to anyone who looks at the arrangement of letters in a line of English text. However, the conformation of proteins is an expression of information in a three-dimensional form. The one-dimensional information has become translated into a three-dimensional form that is the actual operational unit with which the molecule carries out its biological activity. It is clear that difficulties would develop if nature used three-dimensional arrays for retaining information, since it would make the process of copying the information and of translating it extremely difficult. These processes are simplified enormously in the one-dimensional array of nucleotide bases found in DNA or messenger RNA. More generally, in an N-dimensional world a template or mold cannot be more than $(N - 1)$-dimensional. We inhabit a three-dimensional world and structures exist as three-dimensional objects. Information could be two-dimensional—as for example in the reproduction of diagrams and pictures—but in actual fact the one-dimensional form is most frequently used—as in a line of type, in the Morse code, or in a computer tape. We are tempted to ask if biological information is *always* one-dimensional. Are there cases in which two-dimensional molds are used in biological systems? We do not know any examples of this at the present time, but then our knowledge of biological systems is far from complete.

We thus arrive at a concept of the life process as a continuous, ever-repeating alternation between information and conformation, between genotype and phenotype, between DNA and protein. However, we note that there is an element of asymmetry in this system. DNA makes DNA through a replication process. DNA in turn makes protein through a translational process. But protein apparently does not make DNA; thus the informational flow is directional. This classical scheme does not therefore supply a route by which the environment might affect the proteins, and thus at second remove DNA.

There seems to be no mechanism of a Lamarckian kind whereby the environment might have a direct effect upon the progeny of the organism. Thus by the molecular approach we reach the same conclusion as did the classical geneticists working at the observational level.

The only way in which inheritable changes can appear is by alteration of the DNA, and this can be produced only by mutagens, not by the environment in general. These modifications are then subjected to a process of natural selection, which is also an essential part of the scheme. There are of course inadequacies in this description of the scheme; however, these inadequacies are not necessarily criticisms of it. Indeed, the observation of inadequacies and difficulties is exactly what is needed to lead to further progress. We can cite some examples. It is possible that the "information" stage can sometimes be skipped. Sonneborn has shown that some patterns in simple organisms appeared to be self-replicating. Can these occur if they do not involve a portion of the cycle through nucleic acid? Another kind of inadequacy has appeared in consequence of the fact that chromosomal genes are not the unique carriers of information. Many microorganelles such as mitochondria and chloroplasts have DNA in them and we may wonder to what extent they are independently self-reproducing. Another example may be seen in the episomes, which represent fragments of genetic information that are not integrated within the chromosomes. There is in addition a blurred boundary between normal nonchromosomal entities and lysogenic bacteriophage, which can under some conditions be integrated into the chromosome and under other conditions remain independent of it.

We must also note that evolution has seen the emergence of quite new types of information storage and transfer systems. One of these is the central nervous system, which is capable of storing information directly from the environment but which lacks the capacity to pass this information in an inheritable way to the offspring. In this system information transfer is generally what we call the process of education, which provides a new route whereby sense data can be stored and then passed on from one generation to the next.

We may thus describe three different types of the information that are of importance in biological systems: the genetic information, which does not have feedback from the organisms but is passed on from generation to generation; stored sense data, which do have considerable feedback into the storage system of the organisms but are not passed down from generation to generation; and finally, communicated data, which do have feedback and are also passed down to the next generation. It is the possession of the third kind of information in large amounts that makes Homo sapiens unique as a species; even esthetics can be thought of as a complex form of information transfer.

I have tried to show that the subject matter of molecular biology comprises the interplay between information and conformation, between genotype and phenotype, between nucleic acid and protein, and that, because the essence of both information and conformation is molecular structure, the deepest insights can be obtained only if this interplay is examined at the molecular level. It is just because molecular biology is based on techniques for the study and elucidation of complex molecular structures that its impact has been so dramatic during the past decade, when the structures have for the first time been solved. But just as the relationship between genotype and phenotype can be regarded dialectically—being two opposed modes of organization whose transformations determine the nature of the life process, and indeed the very essence of life—so the practitioners of molecular biology have been rather sharply categorized into those who have studied information and those who have studied conformation, into those interested in nucleic acid and genetics and those interested in proteins, enzymes, and metabolism. The most important difference between the approaches of these two kinds of molecular biologist has been in their attitudes to molecular structure. It is true that the recent efflorescence of new knowledge about nucleic acid replication, information transfer, protein synthesis, the nature of the genetic code, and the molecular basis of mutagenesis, is based entirely on a knowledge of structure—the structure of nucleic acid. But the structure has been, as it were, simply a datum. Nearly all the work in this field has depended only on the original determination of DNA structure in 1953 by Watson and Crick, together with the scheme of replication that they propounded at almost the same moment. All the rest has followed without the need for new structural information, simply by using the original concepts as a background to a great range of experimental work, using the techniques of genetic analysis and of classical biochemistry. In the application of molecular biology to problems of conformation it has been quite otherwise. Here there has been no single illuminating generalization; it has rather been a case of slow and tedious progress, of using the lengthy procedures of X-ray analysis to elucidate a series of complex structures. The determination of each structure takes years, and while it is on the road to solution there are few intermediate results of biological interest.

The two parts of the field have attracted different types of mind. Molecular genetics is a subject in which the individual experiment is often rather quick to do; everything depends on the *planning* of the experiment. Ideas, too, are quite easy to come by, and the literature is full of unproved speculations; often the same idea, traveling around the world by way of a highly developed grapevine system, may be tested experimentally in a dozen different laboratories, so that the competition for rapid publication is intense. The successes are scored by those who not only have the ideas and are well-equipped with up-to-date experimental techniques, but who are also capable of a rigorous discipline of mind and are able to think several moves ahead about the implications of their experimental findings. The qualities required are not unlike those needed in other branches of science, but here they are exhibited in an unusually pure form: a capacity for drawing deductions from apparently isolated facts and for setting up hypotheses of the kind that suggests new and decisive experiments. It is not so much the intrinsic probability or esthetic quality of the hypothesis that counts, but rather its potential fruitfulness as a step toward further progress. Two examples were the hypotheses of messenger RNA and transfer RNA. It was correctly deduced, long ahead of direct experimental evidence, first, that a messenger molecule (presumably some kind of nucleic acid) was necessary to carry the information from its permanent home in the nucleus, out to the cytoplasm where it would be transcribed into protein chains. Second, it was realized that because nucleic acid could "recognize" nucleic acid, by the same base-pairing rules that govern the replication of DNA, but could not recognize amino acids, it followed that an adaptor molecule would be needed to recognize on the one hand the code symbols on the nucleic acid of the messenger and on the other the individual amino acids that had to be assembled in the correct order. In both cases the new molecules were hypothesized, and some of their properties predicted, before there was any direct evidence of their existence; and in both cases they were later searched for and found, even though they are present in very small amounts in the cell. Today they are commonplaces of biochemistry; they can be isolated, purified, and directly observed at work. In my opinion the qualities of insight, of rigid self-discipline, and of inductive thinking required and exhibited in this kind of molecular biology at its best are more demanding than in almost any other branch of science.

For the determination of protein structures an entirely different cast of mind is needed: the capacity for tedious experiments in the chemical laboratory, for collecting X-ray data over months and years, and for conducting lengthy computations, all without any immediate reward in the shape of new biological insights; and at the end of it all a complex and irregular structure about which no easy generalizations can be made—at least up to the present time. While the molecular geneticists have been romping ahead, having tremendous fun with new results and new hypotheses coming hot from the press each week or emerging from the grapevine almost each day, the protein crystallographers have immersed themselves for years in some problem, seeing the hoped-for solution perhaps years ahead. Until 1953 there was no doubt in most people's minds that proteins were the most interesting of all biological molecules. It is small wonder that, since the publication of the double-helical structure of DNA, it has been the nucleic acids that have seemed most exciting; proteins seem dull by comparison. During these last ten years the devotees of the two branches of the subject have had rather little to say to one another. Of course the truth is that both proteins and nucleic acids are interesting and exciting. Indeed, during the past year or two the two wings have shown signs of meaningful interconnection, since after all the only *raison d'être* of the complex genetic apparatus of a cell is the production of a functional, three-dimensional protein molecule. Now that the mechanism has been more or less elucidated whereby amino acids are strung together in a long and specific sequence, it becomes relevant to establish the logical connections between this process and the nature and functioning of the completed product. Besides this, it has become apparent that both the genetic apparatus and the metabolic chains require a system of control; genes must be switched on and off in sympathy with the needs of the individual cell for particular proteins, and metabolic pathways must be closed down or rendered more active in accordance with demand for their products. It now transpires that both types of control are exercised by changes in the activities of protein molecules—of genetic repressors in the one case, and of the so-called allosteric enzymes in the other—and these types of control, though still not fully understood, demand subtle alterations in the conformations of protein molecules, which can only be comprehended in the light of a much more intimate knowledge of the principles of protein structure than we possess today. Thus, just as in the living cell nucleic acids are absolutely necessary to proteins and proteins are necessary to nucleic acids, so it is becoming apparent in the day-to-day work of the laboratory that the two kinds of molecular biologists are mutually interdependent and that neither can make progress without the other. So here, too, at the human level we can observe the resolution of a dialectical opposition which is at this moment leading to new insights in the field.

But it is not only within molecular biology that there have been signs of tension. Relations between molecular biologists and their colleagues in other branches of biology have not invariably been smooth. There have been several sources of trouble. One of them derives from a rather irritating tendency on the part of molecular biologists to, so to speak, take over a particular area of biochemistry that happens to interest them and label it "molecular biology" as if it were their own exclusive property. For example, much of the work of the last few years on the mechanism of protein synthesis is actually nothing more nor less than a branch of biochemistry, albeit in a highly sophisticated form, and it does not involve a knowledge of molecular structure to any greater extent than does the rest of biochemistry; but of course protein synthesis is one essential link in the network of relationships between information and conformation and as such can properly be regarded as lying within the purview of molecular biologists. The trouble perhaps lies, as I indicated earlier, in the use of the portmanteau phrase "molecular biology," which like all neat slogans must appear at the same time to claim too little and to claim too much.

But the fact that parts of the field are also simply particular areas of biochemistry should not be allowed to conceal the fact that there is indeed something actually new in the approach of molecular biology. To define what is new about it is something quite difficult to do in any neat phrase, but in the last resort it is a question of geometry and symmetry. Of course these concepts, as ways of looking at the material world, are extremely old, going back to the Greeks, and in more recent times have been developed by classical biologists like D'Arcy Thompson. His *Growth and Form* has perhaps been one of the most stimulating biological texts ever written, to molecular biologists as well as to others. The limitation it suffered was that it perforce stopped short with *visible* structure. In crystallography the study of the modes of aggregation of solid matter could be conducted at an altogether deeper level when it became possible to correlate the external form and symmetry of crystals with the underlying molecular arrangements. Similarly, in biology the decisive event was the acquisition of new tools, especially the X-ray camera and the electron microscope, which made it possible to establish direct connections between *visible* structure and *molecular* structure.

Nevertheless, to some classical biologists the idea of molecular structure as a basis for the study of living organisms has seemed unduly limiting, as leading to a mechanistic interpretation that was a kind of strait jacket, insufficiently flexible to accommodate the infinite variety of life. But of course this is not really the case; one of the most striking features of the biological macromolecules is their plasticity of function. It is rather like the difference between analog computers and digital computers. Any given analog computer has limited application because it can only solve the particular class of problems for which it was designed. Digital computers, however, are not designed to solve particular problems; by their very nature they are adapted to handle any task that can be cast into numerical form. Nucleic acids and proteins similarly have enormous plasticity; just like digital computers, they are made up of a relatively small number of standard parts, but by their nature they are adapted to undertake a great range of tasks. And indeed it has transpired that essentially the same scheme of chemical relationships underlies every one of the cast variety of life forms we know, from the most simple to the most complex.

Perhaps the real basis for the suspicion with which classical biologists and molecular biologists have regarded one another is this question of complexity and, more especially, variety. Classical biologists have always been impressed by the diversity of living matter, which must indeed impress anyone who uses his eyes. Molecular biologists, on the other hand, have concentrated on the universality of cellular behavior at the molecular level. They have taken as their material particular organisms like *E. coli*, which happen to present special advantages for experimental study, and have then tended to assume that the generalizations they have found to apply in those particular systems are universally applicable. Of course it is at the same time true and not true; in one sense the systems *are* common to all living things: essentially the same genetic code, the same scheme for the replication of the genetic material, the same translation mechanism, the same biochemical pathways are apparently to be found in every type of living cell. On the other hand, the infinite diversity of living forms is equally a fact to be reckoned with. It is just because of the diversity and the complexity that molecular biologists, in order to make any progress at all, had to concentrate their efforts first of all on the underlying unity. The question now is, how does that unity engender the variety which we observe? This is the problem of differentiation, to understand the way in which the same genetic information residing in every cell of a multicellular organism can engender the variety of function and structure that go to make up the complete animal or plant, and it is the problem above all problems to which molecular biologists are now addressing themselves. In doing so they will undoubtedly transform their field into something unrecognizably new and different, and when they have done so molecular biology will itself in retrospect seem old-fashioned and classical.

Molecular biologists sometimes accuse classical biologists of looking sentimentally over their shoulders for lingering traces of a vital force, of adopting a mystical outlook simply in order to preserve the mystique of their subject. On the other side of the coin, classical biologists have sometimes felt that molecular biologists, in attempting to interpret the behavior of living organisms in terms of the chemists' molecules and the physicists' forces, are denying the plain fact that the living is different from the dead, that new phenomena emerge at higher levels of complexity. Of course there are phenomena that molecular biologists cannot even begin to tackle—the problems of body and mind, of free will, of ethics. At every level of complexity there is something new, something which in one sense cannot be explained in terms of the level below, and yet which in no way destroys the validity of the conceptions appropriate to that lower level. But here we enter on problems of a philosophical order. These are problems I do not have the time to discuss nor, if I had the time, would I have the capacity, so I must leave them to the professionals.

Speaking as a working scientist I see nothing amiss in the facts that there are certain tensions within my own field of research, nor that molecular biology has been regarded as something of an *enfant terrible* within the family of biological studies, because I think that these tensions are an inevitable accompaniment of real progress—indeed, an indicator that progress is taking place. I think too that it is just criticism to point out that molecular biologists sometimes wear blinkers, but the fact is that blinkers are often of great assistance in seeing the road ahead. Molecular biologists must remember that the *enfant terrible* of today generally turns into the wise traditionalist of tomorrow, and I for one shall have no regrets when molecular biology takes its respectable position as a part of classical biology and gives place to some new impetus in a direction we cannot at this time foresee. The new activity may be even more exciting and fruitful than the studies of information and conformation that I have been discussing.

G. ADAM*
M. DELBRÜCK
Division of Biology
California Institute of Technology
Pasadena, California

Reduction of Dimensionality in Biological Diffusion Processes

Motto: Drunkard: "Will I ever, ever get home again?"
Polya (1921): "You can't miss; just keep going, and stay out of 3D!"

We wish to propose and develop the idea that organisms handle some of the problems of timing and efficiency, in which small numbers of molecules and their diffusion are involved, by reducing the dimensionality in which diffusion takes place from three-dimensional space to two-dimensional surface diffusion.

Consider the problem of bringing an object (a diffusible molecule), from a large or distant area of production (P) to a small target area Q, where it is to be used. The distance between P and Q may be enormously large compared to the distances traveled by the molecule in physiological time intervals by free diffusion. This will be true even in most of the cases in which the molecule is produced and used within the same cell. In other cases (hormone transport), the distances involved may be of the order of meters. In still others, such as the pheromones of insects (the sex attractants), the distance may be measured in miles. These very large distances are certainly not being covered by diffusion only but by various kinds of convection: protoplasmic streaming, blood circulation, wind current.

We wish to suggest that, in addition to these convective devices, in the intermediate range of distances of a few microns a new principle may be involved in that the diffusing molecules reach their destination area not directly by free diffusion in three-dimensional space, but by subdividing the diffusion process into successive stages of lower spatial dimensionality.

Stage I, free diffusion in three dimensions from P to a specially designed surface (interface) in which the target is embedded. Stage II, diffusion to Q on this surface. We imagine the molecule to be held on this surface by forces sufficiently strong to guarantee adsorption but also sufficiently weak to permit diffusion along the surface.

Central to our considerations will be the following theorem concerning diffusion toward a small target of diameter a within a large diffusion space of dimensionality i and diameter b:

Supported by the National Science Foundation.

*Postdoctoral fellow of the Stiftung Volkswagen-Werk. Present address: Institut für physiologische Chemie der Universität, München, Germany.

The mean time of diffusion to the target, $\tau^{(i)}$, can be represented by

$$\tau^{(i)} = \frac{b^2}{D^{(i)}} f^{(i)}\left(\frac{b}{a}\right) \tag{1}$$

The first factor, $b^2/D^{(i)}$, is the time needed to cover a diffusion distance b in any direction. We will call this factor the *distance factor*. Disregarding the material coefficient $D^{(i)}$, it is independent of the dimensionality of the diffusion space.

The second factor, $f^{(i)}(b/a)$, depends strongly on the dimensionality. We will call it the *tracking factor*. If $b/a \gg 1$, it is linear in b/a for three-dimensional space, essentially log b/a for two-dimensional space, and independent of b/a for one-dimensional space. It is this dependence of the tracking factor on the dimensionality which, in the case of small targets, does make reduction of dimensionality in diffusion from *three* to a lower one advantageous. The advantage of a reduction from two to one dimension, on the other hand, is generally negligible.

Two distinct situations may be envisioned in which this reduction of dimensionality may be advantageous for an organism (and possibly also in some chemical engineering situations):

1. If it reduces the mean *time* for the molecule originating at P to reach Q. Intuitively it is clear that this time might be either reduced or extended by dimensional reduction, depending on the value of the diffusion coefficient in the free and in the adsorbed state. The aim of our analysis will be to estimate for simple models the gain (or loss) factor for the diffusion time in its dependence on the ratio of the two diffusion coefficients.

The idea of shortening diffusion times by adsorption to specific surfaces has previously been advanced by Bücher (1953). Bücher pointed out the general possibility that surface diffusion might give high turnover numbers for membrane-bound enzymes, especially in the case of mitochondria. Similarly, Trurnit (1945) remarks that acetyl choline might, by surface diffusion, get more quickly from its site of action to the site where it is destroyed. He also points out the usefulness of surface diffusion for cases of small amounts of activator molecules. Neither of these authors, however, did any quantitative estimate to check the feasibility of their speculation.

2. If it improves the *catch* of molecules from a stream (gaseous or liquid) passing the area Q. Again it is clear intuitively that the catch may be improved or diminished, depending on the quality of the surface in which the target area Q is embedded.

If the molecules hitting the embedding surface stick to it without moving on it, the catch may be diminished; if the surface reflects the molecules, the catch may be the same as that where there was no embedding surface area (except for hydrodynamic effects of the surface on the flow pattern of the passing medium); if the particles stick to the surface sufficiently loosely to permit diffusion along the surface, the catch may be enormously improved.

MEAN DIFFUSION TIMES

In this part, we are interested in a convection-free situation in which a molecule is generated at some point P within a diffusion space and has to reach a target

Q. The main interest, here, is in those cases in which the dimensions of the target are small compared to those of the diffusion space.

In the following, we first outline the general procedure to derive the mean diffusion times. Subsequently, we apply the procedure to different geometric models of the diffusion spaces.

We use the continuum approach and describe the diffusion process by the equation

$$\sum_{r=1}^{i} \frac{\partial^2 c}{\partial x_r^2} = \frac{1}{D^{(i)}} \frac{\partial c}{\partial t} \tag{2}$$

where c is the concentration of the diffusing particles and $i = 1, 2$, or 3, depending on the dimensionality of the diffusion space.

At time zero, we assume a uniform particle concentration c_0 in the diffusion space. In the course of time these particles will diffuse to the target. There is no generation of particles in the diffusion space at time $t > 0$.

Using the proper boundary conditions, we can then derive $c(x_r, t)$, which describes the particle concentration in the diffusion space in dependence of space coordinates and time. If we integrate c over the space coordinates, we find the number of particles which at time t remained in the diffusion space V:

$$P(t) = \int_V dV \cdot c \tag{3}$$

The average lifetime of a particle in the diffusion space—that is, the average time needed by a particle to reach the target—is defined by

$$\tau = \int_0^\infty dt \cdot t \frac{d}{dt} \left\{1 - \frac{P(t)}{P(0)}\right\} \tag{4}$$

where

$$P(0) = c_0 V$$

Our boundary conditions will all be of the form

$$c = 0 \qquad \text{at the target}$$

$$\frac{\partial c}{\partial r} = 0 \qquad \text{at the outer boundary of the diffusion space}$$

With boundary conditions of this kind, the solutions $c(x_r, t)$ of the diffusion equation, and consequently the quantities $P(t)$, can be written as a sum of exponential terms, each decaying with its proper relaxation time τ_n:

$$P(t) = P(0) \sum_n B_n e^{-t/\tau_n} \tag{5}$$

Since

$$\int_0^\infty dt \cdot t \frac{d}{dt} \{1 - e^{-t/\tau_n}\} = \tau_n$$

we find that

$$\tau = \sum_n B_n \tau_n \tag{6}$$

In the following applications the succession of the τ_n is monotonically decreasing, τ_1 representing the longest relaxation time. Moreover, in all cases of interest to us (small targets), the subpopulation decaying with τ_1 represents the bulk of the particles.

One-Dimensional Diffusion in the Linear Interval $a \leq x \leq b$

In this case, the boundary conditions can be written $c = 0$ at $x = a$ and $\partial c/\partial x = 0$ at $x = b$. The initial condition is $c = c_0$ at $t = 0$. The solution under these conditions (Carslaw and Jaeger, 1959, page 101)

$$c(x, t) = c_0 \sum_{n=0}^{\infty} \frac{4}{(2n+1)\pi} \cos \frac{(2n+1)\pi x}{2b(1-k)} \exp \left\{ -\frac{D^{(1)}}{b^2} \left[\frac{(2n+1)\pi}{2(1-k)} \right]^2 t \right\} \tag{7}$$

Here and in the following we use the notation $k = a/b$. Integration over x from 0 to $b(1 - k)$ gives

$$P(t) = P(0) \sum_{n=0}^{\infty} B_n^{(1)} \exp \{-t/\tau_n^{(1)}\} \tag{8}$$

where

$$B_n^{(1)} = (8/\pi^2)[1/(2n+1)]^2 \tag{9}$$

and

$$\tau_n^{(1)} = \frac{b^2}{D^{(1)}} \left[\frac{2(1-k)}{(2n+1)\pi} \right]^2 \tag{10}$$

Furthermore, we have $P(0) = c_0 b(1 - k)$. Thus, Equation (6) yields

$$\tau^{(1)} = \sum_{n=0}^{\infty} \frac{b^2(1-k)^2}{D^{(1)}} 2 \left(\frac{2}{\pi} \right)^4 \frac{1}{(2n+1)^4} \tag{11}$$

or

$$\tau^{(1)} = \frac{b^2}{D^{(1)}} \frac{(1-k)^2}{3} \tag{12}$$

Here we have used [Jolley (1961)]

$$\sum_{n=0}^{\infty} \frac{1}{(2n+1)^4} = \frac{\pi^4}{96}$$

Two-Dimensional Diffusion on the Circular Ring $a \leq r \leq b$

Here the boundary and initial conditions are $r = a$: $c = 0$; $r = b$: $\partial c/\partial r = 0$; $t = 0$; $c = c_0$. In Appendix I is derived the following formula for the number of particles found at $t = 0$ in the circular ring $a \leq r \leq b$:

$$P(t) = P(0) \sum_{n=1}^{\infty} B_n^{(2)} \exp \left\{ -\frac{D^{(2)}}{b^2} y_n^2 t \right\} \tag{13}$$

where

$$B_n^{(2)} = \frac{4}{y_n^2(1-k^2)} \frac{[J_1(y_n)]^2}{[J_0(ky_n)]^2 - [J_1(y_n)]^2} \tag{14}$$

and

$$P(0) = c_0 \pi b^2 (1 - k^2) \tag{15}$$

Here the y_n are the roots of

$$J_0(ky)Y_1(y) - Y_0(ky)J_1(y) = 0 \tag{16}$$

where $J_n(y)$ and $Y_n(y)$ are Bessel functions of order n. The first six roots y_n for $k > 0.05$ are given diagrammatically in Jahnke-Emde-Lösch (1960).

Using Equations (6) and (15), we derive the following for the mean diffusion time to the central circular target:

$$\tau^{(2)} = \frac{b^2}{D^{(2)}} \sum_{n=1}^{\infty} \frac{B_n^{(2)}}{y_n^2} \tag{17}$$

A numerical inspection of Equation (17) shows that for $k \leq 10^{-1}$ the average diffusion time is given within a relative error of less than 6 percent by

$$\tau^{(2)} = \frac{b^2}{D^{(2)} y_1^2} \tag{18}$$

Here, according to Equation (16), the quantities y_1 depend on k. For small k we have evaluated the first root y_1 graphically, by determining the intersection points of the plot of $J_1(y)/Y_1(y)$ with the plots of $J_0(ky)/Y_0(ky)$. The results are given in Table 1.

TABLE 1.

k	10^{-1}	5×10^{-2}	2×10^{-2}	10^{-2}	5×10^{-3}	10^{-3}	10^{-4}
y_1	1.103	0.927	0.792	0.726	0.660	0.568	0.485

Using polynomial approximations (Abramowitz and Stegun, 1965, page 369 ff.) for the Bessel functions in Equation (16), we can derive the following formula for very small k:

$$\frac{1}{y_1^2} = 1.15 \log_{10}(k^{-1}) - 0.250 \tag{19}$$

For $k < 10^{-4}$, this formula yields y_1 with a relative error of less than 2 percent.

Three-Dimensional Diffusion in a Spherical Shell $a \leq r \leq b$

In this case the diffusion space is a large sphere of radius b, and the target is a small sphere of radius a in the center of the large sphere.

The boundary conditions are accordingly at $r = a$: $c = 0$; at $r = b$: $\partial c/\partial r = 0$. In Appendix II it is shown that, with the initial condition $t = 0$: $c = c_0$, $P(t)$ can be written

$$P(t) = P(0) \sum_{n=1}^{\infty} B_n^{(3)} \exp\left\{-\frac{D_1^{(3)}}{b^2} x_n^2 t\right\} \tag{20}$$

where

$$B_n^{(3)} = \frac{6[\sin\{(1-k)x_n\} - x_n \cos\{(1-k)x_n\} + kx_n]^2}{(1-k^3)x_n^4[(1-k) - 1/(1+x_n^2)]} \tag{21}$$

and

$$P(0) = c_0 \frac{4}{3} \pi(b^3 - a^3) \tag{22}$$

In Equation (21) the x_n are the positive roots of

$$x \cot\{(1-k)x\} - 1 = 0 \tag{23}$$

The first six roots for different $k \geq 5 \times 10^{-3}$ are tabulated in Carslaw and Jaeger (1959), page 492.

Using Equations (6) and (22), we find the following for the mean diffusion time to the central target:

$$\tau^{(3)} = \frac{b^2}{D^{(3)}} \sum_{n=1}^{\infty} \frac{B_n^{(3)}}{x_n^2} \tag{24}$$

A numerical inspection, using the roots given by Carslaw and Jaeger (1959, page 492), shows that for $k \leq 10^{-1}$ we can, with a relative error of less than 4 percent, write

$$\tau^{(3)} = \frac{b^2}{D^{(3)} x_1^2} \tag{25}$$

If we expand the cot in Equation (23), drop higher terms, and solve for x, we obtain for the first root

$$x_1 = \frac{\sqrt{3k}}{(1-k)} \tag{26}$$

In this equation the neglect of higher terms introduces into x_1 a relative error of less than 2 percent for $k \leq 10^{-1}$. Thus, for $k \leq 10^{-1}$, within the above approximations for the mean diffusion time to the central target, we have

$$\tau^{(3)} = \frac{[1-(a/b)]^2}{3aD^{(3)}} b^3 \tag{27}$$

For fixed a, the diffusion time varies essentially with the cube of the radius (that is, proportional to the volume) of the diffusion space.

To validate the theorem stated in the introduction we list in the following our resulting $\tau^{(i)}$, the mean diffusion times needed to reach a centrally located, small target of dimension a (sphere, circle, interval) from a random point of a diffusion space of dimension b (sphere, circle, interval).

For $a/b \leq 10^{-1}$	For $a/b \ll 10^{-1}$
$\tau^{(3)} \approx \frac{b^2}{D^{(3)}} \cdot \frac{\left[1-\left(\frac{a}{b}\right)\right]^2}{3a/b}$	$\tau^{(3)} \approx \frac{b^2}{D^{(3)}} \frac{b}{3a}$
$\tau^{(2)} \approx \frac{\tau^2}{D^{(2)}} \frac{1}{y_1^2}$	$\tau^{(2)} \approx \frac{b^2}{D^{(2)}} 1.15 \log \frac{b}{a}$
$\tau^{(1)} = \frac{b^2}{D^{(1)}} \cdot \frac{\left[1-\left(\frac{a}{b}\right)\right]^2}{3}$	$\tau^{(1)} \approx \frac{b^2}{D^{(1)}} \cdot \frac{1}{3}$

Thus we have proved the theorem for the special geometries discussed. We conjecture it is valid for much more general shapes of the diffusion spaces and of the targets. Montroll (1964) has derived similar expressions as limiting cases for random walks on large discrete lattices; see his formulas (26), (28), (29).

Combined Three-Dimensional and Surface Diffusion

We now wish to determine the change in diffusion time, if there are introduced into a spherical system surfaces along which a two-dimensional diffusion can take place. To this end, we compare the two simple models depicted in Figures 1 and 2. In Figure 1 we have the case, treated in the previous section, where the molecules diffuse in a spherical diffusion volume to a small central spherical target.

In Figure 2 is introduced an equatorial plane, which separates the two half-spheres, and on which surface diffusion to the central circular target can occur. To avoid the presentation of involved analysis for this case, we give in the following an upper limit for the mean time necessary for particles diffusing out of the half-spheres to the equatorial plane and from there onto the surface to the central target. If this upper limit is shorter than the mean diffusion time for the case of Figure 1, then the case of the half-sphere with surface diffusion on the equatorial plane certainly gives an improvement in the time of diffusion.

To derive the upper limit of the combined diffusion time, we replace the half-sphere by a circular cylinder of the same radius b as the sphere and of height b. The circular basis of the cylinder is assumed to be perfectly absorbing, and all other surfaces to be reflecting for the diffusing particles. A uniform initial particle distribution in the cylinder has to travel a slightly longer mean diffusion path to the basis than a corresponding distribution in a half-sphere. Thus the time of diffusion to the basis plane will be longer. Furthermore, relatively more particles flow to the outer regions of the circular ring, and therefore need a longer time of surface diffusion to the central target. Both effects tend to increase the mean combined diffusion time in the cylinder as compared to the half-sphere. To derive the diffusion time from the circular cylinder to the basis plane, we observe that this is the one-dimensional problem treated earlier.

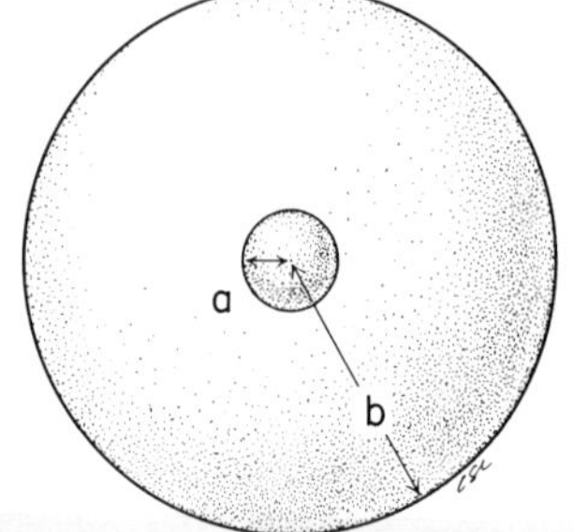

FIGURE 1.
Model for free diffusion in three-dimensional space. The molecules are originally distributed uniformly between the two concentric spheres with radii a and b, respectively. The outer surface of the inner sphere adsorbs the diffusing molecules on first contact [$c(a) = 0$]. The inner surface of the outer sphere totally reflects the molecules.

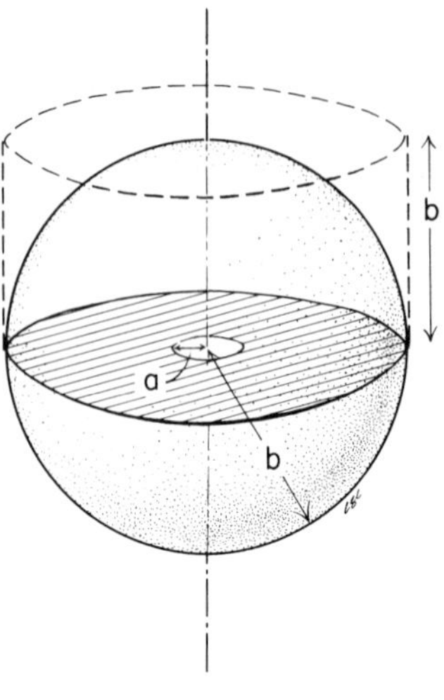

FIGURE 2.
Same as Figure 1, except that molecules contacting the equatorial plane of the other sphere are adsorbed to it and continue to diffuse on it with the new diffusion constant $D^{(2)}$. The mean combined diffusion time, $\tau^{(3,2)}$, is estimated by replacing the upper half-sphere by a cylinder of height b.

We have only to replace $(b - a)$ by b and D_1 by D_3. Then

$$\tau_{\text{cyl}}^{(3)} = \frac{b^2}{3D^{(3)}} \tag{28}$$

A rather lengthy calculation of the precise mean diffusion time τ_{hs} from the half-sphere to the equatorial plane yields a very similar result:

$$\tau_{\text{hs}}^{(3)} = \frac{b^2}{5.40D^{(3)}} \tag{29}$$

This is roughly half the time of Equation (28).

A numerical comparison of Equations (18) and (28), using Table 1, shows that the rate-limiting step in the combined volume and surface diffusion is expressed by Equation (18).

Even in the case $D^{(2)} = D^{(3)}$, we have $\tau_{\text{cyl}} < \tau^{(2)}$. In systems of biological interest we expect $D^{(2)} \ll D^{(3)}$. Thus, for our present purposes, we can neglect the diffusion time to the equatorial plane, and assume that the particles initially in the spherical half-space diffuse instantaneously to the equatorial plane and then diffuse more slowly on its surface to the central target.

In our model the upper limit of $\tau^{(3,2)}$, the combined volume and surface diffusion, is thus approximately

$$\tau^{(3,2)} = \frac{b^2}{y_1^2 D^{(2)}} \tag{30}$$

Biological Implications of the Model Calculations

The ratio of the diffusion times with and without internal surface diffusion in our simple model system, according to Equations (27) and (30), is

$$\frac{\tau^{(3,2)}}{\tau^{(3)}} = \left\{\frac{1}{3} \cdot \frac{D^{(2)}}{D^{(3)}} \left[y_1\left(\frac{b}{a}\right)\right]^2 \frac{b}{a}\left(1 - \frac{a}{b}\right)^2\right\}^{-1} \tag{31}$$

Here the dependence of y_1 on b/a can be taken from Table 1. The ratio of diffusion times depends only on b/a and $D^{(2)}/D^{(3)}$. To gain an insight into possible biological implications of our model calculations, we specify the size of the target to be comparable to the active site of an enzyme molecule. (Another application will be given in the discussion of diffusion of odor molecules.) Thus we apply relation (31) now to the situation in which a molecule is located anywhere in a cell and has to diffuse to a specific enzyme with an active site of size a before it can be processed biochemically and used by the organism. Equation (31), then, represents the factor by which the necessary time of diffusion to the active site is reduced, if by surface diffusion use is made of membranous subdivision of the diffusion space compared to purely three-dimensional diffusion.

Unfortunately, there are no data for surface diffusion coefficients on biological membranes. Yet, from our present knowledge of diffusion coefficients in liquids, gels, and solids, it appears improbable that we will encounter in biological systems any ratios $D^{(2)}/D^{(3)}$ much smaller than 10^{-2}.

If we solve Equation (31) for $D^{(2)}/D^{(3)}$ and consider $\tau^{(3,2)}/\tau^{(3)}$ as a parameter, then the specific value $\tau^{(3,2)}/\tau^{(3)} = 1$ determines for every given $D^{(2)}/D^{(3)}$ the critical ratio $(b/a)_{\text{crit}}$, above which combined three- and two-dimensional diffusion becomes faster than three-dimensional diffusion alone. Hence, in

Figure 3 we have plotted contour lines of $\tau^{(2,3)}/\tau^{(3)}$ in its logarithmic dependence on $(D^{(2)}/D^{(3)})$ and (b/a), as given by Equation (31).

The second scale of the abscissa converts from b/a to the radius b of the diffusion space, assuming the diameter of the active site of the central enzyme molecule to be $2a = 20$ Å. The range of the critical sizes of the diffusion space in Figure 3 is well in the range of sizes of subcompartments in biological systems or of bacteria. The result of our model calculation given in Figure 3 suggests the possibility that the reduction of diffusion times by means of reduction of dimensionality of diffusion contributed to the evolutionary advantage of internal membranes.

One might speculate that this selective advantage of internal membranes contributed greatly to the evolutionary step from small, little-organized, bacteria-like ancestors of cells to fully developed, internally compartmentalized unicellular organisms. For reasonable ratios $D^{(2)}/D^{(3)} > 10^{-2}$, the critical sizes b_{crit} seen in Figure 3 are just of the range of bacterial diameters.

However, besides dimensional reduction, the reduction of molecular transport times by convection, as in protoplasmic streaming, seems to be another way by which nature has overcome the greatly (according to a cubical law) increasing diffusion times for increasing organismal size.

The reduction of diffusion times by compartmentalization and surface diffusion probably is not the only selective advantage of internal membranes. Certainly they render other essential functions to the organism, such as osmotic functions, or provide a hydrophobic environment for certain enzymatic reactions. According to our present knowledge, this is especially likely for the internal membrane system in mitochondria and chloroplasts. There, in all probability, the reduction of diffusion times of substrates or coenzymes from the matrix space to the membrane-bound respiratory or photophosphorylating enzymes is only one of several functions performed by the crista or grana membrane system (see Lehninger, 1965; Mitchell, 1961, 1966; Jagendorf and Uribe, 1966).

Another application of dimensional reduction possibly is in enzyme complexes, which need not necessarily be membrane-bound. Their evolutionary

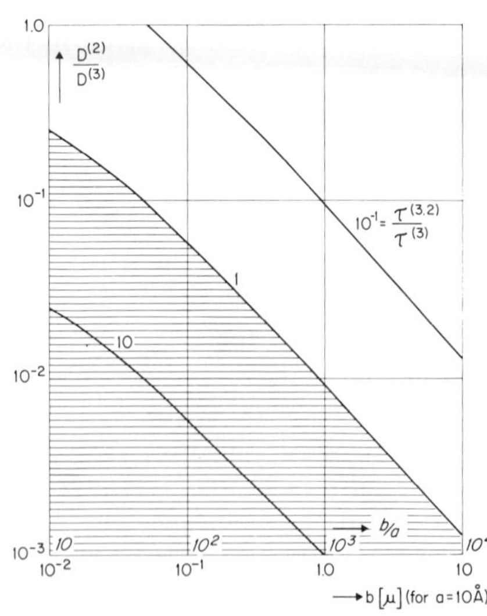

FIGURE 3.
Contour lines for the ratio of the mean diffusion times according to the diffusion regimes illustrated in Figures 2 and 1, respectively, are plotted as a logarithmic function of b/a and $D^{(2)}/D^{(3)}$. The blank area implies a situation in which combined space and surface diffusion is favorable.

advantage, compared to separate soluble enzymes, might rest partially in the vast reduction of diffusion times of substrates in sequential enzyme reactions by diffusion on the surface of the complex.

In the case of *small* amounts of substrates, inhibitors, or regulators, one would expect the effectivity of diffusion on internal cell surfaces to be crucial for survival of the species in evolution, in that it ensures short reaction times upon changes of the environment and an economic use of low concentrations of molecules.

DIFFUSION OF ODOR MOLECULES TO INSECT SENSORY RECEPTORS

We now wish to discuss a rather specialized application of the idea of reduction in dimensionality of diffusion, the chemoreception of a particular odorant by an insect. The insect is the silkworm moth *Bombyx mori*, and the odorant is the sex attractant bombykol, exuded by the female and perceived by the male over great distances. The molecule involved is known precisely: a 16-carbon straight-chain alcohol with two double bonds in a particular configuration. This molecule is perceived by means of very numerous small sense organs, sensilla, each a little hair 100 μ long and 2 μ in diameter. These hairs are arranged in a basket array on the surface of the main stem and the branches of the moth's antennae. Each hair is innervated by the terminal processes of one or two sensory nerve cells, which run up through the length of the hair; however, over most of the length they are shielded from direct contact with the outside air by a cuticular sheath. Direct contact occurs only at minute pores, 150 Å in diameter, distributed along the surface of the sensillum and spaced about 4500 Å apart from each other. Thus the pores constitute only about 1/1000 of the surface of a sensillum. There are about 3200 pores on each hair and about 10,000 sensilla per antenna. We can ignore for the following discussion the presence on the antennae of a minority of other types of sensilla with different functions.

This gigantic setup serves to abstract with a high degree of efficiency bombykol molecules from the passing air and to bring these molecules in some unknown manner to action on the exposed sensitive surfaces of the sensory cells in the pores. We shall not be concerned with the mode of action of the molecules once they reach the pores, but we shall discuss the question of how they get there.

What happens to molecules of bombykol which hit the 99.9 percent portion of the surface of the hairs between the pores? One might assume that they are reflected, or one might assume that they stick where they hit and stay put. These two possibilities have been discussed in an excellent paper on the subject of insect olfaction by Boeckh, Kaissling, and Schneider (1965). We want to recalculate these two cases by procedures that seem to us more appropriate and to discuss a third possibility, suggested by Locke (1965) and (in conversation) by J. R. Platt. This third possibility envisages that bombykol molecules, when hitting a hair's surface anywhere, will stick sufficiently tightly not to come loose, but also sufficiently loosely to be able to diffuse on the surface until they slide into one of the pores. This assumption seems to be eminently reasonable in view of the structure of bombykol and the probable surface nature of the hair (Locke, 1965) and in view of the gain in efficiency of the catch of bombykol molecules that such a regime would entail.

The diffusion problem we have to consider separates into two parts: (1) the diffusion from the passing air to the surface, and (2) the diffusion from a random point on the surface to a pore. The first part determines the catch—that is, the efficiency of the molecular sieve. The second part considers the time needed for diffusion along the sensillum surface to the sensory pores: the average diffusion time should be shorter than the known latency at threshold (<1 sec). It will be treated in Appendix III.

Convective Diffusion to a Cylindrical Surface

We consider one of the nearly cylindrical sensory hairs with its axis at right angles to an air stream containing bombykol molecules (see Figure 4).

To orient ourselves, we first observe that the gas-kinetic mean free path of bombykol in air is about 85 Å, comparable to the pore size but small compared to the sensillum diameter. Further, we compute the Reynold's number for the pertinent aerodynamic situation. The air velocity in the behavioral threshold experiments cited by Boeckh et al. (1965) was $u_0 = 100$ cm/sec. With a hair diameter $d = 2\,\mu$, an air viscosity of $\eta = 1.83 \times 10^{-4}$ poise (Hodgeman, 1961), and an air density of $\rho = 1.29 \times 10^{-3}$ g/cm^3, we find $\mathrm{Re} = u_0\, d\rho/\eta = 0.14$. This means that we have laminar, incompressible flow with little circulation in the wake. The measured streamlines for flow around a circular cylinder at comparably low Reynold's number are given by Tietjens and Prandtl (1931). For the following, we also need Peclet's number $\mathrm{Pe} = u_0 A/D$, where $A = 1\,\mu$ is the radius of the cylinder and D is the diffusion coefficient of bombykol in air. This latter coefficient has not been measured. However, we can arrive at a fairly reliable figure of $D = 0.025$ cm^2/sec by extrapolating from the known figures for several organic vapors in air (Landolt-Börnstein, 1905) to the molecular weight of bombykol. With these numbers, we find $\mathrm{Pe} = 0.4$.

Peclet's number is a measure of the importance of convection and diffusion. The above figure indicates that convection is small but not negligible relative to diffusion. An accurate analysis of the convective diffusion requires first a solution of the aerodynamic problem—that is, finding the detailed velocity distribution of the air around the cylinder. The next step is to calculate the concentration of bombykol near the cylindrical surface, using the differential equation governing convective diffusion,

$$D\left(\frac{\partial^2 c}{\partial x^2} + \frac{\partial^2 c}{\partial y^2}\right) - \left(v_x \frac{\partial c}{\partial x} + v_y \frac{\partial c}{\partial y}\right) = \frac{\partial c}{\partial t} \tag{32}$$

and appropriate initial and boundary conditions.

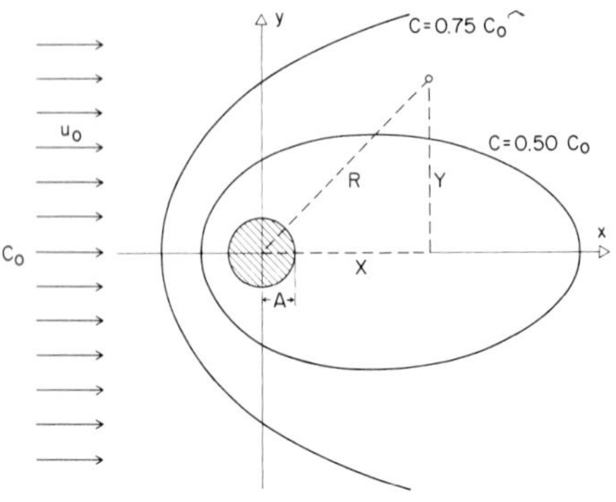

FIGURE 4.
The cross section of a sensillum (hatched) in an airstream of velocity u_0 coming from left, containing bombykol at concentration c_0 before reaching the sensillum. The boundary condition is $c = 0$ at the surface of the sensillum (complete adsorption). The contour lines at which, in the steady state, $c = 0.5\ c_0$ and $0.75\ c_0$ are drawn in.

The final step, then, is to calculate the rate of adsorption to the cylinder for computing the flux:

$$j = +D\left(\frac{\partial c}{\partial r}\right)_{r=A} \tag{33}$$

The exact execution of this program is prohibitively involved, even in the first step on the aerodynamic problem (Lagerstrom, 1964). Fortunately, however, since Peclet's number is small, we can obtain a reasonable approximation by the following asymptotic method.* Let the laminar air stream far from the cylinder be in the positive x-direction (see Figure 4). The approximation of the aerodynamic problem consists in neglecting the perturbation of the stream lines in the vicinity due to the presence of the cylinder. Thus, we assume a coordinate-independent, effective air velocity $\bar{v}$, exclusively in the x-direction. This effective velocity is close to the free stream velocity.

Equation (32) then reduces to the much simpler form

$$\bar{v}\frac{\partial c}{\partial x} = D\left(\frac{\partial^2 c}{\partial x^2} + \frac{\partial^2 c}{\partial y^2}\right) \tag{34}$$

with the boundary conditions

$$c = 0 \qquad \text{for} \quad x^2 + y^2 = A^2 \tag{35}$$

$$c = c_0 \qquad \text{for} \quad x^2 + y^2 \gg A^2 \tag{36}$$

Equation (34) describes the stationary state established within a time that is short compared to the latencies in the biological experiments.

We first introduce dimensionless quantities:

$$C = \frac{c}{c_0}, \qquad X = \frac{x}{A}, \qquad R = \frac{r}{A}, \qquad \bar{V} = \text{Pe}\frac{\bar{v}}{u_0} \tag{37}$$

Equation (34) then reads

$$\bar{V}\frac{\partial C}{\partial X} = \frac{\partial^2 C}{\partial X^2} + \frac{\partial^2 C}{\partial Y^2} \tag{38}$$

with the boundary conditions

$$C = 0 \qquad \text{for} \quad R = 1 \tag{39}$$

$$C = 1 \qquad \text{for} \quad R \rightarrow \infty \tag{40}$$

A solution of Equation (38) is

$$C = 1 - \alpha e^{\bar{V}X/2} K_0\left(\frac{\bar{V}R}{2}\right) \tag{41}$$

where $K_0(z)$ is the modified Bessel function of the second kind of order zero, and α is a constant to be determined from the boundary conditions.

To show that Equation (41) satisfies the boundary condition (40), we use the asymptotic formula (Abramowitz and Stegun, 1965, page 375):

$$K_0(z) = \sqrt{\pi/2z}\, e^{-z}[1 + 0(1/z)] \tag{42}$$

*We are greatly indebted to Professor P. Lagerstrom for guiding us in the handling of this problem.

It follows that the second term in Equation (41) vanishes for large R as $R^{-1/2}$, so that the boundary condition (40) is fulfilled.

To satisfy the boundary condition (39) to the same approximation as represented by the differential equation, we expand the second term of Equation (41), use again the smallness of the Peclet number—that is, reject linear and higher terms in $\bar{V}R/2$—and obtain, for $R \sim 1$,

$$C = 1 + \alpha[\ln(\bar{V}R/4) + \gamma] \tag{43}$$

where $\gamma = 0.577$ is Euler's number.

Here we used for $K_0(z)$ an expansion valid for small z (Abramowitz and Stegun, 1965, page 375):

$$K_0(z) = -[\ln(z/2) + \gamma][1 + 0(z^2)] \tag{44}$$

Equation (43) satisfies the boundary conditions (39) if

$$\alpha = -[\ln(\bar{V}/4) + \gamma]^{-1} \tag{45}$$

Thus our solution is

$$C = 1 - \frac{e^{\bar{V}X/2}K_0(\bar{V}R/2)}{\ln(4/\bar{V}) - \gamma} \tag{46}$$

The flux to the cylindrical surface can be computed from Equations (43) and (45) as follows:

$$J = FD\left(\frac{\partial c}{\partial r}\right)_{r=A} = -\frac{c_0 DF}{A[\ln(\bar{V}/4) + \gamma]} \tag{47}$$

where $F = 2\pi AL$ is the surface area of the cylindrical sensillum. We note that the flux, according to Equation (47), is very insensitive to the effective velocity $\bar{V}$, as can be seen from the following figures.

For $\bar{v} = u_0$ ($\bar{V}$ = Pe), we have $J = (c_0FD/A)0.580$, and for $\bar{v} = u_0/2$ ($\bar{V}$ = Pe/2), we have $J = (c_0FD/A)0.413$.

Thus, in view of the approximate character of our solution, it is sufficient to use $\bar{v} = u_0$.

Application to Threshold Experiments in Olfaction of Bombyx

Equation (47) gives the flux to an isolated hair with its axis at right angles to the air stream. One may wonder whether this result is applicable to the sensilla on the antennae of the moth. Doubts might be entertained on two counts.

First, the antennae offer a great deal of surface, on the main stem and branches, in addition to the sensillar surface. Will not these additional surfaces take out most of the odorant? However, the particle flux per unit area to the main stem and branches of the antennae (stems and branches considered as adsorbing) is by orders of magnitude less than that to the *thin* sensory hairs. In convective diffusion the specific flux to an adsorbing cylindrical surface perpendicular to the air stream is essentially inversely proportional to the radius of the cylinder. Thus the sensory hairs, which are very much thinner than the stem and the branches, are overwhelmingly more efficient in the catch of odorants.

Second, the sensilla stand not at right angles, but in a more or less random orientation to the wind. What contribution to the catch derives from the sensilla standing parallel to the wind? It can be estimated that the flux to a hair parallel to the air stream is not greater than in the perpendicular case. Since, moreover, in a directionally random assembly of vectors the components perpendicular to a chosen direction are greatly preponderant, it is sufficient for our purposes to consider the representative hair to be oriented perpendicular to the air stream.

Thus we can apply Equation (47) to the situation of the threshold experiments of Boeckh et al. (1965). We use $F = 6 \times 10^{-6}\,\text{cm}^2$, $\bar{V} = \text{Pe} = 0.4$, $A = 10^{-4}\,\text{cm}$, $D = 2.5 \times 10^{-2}\,\text{cm}^2/\text{sec}$, and obtain for the flux per sensillum:

$$J = 8.7 \times 10^{-4}\, c_0/\text{sensillum sec} \tag{48}$$

where c_0 is to be measured in molecules per cubic centimeter.

This is the number of molecules reaching the surface of the sensory cells per second, if the diffusion of bombykol on the sensillum surface satisfies the inequality (A21) derived in Appendix III for its coefficient of surface diffusion.

If we use the value $c_0 = 2 \times 10^2$ molecules/cm^3, estimated by the above authors for behavioral threshold, we arrive at 0.17 molecules bombykol reaching a sensillum per second. However, it seems premature to us to discuss the meaning of this result with respect to possible transducer mechanisms of the sensory cells, or integrative mechanisms in the deuterocerebrum (or earlier?). The methods used for estimating c_0 involve estimates of the rates of evaporation of bombykol from filter papers, where it is adsorbed, apparently, in a physically ill-defined manner. The rate of evaporation is not necessarily proportional over a wide range to the amounts of bombykol deposited on the paper. Boeckh et al. mention this difficulty, but have perhaps not sufficiently emphasized the magnitude of error here possible. Extrapolation from weight losses or activity losses at much higher loading of the papers are not really helpful to overcome this difficulty. We feel that it will be indispensable to apply completely different methods for quantitating low stimuli to unravel the mode of response at these levels, a field so very promisingly approached by the investigation of insect olfactory reception by Boeckh et al. (1965).

To obtain an estimate for the effectiveness of the catch by convective diffusion, we have plotted in Figure 4 the contour lines $c = 0.50\, c_0$ and $c = 0.75\, c_0$ for the concentration distribution according to Equation (46). As can be seen from this graph, the sensillum effectively cleans out an air space of several times its diameter. The outline area of one antenna is 0.06 cm^2 (see Figure 14 in Boeckh et al., 1965). The total cross section σ of the 10,000 sensilla specialized for bombykol is roughly three times less ($\sigma = 0.02\,\text{cm}^2$). The sensilla are spaced at distances equal to a low multiple of their diameters (see Figure 15 in Schneider and Kaissling, 1957–1958). The influx to the cross section of one sensillum at our reference air velocity is $c_0 u_0 \sigma = 2 \times 10^{-4}\, c_0$ cm^3/sec. The catch, according to Equation (48) is four times higher; that is, each sensillum, if it were standing alone, would deplete of bombykol an air stream a few times its diameter, and the antenna as a whole should correspondingly clean out the air flow passing its whole outline area (for $u \leq u_0$). This is a conclusion that should be susceptible to direct test, using the antenna of one animal as an absorber in front of another antenna as a testing device for the passing bombykol molecules.

Such an experiment, among others, would give an indication whether the sensillum surface is adsorbing or reflecting. In the case of a reflecting surface, the catch should not be changed much by the presence of another "screening" antenna. As is shown in Appendix IV by Equation (A26), the regime of convective diffusion with surface diffusion on the sensillum predicts a catch of bombykol molecules that is about 40 times larger than can be obtained from an antenna that reflects the molecules everywhere except at the pores, and 1000 times larger than that from an antenna which adsorbs everywhere, without surface diffusion. "Under these circumstances, one would be surprised if Nature had made no use of this possibility" (Dirac, 1931).

APPENDIX I. TWO-DIMENSIONAL DIFFUSION IN THE CIRCULAR RING $a \leq r \leq b$

The initial and boundary conditions are at $t = 0$: $c = c_0$; at $r = a$: $c = 0$; at $r = b$: $\partial c/\partial r = 0$.

For these boundary conditions the solution of the diffusion equation for a unit particle concentration released at $t = 0$ in the circular ring-element $2r'\pi\, dr$ (where $a < r' < b$) is given by Carslaw and Jaeger (1959, page 370) as

$$v(r, r', t) = \frac{\pi}{4} \sum_{n=1}^{\infty} \frac{1}{F(y_n)} \left(\frac{y_n}{b}\right)^2 [J_1(y_n)]^2 C_0(r, y_n) C_0(r', y_n) e^{-D^{(2)} y_n{}^2 t/b^2} \tag{A1}$$

The y_n are the positive roots of

$$J_0(ky)Y_1(y) - Y_0(ky)J_1(y) = 0 \tag{A2}$$

where $k = a/b$.

The other functions in Equation (A1) are

$$F(y_n) = [J_0(ky_n)]^2 - [J_1(y_n)]^2 \tag{A3}$$

and

$$C_0(r, y_n) = Y_0(y_n r/b)J_0(ky_n) - J_0(y_n r/b)Y_0(ky_n) \tag{A4}$$

If at $t = 0$ the particle concentration c_0 is uniform in the area $a \leq r' \leq b$, we have, for the time and coordinate dependence of c,

$$c(r, t) = c_0 \int_a^b dr'\, 2\pi r' v(r, r', t) \tag{A5}$$

Substituting Equation (A1) into (A5), we find that

$$c(r, t) = c_0 \frac{\pi^2}{2} \sum_{n=1}^{\infty} \frac{[J_1(y_n)]^2}{F(y_n)} \left(\frac{y_n}{b}\right)^2 C_0(r, y_n) e^{-D^{(2)} y_n{}^2 t/b^2} I_n \tag{A6}$$

where

$$I_n = \int_a^b C_0(r', y_n)\, r' dr' = \left(\frac{2}{\pi}\right)\left(\frac{b}{y_n}\right)^2 \tag{A7}$$

To evaluate the integral in Equation (A7) we have used (A4) and the relation $\int dz \cdot z \cdot Z_0(z) = z \cdot Z_1(z)$, where $Z_0(z)$ can be $J_0(z)$ or $Y_0(z)$. Since

$$P(t) = \int_a^b dr \cdot 2\pi r \cdot c(r, t) \tag{A8}$$

we use formula (A7) again, and find that

$$P(t) = P(0) \sum_{n=1}^{\infty} B_n^{(2)} e^{-D^{(2)} y_n{}^2 t/b^2} \tag{A9}$$

where

$$B_n^{(2)} = \frac{4[J_1(y_n)]^2}{y_n^2(1 - k^2)\{[J_0(ky_n)]^2 - [J_1(y_n)]^2\}} \tag{A10}$$

and

$$P(0) = c_0\pi(1 - k^2)b^2 \tag{A11}$$

APPENDIX II. THREE-DIMENSIONAL DIFFUSION IN THE SPHERICAL SHELL $a \leq r \leq b$

The boundary conditions are at $r = a$: $c = 0$; at $r = b$: $\partial c/\partial r = 0$.

The solution for a unit particle concentration released at $t = 0$ in a spherical lamina at $r = r'$ is given in Carslaw and Jaeger (1959, page 367) and has the following form:

$$v(r, r', t) = \frac{1}{2\pi rr'}\sum_{n=1}^{\infty} R_n(r)R_n(r')e^{-D^{(3)}x_n^2t/b^2} \tag{A12}$$

where

$$R_n(r) = \frac{(1 + x_n)^{1/2}\sin[(r - a)/b]}{\{b(1 - k)(1 + x_n^2) - b\}^{1/2}} \tag{A13}$$

with $k = a/b$.

The quantities x_n are the positive roots of

$$x\cot[(1 - k)x] - 1 = 0 \tag{A14}$$

If at $t = 0$ the particle concentration c_0 is uniform in the space $a \leq r' \leq b$, we have, for its dependence on time and coordinates,

$$c(r, t) = c_0\int_a^b 4\pi r'^2\,dr'v(r, r', t) \tag{A15}$$

Using Equation (A12), we find that

$$c(r, t) = 2c_0\sum_{n=1}^{\infty} I_n\frac{R_n(r)}{r}\exp\left\{-\frac{D^{(3)}}{b^2}x_n^2t\right\} \tag{A16}$$

where

$$I_n = \int_a^b dr'\,r'^2R_n(r') \tag{A17}$$

or, with Equation (A13),

$$I_n = \frac{b^{3/2}[\sin\{(1 - k)x_n\} - x_n\cos\{(1 - k)x_n\} + kx_n]}{x_n^2\{(1 - k) - [1/(1 + x_n^2)]\}^{1/2}} \tag{A18}$$

In this case Equation (3) reads

$$P(t) = \int_a^b dr\,4\pi r^2c(r, t)$$

or

$$P(t) = P(0)\sum_{n=1}^{\infty} B_n^{(3)}\exp\left\{-\frac{D^{(3)}}{b^2}x_n^2t\right\} \tag{A19}$$

where

$$B_n = \frac{6}{b^3(1 - k^3)}I_n^2 \tag{A20}$$

corresponding to Equations (20) and (21) of the text.

APPENDIX III. TIME OF DIFFUSION ON THE SENSILLUM SURFACE

We are interested in the question of whether the diffusion from an adsorption site anywhere on the sensillum surface to one of the sensory pores is sufficiently fast to render the adsorbed molecule effective within the known latency at threshold (<1 sec).

We can treat this surface diffusion very simply, making use of the results obtained in the section on two-dimensional diffusion. It was shown there that the mean diffusion time, for target areas in the range here encountered, depends only very insensitively on the ratio of radius of diffusion area to radius of target area.

According to Equation (18) we have

$$\tau^{(2)} = \frac{b^2}{D^{(2)}} \frac{1}{y_1^2}$$

In the case of the bombyx sensillum trichodeum we have $b/a \sim 35$, or $y_1 \approx 0.8$.

Thus, to arrive at a latency $\tau^{(2)}$ of less than a second, we need only require that

$$D^{(2)} > 6 \times 10^{-10} \text{ cm}^2/\text{sec} \tag{A21}$$

where we have used $b \sim 2.3 \times 10^{-5}$ cm.

The coefficient of diffusion of bomykol on the surface of a bombyx sensory hair is of course not known. But any reasonable estimate would put it above 10^{-8} cm²/sec. Thus the inequality (A21) in all probability is satisfied by a wide margin.

APPENDIX IV. COMPARISON OF PARTICLE FLUX TO SENSORY PORES FOR DIFFERENT CASES OF SURFACE OF SENSORY HAIRS

If the surface of the sensory hair adsorbs the bombykol molecules from the air stream, we have the situation treated in the section on convective diffusion.

If the particles can diffuse on the surface of the sensory hairs with a diffusion coefficient satisfying inequality (A21), then effectively each of them reaches a sensory pore within the latency period. Under the experimental conditions of Boeckh et al. (1965), the total flux to the sensory pores of one sensillum is then

$$J^{(3,2)} = 8.7 \times 10^{-4}\ c_0/\text{sensillum sec} \tag{A22}$$

Here we have written $J^{(3,2)}$ to indicate that the flux implies three-dimensional and two-dimensional diffusion in succession.

Let us compare this flux with those resulting from the alternative assumptions discussed by Boeckh et al. (1965). If bombykol is adsorbed everywhere on the sensillum but does not diffuse on the surface, we have, with respect to convective diffusion to the hair surface and the concentration distribution in the space around the hair, the same situation as above. We just have to diminish the flux to the pores by the ratio (pore area)/(sensillum area) $\approx 10^{-3}$ and obtain

$$J_{\text{adsorb}} = 8.7 \times 10^{-7}\ c_0/\text{sensillum sec} \tag{A23}$$

where c_0 is to be measured in molecules per cubic centimeter. The corresponding calculation by Boeckh et al. gives a value ten times lower, but their procedure seems unacceptable because it ignores both the aerodynamic aspects and the depletion of the air due to adsorption near the sensillum.

If bombykol is reflected from the sensillum everywhere except at the pores, we cannot use the diffusion equation because the calculated concentration near the pores (which determines the flux) would change rapidly over a distance comparable to the mean free path of bombykol in air. Under these circumstances a gas-kinetic approach, as used by Boechk et al., will yield the better approximation of the flux.

We ignore concentration changes arising from adsorption and use the gas-kinetic number Z of collisions of molecules with a fixed reflecting surface.

If $\bar{w}$ is the average gas-kinetic velocity of bombykol molecules, we have (see, for instance, Kauzmann, 1966)

$$Z = c_0 \bar{w}/4 \tag{A24}$$

collisions per unit time and unit area of surface.

The number of collisions with sensory pores—that is, the flux to these pores—is obtained simply by multiplying the expression (A24) with the total area f of all the sensory pores of one sensillum.

With $f = 5.3 \times 10^{-9}$ cm^2 and $\bar{w} = 1.6 \times 10^4$ cm/sec, we obtain

$$J_{\text{refl}} = 2.1 \times 10^{-5}\ c_0/\text{sensillum sec} \tag{A25}$$

where c_0 is to be measured in molecules per cubic centimeters.

In summary, then, the total fluxes to the pores of one sensillum, according to the three assumptions discussed, stand in the ratio

$$J_{\text{absorb}}:J_{\text{refl}}:J^{(3,2)} = 1:24:1000 \tag{A26}$$

REFERENCES

Abramowitz, M., and I. A. Stegun (1965). *Handbook of Mathematical Functions.* New York: Dover.

Boeckh, J., K. E. Kaissling, and D. Schneider (1965). *Cold Spring Harbor Symposia Quantitative Biology* **30**, 263.

Bücher, T. (1953). *Adv. in Enzymol.* **14**, 1.

Carslaw, H. S., and J. C. Jaeger (1959). *Conduction of Heat in Solids.* Oxford U. P.

Dirac, P. A. M. (1931). *Proc. Roy. Soc.* A **133**, 60. In the preprint of this paper sent to Niels Bohr, the typist had omitted the "no" in the sentence quoted.

Hodgeman, C. D. (1961). *Handbook of Chemistry and Physics.* 44th ed., p. 2264. Cleveland: Chemical Rubber Co.

Jagendorf, A. T., and E. Uribe (1966). *Proc. Nat. Acad. Sci. U.S.* **55**, 170.

Jahnke-Emde-Lösch (1960). *Tables of Higher Functions*, 6th ed., p. 201 ff. New York: McGraw-Hill.

Jolley, L. B. W. (1961). *Summation of Series*, p. 64, New York: Dover.

Kauzmann, W. (1966). *Thermal Properties of Matter: Kinetic Theory of Gases*, p. 178. New York: Benjamin.

Lagerstrom, P. (1964). In Moore, F. K., ed., *Theory of Laminar Flows*, p. 373 ff. Princeton, Berlin.

Landolt-Börnstein (1905). *Physikalisch-Chemische Tabellen*, 3rd ed.

Lehninger, A. L. (1965). *The Mitochondrion*, Chapters 6 and 7. New York: Benjamin.

Locke, M. (1965). *Science* **147**, 295.

Mitchell, P. (1961). *Nature* (London) **191**, 144.

Mitchell, P. (1966). In Tager, J. M., S. Papa, E. Quagliarello, and E. C. Slater, *Regutions of Metabolic Process in Mitochondria*, p. 65. Amsterdam.

Montroll, E. W. (1964). *Proc. Symp. Appl. Math. Am. Math. Soc.* **16**, 193.

Polya, G. (1921). Math Ann. **84**, 149. In this paper it is shown that the probability of a return to the origin of a random walker is 1, 1, <1 in 1, 2, 3 dimensions, respectively. Montroll (1964) showed that in the three-dimensional case this probability is 0.340537330.

Schneider, D., and K. E. Kaissling (1957). *Zool. Jahrb. Abt. f. Anat. u. Ontogenie d. Tiere* **75**, 287.

Schneider, D., and K. E. Kaissling (1957–1958). ibid. **76**, 223.

Tietjens, O., and L. Prandtl (1931). *Hydro- und Aeromechanik*, Band 2, Tafel 23, Abbildung 56.

Trurnit, H. J. (1945). *Fortschr. Chem. organ. Naturst.* **4**, 347.

ARTHUR B. PARDEE
Department of Biology
Princeton University
Princeton, New Jersey

Emphores

It was my good fortune to carry out my graduate studies in Dr. Pauling's group. Such an early exposure to a truly great scientist is, in my opinion, one of the most valuable parts of a young scientist's training. It sets standards and raises ideals toward which one can strive long thereafter. In the research itself, the specificity of antibodies' interactions with haptenic groups, and the kinetics of reaction of purified antibodies with defined haptens was investigated. A third unusual protein was introduced when in these studies we discovered that serum albumin has a strong affinity for haptenic dyes and thus can markedly influence hapten antigen-antibody reactions (Pardee and Pauling, 1949).

Many of Linus Pauling's biochemical investigations have centered on two unusual proteins, hemoglobins and antibodies. Antibodies, hemoglobins, and serum albumin are unusual proteins in that they do not fall into the usual classification of enzymes: they do not catalyze changes in chemical bonding. Nor are they structural proteins. Yet, as is well known, these proteins are extremely important biological entities. In this communication I wish to suggest that they can conveniently be classified as a group of macromolecules which specifically bind small molecules or submolecular groupings, and thereby carry the small molecules into biological activity. The name Emphore is proposed to designate this broad class of compounds.

REGULATORY SUBUNITS OF ENZYMES

Two classical examples of emphores, hemoglobin and antibodies, are discussed by other authors in this volume. We will illustrate the properties of emphores with examples that have arisen from the interests of my own laboratory. The first example is a subunit that modifies an enzyme, making it possible to regulate enzyme activity by end-product inhibition. The second is a protein that appears to play a role in the transport of sulfate through cell membranes.

The regulation of enzyme activities through activation or inhibition by small molecules is now a well-recognized and well-documented phenomenon (Atkinson, 1966). Enzymes can be modified by end products of metabolic pathways, a process commonly known as feedback or end-product inhibition. The important point is that many of these modifiers have no structural resemblance to the substrates of the enzyme, but rather they have a biological role which makes their modifying effects teleologically useful in conserving energy and metabolites.

It is now clear that these regulatory small molecules bind specifically to sites on protein subunits of the enzyme which are distinct from the catalytic sites of the enzyme (Gerhart and Pardee, 1962; Monod, Changeux, and Jacob, 1963). These have been conveniently named allosteric sites. The binding of small molecules causes quaternary structural changes in interactions of the subunits and possible tertiary changes in the internal structures of the subunits, resulting in altered enzyme activity. These changes are similar to the long-recognized influences of hemoglobin subunits upon one another as a consequence of oxygen binding (Monod, Wyman, and Changeux, 1965). With both the enzyme and hemoglobin, one subunit acts as an emphore which binds a small molecule specifically and thereby modifies another subunit; it carries a small molecule into biological action but does not catalyze a change in the molecule's structure, as would an enzyme.

In one case, that of aspartate transcarbamylase, a specific regulatory subunit has been isolated and its properties studied. This subunit binds cytidine triphosphate with a resultant inhibition of the holoenzyme activity (Gerhart and Schachman, 1965). The native enzyme is composed of four subunits, which catalyze the initial reaction of the pyrimidine pathway in *Escherichia coli*, and four regulatory subunits, which modify its activity. The regulatory subunits can combine reversibly with the catalytic subunits. These greatly modify the action of the catalytic subunits, especially in the presence of the inhibitor CTP or activator ATP. The only known function of these regulatory subunits is to cause this modification, the role of which is to regulate pyrimidine metabolism.

The regulatory subunit is a protein of approximately 25,000 molecular weight. It has a high affinity for a single CTP molecule, and no catalytic activity has been discovered. The isolation of this subunit conclusively proves the idea derived from kinetics that the regulatory binding sites are distinct from catalytic binding sites.

Other enzymes appear to be composed of several very similar subunits. Each enzyme subunit presumably contains both catalytic and allosteric sites. Combination of a small molecule with an allosteric site on one subunit modifies the catalytic activities of other subunits. Each subunit serves as a regulatory emphore for the others.

An example recently under study in our laboratory illustrates some of the properties of such an enzyme. The enzyme UTP-aminase carries out the conversion of uridine triphosphate to cytidine triphosphate (Chakraborty and Hurlbert, 1961). This enzyme has been purified approximately 350-fold in our laboratory, and its kinetics and some of its physical properties have been studied (Long, 1966). The results suggest that the enzyme has two subunits, each of which possesses the four catalytic sites required for the four substrates, UTP, ATP, GTP, and glutamine. The enzyme is strongly inhibited by the product, CTP, which appears to combine at the UTP site. The kinetic properties are much more complex than can be explained by action of a monomer. It is seen (Figure 1) that UTP follows a typical classical saturation curve when the other substrates are at high concentrations. However when ATP, another substrate, is at a low concentration the UTP saturation curve becomes dependent upon the third power of this substrate. A variety of other kinetic anomalies of this sort are resolved by the model indicated on the figure, which postulates that two such active subunits are present in each native enzyme molecule. One of them must be combined with the proper substrates of the

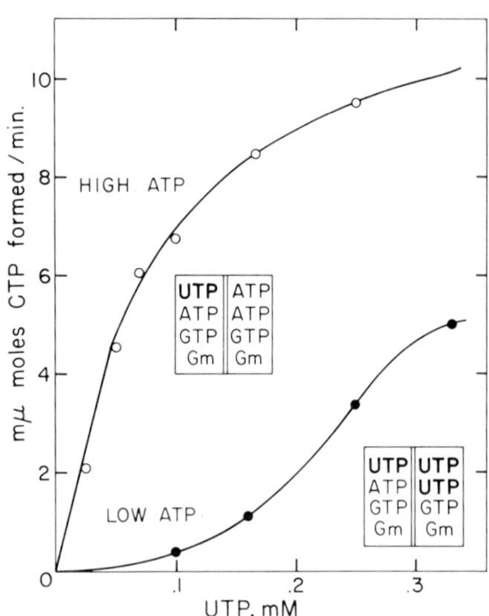

FIGURE 1.
Saturation of UTP aminase with UTP. The rate of production of CTP was measured when the purified enzyme was incubated with saturating concentrations of GTP and glutamine, either a nearly saturating or low concentration of ATP, and variable concentrations of UTP. The inserts indicate the postulated saturation patterns of the enzyme under the two sets of conditions.

reaction; the other must be saturated with a variety of molecules in order to activate the first. When the ATP concentration is high, this compound fills two sites on the second subunit. But when the ATP concentration is low, extra UTP molecules are required to fill these sites. Each subunit is an emphore for the other.

In the last few years, many cases have been noted of nonclassical kinetics, characterized by sinusoidal saturation curves and effects of various modifiers. General schemes to explain these results have been given mathematical formulations (Atkinson, 1966). No doubt the elucidations of the mechanisms of these enzymes in both physical and kinetic terms will occupy many workers for a long time. The point of great importance for biology is that the kind of molecule that modifies catalytic activity is independent of the catalytic site, and is determined only by a regulatory site with its own specificity. The existence of such specific regulatory sites, sometimes built onto special emphores, has far reaching biological importance in many systems.

TRANSPORT EMPHORES

Small molecules can be transported into living cells against concentration gradients. Two recent investigations point to a role of emphores in this active transport process. In one of these, an emphore is involved in the transport of sugars (Kundig, Kundig, Anderson, and Roseman, 1966). In the other, an emphore is involved in the transport of sulfate ions (Pardee, 1966).

In the first system, the transport of sugars is prevented when *E. coli* cells are osmotically shocked. These shocked cells release a special protein (Hpr) into solution. When a purified preparation of Hpr is added to the shocked cells, transport of sugars is sometimes restored. This suggests that Hpr is essential for active transport of sugars. Previous investigations (Kundig, Ghosh, and Roseman, 1964) show that an enzyme is capable of phosphorylating Hpr, and that another enzyme transfers phosphate from Hpr to a sugar. No affinity of Hpr for sugar has been reported. On the basis of these observations, a scheme

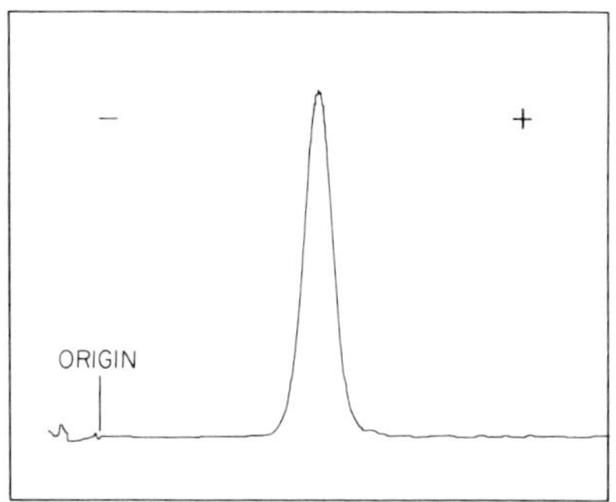

FIGURE 2.
Tracing of a gel electrophoresis run of the sulfate-binding protein. The protein was run on a Canalco gel electrophoresis apparatus, and was stained for protein. A tracing was then made on a Joyce-Loebl densitometer.

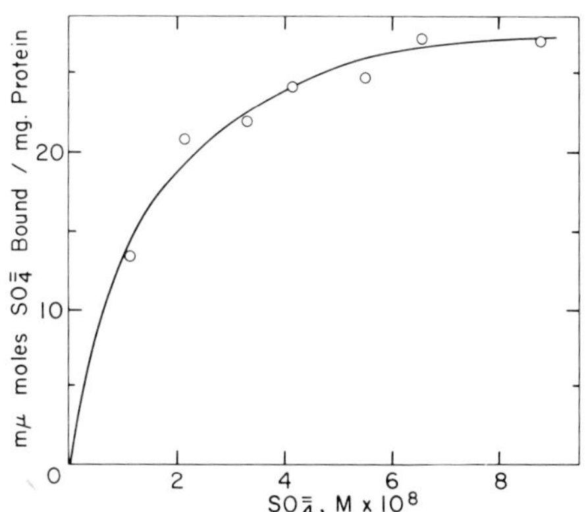

FIGURE 3.
Saturation curve of the sulfate-binding protein. The protein was equilibrated at various concentrations of sulfate in 0.02 *M* phosphate buffer, *p*H 7. Then the amount of bound sulfate was determined by the resin method (Pardee et al., 1966).

can be devised. Possibly Hpr carries high energy phosphate out of the cell where it phosphorylates a sugar. Then the sugar phosphate enters the cell and is dephosphorylated within.

In the second system, transport of sulfate into *Salmonella typhimurium* is prevented by conversion of the bacteria into spheroplasts or by osmotic shock (Pardee, Prestidge, Whipple, and Dreyfuss, 1966). A protein is released into the medium under these circumstances. It has been purified to homogeneity (Figure 2), and it has the following properties (Pardee, 1966). It is of typical amino acid composition and has a molecular weight of approximately 32,000. It binds one sulfate per protein molecule very firmly, and in a manner which suggests that the binding is ionic (Figure 3). No catalytic activity of this protein has yet been discovered. The protein is thus a sulfate-binding emphore.

The participation of this emphore in sulfate transport is deduced from a variety of correlations. Both transport and formation of emphore are repressible, in parallel. Both the emphore and the transport ability are lost upon osmotic shock or spheroplast formation. Both are lost in a class of mutants, and both are restored upon reversion or transduction. Both have similar specificities. The emphore restores transport to shocked cells.

One plausible scheme for the operation of the emphore in sulfate transport is shown in Figure 4. Neither sulfate nor the emphore (B) is postulated to penetrate the cell membrane. But B can carry sulfate through the membrane by diffusion. To be freed of sulfate, B is phosphorylated; it can then return to the exterior of the cell. Phosphate is provided by a high energy donor, either ATP or possibly phosphoenolpyruvate (Kundig et al., 1964). After the phosphorylated emphore diffuses to the exterior of the membrane, a special enzyme splits off phosphate. The emphore is then ready to repeat the transport cycle. The model accounts satisfactorily for a wide variety of properties of this transport system. It may be noted that the transport system appears to be

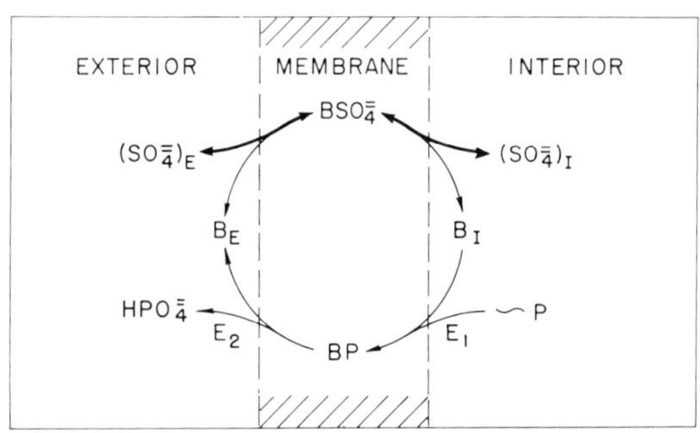

FIGURE 4.
A model for active sulfate transport.

subject to a negative feedback inhibition at high sulfate concentrations. This requires an additional portion of the model which will not be discussed here.

This sulfate-binding emphore would appear very likely to be a carrier, of the sort postulated for many years to serve in the transport of materials into living cells. However further studies will be required to justify the kind of model presented in Figure 4. The name permease was coined some years ago to describe active transport systems of bacteria (Kepes and Cohen, 1962). According to Monod (personal cummunication) this designation "permease" was intended to refer particularly to the specific binding part of a transport system. Thus, by this definition of permease, the emphore described here is apparently an isolated pure permease.

Recently another example of emphores has come to our attention. A factor in wheat germ lipase agglutinates cancer cells but not normal cells (Aub, Tieslau, and Lankester, 1963). This agglutinin has been purified to homogeneity (Burger and Goldberg, 1966). It is a protein of about 25,000 molecular weight with affinity for the N-acetyl-glucosamine group, and with no known enzymatic activity. It has the macromolecular structure and specificity of an emphore; its actual biological function in wheat germ is unknown.

EMPHORES

The work of my laboratory has been so frequently concerned with emphores such as those discussed above that I have been impressed with the desirability of considering these entities as a group. The word emphore means "carrier in," which is a satisfactory description of the biological action of many emphores. The definition, though, of emphore is that of a macromolecule with a specific affinity, but without catalytic activity.

We now can subclassify emphores in terms of their type of activity. These activities fall into four general classes.

The first class of emphores is typified by hemoglobin, whose principal function is to carry oxygen into the interior of the body and release it there. Other such emphores are siderophilin (transferrin) which carries iron through the blood stream, possibly ceruloplasmin which might similarly carry copper, serum albumin which binds and carries a variety of slightly soluble molecules, and alpha and beta globulins which carry lipids through the bloodstream. Other oxygen-carrying pigments occur in other organisms and may be classified under this first group. These include hemocyanins, erythrocrurins, and hemerythrins.

The second class of emphores carry small molecules through membranes. The sulfate-binding protein is an example of this class. The intrinsic factor is a carrier of vitamin B_{12} through the gut wall. Some protein hormones such as insulin and parathyroid hormone, may possibly be emphores, if these are indeed active in the transport of small molecules, glucose and phosphate respectively.

The third class of emphores have the biological function of holding or storing small molecules. One example of this type of emphore is myoglobin, which binds and stores oxygen in muscle. Another is ferritin, which stores iron in the liver. A third is thyroglobulin, which stores thyroxin in the thyroid gland.

The fourth class of emphores are those that bring small molecules into biological action. An example is the regulatory subunit of aspartate transcarbamylase, discussed above. Repressors of enzyme synthesis have been postulated to have properties which would classify them as emphores also (Monod et al., 1963). Antibodies can be classified in this grouping also, since they bind haptenic groups and give them the property of interaction with haptenic groups on other molecules. This makes precipitation and other immunological reactions possible.

Some emphores of this group bind small molecular groupings, generally in an activated form, and bring them into action as substrates of enzymes. Examples include the acyl carrier protein, which is able to provide the acyl group for fatty acid synthesis, ferridoxin, which is an electron transfer protein of low molecular weight, cytochrome C, which also functions as an electron carrier, and thioredoxin, which is active in deoxyribotide synthesis. One can also include here sRNA, which has been long recognized as an adaptor for bringing amino acids into action in protein synthesis.

The relation of these last emphores to coenzymes and enzymes is close. The line distinguishing the emphores from coenzymes can be drawn on the basis of size, although as with all definitions in biology, one soon finds that the delineation between classes is hard to make. Indeed, the most characteristic emphores, such as transferrin, have their analogues in low molecular weight compounds such as ferrichrome, which is an iron-binding hexapeptide.

In spite of the above difficulties, macromolecules with a definite specificity for small molecules or molecular groupings, and with important biological activities, seem worth classifying as a group. This classification should help us to recognize similarities in experimental approaches to the study of otherwise seemingly unrelated molecules.

REFERENCES

Atkinson, D. E. (1966). *Ann. Rev. Biochem.* **35**, 85.

Aub, J. C., C. Tieslau, and A. Lankester (1963). *Proc. Nat. Acad. Sci. U. S.* **50**, 613.

Burger, M. M., and A. R. Goldberg (1967). *Proc. Nat. Acad. Sci.* **57**, 359.

Chakraborty, K. P., and R. B. Hurlbert (1961). *Biochim. Biophys. Acta* **47**, 607.

Gerhart, J. C., and A. B. Pardee (1962). *J. Biol. Chem.* **237**, 891.

Gerhart, J. C., and H. K. Schachman (1965). *Biochemistry* **4**, 1054.

Kepes, A., and G. N. Cohen (1962). In Gunsalus, I. C., and R. Y. Stanier, eds., *The Bacteria*, vol. 4, p. 179. New York: Academic Press.
Kundig, W., S. Ghosh, and S. Roseman (1964). *Proc. Nat. Acad. Sci. U. S.* **52**, 1067.
Kundig, W., F. D. Kundig, B. Anderson, and S. Roseman (1966). *J. Biol. Chem.* **241**, 3243.
Long, C. W. (1966). Thesis. Princeton University.
Monod, J., J. P. Changeux, and F. Jacob (1963). *J. Mol. Biol.* **6**, 306.
Monod, J., J. Wyman, and J. P. Changeux (1965). *J. Mol. Biol.* **12**, 88.
Pardee, A. B., and L. Pauling (1949). *J. Am. Chem. Soc.* **71**, 143.
Pardee, A. B., L. S. Prestidge, M. B. Whipple, and J. Dreyfuss (1966). *J. Biol. Chem.* **241**, 3962.
Pardee, A. B. (1966). *J. Biol. Chem.* **241**, 5886.

ALEXANDER RICH
Department of Biology
Massachusetts Institute of Technology
Cambridge, Massachusetts

On the Assembly of Amino Acids into Proteins

The information for specifying the organization and metabolic activity of a living cell is stored in the sequence of nucleotides found in its DNA. The molecular expression of this information is found in specific proteins which act as enzymes, as structural proteins, or as regulatory proteins. It is the combination of the singular specificity and great heterogeneity of proteins that makes them uniquely capable of carrying out this wide variety of biological activities. At the present time we know a fair amount about the processes which connect the polynucleotide sequence of DNA with the amino acid sequence of proteins. This information has been obtained largely in the past ten or fifteen years, during which time molecular biology has developed as a coherent branch of science. The purpose of this article is to discuss various aspects of the mechanism of protein synthesis and to consider, in a somewhat speculative manner, some of the molecular details concerning the operation of this system. In this discussion, an attempt will be made to correlate molecular structure with function. This is particularly appropriate in the present context because one of the great contributions made by Linus Pauling to molecular science was his clear appreciation of the fact that a detailed understanding of the structure of molecules often leads to considerable insight into the way that they carry out chemical reactions.

THE ACTIVE COMPLEX IN PROTEIN SYNTHESIS

The major pathways underlying the assembly of amino acids into proteins are now understood although the detailed molecular mechanisms have not been determined in most cases.* Beginning at the N-terminal end, amino acids are assembled in an order which is determined by the sequence of codons, or triplets of nucleotides in a ribopolynucleotide strand called messenger RNA. One of the major accomplishments of the past five years has been the elucidation of this mechanism and the establishment of the genetic code which identifies the triplets of nucleotides with particular amino acids. The actual assembly process takes place through the use of a transfer RNA

*In a general article of this nature, it is impossible to cite adequate references for many statements. Excellent summaries of the current understanding of protein synthetic work are found in The Cold Spring Harbor Symposium, Vol. 31, (1966).

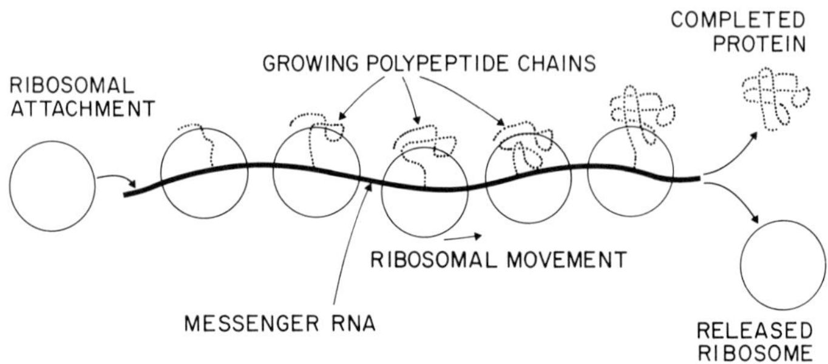

FIGURE 1.
A schematic diagram of polysome function.

molecule, a small polynucleotide containing approximately 75 residues which is believed to be folded together in such a way that three anticodon nucleotides are exposed so that they can combine with the codon triplet on the messenger RNA strand. The transfer RNA is activated for synthesis by forming a chemical link with an amino acid. Formation of the link is due to activating enzymes which apparently have the capacity both to detect a sequence of nucleotides on the transfer RNA and to recognize an individual amino acid. All transfer RNA's terminate in a common nucleotide sequence, cytosine-cytosine-adenine. The amino acid is added to the ribose of the terminal adenosine residue through an ester linkage.

Messenger RNA and the activated transfer RNA molecules condense together on the ribosome, a macromolecular organelle made up of two large fragments which have sedimentation constants of 50S and 30S in bacteria. A great deal about the ribosome is unknown. We have a general understanding of its role in protein synthesis but very little detailed knowledge of its operation and structure. Built of proteins and macromolecular RNA, ribosomes are approximately spherical with a diameter near 200 Å. However, the messenger RNA molecule which specifies the amino acid sequence of proteins is much longer than this. For example, in hemoglobin synthesis the polypeptide chains in the molecule contain about 150 amino acids, and 450 nucleotides are needed to specify such a sequence. If the nucleotides are stacked in their normal configuration with a spacing of 3.4 Å between the planar bases, this will yield a molecule with an overall length of 1500 Å. This is long enough to accommodate several ribosomes, and it is found that hemoglobin synthesis normally occurs on polyribosomal (polysomal) structures in which a single messenger RNA is associated with approximately 5 ribosomes (Warner, Hall, and Rich, 1962). Since the ribosomes are located on different points of the messenger strand, they are assembling different parts of the polypeptide chain as shown in Figure 1. At the left end of the messenger strand in Figure 1 the ribosome attaches, and protein synthesis is initiated at the N-terminal end of the polypeptide chain. Transfer RNA molecules are added to the polysomal structure, and the polypeptide chain elongates. This is associated with a gradual movement of the ribosome down the messenger strand. Each step in this process is believed to involve the reading of a triplet of nucleotides and a translation of approximately 10 Å along the messenger RNA. Electron micrographs of the polysomes active in hemoglobin synthesis are shown in Figures 2, *a* and 2, *b*. Using appropriate staining procedures, it is possible to visualize the messenger RNA strand. If polysomes are engaged in the synthesis of longer polypeptide chains, they are correspondingly larger. The electron micrograph in Figure 2, *c* shows a polysome containing 56 ribosomes that is synthesizing myosin, the structural protein of muscle whose subunits contain more than 1,800 amino acids.

MACROMOLECULAR FLUX AND THE RIBOSOME

Three different types of macromolecules pass through the ribosome. These are a flow of messenger RNA, a flow of transfer RNA molecules, and a flow of amino acids resulting in the production of a growing or nascent polypeptide chain. At the present time very little concerning the structure of the ribosome is known. Electron micrographs show some suggestions of substructure but further information has yet to be obtained that will allow us to describe its internal structure in detail. There are indications that the ribosome binds onto the ribose-phosphate backbone of the messenger RNA rather than on the nucleotide bases. Experiments with synthetic polyribonucleotides as well as with naturally occurring messenger RNA indicate that the 30S or smaller ribosomal subunit associates with the messenger RNA strand, while the transfer RNA and the nascent polypeptide chain is attached to the larger 50S subunit.

Ribosomes have a binding site for transfer RNA molecules (Cannon, Krug, and Gilbert, 1963). However, when ribosomes are actively engaged in protein synthesis in polysomes, they are found to contain two tRNA molecules firmly bound to them (Warner and Rich, 1964). It is known that the nascent or growing polypeptide chain is attached to a tRNA molecule (Gilbert, 1963); thus it is likely that the second tRNA molecule contains the amino acid which is about to be added. This had led to a commonly accepted model involving an amino acyl tRNA site (A) and a peptidyl tRNA site (P), as shown in Figure 3, *a*. The sequence of steps in protein synthesis involves an initial selection of the appropriate activated tRNA onto site (A), and then the transfer of the peptide chain (1) from the tRNA of site (P) to the amino acid on the tRNA of site (A). When the tRNA is removed from site P, the tRNA with the attached peptide chain moves from site A to site P (2), while the messenger RNA moves together with it (3). This may be a 10 Å translation if the three bases in the codon are stacked, or even up to a 20 Å translation if the messenger RNA strand is extended. The process is then ready for a cyclic repetition. It is clear that further experiments will be required to define this process in molecular terms rather than in the schematic terms used here.

Recent experiments involving a step-by-step formation of peptide bonds have strengthened this interpretation of the coordinate action of two tRNA sites (Bretscher and Marcker, 1966). This entire process is normally quite rapid (McQuillan, Roberts, and Britten, 1959). It takes approximately 10 seconds in bacteria to assemble a protein containing 500 to 1,000 amino acids; thus the time required for one step of the cycle is 10 to 20 milliseconds.

Polysomes are found attached to the endoplasmic reticulum in many cells

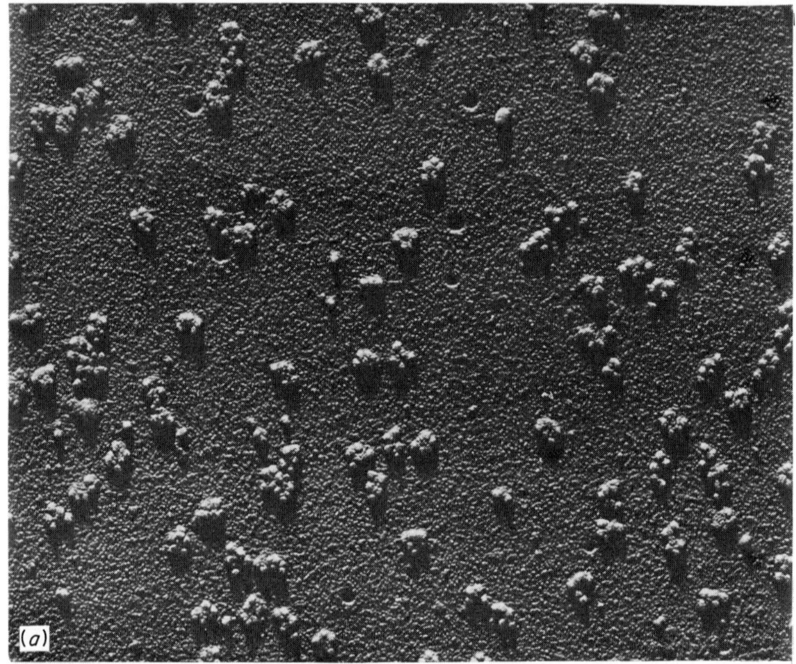

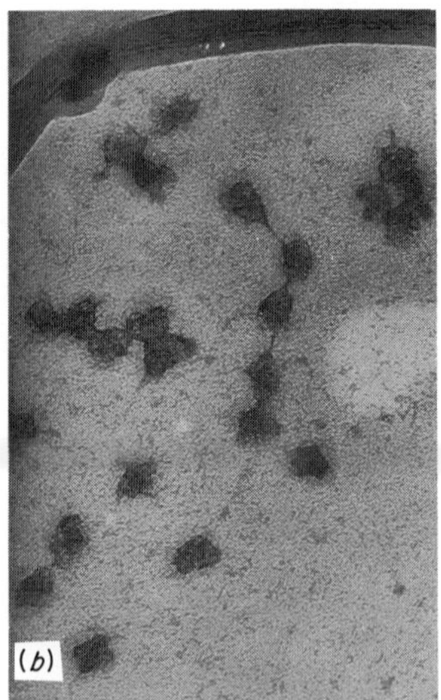

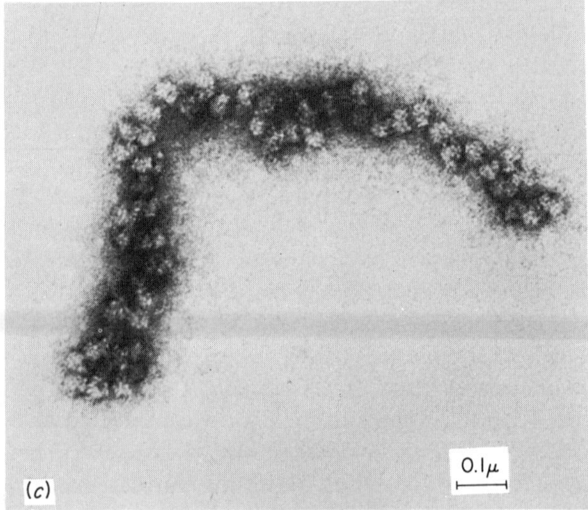

FIGURE 2.
Electron micrograph of polyribosomes. (*a*) and (*b*) Polysomes obtained from reticulocytes synthesizing hemoglobin. The most frequently observed species is a ribosomal pentamer in a somewhat clustered configuration. In (*a*) the polysomes are shadowed with platinum; (*b*) shows the polysomes stained with uranyl acetate which causes RNA to look black. A thin messenger RNA strand is visualized. (Slayter, Warner, Rich, and Hall, 1963). (*c*) Electron micrograph of a large polysome from embryonic chick thigh muscles. The polysome contains 56 ribosomes in a somewhat coiled configuration, and it has been identified as active in myosin synthesis (Heywood, Dowben, and Rich, 1967).

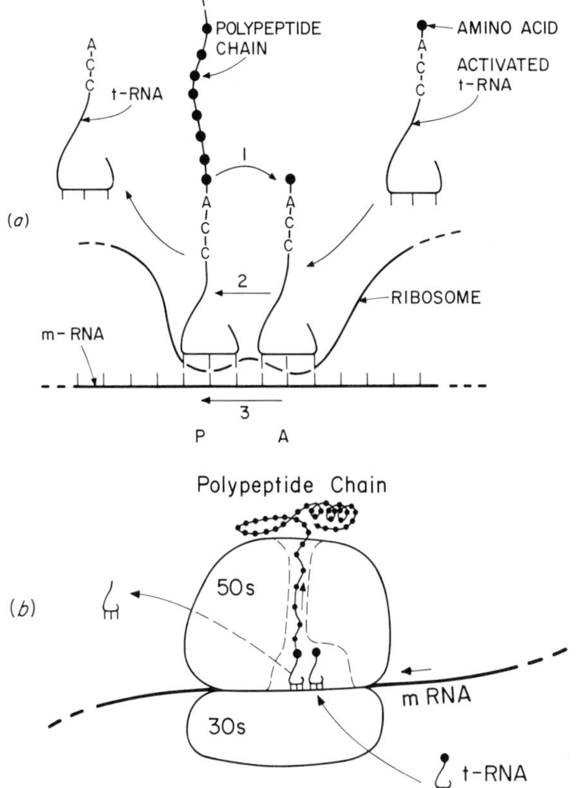

FIGURE 3.
Diagrammatic representation of steps in protein synthesis. (*a*) The mechanism for peptide bond formation involving the use of two transfer RNA sites on the ribosomes. (*b*) A schematic diagram of the ribosome illustrating the three types of macromolecular flow passing through the ribosome. A hypothetical channel is illustrated as an exit guide for the growing polypeptide chain.

and the protein products produced by the polysomes are secreted into the lumen of the endoplasmic reticulum where they are often assembled for export to the outside of the cell. This suggests that the ribosome has a specific exit through which the polypeptide chain emerges. It is not unreasonable to believe that the exit for the emerging polypeptide chain is not the same as the entrance or exit for the messenger RNA strand along which the ribosome is moving, because the growth of a large and often bulky polypeptide chain would be likely to impair messenger RNA movement. Some polypeptide chains are extremely large. For example, the subunit of the enzyme β-galactosidase contains more than 1,100 amino acids and there is good evidence suggesting that the nascent polypeptide chain folds into a globular configuration while it is still attached to the ribosome. Polysomal-bound β-galactosidase activity probably arises from the binding of completed subunits to the folded nascent chain, since only the tetrameric molecule has enzymatic activity (Lederberg, Rotman, and Lederberg, 1964). The ribosomal-bound enzyme has a diameter of almost 100 Å, or roughly half the diameter of the ribosome itself. Thus, it is likely that this molecule would not be found on the ribosome near the entrance or exit for messenger RNA nor near the entrance or exit sites for the flow of transfer RNA molecules.

We may speculate that the three types of macromolecular flux passing through the ribosome are quite distinct from each other in a geometrical sense. Although there is no good evidence for this, it is generally assumed that the messenger RNA lies in the crevice between the two ribosomal subunits.

Perhaps the reason for assuming this is the fact that the crevice itself may then afford a large access region which may be needed for the selection of the appropriate activated transfer RNA. Because the nascent polypeptide chain can be large, it is probably important that it be well removed. This might be brought about by having the peptide chain pass through the 50S subunit, perhaps in a channel. There is no direct evidence regarding the existence of such a channel, but indirect evidence may be found in the fact that a substantial portion of the proximal nascent polypeptide chain is resistant to digestion by externally added proteolytic enzymes (Malkin and Rich, 1967). The resistant fragment contains 30 to 35 amino acids and if the polypeptide chain is in an extended configuration, this would represent a distance of approximately 100 Å. The proteolytic resistant polypeptide chain is the part attached to the tRNA and it may thus be shielded in a channel passing through the 50S ribosomal subunit, as shown in Figure 3, *b*. It should be possible to detect the existence of such a channel in this subunit by careful electron microscopic experiments involving various staining techniques.

The complete elucidation of the molecular machinery for carrying out these operations will probably have to await either the crystallization of the intact 70S ribosomal structure or, more likely, the crystallization of separate 50S or 30S subunits. X-ray analysis of these organelles will provide us with a molecular structure framework which can serve as a basis for interpreting the details of ribosome function. It may also shed some light on the great mystery of the biological role of the ribosomal RNA. This RNA makes up more than half of the mass of the ribosome and it accounts for more than 80 percent of the total RNA of the cell. At the present time, we have no understanding at all of its biological or structural role.

RIBOSE AND DEOXYRIBOSE IN THE NUCLEIC ACIDS

Two kinds of nucleic acids are used in biological systems, those containing ribose and those containing deoxyribose. The differences in their biological role is reasonably well understood. DNA is the major repository of genetic information and it serves as a template for manufacturing the RNA molecules that are involved in the synthesis of proteins. The nucleotide sequence of RNA is based upon sequences in DNA. However, because different sugar residues are used in forming the two polymers, it is therefore possible for a cell to destroy RNA but still not lose the fundamental information required to regenerate the system as long as the DNA remains intact. However, RNA also has the capacity to carry genetic information and to serve as the basis for a system of genetic replication. This is evident in the RNA viruses, which have the capacity to infect a cell, to direct the synthesis of protein, and at the same time to carry out the self replication required to produce more progeny. These observations have served as the basis for the speculation that RNA may have preceded DNA in the course of evolution (Rich, 1959). According to this view, DNA may be a more recently created nucleic acid specializing solely in the reproduction of genetic information. Adding a different sugar in the polymer backbone has allowed the two types of nucleic acids to follow different metabolic pathways.

However, it is pertinent to ask whether the choice of the two sugars reflects different physiological functions. Why is it that RNA can participate in protein synthesis, and DNA cannot? When the structure of DNA was first formulated by Watson and Crick (1953), they suggested that the absence of the hydroxyl group at the C2′ position of deoxyribose may make it possible for DNA to form a double helical complementary molecule whereas RNA would be unable to. However, the discovery of complementary double helical RNA made from synthetic polynucleotides as well as from naturally occurring polynucleotides clearly showed that both types of molecules were able to form a double helical structure. This is presumably the structural basis of RNA replication. We might ask if there is some function in protein synthesis which requires the use of the hydroxyl group on the C2′ of ribose. It has been shown that it is not necessary in messenger RNA. Single-stranded DNA can serve as a messenger RNA in protein synthesis providing the ribosomes are perturbed slightly through the addition of an antibiotic such as neomycin (Holland and McCarthy, 1965). Is it possible that transfer RNA specifically uses the C2′ hydroxyl group? Most of the nucleotides in transfer RNA are involved in forming double helical structures which are similar to those found in DNA. However, a unique event occurs in the attachment of the amino acid to the terminal adenosine of transfer RNA. The carboxyl group of the amino acid is attached to either the 2′ or the 3′ hydroxyl group of the ribose through an ester linkage. It has been shown that there is a very rapid transfer of the acylating amino acid between the 2′ and 3′ positions, both in transfer RNA and in chemically synthesized nucleotides (Griffin, Jarman, Reese, Sulston, and Trentham, 1966). However, is it possible that the presence of two *cis* hydroxyl groups on the same side of the furanose ring may play an important role in peptide bond formation? The usual description of peptide bond formation is as follows: the two amino acyl transfer RNA's are believed to be side by side on the ribosome. Peptide bond formation is believed to involve the transfer of the peptidyl group from the adenosine residue of one transfer RNA (P site, Figure 3, *a*) directly to the amino group of the amino acid on the adjacent amino acyl tRNA (A site). Thus, an ester bond on one tRNA has been broken, and a peptide bond formed on another tRNA. However, we can suggest an alternative formulation. This process could involve the formation of an intermediate involving both hydroxyl groups on adjacent C2′ and C3′ atoms. The peptidyl group could be transferred by transesterification initially to the hydroxyl on C3′ (for example), while the amino acid is still esterified on the hydroxyl at C2′. This intermediate is illustrated diagrammatically in Figure 4. The next step would involve a nucleophilic attack by the nitrogen atom of the amino acid on the carbon atom of the carbonyl group of the peptidyl ester bond. This nucleophilic attack would be accomplished through formation of a cyclic intermediate which would have as its product the formation of a peptide bond and the elongation of the peptidyl chain by one residue. Thus peptide bond formation might utilize the two *cis* hydroxyl groups to form a diester intermediate, which would then react to form a peptide link.

This proposal can be examined in several ways. The stereochemistry shown in Figure 4 is quite reasonable as can be shown with molecular models. Furthermore, intramolecular reactions involving cyclic intermediates are not uncommon. However, an experimental demonstration of this mechanism would involve the preparation of 2′, 3′ diesters of nucleic acid derivatives and the demonstration of their participation in peptide bond formation on ribosomes. Demonstration of such a mechanism would uniquely utilize the ribose residue in peptide bond formation and would thereby explain part of its participation in protein synthesis. Investigations of this mechanism are being actively pursued at the present time in our laboratories by V. Shashona and J. Sheehan.

THE INACTIVATION OF MESSENGER RNA

We are beginning to understand the mechanisms which initiate and terminate the synthesis of polypeptide chains. It has been shown that in *Escherichia coli* the nucleotide triplet AUG codes for N-formyl-methionine transfer-RNA, which appears to be the initiating amino acid for all the proteins of *E. coli* (Capecchi, 1966). This codon is believed to be at the 5′ OH end of the structural part of messenger RNA. Other N-terminal amino acids are produced by cleaving off this amino acid or other amino acids further along the polypeptide chain. Information concerning initiation in other species is not available at the present time. The existence of polypeptide chain terminating nucleotide triplets UAG and UAA have been established in some mutants in *E. coli* (Sarabhai, Stretton, Brenner, and Bolle, 1964). It has been inferred that a modification of this system may actually be used in normal polypeptide chain termination. The important point is that there are definite start and stop symbols on polynucleotide strands which signal the initiation and termination of protein synthesis. These are illustrated diagrammatically in Figure 5, *a* as initiator and terminator blocks.

It is commonly assumed that under normal physiological conditions ribosomes will engage messenger RNA only at an initiator site and nowhere else along the polynucleotide chain. Indeed, if initiation were carried out at random along the messenger RNA, this would result in a series of fragmented polypeptide chains which would have no function in the cell. This assumption

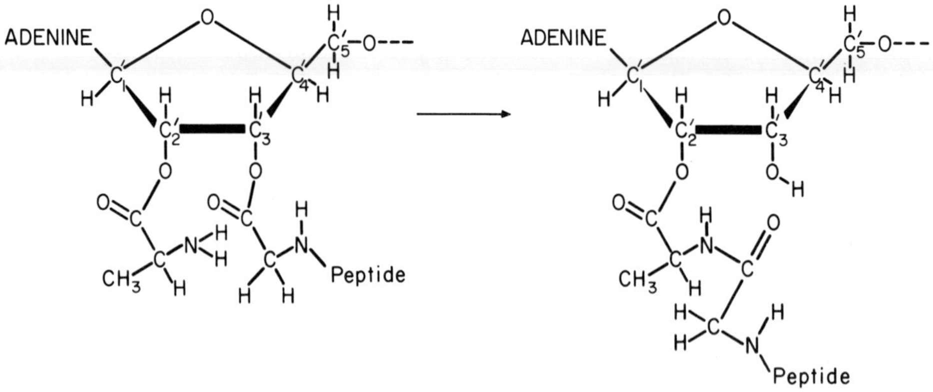

FIGURE 4.
A schematic diagram illustrating a possible intermediate in the formation of the peptide linkage. The ribose residue at the left has an alanine attached to the 2′ hydroxyl group and a peptide on the 3′ hydroxyl group. This intermediate reacts to form the linear peptide shown attached to the 2′ hydroxyl site. An alternative formulation of the mechanism could be drawn with the 2′ and 3′ substituents exchanged.

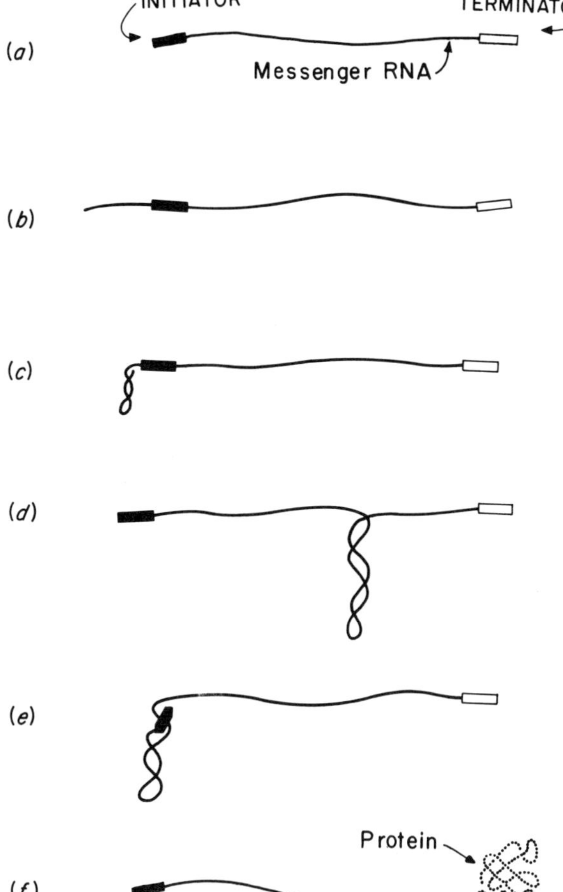

FIGURE 5.
A schematic diagram of various modifications of a single cistron of messenger RNA. (*a*) Parts labeled initiator and terminator indicate the positions where the ribosomes start and finish the synthesis of the single polypeptide chain. (*b*) A small polynucleotide segment is attached on the initiator end of the messenger. (*c*) The attached polynucleotide chain on the initiator end has formed a double helical segment. (*d*) The messenger strand has a large double helical segment in the middle of it, owing to an autocomplementary selection of codons. (*e*) A double helical segment in the messenger RNA involves the initiator codon. (*f*) The messenger RNA has a small polynucleotide chain attached beyond the terminator of the messenger RNA. This can act as a component of a specific regulatory system by binding a completed protein subunit (or molecule). This could even bind the protein synthesized by the cistron. In this position it might block the passage of ribosomes over the terminator portion of the messenger.

of a unique initiator site may be of importance in our thinking about the mechanism of messenger RNA inactivation. It is probable that messenger RNA is inactivated enzymatically by an exonucleolytic attack at the 5′ hydroxyl end of the messenger RNA chain where initiation occurs, because endonuclease action in the center of messenger RNA would result in the production of truncated, unusable polypeptide chains. In addition, since the fragmented messenger RNA would not have termination signals, there would be a gradual inactivation of ribosomes which were unable to disengage from a fragment of a messenger strand. On the other hand, inactivation of messenger RNA by exonucleolytic attack on the initiator end of the messenger strand would bring about a more orderly cessation of protein synthesis. No additional ribosomes would engage the messenger to synthesize protein, but the ribosomes already on the mesesnger strand would complete the synthesis of normal protein. Considerable effort is being directed at the present time toward the isolation of 5′ exonucleases which may serve in this capacity of destroying messenger RNA.

Very little is known at the present time about the detailed interactions which determine the half-life of messenger RNA. For example, in bacterial cells it is known that the average half-life is a few minutes, whereas it is considerably longer in mammalian cells. However, in some cells, such as the reticulocyte, it is apparent that the messenger RNA is very long lived since that cell persists in synthesizing hemoglobin in the absence of a nucleus and in the absence of RNA synthesis. Thus it is not unreasonable to believe that there is a system for regulating the life time of messenger RNA and, thereby, the synthesis of individual proteins.

If we assume the orderly destruction of messenger RNA from the initiator end, we then have an obvious mechanism for modifying the half-life of messenger RNA. This is most easily brought about by the incorporation of additional nucleotides at the 5′ OH end of the messenger RNA strand. These nucleotides are not structural codons to be translated into amino acid sequences but rather represent a type of polynucleotide buffer segment which would serve as a substrate for the messenger inactivating exonuclease (Figure 5, *b*). The length of this buffer nucleotide strand or its composition could determine the rate at which the exonuclease would work through this end of the molecule before it destroyed the initiator sequence. Furthermore, it would then be possible to explain the existence of very long-lived messengers, since these might be obtained by forming an autocomplementary segment in the buffer polynucleotide attached to the end of the messenger strand (Figure 5, *c*). The messenger-destroying nucleases may have a property common to many nucleases in that the rate of hydrolyzing single-stranded polynucleotide chains is much higher than that of hydrolyzing double-stranded segments. Thus, the autocomplementary folding at this end of the messenger would insure the existence of this messenger for a long time period. The experimental resolution of these issues will be obtained when we have purified messenger RNAs and have determined the polynucleotide sequences at the 5′ OH end.

SECONDARY STRUCTURE IN MESSENGER RNA

One of the striking characteristics of the nucleic acids is their capacity to form double-stranded helical structures. This is seen widely in DNA, in RNA, and in the synthetic polynucleotides. It arises from the tendency of the purines and pyrimidines to form specific hydrogen-bonded pairs, leading to helical structures which are geometrically regular and have great stability. In addition to the double helical form of DNA, considerable secondary or helical structure is found in ribosomal RNA as well as in the transfer RNA molecule. Let us consider the consequence of secondary structure in messenger RNA. Suppose we have a messenger RNA which is coding for a particular sequence of 9 amino acids as shown in Figure 6, *a*. From our knowledge of the genetic code, we can say that the messenger RNA for this polypeptide fragment may have the sequence shown in Figure 6, *b*. However, it can be seen that this sequence is capable of forming a double helical segment in which the messenger strand is held together by 12 or 13 pairs of hydrogen-bonded nucleotides. We can call this autocomplementarity, implying the ability of a messenger RNA to fold back upon itself to form large self-complementary segments with double-stranded helical structure and with the (*a*) ···—leu—ser—val—tyr—ala—val—asp—arg—lys—···

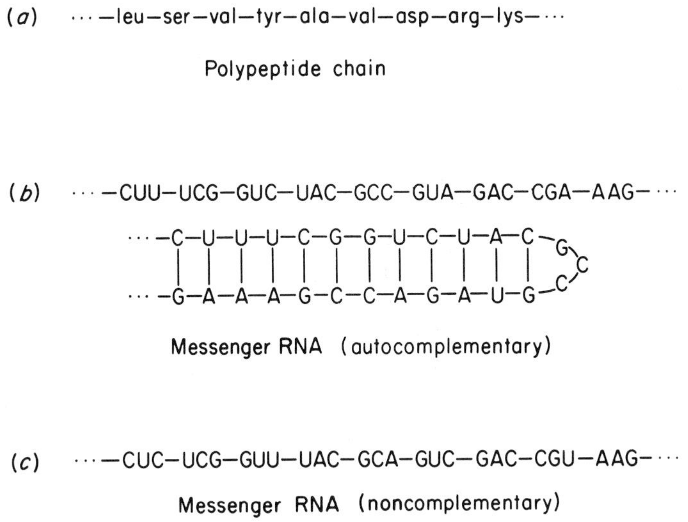

Polypeptide chain

(*b*) ···—CUU—UCG—GUC—UAC—GCC—GUA—GAC—CGA—AAG—···

```
···—C—U—U—U—C—G—G—U—C—U—A—C—G
    |  |  |  |  |  |  |  |  |  |  |  |    C
···—G—A—A—A—G—C—C—A—G—A—U—G—C
```

Messenger RNA (autocomplementary)

(*c*) ···—CUC—UCG—GUU—UAC—GCA—GUC—GAC—CGU—AAG—···

Messenger RNA (noncomplementary)

FIGURE 6.
(*a*) A segment of a hypothetical polypeptide chain containing 9 amino acids. (*b*) The sequence of codons which might be used to synthesize the polypeptide chain in (*a*). This particular set of codons is able to form a large autocomplementary double-stranded helical segment in the messenger RNA. (*c*) An alternative selection of codons which also codes for the polypeptide chain shown in (*a*). This segment is unable to form a large autocomplementary structure.

chains oriented in opposite directions. This is illustrated schematically in Figure 5, *d* in which the double-stranded region is shown in the middle of the messenger RNA. What would be the effect of having a sequence such as this in the messenger RNA? Small autocomplementary segments would have limited stability and it is quite likely that the ribosomes moving along the messenger strand would force the two strands apart without impairment of protein synthesis. However, if this double helical segment were fairly long and if it had a high percentage of guanine and cytosine pairs leading to a higher stability, it is possible that the ribosome would be unable to translate this segment of messenger. An even more extreme case of autocomplementarity is shown in Figure 5, *e* in which the double helical segment includes the initiator component of the messenger RNA. In that case the ribosome would be unable to engage the initiating end of the messenger and translation would be blocked.

One of the characteristics of the amino acid sequences found in proteins is that there appear to be no constraints in terms of allowed sequences; thus they may even include such polypeptides as those in Figure 6, *a*. This suggests that there may be a mechanism for avoiding messenger autocomplementarity. An interesting feature of the genetic code is the fact that there appears to be ambiguity in the third base of the codon triplet. Although the code has sixty-four triplets, there are, in effect, half this number of independent triplets, since in almost all cases the third base is specified either by a purine (adenine or guanine) or by a pyrimidine (cytosine or uracil). This has been interpreted as indicating that in the ribosome the third base does not form the same types of complementary hydrogen bonds as in DNA (Crick, 1966).

However, we suggest that an important functional reason behind this ambiguity is that it allows an element of choice in the selection of the third base such that it can prevent the formation of significant segments of autocomplementary messenger RNA. Thus, for example, the nucleotide sequence in Figure 6, *c* is not able to form significant autocomplementary segments, even though it codes for the same amino acid sequence in Figure 6, *a*. In this form it could not block ribosomal translation owing to significant secondary structure. This process may have been an important component of the selection pressure in the evolution of the particular codon sequences used in messenger RNA.

It is possible to subject this idea to experimental analysis. Synthetic messenger RNAs with autocomplementary segments can be made to show the inhibition. Furthermore, one can take the known amino acid sequence of proteins and write down the possible codons which stand for these amino acids. With the aid of a computer, a program can be made to select those sequences which have significant autocomplementary segments. If the considerations described above have been an important component of evolutionary selection pressure, then we predict that the actual codons used by the natural messenger will differ substantially from the codon sequences which have autocomplementarity. In short, the determination of the nucleotide sequence in naturally occurring messenger RNAs should allow us to decide whether or not this has been a basis for codon selection.

REGULATION OF PROTEIN SYNTHESIS

At the present time the evidence is fairly convincing that the major system for regulating protein synthesis acts at the level of messenger RNA production on DNA. The control system regulating enzymatic induction in microorganisms has been described in detail by Jacob and Monod (1961). The important part of this system is the fact that it operates through negative control in that the presence of a repressor molecule on DNA is used to prevent the synthesis of RNA. It is only through a modification of the environment that the repressor is removed and the synthesis of the RNA ensues. It is not known to what extent this system operates in higher organisms, but there are some indications that analogous regulatory events are occurring in these more complicated systems.

There have been some suggestions that a regulatory system, perhaps a minor component, may also be operating at the level of translation. This is believed to occur by altering the extent to which a particular messenger RNA is active at the polysome level.

One potential mechanism for controlling protein synthesis at the polysome level involves the interaction of proteins, some of which may be nascent chains still attached to ribosomes (Cline and Bock, 1966). For example, it has been shown that the enzyme β-galactosidase is enzymatically active while it is bound to polysomes (Kiho and Rich, 1964). The molecule is a tetramer, even on the ribosome (Lederberg, Rotman, and Lederberg, 1964). It is believed that completed β-galactosidase subunits associate in this case with a nascent or incomplete β-galactosidase polypeptide chain on the ribosomal surface to give rise to active enzyme. The hemoglobin molecule contains 2 α and 2β subunits. Polysomes synthesizing hemoglobin are found to have some completed α-polypeptide chains which may require β chains for their release (Columbo and Baglioni, 1965). Such a mechanism would provide a basis for a stoichiometric production of α and β chains. A further example of this phenomenon is seen in the synthesis of gamma globulin molecules which are found on two classes of polysomes active in the synthesis of heavy and light chains (Becker and Rich, 1966). The heavier polysomes, which synthesize heavy chains, are found to have completed light chains attached to them (Shapiro, Scharf, Maizel, and Uhr, 1966); this has also been cited as a possible regulatory mechanism in that the detachment of the heavy chain may require the presence of the light chain. These are both examples in which completed subunits are believed to associate with incomplete polypeptide chains on the ribosomal surface and this interaction may result in a regulatory role.

The physical basis for this type of mechanism in translational control is related to changes in the conformation of a nascent polypeptide chain when it is associating with another subunit. For example, one may postulate that a particular conformational alteration in the nascent chain may make it impossible for the ribosome to continue traveling down the messenger RNA, thus inhibiting the system. However, other subunits combining with the nascent chains might change the conformation and allow synthesis to continue. In addition, another type of interaction could be postulated for those proteins which contain a prosthetic group, such as heme. There is evidence at the present time that indicates a close coupling between heme synthesis and globin synthesis. In this case, the heme unit might confer upon the nascent polypeptide chain a conformational change which allows the chain to be completed.

In the above we have considered only regulatory mechanisms which depend upon protein interactions. However, there is an additional type of potential regulatory mechanism. Proteins play an important role in regulating the expression of the information found in the nucleic acids. The most striking example of this is in the action of the protein repressor molecule (Gilbert and Müller-Hill, 1966) which has an affinity for a specific site on DNA, presumably a specific nucleotide sequence. When the repressor molecule is bound to the DNA, the synthesis of messenger RNA is inhibited. Another protein-nucleic acid interaction which may be important for regulation is the interaction between the transfer RNA molecule and its activating enzyme. Here we wonder whether it is possible that another interaction between polynucleotides and proteins functions in translational control. Specifically, is it possible that a short polynucleotide segment is appended to messenger RNA, perhaps beyond the terminator sequence, that has a role in regulating translation? If this hypothetical polynucleotide had the capacity to interact specifically with a particular completed polypeptide chain, this could act as a regulatory system by blocking the movement of the ribosome over the terminator sequence (Figure 5,*f*). This mechanism would represent a system of negative feedback. Accumulation of the gene product would automatically turn off the production of this gene, whereas utilization of the gene product would turn on the synthesis of the gene. We are postulating here a system of specialized translation repressors somewhat analogous to the transcription repressors acting at the level of DNA.

The types of interactions described here are similar in character to the interaction which is believed to occur when a repressor molecule inactivates a particular stretch of DNA by preventing the DNA from acting as a template for the synthesis of messenger RNA. Here the messenger strand is inactivated, but it can be reactivated by any process which brings about the removal of the blocking protein unit. It is clear that such a system could also be used as the basis of a mechanism for insuring stoichiometric production of different subunits in a protein.

To a certain extent, RNA viruses may be considered as examples of systems which exhibit translational repression. One of the cistrons found in the viral RNA directs the synthesis of the protein subunit which forms the outer coat of the virus. During the process of viral infection, protein subunits of the viral coat increase in concentration within the cell and eventually build up to form a three-dimensional structure containing the viral RNA. For the bacterial virus MS2, it has already been demonstrated that the addition of coat protein selectively turns off the synthesis of some of the other cistrons which are coded for by the viral strand (Sugiyama and Nakada, 1967). The basis for the inhibition may involve these proposed translational repressors, which may also exist in normal synthetic systems independent of viral infection.

In these last sections we have been describing some hypothetical systems for regulating both the half-life of messenger RNA and the ways of modifying the rate at which translation occurs. It is conceivable that nature employs some of these mechanisms. However, it is unlikely that such mechanisms occur in a consistent fashion in all cases of protein synthesis. It may be more reasonable to believe that nature has evolved a variety of subsidiary methods for influencing the conversion of polynucleotide sequence information into protein structure. Different mechanisms may be utilized for different proteins, and only continued experimentation in a variety of protein synthetic systems will allow us to appreciate the scope of such potential mechanisms. The major reason for presenting speculations such as these is that they can serve as concrete models for designing experiments.

REFERENCES

Becker, M., and A. Rich (1966). *Nature* **212**, 142.

Bretscher, M. S., and K. A. Marcker (1966). *Nature* **211**, 380.

Cannon, M. R., R. Krug, and W. Gilbert (1963). *J. Mol. Biol.* **7**, 360.

Capecchi, M. (1966). *Proc. Nat. Acad. Sci. U.S.* **55**, 1517.

Cline, A. L., and R. M. Bock (1966). *Cold Spring Harbor Symp. Quant. Biol.* **31**.

Columbo, B., and C. Baglioni (1965). *J. Mol. Biol.* **16**, 51.

Crick, F. H. C. (1966). *J. Mol. Biol.* **19**, 548.

Gilbert, W. (1963). *J. Mol. Biol.* **6**, 389.

Gilbert, W., and B. Müller-Hill (1966). *Proc. Nat. Acad. Sci. U.S.* **56**, 1891.

Griffin, B. E., M. Jarman, C. B. Reese, J. E. Sulston, and D. R. Trentham (1966). *Biochem.* **5**, 3638.

Heywood, S. M., R. M. Dowben, and A. Rich (1967), *Proc. Nat. Acad Sci. U.S.* **57**, 1002.

Holland, J. J., and B. J. McCarthy (1965). *Proc. Nat. Acad. Sci. U.S.* **54**, 880.

Jacob, F., and J. Monod (1961). *Cold Spring Harbor Symp. Quant. Biol.* **26**, 193.

Kiho, Y., and A. Rich (1964). *Proc. Nat. Acad. Sci. U.S.* **51**, 111.

Lederberg, S., B. Rotman, and V. Lederberg (1964). *J. Biol. Chem.* **239**, 54.
Malkin, L. M., and A. Rich (1967). *J. Mol. Biol.* **26**, 329.
McQuillan, K., R. B. Roberts, and R. J. Britten (1959). *Proc. Nat. Acad. Sci. U.S.* **45**, 1437.
Rich, A. (1959). Ann. *N. Y. Acad. Sci.* **81**, 709.
Sarabhai, A. S., A. O. W. Stretton, S. Brenner, and A. Bolle (1964). *Nature* **201**, 13.
Shapiro, A. L., M. D. Scharf, J. V. Maizel, and J. W. Uhr (1966). *Proc. Nat. Acad. Sci. U.S.* **56**, 216.
Slayter, H. S., J. R. Warner, A. Rich, and C. E. Hall (1963). *J. Mol. Biol.* **7**, 652.
Sugiyama, T., and D. Nakada (1967). *Proc. Nat. Acad. Sci.* **57**, 1744.
Warner, J. R., C. E. Hall, and A. Rich (1962). *Science* **138**, 1399.
Warner, J. R., and A. Rich (1964). *Proc. Nat. Acad. Sci.* **51**, 1134.
Watson, J. D., and F. H. C. Crick (1953). *Nature* **171**, 964.

W. A. SCHROEDER
W. R. HOLMQUIST
Division of Chemistry and Chemical Engineering
California Institute of Technology
Pasadena, California

A Function for Hemoglobin A_{Ic}?

Hemoglobin A_{Ic} is one of the minor hemoglobin components of the adult erythrocyte. Among the known structures of the several minor components in the adult and fetal erythrocytes, it alone has a particularly unique and interesting structure. For this reason, we have chosen to suggest several possible functions, other than the usual one of oxygen transport, for hemoglobin A_{Ic}. First, however, we shall consider the history and structure of the minor components in general as well as certain facts about the erythrocyte.

MINOR HEMOGLOBINS IN GENERAL

It is already more than a decade since Kunkel and Wallenius (1955) and Morrison and Cook (1955) showed conclusively that hemoglobin from the normal human adult is heterogeneous. The major component, hemoglobin A, which comprises 80–90 percent of the total, is accompanied by several minor components in different amounts. The hemoglobin in the umbilical cord blood of a normal newborn infant is also heterogeneous, not only because of the presence of some adult hemoglobin, but also because of other components. Minor components also accompany the abnormal hemoglobins S and C, which are found in some individuals (Jones, 1961). Whether these are identical to or analogous in structure to the minor components of normal individuals is not known. Although less extensively studied, the hemoglobins of animals are also known to contain minor components (Huisman et al., 1960; Muller, 1961).

Almost all of the heterogeneity in cord blood hemoglobin (other than adult hemoglobin) is attributable to a minor component designated hemoglobin F_I (Allen et al., 1958), which is present to the extent of about 10 percent of the total hemoglobin. Hemoglobin F_I has the amino acid composition, the ultraviolet spectrum, and the stability to denaturation by alkali that is shown by the major fetal hemoglobin F. Nevertheless, there is a distinct chemical difference between the two: an acetyl group on the *N*-terminal glycyl residue of one of the two γ chains distinguishes F_I from F.

One of the human adult minor components, termed hemoglobin A_2, is well separated electrophoretically from the major component, is easily deter-

Contribution No. 3428 from the Gates and Crellin Laboratories of Chemistry. This investigation was supported in part by a grant (HE-02558) from the National Institutes of Health.

mined quantitatively, and has attracted wide attention because of its possible implication in thalassemia. Hemoglobin A_2 has the subunit structure $\alpha_2\delta_2$: it is like normal hemoglobin A, $\alpha_2\beta_2$, in its pair of α chains but the δ chain differs in sequence in 10 places from the sequence of the β chain (see Schroeder and Jones, 1965, for a review). The content of hemoglobin A_2 in the normal individual is 2.5–3.5 percent.

Another minor component(s), A_3, is also electrophoretically apparent in normal adult hemoglobin because of the asymmetry of the spot of the major component. Some of the observations of Kunkel et al. (1957) suggested that this component(s) was the product of aging of the major component. Evidence to support this view came from Muller (1961), who reported that the hemoglobin in this component was combined with glutathione. There appear to be at least five other minor components in addition to A_2 and A_3. There has been a tendency to dismiss these, without real evidence, as transformation or degradation products of the major component.

The difference between A and A_2 as compared to that between F and F_I is one of kind: major differences in sequence differentiate A and A_2, whereas F and F_I have a common structure except for an acetyl group. What may be said about the electrophoretically poorly separated component A_3 to which we have referred above? Although this component separates only poorly from the main component in electrophoretic systems, it is not only completely and easily separated from it chromatographically on the cation exchanger IRC-50, but is further resolved into at least two fractions termed A_{Ia} and A_{Ib} (Schnek and Schroeder, 1961). Laurent et al. (1962) have shown that A_{Ia} itself probably is heterogeneous. Aside from the fact that these components appear to have an amino acid composition identical to hemoglobin A (Jones and Holmquist, unpublished data), nothing else is known about their structure.

Another minor component, which emerges chromatographically (Allen et al., 1958) between A_{Ib} and A, is termed A_{Ic}. It is normally indistinguishable electrophoretically from the major component, is present to the extent of 5–7 percent in the hemoglobin of normal adults, and is the most abundant of the adult human minor components. Let us consider what is known today about A_{Ic}.

HEMOGLOBIN A_{Ic}

Is A_{Ic} a product of the aging of the main component? Does it contain glutathione? Experiments by Huisman and co-workers (Huisman and Dozy, 1962; Huisman and Horton, 1965) have answered both of these questions in the negative. They showed that glutathione indeed will react with hemoglobin A but that the product is not a normal component of the cell. It is, nevertheless, a product that can form in cells in vitro or in aging solutions of hemoglobin. On the other hand, the percentage of A_{Ic} is identical in erythrocytes of different physiological age; even in cells aged in vitro the percentage of A_{Ic} was not altered. The aging of cells or of solutions under a variety of conditions produces components that are readily distinguishable chromatographically from A_{Ic}. However, electrophoretically they are not distinguishable and as a result have led to incorrect conclusions about the effects of aging.

Experiments with radioactive tracers by Holmquist and Schroeder (1966b) have shown conclusively that A_{Ic} is not the product of the physiological aging of the erythrocytes. When A_{Ic} and A were isolated from reticulocyte-rich blood that had been incubated with radioactive valine, both had equal specific activity. Thus A_{Ic} could not be a product solely of older erythrocytes. Other data allowed the conclusions to be drawn that (1) A_{Ic} and A are not rapidly interconvertible, (2) A_{Ic} and A are in rapid equilibrium, or (3) though interconvertible at moderate rates, any formation of one from the other is just offset by the reverse process. That hemoglobin A_{Ic} is a normal and constant component of the erythrocyte has thus been clearly demonstrated. Indeed, the amount of A_{Ic} is largely uninfluenced by a wide variety of hematological diseases except for congenital spherocytosis and the autoimmune hemolytic anemias (Horton and Huisman, 1965).

Chemical studies of the structure of hemoglobin A_{Ic} were begun by Jones (1961), who was able to show that in A_{Ic} the *N*-terminal amino acid of one or both of the β chains had reacted with some group of unknown structure. Superficially, the relationship between A_{Ic} and A resembles that between F_I and F, in which the acetylation of a single γ chain of F results in F_I. Actually, the group at the *N*-terminus of A_{Ic} is very different (Holmquist and Schroeder, 1966a). Attached to the *N*-terminus of one β chain of hemoglobin A_{Ic} through a Schiff's base linkage is an aldehyde or ketone of molecular weight approximately 280. This aldehyde or ketone is not pyridoxal or pyridoxal phosphate; indeed, phosphorus is absent, as are carbohydrates, steroids, amino acid components, and esterified nonketo acyl groups of less than five carbon atoms. Nitrogen and free carboxyl groups also appear to be absent. The molecular weight and chemical properties of the aldehyde or ketone are consistent with those of a long chain aliphatic lipid, though this remains to be verified (Holmquist and Schroeder, 1966a).

The main function of hemoglobin (at least of the main component, and presumably of the minor components) is the obvious one of oxygen transport. But is this the only purpose of the minor components? The erythrocyte is not simply a bag for holding hemoglobin: it is such a beehive of activity with its many active enzyme systems that it is not unreasonable to assume specific functions for the minor hemoglobin components.

For reasons stated in the introduction, we shall focus attention on hemoglobin A_{Ic}. But first, let us consider some of the properties of the erythrocyte, for they bear importantly on the functions we shall consider for A_{Ic}.

GENERAL PROPERTIES OF THE ERYTHROCYTE

Some Properties of the Erythrocyte Interior

On the average, the biconcave erythrocyte is 8.4 μ in diameter. Near the periphery the thickness is 2.4 μ, and at the center it is 1 μ. The volume is 81 μ^3, and the surface area is 163 μ^2 (Dittmer, 1961). Within this volume are packed the hemoglobins, lipids, sugars, salts, and enzyme systems that help to maintain the integrity of this marvelous machine. The hemoglobin in the erythrocyte becomes almost completely oxygenated in one-third of the 780 msec that it spends in a pulmonary capillary (Harris, 1963, page 197). Although the concentration of hemoglobin in the cell is only about 0.0055 *M*, by weight it is 34 percent. The volume of cell is such that the average distance between hemoglobin molecules probably is only about 10 Å (Perutz, 1948; Bateman et al., 1953; Harris, 1963, page 200); it will be remembered that the individual molecules of oxyhemoglobin are globular particles with the over-all dimensions of 50 × 55 × 64 Å (Muirhead and Perutz, 1963). Thus, although the hemoglobin molecules are not in crystalline array in the erythrocytes, their motions are of necessity rather restricted by these geometrical considerations. It is rather interesting and certainly surprising in this connection, that at this concentration the solution of hemoglobin in the erythrocytes behaves in some respects almost as an ideal solution. For example, for the hemoglobin molecule, the Stokes-Einstein radius that is calculated from measured values at 20°C of the diffusion coefficient (Polson, 1939; Keller and Friedlander, 1966) and viscosity (Longmuir and Roughton, 1952) is 31 Å and 35 Å at concentrations of 1 percent and 27 percent hemoglobin respectively, as compared to the X-ray value of 35 Å; furthermore, for an ideal solution, the value for $D_{\text{Hb}} = (28/64{,}450)^{\frac{1}{2}} D_{\text{CO}} = 0.021\ D_{\text{CO}}$ agrees with the experimental value *exactly*. Thermodynamicians and fluid engineers have an interesting problem here!

The Structure of the Cell Membrane

What the exact structure of the cell membrane may be, not only of the erythrocyte but of other cells as well, is the subject of much study and debate. There seems to be little question, however, that the cell membrane contains both protein and lipid. In one model of membranes that has been suggested by Danielli (1958), the membrane consists of a double layer of lipid with the nonpolar groups adjacent. This lipid leaflet is then sandwiched between two layers of protein so that the cell membrane presents protein to the outside world as well as to its interior. Whether or not the cell membrane contains pores or whether entrance and exit through the cell wall occurs by some other means is open to question. These ideas are by no means universally accepted, and other models have been proposed. A stimulating and informative discussion of the known facts and extrapolated conjectures has been given by Kavanau (1963).

The thickness of the erythrocyte cell membrane has been variously estimated to be between 50 and 1000 Å (Prankerd, 1961, page 14). Kavanau (1963) has suggested the possibility that the thickness of the cell membrane may vary, depending on the metabolic state of the cell at any moment. He considers the cell membrane to be composed of a number of accordionlike cylindrical "plaques" oriented perpendicularly to the membrane surface in a hexagonal array. The spaces between these plaques constitute the pores of the cell membrane. As the cell membrane thickens radially, lateral constriction of the plaques increases the pore area, and in this manner the behavior of the membrane toward the entrance or exit of metabolites may be controlled. Experimental evidence for such plaquelike structures exists (Hillier and Hoffman, 1953). A thickness of 1000 Å requires so large a volume for the erythrocyte membrane that the known quantities of lipids and proteins in the membrane would have to be augmented by hemoglobin to make up part of the volume. In the discussions that follow, we shall assume the cell membrane to have an "average" thickness of 200 Å. It thus comprises about 2 percent of the thickness of the erythrocyte.

Whether hemoglobin is part of the membrane has not been definitely answered. Anderson and Turner (1960) have demonstrated that approximately 2 percent of the total hemoglobin in the erythrocyte cannot be removed from the cell membrane by hemolysis with water followed by repeated washing. On the other hand, Hoffman (1958) has concluded rather emphatically that hemoglobin is not covalently and irreversibly bonded to the cell membrane. These experiments of Anderson and Turner, on the one hand, and of Hoffman, on the other, are difficult to reconcile, but taken together they suggest that hemoglobin can be a component of the cell membrane.

The variety of answers to questions about the cell membrane is in part a result of the variety of methods that have been used to examine it.

TRANSPORT OF OXYGEN IN THE ERYTHROCYTE

As Roughton (1959) has pointed out, it has frequently been assumed that oxygenation of the hemoglobin is "instantaneous," after oxygen has penetrated the lung capillaries, diffused through the plasma, and arrived at the erythrocyte membrane. His data, that of Keller and Friedlander (1966), and the calculations to be made below show that this is not the case, but that there are also limiting factors introduced by the membrane and the contents of the cell interior.

Transport of Oxygen Through the Membrane

In principle, the membrane could either hinder the entry of oxygen into the cell (positive resistance) or it could aid this entry by actively transporting oxygen to the inner surface so that the P_{O_2} at the inner surface would be higher than at the outer surface (negative resistance). Since there is no experimental evidence for the active transport of oxygen across the cell membrane, we shall limit our discussion to the possibility that the membrane decreases the flux of oxygen into the cell interior. If, as suggested by Roughton's (1963) theoretical interpretation of the generally concordant data of Roughton and Forster (1957), Gibson et al. (1955), and Staub et al. (1961), $D_{O_2}^{\text{membrane}} \approx 0.04\ D_{O_2}^{\text{cell interior}}$, where D is the diffusion coefficient of oxygen, and if the membrane makes up 2 percent of the over-all thickness of the erythrocyte, then the flux of oxygen into the cell could be reduced by as much as 50 percent over the value it would have in the absence of a membrane. On the other hand, the data of Kreuzer and Yahr (1960) suggest that $D_{O_2}^{\text{membrane}} \approx D_{O_2}^{\text{cell interior}}$, in which case the membrane would have a small ($\sim$2 percent) effect on oxygen transport. One is tempted to place somewhat more confidence in the estimate of Kreuzer and Yahr because Roughton and Forster (1957) and Gibson et al. (1955) used red cells that had been deoxygenated by vacuum boiling. Kreuzer and Yahr (1960) have shown by microscopic examination that a high proportion of such cells were crenated, and, in agreement with the work of Carlsen and Comroe (1958), that such crenated cells required twice as long to take up oxygen. Further, one wonders if the injection of cells through a syringe at high speed into the rapid mixing apparatus of Roughton and his collaborators may not alter the surface properties of these cells. The procedure of Staub et al. (1961), which involved bubbling a gas mixture through the suspension of cells, may also have affected the surface. On the other hand, Kreuzer and Yahr (1960) deoxygenated the cells by flowing a stream of nitrogen over a thin cellular layer: the uptake of oxygen was measured not by a rapid mixing apparatus but by allowing oxygen to diffuse into this thin layer. This procedure seems less likely to damage the cells, and, in fact, the percentage of crenated cells was small in such preparations. An added merit of the work of Kreuzer and Yahr is that it is a direct experimental measurement, which need not be interpreted by a theory whose assumptions, though reasonable, are also uncertain. To settle some of these questions, comparative experiments should be made in the same laboratory by the different methods and with microscopic examination of the cells. Accepting the work of Kreuzer and Yahr, we conclude that the entrance of oxygen into the red cell is not retarded by more than a few percent by the red cell membrane.

Transport of Oxygen Through the Erythrocyte Interior

Under normal physiological conditions it may be expected that each molecule of hemoglobin, which is fully oxygenated when it leaves the lungs will give up about one of its four oxygen molecules by the time that it returns to the lungs (Staub et al., 1961). Within the next 250–300 msec (Harris, 1963, page 198), however, in this efficient machine, about 3×10^8 molecules of oxygen must be transported to 3×10^8 molecules of hemoglobin through a membrane that is probably at least of the order of 200 Å, or 3 or 4 times the average dimensions of the hemoglobin molecule and over 150 times that of the oxygen molecule. Furthermore, in the biconcave erythrocyte, in which the maximum thickness is 2.4 μ, some molecules must be transported a distance of almost 12,000 Å to the innermost point through a volume where the space between hemoglobin molecules is of the order of 10 Å. In doing so, the oxygen molecules must pass 150 to 200 molecules of hemoglobin.* Considering these obstacles, and the affinity of hemoglobin for oxygen, it is at first glance surprising that an oxygen molecule ever reaches the center of the cell at all. The task set for Jason by King Pelias of Iolcus would seem small by comparison. Nevertheless, as in Jason's case, through the propitious marriages of nature, the job does get done. It is therefore worthwhile to inquire into the details of this oxygen transport.

The rate of transport of oxygen through the red cell interior is determined by the partial pressure of oxygen at the inner surface of the cell membrane, the solubility and diffusion coefficients of oxygen in 34 percent w/v hemoglobin solution, the degree of saturation of the hemoglobin with oxygen (that is, $[HbO_2]/[Hb_{total}]$), the kinetics of the over-all reactions $Hb + O_2 \xrightarrow{k'} HbO_2$ and $HbO_2 \xrightarrow{k} Hb + O_2$, the self-diffusion coefficient of hemoglobin,

*During this discussion, we assume that the erythrocyte maintains its biconcave shape at all times. It is, however, probable that there is much distortion (Grant et al., 1963) and consequent internal mixing during the passage through the capillary.

and the volume and shape of the cell interior. Here [Hb] and [HbO_2] refer to the heme and oxyheme concentrations as determined, for example, spectrophotometrically. Fully oxygenated hemoglobin would be represented by Hb_4O_8. More precisely,

$$[Hb] = 4[Hb_4] + 3[Hb_4O_2] + 2[Hb_4O_4] + [Hb_4O_6]$$

$$[HbO_2] = 4[Hb_4O_8] + 3[Hb_4O_6] + 2[Hb_4O_4] + [Hb_4O_2]$$

Somewhat surprisingly, in view of the complexity of the situation, the resistance of the red cell interior to oxygen transport is relatively well understood, and, in fact, the quantitative relationships between the determinants (oxygen solubility, diffusion, kinetics, geometry, and others) of this resistance are known and have been used rather successfully to predict, from in vitro data, the in vivo uptake of oxygen by the red cell in the capillaries of the lung (Roughton, 1959; Staub, 1963).

Briefly, oxygen can be transported through the cell interior in only two ways: (1) the ordinary thermal diffusive motion of the dissolved O_2 molecule, and (2) by riding piggyback on the hemoglobin molecule. The latter phenomenon was discovered by Scholander (1960) and is termed "facilitated" transport. Since the mechanism of the diffusion of dissolved oxygen is by simple Brownian motion, we shall not discuss it further; actually, as will be seen in the following paragraphs, it accounts for approximately 94 percent of the total transport. Facilitated transport, however, is a considerably more complex phenomenon, which has received much attention in recent years (Keller and Friedlander, 1966; Wittenberg, 1966; Wyman, 1966; and LaForce and Fatt, 1962). Because it may be of some importance in the erythrocyte, we shall discuss it in some detail.

One conceivable mode of facilitated oxygen transport is by the rotational motion of hemoglobin molecules. A hemoglobin molecule would pick up an oxygen molecule that is coming from the general direction of the membrane side, rotate by some angle, and then throw the oxygen molecule in the general direction of the center of the red cell. However, according to the calculations of Wyman (1966) on the basis of the experimental data of Oncley (1938), this mechanism can account for only about 0.05 percent of the experimentally observed rate of facilitated transport. This rotational transport of oxygen is, therefore, not of quantitative significance in the erythrocyte.

Another mode of facilitated oxygen transport is by the translational thermal motion of oxyhemoglobin. An earlier, and at the time reasonable, mechanism of this sort was the suggestion of Scholander (1960) that oxygen transport might occur by a "bucket-brigade" type of mechanism in which the diffusing hemoglobin molecules pass oxygen down the line by handing it from one to the other in a region where a gradient in the concentration of oxyhemoglobin exists. This would involve a bimolecular ligand transfer of the type

$$HbO_2 + Hb' \rightleftarrows Hb + Hb'O_2$$

in which the rate of transport should vary as the square of the total hemoglobin concentration. Wittenberg (1966) has provided definitive experimental, as well as theoretical, evidence against this mechanism. In particular, the rate of oxygen transport varies linearly with the total hemoglobin concentration and not as the square.

The main mechanism responsible for facilitated transport appears to be that recently described by Wittenberg (1966), Wyman (1966), and Keller and Friedlander (1966). In this mechanism, free molecular oxygen—that is, oxygen dissolved in solution, but not combined with hemoglobin—is moved through a hemoglobin solution in three steps.

(1) Dissolved oxygen combines with an incompletely oxygenated hemoglobin molecule:

$$\mathrm{Hb}(x, y, z) + \mathrm{O_2}(x, y, z) \xrightarrow{k'} \mathrm{HbO_2}(x, y, z)$$

(2) The oxygenated hemoglobin molecule moves to a region of lower HbO_2 concentration:

$$\mathrm{HbO_2}(x, y, z) \longrightarrow \mathrm{HbO_2}(x', y', z')$$

(3) And finally the oxygen dissociates from the hemoglobin:

$$\mathrm{HbO_2}(x', y', z') \xrightarrow{k'} \mathrm{Hb}(x', y', z') + \mathrm{O_2}(x', y', z')$$

This would be a bucket brigade in which a member of the brigade runs toward the next in line and *throws* rather than hands the bucket to the next fellow. Although facilitated transport plays a dominant role in oxygen transport in thick layers of hemoglobin solutions ($>120\ \mu$) at low oxygen partial pressures under steady-state conditions in which a gradient in HbO_2 can be maintained, its importance in the erythrocyte, in which none of these conditions exists, is unknown. An estimate can be made as follows. From the Einstein relation (Eggers et al., 1964), the approximate time required for an oxygen molecule to diffuse to the center of the red cell at 37°C is

$$t_{O_2} = \frac{\xi^2}{2D_{O_2}} = 0.0007 \text{ sec} \tag{1}$$

where the half-thickness ξ of the erythrocyte has been taken as 1 μ and D_{O_2} as 7.3×10^{-6} cm²/sec; the time required ($t_{\text{hemoglobin}}$) for a hemoglobin molecule to diffuse this same distance is about 0.058 sec because $D_{\text{hemoglobin}} = 1.75 \times 10^{-7}$ cm²/sec under these same conditions.* (Since at 37°C the half-time for an oxygen molecule to dissociate from a heme group may be calculated from data in the literature to be about 0.0078 sec, the oxygen must "hitch a ride" on 7 to 8 hemoglobin molecules in succession.) If the concentration of unoxygenated heme groups in the cell were much greater than that of dissolved oxygen, then, in spite of the relatively long diffusion time of hemoglobin, facilitated transport could play a significant role. In mixed venous blood at a partial pressure of 41.5 mm Hg, 78 percent of the heme groups are fully oxygenated under physiological conditions (Staub et al., 1961). Thus, during the time (780 msec) that the hemoglobin molecule is becoming fully oxygenated in the capillaries of the lung, an average of 11 percent of the heme groups is available for facilitated transport. Since the total heme concentration in the erythrocyte is 0.022 *M*, the average concentration of unoxygenated hemes is $[\mathrm{Hb}]_{\mathrm{Av}} = 0.0024$ *M*. The average concen-

*See the Appendix for the sources of data in this section.

tration of dissolved oxygen at 37°C is 0.000087 *M* at a partial pressure of 70 mm Hg, which is the mean of the mixed venous and capillary oxygen tensions of 41.5 and 100 mm Hg respectively (Staub, 1963). Therefore, the relative importance of facilitated transport to ordinary diffusion can be estimated as

$$\frac{t_{O_2}[\mathrm{Hb}]_{\mathrm{Av}}}{t_{\mathrm{Hb}_4}[\mathrm{O}_2]_{\mathrm{Av}}} \times 100 = 33 \text{ percent}$$

The effectiveness of facilitated transport is probably much less because the oxygen molecule does not combine instantaneously with unoxygenated heme, but requires about 0.0038 sec to do so at *p*H 7.4 at a partial pressure of 70 mm Hg at 37°C. This time of reaction is about six times longer than it takes for the oxygen molecule to get to the center of the erythrocyte by its own thermal motion—Equation (1). *Thus facilitated transport may contribute approximately 6 percent to the total flux of oxygen into the erythrocyte.*

In summary, the role of hemoglobin in transporting oxygen to the interior of the red cell is probably small. Its primary function, with respect to oxygen, is to store this metabolite for later release to other body tissues.

The remarks of this section have covered only a very circumscribed aspect of the activity of the erythrocyte and exclude such important features as carbon dioxide release, ingress and egress of chloride, and so on. However, with the background that we have given here, let us consider possible functions for hemoglobin A_{Ic}.

POSSIBLE FUNCTIONS FOR HEMOGLOBIN A_{Ic}

With the information of the preceding sections as a background, we are now in a better position to consider the possible functional relevance of A_{Ic} to the erythrocyte. Conceivably, A_{Ic} could play a significant role in either the red cell membrane or the red cell interior. We shall consider each of these in turn. In either case, it is necessary to have a reasonable picture of the three-dimensional structure of the A_{Ic} molecule.

The Three-Dimensional Arrangement of Hemoglobin A_{Ic}

It is in the semiordered array of the erythrocyte that hemoglobin A_{Ic} makes up 5–7 percent of the molecules. We assume that A_{Ic}, because it contains α and β chains like hemoglobin A, has the same three-dimensional structure (Muirhead and Perutz, 1963) as A with one important exception: at the *N*-terminus of one of the β chains, the blocking group of about 280 molecular weight is attached in Schiff's base linkage. Present information suggests that this group may be a long chain (perhaps unsaturated or hydroxylated) lipid. The point of attachment is to a residue that is on the surface of the molecule near the dyad axis. A group of this size could extend perpendicularly from the surface of the molecule by as much as 20 Å. Indeed, inasmuch as the *N*-terminal amino acid residue to which the group is attached and also the next two residues are not believed to be in helical configuration (Perutz, 1965), the perpendicular extension might be almost 30 Å. The length of this appendage would then be equal to about half the thickness of the molecule at the point of attachment and could be a significant steric factor in the confined semiordered quarters within the erythrocyte, in which the average distance between hemoglobin molecules is only 10 Å. On the other hand, this appendage, instead of being perpendicular to the surface of the molecule, could more or less be parallel to it in one of many directions. In such a parallel position it would sweep out an area equivalent to one-fourth of the surface area of the hemoglobin molecule and might possibly interfere with the oxygenation or deoxygenation of one of the four heme groups. The oxygen dissociation curve of A_{Ic} has not been measured, so whether such interference actually occurs is not known.

A_{Ic} *as a Component of the Erythrocyte Membrane and Its Possible Functions There*

A simple calculation (Holmquist and Schroeder, 1966b) shows that the amount of hemoglobin A_{Ic} in the red cell is approximately three times that necessary to cover the red cell surface uniformly with a monolayer. If, indeed, A_{Ic} does form such a monolayer, the excess molecules may, by a sort of "mass action," be required to maintain or replenish the monomolecular layer. Although other minor components of the erythrocyte (Kunkel's A_2 and Schnek and Schroeder's A_{Ib}, in particular) are also present in sufficient amount to cover a substantial portion of the cell membrane, we shall limit our discussion to hemoglobin A_{Ic}, because only this component appears to contain a lipidlike group at the *N*-terminus of one of the β chains. The length of the group (estimated at 20 Å) is of the right dimension to participate at the lipoprotein-protein interface that has been postulated at the inner surface of the membrane.

To the best of our knowledge, no one has searched for the presence of A_{Ic} in the erythrocyte membrane. However, the data that do exist with respect to the presence or absence of hemoglobin therein suggest that such a search might well bear fruit. The results of Anderson and Turner (1960) together with those of Hoffman (1958), which were discussed in an earlier section, would indicate that about 2 percent of the total hemoglobin of the erythrocyte is associated, though perhaps not irreversibly so, with the cell membrane. Unfortunately the above investigators did not determine *which* hemoglobins had remained as components of the membrane or within the erythrocyte. Such experiments would be more meaningful if the relative amounts of hemoglobins A, A_{Ic}, and A_2 associated with the membrane were determined.

Assuming that A_{Ic} is a component of the erythrocyte cell wall, we may inquire into its function there. Conceptually, and for convenience of discussion, this may be divided into two categories, chemical and mechanical, though when one begins to think about possible detailed mechanisms for any proposed function, the artificiality of these categories becomes apparent. In the first category, for example, A_{Ic} can be conceived of as a buffer that maintains the cell membrane at the *p*H necessary for its proper functioning. Another possible function of chemical nature would be the facilitation or retardation of oxygen transport through the membrane. In the mechanical category, one can suggest that the presence of A_{Ic} in the membrane is responsible for the "tightness" or "looseness" of the membrane structure, its fluid properties, or elastic strength, for example, each of which may be important in determining the manner in which metabolites enter into or exit from the cell interior. Because of the volume change which the hemoglobin molecule undergoes on oxygenation, it can also act as a molecular pump to bring metabolites into or out of the red cell.

A_{Ic} *as a Membrane Buffer.* One can estimate to what extent a monolayer of hemoglobin on the cell membrane might be responsible for controlling the pH within the membrane. We assume for this calculation that the hemoglobin does not interact with other elements of the membrane. If the thickness of the membrane is 200 Å and its area is 163 μ^2, the membrane volume is about 3 μ^3. Assuming the intact cell membrane to have a pH of about 7.4, there are about 7×10^7 hydronium ions in this volume. The approximately 5×10^6 molecules of hemoglobin in the monolayer each contain 38 histidyl residues. We shall assume half of these are ionized, so that the hemoglobin contributes 9.5×10^7, or virtually all of the hydronium ions needed to maintain the membrane pH. Crude as this calculation is, it shows that a membrane without a "hemoglobin" would be a very different membrane indeed. Although there is no a priori reason why a protein other than hemoglobin could not maintain the pH, the use of hemoglobin would simultaneously serve to maximize the efficiency of the erythrocyte with respect to several other functions to be discussed.

An Oxygen Transport Function for A_{Ic} *in the Membrane?* The question to be considered here is whether A_{Ic} as a component of the cell membrane could play some essential and quantitatively significant role in limiting or augmenting the flux of oxygen into the erythrocyte. Might it act as a sort of "gate" at the membrane surface? We believe the answer to this question is very probably "no." It is nonetheless instructive to consider the facts that lead to this conclusion.

In an earlier section it was estimated by Equation (1) that it requires 0.0007 sec for an oxygen molecule to diffuse from the red cell surface to its center, a distance of about 1 μ. Because oxygen has about the same diffusion coefficient within the membrane as within the cell interior (at worst, $D_{O_2}^{\text{membrane}} = 0.04\, D_{O_2}^{\text{interior}}$), the time required for it to diffuse across the 200 Å or so of the membrane is from Equation (1) only 3×10^{-7} sec. This is *so* much shorter than the 0.004 sec required for the oxygen molecule to react with a heme molecule, that, even allowing for all possible uncertainties in the data and the approximation involved in Equation (1), it is clear that unless hemoglobin A_{Ic} has an oxygenation rate *several* orders of magnitude larger than that of hemoglobin A—an improbable event in view of the similar structures of the two hemoglobins—it plays no role in the transport of oxygen through the membrane.

Although, as shown above, oxygen probably does not react with hemoglobin in passing through the cell membrane, let us suppose that for some reason oxygen cannot simply diffuse through the membrane oblivious of the hemoglobin. Rather, each oxygen molecule can only get into the cell by reacting with hemoglobin A_{Ic} and subsequently dissociating from it: the A_{Ic} molecule is a *mandatory* "gate" through which each oxygen molecule must pass to get through the membrane and into the cell interior. Further transport would require first that the oxyhemoglobin A_{Ic} in the membrane release or transfer its oxygen to another molecule. Following this, the now deoxygenated hemoglobin A_{Ic} molecule or a close neighbor (because of the restricted quarters in the erythrocyte) would be free to accept another oxygen molecule from the outside world. If one-third of the hemoglobin A_{Ic} is actively participating or about 1.8 percent (5×10^6 molecules of A_{Ic}) of the total number of hemoglobin molecules, then, on the average, each molecule of A_{Ic} would have to undergo this reaction about 60 times during the approximately 250 msec that are required to bring 3×10^8 oxygen molecules into the cell during complete oxygenation in the capillaries of the lung. Each cycle of reaction would, therefore, have 5 msec for completion. This time requirement is not unreasonable, since the sum of half-times for association and dissociation of oxyhemoglobin at 37°C is 8–12 msec. The erythrocyte thus seems to have a second moderately effective system for getting oxygen into its interior.

Another manner in which a mandatory gate might function is as follows. If the hemoglobin A_{Ic} is considered to be only a monolayer on the membrane surface, it could act by dissociating the hemoglobin moiety from the attaching group after oxygenation. The attachment through the Schiff base linkage is a more labile one than, for example, that of the acetyl group that has been found in many proteins. Perhaps the conformational change that accompanies oxygenation facilitates hydrolysis of the hemoglobin moiety from the group that remains in attachment to the membrane. Although aliphatic Schiff's bases are relatively labile—and this lability certainly introduced difficulties into the chemical investigation of hemoglobin A_{Ic} as the protein was degraded to small peptides (Holmquist and Schroeder, 1966a)—hemoglobin A_{Ic} itself in vitro does not give evidence of unusual lability (Huisman and Horton, 1965). Isolation of the hemoglobin as well as separation of chains prior to degradation to peptides does not remove the blocking group. These facts would speak against the hydrolysis of the Schiff base imine bond in the erythrocyte membrane, but one can conceive of an enzyme spatially near to membrane-bound A_{Ic} that would make this hydrolysis effective. To bring oxygen effectively into the erythrocyte the activity of this enzyme would have to be at least of the order of 100–300 molecules of A_{Ic} per second or greater.

Having discussed in preceding sections the chemical roles that A_{Ic} might play in the membrane, we now turn our attention toward the possibility that the function of A_{Ic} in the membrane is to confer on the latter certain mechanical properties.

A_{Ic} *as a "Transition" Molecule.* If the properties of a membrane are to be optimal for the passive transport of metabolites such as oxygen, these transport properties should be as close as possible to those of an equivalent thickness of the cell interior and yet the membrane should be as thin as is consistent with the other specialized functions such as restricting the over-all volume. The incorporation of hemoglobin A_{Ic} into the membrane would be an excellent way to accomplish this design because the presence of a monolayer of hemoglobin would tend to minimize the difference between the cell membrane and the cell interior. Such a monolayer would contain 50 percent by weight of hemoglobin and 29 percent by volume.

A_{Ic} *as a Determinant of Membrane Strength.* Might the tensile strength of the red cell membrane be determined in part by the A_{Ic} therein? The surface tension σ of the red cell membrane just at hemolysis is approximately 0.96 dynes per centimeter at 25°C (Hoffman, 1958) and is comparable to that of other cell membranes (Harvey, 1954; Norris, 1939). However, in the case of the erythrocyte membrane it is interesting to speculate that this low surface tension may be caused by A_{Ic}. The structure of A_{Ic} is similar to that of other surface active agents: a long chain aliphatic hydrocarbon (the blocking group in A_{Ic}) to which a polar group (hemoglobin A) is attached. Perhaps some experimental light could be shed on this conjecture by measuring σ in the presence of increasing concentrations of A_{Ic} in the solution that is used to induce hemolysis. A negative result—that is, if σ is independent of $[A_{Ic}]$—would not necessarily be significant, but a positive result—if σ decreased or increased with increasing $[A_{Ic}]$—would support the hypothesis that the tensile strength of the erythrocyte membrane may be in part determined by A_{Ic}. Another experiment that might be done would be to overlayer a homogeneous mixture of A_{Ic} and hemoglobin A with an organic solvent in which the blocking group was soluble. The aqueous phase would be stirred gently for a time, and the interface would be analyzed to see if the A_{Ic} had preferentially collected there. It would also be of interest to determine the difference between the interfacial tensions when solutions of A_{Ic} and hemoglobin A respectively were used for the aqueous phase. Such experiments should be conducted under conditions approximating those in the red cell at 37°C as well as at lower concentrations of hemoglobin and temperature.

It was noted earlier that hemoglobin A_{Ic} is present in the normal range in many hematologically abnormal states but that it has been observed to be lower in congenital spherocytosis and autoimmune hemolytic anemias (Horton and Huisman, 1965). In congenital spherocytosis, the osmotic fragility is increased; that is, the membrane ruptures more easily than normally. This alteration is what one would expect if A_{Ic} strengthens the membrane and if, perhaps, the excess of A_{Ic} over that required for a monolayer is necessary to maintain an adequate monolayer. In the autoimmune hemolytic anemias, the osmotic fragility curve extends both above and below the normal range.

A_{Ic} *as a Pore Size Regulator and Molecular Pump.* If A_{Ic} is attached to the membrane by its appendage, the monolayer should be at least partially oriented. The appendage is near the dyad axis of the molecule, and its interaction with the membrane should produce an orientation of the molecule with the dyad axis perpendicular to the plane of the membrane. From the data of Muirhead and Perutz (1963), it seems likely that the molecular volume of ferrohemoglobin is about 8 percent larger than that of oxyhemoglobin and that a change in linear dimensions of about 5 Å occurs in one of the dimensions perpendicular to the dyad axis. The change, then, would be parallel to the plane of the membrane in such a partially oriented monolayer. Hemoglobin A_{Ic} might act as a pore size regulator in the following way. Suppose the molecules of A_{Ic} were oriented about the rim of a pore not only so that the dyad axis was perpendicular to the membrane but also so that the dimension that alters on oxygenation and deoxygenation was along a radius of the pore. In such an orientation, the pore size could be altered somewhat in the way in which an iris diaphragm operates: oxygenation would cause a decrease in the size of the molecule with opening of the pore, and deoxygenation would bring about a closing. Conceivably, such a mechanism could be correlated with various phases of the "Hamburger shift," with its coordinated features of bicarbonate diffusion, potassium retention, chloride diffusion, and water balance; Harris (1963, page 203) describes the "Hamburger shift" in detail.

From a somewhat different point of view, the changes in volume on oxygenation and deoxygenation may be thought of in terms of a molecular pump. Thus, if 29 percent of the membrane volume is due to a monolayer of hemoglobin, on deoxygenation the volume of the membrane would be increased by 2.3 percent if no mass transfer occurs. Whether or not mass transfer occurs, the volume increase accompanying the deoxygenation of oxyhemoglobin could be used to do mechanical work—for example, in the absence of mass transfer, by stretching the membrane. If mass transfer is permitted and the membrane volume remains constant, some 0.07 μ^3 could be expelled from the membrane either to the cell interior or to the extracellular plasma or, alternatively, material from the cell interior could be passed outward through the membrane into the plasma. In either case about 600,000,000 molecules, if these have a size comparable to the amino acids, would be transferred from one environment to a totally different one. This number is about twice the number of oxygen molecules that enter the cell during its passage through the lung; it is a number whose magnitude is of metabolic significance whatever the actual molecules or ions involved in the transfer may be.

In the erythrocyte, the dominant reaction is not the complete deoxygenation of oxyhemoglobin, but rather the removal of, on the average, only one oxygen molecule from it. There are at present no experimental data about the configurational changes that occur on such partial deoxygenation. However, might not the hemoglobin in the membrane, because it is on the surface of the erythrocyte, lose *more* than the average of one oxygen molecule during the passage through the body? Certainly, it is an open question whether or not A_{Ic} could operate as a molecular pump in vivo.

In passing, it should be pointed out that the total hemoglobin in the erythrocyte is 50 times that of the monolayer. If all acted as a molecular pump, the pumping volume is about 4 percent that of the erythrocyte.

A_{Ic} *as a Component of the Red Cell Interior and Its Possible Functions There*

Only two functions will be considered for A_{Ic} within the red cell interior: (1) that of oxygen transport, and (2) that of a crystallization inhibitor.

Oxygen Transport by A_{Ic} *in the Red Cell Interior*. Earlier in this paper, it was estimated that the total cellular hemoglobin contributes to the transport of perhaps 6 percent of the oxygen through the cell interior. Since hemoglobin A_{Ic} comprises only 5.3 percent of the total hemoglobin, it can contribute at most to 5.3 percent of the total transport or about 0.3 percent. This is not quantitatively significant.

A_{Ic} *as a Crystallization Inhibitor?* Does hemoglobin A_{Ic} function perhaps by inhibiting the crystallization of hemoglobin in the erythrocyte? Does it in a sense depress the "freezing point"? It is somewhat difficult to consider this in the usual way as a "freezing point depression." The crystal of hemoglobin contains about 45 percent water (Perutz, 1948), and the water is not ice. In this ternary system actually, we have what amounts to a 0.001 molal solution of A_{Ic} in A where the mole fraction of A_{Ic} is 0.05. The "freezing point depression constant" would presumably have to be very large. It may be more relevant to consider whether the appendage on A_{Ic} is situated in such a position as to interfere with intermolecular contacts in the crystal. Perutz (1965) has recently discussed the exact position of these contacts in relation to the three-dimensional structure of the molecule. Each hemoglobin molecule in the crystal has six neighbors in octahedral array. The four equatorial neighbors make different contacts than the apical ones. None of these contacts is very near the assumed position of the attached group in A_{Ic}. Indeed, the group is at such a position in A_{Ic} that there would appear to be adequate space between molecules to contain it. Actually, A_{Ic} appears to crystallize with hemoglobin A: hemoglobin that has been crystallized directly from hemolysates may be shown by chromatography to contain minor components (Allen, Schroeder, and Balog, 1958). It seems unlikely, therefore, that hemoglobin A_{Ic} functions to inhibit crystallization.

However, before this suggestion is dismissed entirely, it should be pointed out that deoxygenated hemoglobin S does gel or crystallize under conditions where hemoglobin A is soluble. Yet the alteration in hemoglobin S involves only a single amino acid substitution rather than the presence of an appreciable appendage. The substitution in S is in the same approximate position as the appendage in A_{Ic}. Murayama et al. (1965) has been active in suggesting why hemoglobin S has its own peculiar properties.

What Fulfills the Function of Hemoglobin A_{Ic} *When It is Absent in Erythrocytes?*

Whether hemoglobin A_{Ic} functions in one of the ways in which we have suggested or whether it functions in a way that we have not even guessed at, it is clearly germane to this discussion to ask the above question. The fetal human erythrocyte, for example, does not seem to have a component like A_{Ic}; in addition to hemoglobins F and A, almost the only other component is that termed hemoglobin F_I which, as has already been mentioned, differs from F by the presence of an acetyl group on the *N*-terminus of one of the γ chains. The quantity of F_I in the fetal erythrocyte is about twice that of A_{Ic} in the adult erythrocyte. It is difficult because of the difference in the structures to attribute identical function to F_I and A_{Ic}. There is then no hemoglobin in the fetal human erythrocyte that may be expected to play the role of A_{Ic}. But does the special environment of the fetus eliminate the necessity for a molecule to play the role of A_{Ic} in the fetal erythrocyte? Apparently not. In the homozygote for the condition known as "hereditary persistence of fetal hemoglobin," the blood contains apparently only F_I and F and yet the homozygote is hematologically normal. In the cases of erythrocytes in hemoglobinopathies or from other species, the situation is simply one in which ignorance is bliss. Minor components are present in all of these instances but virtually no chemical investigation has been done. It is entirely possible, therefore, that components analogous to A_{Ic} are present.

CONCLUDING COMMENTS

In the past, the authors of this paper have been largely concerned with the chemical aspects of hemoglobin. The writing of this article has, therefore, even if it serves no other end, been a means of educating them about many phases of hemoglobin in which their prior knowledge was limited or nonexistent. It has pointed out with renewed emphasis what a storehouse of information about hemoglobin is available—much (most?) of it remains to be understood.

APPENDIX. SOURCES OF DATA

Kinetic Constants and Half-times

Association reaction:

$$Hb + O_2 \xrightarrow{k'} HbO_2$$

$$\frac{d[HbO_2]}{dt} = k'[Hb][O_2]$$

which at constant P_{O_2} gives a half-time for the association of O_2 with a heme moiety of

$$^{Assoc}T_{\frac{1}{2}} = \frac{\ln 2}{k'[O_2]}$$

The concentration of dissolved oxygen at 37° was calculated from

$$[O_2] = \alpha P_{O_2} = 1.23 \times 10^{-6} P_{O_2}$$

where $[O_2]$ is in moles/liter and P_{O_2} in mm Hg (Lange, 1961). The coefficient α is for pure water; however, the presence of hemoglobin in aqueous solution has only a negligible effect on this coefficient for nitrogen and carbon monoxide (Longmuir and Roughton, 1952) and, by presumption, for oxygen.

For k' we have taken the average, $2.8 \times 10^6 M^{-1}sec^{-1}$, of the values found at 37°C for sheep hemoglobin, $1.8 \times 10^6 M^{-1}sec^{-1}$ (Gibson and Roughton, 1958), and for human hemoglobin, $3.8 \times 10^6 M^{-1}sec^{-1}$ (Gibson et al., 1955). Possibly, in the erythrocyte, because even in venous blood the hemoglobin is 78 percent saturated, k' should be taken as $k' = k'_4 = 3.2 \times 10^7 M^{-1}sec^{-1}$ (Gibson, 1959). This would reduce the $^{Assoc}T_{1/2}$'s in the text by a factor of 3.5. Between pH 7 to 10, k' is nearly independent of temperature and pH (Hartridge and Roughton, 1925; Roughton, 1959).

Dissociation reaction:

$$HbO_2 \xrightarrow{k} Hb + O_2$$

$$\frac{-d[HbO_2]}{dt} = k[HbO_2]$$

$$^{Diss}T_{\frac{1}{2}} = \frac{\ln 2}{k}$$

We have calculated k from

$$K_e = \frac{k}{k'}$$

taking K_e ($= P_{\frac{1}{2}}$) to be 25.5 mm Hg for human hemoglobin at 37°C (Staub et al., 1962). The $^{Diss}T_{1/2}$ so calculated, 0.0078 sec, is in approximate agreement with those found at 37°C by reduction of HbO_2 with dithionite (0.0045 sec, sheep, Hartridge and Roughton, 1923; 0.005 sec, man, Gibson and Roughton, 1958; 0.003 sec, man, Dalziel and O'Brien, 1961). Since the course of the dithionite reduction is not completely understood, it seems preferable to use the value of k calculated from K_e and k'. Again, perhaps k should be taken to be k_4, the rate constant for the dissociation of the first oxygen in a fully oxygenated molecule. Within the experimental error k_4 does not differ from the other values quoted above (Gibson and Roughton, 1955). Unlike k', k is very strongly temperature dependent; $Q_{10} = 3.1$ (Roughton, 1959), corresponding to an activation energy of about 19,700 kcal/mole. This value was used to convert, where necessary, the literature values to a common basis of 37°C. $^{Diss}T_{1/2} = 0.0078$ sec differs markedly from the value of 0.1 sec quoted by Wyman (1966) as being from Roughton. At 19°C our value would become 0.057 sec, in satisfactory agreement with 0.1 sec.

Diffusion Constants in 34 percent w/v Hemoglobin Solutions at 37°C

Oxygen: $D = 7.1 \times 10^{-6}$ cm²/sec, calculated from D_{N_2} and D_{CO} together with viscosity measurements on hemoglobin solutions (Roughton, 1959)

4.6×10^{-6} cm²/sec, direct measurement (Klug et al., 1956)

10.0×10^{-6} cm²/sec, direct measurement (Polson, 1939)

7.5×10^{-6} cm²/sec, direct measurement with O_2 electrode (Keller and Friedlander, 1966)

Av. = 7.3×10^{-6} cm²/sec

Hemoglobin: $D = 1.75 \times 10^{-7}$ cm²/sec, direct measurement (Keller and Friedlander, 1966)

The earlier calculated estimate by Langmuir and Roughton (1952), $D = 2 \times 10^{-8}$ cm²/sec, can no longer be considered tenable since it contradicts experimental fact (La Force and Fatt, 1962; Keller and Friedlander, 1966).

ACKNOWLEDGMENTS

We take this opportunity to thank Dr. James Bonner, Dr. Richard T. Jones, and Dr. Phillip Sturgeon for their comments and suggestions after reading the manuscript of this paper. Discussions with Dr. S. K. Friedlander and Mr. E. E. Spaeth were most helpful and stimulating.

REFERENCES

Allen, D. W., W. A. Schroeder, and J. Balog (1958). *J. Am. Chem. Soc.* **80**, 1628.

Anderson, H. M., and J. C. Turner (1960). *J. Clin. Invest.* **39**, 1.

Bateman, J. B., S. S. Hsu, J. P. Knudsen, and K. L. Yudovitch (1953). *Arch. Biochem.* **45**, 411.

Carlsen, E., and J. H. Comroe, Jr. (1958). *J. Gen. Physiol.* **42**, 83.

Dalziel, K., and J. R. P. O'Brien (1961). *Biochem. J.* **78**, 236.

Danielli, J. F. (1958). In Danielli, J. F., K. G. A. Pankhurst, and A. C. Riddiford, eds., *Surface Phenomena in Chemistry and Biology*. New York: Pergamon Press.

Dittmer, D., ed. (1961). *Bood and Other Body Fluids.* Washington: Federation of American Societies for Experimental Biology.

Eggers, D. F., Jr., N. W. Gregory, G. D. Halsey, Jr., and D. F. Rabinovitch (1964). *Physical Chemistry*, p. 396. New York: John Wiley & Sons.

Gibson, Q. H., F. Kreuzer, E. Meda, and F. J. W. Roughton (1955). *J. Physiol.* **129**, 65.

Gibson, Q. H., and F. J. W. Roughton (1955). *Proc. Roy. Soc.* (London) B **143**, 310.

Gibson, Q. H., and F. J. W. Roughton (1958). *J. Physiol.* **140**, 37P.

Gibson, Q. H. (1959). *Biochem. J.* **71**, 293.

Grant, M. M., T. P. Bond, R. G. Cooper, and J. R. Derrick (1963). *Science* **142**, 1319.

Harris, J. W. (1963). *The Red Cell.* Cambridge: Harvard Univ. Press.

Hartridge, H., and F. J. W. Roughton (1923). *Proc. Roy. Soc.* (London) A **104**, 395.

Hartridge, H., and F. J. W. Roughton (1925). *Proc. Roy. Soc.* (London) A **107**, 654.

Harvey, E. N. (1954). *Protoplasmatologia* **2**, E5.

Hillier, J., and J. F. Hoffman (1953). *J. Cell. Comp. Physiol.* **42**, 203.

Hoffman, J. F. (1958). *J. Gen. Physiol.* **42**, 9.

Holmquist, W. R., and W. A. Schroeder (1966a). *Biochemistry* **5**, 2489.

Holmquist, W. R., and W. A. Schroeder (1966b). *Biochemistry* **5**, 2504.

Horton, B. F., and T. H. J. Huisman (1965). *Brit. J. Haematol.* **11**, 296.

Huisman, T. H. J., J. v. d. Brande, and C. A. Meyering (1960). *Clin. Chim. Acta* **5**, 375.

Huisman, T. H. J., and A. M. Dozy (1962). *J. Lab. Clin. Med.* **60**, 302.

Huisman, T. H. J., and B. F. Horton (1965). *J. Chromatog.* **18**, 116.

Jones, R. T. (1961). Ph.D. thesis. California Institute of Technology.
Kavanau, J. L. (1963). *Nature* (London) **198**, 525.
Keller, K. H., and S. K. Friedlander (1966). *J. Gen. Physiol.* **49**, 663.
Klug, A., F. Kreuzer, and F. J. W. Roughton (1956). *Helv. Physiol. Pharm. Acta* **14**, 121.
Kreuzer, F., and W. Z. Yahr (1960). *J. Appl. Physiol.* **15** (6), 117.
Kunkel, H. G., R. Ceppellini, O. Muller-Eberhard, and J. Wolf (1957). *J. Clin. Invest.* **36**, 1615.
Kunkel, H. G., and G. Wallenius (1955). *Science* **122**, 288.
La Force, R. C., and I. Fatt (1962). *Trans. Faraday Soc.* **58**, 1451.
Lange, N. A. (1961). *Handbook of Chemistry*. New York: McGraw-Hill.
Laurent, G., M. Charrel, C. Marriq, and Y. Derrien (1962). *Bull. Soc. Chim. Biol.* **44**, 419.
Longmuir, I. S., and F. J. W. Roughton (1952). *J. Physiol.* **118**, 264.
Morrison, M., and J. L. Cook (1955). *Science* **122**, 920.
Muirhead, H., and M. F. Perutz (1963). *Nature* (London) **199**, 633.
Muller, C. J. (1961). *Molecular Evolution*. Assen, the Netherlands: Van Gorcum.
Murayama, M., R. A. Olson, and W. H. Jennings (1965). *Biochim. Biophys. Acta* **94**, 194.
Norris, C. H. (1939). *J. Cell. Comp. Physiol.* **14**, 117.
Oncley, J. L. (1938). *J. Am. Chem. Soc.* **60**, 1115.
Perutz, M. F. (1948). *Nature* (London) **161**, 204.
Perutz, M. F. (1965). *J. Mol. Biol.* **13**, 646.
Polson, O. (1939). *Koll. Z.* **87**, 149.
Prankerd, T. A. J. (1961). *The Red Cell*. Oxford: Blackwell.
Roughton, F. J. W. (1959). *Prog. in Biophysics and Biophysical Chem.* **9**, 55.
Roughton, F. J. W. (1963). *Brit. Med. Bull.* **19**, 80.
Roughton, F. J. W., and R. E. Forster (1957). *J. Appl. Physiol.* **11**, 260.
Schnek, A. G., and W. A. Schroeder (1961). *J. Am. Chem. Soc.* **83**, 1472.
Scholander, P. F. (1960). *Science* **131**, 505.
Schroeder, W. A., and R. T. Jones (1965). *Fortschr. Chem. organ. Naturstoffe* **23**, 113.
Staub, N. C. (1963). *J. Appl. Physiol.* **18**, 673.
Staub, N. C., J. M. Bishop, and R. E. Forster (1961). *J. Appl. Physiol.* **16**, 511.
Staub, N. C., J. M. Bishop, and R. E. Forster (1962). *J. Appl. Physiol.* **17**, 21.
Wittenberg, J. B. (1966). *J. Biol. Chem.* **241**, 104.
Wyman, J. (1966). *J. Biol. Chem.* **241**, 115.

EMILE ZUCKERKANDL
Département de Biochemie Macromoléculaire
Centre National de la Recherche Scientifique
Montpellier, France

Hemoglobins, Haeckel's "Biogenetic Law," and Molecular Aspects of Development

The question has been asked whether the evolution of hemoglobins is in accord with the suggestion that ontogenesis repeats phylogenesis, as Haeckel and Fritz Müller contended (Cuenot, 1951)—more specifically, whether polypeptide chains that function in the embryo are evolutionarily older than their adult counterparts. Haeckel's so-called "biogenetic law" has been severely criticized for its exaggerations, its pretense to rigor and generality, and its erroneous implication that the embryos of contemporary organisms recapitulate the *adult* stage of ancestral forms, whereas recapitulation, to the extent it exists, rather relates to *preadult* stages of the ancestral organism (De Beer, 1964). In spite of its shortcomings, Haeckel's generalization still lurks in the minds of many biologists, because there is a large body of observations that point in its direction. We may wonder what molecules, and especially informational macromolecules, may have to offer in the matter. Informational macromolecules are indeed the fundamental substratum of both constancy and change in organisms and, in conjunction with the environment, the key to the evolutionary process.

Some vertebrate hemoglobin chains are the only type of informational macromolecules whose structurally different "editions," in the embryo and in the adult, have been thoroughly studied with respect to their primary structure (in this case, an amino acid sequence). Only two fetal hemoglobin chains have so far been analyzed in this way—the human fetal non-α chain, called the γ chain (Schroeder et al., 1963), and the cattle fetal chain (Babin et al., 1966). The latter does not appear to be related to the human γ chain by way of the type of homology that makes it advisable to designate two structurally distinct chains by the same Greek letter (Zuckerkandl and Pauling, 1965), but, to conform to usage, it might be called the "γ" chain (with quotation marks). These fetal chains may be compared with the corresponding non-α chains of cattle and human adults, the so-called β chains,* whose primary structure has also been elucidated—completely in the case of the human chain (Braunitzer et al., 1961; Konigsberg et al., 1963) and incom-

This work is part of a series of investigations that are rendered possible, in part, by a grant from N.I.H. No. GM-11272, and by a grant from Délégation Générale à la Recherche Scientifique et Technique, No. 66-00-186.

*In a chain nomenclature oriented toward a phyletic classification, the cattle chain also should be "β" with quotation marks (Zuckerkandl and Pauling, 1965a).

pletely in the case of the cattle chain (Schroeder and Jones, 1965). Other globin chains will also be drawn into the subsequent comparisons. All comparisons are made on the basis of the assumption that polypeptide chains evolve mostly through successive substitutions of individual amino acid residues.

It will become apparent from what follows that the only two structurally known fetal polypeptide chains do not accord with the theory of recapitulation, and therefore many other fetal proteins may be expected to be equally discordant with the theory. After presenting the evidence on which this statement is based, the general situation that the "biogenetic law" tried to describe will be tentatively reformulated as it appears when viewed from the molecular angle.

Structural differences between fetal and adult hemoglobin types are very frequently found, but not universally—at least not when adult organisms are compared with embryos of a relatively advanced developmental stage. For instance there seems to be no specific fetal hemoglobin in the horse (Stockwell et al., 1961). Differences in primary structure between fetal and adult hemoglobin types, or differences in some physicochemical characteristics with probable implications with respect to amino acid sequence, have, however, been detected in vertebrates as primitive as cartilaginous fish (Manwell, 1963) and in the larva of one of the most primitive contemporary vertebrates, the Cyclostome *Petromyzon* (Adinolfi et al., 1959). In some birds several structurally distinct hemoglobins have been found (see, for example, Manwell et al., 1963; Borgese and Bertles, 1965). Thus, in the duck, an early embryonic hemoglobin is followed by a late embryonic hemoglobin, both components being distinct from the adult hemoglobins (Borgese and Bertles, 1965).

It should be emphasized at the outset that the hope of recognizing in a fetal hemoglobin a more "primitive" (in the sense of less highly organized) molecule than in a corresponding adult hemoglobin is, for the most part, bound to be illusory. True, there are some steps of functional advance in hemoglobin evolution, as in the establishment of heme-heme interaction or of a Bohr effect. Such innovations may be termed progress, but they occur very seldom during molecular evolution. In between such events there is no basis on which to consider one globin molecule as more or less highly evolved than another. Perhaps an organism may or may not be "primitive" in the proposed sense, but to say this of a molecule is irrelevant most of the time. Therefore, in inquiring about recapitulation in hemoglobins, the question to be asked must be a different one: Which of two hemoglobins of an organism, the fetal or the adult one, is structurally more similar to their common molecular ancestor?

RATES OF EVOLUTION OF ADULT AND FETAL HEMOGLOBIN CHAINS

According to Ingram's scheme, as published in his classical paper (1961) on the evolution of hemoglobin, the β chain of human adult hemoglobin is derived from the fetal γ chain. The similar scheme that we devised at the California Institute of Technology in 1960 was, likewise, based on the postulates of gene duplication followed by independent mutation of daughter genes and eventually by the translocation of some of these genes; but it differed from Ingram's in one detail of nomenclature, in that we thought fit to consider the common ancestor of any two genes as a gene differing from each of them rather than as identical to one of them, and, accordingly, to give the ancestral gene a name of its own. Thus, the common ancestor of the β, γ, and δ genes was called the β-γ-δ gene. (The δ chain is the non-α chain of a minor hemoglobin component of the human adult.) Figure 1 shows this formal difference between Ingram's and our concept of the evolutionary derivation of hemoglobin genes. The difference is not without some meaning, since Ingram's diagram might be taken to suggest that the γ chain is indeed older than the β chain, as the "biogenetic law" would have it. The phenomenon of duplication implies of course that the two daughter genes have arisen simultaneously from the mother gene, and thus Ingram's scheme amounts to suggesting that the contemporary γ chain is more similar in structure to the ancestral chain than the contemporary β chain—in other words, that, since the time of gene duplication, the β chain has been subjected to evolutionarily effective amino acid substitutions at a higher rate than the γ chain.

Evidence to the contrary—and thus evidence against recapitulation—is contained in the matrix of Table 1, which presents the proportion of differences in amino acid sequence and the minimum average number of corresponding base substitutions per DNA site for various globin chains. The β chains and γ chains seem to have changed at an approximately equal rate, since both show practically the same number of differences, not only in relation to myoglobin (in which case, however, the total number of differences is so great that smaller differences may be blurred on account of a larger proportion of mutational "double hits"), but also in relation to either the carp α chain or the mammalian α chain. This relationship holds whether the figures are expressed in terms of amino acid differences or in terms of the minimum number of base substitutions in the gene that have to be postulated in order to account for the observed differences in amino acid sequence. Thus it seems preferable to adopt the Caltech nomenclature, perhaps as shown in Figure 1b.

The question still remains of whether the ancestral chain functioned as a

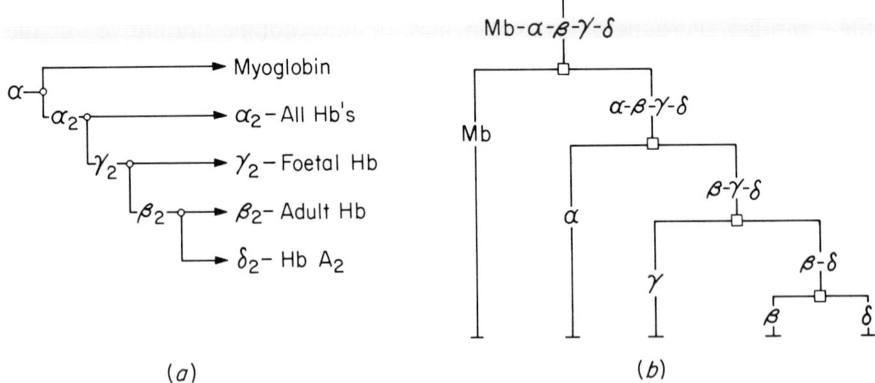

FIGURE 1.
Two diagrams of gene duplication in globins. (Circles or squares represent gene duplication.) (*a*) Ingram's diagram (1961). (*b*) A version of the Caltech diagram.

fetal chain or as an adult chain, or as a chain present in both fetus and adult. Extensive sequence studies on vertebrate hemoglobin should, in the future, give us a clue to the answer in terms of a probability. There is no reason to assume that the β-γ-δ gene has been the first "non-α" gene. It is possible that another non-α gene had arisen at an early time of vertebrate evolution. Then, after the duplication of the β-γ-δ gene to yield the β-δ and the γ gene, this hemoglobin non-α gene, which heretofore had been functional in either the fetus or the adult, may have disappeared from the genome, or become "silent," or been shifted, for its period of synthesis and function, to an as yet incompletely explored period of early ontogenesis.

That the β chain and the γ chain evolve at an approximately equal rate may appear surprising. Perhaps the γ chain has become instated as a fetal chain only in relatively very recent evolutionary times and has in fact been an "adult" hemoglobin chain over most of the time of its existence. This may well not be so. The high resistance to alkali denaturation of hemoglobins from adult prosimians (Buettner-Janusch and Twitchell, 1961)—a typical γ chain characteristic—would be in keeping with the view that the γ chain may have functioned for a long time as an "adult chain." But sequence studies on the adult major component non-α chain of *Lemur fulvus* (Buettner-Janusch and Hill, 1965) show that this chain, although intermediate in character between the human β chain and the human γ chain—in some ways different from both, and, in fact, in some ways different from any other known hemoglobin chain—is still closer in structure to the human β chain than to the human γ chain. Thus an incomplete sequence analysis (Buettner and Hill, 1965) shows, over the known stretches of sequence, that the *Lemur* chain is at 18 positions identical with the human β chain but different from the human γ chain. It is at only 7 positions identical with the human γ chain but different from the human β chain.* Plainly the non-α chain of adult *Lemur* is not a γ chain, and thus the γ chain probably did not play the role of an adult hemoglobin chain in so remote an ancestor as that common to *Lemur* and man.

If, thus, we suppose that the γ chain has functioned as a fetal chain over all or much of the time of its existence, and if, on the other hand, the mammalian embryos, as will be proposed below, can indeed be said to evolve more slowly (to change less drastically in form, function, and behavior) than the adult mammals, then the observation on the equal rate of evolutionarily effective amino acid substitutions in hemoglobin β and γ chains fits in well

*Moreover, there are at least 11 characters of sequence that are either entirely "original" in relation to all known non-α chains (6 of them) or occur (5 of them), not in known β and γ chains, but only in β-δ (Zuckerkandl and Pauling, 1965a) chains (horse, cattle). Although no definite conclusion can be reached at this point, the descent of the *Lemur* adult non-α chain from a particular duplication of the β-δ chain appears possible.

The N-terminal threonine residue of the *Lemur* "β" chain probably has derived from a valine, which is the usual N-terminus. This substitution must however have taken place in two mutational steps (Goldberg and Wittes, 1966), through an intermediary amino acid residue—for instance, methionine, which has already been found as an N-terminus in some non-α chains (Schroeder and Jones, 1965). In view of the relative evolutionary stability of the N-terminal residue (even in non-α chains, in which it is less stable than in the α-chain), this two-step substitution also suggests that the origin of the *Lemur* "β" chain may well be more remote than the common ancestor of *Lemur* and man.

TABLE 1.
Comparison between globin chains.

	α Man	α Horse	α Pig	α Cattle	α Mouse	α Carp	β Man	β Lemur	"β" Horse	"β" Cattle	γ Man	"γ" Cattle	Lamprey	Mb SW
α Man		12 0.15	12.8 0.17	$\sim$ 19.2	$\sim$ 12.8 0.16	$\sim$ 40 0.55	53 0.70	$\approx$ 55 0.72	55 0.74	55 0.71	55 0.72	58 0.76	$\sim$ 60 0.85	72 1.06
α Horse	12 0.15		10.6 0.16	$\sim$ 27	$\sim$ 16.3 0.19	$\sim$ 42 0.57	54 0.70	$\approx$ 53 0.74	53 0.71	52 0.68	53 0.68	56 0.72	$\sim$ 63 0.92	74 1.08
α Pig	12.8 0.17	10.6 0.16		$\sim$ 17	$\sim$ 17.8 0.21	$\sim$ 45 0.59	54 0.71	$\approx$ 53 0.72	53 0.72	55 0.71	55 0.72	57 0.75	$\sim$ 64 0.95	72 1.06
α Cattle	$\sim$ 19.2	$\sim$ 27	$\sim$ 17		$\sim$ 14.2	$\sim$ 42	$\sim$ 56		$\sim$ 58	$\sim$ 57	$\sim$ 56	$\sim$ 57	$\sim$ 64	$\sim$ 72
α Mouse	$\sim$ 12.8 0.16	$\sim$ 16.3 0.19	17.8 0.21	$\sim$ 14.2		$\sim$ 41 0.53	$\sim$ 54 0.69	$\approx$ 58 0.76	$\sim$ 53 0.67	$\sim$ 52 0.68	$\sim$ 50 0.63	$\sim$ 60 0.76	$\sim$ 64 0.89	$\sim$ 74 1.04
α Carp	$\sim$ 40 0.55	$\sim$ 42 0.57	$\sim$ 45 0.59	$\sim$ 42	$\sim$ 41 0.53		$\sim$ 52 0.66	$\approx$ 57 0.71	$\sim$ 56 0.77	$\sim$ 54 0.69	$\sim$ 54 0.71	$\sim$ 57 0.71	$\sim$ 67 0.93	$\sim$ 76 1.07
β Man	53 0.70	54 0.70	54 0.71	$\sim$ 56	$\sim$ 54 0.69	$\sim$ 52 0.66		$\approx$ 20.5 0.25	17.8 0.23	15.8 0.19	27 0.34	22 0.25	$\sim$ 70 1.02	73 1.1

β Lemur	$\approx$ 55 0.72	$\approx$ 53 0.74	$\approx$ 53 0.72		$\approx$ 58 0.76	$\approx$ 57 0.71	$\approx$ 20.5 0.25		$\approx$ 22 0.31	$\approx$ 22 0.25	$\approx$ 26 0.31	$\approx$ 19 0.27	$\approx$ 65 0.89	$\approx$ 66 0.99
"β" Horse	55 0.74	53 0.71	53 0.72	$\sim$ 58	$\sim$ 53 0.67	$\sim$ 56 0.77	17.8 0.23	$\approx$ 22 0.31		21 0.25	29 0.35	21 0.25	$\sim$ 70 1.04	75 1.08
"β" Cattle	55 0.71	52 0.68	55 0.71	$\sim$ 57	$\sim$ 52 0.68	$\sim$ 54 0.69	15.8 0.19	$\approx$ 22 0.25	21 0.25		27 0.34	14.4 0.17	$\sim$ 72 1.03	75 1.08
γ Man	55 0.72	53 0.68	55 0.72	$\sim$ 56	$\sim$ 50 0.63	$\sim$ 54 0.71	27 0.34	$\approx$ 26 0.31	29 0.35	27 0.34		26 0.33	$\sim$ 70 1	76 1.07
"γ" Cattle	58 0.76	56 0.72	57 0.75	$\sim$ 57	$\approx$ 60 0.76	$\sim$ 57 0.71	22 0.25	$\approx$ 19 0.27	21 0.25	14.4 0.17	26 0.33		$\sim$ 68 1.03	74 1.04
Lamprey	$\sim$ 60 0.85	$\sim$ 63 0.92	$\sim$ 64 0.95	$\sim$ 64	$\sim$ 64 0.89	$\sim$ 67 0.93	$\sim$ 70 1.02	$\approx$ 65 0.89	$\sim$ 70 1.04	$\sim$ 72 1.03	$\sim$ 70 1	$\sim$ 68 1.03		$\sim$ 75 1.08
Mb SW	72 1.06	74 1.08	72 1.06	$\sim$ 72	$\sim$ 74 1.04	$\sim$ 76 1.07	73 1.1	$\approx$ 66 0.99	75 1.08	75 1.08	76 1.07	74 1.04	$\sim$ 75 1.08	

NOTE: *Upper figure in each box:* percent differences in amino acid sequences. *Lower figure:* average minimum number of base substitutions per amino acid site. When results are approximate, degree of approximation is indicated by $\sim$ or $\approx$. Figures in boxes not carrying one of these signs may be incorrect by a small amount. The genetic code used for the computation is as in Goldberg and Wittes (1966).

with the idea that the rate of evolution of the organism is largely independent of the rate of amino acid substitution in its proteins (Zuckerkandl and Pauling, 1965a). This is to be expected if most of the evolutionarily effective amino acid substitutions do not entail any important functional change. Thus embryonic proteins may undergo as many evolutionarily effective amino acid substitutions as their adult counterparts, but may tolerate a smaller number of functionally significant ones—that is, a smaller number of mutations that are expressed in striking changes in morphology, physiology, and behavior. Further structural work on fetal and adult hemoglobin chains is needed to substantiate this impression.

THE SUCCESSION OF DIFFERENT FETAL CHAINS DURING HEMOGLOBIN EVOLUTION

That the contemporary γ chain does not "recapitulate" ancient evolution is confirmed by the approximate relation that exists, on the average, between the number of differences between homologous polypeptide chains and the time in the past at which their common molecular ancestor presumably existed in an ancestral organism (Zuckerkandl and Pauling, 1962, 1965a; Margoliash and Smith, 1965). Let us take 7×10^6 years as the mean period of time between evolutionarily effective amino acid substitutions (Zuckerkandl and Pauling, 1965a). This figure, even though its roughness is duly acknowledged, may not apply to a meaningful approximation to certain individual cases, but these cases, if indeed pertinent,* seem to tend to reduce the figure rather than to increase it (Buettner-Janusch and Hill, 1965). With 39 differences between the β chain and the γ chain of man (Schroeder et al., 1963), their common ancestor is thus not expected to date back further than about 140 million years. This figure coincides geologically with the limit between the Jurassic and the Cretaceous eras, a time at which fish had already existed for roughly another 300 million years, and reptiles (the common ancestors of birds and mammals) for about another 150 million years (Newell, 1963). Thus the common ancestor not only of fish and man but also of bird and man—all forms that possess special "editions" of hemoglobin specific for preadult stages—dates back much further than the common ancestor of the human β chain and γ chain. It is probable that if contemporary primitive organisms possess special fetal hemoglobins, their very distant ancestors did so also. Although proper attention must be paid to the possibility that in some cases an interaction of different enzymes might lead to a morphologically similar result, primitive contemporary organisms should have remained more constant with respect to the types of proteins they contain than advanced contemporary organisms since the time of the common ancestry of the two. In all likelihood, then, fetal hemoglobin chains had long been in existence in the ancestry of man at the time the direct ancestor of the present human fetal chain arose by a new gene duplication. Thus *fetal non-α chains of different origin (with respect to gene duplication) probably succeeded each other during hemoglobin evolution.* A fetal chain, rather than recapitulating evolutionary history, may on the contrary be the product of recent evolutionary history.

*They are not pertinent if an apparently "aberrant" chain is in fact not derived from the particular gene duplication that led to the other chains considered in a comparison.

An examination of the second fetal chain whose sequence is known—the cattle fetal ("γ") chain (Babin et al., 1966)—reinforces this point of view. It has been shown elsewhere (Zuckerkandl and Pauling, 1965a) that this cattle fetal chain seems to be more closely related to the cattle and human adult non-α major hemoglobin chains than to the human fetal hemoglobin chain. The number of differences in amino acid sequence between the cattle fetal ("γ") chain and the cattle adult ("β") non-α chain is smaller than between the human fetal (γ) chain and adult non-α chain, being about 21* as against 39. Also the number of differences between the cattle "β" and human β chains (that is, between the adult major component non-α chains of the two animals) is smaller than the number of differences between the cattle fetal and the human fetal non-α chains, being about 23* in the first case as against 39 in the second. Furthermore, some special features of the sequences also point to a particularly close relationship between the cattle fetal chain and the adult non-α chain. It is thus reasonable to assume that the fetal chain found in cattle has arisen as a result of a gene duplication that is much more recent than the one that gave rise to the human β and γ chains. At any rate, the relatively small difference of about 22 characters of sequence between the fetal and the adult non-α chains in cattle points to a common origin of these chains that is well within the time of mammalian evolution. The cattle fetal chain has not been in existence for any impressive stretch of evolutionary history, and it would be meaningless to say that it is "recapitulating" anything.

It may be that similar shifts and replacements of genes with respect to the timing, during the course of ontogenesis, of the synthesis of corresponding polypeptide chains will be found to be a widespread phenomenon in proteins that exist in structurally distinct fetal and adult "editions."

THE RELATIVELY RECENT EVOLUTIONARY ORIGIN OF MANY HEMOGLOBIN GENE DUPLICATES AS FOUND IN ANY ONE ORGANISM

Besides the γ chain, there exists in humans another hemoglobin polypeptide chain, the so-called ϵ chain, normally found only in very young fetuses (Huehns et al., 1964). There is little doubt that this structurally distinct ϵ chain is controlled by a locus different from the loci of the other human hemoglobin chains. The primary structure of this chain has not yet been established. But it is worthwhile to point out that this earlier human fetal hemoglobin chain forms a tetrahemic hemoglobin in combination with α chains, just as the other human non-α chains do. If molecules in very young embryos generally recapitulated very primitive stages of vertebrate evolution, we might have expected to find an early embryonic hemoglobin chain that, at least in the oxygenated state, would not associate with α chains to form a tetrahemic molecule. The most primitive contemporary vertebrates, the Cyclostomes, indeed contain hemoglobins that exist as free single chains in the oxygenated state, as myoglobins do. So far as hemoglobins go, this primitive

*This number, smaller than the approximate one indicated by Zuckerkandl and Pauling (1965a), has been revised according to Schroeder and Jones (1965).

phylogenetic stage and probably subsequent ones as well have apparently been lost for ontogenesis in man.

From the degree of difference between chains, the oldest gene duplication of which there is evidence left among human hemoglobins is the one that gave rise to the α chain and a non-α chain, the latter leading later through three subsequent gene duplications to the β, γ, δ, and ϵ chains. The sequence of these duplications is given in Figure 1, except for the one, whose temporal localization is still undefined, that led to the appearance of the ϵ chain. The α-non-α duplication is somewhat older than the common ancestor of bony fish and man and less old than the common ancestor of lamprey and man. In fact, according to the minimum average number of base substitutions per DNA coding triplet, the α-non-α duplication is just intermediate in age between the ages of these two common ancestors: α_{man}–β_{man}:0.70; α_{man}–α_{carp}:0.55; α_{man}–lamprey chain:0.85; see Table 1. This is a rather old age. In the absence of a show of "recapitulation" performed by non-α chains, does the α chain lend itself to Haeckel's generalization? It precisely does not, since it is the one mammalian hemoglobin chain that is synthesized throughout the life span of the individual, from the earliest embryonic stages that have been investigated (in man).

As to the number of differences between the structurally known human non-α hemoglobin chains, they are smaller than the number of differences between the carp α and the human α chains (Table 1). Since evolutionarily effective amino acid substitutions occur in non-α chains rather faster than they do in α chains, this finding means that the various non-α chains present in contemporary man all have arisen, so far as is known from sequence studies, at a time later than that of the common ancestor of bony fish and man. From a comparative study of tryptic peptide maps (Huehns et al., 1964) it appears very likely that this holds also for the ϵ chain. This early fetal chain may not be more different from the adult β chain than is the later fetal chain, the γ chain. This would be a further blow for "recapitulation." At any rate, the impression is gained that, *among the several "editions" of a polypeptide chain that are controlled by distinct genetic loci and found in the same individual, most are relatively similar in primary structure and thus relatively young.*

What do "old" and "young" exactly mean in the case of an informational macromolecule? The sense is clear enough as long as we speak of the age of a "type" of polypeptide chain—that is, one that has a type of tertiary structure and carries out a type of function. The type of polypeptide chain must have arisen at one period of evolutionary history, and this period defines its age. But what about the age of different polypeptide chains of a given type, such as hemoglobin and myoglobin chains? Clearly all hemoglobin chains are equally young, since they have just been synthesized by a cell, and equally old, since all of them can, in principle, be traced back to the first hemoglobin chain that nature invented, or at least to the first hemoglobin chain that was ancestral to the contemporary vertebrate hemoglobins—surely some invertebrate hemoglobin chain. An old chain, then, may be a chain whose sequence has presumably changed little in comparison with a remote chain ancestor. However, since all chains seem to have been the seat of periodic amino acid substitutions, and since their rate does not appear to differ very considerably for different homologous chains, this definition of old age is presumably of little value much of the time when applied to homologous polypeptide chains and genes. In practice, the only constantly meaningful definition appears to be one that refers to gene duplication. Old or young homologous hemoglobin chains will be chains that derive from a remote or recent gene duplication.

In this sense, the majority of human hemoglobin chains are relatively young, if the time span under consideration is the one of vertebrate evolution since its incipient stages. It may be that this observation on human chains will turn out to be generally applicable and that most organisms lose some very old gene duplicates in the course of evolution and replenish their store of multiple editions of a given polypeptide chain by relatively recent duplications. In this connection it should be mentioned that the fetal and the adult non-α chain of the cartilaginous fish *Squalus suckleyi* appear to be very similar in structure (Manwell, 1963). As to the six hemoglobin polypeptide chains that have been found in the Cyclostome *Petromyzon* (Adinolfi et al., 1959), it will be of interest to find out how similar in sequence they are.

Thus, so far as hemoglobins go, the theory of recapitulation seems to be altogether unsuccessful. Yet most of the critics of this theory acknowledge that there is *something* to it. In what light does that "something" appear when viewed from a more general molecular and genetic level? Phenomena that, happily or unhappily, have been termed "recapitulation" do occur, and they should be controlled by a group of genes that does not include the hemoglobin genes. This, at least, seems a logical conclusion, although it is not as certain as it appears at first, as we shall see in the course of the subsequent discussion.

ONTOGENETIC ACTIVITY GROUPS OF GENES

Part of the contemporary structural genes are likely to be very similar to genes that existed one billion years ago or more (see, for instance, Eck and Dayhoff, 1966; Margoliash and Smith, 1965; Zuckerkandl and Pauling, 1965a). Apart from these structural constancies, there may be a core of structural genes whose *activity distribution*, over the developmental stages of the organisms, from the point of view of messenger RNA and protein synthesis, remains similar throughout evolution. The activity of such genes should be responsible, for instance, for the rudiments of branchial slits, characteristic of fish, at one stage of development of the human embryo, or for the formation of rudimentary teeth in the fetus of whales. Besides groups of genes whose synthetic activity is promoted and/or inhibited simultaneously during normal development of the organism, there are, from the point of view of regulation, two further classes of structural genes: those that are synthetically active throughout the life cycle of a given organism (which does not preclude variations in activity over periods much shorter than a developmental stage through repression or induction), and those that are in general totally inactive throughout this life cycle. Of the latter, some genes may be active in only one type of tissue, others in none. Structural genes for inducible or repressible enzymes may not be inducible by substrate or repressible by product at all times of ontogenetic development. In that case they are to be counted among certain temporal activity groups, even though in the absence of induction or in the presence of repression they will not be active at the same time as the other genes of the activity group.

Gene activity complexes are not expected to be impermeable to "gene flow" any more than are neighboring populations belonging to the same species. *Individual genes, or even groups of genes, may "migrate" from one gene activity complex to another.* In some cases this migration may occur physically by translocation within the genome, in others by mutation, deletion, or insertion of regulatory genes without any displacement of structural genes. The evolutionary history of hemoglobin genes, little as there is at present known about it, already seems to offer at least one example of such intragenome gene flow. Mention was made earlier of a gene duplication that presumably occurred in the ancestry of cattle. The ancestral gene, before duplication, was probably active in the adult. Indeed, its contemporary descendants show about the expected minimum number of base substitutions when compared with the numbers characteristic of other mammalian adult non-α chains (Table 1). Thus a gene of an adult ancestor of cattle apparently gave rise to two daughter genes; one continued in its adult function, and the other migrated into a fetal gene activity complex. The opposite migration should also occur, but no example is known.

There will be many transitions between different activity groups of genes, so that their boundaries can be expected to be blurred and their classification to be complex. Nevertheless, such activity groups undoubtedly exist, in view of the detection of the successive production of different m-RNA's during morphogenesis (Doi and Igarashi, 1964). The principal point to be made here is that, *within limits, gene activity groups should evolve independently.* That they can do so is demonstrated by the cases in which a greatly different embryonic development leads to closely related and homologous adult forms —as for instance, with the higher Crustaceans *Paeneus* and the crayfish (see, for example, Cuenot, 1951). There will thus be several embryonic activity groups of genes that will evolve semiautonomously, as do adult groups of genes.

The genes in the different activity complexes will evolve at different average rates. Sometimes the rate of evolution will be highest at some embryonic stages, as is obviously the case with the Crustaceans just mentioned and in a number of other instances. But in general embryonic and fetal gene groups seem to evolve at slower average rates than gene complexes that are active in the adult. This is to be expected (1) because during considerable evolutionary time-spans embryos generally live in more constant environments than adults, and (2) because certain important driving forces in evolution do not come into the picture in embryos. This second reason refers to the necessity of behavioral adaptation to other organisms that present adaptive challenges, either as predators or as potential prey. More generally, embryos differ from adults in that the requirement for *action,* in the case of the former, is minimal except in the case of free-living larval stages, in which rates of evolution may be more rapid; such cases are, however, to be considered separately. It is almost inconceivable that the fundamental difference between most embryos and adults with respect to the amount and kind of action required for the survival of the individual and of the species should not have a significant effect on rates of evolution. Another factor may independently reduce the rate of evolution in embryos in comparison with that of adults—namely, regulatory interrelations between parts of the total genome that should reduce the degrees of evolutionary freedom in earlier stages of ontogenetic development, since some mutations endowed with a favorable effect in relation to one stage of development might be incompatible with the proper gene activity equilibria at later stages. This is a type of limitation that only the gene complexes active in the adult stage should not have to cope with, provided that mutated or newly introduced genes of the adult remain inactive during the embryonic stages.

We may therefore expect that the situation has to a larger extent remained essentially unaltered in gene activity complexes characteristic of the embryonic stages than in gene activity complexes of the adult. In this light, *"recapitulation" would appear essentially as the expression of a rate of evolution in embryonic gene activity complexes that is slower than the rate in adult gene activity complexes.* The magnitude of the effect will depend on what proportion of structural genes is specifically assigned to particular stages of development.

It will be of interest, though altogether for the future, to evaluate the relative sizes of the following subgenomes: the fraction of the total genome—if any—that is engaged exclusively in regulatory activity and does not control the structure of enzymes or other proteins; the fraction of totally silent genes that are active neither as controller genes nor as structural genes and therefore cease to be "genes" altogether, but remain heritable regions of DNA; the fraction of structural genes that are active at all stages of ontogeny, including the adult; the fraction of structural genes that are active exclusively at some preadult stage; and the fraction of structural genes that are active exclusively in the adult. The evolutionary history of these various proportions should be instructive. These proportions may, in fact, be different for every type of tissue of an organism, although there probably exists a core of genes active at all stages of development and in all tissues. The latter would presumably also be the genes least prone to evolutionary change. The relatively very great evolutionary stability of cytochrome c may be a case in point (Margoliash and Smith, 1965; Zuckerkandl and Pauling, 1965a). One of several possible reasons why this would be so may reside in the restricted number of degrees of evolutionary freedom when a structural gene must remain compatible, on the one hand, with the function (in all tissues and at all stages of ontogeny) of the protein whose synthesis it controls, and, on the other hand, with the stability of gene activity complexes in different tissues and at different stages of development. Indeed, structural genes may also exert a regulator activity on other genes. Furthermore, since amino acid substitutions may alter the rate of synthesis of the polypeptide chain in which they occur (Itano, 1965), and since such a change in rate may not be "tolerable" in a universally active gene, this factor may further promote the constancy in sequence of a protein like cytochrome c.

A very rough and indirect evaluation of the proportion of total DNA representing universally active genes can be proposed in the following way.

From the findings of Hoyer et al. (1965) it can be seen (Figure 2, a reproduction of Figure 4 of Hoyer et al.) that since the time of the origin of the vertebrates about 2 percent of the DNA has apparently remained "similar" in all vertebrates. The meaning of this figure is uncertain, because some features underlying the technique of the authors are not yet clear and, even more importantly, because we must distinguish between the preservation of a polypeptide type and of a polypeptide sequence. Polypeptide sequences, and

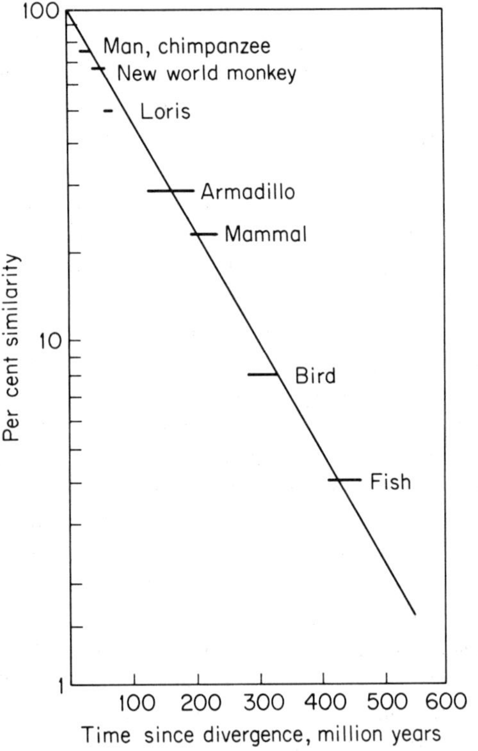

FIGURE 2. Relationship between polynucleotide similarity and time of evolutionary divergence. [From V. Bryson and H. J. Vogel, eds., *Evolving Genes and Proteins*. Academic Press, 1965.]

therefore the corresponding polynucleotide sequences, may be altered profoundly without alteration of the basic functional properties and tertiary structure of the polypeptide, as the example of hemoglobin and myoglobin shows. If 2 percent of all vertebrate DNA remains fairly constant throughout vertebrate evolution, the percentage of gene (polypeptide) *types* that have been preserved may be much larger. It is of course not possible to evaluate this latter figure from the work of Hoyer et al., and the automatic covering of all functional types of DNA (including "inactive" DNA) in the figures of the authors would by itself make any estimate impossible. With respect to this last feature, however, it is striking that the logarithmic relationship found by the authors is compatible with a random decay of DNA similarity throughout vertebrate evolution. The random decay process is exactly what we would expect to find, in keeping with the hypothesis proposed by Zuckerkandl and Pauling (1965a) relative to a fairly constant average rate of evolutionarily effective amino acid substitutions for each type of polypeptide chain, and hence of base substitution in DNA. Not compatible with a random decay process would be the existence of a significant fraction of DNA characterized by rates of base substitution clearly distinct from the rates prevailing in the rest of the genome. The linearity of the figure of Hoyer et al. (Figure 2) suggests that "inactive" stretches of DNA, if quantitatively important, are the seat of evolutionarily effective base substitutions at rates not very different from those prevalent, on the average, in well-characterized genes. If this conclusion were found to be correct, it would be remarkable. It would imply indeed that natural selection works on "inactive" stretches of DNA as well as on "active" ones. This puzzling suggestion would become understandable if much of the "inactive" DNA had in fact highly specific regulatory functions.

Possible reasons for the structural constancy of cytochrome c have been mentioned above: a restricted number of degrees of evolutionary freedom of proteins whose enzymatic function is very widespread in a given organism both in space (different tissues) and time (different phases of ontogeny), and of corresponding structural genes whose regulatory function is very widespread in the same way. For these reasons it does not seem unlikely that the subgenome of genes active at all stages of development and in all tissues of an organism is the one that is most constant—not only with respect to the *types* of polypeptide chains that the genes control, but also in regard to the *sequence* of these polypeptide chains and thus of the sequence of the bases in the corresponding genes. This should be so even in the genes, in spite of the possibility of isosemantic substitutions (= synonymy; Zuckerkandl and Pauling, 1965b; Sonneborn, 1965). On these grounds, the figure drawn from the work of Hoyer et al. may be tentatively used as a basis for the statement that the subgenome of universally active genes probably represents a few percent or less of the total genome.

As to the fraction of structural genes active exclusively at preadult stages, it probably is smaller than the one active exclusively in the adult. Indeed, less complex organisms may contain a smaller amount of total information than more complex organisms (compare Gatlin, 1966), and embryos are, in one sense and with due qualification, simpler organisms. The number of genes subject to evolutionarily effective duplication or to other processes resulting in the increase of genetic material may be smaller at embryonic stages than in the adult organism because in the embryo fewer new genes should in general be needed for adaptive purposes. Of the genes present, a number of those responsible for the specific functions of the different tissues are apparently not yet activated at earlier embryonic stages, when the functions are not yet carried out.

Thus, during evolution toward "higher" organisms, the adult gene activity complex may be expected to increase more rapidly in size and complexity than any embryonic gene activity complex, with the exception of special cases that have already been referred to. As this type of evolution proceeds, one may therefore predict an increasing disproportion between the size of the gene activity complex typical of the adult and the gene activity complexes typical of preadult stages.

One might be tempted to believe that the phenomena that have been interpreted in terms of recapitulation involve but a very small part of the total genome of higher animals. These phenomena might indeed involve only a fraction—as suggested by the case of the hemoglobins—of the group of genes whose activity is limited to preadult stages of development, although genes active throughout development will of course participate in the equilibrium of gene activity complexes at all times. On the other hand, this latter group of genes that are active at all stages and in all tissues, though they presumably repeat in contemporary organisms structures and processes that have existed for billions of years, might be thought unable to participate in any specific way in recapitulation, since by definition their activity is not linked specifically to early phases of ontogeny.

This, at any rate, would be a likely view if it were right to link specific morphological structures of the embryo to a specific group of structural genes that are active in the embryo only. Such an assumption might, however, be partly wrong.

MOLECULAR CHANGE AND MORPHOGENESIS

It was stated above that recapitulation may be simply the expression in embryonic gene activity complexes of a rate of evolution that is slower than it is in adult gene activity complexes. What this difference in rate of evolution exactly means in molecular terms remains to be established. The crux of the matter, however, lies here. Three possibilities present themselves: a lower average rate of evolutionarily effective amino acid substitutions in proteins characteristic of the embryonic gene activity groups; or an average rate of amino acid substitutions in embryonic gene activity groups not significantly different from that found in proteins characteristic of the adult, but a low rate of those substitutions that critically alter the functional properties in the embryonic protein; or finally, whatever the rate of amino acid substitutions in the proteins, an emphasis on the constancy of gene activity in the embryo—that is, fewer alterations in intensity and timing of gene activity. Constancy of regulation might in part be brought about by the elimination of certain base substitutions in the structural gene (Zuckerkandl and Pauling, 1965b; Itano, 1965; Sonneborn, 1965). Sequence studies on embryonic and adult proteins in conjunction with functional studies will decide between the first two possibilities. The third will be effectively examined only as the phylogeny and ontogeny of the *regulation* of the activity of structural genes become better understood.

In this connection, the important possibility should be considered that *reproducible morphogenesis depends on constancy of genic regulation to a larger extent than on constancy of genic structure*. The probable partial dependence of constancy of morphogenesis on the constancy of base sequence in structural genes may be due largely to the fact that changes in structural genes themselves bring about changes in regulation of gene activity—namely, of the synthetic activity (with respect to polypeptide production) of the changed structural gene itself (Itano, 1965; Zuckerkandl and Pauling, 1965b) as well as of the synthetic activity of other structural genes (Lee and Englesberg, 1962; Slonimsky et al., 1963; and "polarity effects" in operons: Jacob and Monod, 1961). Differences in sequence between homologous polypeptide chains (chains that belong to a given "type" and have a common molecular ancestor) often will be responsible for differences in morphogenesis not so much as a result of functional differences between the polypeptide chains as of the influences of these structural differences on the rate and perhaps on the timing of synthesis of the polypeptide chains.

If the same organic forms may thus be controlled by different structural genes, provided the regulatory mechanisms remain constant, different forms may on the other hand be produced by the same structural genes, or the same type of structural genes,* when the ratio of their synthetic activities and/or the timing of their activity are altered.

*The type of a structural gene is defined here by the type of tertiary structure and type of function of the corresponding polypeptide chain.

Thus, during ontogenesis, which is characterized by multiple transformations of an organism's morphology, the same structural genes may contribute to the determination of the different forms that succeed each other. The difference in the successive results of the interaction of the genes may tentatively be attributed to changes in relative rates of synthesis of polypeptide chains under the control of these structural genes and also, of course, to the intervention of other, previously inactive genes and/or to the repression of previously active ones. These latter factors are, however, not distinct in a clear-cut fashion from the first one if phases of development contiguous in time are considered. Indeed, activation and inactivation of genes during ontogenesis may well be mostly a very progressive process, as the example of the hemoglobins shows (for instance, the progressive activation of synthesis of the human β chain and the progressive inhibition of synthesis of the human γ chain during fetal life and in the newborn (for example, see Zuckerkandl, 1965).

Recapitulation, then, may be largely a matter of the temporary reproduction in the embryo of ancestral *ratios of rates* of synthesis, and this may be one reason why, upon examination of the *structures* of individual molecules such as hemoglobins, one does not find any trace of it. Although, as mentioned, mutations in structural genes may alter the rates, and perhaps even the timing, of synthesis for other structural genes, there may be many such mutations that do not have any significant effect in this respect. Therefore a measure of evolutionary change in primary structure of polypeptide chains may be compatible with constancy of the regulatory function of the corresponding gene. Thus, at the moment, the possibility cannot be ruled out that individual informational macromolecules, whose structural features through ontogeny do not recapitulate phylogeny, might in principle still participate in bringing about what there is found of recapitulation at the organismic level.

There may be, to a certain extent, a dissociation in organisms between evolution of form and evolution of function. Both aspects of evolution no doubt imply changes in the sequence of structural genes and of the corresponding polypeptide chains. But evolution of form may depend more directly on the influences of the structural changes in DNA on the regulation of synthesis of polypeptide chains, whereas evolution of function may result more directly from their influences on the physicochemical and steric properties of the polypeptide chains.

This view offers a "handle" for natural selection in a field where, so far, we have lacked an adequate one. Many changes are seen to occur during the evolution of hemoglobin polypeptide chains and of other proteins that seem innocuous from the point of view of protein function and, if one dared to pronounce this word, rather neutral in this respect. How does it come about that such mutations so often spread over the near totality of the individuals that make up a species? The usual answer is that there are no neutral mutations, and that a very small difference in selective value can account for the observed spread in the species of the amino acid substitutions. This easy answer leaves one with some misgivings. First, the statement concerning the absence of neutral mutations is not based on evidence relating to functionally inconspicuous individual amino acid substitutions in polypeptide chains, and the extension of the statement to cover these cases may or may not be warranted. Second, at any one variable molecular site, identical substitutions appear by coincidence in different animals and thus no doubt under different conditions in the external and internal milieu (Zuckerkandl, 1964); it is not obvious why selection pressure should take the same direction under these different conditions. On the other hand, under closely similar conditions, different amino acid residues may be found at a given site. Thus, at β chain site No. 104, the gorilla β chain seems to have lysine instead of arginine present in the human chain (Schroeder and Zuckerkandl, unpublished). Do the differences in internal and external milieu between man and gorilla account for this difference? This seems doubtful in view of the fact that chimpanzees appear to have arginine at this position, like man (Rifkin and Konisberg, 1965). Thus no sooner has man heard the news that his exterior and interior conditions of life, as far as hemoglobins go, are not *quite* exactly the same as in a gorilla, that he learns that they are quite exactly as in a chimpanzee. The postulate of functional nonneutrality of *all* amino acid substitutions from the point of view of protein function thus seems to lead to a somewhat inconsistent picture and is, at any rate, not exempt of arbitrariness at the present time. We may, then, grope for another theory to account for the general spread in the species of "harmless" amino acid substitutions and resort to the theory of random genetic drift. But this interpretation seems to meet with some difficulty—in part because it would be necessary to postulate with disquieting frequency that the species took a new start from a small fraction of the total population.

The intervention of selection pressure in the case of very conservative amino acid substitutions is explainable if one assumes that the majority of evolutionarily effective amino acid substitutions in, say, hemoglobins, has a function with respect to the regulation of the rate of synthesis not only of the hemoglobin itself, but of other proteins as well. Changes of the ratios of these rates, which may account for the larger part of the observed differences between related organisms, would lead to an evolutionarily effective frequent reshuffling of primary structure of proteins with respect to features that are of little or no consequence from the point of view of the function of the protein. One might presume that sometimes an arginine appears instead of a lysine in a hemoglobin chain, not because natural selection favors a change in respiration (even if a very slight such change occurs), but because it favors a change in an apparently unrelated function such as, for instance, in the animal's shape.

The action of amino acid substitutions on morphogenesis may be due to interaction at different levels, including higher levels of organic integration. The latter seems to obtain in the case of an example that can be quoted in partial support of the proposed hypothesis—the effects of sickle-cell disease (homozygosity for HbS) on some morphological features of the affected individual, such as a peculiar structure of the cranial bones (Lehmann and Huntsman (1966). The interaction between the HbS structural gene and the cells of the cranium is in this case no doubt very indirect, through complex physiological effects; nonetheless there may well be here a "regulatory" action that one structural gene exerts on others.

The molecular basis for recapitulation is far from solved. All we can do for the present is to present the limited evidence, to define approaches, and to indicate perspectives.

ACKNOWLEDGMENTS

I thank Mr. Jean Derancourt for preparing Table 1; Dr. Jean-François Pechère, Dr. Louis Thaler, and Mr. Andrew Lebor for their discussions with me; and Dr. Jean-François Pechère for the great pains he has taken in helping me put this manuscript in shape.

REFERENCES

Adinolfi, N., G. Chieffi, and M. Siniscalco (1959). *Nature* (London) **184**, 1325.
Babin, D. R., W. A. Schroeder, J. R. Shelton, J. B. Shelton, and B. Robberson (1966). *Biochemistry* **5**, 1297.
Borgese, T. A., and J. F. Bertles (1965). *Science* **148**, 509.
Braunitzer, G., R. Gehring-Müller, N. Hilschmann, K. Hilse, G. Hobom, V. Rudloff, and B. Wittman-Liebold (1961). *Z. Physiol. Chem.* **325**, 283.
Buettner-Janusch, J., and J. B. Twitchell (1961). *Nature* (London) **192**, 669.
Buettner-Janusch, J., and R. L. Hill (1965). *Science* **147**, 836.
Cuenot, L. (1951). *L'évolution Biologique.* Paris: Masson & Cie.
De Beer, Sir Gavin (1964). *Atlas of Evolution.* London: Th. Nelson & Sons.
Doi, R. H., and R. T. Igarashi (1964). *Proc. Nat. Acad. Sci. U.S.* **52**, 755.
Eck, R. V., and M. O. Dayhoff (1966). *Science* **152**, 363.
Gatlin, L. L. (1966). *J. Theoret. Biol.* **10**, 281.
Goldberg, A. L., and R. E. Wittes (1966). *Science* **153**, 420.
Hoyer, B. H., E. T. Bolton, B. J. McCarthy, and R. B. Roberts (1965). In Bryson, V., and H. J. Vogel, eds., *Evolving Genes and Proteins*, p. 581. New York:Academic Press.
Huehns, E. R., N. Dance, G. H. Beaven, J. V. Keil, F. Hecht, and A. G. Motulsky (1964). *Nature* (London) **201**, 1095.
Ingram, V. M. (1961). *Nature* (London) **189**, 704.
Itano, H. A. (1965). In Jonxix, J. H. P., ed., *Abnormal Hemoglobins in Africa*, p. 3. Oxford: Blackwell Scientific Publications.
Jacob, F., J. Monod (1961). *Cold Spring Harbor Symposia Quantitative Biol.* **26**, 193.
Konigsberg, W., J. Goldstein, and R. J. Hill, *J. Biol. Chem.* **238**, 2028.
Lee, N., and Englesberg, E. (1962). *Proc. Nat. Acad. Sci. U.S.* **48**, 335.
Lehmann, H., and R. G. Huntsman (1966). *Man's Hemoglobins.* Amsterdam: North-Holland Publishing Co.
Manwell, C. (1963). *Arch. Biochem. Biophys.* **101**, 504.
Manwell, C., C. M. Ann Baker, J. D. Roslansky, M. Foght (1963). *Proc. Nat. Acad. Sci. U.S.* **49**, 496.
Margoliash, E., and E. L. Smith (1965). In Bryson, V., and J. Vogel, eds., *Evolving Genes and Proteins*, p. 221. New York: Academic Press.
Newell, N. D. (1963). *Scientific American* **208**, 2.
Rifkin, D., and W. Konigsberg (1965). *Biochim. Biophys. Acta* **104**, 457.
Schroeder, W. A., and R. T. Jones (1965). In *Fortschr. Chem. organ. Naturstoffe* **23**, 113.
Schroeder, W. A., J. R. Shelton, J. B. Shelton, J. Cormick, and R. T. Jones (1963). *Biochemistry* **2**, 992.
Slonimsky, P. P., R. Acher, G. Pere, A. Sels, and M. Somlo (1963). In *Mécanisme de Régulation des Activités Cellulaires chez les Microorganismes* (Symposium CNRS, Centre National de al Recherche Scientifique), p. 435. Paris.
Sonneborn, T. M. (1965). In Bryson, V., and H. J. Vogel, eds., *Evolving Genes and Proteins*, p. 377. New York: Academic Press.
Stockell, A., M. F. Perutz, H. Muirhead, and S. C. Glauser (1961). *J. Mol. Biol.* **3**, 112.

Zuckerkandl, E. (1964). In Peeters, H., ed., *Protides of the Biological Fluids*, p. 120. Amsterdam: Elsevier Publishing Co.

Zuckerkandl, E. (1965). *Scientific American*, **212**, 110.

Zuckerkandl, E., and L. Pauling (1962). In Kasha, M., and B. Pullman, eds., *Horizons in Biochemistry*, p. 189. New York: Academic Press.

Zuckerkandl, E., and L. Pauling (1965a). In Bryson, V., and H. J. Vogel, eds., *Evolving Genes and Proteins*, p. 97. New York: Academic Press.

Zuckerkandl, E., and L. Pauling (1965b). *J. Theoret. Biol.* **8**, 357.

HARVEY A. ITANO
Laboratory of Molecular Biology
National Institute of Arthritis and Metabolic Diseases
National Institutes of Health
Bethesda, Maryland

The Structure-Rate Hypothesis and the Toll Bridge Analogy

The early study of sickle cell anemia at the California Institute of Technology was described by Pauling in his Harvey Lecture (1953–1954). In this paper I shall review the source and development of a hypothesis based on early work at the Institute, and I shall consider the relationship of this hypothesis to current concepts of the genetic control of protein synthesis.

In the summer of 1948, S. J. Singer and I first demonstrated the electrophoretic abnormality of the hemoglobin of sickle cell anemia (Itano and Pauling, 1949). Soon thereafter we found that the hemoglobin in sickle cell trait, an asymptomatic condition associated with sickling cells, has two components, normal hemoglobin (hemoglobin A) and the hemoglobin of sickle cell anemia (hemoglobin S). The qualitative aspect of this observation was consistent with the available genetic data, which indicated that the anemia and the trait were associated with homozygosity and heterozygosity, respectively, for the gene that causes red cells to sickle. However, the quantitative result of our electrophoretic patterns, namely, that there was more normal than abnormal hemoglobin in sickle cell trait, perplexed us. We therefore consulted the geneticists in the Division of Biology at the California Institute of Technology and learned much about the quantitative effects of genes. A discussion of possible genetically controlled mechanisms for the unequal amounts of the two hemoglobins of sickle cell trait in the same cell was part of our paper on sickle cell anemia (Pauling et al., 1949).*

I. C. Wells came to the Institute that fall, and he joined us in the study of hemoglobin S. We carried out a series of analyses on the ratio of the two hemoglobins in 42 unrelated individuals with sickle cell trait and found wide variations (Wells and Itano, 1951). Familial studies of the ratios were conducted in collaboration with Neel, and the results were discussed in terms of a locus independent of the sickle cell locus that controlled the ratios (Neel et

*The title of this paper, "Sickle Cell Anemia, a Molecular Disease," was Pauling's inspiration. The current practice of applying the adjective "molecular" to biological and medical terms began with this paper, although it was at least ten years before the usage became fashionable. Pauling's interest in the molecular approach to biological and medical problems began even earlier with his work on the physical chemistry of hemoglobin and on the chemistry of the antigen-antibody reaction.

al., 1951). Synthesis of hemoglobin A was found to be decreased greatly in thalassemia, another inherited hematologic disorder (Rich, 1952). Individuals who had inherited a gene for hemoglobin S from one parent and a gene for thalassemia from the other parent were found to have more hemoglobin S than A, the reverse of the situation previously observed in sickle cell trait (Sturgeon et al., 1952; Neel et al., 1953).

THE HYPOTHESIS

Consideration of these and additional data involving a second abnormal hemoglobin, hemoglobin C, led to the conclusion that a second locus was not necessary to account for the familial data, and that the ratios could be explained in terms of inherited differences in the rate of synthesis of hemoglobin A. The structure-rate hypothesis in its earliest form was stated as follows: "The simplest genetic hypothesis, which is in accord with the available familial data, is that the sickle cell hemoglobin mechanism, the hemoglobin C mechanism, and the three rate modifications of the normal hemoglobin mechanism depend on alleles" (Itano, 1953). Terms like structural gene, messenger RNA, and codon were not yet part of the vocabulary of genetics.

The discovery of the amino acid substitution of hemoglobin S by Ingram (1957) permitted restatement of the structure-rate hypothesis in terms of structural genes. In a review article that same year I wrote, "For the present we can merely infer on the basis of available evidence [that] a mutation which results in the alteration of the structure of hemoglobin also alters its net rate of synthesis. In other words, *the two functions, determination of structure and determination of net rate of synthesis, may be regarded as properties of the same allele*"* (Itano, 1957). The structure-rate hypothesis as applied to thalassemia reads as follows: "Thalassemia mutants at the *Hb* locus are analogous to the mutants for the abnormal hemoglobins, differing in their failure to alter the net charge of adult hemoglobin and in the greater inhibition that they exert on net rate of synthesis" (Itano, 1957). In the same review I suggested that inherited variations in content of uncharged amino acids may be associated with the observed inherited differences in the rate of synthesis of hemoglobin A relative to that of hemoglobin S in sickle cell trait.

As the body of knowledge concerning hemoglobin structure and genetic mechanisms expanded, restatements of the hypothesis increased in sophistication. Ingram and Stretton (1959) extended the theory of thalassemia to provide for independent genetic control of the α-chain and β-chain of hemoglobin (Itano and Robinson, 1960) by postulating that thalassemia is a mutation of either the α-chain or the β-chain gene, but a mutation that does not affect the electrophoretic behavior of hemoglobin. Next it was suggested that change in net rate of synthesis of a chain without an amino acid substitution could occur because of degeneracy of the nucleic acid code (Itano, 1965). These concepts have since been discussed in greater detail, and a mechanism whereby a change in one RNA codon out of nearly 150 in a

*Italics added.

messenger RNA molecule might greatly decrease the rate of synthesis of a polypeptide chain has been presented (Itano, 1966).

The hypothesis that the defect of thalassemia results from a defective RNA molecule that blocks ribosomes (Ingram, 1964) is equivalent to the structure-rate hypothesis because, by definition, a defective messenger RNA must result from a defective structural gene. Recently G. von Ehrenstein (personal communication, 1966) has suggested that thalassemia mutations might be chain-terminating mutations of the amber type.* A mutation of this type in the structural gene would be consistent with the model of blocked ribosomes on an altered messenger RNA.

THE MOLECULAR MODEL

The assembly of messenger RNA is considered to consist of the sequential addition of four kinds of bases in the order dictated by the sequence of bases on the structural gene. Since each base can add to one of four kinds of bases (including its own kind), there are sixteen possible kinds of addition reactions. In order that a mutation introduce a rate-limiting step, a unique addition reaction—one that does not occur elsewhere in the synthesis of the messenger molecule—must result from the mutation. Moreover, this reaction must be slower than any of the other 450 that are required to assemble the messenger RNA for a hemoglobin chain.

Since there are 20 amino acids, 400 different dipeptide sequences—that is, types of additions—are possible. A set of several transfer RNA molecules bind each amino acid, and one to six codons recognize each set of transfer RNA molecules. The possibility that a mutation in the structural gene will produce a unique addition reaction in peptide synthesis is, therefore, much greater than in messenger RNA synthesis. The molecular model for the structure-rate hypothesis assumes that, in accordance with chemical rate laws, the formation of each peptide bond proceeds at a rate which depends upon the nature, concentration, and activity of reactants and catalysts (enzymes).

A few electrophoretically abnormal hemoglobins with amino acid substitutions are present in heterozygotes in the same proportion as hemoglobin A (Allan et al., 1965). This is taken to mean that the time required to form each of the two peptide bonds that involve the substituted amino acid is no greater than the time required for formation of the rate-limiting bond in the normal chain which, in general, would be elsewhere in the chain.

A subnormal rate of synthesis associated with a mutation is taken to indicate that the mutation has resulted in a step in the assembly of the polypeptide chain that is slower than the rate-limiting step of normal chain assembly. It is immaterial whether the mutation has resulted in the incorporation of an amino acid different from the one normally present, or in the formation of a different codon for the same amino acid. In either instance, formation of a

*Suppression of such a mutation could result in a neutral or a charge-altering amino acid substitution or in incorporation of the amino acid residue normally present at the mutated position.

peptide bond at the mutated site requires the participation of at least one reactant that is different from those taking part in bond formation at the corresponding site of the normal chain. The fact that assembly at the mutated site has become rate-limiting implies that the concentration or reactivity of the new reacting species is abnormally low. A minor leucine transfer RNA is known to incorporate leucine into only one of the leucine positions in the α-chain of rabbit hemoglobin (Weisblum et al., 1965). The hemoglobin S mutation resulted in the incorporation of valine in the position of the β-chain normally occupied by glutamic acid (Ingram, 1957). The subnormal rate of hemoglobin S synthesis suggests the possibility that the new valine site in hemoglobin S requires a minor valine transfer RNA which is not used for valine incorporation elsewhere in the β-chain. Mutation of a codon for an amino acid to another codon, which codes for the same amino acid but which requires a different transfer RNA, could conceivably account for differences in the rates of synthesis of structurally identical polypeptide chains (Itano, 1965).

THE TOLL BRIDGE ANALOGY

I have found a toll bridge analogy useful in explaining the effect of a point mutation on the rate of synthesis of a polypeptide chain with reference to the polyribosomal model of hemoglobin synthesis (Rich et al., 1963). Each messenger RNA molecule is analogous to a lane on a tollbridge, and the automobiles crossing the bridge in a lane are analogous to ribosomes moving along the molecule of messenger RNA. The toll gate is the site of chain initiation. The restriction that automobiles cannot change lanes and cannot pass each other within each lane is imposed. If the rate at which cars pass the toll gate and move onto the bridge is rate-limiting, cars in a given lane will reach the opposite end at the rate at which tolls are collected; consequently a slow lane with fast toll collection will carry more cars across per unit time than a fast lane with slow toll collection. *If chain initiation is rate-limiting, rapid assembly of chains will not increase the number of chains completed per unit time under steady state conditions.* If a lane on the bridge has an obstruction that delays each car longer than the time it takes to collect toll, the delay at the obstruction will become rate-limiting. Cars will be backed up between the toll gate and the obstruction, and will reach the other end at the rate at which they get past the obstruction. Past the obstruction, the cars will be more widely spaced, and the faster they move, the greater will be the distance between them. *If there is a point in the assembly of a chain that is rate-limiting, chains will be completed at the rate at which they pass this point. When proximal to the rate-limiting point, ribosomes are closer together than they are when they are distal to this point.* If a traffic jam at the exit ramp causes a greater delay in movement than does the toll gate or the obstruction, it will cause the cars to back up all the way across the bridge. *If chain termination is rate-limiting, ribosomes, provided there are a sufficient number, will be packed linearly on the messenger RNA at maximum density. The effect on ribosomal distribution of shorter delays at chain initiation or during peptide chain assembly will be obliterated.*

DISCUSSION

The amount of a hemoglobin chain in a mature red cell divided by the time between beginning and completion of hemoglobin synthesis is the mean net rate of synthesis of that chain. The net synthesis is the amount synthesized less the amount destroyed, the amount destroyed being a significant factor when one chain is produced in excess (Bank and Marks, 1966) or when an inherited amino acid substitution results in a structurally unstable chain (Carrell et al., 1966). The relative net rate of synthesis of two chains in the same cell is the same as the ratio of the two chains in the cell at maturation.

The two types of chains of hemoglobin A are controlled by independent genetic loci. In a red cell heterozygous at one of these loci, two chains that differ in their genetic control by one base in about 450 are synthesized in the same environment. Under these conditions, the more complex regulatory mechanisms that have been postulated for the control of the rate of synthesis of microbial enzymes are irrelevant. Any control independent of the structural genes would be exerted equally on both members of such a pair of alleles. For example, the total rate of synthesis of hemoglobin will be subnormal when the supply of iron in a cell is inadequate, but the relative rates of synthesis of two hemoglobins in the same cell will not be affected.

Methods are now available to examine the nature of the rate factor in the structural gene of an abnormal chain. One approach is to use the pulse-labeling technique of Dintzis (1961) to detect points of delay in peptide chain assembly (Winslow and Ingram, 1966); however, the results of this type of experiment should be interpreted with caution. It is evident from the toll bridge analogy that ribosomes will accumulate backward from the delay point; therefore slowing of peptide assembly will begin before the actual delay point on the messenger RNA molecule. An experiment which may now be feasible is to test whether the site of amino acid substitution in an abnormal chain that is synthesized at a subnormal rate requires a special transfer RNA.

The same approaches can be followed to test whether a structural difference in a peptide chain without change in its net charge or a structural change in messenger RNA without change in amino acid sequence can also alter the rate of assembly of a peptide chain. Several workers have recently demonstrated neutral amino acid substitutions in mouse hemoglobin (Rifkin et al., 1966), rabbit hemoglobin (von Ehrenstein, 1966), and horse hemoglobin (Kilmartin and Clegg, 1967). Although no neutral substitutions have been found in hemoglobin A or in thalassemia, the data are as yet too limited to exclude their existence entirely. The experimental approach toward finding neutral substitutions differs somewhat from the established methods of amino acid sequence analysis. The methods used by von Ehrenstein (1966) to detect and quantitate neutral amino acid multiplicities at several positions of rabbit hemoglobin α-chain should prove useful in the re-examination of hemoglobin A and thalassemia hemoglobin.

CONCLUSIONS

The structure-rate hypothesis includes mutations that do not alter amino acid sequence or do not alter net rate of synthesis, because it simply postulates that *the structural gene for a hemoglobin chain exerts its control on both the amino acid sequence and the net rate of synthesis of the chain.* The net effect of this control is evident in the hemoglobin composition of heterozygous cells; however, the mechanism of this control is yet to be elucidated. It appears more likely that the rate effect is imposed at the translational step, rather than at the transcriptional step, in the decoding of genetic information.

REFERENCES

Allan, N. D., D. Beale, D. Irvine, and H. Lehmann (1965). *Nature* **208**, 658.
Bank, A., and P. A. Marks (1966). *Nature* **212**, 1198.
Carrell, R. W., H. Lehmann, and H. E. Hutchinson (1966). *Nature* **210**, 915.
Dintzis, H. M. (1961). *Proc. Nat. Acad. Sci. U. S.* **47**, 247.
von Ehrenstein, G. (1966). *Cold Spring Harbor Symp. Quant. Biol.* **21**, 705.
Ingram, V. M. (1957). *Nature* **180**, 326.
Ingram, V. M. (1964). *Ann. N. Y. Acad. Sci.* **119**, 485.
Ingram, V. M., and A. O. W. Stretton (1959). *Nature* **184**, 1903.
Itano, H. A. (1953). *Am. J. Human Genet.* **5**, 34.
Itano, H. A. (1957). *Adv. in Protein Chem.* **12**, 215.
Itano, H. A. (1965). In Jonxis, J. H. P., ed., *Abnormal Haemoglobins in Africa*, pp. 3–16. Oxford: Blackwell Scientific Publ.
Itano, H. A. (1966). *J. Cell Physiol.* **67**, suppl. 1, 65.
Itano, H. A., and L. Pauling (1949). *Fed. Proc.* **8**, 209.
Itano, H. A., and E. A. Robinson (1960). *Proc. Nat. Acad. Sci. U. S.* **46**, 1492.
Kilmartin, J. V., and J. B. Clegg (1967). *Nature* **213**, 269.
Neel, J. V., H. A. Itano, and J. S. Lawrence (1953). *Blood* **8**, 434.
Neel, J. V. I. C. Wells, and H. A. Itano (1951). *J. Clin. Invest.* **30**, 1120.
Pauling, L. (1953–1954). *Harvey Lectures*, Ser. 49, 216.
Pauling, L., H. A. Itano, S. J. Singer, and I. C. Wells (1949). *Science* **110**, 543.
Rich, A. (1952). *Proc. Nat. Acad. Sci. U. S.* **38**, 187.
Rich, A., J. R. Warner, and H. M. Goodman (1963). *Cold Spring Harbor Symp. Quant. Biol.* **28**, 269.
Rifkin, D. B., M. R. Rifkin, and W. Konigsberg (1966). *Proc. Nat. Acad. Sci. U. S.* **55**, 586.
Sturgeon, P., H. A. Itano, and W. N. Valentine (1952). *Blood* **7**, 350.
Weisblum, B., F. Gonano, G. von Ehrenstein, and S. Benzer (1965). *Proc. Nat. Acad. Sci. U. S.* **53**, 328.
Wells, I. C., and H. A. Itano (1951). *J. Biol. Chem.* **188**, 65.
Winslow, R. M., and V. M. Ingram (1966). *J. Biol. Chem.* **241**, 1144.

ALFRED G. KNUDSON, JR.
Professor of Medicine
State University of New York
Stony Brook, New York

RAY D. OWEN
Division of Biology
California Institute of Technology
Pasadena, California

Molecular Genetics and Disease Resistance

Recognition in 1949 by Pauling, Itano, Singer, and Wells that sickle-cell anemia, an inherited disease, has its basis in a molecular abnormality of hemoglobin constituted a major advance in medicine: formulation of the concept of "molecular disease." Also in 1949, J. B. S. Haldane suggested that the genetic polymorphisms observed in man might have been established largely as a result of the selective influences of infectious disease. The subsequent demonstration by Allison (1954) and others that heterozygosity for the sickle-cell gene confers resistance to falciparum malaria provided, therefore, the first formulation of resistance to infection in molecular terms. Since then several other erythrocyte traits, notably the C and E hemoglobinopathies, the thalassemias, and primaquine sensitivity (glucose-6-phosphate dehydrogenase deficiency), have been similarly associated with resistance to falciparum malaria (Motulsky, 1960).

Evidently a fruitful approach to understanding the molecular bases of resistance to infectious disease, as a guide to treatment or prevention, is via the analysis of genetic polymorphisms. Single-gene polymorphisms give particular promise of revealing relevant molecular mechanisms of resistance and susceptibility. Those most likely to be demonstrably associated with resistance are those causing severe disease in the homozygous individual, such as sickle-cell anemia and thalassemia, because with these the advantage of the heterozygote over the homozygous normal individual is presumably greatest. For example, cystic fibrosis of the pancreas, the most common recessively inherited lethal disorder among Caucasians of European descent, offers a good prospect for investigation (Hallett, Knudson, and Massey, 1965). There is even evidence that the basic abnormality involves some mucosubstance, a class of molecules of potentially great interest in the study of bodily defenses against infectious agents.

In these examples the discovery of polymorphisms has proceeded from clinical findings. With the development of such analytical tools as starch-gel electrophoresis, it has become possible to identify polymorphisms by screening blood samples of large numbers of normal individuals. What do these polymorphisms signify? How did they arise; how and why are they maintained in human populations? So far the answers are not known, but the differences in frequencies that certain polymorphic alleles display in different populations strongly suggest that the polymorphisms do not result simply from balance between forward and back mutation. Some of the species of α and β globulins of serum, especially haptoglobin and transferrin, have been considered in this light. Transferrin has been shown to be an inhibitor of bacterial and viral multiplication, suggesting that its variants may afford resistance to specific infections (Motulsky, 1960). In the quest for these answers one of the most thwarting, yet promising, polymorphisms is the one known for the longest time, that for the ABO blood groups (Mourant, 1954). The suggestions that these antigens may be related to the plague bacillus and smallpox virus (Vogel, Pettenkofer, and Helmbold, 1960) have so far come to naught for lack of supporting evidence.

So far we have taken note of polymorphisms in search of disease resistance, so to speak; by clinical or laboratory means a polymorphism is identified, localized to some molecular species, and, hopefully, then related to resistance to some infectious disease. In this way entirely unsuspected mechanisms of resistance stand to be discovered. On the other hand are some polymorphisms that are related to known mechanisms of immunity and resistance. Here again some are disclosed by clinical disorders. Perhaps the best known of such disorders is allergy. The allergic diathesis is a very common genetic condition that is probably sustained in the population because it confers some selective advantage under certain conditions. The unique feature of the allergic state seems to be the tendency to form reagins, or nonprecipitating antibodies, which could some day be revealed as molecules that neutralize some class of infectious agents. So also with the autoimmune disorders, or diseases of hypersensitivity; it may prove that the hereditary propensity for unusual immune reactions (Hall, Owen, and Smart, 1960) has had a selective value, and will lead us to other new aspects of naturally occurring disease resistance.

In addition to gene-regulated native-resistance mechanisms, the molecular bases for immune responses and reactions, to which Linus Pauling has given important attention, represent another key area in the interplay between disease and survival. There is currently much debate over the origins of specificity of immunoglobulin molecules involved in humoral immune reactions. Facts are accumulating rapidly, and it is possible that by the time this essay appears the evidence will have overwhelmed the proponents of all but one of the presently debated positions. At this writing, however, it is still not clear whether the control of antibody specificity resides entirely in the genetic information of the cell elaborating the antibody, or whether the antigen may play some direct role in the final modulation of antibody specificity, as Pauling theorized many years ago. That genetically determined differences among immunoglobulins exist is demonstrated, for example, by the Gm and Inv specificities of human immunoglobulins, but these differences are not thought to be related to antibody specificity. Nevertheless the important polymorphisms that exist for these specificities may reflect adaptive values (Steinberg, 1962).

The emerging view of the molecular basis of disease and of disease resistance must also include a consideration of interferon production in viral infections. Interferon appears as a first line of defense, and is probably more useful than antibodies during the initial stages of infection. Genetically determined variants of interferon are not known, so no progress has yet been made via the study of polymorphisms. The questions that arise regarding the origin of immunoglobulin specificity may also be asked for interferon. Hopefully, molecular geneticists will begin to effect some separation and characterization of specific interferons.

The whole subject of mechanisms of response to virus infection is incompletely understood. Population differences in susceptibility, exemplified by the severity of measles among Polynesians and of smallpox among American Indians, and the alterations of these differences by natural selection also raise questions about mechanisms. Are these interferon differences, or is some still unknown molecule involved? The study of ethnic differences might of course reveal some new molecular mechanisms. In certain provocative models in the form of animal diseases, such as scrapie in sheep, Aleutian disease in mink, and leukemia in mice, some kind of genetic terrain seems to be implicated. Scrapie was at first identified as a disease of particular breeds of sheep in the British Isles, and there was evidence, through pedigrees, of its vertical transmission; Aleutian disease in mink seemed to be associated with homozygosity for a particular gene affecting coat color; leukemia had a remarkably high incidence in a particular (AK) strain of mice.

In these diseases there is now persuasive evidence that the etiologies rest on virus infection; scrapie has been transmitted to other species including mice, Aleutian disease can be transmitted to previously uninfected mink of different color-genotypes, and various murine leukemia viruses have been transmitted to certain strains under restrictive circumstances. In some instances it may well turn out that early indications of a significant genetic factor were misleading, deriving only from the peculiar vertical transmission of the virus involved. For Aleutian disease in mink, however, there is an apparent difference in susceptibility of the Aleutian strain and that of non-Aleutian mink; not enough data have been reported for mink populations segregating for the Aleutian color gene to permit full confidence that a direct biochemical effect of the gene controlling the color variation may be involved in an important way, either in the infectivity of the virus or in the sequelae of virus infection. For mouse leukemia, detectable strain differences in susceptibility reflect the specificity of the leukemia virus employed. Meanwhile, great interest attaches to these investigations not because the pathological syndromes resulting from infection with these viruses are classical acute diseases but because they do parallel closely certain chronic and degenerative diseases of great concern to mankind (Gajdusek, Gibbs, and Alpers, 1965). Neuropathies, various disorders of the cardiovascular-renal system, aberrations in the lymphoid system, and cancer are among the disorders that may be caused in part by virus infection.

Some of the mild latent viruses that cause chronic disease are transmitted by congenital infection, and their animal hosts are therefore immunologically tolerant. Tolerance for viruses permits a prolonged coexistence, the full implications of which are not yet understood. Although it is true that some fraction of animals so infected falls ill and may die, we do not yet know whether other more subtle effects may occur. It will be of great interest to determine whether tolerant animals are less susceptible to virulent viruses than are nontolerant animals. If so, a kind of polymorphism could be established in which the tolerated, latent virus would play a role similar to that of the sickle-cell anemia mutation (Knudson, 1966). The very agent that causes one kind of disease may be a resistance factor for another.

REFERENCES

Allison, A. C. (1954). Protection afforded by sickle-cell trait against subtertian malarial infection. *Brit. Med. J.* i, 290.

Gajdusek, D. A., C. J. Gibbs, and M. Alpers, ed. (1965). *Slow, Latent, and Temperate Virus Infections.* N.I.N.D.B. Monograph No. 2, U.S. Dept. of Health, Education and Welfare, 1965.

Haldane, J. B. S. (1949). Disease and evolution. *Ricerca Sci.* **19** (suppl.), 68.

Hall, R., S. G. Owen, and G. A. Smart (1960). Evidence for genetic predisposition to formation of thyroid autoantibodies. *Lancet* **2**, 187.

Hallett, W. Y., A. G. Knudson, and F. J. Massey (1965). Absence of detrimental effect of the carrier state for the cystic fibrosis gene. *Am. Rev. Resp. Dis.* **92**, 714.

Knudson, A. G. (1966). Congenital viral infection and human disease. *Am. Naturalist* **100**, 162.

Motulsky, A. G. (1960). Metabolic polymorphisms and the role of infectious disease in human evolution. *Human Biol.* **32**, 28.

Mourant, A. E. (1954). *The Distribution of Human Blood Groups.* Oxford: Blackwell.

Pauling, L., H. A. Itano, S. J. Singer, and I. C. Wells (1949). Sickle-cell anemia, a molecular disease. *Science* **110**, 543.

Steinberg, A. G. (1962). Progress in the study of genetically determined human gamma globulin types (the Gm and Inv groups). *Prog. Med. Genet.* **2**, 1.

Vogel, F., H. J. Pettenkofer, and W. Helmbold (1960). Über die Populationsgenetik der ABO-Blutgruppen: 2. Mitteilung. Gehäufigkeit und epidemische Erkrankungen. *Acta Genet.* (Basel) **10**, 267.

THOMAS L. PERRY, M.D.
Department of Pharmacology
The University of British Columbia
Vancouver, Canada

Homocystinuria: A Challenging Molecular Disease

Homocystinuria is just the sort of human disease which would interest Linus Pauling. I regret that it was not discovered until 1962, at about the time my stimulating association with Professor Pauling as one of his students and colleagues came to an end. Had the disease been known a few years earlier, I might have had the pleasure of discussing some of its intriguing aspects with him, and of exploring in the laboratory some of his incisive speculations about the disease. In this article, I should like first to review what is known about the natural history and clinical manifestations of homocystinuria, as well as about its underlying enzymatic error. Then I wish to describe the results of some experiments on this disease done in our laboratories at the University of British Columbia. I hope this account may stimulate at a distance the conversations about homocystinuria that I should like to have with Linus Pauling, and with other scientists.

Homocystinuria was independently discovered by Carson and Neill (1962) in Ireland, and by Gerritsen, Vaughn, and Waisman (1962) in the United States. From surveys of mentally retarded patients, these two groups of investigators discovered that an abnormal excretion of homocystine in the urine may occasionally accompany mental retardation. Since then, more than 60 patients with homocystinuria have been reported from various parts of the world. What was at first thought to be a very rare disease may prove, when physicians begin to search for it systematically, to be as frequent in occurrence as phenylketonuria. The recent discovery of homocystinuria should make us wonder how many other genetically determined diseases have been plaguing man for centuries or millenia without our yet having recognized them.

CLINICAL MANIFESTATIONS AND PATHOLOGICAL CHANGES IN HOMOCYSTINURIA

A characteristic clinical syndrome has been defined by a number of investigators who have studied patients with homocystinuria (Field, Carson, Cusworth, Dent, and Neill, 1962; Carson et al., 1963; Komrower and Wilson, 1963; Gerritsen and Waisman, 1964a; White, Thompson, Rowland, Cowen,

Supported by grants from the Medical Research Council and the Department of National Health and Welfare of Canada.

and Araki, 1964; Carson, Dent, Field, and Gaull, 1965; Schimke, McKusick, Huang, and Pollack, 1965; Spaeth and Barber, 1965; White, Rowland, Araki, Thompson, and Cowen, 1965; Dunn, Perry, and Dolman, 1966; Werder, Curtius, Tancredi, Anders, and Prader, 1966). Most patients exhibit mental deficiency, but this is usually not as severe as that characteristic of phenylketonuria. An appreciable minority of older individuals with homocystinuria have been found to be of normal intelligence (Schimke et al., 1965; Spaeth and Barber, 1965; Dunn et al., 1966). Dislocation of the ocular lenses is almost always found in homocystinurics, and apparently results from a failure in the normal development of the suspensory ligament of the lens. Skeletal abnormalities are common in homocystinuria. These include dolichostenomelia (long, thin extremities), genu valgum, deformities of the sternum such as pectus excavatum and pectus carinatum, osteoporosis of the vertebrae leading to scoliosis, and abnormalities of the palate with crowding of the teeth. Arterial and venous thromboses commonly occur in homocystinuria, and many patients die in infancy or early childhood of cerebrovascular accidents, due to thromboses of arteries or veins in the brain. Thromboses of coronary arteries, or of other major arteries, and thrombophlebitis in veins leading to pulmonary emboli also cause many deaths. Despite the occurrence of these intravascular thromboses, some homocystinurics survive at least into the fifth decade of life (Schimke et al., 1965).

Autopsy studies have shown, besides the dislocated lenses and skeletal abnormalities observable in living homocystinurics, characteristic changes in various organs caused by old and recent infarcts. In the brain, fresh and old organized thrombi are usually present both in arteries and veins. Throughout the brain are found both acute ischemic necrosis of nerve cells, and areas of encephalomalacia due to previous episodes of cerebrovascular occlusion (Gibson, Carson, and Neill, 1964; White et al., 1965; Dunn et al., 1966).

Schimke et al. (1965) have described pathological changes in the walls of large arteries of autopsied homocystinurics, particularly in the aorta, renal, subclavian, iliac, carotid, and coronary arteries. These investigators found that muscular elements of the media were often widely separated by an inordinate expansion of the intercellular ground substance. The intima of some arteries was greatly thickened by fibroelastic and collagenous material, with reduction of the lumina to very small passages. The intraarterial thromboses characteristic of homocystinuria might well be caused by these structural alterations, but such changes in the media and intima of arteries would not explain why homocystinurics often have venous thromboses as well. An alternative explanation for the intravascular thromboses has been suggested by McDonald, Bray, Field, Love, and Davies (1964). They report that there is increased stickiness of platelets in the blood of homocystinurics, and that the addition of homocystine to normal blood enhances platelet stickiness in vitro.

Homocystinuria has certain clinical manifestations which resemble those encountered in Marfan's syndrome. Dolichostenomelia, arachnodactyly, and tall stature are common in both conditions, as is ectopia lentis. The dilatation of the aorta found in patients with Marfan's syndrome may be mimicked in homocystinuria by dilatation of large arteries secondary to medial changes. Mental deficiency is absent in Marfan's syndrome, however, and the mode of inheritance in the two diseases is different. Marfan's syndrome is inherited as an autosomal dominant; homocystinuria is inherited as an autosomal recessive. A number of patients with so-called Marfan's syndrome who were mentally retarded, and whose antecedents had no similar disease, have proven on more careful investigation to be homocystinurics.

KNOWN BIOCHEMICAL ABNORMALITIES IN HOMOCYSTINURIA, AND THE BASIC METABOLIC ERROR

Characteristic biochemical changes in homocystinuric patients noted by the investigators who have studied the clinical manifestations and pathologic changes are elevation of the concentrations of methionine and of homocystine in blood and the presence of homocystine in the urine. Normally, the concentration of methionine in fasting human plasma is less than 4 μmoles/100 ml, and homocystine is not detectable at all in normal human plasma. Homocystinuric patients may have fasting-plasma methionine concentrations as high as 200 μmoles/100 ml, and plasma homocystine concentrations up to 20 μmoles/100 ml (Brenton, Cusworth, and Gaull, 1965a; Werder et al., 1966; Perry, Hansen, MacDougall, and Warrington, 1967). Homocystine is not detectable in normal urine, but urinary excretion of it may exceed 1000 μmoles per 24 hours in patients with homocystinuria (Werder et al., 1966). Besides homocystine, smaller amounts of methionine and of the mixed disulfide of cysteine and homocysteine are also found in homocystinuric urine.

The basic metabolic error in homocystinuria has been shown to be a deficiency of activity of the enzyme cystathionine synthase (Mudd, Finkelstein, Irreverre, and Laster, 1964). In man, as in other mammals, methionine derived from dietary protein is either synthesized into tissue protein, or is degraded to cysteine and eventually to inorganic sulfate. The major known pathway of methionine degradation is shown in Figure 1. Methionine is first activated to S-adenosylmethionine. A methyl group is then given up to various methyl acceptors to yield S-adenosylhomocysteine, which is hydrolyzed to homocysteine and adenosine. Homocysteine may either be remethylated to yield methionine, or it may be condensed with serine to form cystathionine. The activity of this condensing enzyme, cystathionine synthase, was shown to be absent in a liver biopsy specimen from a homocystinuric patient by Mudd et al., (1964). These investigators later showed that the activity of hepatic cystathionine synthase in known homocystinuric heterozygotes (parents of a homocystinuric child) was only about 40 percent of that found in the livers of control subjects (Finkelstein, Mudd, Irreverre, and Laster, 1964). Since in homocystinurics the major known pathway of methionine degradation is completely or almost completely blocked at the point of cystathionine synthesis (Figure 1), one would expect one or more of the metabolites before the block to accumulate. The homocysteine which must accumulate in tissues is readily oxidized to the disulfide, homocystine, and it is the latter substance which is usually detected in the urine and blood of homocystinurics.

One of the major free amino acids present in normal human brain is cystathionine (Tallan, Moore, and Stein, 1958). Indeed, the concentration of cystathionine in man's brain is higher than in the brains of other primates,

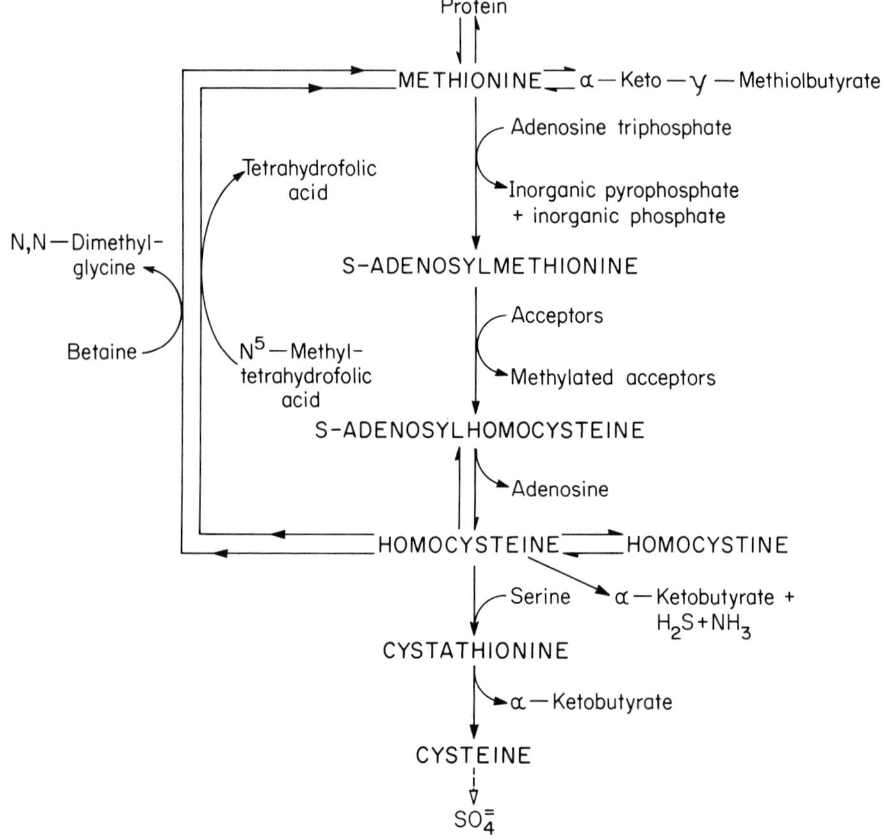

FIGURE 1.
Current concepts of mammalian methionine metabolism.

and very much higher than in the brains of lower mammals and birds. Only low concentrations of cystathionine are found in human liver, kidney, and muscle. Nothing is known yet as to the physiological role played by this amino acid in the central nervous system. There is no evidence that cystathionine introduced into the blood stream can penetrate into brain cells, and cystathionine has been shown to be absent from the developing human brain until late in fetal life (Okumura, Otsuki, and Kameyama, 1960). The cystathionine present in human brain from infancy throughout adult life must be formed in situ, and Mudd, Finkelstein, Irreverre, and Laster (1965) have shown that the enzyme, cystathionine synthase, is present in normal human brain as well as in human liver. As one might expect, homocystinuric patients show at autopsy marked diminution or absence of cystathionine from brain tissue (Gerritsen and Waisman, 1964b; Brenton, Cusworth, and Gaull, 1965b).

A surprising thing about homocystinuric patients is that, although the major known degradative pathway in methionine metabolism is blocked at the step of cystathionine synthesis (Figure 1), the urinary excretion of such known sulfur-containing amino acids as methionine, methionine sulfoxide, homocystine, and the mixed disulfide of cysteine and homocysteine accounts for only a small fraction of the daily dietary intake of methionine. What happens to the considerable proportion of this methionine which is neither synthesized into protein, nor excreted in urine in the form of known sulfur-containing amino acids? Laster, Mudd, Finkelstein, and Irreverre (1965) have shown in interesting loading experiments that patients with homocystinuria have an impaired capacity to convert the sulfur of orally administered L-methionine, but not of L-cysteine, to urinary inorganic sulfate. Their patients were able to convert a small amount of administered L-methionine to inorganic sulfate, which suggested that slight residual cystathionine synthase activity was present in liver cells. The urinary excretion of homocystine was not markedly increased by oral administration of methionine to the homocystinuric subjects, but the amounts of bound methionine and of unidentified neutral sulfur in urine were considerably increased by the methionine supplementation. This and other intriguing findings suggested to us that much is not yet understood about methionine metabolism in man, and led us to attempt in our own laboratory to identify some of the unknown sulfur compounds present in homocystinuric urine.

THE SEARCH FOR UNUSUAL SULFUR-CONTAINING AMINO ACIDS IN HOMOCYSTINURIC URINE

We have been fortunate to have available for study seven children with homocystinuria, belonging to three unrelated families. At the time of this writing, these appear to be the only cases known in Canada. There are certainly many more homocystinurics not yet diagnosed. They will surely be found as physicians begin to seek actively the reasons for delayed mental development in children, and when all patients with dislocated lenses and unexplained intravascular thromboses early in life, and all those thought to have Marfan's syndrome, are scrutinized more carefully. The diagnosis of homocystinuria can readily be excluded in a few minutes by a simple cyanide-nitroprusside screening test on urine. Approximately 0.4 ml of a 5 percent aqueous solution of sodium cyanide is added to 1 ml of urine in a test tube. After a few minutes, one drop of a 5 percent aqueous solution of sodium nitroprusside is added. The immediate development of a beet-red color indicates that the urine contains an unusual amount of cystine or homocystine. Which of the two disulfide amino acids is present can then easily be determined by two-dimensional paper chromatography.

In the course of preparing paper chromatograms on the urines of our patients, we were surprised to find specimens that contained unusual ninhydrin-positive spots, which did not match any of the amino acids known to be present in human urine. We also found in these urines, as we expected, homocystine, the mixed disulfide of cysteine and homocysteine, methionine, and methionine sulfoxide. The unidentified ninhydrin-positive compounds seen on paper chromatograms were shown to be either thioethers or disulfides, based on their characteristic reactions to the chlorplatinate and to the cyanide-nitroprusside spray reagents. Figure 2 shows two of these unusual sulfur-containing ninhydrin-positive compounds. On the left side of the figure is a two-dimensional paper chromatogram prepared from an aliquot of normal human urine containing 25 μg of creatinine. The right side of the figure shows

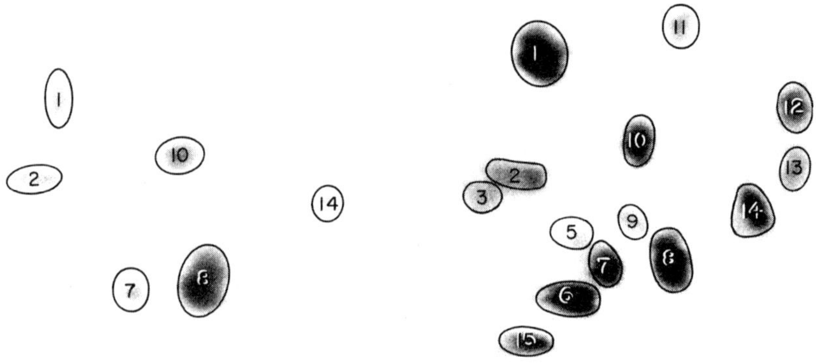

FIGURE 2.
Two-dimensional paper chromatograms prepared from 25 μg creatinine equivalents of the urine of a normal child (*left*), and of a homocystinuric child (*right*). Origins are at the lower left-hand corners of each chromatogram. First solvent (vertical axis): pyridine—acetone—ammonium hydroxide—water (45:30:5:20). Second solvent: isopropanol—formic acid—water (75:12.5:12.5). (1) Taurine, (2) histidine, (3) 5-amino-4-imidazolecarboxamide-5′-S-homocysteinylriboside, (5) lysine, (6) homocystine, (7) glutamine, (8) glycine, (9) methionine sulfoxide, (10) serine, (11) threonine, (12) methionine, (13) β-aminoisobutyric acid, (14) alanine, (15) homolanthionine.

a chromatogram prepared from the same creatinine equivalent of a homocystinuric child's urine. The legend to Figure 2 identifies the expected amino acids, and indicates the position and nature of two very unusual urinary compounds.

Knowing that homocystinuric urine contained unusual sulfur compounds, we then attempted to isolate the ninhydrin-reactive sulfur compounds from large quantities of urine obtained from two homocystinuric children. Urines were concentrated to small volumes by lyophilization, and were then applied to preparative columns of Dowex 50. The columns were developed with a pyridine acetate buffer, and each of many small fractions collected on a fraction collector was examined for the presence of any ninhydrin-positive material, which also reacted positively to either the chlorplatinate or cyanide-nitroprusside reagents. Appropriate fractions were then pooled, and the volatile buffer was removed under reduced pressure on a rotary evaporator. The desired unidentified compounds were then further purified by repeated unidimensional chromatography on paper, using several different solvents. Details of the isolation procedures have been described elsewhere (Perry, Hansen, Bär, and MacDougall, 1966a; Perry, Hansen, and MacDougall, 1966b; Perry et al., 1967). No attempt was made to recover or identify ninhydrin-negative sulfur-containing compounds in homocystinuric urine. We therefore may very possibly have missed other interesting new methionine metabolites.

Five unusual ninhydrin-positive sulfur-containing compounds have been isolated from the urine of homocystinuric children so far. Based on examination of tracings obtained when urines were chromatographed on a Technicon amino acid analyzer, it appears likely that these five compounds are excreted regularly by most patients with homocystinuria. Figure 3 shows a tracing of an amino acid analyzer chromatogram of a 250 μg creatinine equivalent of a homocystinuric child's urine. The amino acids were separated on a 65 cm column using the buffer gradient system recommended for protein hydrolyzate analysis (Technicon Chromatography Corporation, 1965). Figure 4 shows a tracing of a portion of a chromatogram of the same amount of the identical urine, when the amino acids had been separated on a 130 cm column, using a different buffer system recommended by Efron (1965). The legends to the figures indicate the identities of the various common and unusual amino acids present.

One of the unusual amino acids isolated from the urine of one homocystinuric child, and clearly present in the urines of at least three other patients with homocystinuria, has been identified as S-adenosylhomocysteine

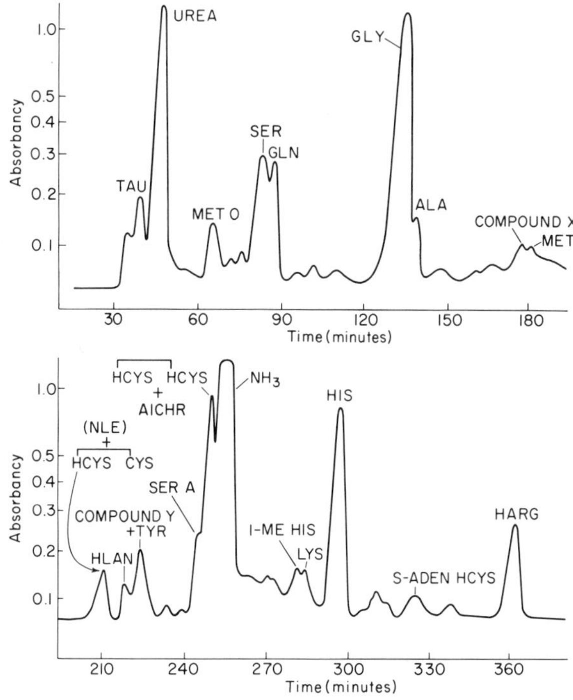

FIGURE 3.
Tracing of amino acid analyzer chromatogram of a homocystinuric child's urine. A 250 μg creatinine equivalent of urine was chromatographed on a 65 cm resin column, as described in text. Abbreviations: TAU = taurine, MET O = methionine sulfoxide, SER = serine, GLN = glutamine, GLY = glycine, ALA = alanine, MET = methionine, HYCS CYS = mixed disulfide of cysteine and homocysteine, HLAN = homolanthionine, TYR = tyrosine, SER A = ethanolamine, HCYS HCYS = homocystine, AICHR = 5-amino-4-imidazolecarboxamide-5′-S-homocysteinylriboside, 1-ME HIS = 1-methylhistidine, LYS = lysine, HIS = histidine, S-ADEN HCYS = S-adenosylhomocysteine, HARG = homoarginine (used as internal standard). Norleucine (NLE) was not used as an internal standard, but its emergence point is indicated.

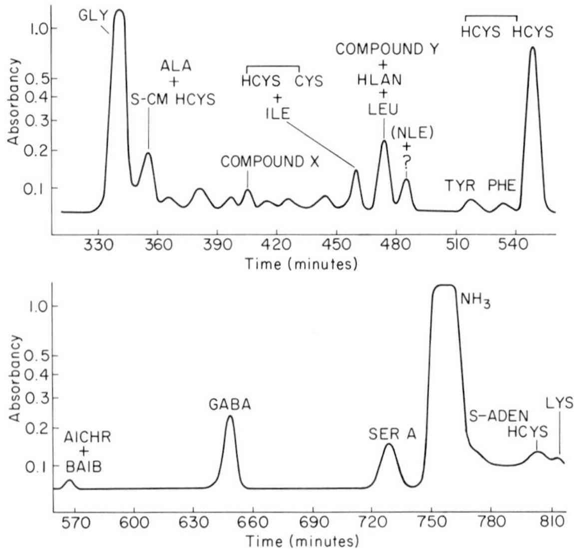

FIGURE 4.
Tracing of part of amino acid analyzer chromatogram of iodoacetate-treated homocystinuric urine. A 250 μg creatinine equivalent of the same urine shown in Figure 3 was chromatographed on a 130 cm resin column, as described in the text, after prior treatment with iodoacetic acid. Only that portion of the total chromatogram from glycine to lysine is reproduced. Abbreviations used are the same as in Figure 3, except for the following: S—CM HCYS = S-carboxymethyl-homocysteine, ILE = isoleucine, LEU = leucine, PHE = phenylalanine, BAIB = β-aminoiso-butyric acid, GABA = γ-aminobutyric acid (used as an internal standard). Norleucine (NLE) was not used as an internal standard, but its emergence point is indicated.

(Perry et al., 1967). This compound gives a positive reaction to the ninhydrin and chlorplatinate reagents, and does not react to nitroprusside, even after prior exposure to NaCN or to methanolic NaOH. The ultraviolet (UV) absorption spectrum of the compound isolated from urine is identical to that of authentic S-adenosylhomocysteine, the maximum at neutral *p*H being 259 to 260mμ. Both the urinary compound and authentic S-adenosylhomo-cysteine co-chromatograph exactly in five different solvents on paper, and in two different buffer gradient systems on the Technicon amino acid analyzer. Mild hydrolysis of the urinary and the authentic compounds in 0.1 *N* HCl at 100°C for 90 minutes in each instance yielded adenine and S-ribosylhomo-cysteine. It was estimated that one of our homocystinuric children excreted approximately 0.13 mmoles of S-adenosylhomocysteine per gram of creatinine.

S-adenosylhomocysteine has not previously been reported in human urine, but one would expect it to be excreted in homocystinuria. As can be seen in Figure 1, it is the metabolite immediately preceding homocysteine in the degradative pathway from methionine to cysteine. De la Haba and Cantoni (1959) have shown that an enzyme is present in mammalian liver which hydrolyzes S-adenosylhomocysteine to adenosine and homocysteine, and which also condenses these two substances to yield S-adenosylhomocysteine. The equilibrium of this reaction lies far in the direction of condensation, so that failure to remove homocysteine in the liver cells of a homocystinuric might well lead to the formation of S-adenosylhomocysteine, and its subsequent excretion in urine.

A second sulfur-containing amino acid that we isolated from the urines of two homocystinuric children has apparently never been encountered in nature, nor has it been synthesized. We have tentatively identified it as 5-amino-4-imidazolecarboxamide-5′-S-homocysteinylriboside (AICHR) (Perry et al., 1966a). The structure of this compound is shown in Figure 5. This compound occupies a position very close to histidine on two-dimensional paper chromatograms of homocystinuric urine (Figure 2), and on paper chromatograms exhibits R_fs almost the same as those of S-adenosylhomocysteine. However, it is eluted from the ion-exchange columns of the amino acid analyzer considerably earlier than S-adenosylhomocysteine (Figures 3 and 4).

AICHR reacts positively to the ninhydrin and chlorplatinate reagents, and fails to react to nitroprusside even after prior exposure to NaCN or methanolic NaOH. It absorbs UV light strongly, and shows a maximum absorption at neutral *p*H between 266 and 268 mμ. The UV absorption spectra of AICHR isolated from urine are identical to those of authentic 5-amino-4-imidazolecarboxamide riboside at *p*H 1 to 2, *p*H 7.5, and *p*H 12 to 13. Both AICHR and the latter compound give a blue color fading to purple-brown when paper chromatograms are sprayed with diazotized sulfanilic acid. Mild hydrolysis of AICHR in 1 *N* HCl at 100°C for 1 hour degrades the urinary compound into two fragments, which have been identified as 5-amino-4-imidazolecarboxamide (AIC) and as S-ribosylhomocysteine. Vigorous acid hydrolysis in 5.7 *N* HCl at 110°C for 16 hours breaks the urinary compound down to homocystine, homocysteine thiolactone, and glycine. The last is an artifact produced by destruction of the AIC moiety. We estimated that one of the two homocystinuric children from whose urine we isolated the compound was excreting as much as 0.3 mmole of AICHR per gram of creatinine.

The metabolic origin of AICHR in homocystinuria needs to be explained. Its presence in the urine of homocystinuric patients suggests that there are alternate pathways of methionine or purine metabolism in man that need exploration. One possibility is that AICHR is synthesized from homocysteine by reaction with AIC riboside, when tissue concentrations of homocysteine become greatly increased. AIC ribotide is a normal intermediate in the biosynthesis of purines, and it can be dephosphorylated to the corresponding

FIGURE 5.
Structure of 5-amino-4-imidazolecarboxamide-5′-S-homocysteinylriboside.

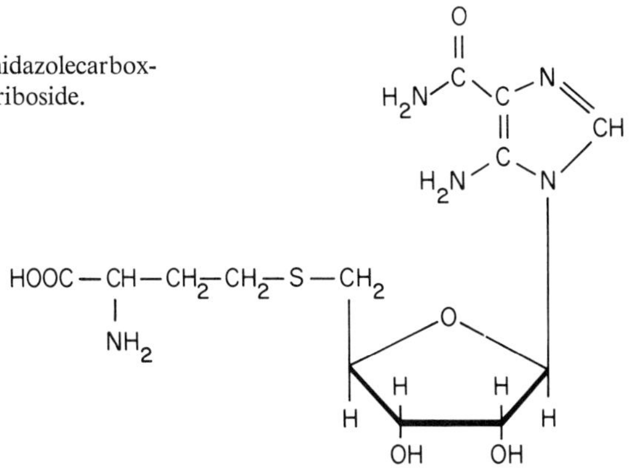

riboside. Although the liver enzyme which condenses adenosine and homocysteine to form S-adenosylhomocysteine is believed to be quite specific with respect to both substrates (de la Haba and Cantoni, 1959), it is possible that this enzyme might also accept AIC riboside as a substrate. A high concentration of homocysteine in tissues might itself favor the accumulation of AIC riboside, and thus promote condensation with the AIC riboside, should such a hypothetical pathway be operative. Warren, Flaks, and Buchanan (1957) have shown that the two biosynthetic steps in purine synthesis leading from AIC ribotide, through 5-formamido-4-imidazole-carboxamide ribotide, to inosinic acid are reversible in vitro. These investigators found that the conversion of inosinic acid back to AIC ribotide by the enzymes that normally synthesize inosinic acid required a reduced folic acid compound, potassium ions, and a reducing substance. Of the reducing substances they tested, homocysteine was far the most effective.

A second possibility for the metabolic origin of AICHR is that this compound is a degradation product arising from S-adenosylhomocysteine, which might first be deaminated to S-inosylhomocysteine, and then, facilitated by excessive tissue concentrations of homocysteine, undergo a series of reactions analogous to those observed by Warren et al. (1957). These would involve an opening of the purine ring and the removal of a formyl group by the same enzymes that convert inosinic acid to AIC ribotide in liver cells in vitro. Figure 6 illustrates the reversible metabolic pathway from AIC ribotide to adenylic acid. If, instead of the phosphate (P) group attached to ribose in each of the structures illustrated in Figure 6, a homocysteine moiety is substituted, it can be seen how AICHR might conceivably be derived from S-adenosylhomocysteine.

The third unusual sulfur-containing amino acid, which we have been able to isolate from the urine of two homocystinuric children and to identify, is homolanthionine (Perry et al., 1966b). This compound is a higher homologue of cystathionine. The structures of these two amino acids are shown in Figure 7. In the urines of each of our homocystinuric patients, we have regularly noted a small peak on amino acid analyzer chromatograms (65 cm column, protein hydrolyzate buffer gradient system) appearing between the mixed disulfide of cysteine and homocysteine and tyrosine (Figure 3). In a few urines, the same amino acid has been detected on paper chromatograms below and to the left of the position occupied by homocystine (Figure 2). This compound reacts positively to the ninhydrin and chlorplatinate spray reagents, but not to nitroprusside—even after prior exposure to NaCN or methanolic NaOH. It is not UV absorbent.

The purified urinary compound has been tentatively identified as homolanthionine on the basis of identical R_fs on paper chromatograms in five different solvents of the urinary compound and of authentic L-homolanthionine, as well as identical elution volumes from the ion exchange columns of the amino acid analyzer, using two different buffer gradient systems. Removal of sulfur from the urinary compound and from authentic homolanthionine with Raney nickel (Huang, 1963) in each case yielded only one amino acid that did not contain sulfur, namely, α-aminobutyric acid. In addition, the optical rotatory dispersion curves of the urinary compound and of authentic L-homolanthionine were very similar in the wavelength range

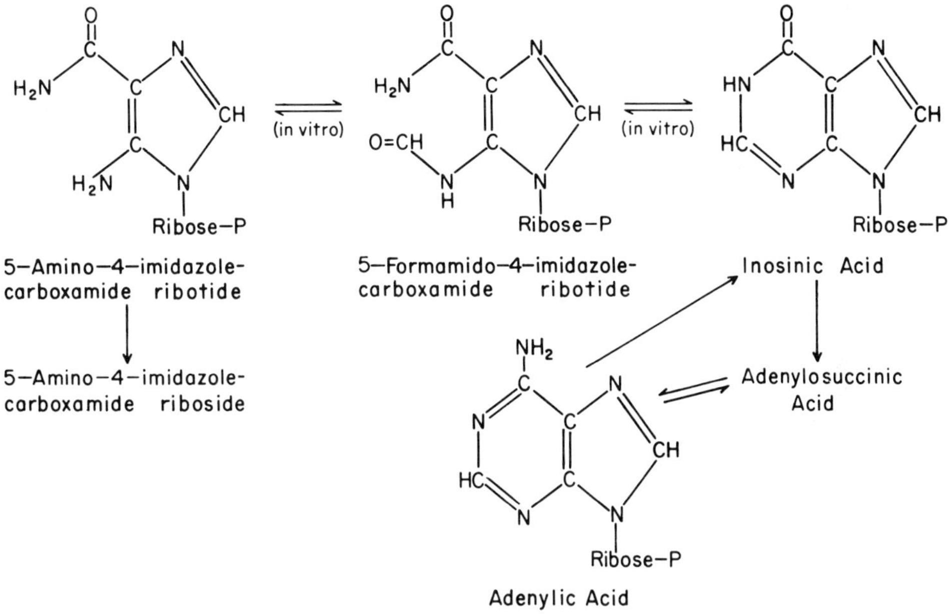

FIGURE 6.
Reversible metabolic pathway from 5-amino-4-imidazolecarboxamide ribotide to adenylic acid.

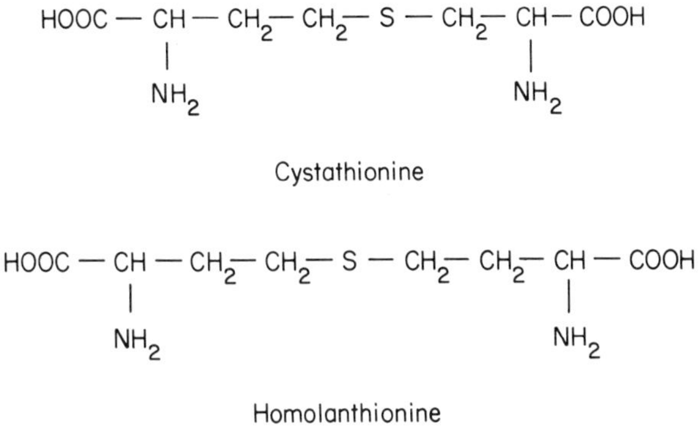

FIGURE 7.
Structures of cystathionine and homolanthionine.

between 208 and 240 mμ. It was estimated that our seven homocystinuric children excreted from 0.07 to 0.14 mmole of homolanthionine per gram of creatinine.

The only other report of homolanthionine in a living system was made by Huang (1963), who found that this amino acid was accumulated by cells of a methionine-requiring mutant of *Escherichia coli.* Presumably the homolanthionine formed by human beings with homocystinuria results from a condensation of homocysteine and homoserine, analogous to the condensation of homocysteine and serine to form cystathionine that occurs in normal individuals. Perhaps patients with homocystinuria do not suffer from an absence of cystathionine synthase, but rather have a slightly altered form of the enzyme, which has greater activity for homoserine than for serine, its normal substrate.

If homolanthionine results from the condensation of homocysteine and homoserine, the source of the latter requires explanation. Homoserine is formed in bacteria and plants—but not in mammals—as an intermediate in the biosynthesis of threonine from aspartic acid. There is evidence, however, for the formation of α-keto-γ-hydroxybutyric acid from pyruvate and formaldehyde in mammalian liver, and transamination of this α-keto acid would result in the production of homoserine (Meister, 1965). Recently, Finkelstein, Mudd, Irreverre, and Laster (1966) have presented direct evidence that extracts of normal human liver are able to catalyze the conversion of L-homoserine to α-keto-γ-hydroxybutyric acid. The presence of this transaminase in liver could provide, from the keto acid, a source of the homoserine required for homolanthionine formation.

We have also isolated and purified from the urines of two homocystinuric children two further homocysteine-containing disulfide compounds, which we have not yet been able to identify. These substances, temporarily designated as Compounds X and Y, appear to be regularly present in homocystinuric urine, based upon the appearance of amino acid analyzer chromatograms (Figures 3 and 4). Both compounds are ninhydrin-positive, and they react strongly to nitroprusside after treatment with NaCN. After hydrolysis in 6 *N* HCl at 100°C for 16 hours, they yield homocystine and homocysteine thiolactone, but no other ninhydrin-positive compounds.

We have also searched for homocysteine itself in the blood plasma and urines of our homocystinuric patients. One would expect to find it present, but the techniques normally used in examining physiological fluids result in the rapid oxidation of cysteine and homocysteine to the corresponding disulfide amino acids. Brenton et al. (1965a) demonstrated the presence of homocysteine in the plasma and urine of one of their homocystinuric patients. We treated blood specimens immediately after venepuncture, and urine specimens immediately after voiding, with iodoacetic acid, using the method developed by Brigham, Stein, and Moore (1960). Any homocysteine present is rapidly converted to the stable thioether, S-carboxymethylhomocysteine, and this compound can then be estimated on the amino acid analyzer. As expected, we found in all our patients that both homocysteine and homocystine are present in freshly drawn plasma and in freshly voided urine. In two of our patients, approximately two-thirds of the homocysteine molecules present in blood were in the —SH form, and only one-third in the —S—S— form.

AN EXPERIMENT IN TREATING HOMOCYSTINURIA

The search for unusual and new sulfur-containing amino acids in our homocystinuric patients has been engrossing, and hopefully may contribute to the uncovering of previously unrecognized pathways of methionine metabolism in man. Equally exciting, and very emotionally satisfying, has been an attempt to prevent mental retardation and the other serious clinical manifestations of homocystinuria in a newborn infant.

Early in 1965 we learned of a new pregnancy in a family that had already two homocystinuric children. One child had died of the disease at the age of ten months, as a result of massive intravascular thromboses. The other child had suffered repeated cerebrovascular accidents, and was seriously mentally retarded. It seemed reasonable that if one could diagnose homocystinuria very early in infancy, we might partially by-pass the biochemical error by administering a diet low in methionine, in the same way that phenylketonuria is alleviated by feeding a low-phenylalanine diet. We were fortunate in being able to persuade a well-known manufacturer of baby foods to provide us in advance with an experimental diet that was unusually low in sulfur-containing amino acids. This infant formula was composed of soy bean protein, to which corn oil, dextrimaltose, and necessary vitamins and minerals had been added. Since in homocystinuria the metabolic pathway from methionine to cysteine is almost completely blocked at the step of cystathionine synthesis, cysteine becomes an essential amino acid for homocystinurics. It therefore seemed wise also to prepare to feed a newborn homocystinuric infant supplemental L-cystine. Armed with the low-methionine formula and a large quantity of L-cystine, we awaited the one in four chance that the expected infant would have homocystinuria.

The baby was born in May 1965. Although he appeared to be in excellent health, a cyanide-nitroprusside screening test carried out on his urine on the fourth day of life was strongly positive, and subsequent chromatography showed that homocystine was present in the urine. The plasma methionine concentration in the cord blood was only 3 μmole/100 ml (normal <4), but by the fifth day it had risen to 129 μmole/100 ml, and on the sixteenth day had reached the extremely high level of 193 μmole/100 ml. At this time the infant was changed from an evaporated-milk formula to the special low-methionine formula, and supplemental L-cystine was administered. He has remained on a low-methionine diet steadily to the present time (age fifteen months). This diet consists of the special low-methionine formula, together with a variety of fruits and vegetables that have a low protein content. The infant throughout this period has had a daily intake of methionine ranging from 25 to 30 mg per kilo body weight (1/4 to 1/3 the usual dietary methionine intake of a normal infant), and has received a daily supplement of L-cystine amounting to 200 mg per kilo. Plasma methionine concentrations have been monitored at regular intervals with an amino acid analyzer, and the urinary content of homocystine has been observed. There has been a marked change in the direction of normal in these biochemical parameters (Perry, Dunn, Hansen, MacDougall, and Warrington, 1966c). Table 1 illustrates the changes in plasma methionine and urinary homocystine in response to the low-methionine diet. The methionine content of the experimental formula available to us has not been sufficiently low to make it possible to keep plasma methionine concentrations as close to normal as we should like. Further reduction of the total amount of formula fed our patient has not proved compatible with normal growth.

The encouraging prospect emerging from the experiment is that the infant so far has remained in excellent health, and has shown normal physical and intellectual development. He walks alone, says single understandable words, and on careful psychometric testing at the age of twelve months had a developmental quotient of 93 (within the normal range). Careful neurological examinations and electroencephalography continue to give no evidence of central nervous system damage. Slit-lamp examination shows no abnormality of the ocular lenses, and X-rays of the long bones show none of the bony changes that are usually apparent in untreated homocystinuria by this age. The baby's parents are gratified at the marked difference in this baby's performance from the unhappy course followed by their two preceding homocystinuric children.

To our knowledge, this is the first homocystinuric infant in North America to be treated specifically for the disease. Westall (1965) in England has had a similar homocystinuric infant under treatment for two years with a low-methionine diet based on gelatin, supplemented with cystine and several other amino acids. His patient is also well and developing normally. At this point we are still guarded in our optimism. We recognize that our patient, although he exhibits all the biochemical stigmata of homocystinuria, may in fact have a mild variant of the disease. We also recognize that he may still develop the life-threatening thrombotic episodes, or that as he becomes older his intellectual performance may be less promising. But perhaps we are not unreasonable in suspecting that the reduction of tissue concentrations of methionine, homocysteine, and many other unusual metabolites, resulting from marked limitation of methionine intake, may actually be responsible for

TABLE 1.
Response of a homocystinuric infant to treatment with a low-methionine diet.

Age (days)	Weight (kg)	Diet	Supplemental L-cystine (mg/kg/day)	Plasma* methionine (μmole/100 ml)	Plasma* homocystine (μmole/100 ml)	Urinary* homocystine
Birth	3.5			3		
4						++
5		Cow's milk formula	None	129		
9				187		++++
16	4.0			193	2.8	++++
22				94		+
29	4.0	Low-methionine	100	33		+
53		formula, ad lib		12		0
63	5.0			10		0
90		Low-methionine formula,		17		0
130		ad lib; plus fruits		48		+
139	7.2	and vegetables		50		+
146		Low-methionine formula,		63		+
153		ad lib; no solids		41		+
160			200	46		++
167	7.2			26		+
194				20		+
240		Low-methionine formula,		18		+
269	8.0	limited intake; fruits		8		+
298		and vegetables, ad lib		8		+
381	9.3			35	2.4	+
389				27		+
430				34		+

*Fasting plasma methionine concentrations are normally not higher than 4 μmole/100 ml. Homocysteine and homocystine (here calculated as homocystine) are normally not detectable in plasma or urine.

our patient's normal progress. We hope that if we can continue the low-methionine diet and the cystine supplement until rapid growth of brain, bones, and arterial walls is complete in this homocystinuric child, he will later be able to survive in reasonable health and with normal mentality on a less restricted diet.

UNSOLVED PROBLEMS IN HOMOCYSTINURIA

Homocystinuria presents a number of intriguing problems, which need imaginative investigators for their solution. One of these, of course, is to explain the exact mechanism of the mental defect present in most homocystinurics. Does cystathionine, one of the major free amino acids in human brain, play a physiological role in the normal functioning of the central nervous system? Why does our treated patient have apparently normal mentation for a fifteen month-old infant, when his brain may well lack cystathionine entirely? We have not given our patient L-cystathionine for a variety of reasons. The amounts of the amino acid required would be prohibitively expensive. Cystathionine is probably not absorbed through the gastrointestinal mucosa, and, if it were administered parenterally, it is likely that it would be rapidly cleared from the circulation by the kidneys. Finally, there is no evidence that cystathionine would penetrate from the circulation into the target area, the brain cells.

Not only does our treated infant appear to have normal intelligence, but one of our untreated homocystinurics, who is now eighteen years old, is also of average intelligence (Dunn et al., 1966). Schimke et al. (1965) have reported a number of adults with homocystinuria who were not mentally defective. It would be surprising to find that man, of all animals, has the highest brain concentrations of cystathionine, but that this amino acid has no essential function in brain!

If lack of cystathionine is not responsible for the mental defect characteristic of most homocystinurics, what is responsible for it? One possibility is that very high tissue concentrations of methionine, homocysteine, or homocystine may interfere with the entrance of other amino acids into developing brain cells, or with the incorporation of other amino acids into protein in rapidly growing brain. Another possibility is that one or more of the unusual sulfur-containing metabolites formed in homocystinuria has a toxic effect on brain. Identification of the remaining unknown sulfur-containing amino acids and ninhydrin-negative sulfur compounds in homocystinuric urine may throw light on this possibility. It also may be that repeated minor thromboses in the cerebral vessels, resulting in recurrent occlusion of the blood supply to many very small areas of brain, is the major cause of the mental defect in homocystinuria.

In our own small group of homocystinuric children, we have the impression that the most severely affected patients, especially with regard to mental defect, have relatively low plasma concentrations of methionine, and relatively high concentrations of homocysteine and homocystine. The less severely affected patients tend to show high plasma methionine concentrations, and relatively low concentrations of homocysteine and homocystine. Does greater efficiency of the enzymatic pathway for remethylation of homocysteine to methionine (Figure 1) in some manner confer partial protection to the patient?

A second puzzling question about homocystinuria is the nature of the defect in collagen formation. A decrease in the degree of cross-linking in collagen might account for the defects in the suspensory ligament of the lens, in the cartilaginous matrix of developing bone, and in the media and intima of arteries. But why should an abnormality in methionine metabolism, or a relative deficiency in cysteine during growth, interfere with normal collagen formation? Collagen is a very unusual animal protein, in that it contains relatively large amounts of glycine, proline, hydroxyproline, and hydroxylysine, but little methionine, and no cysteine. Comparisons need to be made of the chemical composition and physical form of collagen from homocystinurics, with that obtained from normal persons. If differences are found, we must ask which of many sulfur-containing metabolites are responsible for the differences.

The reason, or reasons, for the intravascular thromboses, which are the major cause of death in homocystinurics, urgently needs explanation. If a structural abnormality in the development of collagen in the walls of arteries accounts for the initiation of arterial thromboses, hopefully, use of a low-methionine diet in infancy and childhood might protect the homocystinuric from this complication in later life. But abnormal collagen in the media and intima of arteries does not explain why venous thromboses are also troublesome. The claim by McDonald et al. (1964) that platelet stickiness is increased in homocystinuria could not be confirmed by another group of investigators (Schimke et al., 1965). This claim, however, merits further study. Does homocysteine possibly alter the charge on platelet membranes, or could homocystine molecules link platelets together by disulfide bridges? Certain adenine nucleotides are known to affect the coagulability of blood. Does S-adenosylhomocysteine or AICHR cause an increased tendency toward intravascular coagulation? It will be very important to discover whether there is a continuing chemical cause for hypercoagulability of blood in homocystinuria. If there is, then it may be necessary to continue a low-methionine diet indefinitely for successfully treated homocystinurics.

Another aspect of homocystinuria that merits further exploration is the occurrence of mental illness, especially schizophrenia, among the close relatives of homocystinurics. Carson et al. (1963) reported the pedigree of a homocystinuric family in which six persons with schizophrenia were found in the two generations before the propositus on the mother's side of the family. In our own homocystinuric families, there is also a greater incidence of mental illness among close relatives than might be expected, but it occurs only on one side of each family (Dunn et al., 1966). Two of the homocystinuric patients described by Schimke et al. (1965) had suffered psychotic illnesses.

It would be valuable to have a simple and reliable test for heterozygosity for the abnormal gene that causes homocystinuria. At the present time, heterozygosity can only be determined by assay of cystathionine synthase in liver biopsy specimens (Finkelstein et al., 1964). Unfortunately the enzyme is apparently not present in such readily available tissues as erythrocytes, leucocytes, platelets, and skin. We have tried unsuccessfully to develop both methionine and homocystine loading tests which might differentiate the parents of homocystinuric children from control subjects (Dunn et al., 1966). If a simple and accurate method for determining heterozygosity were available, it would be of interest to study especially the mentally ill relatives of homocystinurics. It is even conceivable that "multiple heterozygosity," as suggested by Lippman (1958), may be a factor in the etiology of the serious mental illnesses of adult life. Perhaps simultaneous heterozygosity, for homocystinuria and for one or more other gene pairs, has a synergistic and harmful effect on a complex metabolic pathway.

LINUS PAULING AND HOMOCYSTINURIA

Although homocystinuria was discovered only at the end of my association with Linus Pauling, any contributions made to the understanding of this disease in our laboratories here in Vancouver are in large measure attributable to him. I do wish to take this opportunity to comment briefly on certain of his personal and scientific characteristics that might well be emulated by others.

I should probably never have entered research in the basic medical sciences had it not been for Linus Pauling's encouragement and practical help. A decade ago, after devoting myself to an active practice of pediatrics for ten years, I was strongly advised by everyone I consulted in the academic or scientific community not to try to enter scientific research. I had been too long away from scientific medicine, and in particular I was too old to be worth retraining, the "experts" told me. Pauling alone gave me different advice. In a very practical way he helped me by offering me the chance to work in his project to investigate the biochemical basis of mental deficiency, then financed by the Ford Foundation. I am most grateful to him for his encouragement, and hope that other scientists will come to acquire the imagination and humanity needed to retool older men and women who genuinely want to embark on a second career of scientific research. I believe the idea, nowadays so prevalent, that only persons in their twenties or thirties can make really significant research contributions, is both fallacious and harmful to the development of science.

It was not only Linus Pauling's attitudes toward older persons interested in science that were unusual a decade ago. He was one of a very small minority of distinguished American scientists courageous enough to give a job to competent persons who had run afoul of the McCarthyite committees of the day. My brother-in-law, the late Dr. Richard W. Lippmann, for being true to himself and to the best American traditions, had been effectively barred from scientific research in his chosen field of renal physiology. After a brilliant investigative career, in which he had made more than seventy published scientific contributions to the study of renal function and kidney disease, no medical school or scientific establishment in the United States could find a place for him to continue the research he most wanted to do. In 1956, Pauling gave Lippman the opportunity, in his laboratories at the California Institute of Technology, to embark on research in the chemistry of mental disease. Had it not been for Richard Lippman's untimely death in 1959, he would certainly now be making contributions in this field as significant as those he made in renal disease. I too found myself barred from a career in research and teaching in the medical schools of my own country because of my refusal to cooperate with a witch-hunting Congressional Committee. Pauling alone gave me a chance. Surely, individual scientists, institutions, agencies administrating grants, and governments need to follow Linus Pauling's example in making honesty and scientific qualifications the only criteria for participation in scientific research.

Finally, I believe that much of value can be learned from observation of Linus Pauling's scientific and humanitarian activities over the last twenty years. His interests have broadened to cover the applications of chemistry to the solution of problems concerning the nature of various pathological conditions which cause extensive human misery. His contributions to molecular biology in the hemoglobinopathies and in mental deficiency, and his theory of general anesthesia, exemplify his growing concern that science should be devoted to enriching men's lives, not to destroying them. His public contributions in alerting the peoples of the world to the radiation hazards of nuclear weapons testing, and to the incomparably worse threat of universal destruction in a nuclear war, are in the best traditions of science and of humanity.

At a time when technological advances and a significant proportion of all scientific research are being perverted for use in the bullying and destruction of poor and helpless people in various parts of the world, we badly need more Linus Paulings. I should like to see an end to all scientific research on methods of chemical and biological warfare. I should like to see, instead, the scientific talent and financial resources of the advanced countries devoted to seeking better methods of preventing or curing the diseases which cause so much human suffering.

I should like to see science master the problems of homograft rejection, for instance, so that we might treat homocystinuria and similar diseases by transplanting into the abdomen an accessory liver, perhaps obtained from a premature infant who died shortly after birth. Such a liver might be able to carry out effectively that one chemical job out of hundreds which the patient's own liver cannot perform. I should like to see chemists, engineers, and surgeons team together to synthesize missing enzymes and to install them in a suitable chamber placed in the circulation of a patient having an enzyme deficiency disease. Molecules of substrate might be able to enter such a chamber, and molecules of reaction product to leave it, while larger antibodies, which might destroy the artificial enzyme, would be excluded.

As we find new ways of keeping people with serious molecular diseases alive and healthy, concern is being expressed about increasing the pool of abnormal genes in the men and women of tomorrow. It seems to me that if we can, as we must, master man's destructiveness and perfect his living relationships with his fellow men, we can surely learn to counter the ill effects produced by abnormal genes in the centuries to come.

ACKNOWLEDGMENTS

I wish to thank Shirley Hansen, Lynne MacDougall, and Patrick D. Warrington, for valuable technical assistance, and Dr. Henry G. Dunn for his help with the patients studied. I am especially grateful to Professor James G. Foulks for many stimulating discussions which have materially aided this research.

REFERENCES

Brenton, D. P., D. C. Cusworth, and G. E. Gaull (1965a). *J. Pediatrics* **67**, 58.
Brenton, D. P., D. C. Cusworth, and G. E. Gaull (1965b). *Pediatrics* **35**, 50.
Brigham, M. P., W. H. Stein, and S. Moore (1960). *J. Clin. Invest.* **39**, 1633.

Carson, N. A. J., and D. W. Neill (1962). *Arch. Dis. Childh.* **37**, 505.

Carson, N. A. J., D. C. Cusworth, C. E. Dent, C. M. B. Field, D. W. Neill, and R. G. Westall (1963). *Arch. Dis. Childh.* **38**, 425.

Carson, N. A. J., C. E. Dent, C. M. B. Field, and G. E. Gaull (1965). *J. Pediatrics* **66**, 565.

de la Haba, G., and G. L. Cantoni (1959). *J. Biol. Chem.* **234**, 603.

Dunn, H. G., T. L. Perry, and C. L. Dolman (1966). *Neurology* **16**, 407.

Efron, M. (1965). Personal communication.

Field, C. M. B., N. A. J. Carson, D. C. Cusworth, C. E. Dent, and D. W. Neill (1962). *Abstr. 10th International Congress of Pediatrics, Lisbon*, p. 274.

Finkelstein, J. D., S. H. Mudd, F. Irreverre, and L. Laster (1964). *Science* **146**, 785.

Finkelstein, J. D., S. H. Mudd, F. Irreverre, and L. Laster (1966). *Proc. Nat. Acad. Sci. U.S.* **55**, 865.

Gerritsen, T., J. G. Vaughn, and H. A. Waisman (1962). *Biochem. Biophys. Res. Commun.* **9**, 493.

Gerritsen, T., and H. A. Waisman (1964a). *Pediatrics* **33**, 413.

Gerritsen, T., and H. A. Waisman (1964b). *Science*, **145**, 588.

Gibson, J. B., N. A. J. Carson, and D. W. Neill (1964). *J. Clin. Path.* **17**, 427.

Huang, H. T. (1963). *Biochemistry* **2**, 296.

Komrower, G. M., and V. K. Wilson (1963). *Proc. Roy. Soc.* (London) **56**, 996.

Laster, L., S. H. Mudd, J. D. Finkelstein, and F. Irreverre (1965). *J. Clin. Invest.* **44**, 1708.

Lippman, R. W. (1958). *Am. J. Ment. Def.* **63**, 320.

McDonald, L., C. Bray, C. Field, F. Love, and B. Davies (1964). *Lancet* **1**, 745.

Meister, A. (1965). *Biochemistry of the Amino Acids*, vol. 2, p. 678. New York: Academic Press.

Mudd, S. H., J. D. Finkelstein, F. Irreverre, and L. Laster (1964). *Science* **143**, 1443.

Mudd, S. H., J. D. Finkelstein, F. Irreverre, and L. Laster (1965). *J. Biol. Chem.* **240**, 4382.

Okumura, N., S. Otsuki, and A. Kameyama (1960). *J. Biochem.* (*Tokyo*) **47**, 315.

Perry, T. L., S. Hansen, H.-P. Bär, and L. MacDougall (1966a). *Science* **152**, 776.

Perry, T. L., S. Hansen, and L. MacDougall (1966b). *Science* **152**, 1750.

Perry, T. L., H. G. Dunn, S. Hansen, L. MacDougall, and P. D. Warrington (1966c). *Pediatrics* **37**, 502.

Perry, T. L., S. Hansen, L. MacDougall, and P. D. Warrington (1967). *Clin. Chim. Acta.* In press.

Schimke, R. N., V. A. McKusick, T. Huang, and A. D. Pollack (1965). *J. Am. Med. Assoc.* **193**, 711.

Spaeth, G. L., and G. W. Barber (1965). *Trans. Am. Acad. Ophth. and Otol.* **69**, 912.

Tallan, H. H., S. Moore, and W. H. Stein (1958). *J. Biol. Chem.* **230**, 707.

Technicon Chromatography Corporation, Ardsley, N. Y. (1965). Technical bulletin.

Warren, L., J. G. Flaks, and J. M. Buchanan (1957). *J. Biol. Chem.* **229**, 627.

Werder, E. A., H.-C. Curtius, F. Tancredi, P. W. Anders, and A. Prader (1966). *Helv. Paediat. Acta* **21**, 1.

Westall, R. G. (1965). Personal communication.

White, H. H., H. L. Thompson, L. P. Rowland, D. Cowen, and S. Araki (1964). *Trans. Am. Neurol. Assoc.* **89**, 24.

White, H. H., L. P. Rowland, S. Araki, H. L. Thompson, and D. Cowen (1965). *Arch. Neurol.* **13**, 455.

RICHARD T. JONES
Department of Biochemistry
University of Oregon Medical School
Portland, Oregon

ROBERT D. KOLER
Division of Hematology and Experimental Medicine
University of Oregon Medical School
Portland, Oregon

Some Contributions of Human Genetics to Molecular Biology

Two of the areas in which Linus Pauling has had continuing interest are human genetics and molecular biology. The concept of "molecular disease," which Pauling was first to recognize and which he and co-workers first elaborated in their studies of sicklemia and sickle-cell anemia (Pauling et al., 1949), served as one of the bridges from classical chemistry and biology to modern molecular biology and molecular medicine. Pauling's long interest in the structure of molecules, including biological macromolecules, has helped set the stage for attempts to understand disease at a molecular level.

This paper will trace how certain studies of human biochemistry and disease, especially hereditary diseases, have contributed fundamental concepts to molecular biology. This will be done by discussing how studies of human genetics have contributed to the development of modern ideas of:

1. the genetic control of enzymes;
2. the genetic control of protein structure and function;
3. mutations and their effects on proteins;
4. multiple subunits of macromolecules;
5. isozymes and the genetic control of metabolism.

As will become evident, hemoglobin has been the prototype system for these studies, and reference to it is made frequently, in part because of contributions that have been made by studies of this system, and in part because we are most familiar with it.

It is interesting to ask why studies of such a complex biological system as the human organism have resulted in any fundamental contributions to molecular biology. Although man is not a convenient experimental system, he is unique because a long and voluminous history of his biologic variation exists. This is partly a reflection of his interest in himself and of his attempt to understand and correct human disease. In his study of diseases, occasional

This work was supported in part by Public Health Service Research Grant No. CA07941 from the National Cancer Institute.

insights into fundamental mechanisms of biology have been gained, which have then been studied in detail in more convenient or simpler systems. Certainly, future insights into underlying causes and mechanisms of human diseases will depend upon application of concepts of molecular biology to the study of man.

For the purposes of our discussion the term human genetics will be used in a broad sense to include many of the biological and chemical studies of normal and diseased states of man. The term molecular biology will be used to include areas of modern biology that are presently being approached from a molecular point of view. Several topics important to this subject, but not discussed in this brief account, include chromosomal genetics, human blood groups, tissue transplantation, growth and development and immunoproteins.

THE GENETIC CONTROL OF ENZYMES

That genes act by regulating chemical events was first deduced by Garrod, and was clearly stated in his Croonian lectures entitled "Inborn Errors of Metabolism." These lectures, delivered before the Royal College of Physicians in 1908, have been reprinted, with a supplement by Harris (1963). Also included is a 1902 paper on "The incidence of alkaptonuria: A study in chemical individuality," in which Garrod describes the excretion of homogentisic acid as "an alternative course of metabolism." He cites the familial incidence, the high frequency of consanguineous matings among parents of affected, and the likelihood, therefore, that the condition is the result of homozygosity for a single recessive character. Garrod attributes this conclusion to Bateson (1902), who had just described the rediscovery of Mendel's work. Moreover, Garrod predicted in his 1902 publication that other and more subtle examples of individual variation in metabolic processes would be found. By 1908, he had added albinism, cystinuria, and pentosuria to his list of inborn errors. He introduces his discussion of these metabolic defects with the prophetic statement that the sources of variation between species, and between individuals within species, should be sought among the highly complex proteins, because the factors that bring about evolution have worked and are working upon chemical variations like alkaptonuria.

Alkaptonuria also served as a model system for investigating metabolic pathways. Neubauer (1928), after doing feeding experiments, was able to identify intermediates that are catabolyzed to homogentisic acid in the alkaptonuric subject, and thus to arrive at the correct scheme of phenylalanine and tyrosine metabolism.

Unfortunately, few of Garrod's contemporaries seemed to appreciate the importance of this relationship between recessive Mendelian traits and blocks at specific steps in normal metabolism; that is, between gene and enzyme. His work was omitted or only briefly presented in texts of genetics or of metabolism. His discouragement is apparent by 1928, when, in a paper entitled "The Lessons of Rare Maladies," he barely mentions inborn errors of metabolism, "a subject with which I fear to weary my colleagues" (Garrod, 1928).

The renaissance that resulted in the one gene-one enzyme hypothesis, and its extension to a general concept of the mechanism of gene action came, not from human genetics, but from a study of mutants purposely produced in Neurospora by Beadle and Tatum (1941). Their nutritional mutants of Neurospora allowed Beadle and Tatum to trace metabolic pathways by methods analogous to the feeding experiments done on alkaptonuric subjects by Neubauer thirty years before. In his lecture given on receiving the Nobel prize in 1958, Beadle (1959) graciously acknowledged the priority of Garrod's ideas.

They, however, accomplished two major steps beyond Garrod's formulation. They extrapolated their observations to *all* organisms, and they convinced their contemporaries, though not immediately (see Delbruck, 1946), that these generalized statements are substantially correct.

The list of inborn errors in man has grown enormously (Harris, 1963; Stanbury, Wyngaarden, and Fredrickson, 1966; Waisman, 1966), stimulated no doubt by the approach introduced by Garrod, and rediscovered and refined by Beadle and Tatum. Biochemical knowledge about such inborn errors, about the enzymes affected and the pathways involved, continues to accumulate; it has saved lives and prevented mental retardation through the development of successful dietary therapy, for example, to counteract the effects of such metabolites as galactose or phenylalanine.

THE GENETIC CONTROL OF PROTEIN STRUCTURE AND FUNCTION

Although the experimental justification for the one gene-one enzyme-one action principle was well established by the late 1940's, the manner in which genes influenced protein function was not clear at that time. The chemical studies of sickle-cell hemoglobin by Pauling and co-workers (1949) produced the first direct evidence that an alteration in the physical chemical properties of a protein is genetically controlled. Their paper entitled, "Sickle Cell Anemia, a Molecular Disease" initiated a number of investigations in molecular biology, which have resulted in many of our present concepts of the genetic control of protein structure. In his Harvey Lecture of April 1954, Pauling (1954) relates how Dr. William B. Castle brought the disease, sickle-cell anemia, to his attention in 1945. Dr. Castle mentioned how the disease had been discovered in 1910 by a physician, J. B. Herrick (1910), who described the change in shape of erythrocytes from such patients to a sickle form and the reversal of this change by oxygen. During his discussion with Castle, Pauling "pointed out that the relation of sickling to the presence of oxygen clearly indicated that the hemoglobin molecules in the red cell are involved in the phenomenon of sickling, and that the difference between sickle-cell-anemia red corpuscles and normal red corpuscles could be explained by postulating that the former contains an abnormal kind of hemoglobin, which, when deoxygenated, has the power of combining with itself into long rigid rods which then twist the red cell out of shape."

In spite of a long history of medical interest in the cause and abnormal physiology of sickle-cell anemia, it was not until Pauling's postulation of an abnormal hemoglobin that a single etiological factor for this disease was considered. Pauling's recognition of this molecular abnormality in sickle-cell anemia must, in part, have resulted from his long-time interest in chemical properties of hemoglobin (Pauling, 1954), his considerations of the chemical structure of proteins, and particularly, his structural approach to chemistry.

It is interesting to note, as did Ingram (1959a), that abnormal human hemoglobins formed the first, most useful system in biochemical genetics in which the chemical changes in a protein due to mutations could be observed. Several reasons account for this. Hemoglobin can be obtained easily in relatively large amounts and in essentially pure form without resorting to elaborate purification procedures. Because it is readily available, it has been studied more completely than any other protein. The recognition of genetic abnormalities of hemoglobin and the worldwide search for these genetic variants have resulted in the collection of the largest number of natural mutants known for a single protein system.

Simultaneous with the chemical studies of sickle-cell hemoglobin by Pauling and co-workers, the genetics of sickle-cell disease was clarified by Neel (1949). Although Emmel (1917) as early as 1917 and later Taliaferro and Huck (1923) recognized a genetic basis for the sickling phenomenon, Neel was the first to demonstrate by "classical" genetic studies that sickle-cell anemia is due to homozygosity for a gene that is responsible for the sickling characteristic. Beet (1949) reached the same conclusion independently, and at about the same time. Pauling and associates (1949) also arrived at the same genetic interpretation by demonstrating the presence of roughly equal amounts of normal adult hemoglobin (Hb A) and sickle-cell hemoglobin (Hb S) in individuals with sickle-cell trait, and only Hb S in individuals with sickle-cell anemia.

Soon after the discovery of Hb S, Itano and Neel (1950) detected a second abnormal hemoglobin, Hb C, which they found in association with Hb S in individuals who did not have Hb A. Examination of the families revealed the presence of Hb A and Hb C in one parent and Hb A and Hb S in the other parent. Although Itano and Neel (1950) were unable to distinguish between a genetic mechanism of two loci and a sytsem of multiple alleles, they did establish that there are different kinds of genetically determined abnormal hemoglobins. In a later paper Itano (1953) hypothesized allelic control of Hb S and Hb C because of the absence of Hb A in Hb S-C disease. Allelic control of Hb S and Hb C was confirmed by Ranney (1954) from studies of the inheritance of Hbs A, S, and C in a single family. Since 1950 more than 100 other variants of human hemoglobins have been discovered and chemically and genetically characterized to various degrees of completion (Schroeder and Jones, 1965).

From physical chemical studies Pauling and co-workers (1949) concluded that Hb S has from two to four more net positive charges per molecule than Hb A. Their preliminary studies of derivatives of the heme group led them to conclude that these differences reside in the globin portion of the hemoglobin molecule. This prompted Schroeder, Kay, and Wells (1950) to determine the amino acid composition of Hb S. They found it to be the same as that of Hb A within the limits of the sensitivity of the methods for amino acid analysis then available. This work demonstrated that structural differences between Hb A and Hb S might be small compared to the similarities.

In spite of continued work in several laboratories it was not until 1956 that Ingram (1956) obtained the first insight into the structural differences between Hb A and Hb S by applying the elegant method of combined paper electrophoresis and paper chromatography to the comparison of peptide fragments of Hb A and Hb S produced by hydrolysis with trypsin. Ingram

(1958, 1959b) found that the only apparent difference in the amino acid sequence between Hb S and Hb A consists in the replacement in each identical half molecule of a valyl residue in Hb S for a glutamyl residue in Hb A. This single difference per half molecule accounts for the difference in charge observed earlier. From the work of several different groups (Hunt and Ingram, 1959; Hill and Schwartz, 1959; Rhinesmith, Schroeder, and Martin, 1958; Vinograd, Hutchinson, and Schroeder, 1959) it was possible to conclude that the amino acid change is located at the sixth residue position from the NH_2-terminal end of the β-chain and that the α-chain was completely normal. Thus it was established that genetic mutation could result in the substitution of only one amino acid out of 267 different amino acid residues in a single protein molecule. A remarkable correlary to this was the conclusion that the remaining portions of the molecules of Hb A and Hb S are apparently identical: This observation contributed significantly to an acceptance of the concept that the primary structure of a protein is quite precise and constant.

Hb C was the next abnormal hemoglobin to be structurally characterized, and was found by Hunt and Ingram (1958, 1960) to have a single substitution of a lysyl residue for the valyl residue number 6 of the β-chain. From a comparison of the structural differences among Hbs A, S and C, these investigators demonstrated that different mutations can alter in different ways the same amino acid residue site in a polypeptide chain. They also proposed that allelism of genes can be deduced from chemical studies of their corresponding protein gene products. Similar chemical evidence has been used since then to conclude allelism of all β-chain variants.

Chemical characterization of other abnormal human hemoglobins led to the recognition of single amino acid substitutions in the α-chain and to the conclusion that the structure of these α-chain variants is probably determined by allelic genes (see Schroeder and Jones, 1965, or Baglioni, 1963, for review of literature).

Following Ingram's discovery of the chemical abnormality in Hb S and the demonstration of two types of polypeptide chains in Hb A (Rhinesmith, Schroeder, and Pauling, 1957), Itano (1957) proposed that there is separate genetic control of the α- and β-chains of human hemoglobin. He postulated that if the two chains were synthesized independently, four hemoglobins would be produced in individuals who are heterozygous for both genetic factors (that is, for both α- and β-chain genes). Although Schwartz et al. (1957) proposed multiple loci to explain their observation in a family with Hb S and Hb G, the first irrefutable evidence of nonallelism of two abnormal hemoglobins was obtained by Smith and Torbert (1958), who observed independent segregation of genes determining Hb S and Hb Hopkins-2. Itano and Robinson (1960), examining individuals studied by Smith and Torbert, found that the α-chain of Hb Hopkins-2 was abnormal, and that four hemoglobins designated $\alpha_2{}^A\beta_2{}^A$, $\alpha_2{}^{Ho-2}\beta_2{}^A$, $\alpha_2{}^A\beta_2{}^S$, and $\alpha_2{}^{Ho-2}\beta_2{}^S$ were present in those individuals who were doubly heterozygous for these genes. Chemical and genetic studies of other hemoglobin combinations (see Schroeder and Jones, 1965, for further details) support the conclusion that two separate genetic loci, namely the α-chain gene and the β-chain gene, control the structure of Hb A. In addition to the family reported by Smith and Torbert (1958) (also reported in Itano, Singer, and Robinson, 1959) four other reports

(Dherte, Lehmann, and Vandepitte, 1959; Cabannes and Portier, 1959; Atwater, Schwartz, and Tocantins, 1960; and Raper et al., 1960) of more than two kinds of adult hemoglobin in the same individual have appeared. All are readily explainable in terms of the occurrence of both mutant and normal alleles at loci controlling the structures of the α- and β-chains. Estimates of linkage between the α and β locus using the method of Morton (1955) and based on the pedigrees described in the above reports give minus lod scores of 4.0 or greater for recombination values less than 0.5. Thus, there is no question of allelism or of pseudoallelism, and human hemoglobin provided the first example of the presence within one functional protein of the products of two genes; the products are clearly separable by genetic tests, and are now known to be different in primary structure. These studies of the genetic control of subunit structure of hemoglobin resulted in a slight revision of the "one gene-one enzyme" statement to a "one gene-one polypeptide chain" concept.

In addition to Hb A two other normal hemoglobins, Hb F and Hb A_2, exist in humans. Chemical studies of these reveal the presence of α-chains in each that are chemically identical to the α-chains of the adult form (Schroeder and Matsuda, 1958; Hunt, 1959; Jones, Schroeder, and Vinograd, 1959; Schroeder et al., 1963; Muller and Jonxis, 1960; Ingram and Stretton, 1962). This suggests that the α-chains of Hb A, F and A_2 are also genetically identical, and that all α-chains are synthesized under the control of a single α-chain locus. Genetic proof of this conclusion has come from the finding of abnormal Hb A_2 and abnormal Hb F associated with α-chain variants of Hb A. The first evidence was obtained from observation of α-chain variants of Hb A_2 that were studied by Huehns and Shooter (1961). Analogous mutants of the α-chain of Hb F were characterized by Minnich and co-workers (1962). (The reader is referred to Baglioni, 1963; and Schroeder and Jones, 1965, for details.) The concept of one gene-one polypeptide chain readily explains these observations of more than one abnormal protein associated with a single genetic mutation if groups of proteins may share common polypeptide chains.

In addition to hemoglobin variants that affect either the β-chain of Hb A or the α-chain of Hbs A, F, and A_2, mutants of the γ-chain of Hb F (Schneider and Jones, 1965) and the δ-chain of Hb A_2 (Jones, Brimhall, Huehns, and Barnicot, 1966) have been chemically characterized and shown to have single amino acid substitutions. Genetic variation in one kind of hemoglobin chain does not, in general, appear to alter the amino acid sequence of the other chains. This indicates that the genetic factors that control the sequence of the γ- and δ-chains are independent of each other and of the genes that determine the structure of the α- and β-chains. Pedigree tests also support the idea of tight linkage between the structure genes which produce human β, γ, and δ chains, though a separate messenger is presumed to be produced by each in contradistinction to the single polycistronic message that characterizes the galactosidase and histidine operons in *E. coli* (Winslow and Ingram, 1966).

Several abnormal human hemoglobins have been observed that, after structural characterization, cannot be classified as differing from their normal counterparts simply by single amino acid substitutions. These include Hb H,

Hb Bart's, Hb Lepore, and Hb Freiburg. The determination of their chemical structure has resulted in further insight into the genetic control of protein structure.

Hemoglobin H was first reported by Rigas, Koler, and Osgood (1955) and independently by Gouttas and associates (1955). Hemoglobin H was found to consist of four normal β-chains and to contain no α-chains (Jones, Schroeder, Balog, and Vinograd, 1959; Jones and Schroeder, 1963). Hemoglobin Bart's was found to be analogous to Hb H and to consist of only four normal γ-chains (Hunt and Lehmann, 1959; Jones and Schroeder, 1963). Recognition of the structure of these two hemoglobins also contributed to the concept of the independent synthesis of the various polypeptide chains of hemoglobin. It has been proposed that these two abnormal hemoglobins result from a deficiency in the production of α-chains (see Jones and Schroeder, 1963, for references to literature). Genetic studies of families with Hb H or Hb Bart's, or both, indicate that double heterozygosity at two independent loci may be interacting at a biochemical level to produce the deficiency of α-chain production (Koler and Rigas, 1961; Wasi, Na-Nakorn, and Suingdumrong, 1964).

Hemoglobin Lepore was first observed by Gerald and Diamond (1958), and was chemically characterized by Baglioni (1962, 1965), who found it to consist of two normal α-chains and two identical chains with the sequence of the δ-chain in their N-terminal halves and a sequence corresponding to the β-chain in their C-terminal halves. Hemoglobin Freiburg is a third kind of unusual abnormal hemoglobin. It has been structurally characterized and found to consist of two normal α-chains and two β-chains that are normal except for the deletion of the valyl residue 23 (Jones, Brimhall, Huisman, Kleihauer, and Betke, 1966). The amino acid sequences on either side of the deletion appear to be normal. The genetic events postulated to have given rise to Hb Lepore and Hb Freiburg are discussed in the following section, which begins on the facing page.

Although most abnormal hemoglobins have been detected through differences in electrophoretic mobility, several have been discovered, at least in part, because of changes in their functional properties. Careful examination of the abnormal functions and correlations of these with their abnormal structures have begun to give insight into the relationships of structure to function of proteins at a molecular (atomic) level. The complex interactions between various residues in the polypeptide chain of hemoglobin, and between these residues and the heme group, are beginning to be analyzed by studies of abnormal hemoglobins. The examination of these mutant hemoglobins for the purpose of obtaining insight into exact relationships of structure to function is comparable to the use of enzyme mutants for the determination of metabolic pathways. The abnormal hemoglobins that have been most useful in these correlation studies can be divided into two groups, those affecting the oxygen-transport function and those affecting the chemical stability.

The Hb M's are prominent among the abnormal hemoglobins that have altered oxygenation properties. These mutants are characterized by having a greater than normal stability of the oxidized or met-form of the heme groups associated with the abnormal chain. Gerald and Efron (1961) were the first to recognize how a single amino acid alteration in Hb M might give rise to an intramolecular coordination complex of ferri heme, which would be more stable to reduction than normal ferri heme. Since their studies, five or more Hb M's have been characterized, and the importance of several residues near the heme group have been recognized (see Baglioni, 1963, for further discussion). Apparently more subtle in their effects are the amino acid substitutions in Hb Chesapeake (Charache, Weatherall, and Clegg, 1966) and Hb Yakima (Osgood et al., 1966) in which there is an increase in oxygen affinity associated with erythrocytosis, and in Hb Kansas (Reissman, Ruth, and Nomura, 1961; Riggs, Personal Communication), which is characterized by a lowered oxygen affinity. It is interesting to note that the amino acid substitution in Hb Kansas is a threonyl residue in place of the asparagyl residue 102 of the β-chain (Riggs, Personal Communication), and in Hb Yakima the aspartyl residue 99 of the β-chain has been replaced by a histidyl.

Hemoglobin Zürich (Muller and Kingma, 1961), Hb Seattle, Hb St. Mary's (Huehns and Shooter, 1965) and Hb Köln (Carrell, Lehmann, and Hutchison, 1966) are examples of unstable abnormal hemoglobins. Instability can be demonstrated under a variety of conditions: these include in vitro exposure of the hemoglobin to heat or alkali and in vitro or in vivo exposure to certain drugs or other chemical compounds. The exact molecular mechanisms to account for the instability of these hemoglobins have not yet been worked out in detail, but may be forthcoming from the accumulating fund of knowledge about the normal structure of hemoglobins.

Although the abnormal hemoglobins provided the first and most useful system for studying the precise effects of genetic changes in protein structure, this aspect of biochemical genetics is now being studied in several other protein systems as well. The one important restriction in studies of the abnormal human hemoglobins is the impossibility of using direct genetic experiments including production of mutations and purposeful matings. The protein of tobacco mosaic virus, tryptophan synthetase of *E. coli*, and several proteins of other microorganisms have become standards for studies of the genetic control of protein structure. These other systems have been selected in part for the ease of genetic investigation. Nevertheless, investigations of human hemoglobins continue to yield important information about the genetic control of protein structure.

TYPES OF MUTATIONS AND THEIR EFFECTS ON PROTEIN

The substitution of a valyl residue for the glutamyl residue number 6 of the β-chain of sickle-cell hemoglobin was the first recognition of the effect of a point mutation on the structure of a protein (Ingram, 1956, 1959b). Substitutions of single amino acid residues, presumably due to single point mutations, appear to account for most of the abnormal human hemoglobins (Schroeder and Jones, 1965), mutant proteins of tobacco mosaic virus, and genetic variants of tryptophan synthetase. Determinations of the structure of abnormal hemoglobins were the first experimental evidence used to postulate that point mutations result from the change of one DNA base for another in the codon that controls the substituted amino acid residue. Ingram (1961b) pointed out that in addition to providing information about point mutations and the genetic code, the results of structural characterization of the first three or four abnormal hemoglobins were consistent with the hypothesis of colinearity of the gene and the amino acid sequence of polypeptide chains. Proof of this colinearity came later from genetic and chemical studies of more favorable systems, including tryptophan synthetase of *E. coli* (Yanofsky et al., 1964) and the head protein of T4 bacteriophage (Sarabhai et al., 1964).

The effect of substitutions of single amino acid residues on the function of a given protein depends upon the role played by the residue in question in the normal protein. With Hb S the effect of the substitution on the solubility of deoxygenated hemoglobins is profound. However, with many other abnormal hemoglobins the substitutions appear to have little or no physiological effect (Huehns and Shooter, 1965). Attempts to determine the chemical stability of mutant proteins may be possible from consideration of energy changes resulting from the substitution of residues on tertiary and quarternary structure of the protein. This may now be possible from the hemoglobin-myoglobin system because of the detailed chemical and structural information which has been presented by Perutz (1965) and Perutz, Kendrew, and Watson (1965). These authors have recognized correlations between the structures of heme proteins from various species which have permitted them to conclude that certain regions of the hemoglobin and myoglobin molecules are invariant in amino acid sequence, whereas other regions can be occupied by a variety of amino acid residues. The degree of variability appears to depend upon the chemical stability and biological activity of the molecule. It should be noted, however, that the differences in structure of hemoglobins and myoglobins from various species reported by these authors are those which have survived biological selection. Only in the biologically unfit carrier of rare mutant hemoglobins has it been possible to study the deleterious result of substitution of certain amino acid residues on molecular function. Examples which affect solubility and oxygen affinity have already been discussed. Other substitutions may occur, but never become available for study if they result in a molecule which is so unstable that it cannot be detected or an organism which is not viable.

In addition to changes in the structure of proteins, point mutations which cause a substitution of single base in a given codon site may alter the rate of synthesis of a mutant protein. This was first postulated by Itano (1953, 1957) to explain the difference in relative amounts of Hb S and Hb A in individuals heterozygous for sickle-cell anemia. Itano later interpreted this "structure-rate" hypothesis in terms of the genetic code, and expanded it by postulating that the degeneracy of the code could permit nonidentical genes to control the rate of synthesis of identical polypeptide chains (Itano, 1965). Thus, mutations that produce only a degenerate change in a codon in a specific gene would not alter the structure of the polypeptide chain determined by that gene, but might result in an alteration in the rate of synthesis of the polypeptide, depending upon the relative concentration or activity of the specific transfer-RNA required for encoding the mutated codon.

The two codons UAG and UAA are unique in that they appear to result in termination of synthesis of polypeptide chains (Stretton, 1965). When either is produced by mutation within a gene only the portion of the polypeptide from the NH_2-terminal end up to the residue designated by the codon just prior to eitherUAG or UAA is produced. Only the NH_2-terminal portion is formed because the polypeptide is assembled in a linear sequence beginning at the NH_2-terminal end (Dintzis, 1961). The "amber" and "ochre" mutants of the head protein of T4 phage have been found to represent such a system (Stretton, 1965). Although no examples of "amber" or "ochre" mutants have yet been recognized in mutants of human proteins, it is interesting to speculate that this type of substitution of a single base may account for genetic defects in which no detectable protein is present.

Evidence of unequal crossing-over as an important mutagenic mechanism has been obtained at the level of protein structure from consideration of the homologies between the α-, β-, γ-, and δ-chains of human hemoglobins and myoglobin, from studies of human haptoglobins, and from structural characterization of two unusual abnormal hemoglobins, Hb Freiburg and Hb Lepore. The first example of gene duplication due to unequal crossing-over was presented by Bridges (1936) in his studies of the bar-eye gene in Drosophila in 1936. Following this, ample biochemical and genetic evidence of gene duplication in several species has been obtained (Lewis, 1951; Wagner and Mitchell, 1964). This genetic evidence undoubtedly served as a background against which Itano (1957), and later Ingram (1961a) and Braunitzer and associates (1961) proposed theories of the origin of the present hemoglobin chains. Because of the striking homology between the primary structure of the various chains of human hemoglobins and myoglobin, it has been suggested that the genes which determine the structure of these polypeptides have been derived from a common precursor by a mechanism of gene duplication. Such duplications may have arisen as the result of unequal crossing-over (Dixon, 1966). In the evolution of hemoglobin, the crossing-over events that are postulated to account for duplication of complete structural genes must have occurred outside the cistron regions for hemoglobin and myoglobin. However, partial gene duplication, probably resulting from intragenic unequal crossing-over, has been proposed by Smithies, Connell, and Dixon (1962) as the mechanism by which certain types of human haptoglobin α-chains arose. Structural analysis of the α-polypeptide of Hp 2α indicates that it is almost double the length of the α-chain of Hp 1α. The Hp 2α-chain appears to consist of one polypeptide chain representing almost a complete duplication of the Hp 1α-chain. A third type of haptoglobin (Johnson type) contains an α-chain which appears to represent almost a triplication of the Hp 1α-chain.

Dixon (1966) and Epstein and Motulsky (1965) have discussed how intragenic (intracistronic) unequal crossing-over theoretically might give rise to additions and deletions of genetic material varying from a single DNA base to one base less than a complete cistron. Addition or deletion of a number of bases not divisible by three would be expected to produce "reading frame shifts" resulting in the formation of "nonsense" polypeptide chains. Such mutants of microorganisms have been induced by acridine in which single- or double-base additions or deletions have occurred (Crick et al., 1961; and Streisinger et al., 1966). Structural characterization of Hb Freiburg, an abnormal variant of the β-chain of adult hemoglobin, revealed the deletion of a valyl residue normally present at position 23 from the N-terminus (Jones,

Brimhall, Huisman, Kleihauer, and Betke, 1966). Because the remaining portions of this abnormal β-chain appear to have a normal sequence, it is postulated that the abnormality arose as the result of a triplet-base deletion. Intragenic unequal crossing-over between two homologous β-chain genes during miosis is postulated as the mutational mechanism giving rise to the deletion.

Molecular evidence of intergenic unequal crossing-over has been postulated by Baglioni (1962, 1965) from chemical studies of two mutant hemoglobins, Hb Lepore Boston and Hb Lepore Hollandia. Both of these hemoglobins have appeared in place of the normal Hb A and of the minor component Hb A_2. Both Lepore hemoglobins contain normal α-chains, but, instead of β- or δ-chains they contain a polypeptide equal in length to the β- and δ-chains, but with an amino acid sequence which begins with the NH_2-terminal part of the δ-chain and ends at the COOH-terminal part of the β-chain. The point at which these unusual polypeptides change from δ to β sequences is different in Hb Lepore Boston and Hb Lepore Hollandia. From this structural characterization Baglioni proposes that the Lepore hemoglobin resulted from intergenic unequal crossing-over between the β-chain locus and a homologous region of the δ-chain locus.

Intragenic and intergenic unequal crossing-over may be important mutational mechanisms in evolution and in the production of heterogeneity in several protein systems, most notably, the immunoproteins. Baglioni (1966) has postulated that one possible advantage of such gene duplication within the same cistron is the production of a polypeptide chain with more than one active site; interaction between two or more active sites may then result in regulatory properties.

MULTIPLE SUBUNIT STRUCTURE AND GENETICS OF PROTEINS

Chemical and genetic studies of proteins that are comprised of more than one polypeptide subunit have begun to reveal interesting features of genetic and metabolic control of the biological functions of these multimeric proteins. This is one of the most active areas of current investigation in molecular biology.

The existence of a protein structure composed of subunits was first appreciated by Svedberg (Svedberg and Pedersen, 1940) from physical chemical measurements. He and his co-workers obtained data which permitted the postulation that "probably the protein molecule is built up by successive aggregation of definite units . . . ." Studies of hemocyanin and several iron heme proteins, including hemoglobin, not only were consistent with the idea that many proteins are aggregates of smaller subunits, but also indicated that reversible dissociation of some proteins into subunits is possible.

Investigations of the structure of hemoglobins, especially human hemoglobins, have been important in understanding the subunit structure of proteins. Physical measurements of hemoglobin in solution made by Svedberg and co-workers (see Svedberg and Pedersen, 1940) and Tiselius and Gross (1934) led to the conclusion that hemoglobin was made up of at least two subunits. Perutz and co-workers (Boyer-Watson, Davidson, and Perutz, 1947) demonstrated by X-ray diffraction studies of crystalline hemoglobin that these subunits were identical half molecules. By chemical analysis of the free amino-terminal groups in hemoglobins of various species, Porter and Sanger (1948) demonstrated the presence of more than one polypeptide chain per molecule. More precise studies of human hemoglobin made by Rhinesmith et al. (1957, 1958) and independently by Braunitzer (1958) established the presence of two different pairs of identical polypeptide chains. From a consideration of all of these structural studies the subunit formula for normal adult hemoglobin was concluded to be $\alpha_2\beta_2$.

Reversible dissociation of hemoglobins in acid solutions and in alkaline solutions was studied by Field and O'Brien (1955) and Hasserodt and Vinograd (1959), respectively. On the basis of these observations Singer and Itano (1959) and Vinograd and Hutchinson (1960) proposed and demonstrated that common subunits of different hemoglobins can be exchanged by dissociating and reassociating the subunits of mixtures of two hemoglobins. The types of molecules that can be isolated following these subunit recombination or hybridization experiments generally consist of symmetrical, four-chain hemoglobins, that is, $\alpha_2\beta_2{}^A$ and $\alpha_2\beta_2{}^S$, not $\alpha_2\beta^A\beta^S$. (The reader is referred to Guidotti, Konigsberg, and Craig, 1963, and Schroeder and Jones, 1965, for a more complete discussion of asymmetrical hemoglobin molecules.) The recombination reactions can be followed by using radioactive labels, chemical modification of the globin chains, or alteration of the oxidation state of the heme group. In an analogous manner hybrid molecules can be formed by subunit recombinations of hemoglobins from different species such as man and dog (Robinson and Itano, 1960), or by recombination of two human hemoglobins that are abnormal in different chains (Itano and Robinson, 1959). These subunit hybridization experiments have been useful for determining the abnormal chain in mutant hemoglobins and in understanding the mechanism of reversible dissociation of hemoglobins (see Schroeder and Jones, 1965, for review).

Two embryonic hemoglobins were first found in embryos of less than three months' gestation by Huehns, Flynn, Butler, and Beaven (1961). These were named Gower I and Gower II hemoglobin. Since only minute amounts of blood are available from such small embryos, complete characterization has not been done. Based on hybridization and peptide map studies (Huehns et al., 1961; Huehns et al., 1964), a fifth hemoglobin subunit, the ϵ-chain, has been recognized; it differs from the previously characterized α-, β-, γ-, and δ-chains by a still undetermined number of amino acid residues. Hecht and co-workers (1966) have recently described hemoglobin types in still smaller embryos, the smallest estimated to be of 37 days gestation, and suggest that the preferential production of hemoglobin subunits occurs in the sequence: ϵ, followed by α, β, and γ, followed by δ. The tetramers, which result at successive stages of development, are ϵ_4 or Gower 1 Hb, $\alpha_2\epsilon_2$ or Gower II Hb, $\alpha_2\gamma_2$ or fetal Hb, and $\alpha_2\beta_2$ and $\alpha_2\delta_2$ or adult, and A_2 Hb. An extension of the hypothesis described for gene duplication to account for all five structure genes of normal human hemoglobin has been presented by Huehns et al. (1964) and Baglioni (1966).

Recognition that more than one kind of hemoglobin is produced during normal human development, that these different kinds are multimeric forms of a few specific polypeptide chains, and that each polypeptide chain is the product of a separate structural gene has provided a rewarding model. The remarkable versatility that such a system confers on functional proteins in diploid organisms is beginning to emerge from studies of enzymes and immunoglobins.

The simplest of such multimeric proteins are those in which the functional multimer is composed of two or more of the same kind of polypeptide chain. Examples include alkaline phosphatase from *E. coli* (Rothman and Byrne, 1963), glyceraldehyde-3-phosphate dehydrogenase (Harris and Perham, 1965), pig-heart fumarase (Hill and Kanarek, 1964) and glutamate dehydrogenase from *Neurospora crassa* (Coddington, Fincham, and Sundaram, 1966). Human glucose-6-phosphate dehydrogenase (Kirkman and Hendrickson, 1962) is probably also a dimer of two identical subunits.

Aggregation of the products of two different structure genes into a functional multimer accounts for the next most complex kind of functional protein. All of the normally occurring hemoglobins, lactic dehydrogenase (Cahn et al., 1962), aspartate transcarbamylase (Gerhart, 1964), and the individual immunoglobulins (Porter, 1963) fall into this category. A still more complex system is known in which a family of multimers, with similar function, results from aggregation of subunits specified by more than two structural genes. The hemoglobins, lactic dehydrogenase, and the immunoglobulines are examples.

Insight into the significance of subunit structure of many diverse proteins has resulted from studies carried out in simpler organisms such as phage, bacteria, and Neurospora. Two phenomena which seem to result from the multimeric structure of proteins deserve mention here. They are complementation and regulation.

Complementation has been used to describe the ability of one defective mutant to compensate for the deficiency of another mutant. Of particular interest have been studies of complementation among mutants that map within the same locus or cistron and that presumably represent allelic forms of the same gene. In Neurospora such mutants can be tested in heterokaryons. The asci produced are haploid, and the production of a functional protein clearly indicates that diffusible gene products from two such complementing mutants mutually correct for the metabolic block that occurs if only one type of mutant is present. From studies of two such mutants for glutamate dehydrogenase, Coddington et al. (1966) have presented evidence that complementation occurs because a conformational defect in the potentially active subunits of enzyme produced by one mutant, am^{19}, is corrected in hybrid polymers with subunits from a second, inactive mutant, am^{1}. The active enzyme is known to consist of eight identical polypeptide chains (Fincham and Coddington, 1963). Active hybrid enzyme was obtained both in vivo, by producing heterokaryons of the two mutants, and in vitro, by treating the purified mutant enzymes with acid or by freezing and thawing.

Such complementation effects may contribute to the understanding of human diversity. It is conceivable that less marked quantitative differences in activity or stability distinguish multimeric proteins produced by heterozygous individuals from those produced by homozygous individuals. As more information becomes available about specific gene products from a sample of normal individuals, heterogeneity of such multimeric proteins indicates the presence of a large number of alleles at structure gene loci in natural populations. Genetic polymorphisms of such enzymes as pseudocholinesterase, acid phosphatase, phosphoglucomutase, and adenylate kinase occur in man (Harris, 1966). It seems possible that allelic genes for such polymorphisms may persist because complementation in the heterozygote confers a selective advantage.

A second kind of property of multimeric proteins which has been recognized largely from studies of simpler forms is the regulatory mechanism effected by a specific metabolite, which induces conformational change, or allosteric transition of a functional multimer, by reacting with a site on the protein other than the active site. In their classic paper on allosteric proteins, Monod, Changeux, and Jacob (1963) cite hemoglobin and six enzymes, including aspartate transcarbamylase, as examples. The oxygen dissociation curves of myoglobin and adult and H hemoglobins, together with X-ray crystallographic evidence by Muirhead and Perutz (1963), are presented as evidence for cooperative binding of oxygen by the four heme groups due to reversible, discrete conformational alterations. Particularly pertinent to this discussion are the observations that the tetramer, hemoglobin, acquires properties distinct from the monomer, myoglobin, and that only tetramers of two different kinds of subunits possess the important functional properties. An extension of this model will be continued in the next section on isozymes.

Aspartate transcarbamylase illustrates other important features of allosteric regulation. It is the first enzyme after the branching point in the metabolic path which leads to the synthesis of cytidine-5′-phosphate. It is not inhibited by intermediates in this pathway but is specifically inhibited by the final product, cytidine-5′-phosphate. The native enzyme has been shown by Gerhart (1964) to consist of two kinds of subunits, one of which carries the active site, and the other, the binding site for the inhibitor. The importance of such regulatory enzymes in many metabolic paths has been reviewed by Changeux (1964). For most of these the molecular explanation for such control remains to be determined. The model, derived from aspartate transcarbamylase, of multimeric proteins with separate subunits bearing sites for different activities may prove to be a general one.

Other factors which affect the timing and rates of synthesis, the specific association, and the stability of subunits of multimeric functional proteins are being studied in many biologic systems. Two examples from human studies will be presented here. Human erythrocyte glucose-6-phosphate dehydrogenase has been shown by Kirkman and Hendrickson (1962) to be a dimer which depends on the presence of TPN for its activity. The cofactor, TPN, is apparently a necessary and specific structural component of the active enzyme. A similar role for cofactors and substrates is under consideration for many multimeric proteins.

An explanation for the occurrence of non-α-chain tetramers of human hemoglobin, such as Hb H and Hb Bart's, has been advanced by Huehns, Flynn, Butler, and Shooter (1961). They propose that with decreased rates of synthesis of α-chains, which seem to occur in α-thalassemia, a relative excess of γ- and β-chains are produced, which aggregate to form γ_4 Hb in the newborn individual and β_4 Hb in the adult. The counterpart of this situation in β-thalassemia would not result in the appearance of significant amounts of α-chains if release of the α-chain from the polysome requires the presence of free non-α-chain. Furthermore, attempts to prepare pure α-chain hemoglobin have resulted in material that has the physical characteristics of a dimer (Huehns, Shooter, and Dance, 1961), suggesting that a tetramer of α-chains may not form for stereochemical reasons, or, if formed, is not stable.

Thalassemia has proved to be a challenge to both the biochemist and the geneticist. Like the abnormal hemoglobins, it is an inherited anemia which affects the rate of synthesis of hemoglobin (Itano, 1957; Ingram and Stretton, 1959). No structural abnormality in the hemoglobins of individuals afflicted with thalassemia has yet been demonstrated. Among the hypotheses that have been advanced to explain this disease are mutations that do not change the amino acid sequence but that depend on a degenerate codon for which the specific transfer RNA is relatively less available and therefore limiting (Itano, 1965). A mammalian counterpart of the microbial operon in which the hemoglobin structure genes are under the control of one or more regulator genes has also been postulated (Neel, 1961; Motulsky, 1962; Cepellini, 1963; Zuckerkandl, 1964). A translation defect at the ribosome that terminates synthesis—a defect caused by an "amber" or "ochre" type of mutation—could also explain certain types of thalassemia. Among patients who have thalassemia as classically defined, genetic differences have been found and it is likely that different mechanisms will be discovered.

ISOZYMES AND THE GENETIC CONTROL OF METABOLISM

Evidence that led Markert and Møller (1959) to coin the term *isozymes* to describe multiple molecular forms of the same enzyme had been accumulating for a number of years. Among the sources they cite are Vesell and Bearn's description (Vesell and Bearn, 1957) of three forms of lactic dehydrogenase in human serum. They were also, no doubt, aware of the evidence for separate genetic control of the different human hemoglobins. Their paper has served as the landmark for a new approach to the investigation of enzymes. In it they anticipated many of the questions which have since been asked about the nature and role of isozymes. All of these reduce to: Why should an organism carry genetic information necessary to fabricate two or more different kinds of molecules if each kind performs the same function?

Partial answers to this question have come from genetic and biochemical studies of a few isozymic systems, and it may be premature to generalize from these findings. Nonetheless, a pattern which is strikingly similar to that for hemoglobin has emerged. It suggests that isozymes differ in kinetic properties and in the control mechanisms which lead to their synthesis. Teleologically, it can be reasoned that an organism or cell would acquire greater selective advantage if it could utilize one isozyme under conditions of low substrate concentration and another isozyme for high substrate concentration. The different isozymic forms of lactic dehydrogenase (Cahn et al., 1962), malic dehydrogenase (Grimm and Doherty, 1961), hexokinase (Katzen and Schimke, 1965), and pyruvate kinase (Campos, Koler, and Bigley, 1965) differ in this way. Other kinds of kinetic differences among isozymes, such as different responses to activators or inhibitors, may also be the bases for selective advantage.

Isozymes have been distinguished by electrophoresis, chromatography, solubility, heat stability, and antigenic specificity, as well as by different kinetic properties. Isozymic forms behave differently when studied by two or more of the methods just mentioned. Such results add confidence to the conclusion that the isozymes are truly different molecules, and not artifacts produced by a certain experimental method.

Evidence that isozymes are the products of two or more genetic loci is more difficult to obtain. Lactic dehydrogenase is the best studied isozymic system. Two kinds of subunits form tetramers of active enzyme, but with one more degree of freedom than for hemoglobin; thus, all possible combinations of the two subunits taken four at a time are represented. Electrophoretically distinguishable mutants affecting one of the subunits have been found in man (Boyer, Fainer, and Watson-Williams, 1963) and in Peromyscus (Shaw and Barto, 1963). Pyruvate kinase (Koler et al., 1964) has been shown to appear in at least two isozymic forms, and the genetically determined deficiency which leads to nonspherocytic hemolytic anemia affects only one of these. A genetic variant affecting the soluble, but not the mitochrondrial, isozyme of isocitrate dehydrogenase in mice has been reported by Henderson (1965).

Most important to the teleologic model is the requirement that different forms of the enzyme be produced under different kinds of genetic control. Differences in distribution of isozymes among subcellular components, for example, malic and isocitrate dehydrogenases, between tissues, for example, lactic dehydrogenase, creatine phosphate, and pyruvic kinase, or between products during sequential periods of embryogenesis, for example, lactic dehydrogenase and creatine phosphokinase, or within in vitro cultures of the same tissue, for example, alkaline phosphatase, are inferential evidence of different control mechanisms. Of more direct bearing on the question of genetic control are studies indicating that synthesis of only one of several types of isozyme may be induced by a specific hormone and that such induction may be blocked by agents known to prevent DNA-dependent synthesis of RNA. Evidence of this sort has been presented for lactic dehydrogenase (Kaplan, 1965), hexokinase (Sols, Sillero, and Salas, 1965; Grossbard, Weksler, and Schimke, 1966), and pyruvate kinase (Tanaka et al., 1965).

The latter two examples, induction of single isozymes of hexokinase and pyruvate kinase by the same hormone, insulin, lead to an expanded concept of the role of isozymes in metabolic control. Coordinate repression of a series of enzymes in a metabolic sequence by the end product has been reviewed by Umbarger (1964), and is now recognized as a control mechanism for several pathways in microorganisms (Ames and Hartman, 1961; Gorini, Gunderson, and Burger, 1961). Such coordinate repression of synthesis of a group of sequential enzymes leads to homeostasis and economy of DNA-directed enzyme synthesis. It is delicately attuned to the cell's environment and has, thus far, been demonstrated only in unicellular organisms. Evolution to a multicellular form places a new demand on the genome: The total genetic information must accommodate metabolic demands that differ between cells and within the same cell at different times. Key steps catalyzed by unidirectional or rate-limiting enzymes are known in many such metabolic pathways.

Weber and co-workers (1965) have presented evidence that three such key steps in the mammalian glycolytic path are bidirectional only because at each step different enzymes catalyze the conversion of substrate in each direction. These steps involve the reactions:

1. glucose ⇄ glucose-6-PO_4
2. fructose-1-PO_4 ⇄ fructose-1,6-diPO_4
3. phosphoenolpyruvate ⟶ pyruvate

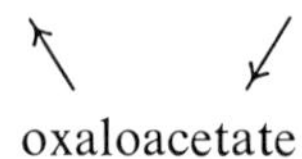

oxaloacetate

Glycolysis is dependent on the action of glucokinase, phosphofructokinase, and pyruvate kinase, which catalyze the three steps from glucose to pyruvate. The reverse steps are catalyzed by glucose-6-phosphatase, fructose-1,6-diphosphatase, pyruvate carboxylase, and PEP carboxykinase, and lead to conversion of pyruvate to glucose. In rat liver, Weber et al. (1965) have demonstrated induction of the three key glycolytic enzymes by insulin, and of the four key gluconeogenic enzymes by glucocorticoids.

A further refinement in the genetic control of the glycolytic path is suggested by the observation of Tanaka and associates (1965), and of Grossbard, Weksler, and Schimke (1966) that only one isozymic form of pyruvate kinase and glucokinase are responsive to insulin induction. Our teleologic model is consistent with evidence suggesting correspondence of the insulin-inducible isozyme to that form which has the highest Km for substrate. The characteristics of the molecular form of pyruvic kinase induced by insulin in rat liver (Tanaka et al., 1965) are identical to those of the human red cell that we have described (Koler et al., 1964; Campos et al., 1965). Similarly, the form of glucokinase which is responsive to insulin is that with the highest Km (Katzen and Schimke, 1965). Weber et al. (1965) have referred to a "functional genome unit," which is activated to produce each set of key enzymes. The similarity of this model to that of the different hemoglobins is apparent, but proof must await purification and structural studies of the subunits of each of these isozymic systems. It is tempting to predict that homologies may be found in the structures—not only of subunits of a series of isozymic forms of the same enzyme—but also of subunits of enzymes with related functions, such as the key glycolytic enzymes.

Whether or not similar genetic and structural subunits are demonstrated in such isozymic forms of the same enzyme or of functionally related enzymes, the contribution which studies of hemoglobin have made to our understanding of the genetic control of functional proteins is a major one. Pauling's role in focusing attention on a molecular basis for these genetic mechanisms has already produced many dividends and will certainly lead to many more.

REFERENCES

Ames, B. N., and P. H. Hartman (1961). In Anderson Hospital, *The Molecular Basis of Neoplasia; Symposium on Fundamental Cancer Research*, p. 322. Austin: Texas Univ. Press.

Atwater, J., I. R. Schwartz, and L. M. Tocantins (1960). *Blood* **15**, 901.

Baglioni, C. (1962). *Proc. Nat. Acad. Sci. U.S.* **48**, 1880.

Baglioni, C. (1963). In Taylor, J. H., ed., *Molecular Genetics*, Part 1, p. 405. New York: Academic Press.
Baglioni, C. (1965). *Biochim. Biophys. Acta* **97**, 37.
Baglioni, C. (1966). Presented at the 3rd Ann. Congress Human Genetics, Chicago, Ill.
Bateson, W. (1902). *Mendel's Principles of Heredity*. Cambridge.
Beadle, G. W. (1959). *Science* **129**, 1715.
Beadle, G. W., and E. L. Tatum (1941). *Proc. Nat. Acad. Sci. U.S.* **27**, 499.
Beet, E. A. (1949). *Ann. Eugen.* **14**, 279.
Boyer, S. H., D. C. Fainer, and E. J. Watson-Williams (1963). *Science* **141**, 642.
Boyer-Watson, J., E. Davidson, and M. F. Perutz (1947). *Proc. Roy. Soc.* (London) A **191**, 83.
Braunitzer, G. (1958). *Z. physiol. chem.* **312**, 72.
Braunitzer, G., N. Hilschmann, V. Rudloff, K. Hilse, B. Liebold, and R. Müller (1961). *Nature* **190**, 480.
Bridges, C. B. (1936). *Science* **83**, 210.
Cabannes, R., and A. Portier (1959). In Jonxis, J. H. P., and J. F. Delafresnaye, eds., *Abnormal Haemoglobins*, p. 51. Oxford: Blackwell Scientific Publ.
Cahn, R. D., N. O. Kaplan, L. Levine, and E. Zwilling (1962). *Science* **136**, 962.
Campos, J. O., R. D. Koler, and R. H. Bigley (1965). *Nature* **208**, 194.
Carrell, R. W., H. Lehmann, and H. E. Hutchison (1966). *Nature* **210**, 915.
Ceppellini, R. (1963). *Proceedings of the Second International Congress of Human Genetics*. Rome: Instituto G. Mendel 1, 526.
Changeux, J. P. (1964). *Brookhaven Symp. Biol.* **17**, 232.
Charache, S., D. J. Weatherall, and J. B. Clegg (1966). *J. Clin. Invest.* **40**, 1826.
Coddington, A., J. R. S. Fincham, and T. K. Sundaram (1966). *J. Mol. Biol.* **17**, 503.
Crick, F. H. C., L. Barnett, S. Brenner, and R. J. Watts-Tobin (1961). *Nature* **192**, 1227.
Delbrück, M. (1946). *Cold Spring Harbor Symp. Quant. Biol.* **11**, 22.
Dherte, P., H. Lehmann, and J. Vandepitte (1959). *Nature* **184**, 1133.
Dintzis, H. M. (1961). *Proc. Nat. Acad. Sci. U.S.* **47**, 247.
Dixon, G. H. (1966). *Essays in Biochemistry* **2**, 147.
Emmel, V. E. (1917). *Arch. Internal Med.*, **20**, 586.
Epstein, C. J., and A. G. Motulsky (1965). In Steinberg, A. G., and A. Bearn, eds., *Progress in Medical Genetics*, vol. 5, p. 85. New York-Grune & Stratton.
Field, E. O., and J. R. P. O'Brien (1955). *Biochem. J.* **60**, 656.
Fincham, J. R. S., and A. Coddington (1963). *Cold Spring Harbor Symp. Quant. Biol.* **28**, 517.
Garrod, A. O. (1928). *Lancet* **1**, 1055.
Gerald, P. S., and L. K. Diamond (1958). *Blood* **13**, 835.
Gerald, P. S., and M. L. Efron (1961). *Proc. Nat. Acad. Sci. U.S.* **47**, 1758.
Gerhart, J. C. (1964). *Brookhaven Symp. Biol.* **17**, 222.
Gorini, L., W. Gunderson, and M. Burger (1961). *Cold Spring Harbor Symp. Quant. Biol.* **26**, 173.
Gouttas, A., P. Fessas, H. Tsevremis, and E. Xefteri (1955). *Sang* **26**, 911.
Grimm, F. C., and D. G. Doherty (1961). *J. Biol. Chem.* **236**, 1980.
Grossbard, L., M. Weksler, and R. T. Schimke (1966). *Biochem. Biophys Res. Commun.* **24**, 32.
Guidotti, G., W. Konigsberg, and L. C. Craig (1963). *Proc. Nat. Acad. Sci. U.S.* **50**, 774.
Harris, H. (1963). *Garrod's Inborn Errors of Metabolism*. New York: Oxford Univ. Press.
Harris, H. (1966). *Proc. Roy. Soc.* (Biol) **164**, 298.
Harris, J. I., and R. N. Perham (1965). *J. Mol. Biol.* **13**, 876, 885.
Hasserodt, U., and J. R. Vinograd (1959). *Proc. Nat. Acad. Sci. U.S.* **45**, 12.
Hecht, F., A. G. Motulsky, R. J. Lemire, and T. E. Shepard (1966). *Science* **152**, 91.

Henderson, N. S. (1965). *J. Exp. Zool.* **158**, 263.

Herrick, J. B. (1910). *Arch. Internal Med.* **6**, 517.

Hill, R. L., and L. Kanarek (1964). *Broohaven Symp. Biol.* **17**, 80.

Hill, R. L., and H. C. Schwartz (1959). *Nature* **184**, 641.

Huehns, E. R., N. Dance, G. H. Beaven, F. Hecht, and A. G. Motulsky (1964). *Cold Spring Harbor Symp. Quant. Biol.* **29**, 327.

Huehns, E. R., F. V. Flynn, E. A. Butler, and G. H. Beaven (1961). *Nature* **189**, 496.

Huehns, E. R., F. V. Flynn, E. A. Butler, and E. M. Shooter (1960). *Brit. J. Haematol.* **6**, 388.

Huehns, E. R., and E. M. Shooter (1961). *J. Mol. Biol.* **3**, 257.

Huehns, E. R., and E. M. Shooter (1965). *J. Med. Genet.* **2**, 48.

Huehns, E. R., E. M. Shooter, and N. Dance (1961). *Biochem. Biophys. Res. Commun.* **5**, 362.

Hunt, J. A. (1959). *Nature* **183**, 1373.

Hunt, J. A., and V. M. Ingram (1958). *Nature* **181**, 1062.

Hunt, J. A., and V. M. Ingram (1959). *Nature* **184**, 640.

Hunt, J. A., and V. M. Ingram (1960). *Biochim. Biophys. Acta* **42**, 409.

Hunt, J. A., and H. Lehmann (1959). *Nature* **184**, 872.

Ingram, V. M. (1956). *Nature* **178**, 792.

Ingram, V. M. (1958). *Biochim. Biophys. Acta* **28**, 539.

Ingram, V. M. (1959a). In Josiah Macy, Jr. Foundations. *Genetics: Genetic Information and Control of Protein Structure and Function*, p. 65. New York.

Ingram, V. M. (1959b). *Biochim. Biophys. Acta* **36**, 402.

Ingram, V. M. (1961a). *Nature* **189**, 704.

Ingram, V. M. (1961b). *Hemoglobin and Its Abnormalities.* Springfield, Ill.: Charles C. Thomas.

Ingram, V. M., and A. O. W. Stretton (1959). *Nature* **184**, 1903.

Ingram, V. M., and A. O. W. Stretton (1962). *Biochim. Biophys. Acta* **62**, 456.

Itano, H. A. (1953). *Am. J. Human Genet.* **5**, 34.

Itano, H. A. (1957). *Adv. in Protein Chem.* **12**, 215.

Itano, H. A. (1965). In Jonxis, J. H. P., ed., *Abnormal Haemoglobins in Africa*, p. 3. Oxford: Blackwell Scientific Publ.

Itano, H. A., and J. V. Neel (1950). *Proc. Nat. Acad. Sci. U.S.* **36**, 613.

Itano, H. A., and E. Robinson (1959). *Nature* **183**, 1799.

Itano, H. A., and E. A. Robinson (1960). *Proc. Nat. Acad. Sci. U.S.* **46**, 1492.

Itano, H. A., S. J. Singer, and E. Robinson (1959). In Wolstenholme, G. E. W., and C. M. O'Connor, eds., *Ciba Symp. Biochemistry of Human Genetics*, p. 96. Boston: Little, Brown.

Jones, R. T., B. Brimhall, E. R. Huehns, and N. A. Barnicot (1966). *Science* **151**, 1406.

Jones, R. T., B. Brimhall, T. H. J. Huisman, E. Kleihauer, and K. Betke (1966). *Science* **154**, 1024.

Jones, R. T., and W. A. Schroeder (1963). *Biochemistry* **2**, 1357.

Jones, R. T., W. A. Schroeder, J. E. Balog, and J. R. Vinograd (1959). *J. Am. Chem. Soc.* **81**, 3161.

Jones, R. T., W. A. Schroeder, and J. R. Vinograd (1959). *J. Am. Chem. Soc.* **81**, 4749.

Kaplan, N. O. (1965). *J. Cell. Comp. Physiol.* **66**, 1.

Katzen, H. M., and R. T. Schimke (1965). *Proc. Nat. Acad. Sci. U.S.* **54**, 1218.

Kirkman, H. N., and E. M. Hendrickson (1962). *J. Biol. Chem.* **237**, 2371.

Koler, R. D., R. H. Bigley, R. T. Jones, D. A. Rigas, P. Vanbellinghen, and P. Thompson (1964). *Cold Spring Harbor Symp. Quant. Biol.* **29**, 213.

Koler, R. D., and D. A. Rigas (1961). *Ann. Human Genet.* **25**, 95.

Lewis, E. B. (1951). *Cold Spring Harbor Symp. Quant. Biol.* **16**, 159.

Markert, C. L., and F. Møller (1959). *Proc. Nat. Acad. Sci. U.S.* **45**, 753.

Minnich, V., J. K. Cordonnier, W. J. Williams, and C. V. Moore (1962). *Blood* **19**, 137.
Monod, J., J. P. Changeux, and F. Jacob (1963). *J. Mol. Biol.* **6**, 306.
Morton, N. (1955). *Am. J. Human Genet.* **7**, 277.
Motulsky, A. G. (1962). *Nature* **194**, 607.
Muirhead, H., and M. F. Perutz (1963). *Nature* **199**, 633.
Muller, C. J., and J. H. P. Jonxis (1960). *Nature* **188**, 949.
Muller, C. J., and S. Kingma (1961). *Biochim. Biophys. Acta* **50**, 595.
Neel, J. V. (1949). *Science* **110**, 64.
Neel, J. V. (1961). *Blood*, **18**, 769.
Neubauer, O. (1928). *Handb. Norm. Path. Physiol.* **5**, 671.
Osgood, E. E., R. T. Jones, B. Brimhall, and R. D. Koler (1966). Unpublished studies.
Pauling, L. (1954). *The Harvey Lectures*, Ser. 49, 216.
Pauling, L., H. A. Itano, S. J. Singer, and I. C. Wells (1949). *Science* **110**, 543.
Perutz, M. F. (1965). *J. Mol. Biol.* **13**, 646.
Perutz, M. F., J. F. Kendrew, and H. C. Watson (1965). *J. Mol. Biol.* **13**, 669.
Porter, R. R. (1963). *Brit. Med. Bull.* **19**, 197.
Porter, R. R., and F. Sanger (1948). *Biochem. J.* **42**, 287.
Ranney, H. M. (1954). *J. Clin. Invest.* **33**, 1634.
Raper, H. B., D. B. Gammack, E. R. Huehns, and E. M. Shooter (1960). *Brit. Med. J.* **2**, 1257.
Reissmann, K. R., W. E. Ruth, and T. Nomura (1961). *J. Clin. Invest.* **40**, 1826.
Rhinesmith, H. S., W. A. Schroeder, and N. Martin (1958). *J. Am. Chem. Soc.* **80**, 3358.
Rhinesmith, H. S., W. A. Schroeder, and L. Pauling (1957). *J. Am. Chem. Soc.* **79**, 4682.
Rigas, D. A., R. D. Koler, and E. E. Osgood (1955). *Science* **121**, 372.
Riggs, A., Personal Communication.
Robinson, E. A., and H. A. Itano (1960). *Nature* **188**, 798.
Rothman, F., and R. Byrne (1963). *J. Mol. Biol.* **6**, 330.
Sarabhai, A. S., A. O. W. Stretton, S. Brenner, and A. Bolle (1964). *Nature* **201**, 13.
Schneider, R. G., and R. T. Jones (1965). *Science* **148**, 240.
Schroeder, W. A., and R. T. Jones (1965). *Forschr. Chem. organ. Naturst.* **23**, 113.
Schroeder, W. A., L. M. Kay, and I. C. Wells (1950). *J. Biol. Chem.* **187**, 221.
Schroeder, W. A., and G. Matsuda (1958). *J. Am. Chem. Soc.* **80**, 1521.
Schroeder, W. A., J. R. Shelton, J. B. Shelton, and J. Cormick (1963). *Biochemistry* **2**, 992.
Schwartz, H. C., T. H. Spaet, W. W. Zuelzer, J. V. Neel, A. R. Robinson, and S. F. Kaufman (1957). *Blood* **12**, 238.
Shaw, C. R., and E. Barto (1963). *Proc. Nat. Acad. Sci. U.S.* **50**, 211.
Singer, S. J., and H. A. Itano (1959). *Proc. Nat. Acad. Sci. U.S.* **45**, 174.
Smith, E. W., and J. V. Torbert (1958). *Bull. Johns Hopkins Hosp.* **101**, 38.
Smithies, O., G. E. Connell, and G. H. Dixon (1962). *Nature* **196**, 232.
Sols, A., A. Sillero, and J. Salas (1965). *J. Cell. Comp. Physiol.* **66**, 23.
Stanbury, J. B., J. B. Wyngaarden, and D. S. Fredricksen, eds. (1966). *The Metabolic Basis of Inherited Disease*, 2nd ed. New York: McGraw-Hill.
Streisinger, G., Y. Okada, J. Emrich, J. Newton, A. Tsugita, E. Terzaghi, and M. Inouye (1966). *Cold Spring Harbor Symp. Quant. Biol.* **31**, 77.
Stretton, A. O. W. (1965). *Brit. Med. Bull.* **21**, 229.
Svedberg, T., and K. O. Pedersen (1940). *The Ultracentrifuge*. Oxford: Clarendon Press.
Taliaferro, W. H., and J. B. Huck (1923). *Genetics* **8**, 594.
Tanaka, T., Y. Harano, H. Morimura, and R. Mori (1965). *Biochem. Biophys. Res. Commun.* **21**, 55.
Tiselius, A., and D. Gross (1934). *Kolloid Z.* **66**, 11.
Umbarger, H. E. (1964). *Science* **145**, 674.
Vesell, E. S., and A. G. Bearn (1957). *Proc. Soc. Exp. Biol. Med.* **94**, 96.

Vinograd, J. R., and W. D. Hutchinson (1960). *Nature* **187**, 216.

Vinograd, J. R., W. D. Hutchinson, and W. A. Schroeder (1959). *J. Am. Chem. Soc.* **81**, 3168.

Wagner, R. P., and H. K. Mitchell (1964). *Genetics and Metabolism*. New York: Wiley.

Waisman, H. A. (1966). *Ped. Clin. N. America* **13**, 469.

Wasi, P., S. Na-Nakorn, and A. Suingdumrong (1964). *Nature* **204**, 907.

Weber, G., R. L. Singhal, N. B. Stamm, and S. K. Strivastava (1965). *Fed. Proc.* **24**, 745.

Winslow, R. M., and V. M. Ingram (1966). *J. Biol. Chem.* **241**, 1144.

Yanofsky, C., B. C. Carleton, J. R. Guest, D. R. Helinski, and U. Henning (1964). *Proc. Nat. Acad. Sci. U.S.* **51**, 266.

Zuckerkandl, E. (1964). *J. Mol. Biol.* **8**, 128.

ARTHUR CHERKIN
Psychobiology Research Laboratory
Veterans Administration Hospital
Sepulveda, California
and Division of Biology
California Institute of Technology
Pasadena, California

Molecules, Anesthesia, and Memory

Since, therefore, the substance of the mind has been found to be extraordinarily mobile, it must consist of particles exceptionally small and smooth and round. This discovery, my dear fellow, will prove a timely aid to you in many problems.

Lucretius, c. 55 B.C.

The thoughts to which I am now giving utterance, and your thoughts regarding them, are the expression of molecular changes in that matter of life (that is, protoplasm) which is the source of our other vital phenomena.

Thomas Henry Huxley, 1865

During the last twenty years much progress has been made in the determination of the molecular structure of living organisms and the understanding of biological phenomena in terms of the structure of molecules and their interaction with one another. The progress that has been made in the field of molecular biology during this period has related in the main to somatic and genetic aspects of physiology, rather than to psychic. We may now have reached the time when a successful molecular attack on psychobiology, including the nature of encephalonic mechanisms, consciousness, memory, narcosis, sedation, and similar phenomena, can be initiated.

Linus Pauling, 1961

Mind is considered as something apart from brain, or as a biological function of brain with molecular correlates accessible to experimentation. In 1895, Maudsley saw a "physics of the mind at the base of all its psychics" and stated that mental disorders "must own similar molecular derangements to those which toxic agents produce"; yet until recently "molecular psychology" (Moore and Mahler, 1965) was an unacceptable term that still evokes frowns. Nevertheless, scientists are turning to the molecular study of mind, as it becomes increasingly evident that mind obeys the same laws as do other biological phenomena, and that facts about the molecular events that occur when the brain secretes memory and behavior are needed for an understanding of the higher-order events associated with those secretions. This is not to say that facts about molecular events alone will suffice; to understand memory, we must clarify the functioning of whole neurons and their complex relationships with other neurons and other cells (Dingman and Sporn, 1964). But the view that brain research must concentrate on complex neuronal patterns is beginning to converge with the view expressed by Huxley and Pauling. Schmitt (1965), summarizing the argument for the molecular approach and for its significance, concluded:

Molecular neurology, which is already on the way to becoming firmly established, together with molecular neuropsychology, which is emerging as a coherent field, seems destined to provide a powerful thrust in modern science. Society may well encourage, indeed demand, full speed ahead in these fields because of their important bearing on mental health, on the understanding of mechanisms of memory, learning, and other psychological parameters basic to science itself, and on man's deep personal concern about the nature of his being.

The molecular theory of general anesthesia (Pauling, 1961) stimulated a series of experiments during which I became impressed with the broad range of biological reactivity, excitatory as well as depressant, of the nonhydrogen-bonding anesthetics, despite their weak chemical reactivity in vivo. Long ago, Claude Bernard (1875, p. 153) clearly recognized that chloroform was a *depressant of all tissues*, animal and vegetable, and it had been recognized since 1797 that anesthetics were *excitants* (Poussel, 1951). Examples of both properties are listed in Table 1.

There are exceptions, of course. For example, Buchheit, Schreiner, and Doebbler (1966) point out that the rank order of the noble gases and nitrogen may differ, depending upon whether their potencies are compared on the basis of inhibition of growth of *Neurospora crassa*, depression of anaerobic glycolysis in tissue slices, depression of oxygen-dependent radiosensitivity, inhibition of insect pupation, or anesthesia of insects. Such differences make difficulties for general theories of nonspecific biological activity. The position taken here, as a first approximation, is to suggest that the numerous differences be ascribed to secondary factors, such as variable rates of uptake and distribution, and that the similarities be ascribed to one or two primary mechanisms of action.

One striking similarity is the uniformity of the anesthetizing partial pressure (P_a) of a given compound in a wide variety of biological preparations (Table 2). Another is the marked dose-dependence of a great variety of compounds, with only a narrow spread between the anesthetic and the lethal partial pressures (Table 3).

This discussion reviews and examines two questions. First, what do structurally nonspecific molecules do to biological systems in general? Secondly, how do such molecules affect memory processes in particular? The examination will first develop the thesis that anesthesia is one aspect of a broad range of interactions of nonhydrogen-bonding molecules with biological systems and that the characterization of a molecule as an "odorant," "convulsant," "amnestic," "anesthetic," or "insecticide" arises from an arbitrary choice based upon practical considerations, rather than from any specific intrinsic property of the molecule. We shall then consider how this thesis applies to the use of anesthetics in the study of memory.

ANESTHESIA

Nonspecific Anesthetics

Discussion will be restricted to the class of general anesthetics considered by Pauling (1961) that operate through London dispersion forces without forming covalent, ionic, or hydrogen bonds in biological systems. The class includes the noble gases, nitrous oxide, hydrocarbons, and halogenated hydrocarbons. A simplifying characteristic of such molecules in the study of anes-

TABLE 1.
Examples of biological activities exhibited by typical nonspecific anesthetic molecules (weak anesthetics are exemplified by N_2, N_2O, *and* Xe; *strong ones, by* $CHCl_3$.)

Activity	Biological system	Anesthetic compound	Relative partial pressure[a]
Excitatory			
Olfaction (threshold)	Man	N_2	0.4[b]
	Man	$CHCl_3$	0.0002[c]
Gustation (threshold)	Man	N_2	0.4[b]
	Man	$CHCl_3$	1.1[d]
Auditory, visual, paresthetic stimulation	Man	N_2O	0.3[e]
Convulsive movements	Man	N_2O	0.3[e]
	Mouse	Xe	0.5[f]
Depressant			
Analgesia	Man	N_2O	0.3[e]
Inhibition of memory fixation	Man	N_2O	0.3[e]
Anesthesia	Mammals	N_2	1.0[g]
	Mammals	N_2O	1.0[g]
	Mammals	Xe	1.0[g]
	Mammals	$CHCl_3$	1.0[g]
Inhibition of protein synthesis	Rat brain	$C_2H_5OC_2H_5$	~1. [h]
Inhibition of growth	Insect pupation	N_2	0.5[i]
	Neurospora	N_2	2.0[j]
	Neurospora	Xe	1.0[j]
	Mouse Sarcoma I	N_2O	0.6[k]
	Mouse Sarcoma I	$CHCl_3$	0.3[k]
Inhibition of enzymatic activity	Rat brain protease	$CHCl_3$	5[l]
Disorganization of mitosis	Onion root tip	N_2	7[m]
	Onion root tip	N_2O	10[m]
	Onion root tip	$CHCl_3$	9[n]
Inhibition of nerve potential	Cat stellate ganglion		
	-synaptic	$CHCl_3$	5[o]
	-axonal	$CHCl_3$	16[o]
Death	Grain weevil	N_2	30[p]
	Grain weevil	$CHCl_3$	17[p]
	Mouse	$CHCl_3$	2[q]

[a]Relative to the following average P_a (anesthetizing partial pressure, in mm Hg) of the same anesthetic compound for mammals: N_2 = 18,000; N_2O = 980; Xe = 930; $CHCl_3$ = 6.0; and $C_2H_5OC_2H_5$ = 25.
[b]Case and Haldane, 1941 (*one* subject detected odor).
[c]Laffort, 1963.
[d]Poussel, 1951.
[e]Parkhouse et al., 1960.
[f]Lawrence et al., 1946.
[g]See footnote a above.
[h]Gaitonde and Richter, 1956.
[i]Frankel and Schneiderman, 1958.
[j]Buchheit et al., 1966.
[k]Fink & Kenny, 1966 (4-day cultures).
[l]Ungar, 1965.
[m]Ferguson, Hawkins and Doxey, 1950.
[n]Östergren, 1951.
[o]Brink and Posternak, 1948.
[p]Ferguson and Hawkins, 1949.
[q]Raventós, 1956.

TABLE 2.
Anesthetizing partial pressures (in mm Hg; partial pressures rounded to 1 or 2 significant figures) in various biological systems, normalized to 37°C.

Anesthetic Compound	Grain Weevil[a]	Brine Shrimp[b]	Gold-fish[c]	Mouse	Dog	Man
N_2	—	—	—	22,000[d]	—	14,000[e]
Xe	—	—	—	650[d]	1600[d] 910[g]	530[f]
N_2O	—	—	—	1100[h] 690[d]	1400[g]	720[d]
$c\text{-}C_3H_6$	—	630	380	130[h] 130[i] 110[d]	130[g]	200[d]
CH_3Cl	170	—	140	110[d] 50[j]	—	110[d]
$n\text{-}C_5H_{12}$	110	290	180	95[k] 78[l]	—	—
C_2H_5Cl	150	—	90	38[j] 36[i]	—	44[d]
$C_2H_5OC_2H_5$	20	16	31	33[i] 26[j] 22[k]	23[g]	20[m]
$CF_3CClBrH$ (halothane)	—	8	14	13[h] 6.5[i] 5.2[n]	6.6[g]	5.6[g]
$CHCl_3$	10	5	11	11[d] 9.9[i] 6.4[h] 5.8[k] 3.8[j]	2.7[o]	2.4[o]
$CHCl_2CF_2OCH_3$ (methoxyflurane)	—	—	2.1	1.5[n]	1.8[g]	0.7[p]

NOTE: The experimental temperatures were the following: grain weevils, 25°C; brine shrimp, 20°C; goldfish, 10° to 30°C. The partial pressures of $c\text{-}C_3H_6$, $C_2H_5OC_2H_5$, $CHCl_3$, CF_3ClBrH, and $CHCl_2CF_2OCH_3$ were normalized to 37°C using the enthalpy changes determined in goldfish[c]. Other compounds were normalized using the average enthalpy change (10.7 kcal/mole) of $C_2H_5OC_2H_5$ and the three halogenated compounds, taken as a group.

[a]Ferguson, 1951.
[b]Robinson et al., 1965.
[c]Cherkin and Catchpool, 1964, and unpublished results.
[d]Miller, Paton, and Smith, 1965.
[e]Haldane, 1951.
[f]Catchpool, 1966.
[g]Eger et al., 1965.
[h]Epstein, Ngai, Brody, and Rittenberg, 1963.
[i]Raventós, 1956.
[j]Meyer and Hopff, 1923.
[k]Fühner, 1921.
[l]Stoughton and Lamson, 1936.
[m]Faulconer, 1952.
[n]Speden, 1964.
[o]Thomas, MacKrell, and Conner, 1961.
[p]Holaday, Garfield, and Ginsberg, 1965.

TABLE 3.
Ratio of lethal dose to anesthetic dose in mice exposed for 30 minutes.

Anesthetic compound	AD_{50} (mm Hg)	LD_{50}/AD_{50}		
		Range	Mean	Reference
Unsaturated hydrocarbons (13)	15–173	1.1–2.1	1.6	Virtue, 1950
Various (6)	6–132	1.5–5.0	2.5	Raventós, 1956
Trichlorethylene	6	5.0		
Halothane	7	3.3		
Chloroform	10	1.5		
Ether	33	1.7		
Ethyl chloride	36	1.7		
Cyclopropane	132	1.5		

thesia or memory is their limited repertoire of interactions with biological systems, restricted to the following: formation of clathrate hydrates (Pauling, 1961) or of structured water (Miller, 1961); binding to proteins (Featherstone and Muehlbaecher, 1963; Schoenborn, Watson and Kendrew, 1965), or dissolution in lipids (Meyer and Hemmi, 1935).

A molecule with a hydrogen-bonding functional group, such as an alcohol, exerts its biological effects through hydrogen bonding as well as through its nonspecific moiety, and expressions have been proposed for calculating the separate contributions (McGowan, 1954; Zahradník, 1962). Molecules with functional groups are excluded here, however, for the reasons given by Pauling (1961) in his statement of his theory of anesthesia. The theory is that nonhydrogen-bonding anesthetics interact with encephalonic water and immobilize it in such a way, in the form of clathrate hydrates, that the overall arrangement of water, anesthetic, proteins, and ions in the brain fluid interferes with nerve function and causes reversible unconsciousness. Pauling's formulation delineates a class of compounds on a rational molecular basis—their independence of hydrogen bonding—that eliminates the complicated pharmacodynamics and metabolic reactivity of specific anesthetics, such as the barbiturates.

The limited and weak chemical reactivity of the nonhydrogen-bonding anesthetics is in contrast to their broad and strong biological activity (Table 1).

Species Uniformity

The statement of any unrestricted theory of anesthesia implies uniformity among species. The conventional view, however, has characterized anesthetic action as showing pronounced species differences. The divergence can be reconciled as follows. The significance of species variation has been magnified because of clinical considerations that demand precise titration of each patient to the required level of depression; values obtained for goldfish or mice are obviously inapplicable to such individual needs. Beyond this, much of the variation ascribed to species differences reflected other experimental differences. Potency values can vary by a factor of 5 because of differences between inspired and alveolar partial pressures (Eger et al., 1965), by a factor of 2 because of different depths of anesthesia, and by factors of 2 to 4 because of different body temperatures (Cherkin and Catchpool, 1964). When normalized for temperature, the anesthetizing partial pressures of a given compound in different species vary from the mean by a factor of less than 2.8 (Table 2). This generalization applies to biological systems as diverse as the grain weevil, brine shrimp, goldfish, mouse, dog, and man, and to compounds varying in anesthetizing partial pressure over a range of 1000-fold. This uniformity supports a unitary mechanism of general anesthesia. From this point of view, the residual "species variation" in anesthetizing partial pressures is ascribed to secondary factors—such as variable rates of uptake and distribution, duration of anesthesia, and toxic effects—and to inter-laboratory differences that appear even in a single species (Table 2).

Dose-Dependence

The profound dependence of the qualitative biological action of a nonspecific compound upon its partial pressure is characteristic. For example, the AD_{50} dose (anesthetizing partial pressure for fifty percent of a group of animals) typically exceeds the AD_1 (minimum anesthetic dose) by a factor of only 2. Similarly, the LD_{50} (lethal partial pressure for fifty percent of a group of animals) typically exceeds the AD_{50} by a factor of 2 (Table 3). Thus, for a given exposure period, a compound could be classified as inert, or anesthetic, or lethal, depending upon the experimental partial pressure, within a factor of 4. Similarly, at a given partial pressure, the biological effect is dependent upon the duration of exposure.

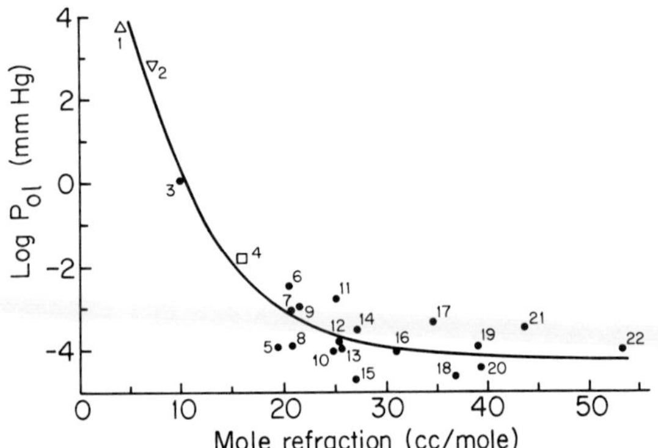

FIGURE 1.
Relationship between mole refraction (R) and olfactory threshold (P_{ol}). $R = (n^2 - 1/n^2 + 2)(M/\rho)$, where n = index of refraction, M = molecular weight, and ρ = density. Related parameters have also been used as a basis of comparison, including these: molecular volume (M/ρ); polarizability, α ($R = [4\pi/3]N\alpha$, where N is Avogadro's number); and parachor, P ($P = M\gamma^{1/4}/(D - d)$, where γ = surface tension, D = liquid density, d = vapor density). R values were calculated from the refractive index and molecular volume, or estimated from the atomic refractions. R of CH_4 = 6.6; R of Xe = 10.2. P_{ol} values: ●, calculated from normalized molar concentrations (Laffort, 1963); △, N_2 (Case and Haldane, 1941); ▽, N_2O, estimated as 1 atm.; □ C_3H_8, estimated as 4 × the Laffort P_{ol} of n-C_4H_{10} (Gerarde, 1963). The key numbers represent: 1, N_2; 2, N_2O; 3, C_2H_6; 4, C_3H_8; 5, $CH_2{=}CH{-}CH{=}CH_2$; 6, n-C_4H_{10}; 7, i-C_4H_{10}; 8, CH_2ClCH_2Cl; 9, $CHCl_3$; 10, Benzene; 11, n-C_5H_{12}; 12, n-C_4H_9Cl; 13, CCl_4; 14, c-C_6H_{12}; 15, Naphthalene; 16, Toluene; 17, n-C_7H_{16}; 18, Xylene; 19, n-C_8H_{18}; 20, n-$C_7H_{15}Cl$; 21, n-C_9H_{20}; and 22, n-$C_{11}H_{24}$.

Stimulatory Effects

When goldfish are placed in an anesthetizing solution, they pass through a phase of hyperactivity during induction, and, with some compounds, during recovery in fresh water. The excitatory properties of anesthetics are well known clinically, and are responsible for the delirium of so-called Stage II anesthesia, that precedes the depression of Stage III, or surgical anesthesia. Wells (1847) held that "stimulation carried to excess always produces complete insensibility of the nervous system," an early anticipation of encephalographic evidence that certain anesthetic compounds cause reversible unconsciousness by creating epileptoid commotions in the brain (Winters and Spooner, 1966). The mechanism of excitation is said to involve depression of inhibitory systems (Esplin, 1965), but there is evidence for direct stimulation of neural activity. Chloroform and ether *increase* the excitability of single nerve fibers before depressing it (Arvanitaki and Chalazonitis, 1951). Paton and Speden (1965) cite experiments that report sensitization or stimulation, by anesthetics, of pulmonary stretch receptors, baroreceptors, chemoreceptors, muscle spindles, frog sciatic nerve, ganglion cells of *Limulus*, the mammalian sympathoadrenal system, guinea pig intestinal nerve and smooth muscle, and even germination of seeds and movement of spermatozoa.

Olfaction

The direct stimulation of olfactory receptors by volatile anesthetics is a common experience, and the parallelisms between olfaction and anesthesia have attracted frequent comment (Poussel, 1951; Mullins, 1955). Research on olfaction reveals a bewildering wealth of detailed complexity (Lettvin and Gesteland, 1965; Wenzel and Sieck, 1966). A simplified approach is to restrict attention to the nonhydrogen-bonding compounds.

The importance of the shape of a molecule to its ability to stimulate gustatory or olfactory receptors was emphasized by Pauling (1946) in his statement that "even the senses of taste and odor are based upon molecular configuration rather than upon ordinary chemical properties—a molecule which has the same shape as a camphor molecule will smell like camphor even though it may be quite unrelated to camphor chemically." The stereochemical theory of olfaction has been developed in detail by Amoore (1965), and Pauling's theory of general anesthesia is also based upon configurational properties.

All nonhydrogen-bonding compounds that anesthetize at pressures below 1 atm stimulate the olfactory receptors. Mullins (1955) considered xenon to be an exception but recent data suggest that its partial pressure for Stage III anesthesia is above 1 atm (Table 2); consequently, the olfactory threshold may also be above 1 atm. Nitrogen at 0.8 atm is tasteless and odorless, and inert as an anesthetic, but nitrogen at 10 atm has been reported to have taste and odor (Case and Haldane, 1941), and at 16 to 20 atm it causes general anesthesia (Haldane, 1951). The Case and Haldane experiments need repeating on more subjects and with nitrogen of assured purity. Nevertheless, the rough correlation between olfactory threshold and mole refraction (Figure 1) suggests that olfactory thresholds should be found for xenon, methane, nitrogen, and argon in the range of 1 to 20 atm. The conclusion that small molecules, such as Xe, CH_4, N_2, and Ar, are incapable of stimulating olfactory receptors (Mullins, 1955) cannot be considered firm until these gases are tested under pressure, as is now feasible in hyperbaric chambers. Dogs have been exposed to Xe at ~2 atm, with recordings from electrodes implanted in the olfactory bulb (Domino et al., 1964), but the Xe was administered through a tracheotomy tube and the hypersynchronous activity recorded was probably a central effect.

The rough correlation between anesthetizing partial pressure (P_a) and olfactory threshold (P_{ol}) is shown in Figure 2. As a generalization $P_a/P_{ol} \approx 10^5$; for small molecules, with mole refractions below 15 cc/mole, the ratio P_a/P_{ol} becomes much smaller, perhaps even less than 1. The usefulness of this correlation is limited by the uncertainty of reported olfactory threshold values that typically vary by factors of 2 to 60 (Laffort, 1963).

In considering chemically inert molecules, Mullins (1955) concluded that "very small and very large molecules do not have odors." He suggested that a minimum molecular volume of about 40 cc/mole was required to initiate olfaction, that 100 cc/mole appeared optimal, and that high molecular volumes (>200 cc/mole) reduced olfactory potency. The C_3 and C_4 fluorocarbons deviate from Mullins' conclusion, because their molecular volumes of 117 to 147 cc/mole are near the optimal range, yet they are "either odorless or very nearly so" (Simons and Block, 1939). It is interesting that the aliphatic fluorocarbons also have a low anesthetic potency (P_a > 600 mm Hg); in contrast, the aromatic perfluorobenzene has a strong odor and is a potent anesthetic (P_a = 6 mm Hg). A quantitative comparison of the olfactory thresholds and anesthetizing partial pressures of pure perfluorocompounds would be of interest.

There are of course differences, as well as similarities, between anesthesia and olfaction. Olfaction is stimulated at such low partial pressures that the formation of clathrate hydrates cannot be involved. Mullins (1955) suggested that olfaction required specific size and shape correspondence between a molecule and various receptors in the olfactory membrane, whereas narcosis involved nonspecific occupation of membrane interspaces. His results with the isomeric 2-butenes tend to support this view. The olfactory threshold for *trans*-2-butene was 95 times that for *cis*-2-butene; the corresponding factor for anesthesia is only 1.1 (Virtue, 1950).

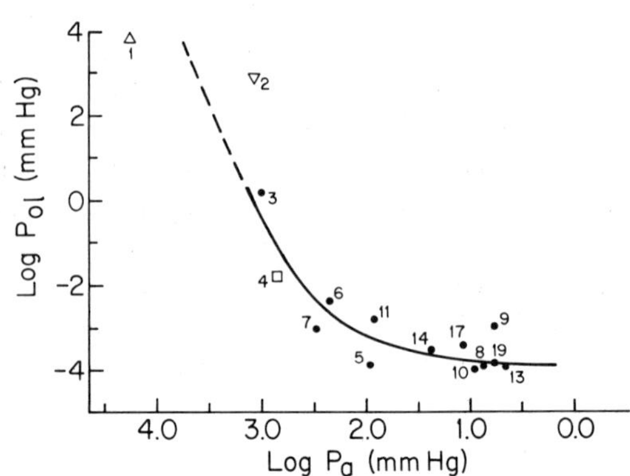

FIGURE 2.
Relationship between anesthetizing partial pressure (P_a) and olfactory threshold (P_{ol}). The key numbers and symbols are as in Figure 1. Those shown represent the nonhydrogen-bonding compounds for which P_a values in mammals, and normalized olfactory thresholds (Laffort, 1963), are available. P_a values: 1, 2, 9, 11 (averages from Table 2); 3, 4 (Miller et al., 1965); 5 (Adriani, 1962); 6, 7 (Stoughton and Lamson, 1936); 8 (Meyer and Hopff, 1923); 10, 17, 19 (Führer, 1921); 13 (Meyer, 1951); 14 (Virtue, 1949).

Stimulation of Other Senses; Convulsions; Toxicity

The evidence of interaction of nonspecific anesthetics with sensory systems other than olfaction is more indirect. On the theoretical level, Hodgson (1965) concluded, "granting the need for caution in extending or modifying basic concepts, the time has obviously arrived when the one cell—one function type of analysis is sometimes unwarranted. More than one kind of sensory mechanism can be built into a single cell." He suggests that the receptor *cell* be replaced by the receptor *site* as the unit of analysis, pointing out that if heterogeneous sites are involved in single cells, strict purity of stimulus modalities need not be postulated. The receptor site concept provides a theoretical accommodation for nonspecific chemosensitivity of all sensory receptors.

Further theoretical support is found in the 1826 precept of Johannes Müller, that sensation is a function of the organ, not of the stimulating agent. Bayliss (1960) states that "one of the best proofs of this is afforded by the fact that mechanical, chemical, or electrical stimulation of the chorda tympani nerve as it passes through the tympanic cavity of the ear produces equally a sensation of taste . . . any kind of stimulus which will excite a nerve will also excite a receptor organ; but the important point is that specialized receptors are much more sensitive to their appropriate type of stimulus than to any other. By whatever means they are excited, the sensation is the same." Thus, there is no theoretical bar to stimulation of receptor systems by nonhydrogen-bonding anesthetic agents, in view of their ability to stimulate nerves.

On the experimental side, Ngai (1963) has discussed the effects of nonspecific anesthetics upon neurostructures, including peripheral receptors; the concentrations that sensitized the peripheral receptors were approximately one-third the anesthetizing concentrations for the same receptors. Numerous such reports have appeared, and Mullins (1955) has discussed the relationship of excitatory, olfactory, and convulsant action. Lawrence et al. (1946) noted convulsant activity in mice during induction with xenon at 440 to 580 mm Hg. Convulsant activity has also been observed during recovery from various anesthetics (Virtue, 1950). We have observed convulsions in goldfish during both induction and recovery; further, in subanesthetic concentrations of the clinical convulsant hexafluorodiethyl ether (Indoklon®, $CF_3CH_2OCH_2CF_3$), goldfish showed sustained convulsive activity throughout one-hour experiments (Cherkin et al., 1967). Bees exposed to $CHCl_3$ or $C_2H_5OC_2H_5$ also showed initial excitation, which was sometimes prolonged for an hour (Schmid, 1964).

It is of interest that the excitatory properties of nonspecific compounds were observed even before their depressant effects. Sir Humphry Davy (1799) recorded the following:

> I began to respire twenty quarts of unmingled nitrous oxide: a thrilling, extending from the chest to the extremities was almost immediately produced. I felt a sense of tangible extension, highly pleasurable, in every limb; my visible impressions were dazzling, and apparently magnified; I heard distinctly every sound in the room, and was perfectly aware of my situation. By degrees, as the pleasurable sensations increased, I lost all connection with external things; trains of vivid visible images rapidly passed through my mind, and were connected with words in such a manner as to produce perceptions

perfectly novel. I existed in a world of newly-connected and newly-modified ideas; I theorised, I imagined that I made discoveries. When I was awakened from this semi-delirious trance by Dr. Kinglake, who took the bag from my mouth, indignation and pride were the first feelings produced by the sight of the persons about me. My emotions were enthusiastic and sublime, and for a minute I walked around the room, perfectly regardless of what was said to me. As I recovered my former state of mind, I felt an inclination to communicate the discoveries I had made during the experiment. I endeavoured to recall the ideas: they were feeble and indistinct; one collection of terms, however, presented itself; and, with the most intense belief and prophetic manner, I exclaimed to Dr. Kinglake, "*Nothing exists but thoughts! The universe is composed of impressions, ideas, pleasures, and pains.*"

The similarity of Davy's experience to those reported by subjects taking modern hallucinogenic drugs is striking! As would be expected, friends of Davy tried N_2O inhalations, and Davy became the first to observe and report the convulsant action of a non-specific molecule—as well as the first to anticipate the "ether frolics" and "LSD freak-outs" that had to await the growth of synthetic organic chemistry.

Heightened auditory and visual acuity, hallucinations, and mental dissociation have been confirmed by anecdotal reports from patients undergoing anesthesia and by controlled studies. Parkhouse et al. (1960) reported observations on seventeen subjects inhaling N_2O, with the stimulatory effects upon the sensory systems shown in Table 4, as well as evidence of convulsant action in some subjects.

Lethal effects of anesthetics have been observed in bacteria, plants, insects, fish, amphibia, birds, and mammals. Stage IV anesthesia, characterized by irreversible depression and death, occurs in mice at partial pressures only 1.1 to 5 times the anesthetizing partial pressures (Table 3).

MEMORY

In view of the broad range of interactions of nonhydrogen-bonding molecules with nervous tissue, it is no surprise that they interact with a key function of brain—the input, storage, and recall of memories. The art of anesthesia and the mystery of memory have been closely linked since the original sug-

TABLE 4.
Percentage of 17 subjects reporting sensory effects during nitrous oxide inhalation.

Sensation reported	Partial pressure of N_2O (in mm Hg)			
	0	150	230	300
Paresthetic	6	59	100	100
Auditory	0	12	24	59
Cold	6	6	24	47
Visual	0	12	18	24

SOURCE: Parkhouse et al., 1960.

gestion of general anesthesia for surgery and the original report of amnesia caused by an anesthetic molecule, both by Davy (1799). Despite the dramatic sensory effects he experienced while inhaling N_2O, he could not remember them later. He stated, "The next morning the recollections of the effects of the gas were very indistinct; and had not remarks, written immediately after the experiments, recalled them to my mind, I should have even doubted of their reality."

Pauling (1961) introduced the hydrate theory of anesthesia with a discussion of ephemeral and permanent memory. Memory has to do with retaining information in the brain for a period of time. When the time is less than a few minutes, it is referred to as ephemeral or short-term memory; this is what we use to hold a new telephone number in mind while we dial it. When the hold time exceeds a few minutes, and endures for hours or days or years, it is called permanent or long-term memory; this is what we use for retaining our home telephone number. There is a "two-store" theory (Brown, 1964), which states that short-term memory is retained in some kind of unstable active reverberating pattern, without any material trace, while long-term memory has a stable physical trace in the brain—the "engram." "Tri-trace" theories have also been proposed (McGaugh, 1966). It has been theorized that the engram consists of modified synapses, facilitated neural loops, specific memory molecules, induced neuronal enzymes, or some other physical change or changes. The "mnemon" concept (Cherkin, 1966) suggests that memory has a quantal nature, whatever its physical embodiment may be. A "weak" memory reflects the presence of a few mnemons and a "strong" memory reflects the presence of many. According to this concept, the effect of graded partial anesthesia would be to depress, in a graded way, mnemon production induced by a standardized learning experience of a given learning strength.

We may represent short-term, labile-store memory, and long-term, stable-store memory, as follows (with representative half-times, in seconds):

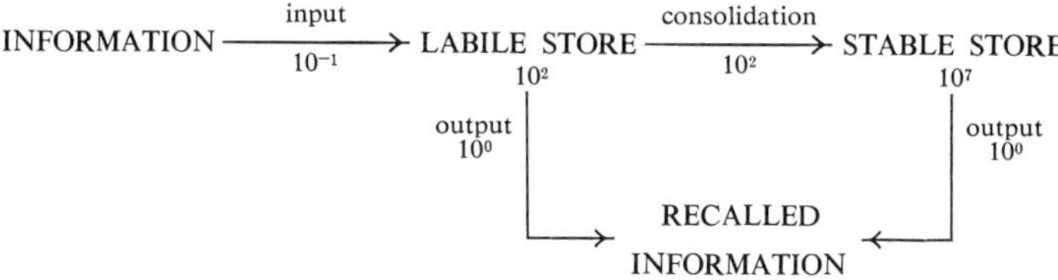

Each step has been explored with the aid of anesthetic compounds, as well as by other techniques. The findings detailed below indicate that the input, labile store, stable store, and recall are relatively resistant to levels of anesthetic compounds that block consolidation. Evidence for the relative vulnerability of the consolidation process also comes from clinical studies of memory impairment in patients with senile amnesia (Kay, 1959), Korsakoff's psychosis, which involves brain damage from prolonged nutritional deficiencies (Victor, 1964), or bilateral lesions of the hippocampal complexes (Drachman and Arbit, 1966). In each of these conditions, short-term memory and recall of old memories are reasonably functional, but the ability to lay down a new memory trace is severely restricted or completely lacking. Presumably, the problem is faulty consolidation, because input, reverberation (labile store), storage (stable store), and recall are all operating. The apparent absence of memory fixation in infants, and its deterioration in advanced age, suggest that modern searchers for the engram might well hunt for molecules that wax in the young brain and wane in the senile brain.

The consolidation theory finds support in a variety of experimental evidence that fixing a memory trace requires time, on the order of 10^1 to 10^3 seconds or more (Kopp, Bohdanecky and Jarvik, 1966). When the fixing process is interrupted shortly *after* an event, the memory of the preceding event is impaired. Clinically, such "retrograde amnesia" is a well-known consequence of concussion, convulsions, or anesthesia. There is a body of research that disputes the consolidation interpretation of retrograde-amnesia experiments and proposes instead inhibitory behavioral mechanisms (McGaugh, 1966, Reference 8), but the view advanced here is our working hypothesis.

Patients entering anesthesia, or emerging from it, pass through a conscious stage, during which they have short-term memory and can recall prior memories but cannot form new memories that they can recall later. On the other hand, there are reports that, even under deep surgical anesthesia, patients form memory traces that can later be recalled *under hypnosis* (Cheek, 1964). This discrepancy, too, has yet to be resolved. The kinetics and temperature-dependence of memory fixation should prove instructive in testing conflicting hypotheses and in ascertaining the correct mechanisms.

The marked dependence of biological functions in general upon the partial pressure of anesthetic agents suggests that a similar relationship applies to the memory process. The quantitative aspect is likely to be of particular importance because of the aforementioned capability of anesthetic compounds to cause excitation. Because memory formation is enhanced by stimulant compounds, like strychnine and picrotoxin (McGaugh and Petrinovich, 1966), "stimulant" levels of anesthetic molecules may enhance memory formation. More typically, subanesthetic levels depress memory formation (Artusio, 1954; Robson, Burns, and Welt, 1960; Parkhouse et al., 1960). Information concerning the quantitative relationship between the partial pressure of inhaled anesthetic compounds and the ability to lay down a memory trace is becoming available. Evidence in man for the resistance to anesthetics of the input and reverberation mechanisms is found in experiments with diethyl ether (Artusio, 1954) and with N_2O (Robson et al., 1960; Parkhouse et al., 1960). Artusio's findings in 115 surgical patients are summarized in Table 5. Of particular interest is Stage I, Plane 2, that starts with the onset of partial analgesia and ends with complete analgesia. At this plane, sensory perception was qualitatively unimpaired. Painful stimuli were perceived, objects and individuals were recognized, colors were distinguished, hearing and taste remained intact. The subjects gave correct answers to questions that required recall of old memories, such as birth date, and of current matters, such as events of the preceding day. The subjects readily detected incorrect content inserted into the questions. In short, the input, reverberation, and recall steps were functional. But when the subjects were questioned immediately after completion of their operation, or on postoperative days 1, 3, 5, 7, they had no recall at all for the events that occurred during Plane 2 analgesia. The conclusion is that no consolidation of memory traces occurred, at a partial pressure of ether that did not impair cerebration, short-term memory, or sensory perception.

The Artusio study was complicated by the use of surgical patients as subjects, and by the administration of atropine, thiopental, N_2O, and a high initial level of ether. Furthermore, the findings are subject to an alternative interpretation, namely, that they reflect "state-dependent learning" (Overton, 1966); in this case the patients would be shown to have unimpaired memory by a test of their recall while they were held in the same state as that during their learning experience, that is, under Plane 2 analgesia. With Pauling's encouragement, I am preparing to repeat the study, using volunteer subjects and experimental animals, which are held at Plane 2 with known partial pressures of nonspecific anesthetics, without additional medication, and with controlled sensory input and quantitative measures of short-term and long-term memory.

The relative effect of N_2O (0.3 P_a) upon the input and the recall steps of short-term memory was studied by Steinberg and Summerfield (1957) in subjects who were memorizing series of nonsense syllables; the results indicated that the input or reverberatory phase, or both, were impaired, whereas recall was not.

Robson et al. (1960) made a more quantitative study, using a range of N_2O partial pressures (0.2 to 0.5 P_a). The test was memorization of visually presented numbers, tested after a lapse of 30, 120, or 300 seconds. The results indicated no impairment of input, reverberation, or recall, since memory scores were high 30 seconds after presentation. As the partial pressure of N_2O was gradually raised, recall after 30 seconds remained correct but errors were made when 120 or 300 seconds elapsed. The significant feature of the Artusio and the Robson experiments is the finding of anesthetic conditions that dissected short-term from long-term memory, by demonstrating a level at which short-term memory was not impaired while long-term memory apparently was.

TABLE 5.
Sensation, cerebration, and memory of 115 patients under ether, at the subanesthetic levels of Stage I analgesia.

	Plane 1	Plane 2	Plane 3	
			Start	End
Pain perception	++	+	0	0
Color discrimination	++	++	+	0
Taste	++	++	++	0
Eye focus	++	++	+	0
Hearing	++	++	++	0
Cerebration	++	++	++	0
Memory recall				
of distant past[a]	++	++	++	0
of recent past[a]	++	++	+	0
Memory fixation during test period[b]	++	0	0	0

NOTE: ++ = normal; + = impaired; 0 = absent.
[a]Recall tested during ether inhalation.
[b]Recall tested after operation, and after lapse of 1, 3, 5, and 7 days.
SOURCE: Artusio, 1954.

Burns, Robson, and Welt (1960) observed that N_2O (0.3 P_a) approximately doubled the sensory thresholds for vision, touch, skin pain, and warmth, and raised the threshold for hearing about ten-fold. Both in man and in the cat, N_2O (0.3 P_a) changed the biological time sense in such a way that estimates of a clocked 15-second interval rose to about 30 seconds or more (Robson et al., 1960).

Parkhouse et al. (1960) used a battery of memory tests, both immediate and delayed, to study the effect upon memory of N_2O at partial pressures of 0.2, 0.3, or 0.4 P_a. The results may be interpreted as follows:

1. The influence of N_2O upon memory was highly dose-dependent. The effective partial pressure (0.4 P_a) exceeded the ineffective partial pressure (0.2 P_a) by a factor of only 2.

2. At 0.3 P_a, N_2O impaired short-term memory (0.5-minute) and long-term memory (30-minute) equally; input and reverberation were evidently affected, whereas consolidation was not.

3. N_2O at 0.4 P_a impaired short-term memory only slightly more than it did at 0.3 P_a, but it abolished long-term memory. Input, reverberation, *and* consolidation were therefore disrupted.

In the laboratory, experiments with goldfish, chicks, hamsters, mice, and rats confirm that interruption of brain activity shortly *after* a learning experience impairs the memory of that experience. The anesthetics shown to be effective include pentobarbital and ether (Pearlman, Sharpless and Jarvik, 1961), halothane (Cherkin and Lee-Teng, 1965), carbon dioxide (Leukel, 1957; Taber and Banuazizi, 1965; Quinton, 1966), ether (Herz, Peeke and Wyers, 1966), and N_2O (Bovet, McGaugh and Oliverio, 1966). Conflicting results with ether (Chorover, 1965) demonstrate the dependence of consolidation disruption upon the partial pressure and duration of application of the anesthetic agent. For example, Herz et al. (1966) studied the ability of ether to cause retrograde amnesia in mice, when it was administered immediately after a one-trial learning experience. Retrograde amnesia was produced by exposing the mice to the vapor of the liquid ether at 31°C (P_o = 676 mm Hg) for 70 seconds, but it was not produced at 31°C for 40 seconds, nor at 22°C (P_o = 480 mm Hg) for 120 seconds.

Quinton (1966) recently reported that 2-minute exposure to carbon dioxide at 380 mm Hg, or 15-minute exposure at 220 mm Hg, produced marked retrograde amnesia in rats. Increasing the exposure at the higher partial pressure did not increase the amnesia, whereas memory impairment at the lower pressure was a function of the time of exposure (2 to 15 minutes). The lower pressure is close to the anesthetizing partial pressure of CO_2.

In an experiment with halothane ($CF_3CBrClH$), one-day-old chicks were subjected to a simple one-trial learning experience (Cherkin and Lee-Teng, 1965). The experiment consisted of permitting a chick to peck spontaneously at a small porous lure saturated with *n*-propanol. Typically, the chick refused to peck again, when the lure was re-presented 18 hours later. When the first peck was followed by immediate anesthesia by halothane at 15 mm Hg for 5 minutes, the chick did peck at the lure upon re-presentation 18 hours later; evidently, it did not remember the disagreeable effect. When anesthesia was delayed for 90 minutes after the pecking, avoidance was not impaired. Control chicks, tested with water instead of *n*-propanol, showed no increased tendency later to avoid the lure. The results are shown in Figure 3.

A disadvantage of the use of anesthesia for disrupting memory is that induction requires 10 to 20 seconds, limiting its quantitative application to learning situations with consolidation half-times of 100 seconds or more. For shorter interruption times, down to 0.5 seconds, electroconvulsive shock can be applied to the chick (Lee-Teng and Sherman, 1966).

There appears to be no evidence against the resistance to anesthesia of stored long-term memory. Early reports that stored homing memory in bees was destroyed by N_2O, $CHCl_3$, and other compounds have been refuted by the demonstration that the disruption was transitory, and that the homing memory returned within a few days after anesthesia (Schmid, 1964).

Correlation of the effects upon memory consolidation with the simultaneous changes in other neurologic functions, brought about by graded partial pressures and graded exposure times to various non-specific anesthetics, may be expected to shed light upon the relationships between memory formation and other functions of the brain.

As an aside, it is curious that recall of distant memory in man is *facilitated* by sub-anesthetic levels of barbiturates, or by hypnosis, but this phenomenon seems to involve state-dependent learning (Overton, 1966) or a release of psychological repression, and will not be dealt with here. The controversial role of RNA in memory (Dingman and Sporn, 1964), currently the subject of active research in many laboratories, is also outside the scope of this discussion.

Today, opinion is still divided about the timeliness of the molecular attack upon psychobiology proposed by Pauling in 1961. But the winds of change are blowing. With full respect for the formidable complexities of brain function, increasing numbers of research workers are beginning to share his belief and to interest themselves in brain research. For them, another Lucretian "timely aid in many problems" is a Paulingian word-of-wisdom that came to me in the following way. When Pauling's goldfishers (Cherkin and Catchpool, 1964) compared P_a values with his brine-shrimpers (Robinson et al., 1965), the figures agreed within a factor of 2 for cyclopropane, ether, halothane, and chloroform (Table 2). But because the values for *n*-pentane differed by a factor of 12, confidence in the theory that anesthetic potency is essentially species-independent, was shaken. Careful rechecking of our notebooks disclosed no errors, but revealed that for technical reasons the literature value for the aqueous solubility of *n*-pentane entered into the calculation of P_a for brine shrimp, but not for goldfish. If the correct solubility were 36

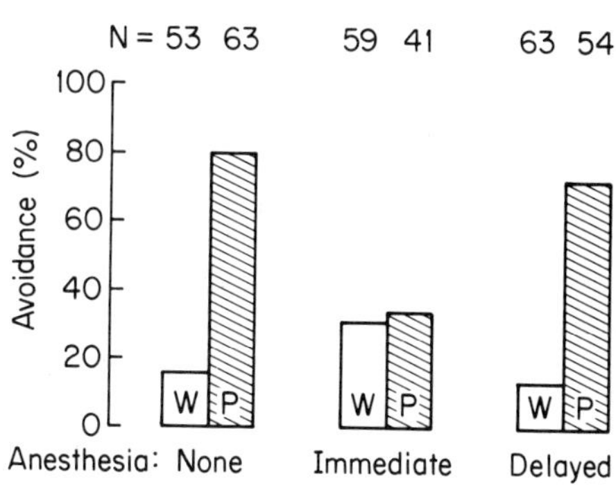

FIGURE 3.
Interruption of memory consolidation by halothane anesthesia in the one-day-old male chick. N = number of chicks per group; P = *n*-propanol lure; W = water lure. The lures were identical in appearance. Avoidance is the measure of memory retention; it represents the percent of chicks that avoided pecking the lure when it was presented to them 18 hours after their initial peck at the same lure. Anesthesia was in halothane vapor (15 mm Hg) for 5 minutes, starting immediately after the first peck, or after a 1.5-hour delay. Immediate anesthesia abolished the increased avoidance of the lure by chicks that had pecked the *n*-propanol lure.

mg/liter, instead of the reported 360 mg/liter (Fühner, 1924), the discrepancy would be resolved. At that moment, McAuliffe (1963) published his modern determinations of the solubility of hydrocarbons in water; his value for *n*-pentane was 38.5 $\pm$ 2.0 mg/liter! Then all the P_a values for goldfish and brine-shrimp agreed, faith in our theory was restored, and Pauling's joking comment was, "Hmm, perhaps you should use goldfish to measure solubilities."

Pauling's word-of-wisdom was this: "You know, you never want to let a 'fact' stand in the way of a good theory." The molecular attack on psychobiology, like the other attacks on this tough stronghold of human ignorance, has many "facts" to correct and many facts to discover, before it can contribute much to a good theory of mind. The need and the opportunities present an irresistible challenge.

ACKNOWLEDGMENTS

I am grateful to Linus Pauling for opening the door to a meaningful, timely, and exciting area of research, and for his continuing interest and encouragement. I thank Professor Roger W. Sperry for the generous hospitality of his laboratories and associates, for his suggestion to apply anesthesia to memory consolidation, and for the instructive opportunity to collaborate with Dr. Evelyn Lee-Teng in developing the one-trial learning model in chicks. The comments of Professor Seymour Benzer, Dr. W. G. Clark, and Mr. C. F. Leonard on the manuscript are acknowledged with thanks. This work received support from the Ford Foundation and from USPHS fellowship 1-F3-MH-25, 443-01, NIMH.

REFERENCES

Adriani, J. (1962). *The Chemistry and Physics of Anesthesia*, 2nd ed. Springfield: Charles C. Thomas.

Amoore, J. E. (1965). *Cold Spring Harbor Symposia Quantitative Biology* **30**, 623.

Artusio, J. F. (1954). *J. Pharm. Exptl. Therap.* **111**, 343.

Arvanitaki, A., and N. Chalazonitis (1951). *Colloq. Intern. Centre Natl. Recherche Sci. Paris*, p. 195.

Bayliss, L. E. (1960). *Principles of General Physiology*, vol. 2, p. 417. London: Longmans, Green.

Bernard, C. (1875). *Leçons sur les anesthésiques et sur l'asphyxie*. Paris: Baillière.

Bovet, D., J. L. McGaugh, and A. Oliverio (1966). *Life Sci.* **5**, 1309.

Brink, F., and J. M. Posternak (1948). *J. Cell. Comp. Physiol.* **32**, 211.

Brown, J. (1964). *Brit. Med. Bull.* **20**, 8.

Buchheit, R. G., H. R. Schreiner, and G. F. Doebbler (1966). *J. Bacteriol.* **91**, 622.

Burns, B. D., J. G. Robson, and P. J. L. Welt (1960). *Can. Anaesth. Soc. J.* **7**, 411.

Case, E. M., and J. B. S. Haldane (1941). *J. Hyg.* **41**, 225.

Catchpool, J. F. (1966). *Fed. Proc.* **25**, 979.

Cheek, D. B. (1964). *Am. J. Clin. Hypnosis* **6**, 237.

Cherkin, A. (1966). *Proc. Nat. Acad. Sci. U.S.* **55**, 88.

Cherkin, A., and J. F. Catchpool (1964). *Science* **144**, 1960.

Cherkin, A., and E. Lee-Teng (1965). *Fed. Proc.* **24**, 328.

Cherkin, A., J. F. Catchpool, C. F. Leonard, and L. Pauling (1967). In preparation.

Chorover, S. L. (1965). In Kimble, D. P., ed., *The Anatomy of Memory*, vol. 1, p. 252. Palo Alto, Calif.: Science and Behavior Books, Inc.

Davy, H. (1799). Cited by J. Davy, *Memoirs of the Life of Sir Humphry Davy, Bart.*, pp. 96, 98. London: Longman, Rees, Orme, Brown, Green & Longman, 1836.
Dingman, W., and M. B. Sporn (1964). *Science* **144**, 26.
Domino, E. F., S. F. Gottlieb, R. W. Brauer, S. C. Cullen, and R. M. Featherstone (1964). *Anesthesiology* **25**, 43.
Drachman, D. A., and J. Arbit (1966). *Arch. Neurol.* **15**, 52.
Eger, E. I., B. Brandstater, L. J. Saidman, M. J. Regan, J. W. Severinghaus, and E. S. Munson (1965). *Anesthesiology* **26**, 771.
Epstein, R. M., S. H. Ngai, D. C. Brody, and D. M. Rittenberg (1963). *Anesthesiology* **24**, 130.
Esplin, D. W. (1965). In Goodman, L. S., and A. Gilman, eds., *The Pharmacological Basis of Therapeutics*, p. 40. New York: Macmillan.
Faulconer, A. Jr. (1952). *Anesthesiology* **13**, 361.
Featherstone, R. M., and C. A. Muehlebaecher (1963). *Pharm. Rev.* **15**, 97.
Ferguson, J. (1951). *Colloq. Intern. Centre Natl. Recherche Sci. Paris*, p. 25.
Ferguson, J., and S. W. Hawkins (1949). *Nature* (London) **164**, 963.
Ferguson, J., S. W. Hawkins, and D. Doxey (1950). *Nature* (London) **165**, 1021.
Fink, B. R., and G. E. Kenny (1966). *Fed. Proc.* **25**, 561.
Frankel, J. and H. A. Schneiderman (1958). *J. Cell. Comp. Physiol.* **52**, 431.
Führer, H. (1921). *Biochem. Z.* **115**, 235.
Führer, H. (1924). *Ber.* **57**, 510.
Gaitonde, M. K., and D. Richter (1956). *Proc. Roy. Soc.* (London) B **145**, 83.
Gerarde, H. W. (1963). In Patty, F. A., ed., *Industrial Hygiene and Toxicology*, vol. 2. New York: Interscience.
Haldane, J. B. S. (1951). *Colloq. Intern. Centre Natl. Recherche Sci. Paris*, p. 47.
Herz, M. J., H. V. S. Peeke, and E. J. Wyers (1966). *Psychon. Sci.* **4**, 375.
Hodgson, E. S. (1965). In Carthy, J. D., and C. L. Duddington, eds., *Viewpoints in Biology*, p. 104. London: Butterworth.
Holaday, D. A., J. Garfield, and D. Ginsberg (1965). *Anesthesiology* **26**, 251.
Huxley, T. H. (1865). Cited by F. O. Schmitt (1965). *Science*, **149**, 931.
Kay, H. (1959). In Birren, J. E., ed., *Handbook of Aging and the Individual*, Chicago: Univ. of Chicago Press.
Kopp, R., Z. Bohdanecky, and M. E. Jarvik (1966). *Science* **153**, 1547.
Laffort, P. (1963). *Arch. Sci. Physiol.* **17**, 75.
Lawrence, J. H., W. F. Loomis, C. A. Tobias, and F. H. Turpin (1946). *J. Physiol.* **105**, 197.
Lee-Teng, E., and S. M. Sherman (1966). *Proc. Nat. Acad. Sci. U.S.* **56**, 926.
Lettvin, J. Y., and R. C. Gesteland (1965). *Cold Spring Harbor Symposia Quantitative Biology* **30**, 217.
Leukel, F. (1957). *J. Comp. Physiol. Psychol.* **50**, 300.
Lucretius (c. 55 B.C.). *Nature of the Universe*, trans. by R. Latham. Baltimore: Penguin Books, 1951.
Maudsley, H. (1895). Cited by H. McIlwain, *Maudsley, Mott and Mann on the Chemical Physiology and Pathology of the Mind.* London: H. K. Lewis & Co., 1955.
McAuliffe, C. (1963). *Nature* (London) **200**, 1092.
McGaugh, J. L. (1966). *Science* **153**, 1351.
McGaugh, J. L., and L. Petrinovich (1966). *Psychol. Rev.* **7**, 382.
McGowan, J. C. (1954). *J. Appl. Chem.* **4**, 41.
Meyer, H. (1951). *Colloq. Intern. Centre Natl. Recherche Sci. Paris*, p. 17.
Meyer, H., and H. Hemmi (1935). *Biochem. Z.* **277**, 39.
Meyer, H., and H. Hopff (1923). *Z. Physiol. Chem.* **126**, 281.
Miller, K. W., W. D. M. Paton, and E. B. Smith (1965). *Nature* (London) **206**, 574.
Miller, S. L. (1961). *Proc. Nat. Acad. Sci. U.S.* **47**, 1515.
Moore, W J., and H. R. Mahler (1965). *J. Chem. Educ.* **42**, 49.

Mullins, L. J. (1955). *Ann. N. Y. Acad. Sci.* **62**, 247.

Ngai, S. H. (1963). In Root, W. S., and F. G. Hofmann, eds., *Physiological Pharmacology*, vol. 1. New York: Academic Press.

Östergren, G. (1951). *Colloq. Intern. Centre Natl. Recherche Sci. Paris*, p. 77.

Overton, D. A. (1966). *Psychopharmacologia* **10**, 6.

Parkhouse, J., J. R. Henrie, G. M. Duncan, and H. P. Rome (1960). *J. Pharm. Exptl. Therap.* **128**, 44.

Paton, W. D. M., and R. N. Speden (1965). *Brit. J. Pharm. Chemotherapy* **25**, 88.

Pauling, L. (1946). *Chem. Eng. News* **24**, 1064.

Pauling, L. (1961). *Science* **134**, 15.

Pearlman, C., S. K. Sharpless, and M. E. Jarvik (1961). *J. Comp. Physiol. Psychol.* **54**, 109.

Poussel, H. (1951). *Colloq. Intern. Centre Natl. Recherche Sci. Paris*, p. 157.

Quinton, E. E. (1966). *Psychon. Sci.* **5**, 417.

Raventòs, J. (1956). *Brit. J. Pharm.* **11**, 394.

Robinson, A. B., K. F. Manly, M. P. Anthony, J. F. Catchpool, and L. Pauling (1965). *Science* **149**, 1255.

Robson, J. G., B. D. Burns, and P. J. L. Welt (1960). *Can. Anaesth. Soc. J.* **7**, 399.

Schmid, J. (1964). *Z. Vergleich. Physiol.* **47**, 559.

Schmitt, F. O. (1965). *Science* **149**, 931.

Schoenborn, B. P., H. C. Watson, and J. C. Kendrew (1965). *Nature* (London) **207**, 28.

Simons, J. H., and L. P. Block (1939). *J. Am. Chem. Soc.* **61**, 2962.

Speden, R. N. (1964). Personal communication.

Steinberg, H., and A. Summerfield (1957). *Quart. J. Exptl. Psychol.* **9**, 138.

Stoughton, R. W., and P. D. Lamson (1936). *J. Pharm. Exptl. Therap.* **58**, 74.

Taber, R. I., and A. Banuazizi (1965). *Fed. Proc.* **24**, 329.

Thomas, D. M., T. N. MacKrell, and E. H. Conner (1961). *Anesthesiology* **22**, 542.

Ungar, G. (1965). *Nature* **207**, 419.

Victor, M. (1964). In M. A. B. Brazier, ed., *Brain Function*, vol. 2. Berkeley: Univ. of Calif. Press.

Virtue, R. W. (1949). *Anesthesiology* **10**, 318.

Virtue, R. W. (1950). *Proc. Soc. Exptl. Biol. Med.* **73**, 259.

Wells, H. (1847). Cited by F. Cole, *Milestones in Anesthesia.* Lincoln: Univ. of Nebraska Press, 1965.

Wenzel, B. M., and M. H. Sieck (1966). *Ann. Rev. Physiol.* **28**, 381.

Winters, W. D., and C. E. Spooner (1966). *Electroenceph. Clin. Neurophysiol.* **20**, 83.

Zahradník, R. (1962). *Arch. Int. Pharmacodyn.* **135**, 311.

J. F. CATCHPOOL
Department of Pharmacology
San Francisco Medical Center
University of California
San Francisco, California

The Pauling Theory of General Anesthesia

On January 4, 1960, I arrived in Pasadena to start a postdoctoral fellowship at the California Institute of Technology, and it was after five when I knocked on the door of Dr. Pauling's office. I half expected him to have already gone home. To my surprise, he opened the door himself; and a few moments later, after assuring himself that I was provided with accommodation for the night, he launched into a discussion of what research I might undertake. "Nearly all the people that come to work with me want to work on their own projects, but sometimes I have ideas of my own and it is difficult for me to find time to conduct experiments, to indicate the correctness or incorrectness of these ideas, and I must admit I do not have enough patience for experimental work." I made a mental note that a good place to start my career in science would be to work on one of Pauling's ideas. I told him of my very modest clinical investigations of the effects of hypothermia on the course of acute tetanus, and of some other work I had done on local anesthetics. "I have an idea how anesthetics may be exerting their effects," Pauling continued. "It may be that anesthesia is produced by tiny microcrystals of ice forming in the brain and interfering with the passage of electrical impulses. How would you like to work on this problem?"

Good grief, I thought, he is inventing a theory of anesthesia, just for my benefit! I found myself wondering if Pauling had ever been in an operating theater, but quickly recalled that the conventional explanation of the mechanism of anesthesia, originated by Meyer and Overton at the turn of the century, is unsatisfactory from many points of view. The old "theory," based on a correlation between anesthetic potency and the solubility of an anesthetic agent in olive oil, states that narcosis commences when any chemically indifferent substance has attained a certain molar concentration in the lipids of the cell (Meyer, 1899).

I remembered that many great scientists sought a rational explanation for the complex and dramatic clinical effects of the enormous number of diverse substances with anesthetic properties. As long ago as 1799 Sir Humphry Davy noticed the anesthesia produced by breathing nitrous oxide, and later Michael Faraday observed the effects of inhaling ethyl ether. They both sought to explain the phenomenon by postulating a reversible asphyxiation of the vital centers.

Pauling recalled that his interest in the mechanism of anesthesia had begun in 1950 when he heard Professor Henry K. Beecher of the Massachusetts

General Hospital report the discovery (Cullen and Gross, 1951) that the noble gas xenon could cause anesthesia in man and animals. He wondered how these chemically unreactive atoms of xenon could affect the brain, because they could not possibly take part in reactions involving covalent or ionic bonds. Apart from the recently discovered compounds of xenon and fluorine (predicted by Pauling in 1933), xenon can form only hydrates of the clathrate type, which melt near 0°C, far below the temperature of the body.

Pauling is able to recall the exact moment that he formulated the hydrate microcrystal theory of general anesthesia. On a morning in April, 1959, he was reading a paper (McMullan and Jeffrey, 1959) describing the determination of the structure of a crystalline hydrate of an alkylammonium salt; it was found to be a clathrate structure, resembling that of xenon hydrate. As he later wrote,

> It is my memory that my thoughts during the next few seconds were the following: 'This hydrate crystal decomposes (melts) at about 25°C. It contains dodecahedral chambers that might be occupied by xenon molecules, which would stabilize it enough to raise the decomposition temperature to above 37°C, the temperature of the human body. Alkylammonium ions resemble substances normally present in the brain—amino acids, the side chains of protein molecules. Hydrate microcrystals involving these substances in the brain might form if the brain were cooled, or might interfere with the motion of ions or electrically charged protein side chains that normally contribute to the electric oscillations in the brain that are involved in consciousness and ephemeral memory, reducing their amplitude enough to cause loss of consciousness; or they might effect this result by interfering with some chemical reaction involved in supporting the electric oscillations, as by masking the active region of an enzyme molecule. The activity of anesthetic agents should be proportional to the polarizabilities of their molecules, which determine their effectiveness in stabilizing the hydrate crystals by the London electron-correlation intermolecular interactions; hence xenon should be more effective than argon, chloroform more effective than methyl chloride, as observed. This is a molecular theory of anesthesia.' (Pauling, 1964).

Many of the commonly used general anesthetics are known to form hydrates. The hydrate of nitrous oxide was first reported in 1799 by Humphry Davy, who also suggested that the gas be used for the relief of pain during surgical operations. The hydrate of chloroform was also well known long before chloroform was ever used as an anesthetic agent.

Hydrates of anesthetic agents are usually of the gas hydrate type, in which the anesthetic molecule is encaged in a distorted ice matrix. These crystalline compounds usually occur in two forms. In the type I gas hydrate the smaller molecules—such as nitrogen, methane, nitrous oxide, xenon, and cyclopropane—form a structure with the ratio of $5\frac{3}{4}$ water molecules to each anesthetic molecule. The exact arrangement of the molecules of this type of hydrate was worked out by Pauling and Marsh (1952). They studied the structure of chlorine hydrate by X-ray crystallography, and found that the chlorine molecules are held in cages formed by twenty water molecules joined together by tetrahedrally placed hydrogen bonds to form regular dodecahedrons. At every corner of each dodecahedron another water molecule is attached by a hydrogen bond, so that other dodecahedra can be formed. In this manner the oxygen atom of every water molecule is tetrahedrally bonded via hydrogen atoms to four other water molecules.

Pentagonal dodecahedra, however, cannot pack together to fill space com-

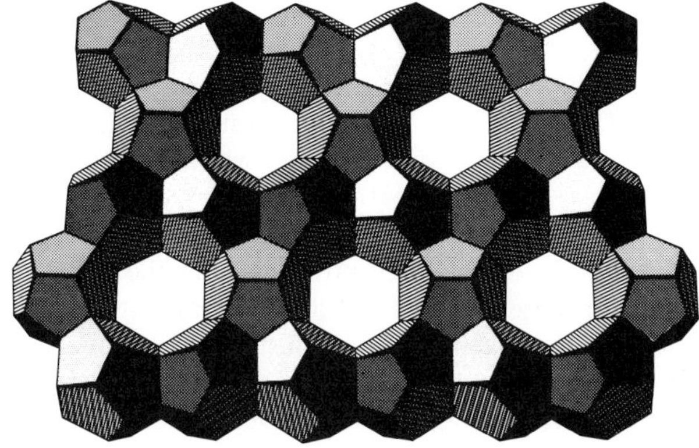

FIGURE 1.
The structure of the type I hydrate. The tetrakaidecahedral cavities (12 pentagonal plus 2 hexagonal faces) are formed when layers of dodecahedra, arranged as shown, are stacked on top of each other so that the hexagonal openings line up perpendicular to the plane of the drawing.

pletely, so that holes are left between the dodecahedra; in the type I gas hydrate structure, these holes occur between two layers of dodecahedra, the dodecahedra of each layer being arranged in rings of 6. These holes have 14 sides: 12 pentagonal faces, each contributed by one dodecahedron, and 2 hexagonal faces, formed by the ring of 6 dodecahedra in each layer. The cavities in the dodecahedra can be occupied by molecules with diameters of less than 5.1 Å, and the 14-sided cavities (tetrakaidecahedra) can encage a molecule less than 5.8 Å in diameter (Figure 1).

Slightly larger anesthetic molecules, such as chloroform, propane, and ethyl chloride, too large to fit in either the dodecahedra or the tetrakaidecahedra, form the type II gas hydrates, which have a slightly different arrangement of dodecahedra in which the ratio of gas molecules to water molecules is 1 to 17. In this structure, first described by von Stackelberg in 1952, the water molecules are all tetrahedrally bonded to each other to form dodecahedra, but the spaces formed between the dodecahedra have 16 sides (hexakaidecahedra) and are slightly larger than the type I tetrakaidecahedral cavities. These spaces can contain guest molecules up to 6.7 Å in diameter. The 16-sided cavities are formed by 12 dodecahedra framing 4 hexagonal faces lying on the 4 sides of a tetrahedron. In the type II structure the cavities do not stack up as in the type I structure, but honeycomb the crystal in a zigzag manner (Figure 2).

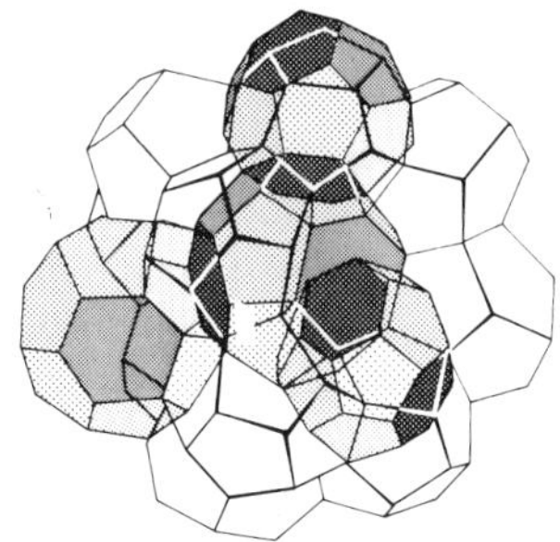

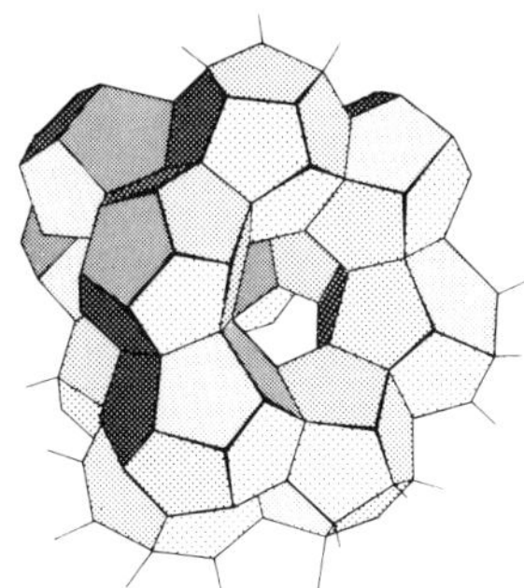

FIGURE 2.
Two drawings of a folded-paper model of the hexakaidecahedral cavity of the type II hydrate structure. The 16-sided cavities are formed between 16 dodecahedra framing 4 hexagonal faces. The plane of each hexagonal face lies on the side of a regular tetrahedron. Each hexagonal face is shared by 2 hexakaidecahedra.

In both these hydrate structures, no ordinary chemical bonds are needed between the atoms of the guest anesthetic molecules and the host water molecules. The strength or stability of the structure largely depends on the percentage of cavities occupied, the size and fit of the guest molecules lying in the cavities, and the symmetry of electrical charges within each cage's molecule. Strongly polar molecules would distort the symmetry of the cages, so that neither the type I or type II hydrate is likely to form. Molecules that are not spherically symmetrical and cannot fit comfortably into the 12-, 14-, or 16-sided cavities will not form these types of hydrates. Hydrate structures, however, have been reported for hydrogen-bonding substances; acetone, for instance, is known to form the type II hydrate (Quist and Frank, 1961).

Starting with the paper of McMullan and Jeffrey (1959), describing the structure of the hydrates of quaternary ammonium salts, G. A. Jeffrey and his co-workers have published a series of six papers (Feil and Jeffrey, 1961; Bonamico et al., 1962; Jeffrey and McMullan, 1962; McMullan et al., 1963; Beurskens et al., 1963) describing polyhedral clathrate hydrate structures in which the idealized water structure takes the form of layers of $H_{40}O_{20}$ pentagonal dodecahedra, interlinked in various arrangements by tetrakaidecahedra and pentakaidecahedra. The anions of these salts are enclosed in the tetrakaidecahedra and pentakaidecahedra. The hydrogen-bonding atoms of these large anions replace oxygen atoms in the water lattice, so that the anions can each occupy four adjacent pentakaidecahedra or four adjacent tetrakaidecahedra. In all these structures there was evidence that some of the dodecahedra were occupied by guest water molecules when distorted by the presence of cations. All six structures reported were analogues of the type I hydrate and could be visualized as alternating layers of polyhedra. No counterpart of the type II hydrate was reported in this series.

The recently reported hydrate of hexamethylenetetramine that melts at 13.5°C is of interest because the guest molecule is linked to the water framework by hydrogen bonds from three of its four nitrogen atoms and hangs batlike by its hydrogen bonds in the water cage. The water cage in this structure is slightly different from the other hydrate cages and consists of columns of slightly puckered and staggered six-member rings of water molecules. Jeffrey suggests that "we may expect to find a rational and continuous series of hydrated crystal structures ranging from the pure clathrates of the gas hydrate type through intermediate framework structures, to structures where the water molecules are arranged in hydrogen-bonded sheets, ribbons and chains, and finally structures where the water molecules are isolated from each other as in many of the simpler hydrated salts" (Jeffrey and Mak, 1965). Hydrate structures have also been reported for ethyl alcohol (Glew, 1962), and for dimethyl ether, ethylene oxide, and acetone (von Stackelberg et al., 1947). X-ray diffraction studies of a low-density form of silica containing hydrocarbon impurities reveal a structure analogous to the type I hydrate. In this the tetrahedrally bonded silicate framework forms dodecahedra and tetrakaidecahedra, with the straight-chain hydrocarbon impurities occupying a stack of tetrakaidecahedral cavities. The hexagonal openings between adjacent tetrakaidecahedra are just large enough to accommodate the waists of the carbon-carbon bonds of the hydrocarbon chains (Kamb, 1965). Many substances have already been shown to exist in hydrated forms when dissolved in water, and many more hydrates will be found. Proteins themselves form highly hydrated crystals and regiment many thousands of surrounding water molecules when dissolved in water. Gas hydrate formation has been shown to be possible in highly concentrated aqueous solutions of proteins, lipids, and carbohydrates (Huang et al., 1965). Hydrates that can remain solid up to 31°C are known. When a protein melts or denatures, disordering of its ordered water shell may be considered somewhat analogous to the melting of the simple gas hydrates.

Local solvent structure in the neighborhood of macromolecules seems to influence greatly the stability of such large biopolymers as proteins, collagen, RNA, and DNA. X-ray patterns of collagen have shown water molecules, immobilized close to the chains of collagen, that appear to play an important role in the stabilization of their macromolecular configuration. The helical configuration of DNA fibers is dependent on the presence of water at high humidities. $(NH_4)_2SO_4$ and sucrose are known to stabilize proteins against denaturation; urea, LiBr, and KSCN, to destabilize them. These molecules exert their effects on biopolymers by modifying the structure of the solvent water, thereby affecting the interactions between the macromolecules and water and changing the conformation and behavior of these large molecules. Jeffrey and his co-workers have shown that a surprisingly wide variation in the size and shape of hydrate cages can occur, and a much more general form of enclathration of large molecules with exposed apolar groups is possible. Hydrate cages with volumes of 1550 $Å^3$ are described in which the hydrogen-bonding atoms of the guest molecules sometimes take the place of oxygen atoms in the idealized water cage. In the peralkylated ammonium salt hydrates the apolar side chains, lying in contiguous polyhedra, almost completely shield their hydrophobic atoms from the aqueous environment. Klotz, like Pauling, believes that hydrate structures analogous to those proposed by Jeffrey are especially likely to form around proteins in solution, because the high local concentration of apolar side chains produces cooperative effects to induce a stabilized arrangement of water in hydrate forms. Thus temperature denaturation of proteins can be understood in terms of the melting of ordered water regions, or "hydrotactoids," as Klotz calls water structures around apolar side chains of proteins (Klotz, 1965).

As mentioned before, Pauling's theory of anesthesia was partly inspired by Jeffrey's work. Pauling surmised in 1959 that the structure of water around the proteins was likely to be an arrangement of water molecules tetrahedrally bonded to each other to form dodecahedra with polyhedral cavities accommodating the apolar side chains. It was also apparent to him that the addition of apolar anesthetic molecules might cooperatively stabilize the hydrate structure by forming a mixed hydrate in this "hydrotactoid" water shell. In 1960, in an attempt to measure in brain tissue the cooperative stabilization of these hydrate shells by anesthetic agents, I built a calorimeter to see if there was any measurable change in the heat capacity of brain tissue between 0° and 40°C. I was looking for a small plateau in the heat capacity that would indicate a partial melting of hydrate structures. I was hoping that the addition of anesthetic agents might shift the plateau to a higher temperature, but the small plateau, if any, was too small to measure.

The first published account of Pauling's "molecular theory of general

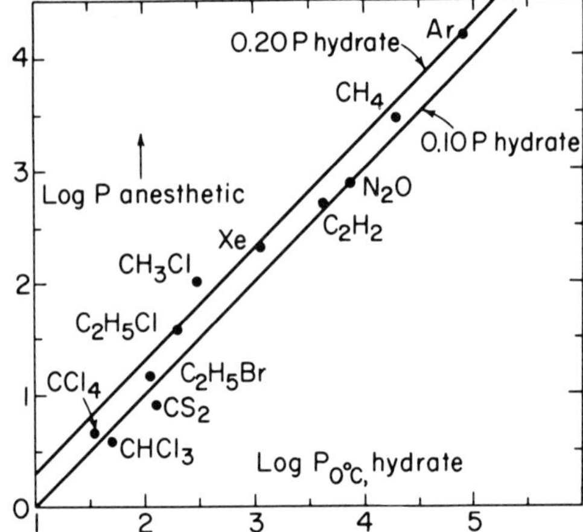

FIGURE 3.
The logarithm of the anesthetizing partial pressure of nonhydrogen-bonding anesthetic agents plotted against the equilibrium partial pressure of their hydrate crystals. [From Pauling, 1961.]

anesthesia" appeared in *Science* on July 7, 1961. The theory had already been presented in lectures at Stanford Medical School in Palo Alto, and at a meeting of the American Chemical Society in Hawaii, as well as to other small groups. By 1961, the state of the art had evidently reached a point where, as Pauling said, "this theory was forced upon us by the facts about anesthesia." The inevitability became almost immediately evident when in September 1961 Professor Stanley Miller of the Department of Chemistry of the University of California at San Diego quite independently published an almost identical description of "A Theory of Gaseous Anesthetics" (Miller, 1961). Miller arrived at these conclusions while investigating the occurrence and formation of gas hydrates at low temperatures in the solar system. He noticed that many gases that form hydrates are also gaseous anesthetics. Both Miller and Pauling reported the remarkable correlation between the partial pressure of an anesthetic gas required for anesthesia and the equilibrium partial pressure of the hydrate crystals at 0°C (Figure 3).

Both Pauling and Miller recognized that the mechanism of narcosis could not simply be due to the formation of crystals of anesthetic gas and water with the same ratio of gas to water found in the pure hydrates, because obviously the dissociation pressure of the hydrates at mammalian body temperature would equal many atmospheres, whereas the partial pressure of most anesthetic agents needed for narcosis would be only a fraction of an atmosphere. Methyl chloride, for instance, is narcotic for mammals at a partial pressure of about 0.14 atmospheres but, as Pauling pointed out, the crystals of the hydrate are not stable at 37°C until a pressure of 40 atmospheres has been reached. Even atmospheric nitrogen becomes a narcotic if breathed by deep sea divers (Behnke et al., 1935). At great pressures the small apolar molecules of nitrogen may be forced into solution and into the dodecahedral cavities in the hydrate structures forming around the side chains of proteins.

It is known, however, that two hydrate-forming agents acting together can increase the stability of the hydrate framework by filling a greater percentage of the cavities in the hydrate lattice. For example, chloroform hydrate that normally melts at 1.4°C melts at over 14.7°C when stabilized by the addition of xenon gas at 1 atmosphere pressure. The small xenon atoms enter the dodecahedra surrounding the hexakaidecahedral cavities occupied by the chloroform molecules, so that the resulting double hydrate has extra internal bracing and does not fall apart until a higher temperature is reached. When the ratio of chloroform to xenon to water molecules is 1:2:17, all the cavities in the type II hydrate are occupied.

Investigations by nuclear magnetic resonance of the structure of water and of solutions of xenon in O^{17} water (Glasel, 1966) show that in solid xenon hydrate the water molecules behave as if they were in a loosely ordered state, free to undergo individual reorientation but not tightly bound as in the ice lattice. Nuclear magnetic resonance spectra of liquid xenon solutions are similar, showing that the solvated xenon also increases the activation energy of the water molecule; "in essence, the xenon lowers the effective temperature of the water." Glasel concludes that water has a quasi-lattice structure from 4 to 30°C, but above 30°C an abrupt change takes place in the structure. Above this temperature any addition to the water leads to ordering of the liquid structure and "must therefore have a profound effect on the mechanisms of reactions involving water as a physical entity."

There are many molecules normally present in the aqueous medium of the brain, such as the alkyl carboxylate ion side chains of the aspartate and glutamate residues and other side chains of proteins, as well as other solute molecules that have the properties of hydrate-forming agents. Water in the neighborhood of these molecules, with distributions of electrostatic charges and geometry favorable to hydrate formation, is in a more highly ordered state and contains cavities that can be filled by small gas-hydrate-forming molecules. Occupancy of these cavities will contribute to the stability of the water structure, extending it further into the surrounding unstructured water, and increasing the "iceberg cover," as Miller calls it. Anesthetic gases in aqueous solution have low solubilities, large entropies of solution, and large partial molal heat capacities, indicating that the water surrounding the dissolved gas molecules is in a more highly structured state than the water not affected by the apolar gas molecules.

Hydrate structures are held together in part by the energy of the hydrogen bonds of the tetrahedrally bonded water-molecule framework and partly by the interaction of the intermolecular attractive forces between the encaged molecules and the water framework. The empty hydrogen-bonded framework of a hydrate would be expected to be as stable as that of ordinary ice, but it is in fact less stable because the framework is more open because of the voids between the dodecahedra. Van der Waals and Platteeuv (1958) calculated the free energy per water molecule of the hydrate framework, and found it to be greater than the free energy of the ice at 0°C by 0.167 kcal/mole for the type I framework and by 0.19 kcal/mole for the type II framework.

The extra stability given to these structures by the encaged molecules can be calculated by applying the London equation for the electronic dispersion interaction between two molecules. Pauling calculated that $Xe \cdot 5\frac{3}{4}\ H_2O$ hydrate would have an enthalpy of formation from gaseous xenon and ice of 9.4 kcal/mole, from which should be subtracted a small correction for the van der Waals repulsion of the xenon atoms and the surrounding water molecules. The enthalpy of formation of xenon hydrate was found by experiment to be 8.4 kcal/mole. The stabilization of hydrate crystals can therefore be understood in terms of the interactions of van der Waals intermolecular forces. These intermolecular forces are significant only when the molecules are very close together. At distances greater than a few molecular diameters the strength of these forces, decreasing by the inverse seventh power, falls to essentially nothing. If two molecules approach each other to a point at which their electron shells start to penetrate each other, then the positive nuclei of the atoms will begin to repel each other.

The intermolecular forces responsible for hydrate formation and stability can be correlated with the index of refraction of the hydrate-forming agent. The index of refraction is a measure of the interaction of light with the substance it is traversing. The molar refraction of a substance is a monotonically increasing function of its molecular polarizability, and the magnitude of the molecular interactions between two molecules is proportional to the product of their electric polarizabilities. Therefore, the strength of intermolecular forces should increase with increasing molar refraction. Pauling (1961) plotted the partial pressure of anesthetic agents in equilibrium with their hydrate crystals and water and ice versus their mole refraction, and showed that the same curve could be obtained by plotting the mole refraction against the anesthetizing partial pressure for mice (Figure 4).

Both Miller and Pauling observed that decreasing the temperature of an aqueous solution of anesthetic agents would tend to stabilize and extend the water structure. Narcosis is observed to occur at about 27°C in human beings. The narcotic effects of hypothermia could be explained as the result of the formation of hydrate structures in the synaptic regions of the brain.

To see if anesthetic potency does in fact become greater at a lower tempera-

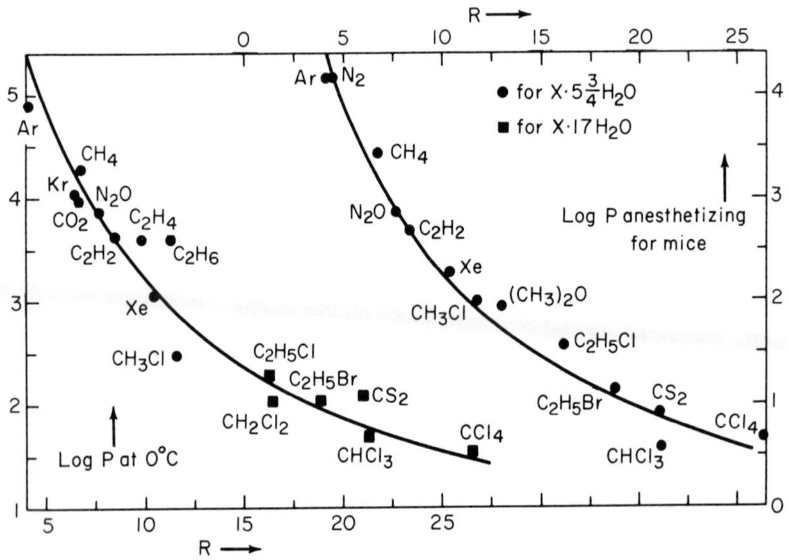

FIGURE 4.
Left curve: Logarithm of the partial pressure of anesthetic agents in equilibrium with their hydrate crystals and water and ice at 0°C (left scale) plotted against the mole refraction of the anesthetic agent (left bottom scale). *Right curve:* Logarithm of the anesthetizing partial pressure for mice (right scale) plotted against the mole refraction of the anesthetizing agent (top scale). Both curves are identical. [From Pauling, 1961.]

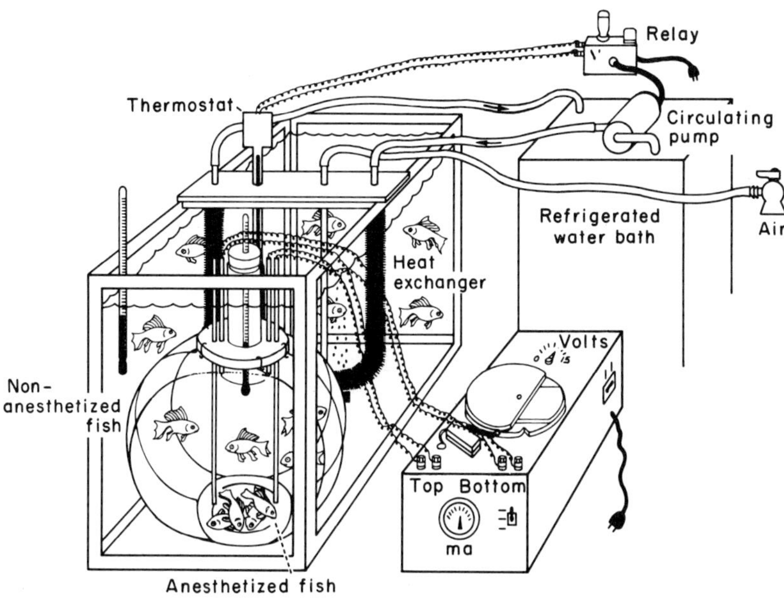

FIGURE 5.
The experimental setup used to determine the relationship between body temperature and the partial pressure that anesthetizes 50 percent of the goldfish. [From Cherkin, Catchpool, Leonard, and Pauling, unpublished.]

ture, my colleague, Arthur Cherkin, and I carried out a series of experiments with goldfish to determine the potency of various anesthetic agents (Cherkin and Catchpool, 1964). The advantage of using the intact animal was that we could observe true reversible general anesthesia. The fish were obliged to perform a coordinated task—specifically to keep out of a region in a bowl of water where they would receive an unpleasant shock. We used several fish at a time so that biological variations were minimized, ran replicate experiments to reduce experimental error, and used a range of doses to determine the AD/50, the dose at which half of the goldfish were, by our criteria, anesthetized.

To determine anesthetic potency at the site of action in the nervous system, it is necessary that the thermodynamic activity of the anesthetic agent be the same in all tissues of the body and in the external environment in which the concentration is measured. By adding measured quantities of the anesthetic to the water in which the fish were swimming and allowing sufficient time (at least 20 minutes) for the anesthetic to saturate all the tissues equally, we determined the AD/50 under conditions approaching true thermodynamic equilibrium. By acclimating the goldfish in tanks of water held at various temperatures for several days, we were able to measure the effect of temperature on the potency of an anesthetic agent without also measuring the effects of hypothermia on the fish's metabolism (Fig. 5). We used ten fish at a time, and added an exact amount of a concentrated solution of anesthetic agent to 4 liters of water saturated with air. The fish were kicked with a small electric shock every 6 seconds to see if they were truly unconscious. By our criteria those fish not responding to the stimulus were considered to be anesthetized. After using about 1500 15-cent fish we arrived at the conclusion that anesthetic potency nearly doubles for every 10°C *decrease* in temperature —a result in accord with the hydrate theory of anesthesia, but contrary to lipid solubility theories.

The stability of a hydrate also depends on the fit of the encaged molecule. A molecule too small for the cavity it occupies contributes little stability to the structure. Molecules larger than the optimum size have a slight repulsing effect on the water cage, and the hydrates will have a higher dissociation pressure. Hydrate cages may form around molecules too large to fit into the hexakaidecahedral cavities, but the cages will be irregularly shaped and there will be greater distortions of the tetrahedral hydrogen bond angles. Slight differences in anesthetic potencies between anesthetic molecules of the same composition and molecular weight (but with differing steric arrangements of their atoms) might be demonstrable if a technique could be devised for measuring anesthetic potency accurately enough to show these small differences.

In the summer of 1963 a technique for measuring anesthetic potency under conditions approaching thermodynamic equilibrium, using 100,000 animals in 12 different anesthetic concentrations, was devised by A. Robinson et al., (1965). Pooled, dry, brine shrimp eggs were hatched in artificial seawater at 20°C and a population of uniform viability was obtained by using the phototactic instinct of the microscopic 3-day-old larvae. Only those shrimp able to swim an obstacle course toward a light shining through the water were used. Equal portions of the seawater containing the swimming shrimp were exposed to twelve different concentrations of anesthetic agents dissolved in seawater contained in separatory funnels in a thermostated water bath. After 10 hours' exposure, the percentage of shrimp affected by each dose was determined. This was done by means of a photocell counter to tally the number of shrimp still swimming toward the light at the top and the number of shrimp so overcome that they had fallen through the stopcock of the separatory funnel into a darkened container (Figure 6).

In the time available, it was shown that the anesthetic potencies of four agents with similar electric polarizabilities appeared to show a correlation with molecular size and shape. Chloroform, the best fit for the hexakaidecahedral cavities, is the most potent. Halothane, the next most potent, is larger and not so spherically symmetrical. Diethyl ether is a linear molecule and a poor fit, but it might make use of its hydrogen-bonding capability in the formation of a hydrate. The linear molecule *n*-pentane has no hydrogen-bonding capability. But cyclopropane, with an electric polarizability half as great as that of each of the other four compounds, is the least potent, although it has a spherical molecule and is known to form hydrates.

To propose a theory of general anesthesia it is necessary to have a definition of general anesthesia. The state of anesthesia is usually described as a reversible depression of the level of consciousness. Anesthesia can be brought about by either physical or chemical means, and all anesthetic agents depress all biochemical, motor, and sensory activity to some extent, but of course the most noticeable effect is a depression of the higher centers of the central nervous system. General anesthetic agents are observed to depress the peripheral nerves only when applied to nerve-muscle preparations in concentrations many times the concentration needed to bring about a lethal effect in the intact animal. Certain functions in certain parts of the central nervous system may reasonably be expected to be more susceptible than others to interruption of function. The phenomenon of general anesthesia can therefore be accurately quantitated only in the intact animal.

If anesthesia is the reversible depression of the level of consciousness, how then can consciousness be disturbed by these chemically inert hydrate-forming molecules? If consciousness and ephemeral memory are considered in terms of a certain level of electrical activity in the brain and if permanent memory involves the laying down of nucleoproteins to provide electrical circuits over which a sequence of oscillations can retrace an electrical pattern, then anything that disturbs the pattern, or offers impedance to the electrical oscillations, will tend to alter the state of consciousness. Loss of consciousness, as in sleep or barbiturate anesthesia, is manifested by a diminished output of electrical energy by the exciting mechanism. Some drugs and narcotics may exert their effects by depressing centers that initiate the electrical activity, but other depressants (such as the general anesthetics, because of their extremely simple chemical nature) do not affect specific centers only but every tissue of the body. That general anesthesia in humans can be produced by xenon demonstrates that anesthesia need not depend on the specific effects of any structural grouping. Some parts of the body, particularly some segments of the neural electrical circuits, are much more affected than, say, the peripheral motor circuits and reflex arcs. Other circuits, such as the autonomic nervous system, need more impedance thrown into the circuit before they are functionally disturbed.

Pauling has suggested that hydrates forming in the water contained in the neurons and around the neural network, both in the cells of the neurons and in the synaptic regions between the neurons, may trap some of the electrically charged side chains of neuronal proteins and interfere with the move-

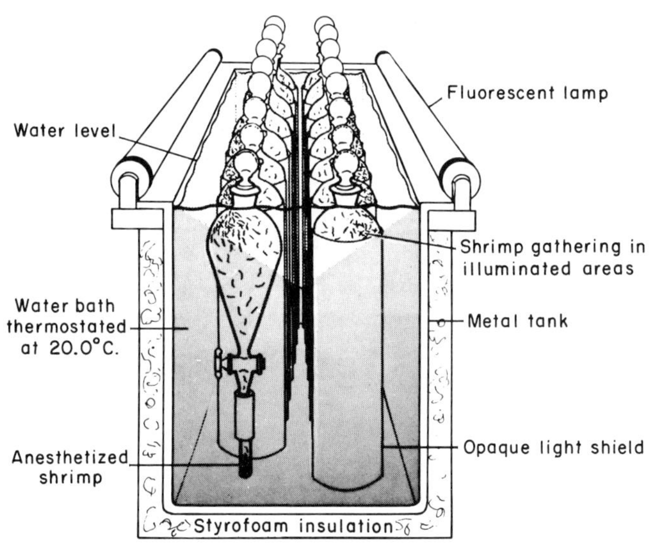

FIGURE 6.
The determination of anesthetic potency, using 100,000 brine shrimp in 12 different anesthetic concentrations and 4 controls. Thermodynamic equilibrium may be approached by exposing the shrimp to an unvarying environment for 10 hours. [From Robinson, Manly, Anthony, Catchpool, and Pauling, 1965.]

ment of ions in the synapses and in the neurons, in such a way as to increase the impedance to the reverberating electric oscillations, thus reducing the electrical activity to the point at which consciousness is lost. Hydrates may also inertly lodge in key enzyme sites on proteins, preventing reactant molecules from coming close enough to react with them.

Both Pauling and Miller base their theories of anesthesia on the excellent correlation between anesthetic potency and ability to form hydrates. But, as Pauling pointed out, such correlations can be made between narcotic potency and almost any other physicochemical property involving van der Waals intermolecular forces. The Meyer-Overton theory of anesthesia is based on the empirical correlation between narcotic potency and solubility in olive oil. Ferguson's thermodynamic theory of anesthesia (1939) is based on the constancy of the ratio of anesthetizing partial pressure to the thermodynamic activity of the pure nonhydrogen-bonding liquids measured at a standard temperature. Pauling showed another good correlation between narcotic potency and the hydrate dissociation pressure. Other workers have reported different correlations, related to the energy of intermolecular attraction. By sticking his neck out a considerable distance, Pauling suggested a mechanism whereby consciousness may be lost in a reversible manner, and it can be said that his theory is not just another empirical correlation, but the first explicit theory of general anesthesia. The theory has the additional attribute that it is extremely difficult to prove or disprove and, by adding to the controversy, has stimulated new research into the old unsolved problem of the mechanism of anesthesia.

ACKNOWLEDGMENT

My thanks to the George Williams Hooper Foundation, University of California San Francisco Medical Center, for help with the preparation of this manuscript.

REFERENCES

Behnke, A. R., R. M. Thompson, and E. P. Motley (1935). *Am. J. Physiol.* **112**, 554.

Beurskens, G., G. A. Jeffrey, and R. K. McMullan (1963). *J. Chem. Phys.* **39**, 3311.

Bonamico, M., G. A. Jeffrey, and R. K. McMullan (1962). *J. Chem. Phys.* **37**, 2219.

Cherkin, A., and J. F. Catchpool (1964). *Science* **144**, 1460.

Cullen, S. C., and E. G. Gross (1951). *Science*, **113**, 580.

Davy, H. (1799). Cited by J. Davy (1836), in *Memoirs of the Life of Sir Humphry Davy, Bart.* London: Longman, Rees, Orme, Brown, Green and Longman.

Faraday, M. (1818). *Quart. J. Sci. Arts* **4**, 158.

Feil, D., and G. A. Jeffrey (1961). *J. Chem. Phys.* **36**, 1863.

Ferguson, J. (1939). *Proc. Roy. Soc.* (London) B **127**, 387.

Glasel, J. A. (1966). *Proc. Nat. Acad. Sci. U.S.* **55**, 479.

Glew, D. N. (1962). *Nature* (London) **195**, 698.

Huang, C. P., O. Fennema, and W. D. Powrie (1965). *Cryobiology* **2**, 109.

Jeffrey, G. A., and T. C. W. Mak (1965). *Science* **149**, 178.

Jeffrey, G. A., and R. K. McMullan (1962). *J. Chem. Phys.* **37**, 2231.

Kamb, B. (1965). *Science* **148**, 232.

Klotz, I. M. (1965). *Fed. Proc.* 24, part III, suppl. **15**, S-24.

McMullan, R. K., and G. A. Jeffrey (1959). *J. Chem. Phys.* **31**, 1231.

McMullan, R. K., M. Bonamico, and G. A. Jeffrey (1963). *J. Chem. Phys.* **39**, 3295.

Meyer, H. (1899). *Arch. Exp. Path. Pharmak.* **42**, 109.
Miller, S. L. (1961). *Proc. Nat. Acad. Sci. U.S.* **47**, 1515.
Pauling, L. (1933). *J. Am. Chem. Soc.* **55**, 1895.
Pauling, L. (1961). *Science* **134**, 15.
Pauling, L. (1964). *Anesth. Analg.* **43**, 1.
Pauling, L., and R. E. Marsh (1952). *Proc. Nat. Acad. Sci. U.S.* **38**, 112.
Quist, A. S., and H. S. Frank (1961). *J. Phys. Chem.* **65**, 650.
Robinson, A. B., K. F. Manly, M. P. Anthony, J. F. Catchpool, and L. Pauling (1965). *Science* **149**, 1255.
Van der Waals, J. H., and J. C. Platteeuv (1958). *Mol. Phys.* **1**, 91.
Von Stackelberg, M., et al. (1947). *Fortschr. Mineral.* **26**, 122.

LUDVIK BASS
WALTER J. MOORE
Department of Mathematics
University of Queensland
Brisbane, Australia

A Model of Nervous Excitation Based on the Wien Dissociation Effect

In 1850 Helmholtz successfully completed an experiment then generally regarded as impossible: he measured the conduction velocity in frog nerve, finding it to be about 30 m/sec. It is surprising to realize that it was only sixteen years later that Bernstein began his researches on the origin of the nerve impulse. This was well before the advent of the ionization theory of Arrhenius in 1883 and the brilliant advances in electrochemistry that marked the last years of the nineteenth century. Nernst's paper on electrochemical cells appeared in 1888 and Planck's beautiful analysis of the electrodiffusion problem in 1890. In 1902 Bernstein published a definitive statement of his membrane theory of the nervous impulse, based on the depolarization of an electrical potential across a membrane selectively permeable to K^+ ions, but in the same year Overton showed that Na^+ ions played an essential role in the excitation of the action potential. The effect by which a depolarization of about 20 mV from the resting potential of -70 mV initiates the radical changes in membrane permeability responsible for the action potential is one of the most remarkable of all natural phenomena. The elucidation of the molecular and ionic mechanism of this effect may well be the most exciting problem that faces biological electrochemistry today.

During the past fifty years advances in our understanding of nerve conduction have been closely coupled with the invention of new experimental techniques. Much important work arose from the discovery of Young (1936) that certain cephalopods possess giant unmyelinated axons, with diameters as large as 1000 μ, compared to the usual 0.1 to 20 μ. Such preparations made it possible to introduce long potential electrodes into the axoplasm from one end of the axon. When the necessary squids became available at Plymouth (Hodgkin and Huxley, 1939) and Woods Hole (Curtis and Cole, 1940) it was found that the action potential across the membrane was not limited, as Bernstein had suggested, to a depolarization pulse to zero, but continued with an overshoot of as much as $+50$ mV.

In 1949 Cole invented the voltage-clamp method. A metal wire was inserted axially through the axon and the membrane potential was held fixed by an

Walter J. Moore is Research Professor, Australian-American Educational Foundation. Permanent address: Chemical Laboratory, Indiana University, Bloomington.

electronic potentiostat. With this device it was possible to measure precisely the electrical characteristics of axons. In the hands of Hodgkin and Huxley, the new technique led to a series of important papers, culminating in 1952 in the masterly "A Quantitative Description of Membrane Current and its Application to Conduction and Excitation in Nerve." This paper marked the essential completion of a first stage in our understanding of nerve conduction, the provision of a phenomenological theory adequate to describe electrical properties measured in voltage-clamp experiments as well as propagation of the action potential in the natural axon.

In the present paper we first give a concise summary of the Hodgkin-Huxley theory in a form relevant to the subsequent discussion. In the second section we consider the clockwork reliability of nervous excitation, which imposes an important condition on any microscopic model of the process. In the third section we develop the idea that local alkalosis due to the Wien effect on acid dissociation is the universal physicochemical link between electrical depolarization and the chemical or conformational change leading to action potential. In the final section we consider several molecular and ionic mechanisms through which local alkalosis may control membrane permeability.

THE PHENOMENOLOGICAL FRAMEWORK

The macroscopic description of the nerve impulse is greatly simplified by the relative unimportance of magnetic effects. The velocity of propagation of the standard impulse is determined by the rate of release of stored-up electrochemical energy, not (as in ordinary electromagnetic propagation) by the interplay of electric and magnetic field energy. In the observed range (Hodgkin, 1964) of electric current density (mA/cm^2) and its time-change (in msec) magnetic effects appear to play no appreciable role in axonal transmission. The time-change of the current density is, on the other hand, rapid enough to eliminate concentration polarization in the axoplasm, so that Ohm's law is valid for axial currents. The basic partial differential equation connecting current density with membrane potential arises then from the combination of charge conservation with Ohm's law.

Consider a long cylindrical axon of radius a and resistivity R of axoplasm. In cylindrical coordinates (r, z) it occupies the domain $r \leq a$, with the outer surface of the thin axonal membrane at $r = a$. If the resistance of the external medium is neglected, the electrical potential may be put to zero for $r > a$ and, for sufficiently low R, independent of $r < a$ except within the membrane. Thus $V(z, t)$ is the potential difference across the membrane. The outflow of total (displacement and conduction) current from any segment of the axon of thickness Δz must vanish: $\pi a^2 \, \Delta i_z + 2\pi a i_r \, \Delta z = 0$, where i_z, i_r are components of the total current density. Thus

$$\frac{a}{2}\frac{\partial i_z}{\partial z} + i_r = 0 \tag{1}$$

We can eliminate i_z by using Ohm's law, $Ri_z = -\partial V/\partial z$:

$$\frac{a}{2R}\frac{\partial^2 V}{\partial z^2} = i_r \tag{2}$$

Here i_r consists of the displacement (charging) current density $C(\partial V/\partial t)$ (C being the membrane capacitance per unit area), and a conduction c.d. carried by ions, predominantly Na^+ and K^+, to *both* of which the membrane is permeable.

From the composition of the solutions at $r < a$, $r > a$, we can compute the Nernst equilibrium potentials V_{Na}, V_K for sodium and potassium alone: $+55$ mV and -75 mV, respectively, for squid giant axon. If $V = V_K$, potassium would not contribute to the conduction current despite the permeability of the membrane, and the same consideration applies to sodium. This may be expressed by writing

$$i_r = C\frac{\partial V}{\partial t} + g_{Na}(V - V_{Na}) + g_K(V - V_K) \tag{3}$$

where the positive generalized conductances g_{Na}, g_K are bounded. A less important leakage term $g_L(V - V_L)$ may be added to account for the small permeability of the membrane to other ionic species. The nerve impulse arises from the strong dependence of the g's on V. Consider first the formal solution

$$V = \frac{g_{Na}V_{Na} + g_K V_K + i_r - C\dfrac{\partial V}{\partial t}}{g_{Na} + g_K}$$

In the resting state, i_r and $C(\partial V/\partial t)$ vanish, and it is found empirically that $g_K \gg g_{Na}$. Hence, regardless of the individual values of the g's, $V \approx V_K$ or, more exactly, $V \gtrsim V_K$, since V_{Na} is positive. Next, a metal wire may be inserted into the axon, rendering $\partial V/\partial z = 0$ and thus, except during stimulation by an external e.m.f., $i_r = 0$ from (2). At the observed peak of the (nonpropagating) membrane action potential after stimulation, $\partial V/\partial t = 0$ instantaneously. It is then found empirically that $g_{Na} \gg g_K$ at that time, so that $V \lesssim V_{Na}$ at the peak of the membrane action potential. A similar analysis applies to a sufficiently small patch of the membrane without the internal metal wire. Thus $V_K \leq V \leq V_{Na}$. Higher approximations to the resting potential depend on the essentially nonequilibrium character of the potential difference and on its spatial distribution within the membrane.

The deeper reason for the splitting-off of *linear* factors in (3) is in the observation that, on a *sudden* change of V induced from an external source of e.m.f., the g's change gradually. More generally, although V and its derivatives, appearing explicitly in (2) and (3), are taken at any one time t, the g's at that time depend on the *history* of V. This dependence expresses the *time-delay* involved in activating, by changes in potential, the unknown electrochemical mechanism of the (absolute and relative) permeability change of the membrane:

$$g_i = g_i[t, V(t' \leq t)], \qquad i = Na^+, K^+ \tag{4}$$

Equations (2) through (4) form a natural framework for a theory of the nerve impulse. It remains to construct a mathematical description of (4) derivable from microscopic foundations.

Hodgkin and Huxley (1952) chose to construct the g's as powers and products of quantities which satisfy *first-order* differential equations with potential-dependent coefficients. For example,

$$g_K = \text{const} \cdot n^4 \tag{5}$$

$$\frac{dn}{dt} = \alpha(V)(1 - n) - \beta(V)n \tag{6}$$

where α and β depend on time through V:

$$\alpha_t \equiv \alpha[V(t)], \qquad \beta_t \equiv \beta[V(t)]$$

The solution of (6) for initial values n_0, V_0 is

$$n = n_0 \exp\left[-\int_0^t (\alpha_{t'} + \beta_{t'})dt'\right] + \int_0^t \alpha_{t'} \exp\left[-\int_{t'}^t (\alpha_{t''} + \beta_{t''})dt''\right] dt' \tag{6a}$$

expressing the dependence of n (and hence of g_K) on the history of $V(t' \leq t)$ involved in the time integrals of $\alpha + \beta$.

The dependence of α and β on instantaneous values of V is found empirically by inserting a wire into the axon, connecting it to a potentiostat and then holding V at an imposed constant value (voltage-clamp technique). In this case $i_r \neq 0$ continues to flow from the external source of e.m.f., α and β become constant, and n is reduced to

$$n = n_0\, e^{-(\alpha+\beta)t} + \frac{\alpha}{\alpha + \beta}\left(1 - e^{-(\alpha+\beta)t}\right)$$

for each V = const. Since ratios of n^4 at different times are observable, α and β can be deduced. The results of a series of such experiments at different constant V are summarized in the equations

$$\alpha = 0.01\, \frac{\psi + 10}{e^{(\psi+10)/10} - 1}, \qquad \beta = 0.125\, e^{\psi/80} \tag{7}$$

where $\psi = V - V_{\text{rest}}$ is the deviation of the potential from its resting value (measured in mV). A similar though less simple analysis of g_{Na} yields similar empirical functions. The functions (7) are then used in (6a) and (5) for the natural axon (without potentiostat or internal wire) to give the desired relation (4) for g_K, and the corresponding functions for sodium yield g_{Na}. Thus the mathematical specification of the membrane excitation is completed. This involves at least six empirical functions, such as appear in (7), and some twenty-one empirical parameters. The specifications (4) may then be viewed as an admirable summary of the voltage-clamp measurements in analytical form, or else interpreted as the kinetics of several charged particles whose simultaneous positions, influenced by the membrane potential, control the permeability of the membrane.

Independently of the plausibility of their physical interpretation, the analytically summarized data, obtained exclusively from local and nonpropagating excitations, suffice to describe correctly the *propagation* of the nerve impulse. Use is made only of the further basic observation that the time course and propagation velocity θ of the nerve impulse is always the same on repetition or propagation, regardless of the considerable dissipation of energy in the resisting medium:

$$V(z, t) = V(z \pm \theta t) \tag{8}$$

Thus we may insist on replacing $\partial^2 V/\partial z^2$ by $(1/\theta^2)(\partial^2 V/\partial t^2)$ in (2), taken with (3) and (4), provided that θ is the actual velocity of the physiological impulse, whose amplitude is continuously maintained by the release of stored electrochemical energy through the variation of the g's. If we now choose the "wrong" θ, the amplification is lost and no *finite* wave form can be maintained unchanged; accordingly, the solutions $V(t)$ are found to grow out of all bounds. Only one velocity θ gives a finite (and fixed) wave form, and both that velocity and that wave form are in good agreement with observations. It is to be noted that the large number of empirical functions and parameters used to summarize local *nonpropagating* membrane excitations is in no way extended to bring about agreement between calculated and observed *propagating* excitations. It follows that any theory which can reproduce the voltage-clamp data will also yield the correct propagation features of the nerve impulse.

CLOCKWORK REGULARITY OF THE ACTION POTENTIAL

In Chapter VII of his quintessential book *What Is Life?* Schrödinger analyzed the way in which living cells, subject to the laws of statistical thermodynamics, can nevertheless operate with the clockwork regularity characteristic of dynamical systems. He pointed out that "clockworks made of real physical matter (in contrast to imagination) are not true 'clock-works.' The element of chance may be more or less reduced. The likelihood of the clock suddenly going altogether wrong may be infinitesimal, but it always remains in the background." He went on to show that a system can display clockwork regularity when the molecular disorder due to thermal fluctuations has no appreciable effect on its operation. Not the least remarkable feature of the action potential is the clockwork regularity with which it fires consequent upon a depolarization pulse of amplitude greater than the threshhold value. This regularity becomes highly significant as soon as we realize that most of the direct physical effects of a depolarization pulse can only be due to energy changes of less than $\mathbf{k}T$ per particle (about 1/40 eV at room temperature). We must always demand of any proposed mechanism, therefore, not only that it explain the triggering of the action potential, but also that it does not permit random firing of the action potential as a result of thermal fluctuations.

Let us first outline the theory (Landau and Lifshitz, 1959) of such spontaneous (thermal) firings. Consider a small subsystem of a large system of fixed energy—in our case a small element of the entire *resting* membrane. The probability w' of a fluctuation of the large system to a nonequilibrium state of entropy S' is

$$\frac{w'}{w} = e^{(S'-S)/\mathbf{k}} \tag{9}$$

where w is the probability and S the entropy of the equilibrium state. The fixed energy U of the large system is unchanged by the fluctuation, $U = U'$, but its entropy is modified ($S \neq S'$). At constant temperature,

$$S' - S = -\frac{1}{T}(U' - TS' - U + TS) = -\frac{F' - F}{T} = -\frac{A_{\min}}{T}$$

where $A_{\min}$, the free energy increment, is the minimum external work that would need to be done on the system to produce the same fluctuation reversibly. This is the result given by Einstein (1910). In accord with the ergodic hypothesis, the times t, t' spent in states of probabilities w, w' are related by

$$\frac{t'}{t} = \frac{w'}{w} = e^{-A_{\min}/\mathbf{k}T} \tag{10}$$

If a fluctuation lasts a time τ, then $t' = n\tau$ gives $t'/t = n\tau/t = \nu\tau$, where $\nu = n/t$ is the frequency of the fluctuation, and we have

$$\nu = \frac{1}{\tau}\, e^{-A_{\min}/\mathbf{k}T} \tag{11}$$

The duration of the fluctuation which would bring about an action potential will not be longer than the rise-time of the latter: $\tau \leq 1$ msec. We may use (11) to test various microscopic models of initiation of the action potential by requiring that the computed ν should not be noticeable, in accordance with the observed clockwork regularity of the actual firing mechanism.

As an example of this kind of calculation, consider the effect of depolarization on dipoles, which may be capable of acting as gates in channels for ion transport (Goldman, 1964). The polarization due to N dipoles of electric moment μ in a field E is given by the Langevin-Debye theory as

$$P = N\bar{\mu} \approx \frac{N\mu^2 E}{3\mathbf{k}T}$$

if $\mu E \ll \mathbf{k}T$. This inequality is satisfied in the membrane ($E \approx 10^5$ V/cm) even for dipole moments $\mu = 10\mu_{\mathrm{H_2O}} = 18$ Debyes, which we shall assume as the upper limit for dipoles with a rotational degree of freedom in the membrane. The minimum work corresponding to depolarization by ΔE is

$$A_{\min} = \int_{\Delta E} E\,dP = \frac{N\mu^2}{6\mathbf{k}T}\,\Delta(E^2) \approx \frac{N\mu^2}{3\mathbf{k}T}\,E\,\Delta E \tag{12}$$

This expression can also be obtained from the change of free energy expressed in terms of the partition function of the dipoles. It remains to estimate the number N of dipoles (or channels, each controlled by one dipole) which, when subjected to critical depolarization, suffice to initiate an action potential. Fortunately, myelinated nerve fibers provide us with a direct experimental measure of a small area sufficient for excitation: the area of the node of Ranvier A_R is about 2.2×10^{-7} cm^2 in *frog* (Hodgkin, 1964) and saltatory nerve conduction can be initiated at *any* node, so that the frequency (11) is to be multiplied by the number of nodes on the nerve.

If A_c is the effective area of a channel, the peak inward flux of Na^+ ions at the node is at least

$$jA_R = \mathbf{e}\, ND\, [\partial c/\partial x]_{\mathrm{Av}}\, A_c \tag{13}$$

The observed peak inward current density is 20 mA/cm^2; A_c can be estimated from the value 4.5 Å for the radius of a channel obtained from transport of nonelectrolytes across nerve membranes (Villegas, Caputo, and Villegas, 1962), and the concentration gradient of Na^+ is known. If we take $D = 1.25 \times 10^{-5}$ cm^2/sec (for the free ion in an open channel), from (13) we obtain $N = 1.96 \times 10^3$. For $E = 70$ mV/70 Å $= 10^5$ V/cm and $\Delta E = 20$ mV/70 Å $\approx 3 \times 10^4$ V/cm, from (11) and (12) we obtain

$$\nu \geq 10^3\, e^{-4.2} = 15\ \mathrm{sec}^{-1}$$

It would thus appear that the dipole orientation mechanism could hardly provide the requisite clockwork operation of the action potential.

On the other hand, the node of Ranvier is too large to be brought to bear on a mechanism in which gates for ions are operated by the transfer of ions across the *whole* membrane width δ; an analogous estimate of $A_{\min}$ for this case* gives $\nu \geq 10^3 e^{-91}$ (note that $e\delta$ corresponds to 336 Debyes). However, smaller areas capable of initiating action potentials might be observed in the future (for example, by using a capillary electrode penetrating a myelin sheath), and the number of such areas on an *unmyelinated* axon will be very large.

THE ROLE OF THE WIEN DISSOCIATION EFFECT

Action potentials are initiated in the natural axon by critical depolarizations of some 20 mV, corresponding to a reduction of the resting field of about 10^5 V/cm by a factor of 5/7. Through some extraordinarily nonlinear link this causes a selective increase in the membrane permeability by a factor of 40. It is therefore necessary to examine calculable effects of the critical depolarization in an attempt to find some essential physicochemical parameter of the state of the membrane that possesses the following properties. (a) The critical depolarization is calculated to cause in this membrane parameter a critical increment that is known to initiate an action potential if brought about by nonelectrical means. (b) Within its critical increment, this parameter is stable with respect to thermal fluctuations in membrane patches whose areas are small compared to that of a node of Ranvier.

We shall show that such a parameter of the membrane is its pH (a measure of the concentration of free protons), and the corresponding effect of depolarization is the reduction of the Wien effect on the dissociation of the weak-acid-base system controlling the pH (Bass and Moore, 1966). The observed nervous activity brought about by equivalent nonelectrical changes of pH is the well-known physiological condition of alkalosis. The observed critical increment of pH in the axoplasm (as determined by experiments with perfused axons) is 0.1 to 0.2 pH units to the positive of its normal value (pH 7.2–7.4). Such a ΔpH causes spontaneous repetitive action potentials (Tasaki, Singer, and Takenaka, 1965).

We assume that the proton concentration, $c = 10^{-p\mathrm{H}}$ mole/liter, in the axolemma is related to its counterpart $\bar{c}$ in the axoplasm through a fixed distribution coefficient,

$$c = K_d \bar{c} \tag{14}$$

Hence, taking logarithms of both sides of (14), we see that, at equilibrium, increments of pH are the same in the membrane and in the axoplasm. The acid-base buffer systems in the membrane need not be the same as those in the axoplasm, since identical effects of internal pH can be achieved by using different buffers as perfusion media. In such experiments, the axoplasm is gently

*Suppose that ions have only two energy states, separated by an energy gap $\mathbf{e}\psi = 70$ mV. Then

$$F = -N\mathbf{k}T \ln(1 + e^{-\mathbf{e}\psi/\mathbf{k}T}), \qquad A_{\min} = \Delta F \approx \frac{N\mathbf{e}\Delta\psi}{e^{\mathbf{e}\psi/\mathbf{k}T} + 1}$$

so that $A_{\min}/\mathbf{k}T \approx 91$.

extruded and an artificial internal medium is continuously perfused through the axon.

We need to show that a critical membrane depolarization can cause a critical positive increment $\Delta p\mathrm{H}$ in the membrane, and also that the time course of $\Delta p\mathrm{H}$ is consistent with data for the action potential:

$$c = c[V(t' \leq t), t] \tag{15}$$

We leave aside at present the second part of the problem, the effect of $\Delta p\mathrm{H}$ on permeability, which would complete the theory:

$$g_i = g_i[c(t' \leq t), t], \qquad i = \mathrm{Na}^+, \mathrm{K}^+ \tag{16}$$

Let the membrane acid-base system (for example, the substituted phosphoric acids of the phospholipids) consist of a weak acid A of concentration c_A and its conjugate base B of concentration c_B:

$$A \underset{k}{\overset{kK}{\rightleftarrows}} B + \mathrm{H}^+$$
$$c_A + c_B = \mathrm{const} = M \tag{17}$$

Here k is the rate constant for association and K is the acid dissociation constant. At equilibrium $c = Kc_A/c_B$. The dielectric constant ϵ of the medium, from capacitance measurements, is between 5.7 and 8, corresponding to capacitor spacings of 50 Å and 70 Å, respectively. In a medium of such low ϵ even an acid that would be moderately strong in aqueous solution ($K \sim 10^{-3}$) will necessarily become a weak acid ($K \sim 10^{-9}$) (Gurney, 1953). The low dielectric constant also indicates that relatively few free water molecules can exist in the membrane as compared to the surrounding media. Therefore H^+ need not be coupled to OH^- as in an aqueous solution, and the Grotthuss mechanism for proton conduction will probably be lost within the membrane.

The theory for the effect of an electric field E on the dissociation constant of a weak electrolyte (Wien dissociation effect) was developed in 1934 by Onsager. For the case of a strong field, his relation for a 1:1 electrolyte is

$$\frac{K(E)}{K(0)} \approx \left(\frac{2}{\pi}\right)^{1/2} (8b)^{-3/4} \exp{(8b)^{1/2}}, \qquad b = \frac{\mathbf{e}^3|E|}{2\epsilon \mathbf{k}^2 T^2} \tag{18}$$

Here $K(E)$ is the dissociation constant in the presence of the field E, and $K(0)$ is the constant when $E = 0$. For the case of polyvalent ions, $K(E)$ is a somewhat more complicated function of E. In tests of this equation with various weak electrolytes, experimental agreements have been excellent.

We may use (18) to calculate the $\Delta p\mathrm{H}$ caused by a depolarization from E to E' so rapid that c_A/c_B remains unchanged:

$$\Delta p\mathrm{H} = -\log K(E') + \log K(E) = \log \frac{K(E)/K(0)}{K(E')/K(0)} \tag{19}$$

Since $E' < E$ in absolute value, $\Delta p\mathrm{H} > 0$. The results of calculations of $\Delta p\mathrm{H}$ on critical depolarization for a 50 Å and a 70 Å membrane and for 1:1 and 1:2 electrolytes (the latter for the case of the double charge residing on a relatively immobile base) are summarized in Table 1. The increments of $p\mathrm{H}$ are seen to be within the critical range of $\Delta p\mathrm{H}$ observed to initiate action potentials in perfusion experiments.

TABLE 1.
Computed Wien dissociation factors in axonal membranes and resulting changes in pH *on depolarization.*

Factor		Membrane and electrolyte			
		50 Å, $\epsilon = 5.7$		70 Å, $\epsilon = 8.0$	
		1:1	1:2	1:1	1:2
$K(E)/K(0)$	(70 mV)	8.9	37.3	3.4	9.2
$K(E)/K(0)$	(50 mV)	5.5	17.3	2.5	5.7
ΔpH		0.21	0.33	0.13	0.21

We now consider whether the kinetics of the changes in pH are consistent with the time course of action potentials. For the external buffer,

$$\bar{A} \underset{\bar{k}}{\overset{\bar{k}\bar{K}}{\rightleftarrows}} \bar{B} + \mathrm{H}^+ \tag{20}$$

with equilibrium concentrations $\bar{c} = \bar{K}c_{\bar{A}}/c_{\bar{B}}$. There is also diffusional proton transfer into and out of the membrane,

$$\bar{c} \underset{1/\tau_{\mathrm{diff}}}{\overset{K_d/\tau_{\mathrm{diff}}}{\rightleftarrows}} c \tag{21}$$

with the equilibrium distribution (14) and the diffusional relaxation time $\tau_{\mathrm{diff}} \approx \delta^2/2D_{\mathrm{H}^+}$; with $D_{\mathrm{H}^+} \approx 2.5 \times 10^{-10}\ \mathrm{cm}^2/\mathrm{sec}$ we obtain $\tau_{\mathrm{diff}} \approx 1$ msec. The D of H^+ in the membrane has been taken to be the same as that of K^+ (Cole, 1965). Finally, the buffer character of the acid-base systems is expressed by

$$c_A,\ c_B,\ c_{\bar{A}},\ c_{\bar{B}} \gg c \tag{22}$$

It is clear from the discussion in the section on the clockwork regularity of the action potential that the number of buffer ions in the vicinity of even a single pore would be much too large for appreciable thermal fluctuations of pH.

From (17), (20), and (21), we obtain the kinetic equations

$$\frac{dc}{dt} = \frac{K_d\bar{c} - c}{\tau_{\mathrm{diff}}} + kKc_A - kc_Bc$$

$$\frac{dc_B}{dt} = kKc_A - kc_Bc, \qquad c_A + c_B = M$$

or equivalently,

$$\begin{aligned} \frac{d}{dt}(c - c_B) &= \frac{d}{dt}(c + c_A) = \frac{K_d\bar{c} - c}{\tau_{\mathrm{diff}}} \\ \frac{dc_B}{dt} &= kK(E)(M - c_B) - kc_Bc \end{aligned} \tag{23}$$

determining $c(E, t)$ and, if $V \approx -E\delta$ is used, $c(V, t)$; c is coupled to E through

(18). In equilibrium the proton concentration c is subject to *two* relations:

$$c = K_d\bar{c} = K\frac{c_A}{c_B} \tag{24}$$

If $\bar{c} = \bar{K}c_{\bar{A}}/c_{\bar{B}}$ is changed by perfusion, both c and c_A/c_B are adjusted, the membrane acid-base balance being modified by the large external buffer reservoir, as indicated by (14). On the other hand, depolarization modifies $K(E)$. If the resting field E is suddenly reduced to a smaller constant field E' (voltage clamp), c will be reduced and protons will diffuse into the membrane. Since c is not directly observable, and its functional connection (16) with the observable g remains to be found, it will suffice to deduce from (23) and (18) the main features of the electrically induced alkalosis.

The rate constant for association (Onsager, 1934) is

$$k = 4\pi D_{H^+}\frac{\mathbf{e}^2}{\epsilon\mathbf{k}T}$$

and the "chemical" relaxation time of a new association state of the relatively few H^+ ions with the numerous base ions, according to Equations (22) and (23), is $\tau_{chem} \approx (kc_B)^{-1}$:

$$\tau_{chem} = \frac{\epsilon\mathbf{k}T}{4\pi D_{H^+}\mathbf{e}^2 c_B} \approx \frac{0.73}{c_B}\,10^{-6}\text{ sec} \tag{25}$$

for $\epsilon = 8$ at room temperature and c_B in mol/liter. Since the phospholipid acid-base ions are part of the building material of the membrane itself, c_B can hardly be much less than of the order of unity. Thus $\tau_{chem} \ll \tau_{diff}$ and, for a time of the order of τ_{chem} following the depolarization, the membrane is effectively cut off from the diffusional proton transfer. Thus an internal chemical equilibrium $c - c_B \approx \text{const} = c(0) - c_B(0)$, $dc_B/dt \approx 0$, is effectively reached before the outside buffer system can react. From (23)

$$c \approx K(E')\frac{M - c_B}{c_B} = K(E')\frac{c_A(0) + c(0) - c}{c_B(0) - c(0) + c} \approx K(E')\frac{c_A(0)}{c_B(0)} \tag{26}$$

the last equality arising from (22). The process (23) is thus split into an association step so rapid that protons (bound and free) are conserved in the membrane, and a subsequent slow step of proton diffusion, for which (26) serves as an initial condition. It should be noted, however, that τ_{chem} is merely a lower limit for the time rise of the g's, since the effect of ΔpH on the membrane may introduce a further time delay.

The diffusion step restores the original pH in the membrane within $\tau_{diff} \approx$ 1 msec. Alternatively, only the inner edge of the membrane is involved in the activation and inactivation of the membrane by pH changes (Bass and Moore, 1967). The diffusion relaxation time is then short as compared to the time taken (at constant proton flux) to override the membrane buffer in the vicinity of the inner membrane edge. Again, the effect of the pH restoration on the g's may be subject to some delay. If E' is held by a voltage clamp, the restoration of pH entails a shift of the internal buffer ratio c_A/c_B according to (24) with $K = K(E')$, whereas in the natural axon the internal buffer returns to its initial state. We therefore conjecture that although g_{Na} (which is rapidly reduced to initial value in *either* case) depends on pH alone, g_K depends also on the state of the buffer. If the field is reduced slowly (in terms of τ_{diff}), proton diffusion can keep up with the change of the equilibrium constant and only c_A/c_B is changed; this corresponds to the phenomenon of *accommodation*, the failure of a slowly rising depolarization to set up an action potential.

It is possible to construct a phenomenological form of the missing relation (16) somewhat along the lines of (5) and (6a), but with dependence on c and c_A/c_B replacing dependence on V. Agreement with voltage-clamp data would again require the introduction of purely empirical functions and parameters, which would not be in keeping with the microscopic level of the foregoing theory of the relation (15). We therefore defer the mathematical treatment of (16) until it can be given a quantitative physicochemical basis. However, even the present preliminary considerations suggest strongly that the link beween depolarization and permeability change is alkalosis, not in its pathological form, but as a transient and local condition of the normal axonal membrane when engaged in transmitting the nerve impulse.

The Wien coupling between depolarization and pH in thin membranes under a high field is likely to have other applications. For example, the enzyme acetylcholine esterase plays an important part in the polarized postsynaptic membranes of cholinergic synapses; the activity of the esterase depends strongly on pH (Nachmansohn, 1959), and the membrane is depolarized whenever the synapse fires. Depending upon the (unknown) pH in the membrane, depolarization must be equivalent to a substantial increase or decrease in the activity of esterase in the membrane.

The mechanism based on the Wien effect as well as the other microscopic models discussed above all presuppose a high electric field in the resting membrane. If action potentials could be obtained by the reduction of a *low* resting field, such models would be inadequate. The latter possibility appears to arise from recent perfusion experiments in squid giant axons (Narahashi, 1963). A range of low resting membrane potential differences (between -25 mV and 0) was obtained by reducing the internal potassium concentration [K^+], in which action potentials were abolished when [K^+] *alone* was reduced but were still obtained when the internal ionic strength was reduced proportionately.

At least two models can account for this effect of low internal ionic strength while preserving a high field in the membrane. Chandler, Hodgkin, and Meves (1965) postulate excess surface charges adsorbed at the inner surface of the membrane. The theoretical consequences of the model are satisfactory, but the postulated density of adsorbed charges and the resulting dependence of the membrane capacitance on ionic strength remain to be detected experimentally. Another model (Bass and Moore, 1967) is based only on Planck's electrodiffusion field. That field is uniform across the membrane for equal internal and external ionic strengths (as in the natural axon), but, when the internal ionic strength is reduced, the field becomes so distributed that a *locally* high field strength is maintained within the membrane in regions of low ionic strength. This model would be valid if the (as yet unmeasured) concentrations of ions *in* the membrane were high enough to render the Debye length short as compared to the membrane thickness (Bass, 1964). Some combination of the two models is likely to account fully for the observed perfusion phenomena.

CONJECTURES ON MECHANISMS OF NERVE EXCITATION BY ALKALOSIS

One of the basic facts of physiology is that alkalosis greatly increases the excitability of nerve cells. For instance, hyperventilation may elevate the blood *p*H only momentarily, but in a person subject to epilepsy this slight alkalosis suffices usually to initiate a seizure. On the other hand, acidosis depresses neuronal excitability, even leading eventually to coma, as in diabetes or uremia. Ketogenic diets have had some success in treatment of epilepsy. Paralleling these well-established effects arising from changes in concentration of hydrogen ions are similar effects due to changes in concentration of calcium ions. If serum calcium in man is lowered by only about 30 percent from the normal 2.5 mM, tetany supervenes. Latent tetany can be maintained by keeping the lowered serum calcium just above the critical level. In a subject in this condition, hyperventilation causing a slight alkalosis will immediately induce tetanic seizures. In such alkalosis the measured serum calcium is not actually lowered. It is not certain, therefore, whether the alkalosis acts in parallel with a previous lowering of calcium, or whether the ionized calcium level is actually lowered even though the total calcium remains steady. An effective equilibrium constant such as the following has been suggested (Best and Taylor, 1961):

$$K = \frac{[Ca^{++}][HCO_3^-][HPO_4^{--}]}{[H^+]}$$

In this case alkalosis would decrease the Ca^{++} concentration.

Such physiological data lead us to consider whether alkalosis in the axolemma may exert an effect on membrane permeability by causing shifts in bound calcium ions. The affinity of anionic sites for protons is much greater than that for other univalent ions and indeed usually greater than that for bivalent ions such as Ca^{++} or Mg^{++}. The intense electrostatic field of the proton, due to its close interaction range, causes it to behave more like a multivalent cation in its affinity for anions. For example, a study of cation binding by rat-liver microsomes and erythrocyte ghosts (Carvalho, Sanui, and Pace, 1963) showed that their binding affinity for H^+ was some hundred times that for Ca^{++} or Mg^{++}, and these doubly charged ions were about one hundred times more tightly bound than Na^+ or K^+. The *p*K values of the binding sites are consistent with those of secondary phosphate and imidazolium groups. Eisenman (1960) has outlined quantitatively mechanisms of ionic specificity based on the relative binding of cations by fixed anions, showing that only strong acid groups have a field strength sufficient to bind alkali ions in preference to protons. Inside the axonal membrane, therefore, available protons would invariably displace Na^+ or K^+ ions from anionic sites.

Several theories of the permeability changes associated with the action potential have been based on displacements of Ca^{++} ions, but no quantitative conclusions have yet been drawn from such models. An example is the discussion by Lettvin and co-workers (1964), based on postulating two different channels for Na^+ and K^+, with Ca^{++} blocking the Na^+ channel internally and the K^+ channel at the outer end. Such models have the virtue of qualitative consistency with (a) the small observed Ca^{++} influx associated with the Na^+ influx during action potential, and (b) the increase in rate of recovery of the Na^+ inactivation with increase in external Ca^{++} concentration, following a pulse depolarization.

In our mechanism, the effect of alkalosis can occur either as a result of depolarization or as a result of perfusion of the axon with a more alkaline buffer. The depolarization produces alkalosis in the membrane by increasing the association of protons with anionic basic sites, whereas perfusion draws protons out of the membrane into the axoplasmic region. It is evident, therefore, that competition between H^+ and Ca^{++} for a *single* anionic site could not account for both cases. It would be necessary to assume at least two different sites, α and β, such that removal of H^+ from site β is followed by shift of Ca^{++} from the α to the β site, thus unblocking a channel. A further complexity in the ionic competition model is that the axon is sensitive to changes in Ca^{++} only at its outer surface. Effects due to pH changes have so far been observed relative only to the inner surface (Tasaki, Singer, and Takenaka, 1965). Such studies, however, suffice to indicate that the whole membrane must be involved in permeability changes, since the excitatory effect due to $\Delta p\text{H} > 0$ inside the axon can be counteracted by $\Delta[Ca^{++}] > 0$ outside. We can confidently predict that a decrease in excitability due to $\Delta p\text{H} < 0$ inside would be counteracted by lowering $[Ca^{++}]$ outside.

The second category of mechanisms for excitation via alkalosis includes effects of pH on conformations of molecules, especially proteins or protein segments, in the membrane. Such conformational changes with pH have been observed in certain proteins. For example, the solubility of paramyosin in aqueous media undergoes an abrupt change between pH 6 and 7 (Johnson, Kohn, and Szent-Gyorgi, 1959). At any given ionic strength the change may be complete within 0.3 pH units, the protein passing from virtual insolubility at pH 6.4 to complete solubility at p*H* 6.7. Presumably this is an example of a coil to helix transition effected by changes in net charges on side chains. Somewhat similar effects have been observed in synthetic polypeptides. For instance, poly-L-lysine was found to have a helical conformation at neutral pH but to undergo transition to a random coil on conversion to a more ionized form (Applequist and Doty, 1962). Such changes due to addition or subtraction of protons might also be effected by binding or release of Ca^{++} ions. The conformational changes of proteins in or near channels may act to clear or to block them.

At present we do not have sufficient data to decide between the two possible modes of action of H^+ and Ca^{++} that have been outlined: either (a) Ca^{++} and H^+ act sequentially, so that alkalosis changes the Ca^{++} binding on certain sites, or (b) Ca^{++} and H^+ act in parallel, each ion independently being capable of producing certain conformational changes in the axolemma.

REFERENCES

Applequist, J., and P. Doty (1962). *In* Stahmann, M. A., ed., *Polyamino Acids, Polypeptides and Proteins*, p. 161. Madison: Univ. of Wisconsin Press.

Arrhenius, S. (1887). *Z. physik. Chem.* **1**, 631.

Bass, L. (1964) *Trans. Faraday Soc.* **60**, 1914.

Bass, L., and W. J. Moore (1966). *Proc. Australian Physiol. Soc.*, 9th meeting.

Bass, L., and W. J. Moore (1967). *Nature* (London) **214**, 393.
Bernstein, J. (1866). *Arch. Anat. Physiol. u. wissensch. Med.*, 596.
Bernstein, J. (1902). *Pflüg. Arch. Ges. Physiol.* **92**, 521.
Best, C. H., and N. D. Taylor (1961). *Physiological Basis of Medical Practice*, p. 1052. Baltimore: Williams & Wilkins.
Carvalho, A. P., H. Sanui, and N. Pace (1963). *J. Cell Comp. Physiol.* **62**, 311.
Chandler, W. K., A. L. Hodgkin, and H. Meves (1965). *J. Physiol.* **180**, 821.
Cole, K. S. (1949). *Arch. Sci. Physiol.* **3**, 253.
Cole, K. S. (1965). *Physiol. Rev.* **45**, 340.
Curtis, H. J., and K. S. Cole (1940). *J. Cell. Comp. Physiol.* **15**, 147.
Einstein, A. (1910). *Ann. Physik* **33**, 1275.
Eisenman, G. (1961). *Membrane Transport and Metabolism*, p. 163. New York: Academic Press (Proc. Symposium Prague 1960).
Goldman, D. E. (1964). *Biophys. J.* **4**, 167.
Gurney, R. W. (1953). *Ionic Processes in Solution*, p. 64. New York: McGraw-Hill.
Helmholtz, H. von (1850). *Arch. Anat. Physiol.*, 277.
Hodgkin, A. L. (1964). *The Conduction of the Nervous Impulse*. Liverpool University Press.
Hodgkin, A. L., and A. F. Huxley (1939). *Nature* (London) **144**, 710.
Hodgkin, A. L., and A. F. Huxley (1952). *J. Physiol.* **117**, 500.
Johnson, W. H., J. S. Kohn, and A. F. Szent-Gyorgi (1959). *Science* **130**, 160.
Landau, L. D., and E. M. Lifshits (1959). *Statistical Physics*, p. 351. London: Pergamon Press.
Lettvin, J. Y., W. F. Pickard, W. S. McCulloch, and W. Pitts (1964). *Nature* (London) **202**, 1338.
Nachmansohn, D. (1959). *Chemical and Molecular Basis of Nerve Activity*, p. 112. New York: Academic Press.
Narahashi, T. (1963). *J. Physiol.* **169**, 91.
Nernst, W. (1888). *Z. physik. Chem.* **2**, 613.
Onsager, L. (1934). *J. Chem. Phys.* **2**, 599.
Overton, E. (1902). *Pffüg. Arch. Ges. Physiol.* **92**, 346.
Planck, M. (1890). *Ann. Phys. Chem.* **39**, 161.
Schrödinger, E. (1944). *What is Life*?, p. 84. Cambridge Univ. Press.
Tasaki, I., I. Singer, and T. Takenaka (1965). *J. Gen. Physiol.* **48**, 1095.
Villegas, R., C. Caputo, and L. Villegas (1962). *J. Gen. Physiol.* **46**, 245.
Young, J. Z. (1936). *Quart, J. Micr. Sci.* **78**, 367.

J. D. BERNAL
Birbeck College
University of London
London, England

The Pattern of Linus Pauling's Work in Relation to Molecular Biology

Pauling must be considered as one—in many ways the principal one—of the originators of modern molecular biology. This stems from a number of different aspects of his work, one of which, the structure and transformation of DNA, turns out to be of crucial importance to the modern understanding of molecular structures of living organisms. To see what Pauling has been doing, it is necessary to go right back into the history of his work, and this brings out a leading feature of his scientific mentality. I would call it a combination of a deep appreciation of quantum quantitative chemistry and a geometrical understanding of crystal structures.

Pauling was the man who more than anyone else spread the knowledge of quantum theory in the fields of classical chemistry. Some of the theories he embodied in his book *The Nature of the Chemical Bond* have been criticized on account of their lack of mathematical rigor but they have had an enormous effect both on chemists and crystallographers because of their comprehensibility. Pauling could afford to leave out some of the niceties of quantum mechanics because he had already proved himself in the early years to have such an ingrown sense of the realities of the quantum as applied to chemistry that he did not need to think about detailed derivations but thought automatically in quantum terms. He occupied himself very early with the boundaries of the subject, replacing to a large extent his former concern with metallurgy and mineralogy with one of biology and especially with the nature and role of protein structures, as is shown in various essays in this volume.

His greatest triumph, however, was the elucidation of the fundamental nature of the secondary structure of proteins, the helical peptide-linked polymer molecule. That the protein structure was the key to many of the problems of biology was already sensed in the nineteenth century. In fact, it occurs in Engel's definition of life as "the mode of existence of proteins" (albumins, egg white). This was to be verified by the biochemical realization that the enzymes that make possible nearly all reactions in living systems are themselves protein molecules.

However, it was only in the twentieth century that serious study of that structure could be undertaken. This was achieved by a mixture of chemistry and crystallography. The first stages were necessarily highly inexact, stemming from Fischer's illuminating and correct generalization of proteins as polypeptides, in turn reducing the problem to the structure of the constituent amino acids and their mode of attachment to each other.

Proteins were known in two major states: as natural fibers responsible for the structure of wool, silk, muscle, and connective tissue, and also as soluble molecules occurring either as natural crystals in certain living structures (for instance, in the Bence Jones protein) or relatively easily prepared as the hemoglobin crystals produced in a variety of forms in the preparations of blood. Both these forms were, therefore, open to attack by X-ray methods. The fibrous proteins, particularly wool, were studied by X-rays by Astbury as far back as 1926. From wool and other elastic tissues he produced rather vague fiber diagrams, which he proceeded to analyze with great ingenuity but without full accuracy. He was already able to establish, however, that these proteins existed in two forms, α and β. The α form, having a smaller unit fiber length, could be transformed reversibly by stretching into the β, which was longer and more regular and proved to be identical with the structure of silk.

After a number of fruitless attempts I was, myself, in 1934 (Bernal, 1934) able to establish the regular crystal nature of the soluble proteins and thereby set going the study of their detailed structure by classical X-ray crystallographic methods, first successfully achieved by Kendrew in 1962 in the structure of myoglobin. Pauling had a clear idea of the thermodynamic nature of protein structure and understood the phenomenon of denaturation, the turning of a regular pile into an irregular tangled skein that would, of its nature, be insoluble. This was the basis of his work on denaturation and also on the antigen-antibody complexes of immunology (Pauling, 1940) which could be taken as a higher degree of compounding than that of protein molecules such as are found naturally to occur in the molecule of hemoglobin, itself really a double pair of single-chain protein molecules (Perutz et al., 1960). Pauling had shown the relation of this structure to the kinetics of oxygen control by hemoglobin, which has to be very delicately balanced so that oxygen can be taken up from the atmosphere by a strong combination and released by an even stronger combination with the tissue cells (Pauling, 1949; St. George and Pauling, 1951; Lein and Pauling, 1956). The same idea proved to be the clue to the action of enzymes, one in which the substrate molecule is fitted into a cleft of the enzyme and was first shown to be so by Phillips and co-workers (Black et al., 1962) in the case of lysozyme. It now appears to hold also for ribonuclease (Avey et al., 1967; Kartha et al., 1967).

Even one of the clues to the specific nature of biochemistry is really due to Pauling. The so-called energy-rich phosphate bond, adenosine triphosphate, active in enzymes, is itself an example of the energy level change due to a slightly displaced oxygen atom, which changes its state according to its position in the crystal cell or in the solution.

At about the same time an advance was made on the physical side by Svedberg in Uppsala (Svedberg and Pedersen, 1940) in the development of the ultracentrifuge. This enabled him to measure the sedimentation properties of single protein molecules in solution. His work, in fact, implied that a pure protein solution is monodispersed and the molecules are all of predominantly the same molecular weight, and therefore the protein itself is a specific type of molecule, a fact that had long been suspected and was really implied by the existence of crystals. It is true that these crystals were of a peculiar kind, their size and angles depending on the constitution of the liquid from which they were crystallized and, consequently, their true nature remained in doubt until X-ray studies could be carried out on them. Svedberg thus arrived at a generalization, which was at first taken too literally, that the molecular weight of all proteins are multiples of 17,000 and that therefore they had a certain structure in common. This was altogether too simple to be true. In fact, there does seem to be a rough equality of chain length of the peptides of similar proteins and, consequently, of the submultiples of their combined formation. For instance, the molecular weight of hemoglobin is 65,000 and that of myoglobin corresponds to the second one of the four heme-containing units of hemoglobin, which is 17,000. Lactoglobin, which has a molecular weight of 35,000, has proved to consist of two equivalent halves of 17,500 each. But it is altogether too much to expect this to hold for all proteins. Indeed, we know many exceptions.

During this time the chemical approach was beginning to yield results. The analysis of proteins at first required a large amount of material and could only give the relative proportions of different amino acids, with no hint of their order of arrangement. This was made possible by the work of Martin and Synge (1941) who introduced the technique of paper chromatographic analysis, enabling a very much smaller quantity of substance to be used. In the hands of Sanger (Sanger and Tuppy, 1951; Sanger and Thompson, 1953) the technique established for the first time the order of amino acids in a protein chain, the so-called primary or chemically established structures.

By 1940 it was clear that a successful attack on the complete protein structure could be made, but there were still many difficulties. Two modes of attack suggested themselves: the first was a straightforward X-ray crystallographic study of crystalline protein, using all the techniques of an advanced crystal analysis. Computers were not available for this until much later, in the mid-1950's. The second was a model building method based on an exact knowledge of the structure of amino acids and smaller peptides themselves and an attempt to build up the protein a priori and then check the structure by X-ray methods. I remember very well discussing the problem with Pauling just before the war. He was in favor of the second method, which I thought indirect and liable to take a very long time. Nevertheless, it was Pauling's ideas that were to have a decisive effect on the result. The series of structures of the amino acids came from R. B. Corey and his colleagues at the Gates and Crellin Laboratories at the California Institute of Technology (Donohue, 1950; Shoemaker et al., 1950, 1953; Donohue and Trueblood, 1952). These workers were able to establish far more accurately than I had been able to do (Bernal, 1931) the structure of these molecules (*see* Hodgkin and Riley in this volume). The key to the configuration of the polypeptide structure lay in the X-ray analysis of the di- and tripeptides. Here Pauling's chemical intuition enabled him to introduce the decisive formulations that were to help to fix the configuration. The elementary chemical view was that the configuration about the bond between the first carbon atom and the succeeding amide nitrogen atom was to be a simple, single bond capable of any orientation (see Figure 1). Pauling, however, pointed out that there would be some resonance between the planar carboxy group and the carbon-nitrogen bond, resulting in a shortening of the carbon-nitrogen bond and the introduction of a partial resonance of 50 percent double-bond character that would prevent rotation about the bond. The shortening of the bond from 1.47 to 1.32 Å was dem-

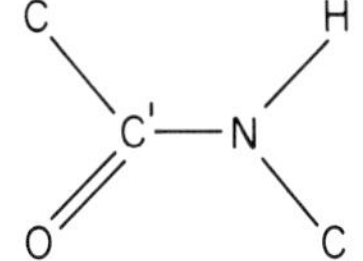

FIGURE 1.
Formula of planar amide group. According to Pauling there is resonance between the C′ and the N single bond and no possibility of free rotation. [Corey and Pauling, 1953.]

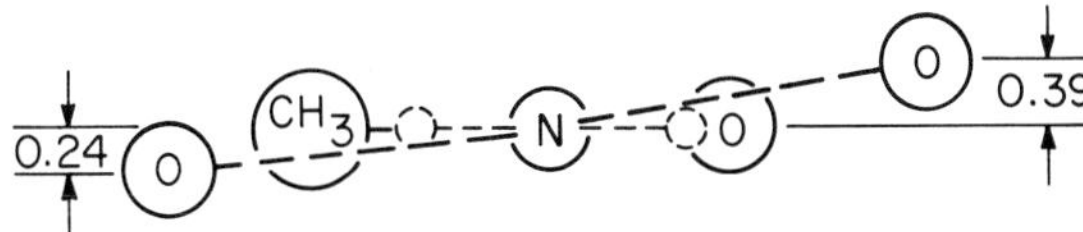

FIGURE 2.
Actual projection of amide grouping from acetamide (Senti and Harker, 1940) showing its planar character. The outlying oxygen atoms are only 0.39 Å to 0.24 Å respectively from the plane. [Corey and Pauling, 1953.]

onstrated by Hughes and his collaborators for a number of dipeptides, showing that the planar amide form was a general feature in peptides (see Figure 2).

These ideas straight away very much simplified the problem of protein structure but they required very large modifications of the basic ideas of crystallography. These ideas had contained the restriction that a helix was possible in crystals only with a helicity of 2-, 3-, 4-, and 6-fold symmetry, the screw axes of elementary space-group theory. It was not a new idea, by any means, that the peptide chain could be helical, but this limitation appeared much too stringent to account for the variety of protein structures. In fact, there was no real reason why the crystallographic limitation of symmetry should apply to the internal structure of a molecule. It only strictly applied to relations of separate molecules in the same cell. The stroke of genius on the part of Pauling was to abandon the idea of integral repeats along a helix and to substitute a helix of peptides with an irrational and, therefore, not exactly repeating structure. In the classical series of papers entitled "The Structures of Proteins," published by the National Academy of Sciences in 1951 (Pauling, Corey, and Branson, 1951), Pauling put forward "two hydrogen-bonded helical configurations of the polypeptide chain." These papers contained two fundamental chemical ideas. One, already mentioned, is the negative notion of the nonintegral character of the helical twist. The other idea is that the resulting polymer is stabilized by hydrogen bonds between the oxygen of the carboxy group and the nitrogen of the amide group *in the same chain*. This was the leading idea of the so-called *secondary* structure of proteins. He did this by attempting to make models on the basis of the exact measurements of the residues and linking them by hydrogen bonds. As Professor Hodgkin tells the story, he built models by drawing planar extended sheets of peptides on the correct scale (see Figure 3) and then rolled them up somewhat askew to form tubes, joining the edges so as to obtain a perfect fit of the atoms in the peptide chain. Thus he arrived at two helices, both nonintegral, the α or 3.7 residue helix and the γ or 5.1 residue helix (Figure 4). In the α helix he found that each planar amide group could be hydrogen bonded to the third amide group beyond it all along the helix, and that in the γ helix each is bonded to the fifth amide group beyond it. He favored the structure of the α helix for the normal fibrous proteins and suggested that

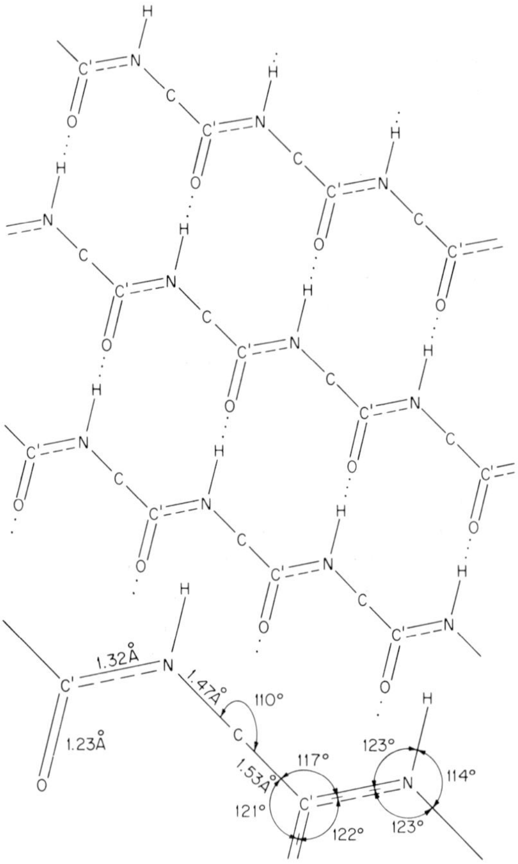

FIGURE 3.
Peptide chains in the α-helix conformation, showing the unrolled planar net from which they can be derived. [Pauling, Corey, and Branson, 1951.]

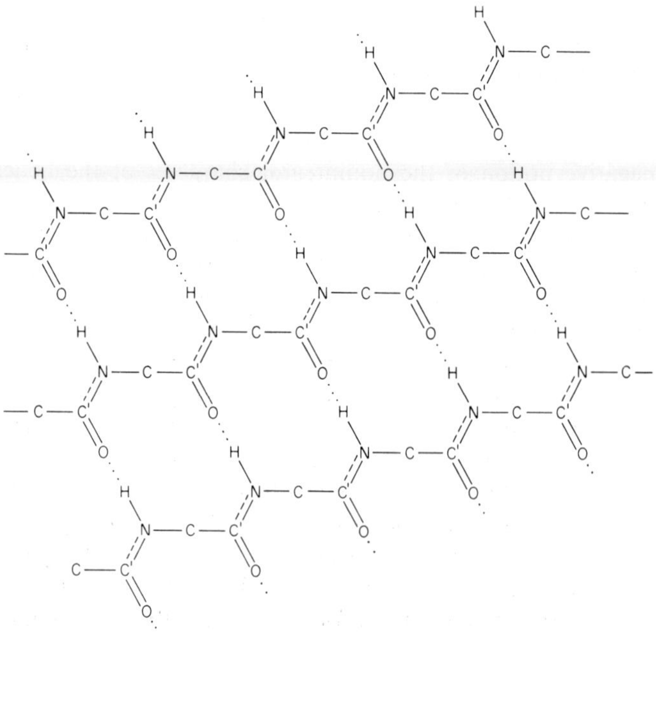

FIGURE 4.
Peptide chains in the γ-helix arrangement, showing the net from which it can be derived. [Pauling, Corey, and Branson, 1951.]

both helices might appear in some of the crystalline proteins. In his own words (1951), "It is our opinion that the structure of α-keratin, α-myosin and similar fibrous proteins is closely represented by our 3.7 residue helix, and that this helix also constitutes an important structural feature in hemoglobin, and other globular proteins, as well as of synthetic polypeptides."

The α helix was, therefore, quite correctly seen by Pauling as a clue to the structure of all proteins and so, with the appropriate modifications, it was found to be its best confirmation by X-rays in the study of the synthetic polypeptides such as alanine, which was then coming into use (*see* Hodgkin and Riley in this volume). This was particularly so in the strong reflection that occurred at a spacing of 1.6 Å perpendicular to the axis of the fiber, the pitch of the screw in which every residue contributed.

Pauling proved to be right in his intuition but the details turned out to be more complicated than he expected. Pauling's chemical hypothesis was presented to crystallographers at the meeting of the International Union of Crystallography in Stockholm in 1951. It was so comprehensible that it won immediate acceptance. I remember how Perutz, who had spent years vainly trying to solve the crystal structure of hemoglobin by orthodox methods, seized upon it as a solution and verified the presence of a strong 1.5 Å reflection and the existence of the α helix in the structure. I, myself, remained sceptical, for there was no such diffraction to be seen in the structure of ribonuclease, on which Carlisle had already started work. Here, also, Pauling made the same simplifying but incorrect assumption that the structure of the globular proteins consisted of rods of polypeptides arranged parallel to each other in different kinds of order. Crick (1952, 1953) was able to show that this was incompatible with the intensities of the X-ray reflections, which ought to be, on this hypothesis, much stronger than those observed. I had said that if the structure of globular protein was simple, we should be able to find it out relatively quickly, but, in fact, it took years and the structure was not a simple one. It now appears that the only thing that was wrong in Pauling's hypothesis, but carefully not stated, was the implication that the α helix was an important structural feature of *all* globular proteins. If it had been stated as *some* globular proteins, it would have been correct as well as illuminating.

In fact, we know now that globular proteins are characterized by the amount of α helix their peptide chains contain. Obviously, the possibility of working out the structure depended on this being as large as possible, up to 70 percent in hemoglobin and as little as 20 percent in the much smaller insulin, which obstinately refused to yield to this type of analysis. Nevertheless while the Pauling hypothesis put heart into those who were trying to interpret crystalline protein structures, it was to prove the way in which these structures were finally analyzed. Here the strict crystallographic methods and the well-established idea of heavy-atom isomorphous replacement proved in the end to provide the *phases* of reflection, which enabled Kendrew in 1958 to work out the structure of the simplest of the crystalline proteins, myoglobin, with one single chain of peptides (Kendrew et al., 1960). It is true that he found there long stretches that he recognized as α helix but he found as well that the peptide chain consisted of a number of stretches of straight helix connected by a number of bends in which no helix can be detected. This is the now-recognized *tertiary* structure of a protein, which was not foreseen in Pauling's work but is the logical pendant of it.

It can be said that, by and large, Pauling's idea played an essential role in the working out of protein structure. But it did far more. It broke away from the limitation imposed by crystallographers on the integral nature of the turns of a helix. It eventually led to a new generalization of crystallography that was to have immense repercussions. It might be said, "Only a crystallographer could have predicted this development, but if they were good crystallographers they would have been bound to reject it." Indeed, Pauling's generalization opened the field to a new and much more wide-sweeping account of semi-regular structures that are similar to the helical. Its implications in diffraction were worked out largely by Cochran, Crick, and Vand (1952). It appeared that such a structure in a fibrous form had a transform in X-ray diffraction, which showed definite layer lines with a characteristic butterfly pattern on the higher layer lines. Such patterns had been seen but not properly interpreted before. They were particularly noticeable in some fiber photographs obtained, for instance, in tobacco mosaic virus. When I first saw these photographs, I recognized that they were due to the existence of subunits (Bernal and Fankuchen, 1941) but had to face the contradiction that, to account for certain reflections, I had to assume a unit cell that was larger than the isolated virus particles I was observing in solution, or, alternatively, to accept the presence of fractional indices. The reflections I was looking at were, in fact not those from planes with any kind of index but the maxima of certain Bessel functions, as first explained by Watson (1954). The same type of photograph with a butterfly pattern was also seen in the X-ray photographs of fibers of nucleic acid.

It was to this vitally important subject that Pauling was next to turn his attention. It was already evident from the work of Caspersson (1947) that nucleic acids were intimately connected with the production of protein in actual cells. Attempts to purify or to crystallize them had proved unavailing, and it was evident that their structure must be that of a polymer, far more complicated than that of a protein. Nevertheless, it proved actually easier to analyze though not to the same degree of detail. Astbury, who first studied the fibers of thymus nucleic acid (now DNA), had made the essential observation that it had a marked *negative* birefringence, indicating that it contained strongly birefringent molecules arranged perpendicular to the axis of the fiber (Astbury and Bell, 1938). These could be the purines and the pyrimidines, which were known chemically to exist as its main constituents, the phosphate and the sugar being essentially isotropic.

An attempt at the crystallographic solution of the structure, though a partial one, was made by Furberg (1950a, 1950b, 1952, 1959), working on the structure of two of its components, cytidine and cytidylic acid, in which he showed that the ribose phosphate polymer was arranged so that the plane of the five carbon, ribose, sugar molecule was parallel to the axis and, therefore, at right angle to the bases by the turn of the helix. By building a model Furberg was able to show that the 3-fold polymer molecule of sugar-base phosphoric acid could be arranged in a helix in two different ways (see Figure 5).

Pauling, who had an intense interest in nucleic acids, chose the first of these ways, described by him (1953a, 1953b) on the basis of his protein hypothesis, the central chain containing the polymer with the bases projecting at the side: "The phosphate groups are closely packed about the axis of the molecule, with the pentose residues surrounding them, and the purine and pyrimidine groups projecting radially, their planes being approximately perpendicular to the molecular axis" (Pauling, 1953a). He proposed that these phosphate centered chains were themselves, in turn, composed of helices and proposed a model that contained three of these helices arranged around the main axis of the molecule.

This suggestion was not to lead anywhere because it missed the essential double nature of the nucleic acid chain, which contains the secret of its genetic importance. Here was an exceptional case in which Pauling's chemical intuition got ahead of his crystallographic dimensional approach. The single-chain structure that he favored would have been geometrically unstable and gave no answer to why the base groups should be in parallel sheets 3.5 Å apart. The double-chain idea was to come essentially from chemical considerations in the discovery by Chargaff (1951) that the sum of the purine (double-ring) bases and of the pyrimidine (single-ring) was always equal, implying, as Watson and Crick saw, that the nucleic acids, or at least DNA, could be considered as a strictly duplex molecule in which the pile along the center did not contain the sugar and the phosphate but only the base pairs. Actually, this arrangement was predicted on topological and genetic grounds by Pauling (Pauling and Delbrück, 1940) long before, but not applied par-

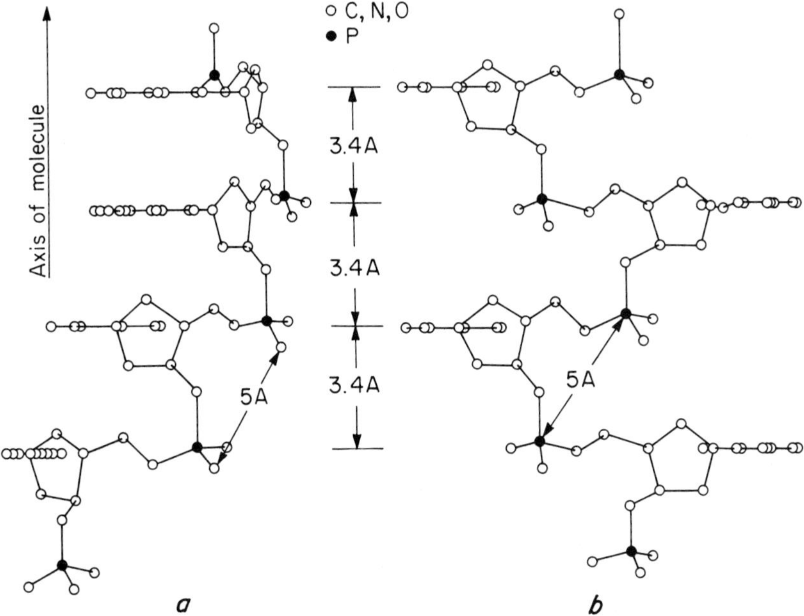

FIGURE 5.
(*a*) The model showing piling of bases of phosphate-sugar groups on the outside of the helix. This is the arrangement favored by Furberg and corresponds to the Watson-Crick arrangement but has a single helical chain only. (*b*) Alternative ways for deriving a structure of nucleic acid according to Furberg, in which the phosphate-sugar groups are the central axes. This was the arrangement favored by Pauling. [Furberg, 1952.]

ticularly to nucleic acids. The particular application by Watson and Crick (1954) led to the celebrated Watson-Crick structure, consisting of two helices unequally spaced along the axis with phosphate and sugar residues on the outside of the cylinders and coiled in opposite directions so that the order of the four nucleotides could be preserved and duplication made possible. The resulting structure was essentially one derived from a priori notions of the arrangements of the elements in the polymer and was confirmed soon after by X-ray analysis by Wilkins, Stokes, and Wilson (1953), Rosalind Franklin and Gosling (1953a, 1953b). The complexity that enables a single DNA molecule to assume its structure applied not to the "information-containing" part or the order of the nucleotides, which is of interest to geneticists, but only to the general structure of the molecule, which turns out to be much simpler than that of the proteins themselves.

Thus Pauling himself did not contribute immediately to the Watson-Crick hypothesis. However, his helical hypothesis was essential to its formulation because it made it possible to predict such a structure and to predict it accurately. Pauling's work in molecular biology can thus be seen to be absolutely fundamental in nature and we ought to inquire how it came itself to be based on leading ideas that Pauling held from the very start. We can distinguish two of these ideas. The first is his intuitive hold of quantum chemistry, which enabled him to think in terms of quantized energy levels from the outset. As hinted before, in his later years Pauling could afford to be very free with his quantum theory.

The second leading idea is Pauling's appreciation of quantitative geometrical linking as found in all kinds of substances from metals to nucleic acids. He knew his atoms and their various states and binding conditions so well that he was prepared to break with what are after all only conventions—such as the regularities of classical crystallography—if they could not be fitted into these regularities. He thus opened wide the possibility of extending crystallography and made possible an understanding of structures of such complexity as the proteins and the nucleic acids. In this way we owe to him the enormous emancipation of modern biomolecular structures.

It should be apparent that Pauling's work already held the key to much of the problem of biochemistry proper as well as of molecular biology, but simple and convincing as the key is, it will require many minds and hands to use it to solve the innumerable problems of the inner workings of life.

REFERENCES

Astbury, W. T., and F. O. Bell (1938). *Nature* (London) **141**, 747.

Avey, H. P., M. O. Boles, C. H. Carlisle, S. A. Evans, S. J. Morris, R. A. Palmer, B. A. Woolhouse (1967). *Nature* (London) **213**, 557.

Bernal, J. D. (1931). *Z. für Krist* **48**, 363.

Bernal, J. D. (1934). *Nature* (London) **133**, 794.

Bernal, J. D., and I. Fankuchen (1941). *J. Gen. Physiol.* **25**, 111.

Black, C. C. F., R. H. Fen, A. C. T. North, D. C. Phillips, and R. J. Poljak (1962). *Nature* (London) **196**, 1173.

Caspersson, T. (1947). *Symp. Soc. Exp. Biol.* No. 1, 127.

Chargaff, E. (1951). *J. Cell and Comp. Physiol.* **38**, suppl. 1, 41.

Cochran, W., F. H. C. Crick, and V. Vand (1952). *Acta Cryst.* **5**, 581.

Corey, R. B., and L. Pauling (1953). *Proc. Roy. Soc. B.* **141**, 10–20.

Crick, F. H. C. (1952). *Acta Cryst.* **5**, 381.
Crick, F. H. C. (1953). *Acta Cryst.* **6**, 600.
Donohue, J. (1950) *J. Am. Chem. Soc.* **72**, 949.
Donohue, J., and K. N. Trueblood (1952). *Acta Cryst.* **5**, 414, 419.
Franklin, R. E., and R. G. Gosling (1953a). *Nature* (London) **171**, 740.
Franklin, R. E., and R. G. Gosling (1953b). *Nature* (London) **172**, 156.
Furburg, S. (1950a). *Acta Cryst.* **3**, 325.
Furberg, S. (1950b) *Trans. Faraday Soc.* **46**, 791.
Furberg, S. (1952) *Acta Chem. Scand.* **6**, 634.
Furberg, S. (1959). *Acta Chem. Scand.* **13**, 910.
Kartha, G., J. Bello, D. Harker (1967). *Nature* (London) **213**, 862.
Kendrew, J. C., R. E. Dickerson, B. E. Strandberg, R. G. Hart, D. R. Davies, D. C. Phillips, and V. C. Shore (1960). *Nature* (London) **185**, 422.
Lein, A., and L. Pauling (1956). *Proc. Nat. Acad. Sci. U.S.* **42**, 51.
Martin, A. J. P., and R. L. M. Synge (1941). *Biochem. J.* **35**, 1358.
Pauling, L. (1940). *J. Am. Chem. Soc.* **62**, 2643.
Pauling, L. (1949). In *Haemoglobin* (*Symposium on Conf. Cambridge in Mem. Joseph Barcroft*), p. 57. London: Butterworth's Scientific Publications.
Pauling, L. (1953a). *Nature* (London) **171**, 346.
Pauling, L. (1953b). *Proc. Nat. Acad. Sci. U.S.* **39**, 84.
Pauling, L., R. B. Corey, and H. R. Branson (1951). *Proc. Nat. Acad. Sci. U.S.* **37**, 205.
Pauling, L., and M. Delbrück (1940). *Science* **92**, 77.
Perutz, M. F., M. G. Rossmann, A. F. Cullis, H. Muirhead, G. Will, and A. C. T. North (1960). *Nature* (London) **185**, 416.
St. George, R. C. C., and L. Pauling (1951). *Science* **114**, 629.
Sanger, F., and E. O. P. Thompson (1953). *Biochem. J.* **53**, 353, 366.
Sanger, F., and H. Tuppy (1951). *Biochem. J.* **49**, 463, 481.
Shoemaker, D. P., J. Donohue, V. Schomaker, and R. B. Corey (1950). *J. Am. Chem. Soc.* **72**, 2328.
Shoemaker, D. P., J. Donohue, R. Barieau, and C. S. Lu (1953). *Acta Cryst.* **6**, 241.
Svedberg, T., and K. O. Pedersen (1940). *The Ultracentrifuge*. Oxford: Clarendon Press.
Watson, J. D. (1954). *Biochem. Biophys. Acta.* **13**, 10.
Watson, J. D., and F. H. C. Crick (1954). *Proc. Roy. Soc.* (London) A **223**, 80.
Wilkins, M. H. F., A. R. Stokes, and H. R. Wilson (1953). *Nature* (London) **171**, 738.

NUCLEIC ACIDS

CRELLIN PAULING

Department of Life Sciences
University of California, Riverside
Riverside, California

The Specificity of Thymineless Mutagenesis

Certain bacteria, upon incubation under conditions in which the synthesis of DNA has been specifically inhibited, lose the ability to divide and produce a colony. This phenomenon, first described in 1954 by Barner and Cohen (1954), has been termed "thymineless death," in as much as it has been observed in thymine auxotrophs of *Escherichia coli* (Barner and Cohen, 1954), *Bacillus subtilis* (Brabander and Romig, 1960), and *B. megaterium* (Wachsman, Kemp, and Hogg, 1964). A similar phenomenon has been observed upon chemical inhibition of DNA synthesis in wild type *E. coli* (Cohen and Barner, 1956), *B. subtilis* (Mennigmann and Szybalski, 1962), and *B. megaterium* (Wachsman and Hogg, 1964). However, thymine auxotrophs of *E. coli* that under some conditions do not undergo thymineless death have been isolated; hence thymine auxotrophy is a necessary but not sufficient condition for thymineless death.

In the original description of the phenomenon, Barner and Cohen (1954) reported that lethality in the absence of thymine is contingent on growth, in as much as no loss of viability was observed when the cells were also deprived of a carbon source. Subsequently the same authors (Barner and Cohen, 1957, 1958) have reported that in amino acid or uracil auxotrophs derived from *E. coli* 15T−, thymineless death is somewhat mitigated by deprivation of the required growth factor. These observations led Barner and Cohen to refer to thymineless death as "death through unbalanced growth."

Gallant and Suskind (1962) have examined macromolecular synthesis during thymine starvation in *E. coli* B3. These authors absolve gross protein synthesis of responsibility for thymineless death; however, the net synthesis of RNA is closely correlated with lethality. They conclude that RNA synthesis in the absence of concomitant DNA synthesis is the basis for the observed loss of viability. Hanawalt (1963) has presented evidence that messenger RNA synthesis in the absence of DNA synthesis is specifically involved in the killing process in thymineless death.

The work reported in this paper was submitted to the Graduate School, University of Washington, as partial fulfillment of the requirements for the Ph.D. degree. The author acknowledges the guidance of Dr. J. Gallant and Dr. H. L. Roman, and the support of USPHS Training Grant 5-TI GM-182 and USPHS Predoctoral Fellowship 5-TI GM-14, 650.

Interest in the mechanism of thymineless death is stimulated by the effects, which, in addition to lethality, have been correlated with thymine deficiency. These effects include the induction (or diversion) of prophage in lysogenic bacteria (Korn and Weissbach, 1962; Melechen and Skaar, 1962); stimulation of genetic recombination (Gallant and Spottswood, 1964; Gallant and Spottswood, 1965; Weigle and D'Ari, personal communication); induction of colicin production in colicinogenic strains (Luzzati and Chevallier, 1964); the premature initiation of the DNA replication cycle (Pritchard and Lark, 1964); the stimulation of dark repair of DNA (Pauling and Hanawalt, 1965); and the induction of mutations (Coughlin and Adelberg, 1956; Weinberg and Latham, 1956; Kanazir, 1958; Pauling, 1964).

Extensive investigation in recent years has given much insight into mutation as a molecular event, and into the mode of action of specific chemical mutagens (for review, see Krieg, 1963). In an attempt to determine the nature of the mutagenicity of thymine deprivation at the molecular level, and possibly to gain insight into the molecular mechanism of thymineless death, we have examined in some detail the induction of mutations by thymine deprivation.

In the initial description of the mutagenicity of thymine starvation, Coughlin and Adelberg (1956) showed that thymine starvation induced reversion to prototrophy of a histidine auxotroph of *E. coli* 15T−. However, they observed no net increase in the number of revertants, and a reconstruction experiment was necessary to distinguish between *selection* of revertants preexisting in the population, and bona fide *induction* of revertants. They concluded that the increase in revertant frequency could not be explained by selection, and that thymine starvation was in fact mutagenic. Kanazir (1958) was able to demonstrate an increase in the absolute number of revertants to prototrophy of a uracil auxotroph of 15T− during the lag period prior to the onset of loss of viability.

Both Coughlin and Adelberg, and Kanazir, concluded that thymine starvation was in fact mutagenic for derivatives of *E. coli* 15T−. We first examined derivatives of *E. coli* B3 for mutagenic response to thymine starvation, in order to (1) determine if the apparent mutagenicity of thymine starvation of *E. coli* 15T− was a property of thymine starvation itself, and not unique to that strain, and (2) attempt to obtain further evidence on the question of selection versus induction of revertants.

Extensive investigation in recent years has demonstrated that agents which induce mutations affect a unique, nonrandom distribution of genetic sites (Benzer, 1961). In some cases an analysis of the distribution of both forward and reverse mutation induced by particular mutagens and combinations thereof has led to the development of hypotheses concerning the mechanism of action of these agents. Mutations leading to a mutant phenotype of an organism can be divided into two broad classes: revertible mutants and nonrevertible mutants. Nonrevertible mutants are generally assumed to be the result of reasonably large deletions of the genetic material, and will not be considered further in this discussion.

Freese (1959a, 1959b) has proposed that point mutations involve one of two distinct types of base pair substitutions, *transitions* or *transversions*. Transition mutations represent the substitution of a purine for a purine, and a pyrimidine for a pyrimidine. Transition mutations can be represented by the scheme:

$$\begin{array}{ccc} A & & G \\ | & \longleftrightarrow & | \\ T & & C \end{array}$$

Transversion mutations represent the substitution of a pyrimidine for a purine, and a purine for a pyrimidine. Likewise, transversion mutations can be represented by:

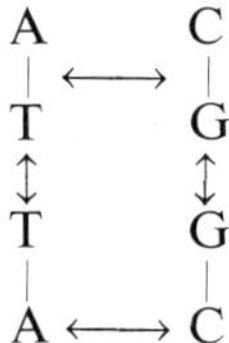

Brenner et al. (1961) have proposed an additional type of point mutation, in which the insertion or deletion of a single base pair is the molecular basis for the mutation.

The elegant experiments of Crick et al. (1961) represent strong support for the existence of mutations which result from the insertion or deletion of a single base pair. The argument is based on considerations of the genetic code, and on an analysis of a series of revertants of a proflavin-induced rII mutant of bacteriophage T4. Crick et al. (1961) observed that a large number of revertants of this rII mutant were in fact due to the existence of suppressor mutations, which, when separated from the original mutant by crossing, proved themselves to be nonleaky rII mutations. Additional mutants which functioned as suppressors of selected mutants of this group were then isolated. The original proflavin-induced mutant is arbitrarily designated as a (+) mutant, and the suppressors of it are designated as (−) mutants; suppressors of the (−) mutants are then *new* (+) mutants. Combinations of the various types of mutants were constructed by genetic means, and, in general, combinations of the (++) or (−−) types had the mutant phenotype, whereas (+−) or (−+) combinations had a phenotype approaching that of the wild type, providing that the mutants were not too far apart. Certain combinations of the (+++) and (−−−) type also showed wild type function, supporting the assumption of a triplet code.

On the basis of this evidence, Brenner et al. (1961) conclude that the mutagenic action of acridines, of which proflavin is an example, is due to the insertion or deletion of a single base pair. The molecular mechanism of this action is not clear; acridines bind to DNA by intercalation between adjacent base pairs (Lerman, 1961, 1963; Luzzati, Masson, and Lerman, 1961).

Freese (1959a, 1959b) postulated that the thymine analog 5-bromouracil (5-BU) and the adenine analog 2-aminopurine (2-AP) specifically induce transitions. The mutagenic action of these agents is presumed to be due to the possibility of two alternative base pairings as the analog is incorporated into the DNA, for example, 5-BU pairing usually with adenine but occasionally with guanine. The observation that base analog mutants were generally induced to revert with base analogs, but that proflavin-induced mutants and most spontaneous mutants were not induced to revert with base analogs, supports the theory for the mutagenic action of base analogs.

The ability of a base analog to induce both kinds of transitions is inherent in its capacity for alternative pairings, since it could make the less likely pairing, or "pairing error," either as an error of incorporation into the DNA or as an error of replication during subsequent DNA synthesis. Errors of incorporation clearly must occur during growth in the presence of the base analog, and hence the possibility remains that the mutagenic action may be due to a physiological effect of the base analog, rather than to actual incorporation of the base analog into the DNA. Errors of replication, on the other hand, would be expected both during growth in the presence of the analog, and during "clean growth" in the absence of the base analog following a period of "dirty growth" in the presence of the base analog. Mutations induced during clean growth following a period of dirty growth would of necessity be errors of replication, and may more likely result from pairing errors than from some physiological effect of base analog.

The powerful mutagenic effect of growth in 5-BU or 2-AP has been amply documented (see Krieg, 1963). Clean growth mutagenesis following growth in 5-BU has been demonstrated for phage T4 by Terzaghi, Streisinger, and Stahl (1962), and for bacteria by Strelzoff (1961, 1962). These authors conclude that the mutagenicity of 5-BU is due to incorporation at the site of an A-T pair, and to pairing with G during a subsequent division, thus leading to an A-T to G-C transition. The remaining mutations, requiring the presence of 5-BU during DNA synthesis, represent the incorporation of 5-BU opposite G, but the pairing with A during subsequent divisions, and thus lead to G-C to A-T transitions. Similar results have been obtained for clean growth following a pulse of 2-AP by Rudner (1961a), Strelzoff (1961, 1962), and Margolin and Mukai (1961), who conclude that 2-AP induces G-C to A-T transitions by incorporation errors, and A-T to G-C transitions by replication errors.

An additional mutagen that appears to be specific for the induction of transition mutations is hydroxylamine. Freese, Bautz, and Freese (1961) have examined the chemical and mutagenic activity of hydroxylamine, and they assign the mutagenic specificity of that agent to its effect on cytosine. Treatment of d-cytidine with hydroxylamine results in the production of N-6-hydroxycytidine; the *p*H dependence of hydroxylamine-induced mutation of phage T4 resembles that of the reaction with cytosine. Freese concludes that hydroxylamine preferentially inudces G-C to A-T transitions; this conclusion has been supported by Tessman, Poddar, and Kumar (1964).

The assignment of base analog-induced mutants to the transition class, and of acridine-induced mutants to the insertion-deletion class, predicts that these classes should be mutually exclusive, that is, that base-analog-induced mutants should revert with base analogs, but not with acridines, and vice versa. This prediction is generally borne out, but not without exceptions. Krieg (1963) has reported that a fraction of phage T4 rII mutants which are highly base-analog-revertible will also revert with 5-amino-acridine; thus the exclusion between sensitivity to base-analog and acridine induction of revertants is not absolute. The prediction that base-analog mutants should revert with base analogs is supported by the results of Freese (1959b) and of Champe and Benzer (1962), although the latter authors find that 10–40 percent of the mutants are possible exceptions to this rule.

In brief, the base analogs 5-BU and 2-AP, as well as hydroxylamine, are considered to induce transition mutations, whereas acridines are considered to induce mutations which result from insertion-deletion of a single base pair. Owing to the inherent ability of the base analogs to be effective by either errors of incorporation or errors of replication, neither 5-BU nor 2-AP is specific for transitions of one direction. Hydroxylamine, however, apparently is specific for transitions of the G-C to A-T variety. The existence of base-analog-induced mutants which fail to revert with base analogs, as well as the existence of mutants which revert with both base analogs and acridines, implies that the suggested mechanisms for these agents are not absolutely specific.

We have performed experiments designed to indicate the specificity of the induction of mutations by thymine starvation, and we present results which suggest that thymine starvation specifically induces transition mutations.

MATERIALS AND METHODS

Escherichia coli B3, a thymine auxotroph of *E. coli* B, and several derivatives thereof, were used in all experiments. The bacteria were grown in tris-glucose medium (Gallant and Suskind, 1962). The medium was supplemented with thymine (10 μg/ml) and amino acids (20 μg/ml) as required. Viable count determinations were performed by the pour-plate method (Gallant, 1962). All data for viability are presented relative to the value at time zero. Selective platings were performed by the pour-plate method on tris-glucose agar, supplemented with growth factors as required for the particular experiment. Media transfers were accomplished by centrifugation.

RESULTS

The results of a typical experiment with B326, an ultraviolet (uv) light induced lactose− (lac−) mutant of B3, are given in Figure 1. The survival data show a typical curve for thymineless death at 37°C, with a lag period of approximately 60 minutes, followed by exponential kill with a decade time of approximately 50 minutes. These kinetics are characteristic of B3 for thymineless death at 37°C; the lag period varies from 50 to 70 minutes from experiment to experiment. The reversion data, plotted as frequency of lac+ per viable bacterium, show an increase of more than two logs in the frequency of lac+ revertants among the survivors in the 150-minute duration of the experiment, and no significant change in the frequency of lac+ revertants in the control. At 50 minutes, before there has been any significant loss of viability, the revertant frequency has increased six-fold; hence the increase in revertant frequency cannot be explained by selection of preexisting revertants in the population. It can also be seen in Figure 1 that the lac+ revertants appear with apparent exponential kinetics, indicating that the culture is homogeneous with respect to mutability during both the lag period and the period of exponential death.

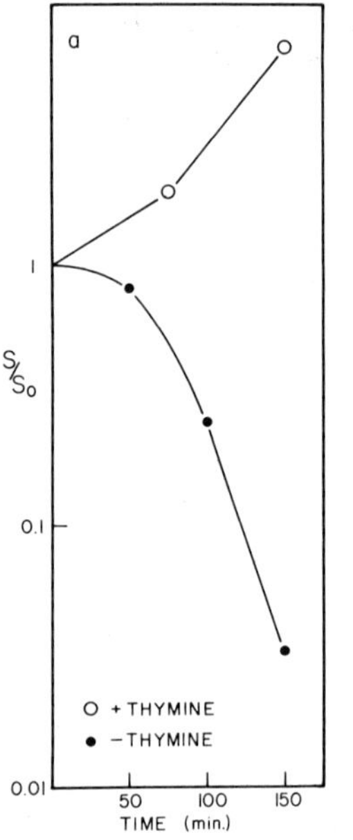

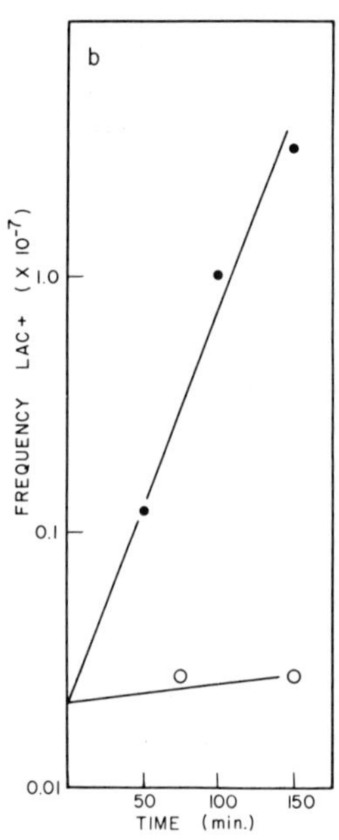

FIGURE 1.
Induction of lac+ revertants of B326 during thymineless growth. An exponentially growing B326 culture was harvested, washed, and resuspended in warmed tris-glucose medium. To one half, thymine was added (light circles); the other half was maintained without thymine (dark circles). At the indicated times survival (*a*) and the frequency of lac+ revertants (*b*) were determined.

Figure 2 shows the results of a similar experiment with B301, a methionine (met) requiring derivative of B3. Qualitatively the results are the same as for B326: an apparent exponential increase in the frequency of met+ revertants, and a clear increase in the frequency of revertants before any significant loss of viability. In both the lac− strain B326 and the met− strain B301, thymine starvation proved effective in the induction of revertants; in both cases a clear increase in revertant frequency was observed before the cultures had sustained significant loss of viability. These results seem to warrant the conclusion that the apparent mutagenicity of thymine starvation is in fact a property of thymine starvation, and they demonstrate that reconstruction experiments to distinguish between mutation induction and selection of preexisting revertants are unnecessary.

The results for survival and induction of lac+ revertants for B326 during thymine starvation following 60 minutes of growth in medium containing 0.5 mg/ml of 2-AP are given in Figure 3. The spontaneous revertant frequency prior to growth in 2-AP was 0.037×10^{-7}. Comparison with the control (Figure 1) shows that pretreatment with 2-AP has a significant effect both on the kinetics of thymineless death and on the induction of revertants. The lag period prior to the onset of kill is much reduced; the back extrapolate of the exponential portion of the survival curve is reduced to 25 minutes, compared with 60 minutes for normal conditions. At 120 minutes survival is 0.2 percent, compared with 10 percent for normal conditions. As a result of growth in 2-AP, the frequency of lac+ revertants at the beginning of thymine starvation was 1.2×10^{-7}; after 120 minute-thymine starvation, the frequency had increased to 390×10^{-7}. Figure 1 shows an increase under normal conditions from 0.022×10^{-7} to 1.6×10^{-7} in the same period of time. At t = 40 minutes, thymine was added to a sample removed from the experiment; both loss of viability and induction of revertants cease immediately upon addition

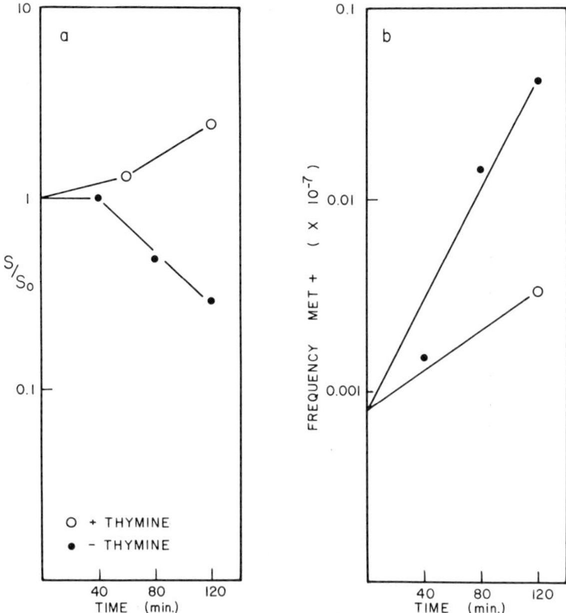

FIGURE 2.
Induction of met+ revertants of B301 during thymineless growth. An exponentially growing B301 culture was harvested, washed, and resuspended in warmed tris-glucose medium. To one half, thymine was added (light circles); the other half was maintained without thymine (dark circles). Methionine was added to both fractions. At the indicated times survival (*a*) and the frequency of met+ revertants (*b*) were determined.

of thymine. We interpret the reduction in frequency of lac+ revertants after readdition of thymine to be the consequence of nuclear segregation. The control grown in the presence of exogenous thymine shows exponential growth after a short lag; there was at most a doubling in frequency of revertants during normal growth after the removal of 2-AP. This result is consistent with the hypothesis that for this mutant, B326, 2-AP is effective in inducing reversion primarily by error of incorporation into the DNA, and not detectably by error of replication following incorporation of the base analog in the DNA.

The results of a similar experiment, showing survival and induction of lac+ revertants during thymine starvation following 90-minute growth in medium containing 0.02 mg/ml 5-BU, are given in Figure 4. The spontaneous revertant frequency prior to growth in 5-BU was 0.034×10^{-7}. Comparison with the control (Figure 1), which was thymine starved without prior treatment, shows that pregrowth in 5-BU has some effect on the survival, but not a marked effect on revertant frequency. The lag period prior to the onset of kill is absent in this experiment; the culture sustained exponential loss of viability from time 0. At 120 minutes the survival was 5 percent, compared with 10 percent for normal conditions. As a result of the period of growth in 5-BU, the revertant frequency at the onset of thymine starvation was 0.31×10^{-7}; after 120 minutes of thymine starvation the frequency had increased to 3.5×10^{-7}. The control, grown in the presence of thymine, paralleled the rise in revertant frequency observed for thymineless growth for 60 minutes, and then fell slightly. It must be emphasized that thymine was absent during the period of growth in 5-BU, and that the increase in revertant frequency observed cannot be attributed solely to the mutagenic action of 5-BU, in as much as 90 minutes growth in the absence of thymine is itself sufficient to increase the revertant to 0.5×10^{-7}. The marked divergence in revertant

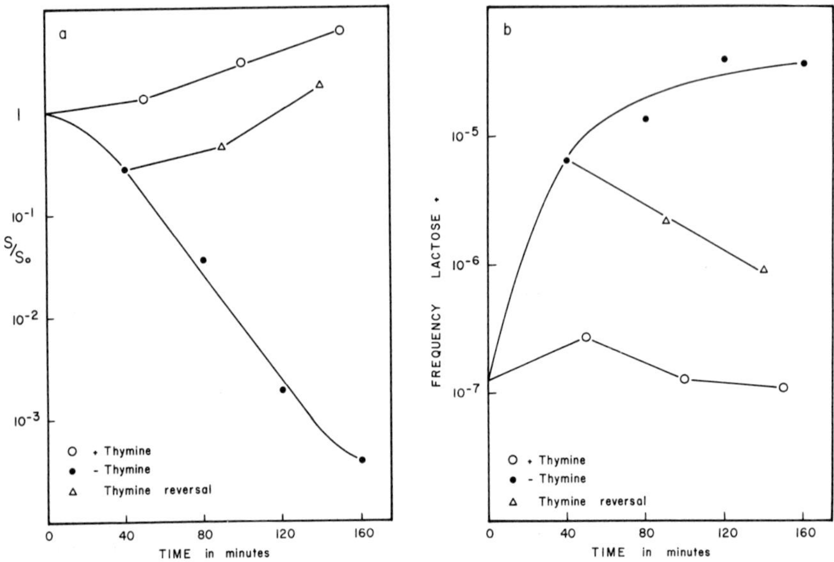

FIGURE 3.
Effect of pulse of 2-AP on thymineless induction of lac+ revertants of B326. 2-AP (final concentration of 0.5 mg/ml) was added to an exponentially growing B326 culture. After 60 minutes, the culture was harvested, washed, and resuspended in warmed tris-glucose medium. To one half, thymine was added (light circles); the other half was maintained without thymine (dark circles). After 40 minutes thymine was added to a portion of the thymineless culture (triangles). At the indicated times survival (*a*) and the frequency of lac+ revertants (*b*) were determined.

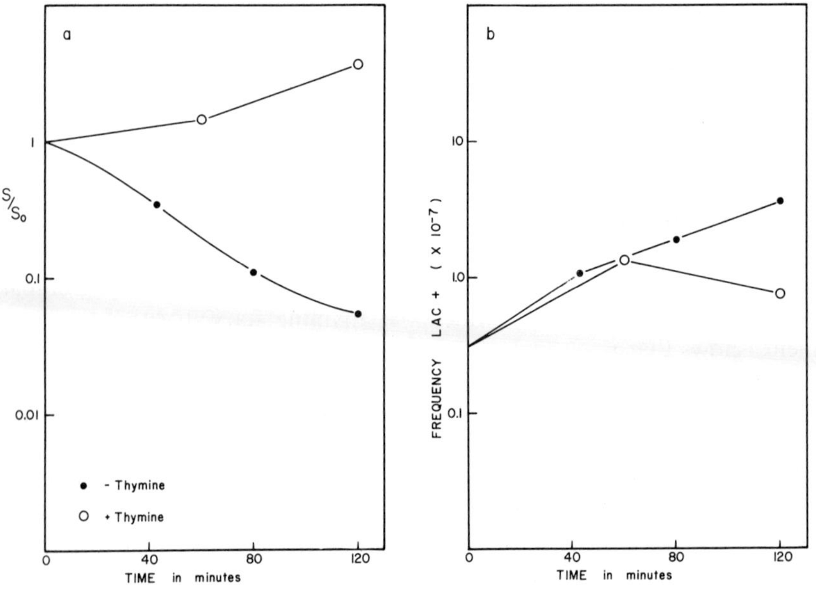

FIGURE 4.
Effect of pulse of 5-BU on thymineless induction of lac+ revertants of B326. An exponentially growing B326 culture was harvested, washed, and resuspended in warmed tris-glucose medium containing 0.02 mg/ml 5-BU. After 90 minutes of growth in this medium, the culture was again harvested, washed, and resuspended in warmed tris-glucose medium. To one half, thymine was added (light circles); the other half was maintained without thymine (dark circles). At the indicated times, survival (*a*) and the frequency of lac+ revertants (*b*) were determined.

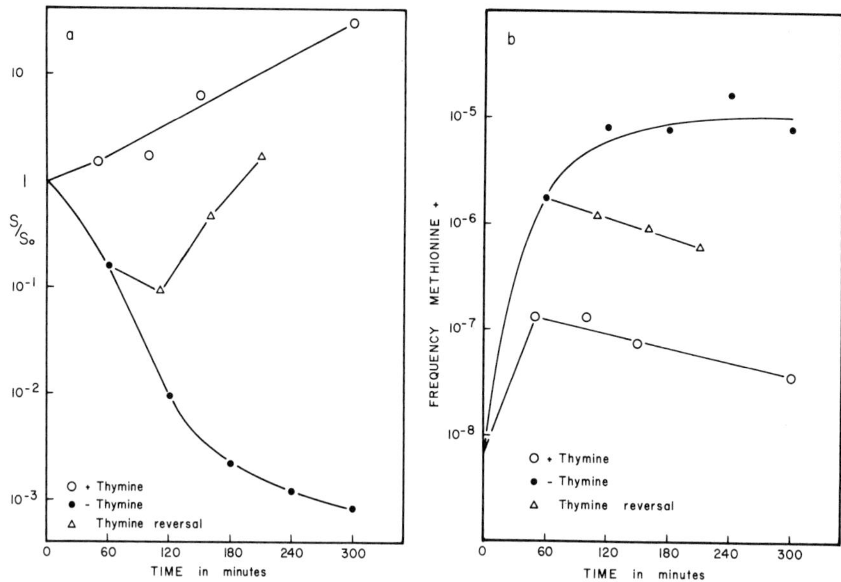

FIGURE 5.
Effect of pulse of 2-AP on thymineless induction of met+ revertants of B301. 2-AP (final concentration of 0.5 mg/ml) was added to an exponentially growing B301 culture. After 60 minutes, the culture was harvested, washed, and resuspended in warmed tris-glucose medium. To one half, thymine was added (light circles); the other half was maintained without thymine (closed circles). Methionine was added to both fractions. After 60 minutes thymine was added to a portion of the thymineless culture (triangles). At the indicated times, survival (*a*) and the frequency of met+ revertants (*b*) were determined.

frequency between plus and minus thymine growth noted in the 2-AP experiment is not observed in the case of pregrowth in 5-BU, suggesting that the synergism may be specific to 2-AP.

Results of a similar experiment, showing the survival and induction of met+ revertants of B301 during thymine starvation following growth in 2-AP, are given in Figure 5. The spontaneous frequency of met+ revertants prior to growth in 2-AP was $0.0023 \times {}^{-7}$. These results show an effect similar to that described above for B326. The lag period prior to the onset of exponential kill is reduced, and the survival after 120 minutes is substantially less than that observed under normal conditions. The induction of met+ revertants is similarly affected, with an increase from $0.07 \times {}^{-7}$ to 62×10^{-7} in 120 minutes, compared with an increase from 0.00078×10^{-7} to 0.0042×10^{-7} in 120 minutes under normal conditions (Figure 2). In this experiment loss of viability did not cease immediately upon thymine reversal, but the induction of revertants was stopped immediately upon thymine reversal. For this strain the control grown in the presence of thymine showed an increase in the frequency of met+ revertants from 0.07×10^{-7} to 1.3×10^{-7} during the first division; thereafter the frequency of revertants fell, presumably because of nuclear segregation. We interpret this observed increase in frequency of revertants during clean growth following removal of 2-AP to be the result of phenotypic lag in the expression of reversion, because of the cessation of mutagenesis after the first division. Hence, for this mutant as well as for B326, the mutagenicity of 2-AP is due primarily to errors of incorporation.

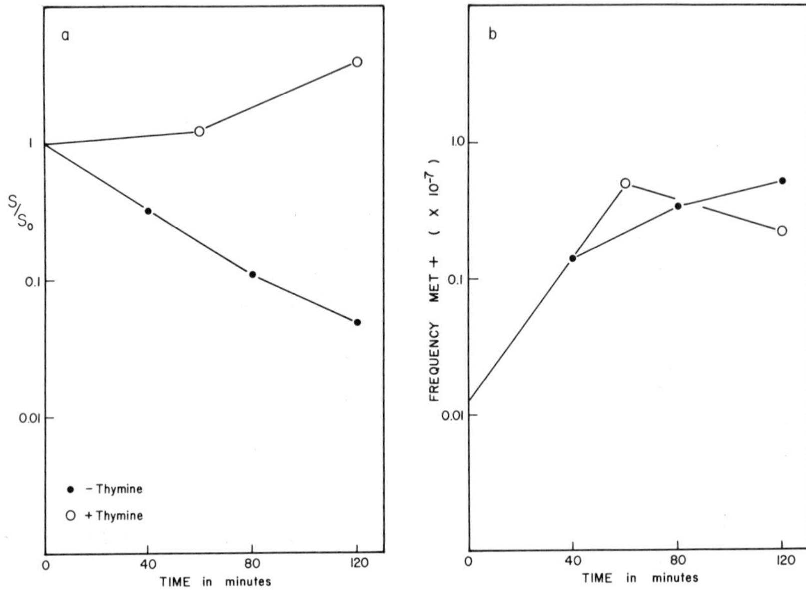

FIGURE 6.
Effect of pulse of 5-BU on thymineless induction of met+ revertants of B301. An exponentially growing B301 culture was harvested, washed, and resuspended in warmed tris-glucose medium containing 0.02 mg/ml of 5-BU. After 100 minutes growth in this medium, the culture was again harvested, washed, and resuspended in warmed tris-glucose medium. To one half, thymine was added (light circles); the other half was maintained without thymine (dark circles). Methionine was added to both fractions. At the indicated times, survival (*a*) and the frequency of met+ revertants (*b*) were determined.

The effect of pregrowth in 5-BU on survival and reversion of this mutant, B301, is shown in Figure 6. The frequency of met+ revertants prior to growth in 5-BU was 0.0021×10^{-7}. These results show an effect similar to that described for B326. The lag period prior to the onset of exponential death is absent, but the induction of revertants is not markedly affected. As a result of 100-minute growth in medium containing, in place of thymine, 0.02 mg/ml 5-BU, the revertant frequency rose to 0.013×10^{-7}; 100 minutes thymine starvation without 5-BU would increase the revertant frequency to 0.02×10^{-7}. Again, the rise in revertant frequency was similar for both plus and minus thymine; the divergence in revertant frequency noted in the pregrowth in 2-AP is not observed with 5-BU. The increase in revertant frequency plus and minus thymine may be the result of phenotypic lag in the expression of revertants induced during the period of growth in 5-BU.

An experiment was performed to determine the effect of prior growth in 2-AP on survival and thymineless induction of revertants as a function of time in 2-AP. Aliquots of a culture of B326 were grown in medium containing 0.5 mg/ml of 2-AP for 20, 40, and 60 minutes, and then subjected to thymine starvation as usual. The data are presented in Figure 7; the data from Figure 1 are repeated for comparison. The data for survival indicate that the lag period prior to the onset of exponential death is reduced roughly proportionally to the period of growth in 2-AP; however the rate of kill during the exponential phase is apparently independent of pregrowth in 2-AP.

The induction of lac+ revertants is also affected by the length of the pulse

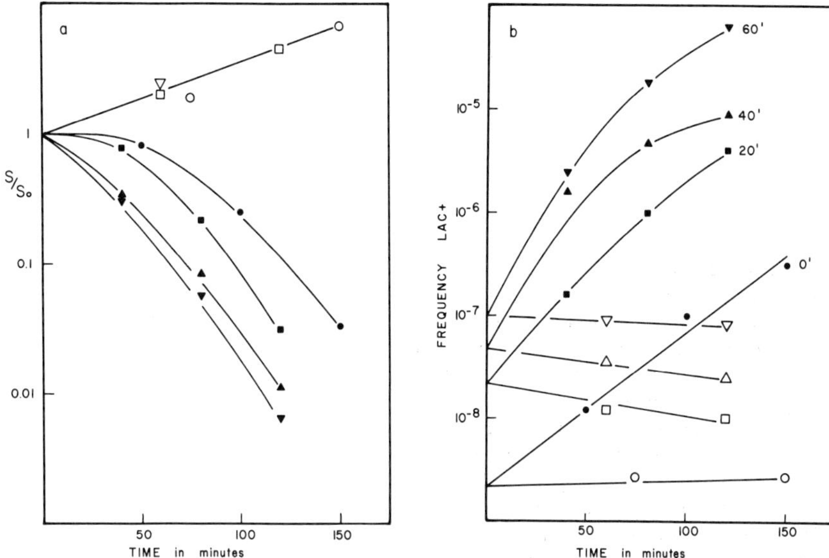

FIGURE 7.
Effect of length of pulse of 2-AP on thymineless induction of lac+ revertants of B326. 2-AP (final concentration of 0.5 mg/ml) was added to an exponentially growing B326 culture. After 20 minutes (squares), 40 minutes (triangles), and 60 minutes (inverted triangles), growth in this medium samples were harvested, washed, and resuspended in warmed tris-glucose medium. To one half of each culture, thymine was added (light symbols); the other half of each culture was maintained without thymine (dark symbols). At the indicated times, survival (*a*) and the frequency of lac+ revertants (*b*) were determined. The data presented in Figure 1 are replotted for comparison (circles).

of 2-AP prior to the period of thymine starvation. The increase in initial revertant frequency reflects the mutagenicity of 2-AP itself for this strain. It is also seen that both the slope and the number of revertants induced during thymine starvation are positively correlated with time in 2-AP.

The data for revertant frequency presented in Figure 7 have been replotted in Figure 8. In this figure the frequency of revertants induced during growth in 2-AP is represented by the dark circles. The frequency of revertants induced by 80 minutes thymine starvation following the pulse of 2-AP are represented by the light circles, corrected for revertants present at the onset of thymine starvation. Note that the shapes of the two curves are similar. The increase in frequency of revertants during dirty growth in the presence of 2-AP is a measure of the extent of incorporation of 2-AP in the DNA (Rudner, 1961b); hence this treatment of the data emphasizes the positive correlation between the extent of substitution of 2-AP in the DNA and the efficiency of subsequent clean thymineless-induced mutagenesis.

Figure 9 gives the results of an experiment in which 2-AP was present during the period of thymine starvation. The survival and frequency of lac+ revertants is given; the strain used was B326. In this case the presence of 2-AP has little effect on survival; the loss of viability during thymineless growth is normal. For the control in the presence of thymine, growth is somewhat impaired; by 120 minutes the survival is 1.6, whereas for normal growth under similar conditions the doubling time is approximately 50 minutes. The frequency of lac+ revertants increases essentially in a parallel

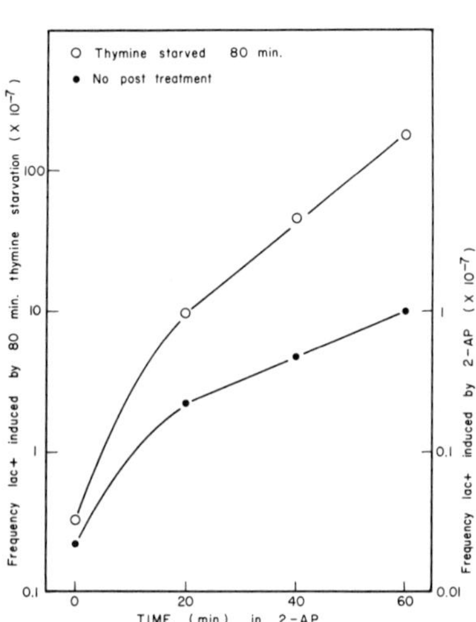

FIGURE 8.
Relationship between time in 2-AP and thymineless induction of lac+ revertants of B326. The data presented in Figure 7 are replotted in this figure to compare directly the effect of the length of the pulse of 2-AP on thymineless induction of lac+ revertants. The dark circles represent the frequency of lac+ revertants following growth of B326 for the indicated times in medium containing 0.5 mg/ml of 2-AP; the ordinate is to the right. The light circles indicate the frequency of lac+ revertants induced during 80 minutes of thymine starvation following growth for the indicated times in medium containing 2-AP; the ordinate is to the left.

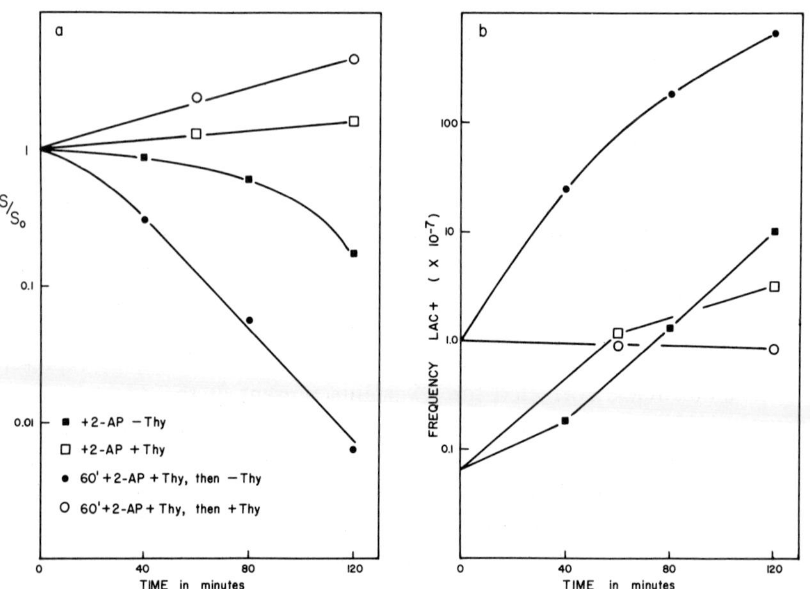

FIGURE 9.
Effect of 2-AP during thymine starvation on thymineless induction of lac+ revertants of B326. An exponentially growing B326 culture was harvested, washed, and resuspended in warmed tris-glucose medium containing 0.5 mg/ml of 2-AP. To one half, thymine was added (light squares); the other half was maintained without thymine (dark squares). At the indicated times, survival (*a*) and the frequency of lac+ revertants (*b*) were determined. The data presented in Figure 7, for the sample given a 60 minute pulse of 2-AP, are replotted for comparison (light circles represent + thymine; dark circles represent − thymine).

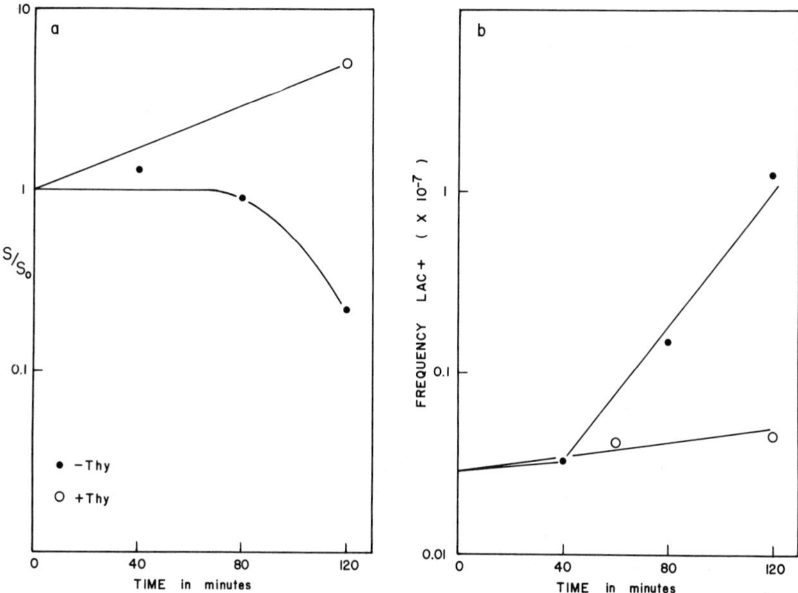

FIGURE 10.
Effect of pulse of adenine on thymineless induction of lac+ revertants of B326. Adenine (final concentration of 0.5 mg/ml) was added to an exponentially growing B326 culture. After 60 minutes, the culture was harvested, washed, and resuspended in warmed tris-glucose medium. To one half, thymine was added (light circles); the other half was maintained without thymine (dark circles). At the indicated times, survival (*a*) and the frequency of lac+ revertants (*b*) were determined.

manner plus and minus thymine; after 120 minutes growth the frequency of lac+ revertants is 10×10^{-7} and 3.3×10^{-7} for the minus and plus thymine experiments respectively. As seen in Figure 1, 120 minutes of growth in thymineless medium without the addition of 2-AP would increase the revertant frequency to 1.2×10^{-7}; hence the presence of 2-AP during thymine starvation does affect the efficiency of thymineless mutagenesis. The results for thymineless induction of revertants following a 60-minute pulse of 2-AP, as given in Figure 7, are replotted in Figure 9 for comparison. It can be seen that the effect of 2-AP present during thymine starvation is not as marked as in the case of a pulse of 2-AP prior to thymine starvation.

Figure 10 gives the results of the effect of a period of growth in medium containing 0.5 mg/ml adenine prior to thymine starvation. Growth in adenine has no apparent effect on either survival or thymineless induction of lac+ revertants for this mutant, B326.

DISCUSSION

Data presented above show that loading the DNA of the bacteria with the base analog 2-AP prior to the period of thymine starvation strongly increases the efficiency of thymineless mutagenesis. If we assume that the increase in revertant frequency during dirty growth is a measure of the extent of incorporation of 2-AP in the DNA, the increase in efficiency is positively correlated with the extent of substitution of 2-AP in the DNA. This strong synergism between base analog substitution and thymine starvation is specific for 2-AP; substitution with 5-BU prior to thymine starvation has no apparent effect.

In the case of the pulse of 5-BU experiments, the entire increase in revertant frequency can be accounted for by thymineless mutagenesis alone, assuming that the cells sustain thymineless damage during the period of growth in 5-BU. For both strains tested, the increase in revertant frequency during growth in 5-BU is no greater than that observed during an equal or shorter period of thymine starvation, and the additional increase in revertant frequency may represent a continuation of that effect. In addition, no divergence in revertant frequency between plus and minus thymine growth is observed in the pulse of 5-BU experiments.

In the case of the 2-AP experiments, however, the increase in revertant frequency during thymine starvation is far too extensive to be accounted for by thymine starvation alone; it is clear that there is a strong interaction between thymine starvation and 2-AP substituted in the DNA of the bacteria. A pulse of adenine had no effect; the effect of 2-AP is therefore specific for the adenine analog. Furthermore, there is a marked divergence in revertant frequency between plus and minus thymine growth following a pulse of 2-AP. This marked divergence indicates that a process is occurring during thymine starvation that is not occurring during growth in the presence of thymine, and that the substitution of 2-AP in the DNA potentiates this process.

It has also been seen that there is an effect on induction of revertants when 2-AP is present during the period of thymine starvation; there is no large divergence between plus and minus thymine growth under these conditions. We interpret the increase in revertant frequency to be due to 2-AP pairing errors occurring during residual DNA synthesis, coupled with thymineless induction potentiated by the 2-AP so incorporated.

Many of the consequences of thymine deprivation listed in the introduction parallel those of uv irradiation. Of particular interest in this discussion is the observation that thymine deprivation stimulates the dark repair of DNA (Pauling and Hanawalt, 1965). The stimulation of the dark repair of DNA by thymine deprivation suggests a mechanism for the observed mutagenicity. Pauling and Hanawalt (1965) discuss the implications of the dark repair of DNA on the phenomena associated with thymine deprivation, and in particular suggest that thymine deprivation induced single strand breaks in the DNA. Hence the (attempted) repair of such breaks during and after a period of thymine starvation could account for the patterns of mutation induction reported here. In addition, Pauling and Hanawalt (1965) predict the occurrence of radiation-sensitive mutants of *E. coli* that die thymineless death without the customary lag period prior to the onset of exponential loss of viability; Cummings and Taylor (1966) have reported the isolation of a strain of *E. coli* that fulfills this prediction.

On the basis of the evidence presented above, we submit the following model for the molecular mechanism of thymineless mutagenesis.

We propose that thymine starvation induces transition mutations, preferentially in the A-T to G-C direction. The reasoning for this model is presented on the following page.

1. The mutagenicity of 2-AP has been ascribed to its ability for alternative base substitutions (Freese, 1959a, 1959b). The marked potentiation of thymineless mutagenesis observed upon loading the DNA of the bacteria with 2-AP prior to thymine starvation suggests that thymineless induction of mutants is the result of base-pairing errors leading to the substitution of cytosine for thymine; the increase in efficiency following incorporation of 2-AP in the DNA reflects the greater ability of cytosine to pair with 2-AP than with adenine.

2. The synergism observed between 2-AP and thymine starvation is specific for 2-AP, and is not observed for 5-BU. Since the mutagenicity of 2-AP and of 5-BU is presumed to be due to their inherent capacity for alternative base pairings, leading to transition type base substitutions, the observation that pregrowth in 2-AP strongly potentiates thymineless mutagenesis, whereas pregrowth in 5-BU does not, suggests that the specific base-pairing error allowed by 2-AP is involved in thymineless mutagenesis. Because 2-AP is an adenine analog, presumably the "normal" pairing is with thymine, and the "mistake" is with cytosine.

3. Thymine starvation of a thymine auxotroph would be expected to result in a reduction of the internal concentration of deoxythymidylate, leading to an increase in the ratio of deoxycytidylate to deoxythymidylate. This increase in the ratio of deoxycytidylate to deoxythymidylate during thymine starvation might be expected to enhance pairing errors resulting in incorporation of cytosine in place of thymine, leading to A-T to G-C transitions.

Previously described techniques for the induction of transitions either have no directional specificity, or have been specific for G-C to A-T transitions (Krieg, 1963). Thymine starvation may supply a technique for the preferential induction of transition mutations of the A-T to G-C variety.

It must be pointed out that, although the data are consistent with this model, there is no evidence to rule out alternative or additional classes of mutation induction by thymine starvation. The model has a clear prediction, however, which allows a test of its validity.

The one agent that appears to be specific with respect to direction of transition induced is hydroxylamine. The evidence is quite strong that hydroxylamine specifically induces G-C to A-T transitions (Tessman, Poddar, and Kumar, 1964). The assertion that thymine starvation induces A-T to G-C transitions leads to a specific prediction. Mutants induced with hydroxylamine would be expected to be strongly revertible by thymine starvation, and mutants induced by thymine starvation would be expected to be strongly revertible with hydroxylamine. Furthermore, mutants which do not revert with hydroxylamine would be expected to revert during thymine starvation, and mutants which do not revert during thymine starvation would be expected to revert in response to hydroxylamine. These predictions remain to be tested.

REFERENCES

Barner, H. D., and S. S. Cohen (1954). *J. Bacteriol.* **68**, 80.
Barner, H. D., and S. S. Cohen (1957). *J. Bacteriol.* **74**, 350.
Barner, H. D., and S. S. Cohen (1958). *Biochim. Biophys. Acta* **30**, 12.
Benzer, S. (1961). *Proc. Nat. Acad. Sci. U.S.* **47**, 403.
Brabander, W. J., and W. R. Romig (1960). *Bacteriol. Proc.*, 187.

Brenner, S., L. Barnett, F. H. C. Crick, and A. Orgel (1961). *J. Mol. Biol.* **3**, 121.
Champe, S. P., and S. Benzer (1962). *Proc. Nat. Acad. Sci. U.S.* **48**, 532.
Cohen, S. S., and H. D. Barner (1956). *J. Bacteriol.* **72**, 588.
Coughlin, C. A., and E. A. Adelberg (1956). *Nature* **178**, 531.
Crick, H. F. C., L. Barnett, S. Brenner, and R. J. Watts-Tobin (1961). *Nature* **192**, 1227.
Cummings, D. J., and A. L. Taylor (1966). *Proc. Nat. Acad. Sci. U.S.* **56**, 171.
Freese, E. (1959a). *J. Mol. Biol.* **1**, 87.
Freese, E. (1959b). *Proc. Nat. Acad. Sci. U.S.* **45**, 622.
Freese, E., E. Bautz, and E. B. Freese (1961). *Proc. Nat. Acad. Sci. U.S.* **47**, 845.
Gallant, J. (1962). *Biochim. Biophys. Acta* **61**, 302.
Gallant, J., and T. Spottswood (1964). *Proc. Nat. Acad. Sci. U.S.* **52**, 1591.
Gallant, J., and T. Spottswood (1965). *Genetics* **52**, 107.
Gallant, J., and S. R. Suskind (1962). *Biochim. Biophys. Acta* **55**, 627.
Hanawalt, P. C. (1963). *Nature* **198**, 286.
Kanazir, D. (1958). *Biochim. Biophys. Acta* **30**, 20.
Korn, D., and A. Weissbach (1962). *Biochim. Biophys. Acta* **61**, 775.
Krieg, D. R. (1963). *Progress in Nucleic Acid Research*, vol. 2, p. 125. New York: Academic Press.
Lerman, L. S. (1961). *J. Mol. Biol.* **3**, 18.
Lerman, L. S. (1963). *Proc. Nat. Acad. Sci. U.S.* **49**, 94.
Luzzati, D., and M. R. Chevallier (1964). *Ann. Inst. Pasteur* **107**, 152.
Luzzati, V., F. Masson, and L. S. Lerman (1961). *J. Mol. Biol.* **3**, 634.
Margolin, P., and F. Mukai (1961). *Z. Vererbungslehre* **92**, 330.
Melechen, N. E., and P. D. Skaar (1962). *Virology* **16**, 21.
Menningmann, H. D., and W. Szybalski (1962). *Biochem. Biophys. Res. Comm.* **9**, 398.
Pauling, C. (1964). Ph.D. thesis, University of Washington.
Pauling, C., and P. Hanawalt (1965). *Proc. Nat. Acad. Sci. U.S.* **54**, 1728.
Pritchard, R. H., and K. G. Lark (1964). *J. Mol. Biol.* **9**, 288.
Rudner, R. (1961a). *Z. Vererbungslehre* **92**, 361.
Rudner, R. (1961b). *Z. Vererbunglehre* **92**, 336.
Strelzoff, E. (1961). *Biochem. Biophys. Res. Comm.* **5**, 384.
Strelzoff, E. (1962). *Z. Vererbunglehre* **93**, 301.
Terzaghi, B. E., G. Streisinger, and F. W. Stahl (1962). *Proc. Nat. Acad. Sci. U.S.* **48**, 1519.
Tessman, I., R. K. Poddar, and S. Kumar (1964). *J. Mol. Biol.* **9**, 352.
Wachsman, J. T., and L. Hogg (1964). *J. Bacteriol.* **87**, 1137.
Wachsman, J. T., S. Kemp, and L. Hogg (1964). *J. Bacteriol.* **87**, 1079.
Weinberg, R., and A. B. Latham (1956). *J. Bacteriol.* **72**, 570.

HERBERT JEHLE
WILLIAM C. PARKE
Physics Department
George Washington University
Washington, D. C.

Nucleic Acid Replication and Transcription

The structural aspects of the synthesis of biological macromolecules are an important and fundamental part of modern molecular biology (Pauling, 1955, 1957, 1960; Pauling and Corey, 1956; Pauling and Hayward, 1964). The structural aspects of synthesis of nucleic acids are better understood than other aspects of the biosynthesis of macromolecules (Langridge, Wilkins, et al., 1960; Meselson, Stahl, and Vinograd, 1957; Meselson and Stahl, 1958; Meselson and Weigle, 1961).

The conventional scheme of replication, however, did not convincingly account for the observed high accuracy in the correct selection of filial nucleotides. It did not provide for an adequate lock-and-key operation in the selection process, nor did it give an indication of the ways in which the polymerizing enzymes could be structurally involved in the process. We have, therefore, suggested some modifications (Jehle, Ingerman, Shirven, Parke and Salyers, 1963; Jehle, 1965) in the structure of the replicating-fork region in DNA. This proposal relegates the actual replication process to the stem region of the Y, ahead of the replicating fork. It postulates that the stem region is stabilized by groove-filling molecules, presumably proteins—perhaps even the polymerases themselves (Kozinski, 1966; Reich, 1964; Farmer, 1966; Rothman, 1965; Hamilton, Fuller, and Reich, 1963; Tabor and Tabor, 1964). These molecules snugly fill out the grooves of a double-strand nucleic-acid helix in a structurally and charge-complementary manner (Wilkins, 1956; Zubay and Wilkins, 1964; Steiner and Beers, 1961).

The firmly structured stable helix which ensues is then admirably suited to act as a template after one parental nucleic-acid strand is gently pried out (owing perhaps to a small local change in the ionic conditions of the surroundings). Filial nucleotide triphosphates which are identical with those of the displaced parental chain fill the cavities immediately and accurately as in a lock and key situation (Pauling, 1957). We do not need to depend exclusively on the hydrogen bond complementarity (Hotchkiss, 1962; Rosenberg and

Professor Parke is now at the Division of Radiation Physics, National Bureau of Standards, Gaithersburg, Maryland.

This work received support from USPH research grants CA 04989 and GM 12054 and from the George Washington University Committee on Research. It was performed at the Hansen Laboratory of Biophysics, Stanford University, and at the National Institutes of Health, Bethesda, Maryland.

Cavalieri, 1963, 1964) between bases since it is not a very reliable guide for correct nucleotide selection at the time when the helix opens up.

Since this nucleic-acid replication process has been proposed in earlier papers, we limit its description to the legends of Figures 1, 2, and 3, and proceed to a consideration of additional questions that arise in the discussion of nucleic acid synthesis.

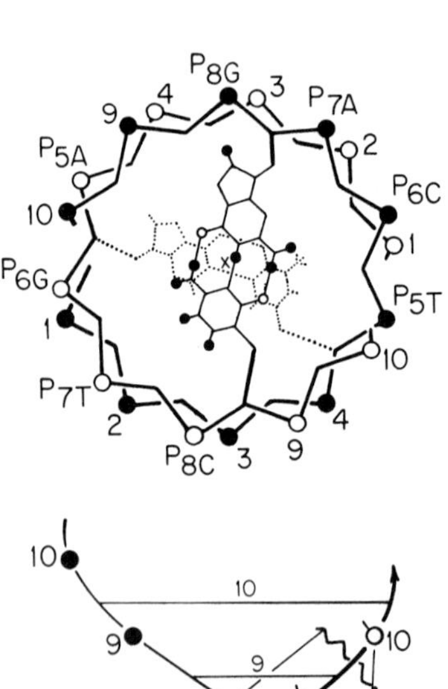

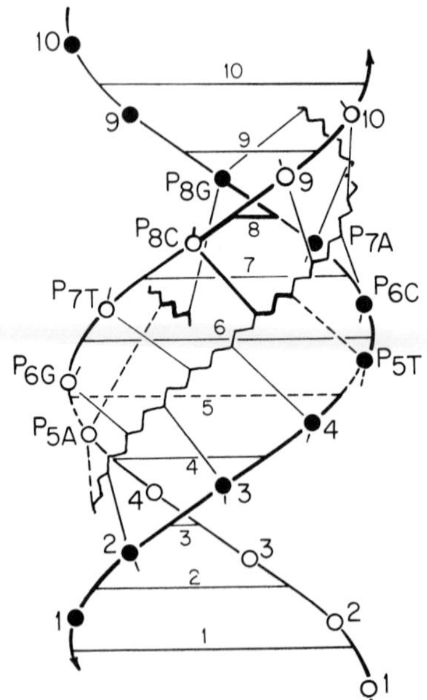

FIGURE 1.
The upper diagram is a cross section on base-pair levels 5–5 (· · · ·) and 8–8 (——) of a DNA double-strand helix which has, as first suggested by Wilkins, its groove snugly filled with protein strands. This "Watson-Crick-Wilkins" helix (of the B structure type of Langridge and Wilkins) is shown in side view in the lower diagram. The location of the phosphates is indicated by small circles along both helical backbone chains (strands). The fifth through the eighth base layers have their phosphates marked P, followed by a number indicating the layer and a letter referring to the attached base. These are the layers principally involved in the state of replication shown in Figure 2. The pentoses are not specially marked: they are situated where the heavy zig-zag line shows inner corners; the outer corners represent the phosphates. The two grooves of the nucleic acid are filled with two protein chains, whose backbones are indicated by zig-zag lines; side chains are drawn as lines reaching from the protein backbone towards the phosphates. For greater legibility, the protein in the wider groove is indicated by a short zig-zag line, and only the two side chains that reach to the phosphates P_{5A} and P_{8G} are shown. This makes the subunits (for example, 5, 6, 7, 8 or 6, 7, 8, 9) stable and compact. For the narrow-groove protein, two corresponding side chains, reaching to P_{5T} and P_{8C}, are shown. The groove-filling proteins are omitted from the upper diagram.

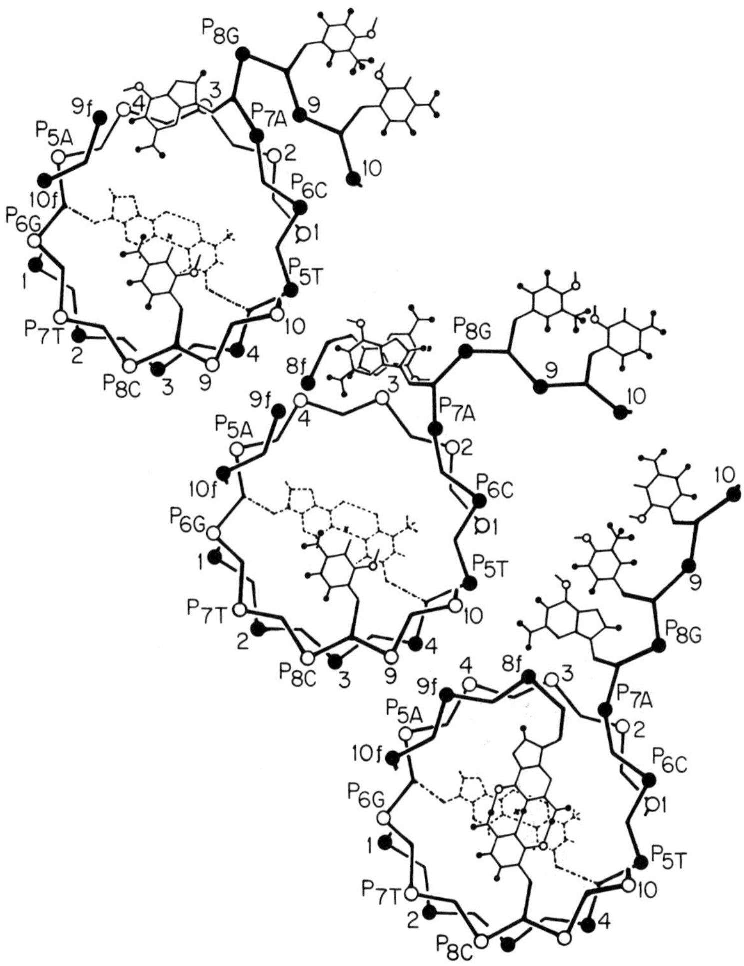

FIGURE 2.
The proposed replicative process starts with the intact double-strand Watson-Crick-Wilkins helical structure pictured in Figure 1. A local change in the ionic condition of the medium might loosen the hydrogen bonding between complementary bases and make the P_{8G} phosphate break loose from the wide-groove protein side chain (upper diagram of Figure 2), the guanine 8G thereupon being pried out. This change presumably does not need be specific. A filial guanine 8G*f* (*f* standing for filial), brought on by Brownian motion and specifically held by London forces to guanine 8G, might attach itself to the wide-groove protein side chain at P_{8Gf} (middle diagram) and thereafter snugly fit into the open slot (lower diagram); the process repeats itself at the next and lower levels. The cavity (open slot) provides for a highly specific lock-and-key situation; only the correct filial nucleotide may enter there. Whereas during replication the scarcity of protons lessens the influence of hydrogen bonding, the re-establishment of normal ionic conditions thereafter brings the Watson-Crick complementarity condition in force again. The nucleotide 8G is a monomer of one of the parental nucleic acid strands which peels off eventually. The diagrams are so drawn that the conservative half of the nucleic acid and the two groove proteins lose their shape but imperceptibly, to make the drawing readable. The proteins are omitted to keep the diagrams from becoming overloaded. The pyrophosphate of the incoming nucleotide triphosphate has also been omitted in the middle diagram.

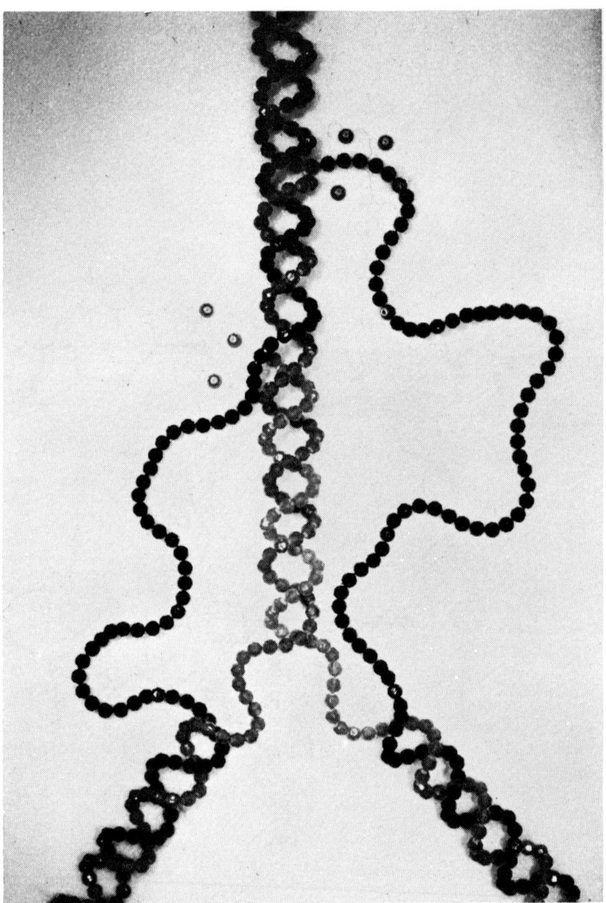

FIGURE 3.
Stem replication scheme for DNA. The strand-synthesis process described in Figures 1 and 2 occurs twice, one after the other on both strands. Both parental strands (represented by dark beads) peel off from the stem region of the Y-shaped fork, while filial nucleotide triphosphates (lighter beads, three of them shown as individual beads approaching each replicating region) are incorporated into the helix, replacing the parental nucleotides, and polymerization occurs. The stem region of this WCW helix remains structurally intact because of the groove-stabilizing molecules (presumably proteins or even polymerases) which are omitted in this picture. The parental strand which peels off nearest the Y juncture of the fork is called the masterstrand (see Figure 4) because both filial strands are formed in its image. The peel-off process implies a torque on the stem which therefore exhibits a screwlike motion (because the motion of the parental strand is impeded by the viscosity of the medium). The stem is thus eventually opened at its filial-filial end, that is, the Y juncture. It is assumed that two filial single-strand nucleic acids result, each one carrying one groove molecule with it. The peeled-off parental strands base-pair with the complementary filial strands, aided by the attached groove molecules. This suggested scheme differs from the conventional one in that the synthesis region is relegated to the stem of the Y (rather than to the Y juncture), where a well-defined Watson-Crick helix may assure exact replica formation.

MASTERSTRAND (FIGURE 4)

The essential point for genetics is that, in the present scheme of duplication of a double-strand DNA, WC (where W is one strand and C is the other strand), there is a masterstrand W which serves as a template for the new complementary strand C′; then a little further up toward the replicating fork, W separates from the stem—that is, from C′—which then provides the template on which W′, the copy of the masterstrand, is synthesized. Thus any mutation in W is copied in C′ and ends up in both strands C′ and W′ and thus in both arms of the Y.

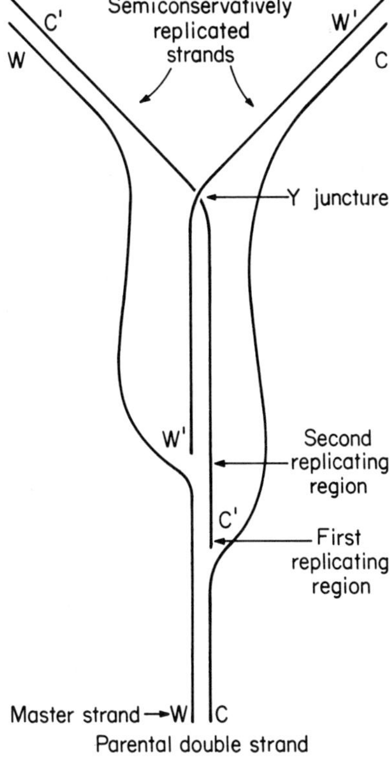

FIGURE 4.
Masterstrand. Diagrammatic sketch of the replication scheme of Figure 3, illustrating the concept of a masterstrand that employs only one double-strand parental DNA. The masterstrand W directs the assembly of a complementary filial strand C′ which, further up, directs the assembly of the filial strand W′. At the two replicating regions filial nucleotide triphosphates are supplied from the medium; all materials stream upward; but the picture does not change.

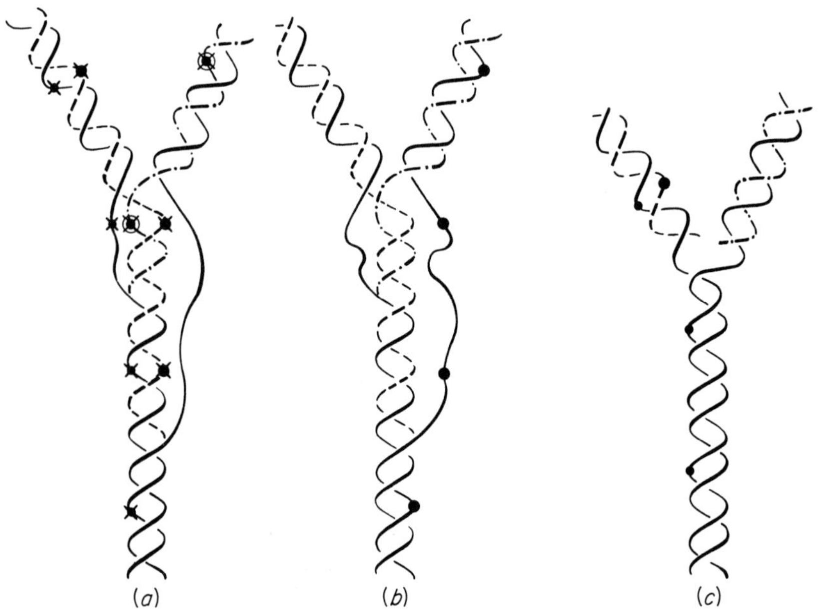

FIGURE 5.
The result of a mutation (indicated by ✖ or ●) of only one of the parental strands. The process shown in (*a*) is initiated by a mutation of the nucleotide sequence on the masterstrand, that is, base(s) which are not complementary to the base(s) on the other, nonmutated strand. When the mutation reaches the first replication region (see Figure 4) it causes incorporation of bases that are exactly complementary to the mutated bases, and these newly incorporated bases in turn cause, at the second replication region, incorporation of exactly complementary bases that are identical with the bases of the mutated parental strand. In (*a*) the mutated strand becomes the masterstrand; in (*b*) the other strand carries the mutation; in (*c*) is shown the process in the conventional replication scheme. The drawings show what can happen to a single mutation as it flows up towards the filial arms; accordingly the markings show the position of the mutation at the subsequent levels on the same Y structure.

MUTATIONS WITHOUT SEGREGATION (FIGURE 5)

A series of investigations (Kubitschek, 1964; Kubitschek and Henderson, 1966; Witkin and Sicurella, 1964; Goodgal and Herriott, 1961; Stoker; Freese, 1963) have shown the importance of events which indicate the possibility of mutation without segregation. We will not go into a discussion of the molecular scheme of replication proposed by Kubtischek to explain this but rather concentrate on the genetic phenomenon itself. The experiments indicate that a mutation on one strand may not cause a mutation of the complementary strand before the material reaches the region of replication, but it may nevertheless give rise to mutations of both filial arms, with prospects of an admixed percentage of single-arm mutations.

Parts *a* and *b* of Figure 5 illustrate how a mutation at one or at the other parental strand, in its flow through the replication fork, is propagated to the filial strand. The strand which peels off closer to the fork, that is, which is behind in the peel-off process, acts as a masterstrand, forming a comple-

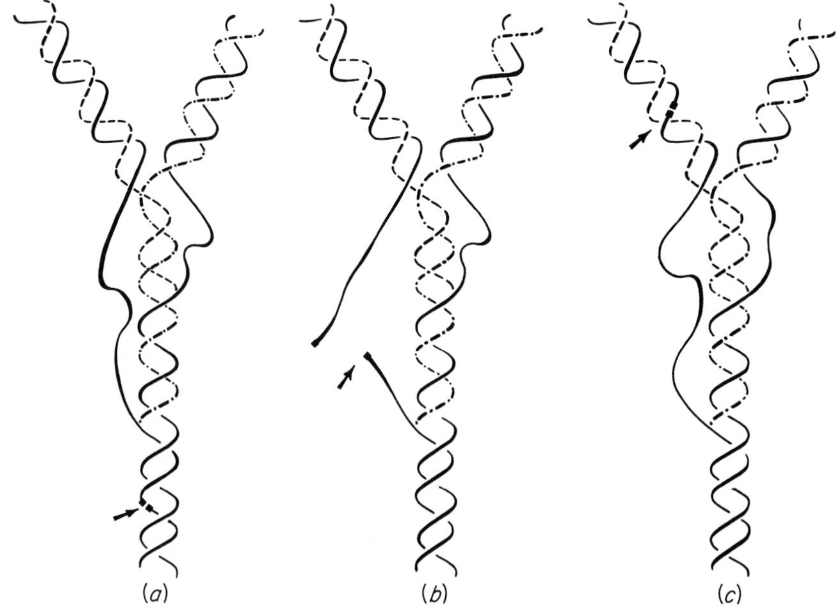

FIGURE 6.
Since it is expected that the parental double-strand helix will occasionally show a break in one of its strands, we indicate here the sequence of events (three sequential pictures) that show the possible occurrences if the masterstrand is not broken. Note that in the conventional replication scheme (Figure 5, *c*) the parental single-strand break is likely to cause a break of the ensuing filial arm at that spot.

mentary filial strand in its image. This filial strand forms a second filial strand complementary to itself, in its image; this second filial strand is thus identical with the parental masterstrand. If the masterstrand was mutated, three mutated final strands result. If the other parental strand was mutated, it remains the only mutated one.

INTERNAL HETEROZYGOTES ON T_4

Of even more direct importance for the issue are the experiments on internal HETS (Stahl and Green; Doermann and Boehner, 1963; Streisinger, Edgar, and Denhardt, 1964; Séchaud et al., 1965; Shalitin and Stahl, 1965; Stahl et al. 1965; Barricelli, 1955, 1956, 1960, 1965; Barricelli and Doermann, 1960, 1961; Barricelli and Womack, 1965). The question at stake is whether heterzygosity is lost, partly lost, or retained, in a round of replication. In the proposed stem-replication scheme this question again refers to the situation represented in Figure 5. In either case (Figure 5*a* and *b*) one filial arm will be heterozygous, the other homozygous.

SINGLE-STRAND BREAKS AND REPLICATION (FIGURE 6)

In the same manner as is assumed in the conventional replication scheme, we imagine the nucleic-acid material to flow through the fork region in a helical right-handed screw motion. If a movie were taken of the process pictured in Figure 3, its appearance would not alter when the individual nucleotides flow along the helixes (strands of beads). This means, of course, that the parental material spins around its helix axis, and the problem that arises is how far ahead of the fork this spinning occurs in a synchronous fashion. In a recent lecture Phil Hanawalt illustrated the subtlety of this problem by using twine to represent a scale model of an actual nucleic-acid double-strand helix. If we try to twist the twine (which, to represent *E. coli* DNA, might be stiffened by starching it, but would be more than a thousand feet long) on one end, we notice that it tangles up immediately, forming hairpin loops; the twine becomes particularly entangled if the twisting motion is rapid or if the twine is coiled together as DNA is in its natural habitat. In short, it may only be because of single-strand breaks that an orderly spinning motion becomes possible—at such loci the half-broken nucleic-acid double-strand helix can rotate freely around its helix axis by virtue of the free rotation about the several single bonds of the one intact strand.

The conventional replication scheme (Figure 5,*c*) does not account satisfactorily for the postulated single-strand breaks; it implies a double-strand break of the ensuing filial arm unless, by an ad hoc assumed mechanism, every single-strand break heals before reaching the fork.

The proposed stem-replication scheme of Figures 1, 2, 3 and 4, however, seems to permit filial arms to be synthesized (they contain nothing more than the inherited single-strand break). When the broken end of the parental strand in question peels off from the stem (Figure 6,*b*), the replacing filial nucleotide triphosphates may form a filial chain just as well as when there was no break in the parental chain.

If the broken strand replicates before the unbroken strand does (as in Figure 6), the stem's rigidity is never impaired by the single-strand break.

If, on the other hand, the intact strand peels off first (not shown in Figure 6), the locus of break (on the broken strand, which stays in the stem) may not necessarily cause a break of the stem because the proposed groove-filling molecules provide strength and rigidity to the entire structure.

TRANSCRIPTION

Some comments relating to transcription may be made here (see Figure 7). Spiegelman's work (Haruna and Spiegelman, 1965; Shipp and Haselkorn, 1964; Haselkorn and Fox, 1965) on in vitro synthesis of biologically active RNA from a single-strand primer to which the synthesized RNA is identical is of paramount importance here. We would like to think of this process in terms of the aforementioned model of transcription. It is assumed that the single-strand primer collects so many complementary pieces of nucleic acid as well as groove molecules that a stable, structure of the type shown in Figure 1 results. The complement to the primer consists of short pieces and thus may not be identifiable as forming a "replicative phase." If a transcription process should start with some of these complementary nucleic-acid pieces peeling off, the process would then yield only short pieces of nucleic acid—not a complete single strand. If, on the other hand, the transciption process

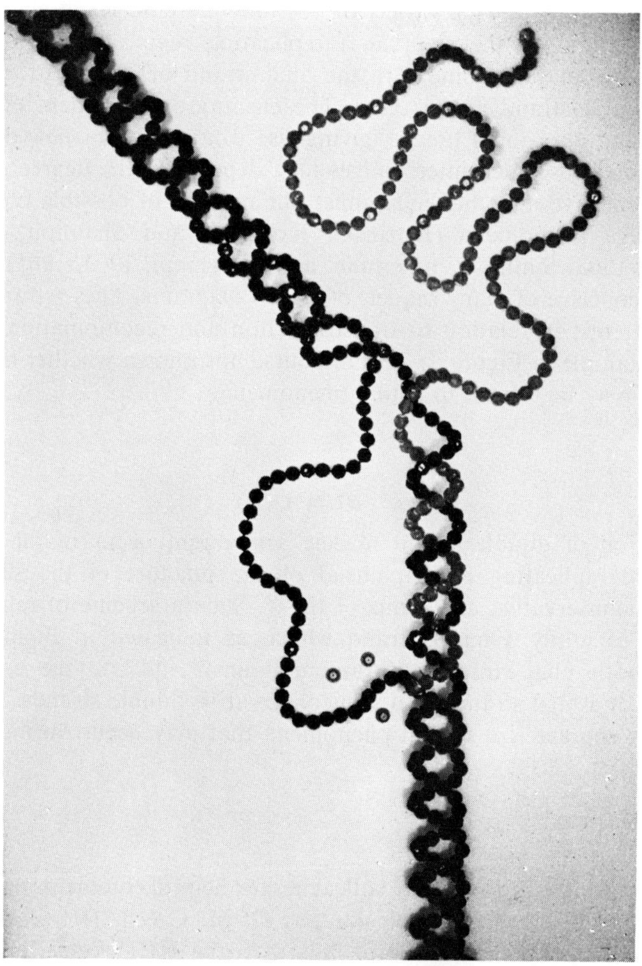

FIGURE 7.
Transcription of an RNA, from a double-strand DNA functioning as a template. This process shown here, differs from the DNA replication process of Figure 3 in that only one of the parental nucleic-acid strands is peeled off and replaced, here by filial ribonucleic-acid monomers. Along the stem a hybrid double-strand helix is being formed. The stem is presumably reinforced by tightly fitting polymerases laid into the grooves. The off-peeling parental DNA strand exerts a torque at the synthesis site which eventually causes untwining of the stem at the hybrid site, thus giving the newly formed RNA a chance to peel off and perhaps be replaced on the helix by the formerly peeled-off parental DNA strand. The motion is, in both Figures 3 and 7, a helical flow channeled along the backbone helix; the conformation of the replicating region remains essentially unchanged as time goes on.

involves peeling off the intact single-strand primer, an identical replica of it may be formed in the manner indicated in Figures 1 and 2. In this way it may become understandable how an exact replica could be synthesized from a single-strand primer.

STRAND EXCHANGES

There seems to be a wide variety of possible interchanges in which a parental strand, peeling off the stem at a replicating region of the double-strand nucleic acid, may combine with the filial strand belonging to another replicating double-strand nucleic acid. The ensuing entanglement of the strands may lead to occasional breaks, giving rise to a variety of new double-strand helix formations. Such interchanges may depend on the degree of homology of the respective interchange regions. The number of possible types of strand interchange phenomena (Kozinski, Kozinski, and Shannon, 1963, 1965; Bodmer, 1965; Simha, Zimmerman, and Moacanin, 1963, 1965) is large and we may conceive of many sequels of strand diagrams. These strand exchanges are of interest in relation to transformation and recombination (we show a simple example in Figure 8). It is of course not known whether these possible schemes may be related to actual phenomena.

SUMMARY

Replication of double strand nucleic acid might occur in the stem of the Y shaped replicating region, ahead of the juncture of the stem with the two semiconservative filial arms of the Y. Such a scheme of replication may be seen to imply a masterstrand which, as indicated in Figures 3 and 4, directs both filial strands. "Strand diagrams" indicate the association of parental and filial strands into semiconservative double strands. These strand diagrams characterize several phenomena that may occur during replication.

ACKNOWLEDGMENT

We would like to thank our colleagues for helpful comments and for reports on work in progress, in particular Dr. D. M. Green, Dr. F. W. Stahl, Dr. M. Meselson, Dr. C. Pauling in regard to the HETS issue, and to Dr. G. W. Bazill, Dr. D. Bradley, Dr. N. Davidson, Dr. H. de Voe, Dr. J. Farmer, Dr. M. Fox, Dr. P. Hanawalt, Dr. L. Heppel, Dr. R. D. Hotchkiss, Dr. E. W. Kozinski, Dr. M. Nirenberg and Dr. E. Trucco, in regard to related issues of replication and transcription.

We have the great pleasure to give thanks for the year (1965–66) of hospitality and inspiration at the laboratories of Dr. M. Nirenberg at the National Institutes of Health, and Dr. M. Weissbluth at Stanford, while we were on a fellowship from the National Cancer Institute; and for the year (1956–57) we spent at Dr. Linus Pauling's laboratory at the California Institute of Technology, where this work was started.

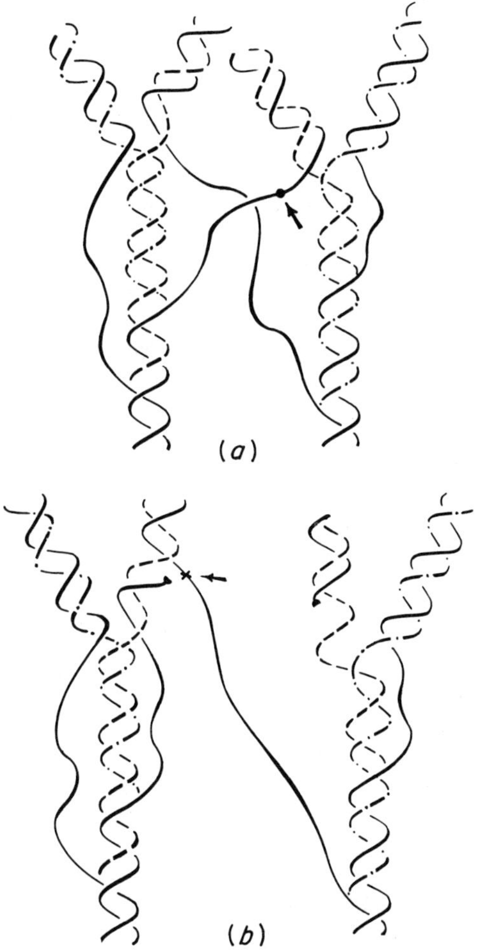

FIGURE 8.
An example showing one of the many different strand diagrams that may result when strands from different parental double-strand helixes combine. The arrow in the upper diagram (*a*) indicates what may be either a new single-strand break to be, or simply a preexisting break on that parental strand. The arrow in the lower diagram (*b*), which represents a subsequently possible strand arrangement, indicates a new break arising because the foreign parental strand encounters, at the locus marked by X, a double-strand region of that arm. The loose parental strand is likely to pair with the filial complementary strand of the same Y structure. These diagrams serve only as an example; the real issue is to find out by which strand diagrams a particular observable genetic effect may be represented.

REFERENCES

Barricelli, N. A. (1965). *Z. Vererbungslehre* **97**, 79.

Barricelli, N. A., and A. H. Doermann (1960). *Virology* **11**, 136.

Barricelli, N. A., and A. H. Doermann (1961). *Virology* **13**, 460.

Barricelli, N. A., and F. C. Womack (1965). *Virology* **27**, 589.

Barricelli, N. A., and F. C. Womack (1965). *Virology* **27**, 600.

Bodmer, W. F. (1965). *J. Mol. Biol.* **14**, 534.

Doermann, A. H., and L. Boehner (1963). *Virology* **21**, 551.

Farmer, J. L. (1966). Ph.D. thesis, Brown University.

Freese, E. (1963). In Taylor, J. H., ed., *Molecular Genetics* part I, p. 207. New York: Academic Press.

Goodgal, S. H., and R. M. Herriott (1961). *J. Gen. Phys.* **44**, 1201.

Goodgal, S. H., and R. M. Herriott (1961). *J. Gen. Phys.* **44**, 1229.

Hamilton, L. D., W. Fuller, and E. Reich (1963). *Nature* **198**, 538.

Haruna, I., and S. Spiegelman (1965). *Science* **150**, 884.

Haselkorn, R., and C. F. Fox (1965). *J. Mol. Biol.* **13**, 780.

Hotchkiss, R. D. (1962). Dyer lecture.

Jehle, H., M. L. Ingerman, R. M. Shirven, Wm. C. Parke, and A. A. Salyers (1963). *Proc. Nat. Acad. Sci. U.S.* **50**, 738.

Jehle, H. (1965). *Proc. Nat. Acad. Sci. U.S.* **53**, 1451. Reprinted in Pullman, B., and M. Weissmuth, eds., *Molecular Biophysics*, p. 359. New York: Academic Press.

Kozinski, A. W., P. B. Kozinski, and B. Shannon (1963). *Proc. Nat. Acad. Sci. U.S.* **50**, 746.

Kozinski, A. W., P. B. Kozinski, and B. Shannon (1963). *Virology* **20**, 213.

Kozinski, A. W., P. B. Kozinski, and B. Shannon (1965). *Proc. Nat. Acad. Sci. U.S.* **54**, 634.

Kozinski, A. W. (1966). Personal communication.

Kubitschek, H. E. (1964). *Proc. Nat. Acad. Sci. U.S.* **52**, 1374.

Kubitschek, H. E. (1966). *Proc. Nat. Acad. Sci. U.S.* **55**, 269.

Kubitschek, H. E., and T. R. Henderson (1966). *Proc. Nat. Acad. Sci. U.S.* **55**, 512.

Langridge, R., M. H. F. Wilkins, et al. (1960). *J. Mol. Biol.* **2**, 19.

Langridge, R., M. H. F. Wilkins, et al. (1960). *J. Mol. Biol.* **2**, 38.

Meselson, M., F. W. Stahl, and J. Vinograd (1957). *Proc. Nat. Acad. Sci. U.S.* **43**, 581.

Meselson, M., and F. W. Stahl (1958). *Proc. Nat. Acad. Sci. U.S.* **44**, 671.

Meselson, M., and J. J. Weigle (1961). *Proc. Nat. Acad. Sci. U.S.* **47**, 857.

Pauling, L. (1960). In Florkin, M., ed., *Aspects of the Origin of Life*, p. 132. New York: Pergamon Press.

Pauling, L., and H. A. Itano (1957). *Molecular Structure and Biological Specificity*. Am. Inst. Biol. Sci.

Pauling, L., and R. B. Corey (1956). *Arch. Biochem. Biophys.* **65**, 164.

Pauling, L., and R. Hayward (1964). *The Architecture of Molecules*. San Francisco: W. H. Freeman and Company.

Pauling, L. (1955). *Aspects of Synthesis and Order in Growth*, p. 3. Prinecton Univ. Press.

Pauling, L. (1957). *Festschrift f. Arthur Stoll*, p. 597. Basel: Birkhäuser.

Reich, E. (1964). *Science* **143**, 684.

Rosenberg, B. H., and L. Cavalieri (1963). *Prog. Nucleic Acid Research* **2**, 2.

Rosenberg, B. H., and L. Cavaliere (1964). *Proc. Nat. Acad. Sci. U.S.* **51**, 826.

Rothman, F. (1965). Personal communication.

Sechaud, J., G. Streisinger, J. Emrich, J. Newton, H. Lanford, H. Reinhold, and M. M. Stahl (1965). *Proc. Nat. Acad. Sci. U.S.* **54**, 1333.

Shalitin, C., and F. W. Stahl (1965). *Proc. Nat. Acad. Sci. U.S.* **54**, 1340.

Shipp, W., and R. Haselkorn (1964). *Proc. Nat. Acad. Sci. U.S.* **52**, 401.
Simha, R., J. M. Zimmerman, and J. Moacanin (1963). *J. Chem. Phys.* **39**, 1339.
Simha, R., J. M. Zimmerman, and J. Moacanin (1965). *J. Theor. Biol.* **8**, 81.
Simha, R., J. M. Zimmerman, and J. Moacanin (1965). *J. Theor. Biol.* **9**, 156.
Stahl, F., H. Modersohn, B. E. Terzaghi, and J. M. Crasemann (1965). *Proc. Nat. Acad. Sci. U.S.* **54**, 1342.
Stahl, F. W., and D. M. Green (1966). Personal communication.
Steiner, R. F., and R. F. Beers (1961). *Polynucleotides.* Amsterdam: Elsevier.
Streisinger, G., R. S. Edgar, and G. H. Denhardt (1964). *Proc. Nat. Acad. Sci. U.S.* **51**, 775.
Tabor, H., and C. W. Tabor (1964). *Pharmacol. Rev.* **16**, 245.
Wilkins, M. H. F. (1956). *Cold Spring Harbor Symp. Quant. Biol.* **21**, 75.
Witkin, E. M., and N. A. Sicurella (1964). *J. Mol. Biol.* **8**, 610.
Zubay, G., and M. H. L. Wilkins (1964). *J. Mol. Biol.* **9**, 246.

JAMES BONNER
DOROTHY Y. H. TUAN
Division of Biology
California Institute of Technology
Pasadena, California

On the Structure of Chromosomal Nucleohistone

INTRODUCTION

The genetic material of organisms does not occur in the cell as chemically pure DNA, but rather occurs complexed with one or another kind of polycation. In the higher organisms—those that have an organized nucleus and that include all organisms known today other than bacteria and blue-green algae—the polycations that are complexed to the chromosomal DNA are a characteristic group of small proteins, the histones. Thus far, histones have been found in the cell only in combination with DNA, and no nucleated cell has been found in which DNA is not complexed with histones or the related protamines. Although the structure of DNA itself is well known through the work of Watson, Crick, and Wilkins and their successors, and although the structures of proteins are even better known through the pioneering work of Pauling and his associates, and of Perutz and Kendrew and their followers, the structure of the nucleohistone complex is less well known. Since it is nucleohistone that forms the fundamental genetic material, and since histones play a role not only as cation for the anionic DNA but also in the regulation of the genetic activity of the DNA to which they are complexed (Huang and Bonner, 1962), it is of interest to consider their structure. It is the purpose of this paper to discuss the alterations in structure of DNA that result from its complex formations with histone, to discuss the structure of histones and the changes in their conformations that result from their complexing with DNA, to consider also the geometry with which proteins are disposed around the DNA molecule, and, finally, to consider how our knowledge of nucleohistone structure might be used in the elucidation of the structure of chromosomes themselves.

COMPONENTS OF CHROMOSOMAL NUCLEOHISTONE

In the course of the past five years or so, methodology for the isolation of pure chromosomes in the so-called interphase state—the extended state in which chromosomes occur between cell divisions—has been worked out in some detail. It is now possible, therefore, to prepare relatively large masses of pure chromatin, and to use such chromatin for chemical, physical, and biochemical studies (Bonner et al., 1967). Chromatin, as it is isolated by the method of Huang and Bonner (1962), is a clear colorless material, soluble in low ionic strength buffer, and dissociable by dilution into individual DNA strands, which are very long and each complexed with an appropriate array of proteins. Since chromosomes contain large amounts of DNA (the molecular weight of DNA in an individual chromosome of an individual higher organism may be 100 or more times greater than the 2,000 million molecular weight of the *E. coli* chromosome), chromatin has a very large sedimentation coefficient and is totally sedimented in centrifugal fields as low as 2,000 *g*. It is therefore helpful for many studies of chromosomal properties to subject chromatin to mechanical shearing, which reproducibly decreases the material into fragments possessing a sedimentation coefficient of approximately 30S, and a molecular weight of approximately 20 million. Such sheared chromatin is known as soluble nucleohistone, and the method of its preparation was first described by Zubay and Doty (1959). Application of the method of Zubay and Doty to chromatin purified by the methods of Huang and Bonner yields molecularly dispersed, completely soluble material (Bonner and Huang, 1963) suitable for studies such as those described below.

Chromosomal nucleohistone is composed of DNA, histone proteins, some nonhistone protein, and a small amount of RNA. The DNA of a nucleohistone particle of molecular weight 20 million is a single DNA molecule of an approximate molecular weight of 8 million. The histone to DNA mass ratio of a typical nucleohistone is approximately 1:1.35, the ratio required to provide one basic amino acid group for each phosphate group of the DNA. In a typical nucleohistone, too, the content of nonhistone protein is low, less than 0.1 that of the mass of DNA. DNA that is stoichiometrically complexed with histone does not support the formation of RNA by RNA polymerase (Huang and Bonner, 1962; Bonner and Huang, 1963). The chromatin from which nucleohistone is prepared by shearing does support such synthesis. This is due to the fact that chromatin contains not only DNA complexed with histones but also DNA complexed primarily with nonhistone proteins (Marushige, 1967; Bonner, 1967). The latter appears to constitute the template-active portion of the genome. Subsequent discussion of nucleohistone structure in this paper will not include the nonhistone proteins.

Chromatin contains about 5 percent as much RNA as DNA. This RNA is of a special class characterized by short chain length [approximately 40 nucleotides (Huang and Bonner, 1965)] and extensive homology with the genomal DNA [about 5 percent of DNA hybridized (Bonner and Widholm, 1967)]. This class of RNA, which appears to be concerned with control of gene activity, will also be omitted from further discussion in this paper.

THE HISTONES

The properties of histones have been described (Bonner and Ts'o, 1964; Murray, 1965). Their composition is such that one amino acid in four is a basic lysine or arginine. They are resolvable by column chromatography on Amberlite CG-50, using a gradient of guanidinium chloride as the eluant, into a characteristic series of six components, the first two of which are known as histones Ia and Ib, the second two, as histones IIa and IIb, and the third two as histones III-IV. Histones I are the "lysine-rich" histones, possessing a lysine to arginine ratio of approximately 10:1, and a proline content of

10 mole percent. Histones II are known as the "slightly lysine-rich" histones and have a lysine to arginine ratio of about 1.5:1. Histones III-IV, the "arginine-rich" histones, have a ratio of lysine to arginine of approximately 0.7:1. Both histones II and III-IV are lean in proline. Rechromatography and examination by disc electrophoresis in polyacrylamide gel, all conducted in 10 *M* urea (Bonner et al., 1967), reveal that the histone fractions thus prepared are essentially pure entities, and this is confirmed by tryptic digestion and fingerprinting (Fambrough, 1967). Similarly, the histones III-IV fraction may be separated into two pure components by preparative polyacrylamide gel electrophoresis. There are, then, six major histones, and although examples have been found in which a particular organ of a particular creature may lack one or another histone (for example, pea cotyledons lack histones Ia and Ib, and calf thymus lacks histone IIa), it may be stated as a general proposition that all organs of a given organism possess the same spectrum of histones. More surprisingly, all organisms possess histones that are remarkably alike. For example, although there are amino acid sequence differences between histones I of peas and histones I of cows, or between histones II of peas and histones II of cows, they share many peptides in common: their end terminal groups are identical, their electrophoretic mobilities are identical, and their amino acid analyses are identical within the error of measurement (Fambrough and Bonner, 1966). It is clear that histones perform a basic and important function, and that this function (or functions) places restriction upon the modifications to which histones may be subjected by evolution.

PARTITION OF HISTONE CLASSES

We first consider the topographical distribution of the several species of histone molecules along the DNA molecule. Are the different species intermixed among one another or are long segments of DNA covered by repeating units of a single species? This matter has been studied by Ohlenbusch, Olivera, Tuan, and Davidson (1967), who used techniques for the selective removal of individual histones followed by determination of the melting profile of the partially dehistonized nucleoprotein. Thus, thymus nucleohistone is pelleted from 0.6 *M* NaCl or 0.25 *M* $NaClO_4$. Histones Ia and Ib remain in solution. The pellet containing DNA and histones IIb and III-IV is redissolved in a medium of low ionic strength (2.5×10^{-4} *M* EDTA) and melted. If histone I occupied long stretches of the DNA, these stretches should now melt as deproteinized DNA rather than at the higher temperature characteristic of the melting of DNA of nucleohistone. However, nucleohistone lacking histone I exhibits no melting characteristic of deproteinized DNA.

According to Olivera (1966) a stretch of approximately 200 consecutive base pairs is required if the stretch is to exhibit the cooperative melting characteristic of high molecular weight DNA. Such a stretch would require about eight histone I molecules to complex completely with it. We conclude that infrequently, if ever, are eight histone I molecules adjacent to one another in thymus nucleohistone.

Histone II cannot at present be removed from nucleohistone without simultaneous removal of histone I. Therefore, as increasing amounts of histone II are removed, the bare patches thus formed are added to those generated

TABLE 1.
Molar extinction coefficients of calf thymus DNA as function of histone-DNA interactions.

State of DNA	Molar extinction coefficient E(p)
Native Na DNA (Sigma)	$(6.815 \pm .05) \times 10^3$
Denatured Na DNA (Sigma)	$(9.280 \pm .05) \times 10^3$
Native nucleohistone	$(7.561 \pm .099) \times 10^3$
Nucleohistone extracted with 0.6 F NaCl	$(7.635 \pm .149) \times 10^3$
Nucleohistone extracted with 0.8 F NaCl	$(7.110 \pm .174) \times 10^3$
Nucleohistone extracted with 0.9 F NaCl	$(7.083 \pm .075) \times 10^3$
Nucleohistone extracted with 1.2 F NaCl	$(6.989 \pm .050) \times 10^3$
Nucleohistone extracted with 1.6 F NaCl	$(6.815 \pm .149) \times 10^3$
Nucleohistone extracted with 4.0 F NaCl	$(6.815 \pm .050) \times 10^3$
Reconstituted nucleohistone I	$(6.715 \pm .025) \times 10^3$
Reconstituted nucleohistone II	$(6.815 \pm .075) \times 10^3$
Reconstituted nucleohistone III–IV	$(6.690 \pm .024) \times 10^3$

by removal of histone I. Extraction of nucleohistone with 1.2 *M* NaCl or 0.45 *M* $NaClO_4$, which removes essentially all of histones I and II (approximately 84 percent of total histone), does result in a preparation that yields a two-step melting profile, although even in this instance the T_M of the lower step is somewhat higher than that of pure DNA. Histones of all three categories are therefore scattered along the DNA molecule, with no long stretches covered by histones of any single kind. There are no gene-length stretches covered by one histone species. Whether they are ordered in a logical or meaningful manner we cannot yet say. The presence of a histone code, one written in a 5-symbol alphabet in the form of sequences of histone species, is not rigorously excluded.

STATE OF DNA IN NUCLEOHISTONE

DNA is present in nucleohistone in a state that yields, by deproteinization, the material we know as DNA. Nonetheless, the attributes of DNA in nucleohistone are not identical with those of pure DNA. For example, the DNA of nucleohistone is hyperchromic with respect to pure native DNA as shown in Table 1 (Tuan, 1967). Approximately one-third of the hypochromicity of DNA is lost by association of DNA with histone. Thus formation of the complex results in a distortion of the base-stacking interaction, a distortion in the direction of melting. A DNA molecule in native nucleohistone is also less asymmetrical than the same DNA freed of histone. This has been indicated qualitatively by light scattering. The radius of gyration of native DNA is Pq = 2500 Å (Scheraga and Mandelkern, 1953), and that of calf thymus nucleohistone is 2000 Å (Zubay and Doty, 1959; Bayley et al., 1962). A lesser asymmetry of DNA in nucleohistone is indicated too by the results of analytical sedimentation and by the viscosity. Gianonni and Peacocke (1963), using experimental values for molecular weight, M_L, viscosity [n] and S in the Scheraga-Mandelkern (1953) equation, have calculated the parameter β, which is a measure of effective hydrodynamic ellipsoid axial ratio of a molecule, or alternately, in terms of a coil model, a measure of the stiffness of the coil. The value of β for DNA of molecular weight 8×10^6 is 3.3, and that for nucleohistone containing DNA of the same molecular weight is 2.6. DNA in nucleohistone is therefore less asymmetric than DNA alone, or alternatively, less stiff. In general, we conclude that DNA in nucleohistone would appear to be shortened and fattened as compared with pure DNA.

In nucleohistone the base pairs are less fully oriented normal to the long axis of the molecule than are the base pairs in DNA, as has been shown by Ohba (1966) through study of flow birefringence and flow dichroism. DNA in the form of nucleohistone exhibits a greater rotary diffusion constant than pure DNA from the same preparation. Nucleohistone molecules are therefore less oriented at any given shear gradient than are the DNA molecules derived from it. In order to compare magnitudes of flow dichroism and flow birefringence of DNA and nucleohistone, Ohba (1966) therefore sheared the DNA until its rotary diffusion constant matched that of nucleohistone. Under these conditions both the flow dichroism and the flow birefringence of DNA in nucleohistone are approximately 0.4 that of DNA. The birefringence of DNA is due to the orientation of base planes normal to the DNA molecular axis. Therefore in nucleohistone the degree of such orientation normal to the long axis of the molecule is about 0.4 that of DNA. That base pairs are differently and less completely stacked in nucleohistone than are those in DNA is also shown by the hyperchromicity of DNA in nucleohistone, discussed above (Tuan, 1967).

The findings summarized above indicate that when histone is complexed to DNA, the DNA is shortened, the base pairs are partially unstacked (hyperchromic effect), and the base pairs are disoriented relative to the long axis of the molecule. These facts all point to, among other things, some sort of supercoiling of DNA as a result of its association with histone. The presence or absence of such supercoiling in nucleohistone has been studied by X-ray diffraction and directly by electron microscopy.

The X-ray diffraction pattern exhibited by pure DNA fibers at high (92 to 98 percent) humidity is characterized by (1) a strong meridional 3.4 Å reflection generated by the base pair-separation, (2) a strong equatorial 20 Å reflection generated by the lateral separation of DNA molecules, and (3) a series of layer lines of spacing corresponding to the helix pitch (Wilkins, Stokes, and Wilson, 1953). The X-ray diffraction pattern exhibited by nucleohistone fibers exhibits features (1) and (3) although in both features the diffraction patterns are much broader, indicating less orientation of the molecules in nucleohistone fibers than of those in DNA. The equatorial reflection generated by lateral separation is altered from the 20 Å characteristic of DNA to about 35 Å, as might be expected for DNA molecules clothed with a more than equal mass of protein. In addition, however, nucleohistone fibers generate a unique set of reflections absent from DNA patterns. These consist of a set of semimeridional spacings corresponding to separations of 22, 27, 37, 55, and 110 Å. These spacings were first noted by Wilkins, Zubay, and Wilson (1959) and have been studied in more detail by Bonner and Richards (1964), Palau, Pardon, and Richards (1967), Pardon and Wilkins (1967), and Pardon and Richards (1967). Pardon and Richards have now found a curious and entertaining fact: the semimeridional spacings characteristic of nucleohistone disappear when the fiber is severely stretched and reappear when the fiber is again relaxed. This immediately suggests that they are due to longitudinal spacings along the nucleohistone molecule that are generated by a form of supercoiling. These spacings are in accord with a model of nucleohistone in which the DNA is supercoiled with a pitch of 110 Å.

Although it is possible to reconstitute nucleohistones that possess the melting and template properties of native nucleohistone from purified DNA and histone (Huang, Bonner, and Murray, 1964), such preparations yield fibers that do not generate any of the semimeridional spacings (Bonner and Richards, 1964). Such reconstituted nucleohistones appear to lack any superstructure. It is of interest, therefore, that nucleohistone dissolved in, say 2.0 *M* NaCl [in which histone is known to be dissociated from DNA (Ohlenbusch, et al., 1967)] and then dialyzed to low ionic strength, yields fibers that exhibit the semimeridional spacings we have attributed to supercoiling. Apparently salt-dissociated histone retains properties that are lost when histone is extracted from nucleohistone by 0.2 *N* H_2SO_4 and purified by column chromatography (Fambrough and Bonner, 1966).

Finally, that the DNA of native nucleohistone is wound into a supercoil of pitch approximately 110Å can be visualized directly by electron microscopy. The nucleohistone molecule of width approximately 35 Å is seen to lie upon the electron microscope grid as a uniformly supertwisted strand in which three turns of the DNA helix occupy one turn of the supercoil (Moudrianakis, 1967).

The supercoil model of native nucleohistone accounts qualitatively at least for all of the aspects noted above in which DNA of nucleohistone differs from pure DNA. Thus the deformation of the DNA molecule required to transform it into a rather steep supercoil requires some sidewise sliding of adjacent base pairs; this accounts both for the hyperchromicity of DNA in nucleohistone and for its altered optical rotary dispersion spectrum. The fact that the base pairs of the supercoil are only statistically normal to the longitudinal axis of the molecule, whereas many have a component parallel to it, accounts for the decreased dichroism of nucleohistone as compared to DNA. Most satisfyingly, the model accounts for the shortening of the DNA molecule in nucleohistone. The required shortening of about 35 percent would be achieved by a supercoil of pitch 110 Å, making an angle with the screw axis of about 35°, and of total width (including that of nucleohistone molecule) of about 75 Å.

ROLE OF INDIVIDUAL HISTONES

Since native nucleohistone contains a finite, though small, number of histone species, it is worthwhile to inquire whether any particular species is charged with responsibility for the physical alterations that histones wreak on DNA. This question has been answered by means of selective dissociation of histones from nucleohistone as discussed above. Removal of lysine-rich histone I does not alter nucleohistone with respect to hyperchromicity (Tuan, 1967), X-ray diffraction pattern (Pardon and Richards, 1967), flow dichroism (Ohba, 1966), or ORD spectrum (Tuan, 1967). Removal of both histones I and II causes all of these properties to revert to the values more nearly characteristic of pure DNA. Histone II (or histone II in association with histone III-IV) is responsible for the principal physical properties conferred upon DNA in the nucleohistone complex.

CONFORMATION OF PROTEIN IN NUCLEOHISTONE

Purified histones in aqueous solution of low ionic strength exhibit a low proportion of α-helical structure. This has been determined by following rate of equilibration of amide protons with deuterons, using the N-D stretching frequency in the infrared (Bradbury and Crane-Robinson, 1964), and by optical rotatory dispersion (Tuan, 1967). However, histones, as they are present in native nucleohistone, exhibit abundant α-helical structure. Bradbury and Crane-Robinson conclude that approximately 60 percent of the total histone of native nucleohistone is present in α-helical conformation. The ORD data of Tuan in the UV region suggest approximately 40 percent α-helical content. This, however, is distributed among the different histones unequally. Thus, histone I possesses low α-helical content whether it is free or DNA-bound (Tuan, 1967). This is as it should be, since histone I, because of its high content of proline [10 mole percent (Murray, 1965)] cannot assume the α-helical form. Both histone II and histone III-IV have been shown by the ORD studies of Tuan (1967) to possess 40 to 60 percent α-helical conformation in nucleohistone. That these histones lack α-helical form when dissolved in low ionic strength media in the absence of DNA is apparently due to disruption of structure by internal charge repulsion. Thus histone II in particular, when dissolved in 3–4 *M* NaCl, exhibits an α-helical content even larger than that which it exhibits when bound to DNA.

ARRANGEMENT OF PROTEINS IN NUCLEOHISTONE

No experimental method has yet been devised to determine the geometry with which histones are combined with DNA in nucleohistone. Reflections due to α-helical structure are not detectable in X-ray diffraction patterns of nucleohistone fibers. This is in part due to the poor orientation of DNA in such fibers as described above, the diffraction patterns being of corresponding diffuseness. The principal contributions to the elucidation of the structure of nucleohistone by X-ray diffraction has been made by studies on the structure of the complex between a random alanine-lysine (ratio 4:1) copolymer and DNA (Zubay, Wilkins, and Blout, 1962). Fibers of this complex yield fiber patterns that include the 7.5 Å and 4.4 Å spacings of the polyalanine α-helix. The 7.5 Å spacing appears as an equatorial one, the 4.4 Å spacing as a meridional one. Zubay et al. interpret this structure as consisting of polypeptide α-helices generally parallel to the fiber axis. The complex generates, however, none of the meridional spacings interpreted above as due to supercoiling, and the structure is not clearly identifiable as closely related to that of native nucleohistone. The same may be said of the protamine-DNA complexes studied by Feughelman et al. (1955). In this case, in which the native structure of the DNA is well preserved, the protamine (in coil form) appears to lie along the narrow groove of the DNA.

Bradbury and Crane-Robinson (1964) have approached the matter by examining, with polarized infrared radiation, the peptide bands of the histone of stretched nucleohistone films. They find complete absence of dichroism in all peptide bands. This is despite the fact that all DNA bands examined revealed strong dichroism, showing that the DNA was in fact oriented. It appears that the histone component is either totally disordered or disposed in an ordered but statistically isotropic manner. Such would be the case, for example, if nucleohistone molecules paralleled the grooves of the DNA. New approaches are needed for the study of this matter. One such approach might be the labeling of selected amino acids with electron opaque derivatives for inspection by electron microscopy.

One interesting facet of nucleohistone structure relates to histone I. Although histone I has a higher positive charge density than histones of the other two classes, it is nonetheless the histone most readily dissociated from the nucleohistone complex. This behavior may be due to the conformational differences between histone I, which possesses coil structure even when associated with DNA, and other histones which, as described above, possess substantial α-helical content when associated with DNA. It might be, for example, that histone I is fitted into one of the DNA grooves and that as a consequence not all basic groups are able to associate intimately with DNA counter ion. Walker (1965) has in fact shown by spectrophotometric and electrometic titration studies that although approximately 80 percent of the basic groups of the histone of native nucleohistone are masked and not available for titration, 20 percent of such groups are available. It would be of interest to extend this work and to determine whether the titration-available basic groups are predominantly contained in histone I. In any case, histones II and III-IV must, because of the large diameter of the α-helix, lie largely along the outer surface of the DNA helix. It may be further noted that histone I, even though it is most readily dissociated, is the most effective histone in lowering template activity of DNA for RNA synthesis, as well as the most effective histone in stabilization of DNA against melting. Clearly the geometry of the histone-DNA complex and the differences in this geometry among the several histones are important matters for further study.

RELATION OF NUCLEOHISTONE TO CHROMOSOME STRUCTURE

The nucleohistone molecules whose properties have been considered above are of course but fragments of chromosomes. They are in fact fragments of the size to which DNA is readily reduced by mechanical shear. The interphase chromosome is presumably made up of a near-continuum of such nucleohistone, regions of DNA stoichiometrically complexed with histone protein [template inactive (Huang and Bonner, 1962; Huang, Bonner and Murray, 1964)], being interspersed with shorter regions of DNA complexed with nonhistone protein [template active (Bonner and Huang, 1963; Marushige, 1967; Bonner, 1967)]. Among the principal present questions of chromosome structure are these: (1) How long are the DNA molecules of a chromosome; is a chromosome made up of one DNA molecule or many? (2) Is the DNA of a chromosome circular, as is that of *E. coli*, mitochrondria, and many viruses; if so, does a chromosome consist of one circle or many? (3) What is the relation of the interphase chromatin to the compact metaphase chromosome of classical cytology (DuPraw, 1965)? That the DNA molecules of interphase chromosomes are of exceedingly large size has been shown by Huberman and Riggs (1966), who used the Cairns (1963) autoradiographic technique, which minimizes both shear and nuclease action. They have found, in preparations from Chinese hamster cells, occasional DNA molecules that are 1.6–1.8 mm long, and many that are more than 0.8 mm long. These molecules are shorter than the total length of DNA (about 9 cm) in the Chinese hamster chromosome. There is, however, evidence to suggest that chromosomes may be made up of several DNA molecules. Thus, in the salivary gland chromosomes of Drosophila (in which about 2,000 chromatin strands lie side by side in register) there are fifty separate points at which replication (followed by incorporation of labeled thymidine) takes place simultaneously (Plaut, 1964). On the reasonable assumption that each replication point represents a separate DNA molecule, these molecules would be on the average about 0.9 mm long, which is the general size found by Huberman and Riggs. The available facts suggest then that a chromosome may be made up of numerous (50–100) separate DNA molecules, each about the same size as the *E. coli* chromosome, and each of the conformation discussed above. How these individual pieces are held together and whether or not they are circular, as are *E. coli* chromosomes, are important questions for the future.

REFERENCES

Bayley, P., B. Preston, and R. Peacocke (1962). *Biochim. Biophys. Acta* **66**, 943.

Bonner, J. (1967). *Biochim. Biophys. Acta.* In press.

Bonner, J., G. R. Chalkley, M. Dahmus, D. Fambrough, F. Fujimura, R. C. Huang, J. Huberman, R. Jensen, K. Marushige, H. Ohlenbusch, B. Olivera, and J. Widholm (1967). In Colowick, S. P., and N. O. Kaplan, eds., *Methods in Enzymology*, Vol. 12. New York: Academic Press. In press.

Bonner, J., and R. C. Huang (1963). *J. Mol. Biol.* **6**, 169.

Bonner, J., and B. Richards (1964). Unpublished work.

Bonner, J., and P. O. P. Ts'o (1964). In Bonner, J., and P. O. P. Ts'o, eds., *The Nucleohistones*, p. 398. San Francisco: Holden-Day.

Bonner, J., and J. Widholm (1967). *Proc. Nat. Acad. Sci. U.S.* In press.

Bradbury, E. M., and C. Crane-Robinson (1964). In Bonner, J. and P. O. P. Ts'o, eds., *The Nucleohistones*, p. 117. San Francisco: Holden-Day.

Cairns, J. (1963). *J. Mol. Biol.* **6**, 208.

DuPraw, E. J. (1965). *Nature* **206**, 338.

Fambrough, D., and J. Bonner (1966). *Biochemistry* **5**, 2563.

Fambrough, D. (1967). Ph.D. Thesis, California Institute of Technology.

Feughelman, M., R. Langridge, W. Seeds, A. Stokes, H. Wilson, C. Hooper, M. H. F. Wilkins, R. Barclay, and L. Hamilton (1955). *Nature* **175**, 834.

Gianonni, R., and A. Peacocke (1963). *Biochem. Biophys. Acta* **68**, 157.

Huang, R. C., and J. Bonner (1962). *Proc. Nat. Acad. Sci. U.S.* **48**, 1216.

Huang, R. C., and J. Bonner (1965). *Proc. Nat. Acad. Sci. U.S.* **54**, 960.

Huang, R. C., J. Bonner and K. Murray (1964). *J. Mol. Biol.* **8**, 54.

Huberman, J., and A. Riggs (1966). *Proc. Nat. Acad. Sci. U.S.* **55**, 599.

Marushige, K. (1967). In preparation.

Marushige, K., and J. Bonner (1966). *J. Mol. Biol.* **15**, 160.

Moudrianakis, E. N. (1967). In preparation.
Murray, K. (1965). *Ann. Rev. Biochem.* **34**, 209.
Ohba, Y. (1966). *Biochim. Biophys. Acta* **123**, 76.
Ohlenbusch, H., B. Olivera, D. Tuan, and N. Davidson (1967). *J. Mol. Biol.* In press.
Olivera, B. M. (1966). Ph.D. Thesis, California Institute of Technology.
Palau, J., J. Pardon, and B. Richards (1966). *Biochim. Biophys. Acta* **129**, 633.
Pardon, J., and B. Richards (1967). In preparation.
Pardon, J., and M. H. F. Wilkins (1967). In preparation.
Plaut, W. J. (1964). *J. Mol. Biol.* **7**, 632.
Scheraga, H., and L. Mandelkern (1953). *J. Am. Chem. Soc.* **75**, 179.
Tuan, D. Y. (1967). Ph.D. Thesis, California Institute of Technology.
Walker, I. O. (1965). *J. Mol. Biol.* **14**, 381.
Wilkins, M. H. F., A. R. Stokes, and H. R. Wilson (1953). *Nature* **171**, 4.
Wilkins, M. H. F., G. Zubay, and H. R. Wilson (1959). *J. Mol. Biol.* **1**, 179.
Zubay, G., and P. Doty (1959). *J. Mol. Biol.* **1**, 1.
Zubay, G., M. H. F. Wilkins, and E. R. Blout (1962). *J. Mol. Biol.* **4**, 69.

DAVID R. DAVIES
GARY FELSENFELD
National Institute of Arthritis and Metabolic Diseases
National Institutes of Health
Bethesda, Maryland

The Structure of RNA

When the structure of DNA was proposed by Watson and Crick in 1953, it appeared reasonable to assume that the structure of RNA would soon yield to a similar type of investigation. This assumption was incorrect, and it is only in recent years that acceptable structures for RNA have been proposed. In this brief note we shall review some of the steps which have led to the present views of RNA structure. We shall consider two main bodies of evidence: that based on X-ray diffraction work on fibers of RNA and polynucleotides, and that based on solution studies.

In any discussion of the structure of RNA it is necessary to define clearly the class of RNA being discussed. We shall recognize four different classes of RNA.

1. Messenger RNA: The expression of its function presumably depends only on its linear sequence of bases, and on its translation into a polypeptide chain. In this class we would also include single-stranded viral RNA, although this may also carry with it some structural information that enables it to ball up so that it may be encapsulated by the virus protein.

2. Ribosomal RNA: Together with protein, it composes the particles on which protein synthesis takes place. Although it is an essential ingredient of the ribosome, it is not clear whether it is there only to provide structural stability or whether it has other functions, such as providing direct attachment sites for the tRNA.

3. Transfer RNA: This relatively small nucleic acid containing approximately 75 bases has several functions to perform. It must be recognized by the correct amino acid transfer enzyme and it must bind to the ribosome only when the site is defined by the appropriate codon of the messenger RNA. It must attach to this site when it is charged with the appropriate amino acid, and it is released upon transfer of the growing polypeptide chain to the next tRNA molecule. Because of its low molecular weight, its isolation in pure form, and the sequence data for it that now exist, it offers the most promising opportunity for defining a complete nucleic acid three-dimensional structure.

4. Double-stranded viral RNA: Examples of these are the reovirus, the wound tumor virus, the rice dwarf virus RNA and the replicating forms of the RNA of viruses such as MS2. These RNA's form a class of regular helices

with Watson and Crick base pairing, giving rise to a well-defined structure. The work of Langridge and Gomatos (1963), Langridge et al. (1964), Tomita and Rich (1964), and Sato et al (1966), together with the complementarity of the base composition, has established beyond a reasonable doubt that the structure is a double-stranded helix, although the exact dimensions of the helix (that is, whether it has ten or eleven base pairs per turn) remain uncertain. The structure, therefore, resembles that of DNA in that it is double-stranded and regular throughout almost its entire length. In this respect these RNA's are quite different from those of the other three classes: however, their structures have a direct bearing on the conclusions that may be drawn for other RNA's, with regard both to their internal structures and to intermolecular contacts between adjacent helical regions (Arnott et al., 1966).

X-RAY DIFFRACTION

The earliest studies on the X-ray diffraction of RNA (Rich and Watson, 1954) established that the diffraction patterns were of the same general nature as those of DNA, having intensity on the meridian in the region of 3 to 4 Å spacing, and having an intensity distribution appropriate to a helical structure. However, at that time the structure could not be deduced because of the lack of good orientation in the fiber and the diffuseness of the diffraction pattern. The absence of complementary base ratios in RNA was also incompatible with a simple DNA-like double helix.

Studies on the polynucleotide interactions revealed an increasingly complex number of structural possibilities with many base-pairing arrangements other than those proposed by Watson and Crick. In addition it became clear that structures with more than two strands were possible (Felsenfeld et al., 1957). Suggestive evidence for the RNA structure came from the fact that the copolymer rAU gave diffraction patterns very similar to those of RNA (Rich, 1956). Also, the double-stranded complex of rI:rC was observed to have two kinds of diffraction patterns: one was a highly crystalline pattern which was clearly that of a double-stranded helix with Watson and Crick base pairing; the other pattern was noncrystalline and was very similar to that of RNA (Davies and Rich, 1958; Davies, 1960), and it has occasionally been possible to transform the material from this form into the crystalline form in the same fiber (Davies, unpublished). This finding, combined with the observation that only a 1-1 complex of rI:rC is formed in solution, and with the fact that the noncrystalline pattern is unlike that of rI or rC alone, suggested that the predominant structure giving rise to the RNA pattern is one with Watson and Crick base pairing. A considerable degree of disorder in the structures of RNA from viruses and ribosomes is indicated by the diffuseness of the diffraction pattern and its lack of intensity (exposure times necessary to obtain diffraction patterns of these RNA's are longer than the times necessary to produce DNA patterns of comparable intensity).

More recent work on partially degraded ribosomal RNA, originally believed to be transfer RNA (Spencer et al., 1962; Spencer and Poole, 1965), showed that this material could form liquid crystals of the spherulite type, and that it could give quite well-oriented, partially crystalline X-ray diffraction patterns which resembled those of the A form of DNA. The intensity distribution of these crystalline patterns closely corresponds to the intensity of patterns from noncrystalline fibers of the same material. The latter patterns resemble those previously obtained from ribosomal RNA or from noncomplementary viral RNA. It has been inferred from these observations that the structure giving rise to the noncrystalline diffraction pattern consists principally of Watson and Crick base pairs. Subsequent examination of reovirus, wound tumor virus, rice virus, and MS2RF RNA patterns have confirmed that well-ordered double-stranded RNA will crystallize in a form like that of the A form of DNA. The infrared dichroism of oriented films of the rice dwarf virus indicates however that the phosphates have an orientation that is different from that in either A or B forms of DNA (Sato et al., 1966).

Sasisekharan and Sigler (1965) have shown that if conditions are carefully controlled, the diffraction pattern obtained from the polymer complex rA:rU resembles that of reovirus RNA. Under other conditions the material forms a triple-stranded rA:rU:rU complex with excess uncombined rA. The close similarity of the reovirus RNA pattern to that of the rA:rU suggests that the latter forms preferentially a Watson and Crick base pair, and not one of the other kinds of pairing possible for this material (for example, Hoogsteen, 1963).

The X-ray evidence, in general, therefore supports the Watson and Crick type of base pairing as the predominant pairing to be found in the first three classes of RNA. Other possible base-pairing arrangements cannot be ruled out on the basis of the available X-ray diffraction patterns. However, because these diffraction patterns resemble those of noncrystalline base-paired RNA more closely than they resemble any of the other polynucleotide patterns, there is little support for extensive formation of these other forms of base pairing.

The disorder reflected in the diffraction pattern may then be accounted for by the fact that only relatively small regions of these RNA's exist in the form of regular double helices, with unpaired bases either being looped out or forming irregular links with the helical regions. It is probably not possible to provide from the X-ray data alone an estimate of the percentage of bases in a complementary base-paired configuration.

SOLUTION STUDIES

The configuration of naturally occurring single-strand ribonucleic acids in solution has been the subject of much debate. Doty et al. (1959) suggested some years ago that extensive regions of intramolecular hydrogen bonding could form within the RNA strand, provided that mismatches "looped out." Further work by Fresco and Alberts (1960) with synthetic polynucleotides showed that mismatched bases were indeed excluded from an otherwise base-paired structure. The picture of RNA that emerged involved hairpin-shaped regions of intramolecular double-strand hydrogen bonded base pairs, interspersed with single-strand unbonded regions. Calculations showed that a large fraction of nucleotides could be accommodated on a random basis in the hydrogen bonded helical regions. This model for RNA structure appeared to be consistent with a number of physical properties of RNA in solution: (1) RNA behaves hydrodynamically as a random coil, unlike DNA of equivalent molecular weight. (2) The optical properties of RNA are consistent with the existence of some kind of ordered structure at the local level. The optical rotatory dispersion and the hypochromism of RNA (relative to the nucleotides) show that there are restrictions on the relative motions of the chromophores. Furthermore, these properties are temperature-dependent; it is possible to "melt out" the ordered structure. (3) The rate of reaction of RNA with formaldehyde is temperature-dependent; that is, at each temperature the rate of reaction is correlated with the amount of structure estimated from the parameters mentioned in (2).

This model of RNA structure was unquestioned for several years; it appeared to explain, qualitatively at least, the properties of such diverse species as the RNA of tobacco mosaic virus (Doty et al., 1959) and the low molecular weight transfer RNA's (Fresco et al., 1963; Felsenfeld and Cantoni, 1964). The end of the period of general acceptance coincides with the renewal of interest in the structure of the single-strand synthetic polynucleotides. Since 1964, a number of workers have studied the effect of modifying polynucleotides to alter drastically or destroy completely the hydrogen bonding capability of the bases. Fasman et al. (1964) have formylated poly C; Van Holde et al. (1965) have studied the polymer of N6-hydroxyethyl adenine; and Griffin et al. (1964) have examined poly 2,6-diaminopurine-ribotide. The optical rotatory dispersion and hypochromism of these polymers are quite similar to those of the unsubstituted polynucleotides, poly C and poly A at neutral *p*H. These facts, together with the observation of temperature-dependent optical rotation and hypochromism in the dimer ApA, led to a model of homopolynucleotide structure in which hydrogen bonding between bases is not involved. The bases are held in a local and noncooperatively formed "stacked" configuration by the action of electrostatic forces and solvent interaction effects (Holcomb and Tinoco, 1965; Van Holde et al., 1965; Leng and Felsenfeld, 1966).

Since poly rC and poly rA had been thought of as prototypes of RNA structure (that is, they were at first supposed to contain internal hydrogen bonded hairpin-shaped regions), the discoveries concerning the synthetic polymers raised grave doubts about the structure of naturally occurring RNA. Since poly rC and poly rA are optically active and hypochromic, some workers have suggested that the optical properties of the transfer RNA's (for example) can also be explained by the formation of single-strand "stacks," and that internal hydrogen bonding may be of only minor importance (Michelson et al., 1966; Fasman, et al., 1965). This point of view does not, however, appear to us to be consistent with a detailed examination of the optical and other physical properties of RNA in solution. The comments that follow present the evidence supporting the existence of a considerable amount of interbase hydrogen bonding in RNA. Though the measurements cited were made on transfer RNA (tRNA) there is at present no reason to believe that the *secondary* structure of other RNA's is different.

In trying to understand the significance of the single-strand stacked structure of neutral poly rA or poly rC for the structure of RNA, one should abandon the early tenet that these polymers are prototypes of RNA structure. Both of the homopolymers obviously differ from RNA in that they are incapable of forming hydrogen bonded base pairs of the Watson and Crick type. Furthermore, the fact that poly rA and poly rU combine, even when one of the strands contains "imperfections" in the form of bases that cannot pair (Fresco and Alberts, 1960; Bautz and Bautz, 1964), shows that hydrogen bonded structures often form at the expense of single-strand structures even when perfect helices cannot be made.

There is also a considerable amount of direct experimental evidence suggesting that base pairing occurs in RNA. The optical properties of tRNA, for example, resemble much more closely those of the double-strand polyribonucleotides, poly rA:rU or poly rI:rC, than they do those of single-strand poly rA. The magnitude of the hypochromism of tRNA at temperatures near 0°C is greater than that expected from single-strand structures. In 1 *M* NaCl solutions, there is a plateau region at the low-temperature end of the absorbance "melting" curve of tRNA, whereas none of the single-strand polymer structures is completely ordered at temperatires above 0°, and consequently no plateau is observed at these temperatures for single-strand structures. Furthermore, the extent of formation of the single-strand poly rA structure is nearly independent of a variation in ionic strength from 0.01 to 1, whereas the extent of structure formation in tRNA (as judged by optical properties) is strongly dependent upon ionic strength in a manner resembling the ionic strength dependence of the stability of poly rA:rU or poly rI:rC. Recently, Cantor et al. (1966) have attempted to correlate the optical rotatory dispersion (ORD) of yeast alanine tRNA with the ORD to be expected from a single-strand stack of that base sequence. The ORD is calculated by summing the contributions to ORD from the interactions of all nearest neighbor dinucleotide pairs in the chain. These workers concluded that they could not account for the observed ORD of alanine tRNA at low temperature without introducing a large amount of hydrogen bonded base pairing. They have reached similar conclusions about other kinds of RNA. Fasman et al. (1965) have shown that formylation of tRNA results in a marked change in the shape of the thermal denaturation curve, while the curve of poly rC is almost unchanged by formylation. This finding suggests that a cooperatively formed, hydrogen bonded tRNA structure is being disrupted by the chemical reaction.

Unfortunately, it is not yet possible to make a rigorous statement about the nature of the hydrogen bonding, or its exact extent. The most recent estimate of the number of hydrogen bonded bases is that of Cantor et al. (1966) who, as a result of their analysis of optical rotatory dispersion curves, found 64 percent of the bases paired in alanine tRNA. Englander and Englander (1965) have measured the number of slowly exchanging protons in unfractionated tRNA, and conclude that 82 percent of the bases are hydrogen bonded, but this may be an upper limit because of possible contributions from 2′OH hydrogen bonds. The existence of such intramolecular hydrogen bonds involving 2′-OH groups have also been suggested by a number of workers. Among the receptor groups which have been proposed are the phosphate oxygen atom, the ribose ring oxygen atom, and various sites on the bases. There is at present very little experimental evidence to support any one model.

It is clear from the work of Chamberlin (1965), and Ts'o et al. (1966) that there are differences between the stabilities of ribose polymers and their deoxyribose homologues. The double-strand ribose polymers are always more resistant to thermal denaturation than the corresponding deoxyribose polymers, or hybrids consisting of one ribose and one deoxyribose strand (Chamberlin, 1965) but that is the only generalization that can be made.

The principal difficulty in determining the extent of hydrogen bonding is in estimating the number of unorthodox base pairs which may form, because there is insufficient information about their stability and their optical properties; this is a particularly serious problem in the study of tRNA, which has a large number of unusual bases. Some of these bases can form the Watson and Crick type of base pairing, whereas others clearly cannot do so because of methyl substitution for hydrogens in those positions which are required for hydrogen bonding. We shall consider several special cases below.

1. Hypoxanthine has been found in several tRNA's. It is believed to be the first base of the anticodon in yeast tRNA specific for alanine, valine, and serine (Crick, 1966). It can clearly form a base pair with cytosine (Davies and Rich, 1958; Davies, 1960), and can in addition pair in two ways with adenine (Rich, 1958). Crick has provided a tentative explanation for its presence in terms of the "wobble" hypothesis in which the hypoxanthine might pair with adenine instead of with cytosine. He pointed out that guanine, because of its additional amino group, seems unlikely to pair with adenine in the same way.

2. Pseudo-uridine is able to form the same hydrogen bonds with adenine as can uracil. However, it also has the hydrogen bonding potential to acquire an additional adenine.

3. Dihydrouracil is a nonplanar molecule; at the very least its C6 and C5 hydrogens project above and below an otherwise planar molecule. Staggering of the hydrogens about the single bond may also result in further loss of planarity. Cerutti et al. (1966) have concluded that polydihydrouridylate will probably not combine with poly rA, and they suggest the lack of planarity as a reason for this. Similarly, in tRNA, it could not be incorporated into a regular helical region without distortion of the helix.

4. Thymine and 5-methylcytosine can form Watson and Crick base pairs.

5. 2-thiouracil and 4-thiouracil can probably form Watson and Crick base pairs. In addition, however, they provide an opportunity for cross-linking the RNA through the formation of disulfide bridges (Carbon et al., 1965; Lipsett, 1965).

6. Most of the other methylated species would be unable to form Watson and Crick base pairs. However, they would have acquired an increased capacity for hydrophobic interactions.

The sequences of yeast tRNA specific for alanine and for tyrosine have been determined (Holley et al., 1965; Madison et al., 1966). It is perhaps significant that several of the unusual bases occur at identical distances from the anticodon in both molecules, and that similar "clover leaf" models can be proposed for the molecules; these findings suggest that perhaps these molecules have similar tertiary structures.

These models consist of three "arms" of double-strand base-paired regions; tertiary structure could arise by interactions among the arms. Unlike the α-helical regions of proteins, which may tend to approach each other because of hydrophobic interactions between nonpolar side chains, the helical regions of tRNA would probably have to be brought into close association, either by specific interaction between some of the looped-out bases not already involved in hydrogen bonding, or by salt linkages involving metal ion bridges between phosphate groups. The possible role of ions such as Mg^{++} in bridging is clearly suggested in the studies of tertiary structure to be mentioned below. It may be that the unusual effect of Mg^{++} in increasing the sharpness of the thermal transition of tRNA arises from its role in stabilizing tertiary structure. In any case, the close approach of different helical regions of RNA to one another probably requires the rather complete screening of the backbone negative charge.

It is possible that these regions may also be joined together by cross-links of the kind postulated for reovirus RNA (Arnott et al., 1966). However cross-links involving hydrogen bonding between the 2′ OH and the phosphate oxygens seem unlikely to contribute much to the stability of the structure in an aqueous environment because of competition with water and because of the infrequency of their occurrence (two per eleven base pairs).

The arguments concerning secondary structure of tRNA apply equally well to ribosomal RNA (Spencer et al., 1962), and to the single-strand viral RNA's, such as TMV RNA. There may be considerable variation among RNA's however, in tertiary structure. Though very little is known about the folding of RNA, recent work (Henley et al., 1966; Millar and Steiner, 1966) shows that the hydrodynamic properties of tRNA at low temperature are those corresponding to a rather asymmetric molecule, and that as the temperature is raised, shape changes corresponding to a decrease in asymmetry occur which are not accompanied by significant changes in optical absorbance. This is the most direct evidence of existence of tertiary structure in RNA. Recently, it has been shown that certain tRNA molecules can be reversibly inactivated as amino acid acceptors by removing Mg^{++} (Lindahl et al., 1966; Gartland and Sueoka, 1966). The reactivation process requires that a divalent ion be reintroduced while the structure is partly disrupted by heating, by low *p*H, or by low ionic strength of the solvent. It has been suggested that the binding of Mg^{++} or other divalent ions to specific sites on tRNA may be required to hold the molecules in an active configuration. The compact tertiary structures of tRNA suggest that these molecules may be crystallizable as single crystals, and thus amenable to methods such as those used for protein crystal structure determination.

REFERENCES

Arnott, S., F. Hutchinson, M. Spencer, M. H. F. Wilkins, W. Fuller, and R. Langridge (1966). *Nature* **211**, 227.

Bautz, E. K. F., and E. F. Bautz (1964). *Proc. Nat. Acad. Sci. U.S.* **52**, 1476.

Cantor, C. R., S. R. Jaskunas, and I. Tinoco, Jr. (1966). *J. Mol. Biol.* **20**, 39.

Carbon, J. A., L. Hung, and D. S. Jones (1965). *Proc. Nat. Acad. Sci. U.S.* **53**, 979.

Cerutti, P., H. T. Miles, and J. Frazier (1966). *Biochem. Biophys. Research Commun.* **22**, 466.

Chamberlin, M. J. (1965). *Federation Proc.* **24**, 1446.

Crick, F. H. C. (1966). *J. Mol. Biol.* **19**, 548.

Davies, D. R. (1960). *Nature* **186**, 1030.
Davies, D. R., and A. Rich (1958). *J. Am. Chem. Soc.* **80**, 1003.
Doty, P., H. Boedtker, J. Fresco, R. Haselkorn, and M. Litt (1959). *Proc. Nat. Acad. Sci. U.S.* **45**, 482.
Englander, S. W., and J. J. Englander (1965). *Proc. Nat. Acad. Sci. U.S.* **53**, 370.
Fasman, G. D., C. Lindblow, and L. Grossman (1964). *Biochemistry* **3**, 1015.
Fasman, G. D., C. Lindblow, and E. Seaman (1965). *J. Mol. Biol.* **12**, 630.
Felsenfeld, G., and G. L. Cantoni (1964). *Proc. Nat. Acad. Sci. U.S.* **51**, 818.
Felsenfeld, G., D. R. Davies, and A. Rich (1957). *J. Am. Chem. Soc.* **79**, 2023.
Fresco, J. R., and B. Alberts (1960). *Proc. Nat. Acad. Sci. U.S.* **46**, 311.
Fresco, J. R., L. C. Klotz, and E. G. Richards (1963). *Cold Spring Harbor Symp. Quant. Biol.* **28**, 83.
Gartland, W. J., and N. Sueoka (1966). *Proc. Nat. Acad. Sci. U.S.* **55**, 948.
Griffin, B. E., W. J. Haslam, and C. B. Reese (1964). *J. Mol. Biol.* **10**, 353.
Henley, D. D., T. Lindahl, and J. R. Fresco (1966). *Proc. Nat. Acad. Sci. U.S.* **55**, 191.
Holcomb, D. N., and I. Tinoco (1965). *Biopolymers* **3**, 121.
Holley, R. W., J. Apgar, G. A. Everett, J. T. Madison, M. Marquisee, S. H. Merrill, J.R. Penswick, and A. Zamir (1965). *Science* **147**, 1462.
Hoogsteen, K. (1963). *Acta Cryst.* **16**, 907.
Langridge, R., M. A. Billeter, P. Boost, H. R. Burdon, and C. Weissmann (1964). *Proc. Nat. Acad. Sci. U.S.* **52**, 114.
Langridge, R., and P. J. Gomatos (1963). *Science* **141**, 694.
Leng, M., and G. Felsenfeld (1966). *J. Mol. Biol.* **15**, 455.
Lindahl, T., A. Adams, and J. R. Fresco (1966). *Proc. Nat. Acad. Sci. U.S.* **55**, 940.
Lipsett, M. N. (1965). *J. Biol. Chem.* **240**, 3975.
Madison, J. T., G. A. Everett, and H. Kung (1966). *Science* **153**, 531.
Michelson, A. M., T. L. V. Ulbricht, T. R. Emerson, and R. J. Swan (1966). *Nature* **209**, 873.
Millar, D. B., and R. F. Steiner (1966). *Biochemistry* **5**, 2289.
Rich, A. (1956). In McElroy, W., and B. Glass, eds., *The Chemical Basis of Heredity*. Baltimore: Johns Hopkins Press.
Rich, A. (1958). *Nature* **181**, 521.
Rich, A., and J. D. Watson (1954). *Proc. Nat. Acad. Sci. U.S.* **40**, 759.
Sasisekharan, V., and P. B. Sigler (1965). *J. Mol. Biol.* **12**, 296.
Sato, T., Y. Kyogoku, S. Higuchi, Y. Mitsui, Y. Iitaka, M. Tsuboi, and K. Miura (1966). *J. Mol. Biol.* **16**, 180.
Spencer, M., W. Fuller, M. H. F. Wilkins, and G. L. Brown (1962). *Nature* **194**, 1014.
Spencer, M., and F. Poole (1965). *J. Mol. Biol.* **11**, 314.
Tomita, K., and A. Rich (1964). *Nature* **201**, 1160.
Ts'o, P. O. P., S. A. Rapaport, and F. J. Bollum (1966). *Biochemistry* **5**, 4153.
Van Holde, K. E., J. Brahms, and A. M. Michelson (1965). *J. Mol. Biol.* **12**, 726.

NORMAN DAVIDSON
Division of Chemistry and Chemical Engineering
California Institute of Technology
Pasadena, California

JAMES C. WANG
Department of Chemistry
University of California
Berkeley, California

Effects of Density-Gradient Centrifugation on Equilibria Involving Macromolecules

It is well known that, because the total mass of the products is equal to the total mass of the reactants, an acceleration field (gravitational or centrifugal) does not change the equilibrium constant of a chemical reaction (Guggenheim, 1957). However, a sufficiently strong acceleration field will cause a spatial variation of the concentration for each of the various components, and the degree of dissociation or of other reaction may vary with position in space. These effects are particularly marked and interesting for the case of density-gradient centrifugation, because macromolecular species concentrate in narrow bands at different points in the centrifuge cell corresponding to their different buoyant densities.

It is our purpose to examine the effects of density-gradient centrifugation on several types of equilibria involving the dissociation of a macromolecule into half-molecules or the binding of a small ligand to a macromolecule. The formal structure of the arguments and the physical content of the results are rather similar to those obtained by Hearst and Vinograd (1961) in their study of the effects of water activity on the buoyant density of nucleic acids in density-gradient centrifugation. However, the particular examples considered here are important enough to warrant explicit treatment. Furthermore, we shall consider a greatly simplified model, neglecting certain three-component effects and neglecting deviations from ideality, in order to expose the basic phenomena in their simplest forms.

In a centrifuge cell there is a pressure gradient down the cell as well as a variation in gravitational potential. We consider only reactions with zero volume change so that the equilibrium constant is not affected by pressure.

The specific problems to be considered are these:

1. The effect of a density gradient on the association reaction between the right half-molecule of λ-bacteriophage DNA and the left half-molecule (RH and LH) to give the end-to-end joined molecule (JH), in view of the fact that the two halves have slightly different buoyant densities (Hogness and Simmons, 1964; Hershey and Burgi, 1965; Wang and Davidson, 1966).

Contribution No. 3453 from the Gates and Crellin Laboratories of Chemistry.

2. The effect of a density gradient on the binding of a ligand, such as mercuric ion (Hg^{++}), silver ion (Ag^{+}), or methylmercuric cation (CH_3Hg^{+}), to DNA, since these ions are sufficiently dense and massive so that their concentration is not uniform throughout the cell in density-gradient centrifugal experiments.

DEFINITIONS, NOTATION, AND SEDIMENTATION EQUILIBRIUM FOR A SINGLE SPECIES A

Let the acceleration field be a constant g. To a first approximation, if a two-component system such as a CsCl solution, is centrifuged, there is a constant density gradient which is proportional to g. Therefore, we assume that the density gradient is given by

$$\rho = \rho_0 + \alpha g x \tag{1}$$

where x is the distance from an arbitrary point in the cell where the density is ρ_0 and α is a coefficient characteristic of the salt (Vinograd and Hearst, 1962). We make the rather drastic assumption that the various chemical equilibrium constants are not affected by the changing ratio of cesium salt to water through the centrifuge cell. Let v_A, θ_A, M_A, and V_A be the partial specific volume, buoyant density, molecular weight, and partial molar volume of macromolecule A.

$$(1/v_A) = \theta_A = M_A/V_A \tag{2}$$

Then it is well known that the position (x_A) of the center of the buoyant band of the macromolecule occurs when

$$\rho = \theta_A = 1/v_A \qquad \text{or} \qquad x_A = \frac{\theta_A - \rho_0}{\alpha g} \tag{3}$$

Furthermore, the concentration distribution of A is given by

$$[A] = [A_0] \exp\left[-(M_A v_A g^2 \alpha/2RT)(x - x_A)^2\right] \tag{4}$$

$$= [A_0] \exp\left[-\beta_A (x - x_A)^2\right] \tag{5}$$

where

$$\beta_A = M_A v_A g^2 \alpha/2RT \tag{6}$$

The root mean square bandwidth is given by

$$\langle (x - x_A)^2 \rangle^{1/2} = (\pi/2\beta_A)^{1/2} = (RT/M_A v_A g^2 \alpha)^{1/2} \tag{7}$$

The total amount of material in the band, Σ_A (assuming a cross-sectional area of 1 cm^2 for the centrifuge cell) is given by

$$\Sigma_A = \int [A] dx = [A_0](\pi/\beta_A)^{1/2} \tag{8}$$

ASSOCIATION OF TWO MACROMOLECULES

Consider now the equilibrium

$$A + C \longleftrightarrow AC \tag{9}$$

where A and C are macromolecules. Let the molar volumes be additive so that

$$(M_A + M_C)v_{AC} = M_A v_A + M_C v_C \tag{10}$$

First, consider the distribution of each component through the cell, ignoring the effects of chemical equilibrium. It follows from Equations (6) and (10) that the β coefficients for the concentration distribution satisfy the relation

$$\beta_{AC} = \beta_A + \beta_C \tag{11}$$

Furthermore, the positions of the centers of the buoyant bands, x_A, x_C, and x_{AC}, as given by equation (3), satisfy the relations,

$$x_{AC} = \frac{\beta_A x_A + \beta_C x_C}{\beta_{AC}} = \frac{\beta_A x_A + \beta_C x_C}{\beta_A + \beta_C} \tag{12}$$

$$x_A - x_C = \frac{(\theta_A - \theta_C)}{\alpha g} \tag{13}$$

Thus, the concentration distributions of the several components are given by

$$[A] = [A_0] \exp[-\beta_A(x - x_A)^2] \tag{14}$$

$$[C] = [C_0] \exp[-\beta_C(x - x_C)^2] \tag{15}$$

$$[AC] = [AC_0] \exp[-(\beta_A + \beta_C)(x - x_{AC})^2] \tag{16}$$

Note that $[A_0]$, $[C_0]$, and $[AC_0]$ are concentrations at the respective band maxima, and therefore at three different places in the cell.

The concentration distribution for a macromolecular component as given, for example, by Equation (14) is derived from the requirement that the chemical potential of that component should be constant through the cell (Meselson, Stahl, and Vinograd, 1957). The condition for chemical equilibrium at any point in the cell is

$$\mu_A + \mu_C = \mu_{AC} \tag{17}$$

Then Equations (14), (15), and (16) should assure that if Equation (17) is satisfied at one point in the cell by an appropriate choice of $[A_0]$, $[C_0]$, and $[AC_0]$, it will be satisfied throughout the cell. We shall show by an explicit calculation that this is indeed the case. The condition for chemical equilibrium at any point in the cell is that

$$\frac{[AC]}{[A][C]} = K \tag{18}$$

where K is the equilibrium constant for the chemical equation (9). By substitution of Equations (14), (15), and (16) in (18), we obtain

$$\ln \frac{[AC_0]}{[A_0][C_0]K} = -\beta_A(x - x_A)^2 - \beta_C(x - x_C)^2 + \beta_{AC}(x - x_{AC})^2 \tag{19}$$

According to the arguments given above, the right-hand side of Equation (19) should be independent of x. Using Equations (11), (12), and (13), this is indeed found to be true, and the final result is

$$\frac{[AC_0]}{[A_0][C_0]} = K \exp\left[-\frac{\beta_A\beta_C}{(\beta_A + \beta_C)}(x_A - x_C)^2\right] \tag{20}$$

For the special case that $M_A = M_C$, this becomes, with the use of (13) and (6),

$$\frac{[AC_0]}{[A_0][C_0]} = K \exp\left[-\frac{M(\theta_C - \theta_A)^2}{2\alpha RT(\theta_C + \theta_A)}\right] \tag{21}$$

We are dealing with an association equilibrium for which the degree of reaction depends on the concentration of reactants. Therefore, there is no really objective way of comparing the degree of reaction in the presence of the field with that at zero field. But a reasonable, practical way to put the matter is as follows. The bandwidths of the several bands are all of the same order of magnitude; that is, of the order of $1/\beta^{1/2}$. Thus, the mean concentration of each component in its band is of the order of $[A_0]$, $[C_0]$, and $[AC_0]$, respectively. Thus, the value of the exponential factor in Equation (21) is a reasonable measure of the shift of the degree of reaction due to the field.

For the equilibrium between the right half-molecules and the left half-molecules of λb^+ DNA in CsCl, $M \approx 1.5 \times 10^7$, $\theta_A - \theta_C = 0.009$ g cm^{-3} (Hershey and Burgi, 1965), $\alpha = 8.4 \times 10^{-10}$ g cm^{-5} sec^2, $(\theta_A + \theta_C)/2 = 1.70$. By substitution of these parameters in Equation (21),

$$\frac{[AC_0]}{[A_0][C_0]} = Ke^{-8.5} = 2 \times 10^{-4}K \tag{22}$$

Thus, it is predicted that the equilibrium is very considerably shifted towards dissociation by the acceleration field. The observed ΔH for this reaction is 100 ($\pm$10 kcal), and at an $A_{260} = 1.0$, the degree of dissociation (in 2 M NaCl) is about unity at 65° C in the absence of a gravitational field (Wang and Davidson, 1966). To compensate for a decrease in the equilibrium constant by a factor of 5×10^3, the temperature would have to be lowered by 17°.

A still more spectacular shift of the equilibrium is predicted for the joining of linear molecules of λ (BU) DNA (that is, λ DNA labeled with 5-bromuridine) and λ DNA to give the hybrid linear dimer (Baldwin, Barrand, Fritsch, Goldthwait, and Jacob, 1966) or for the joining of half-molecules of λ (BU) and λ to give hybrid "joined-halves" molecules. In these examples, the density difference is larger ($b_A - b_C = 0.088$ g ml^{-3}), and the predicted shifts in the equilibrium constants are by factors of 10^{-345} and 10^{-690} for the joining of half-molecules and whole molecules, respectively.

Thus we predict that under some circumstances, the buoyant band of AC is unstable relative to dissociation into separated buoyant bands of A and C. The next question is whether the rate of disappearance of the AC band and formation of the A and C bands will be fast enough to be observable. An analysis which will not be presented here shows that for the λ (BU): λ linear hybrid, the rate is limited by the rate of transport by the centrifugal field of the components A and C away from the AC buoyant band toward the A and C bands, respectively, and that in typical centrifugal fields the dissociation will take place in a day or so at 10° below the T_m in the absence of a field (T_m is the temperature at which the degree of dissociation is unity), but the dissociation is predicted to be unobservably slow, 20° or 30° below T_m, for those cases in which the equilibrium calculation asserts that dissociation should occur.

A remarkable feature of these results is the fact that the predicted shift in the chemical equilibrium as expressed by Equation (21) is independent of the magnitude of the acceleration field, *g*. The weaker the field, the shallower the density gradient [Equation (1)] and the greater the separation between band centers; however, at the same time, the bandwidths increase [Equation (7)] and there is no over-all effect on the degree of reaction. In principle, DNA solutions in cesium chloride, if they are standing in a beaker on a laboratory

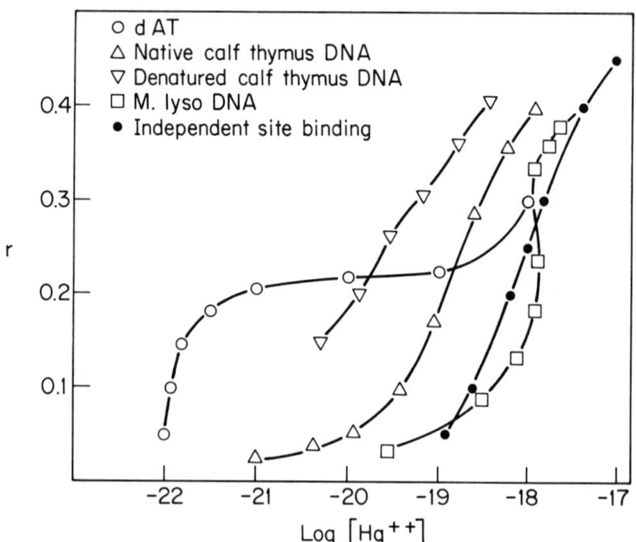

FIGURE 1.
Experimental data on the binding of mercuric ion by different DNA's at *p*H 9.0 from Nandi, Wang, and Davidson (1965). The quantity r is the ratio of metal ions bound in nucleotides. The concentration of $Hg(OH)_2$ can be calculated from the equation, $[Hg(OH)_2] = 10^{11.7}$ $[Hg^{++}]$ (at *p*H 9.0). Please note that the horizontal coordinate scale in the original publication (Nandi, et al., 1965) is incorrect by one log $[Hg^{++}]$ unit.

bench in the earth's gravitational field, should be subject to the formation of a shallow density gradient and to the same shifts in the chemical equilibrium. In practice, convection is sufficient to overcome the very slow formation of the density gradient and the bands under these circumstances. [Note that Equation (21) deals with concentrations at band center; the bandwidth and the total amount of material that has reacted (Σ_{AC}) does depend on the field.]

HEAVY METAL BINDING. GENERAL CONSIDERATIONS

We now consider a second class of problems associated with the effect of density-gradient centrifugation on the binding of a heavy metal ion by a nucleic acid. It may be recalled that DNA reversibly binds mercury and silver ions and that this binding greatly increases the buoyant density of the nucleic acid in a Cs_2SO_4 density gradient (Nandi, Wang, and Davidson, 1965; Jensen and Davidson, 1966).

As shown in Figure 1 for the case of mercury ions, the binding is expressed as a ratio (r) of moles of metal ion to moles of DNA nucleotide. The quantity r is a function of the concentration (M), of metal ion, that is,

$$r = r([M]) \tag{23}$$

Note that in Equation (23), parentheses indicate functional dependence and square brackets indicate a concentration. In the binding curves of Figure 1, dAT shows a saturation degree of binding at 0.25 mole of mercury per nucleotide. For the other DNA's, the saturation degree of binding is 0.5 mole of mercury per nucleotide. Most of the binding curves shown in Figure 1 are somewhat steeper (more cooperative) than expected for independent site binding. For independent site binding, the law of mass action applies

$$\frac{r}{0.5 - r} = K[M] \tag{24}$$

The plot of this equation for an arbitrary value of K is also shown in Figure 1. The slope of the curve $dr/d\log_{10}[M]$ for Equation (24) is given by $2(2.3)r(0.5 - r)$, with a maximum value of 0.29 at $r = 0.25$. The slope of the *M. lysodeiktikus* DNA titration curve around $r = 0.25$ is about 1.9; the average value between $r = 0.12$ and $r = 0.32$ is $\Delta r/\Delta\log_{10}[M] = 0.48$. The binding curves for silver ion at *p*H 8 (Jensen and Davidson, 1966) are still steeper and more cooperative than are those for mercuric ion.

The main species of inorganic mercury under the conditions of these experiments is $Hg(OH)_2$. The overall reaction with DNA at *p*H 9 may be written as

$$Hg(OH)_2 + DNA \rightleftharpoons (DNA{-}2H^+)Hg^{++} + 2H_2O \tag{25}$$

The notation $(DNA{-}2H^+)$ indicates that two protons are released from the DNA per Hg^{++} bound. The equilibrium condition for Equation (25) is independent of *p*H. We assume that the volume change in the reaction is zero so that the equilibrium is unaffected by pressure. Thus, the equilibrium in the centrifuge cell is governed by an equation of the type of (23), where by [M] we mean the concentration of $Hg(OH)_2$. It is clear from Figure 1 that the function $r([M])$ is different for different DNA's.

There are two limiting classes of problems. In one class, the binding is sufficiently strong so that almost all of the metal ion added is bound. Thus, if only a single DNA is present, r is determined by the stoichiometric amounts of DNA and $Hg(OH)_2$ added. The concentration of free $Hg(OH)_2$ is then read off from a curve, which is like those in Figure 1, for the appropriate DNA. In the second class, the binding is sufficiently weak so that most of the added metal ion is free. Therefore, the concentration of metal ion, [M], is known and the amount bound is determined from Figure 1.

We assume that the species M is sufficiently dense so that the product of its specific volume, v_M, and the solution density, ρ, is sufficiently small that it may be approximated as a constant throughout the density gradient in evaluating $1 - v_M\rho$. Then the concentration distribution of M is given by

$$\frac{1}{[M]}\frac{d[M]}{dx} = \frac{M_M g}{RT}(1 - v_M\rho) = \phi g \tag{26}$$

where M_M is the molecular weight and

$$\phi = M_M(1 - v_M\rho)/RT \tag{27}$$

Then

$$\ln [M]/[M_0] = \phi g x \tag{28}$$

where $[M_0]$ is the concentration at the point where $x = 0$. For $Hg(OH)_2$ in Cs_2SO_4, we estimate $v_M \approx 0.16$ cc gram^{-1}; if $\rho = 1.50$, $1 - v_M\rho = 0.75$, and

$$\phi_{Hg(OH)2} = 7.0 \times 10^{-9} \text{ sec}^2 \text{ cm}^{-2} \tag{29}$$

In a typical experiment, $g \approx 1.3 \times 10^8$ cm sec^{-2}, so that the concentration ratio of $Hg(OH)_2$ across a 1-cm centrifuge column is about $e^{0.9}$.

Another point of general interest is that the buoyant density of the DNA increases approximately linearly with the amount of metal ion bound,

$$\theta = \theta_0 + m\,r \tag{30}$$

where θ_0 is the buoyant density for zero metal ion bound. For the silver-DNA complexes, $m = 0.30$; for the mercury-DNA complexes, $m = 0.40$.

The basic point then is that since the metal ion concentration is greater at the bottom of the cell than at the top, a DNA, which, for one reason or another, moves to the bottom of the cell will bind more metal ion than it would at the top.

In the problems that follow we shall be interested only in the position of the center of the band and shall not study the effect of the metal ion concentration gradient on the band profile. (For a consideration of the analogous problem in connection with the varying hydration of DNA in a density gradient, see Hearst and Vinogard, 1961.)

EFFECT OF METAL ION BINDING ON THE SEPARATION BETWEEN ISOTOPICALLY LABELED MOLECULES

We consider that DNA(1) and DNA(2) are chemically the same, but DNA(2) is labeled with a heavy isotope, thus increasing its buoyant density. Assume that the strong binding approximation applies. Let DNA(1) be present in large quantities and let the amount of metal ion added be such that the amount bound to DNA(1) is known to be r_1. Let DNA(2) be present in tracer quantities. In the absence of an acceleration field, the value of r_2 is the same as r_1, since they are chemically identical. But since DNA(2) bands at a larger value of x, it will be exposed to a higher concentration of M, and hence $r_2 > r_1$.

DNA(1) with r_1 metal ions bound per nucleotide has a buoyant density θ_1 and bands at x_1; that is,

$$\theta_1 = \theta_{10} + mr = \rho_0 + \alpha g x_1 \tag{31}$$

The concentration of M at x_1 is determined implicitly by

$$r_1 = r\{[\mathrm{M}(x_1)]\} \tag{32}$$

where the functional dependence of r on [M] has been determined from a titration curve.

Then [M] at any other point x_2 is given by Equation (28), that is

$$\ln[\mathrm{M}(x_2)] = \ln[\mathrm{M}(x_1)] + \phi g(x_2 - x_1) \tag{33}$$

DNA(2) bands at x_2; its buoyancy density is given by

$$\theta_2 = \theta_{20} + mr_2 = \rho_0 + \alpha g x_2 \tag{34}$$

At this point,

$$r_2 = r\{[\mathrm{M}(x_2)]\} \tag{35}$$

r_2 and x_2 are determined by solving Equations (33), (34), and (35).

We can obtain an explicit solution for the case that r_2 and r_1 differ only by a small amount, so that we can approximate (35) by

$$r_2 = r_1 + \frac{\mathrm{d}r}{\mathrm{d}\ln[\mathrm{M}]}\frac{\mathrm{d}\ln[\mathrm{M}]}{\mathrm{d}x}(x_2 - x_1) \tag{36}$$

The quantity $\mathrm{d}r/\mathrm{d}\ln[\mathrm{M}]$ is to be taken from the binding curve (Figure 1), and $\mathrm{d}\ln[\mathrm{M}]/\mathrm{d}x$ is given by Equation (26). Then, from (31), (34), and (36),

$$r_2 - r_1 = \frac{m\phi(\mathrm{d}r/\mathrm{d}\ln[\mathrm{M}])(\theta_{20} - \theta_{10})}{\alpha - m\phi(\mathrm{d}r/\mathrm{d}\ln[\mathrm{M}])} \tag{37}$$

The function in the denominator of Equation (37) is an "effective density gradient" at unit field. It appears in many problems of this type. It is fairly easy to see that a condition for a stable band is that this function be positive (see Hearst and Vinograd, 1961). If the effective density gradient is negative, a single band should split into two bands of high and low r. For example, in an extreme case of a very steep titration curve, a single DNA band with metal ion added so that $r = 0.25$, should disproportionate into two bands: one toward the bottom of the cell with $r = 0.50$, and one toward the top with $r = 0.0$. For Cs_2SO_4, $\alpha = 1.40 \times 10^{-9}$ g cm^{-5} sec^2; if $m = 0.40$, and, as we estimate, $\phi = 7.0 \times 10^{-9}$, the condition for stability is that $dr/d \ln[M] < 0.50$, or $dr/d \log_{10}[M] < 1.15$. As already mentioned, for the mass action law [Equation (24)], $dr/d \log_{10}[M] < 0.29$ and the bands are stable. In Figure 1 it appears that for $r \approx 0.25$ for the *M. lysodeikticus* band in DNA, the condition for stability is not satisfied because the slope is about 1.9, and a centrifugal instability is expected. However, the average slope between $r = 0.12$ and $r = 0.32$ is about $dr/d \log_{10}[M] = 0.48$ and $\alpha - m\phi(dr/d \ln[M]) = 0.9 \times 10^{-9}$. Our estimate of ϕ may not be accurate, and there is considerable uncertainty about the slopes of the titration curve. Thus it is difficult to estimate how accurately the important quantity $\alpha - m\phi(dr/d \ln[M])$ is known. However, it seems probable that some of the mercury-DNA bands and the silver-DNA bands (for which the titration curve is very steep) are predicted to be unstable by the criterion of the sign of the quantity $\alpha - m\phi(dr/d \ln[M])$. From a practical point of view, the bands are observed to be stable for several days. The discrepancy may be a rate effect and due to the slow rate of transfer of metal ions from the low field region of the band to the high field region. Our efforts (Nandi, et al., 1965) to observe Hg-DNA bands at lower pH's (below 8) gave nonreproducible results and, frequently, band spreading. This may be an example of the instability discussed here.

Assume that, for one reason or another, the bands are stable. We return to the original question of the effect of the field in increasing the relative amount of metal ion bound by the DNA with the heavy isotope label. For the quantities estimated above, with $dr/d \log_{10}[M] = 0.45$, $(r_2 - r_1)/(\theta_{20} - \theta_{10}) = 0.56$. If $\theta_{20} - \theta_{10} = 0.02$, $r_2 - r_1 = 0.01$, and the buoyant density difference of the isotopic DNA's would be increased by 0.004 or 20 percent more than that expected without consideration of these field effects.

COMPETITIVE BINDING BY TWO DNA'S

Finally, we consider the problem of competition by two different DNA's for a nonsaturating amount of metal ion. Again, it is assumed that the strong binding approximation applies. The situation in zero acceleration field is shown in Figure 2. In general, at zero field, the metal ion distributes between the two DNA's so that both r_2 and r_1 correspond to the same value of [M]. The actual values of r_1 and r_2 depend on the amount of each DNA and the amount of M added. For simplicity, we assume that DNA(1) is present in large quantity and DNA(2) in tracer quantity. The value of r_1 is then fixed by the amount of metal ion added, and the value of r_2 is determined by the binding curves, plus the effects due to the centrifugal field, which are discussed on the next page.

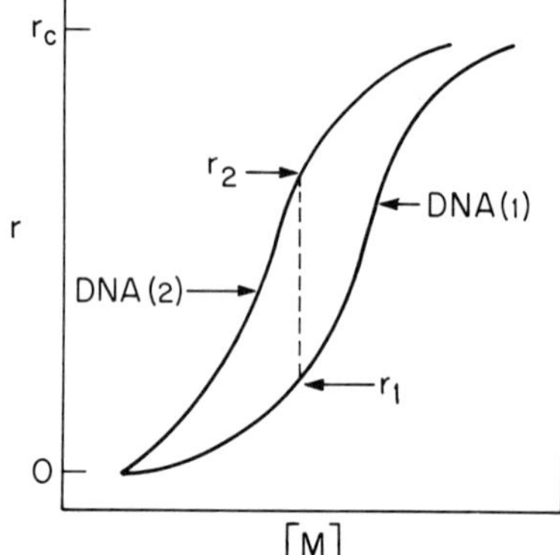

FIGURE 2.
Schematic diagram of competitive binding between DNA(1) and DNA(2) at zero field. At the value of [M] for which DNA(1) binds r_1 moles of Hg^{++}, DNA(2) binds r_2 moles of Hg^{++}.

The principles and the results are not significantly different for the general case of comparable amounts of DNA(1) and DNA(2).

We assume that the buoyant densities with zero metal ion bound are the same, $\theta_{01} = \theta_{02}$. This is approximately true for DNA's of different base composition in Cs_2SO_4. DNA(1), with r_1 moles of metal ion bound and density $\theta_0 + mr_1$, bands at x_1, and so on. The equations in this case are

$$\ln [\mathrm{M}(x_2)] = \ln [\mathrm{M}(x_1)] + \phi g(x_2 - x_1) \tag{38}$$

$$r_2 = r_2\{[\mathrm{M}(x_2)]\} \tag{39}$$

$$\theta_2 = \theta_0 + mr_2 = \rho_0 + \alpha g x_2 \tag{40}$$

$$\theta_1 = \theta_0 + mr_1 = \rho_0 + \alpha g x_1 \tag{41}$$

Then

$$m(r_2 - r_1) = \alpha g(x_2 - x_1) \tag{42}$$

The crucial equation is (39), which asserts that DNA(2) binds an amount of metal ion determined by the concentration of M at x_2, whereas $[\mathrm{M}(x_1)]$ is known from the prescribed value of r_1 and the dissociation curve of DNA(1). If x_2 is not too far from x_1, we expand (39),

$$r_2\{[\mathrm{M}(x_2)]\} = r_2\{[\mathrm{M}(x_1)]\} + \frac{dr_2}{d \ln \mathrm{M}} \frac{d \ln \mathrm{M}}{dx}(x_2 - x_1) \tag{43}$$

By solution of these equations we obtain

$$\frac{r_2\{[\mathrm{M}(x_2)]\} - r_2\{[\mathrm{M}(x_1)]\}}{r_2\{[\mathrm{M}(x_1)]\} - r_1\{[\mathrm{M}(x_1)]\}} = \frac{\phi(dr_2/d \ln [\mathrm{M}])}{\alpha - m\phi(dr_2/d \ln [\mathrm{M}])} \tag{44}$$

Recall that $r_2\{[\mathrm{M}(x_1)]\}$, which can be read off from the binding curves, is the value of r_2 in the absence of field effects. Thus, the numerator of the left-hand side of (44) is the quantity desired, the change in r_2 due to field effects, and the denominator is the difference between r_2 and r_1 at zero field.

Again, the predicted change in r_2 is independent of field. With the numerical values assumed in the previous section, including $dr_2/d \ln [\mathrm{M}] = 0.19$,

$$\frac{r_2\{[\mathrm{M}(x_2)]\} - r_2\{[\mathrm{M}(x_1)]\}}{r_2\{[\mathrm{M}(x_1)]\} - r_1\{[\mathrm{M}(x_1)]\}} = 1.5 \tag{45}$$

Thus, there is a predicted increase of 50 percent in r_2 due to centrifugal field effects.

CONCLUSION

Other problems of a related nature can be treated similarly. In particular, the case of weak binding applies to methylmercury and DNA, and calculations show that the amount of binding is predicted to vary considerably with the position of the DNA in the centrifuge cell.

It seems likely that the most damaging simplifying approximation in the preceding treatment is the neglect of activity coefficient effects that are associated with the varying concentration of cesium salt in the gradient. This affects the net hydration of the DNA and probably also the equilibrium conditions for metal ion binding at zero field.

The available experimental data are not sufficient for a test of the various predictions made here. The general prediction that there are significant effects of the centrifugal field on chemical equilibria in a density gradient is firmly founded, however, and deserving of experimental study.

ACKNOWLEDGMENT

This research has been supported by grant GM 10991 from the United States Public Health Service. We gratefully acknowledge the advice and criticism of Professor Jerome Vinograd.

REFERENCES

Baldwin, R. L., P. Barrand, A. Fritsch, D. A. Goldthwait, and F. Jacob (1966). *J. Mol. Biol.* **17**, 343.

Guggenheim, E. A. (1957). *Thermodynamics*, 3rd ed., p. 409. Amsterdam: North-Holland Publishing Company.

Hearst, J. E., and J. Vinograd (1961). *Proc. Nat. Acad. Sci. U.S.* **47**, 825; **47**, 1005, 1015.

Hershey, A. D., and E. Burgi (1965). *Proc. Nat. Acad. Sci. U.S.* **53**, 325.

Hogness, D. S., and J. P. Simmons (1964). *J. Mol. Biol.* **9**, 711.

Jensen, R. H., and N. Davidson (1966). *Biopolymers* **4**, 17.

Meselson, M., F. W. Stahl, and J. Vinograd (1957). *Proc. Nat. Acad. Sci. U.S.* **43**, 581.

Nandi, U. S., J. C. Wang, and N. Davidson (1965). *Biochemistry* **4**, 1687.

Vinograd, J., and J. E. Hearst (1962). *Progress Chem. Org. Nat. Prod.* **20**, 395.

Wang, J. C., and N. Davidson (1966). *J. Mol. Biol.* **19**, 469.

HYDROGEN BONDING, WATER, AND ICE

JERRY DONOHUE

Department of Chemistry
University of Pennsylvania
Philadelphia, Pennsylvania

Selected Topics in Hydrogen Bonding

My first introduction to the topic of hydrogen bonding occurred more than twenty-five years ago when I was an undergraduate at Dartmouth College. A new addition to the faculty was Professor Lindsay Helmholz, who had just arrived after a year or two as a postdoctoral fellow in Pasadena. Some of the crystal structures that Helmholz had determined while at Cal Tech were those of potassium bifluoride (Helmholz and Rogers, 1939), ammonium bifluoride (Rogers and Helmholz, 1940), diammonium trihydrogen paraperiodate (Helmholz, 1937), and neodymium bromate enneahydrate (Helmholz, 1939); all of these are examples of structures in which hydrogen bonding is a dominant feature. Although Helmholz had little difficulty in imparting the reality of this feature to his students, there was some opposition from some of the more conservative faculty members, partly, I suppose, because the first edition of *The Nature of the Chemical Bond* (Pauling, 1939) had not yet penetrated into New Hampshire.

Needless to say, much has happened since that time.

The present discussion—I do not wish to call it a review—is intended to be more selective than comprehensive; for the topics I have chosen to discuss, the literature search was terminated when only a dozen or so examples were discovered.

An early review of hydrogen bonding is that of Pauling (1939, Chapter IX), and since then there have been a number of somewhat extensive reviews (for example: Donohue, 1952; Robertson, 1953; Fuller, 1959; Pimentel and McClellan, 1960; Nyburg, 1961; Wells, 1962), as well as treatments of some of the more specialized aspects (for example: Lipscomb, 1954; Ubbelohde and Gallagher, 1955; Cannon, 1958; Hamilton, 1962; Chidambaram, 1962; Clark, 1963), and there seems little point in redigesting all of this material. I have, accordingly, chosen for discussion a relatively small number of features that I feel are still interesting enough to warrant reairing, even though there has been a recent tendency to sweep some of these under the rug, and to discuss others in terms that are misleading, if not patently false.

THE GEOMETRIC ROLE OF THE ACCEPTOR ATOM

The most commonly encounted hydrogen bonds are those of the type O—H···O and N—H···O. I shall first discuss the geometry of the oxygen-atom acceptors.

Water Molecules as Hydrogen-Bond Acceptors

A popular current concept of the water molecule is shown in Figure 1: the two H—O bonds occupy two sp^3 hybrid orbitals of the oxygen atom, and the two unshared pairs occupy the remaining two orbitals. If, as is commonly stated, the H—O covalent bonds contain more than 75 percent p character, then the angle between the two lobes occupied by the unshared pairs would be expected to increase to somewhat more than the tetrahedral value of 109°28′, but such fine points need not concern us here. The real point is that this picture "explains" the structure of ice, and is so ingrained that in many quarters the tetrahedral water molecule, in terms of two hydrogen-bond donors and two hydrogen-bond acceptors, all four at tetrahedral angles, as in ice, is *the* water molecule. Although the notion that in O—H···O bonds the O—H vector is directed towards an unshared pair on the oxygen atom is not unattractive, some experimental results suggest otherwise. Let us therefore balance the quantum-mechanical picture of Figure 1 against the observations of X-ray and neutron diffraction.

In Figure 2, *a*–l, are shown the environments* of one dozen water molecules in various crystals. In all of these the disposition of the water oxygen, the two acceptor atoms, and the one donor hydrogen-bonded atom is coplanar, or very nearly so, and the situation of the water molecule can be described as being roughly plane-trigonal. (The 12 angles O···HOH···O average 119°, average deviation 10°; the 24 angles O···HO···O or O···HO···N average 120°, average deviation 12°.)

*If the appropriate structural data were not presented in the original papers, these were calculated from the published parameters and lattice constants.

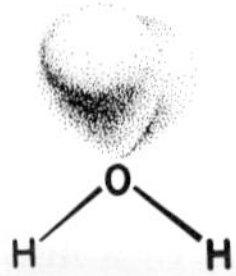

FIGURE 1.
Representation of a water molecule, with the two unshared pairs of the oxygen atom occupying two of the sp^3 hybrid orbitals.

FIGURE 2.
Environments of water molecules in twelve different hydrates. The compounds chosen are those in which the sum of the angles about the oxygen atom of the water molecule is close to 360°, and the coordination is thus approximately planar-trigonal. (*a*) Glycyltryptophan dihydrate (Pasternak, 1956). (*b*) Asparagine monohydrate (Kartha and de Vries, 1961). (*c*) Nitric acid monohydrate (Luzzati, 1951). (*d*) (+)-Demethanolaconinone hydriodide hydrate (Donohue, 1964). (*e*) Cytosine monohydrate (Jeffrey and Kinoshita, 1963). (*f*) Barbituric acid dihydrate, first water molecule (Jeffrey, Ghose, and Warwicker, 1961). (*g*) Barbituric acid dihydrate, second water molecule. (*h*) Histidine hydrochloride monohydrate (Donohue, Lavine, and Rollett, 1956; Donohue and Caron, 1964). (*i*) Shelloic bromolactone hydrate (Gabe, 1962). (*j*) L-Arginine dihydrate (Karle and Karle, 1964). (*k*) Violuric acid monohydrate (the distances shown are all O···O). (Craven and Mascarenhas, 1964; Craven and Takei, 1964). (*l*) Xanthazole monohydrate (Nowacki and Bürki, 1955; Mez and Donohue, 1966).

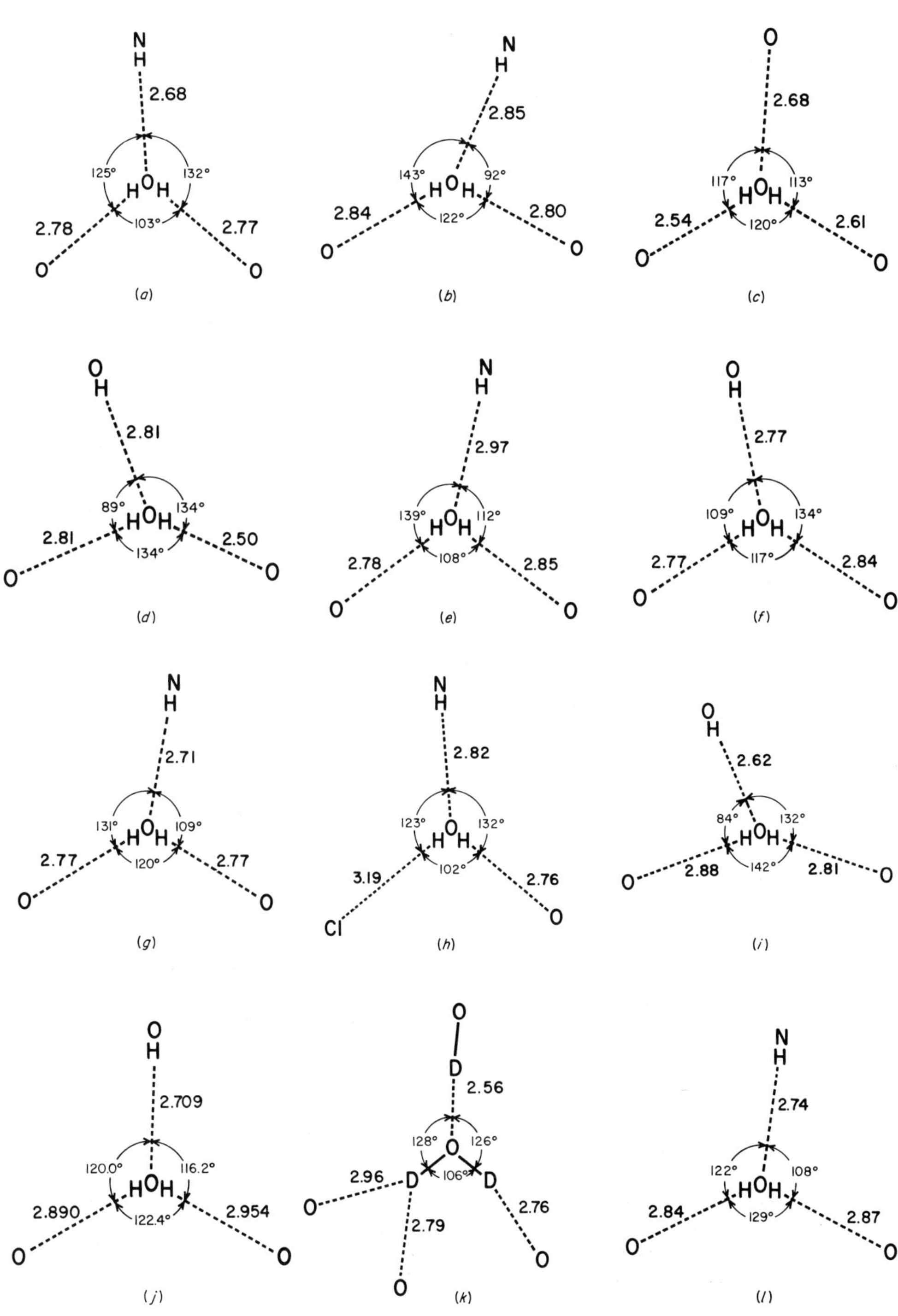

N
H
2.68
125°
132°
H O H
2.78
103°
2.77
O
O
(a)
N
H
2.85
143°
92°
H O H
2.84
122°
2.80
O
O
(b)
O
2.68
117°
113°
H O H
2.54
120°
2.61
O
O
(c)
O
H
2.81
89°
134°
H O H
2.81
134°
2.50
O
O
(d)
N
H
2.97
139°
112°
H O H
2.78
108°
2.85
O
O
(e)
O
H
2.77
109°
134°
H O H
2.77
117°
2.84
O
O
(f)
N
H
2.71
131°
109°
H O H
2.77
120°
2.77
O
O
(g)
N
H
2.82
123°
132°
H O H
3.19
102°
2.76
Cl
O
(h)
O
H
2.62
84°
132°
H O H
O
2.88
142°
2.81
O
(i)
O
H
2.709
120.0°
116.2°
H O H
2.890
122.4°
2.954
O
O
(j)
O
D
2.56
128°
126°
O
2.96
D
106°
D
O
2.79
2.76
O
O
(k)
N
H
2.74
122°
108°
H O H
2.84
129°
2.87
O
O
(l)

It would thus appear that the fuzzy picture of the water molecule, Figure 1, although attractive from a theoretical point of view, is, in fact, quite unimportant as far as hydrogen bonding is concerned. The fact that many structures contain tetrahedral water molecules is not germane, since these could result from other requirements of these structures; surely it is obvious that the tetrahedral water molecule is *not* a stringent structural feature, and, in view of the examples which are shown above, it appears doubtful if it should be seriously considered.*

Carbonyl Groups as Hydrogen-Bond Acceptors

Another popular current concept is that of a carbonyl group as shown in Figure 3: the C—O bond consists of one "σ" bond and one "π" bond (not shown), with the two unshared pairs of electrons on the oxygen atom lying in the remaining two sp^2 lobes, coplanar with the carbon ligands. It is thus predicted that in the OH···O=C or NH···O=C situation the angle OH···O=C or NH···O=C should be close to 120°, and that the OH or NH donors should lie very nearly in the plane defined by the carbon atom and its three ligands. This picture has been emphasized recently (Robertson, 1964) with respect to carboxyl groups.

Again, let us look at some experimental results. (To be sure, the basis of any theory should be grounded on experiment, even though there has been a recent tendency for some theoreticians to consider calculation superior to observation.) In Figure 4, *a*–l, are shown one dozen >C=O groups that are

*It might be argued that in these trigonally coordinated water molecules a bifurcated system (see the following section on bifurcated hydrogen bonds) is present, with the two lone pairs interacting equally with the donor hydrogen atom. I find this suggestion singularly unattractive

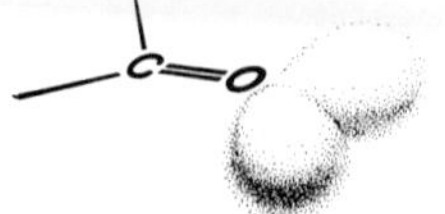

FIGURE 3.
Representation of a carbonyl group, with the two unshared pairs of the oxygen atom occupying two of the sp^2 hybrid orbitals. The πp orbitals of the carbon and oxygen atoms are not shown.

FIGURE 4.
Environments of twelve different carbonyl groups in various crystals. In each case the central carbon atom and its three ligands are closely coplanar; the distance of the hydrogen-bonding oxygen or nitrogen atom from the mean plane is given (in Å) as an underlined number. (*a*) Cytosine (Baker and Marsh, 1964). (*b*) β-Glycylglycine (Hughes and Moore, 1949). (*c*) 5-Fluoro-2′-deoxy-β-uridine (Harris and Macintyre, 1964). (*d*) Benzamide (Penfold and White, 1959). (*e*) Calcium thymidylate (Trueblood, Horn, and Luzzati, 1961). (*f*) Creatinine (du Pré and Mendel, 1955). (*g*) Nicotinamide (Wright and King, 1954). (*h*) 1-Methylthymine: 9-Methyladenine (Hoogsteen, 1963b). (*i*) Glycylphenylalanylglycine, alanyl carbonyl (Marsh and Glusker, 1961). (*j*) Glycylphenylalanylglycine, glycyl carbonyl. (*k*) α-Helix (Pauling and Corey, 1951). (*l*) N-Acetylglycine (Donohue and Marsh, 1962).

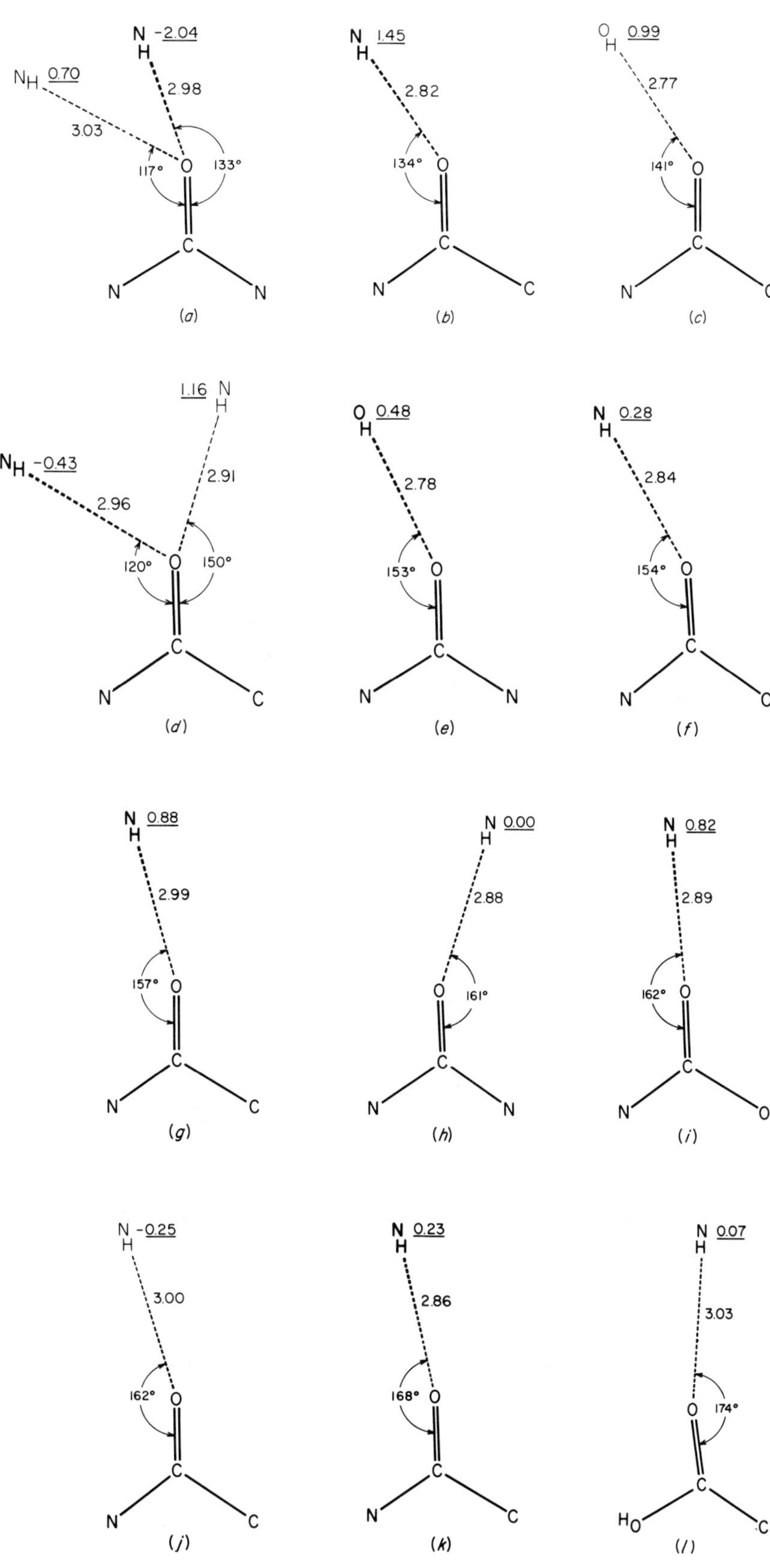

N -2.04
H
NH 0.70
2.98
3.03
117°
133°
O
C
N
N
(a)
N 1.45
H
2.82
134°
O
C
N
C
(b)
O 0.99
H
2.77
141°
O
C
N
C
(c)
1.16 N
H
NH -0.43
2.91
2.96
120°
150°
O
C
N
C
(d)
O 0.48
H
2.78
153°
O
C
N
N
(e)
N 0.28
H
2.84
154°
O
C
N
C
(f)
N 0.88
H
2.99
157°
O
C
N
C
(g)
N 0.00
H
2.88
161°
O
C
N
N
(h)
N 0.82
H
2.89
162°
O
C
N
O
(i)
N -0.25
H
3.00
162°
O
C
N
C
(j)
N 0.23
H
2.86
168°
O
C
N
C
(k)
N 0.07
H
3.03
174°
O
C
HO
C
(l)

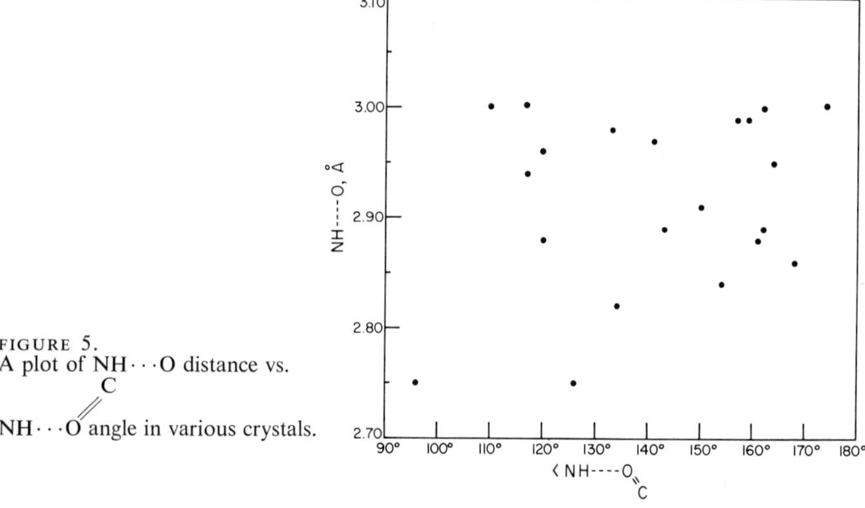

FIGURE 5.
A plot of NH···O distance vs. NH···O=C angle in various crystals.

accepting hydrogen bonds. This handful does not represent a random sampling; not included are examples having the familiar situation of a hydrogen-bond system forming an elongated hexagon, in which all of the angles are necessarily close to 120°. These include carboxylic acid dimers,

```
        O···HO
      //      \
   —C          C—
      \      //
        OH···O
```

and the sterically similar situation possible in many purines and pyrimidines,

```
                 /
        O···HN
      //      \
   —C          C—
      \      //
        NH···O
       /
```

It is obvious from Figure 4 that the notion of the 120° angle is also quite unimportant in hydrogen bonds of this type. Not only is there wide variation of the angle NH···O=C from 120°, but also there are large deviations of the NH (or OH) donors from the plane of the carbonyl system. Moreover, there does not appear to be any weakening of the NH···O bonds as the angle deviates from 120°, as is apparent from Figure 5, in which some of the data of Figure 4, together with additional examples taken from Table 10-I of Pimentel and McClellan (1960), have been plotted. These data give an array of points through which even a connoisseur of solvolysis rates would hesitate to draw a straight line.

The same situation obtains with respect to OH···O=C bonds, but for these I give but one short series of compounds, the first two fatty acids. In crystalline formic acid (Holtzberg, Post, and Fankuchen, 1953) the OH···O distance is 2.58 ± 0.03 Å, and the $\mathrm{OH\cdots O{=}C}$ angle is 122°, and in acetic acid (Jones and Templeton, 1958) the corresponding values are 2.61 ± 0.02 Å and 144°. The absence of a marked difference in this particular case has been noted previously by Pimentel and McClellan (1960, page 267) who remarked, "This suggests that the H-bond interaction is relatively insensitive to the angle between the carbonyl and O···O bonds." I would be inclined to emend this statement by deleting the word "relatively." (However, they also report the opposite view, because they had previously noted (page 232) that "any prediction of the orientation of the H bond must be assigned an uncertainty of about 20°. On the other hand, the bond angles [i.e., $\mathrm{NH\cdots O{=}C}$ and $\mathrm{OH\cdots O{=}C}$] display a tendency which reassures us in using the concept of orbital hybridization. . . ." In view of the data presented in Figures 2, 4, and 5 above, this reassurance is clearly no longer very reassuring.

Discussion

Until recently, the hydrogen bond was most often discussed in terms of the electrostatic model first proposed by Pauling in 1928. This model attributes the properties of the hydrogen bond to ionic forces only, and apparently places no restriction on the geometry of the acceptor atom. This point charge approach has been widely used in calculations of various properties of hydrogen bonds (see, for example, some of the reviews cited near the beginning of this paper). The more recent situation, on the other hand, has been summarized by Pimentel and McClellan (1960, page 230): "Modern discussions of an electrostatic model of the H bond are more sophisticated than the earliest point charge calculations. Theorists believe that the charge distribution of the nonbonded electrons must be explicitly included in the calculation. Furthermore, the concept of orbital hybridization provides a basis for deciding this charge distribution." Now, my dictionary* gives, for the definition of sophisticated, "Deprived of native or original simplicity; made artificial, or, more narrowly, highly complicated . . .", and it seems to me that these "modern discussions" are just that. This seems to have become a general trend lately, and, unfortunately, the artificialities and complications are too often presented as "understandings" which follow from an artificial and complicated display of wave functions or other mathematical sophistications. A current pervasive idea appears to be that a discussion, of no matter what, somehow is not respectable unless it is made in terms of a set of ψ_i. It cannot be denied, of course, that this approach has been highly successful in a wide diversity of problems, but it also cannot be denied that, when hydrogen bonding is concerned, the appropriate approximations and assumptions have not yet been discovered. The theory should cover the situation in general; at the present time it does not.

*Webster's New International Dictionary, Second Edition.

BIFURCATED HYDROGEN BONDS

α-Glycine

Albrecht and Corey (1939) were the first to propose a situation in which one hydrogen atom was shared between two hydrogen-bond acceptors, in crystals of α-glycine. They found that the $-NH_3^+$ group has four near oxygen neighbors, at 2.76, 2.88, 2.93, and 3.05 Å. Although they had no direct evidence for the location of the hydrogen atoms, they also found that it was possible to orient a tetrahedrally disposed $-NH_3^+$ group such that one hydrogen atom was very nearly directed towards the oxygen atom at 2.76 Å, the second hydrogen atom was very nearly directed towards the oxygen atom at 2.88 Å, and the third hydrogen atom was "sharing its bond-forming capacity nearly equally between two nearest oxygen atoms [at 2.93 and 3.05 Å] in the adjacent layer." This situation was pointed out by Pauling (1939, pages 266–267) to represent an exception to the condition that "the coordination of hydrogen does not exceed two," and that "bonds of this type probably occur only rarely."

The positions of the hydrogen atoms were then verified by Marsh (1958), in an exceptionally thorough X-ray investigation. Some structural details of the unusual hydrogen bond are presented in Figure 6, in which it may be seen that although the nitrogen is closer to one oxygen, the hydrogen is closer to the other, because, as pointed out by Marsh, of the more favorable angle of attack. The revised lengths of the two "normal" NH···O bonds are 2.77 and 2.85 Å.

β-Glycine

In a second form of glycine, Iitaka (1960) found that the $-NH_3^+$ group also has four near oxygen neighbors, at 2.76, 2.83, 3.00, and 3.02Å. Assuming the hydrogen atoms to be tetrahedrally disposed, he found that the first two of these distances corresponded to normal NH···O bonds and the other two led to a bifurcated situation as shown in Figure 7, in which it may be seen that this detail of the structure closely resembles that in α-glycine.

γ-Glycine

In a third form of glycine Iitaka (1961) found that the $-NH_3^+$ group has five near oxygen neighbors, at 2.80, 2.82, 2.97, 2.90, and 3.06 Å. According to Iitaka, the first three of these are arranged approximately in the tetrahedral directions with maximum deviation of about 20° from the tetrahedral angle. He therefore concluded that these three oxygens take part in hydrogen bonds, and that the other two (for which the C—N···O angles are 166° and 86°) are linked by electrostatic attraction to the $-NH_3^+$ group. No bifurcation was postulated.

Glycine Hemihydrochloride

One of the nitrogen atoms in glycine hemihydrochloride was found by Hahn and Buerger (1957) to have four close neighbors, an oxygen at 2.90 Å, and three chloride ions at 3.13, 3.23, and 3.32 Å. The angles C—N···O,

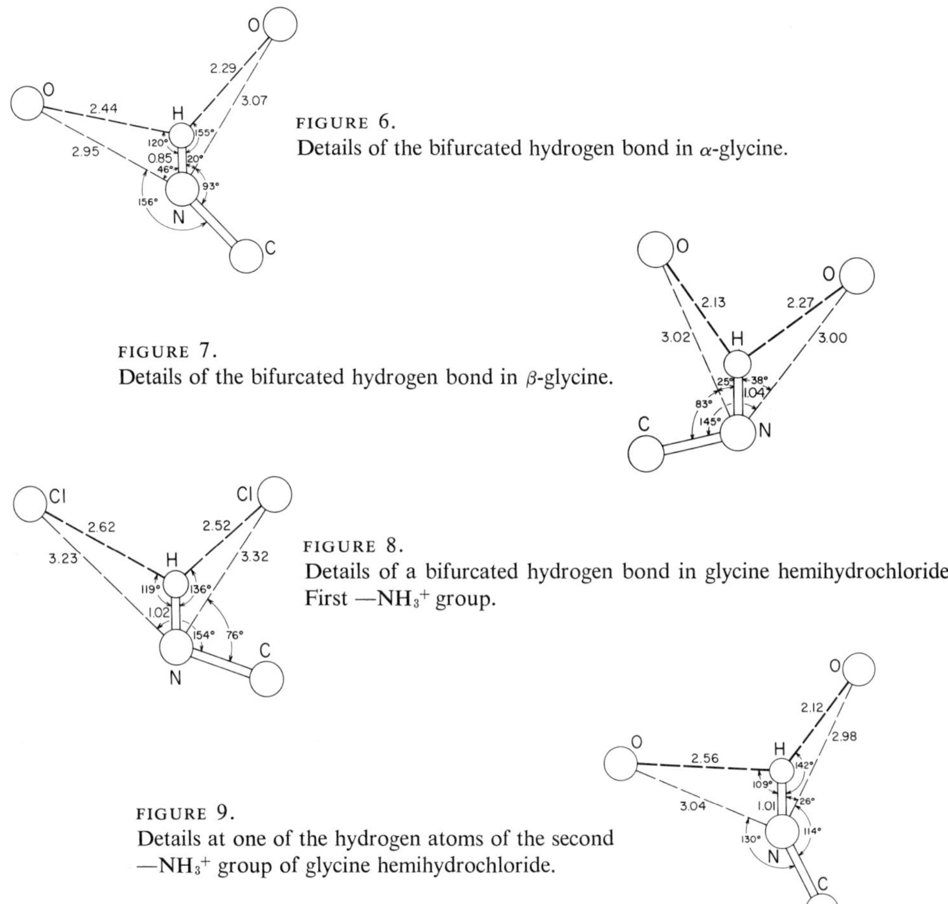

FIGURE 6.
Details of the bifurcated hydrogen bond in α-glycine.

FIGURE 7.
Details of the bifurcated hydrogen bond in β-glycine.

FIGURE 8.
Details of a bifurcated hydrogen bond in glycine hemihydrochloride. First —NH_3^+ group.

FIGURE 9.
Details at one of the hydrogen atoms of the second —NH_3^+ group of glycine hemihydrochloride.

C—N···Cl, and O···N···Cl for the chloride at 3.13 Å are all within 5° of tetrahedral, so the authors positioned two hydrogen atoms accordingly, thus fixing the position of the third hydrogen of this —NH_3^+ group. The third hydrogen is about equally close to the two remaining chloride ions, as shown in Figure 8. Hahn and Buerger, after noting that both of the H···Cl distances are shorter than the van der Waals approach of 3.0 Å, stated that "this indicates that some sort of weak hydrogen bond exists between H and the Cl's."

The second nitrogen atom in this crystal has five close neighbors, four oxygens at 2.93, 2.98, 3.04, and 3.09 Å, and a chloride at 3.22 Å. Hahn and Buerger noted that the relevant angles involving the nitrogen, the chloride at 3.22 Å, and the oxygen at 2.93 Å are all close to the tetrahedral value, and placed two hydrogen atoms accordingly, with the third hydrogen atom in the remaining tetrahedral position. The situation regarding it is shown in Figure 9, and Hahn and Buerger suggested that the hydrogen is forming a weak hydrogen bond with the oxygen at 2.98 Å from the nitrogen, and "probably to a slight degree" with the one at 3.04 Å, even though, as they point out, the H···O distance of 2.56 Å in the latter is almost the van der Waals H···O distance of 2.6 Å (Pauling, 1939). Obviously, the case for bifurcation here is weaker than for the examples given above.

The locating of the hydrogen atoms by Hahn and Buerger was not done solely on the basis of presumed hydrogen bonds. There was additional evidence for them in electron-density difference maps, and in the fact that the conformations about the C—N bonds are the expected staggered ones.

Nitramide

The $-NH_2$ group in nitramide was reported by Beevers and Trotman-Dickenson (1957) to have four close oxygen neighbors, two each at 3.09 Å and 3.12 Å. These correspond to two bifurcated hydrogen bonds, as shown in Figure 10.

Violuric Acid Monohydrate

The crystal structure of perdeuterated violuric acid monohydrate has been examined by means of X-ray diffraction for location of the carbon, nitrogen, and oxygen atoms (Craven and Mascarenhas, 1964) and by means of neutron diffraction for location of the deuterium atoms (Craven and Takei, 1964). The water molecule was found to have four close oxygen neighbors, at 2.56, 2.76, 2.79, and 2.96 Å. The first of these represents the acceptance of a hydrogen bond from the =N—OH group of the acid molecule, the second, a normal OH···O bond, and the last two, a bifurcated system, as shown in Figure 11. As pointed out by Craven and Takei, a rotation of the water molecule by about 16° would lead to two very nearly linear hydrogen bonds (this was, in fact, a structure they tested in the first stages of the refinement), but that "more detailed stereochemical considerations suggest reasons for the observed bifurcated configuration having a lower energy. . . ." These reasons include the specific location of the lone pairs of electrons on the "trigonally hybridized acceptor oxygen atoms" as well as those on the "tetrahedrally hybridized" water molecule, but, in the light of the section above on the geometry of the acceptor atoms, are probably not very compelling.

Magnesium Sulfate Tetrahydrate

Both X-ray (Baur, 1962) and neutron (Baur, 1964) studies have been carried out on $MgSO_4 \cdot 4H_2O$. Seven of the eight hydrogen atoms form normal OH···O hydrogen bonds, with O···O distances of from 2.73 to 2.88 Å, and H—O···O angles of from 4° to 25°. Baur considers that the "O_w—H—O bonds are bent considerably;" the question of the linearity of hydrogen bonds is discussed in the next section of this paper, but it seems worthwhile to point out here that a deviation from linearity of 25° in a complicated structure which contains a large number of hydrogen bonds is by no means exceptional. The situation at the eighth hydrogen atom is shown in Figure 12, which was adapted from Figure 3 of Baur. A too casual perusal of these figures might suggest that this is another example of a bifurcated hydrogen bond, but examination of the H···O distances shows that it is not: one of the H···O distances is the value expected for a normal van der Waals contact of 2.6 Å; the other H···O distance for that same hydrogen atom is 2.39 Å, or 0.3 to 0.6 Å longer than expected in a normal hydrogen bond formed by a water molecule. Baur concludes that this hydrogen atom is not involved in a

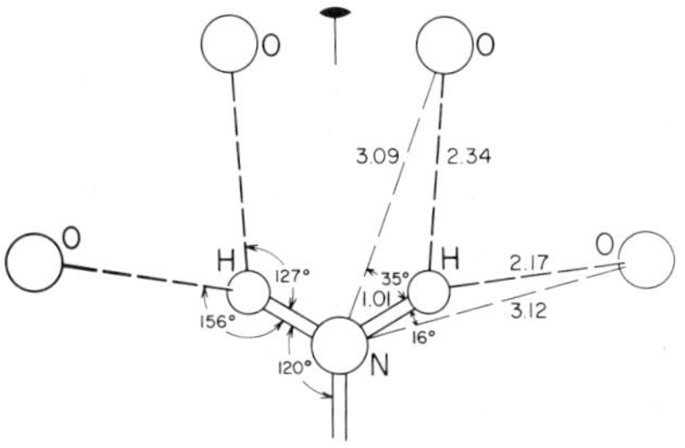

O O O O H H N
3.09 2.34 2.17 3.12 1.01 127° 35° 156° 120° 16°

FIGURE 10.
Details of the bifurcated hydrogen bonds in nitramide. The molecule is required by the crystal symmetry to have a twofold axis.

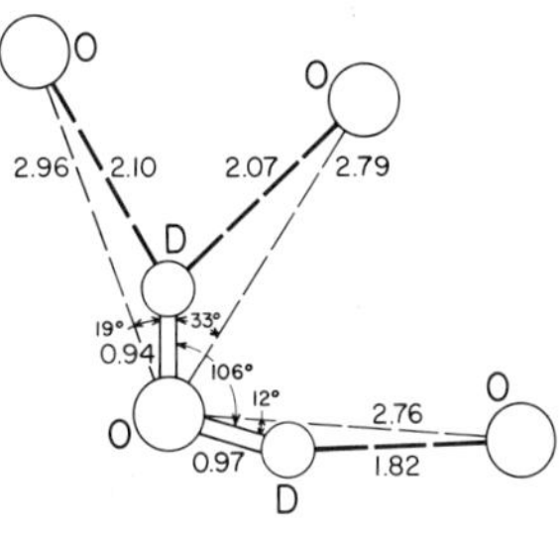

FIGURE 11.
Details of the bifurcated hydrogen bond in perdeuterated violuric acid monohydrate.

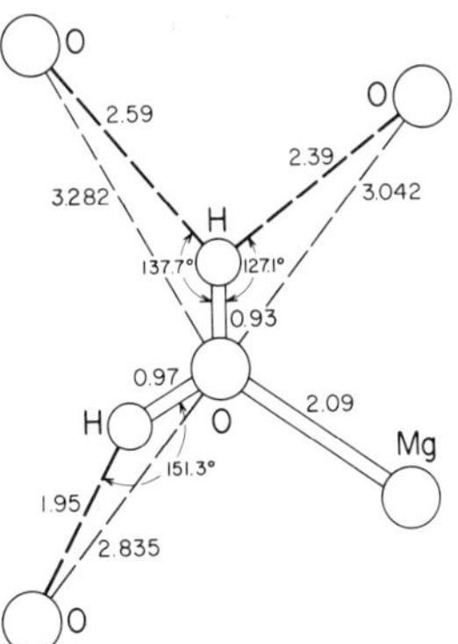

FIGURE 12.
The situation at water molecule II in magnesium sulfate tetrahydrate. [After Baur, 1964.]

hydrogen bond, and gives as supporting evidence the unusual thermal motion of that atom as compared with the other hydrogen atoms. On the other hand comparison of the distances in Figure 12 with those in Figures 6 and 10 suggests that this hydrogen atom might more properly be considered to be involved in one-half of a bifurcated hydrogen bond.

Dicyandiamide

Hughes (1940) found that in the crystal structure of dicyandiamide the two amino groups together have five close N···N neighbors, at 2.94, 3.02, 3.04, 3.15, and 3.16 Å. Because this molecule

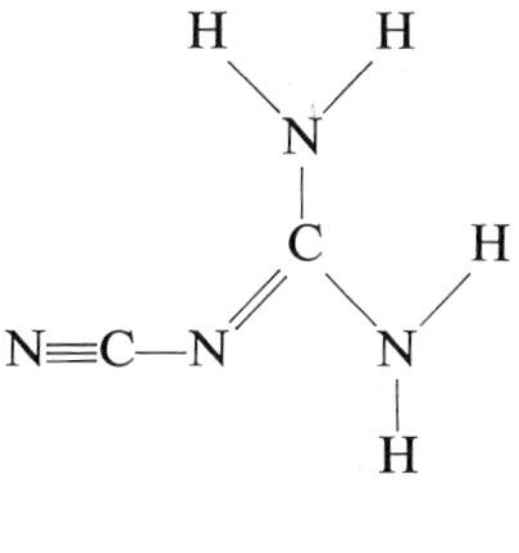

H H N C H N≡C—N N H

has only four hydrogen atoms available as hydrogen-bond donors, either one of the above distances is not a hydrogen bond, or one of the hydrogen atoms must form a bifurcated hydrogen bond. Hughes took the latter interpretation which involves rotation of one of the —NH_2 groups by 30° out of the molecular plane. As has been previously pointed out (Donohue, 1952), it is quite probable that the entire molecule is planar: this situation would rule out the 3.15 Å distance as being part of a bifurcated hydrogen bond, because the presumed acceptor atom is about 3 Å from the molecular plane. An N···N distance of 3.15 Å is slightly larger than twice the van der Waals radius of nitrogen. It would be interesting to have the locations of the hydrogen atoms determined by a direct method.

Maleic Acid

In maleic acid (Shahat, 1952), one of the carboxylic hydrogen atoms is involved in an intramolecular hydrogen bond. The closest oxygen neighbors of the other carboxylic OH group are at 2.75 and 2.92 Å (not 2.98 Å, as given by Shahat), and Shahat calls this grouping a bifurcated hydrogen bond. However, the unsatisfactory geometry negates the validity of this interpretation. In the example of the 2.75 Å distance, because the angle C—O···O is 112°, it is probable that the hydrogen atom lies very nearly on the O···O line: this places it more than 3 Å from the second oxygen atom at 2.92 Å, or much too far for any significant interaction. Since the van der Waals diameter of oxygen is 2.8 Å, no special discussion of this distance is necessary.

Sulfamic Acid

The crystal structure of sulfamic acid was first determined by Kanda and King (1951). They found that the $-NH_3^+$ group has five close oxygen neighbors, at 2.82 to 3.07 Å, and, referring to them, they stated "These five hydrogen bonds can be accounted for only by postulating that two of the three hydrogens are involved in bifurcated bond systems in which each is offered equally by the nitrogen to two oxygen atoms."

This same interpretation was made by Osaki et al. (1955) in a later refinement that gave the N···O distances of 2.93 to 3.01 Å. These were used to postulate the "tentative location of the three hydrogen atoms, which leads to one single and two bifurcated hydrogen bonds . . . (This location turned out to be similar to that obtained by Kanda and King.)." Osaki et al. also found a sixth oxygen, closer than the other five, at 2.84 Å (apparently overlooked by Kanda and King), but stated that this one "did not seem to be concerned with the hydrogen-bond formation substantially because it lies in the extended direction of the sulfur-nitrogen axis of the molecule."

This structure was then studied by neutron diffraction by Sass (1960). In this work, Sass located the hydrogen atoms directly without making any assumptions concerning the hydrogen bonding, taking the positions of the sulfur, nitrogen, and oxygen atoms as given by Osaki et al. as the starting point of his structure refinement. This structure refined to one in which the positions of the hydrogen atoms were quite different from those proposed previously: there are three normal N—H···O bonds, and *no* bifurcated hydrogen bonds. The situation around the $-NH_3^+$ group is shown in Figure 13.

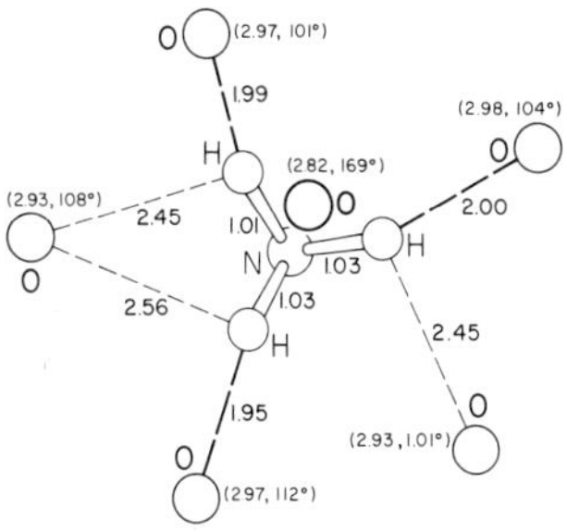

FIGURE 13.
The projection down the N—S bond of sulfamic acid, showing the hydrogen atoms and the six nearest neighbors of the nitrogen atom. The N···O distances and S—N···O angles are given in parentheses at each oxygen atom. [After Sass, 1960.]

The moral here, I suppose, is that distances and bond angles alone may be insufficient for the deduction of a hydrogen-bond scheme. In the case of sulfamic acid the fact that the conformation about the N—S bond in the bifurcated scheme is eclipsed might possibly have aroused some suspicions. Although the correct conformation found by Sass is very nearly staggered, examination of Figure 13 suggests to me that it would have been quite difficult, in the absence of direct evidence for the hydrogen positions, to have arrived at the correct hydrogen-bonding scheme with any degree of confidence. In fact, an incorrect hydrogen-bonding scheme, which was proposed in order to avoid bifurcated hydrogen bonds, was published some years prior to the work of Sass (Donohue, 1957b).

Cycloserine Hydrochloride

Regarding the crystal structure of cycloserine hydrochloride,

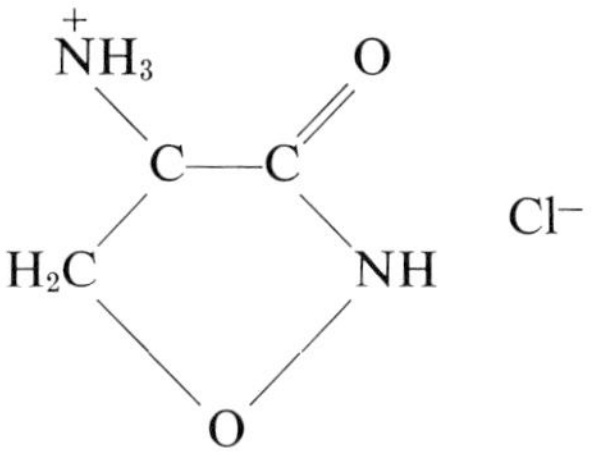

Turley and Pepinsky (1956), upon finding that the carbonyl oxygen has two neighbors, a ring nitrogen at 2.89 and a ring oxygen at 2.92 Å, suggested that "it seems at least possible that a bifurcated hydrogen bond exists here." It was then pointed out (Donohue, 1957a) that this interpretation was impossible, and that the 2.89 Å distance corresponded to an NH···O hydrogen bond and that the 2.92 Å distance was a normal van der Waals contact.

Iodic Acid

The crystal structure of iodic acid was first determined by Rogers and Helmholz (1941). They found that each iodine forms three covalent bonds to oxygen, giving discrete HIO_3 molecules, and, in addition, three secondary bonds to oxygen atoms of neighboring molecules, giving a distorted octahedron about the iodine atom. The oxygen atom having the longest covalent I—O distance, and therefore the hydroxyl oxygen, was found to have two close oxygen neighbors, at 2.76 and 2.78 Å, in different HIO_3 molecules, and Rogers and Helmholz assumed that this corresponded to a bifurcated hydrogen bond system.

Wells (1949) then gave an alternative interpretation. He noted that one of these short O···O distances was within the distorted octahedron, and that it was "not justifiable to regard the short length of any *edge* of this distorted IO_6 group an evidence of hydrogen bonding." He described the structure as a system of linked octahedra, and with no bifurcated hydrogen bonds.

Wells' hypothesis was then verified by Garrett (1954b) in a neutron diffraction study which included locating the hydrogen atom directly. Garrett found HIO_3 molecules were joined by normal hydrogen bonds, OH···O, having an O···O distance of 2.69 Å, and forming infinite chains as described by Wells.

Summary

The conclusions to be drawn for each of the compounds included in this section appears, at present, to be as follows:

1. Clearly bifurcated hydrogen bonds, that is, those in which the donor hydrogen atom is about equidistant from the two acceptor atoms, and in which the H···X distances are both shorter by 0.3 Å or more than the sum of the Pauling van der Waals radii: β-glycine, glycine hemihydrochloride (first $-NH_3^+$ group), and violuric acid monohydrate.

2. Asymmetrically bifurcated hydrogen bonds, that is, those in which the donor hydrogen atom is markedly closer to one of the two acceptor atoms, and in which the H···X distances are both shorter by 0.2 Å, or more, than the van der Waals radius sum: α-glycine and nitramide.

3. Doubtful bifurcated hydrogen bonds, that is, those in which one of H···X distances is 0.2 Å, or more, less, but in which the other is within 0.1 Å of the van der Waals distance: glycine hemihydrochloride (second $-NH_3^+$ group) and magnesium sulfate tetrahydrate.

4. No bifurcated hydrogen bonds, that is, compounds which have previously been suggested to contain such bonds, but which, upon closer examination, are found not to contain any: dicyandiamide, maleic acid, sulfamic acid, cycloserine hydrochloride, and iodic acid.

It may be concluded that bifurcated hydrogen bonds are rare, but not extinct; they are probably more common than the bifurcated yucca described by Albrecht (1952). More examples will doubtless be discovered.

THE LINEARITY OF HYDROGEN BONDS

The first considerations of the collinearity of hydrogen-bond systems such as O—H···O or N—H···O were usually based on the observed angles C—OH···O or C—NH···O because the hydrogen atoms had not been located directly. An early suggestion (Donohue, 1952)—that the length of NH···O hydrogen bonds formed by $-NH_3^+$ groups increased as the deviation of the angle C—NH···O from the tetrahedral increased—was based on rather meager data, but in the light of later and more extensive data (Fuller, 1959) it was shown to be untenable. At the present time, with the direct locating of hydrogen, it is now possible to examine the linearity directly. In Table 1 are collected data for fourteen different compounds, and data selected from Table 1 are presented in Figure 14; data for the O—H···O bonds are not included because the lengths of these bonds is highly dependent on the nature of both

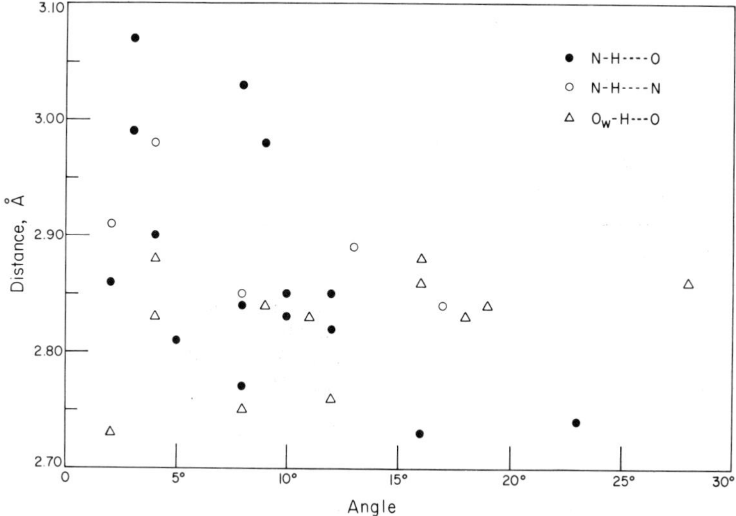

FIGURE 14.
A plot of NH···O (filled circles), NH···N (open circles) and O water-H···O (triangles) distance vs. the angle between the N—H or O—H vector and the N···O, N···N, or O···O vector.

the donor and the acceptor oxygen atoms. It is apparent from Table 1 and Figure 14 that there is little correlation between the length of a hydrogen bond and its deviation from strict linearity.

The fact that hydrogen bonds are not, in general, strictly linear should really not be cause for the surprise that has been expressed in some quarters. The surprising thing to me is that these complicated molecules develop crystal structures in which so many of the requirements of the principles of structural chemistry are simultaneously satisfied, including not only the distances and angular properties, but also the formation of hydrogen bonds by all hydrogen atoms able to do so, that is, those bonded to oxygen or nitrogen. Even more remarkable is the fact that there are *three* different crystal structures for glycine, each with a satisfactory N—H···O hydrogen-bond system; also in each the conformation about the C—N bond is very close to the expected staggered conformation, as shown in Figure 15. This result suggests to me that when "unusual" structural features are reported, such as an abnormal

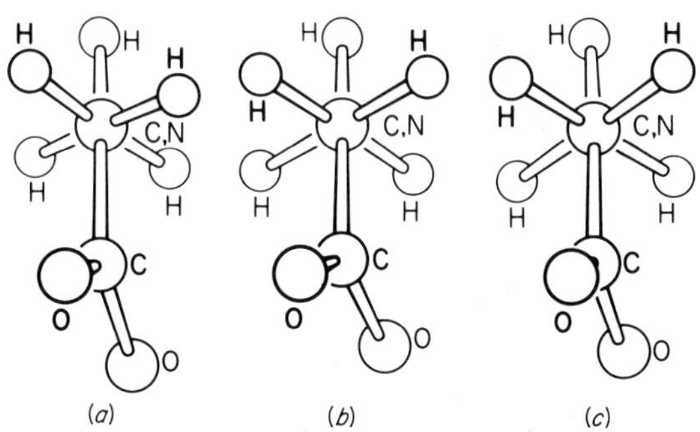

FIGURE 15.
Projection of the glycine molecule down the Cα—N bond in crystals of (*a*) α-glycine, (*b*) β-glycine, and (*c*) γ-glycine.

TABLE 1.
Some distances and angles in assorted hydrogen-bonded crystals.

O—H···O								
Compound	O···O	O—H	H···O	∠H—O···O	Donor	Acceptor	Method*	Ref.
Boric acid	2.69	0.95	1.75	0°	acid OH	acid OH	N	*a*
	2.69	0.97	1.72	0°	acid OH	acid OH		
	2.70	0.99	1.71	2°	acid OH	acid OH		
	2.72	0.95	1.78	3°	acid OH	acid OH		
	2.73	0.99	1.74	1°	acid OH	acid OH		
	2.74	0.98	1.77	1°	acid OH	acid OH		
Oxalic acid dihydrate	2.52	1.06	1.46	3°	carboyxl OH	H_2O	N	*b*
	2.84	0.95	1.91	9°	H_2O	carboxyl O		
	2.86	0.97	1.94	16°	H_2O	carboxyl O		
Acetylglycine	2.56	0.94	1.63	2°	carboxyl OH	carbonyl O	X	*c*
Violuric acid monohydrate	2.56	1.02	1.54	3°	acid OH	H_2O	N, X	*d*
	2.76	0.97	1.82	12°	H_2O	carbonyl O		
Iodic acid	2.69	0.99	1.70	5°	acid OH	acid O	N	*e*
Xanthazole monohydrate	2.83	1.02	2.02	18°	H_2O	carbonyl O	X	*f*
	2.88	0.87	1.92	16°	H_2O	carbonyl O		
Paraperiodic acid	2.59	0.98	1.62	2°	acid OH	acid O	N	*g*
	2.61	0.95	1.66	3°	acid OH	acid O		
	2.75	0.94	1.81	3°	acid OH	acid OH		
	2.77	0.95	1.82	5°	acid OH	acid OH		
	2.81	0.98	1.84	6°	acid OH	acid OH		
$MgSO_4 \cdot 4H_2O$	2.73	0.98	1.75	2°	H_2O	sulfate O	N, X	*n*
	2.75	0.95	1.82	8°	H_2O	sulfate O		
	2.83	0.96	1.90	11°	H_2O	sulfate O		
	2.83	0.99	1.85	4°	H_2O	sulfate O		
	2.84	0.97	1.95	19°	H_2O	sulfate O		
	2.86	0.95	2.06	28°	H_2O	sulfate O		
	2.88	0.97	1.92	4°	H_2O	sulfate O		

N—H···O								
Compound	N···O	N—H	H···O	∠H—N···O	Donor	Acceptor	Method*	Ref.
Cytosine	2.98	0.86	2.14	9°	$-NH_2$	carbonyl O	X	*h*
	3.03	0.88	2.17	8°	$-NH_2$	carbonyl O		
L-alanine	2.81	0.90	1.91	5°	$-NH_3^+$	carboxyl O	X	*i*
	2.83	0.88	2.00	16°	$-NH_3^+$	carboxyl O		
	2.85	0.95	1.92	10°	$-NH_3^+$	carboxyl O		
Isocytosine	2.73	0.96	1.83	16°	ring NH	carbonyl O	X	*j*
	2.82	0.89	1.95	12°	$-NH_2$	carbonyl O		
	2.86	0.89	1.97	4°	$-NH_2$	carbonyl O		
	2.90	0.95	1.95	4°	$-NH_2$	carbonyl O		
Violuric acid monohydrate	2.99	0.97	2.03	3°	ring NH	carbonyl O	N, X	*d*
	3.07	1.06	2.01	3°	ring NH	carbonyl O		
Glycine	2.77	0.92	1.87	8°	$-NH_3^+$	carboxyl O	X	*k*
	2.85	0.85	2.03	12°	$-NH_3^+$	carboxyl O		
Xanthazole monohydrate	2.74	0.83	2.01	23°	ring NH	H_2O	X	*f*
	2.84	0.89	1.96	8°	ring NH	carbonyl O		

			N—H···N					
Compound	N···N	N—H	H···N	∠H—N···N	Donor	Acceptor	Method*	Ref.
Isocytosine	2.91	0.95	1.96	2°	ring NH	ring N	X	*j*
	2.98	1.02	1.96	4°	$-NH_2$	ring N		
Cytosine	2.84	0.88	2.02	17°	ring NH	ring N	X	*h*
Purine	2.85	0.86	2.00	8°	ring NH	ring N	X	*l*
Xanthazole monohydrate	2.89	0.97	1.96	13°	ring NH	ring N	X	*f*

			O—H···N					
Compound	O···N	O—H	H···N	∠H—O···N	Donor	Acceptor	Method*	Ref.
Dimethylglyoxime	2.77	1.02	1.91	26°	oxime OH	Oxime N	N	*m*

*N = neturon diffraction, X = X-ray diffraction. (The X-ray method often gives O—H and N—H distances which are up to 0.3 Å shorter than the true internuclear distances; neutron and X-ray results should thus be compared with caution.)

REFERENCES: *a*, Craven and Sabine (1966); *b*, Garrett (1954a); *c*, Donohue and Marsh (1962); *d*, Craven and Mascarenhas (1964), Craven and Takei (1964); *e*, Garrett (1954b); *f*, Mez and Donohue (1966); *g*, Feikema (1966); *h*, Barker and Marsh (1964); *i*, Simpson and Marsh (1966); *j*, Sharma and McConnell (1965); *k*, Marsh (1958); *l*, Watson, Sweet, and Marsh (1965); *m*, Hamilton (1961), *n*, Baur (1964).

bond length or angle, an unexpectedly large departure from an expected planarity, and so on, which are then attributed to "crystal forces" or "packing distortions," such suggestions should be taken *cum grano*, because it seems highly unlikely that such effects could cause molecules in crystals to take up a geometry which is very far from that corresponding to minimum energy in the isolated molecule.

One further remark may be made concerning the data of Table 1: it should by now be obvious that a bent hydrogen bond is the rule, rather than the exception, and in view of some of the rather large deviations from zero of the angle H—O···O, and so on, which have been reported, it appears that the maximum allowable bend is near 30°, a value somewhat larger than had been customarily used previously. On the other hand, I believe that a structure in which *all* of the hydrogen bonds are that strained is highly unlikely, with the exception of molecules containing a restraint such as that occurring in dimethylglyoxime.

THE C—H···O HYDROGEN BOND: WHAT IS IT?

Some of the early evidence for the formation of C—H···O hydrogen bonds was discussed by Hunter (1946) who remarked, "so weak is the tendency of the CH group to form a hydrogen bond that it is manifested only under the influence of activating atoms or groups tending to promote the ionisation of the hydrogen atom, as in chloroform, hydrogen cyanide, and phenylacetylene." Other compounds mentioned by Hunter as exhibiting a C—H···O interaction include o-nitrotoluene, o-toluic acid, o-nitrophenylacetic acid, and ethyl o-toluate. At that time the principal evidence for the presence of this interaction consisted of one or more abnormal physical properties.

Somewhat later, Sutor (1962) collected data concerning abnormally short C···O distances in eight different compounds. These compounds were: caffeine 3.18 and 3.24 Å (Sutor, 1958b); theophylline, 3.22 Å (Sutor, 1958a); glycyl-L-tyrosine hydrochloride, 3.07 Å (Smits and Wiebenga, 1953); ethylene carbonate, 3.11 Å (Brown, 1954); 1,3,7,9-tetramethyluric acid, 3.00 Å (Sutor, 1963); acetylcholine bromide, 3.00 Å (Sörum, 1959); muscarine iodide, 2.87 Å (Jellinek, 1957); indirubine, 3.01 Å (Pandraud, 1961).

In the longer of the two short C···O distances in caffeine, and in ethylene carbonate, Sutor (1962) showed that the H···O distances were longer than the expected van der Waals contact of 2.6 Å. On the other hand, in the cases of the shorter C···O distance in caffeine, of theophylline, and of 1,3,7,9-tetramethyluric acid, the H···O distances were stated to be 2.25, 2.07, and 2.26 Å, respectively, or less than the van der Waals sum, and Sutor stated "These interactions are therefore hydrogen bonds." In regard to the short C···O distances in muscarine iodide and indirubine, she commented, "These probably represent intramolecular C—H···O hydrogen bonds." She did not discuss further the examples of glycyl-L-tyrosine hydrochloride or acetylcholine bromide.

Regarding these last two compounds, it has since been shown by Dunitz (1963) that Sörum chose an incorrect space group for acetylcholine bromide and that the published structure would therefore have to be revised. Furthermore, the structure of muscarine iodide is based on three projections only, two of which are very poorly resolved; that structure therefore needs further confirmation before it can be accepted. Finally, in glycyl-L-tyrosine hydrochloride the oxygen atom of the C—H···O system is a water molecule, which, according to Smit and Wiebenga, shows abnormal thermal motion [a root mean square amplitude of vibration of 0.42 Å in a direction 29° to the *a*-axis in the (h0l) projection]; its exact situation is accordingly not settled.

The foregoing discussion reduces to four the number of short C···O distances that are candidates for C—H···O hydrogen bonds.

Meanwhile, seven additional short C···O distances may be noted. The compounds in which these occur are: L-threonine, 3.28 Å (Shoemaker, et al., 1950); cytidine, 3.24 Å (Furberg, 1950); hydroxy-L-proline, 3.22 Å (Donohue and Trueblood, 1952); uracil, 3.19 and 3.28 Å (Parry, 1954); l-methylthymine, 3.11 Å (Hoogsteen, 1963a); and cytidylic acid, 3.25 Å (Sundaralingam and Jensen, 1965). Details of the C—H···O systems in these ten compounds are shown in Figure 16, *a–j*.

Before discussing the nature of these interactions, I wish to make a comparison between two molecules that have superficially similar geometries, namely, salicylic acid and o-nitrobenzaldehyde. Cochran (1952) found that the entire molecule of salicylic acid is coplanar. Details of the O—H···O hydrogen bond are presented in Figure 17. The H···O interaction is obviously one of attraction, and, the geometrical restraint in this molecule notwithstanding, inclusion of the appropriate data for it in Table 1 would not have been out of place. In o-nitrobenzaldehyde (Coppens and Schmidt, 1964; Coppens, 1964), however, the situation is quite different, as shown in Figure 18; Coppens (1964) presented evidence that the H···O interaction here is one

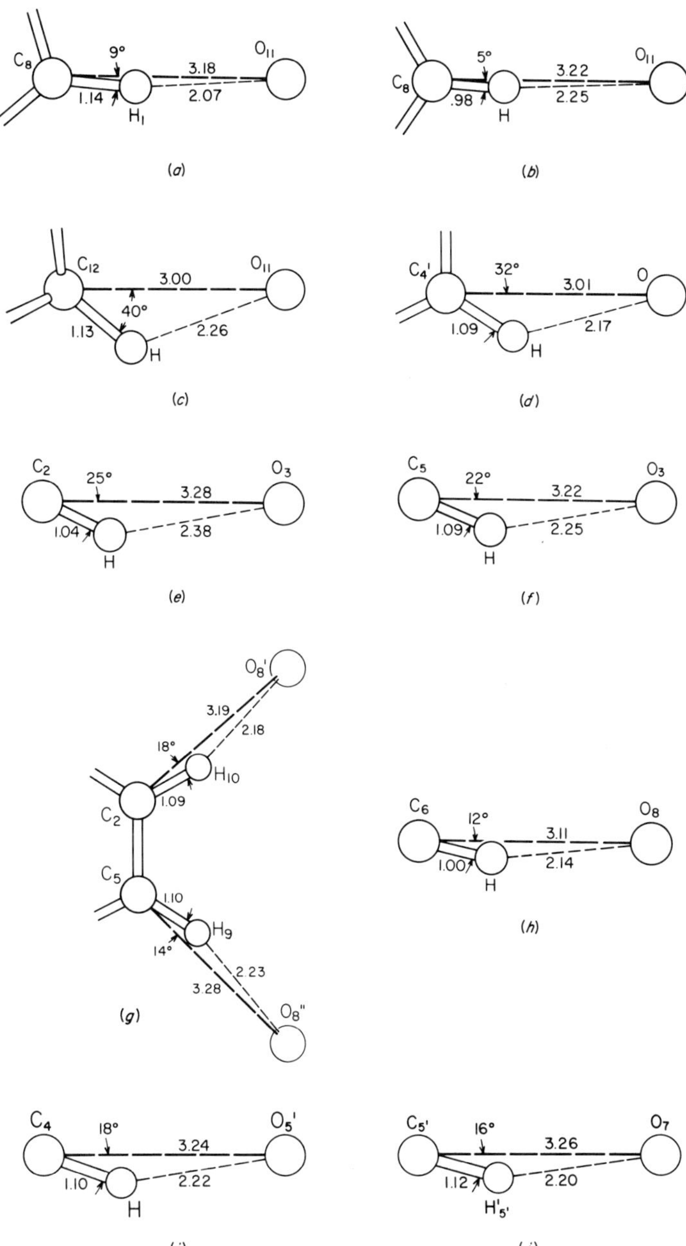

FIGURE 16.
Some details of H· · ·O interactions in short C· · ·O situations in:
(*a*) caffeine, (*b*) theophylline, (*c*) 1,3,7,9-tetramethyluric acid, (*d*) indirubine (hydrogen position assumed), (*e*) L-threonine, (*f*) hydroxy-L-proline, (*g*) uracil, (*h*) l-methylthymine, (*i*) cytidine (hydrogen position assumed), (*j*) cytidylic acid.

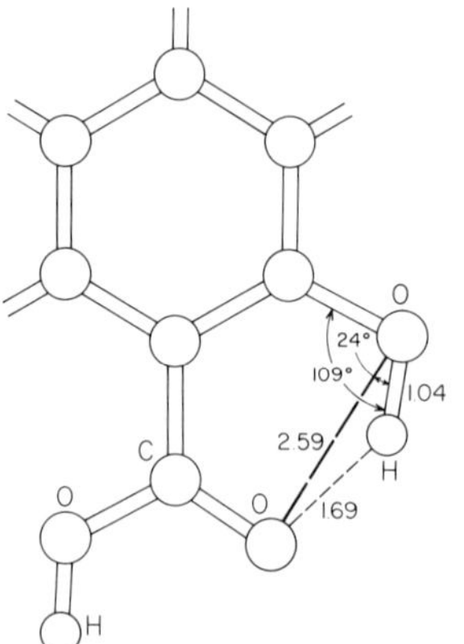

FIGURE 17.
Some details of the molecular structure of salicylic acid.

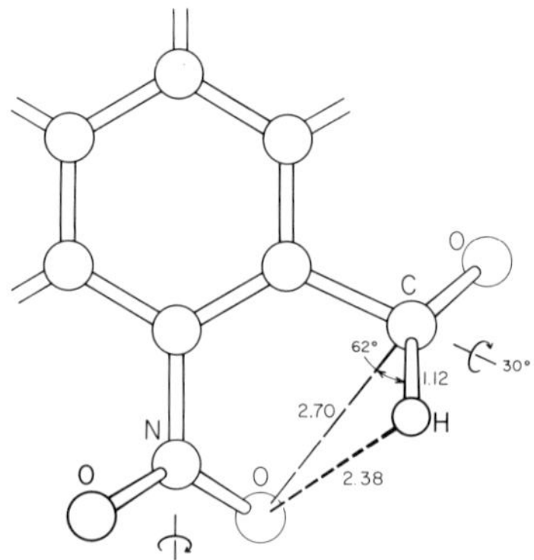

FIGURE 18.
Some details of the molecular structure of o-nitrobenzaldehyde.

of repulsion, demonstrating that the molecule assumes a conformation in which the H···O repulsion is balanced by the restoring force that makes the aldehyde group coplanar with the ring. (Apparently Coppens did not take into account the restoring force that makes the nitro group coplanar with the ring.) Coppens stated: "This definitely rules out the existence of an O···H—C hydrogen bond in this molecule."

It therefore seems reasonable to assume that an O···H distance of 2.38 Å is not presumptive evidence that hydrogen-bond formation is occurring; it should be noted, moreover, that all of the O···H distances in Table 1 are considerably shorter than this. It is also worth noting that after a "detailed analysis of the available structural data on various organic compounds" (details not given), Ramachandran et al. (1963) settled on a set of minimum contact distances which included a "normally allowed" H···O distance of 2.40 Å, and an "outerlimit" H···O distance of 2.20 Å. These distances, which are to be used in nonhydrogen-bonding situations, are both considerably smaller than the value of 2.60 Å, which was obtained from the van der Waals radii of Pauling, and which has been generally used previously. Serious doubt is thus cast upon the interactions that are depicted in Figure 16 as representing C—H···O "hydrogen bonds" in the same sense that the term is used for O—H···O, N—H···O, and N—H···N systems. Only two of the H···O distances are less than 2.2 Å, but in the example of l-methylthymine (2.14 Å) the standard error in the position of a hydrogen atom is 0.06 Å, and in the example of caffeine (2.07 Å), the position of H_1 is probably in error by as much as 0.2 Å, a condition strongly suggested by the published bond angles involving that atom.

Furthermore, there appears to be very little in common among the situations of these hydrogen atoms: some of them are bonded directly to a purine or pyrimidine ring, and others are part of methyl or methylene groups. Moreover, if it is argued that the hydrogen atom on C_4 of cytidine is somehow activated to form a C—H···O hydrogen bond, it should be expected that the same hydrogen atom in cytidylic acid would be part of a similar system, but this is not observed: the distance from H on C_6* to O_5, is 2.72 Å, as opposed to 2.22 Å in cytidine. The short H···O distance (of 2.20 Å) in cytidylic acid (Figure 16, *j*) does not involve a pyrimidine hydrogen atom, but one that is part of the ribose moiety.

It appears to me that the answer to the question in the title of this section is: "It isn't."

CONCLUSION

Although I have discussed only four topics which are a part of the hydrogen-bond problem, there are, obviously, many others that are equally interesting. Among these might be included the ice question, concerning which general opinion does not appear to have solidified; other residual entropy questions, symmetrical hydrogen bonds, isotope effects, and infrared frequency shifts.

Application of knowledge of the geometrical properties of hydrogen bonding has led to some notable successes, such as the construction of the polypeptide α-helix (Pauling and Corey, 1950; Pauling, Corey, and Branson, 1951) and the DNA double helix (Watson and Crick, 1953; Crick and Watson, 1954). It has also led to some notable, but quite interesting failures, such as polypeptide helices having only integral screw symmetry and consequent relaxation of certain geometrical requirements (Bragg, Kendrew, and Perutz, 1950), and a DNA triple helix held together by hydrogen bonds the hydrogen atoms of which, unfortunately, are not present in DNA (Pauling and Corey, 1953; see also Watson and Crick, 1953). I expect the future to bring forth more of both successes and failures.

ACKNOWLEDGMENTS

I wish to thank Dr. R. E. Marsh for reading the manuscript and pointing out numerous ways in which it could be improved (but any errors remaining are my responsibility); Mrs. Maryellin Reinecke who translated my rough sketches into drawings of her accustomed excellence; and, last, but not least, Professor Pauling, who, although he did not read the manuscript, has been a never failing source of advice and encouragement during a period of many years. This work was supported in part by the United States Atomic Energy Commission.

REFERENCES

Albrecht, G. (1952). *Science* **115**, 219.
Albrecht, G., and R. B. Corey (1939). *J. Am. Chem. Soc.* **61**, 1087.
Barker, D. L., and R. E. Marsh (1964). *Acta Cryst.* **17**, 1581.
Baur, W. H. (1962). *Acta Cryst.* **15**, 815.
Baur, W. H. (1964). *Acta Cryst.* **17**, 863.
Beevers, C. A., and A. F. Trotman-Dickenson (1957). *Acta Cryst.* **10**, 34.
Bragg, L., J. C. Kendrew, and M. F. Perutz (1950). *Proc. Roy. Soc.* (London) A **203**, 321.
Brown, C. J. (1954). *Acta Cryst.* **7**, 92.

*Different numbering systems are used in the two papers.

Cannon, C. G. (1958). *Spectrochim. Acta* **10**, 341.
Chidambaram, R. (1962). *J. Chem. Phys.* **36**, 2361.
Clark, J. R. (1963). *Revs. Pure and Appl. Chem.* **13**, 50.
Cochran, W. (1952). *Acta Cryst.* **6**, 260.
Coppens, P. (1964). *Acta Cryst.* **17**, 573.
Coppens, P., and G. M. J. Schmidt (1964). *Acta Cryst.* **17**, 222.
Craven, B. M., and Y. Mascarenhas (1964). *Acta Cryst.* **17**, 407.
Craven, B. M., and T. M. Sabine (1966). *Acta Cryst.* **20**, 214.
Craven, B. M., and W. J. Takei (1964). *Acta Cryst.* **17**, 415.
Crick, F. H. C., and J. D. Watson (1954). *Proc. Roy. Soc.* (London) A **223**, 80.
Donohue, J. (1952). *J. Phys. Chem.* **56**, 502.
Donohue, J. (1957). *Acta Cryst.* **10**, 383.
Donohue, J. (1957). In Pauling, L., and H. A. Itano, eds., *Molecular Structure and Biological Specificity*, Chapter 5. Baltimore: Waverly Press.
Donohue, J. (1964). *Acta Cryst.* **17**, 771.
Donohue, J., and A. Caron (1964). *Acta Cryst.* **17**, 1178.
Donohue, J., L. R. Lavine, and J. S. Rollett (1956). *Acta Cryst.* **9**, 655.
Donohue, J., and R. E. Marsh (1962). *Acta Cryst.* **15**, 941.
Donohue, J., and K. N. Trueblood (1952). *Acta Cryst.* **5**, 419.
Dunitz, J. D. (1963). *Acta Chem. Scand.* **17**, 1471.
Feikema, Y. D. (1966). *Acta Cryst.* **20**, 765.
Fuller, W. (1959). *J. Phys. Chem.* **63**, 1705.
Furberg, S. (1950). *Acta Cryst.* **3**, 325.
Gabe, E. J. (1962). *Acta Cryst.* **15**, 759.
Garrett, B. S. (1954a). Oak Ridge Nat. Lab., Rep. 1745, p. 13; *Structure Reports* **19**, 519.
Garrett, B. S. (1954b). Oak Ridge Nat. Lab., Rep. 1745, p. 97; *Structure Reports* **18**, 393.
Hamilton, W. C. (1961). *Acta Cryst.* **14**, 95.
Hamilton, W. C. (1962). *Ann. Rev. Phys. Chem.* **13**, 19.
Hahn, T., and M. J. Buerger (1957). *Z. Krist.* **108**, 419.
Harris, D. R., and W. M. Macintyre (1964). *Biophys. J.* **4**, 203.
Helmholz, L. (1937). *J. Am. Chem. Soc.* **59**, 2036.
Helmholz, L. (1939). *J. Am. Chem. Soc.* **61**, 1544.
Helmholz, L., and M. T. Rogers (1939). *J. Am. Chem. Soc.* **61**, 2590.
Holtzberg, F., B. Post, and I. Fankuchen (1953). *Acta Cryst.* **6**, 127.
Hoogsteen, K. (1963a). *Acta Cryst.* **16**, 28.
Hoogsteen, K. (1963b). *Acta Cryst.* **16**, 907.
Hughes, E. W. (1940). *J. Am. Chem. Soc.* **62**, 1258.
Hughes, E. W., and W. J. Moore (1949). *J. Am. Chem. Soc.* **71**, 2618.
Hunter, L. (1946). *Ann. Rep. Progr. Chem.* **43**, 153.
Iitaka, Y. (1960). *Acta Cryst.* **13**, 35.
Iitaka, Y. (1961). *Acta Cryst.* **14**, 1.
Jeffrey, G. A., S. Ghose, and J. O. Warwicker (1961). *Acta Cryst.* **14**, 881.
Jeffrey, G. A., and Y. Kinoshita (1963). *Acta Cryst.* **16**, 20.
Jellinek, F. (1957). *Acta Cryst.* **10**, 277.
Jones, R. E., and D. H. Templeton (1958). *Acta Cryst.* **11**, 484.
Kanda, F. A., and A. J. King (1951). *J. Am. Chem. Soc.* **73**, 2315.
Karle, I. L., and J. Karle (1964). *Acta Cryst.* **17**, 835.
Kartha, G., and A. de Vries (1961). *Nature* (London) **192**, 862.
Lipscomb, W. N. (1954). Seminar, University of Minnesota, Jan. 13.
Luzzati, V. (1951). *Acta Cryst.* **4**, 239.
Marsh, R. E. (1958). *Acta Cryst.* **11**, 654.
Marsh, R. E., and J. P. Glusker (1961). *Acta Cryst.* **14**, 1110.
Mez, H.-C., and J. Donohue (1968). *Z. Krist.* To be published.

Nowacki, W., and H. Bürki (1955). *Z. Krist.* **106**, 339.
Nyburg, S. C. (1961). *X-Ray Analysis of Organic Structures.* New York: Academic Press.
Osaki, K., H. Tadokoro, and I. Nitta (1955). *Bull. Chem. Soc. Japan* **28**, 524.
Pandraud, H. (1961). *Acta Cryst.* **14**, 901.
Parry, G. S. (1954). *Acta Cryst.* **7**, 313.
Pasternak, R. A. (1956). *Acta Cryst.* **9**, 341.
Pauling, L. (1928). *Proc. Nat. Acad. Sci. U.S.* **14**, 359.
Pauling, L. (1939). *The Nature of the Chemical Bond.* Ithaca: Cornell Univ. Press.
Pauling, L., and R. B. Corey (1950). *J. Am. Chem. Soc.* **72**, 5349.
Pauling, L., and R. B. Corey (1951). *Proc. Nat. Acad. Sci. U.S.* **37**, 235.
Pauling, L., and R. B. Corey (1953). *Proc. Nat. Acad. Sci. U.S.* **39**, 84.
Pauling, L., R. B. Corey and H. R. Branson (1951). *Proc. Nat. Acad. Sci. U.S.* **37**, 205.
Penfold, B. R., and J. C. B. White (1959). *Acta Cryst.* **12**, 130.
Pimentel, G. C., and A. L. McClellan (1960). *The Hydrogen Bond.* San Francisco: W. H. Freeman and Company.
du Pré, S., and H. Mendel (1955). *Acta Cryst.* **8**, 311.
Ramachandran, G. N., C. Ramakrishnan, and V. Sasisekharan (1963). *J. Mol. Biol.* **7**, 95.
Robertson, J. H. (1964). *Acta Cryst.* **17**, 316.
Robertson, J. M. (1953). *Organic Crystals and Molecules.* Ithaca: Cornell Univ. Press.
Rogers, M. T., and L. Helmholz (1940). *J. Am. Chem. Soc.* **62**, 1533.
Rogers, M. T., and L. Helmholz (1941). *J. Am. Chem. Soc.* **63**, 278.
Sass, R. L. (1960). *Acta Cryst.* **13**, 320.
Shahat, M. (1952). *Acta Cryst.* **5**, 763.
Sharma, B. D., and J. F. McConnell (1965). *Acta Cryst.* **19**, 797.
Shoemaker, D. P., J. Donohue, V. Schomaker, and R. B. Corey (1950). *J. Am. Chem. Soc.* **72**, 2328.
Simpson, H. J., and R. E. Marsh (1966). *Acta Cryst.* **20**, 550.
Smits, D. W., and E. H. Wiebenga (1953). *Acta Cryst.* **6**, 531.
Sörum, H. (1959). *Acta Chem. Scand.* **13**, 345.
Sundaralingam, R., and L. H. Jensen (1965). *J. Mol. Biol.* **13**, 914.
Sutor, D. J. (1958a). *Acta Cryst.* **11**, 83.
Sutor, D. J. (1958b). *Acta Cryst.* **11**, 453.
Sutor, D. J. (1962). *Nature* (London) **195**, 68.
Sutor, D. J. (1963). *Acta Cryst.* **16**, 97.
Trueblood, K. N., P. Horn, and V. Luzzati (1961). *Acta Cryst.* **14**, 965.
Turley, J. W., and R. Pepinsky (1956). *Acta Cryst.* **9**, 948.
Ubbelohde, A. R., and K. J. Gallagher (1955). *Acta Cryst.* **8**, 71.
Watson, D. G., R. M. Sweet, and R. E. Marsh (1965). *Acta Cryst.* **19**, 573.
Watson, J. D., and F. H. C. Crick (1953). *Nature* (London) **171**, 737.
Wells, A. F. (1949). *Acta Cryst.* **2**, 128.
Wells, A. F. (1962). *Structural Inorganic Chemistry*, 3rd ed. Oxford: Clarendon Press.
Wright, W. B., and G. S. D. King (1954). *Acta Cryst.* **7**, 283.

WALTER C. HAMILTON
Chemistry Department
Brookhaven National Laboratory
Upton, New York

On Hydrogen Bonding in Inorganic Crystals: Some Generalizations, Some Recent Results, and Some New Techniques

The importance of the hydrogen bond was recognized by Pauling (1928) early in his studies of the chemical bond, and the subject pervades much of his work, particularly that which deals with the structure and function of biological molecules. The recognition of the importance of the restraints on molecular structure imposed by hydrogen bonding led to the first successful prediction of an important feature of protein structure—the α-helix (Pauling, Corey, and Branson, 1951).

Hydrogen bonds are also common in inorganic crystals, and consideration of the large amount of information about the geometry of the hydrogen bond that has been made available through the use of modern tools of structural chemistry in the course of the past several years leads to the following conclusion: The hydrogen bond is important in chemistry, not only because of the restraints it imposes on the geometry of molecular complexes but also because of the flexibility of the geometrical requirements of hydrogen bonding. This flexibility is illustrated in the many structures of ice and in the polymorphism of the ammonium halides.

Consideration of the latter group of compounds brings us to a discussion of recent experimental data on the rotation of the ammonium ion in crystalline solids, a subject to which Pauling addressed himself many years ago in one of the early applications of quantum mechanics to molecular structure (Pauling, 1930).

One of the most fruitful methods of study of the hydrogen bond has been that of neutron inelastic scattering. Inasmuch as this technique promises to become one of the most important contributions of nuclear science to the study of molecular and crystal structures, it seems useful to conclude this paper with a digression suggesting the implications that this new technique may have for the study of the solid state.

Research performed under the auspices of the United States Atomic Energy Commission.

DEFINITION OF THE HYDROGEN BOND

We say that a hydrogen bond exists when a hydrogen atom is bonded to at least two atoms at the same time. A hydrogen bond may thus be denoted

$$A—H—B$$

Although there are many well-studied systems in which the bond A—H is as strong as the bond H—B, in most hydrogen bonds one bond is much stronger than the other. Such a hydrogen bond may be denoted

$$A—H \cdots B$$

Although many criteria have been used to establish the existence of the hydrogen bond, our discussion here will be largely devoted to the hydrogen bond in the solid state, in which the relative positions of all the atoms in the crystal are known or may be reasonably inferred. We adopt the criterion that a hydrogen bond exists when the atoms H and B, usually (but not necessarily) in different molecules, are considerably closer together than the accepted van der Waals contact distance for nonbonded atoms.

Table 1 gives some values for $H \cdots B$ distances below which a hydrogen bond may be said to exist. The criterion used is that the distance $R(H \cdots B)$ shall be less than $W_H + W_B - 0.2$ Å, where W_H and W_B are the van der Waals radii for atoms H and B. What are the conditions necessary for the formation of a hydrogen bond? In most hydrogen bonds of importance, the atoms A and B must be electronegative, so that the system can be represented *before* the formation of the hydrogen bond as

$$A^{-\delta}—H^{-\delta} \qquad B^{-\epsilon}$$

The atom A must be electronegative to insure the presence of an acidic hydrogen; the atom B must be electronegative to be capable of retaining the extra electronic charge that makes it a strong Lewis base. In most hydrogen bonds, the acceptor atom B possesses lone-pair electrons, which may be largely responsible for the formation of the hydrogen bond.

O—H···O HYDROGEN BONDS IN HYDRATES

Among the hydrogen bonds most common in nature are those which occur in the hydrated crystals of inorganic salts. The structures of these crytsals provide us with excellent examples of the flexibility of the hydrogen bond, illustrating that the geometrical requirements for hydrogen bonding are not nearly as severe as those for covalent bonding. This flexibility is perhaps largely responsible for the ubiquitous appearance of the hydrogen bond. Its ability to tie large ions together in a variety of ways in complex crystal structures leads to great stability of these structures.

If we consider the water molecules in inorganic crystals, we find that they are primarily of two types. In one of these, the oxygen atom of the water molecule may be considered to be planar-trigonal coordinated. There are two hydrogen atoms, of course, each usually participating in a hydrogen

TABLE 1.
H···B *distances below which a hydrogen bond may be said to exist.*

H···F	2.3
H···O	2.4
H···N	2.5
H···Cl	2.8
H···S	2.9
H···Br	3.0
H···I	3.2

bond. A third neighbor atom is situated exactly or approximately on an axis that coincides with the two-fold axis of the water molecule. Two such configurations are shown in Figure 1. The ligand L in Figure 1a is typically a metal ion; in Figure 1b, the oxygen atom acts as the acceptor atom in another hydrogen bond.

In the second type, the oxygen atom of the water molecule is tetrahedrally coordinated. Again, two hydrogen atoms form hydrogen bonds from the water molecule to other parts of the crystal. The isolated water molecule may be considered to be tetrahedral, with two O—H bonds directed toward two corners of a tetrahedron and two lone-pair electron orbitals directed toward the other two corners. The last two sites may be occupied either by positive ions or by hydrogen bond donors. Three types of such coordination are illustrated in Figure 2. The ligands L are typically singly or doubly charged positive ions.

The various geometries of water molecules in crystals have been classified and examples given of each by Chidambaram, Sequeira, and Sikka (1964). These authors give some examples of threefold coordination, in which a single ligand occupies a site at one of the corners of the tetrahedron but the fourth site is empty.

A final type of coordination has been shown to exist in at least one compound, sodium perxenate octahydrate (Ibers, Hamilton, and MacKenzie, 1964), whose crystals contain some tetrahedrally coordinated water molecules. More interestingly, two different types of five-fold coordination are exhibited, namely those shown in Figure 3. Each starts with the planar trigonal coordination shown in Figure 1a and adds either a metal ligand or another hydrogen bond at each of the apices of a trigonal bipyramid.

It should also be remarked that the ideal geometrical configurations implied by the use of the words "trigonal," "tetrahedral," and "trigonal bipyramidal" to describe hydrogen-bond coordination are seldom observed in real crystals. Rather the angles are distorted from the ideal values that would obtain in perfect geometrical figures. Many of these distortions are severe. For example, the angles in the coordination polyhedra described as trigonal bipyramids in sodium perxenate octahydrate range from 70° to 105°, 85° to 150°, and 170° to 172° rather than having the ideal values of 90°, 120°, and 180°.* This variability in the types of coordination and in their departures

*These are angles involving the water molecule oxygen atoms—not the hydrogen atoms. As shown in Table 2, corresponding angles involving the hydrogen atoms can be quite different.

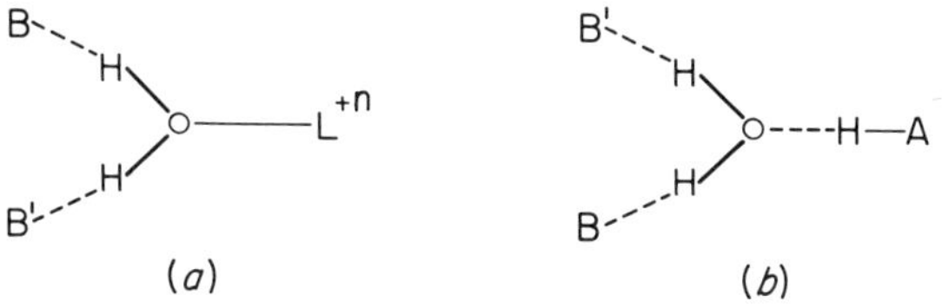

FIGURE 1. Trigonally coordinated oxygen atoms in hydrates. L^{+n} is a metal ion.

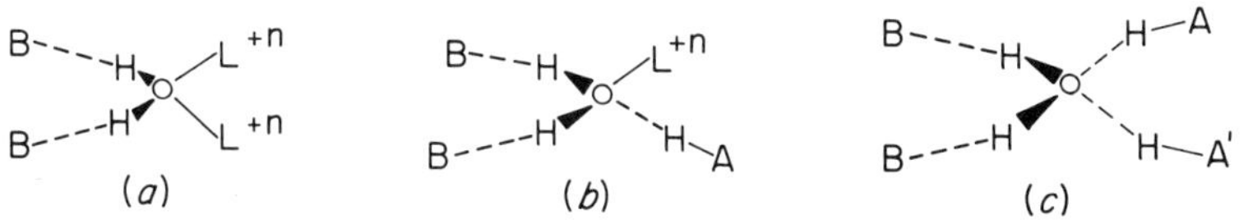

FIGURE 2. Tetrahedrally coordinated oxygen atoms in hydrates.

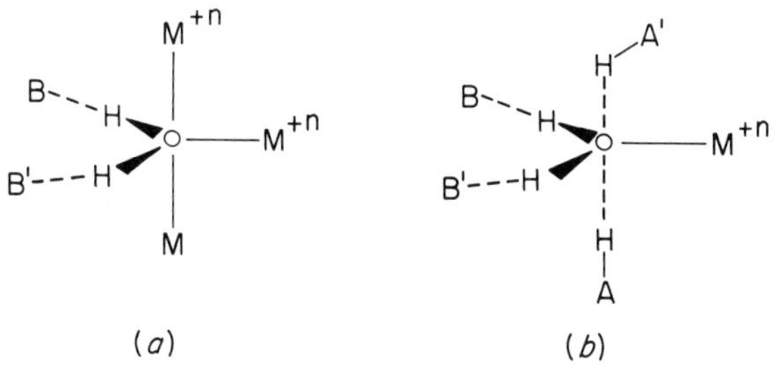

FIGURE 3. Trigonal bipyramidal coordination of oxygen atoms in hydrates.

from regularity is but one example of the enormous flexibility of the hydrogen bond, which allows it to compromise effectively with other geometrical constraints in the formation of ionic crystal structures.

As a further example of the flexibility of the hydrogen bond, we need only look at a comparison between the H—O—H angle of the water molecule and the O···O···O angle in systems which may be depicted as

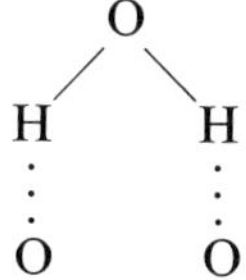

The values of these angles for a number of crystals are presented in Table 2. It may be seen that the variation of the O···O···O angles is much greater than that of the H—O—H angles, which are relatively constant. (The probable error in some of these quoted angles may be 2 to 3°.) This variation in what we may call the *acceptor angle* is evidence that the A—H···B bond need not be a linear bond, although the assumption of linearity has often been used in guessing structures for which the hydrogen-atom positon has not been determined.

Further evidence of the same type is presented in Figure 4, which illustrates the deviations of the O—H···O bond from linearity in all compounds that have been studied by neutron diffraction: As the figure shows, very large deviations frequently occur. The mean value of 165° is not to be interpreted as an energetically preferred value for the hydrogen-bond angle. Rather it provides evidence that, although the energetically preferred situation may be

TABLE 2.
H—O—H *angles and* O···O···O *acceptor angles in some hydrogen-bonded hydrates.*

Compound	∠H—O—H	∠O···O···O	Reference
$CuSO_4 \cdot 5H_2O$	114	119	Bacon and Curry, 1962
	111	121	
	109	130	
	109	105	
	106	122	
$UO_2(NO_3)_2 \cdot 6H_2O$	107	113	Taylor and Mueller, 1965
	107	110	
	115	81	
$Th(NO_3)_4 \cdot 5H_2O$	110	100	Taylor, Mueller, and Hitterman, 1966
	111	109	
	107	122	
$Mg(SO_4) \cdot 4H_2O$	110	105	Baur, 1964
	111	147	
	111	92	
	109	138	
	109	114	
$NaCO_3 \cdot NaHCO_3 \cdot 2H_2O$	107	114	Bacon and Curry, 1956

the linear hydrogen bond, the potential energy well that describes this energetically preferred configuration is a shallow one and is easily distorted or made entirely different in nature when other forces that may be energetically important in the crystal are brought into consideration.

As a final piece of evidence for the great flexibility of the hydrogen bond, we cite the existence of many bifurcated hydrogen bonds—bonds in which a covalently bound hydrogen atom participates in two weaker hydrogen bonds:

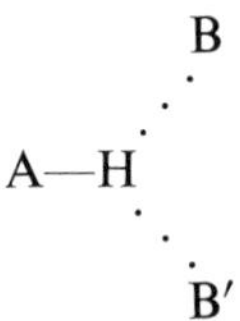

A good example of such a bond is provided by one of the water molecules in $MgSO_4 \cdot 4H_2O$ (Baur, 1964).

The fact that the water molecule, by forming hydrogen bonds, can enter into the crystal structures of inorganic salts in many different ways—and without severe geometrical constraints—is undoubtedly one of the main reasons for the widespread appearance of hydrated crystals in nature. The water molecule assists the formation of stable crystals of many compounds by isolating highly charged complex ions from each other. In particular, the stability of hydrated structures containing oxyanions is probably due to this isolation of the anions by hydrogen-bonded water molecules. Thus, we may expect that many sulfates and phosphates will preferentially crystallize as hydrates: the XeO_6^{-4} ion, for example, forms highly hydrated salts of several kinds, and the anhydrous sodium salt is difficult to obtain in crystalline form.

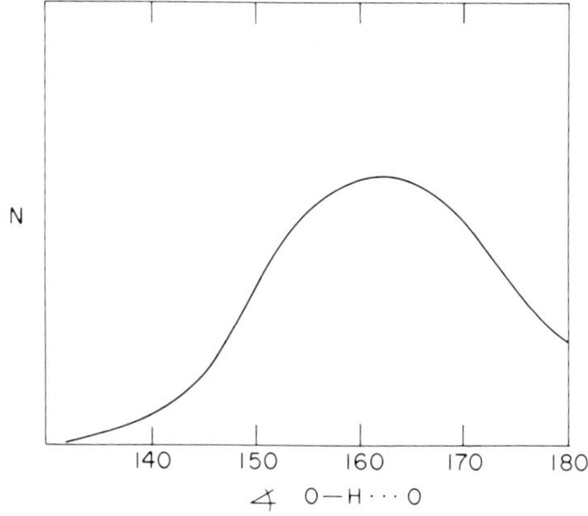

FIGURE 4.
Distribution of O—H···O angles that have been observed in neutron diffraction studies.

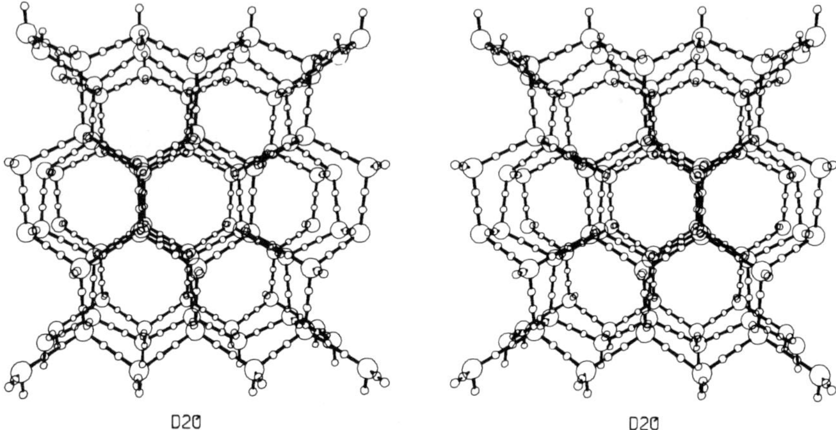

FIGURE 5.
The structure of Ice I viewed down the hexagonal axis. The large circles are oxygen atoms; the small circles represent the half-hydrogens in the disordered model. Note the large voids in the structure. This is a stereo pair and may be viewed by use of a small hand-held stereoscope.

We thus conclude that many ionic structures are stabilized by the hydration that is made possible by the flexibility of the hydrogen bond.

The stabilization of ionic structures by the presence of water molecules is paralleled by the stabilization of some of the structures of water (solid) by the presence of foreign molecules. The structures of all the polymorphs of ice apparently involve an oxygen framework in which each oxygen atom is surrounded by four other oxygen atoms at the corners of a possibly distorted tetrahedron. That ordinary ice has the proton disordered structure suggested by Pauling (1935) has been confirmed by the neutron diffraction work of Peterson and Levy (1957). Figure 5, which shows the structure of Ice I, makes it evident that there are fairly large holes in the structure. The loss in energy due to this non-space-filling arrangement must be made up for by the gain in energy due to the hydrogen bonds. If this structure is slightly expanded

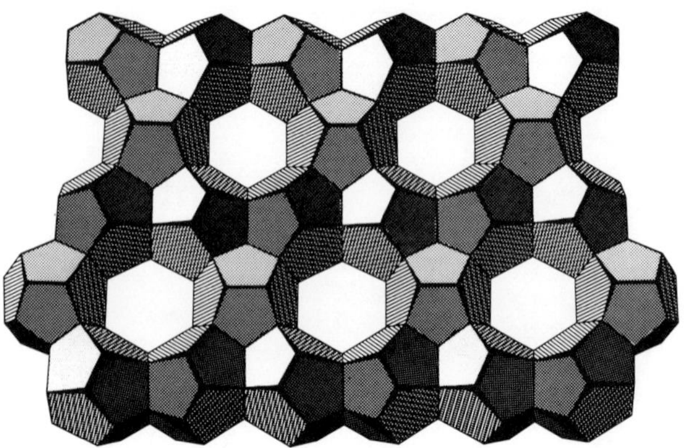

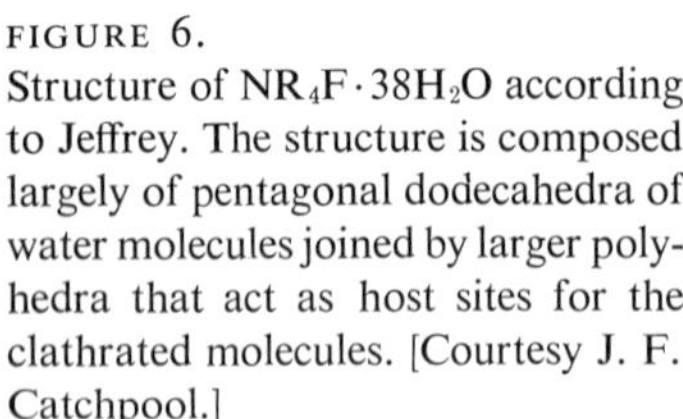

FIGURE 6.
Structure of $NR_4F \cdot 38H_2O$ according to Jeffrey. The structure is composed largely of pentagonal dodecahedra of water molecules joined by larger polyhedra that act as host sites for the clathrated molecules. [Courtesy J. F. Catchpool.]

additional water molecules can be accommodated in the voids, thus increasing the density; such a model has been proposed by Danford and Levy (1962) as being the one most compatible with the X-ray scattering data from liquid water. These authors have rejected the pentagonal dodecahedral model of water suggested by Pauling (1959).

Studies of the clathrate hydrates (see, for example, McMullan and Jeffrey, 1965) show that the crystals of many such substances consist mainly of pentagonal dodecahedral arrangements of water molecules, with occasional larger polyhedra of water molecules acting as the host sites for the solute molecules. Such an arrangement is shown in Figure 6, which illustrates the structure of a highly hydrated tetra(alkyl)ammonium fluoride: $NR_4F \cdot 38H_2O$. The presence of 1 part in 39 of a foreign salt has caused most of the water to crystallize in a pentagonal dodecahedral arrangement. One wonders whether liquid solutions of tetraalkyl ammonium salts in this concentration might not also have a pentagonal dodecahedral structure—distinct from the approximate hexagonal structure proposed by Danford and Levy. X-ray scattering experiments on such solutions should prove to be extremely interesting.

STRUCTURE AND MOTION IN HYDROGEN-BONDED CRYSTALS

In the determination of molecular structure, we are interested not only in the mean positions of atoms in the structure but in the variations of the atomic positions from these mean positions with time. More fundamentally, we are interested in the forces between atoms and molecules and the resulting potential energy surfaces on which the atoms must move. There are two classical methods of structural chemistry, each of which can give information on the nature of these potential surfaces. First, the determination of the mean amplitudes of motion of individual atoms by *diffraction methods* is pertinent to the determination of parameters concerning the potential energy surface. Second, the determination of the energy levels of the system by *spectroscopic methods*, gives us different information, from which we can also infer information regarding the potential energy surface. (The effect of motion on the shape of nuclear magnetic resonance spectral lines can also be an important source of information.) In addition to these two well-tested experimental methods for the investigation of potential functions, there is of course always theory, and it is indeed possible that in some simple cases theoretical calculations of potential energy surfaces—if not perhaps quantitatively realistic—can provide at least some help in the interpretation of the experiments.

To obtain pertinent information concerning potential surfaces from either the diffraction-determined moments of the atomic distributions or from the spectroscopically determined energy levels, it is always necessary to assume a mathematical model for the potential energy surface, the parameters of the model being determined by fitting them to the experimental data. If the model is correct, one should derive the same parameters from the diffraction experiment and the spectroscopic experiment. (The spectroscopic experiment will usually produce more precise results in this kind of study.) As McGaw and Ibers (1963) have shown in their treatment of the strongly hydrogen-bonded F—H—F$^-$ ion, the use of both spectroscopic and diffraction results in a complementary way can produce extremely useful and almost unambiguous results. A disagreement between the parameters derived by diffraction and spectroscopic methods suggests that the model is probably wrong and may indeed lead to the discovery of a more suitable and perhaps correct model.*

As examples of the importance of models and the ambiguities that result from the necessity for using them, we will quote some results from our recent neutron diffraction results at Brookhaven and the very important neutron inelastic scattering results of Rush and Taylor (1965, 1966; Rush, 1966; Rush, Taylor, and Havens, 1960).

AMMONIUM FLUOROSILICATE

Next to the hydrates, salts that include the ammonium cation are probably the most common and the most thoroughly studied of any hydrogen-bonded inorganic compounds. The ammonium ion is capable of contributing to four hydrogen bonds at the corners of a regular tetrahedron. The flexibility of the hydrogen bond and of hydrogen-bonded species in forming many different types of structures is again in evidence. Most ammonium salts exhibit a high degree of polymorphism, many of the polymorphs showing disorder—either static or dynamic.† One such salt is ammonium fluorosilicate—$(NH_4)_2SiF_6$. First of all, this substance crystallizes in both a cubic and a hexagonal phase, which has a transition at 38°K that is probably second-order. Although we have studied the structures of both phases by neutron diffraction (Schlemper, Hamilton, and Rush, 1966; Schlemper and Hamilton, 1966b) we will discuss here only the rather interesting cubic phase.

In this structure (Figure 7), the ammonium ion sits in a site of tetrahedral symmetry. Each corner of the tetrahedron is occupied by a triangle of three

*Lest this discussion sound too pessimistic, we hasten to remark that the model of harmonic motion for many of the molecular vibrations is close enough to the truth to provide much valuable information. It is mainly in the area of low-frequency hindered rotations and in the intermolecular coupling of otherwise harmonic motions that it breaks down severely.

†By dynamic disorder we mean a rapid motion of an atom from one point to another which is well separated in space from the first.

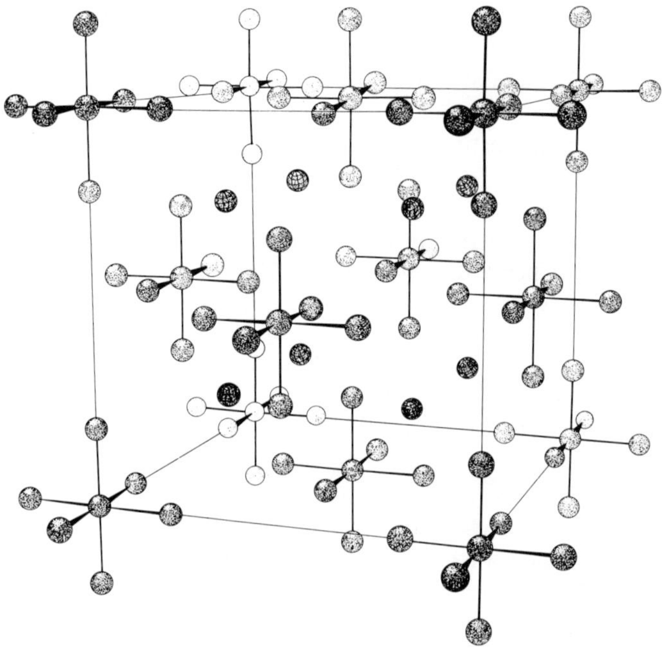

FIGURE 7.
The structure of cubic ammonium fluoroscilicate. Only the nitrogen atom of the ammonium group is shown. (It is indicated by tweedy shading.) Note that it has 12 equidistant fluorine neighbors.

F atoms which belong to the SiF_6^{--} ions. Thus each nitrogen atom has twelve F atoms at identical distances. What position does the ammonium ion assume under these circumstances? The nuclear scattering density due to hydrogen in a plane perpendicular to the threefold axis of the crystal is shown in Figure 8. The triangular figure is thus a probability density function for the position of a single hydrogen atom of the ammonium group. One model that can explain this structure is a static disordered model, with equal probabilities that the hydrogen atom is at each of the positions indicated by the small bumps on the periphery of the triangle. These positions are only 0.75 Å apart and it must be extremely easy for the atom to move from one position to the next. In fact, the shape of the scattering density plot suggests that there is a very large region—corresponding to the entire triangle—in which the hydrogen atom is relatively free to move. The scattering density contours may very well closely approximate the contours of the potential energy surface governing the motion of the hydrogen atom. From this point of view, we consider that we are dealing with a dynamic disorder—that is, a situation in which the motion of the hydrogen atom is quite different from the ellipsoidal motion characteristic of harmonic motion in three dimensions.

The inelastic neutron spectrum* of cubic ammonium fluorosilicate shows a peak at 168 ± 8 cm^{-1} which has been assigned to a hindered rotational motion of the ammonium group. By assuming a threefold cosine-shaped potential barrier to rotation, Schlemper, Hamilton, and Rush (1966) estimated a barrier height *for this model* of 2.1 kcal/mole.

The model suggested by the diffraction evidence, however, provides another interpretation of the spectroscopic results. We consider that the hydrogen

*See the final section of this paper.

FIGURE 8.
Nuclear scattering density in a plane perpendicular to the three-fold axis in cubic ammonium fluorosilicate. The disorder of the hydrogen atom is evident, and there is support for a model of dynamic rather than static disorder.

atom is free to move in a more or less uniform two-dimensional potential of the shape indicated in Figure 8, but that the potential barrier separating one such area of uniform potential from another of the same type is very high. As an approximation to this potential, we may consider a square two-dimensional box with area 1 Å^2. This is the area of appreciable scattering density seen in Figure 8. The potential separating such regions of uniform density is thus infinitely high. Elementary calculation of the energy levels for such a box for a particle with a mass of four hydrogen atoms shows that the lowest transition will be between states separated by 124 cm^{-1}. This is perhaps the transition that is seen in the neutron spectrum.

Thus, depending on the model we have chosen, the potential barrier separating the equivalent tetrahedral sites for the hydrogen atoms of the ammonium group may be found to be as low as 2.1 kcal/mole or to be infinitely high. This rather extreme but nevertheless realistic example emphasizes that the heights often given for the barriers limiting rotation in molecules or crystals are seldom quantitatively meaningful unless there is firm and independent evidence of the shape of the potential function. Thus, presumed heights of barriers to rotation of ammonium groups have often been obtained in nuclear magnetic resonance experiments by the measurement of spin-lattice relaxation times or line-width transition temperatures. Although a comparison of such results for several related compounds will give information on the relative heights of potential barriers, it is clear from the comparison of these results with those from experiments measuring vibrational spectra and also those from theoretical considerations that the models being used for the interpretation of the results are not always analogous.

PHOSPHONIUM IODIDE

A somewhat different situation exists in phosphonium iodide, but again we would like to illustrate the importance of a choice of a model and also to indicate the help that theory may provide in choosing such a model.

Phosphonium iodide (PH_4I) has, like one of the phases of ammonium bromide, a distorted CsCl structure, but the distortions are greater in phosphonium iodide. Each iodide ion has eight phosphonium neighbors, and each phosphonium ion has eight iodide neighbors; these neighbors are not, however, equidistant.

Rush (1966) has measured the inelastically scattered neutron spectrum of phosphonium iodide and has assigned a band with a maximum at 335 cm^{-1}

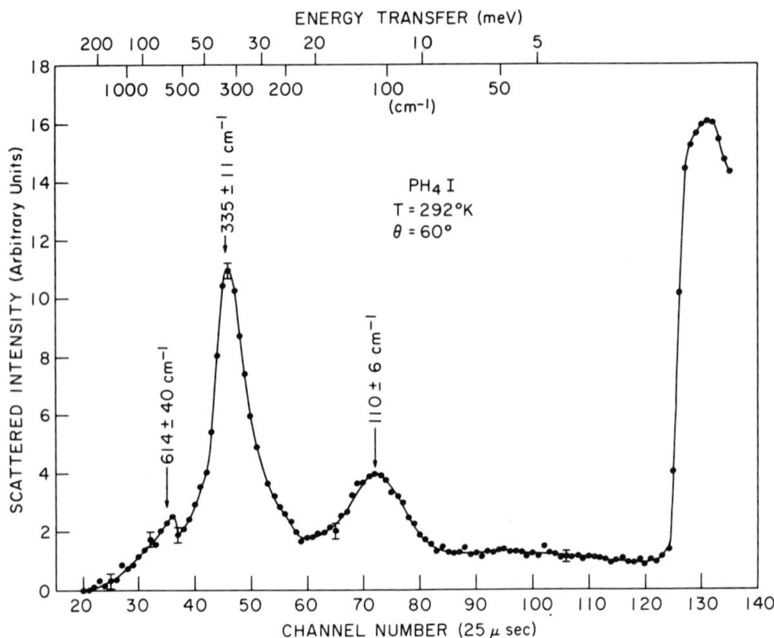

FIGURE 9.
Inelastic neutron spectrum of phosphonium iodide. The peak at 335 cm^{-1} is assigned to the 1–0 transition of the hindered rotational mode of the ammonium ion. [Reproduced from a drawing of J. J. Rush (1966).]

to the torsional oscillation of the ammonium ion (see Figure 9). He has derived from this frequency a barrier height of 7 kcal/mole by using an expression given by Gutowsky, Pake, and Bersohn (1954) for the rotation of a tetrahedral ion in a CsCl lattice. Acting on the assumption that the structure of phosphonium iodide is the same as that of ammonium bromide (Phase III), he used the electrostatic model of Gutowsky, Pake, and Bersohn to calculate the electrostatic potential in PH_4I, and found it to be 7.5 kcal/mole. This good agreement led him to believe that the model is correct and the barrier height well established.

Now, the barrier height as estimated from the torsional frequency for ammonium bromide in the analogous phase is 3.7 kcal/mole. Why, then, does the barrier appear to be so much higher in phosphonium iodide? Is hydrogen bonding perhaps responsible?

We have recently carried out a neutron diffraction study of the structure of phosphonium iodide (Sequeira and Hamilton, 1967). We find that the structure is not identical to that of ammonium bromide, although in both structures the tetrahedral ion has four close anion neighbors and four somewhat farther away. (In the ideal CsCl structure, all eight would be at the same distance.) In ammonium bromide, the four hydrogen atoms of the ammonium ion are directed toward the nearest bromide ions, but in phosphonium iodide, the hydrogen atoms are directed toward the four iodide ions that are somewhat farther away (see Figure 10). The H···I distances are 2.87 and 3.35 Å. The latter is very close to the sum of the van der Waals radii for H and I, whereas the former is considerably shorter. Thus, by our definition, a hydrogen bond exists.

We then asked ourselves whether the structure could be predicted on the basis of electrostatic interactions between the ions. A calculation of the electrostatic energy of the crystal as a function of the rotation of the phosphonium ion about the tetragonal ($\bar{4}$) axis was carried out, using a program written by Baur (1965). Baur, by the way, has found that the orientation of the water molecules in a number of hydrates can be calculated quite accurately by minimizing the electrostatic energy. We found, in agreement with the assumption of Rush, that the *electrostatic* energy was minimized when the orientation of the phosphonium ion was identical to that in ammonium bromide, that is, rotated by 90° from the position we found from the neutron diffraction study. There was a subsidiary minimum in the electrostatic energy at the experimental position.

Since the P—H bond length is greater than the N—H bond length, and since the iodide ion is larger than the bromide ion, the reasonable assumption is that the structure must be determined more by the repulsive terms in the potential-energy expression than by the electrostatic energy. Accordingly, a potential function of the Lennard-Jones type,

$$V(r) = 4\epsilon\left[\left(\frac{\sigma}{r}\right)^{12} - \left(\frac{\sigma}{r}\right)^{6}\right]$$

was assumed to be operative in addition to the purely electrostatic terms. The parameters in the Lennard-Jones potential were chosen to give a minimum at the expected van der Waals distances with the depth of the minimum the same as that for the interaction of the nearest noble gas atoms (Hirschfelder, Curtiss, and Bird, 1954). The total energy of interaction, including the electrostatic, is shown in Figure 11. There is now a minimum at the experimentally determined orientation.*

*A similar calculation on the ammonium bromide system shows that the electrostatic term again produces a minimum in the experimental position and that the repulsive term is small. The low barrier is, however, quite sensitive to the unknown parameters of the L-J potential

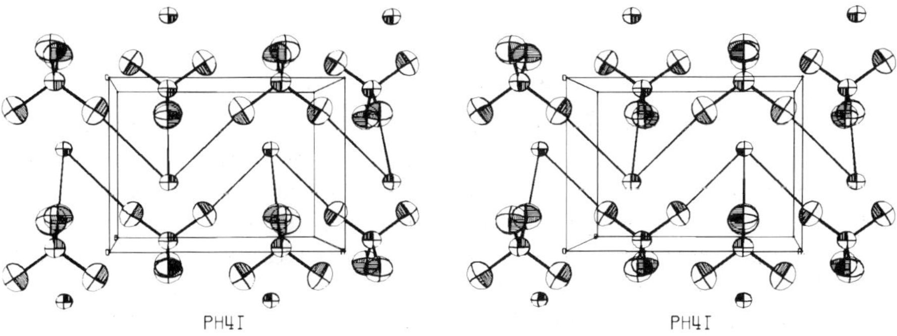

FIGURE 10.
The structure of phosphonium iodide, showing the hydrogen bonding. In ammonium bromide, the tetrahedral ions are rotated about the *c* axis by 90° from the position shown in this figure.

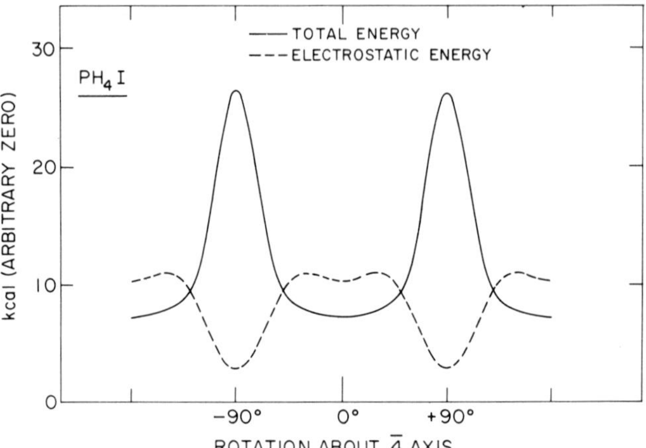

FIGURE 11.
Potential function for rotation of PH_4^+ ion in PH_4I. The broken line is the electrostatic contribution. The solid line is the sum of the electrostatic and the Lennard-Jones energies.

More interesting perhaps is the shape of the potential function. Again a broad flat potential is found with a barrier rather higher than that predicted on the assumption of a cosine potential. Although the height of the barrier is quite sensitive to the parameters used in the Lennard-Jones potential, it seems reasonable to assume that the shape of the potential function is essentially correct, and the obvious next step is to determine the parameters in the Lennard-Jones potential by carrying out a numerical solution of the wave equation for the calculated potentials to obtain a fit to the data for the energy levels.*

In phosphonium iodide, the mean-square amplitudes of oscillation ϕ^2 can be derived from the diffraction data. For small oscillations in a harmonic potential around an axis with moment of inertia I, the librational frequency is given by a formula quoted by Cruickshank (1956):

$$\nu^2 = kT/4\pi^2 I\phi^2.$$

For PH_4I, the frequencies calculated are 250 cm^{-1} for rotation around the tetragonal axis and 800 cm^{-1} for rotation around axes perpendicular to the tetragonal axis. This is in qualitative agreement with the calculated potentials, which indicate a much higher barrier to rotation around the nontetragonal axes.

AMMONIUM SULFATE

The structure of ammonium sulfate has been studied by neutron diffraction (Schlemper and Hamilton, 1966a) in both the ferroelectric and the paraelectric phase. The difference between the two is that there is a movement of the two ammonium groups from mirror planes as the temperature is lowered through the ferroelectric Curie point. The result is a less symmetrical structure in which shorter, and presumably stronger, hydrogen bonds are formed. There is also less distortion of the valence angles in the low-temperature

*Such an approach to obtaining good intermolecular potentials from crystallographic data has been proposed in a somewhat different context by Williams (1965). We propose to report in a future communication the results of such calculations for a number of ammonium salts.

phase than there is in the high-temperature phase. Again, the librational frequencies can be derived from the diffraction-determined amplitudes of vibration if a harmonic potential is assumed. For the two independent ammonium groups in the structure, the calculated mean librational frequencies are 168 and 177 cm^{-1} above the transition and 162 and 158 cm^{-1} below the transition. These differences are not considered to be significantly different, nor are the values between the two ammonium groups significantly different. Rush and Taylor (1965) have observed two peaks in the neutron spectrum, at 200 and 335 cm^{-1}; on the basis of the diffraction data, it seems reasonable to assign the 200 cm^{-1} peak to the librational modes of the ammonium ions. In any case, the evidence seems to be conclusive that the ferroelectric transition is not associated with any change in the motional freedom of the ammonium ions. Subtle differences in the environments of the two ammonium groups are indicated by the nuclear magnetic resonance data of Blinc and Levstek (1960), who interpret their results in terms of two ammonium groups, one of which persists in its reorientation down to −180°C, while the other shows a line width transition—although not at the ferroelectric Curie point. The shape of the potential-energy barrier in this crystal is certainly not simple, and any interpretation of the librational frequencies in terms of a barrier height is highly uncertain.

NEUTRON INELASTIC SCATTERING

In the sections above, we have quoted some results of vibrational frequencies obtained from neutron inelastic scattering experiments. In view of the newness of this technique, particularly as applied to systems of chemical interest, it seems appropriate to present here some elements of the theory, to quote some further results that have been obtained for hydrogen-bonded systems, and to offer a prognostication of the impact that such studies may have on the investigation of molecular structure in general.

A neutron at room temperature has an energy of 0.025 electron volts, an amount of energy equivalent to a frequency of 200 cm^{-1}. A neutron may be scattered by a molecular system with the loss or gain of energy equal to the difference in the energy levels of two states of the system. By measuring the difference in energies between the incident and scattered neutron, one has available a method of carrying out spectroscopic measurements. Many interesting vibrational modes in solids, particularly those involving intermolecular motions and molecular or submolecular rotations, have frequencies of the order of a few hundred cm^{-1} or less. Thus the energy change of the neutron in the inelastic scattering process is comparable to the energy of the incident neutron and is easily measured.

In one type of experiment, which has been extensively used for the study of group motions in hydrogenous materials (Boutin, Safford, and Brajovic, 1963; Rush, 1966; Rush and Taylor, 1965; Janik et al., 1964), neutrons of very low energy are scattered by the system that is under investigation when they absorb energy from the scattering system. The energies of the scattered neutrons are measured by a time-of-flight apparatus. A typical curve, that for phosphonium iodide, is shown in Figure 9. The method is particularly applicable to hydrogen-containing systems, as hydrogen has an extremely high incoherent scattering cross section. The transitions for vibrational modes that involve hydrogen atom motions thus dominate the spectrum. As with any spectroscopic method, the assignment of the bands to individual modes is not always unambiguous, so that again one must proceed with due caution and attempt to confirm the assignments by comparison with the results of other structural studies.

The basic relations governing the intensity of neutron inelastic scattering are simple, as we shall see below. The application to real systems may be complex, because of the necessity of summing over the Boltzmann distribution of states, degeneracies of energy levels, and poor resolution in the usually low-intensity experiments (see, for example, Egelstaff, 1965).

Each nucleus has associated with it a coherent scattering amplitude a^{coh} and an incoherent scattering amplitude a^{inc}. Consider a crystal to be composed of atoms with mass M_j located at positions $\mathbf{r}_j$ relative to the unit-cell origin. Furthermore, let the coordinate of the normal mode excited or de-excited by interchange of energy with the neutron be denoted by q. The atomic displacement vectors $\triangle\mathbf{r}_i$ are related to q by vector coefficients $\mathbf{C}_i$ such that

$$\triangle\mathbf{r}_i = \mathbf{C}_i M_i^{-\frac{1}{2}} q$$

where q is the normal mode coordinate for a single cell with

$$\Sigma\, \mathbf{C}_i^2 = 1$$

The incident and scattered neutron beams may be characterized by wave vectors $\mathbf{k}_o$ and $\mathbf{k}_s$ in the directions of the corresponding beams, with magnitudes $2\pi/\lambda_o$ and $2\pi/\lambda_s$, where λ denotes neutron wavelength. The difference between $\mathbf{k}_o$ and $\mathbf{k}_s$ defines the scattering vector

$$\boldsymbol{\kappa} = \mathbf{k}_s - \mathbf{k}_o$$

For complete generality, we assume that each atom has associated with it a Debye-Waller temperature factor $T_i(\boldsymbol{\kappa})$. With this notation, we may write the ordinary geometric structure factor for elastic (Bragg) scattering as

$$\mathrm{F}(\boldsymbol{\kappa}) = \sum_{\text{cell}} \exp(i\boldsymbol{\kappa}\cdot\mathbf{r}_j) a_j^{\text{coh}} T_j$$

The intensity of the Bragg scattering is proportional to the square of this structure factor; a further condition is that the scattering vector $\boldsymbol{\kappa}$ must be equal to a reciprocal lattice vector $2\pi\mathbf{h}$.

Similarly we can define a normal mode structure factor* as

$$H(\boldsymbol{\kappa}) = \sum_{\text{cell}} \exp(i\boldsymbol{\kappa}\cdot\mathbf{r}_j) a_j^{\text{coh}} T_j M_j^{-\frac{1}{2}} (\boldsymbol{\kappa}\cdot\mathbf{C}_j). \tag{1}$$

The intensity of the one-phonon† coherent scattering from a crystal is proportional to the square of this structure-factor:

*The exponential term in the definition of H indicates that the coherent scattering is dependent on the crystal structure, and I have previously suggested (Hamilton, 1966) that this dependence plus the possibility of observing the inelastic scattering throughout reciprocal space may provide a partial solution to the classical phase problem in crystallography.

†A one-phonon process is one in which a single vibrational mode changes its quantum number by ± 1.

$$I = \text{constants}\,\frac{\lambda_0}{\lambda_s \triangle E}\,|H(\kappa)|^2$$

where $\triangle E$ is the energy involved in the transition.

The incoherent scattering is given by a similar expression, which, however, does not contain an explicit dependence on the crystal structure:

$$I = \text{constants}\,\frac{\lambda_0}{\lambda_s \triangle E}\sum_{\text{cell}}\left[a_j^{\text{inc}} T_j\,(\kappa\cdot\mathbf{C}_j) M_j^{-\frac{1}{2}}\right]^2$$

It is important in both coherent and incoherent scattering to note the following facts:

1. The contribution to the intensity from a particular atom is inversely proportional to the mass of the atom.
2. The contribution to the intensity from a particular atom is proportional to the amplitude of vibration of this atom in the normal mode for which the transition takes place.
3. The intensity is inversely proportional to the amount of energy transferred $\triangle E$.
4. The intensity is also proportional to the wavelength of the incident neutron.

All of these factors suggest that the scattering from low-frequency modes involving hydrogen atoms will be important; coupled with the high incoherent scattering amplitude for hydrogen, these factors provide an explanation for the important contributions that neutron inelastic scattering has made to the study of hydrogen-bonded systems.

The form of equation (1) suggests the possibility of structure determination by use of the coherent inelastic scattering (see, for example, Hamilton, 1966). By measurement of normal modes to which only certain atoms contribute strongly, one essentially makes these atoms *heavy atoms* in the structure determination. That the scattering is dominated by these atoms may perhaps make the structure solution easier. There is the possibility that one may thus arrive at partial structure solutions by measurement of the intensity of the inelastic scattering. Refinement of the structure could proceed by ordinary Bragg scattering measurements. Isotopic substitution has an obvious effect on the inelastic structure factor expression and can be used as it is in elastic neutron diffraction as an isomorphous replacement method for crystal structure determination. Furthermore, the effects of isotopic substitution might facilitate the proper assignment of the normal mode frequencies.

Depending on the experimental arrangements, one must make sums over all contributions to the scattered intensity. As an extreme example, we consider the very fruitful total cross-section measurements exemplified by the work of Rush, Taylor, and Havens (1960). Here, a hydrogenous material is placed in a beam of long wavelength neutrons, and the fraction of the incident intensity that is transmitted is measured. That intensity which is not transmitted is removed from the beam by pure absorption or by scattering out of the beam. In the case of hydrogen, the latter process is the only important one. The scattered intensity is the sum of contributions from the scattering from all normal modes excited in the sample. Furthermore there is averaging

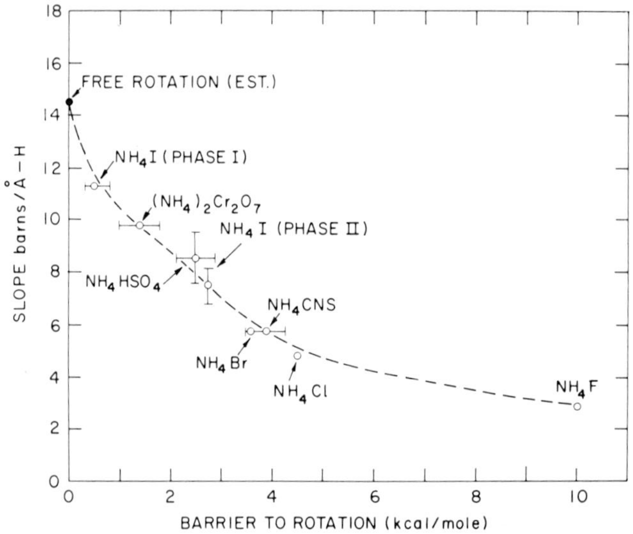

FIGURE 12.
Cross-section slope from total neutron scattering data, plotted against rotational barrier as estimated from nuclear magnetic resonance or spectroscopic data. Barriers depend on the model assumed for the potential energy surface. [Data supplied by J. J. Rush.]

over all orientations of the crystallites in a polycrystalline material. One would think perhaps that such an experiment would give little information. However, if one performs the necessary sums, one finds that the scattering cross section is directly proportional to the wave length of the incident neutron,

$$\sigma = C\lambda_o$$

with a slope C which is intimately connected with the freedom of motion of the hydrogen atoms in the material. Theoretical calculations (Leung, Rush, and Taylor, 1966) agree well with the experimental results and show that a high slope is associated with much motional freedom. For example, in a number of ammonium salts, the cross-section slope is nicely correlated with the barrier hindering rotation that has been determined from nuclear magnetic resonance or spectroscopic experiments (see Figure 12). Similar results hold for methyl group rotation (Rush and Taylor, 1966).

One factor that has not been mentioned above is that there is a conservation condition on the one-phonon scattering analogous to that for the Bragg scattering: namely, there must be a relationship between the scattering vector $\boldsymbol{\kappa}$, the reciprocal lattice vector $2\pi\mathbf{h}$ and the wave vector of the lattice vibration $\mathbf{f}$, such that

$$\boldsymbol{\kappa} = 2\pi\mathbf{h} + \mathbf{f}$$

The wave vector $\mathbf{f}$ takes on all values in the first Brillouin zone of the reciprocal lattice, and measurement of the transition frequency as a function of $\mathbf{f}$ (the dispersion surface) makes possible a determination of the intermolecular forces in a molecular crystal. Infrared spectroscopy and Raman spectroscopy are limited to transitions with $\triangle \mathbf{f} = 0$. Although neutron inelastic scattering has not yet been extensively applied in this context to molecular crystals, one may expect important developments in this field in the next few years.

ACKNOWLEDGMENTS

Much of the work quoted in this paper has been carried out through the diligent efforts of Elmer Schlemper and Anisbert Sequeira. In addition, I am grateful to J. J. Rush for many stimulating conversations regarding the applications of neutron inelastic scattering to hydrogen-bonded systems. I consider myself most fortunate to have received my early training in structural chemistry as one of the enthusiastic group that surrounded Linus Pauling at the California Institute of Technology.

REFERENCES

Bacon, G. E., and N. A. Curry (1956). *Acta Cryst.* **9**, 82.
Bacon, G. E., and N. A. Curry (1962). *Proc. Roy. Soc.* (London) A **266**, 95.
Baur, W. (1964). *Acta Cryst.* **17**, 863.
Baur, W. (1965). *Acta Cryst.* **19**, 209.
Blinc, R., and I. Levstek (1960). *J. Phys. Chem. Solids* **12**, 295.
Boutin, H., G. J. Safford, and V. Brajovic (1963). *J. Chem. Phys.* **39**, 3135.
Chidambaram, R., A. Sequeira, and S. K. Sikka (1964). *J. Chem. Phys.* **41**, 3616.
Cruickshank, D. W. J. (1956). *Acta Cryst.* **9**, 1005.
Danford, M. D., and H. A. Levy (1962). *J. Am. Chem. Soc.* **84**, 3965.
Egelstaff, P. A. (1965). *Thermal Neutron Scattering*. London: Academic Press.
Gutowsky, H. S., G. E. Pake, and R. Bersohn (1954). *J. Chem. Phys.* **22**, 643.
Hamilton, W. C. (1966). *Trans. Am. Cryst. Assoc.* **2**, 53.
Hirschfelder, J. O., C. F. Curtiss, and R. B. Bird (1954). *Molecular Theory of Gases and Liquids*, p. 1110. New York: Wiley.
Ibers, J. A., W. C. Hamilton, and D. R. MacKenzie (1964). *Inorganic Chem.* **3**, 1412.
Janik, J. A., J. M. Janik, J. Mellow, and H. Palevsky (1964). *J. Phys. Chem. Solids*, **25**, 1091.
Leung, P., J. J. Rush, and T. I. Taylor (1966). Unpublished work.
McGaw, B. L., and J. A. Ibers (1963). *J. Chem. Phys.* **39**, 2677.
McMullan, R. K., and G. A. Jeffrey (1965). *J. Chem. Phys.* **42**, 2725.
Pauling, L. (1928). *Proc. Nat. Acad. Sci.* **14**, 359.
Pauling, L. (1930). *Phys. Rev.* **36**, 430.
Pauling, L. (1935). *J. Am. Chem. Soc.* **57**, 2680.
Pauling, L. (1959). In. Hadzi, D., ed., *Hydrogen Bonding*, p. 1. London: Pergamon Press.
Pauling, L., R. B. Corey, and H. R. Branson (1951). *Proc. Nat. Acad. Sci.* **37**, 205.
Peterson, S. W., and H. A. Levy (1957). *Acta Cryst.* **10**, 70.
Rush, J. J. (1966). *J. Chem. Phys.* **44**, 1722.
Rush, J. J., and T. I. Taylor (1965). *Inelastic Scattering of Neutrons*, p. 333. Vienna: International Atomic Energy Agency.
Rush, J. J., and T. I. Taylor (1966). *J. Chem. Phys.* **44**, 2749.
Rush, J. J., T. I. Taylor, and W. W. Havens, Jr. (1960). *Phys. Rev. Letters* **5**, 507.
Schlemper, E. O., and W. C. Hamilton (1966a). *J. Chem. Phys.* **44**, 4498.
Schlemper, E. O., and W. C. Hamilton (1966b). *J. Chem. Phys.* **45**, 408.
Schlemper, E. O., W. C. Hamilton, and J. J. Rush (1966). *J. Chem. Phys.* **44**, 2499.
Sequeira, A., and W. C. Hamilton (1967). *J. Chem. Phys.* In press.
Taylor, J. C., and M. H. Mueller (1965). *Acta Cryst.* **19**, 536.
Taylor, J. C., M. H. Mueller, and R. L. Hitterman (1966). *Acta Cryst.* **20**, 842.
Williams, D. E. (1965). *Science* **147**, 605.

RICHARD E. MARSH

Gates and Crellin Laboratories of Chemistry
California Institute of Technology
Pasadena, California

Some Comments on Hydrogen Bonding in Purine and Pyrimidine Bases

It is perhaps not fully remembered today that the first proposal for the detailed structure of DNA was made by Pauling and Corey (1953). Although this structure was incorrect, it was based on the valid concept of a nonintegral, multiple helix comprising repeating units with predictable dimensions. The crucial concept of hydrogen-bonded base pairs was recognized later that same year by Watson and Crick (1953), and their classical work was the result.

Since that time, there has been much work and many discussions on the structural features of the purine and pyrimidine bases and on their relationship to the properties of DNA. However, there is one feature that has not received much attention, and I should like to take this opportunity to discuss it briefly. This feature might be called the amphoteric nature of the bases—specifically, the ability of a ring nitrogen atom to act as both a hydrogen-bond donor and an acceptor.

Let us take as an example the purine molecule, $C_5H_4N_4$. One of the four hydrogen atoms must be attached to a nitrogen atom; moreover, in order that benzene-like resonance in the six-membered ring be retained, this hydrogen atom must be attached either to N(7) or to N(9) of the five-membered ring. Crystal-structure analysis has shown that, in the solid state at least, the proton bonds to N(7) (Watson, Sweet, and Marsh, 1965), and the standard valence bond representations of the molecule are structures (*a*) and (*b*) in Figure 1. However, the N(7)—C(8) and C(8)—N(9) bond distances are nearly equal in length (1.33 and 1.31 Å), suggesting that structures (*c*) and (*d*) are very nearly as important as (*a*) and (*b*) in describing the overall resonance hybrid. (Structures (*e*) and (*f*), as well as those involving formal charges on the carbon atoms, appear to be less important.) Thus, there is a relatively large positive charge—perhaps about +0.4 units—on N(7), making it an excellent hydrogen-bond donor; concomitantly, N(9) has a negative charge and is an excellent hydrogen-bond acceptor. As a result, a feature of the crystal structure of purine is the formation of strong intermolecular hydrogen bonds N(7)—H···N(9); the N(7)···N(9) distance is quite short, 2.85 Å, and the melting point of the compound is abnormally high, 213°.

Contribution No. 3456 from the Gates and Crellin Laboratories of Chemistry. This work was supported, in part, by Research Grant HE-02143 from the National Heart Institute of the National Institutes of Health, U.S. Public Health Service, to the California Institute of Technology.

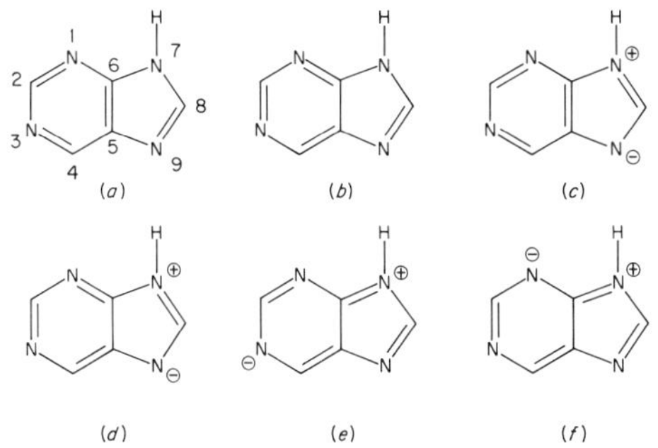

FIGURE 1.
Valence-bond structures of the purine molecule.

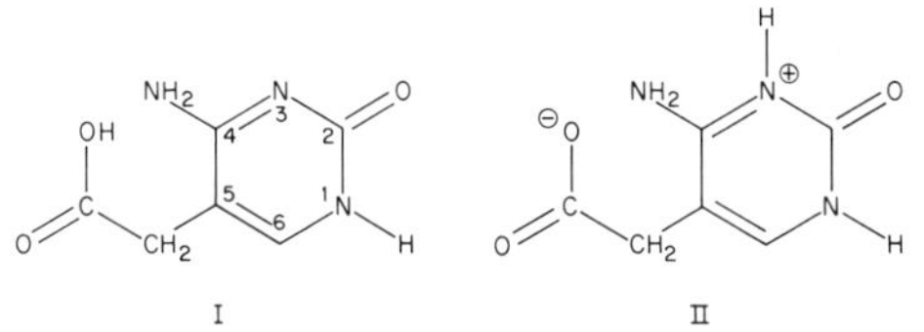

FIGURE 2.
The two tautomeric structures of cytosine-5-acetic acid.

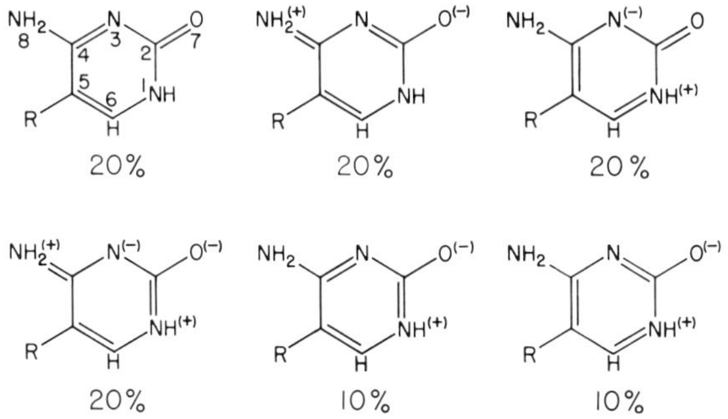

FIGURE 3.
Valence-bond structures of the cytosine residue in cytosine-5-acetic acid, tautomer I, and their estimated contributions to the resonance hybrid.

The point here is that, of the two nitrogen atoms in the five-membered ring, one—that covalently bonded to the proton—is an excellent hydrogen-bond donor, and the other is an excellent acceptor. If it had turned out that N(9) rather than N(7) was protonated, strong N(9)—H···N(7) hydrogen bonds would no doubt have been formed. Thus, a nitrogen atom in an unsaturated ring can be an excellent hydrogen-bond donor or an equally good acceptor, depending upon whether or not it has a proton attached to it. This is not true of normal amine groups: although the group $R—NH_3^+$, for example, is an excellent donor, the $R—NH_2$ group is a rather poor acceptor, and N—H···N distances involving amino groups are typically greater than 3.0 Å.

An even more striking example of the amphoteric nature of a ring nitrogen atom is found in crystals of cytosine-5-acetic acid (Marsh, Bierstedt, and Eichhorn, 1962). These crystals contain, in equal quantity, two different moieties, shown in Figure 2; I is the standard formulation, and II is a zwitterion, the proton of the carboxyl group being transferred to the ring nitrogen atom N(3). The contributions of the resonance structures shown in Figure 3 lead to an estimated charge on N(3) of approximately -0.4 in molecule I and $+0.6$ in

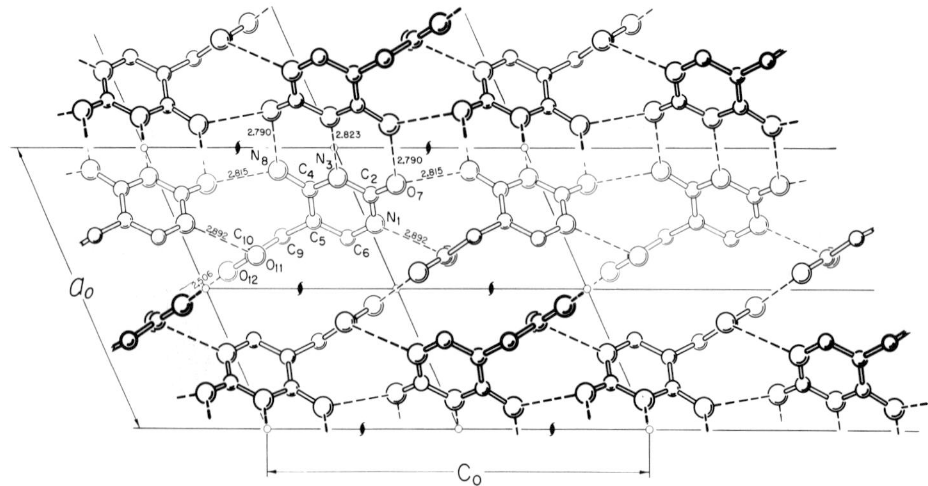

FIGURE 4.

The crystal structure of cytosine-5-acetic acid, viewed down the *b*-axis. The dashed lines represent hydrogen bonds.

molecule II. Accordingly, this atom can act as a hydrogen-bond acceptor in one molecule and as a donor in the other. The resulting structure is shown in Figure 4. The two types of molecules form a "base pair" exactly analogous to the cytosine-guanine pairing in DNA (Watson and Crick, 1953). (Actually, the two types of molecules of cytosine-5-acetic acid are indistinguishable. Disorder in the positions of the hydrogen atoms gives rise to a crystallographic center of symmetry midway between the two N(3) atoms, and to another center between O(12) atoms of two carboxyl groups. These are only statistical centers of symmetry; the hydrogen atoms lie not at the centers, but half the time on one N(3), or O(12), atom and half the time on the other.)

A similar situation obtains in crystals of isocytosine (Sharma and McConnell, 1965). Again, two different tautomers, A and B, occur, and again a ring nitrogen atom N(3) is protonated in one form but not in the other; in this case, the two molecules are structurally distinct. The resulting base pairing is shown in Figure 5. Protonation of N(3) in half of the molecules permits hydrogen bonding between A and B that is similar to the C-G pairing in DNA. (In addition, molecules of type B are bonded together, around a center of symmetry, by N—H···N hydrogen bonds from the amino group to the ring atom N(1). A planar tetramer is thus formed. The complete crystal structure contains additional N—H···O hydrogen bonds which, however, do not lie in the plane of the tetramer.)

In the two preceding examples, a particular atom—a ring nitrogen atom—acts either as a hydrogen-bond donor or as an acceptor, depending upon which tautomeric form the molecule assumes. In cytosine-5-acetic acid one of the tautomers is a zwitterion and might be expected to have a significantly different energy than the neutral form; in isocytosine, however, there appears to be very little reason to choose between the two tautomeric structures.

In both cases the choice of tautomeric form appears to be dictated by the environment of the molecule, and in particular by the hydrogen bonding.

The role of a ring nitrogen atom can also be changed from acceptor to donor, or vice versa, by a change in the *p*H of the system as a whole. As the *p*H is increased, resulting in a net negative charge on the base, an additional acceptor site will usually be formed; as the *p*H is lowered, an additional donor may be formed. The site of this additional acceptor or donor will depend upon the particular tautomeric form the molecule assumes, which in turn may be dictated by the local environment of the molecules. In Figure 6 are shown the various reasonable tautomeric structures the four bases—adenine, cytosine, guanine, and thymine—may assume in neutral, acidic, and basic media; for each of the structures the hydrogen-bond donor (D) and acceptor (A) sites are labeled. (I have not shown forms involving substituent imino (=NH) or hydroxyl groups, because they do not seem to be structurally important.)

Included in Figure 6 are three possible structures for the neutral guanine residue, which differ from one another in the position of the proton. The third structure—the zwitterion, with a positive charge on N(9) and a negative charge on N(3)—is particularly interesting because it permits a pairing with a cytosine residue exactly analogous to the adenine-thymine pairing found some years ago by Hoogsteen (1959). For many years I have been attracted by the thought that such a base-pair scheme might be important either in the structure of DNA itself or as a mechanism for replication. However, no evidence of this pairing has yet been found; crystals of base pairs invariably show the Watson-Crick-type pairing between cytosine and guanine, although the Hoogsteen arrangement has now been found in several A-T or A-U pairs (Sobell, 1966).

As is apparent from Figure 6, there are many reasonable structures that may be written for the nucleic acid bases, depending upon both *p*H and the more subtle effects that influence tautomerism. These structures can give rise to a large variety of hydrogen-bonding arrangements. The point of this discussion is to suggest that these arrangements and their base-pair specificities should be carefully considered in investigations of the properties and functions of DNA.

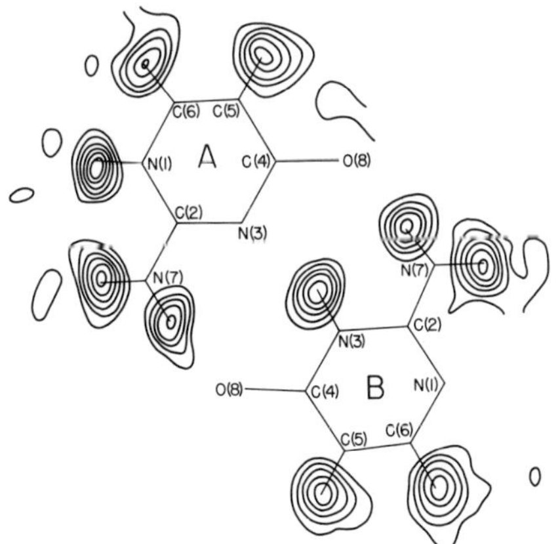

FIGURE 5.
An electron-density map showing the locations of the hydrogen atoms in crystals of isocystosine. Molecules A and B are tautomeric, differing from one another in the location of the hydrogen atom bonded to a ring nitrogen atom. The two molecules are held together, in the crystals, by three hydrogen bonds: N(7A)···O(8B), N(3B)···N(3A), and N(7B)···O(8A).

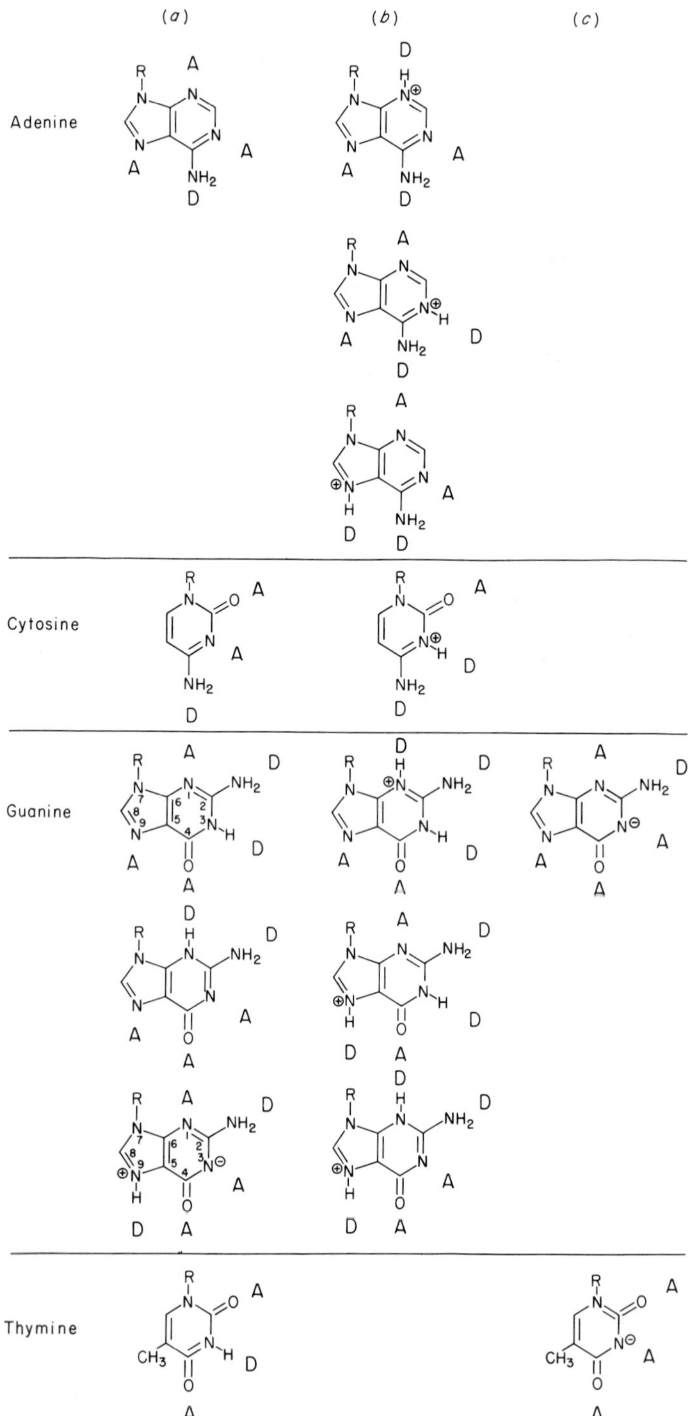

FIGURE 6.
Tautomeric forms which may be assumed by the DNA bases as neutral (column *a*), positively charged (column *b*), and negatively charged (column *c*) species. The sites of potential hydrogen bonding are designated A (acceptor) and D (donor).

REFERENCES

Hoogsteen, K. (1959). *Acta Cryst.* **12**, 822.
Marsh, R. E., R. Bierstedt, and E. L. Eichhorn (1962). *Acta Cryst.* **15**, 510.
Pauling, L., and R. B. Corey (1953). *Proc. Nat. Acad. Sci. U.S.* **39**, 84.
Sharma, B. D., and J. F. McConnell (1965). *Acta Cryst.* **19**, 797.
Sobell, H. M. (1966). *J. Mol. Biol.* **18**, 1.
Watson, D. G., R. M. Sweet, and R. E. Marsh (1965). *Acta Cryst.* **19**, 573.
Watson, J. D., and F. H. C. Crick (1953). *Nature* (London) **171**, 737.

D. P. STEVENSON
Shell Development Company
Emeryville, California

Molecular Species in Liquid Water

In recent years there have been two divergent views of the manner in which the energy of hydrogen bonds in liquid water depends on the O—H—O angle. Lennard-Jones and Pople (1951) have advocated a flexible hydrogen bond model in which the "bond energy" decreases but slowly with deviation of the angle (O—H—O) from 180°. Franck (1958) and Pauling (1959, 1966), on the other hand, postulate a "bond energy" dependence on the angle (O—H—O) with a rather narrow minimum. As has been pointed out by Franck (1958), these two points of view lead to quite different descriptions of the relations of a particular water molecule in the liquid to its nearest neighbors. In the Lennard-Jones and Pople (1951) model each molecule is viewed as hydrogen-bonded to four nearest neighbors with the possibility of a virtually continuous distribution of "bond energies," not only from molecule to molecule but also from "bond" to "bond" for a particular molecule. From the second point of view (Franck, 1958; Pauling, 1959) a particular molecule is to be described as forming 4, 3, 2, 1, or 0 hydrogen bonds with its nearest neighbor molecules. This description, involving a discrete number (5) of molecular species, has been frequently employed in constructing models to account for the physical properties of liquid water (Pauling, 1959; Franck and Quist, 1961; Nemethy and Scheraga, 1962; Buijs and Choppin, 1963; Marchi and Eyring, 1964).

To the extent that it is possible to describe the nearest neighbor relations of water molecules in the liquid in terms of molecular species characterized by a discrete number of hydrogen bonds from a water molecule to its nearest neighbors, it should be possible to ascribe reasonably definite observable properties to these species, and thus make possible measurements of their concentrations. In a recent paper, I (Stevenson, 1965) made a start on such a program by discussing the expected observable properties of the free or non-hydrogen-bonded water molecules that play an important role in the models based on essential linearity of the hydrogen bond (Pauling, 1959; Nemethy and Scheraga, 1962; Marchi and Eyring, 1964). In the present paper, the previous discussion is extended to include the spectroscopic properties to be expected of the other eight, nominally distinguishable molecular species, included in the designation of water molecules as forming 1, 2, 3, and 4 hydrogen bonds with nearest neighbors. The spectroscopic properties of water are then exmained in terms of these expected properties of the hypothetical molecular species. Contrary to the conclusions of Buijs and Choppin (1963), Goldstein and Penner (1964), and Thomas, Scheraga, and Schrier (1965), but in agreement with those of Wall and Hornig (1965), we conclude that the description of liquid water in terms of a discrete number of sharply defined hydrogen-bonded species is not consistent with its spectroscopic properties.

MOLECULAR SPECIES IN LIQUID WATER

As indicated above, when it is assumed that the hydrogen bonds in liquid water are essentially linear, it is appropriate to describe the molecules as forming 0, 1, 2, 3, or 4 hydrogen bonds. The prototype of the first of these classifications is water in the dilute vapor state or in solution in a saturated hydrocarbon; the prototype of the last is water in ice. Because water molecules tend to be in total violation of the Polonian Precept*—that is, they act as either donors (D) or acceptors (A) of protons in hydrogen bond formation —the statement that a molecule forms 1, 2, or 3 hydrogen bonds is not a unique specification of the species. It is convenient to use the notation D_iA_j to describe the possible molecular species, where the indices i and j specify the number of times the particular water molecule acts respectively as a donor or an acceptor of a proton. The indices i and j can take independently the values 0, 1, and 2, and the sum, $i + j = n$, is the number of hydrogen bonds formed by the species. For $n = 1$, 2, or 3, there are respectively two, three, and two different species, D_iA_j, and thus a total of nine molecular species to be considered as "significant structures."

In their application of the Eyring significant structure theory (Eyring and Ree, 1961; Eyring and Marchi, 1962) to the discussion of the macroscopic properties of liquid water, Nemethy and Scheraga (1962) considered the significant structures to be only those specified by the summation index, n—that is, five species, ignoring the distinction between different species with common n, such as $n = 2$, D_2A_0, D_1A_1, and D_0A_2. Marchi and Eyring (1964), in their application of the significant structure theory to a similar discussion of the properties of liquid water, consider $n = 0$ and $n = 4$ to be the only significant structures.

The pentagonal dodecahedra of water molecules that Claussen (1951) proposed as the basic unit for the crystal structure of the cubic hydrates, which von Stackelberg and Müller (1951) showed to be cubic crystals with edge of unit cell equal either to about 17 Å or about 12 Å, and which Pauling and Marsh (1952) showed to be the characteristic units of the hydrates with cubic cells with 12 Å edges (such as $Cl_2 \cdot 8H_2O$), consist of twenty water molecules, forming thirty hydrogen bonds distributed among our species: 20 D_2A_1 and 10 D_1A_2. Pauling's (1959) model for liquid water emphasizes the importance of the D_2A_1 and D_1A_2 species, with a greater or lesser proportion of D_2A_2 and D_0A_0 also contributing to the structure.

In the recent discussions of the changes with temperature of the envelope of the near infrared absorption bands of liquid water, H_2O and D_2O, Buijs and Choppin (1963), Goldstein and Penner (1964), and Thomas, Scheraga, and Schrier (1965) have treated as single entities the species with common values of the donor index, i. That is, they ascribe coincidence to the combination bands of the sets of species: [D_2A_0, D_2A_1, D_2A_2], [D_1A_0, D_1A_1, D_1A_2], and [D_0A_0, D_0A_1, D_0A_2].

*"Neither a borrower, nor a lender be" (*Hamlet*, Act I, Scene III).

In the following paragraphs we shall treat the question of our ability to provide operational characteristics (that is, spectroscopic observables) for the nine possible molecular species of liquid water, D_iA_j ($i = 0, 1, 2; j = 0, 1, 2$), and try to find evidence in the spectroscopic properties of liquid water to show which species, if any, contribute to the structure of the liquid.

THE MOLECULAR SPECIES D_iA_0 AND D_0A_j

Of the five molecular species represented by the symbols D_iA_0 and D_0A_j, the particular one that represents non-hydrogen-bonded or monomeric water molecules, D_0A_0, has played an important part in the descriptions of the structure and properties of liquid water. For example, in Pauling's description of the structure of the liquid, such molecules occupy the centers of the pentagonal dodecahedra and other positions, so that the fraction of D_0A_0 species lies in the range 1/21 to 4/23. In a recent paper I (Stevenson, 1965) proposed two spectroscopic criteria or operational models for monomeric, non-hydrogen-bonded water molecules in liquid water. (1) The vibration-rotation spectra (infrared absorption or Raman) of monomeric water molecules in liquid water should be very similar to those spectra of water in such more or less inert solvents as carbon tetrachloride and/or chloroform. (2) The long wavelength side of the $n \longrightarrow \sigma^*$ absorption band of monomeric water molecules in liquid water should be more or less coincident with that of this absorption band of water vapor and/or water in an inert solvent such as saturated hydrocarbon (*n*-hexane, isooctane, and others). I should like to propose extensions of these operational models for the D_0A_0 species.

(a) The vibration-rotation spectra of water in solution in such solvents as CCl_4 or $CHCl_3$ is a model for the vibration-rotation spectra to be expected of the D_0A_j ($j = 0$, 1, and 2) molecular species of water in its liquid, and (b) the long wavelength side of the $n \longrightarrow \sigma^*$ absorption band of water vapor, or water in isooctane solution, is a model for the longest wavelength portions of the ultraviolet absorption spectra to be expected for the D_iA_0 ($i = 0$, 1, and 2) molecular species. The observational basis for these proposed models for the spectroscopic behavior of the D_iA_0 and D_0A_j species follows.

The absence of monomerlike bands from both the infrared absorption spectrum and Raman spectrum of water has plagued proponents of theories of the structure of the liquid that require a significant (≥ 10 percent) fraction of the non-hydrogen-bonded species. It has usually been suggested that red shifts caused by the effect of dielectric constant on vibrational frequencies shift the monomer bands under the high-frequency limb of the broad, intense band characteristic of the hydrogen-bonded OH modes (Franck and Quist, 1961). For this explanation to be valid it would be necessary that the increased dielectric constant of liquid water over that of CCl_4, for example, cause the ν_3 mode of monomeric H_2O to shift to the red by 100 to 150 cm^{-1} in order that ≥ 10 percent monomeric water in the liquid should not give easily recognized structure to the high-frequency side of the 3000–3700 cm^{-1} band of liquid water.

Bellamy and Williams (1960), on the other hand, have suggested that though the red shifts of group vibrational frequencies characteristic of the change in state (vapor to solution) are largely due to the change in dielectric con-

TABLE 1.
Frequencies of fundamental and first overtone of free OH of alcohols in vapors and solvents.

	ν(OH) (cm^{-1})				2ν(OH) (cm^{-1})			
	Vapor	Isooctane	CCl_4	Liquid	Vapor	Isooctane	CCl_4	Liquid
Dielectric Constant	1.00	1.94	2.23	11–20	1.00	1.94	2.24	11–20
n-C_3H_7OH	3683	—	3635	*	7190	7126	7106	7096
s-C_3H_7OH	3662	—	3630	*	7145	7099	7083	7094
n-C_4H_9OH	—	3649	3641	*	—	7122	7105	7096
s-C_4H_9OH	3660	3638	3632	*	7146	7100	7084	7096
iso-C_4H_9OH	3683	3649	3642	3635	—	7124	7109	7101
tert-C_4H_9OH	3646	3625	3620	3618	7111	7070	7060	7072

SOURCE: All data from Shell Development Company Laboratory. The ν(OH) from spectra obtained with a Beckman IR-7 and the 2ν(OH) from spectra obtained with an Applied Physics Corp. Model 14. The author wishes to acknowledge the assistance of variously Drs. A. C. Jones, F. S. Mortimer, and D. O. Schissler in obtaining a large fraction of these data. In general the frequencies reported here for the locations of the bands in isooctane and CCl_4 solution agree within experimental error with the values to be found in the literature. See, for example, R. F. Goddu, "Near infrared spectrophotometry," in *Recent Adv. Anal. Chem.*, Interscience, New York (1960); R. Moccia and H. W. Thompson, *Proc. Roy. Soc.* **243A**, 154 (1957); J. M. Goldman and R. A. Crisler, *J. Org. Chem.* **23**, 751 (1958).
*Not observable

stant, the further shifts in frequency that accompany changes in solvent are primarily caused by specific solute-solvent interactions and not by changes in the dielectric constant of the solvent. Bellamy and Williams cite the fact that the free OH vibrational bands of the hindered phenol, 2,6-ditertiary butyl-4-methyl phenol in carbon tetrachloride (dielectric constant = 2.24) and diethylether ($D = 4.34$) differ by but 1 cm^{-1}.

The data assembled in Table 1 on the frequencies of the fundamental and first overtone of the free OH mode of some propyl and butyl alcohols in the vapor, isooctane, CCl_4, and in the pure liquid, appear to provide impressive evidence for the proposal of Bellamy and Williams (1960) that solvent shifts are associated with specific solvent-solute interaction rather than a dielectric effect. Note particularly the changes in sign of the shift of 2ν(OH) between CCl_4 solution and the pure alcohol, a red shift for the primary alcohols and a blue shift for the secondary and the tertiary alcohols.

With respect to our hypothesis that the vibration-rotation spectra (IR and/or Raman) of monomeric water in solvents such as CCl_4 and $CHCl_3$ provides a spectroscopic model for the species D_0A_1 and D_0A_2, as well as D_0A_0 in liquid water, we present data on the location of the vibration-rotation bands of ammonia in the vapor, and in CCl_4, $CHCl_3$, and water solutions in Table 2. Here it is seen that of the normal modes of vibration of the ammonia molecule the frequency of only the symmetric deformation (umbrella or incipient inversion) mode is subject to significant "solvent shifts," and these shifts are most certainly the consequence of interaction between the nitrogen atom unshared electron pair and the solvents (Datta and Barrow, 1965).

Since ammonia, in water, is rather strongly and completely hydrogen-bonded to water protons, the changes and similarities of the spectra of ammonia in CCl_4 and water should provide a sound basis for predicting the

TABLE 2.
The frequencies of the fundamental, overtone, and combination bands of "NH_3" in the vapor and various solvents (cm^{-1})

Mode	Vapor[a]	CCl_4[b]	$CHCl_3$[b]	Water[b]	
				(IR)	(Raman)
$[\nu_2]$ (a_1)[d]	950(ave)	994	1037	1114	1115
$[\nu_4]$ (e)	1627	1621	1626	(1620–1650)[c]	
$[2\nu_4]$	3218	3232	3239	—	3229
$[\nu_1]$ (a_1)	3337	3314	3313	3310	3313
$[\nu_3]$ (e)	3443	3417	3413	3413	3400
$[\nu_1 + \nu_2]$	4307	4329	4365	4415	
$[\nu_2 + \nu_3]$	4426	4439	4469	4523	
$[\nu_1 + \nu_4]$	4956				
$[\nu_3 + \nu_4]$	5053	5021	5018	5018	
$[\nu_2 + \nu_3 + \nu_4]$	6025	6034	6068	6138	
$[\nu_1 + \nu_3]$	6609	6570	6565	6563	
$[\nu_3 + 2\nu_4]$		6628	6627	6635	
$[\nu_1 + \nu_2 + \nu_3]$			7627	7683	
$[\nu_1 + \nu_3 + \nu_4]$		8137	8132	8146	

[a]Benedict, Plyler, and Tidewell, *Can. J. Phys.* **35**, 1235 (1957).
[b]Shell Development Company Laboratory (unpublished), D. P. Stevenson with Drs. A. C. Jones and D. O. Schissler. For water our Raman results are in good agreement with the literature as summarized by Plint, Small, and Walsh, *ibid.* **32**, 653 (1954), who discuss the assignments of the fundamentals.
[c]From the relation $2[\nu_3 + \nu_4] + 2[\nu_4] - [\nu_3] - [\nu_3 + 2\nu_4] = 2[\nu_4]$, we calculate $[\nu_4] = 1615$ cm^{-1}.
[d]Datta and Barrow, *J. Am. Chem. Soc.*

TABLE 3.
The frequencies of fundamental, overtone, and combination bands of monomeric waters in the vapor and various solvents (cm^{-1}).

Mode	Vapor[a]			CCl_4[b]			$CHCl_3$[b]	DEE[b]	Dioxane[b]
	H_2O	HDO	D_2O	H_2O	HDO	D_2O	H_2O	H_2O	H_2O
$[\nu_2]$ a_1	1595	1402	1179	(1596)[c]	(1399)[d]	(1180)[e]	1608	1631	1637
$[2\nu_2]$	3151	2809		3164					3280
$[\nu_1]$ a_1	3652	2719	2660	3631	2694	2643	3610	3525	3520
$[\nu_3]$ b_1	3756		2789	3712	3667	2757	3696	3585	3585
$[\nu_1 + \nu_2]$					4086				
$[\nu_2 + \nu_3]$	5332			5285	5047	3923	5276		
$[2\nu_2 + \nu_3]$	6874			6830		5076	6825		
$[2\nu_1]$				7124		5242	7082		
$[\nu_1 + \nu_3]$	7252			7167		5317	7148		
$[\nu_1 + \nu_2 + \nu_3]$	8807		6538	8726		6485	8705		
$[2\nu_2 + \nu_3]$	10,613			10,500					

[a]G. Herzberg, *Infrared and Raman Spectra of Polyatomic Molecules*, Van Nostrand and Co., New York (1945), pages 281 and 282.
[b]Shell Development Company Laboratory, unpublished. The present author with Drs. A. C. Jones and D. O. Schissler.
[c]From the relation, $2[\nu_2\ \nu_3] + [2\nu_2] - [\nu_3] - [2\nu_2 + \nu_3] = 2[\nu_2]$.
[d]From the relations, $[\nu_2 + \nu_3] - [\nu_3] = \nu_2 - x_{23}$ and $[\nu_1 + \nu_2] - [\nu_1] = \nu_2 + x_{12}$, and the values of x_{12} and x_{23}, $-14cm^{-1}$ and -23 cm^{-1}, respectively, and the Dennison [*Rev. Mod. Phys.* **12**, 175 (1950)] isotope relation, $x_{ij}'/x_{ij} = \nu_i'\nu_j'/\nu_i\nu_j$.
[e]From $[\nu_2 + \nu_3] - [\nu_3] = \nu_2 + x_{23}$, etc.

relations to be expected between the corresponding spectrum of water in CCl_4 and of the D_0A_1 and D_0A_2 species of water in water. We would expect the bending frequency, ν_2, of water to undergo a blue shift from CCl_4 solution to water, but smaller than that of ammonia undergoes, because motion of the oxygen atom should contribute less to the normal coordinate for this mode than does the nitrogen atom to the ν_2 mode of ammonia. The stretching frequencies, ν_1 and ν_3, of water would be expected to show virtually no shifts in frequency between CCl_4 and liquid water. Table 3 summarizes the observed frequencies of the bands assignable to the various modes of vibration of H_2O, HDO, and D_2O in CCl_4 and $CHCl_3$ solution as well as in the vapor phase.

Buijs and Choppin (1963) have suggested that for the species D_0A_0, D_0A_1, and D_0A_2 there would be expected a blue shift of $+30$ cm^{-1} for the bending mode, ν_2, and red shifts of -60 cm^{-1} and -150 cm^{-1} for the symmetric and the antisymmetric stretching modes, ν_1 and ν_3, respectively, between the vapor and water solution. However, the similarity to be expected between the behavior of ammonia and water molecules does not admit such large shifts, and thus the Buijs and Choppin (1963) suggestion is based on error in assignment of positions of near infrared band maxima to species. Further evidence for their overestimation of the vapor to liquid water shift to be expected in the bend frequency, ν_2, of the D_0A_j species is the fact that in dilute solutions in diethylether and dioxane, in which the water molecules certainly form reasonably strong hydrogen bonds to the solvent molecules (that is, are probably D_2A_0 species), the bend frequency ν_2 is found to be 1631 and 1637, respectively, shifted but 39 ± 3 cm^{-1} from its vapor and/or CCl_4 solution positions (see Table 3).

In a recent paper (Stevenson, 1965) I suggested that the absence of structure on the high-frequency side of the absorption band of liquid water (3000–3600 cm^{-1}) indicated the concentration of monomeric water molecules (D_0A_0 species) to be less than 5 percent at about 25°C. In Figure 1 are shown the ν_2 bending bands of water in $CHCl_3$ solution and the 25°C absorption spectrum of water (1400–1800 cm^{-1}). These spectra suggest that 5 percent is a very conservative estimate of the upper limit to the sum of the concentration of the three molecular species, D_0A_0, D_0A_1, plus D_0A_2, that would be expected to have a bending mode absorption band very like that of water in CCl_4 or $CHCl_3$.

In the previous paper (Stevenson, 1965) the suggestion that the long wavelength portion of the $n \longrightarrow \sigma^*$ band of monomeric water molecules (D_0A_0 species) in liquid water should be well approximated by the vapor spectrum of water was based on the previous finding that the long wavelength portions of the $n \longrightarrow \sigma^*$ bands of ammonia and the alkylamines in the vapor, and in solution in isooctane, diethylether, and acetonitrile in the case of triethylamine, are essentially coincident (Stevenson, Coppinger, and Forbes, 1961). The results of later experiments, presented below, provide much more direct evidence for the validity of the hypothesis.

It is well known that the monomeric alcohols in their vapor phase or in dilute solution in inert solvents have sharp bands at about 2.8 μ and 1.4 μ in their absorption spectra; these bands can be unambiguously assigned to the non-hydrogen-bonded OH stretch vibrational mode fundamental and first overtone, respectively. Numerous workers have employed the variation of

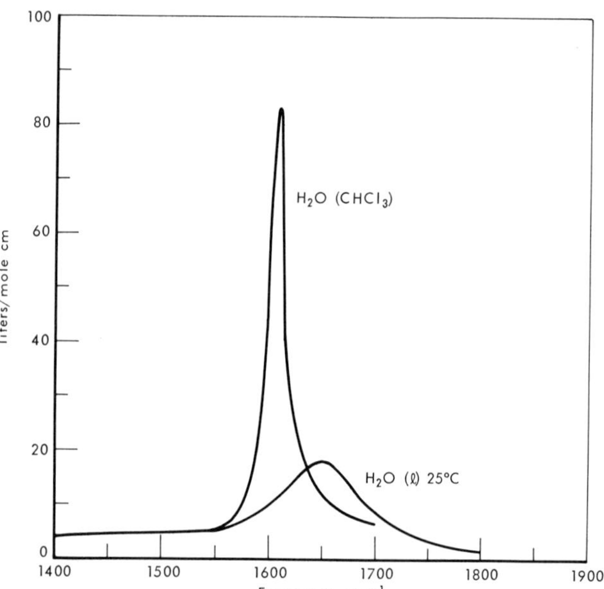

FIGURE 1.
The absorption spectrum of H_2O in $CHCl_3$ and that of H_2O liquid at 25°C as measured with a Beckman IR-7 spectrophotometer. For liquid water the measurements were made in an 0.0014-cm cell for which the cell blank had been determined with carefully dried $CHCl_3$. An approximately saturated solution of water in $CHCl_3$ was employed, the H_2O concentration being determined from the integrated proton magnetic resonance spectra of the solution.

specific intensity of one or the other of these free OH bands with dilution in dilute solution to estimate the various association constants, dimerization, trimerization, and so on, of the alcohols. As indicated above, in some of the propyl and butyl alcohols these bands are observable in the pure liquids at 25°C, and thus in principle it is possible to trace for these alcohols the concentration of "free OH" from infinite dilution in an inert solvent such as isooctane to the pure liquid. Methanol, which we might expect to be more like water in its behavior than the higher aliphatic alcohols, differs from higher alcohols in that, for methyl alcohol mole fractions greater than about 0.88,* neither the free OH band at 2.8 nor the overtone band at 1.4 μ can be seen above the background of the short wavelength side of the bands because of hydrogen-bonded OH.

Like water, the liquid alcohols are much more transparent in the wavelength range of the long wavelength edge of their $n \longrightarrow \sigma^*$ absorption bands in the ultraviolet than in the vapor phase. Because of their greater or lesser mutual miscibility with isooctane, it is possible to observe the changes of apparent absorptivity (specific absorption) with dilution from pure alcohol to infinite dilution in isooctane. Shown in Figures 2 and 3 are the relative absorptivities of methanol and tertiary butyl alcohol as a function of alcohol concentration in isooctane ($0.001 \leq X_{\mathrm{ROH}} \leq 1$) for their 1.4 μ OH overtone bands and for an arbitrary wavelength on the side of the $n \longrightarrow \sigma^*$ absorption bands.

*Because of partial immiscibility of methanol and isooctane at about 25°C, solutions with alcohol mole fractions, X_a ($0.22 \leq X_a \leq 0.88$) are not accessible. In CCl_4 solution the free OH band of CH_3OH can be observed for $X_a \leq 0.7$. Note in Table 1 that although the overtone of the OH stretch mode is observable in all of the liquid propyl and butyl alcohols, the fundamental is only observable in liquid and tertiary butyl alcohols.

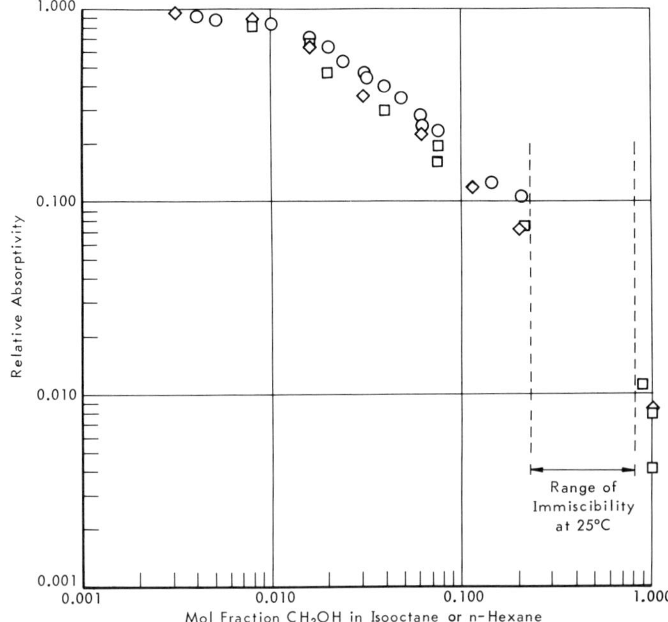

FIGURE 2.
The relative absorptivities of CH_3OH in isooctane solution at about 25°C at $1.401_0 \mu$ (*circles*) and 1950 Å (*squares*). The diamonds are 1950 Å data for methanol in *n*-hexane from Kaye and Poulson (*Nature*, 193: 675 (1962).

The ultraviolet wavelength chosen was one for which it was possible within the limits of the cell lengths available and the transparency of the solvent and alcohol, to make reliable absorbance measurements over the entire concentration range.* The essential one-to-oneness of the decreases of the near infrared and the ultraviolet relative absorptivities is the exact relationship needed to ensure the validity of the following hypotheses. (1) The relative absorptivity at 1.4 μ measures the fraction of alcohol OH not hydrogen-bonded as a *donor.* (2) The relative absorptivity in the ultraviolet measures the fraction of the alcohol not hydrogen-bonded as an *acceptor.* These results seem to be strong confirmation of the previous postulate that the absorptivity of liquid water relative to that of water vapor in the 1800 Å region of their absorption spectra provides a good measure of the upper limit to the relative concentration of monomeric water molecules.

In Figures 2 and 3 the pure alcohol data indicate a very marked reduction in the fraction of non-hydrogen-bonded alcohol molecules as the size of the alkyl group decreases from butyl to methyl. We might infer that in water there should be found an even smaller relative concentration of non-hydrogen-bonded molecules than are found in methanol.

Although not evident in the data presented in Figures 2 and 3, the absorptivities of the alcohols in isooctane (infinite dilution) at the ultraviolet wavelengths there employed are 25 to 50 percent lower than the absorptivities of the vapors at the same wavelengths. As implied in a previous paper (Stevenson, 1965), these facts introduce an uncertainty of the order of a factor of 2

*In the case of methanol there was literature data on the absorptivity of hexane solutions at only 1950 Å.

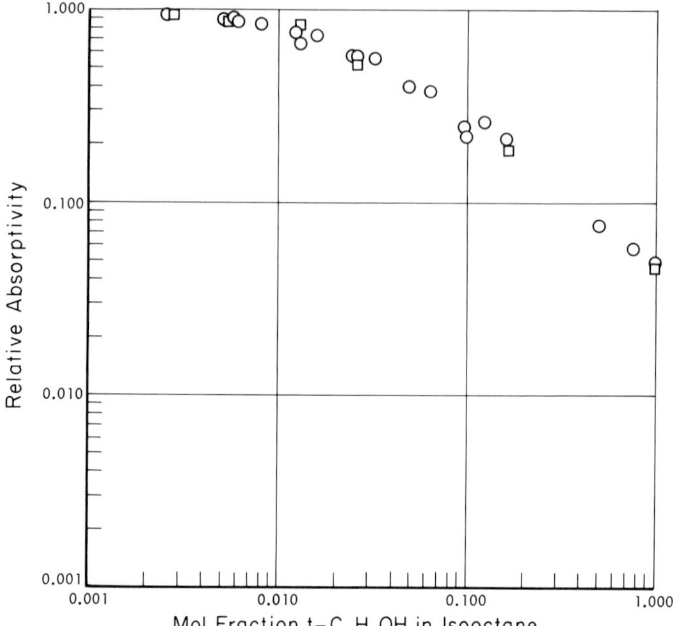

FIGURE 3.
The relative absorptivities of *tert*-C_4H_9OH in isooctane solution at about 25°C at $1.416_5\mu$ (*circles*) and 1900 Å (*squares*).

in the use of the water vapor ultraviolet absorption spectrum in the estimation of the free water molecule concentration in the liquid. In an effort to eliminate this ambiguity we measured the absorptivity of water in isooctane solution in a narrow range of wavelengths near 1800 Å. The narrow range of wavelengths within which we succeeded in making measurements was limited on the short wavelength side by the solvent cut off in transparency and, on the long wavelength side, by the small value of the product of absorptivity by concentration, resulting from the low solubility of water in isooctane (Joris, Black, and Taylor, 1948).* The results of these measurements are shown in Figure 4 as circle points along the smooth curves, which show the absorption spectra of water and D_2O vapors and liquid (25°C) for wavelengths greater than 1700 Å. The water data are from previous work (Stevenson, 1965); the D_2O data are hitherto unpublished results.

Near 1800 Å the absorptivity of water in isooctane is essentially equal to that of the vapor, and thus our previous use of the vapor spectrum as a model for monomeric water in solution is probably more justified than the results of studies of the methanol and tertiary butyl alcohol spectra would indicate.

Previous arguments led to the expectation that the $n \longrightarrow \sigma^*$ band of an acceptor species D_0A_1 or D_0A_2 should undergo a blue shift relative to that of the $n \longrightarrow \sigma^*$ band of the nonacceptor species, D_0A_0. The same arguments lead to the further expectation that acting as a donor should cause a red shift of this electronic band of a nonacceptor molecular species. Thus the $n \longrightarrow \sigma^*$ absorption curve of water vapor (or water in isooctane) that we

*This solubility was taken as the mean of the values given for water in *n*-heptane and *n*-octane at 20–25°C.

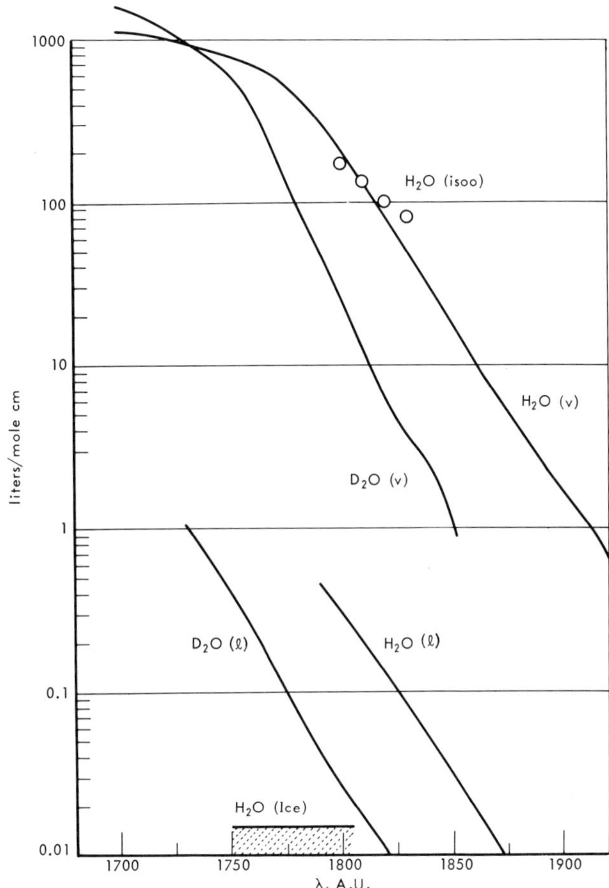

FIGURE 4.
The vapor and liquid (25°C) spectra of H_2O and D_2O for $\lambda \geq 1700$ Å and those of H_2O in isooctane solution (*circles*). The cross-hatched region $1750 \leq \lambda \leq 1800$ Å represents the upper limit to the absorptivity of H_2O-ice at -10°C. The H_2O vapor spectrum here shown is in good agreement with the data of Wilkinson and Johnson [*J. Chem. Phys.* **18**:190 (1950)] in the wavelength range of overlap of measurement, $1700 \leq \lambda \leq 1800$. No evidence was found of the deviation from Beer's law of the vapor spectra of H_2O and D_2O that Johannin-Gilles reported [*Compt. Rend.* **236**:676 (1953), and **250**:1523 (1955)]. For H_2O-ice ($\lambda < 1700$), see Dressler and Schnepp [*J. Chem. Phys.* **33**:270 (1960)].

propose as a model for the D_0A_0 species in liquid water becomes a lower limit for the band of the D_1A_0 and D_2A_0 species. Hence the absorptivity of liquid water relative to the vapor, which we have taken as a measure of the fraction of the D_0A_0 species, becomes a firm upper limit to the sum of fractions of $D_0A_0 + D_1A_0 + D_2A_0$ species.

The D_2O vapor absorption curve shown in Figure 4 provides quantitative verification of the previous qualitative to semiquantitative evidence of an isotopic blue shift relative to the water vapor spectrum. The manner in which the D_2O curve starts to the blue of the water curve and then crosses at about 1725 Å indicates that the water absorption at $\lambda > 1700$ is exclusively due to the $n \longrightarrow \sigma^*$ transition, as has been previously supposed. The vibrational structure found at shorter wavelengths (1650–1700 Å) must arise from overlap by an $n \longrightarrow$ R band.

The opacity of liquid D_2O at 1800 Å is about 40 percent less than would be expected from the magnitude of the water to D_2O blue shift in the vapor spectra. This greater transparency of liquid D_2O relative to its vapor over liquid water relative to its vapor corresponds to a D_iA_0 species concentration in liquid D_2O that is 0.72 times that in liquid water. The difference in H-bond and D-bond energies, corresponding to the difference in heats of vaporization of H_2O and D_2O (256 cal/mole; Lewis and MacDonald, 1933), would suggest that D_iA_0 in liquid D_2O is 0.65 times that in liquid H_2O at about 25°C. The magnitudes of the absorptivities of liquid H_2O and D_2O relative to their vapors at 1800 Å indicate that the sums of the mol fractions of the species $D_0A_0 + D_1A_0 + D_2A_0$ are less than 1.5×10^{-3} and 1.0×10^{-3}, respectively, at about 25°C.

MOLECULAR SPECIES D_1A_1, D_1A_2, D_2A_1 AND D_2A_2

The four species, D_1A_1, D_1A_2, D_2A_1, and D_2A_2, form a single set with respect to their expected electronic absorption spectra, and form two sets of two species each with respect to their vibration-rotation spectra. As indicated above, when an oxygen atom acts as a proton acceptor, the ground electronic state is markedly stabilized relative to the lowest accessible excited state, and as a consequence has negligible absorption in the energy range corresponding to the long wavelength side of the $n \longrightarrow \sigma^*$ band of water molecules, whose oxygen atoms do not act as proton acceptors. It would be expected that interaction with a second proton would even further stabilize the electronic ground state relative to the lowest excited state of an oxygen atom of a D_iA_2 species. However, this would not permit the distinction of a D_iA_2 species from a D_iA_1 species in the presence of D_iA_0 species, which have a longer long wavelength edge to this $n \longrightarrow \sigma^*$ band. Hence the vacuum ultraviolet absorption spectrum of liquid water is not expected to provide information that would allow distinction between the species D_iA_1 and D_iA_2.

The fundamental stretching frequencies of the D_2A_1 and D_2A_2 species would be expected to be essentially identical, as would the fundamental stretching frequencies of the species, D_1A_1 and D_1A_2. Several lines of evidence indicate that when the oxygen atom acts as a proton acceptor in hydrogen bond formation the OH stretching modes are essentially unaffected; one line of evidence cited above, has shown that the stretching frequencies of ammonia in CCl_4, $CHCl_3$, and water solution are essentially the same. Evidence that the behavior of N—H bonds provides a good model for the expected behavior of OH bonds is provided in the infrared absorption spectra of various diols in dilute solution in CCl_4. These molecules tend to form intramolecular hydrogen bonds (when geometry permits and the dilution is sufficiently great that there is no competition by intermolecular hydrogen-bond formation), so that one hydroxyl group is of the species D_1A_0 and the other D_0A_1. As a consequence two bands are observed in the infrared absorption spectrum, usually of about equal intensity, in the 2.8–3.0 μ region characteristic of the OH stretch. Table 4 summarizes the observed positions of these absorption bands in the spectra of diols with varying geometric capability of forming intramolecular hydrogen bonds. It is seen that even in 1,4-butanediol, in which the hydrogen bond OH frequency is shifted 169 cm^{-1} from that of the free OH frequency, the free OH frequency is identical with that found in normal butyl alcohol (Table 1).

From the above it is concluded that the D_1A_j species should have one stretching frequency about equal to the average of the symmetric and antisymmetric stretching frequencies of the D_0A_j species, and another stretching frequency about equal to the average of the symmetric and antisymmetric

TABLE 4.
OH *stretch frequencies of diols in dilute* CCl_4 *solution* (cm^{-1}).

	ν_f	$(\Delta\nu)_{\frac{1}{2}}$	ν_b	$(\Delta\nu_b)_{\frac{1}{2}}$
Resorcinol[a]	3614	16	—	—
Catechol[a]	3618	20	3574	24
Ethylene glycol[b]	3644	—	3604	32
1,3- Propane diol[c]	3636	—	3558	60
1,4-Butane diol[c]	3636	28	3477	92

[a]Resorcinol and catechol, A. C. Jones, unpublished.
[b]$C_2H_4(OH)_2$, Kruger and Metler, *J. Mol. Spec.* **18**, 131 (1965).
[c]$C_3H_6(OH)_2$, $C_4H_8(OH)_2$, Kuhn et al, *J. Am. Chem. Soc.* **86**, 650, 2161 (1964).

stretching frequencies of the D_2A_j species. For the free OH frequency characteristic of the D_1A_j species we expect 3660 $\pm$ 20 cm^{-1} for OH and 2690 $\pm$ 20 cm^{-1} for OD.

Various authors have noted that the absorption spectrum of water in ice should form a good model for the expected absorption spectrum of the D_2A_j species in water. These authors (Buijs and Choppin, 1963; Goldstein and Penner, 1964; Thomas, Scheraga, and Schrier, 1965) have attempted to calculate for various temperatures the fractions of water molecules in the sets of D_0A_j, D_1A_j, and D_2A_j, from the absorptivities of liquid water at various wavelengths across several of the near infrared bands of liquid water. They assumed that the D_2A_j species of the liquid has the absorption characteristics of ice (at about 0°C) and made certain other ad hoc assumptions about the absorption characteristics of the other two sets of species. It has, however, been noted by Haas and Hornig (1960) that there are difficulties in the interpretation of the absorption spectrum of ice (H_2O or D_2O), which can be partially circumvented in the case of the spectrum of HDO in dilute solution in H_2O-ice and D_2O-ice (van Eck, Mendel, and Fahrenfort, 1958).

There are two separate difficulties in the interpretation of the absorption spectra of H_2O-ice and D_2O-ice. The first arises from the broadening of all bands due to mechanical resonance between coupled identical oscillators. The second difficulty results from Fermi resonance between the first overtone of the bending mode and the fundamental of the symmetric stretch mode. There is little or no perturbation of the absorption spectrum of free or monomeric H_2O in the 3 μ region from Fermi resonance between $2\nu_2$ and ν_1. However, hydrogen-bond formation tends to increase ν_2 (and thus $2\nu_2$) while decreasing the stretch frequency (compare the dioxane solution of H_2O with CCl_4 solution in Table 3). As a consequence, $2\nu_2$ approaches ν_1 and the Fermi resonance causes ν_1 to increase, thus decreasing the separation between the symmetric and antisymmetric (ν_3) stretch frequencies. Because of all these complicating effects, no really satisfactory analysis of the absorption spectra of H_2O-ice and D_2O-ice has been achieved to date.

Neither of the difficulties indicated is found in HDO in dilute solution in H_2O-ice or D_2O-ice. In D_2O-ice the OH stretch mode of HDO is isolated from mechanical resonance as is the OD stretch mode in H_2O-ice. Since the frequency of the bend mode of HDO is sufficiently greater than half that of

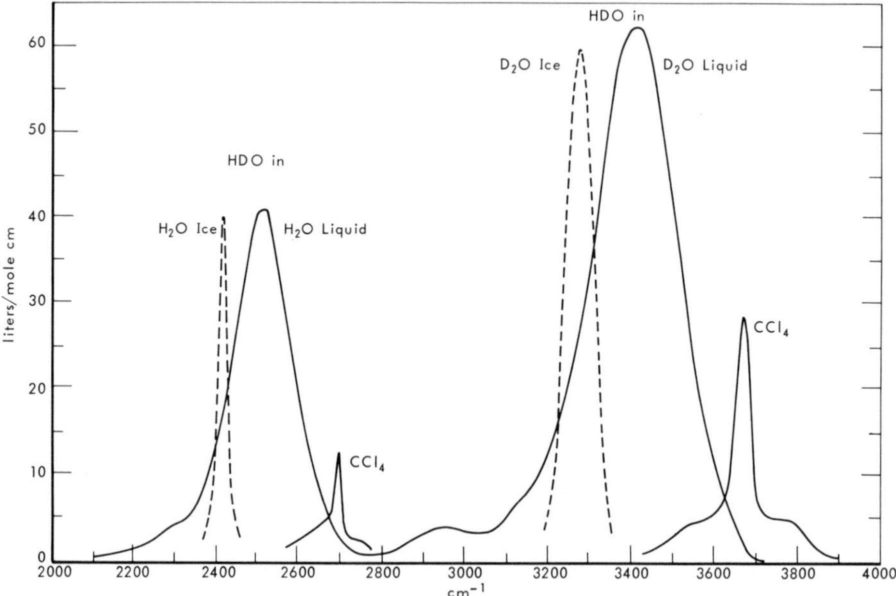

FIGURE 5.
The OD and OH stretch bands of HDO in H_2O and D_2O, as measured at 23°C in the Shell Development Company Laboratories. Also shown are the results of measurements of these bands of HDO in CCl_4 solution. The dashed line curves are the OD and OH bands of HDO in H_2O-ice and D_2O-ice, respectively, constructed from data of Wall and Hornig (1965). See also van Eck, Mendel, and Fahrenfort (1958).

the OD stretch mode and sufficiently less than half that of the OH stretch mode, there is little or no Fermi resonance of the bend overtone with either stretch mode fundamental. Thus the stretch frequencies found by Haas and Hornig (1960) for dilute HDO in ices at -195°C may be taken as the stretch frequencies characteristic of the D_2A_j species of HDO in the liquid. Haas and Hornig's value for the OD stretch was 2416 cm^{-1}, and for the OH stretch, 3275 cm^{-2}; the band widths at half-height were equal to 20 cm^{-1} for the OD, and 80 cm^{-1} for the OH stretch.

To circumvent the interpretational difficulties resulting from the effects of mechanical resonance broadening and Fermi resonance in the spectra of H_2O and D_2O, this paper will be limited to consideration of the spectral characteristics of HDO in dilute solution in H_2O (for the OD stretch band) and D_2O (for the OH stretch band). The stretching frequency bands of HDO under these conditions should consist of the superposition of two component bands. These component bands would be expected to be a relatively narrow high frequency band ($\sim$2690 cm^{-1} for OD and $\sim$3660 cm^{-1} for OH) and a broader low-frequency band ($\sim$2420 cm^{-1} for OD and 3280 cm^{-1} for OH). The high-frequency component would be expected to arise from the D_0A_j species and one of the two bands of the D_1A_j species; the low-frequency component would be expected to arise from the second band of the D_1A_j species and the D_2A_j species.

Shown in Figure 5 are the OD and OH absorption bands of dilute HDO in H_2O and D_2O respectively characteristic of water at $\sim$25°C. These absorp-

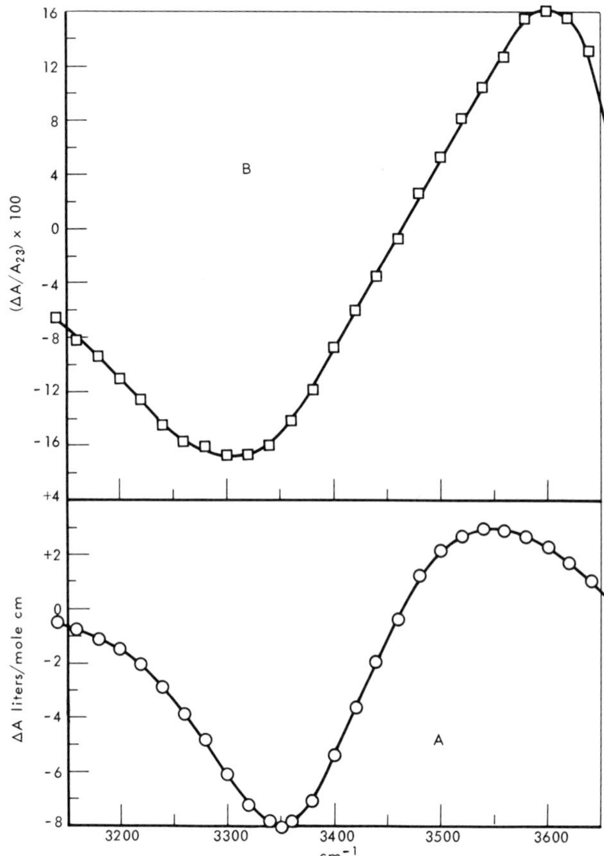

FIGURE 6.
The difference between the OH band of HDO in D_2O at 60°C and that at 23°C. A: The difference spectrum in absorptivity units. B: The difference spectrum as percentage change in absorptivity.

tions bands are of virtually the same shape that Wall and Hornig (1965) found for the corresponding Raman bands. It is apparent that neither of these symmetrical bands can be the resultant of the superposition of two bands with absorption maxima at 2420 and 2690 cm^{-1} in the case of OD and 3280 and 3660 cm^{-1} in the case of OH, since the absorptivity at the frequencies of the hypothetical component bands is but a small fraction of the maximum absorptivity of the observed band in each case.

When the temperature of the "water" is increased, the OD and OH stretch bands of HDO change in a systematic fashion (van Eck, Mendel, and Fahrenfort, 1958). While remaining essentially symmetrical, the absorption maximum shifts to higher frequency with increasing temperature. The nature of the change in band shape with temperature is shown in two fashions for the OH band of HDO in Figure 6, A and B. Figure 6, A shows the difference between the spectrum of HDO at 60°C and its spectrum at 23°C; the difference is shown in absorptivity units, but in Figure 6,B the ordinate is the percentage difference, 100 $(A_{60}\text{-}A_{23})/A_{23}$. The simple difference spectrum (Figure 6, A) suggests that, as the temperature of the water increases, a species with OH stretch absorption maximum at 3350 cm^{-1} is replaced by one with the OH stretch absorption maximum at 3550 cm^{-1}. However, from the lack of constancy of $\Delta A/A_{23}$ at the extremes of the sigmoidal $\Delta A/A_{23}$ versus cm^{-1} curve (Figure 6, B), it is apparent that the absorption band cannot be represented as the superposition of two bands with maxima at 3350 and 3550 cm^{-1}.

Thus, as was concluded by Wall and Hornig (1965) from their examination of the Raman bands of HDO, we may conclude from the infrared bands of this substance that there is no evidence for a discrete, small number of molecular species of the type defined by the symbols D_iA_j ($i = 1, 2; j = 1, 2$) in liquid water.

DISCUSSION

The ultraviolet absorption spectrum of liquid water in the temperature range 0–100°C (Stevenson, 1965) admits the presence of negligibly small (<1 percent) concentration of water molecules, with neither electron pair (D_iA_0 species) engaged in hydrogen-bond formation. The nature of the infrared absorption of liquid water in the 1600–1700 cm^{-1} region similarly admits negligibly small concentrations of water molecules, with neither hydrogen (D_0A_j species) engaged in hydrogen-bond formation. Finally, the nature of the 2500 and 3400 cm^{-1} region absorption bands of HDO in dilute solution in H_2O (about 2500) and D_2O (about 3400) cannot be accounted for in terms of a mixture of free and hydrogen bonded H's (D's)—that is, the expected absorption spectra of D_2A_1, D_1A_2, D_2A_1, D_2A_2 species. Thus we are forced to the conclusion that the proposed description of nearest neighbor relations of water molecules in the liquid in terms of a discrete number of species, defined by the number and donor-acceptor quality of hydrogen bonds, is not a suitable one. Rather, the spectroscopic (vibration-rotation and electronic absorption) spectral properties of liquid water indicate that the nearest neighbor relations are best described by a model in which each molecule forms four hydrogen bonds (twice donor and twice acceptor), with a continuous distribution of these hydrogen bonds in energy. The energies range from strong ones, such as those in ice, to very weak ones or an energy of zero strength.

According to this model, in describing the energy absorption accompanying the fusion of ice to water it is better to say that the average strength of the hydrogen bonds in the liquid is 15 percent weaker than those of ice than to say that 15 percent of the bonds are broken (Pauling, 1960, page 472).

According to the model of the hydrogen bond proposed by Lennard-Jones and Pople (1951), the strength-weakness character of a particular hydrogen bond depends primarily on the oxygen-oxygen distance, not on the (O—H—O) angle. Thus the proposal of Wall and Hornig (1965)—that the vibrational Raman bands of HDO (OH stretch and OD stretch) can be transformed into a nearest neighbor O to O distance distribution function with the aid of the (O—H—O) distance versus OH stretch frequency correlation curve (Nakamoto, Margoshes, and Rundle, 1955)—is completely reasonable, even though the correlation curve, ν(O—H—O) versus R(O—H—O), is based on data for crystals in which the hydrogen bond is believed to be linear.

The ultraviolet absorption spectrum of liquid water indicates that the fraction of water molecules that have neither unshared electron pair engaged

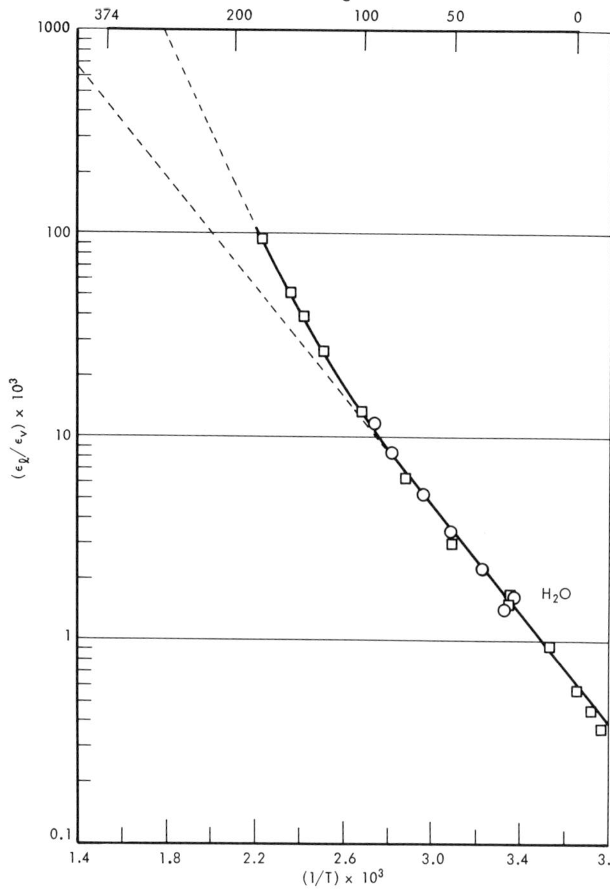

FIGURE 7.
The absorptivity of liquid water (1770–1850 Å) relative to that of the vapor as a function of the temperature of the liquid. The circle points for H_2O are from Stevenson (1965). The square points are heretofore unpublished measurements.

in hydrogen-bond formation approach the magnitude of 0.01 at about 90°C and that this fraction increases rapidly with temperature above 100°C, as may be seen in Figure 7.* The fraction exceeds 0.1 at 200°C, and extrapolation suggests that the fraction becomes unity at a temperature below the critical point. In this connection it should be noted that Luck (1965) reports that the maxima of the near-infrared absorption bands of liquid water approach the values of the frequencies of water in CCl_4 as the critical temperature of liquid water is approached. Thus, at 375°C, Luck (1965) finds 5291, 7153, 8749 and 10,499 cm^{-1} for the band maxima respectively associated with the excitation of $[\nu_2 + \nu_3]$, $[\nu_1 + \nu_3]$, $[\nu_1 + \nu_2 + \nu_3]$, and $[2\nu_1 + \nu_3]$ modes of H_2O. These frequencies are within ± 25 cm^{-1} of those shown for H_2O in CCl_4 in Table 3.

ACKNOWLEDGMENTS

I was assisted in various phases of the experimental work upon which this paper is based by colleagues at Shell Development Company. I wish to express my thanks explicitly to Mr. S. J. Rehfeld for assistance in making the meas-

*Figure 7 is from my own data (Stevenson, 1965) supplemented by data on lower (to −8°C) and higher (to +175°C) temperatures, supplied by Mr. Rehfeld of the Shell Development Company.

urements of the vacuum ultraviolet absorption spectrum of water above its normal boiling point, and to Drs. A. C. Jones, D. O. Schissler, and R. G. Snyder for various infrared spectral measurements with either the Beckman IR-7 or the Applied Physics Corporation Model 90 spectrophotometer.

REFERENCES

Bellamy, L. J., and R. L. Williams (1960). *Proc. Roy. Soc.* (London) A **258**, 22.
Buijs, K., and G. R. Choppin (1963). *J. Chem. Phys.* **39**, 2035.
Claussen, W. F. (1951). *J. Chem. Phys.* **19**, 259, 1425.
Datta, P., and G. M. Barrow (1965). *J. Am. Chem. Soc.* **87**, 3058.
Eyring, H., and R. P. Marchi (1962). *J. Chem. Educ.* **40**, 562.
Eyring, H., and T. Ree (1961). *Proc. Nat. Acad. Sci. U.S.* **47**, 526.
Franck, H. S., and A. S. Quist (1961). *J. Chem. Phys.* **34**, 604.
Franck, H. S. (1958). *Proc. Roy. Soc.* (London) A **247**, 481.
Goldstein, R., and S. S. Penner (1964). *J. Quart. Spec. Rad. Trans.* **4**, 441.
Haas, C., and D. F. Hornig (1960). *J. Chem. Phys.* **32**, 1763.
Joris, Black, and Taylor (1948). *J. Chem. Phys.* **16**, 537.
Lennard-Jones, J., and J. A. Pople (1951). *Proc. Roy. Soc.* (London) A **205**, 155.
Lewis, G. N., and R. MacDonald (1933). *J. Am. Chem. Soc.* **55**, 3057.
Luck, W. (1965). *Ber. Bunsenges physick Chem.* **69**.
Marchi, R. P., and H. Eyring (1964). *J. Phys. Chem.* **68**, 221.
Nakamoto, Margoshes, and Rundle (1955). *J. Am. Chem. Soc.* **77**, 6480.
Nemethy, G., and H. A. Scheraga (1962). *J. Chem. Phys.* **36**, 3382.
Pauling, L. (1959). *Hydrogen Bonding*, P. Hadzi, Pergamon Press, London.
Pauling, L. (1960). *Nature of the Chemical Bond.* Ithaca: Cornell Univ. Press.
Pauling, L., and R. E. Marsh (1952). *Proc. Nat. Acad. Sci. U.S.* **38**, 112.
Pople, J. A. (1951). *Proc. Roy. Soc.* (London) A **205**, 163.
Shakespeare, W., Hamlet, Act. I, Scene II.
Stackleberg, M. V., and H. R. Müller (1951). *J. Chem. Phys.* **19**, 1319.
Stevenson, D. P. (1965). *J. Phys. Chem.* **69**, 2145.
Stevenson, Coppinger, and Forbes (1961). *J. Am Chem. Soc.* **83**, 4350.
Thomas, Scheraga, and Schrier (1965). *J. Phys. Chem.* **69**, 3722.
van Eck, Mendel, and Fahrenfort (1958). *Proc. Roy Soc.* (London) A **247**, 472.
Wall, T. T., and D. F. Hornig (1965). *J. Chem. Phys.* **43**, 2079.

BARCLAY KAMB
California Institute of Technology
Pasadena, California

Ice Polymorphism and the Structure of Water

In an effort to explain the unusual physical and chemical properties of water, a variety of structural models has been proposed: many of the proposals are conflicting, and wide disagreement remains about even the most basic features, such as the coordination number of the water molecules and the extent of hydrogen bonding (Conway, 1966; Wicke, 1966; Kavanau, 1964).

The idea that the molecular arrangement in water is somehow related to the structure of ice has been widely considered since the publication of the classic paper of Bernal and Fowler (1933). Even earlier, Tammann (1926), who discovered the high-pressure polymorphism of ice, noted that the arrangements in the high-pressure solid forms should have their reflection in the structure of the liquid, and he suggested that there ought to be as many "kinds of molecules" in water as there are forms of ice. Bernal and Fowler (1933) similarly took note of ice polymorphism in proposing a quartzlike structure for water: they pointed out that the quartzlike structure should occur among the denser solid polymorphs.

Probably the most celebrated property of water that requires a structural explanation is the large decrease of volume on melting. For most and probably for all substances that melt with negative volume change ΔV_f, denser solid phases appear at high pressure, above which ΔV_f becomes positive (Klement and Jayaraman, 1967). A striking example is cesium, as seen in the curve of melting point versus pressure (Figure 1), measured by Kennedy, Jayaraman, and Newton (1962). The melting curve has two maxima, and there are two intervals of pressure over which the slope of the curve is negative, indicating negative ΔV_f. At the upper end of each of these intervals, the solid transforms to a denser phase, whereupon ΔV_f becomes positive. This evidence shows that the liquid is in some sense able to "anticipate" the structure of a denser solid phase at a pressure some 5 to 15 kb below that at which the denser solid actually becomes stable. In the structure of the liquid, the structure of the solid is in effect averaged over a range of pressures. If transitions to dense enough solid phases lie within the pressure range effectively averaged, the liquid can show anomalously high density relative to the solid stable at the lower pressure.

Contribution No. 1443 from the Division of the Geological Sciences, California Institute of Technology.

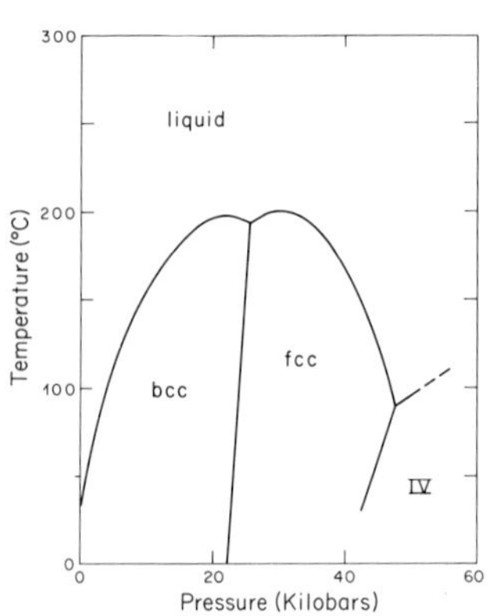

FIGURE 1.
Phase diagram of Cs (Kennedy, Jayaraman, and Newton, 1962), showing the melting curve as a function of pressure. The solid phases are marked "bcc," "fcc," and "IV" (structure unknown).

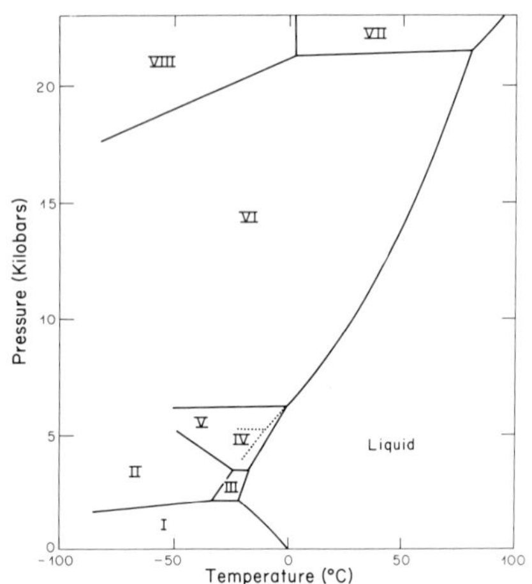

FIGURE 2.
Phase diagram of the H_2O system (Bridgman, 1912, 1937; Brown and Whalley, 1966). The field of ice IV is inferred from the D_2O system (Bridgman, 1935).

Solids that melt with negative ΔV_f exist at low pressure as phases of relatively open structure and of low coordination number, and the structural change on melting involves an increase in coordination number. For the group IV elements (Si, Ge, Sn) and the related III–V and II–VI compounds, the basic structural change is of the gray tin $\longrightarrow$ white tin type, involving an increase in the number of bonded neighbors from four to six. The low-pressure phases have tetrahedrally bonded structures that are analogous, in their framework of tetrahedral linkage, to ordinary ice I. The question thus arises whether the high-pressure forms of ice involve, similarly, an increase in number of bonded neighbors, or whether the increased density is achieved in some other way. We may expect that the increase in density on melting of ice to water is accomplished primarily by a structural change similar to the change that allows the closer molecular packing of the denser solid forms.

The present paper examines this idea in the light of recent information on the structures and physical properties of the dense forms of ice.

THE POLYMORPHS OF ICE

The crystalline polymorphs of H_2O, nine in number, are listed in Table 1, together with characterizing physical and structural information. Ices I and Ic are low-pressure forms, while ices II–VIII are produced only under pressure. The fields of stability are shown in Figure 2. Ice Ic has no known field of actual stability, being formed only as a metastable phase. Ice IV is unstable relative to ice V, but its metastable equilibrium with ice VI and with the liquid was observed by Bridgman (1935).

The low-pressure forms I and Ic have the structures shown in Figure 3, where the convention used depicts the oxygen atom of each water molecule as a ball, and the hydrogen bonds formed by the molecule as rods, the hydrogen atoms themselves not being shown. In these structures, the individual water molecules retain their identity and are little altered from their configuration in the vapor. Each water molecule is hydrogen-bonded to four near neighbors, in perfect (ice Ic) or nearly perfect (ice I) tetrahedral coordination, with an O—H···O distance of 2.75 Å (at 100°K). The H—O—H angle of the isolated water molecule, 104.6°, matches fairly closely the ideal tetrahedral coordination angle of 109.5°. The protons lie asymmetrically in each hydrogen bond, about 1.01 Å from one oxygen atom and 1.74 Å from the other, and each oxygen atom has only two nearby protons, so that each oxygen site corresponds to an intact water molecule. At each oxygen site there are six possible hydrogen-bonded orientations for the water molecule, and these occur statistically, with equal probability (Pauling, 1935; Peterson and Levy, 1957; Shimaoka, 1960), but in such a way that hydrogen bonds are not broken—that is, such that there is one and only one proton along each bond axis. This type of proton disorder (water-molecule orientation disorder) results in a configurational entropy amounting to $R \ln \frac{3}{2} = 0.81$ e.u., and, since the proton arrangement does not become ordered on cooling, the proton-disorder entropy is measured as zero-point entropy of ice I (Pauling, 1935). It is because of this disorder that the protons are not depicted in Figure 3.

The prevalent concept of H_2O as a tetrahedral molecule stems primarily from the essentially ideal tetrahedral H-bond configuration in the rather open structures of ice I and Ic. We now inquire how this tetrahedral character is altered upon packing water molecules more closely together, under pressure.

The known structures of the high-pressure forms of ice are shown in Figures 4 and 5. In all of these structures, with the exception of ice VII, each water molecule has four near neighbors at distances of 2.75–2.87 Å, followed

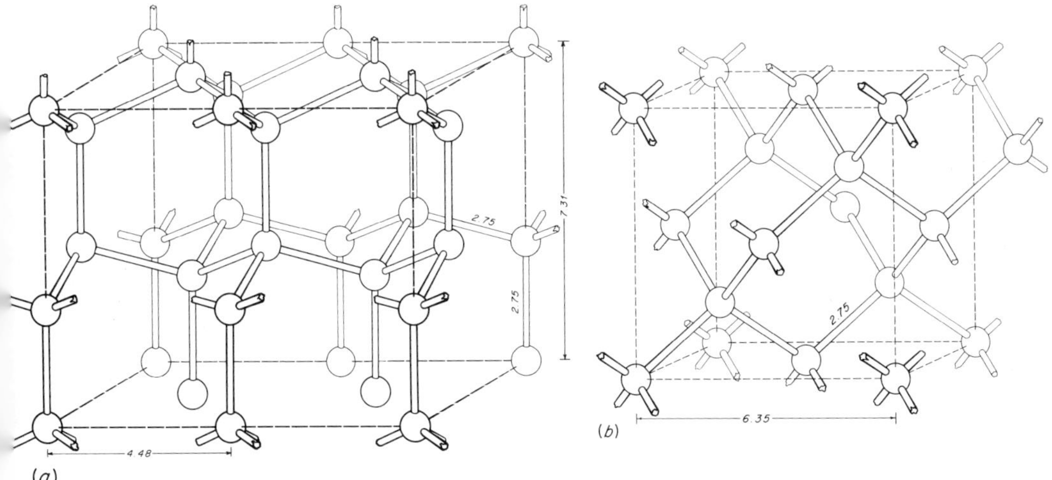

FIGURE 3.
Structure of low-pressure ice phases: (a) ice I, (b) ice Ic. Oxygen atom of each water molecule is shown by a ball, and hydrogen bonds by rods. The hydrogen atoms are omitted; they lie essentially on the hydrogen bond lines. In (b), the (cubic) unit cell is outlined with dashed lines; dimensions are in Å. In (a), the cell outlined is a rectangular (orthohexagonal) cell larger than the true hexagonal unit cell. The hexagonal c axis is labeled 7.31 Å, and one of the hexagonal a axes is labeled 4.48 Å. Bond lengths, in Å, are indicated. All dimensions are at 100°K. (Peterson and Levy, 1957; Shimaoka, 1960; Lonsdale, 1958).

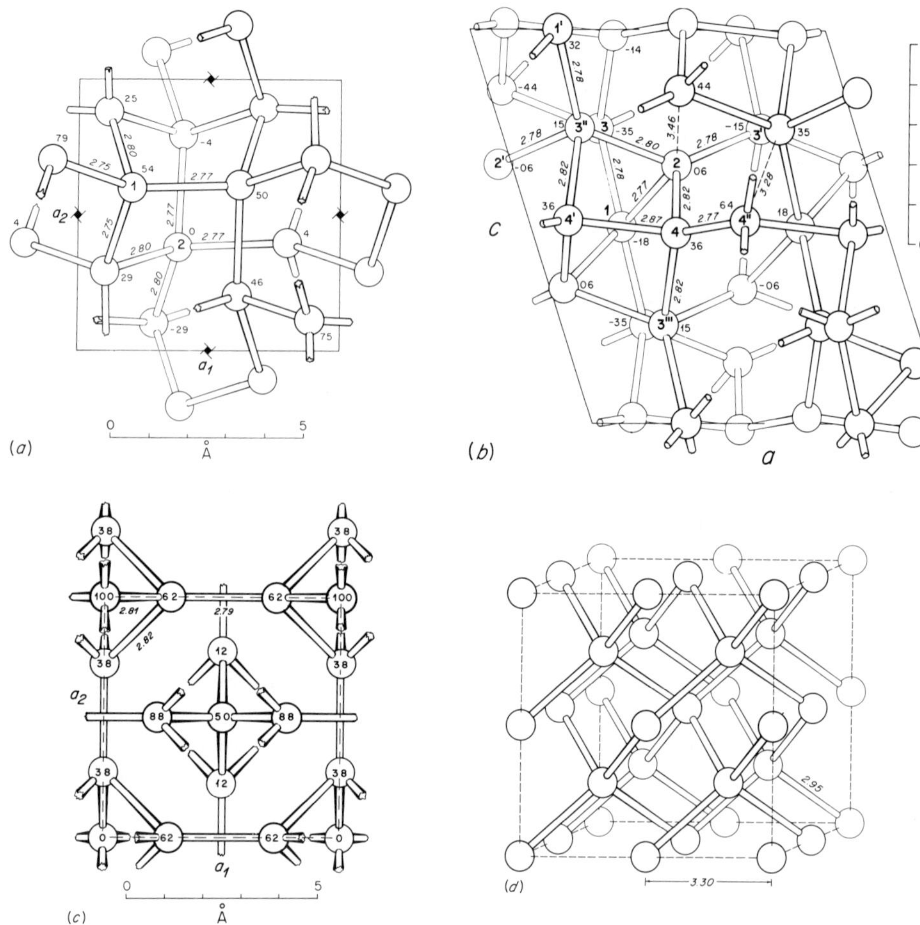

FIGURE 4.
Structures of high-pressure ice phases, shown with the same conventions as in Figure 3. Where drawings are projections along a crystallographic axis, the coordinate of each atom along that axis (height above the plane of projection) is given in hundredths of the axial length. Dimensions are at atmospheric pressure and 100°K. (a) ice III, projected along the c axis (= 6.83 Å); fourfold screw axes (parallel to c) are indicated; nonequivalent water molecules labeled "1" and "2." (b) ice V, projected along b axis (= 7.54 Å); nonequivalent water molecules labeled "1" to "4"; short nonbonded oxygen-oxygen distances indicated with dashed lines. (c) ice VI, projected along c axis (= 5.79 Å); molecular groups around cell corners link to form one bonded framework, group around center of cell is part of second framework. (d) ice VII, clinographic projection of cubic cell (a = 3.30 Å); cell outlined is a cube twice as large, to allow direct comparison with ice Ic (Figure, 3, *b*). (a: Kamb and Prakash, unpublished work; Kamb and Datta, 1960. b: Kamb, Prakash, and Knobler 1967. c: Kamb, 1965a (note error in Figure 3 of this reference); Kamb and Prakash, unpublished work. d: Kamb and Davis, 1964; Weir, Bloch, and Piermarini, 1965.)

by a variable number of more distant neighbors at distances of 3.24 Å or more. The four near neighbors form a distorted tetrahedron around the central molecule. The clear distinction between these four neighbors, at typical hydrogen-bond distances, and the more distant neighbors, at distances typical of non-hydrogen-bonded contacts between OH groups in crystals, shows that a tetrahedral network of hydrogen bonds remains the basic structural feature of these dense forms of ice.

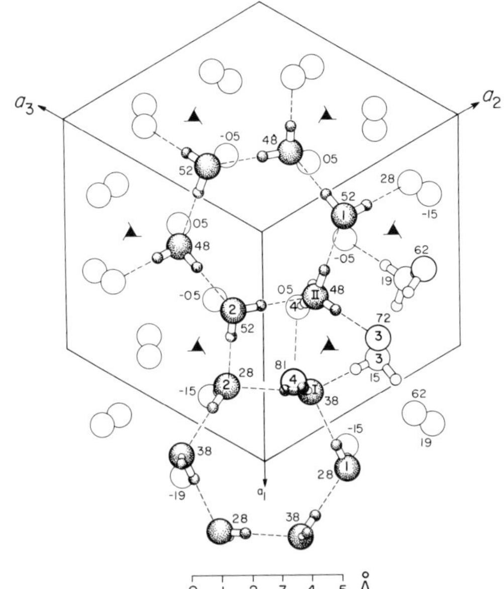

FIGURE 5.
Structure of ice II, viewed in projection along the hexagonal c_{H} axis. Lower edges of the rhombohedral cell are outlined with solid lines, and threefold screw axes are indicated. Some of the water molecules are here portrayed complete with protons, in the ordered arrangement deduced crystallographically. H-bonds are shown by dashed lines. Nonequivalent water molecules are labeled "I" and "II," other numbering is for reference to neighbors. Bond lengths are: I-1, 2.81 Å; I-3, 2.84 Å; I-4, 2.80 Å; II-1, 2.75 Å. (Kamb, 1964.)

The increase in density is thus achieved not by an increase in the coordination number of bonded neighbors, as in the gray tin ⟶ white tin transformation, but instead by an increase in non-nearest-neighbor coordination. In ice I and Ic, the nearest nonbonded neighbors lie at a distance of 4.50 Å, whereas in the high-pressure forms of ice, some 5 to 10 neighbors appear at distances in the range 3.2–4.0 Å (Figure 6). This increase in non-nearest-neighbor coordination is accomplished by a bending of the hydrogen bonds—that is, by a distortion from the angular geometry of ideal tetrahedral coordination. In achieving a close approach of nonbonded neighbors, bending distortion is minimized by a "doubling back" in the bond network, such that the near nonbonded neighbors are topologically distant, in the sense of the network topology. Thus, the 3.24 Å neighbor in ice II is topologically a third neighbor, and the 3.28 Å neighbor in ice V is topologically a fourth neighbor.

In the less-dense high-pressure phases II, III, and V, the network of hydrogen bonds is a single, complete tetrahedral framework, as is the case also in ice I and Ic. With the denser form ice VI, a new feature of molecular packing makes its appearance; a "self-clathrate" structure is built by incorporation of two identical but independent tetrahedral frameworks. The two independent frameworks interpenetrate, but are not interconnected by hydrogen bonds. Each framework fills void space in the other, hence the designation "self-clathrate." The self-clathrate feature is apparently favored because it allows a relatively large number (eight) of nonbonded near neighbors, without requiring excessive bond-bending strain.

The self-clathrate feature appears again in the densest form, ice VII, which consists of the interpenetration of two frameworks of ice Ic type. This arrangement places each water molecule equidistant from eight neighbors, so that here, for the first time among the dense forms of ice, the nearest-neighbor coordination number rises above four. The eight neighbors lie at the corners of a cube. The water molecule at the center is hydrogen bonded to four of these, in perfect tetrahedral coordination, and is in repulsive (nonbonded) contact

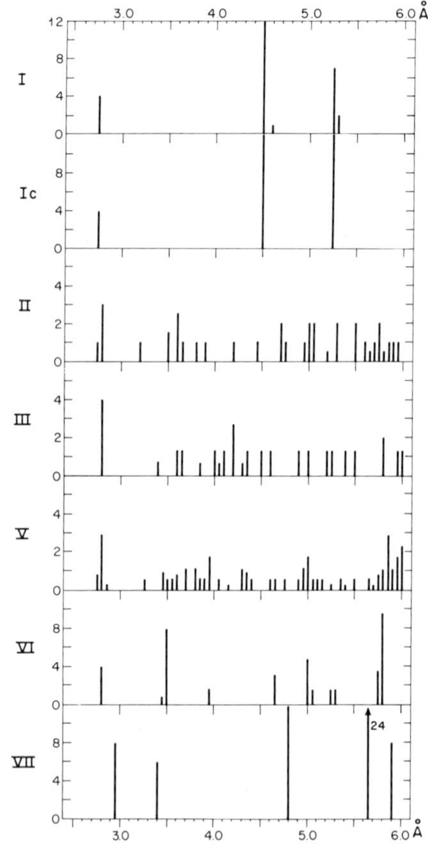

FIGURE 6.
Oxygen-oxygen interatomic distances in the ice structures, shown in histogram form.

with the other four. Because of the self-clathrate feature, the density of ice VII would be just twice that of ice Ic, were it not for the fact that the repulsion between each molecule and its four nonbonded nearest neighbors causes a lengthening of the bond lengths from the 2.75 Å in ice Ic to 2.86 Å in ice VII at 25 kb, or to 2.95 Å at atmospheric pressure. By accommodating the four nonbonded nearest neighbors, the ice VII structure is able to return to the ideal tetrahedral bond geometry of the low-pressure phases and thus to reduce the bond-bending strain essentially to zero.

Proton disorder (water-molecule orientation disorder) is present again in the structures of ice III, V, VI, and VII, as indicated by the entropies of these phases (see below), and by their dielectric properties (Wilson et al., 1965; Whalley, Davidson, and Heath, 1966). In contrast, ice II and VIII have proton-ordered structures. Crystallographic evidence allows the molecular orientations in the ice II structure to be inferred, and they are shown in Figure 5, (which thus departs from the ball-and-stick convention used in Figures 3 and 4). The detailed structure of ice VIII is not yet known; it apparently involves a distortion of the ice VII structure, the protons becoming ordered within the structural framework provided by the basic oxygen arrangement of ice VII (Kamb and Davis, 1964).

The entropies of ice I, III, V, VI, and VII (Table 1) differ relatively little

TABLE 1.

The ice polymorphs: crystallographic[a] and physical[b] data.

Ice	Crystal System	Space Group	Z[c]	Cell constants[d] (Å)	P (kb)	T (°C)	Density[e] (g cm^{-3})	T range (°C)	E(P)[f] $\left(\frac{\text{kcal}}{\text{mole}}\right)$	S(0)[g] (e.u.)	g[h]	τ[h] (μ sec)
I	Hexag.	$P6_3/mmc$	4	a 4.48, c 7.31	0.0	0°	0.92	—	0	0	3.4	18
Ic	Cubic	$Fd3m$	8	a 6.35	0.0	−87[i]	0.93[j]	—	0.04[i]	<0.2[i]	—	—
II	Rhombo.	$R\bar{3}$	12	a 7.78, α 113.1°	2.1	−35	1.18	−75, −35	0.01	−0.77	−0	>4 × 10^4
III	Tetrag.	$P4_12_12$	12	a 6.73, c 6.83	2.1	−30	1.15	−35, −22	0.21[k]	+0.26[k]	2.7	0.2
IV[l]	—	—	—	—	5.4	−6	1.29	−14, −6	0.34	0.20	—	—
V	Monocl.	$A2/a$	28	a 9.22, b 7.54, c 10.35, β 109.2°	3.4	−20	1.24	−17, −24	0.27	0.20	2.9	0.3
VI	Tetrag.	$P4_2/nmc$	10	a 6.27, c 5.79	6.2	0	1.33	−20, 0	0.37	0.19	3.4	0.3
VII	Cubic	$Pn3m$	2	a 3.41	21.5	0	1.56[m]	0, 82	1.13	0.11	2.4	0.8
VIII[n]	—	—	—	—	21.5	0	1.56[m]	−40, 0	0.87[p]	−0.82[p]		—

[a]Sources of crystallographic data are given in captions to Figures 3–5.

[b]Density, energy, and entropy are from Bridgman (1912, 1935, 1937, 1942) except as noted. Dielectric correlation factor *g* and relaxation time *τ* are from Auty and Cole (1952), Wilson et al. (1965), and Whalley, Davidson, and Heath (1966).

[c]Number of molecules in the unit cell.

[d]Cell dimensions are at 100°K and atmospheric pressure (quenched). For dimensions under pressure see Kamb, Prakash, and Knobler (1967), and Kamb and Davis (1964).

[e]Density is at the pressure and temperature listed. A small thermal-expansion correction (Kamb, 1965b) has been applied to the density values measured by Bridgman (1912, 1935, 1937, 1942) on the ice-water equilibrium curve.

[f]*E*(*P*) is energy at pressure *P*, relative to ice I at atmospheric pressure. Values are approximate, and are derived from Bridgman's data in the way described by Kamb (1965b). ΔE values for the ice-ice transitions are taken as averages over the equilibrium curves, which extend over the *T* ranges indicated for each phase (in equilibrium with the phase at immediately lower pressure).

[g]*S*(0) is entropy relative to ice I, computed by summing the average ΔS values over the equilibrium curves, and omitting variations of entropy within the individual phase fields. The *S*(0) values computed in this way are an approximation to the entropies of the phases when hypothetically decompressed to atmospheric pressure and compared with Ice I at the same temperature.

[h]The Kirkwood correlation factor, *g*, is derived from the dielectric constant measured under conditions of stability for the individual phases; *τ* is the dielectric relaxation time at pressure *P*, interpolated or extrapolated (using measured activation energies) to 0°C.

[i]Inversion temperature of ice Ic at warming rate of 16°C min^{-1} (McMillan and Los, 1965). *E* and *S* values for ice Ic are from measured heat of inversion.

[j]Density from X-ray measurements (Shimaoka, 1960; Lonsdale, 1958).

[k]For the transition ice I → III, ΔE and ΔS increase with temperature, which suggests ordering of the protons at lower temperatures, as inferred also from spectroscopic data (Bertie and Whalley, 1964b; Whalley and Davidson, 1965). Crystallographic evidence suggests, however, that any proton ordering in ice III at 100°K is only partial (Kamb and Prakash, unpublished work).

[l]Values for ice IV are estimates based on Bridgman's (1935) measurements on D_2O ice. The differences in molar volume, energy, and entropy between ices IV and VI are assumed to be the same for the D_2O and H_2O ices.

[m]The X-ray-determined density at 25 kb, −50°C, is 1.66 g cm^{-3} (Kamb and Davis, 1964). Ice densities determined by X-rays under pressure are consistently somewhat higher than Bridgman's volumetrically measured densities (Kamb, Prakash, and Knobler, 1967). The density difference between ices VII and VIII is too small to measure by standard volumetric techniques (Brown and Whalley, 1966).

[n]Ice VIII as designated here (Whalley, Davidson, and Heath, 1966) is not to be confused with the icelike substance called "ice VIII" by Cohen and van der Horst (1938) (also Gränicher, 1958, p. 434), which has been shown to be a clathrate hydrate (Quist and Frank, 1961; Wilson and Davidson, 1963).

[p]*E* and *S* values for ice VIII are obtained from those for ice VII by using the transition ΔS and ΔV values of Brown and Whalley (1966).

among one another ($\leq$0.3 e.u.). This near equality reflects the fact that all of these phases have proton-disordered structures within their fields of stability (Kamb, 1964), so that the proton configurational contribution to the entropies is the same as in ice I, 0.8 e.u. The small entropy differences among the phases are differences in the lattice vibrational entropies and inter- and intramolecular vibrational entropies of the different structures. The entropies of ice II and VIII, on the other hand, are about 0.8 e.u. lower than those of the other forms. This entropy difference, essentially equal to the configurational entropy of proton disorder in the disordered forms, is a direct indication of the proton order in ice II and VIII (Kamb, 1964; Brown and Whalley, 1966).

ENERGY AND HYDROGEN BOND STRENGTH

Energies of the ice phases are compared with structural information in Table 2. The structures are characterized by density, mean bond length $\langle d \rangle$, and by $\langle (\delta\theta)^2 \rangle$, the mean square departure of the O$\cdots$O$\cdots$O bond angles from the ideal tetrahedral value of 109.5°. $\langle (\delta\theta)^2 \rangle$ is a measure of H-bond bending, in terms of the angular distortion of the coordination away from ideal tetrahedral geometry. The increasing density of molecular packing in the phases I through VI is accomplished at the expense of a progressive increase in bond-bending distortion $\langle (\delta\theta)^2 \rangle$. (Ice VII does not follow this trend, since it has no bond-bending strain.)

The energies of the ice phases increase in the sequence I, Ic, II, III, V, VI, VIII, VII, that is, essentially in the sequence of increasing density of molecular packing. For ices III to VI, whose energies are dominated by H-bond-bending distortion, the energies increase roughly proportionally to $\langle (\delta\theta)^2 \rangle$ (Figure 7), so that an effective bond-bending "force constant" K_θ can be associated with distortions of the O$\cdots$O$\cdots$O bond angles, via the relation

$$E - E_{\mathrm{I}} = K_\theta \, \langle (\delta\theta)^2 \rangle \tag{1}$$

K_θ has the approximate value 0.9 cal deg^{-2} mole^{-1}. This value can be compared with force constants for H-bond bending based on the elastic shear modulus c_{44} of ice I, and on the molecular libration frequencies in ice, which correspond to $K_\theta \cong 5.2$ cal mole^{-1} deg^{-2}. Two factors may contribute to the discrepancy between the values 0.9 and 5.2 cal mole^{-1} deg^{-2} for K_θ:

1. Increased van der Waals energy in the denser ice forms will tend to offset the bond-bending energy. The estimated magnitude of the increased van der Waals stabilization (taking account of both dispersion and overlap interactions) is 0.4–0.6 kcal mole^{-1} for ice II (Kamb, 1965b) and 0.3–0.5 kcal mole^{-1} for ice VI. Although significant, this is not large enough to explain fully the stability of these phases in relation to the upper, dashed line in Figure 7. The lack of an offsetting energy stabilization in less-dense icelike structures is indicated by the energy of the type I clathrate-hydrate framework (plotted as an open triangle in Figure 7).

2. A covalent contribution to the H bonds might drop off rapidly with distortion from ideal tetrahedral geometry, so that the bond energy as a function of $\langle (\delta\theta)^2 \rangle$ might have a sharp narrow minimum near $\langle (\delta\theta)^2 \rangle = 0$ and only a shallow variation at larger $\langle (\delta\theta)^2 \rangle$ (Frank, 1958). This is suggested by the dotted curve in Figure 7. However, such a large change in the bond-bending force

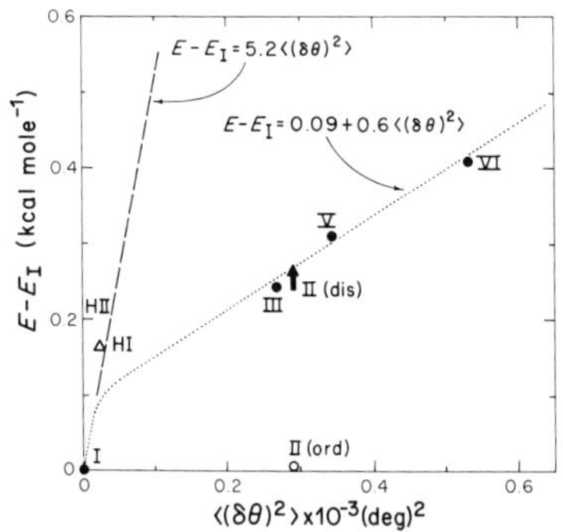

FIGURE 7.
Energies of ice phases relative to ice I, $E - E_I$ (Table 2), as a function of mean coordination distortion $\langle(\delta\theta)^2\rangle$. Arrow indicates energy for hypothetical proton-disordered ice II structure. Upper (dashed) curve is bond-bending energy relationship based on c_{44} of ice I; lower (dotted) curve is drawn to conform to the high-pressure ice phases. Open triangle ("HI") is for the icelike framework of the type I clathrate-hydrate structure (van der Waals and Platteeuw, 1959; McMullan and Jeffrey, 1965).

constant as a function of $\langle(\delta\theta)^2\rangle$ would require a much larger decrease in the molecular libration frequencies than is observed (Table 2).

The difference between the values 0.9 and 5.2 cal mole^{-1} deg^{-2} for K_θ is significant in relation to the energies of possible molecular configurations in liquid water. From this difference between effective bond-bending force constants under different types of structural distortion, it follows that there is wide latitude for other hypothetical ice polymorphs having coordination distortions comparable to those of the stable high-pressure forms, but energies that may be considerably larger. The elusive ice IV, unstable relative to ice V, seems to be an example. Analogues of quartz and coesite (high-pressure SiO_2) are obviously possible structures for dense ice phases, but are unfavored energetically.

The average hydrogen bond length (Table 2), which is an inverse measure of bond strength, increases progressively from 2.75 Å in ice I and Ic to 2.81 Å in ice VI. Bond length thus works against the effect of bond bending in achieving denser molecular packing in these high-pressure forms. In ice VII and VIII the H-bond length increases to 2.95 Å (at atmospheric pressure). As noted earlier, this marked expansion of the ice VII structure is due to repulsion between each water molecule and its four nonbonded near neighbors. Likewise, the high energy of ice VII is the direct consequence of this repulsion (there is no bond bending): the repulsive energy contribution combines with the energy stored in the stretched H-bonds to give the observed destabilization of 0.92 kcal mole^{-1}. From the dimensions and energy of the ice VII structure, it is possible to derive values of parameters in the repulsion potential between non-hydrogen-bonded water molecules (Kamb, 1965b).

Two possible sources for the increased bond lengths of 2.77–2.81 Å in ices II–VI may be distinguished: (1) repulsion of close nonbonded neighbors, causing a general expansion of the structures as in ice VII; (2) a dependence of equilibrium bond length on bond bending.

An increase of bond distance with increase in coordination number is a general effect in crystals, although it can be given different interpretations for different types of bonding (Pauling, 1960, pp. 253, 401, 537). Whether a similar effect is to be expected for an increase in non-nearest-neighbor coordination, as in the dense ice phases, depends on the intermolecular potential function and on the detailed distribution of interatomic distances. When evaluated on the basis of the repulsion potential derived from ice VII, no effect of this kind

TABLE 2.
Structural and spectroscopic information for H_2O phases.

Phase	Density[a] ρ (g cm^{-3})	Heat of inversion[b] ΔH_{inv} $\left(\frac{kcal}{mole}\right)$	Energy[c] $E - E_I$ $\left(\frac{kcal}{mole}\right)$	Bond bending[d] $\langle(\delta\theta)^2\rangle$ (deg$^2 \times 10^3$)	Av. bond length[e] $\langle d \rangle$ (Å)	Bond-length range[f] (Å)
Ice (at 77°K)						
I	0.94	0	0	0	2.75	.01
Ic	0.94	.03	—	0	2.75	.00
II	1.18	<.04	.01	.29	2.80	.09
III	1.16	.10	.24	.27	2.775	.03
V	1.23	.23	.31	.34	2.80	.10
VI	1.31	.32	.41	.53	2.81	.02
VIII	1.50	.59	.66	~0	2.95	—
VII	1.50	—	.92	0	2.95	.00
Water						
l(0°C)	1.00	—	1.44	(.66)[k]	2.84	.28
l(26°C)	1.00	—	1.91	(.73)[k]	2.86	.32
l(65°C)	0.98	—	2.61	(.81)[k]	2.90	.38
(CCl_4)[n]	—	—	—	—	—	—
vapor	—	—	11.65	—	—	—

[a]Calculated from the cell size and contents as measured by X-rays at 100°K. The thermal contraction to 77°K is probably small, and is neglected.

[b]The heat of inversion of ice Ic to ice I was measured as 0.036 kcal mole^{-1} by McMillan and Los (1965), but an upper limit of 0.027 kcal mole^{-1} was set in the measurements of Beaumont, Chihara, and Morrison (1961). The ΔH_{inv} value 0.03 kcal mole^{-1} listed here for ice Ic has been added to the heats of inversion of the dense ice phases to ice Ic, measured by Bertie, Calvert, and Whilley (1964), to obtain the ΔH_{inv} values listed.

[c]Energy values E are given relative to ice I, and are derived by summing the transition ΔE values measured by Bridgman (1912, 1937), averaged along the individual phase boundaries. The values derived in this way, with omission of energy changes due to compression within the individual phase fields, are approximately the energies of the ice phases when hypothetically decompressed to atmospheric pressure. This is supported by a detailed computation of the energy of ice VII at atmospheric pressure (Kamb, 1965b). The heats of inversion at atmospheric pressure, ΔH_{inv}, confirm the general pattern and magnitude of the E values, but suggest some additional energy stabilization, particularly in ice III, in which it may be due to proton ordering at low temperature (Table I, footnote *k*).

[d]Values for ice phases are from structural information cited in Figures 3–5, and are assummed not to change from 100°K to 77°K. For water, see footnote *k*.

[e]For ice phases, data from Figures 3–5. Change in dimensions from 100°K to 77°K is neglected. For water, the values listed are r_{max}, the first maximum in the r.d.f., as read from the straight line in Figure 11.

[f]The bond-length range for ice structures is from sources for Figures 3–5, and does not include the effects of thermal motion. For water, the values quoted are the width of the nearest-neighbor peak of the r.d.f. $\rho(r)$ (Narten, Danford, and Levy, 1967) at a height halfway from $\rho = 1$ to the crest of the peak; the contribution of thermal motion to this halfwidth is uncertain, but might be as large as the 0.25 Å r.m.s. thermal displacement in ice I near 0°C (Owston, 1958; Lonsdale, 1958).

[g]Stretching frequencies are for HDO in dilute concentration, for ice phases from Bertie and Whalley (1964a, b), for water from Falk and Ford (1966). Frequencies of resolved lines in ices II and III are averaged to give a mean frequency $\bar{\upsilon}_{OH}$ or $\bar{\upsilon}_{OD}$. For the unresolved lines of the other phases, $\bar{\upsilon}$ is taken as the peak frequency of the absorption band. Frequency shifts for the ice phases (at 77°K) are relative to ice I at 77°K ($\bar{\upsilon}_{OH}$ =

OH or OD stretching frequency (shift relative to ice I, cm^{-1})				Libration frequency[l] (cm^{-1})		Hindered translation[m] (cm^{-1})	
Infrared[g]		Raman[h]					
$\Delta\bar{\upsilon}_{OH}$	$K\Delta\bar{\upsilon}_{OD}$	$\Delta\bar{\upsilon}_{OH}$	$K\Delta\bar{\upsilon}_{OD}$	$\bar{\upsilon}_L(H_2O)$	$\bar{\upsilon}_L(D_2O)$	$\bar{\upsilon}_T(H_2O)$	$\bar{\upsilon}_T(D_2O)$
0	0	0	0	840	640	225	217
0	—	1	—	840	640	—	—
73	76	106	97	720	550	151	146
44	54	72	57	770	590	—	166
72	73	99	88	730	540	169	159
(79)[j]	(98)[j]	119	117	—	—	—	—
(165)[j]	(183)[j]	248	220	—	—	—	—
—	—	—	—	—	—	—	—
88	70	—	—	705	525	175	175
104	88	139	103	—	—	—	—
130	115	—	120	—	—	—	—
370	345	—	—	—	—	—	—
404	387	536[i]	483[i]	—	—	—	—

3277 cm^{-1}, $\bar{\upsilon}_{OD}$ = 2421 cm^{-1}); shifts for water are given relative to ice I at 0°C ($\bar{\upsilon}_{OH}$ = 3300 cm^{-1}, $\bar{\upsilon}_{OD}$ = 2440 cm^{-1}: Falk and Ford, 1966). This tends to remove the effects of ice thermal expansion between 77° and 273°K in comparing the ice and water frequencies. The shifts $\Delta\bar{\upsilon}_{OD}$ are multiplied by the scaling factor $K = 1.35$ to facilitate comparison with $\Delta\bar{\upsilon}_{OH}$, 1.35 being the average observed ratio $\upsilon_{OH}/\upsilon_{OD}$ for individual HDO stretching frequencies in ices I–V (Bertie and Whalley, 1964b).

[h]Raman frequencies listed are for the pure H_2O and D_2O ices (Taylor and Whalley, 1964; Marckmann and Whalley, 1964), and are given as average frequency shifts relative to the Raman lines 3080 and 3210 cm^{-1} of H_2O ice I and 2283 and 2416 cm^{-1} of D_2O ice I at 77°K. For liquid water, the frequencies listed are the peak frequencies of the Raman stretching vibration bands of HDO in dilute concentration (Wall and Hornig, 1965), and are given relative to the HDO frequencies for ice I at 0°C. OD frequency shifts are multiplied by the scaling factor $K = 1.35$ (see footnote *g*).

[i]The water vapor frequency shifts are for H_2O and D_2O, and are given relative to the average Raman stretching frequency for H_2O and D_2O ice I at 0°C, estimated from the HDO frequencies by assuming that the H_2O and D_2O Raman frequencies are depressed below the HDO frequencies by the same amount at 0°C as at 77°K, where both have been measured (Bertie and Whalley, 1964a).

[j]HDO frequency estimated from average Raman frequency shift of the pure H_2O or D_2O phases by applying the average ratio $\Delta\bar{\upsilon}_{OH}(HDO)/\Delta\bar{\upsilon}_{OH}(H_2O: Raman)$ or $\Delta\bar{\upsilon}_{OD}(HDO)/\Delta\bar{\upsilon}_{OD}(D_2O: Raman)$.

[k]Bond-bending values for water are purely theoretical, as calculated by Pople (1951) in a theory of water structure based on bond bending.

[l]Libration frequencies $\bar{\upsilon}_L$ given are an estimate of the mean frequency of the libration bands of the pure H_2O and D_2O ices as observed by infrared absorption (Bertie and Whalley, 1964a, b). For bands without resolution the frequency $\bar{\upsilon}_L$ is taken to be the peak frequency, whereas for bands showing some resolution, an attempt is made to estimate a corresponding peak frequency for a "smeared" band. For water, the figures listed are the frequencies designated "υ_{l_2}" by Walrafen (1964), who summarized earlier data.

[m]Ice frequencies from Taylor and Whalley (1964), water frequencies from Walrafen (1964).

[n]Water in (dilute) solution in CCl_4 (Falk and Ford, 1966).

sufficient to account for the observed expansion in bond lengths is found.

Conceptually, a continuous change in relative molecular orientation from the ideal bonded configuration in ice I to the nonbonded configuration in ice VII is possible. The H_2O—H_2O equilibrium distance increases correspondingly from 2.75 Å (ice I) to about 3.5 Å (nonbonded pair).* Between these extremes one may expect a continuous variation of equilibrium distance with relative molecular orientation, so that the equilibrium H-bond length must be dependent on H-bond bending.

The mean bond length does not increase abruptly to much higher values as we go from ice I to the denser forms of ice, with distorted tetrahedral coordination. An abrupt increase in bond length would, however, be expected if, as proposed by Frank (1958), an abrupt weakening of the H-bonds were to take place as soon as the coordination is distorted significantly from ideal tetrahedral. Also, the unique energetic stabilization hypothesized by Frank (1958) for the ideal tetrahedral geometry seems ruled out by the relatively low energies of the dense ice structures (Table 2).

The O—H stretching vibration is a sensitive indicator of the strength of hydrogen bonding. Average frequencies of uncoupled O—H and O—D stretching vibrations in the phases of ice, measured at atmospheric pressure and 77°K by E. Whalley and collaborators, are summarized in Table 2 for comparison with the structural and thermodynamic information. The stretching frequencies are given as shifts relative to the frequencies in ice I, and the shifts are in all cases positive (shifts to higher frequency), indicating weakening of the hydrogen bonds relative to those in ice I. The mean O—H stretching frequencies of the forms of ice increase in the sequence I, III, V, II, VI, VIII. For ices II–VI, the mean frequency shifts relative to ice I are less than one-fifth the shifts that occur on breaking the hydrogen bonds of ice I, that is, on passing from ice I to water vapor (last row in Table 2). Although individual frequencies deviate somewhat, the individual shifts are in all cases comparable to the means, hence it follows that the high-pressure polymorphs are fully hydrogen bonded, with hydrogen bonds only slightly weaker than those in ice I (Bertie and Whalley, 1964b). This coincides with the structural conclusions presented earlier. For ice VIII, the frequency shifts are about double those for ice VI. Thus ice VII and ice VIII stand noticeably apart from the other phases, both spectroscopically and structurally, their unique structural feature being their coordination number of 8. But even in this case, the frequency shifts are less than half the shift on breaking the H-bonds of ice I, and the bonding remains strong (Marckmann and Whalley, 1964).

The mean O—H and O—D stretching frequencies in Table 2 increase in a systematic way with increasing mean bond length $\langle d \rangle$ (Figure 8). The observed variation of $\bar{\nu}_{OH}$ with $\langle d \rangle$ differs slightly from the curve given by Wall and Hornig (1965) (Figure 8, dotted curve). Since mean bond length $\langle d \rangle$ tends to increase with bond-bending distortion $\langle (\delta\theta)^2 \rangle$, there is also a correlation between $\bar{\nu}_{OH}$ and $\langle (\delta\theta)^2 \rangle$ for ices I–VI. An influence of bond bending is seen in the fact that, although there are bonds of length 2.75 Å in ices II and III, the lowest O—H frequencies of these phases are nevertheless shifted about 40 cm^{-1}

*The figure 3.5 Å is based on the repulsion potential for nonbonded neighbors derived from ice VII (Kamb, 1965b).

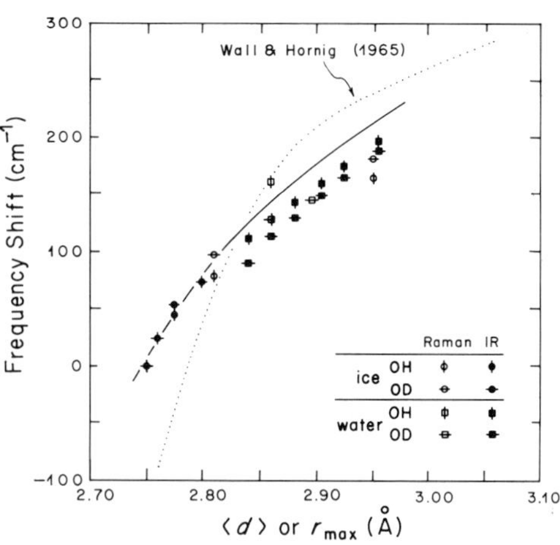

FIGURE 8.
Relationship between mean OH/OD stretching frequency and mean bond length $\langle d \rangle$ for ice phases (circles), and between peak OH/OD stretching-band frequency (Raman/IR) and peak H-bond length r_{max} for water (squares). Frequencies are plotted as shifts relative to the mean frequency for ice I at 77°K (OH: 3277 cm^{-1}; OD: 2421 cm^{-1}). OD frequency shifts are multiplied by the scaling factor 1.35. Data points are, from left to right: ice I (100°K); ice I (0°C); ices III, II, V, VI (100°K); water at 0°, 25°, 50°, 75°, 100°; ice VII (90°K); water at 129°C. Points for water are plotted on the basis of r_{max} values read from the straight line in Figure 11. Dotted curve is from Wall and Hornig (1965, Figure 3). Solid line is estimated mean curve of bond length versus shift of HDO stretching frequency, for bond distortions of the type that occur in the dense ice phases (except ice VII) and in water. The curve is drawn taking into account the observed difference in peak frequency in the infrared and Raman HDO bands of water, and represents the estimated relation between peak frequency and r_{max} for the Raman band. References to data sources are given in footnotes to Table 2.

with respect to the frequency of ice I. For ice VII, conversely, the return to unbent bonds may be responsible for the fact that the frequency shifts (open circles at $d = 2.95$ Å in Figure 8) are somewhat smaller than might be expected on the basis of bond length by extrapolation from the less dense phases and from liquid water (solid curve in Figure 8).

The energies of the ice phases (Table 2) increase systematically with the mean O—H frequencies, except for ice II. Ice II also presents a distinct anomaly in the relation between energy and tetrahedral distortion $\langle(\delta\theta)^2\rangle$ (open circle in Figure 7). Remarkably, the energy of ice II is only 0.01 kcal mole^{-1} greater than that of ice I. Ice II would be widespread in nature, were it not for the destabilizing effect of its low entropy. From the fact that ice II does not invert at higher temperatures to a proton-disordered form with the same tetrahedral framework, but instead transforms to ice III, it follows that the energy of a hypothetical, proton-disordered ice II phase must be greater than about 0.25 kcal mole^{-1}, as would fit the variation of energy with $\langle(\delta\theta)^2\rangle$ for the other forms (arrow in Figure 7). The extra energy stabilization of the proton-ordered ice II structure is not achieved primarily by an increase in H-bond strength, as measured by the mean O—H frequency.

The libration frequencies, which are a direct measure of the bond-bending force constants, present a very similar picture to the O—H stretching frequencies in their ranking of the bond strengths for the different ice phases (Table 2).

PROPERTIES AND STRUCTURE OF WATER IN RELATION TO ICE

The concept that liquids near the melting point have many basic structural features in common with their solid counterparts has been widely considered (see Ubbelohde, 1965). A structure for water based on solid-like arrangements of water molecules would require contributions from dense ice structures, simply because of the increase of density on melting. As already pointed out, such contributions are expected for a substance showing extensive polymorphism at relatively modest pressures. Because of the labile character of the liquid, due to its relatively large thermal motions and the rapid changes in molecular configuration taking place, one would not expect that a single molecular arrangement would be rigidly selected for representation, but instead that many configurations would occur, particularly since the corresponding solid shows, by its rich polymorphism, that a variety of molecular arrangements with roughly comparable energies can be constructed. It is therefore of interest to examine some structurally controlled properties of water from the standpoint of their evidence for or compatibility with the presence of icelike molecular configurations in the liquid.

RADIAL DISTRIBUTION FUNCTION

X-ray diffraction provides geometrical information on water structure, in the form of the average distribution of oxygen-oxygen distances r occurring in the liquid (radial distribution function, r.d.f.). The r.d.f. determined by Morgan and Warren (1938) for water at 1.5°C is shown as the lowermost curve of

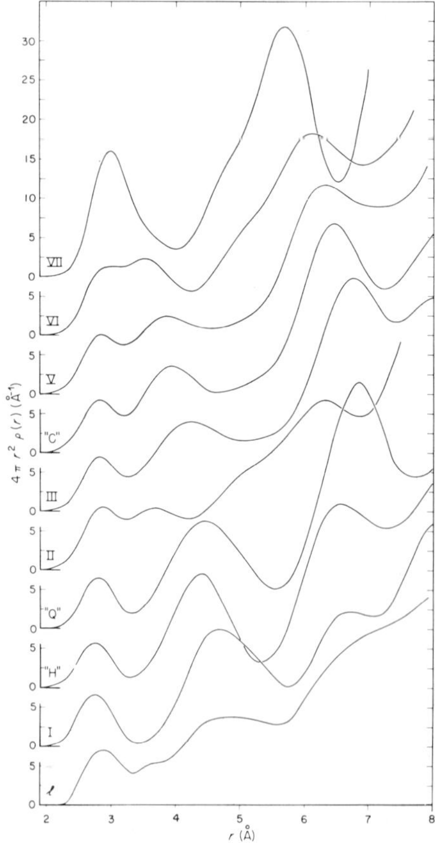

FIGURE 9.
Comparison of r.d.f. for water at 4°C (bottom curve, "l"; Morgan and Warren, 1938) with r.d.f.'s for ice polymorphs and hypothetical ice structures. Ice polymorphs are identified with roman numerals; hypothetical structures are as follows: "H": clathrate-hydrate framework of type I (Pauling and Marsh, 1952); "Q": β-quartz-type framework; "C": coesite-type framework. R.d.f.'s for the solid structures are computed from the discrete interatomic distances (Figure 6) by applying a Gaussian "smearing" (convolution) to represent thermal motions; r.m.s. motions are taken to be 0.25 Å for the bonded distances, and 0.4 Å for longer distances.

Figure 9; recent X-ray data (Narten, Danford, and Levy, 1966, 1967) confirm the main features of Morgan and Warren's curve. For comparison with water in Figure 9 are theoretical r.d.f.'s for known and hypothetical forms of ice, calculated from the discrete distributions of mean oxygen-oxygen distances in the crystals (Figure 6) by applying a Gaussian smoothing to represent the effects of thermal motion.

The primary feature of the water r.d.f. is the presence of about four oxygen neighbors at distances near $r = 2.8$ Å. Morgan and Warren (1938) interpreted this as indicating a basically tetrahedral coordination for the water molecules, similar to the coordination in ice. In comparing the observed r.d.f. for water with calculated r.d.f.'s based on the ice I structure, Morgan and Warren (1938) noted that (1) the nearest neighbor peak is shifted from 2.76 Å in ice I to about 2.9 Å in water, (2) the next-nearest-neighbor peak near 4.5–5.0 Å is broader and lower in water than in ice I, and (3) in the interval between 3.0 and 4.0 Å in the water r.d.f. there is a considerable excess of oxygen-oxygen distances, corresponding to a peak containing about 3.4 extra distances centered at about 3.6 Å. All of these features correlate in a general way with what is found in the denser ice phases: the increase in H-bond lengths, the variable distances of next-nearest neighbors, and the presence of numerous non-nearest neighbors in the interval 3.0–4.0 Å (Figure 6).

In a quartzlike structure for liquid water, as proposed by Bernal and Fowler (1933), the next-nearest peak in the r.d.f. should occur at $r \approx 4.24$ Å, distinctly closer than the $r \approx 4.6$ Å for a structure like that of ice I (curves Q and I, Figure 9). Instead, the observed peak tends to broaden and shift toward larger distances, the maximum occurring at about $r = 5.1$ Å. This contrary behavior was cited by Morgan and Warren (1938) as evidence against the proposed quartzlike structure for water, and it is noteworthy that a structure analogous to quartz does not occur among the dense forms of ice, contrary to the expectation of Bernal and Fowler (1933). Among the known dense ice phases, there is also a general tendency for the main next-nearest neighbor peak to shift toward distances shorter than 4.6 Å, but ice II is an exception, its r.d.f. showing a strong slope toward larger r near $r = 4.6$ Å. It also has a partially resolved peak near $r = 3.5$ Å, as required in the water r.d.f.

For the calculated r.d.f. in Figure 10, ices I, II, and III have been combined

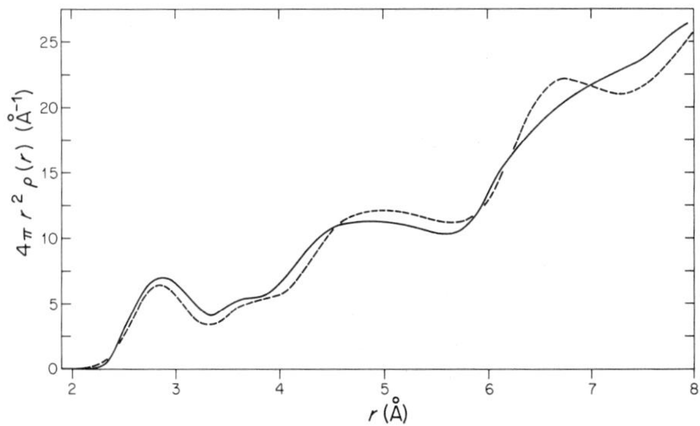

FIGURE 10.
Comparison of r.d.f. for water at 4°C (solid curve) with theoretical r.d.f. for a structural mixture consisting of 50 percent ice I, 33 percent ice II, and 17 percent ice III (dashed curve). A linear thermal expansion of 1.02 relative to the structures at 100°K has been assumed for the ice components. Assumed r.m.s. Gaussian thermal motion displacements $\sigma(d)$ are as follows:
$0 \leq d < 3.65$, $\sigma = 0.25$; $3.65 \leq d < 6.6$, $\sigma = 0.4$; $6.6 \leq d \leq 7.0$, $\sigma = 0.5$; $d > 7.0$; $\sigma = 0.7$ Å.

in proportions of 50, 33, and 17 percent to give a curve that matches the observed r.d.f. for water fairly well (probably to within the experimental uncertainty, except at the largest distances, where the theoretical curve has too little thermal smoothing). It appears likely that with further adjustment of available parameters, and particularly with the inclusion of ice phases other than I, II, and III, an even better fit to the observed curve could be obtained, but such refinement does not seem justified, because of the experimental uncertainties and because structural interpretations of the r.d.f. are not unique anyway. Figures 9 and 10 do, however, indicate that the r.d.f. for water can be accounted for by the presence in the liquid of icelike molecular arrangements.

The most-abundant H-bond length in water at 0°C is about 2.84 Å, according to the interpretation in Figure 11 of the r.d.f. data of Narten, Danford, and Levy (1966).* This is somewhat longer than the mean bond length 2.79 Å for the ice I-II-III mixture described, taking into account an 0.02 Å thermal expansion between 77° and 273°K. Some increase in apparent bond length can be expected as a result of overlap of the bonded-nearest-neighbor peak with the tail of the nonbonded neighbor distribution at larger r; this increase amounts to about 0.02 Å, as judged from the shift in nearest-neighbor maximum for the ice III r.d.f. in Figure 9, where the non-nearest-neighbor density at $r \cong 3.2$ Å is about the same as it is in water. The H-bond length in water at 0°C appears comparable to that expected for ice VI at this temperature; this probably reflects a significant concentration of molecules having H-bonds about as distorted as those in ice VI.

Other properties of water require the presence of some severely weakened or broken bonds, and these must have a proportionate contribution to the r.d.f. that cannot be included in a representation containing contributions only from the fully bonded structures of the solids, such as that in Figure 10.

DENSITY, THERMAL EXPANSION, AND COMPRESSIBILITY

Since ice II and ice III have densities about 15 percent greater than water, and ice I is about 10 percent less dense, a mixture of ice I with ice II or III in roughly equal proportions will account for the density of water. As observed in the melting of solids generally, a decrease in density from the contributing solid structures would be expected owing to incorporation of structural defects and disorder. It is thus not unreasonable that the density represented by the combination of solids in Figure 10 is 4 percent too great as it stands. Denser molecular arrangements, such as those of the denser ice phases, could contribute significantly to the liquid structure if there were a correspondingly greater general loosening of the structure through incorporation of defects. There could also be contributions from structures less dense than ice I, such as the ice frameworks of the clathrate hydrates.

*The nearest-neighbor maxima at 2.82 Å (4°C) and 2.95 Å (200°C) reported by Narten, Danford, and Levy (1967) are probably the maxima in $\rho(r)$ rather than in $D(r) = 4\pi r^2\rho(r)$. The maximum at 2.88 Å (25°C) quoted by Narten, Danford, and Levy (1967) agrees with the value given in their earlier report (Danford and Levy, 1962) but does not quite agree with the values 2.85 Å and 2.87 Å that I calculate from the $\rho(r)$ and $D(r)$ data of the more recent papers. As shown by Narten, Danford, and Levy (1966), the larger r_{max} values obtained by other authors in previous X-ray work are the result of the lower resolution of the earlier X-ray data.

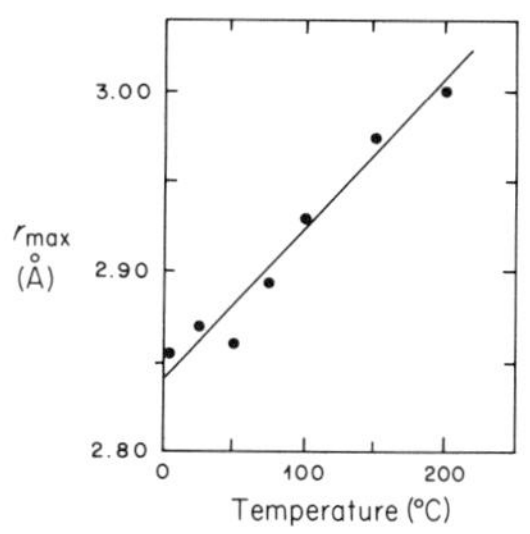

FIGURE 11.
Most-abundant H-bond length in water as a function of temperature, as indicated by the X-ray r.d.f. The ordinate r_{max} is the position of the first maximum in $D(r) = 4\pi r^2 \rho(r)$, determined by fitting a cubic polynomial to r.d.f. values D2(r) of Narten, Danford, and Levy (1966). The scatter of r_{max} values is here interpreted as due simply to random experimental errors, and a straight line is drawn as the most reasonable representation of the dependence of r_{max} on T, according to the experimental data. The values of r_{max} used in the text and tables are read from this straight line.

The celebrated negative thermal expansion of water up to 4°C, leading to a minimum in specific volume at this temperature, is an indication that the process of densification that occurs abruptly on melting continues as the temperature rises to have a further densifying effect strong enough to offset the normal thermal expansion up to 4°C. This implies an increase with temperature in the contributions of the denser molecular arrangements, relative to the less dense ones. Such an increase will tend to occur if the denser structures have somewhat higher energy, as do the denser forms of ice in relation to ice I (Table 2). Most theories of water structure that visualize two or more molecular arrangements in the liquid have hypothesized an energy relationship of this type, to explain the negative thermal expansion. At higher temperatures, the "structural contraction" gradually dies out, presumably because the proportions of the different possible molecular arrangements change less and less as they come to be more or less equally represented. A "normal" thermal expansion then becomes evident. This "normal" thermal expansion for the liquid is much larger than the expansions of the corresponding solid phases, because it involves the progressive incorporation of significant numbers of "defects" or "holes" into the liquid structure.

The compressibility of water, 51×10^{-12} dyne^{-1} cm^2 at 0°C, is four times that of ice I (13×10^{-12} dyne^{-1} cm^2), whereas liquid compressibilities near the melting point are generally no more than twice those of the corresponding solids (see Bridgman, 1942). Compressibilities of the dense forms of ice can be estimated from the PVT data of Bridgman (1912; 1935; 1937); they are less than the compressibility of ice I, and decrease progressively with decreasing specific volume, as is observed in other polymorph systems. Liquid water is in conspicuous violation of this correlation between density and compressibility. Its abnormally large compressibility is in part "structural," in the sense that it reflects a basic change in structure or molecular arrangement under pressure, in addition to the normal compressibility due to elimination of volume-consuming defects ("holes") and to squeezing of the molecules uniformly closer together under pressure. This structural compressibility (Frank and Quist, 1961; Lawson and Hughes, 1963, p. 217) must arise from the increased stabilization under pressure of the denser molecular configurations. The proportion of ice-I-like configurations must progressively decrease and of denser icelike configurations progressively increase under pressure, so that at any particular pressure, molecular groupings similar to those occurring in the solid phase at about the same pressure tend to be favored.

Since the structural compressibility involves a molecular rearrangement in which bonds must be broken and reformed, it can be expected that at very high frequencies of oscillatory compression, the structural changes will be too sluggish to respond and the compressibility will drop to an "unrelaxed" value characteristic of the individual structures present. The required frequency is too high for direct observation of this effect in pure water by present ultrasonic techniques. However, by an extrapolation from glycerol-water mixtures, Slie, Donfor, and Litovitz (1966) arrive at an unrelaxed compressibility for water of 18×10^{-12} dyne^{-1} cm^2 (see also Hall, 1948). This is somewhat larger than the compressibility of ice I, but not unreasonably so in view of the necessary presence of defects in the liquid structure. The remaining 33×10^{-12} dyne^{-1} cm^2 is then the structural compressibility of liquid water (at 0°C).

DENSITY FLUCTUATIONS

Microscopic fluctuations in thermal energy lead to fluctuations in local density and pressure. Applied to water at 0°C, the elementary theory of fluctuations (Davidson, 1962, p. 267) indicates that the r.m.s. density fluctuations $\langle(\delta\rho)^2\rangle^{1/2}$ in a volume containing on the average n molecules are given by

$$\langle(\delta\rho)^2\rangle^{1/2} = \frac{0.25}{\sqrt{n}} \text{ g cm}^{-3}$$

For $n = 6$ to 18, this relation gives $\langle(\delta\rho)^2\rangle^{1/2} = 0.10$ to 0.06 g cm^{-3}. These are density fluctuations of about the size required for mixtures of ice-I-, ice-II-, and ice-III-like structures. The choice $n = 6$ represents the smallest compact molecular grouping around a given central molecule (in 4-coordination with fixed bond length) for which a marked variation in local packing density can be recognized. When $n = 18$, a bonded coordination group is allowed in which all second-nearest neighbors of a central molecule are represented, and one topologically third-nearest neighbor is also included; it thus allows a short nonbonded distance in the range 3.2–4.0 Å, such as occurs in ice II, by a doubling back of the bond framework as described earlier. The features of coordination mentioned for $n = 6$ to 18 are the minimal ones necessary in a characterization of the local coordination as typical of one particular form of ice or another. Thus a representation of water in terms of regions of various icelike structures is only marginally permitted by the theory of fluctuations: molecular groupings large enough to be clearly recognized as representative of particular ice structures tend to be somewhat larger than should be common, according to the theory, at the density contrasts needed for the different ice structures.

The theory of fluctuations as represented above takes no account of any actual peculiarities of microscopic structure that might result in a tendency for the local density in the liquid to assume preferentially certain values larger or smaller than the mean. Any such tendency to "clustering," which in water might arise from the fact that the known icelike structures are energetically more stable than arrangements of intermediate density, should result in density fluctuations somewhat larger than predicted above.

Experimental information on density fluctuations can be obtained by means of light scattering (Mysels, 1964) or small-angle X-ray scattering (Weinberg, 1963). In both cases, the scattered intensity is found to be about 10 percent larger than predicted on the basis of fluctuation theory. The data thus suggest somewhat larger density fluctuations, but apparently rule out large clusters ($n \sim 100$) with density contrast $\langle(\delta\rho)^2\rangle^{1/2} \sim 0.2$ g cm^{-3}.

Further evidence of clustering is provided by homogeneous nucleation in the freezing of water (Langham and Mason, 1958). At the homogeneous nucleation temperature (-41°C for water drops 1μ in size), ice clusters that form in the course of thermal fluctuations are large enough to continue to grow and hence nucleate freezing. From the well-known relation between cluster surface energy σ_{SL} and critical cluster radius r for spontaneous growth,

$$r = \frac{2\sigma_{SL}}{t\rho\Delta S_f}$$

where $-t$ is the homogeneous nucleation temperature in °C and ΔS_f is the entropy of fusion, the critical cluster radius is $r = 7.5$ to 14.2 Å (based on different published estimates of σ_{SL}), corresponding to $n = 50$ to 250. Clusters of this size must therefore occur with reasonable frequency at -41°C in a 10^{-12} cm^3 volume of water; smaller clusters will be more abundant. According to the theory of fluctuations, the possible cluster size for given density contrast increases with temperature, but other structural considerations suggest instead a breakdown of cluster size with rising temperature.

The homogeneous nucleation behavior of water is similar to that of molten metals and salts (Ubbelohde, 1965, p. 284), indicating that solid-like clusters are not anomalously large in water, in relation to these other liquids. This similarity is in agreement with the fact that for other substances showing high-pressure polymorphism, a "premonitory" decrease in liquid volume often occurs, just as for ice I, leading to negative ΔV_f at pressures well below the solid-solid transformation. The effective averaging of the liquid structure over a range of pressures must be realized physically in the existence of clusters or fluctuations of different structure and density.

DIELECTRIC CONSTANT

The dielectric constant provides information on the hydrogen-bonded structure of water when analyzed in terms of the Kirkwood correlation factor g (Kirkwood, 1939; Cole, 1963). The value of g is determined by correlations imposed, via the bond framework, on the relative orientations of nearby molecular dipoles. For tetrahedral H-bonded frameworks it is near 3, as estimated theoretically for the ice I structure (Powles, 1952; Haggis, Hasted, and Buchanan, 1952). The measured g value for water at 0°C, 2.9 (Hasted, 1961), may be compared with values for the ice polymorphs (Wilson et al., 1965; Whalley, Davidson, and Heath, 1966) given in Table 1.

These data indicate that hydrogen bonding remains an important feature of liquid water. The modest decrease in g on melting is in accord with the appearance in the liquid of denser icelike structures, whose g values are generally lower than that of ice I (Table 1). The value $g = 2.9$ for water at 0°C could be accounted for by mixing ice I, II, and III, provided that the ice-II-like regions are mainly proton-disordered, rather than ordered as in the stable solid phase at lower temperatures. It is likely that, because of the tendency to energy stabilization of the proton-ordered arrangement in the ice II structure, a proton-disordered ice II phase would have an effective g value somewhat less than that of ice III.

This interpretation is not unique, for the observed g value of water could be achieved by various mixtures of known or hypothetical icelike structures. However, the observed g in relation to the known g values of ice structures rules out as major constituents any structures having g values near unity. In particular, a hypothetical "non-hydrogen-bonded liquid" in which the molecules interact only by dipole-dipole forces in addition to the spherically symmetrical interactions (Némethy and Scheraga, 1962) is ruled out as a major constituent, since g for such a liquid would be 1 or approximately 1 (Cole, 1963, p. 2609).

The decrease of g with temperature for water (to 2.6 to 100°C) contrasts with the temperature-independence of g shown by the forms of ice, and is evidence that the hydrogen-bonded structure of water changes progressively with temperature. Insofar as the relative proportions of the various icelike, hydrogen-bonded configurations in water should tend to change less and less at higher temperatures, as noted above in relation to the thermal expansion, and insofar as g for water at 70°C is already about as low as can be explained by combinations of ice phases, the further decrease in g up to 100°C and beyond (see Quist and Marshall, 1965) requires an additional type of structural change. The obvious change is a progressive breakdown by incorporation of structural defects, which will tend to lower the correlations imposed on the relative molecular orientations by the bond framework, and hence will lower g.

MOLECULAR VIBRATIONS

Spectra of the O—H or O—D stretching vibrations for HDO in dilute solution in D_2O or H_2O are highly informative of hydrogen bonding in liquid water (Wall and Hornig, 1965). Peak frequencies of the infrared absorption bands (Falk and Ford, 1966) and Raman bands (Wall and Hornig, 1965) are given in Table 2, for comparison with the corresponding data for ice phases. On the basis of these frequencies, the strength of hydrogen bonding in water is comparable to that in dense forms of ice, and there is a marked contrast with non-hydrogen-bonded water, either as water vapor or as water in CCl_4 solution.

In Figure 12 are plotted the infrared and Raman bands for the OH vibration of HDO in water, at 25°C. While the two bands are very similar, the infrared band is shifted to slightly lower frequencies, as a result of the increase in infrared absorption intensity with H-bond strength; since the Raman band is free from this complication, it is the better curve to compare with ice vibration frequencies in considering the distribution of H-bond strengths in water (Wall and Hornig, 1965).

The range of OH frequencies in water is much greater than in any of the individual ice phases, as shown in Figure 12, where the ice HDO absorption bands are compared with the water band. This indicates that a much wider range of molecular configurations is present in water than in any individual ice phase.

A combination of ice structures could account in a general way for the shifted peak frequency of the water band, and could account for a considerable

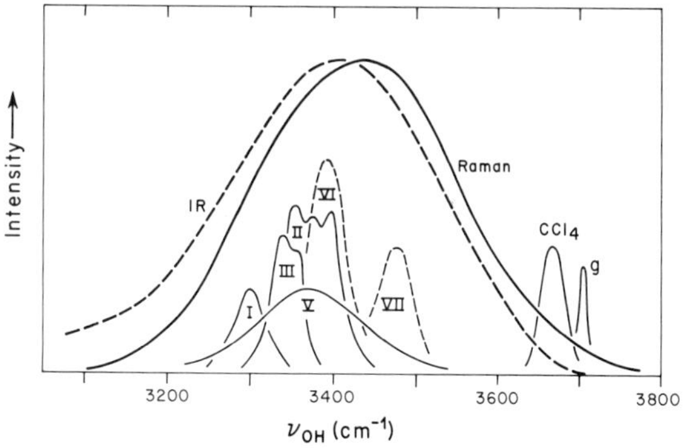

FIGURE 12.
Comparison of OH stretching vibration bands for HDO in water and in the ice polymorphs. The band for ice I is plotted in its position at 0°C (Falk and Ford, 1966), and the bands for the high-pressure ice phases are plotted relative to ice I, from measurements at 77°K (Bertie and Whalley, 1964 a, b); this tends to remove the effects of thermal expansion between 77° and 273°K in comparing the ice and water bands. The bands for ice VI and VII (dashed) are schematic only, being derived from the Raman frequencies for the pure H_2O and D_2O phases (Marckmann and Whalley, 1964) by empirical correction (Table 2); the halfwidths of these bands have not yet been reported, and are probably larger than shown. Band intensity is plotted in arbitrary units, the vertical scale not being the same for the different H_2O phases. The infrared band (Falk and Ford, 1966) and Raman band (Wall and Hornig, 1965) for HDO in water are at 26°C and 27°C, respectively. The frequency shown for HDO in CCl_4 solution is from Falk and Ford (1966), and for HDO vapor (labeled "*g*") from Benedict, Gailar, and Plyler (1956).

spread of frequencies within the band, but the needed combination would have a density much greater than that of liquid water, even allowing for a normal expansion of about 10 percent on melting. A simple combination of ice I, II, and III that has about the correct density, such as the one used in Figure 10, would give an absorption band with a peak frequency that is too low by about 100 cm^{-1}, and a halfwidth too small by about 150 cm^{-1}. The presence of some low-density H-bonded structures would, however, make it possible to satisfy the density requirement in icelike mixtures capable of accounting for the main OH frequencies of liquid water. Possible icelike structures less dense than ice I are the ice VI structure with one of the two independently bonded frameworks omitted, and the frameworks that occur in the various clathrate hydrates (Pauling, 1960, p. 469; McMullan and Jeffrey, 1965; Mak and McMullan, 1965; Jeffrey, 1967). The clathrate hydrate frameworks have hydrogen bond lengths in the range 2.77–2.84 Å and O···O···O angles in the range 89–128°; the H-bond distortion in these frameworks is thus no larger than in ices II–VI, and the O—H frequencies should accordingly lie in the same range.

Neither the hydrate structures nor ices I–VI can be expected to contribute to the higher frequencies sufficiently to account for the high-frequency flank of the water Raman band, where ice VII is the only available contributor among known forms of ice or ice-related hydrate frameworks. It is evident in

Figure 12 that contributions from ice VII, with frequencies as there depicted, are unable to explain adequately the high-frequency flank of the Raman band of water; even if the halfwidth of the ice VII absorption band were greatly increased, the contribution from ice VII necessary to account for the upper flank of the water band would be inordinately large in relation to the other forms of ice. It therefore appears that a large part, if not all, of the high-frequency flank of the Raman band represents water molecules in arrangements such that the hydrogen bonds are more distorted by bending than they are in any form of ice or icelike hydrate, or are more distended by stretching than they are in any ice form except ice VII or VIII.

The extent in liquid water of a quasi-crystalline, icelike structure, in the sense that it can be referred to known ice structures or to known icelike hydrate structures, is thus limited to roughly the lower half of the Raman band, and hence to about 50 percent of the complete structure of the liquid. The most abundant hydrogen bonds in water at 0°C appear to be about as distorted as the bonds in ice VI, in agreement with the mean bond length as discussed earlier. Many of the more-distorted molecular arrangements are probably still icelike in the sense that they involve tetrahedrally linked H-bond frameworks, because the Kirkwood g factor (discussed earlier) is too large to allow an absence of tetrahedral bonding in 50 percent of the water structure.

The peak infrared and Raman frequencies for water at temperatures from 0° to 130°C, in relation to the peak nearest-neighbor distances from the r.d.f. (Figure 11), follow a continuation of the curve of frequency shift versus bond length defined by ices I–VI (Figure 8). This suggests that the bond distortions in water at these higher temperatures are simply accentuated versions of the ice-VI-like distortions common in water at 0°C.

As bond distortions increase, the distorted hydrogen bonds tend toward broken bonds, for which the expected O—H stretching frequencies are those of unbonded water molecules in water vapor or in solution in nonpolar solvents such as CCl_4. As can be seen from Figure 12, the proportion of hydrogen bonds actually broken in this sense in liquid water at 25°C is small (Stevenson, 1965; Wall and Hornig, 1965).

The smooth contour of the HDO Raman and infrared absorption bands for water (Figure 12) has been emphasized by Wall and Hornig (1965) and by Falk and Ford (1966) as evidence that the water structure cannot be built up from a small number of discrete "molecular species" or "well-structured" regions. A mixture of ice phases and monomers, with OH stretching bands as indicated in Figure 12, would clearly result in a composite stretching band showing some resolved peaks or shoulders.

The actual amount of resolution to be expected in a mixture of ice structures is, however, less than would appear from the relatively narrow O—H frequency band of ice I, with halfwidth $\sim$50 cm^{-1} (Haas and Hornig, 1960; Bertie and Whalley, 1964a). Thus, ice V has a broad and featureless OH stretching band with halfwidth $\sim$150 cm^{-1}, more than half the halfwidth of the water band (see Figure 12). This occurs in spite of the fact that ice V is "well-structured" in the sense of long-range order of the oxygen atoms, and that there are only four crystallographically distinct types of water molecules in the structure. The broadening and smoothing of the frequency band is due to proton disorder, which results in the presence of many more distinguishably different nearest-neighbor relationships between water molecules than would occur under proton order (Bertie and Whalley, 1964b; Kamb, Prakash, and Knobler, 1967). Conversely, the resolution of peaks in the frequency band for ice II is an indication of the proton order in this phase (Bertie and Whalley, 1964b). As a potential contributor to the water OH band at much higher temperatures, a proton-disordered ice II structure can be expected to show a broad and featureless band similar to that of ice V. Apparently a similar effect is observed in ice VI (Wilson et al., 1965, p. 2390). (The frequency band suggested for ice VI in Figure 12, on the basis of Raman spectra of the pure H_2O and D_2O solids, is probably too narrow.) Similar expectations apply to other H-bonded structures also, such as the highly distorted icelike structures that would contribute to the high-frequency flank of the water band.

It therefore appears that a mixture of several icelike structures can account for the featurelessness of the main part of the HDO Raman and infrared bands of water, as the consequence of an overlapping of broad bands for the individual structures. Wall and Hornig (1965, pp. 2079, 2086) indicate that the individual bands will also be broadened if the local regions are "poorly defined structurally," and that the bands of "well-structured" regions will be broadened to the extent that the structure is "broken down." However, there seems to be no quantitative information on the magnitude of these possible effects in relation to parameters of the actual short-range order of the liquid, such as the cluster size or fluctuation size n, representing the number of molecules in the typical icelike region. A cluster of size $n \sim 20$ to 30 is barely large enough to be recognizable as representing one particular solid phase or another, and in this sense the structure at this level of organization is inherently poorly defined, and the resulting bands presumably should therefore be broad. Broadening of this type seems required if the water structure contains a significant proportion of ice-I-like regions, which otherwise would necessarily give a resolvable peak or shoulder near 3300 cm^{-1} in the Raman and infrared bands of water. Such broadening would also help to explain the extension of the bands to even lower frequencies (Figure 12).

The observation of a well-defined libration band in water (Walrafen, 1964) gives independent evidence for hydrogen bonding. Although the average libration frequencies listed in Table 2 are necessarily rough, they show clearly that the H-bond weakening in water is comparable to that in the high-pressure phases of ice (data for ice VI are not available).

ENERGY

The energy added upon fusion of ice I, 1.44 kcal $mole^{-1}$ at atmospheric pressure, must be stored in the liquid in the form of well-defined structural features distinguishing the liquid from the solid. In molecular clusters of icelike type in which intermolecular force constants are about the same as those of ice I, energy storage relative to ice I is primarily in the form of potential energy of hydrogen-bond bending and stretching, and of intermolecular repulsion. If broken hydrogen bonds occur extensively enough to allow water molecules to undergo partial or complete free rotation or translation, the associated energy is held primarily in kinetic form, in addition to the potential energy of H-bond breakage. Between these extreme cases are molecular arrangements in which free rotation or translation does not occur, but in which the intermolecular force constants are significantly reduced; in this case there is increased vibrational energy in addition to any potential energy of bond distortion or breakage.

Spectroscopic data on the intermolecular vibration frequencies (Zimmerman and Pimentel, 1962; Ockman, 1958; Bertie and Whalley, 1964a; Walrafen, 1964) indicate little change from ice I to water. The peak frequency ν_R of the libration band of H_2O decreases from 795 cm^{-1} for ice I (at 0°C) to 705 cm^{-1} for water, and the prominent hindered translation frequency ν_T decreases from about 215 cm^{-1} to 175 cm^{-1}. If ν_R is interpreted in terms of a collection of Einstein oscillators, and if ν_T is treated as the maximum frequency of a Debye spectrum for the lattice vibrations, the energy increase resulting from the above frequency changes is only 0.15 kcal $mole^{-1}$.

There remains an energy of 1.3 kcal $mole^{-1}$ to be stored in the form of other structural changes. Since this energy exceeds the energies of any of the ice phases, the various icelike molecular arrangements can be abundantly present in water, as far as available energy is concerned. But by the same token, icelike structures alone are insufficient to account for the energy storage in water. A mixture of ice I, II, and III that accounts for the density of water would have an energy of only about 0.1 kcal $mole^{-1}$ relative to ice I, and an equal mixture of all of the ice phases would have an energy of only about 0.3 kcal $mole^{-1}$.

The structural possibilities for storing an appreciable part of the energy of fusion as potential energy of H-bond bending depend on the value of the effective bond-bending force constant K_θ (Eq. 1).

If the low force constant $K_\theta = 0.9$ cal $mole^{-1}$ deg^{-2} were appropriate to most of the bonded molecular configurations occuring in liquid water, then very large average distortions ($\langle(\delta\theta)^2\rangle \sim 1.4 \times 10^3$ deg^2) would be necessary to account for storage of the fusion energy by bond-bending. Such extreme distortions involve r.m.s. bond-angle departures of more than 30° from the tetrahedral, and hence require the frequent occurrence of configurations in which the distance between a water molecule's bonded neighbors is about as short as the H-bonds themselves. This situation would arise for bifurcated hydrogen bonds, and for three-ring connections in an H-bonded framework; neither of these features occur in forms of ice, although bifurcated H-bonds are known in other substances. An extreme case is the situation in ice VII, where O—O—O angles of 70° and 180° are presented to each water molecule by some of its eight near neighbors; however, as noted earlier, these angles do not qualify as distorted bond angles, and the increased energy in this case is due to bond stretching and intermolecular repulsion, rather than to bond bending.

If the value $K_\theta = 5.2$ cal $mole^{-1}$ deg^{-2} in Eq. (1) were appropriate to most of the distorted H-bond configurations that occur in the liquid, then an average coordination distortion in the liquid of only $\langle(\delta\theta)^2\rangle = 0.25 \times 10^3$ deg^2, about the same as that in ice II and III, would be sufficient to account for the energy of fusion. However, this value of K_θ corresponds to the shear elasticity of ice I. Since the energy storage contributed by elastic distortion is already contained in the vibrational energy discussed previously, it must be assumed that the distorted molecular configurations that contribute large H-bond strain energies to the liquid structures are not simply shear distortion of ice-I-like arrangements, but are instead distorted configurations in local mechanical equilibrium (local minima in potential energy), which lack the special energy-stabilization features of the known dense ice phases, reflected in their much lower K_θ of approximately 0.9 cal mole^{-1} deg^{-2}.

It is likely that many of the molecular configurations in liquid water are intermediate in character between the two extremes just discussed. Such configurations would be represented by points in Figure 7 lying between the upper, dashed line and the lower, dotted line. Because of the increase in van der Waals energy stabilization with increasing density, it can be assumed that most of these configurations are less dense than the ice phases of comparable distortion. The clathrate hydrate frameworks are possible examples, although the energies that have been estimated for the best known of these structures are small (0.16 kcal mole^{-1} for Structure I, 0.19 kcal mole^{-1} for Structure II: van der Waals and Platteeuw, 1959), the H-bond distortions being small also ($\langle(\delta\theta)^2\rangle = 0.02 \times 10^3$ deg^2 for Structures I and II).

With regard to conjectured molecular arrangements of higher energy, little that is definite can be said about actual structures or possible energies, since the corresponding solid phases are hypothetical, being precluded by the existence of the energetically more stable ice polymorphs. Possible examples are quartzlike and coesitelike frameworks. A structure derived from ice VI by omitting one of the two interpenetrating H-bonded frameworks will have an energy greater than that of ice VI by at least the amount of the stabilizing van der Waals interaction between each water molecule in one of the ice VI frameworks and its eight neighbors at 3.51 Å in the other framework. On the basis of dispersion and overlap parameters derived from the ice VII structure (Kamb, 1965b), this interaction amounts to about 0.7 kcal mole^{-1}, hence the "single-framework" ice-VI structure would have an energy of $0.4 + 0.7 = 1.1$ kcal mole^{-1}. This structure thus exemplifies the type of molecular arrangement that could store the fusion energy of water in the form of potential energy of H-bond distortion.

This is consistent with the Raman and infrared spectra of water, discussed earlier, according to which the typical H-bond distortion in liquid water at 0°C is comparable to that in ice VI. If we make the simple assumption that the average energy of bond distortion varies linearly with the shift in OH stretching frequency over the Raman band of water, then a proportionality factor of 11 cal mole^{-1} cm accounts for the energy of fusion. On this basis, the icelike configurations represented by the low-frequency flank of the Raman band contribute only modestly to the energy storage in water ($\sim$0.3 kcal per mole of water in such configurations), while configurations near the peak of the band contribute $\sim$1 kcal mole^{-1}, as expected for the "single-framework" ice-VI structure just mentioned. If the bond distortions in the local molecular arrangements in liquid water are about the same as those in ice phases at the same OH frequency shift, then the foregoing energies correspond to an effective bending force constant of $K_\theta = 2.3$ cal mole^{-1} deg^{-2}, intermediate between the limiting values 0.9 and 5.2 cal mole^{-1} deg^{-2} discussed earlier, and in rather good agreement with the bending force constant $K_\theta = 2.1$ cal mole^{-1} deg^{-2} used by Pople (1951) in a statistical-mechanical analysis of water structure from the "continuum" point of view.

The low energy contributions from the low-frequency flank of the Raman band must be balanced by large contributions ($\sim$2.5 kcal mole^{-1}) from the molecular arrangements represented by the high-frequency flank. The largest of these contributions, corresponding to the uppermost tail of the Raman band, should approach the energy for breakage of the H-bond in water, estimated as 4 kcal mole^{-1} (per H-bond) from the sublimation energy of ice (11.6 kcal mole^{-1}) in relation to that of "non-hydrogen-bonded water" (3.5 kcal mole^{-1}, by extrapolation of the heats of sublimation of H_2Te, H_2Se, H_2S). This limiting energy is rather higher than indicated by the linear relation between E and ν_{OH} assumed above, which is doubtless an oversimplification.

In molecular arrangements involving broken H-bonds, kinetic energy can be stored in free molecular rotation or translation. This is a fairly effective energy storage mechanism in the case of free rotation (0.66 kcal per mole of freely rotating molecules), because the librational energy of ice is low (0.16 kcal mole^{-1}), but for free translation there is actually a net energy loss of 0.34 kcal mole^{-1}, owing to the fact that the lattice vibrational modes in ice are extensively excited and the vibrational potential energies are lost in going over to free translation. The net energy storage in the free motions, 0.32 kcal mole^{-1}, is too low to make this possible as the main source for the fusion energy. A more effective energy storage may be achieved in arrangements with highly distorted but not broken bonds, or with too few broken bonds to allow free motions, such that the motions remain vibrational but the intermolecular force constants are significantly reduced. In the extreme case in which librational and hindered translational force constants are reduced practically to 0, an energy of 1.96 kcal mole^{-1} could be stored in this way. These effects are, however, limited to molecular arrangements represented by the upper tail of the Raman band, where the increase in potential energy is already large. As contributors to the fusion energy, these arrangements are also limited by the fact, noted earlier, that strong librational and hindered rotational frequencies, typical of hydrogen-bonded structures, are observed in liquid water. The highly weakened structures could be represented by broad bands extending to lower frequencies, but if so, these bands are not conspicuous enough to have been yet recognized as such in the spectra of water.

In summary, the fusion energy of water may be roughly pictured as follows. A general vibrational loosening of the structure (relative to ice I) contributes 0.15 kcal mole^{-1}. Molecular arrangements of the various icelike types, represented by the lower third of the Raman band, contribute on the average about $\frac{1}{3} \times 0.15 = 0.05$ kcal mole^{-1}. The central third of the Raman band represents the most distorted of the icelike arrangements (ices VI, VII), and other hydrogen-bonded arrangements that do not occur as solid phases because their H-bond strain energies are too large; they contribute roughly $\frac{1}{3} \times 1.1 = 0.4$ kcal mole^{-1}. To the upper third of the Raman band correspond arrangements in which the H-bond distortion becomes severe, approaching rupture of the hydrogen bonds. An average energy of 2.5 kcal per mole of molecules in these arrangements is the combined result of high bond-strain energy (approaching the H-bond-breaking energy of 4 kcal mole^{-1}), increased vibrational energy for structures with weakened bonds, and rotational or translational kinetic energy, for molecules with more than two broken bonds. These arrangements contribute about $\frac{1}{3} \times 2.5 = 0.8$ kcal mole^{-1}, for a total energy storage of about 1.4 kcal mole^{-1} relative to ice I.

ENTROPY

The entropy of fusion is a manifestation of structural change essentially distinct from the energy. The entropy of fusion for water, 5.26 e.u., exceeds by about 2–3 e.u. the entropy change that is normal for "translational" melting, as indicated by the fusion entropies of most monatomic solids. This normal "translational" entropy of fusion (1.7–3.3 e.u.) includes the entropy of positional disorder related to three translational degrees of freedom ("communal entropy"), any mixing entropy associated with "holes" or other defects in the structure of the monatomic liquids, and vibrational entropy associated with a general "loosening" of the structure and a decrease in the hindered translation frequencies.

The observed change in hindered translation frequency from ice I to water, noted earlier, corresponds to an entropy increase of 1.16 e.u.,* which is probably contained in the normal "translational" fusion entropy component. The entropy increase due to change in the libration frequencies is, on the other hand, an extra contribution, but it amounts to only 0.2 e.u. The remaining excess entropy, while quantitatively uncertain owing to the range of possible translational entropies, is nevertheless a large and prominent quantity requiring a structural explanation in terms of any model of the water structure.

In a structural-mixture model of the liquid, as considered here, the sources of excess entropy are of two basic types: (1) excess entropy of the individual component structures; (2) an overall configurational entropy associated with the composite nature of the entire structure. Since all of the proton-disordered forms of ice have essentially equal entropies, no excess entropy of fusion can be associated with the icelike components themselves. Excess entropy in non-ice-like components can arise from molecular free rotation, or from a significant reduction of librational force constants below those of the icelike portions. The maximum contribution is for completely free rotation, and amounts to 9.4 e.u. (relative to the librational entropy of ice I); in case of free rotation around one axis, the contribution would be about 3 e.u. To account for the excess entropy of fusion from contributions of this kind would require 15–25 percent of the molecules to be in completely free rotation, or larger percentages ($\sim$50 percent) to be in partial free rotation or in highly distorted configurations with greatly reduced librational force constants. Such large percentages are incompatible with the infrared, ultraviolet absorption, and Raman spectra, which allow only about 5 percent (Wall and Hornig, 1965) or less (Stevenson, 1965) of the water molecules to be vapor-like in their freedom from H-bond interactions, and hence able to rotate freely.

It therefore appears necessary, in explaining the excess entropy of fusion, to appeal to an entropy associated with the composite structure of the liquid, a configurational entropy that arises from the multiplicity of local molecular

*This large entropy change must be regarded with some suspicion. Since the decrease in ν_T for the dense ice phases is comparable to that for water (Table 2), a comparably large vibrational entropy increase for the dense ice phases would be predicted on this basis, whereas the observed increase is only about 0.3 e.u. (ice I $\rightarrow$ III). For the solid phases, decreased compressibility at the higher densities tends to have an effect on the vibrational entropy opposite to that of the change in ν_T.

arrangements that can occur in the liquid structure. If the liquid structure could be viewed in terms of a lattice consisting of definite regions, each containing n molecules, and each capable of being organized into m different, equally probable, local structural arrangements, then the configurational entropy would be $(R/n)\ln m$. In fact the configurational entropy must be larger than this by some unknown amount, because the individual "structurable" regions are not defined in advance but can be chosen in different ways, the sizes and shapes of the "structurable" regions are not fixed, and various orientations for the structures within each region are possible. The number of structural arrangements m required to account for a particular excess entropy is highly sensitive to the various unknown and uncertain factors, so that a quantitative discussion of the configurational entropy cannot be given. It nevertheless seems apparent from the above form of the entropy expression that significant contributions from this source can arise only if n is small ($\sim$10), and if m is large ($\sim$100). Such an m is almost an order of magnitude larger than the approximately fifteen known or hypothetical ice and icelike (hydrate) structures, but this is not particularly troublesome, since the solid phases represent only the most stable of an imaginably great number of possible molecular arrangements. However, when n is small, the enumeration of structures as crystallographically distinguishable entities becomes dubious. It thus appears that a composite structure made up of definite icelike regions is only marginally capable of providing the needed configurational entropy: as in the previously discussed case of cluster size in relation to fluctuation theory, the n needed to account for the entropy seems somewhat too small to allow clear identification of the local structure as that of particular ice phases.

In spite of this, one remains tempted to speculate that the excess entropy of fusion is primarily configurational rather than vibrational or rotational, and arises from a multiplicity of possible molecular arrangements that, on a much-restricted scale, is reflected in the extensive polymorphism of the solid phases.

The prominence of ice-VI-like structures relative to ice-I-like ones, as seen in the Raman band of water, implies a relatively much greater number of available structures with distorted bonds, giving effectively a stabilizing configurational entropy that offsets the higher energy of these structures. This is perhaps reminiscent of the observation by Bell and Dean (1966) that the building of random tetrahedral models (for SiO_2) becomes more difficult when the available bond-bending is small. As it stands, this observation applies to a "random network" or "continuum" view of the water structure, rather than to a "quasi-crystalline" view based on solid-like molecular arrangements, but the greater freedom in building random models may well be reflected in a greater diversity in possible nonrandom (crystalline) structures with the same extent of bond distortion.

DEFECT STRUCTURE OF LIQUID WATER

A necessary feature of composite liquid structures built up of "crystalline" molecular arrangements is the existence of surfaces or zones of discontinuity between the solid-like, structured regions. The discontinuities are due to incompatibilities in the schemes of bond connectivity between adjacent structures of different types, and also between adjacent structures of the same type in different orientations. Their existence is thus necessary for the same reason that grain boundaries are present as intercrystalline discontinuities in polycrystalline solids. The composite structure of the liquid might therefore be viewed as a "micropolycrystalline" solid structure with "grain size" of the order of lattice dimensions of the solid phases, this arrangement being stabilized relative to the crystalline solids primarily by the configurational entropy of the composite structure. In this view of the liquid structure, the fusion energy is primarily stored in the bond distortions of the "grain boundary regions," whereas the stabilizing entropy arises primarily from the multiplicity of solid-like bonded structures represented within the regions between the "grain boundaries."

The "grain boundary regions" or zones of discontinuity between icelike molecular clusters in the liquid are regions where there must be considerable bending and stretching distortion of the hydrogen bonds, and where broken bonds must occasionally occur. It is natural to equate these distorted regions with the "defects" in the liquid structure, which have been mentioned from time to time in the preceding discussion. Defects have a major influence on many of the physical properties of water, particularly on the various relaxation and transport properties. Indeed, relaxation behavior is the structural manifestation that typifies water as a *liquid* rather than a glass.

The solid phases of H_2O have a well-defined, stable defect structure, which is responsible for their dielectric relaxation, mechanical relaxation, and self diffusion. The decrease by many orders of magnitude in the relaxation times for these properties in going from the solids to liquid water* shows that there is a vast quantitative if not qualitative change in the nature of the defect structure. Because of this, the detailed picture that has been developed of the defect structure in ice (Gränicher, 1963; Wilson et al., 1965) cannot be directly utilized in interpreting relaxation phenomena in water. For the solids, grain boundaries are thermodynamically unstable; the stable defect structure consists of point defects (vacancies, L and D defects, ion states, and perhaps interstitial molecules) and line defects (dislocations). If the predominating defects in water are the "grain boundary" discontinuity zones between icelike molecular clusters, as suggested above, then the defect structure of the liquid is needed qualitatively different from that of the solids. In this view, relaxation processes and structural reorganization in water take place by propagation of the cluster boundary zones through the structure.

The implied dynamic nature of the cluster boundaries is in harmony with the extensive weakening or breakage of hydrogen bonds hypothesized in these zones, which makes possible the molecular rotations needed to reorganize the structure from that of one cluster to the adjacent one, as the boundary zone propagates. In this respect the cluster boundaries do not resemble closely the actual grain boundaries in polycrystalline ice I, which, though they do migrate, are under normal conditions vastly less mobile than the cluster boundaries must be in water.

If the cluster boundaries have a surface energy of 80 erg cm^{-2}, corresponding to breakage of about half the bonds across arbitrary surfaces in ice I or to less severe distortions of a larger fraction of the bonds (compare surface tension of water, 76 erg cm^{-2}), then the cluster size implied by an 0.8 kcal mole^{-1} defect contribution to the energy of fusion (see earlier discussion) is $n \approx 100$. This

*Compare the dielectric relaxation times for ice phases (Table 1) with the value 17×10^{-12} sec for water at 0°C (Hasted, 1961).

is probably an upper limit, and is sensitive to the assumed surface energy.

Of fundamental importance in the relationship between the structures of solid and liquid H_2O is the question of the average density of the defect regions in the liquid structure. Many theories of water structure have assumed a priori that breakage of the hydrogen bonds in ice I will lead to an increased density for a resulting "unbonded" or "broken-down" structure, so that the increase of density on melting can be achieved by mixing ice-I-like and "unbonded" regions. The structures of the solid phases cannot provide a direct picture of the detailed molecular organization (and hence of the density) in the defect regions, because the required extent of bond distortion or breakage does not occur in any of the solids. A discussion must therefore depend on hypothetical structural models and on calculations from intermolecular forces.

In speaking of broken hydrogen bonds, it is best to distinguish two general types of relationships between water molecules: (1) nonbonded contacts, of the general type present in the relationship between nonbonded nearest neighbors in ice VII; (2) antibonded contacts, which occur where protons of adjacent water molecules tend to oppose one another, or where acceptor electron pairs tend to oppose one another, as in the D and L defects in ice. From the intermolecular potentials derived for ice VII (Kamb, 1965b), the equilibrium distance for pairs of molecules in nonbonded relative orientation is found to be 3.5–3.7 Å. For molecules in antibonding orientation, the equilibrium distance will be at least the van der Waals contact distances of 2.8 Å for L contacts and 4.4 Å for D contacts, but will probably be considerably larger because the dipole-dipole interaction is always repulsive for the antibonding molecular configurations. In the extreme case in which the dipole moment vectors oppose one another, the repulsive dipole-dipole interaction outweighs the van der Waals attraction at all distances, so that there is no finite equilibrium distance. It therefore seems conservative to estimate that the average equilibrium separation among water molecules whose hydrogen bonds have been broken will be at least 3.6 Å. The actual arranging of water molecules into a crystal structure so as to form no hydrogen bonds is only a moot possibility, but if this difficulty is overlooked and the molecules are simply stacked in closest packing at separation 3.6 Å, the resulting density is 0.91 g cm^{-3}, slightly less than the density of ice I. The density of an "unbonded liquid" derived from this "unbonded solid" would presumably decrease by about 10 percent due to incorporation of "holes," as in melting of the (close-packed) noble gases and molecular crystals generally, hence giving a density of about 0.8 g cm^{-3}. There is thus no indication that a "broken-down" structure considerably denser than ice I can result from breakage of the hydrogen bonds among the water molecules; the defect regions of the water structure are probably regions of somewhat lower density than the average, rather than higher.

It is also pertinent to estimate the equilibrium intermolecular distance if the H-bonds are broken dynamically, by free rotation of the water molecules, rather than statically, as considered above. The density of a "dynamically unbonded" H_2O liquid, with the molecules in essentially free rotation, can be estimated as 0.8 g cm^{-3} by extrapolating the packing efficiencies (in relation to molecular size as estimated from the van der Waals radii) of liquid H_2Te, H_2Se, and H_2S. Working in the direction of greater intermolecular attraction and greater density for a dynamically unbonded H_2O liquid is the possible effect of dipole-dipole interactions among the water molecules, either by dipole-dipole correlation, in the case of free rotation (Keesom force), or by static dipole-dipole interaction, in the case of partial free rotation about the dipole axis (Némethy and Scheraga, 1962). These effects would be difficult to analyze adequately, and there has been no demonstration of the reasonableness of an equilibrium distance of about 3.2 Å, which is needed in an unbonded structure dense enough to account for the density of water by mixing with equal quantities of ice I (Némethy and Scheraga, 1962.) The Keesom force is open to suspicion anyway, since the dipole-dipole interaction for water molecules at 3.2 Å (2.9 kcal mole^{-1}) is so much larger than kT (0.54 kcal mole^{-1} at 0°C) as to prevent free rotation except as a rare event. But even if a "dynamically unbonded" distance of 3.2 Å were hypothetically reasonable, the peak nearest-neighbor distance of only 2.84 Å in water at 0°C (Figure 11) rules out the presence of anything like 50 percent of the molecules in a structure with about eight nearest neighbors at 3.2 Å.

The effect of pressure on the water structure should provide a test of whether the defect regions are denser or less dense than the remainder of the structure, but, as with other properties of water, many complicated factors enter, and clear-cut experimental indications are difficult to obtain. The celebrated decrease in viscosity with pressure (up to about 1 kb), discovered by Bridgman (1949), has often been cited as evidence of a "breakdown" of the water structure under pressure, implying that the "broken-down" defect structure is denser than 1.0 g cm^{-2}. However, the dielectric relaxation times of the dense forms of ice are two orders of magnitude lower than that of ice I (Table 1), and ice III is much more plastic than ice I (Brace and Kamb, unpublished work). One can therefore expect a general quickening of relaxation phenomena, as denser H-bonded structures tend to replace less dense ones under pressure.

A rather different type of defect structure involving molecular free rotation consists of freely rotating water "monomers" located within clathrate-type cages of hydrogen-bonded water molecules (Pauling, 1960, p. 473; Frank and Quist, 1961; Danford and Levy, 1962). Such arrangements would be as stabilizing for possible dense solid phases as for the liquid structure: hence the fact that they do not occur among the dense polymorphs of ice makes them unattractive as structural features for liquid water. Moreover, as defect contributions to the water structure, they have the added drawback of providing in themselves no mechanism for rapid relaxation of the bulk of the structure, contained in the clathrate frameworks.

Closely tied to the defect structure of water is the remarkable doubling of the heat capacity from ice to water, which contrasts greatly with the modest increase (less than 30 percent) for most substances on melting. The change in intermolecular vibration frequencies accounts for only 0.6 cal deg^{-1} mole^{-1} of the increase from 9.1 (ice I) to 18.0 cal deg^{-1} mole^{-1} (water). The remaining 8.3 cal deg^{-1} mole^{-1} must be absorbed in structural change in the liquid as a function of temperature. Some of this can be accommodated in an increasing proportion of the more energetic, denser icelike structures, as necessary to explain the thermal contraction up to 4°C. However, this contribution can be only a modest one, because of the relatively low energies of the ice phases. As with the energy of fusion, the bulk of the extra energy must be used to generate molecular arrangements more distorted and more energetic than any of the ice structures (except ice VII). This is seen in the behavior of the infrared and

Raman bands (Falk and Ford, 1966; Wall and Hornig, 1965), which shift slowly to higher frequency, with little change in halfwidth, as the temperature is raised. This implies that the nature of the structural organization remains basically the same, but the proportions of the different types of structures change. The greatest increase occurs in the upper flank of the Raman band, the part which we have attributed to the defect structures in the previous discussion, hence it follows that the high excess heat capacity of water is mainly a consequence of the generation of additional defects in the structure as the temperature rises.

The estimated excess heat capacity integrated from 25°C to 120°C is accounted for, in relation to the observed shift in the IR bands over this same temperature interval (Falk and Ford, 1966), by a proportionality factor of 9 cal $mole^{-1}$ cm in the simple linear relation assumed earlier between bond distortion energy and OH frequency. The approximate agreement of this figure with the 11 cal $mole^{-1}$ cm appropriate for the fusion energy supports this rough assignment of bond distortion energy over the vibration band and implies, as before, an energy of about 2 to 4 kcal $mole^{-1}$ (relative to ice I) for the "defect" molecular arrangements represented by the upper flank of the band.

SUMMARY AND CONCLUSION

The long-considered idea that liquid water is in many respects icelike in structure is here reexamined in the light of recent information on the structures and physical properties of the several ice polymorphs. No attempt is made here to relate this approach to the many and diverse theories of water structure that have been proposed, but a relationship is implicit in the structural considerations involved. A number of theories have postulated icelike contributions to the water structure, but have considered only the ice I structure as a possible contributor. This restriction immediately forces the postulation, in such theories, of a "broken-down" structural component that must be significantly denser than 1.0 g cm^{-3} in order to account for the density of liquid water. In two recent theories (Marchi and Eyring, 1964; Jhon et al., 1966), a dense, hydrogen-bonded component introduced into statistical-mechanical calculations is designated as an "ice-III-like structure" or as "perhaps analogous to ice III," but some of the properties assumed or calculated for this component are unlike those of actual ice III. None of the theories have yet given close consideration to the particular structural features shown by the dense ice polymorphs.

The low-pressure ice phases (ices I and Ic) exhibit hydrogen bonding in classic form, with essentially perfect tetrahedral coordination of the water molecules. In the dense phases ice II-VI, tetrahedral hydrogen bonding is retained, but the coordination is distorted from the ideal tetrahedral, allowing an increase in the non-nearest-neighbor coordination and thus in the density. The distortion, measured by the departure of the coordination angles from 109.5°, increases progressively with increasing density, and is accompanied by a corresponding progressive lengthening of the hydrogen bonds. Energies and O—H stretching frequencies generally increase progressively with bond distortion. The observed energy increase is, however, much less than would be expected on the basis of the H-bond-bending force constant alone, indicating that the bond distortion energy is offset by stabilizing energy contributions, in particular by the increased van der Waals interaction at higher densities. Depending on the extent of this energy offsetting, there is wide latitude for other icelike, H-bonded structures with higher energies. Among these are structures less dense than ice I, such as the icelike frameworks of the clathrate hydrates.

The energy of fusion of ice I is larger than the energies of the dense ice phases relative to ice I, so that, as far as available energy is concerned, all of the ice structures can be abundantly represented in liquid water even at atmospheric pressure. Consistent with contributions of the denser ice structures to the liquid are the decrease of volume on melting, the increase in nearest-neighbor distance, the increase in number of non-nearest neighbors in the distance range 3.0–4.0 Å, and the modest decrease in dielectric response as reflected in the Kirkwood correlation factor *g*. The negative thermal expansion up to 4°C suggests continued increase in the proportion of denser bonded structures, which more than compensates the normal expansion due to incorporation of volume-consuming defects ("holes"). The high structural compressibility reflects an increase in proportion of the denser icelike structures under pressure, without any implication that the water structure under pressure becomes "broken down" or less basically icelike, as required in theories that overlook the contributions of the dense ice structures.

The peak O—H and O—D stretching frequencies in water at 0°C, and the peak H-bond length, correspond approximately to those of ice VI. The ice VI structure thus gives an idea of the typical extent of H-bond distortion in water. However, the lower density of water indicates a significant proportion of more open structures with comparable distortion, such as the structure obtained by omitting one of the two H-bonded frameworks in the actual self-clathrate structure of ice VI. Such structures are icelike in having completely H-bonded frameworks, although they do not occur as stable solid phases because of their higher energies. The higher energy contribution of such structures also helps to explain why the fusion energy (1.4 kcal mole^{-1}) is much larger than the energy of ice VI (0.4 kcal mole^{-1}).

The width of the O—H stretching vibration band shows that a considerably greater range of H-bond distortions is present in water than in any individual form of ice. The lower half of the stretching band of water corresponds to the range of ice structures (omitting ice VII), while the upper half of the band reveals the presence of molecular configurations with H-bonds more severely distorted than in any form of ice (except ice VII). The distortions extend to broken H-bonds, present in small amount. The energies of these severely distorted configurations, extending in the limit to 4 kcal per mole of broken bonds, make the major contribution to the energy of fusion.

In a quasi-crystalline structure for water built from icelike molecular clusters, the most severely distorted molecular configurations (corresponding to roughly the upper third of the O—H vibration band) must occur in the boundary zones between the icelike clusters. These zones are the predominating "defect" features of such a liquid structure. Relaxation and transport processes, such as dielectric relaxation and viscous flow, must take place mainly by the propagation of these defect zones through the structure. The known ice phases do not give a direct structural picture of the defects, but do provide indirect information indicating that the defect zones should be regions of relatively low molecular packing density. The high heat capacity of water, like the high fusion energy, is mainly due to energy storage in the defect regions, which grow with increasing temperature.

The excess entropy of fusion, on the other hand, is not mainly associated with the defects, as rotational or vibrational entropy. Instead, as speculated here, it is primarily a configurational entropy of the entire structure, arising from a multiplicity of possible icelike molecular arrangements that, on a restricted scale, is reflected in the extensive polymorphism of the solid phases. No attempt is made here to express this quasi-crystalline model of water structure in terms of a partition function, primarily because a method for properly evaluating the configurational entropy seems to be lacking. Existing statistical-mechanical theories of water structure have avoided this difficulty by introducing in effect only simple entropy contributions from rotation, translation (including "communal entropy"), and vibration, plus a questionable entropy of mixing among various molecular "species," treated as though they were distinguishable.

The meaningfulness of a quasi-crystalline model of water structure depends on the size of the "crystalline" regions or solid-like molecular clusters that have physical reality. The characteristic coordination features of most of the ice phases can be recognized and distinguished within regions the size of their unit cells, containing from 8 to 28 molecules, but below this size, molecular arrangements rapidly lose their identity as particular icelike structures. Indications of cluster size from fluctuation theory, light scattering, small-angle X-ray scattering, homogeneous nucleation, contour of the infrared and Raman bands, configurational entropy, and fusion energy in relation to cluster boundary energies, suggest consistently that the cluster size must be small but might lie in the range 10–100 molecules. Thus a quasi-crystalline discription appears only marginally possible.

Although the merits of the alternative "uniformistic" or "continuum" description (Pople, 1951) are not explored here, one expects that the local molecular structure will, in this description, continue to be governed by the same factors that determine the coordination relations of the solid phases, so that the bonded portions of the structure will have features often recognizable as fragments of the solid phases. Ice polymorphism remains indirectly significant in showing the extent of flexibility of the water molecule coordination and, in particular, the nature of coordination distortions that allow an increased density. The polymorphism also provides information on the energetics of bond distortion, which is an important factor governing the local molecular configuration. The low energies of the high-pressure ice phases, in spite of their large bond distortions, suggest a greater flexibility in hydrogen-bonded networks than seems to be visualized in the most widely quoted versions of Frank's (1958) "flickering cluster" concept of water structure, which presupposes a rather unique energy stabilization for icelike structures with perfect tetrahedral coordination.

An important aspect of water structure not considered here is the modification of the structure by solute molecules or ions. To the extent that one may hope to understand the structure of the pure liquid in terms of its relationship to the solid phases of H_2O, one may also hope to understand the modifying effects of solutes in terms of the structural features shown by crystalline hydrates (Clark, 1963; Baur, 1965; Jeffrey, 1967).

ACKNOWLEDGMENTS

This work was supported by grants from the National Science Foundation and from the Sloan Foundation, and was done in part during tenure of the Crosby visiting professorship at the Massachusetts Institute of Technology. I am grateful also for informative and stimulating discussions with Linus Pauling, who awakened my interests in ice polymorphism and water structure in the first place.

REFERENCES

Auty, R. P., and R. H. Cole (1952). *J. Chem. Phys.* **20**, 1309.
Baur, W. H. (1965). *Acta Cryst.* **19**, 909.
Beaumont, R. H., H. Chihara, and J. A. Morrison (1961). *J. Chem. Phys.* **34**, 1456.
Bell, R. S., and P. Dean (1966). *Nature* **212**, 1354.
Benedict, W. S., N. Gailar, and E. K. Plyler (1956). *J. Chem. Phys.* **24**, 1139.
Bernal, J. D., and R. H. Fowler (1933). *J. Chem. Phys.* **1**, 515.
Bertie, J. E., and E. Whalley (1964a). *J. Chem. Phys.* **40**, 1637.
Bertie, J. E., and E. Whalley (1964b). *J. Chem. Phys.* **40**, 1646.
Bertie, J. E., L. Calvert, and E. Whalley (1964). *Can. J. Chem.* **42**, 1373.
Bridgman, P. W. (1912). *Proc. Amer. Acad. Arts Sci.* **47**, 441.
Bridgman, P. W. (1935). *J. Chem. Phys.* **3**, 597.
Bridgman, P. W. (1937). *J. Chem. Phys.* **5**, 964.
Bridgman, P. W. (1942). *Proc. Am. Acad. Arts Sci.* **74**, 399.
Bridgman, P. W. (1949). *The Physics of High Pressure.* London: G. Bell and Sons.
Brown, A. J., and E. Whalley (1966). *J. Chem. Phys.* **45**, 4360.
Clark, J. R. (1963). *Aust. J. Pure Appl. Phys.* **13**, 50.
Cohen, E., and C. J. G. van der Horst (1938). *Z. phys. Chem.* B **40**, 231.
Cole, R. H. (1963). *J. Chem. Phys.* **39**, 2602.
Conway, B. E. (1966). *Ann. Rev. Phys. Chem.* **17**, 481.
Danford, M. D., and H. A. Levy (1962). *J. Am. Chem. Soc.* **84**, 3965.
Davidson, N. (1962). *Statistical Mechanics.* New York: McGraw-Hill.
Falk, M., and T. A. Ford (1966). *Can. J. Chem.* **44**, 1699.
Frank, H. S. (1958). *Proc. Roy. Soc.* (London) A **247**, 481.
Frank, H. S., and A. S. Quist (1961). *J. Chem. Phys.* **34**, 604.
Gränicher, H. (1958). *Z. Krist.* **110**, 432.
Gränicher, H. (1963). *Phys. kondens. Mat.* **1**, 1.
Haas, C., and D. F. Hornig (1960). *J, Chem. Phys.* **32**, 1763.
Haggis, G. H., J. B. Hasted, and T. J. Buchanan (1952). *J. Chem. Phys.* **20**, 1452.
Hall, L. (1948). *Phys. Rev.* **73**, 775.
Hasted, J. B. (1961). *Progr. Diel.* **3**, 103.
Jeffrey, G. A. (1967). *Progress Inorg. Chem.* In press.
Jhon, M. S., J. Grosh, R. Ree, and H. Eyring (1966). *J. Chem. Phys.* **44**, 1465.
Kamb, B. (1964). *Acta Cryst.* **17**, 1437.
Kamb, B. (1965a). *Science* **150**, 205.
Kamb, B. (1965b). *J. Chem. Phys.* **43**, 3917.
Kamb, B., and S. K. Datta (1960). *Nature* **187**, 140.
Kamb, B., and B. L. Davis (1964). *Proc. Nat. Acad. Sci. U.S.* **52**, 1433.
Kamb, B., A. Prakash, and C. Knobler (1967). *Acta Cryst.* **22**.
Kavanau, J. L. (1964). *Water and Solute-Water Interactions.* San Francisco: Holden-Day.
Kennedy, G. C., A. Jayaraman, and R. C. Newton (1962). *Phys. Rev.* **126**, 1363.
Kirkwood, J. G. (1939). *J. Chem. Phys.* **7**, 911.
Klement, W., and A. Jayaraman (1967). *J. Phys. Chem. Solids.* In press.

Langham, E. J., and B. J. Mason (1958). *Proc. Roy. Soc.* (London) A **247**, 793.
Lawson, A. W., and A. J. Hughes (1963). In R. S. Bradley, ed., *High Pressure Physics and Chemistry*. New York: Academic Press, p. 207.
Lonsdale, K. (1958). *Proc. Roy. Soc.* (London) A **247**, 424.
Mak, T. C. W., and R. K. McMullan (1965). *J. Chem. Phys.* **42**, 2732.
Marchi, R. P., and H. Eyring (1964). *J. Phys. Chem.* **68**, 221.
Marckmann, J. P., and E. Whalley (1964). *J. Chem. Phys.* **41**, 1450.
McMillan, J. A., and S. C. Los (1965). *Nature* **206**, 806.
McMullan, R. K., and G. A. Jeffrey (1965). *J. Chem. Phys.* **42**, 2725.
Morgan, J., and B. E. Warren (1938). *J. Chem. Phys.* **6**, 666.
Mysels, K. J. (1964). *J. Am. Chem. Soc.* **86**, 3503.
Narten, A. H., M. D. Danford, and H. A. Levy (1966). Oak Ridge Nat. Lab. Tech. Rept. ORNL-3997.
Narten, A. H., M. D. Danford and H. A. Levy (1967). *Disc. Faraday Soc.* In press.
Némethy, G., and H. A. Scheraga (1962). *J. Chem. Phys.* **36**, 3382.
Ockman, N. (1958). *Adv. Phys.* **7**, 199.
Owston, P. G. (1958). *Adv. Phys.* **7**, 171.
Pauling, L. (1935). *J. Am. Chem. Soc.* **57**, 2680.
Pauling, L. (1960). *The Nature of the Chemical Bond*, 3rd ed. Ithaca: Cornell Univ. Press.
Pauling, L., and R. E. Marsh (1952). *Proc. Nat. Acad. Sci. U.S.* **36**, 112.
Peterson, D. W., and H. A. Levy (1957). *Acta Cryst.* **10**, 70.
Pople, J. A. (1951). *Proc. Roy. Soc.* (London) A **205**, 155.
Powles, J. G. (1952). *J. Chem. Phys.* **20**, 1302.
Quist, A. S., and H. S. Frank (1961). *J. Phys. Chem.* **65**, 560.
Quist, A. S., and W. L. Marshall (1965). *J. Phys. Chem.* **69**, 3165.
Shimaoka, K. (1960). *J. Phys. Soc. Japan* **15**, 106.
Slie, W. M., A. R. Donfor, and T. A. Litovitz (1966). *J. Chem. Phys.* **44**, 3712.
Stevenson, D. P. (1965). *J. Phys. Chem.* **69**, 2145.
Tammann, G. (1926). *Z. anorg. Chem.* **158**, 1.
Taylor, M. J., and E. Whalley (1964). *J. Chem. Phys.* **40**, 1660.
Ubbelohde, A. R. (1965). *Melting and Crystal Structure*. New York: Oxford Univ. Press.
van der Waals, J. H., and J. C. Platteeuw (1959). *Adv. Chem. Phys.* **2**, 1.
Wall, T. T., and D. F. Hornig (1965). *J. Chem. Phys.* **43**, 2079.
Walrafen, G. E. (1964). *J. Chem. Phys.* **40**, 3249.
Weinberg, D. L. (1963). *Rev. Sci. Instr.* **34**, 691.
Weir, C., S. Bloch, and G. Piermarini (1965). *J. Res. Nat. Bur. Std.* **69C**, 275.
Whalley, E., and D. W. Davidson (1965). *J. Chem. Phys.* **43**, 2148.
Whalley, E., D. W. Davidson, and J. B. R. Heath (1966). *J. Chem. Phys.* **45**, 3976.
Wicke, E. (1966). *Angew. Chem.* (int. ed) **5**, 106.
Wilson, G. J., and D. W. Davidson (1963). *Can. J. Chem.* **41**, 264.
Wilson, G. J., R. K. Chan, D. W. Davidson, and E. Whalley (1965). *J. Chem. Phys.* **43**, 2384.
Zimmerman, R., and G. C. Pimentel (1962). In A. Mangini, ed., *Advances in Molecular Spectroscopy*. London: Pergamon Press, p. 726.

THE CHEMISTRY AND STRUCTURE OF SMALLER MOLECULES

RUTH HUBBARD
GEORGE WALD
Biological Laboratories of Harvard University
Cambridge, Massachusetts

Pauling and Carotenoid Stereochemistry

One of the admirable things about Linus Pauling's thinking is that he pursues it always to the level of numbers. As a result, there is usually no doubt of exactly what he means. Sometimes his initial thought is tentative because the data are not yet adequate, and then it may require some later elaboration or revision. But it is frequently he who refines the first formulation.

So it is with Pauling's "rule" concerning which of the double bonds in carotenoids can assume the *cis* configuration. This rule was a logical application of resonance theory and did much to stimulate progress in carotenoid chemistry, yet it aroused a certain amount of controversy in the late forties and early fifties. It raised a fundamental issue, which we wish to explore in this paper: the significance of planarity and the tolerance of aplanarity in conjugated systems.

In 1937, Zechmeister and his colleagues began the systematic isolation and identification of carotenoids from natural sources. As they tried to purify these molecules, they soon saw, as had Gillam and El Ridi (1935, 1936), that starting with a pure pigment they sometimes ended with several interconvertible forms. After considering a number of different explanations, Zechmeister and Tuzson (1938) concluded that these were *cis-trans* isomers about one or several of the conjugated double bonds.

Zechmeister and his colleagues tried next to identify these isomers (for reviews, see Zechmeister, 1944, 1962). It proved relatively easy to recognize the all-*trans* isomer, since in carotenoids the absorption maximum (λ_{max}) of this isomer lies at a longer wavelength and has a higher specific absorbance than those of the *cis* isomers. The all-*trans* isomer also is usually adsorbed most strongly in chromatograms. But the problem of assigning specific *cis-trans* configurations to the various other isomers present in the mixtures was at first sight overwhelming.

The usual C_{40} carotenoids contain 9, 10, or 11 conjugated double bonds in a straight chain, sometimes in addition to others whose steric configuration is fixed by their location in rings (see Figure 1). If all the straight-chain double bonds (n) were equally ready to assume the *cis* configuration, we would expect to find 2^n *cis* isomers—that is, 512, 1024, or 2048, depending on whether n is 9, 10, or 11. This leaves little hope of identifying the configura-

This work was supported in part by grants to G. W. from the Rockefeller Foundation, the Office of Naval Research, and the National Science Foundation, and to R. H. from the U.S. Public Health Service (N.I.N.D.B. Grant No. B-568).

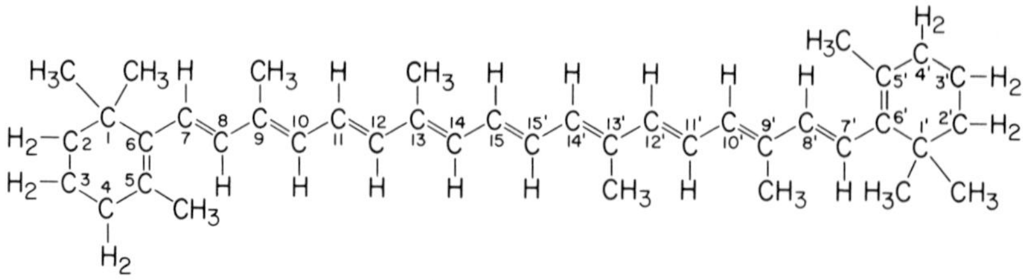

β-carotene

γ-carotene

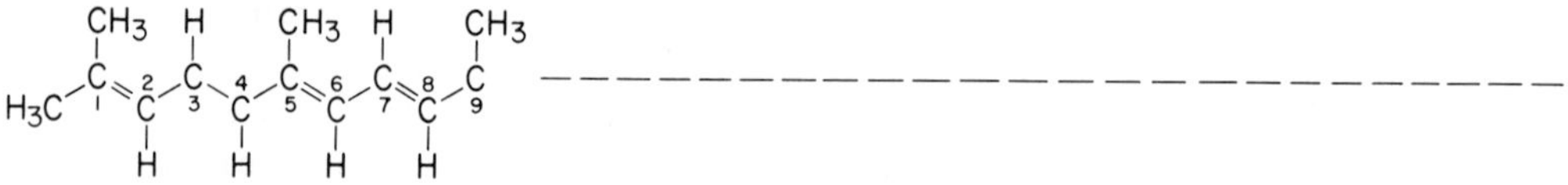

Lycopene

FIGURE 1.
Structural formulas of the all-*trans* isomers of β- and γ-carotene and of lycopene, showing the system of numbering carbon atoms. When adjacent to β-ionone rings, the 6,7 single bond is drawn in the s-*cis* configuration, an arbitrary choice since its configuration in β- and γ-carotene has not been determined. It is, however, known to be s-*cis* in the closely related molecule, all-*trans* 15,15′-dehydro β-carotene (Sly, 1964; see also Table II and text below).

tions of the individual isomers. The only chance for plausible assignments of configuration lay in the development of a rationale for rating this large number of possible isomers in terms of their relative probabilities of occurrence.

We can say immediately that even if all the double bonds were equally likely (on thermodynamic grounds) to assume the *cis* configuration, the *monocis* isomers would be much more probable than the *di-* or *polycis* forms. If, for example, we arbitrarily assign a probability of 0.1 to each *trans*⟶*cis* rotation, each of the *monocis* isomers would have a probability of 0.1, the *dicis* isomers a probability of 0.01, and so on.

Pauling introduced a further simplification by estimating the relative thermodynamic stabilities of the *monocis* forms on configurational grounds. His argument ran as follows (Pauling, 1939):

> As the result of resonance of the type
>
> $$\mathrm{C{=}C{-}C{=}C,\ \dot{C}{-}C{=}C{-}\dot{C}}$$
>
> each of the single bonds in the conjugated system has sufficient double-bond character to keep the entire system coplanar. It is this resonance stabilization of configuration about the single bonds which permits the easy crystallization of the carotenoids. The bond distances have average value 1.40 Å, with the double bonds somewhat shorter (1.35 to 1.39 Å) and the single bonds somewhat longer (1.42 to 1.40 Å), as shown by the

calculations of Coulson (32) quoted above. The bond angles along the chain can be confidently assumed to be close to 125° 16′, the value given by the theory of the tetrahedral carbon atom and supported by electron-diffraction results for simpler substances.

The configuration about the double bonds 5,6 and 5′,6′ is determined by the ring. The remaining double bonds are of several different kinds: (*a*) 15–15′; (*b*) 13–14 and 13′–14′; (*c*) 9–10 and 9′–10′; (*d*) 11–12 and 11′–12′; and (*e*) 7–8 and 7′–8′. [Pauling uses the same system of numbering carbon atoms as is shown in Figure 1.] Now let us consider a chain

—CX—CR═CR′—CX′—

with the cis configuration:

```
     R           R'
      \         /
       C═══════C
      /2       3\
  —C1            4C—
      \         /
        X     X'
```

From the structural parameters it can be calculated that if X and X′ are both hydrogen atoms they will be 1.7 Å apart. This is somewhat less than the usual distance of van der Waals contact (2.0 to 2.4 Å), and will tend to make the molecule less stable than for the trans configuration (with R, R′ = H); but the strain could be removed in large part by small rotations about the bonds 1,2 and 3,4. On the other hand, with X = methyl or some similar group and X′ = H the distance is only 1.6 Å; this is so very small compared with the expected distance of van der Waals contact, 3.2 Å, that the cis configuration would surely be unstable. Hence we conclude that *the cis configuration will be assumed only by those double bonds which are of the type* —CH—CR═CR′—CH—; that is, which are adjoined by two CH groups. Hence the bonds of types *d* and *e* must have the trans configuration.

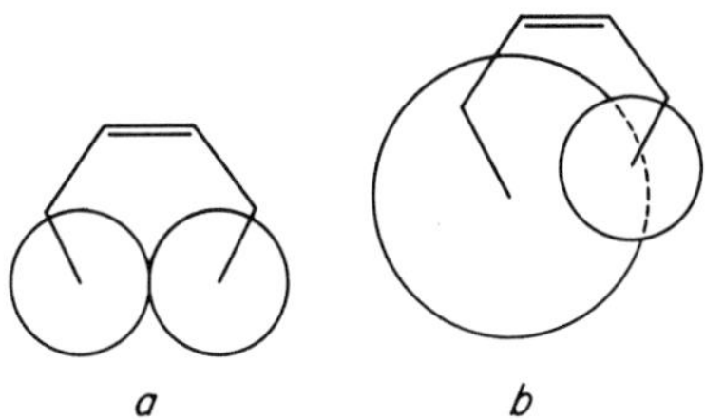

FIGURE 10.
Diagrams showing steric interactions of (*a*) —CH—CR═CR—CH— and (*b*) —C(CH_3)—CR═CR—CH groups for the *cis*-configuration about the double bond.

Guided by this viewpoint, Zechmeister and his colleagues began to isolate and identify the stereoisomers of a large number of naturally occurring carotenoids and synthetic polyenes (for a recent summary, see Zechmeister, 1962).

Pauling's "rule" was first challenged by Karrer et al. (1948) with the synthesis of the sterically hindered *dicis* molecule *cis,cis*-β-muconic acid. In answer, Pauling (1949) pointed out that he had not intended an absolute statement, but rather to define a scale of thermodynamic probabilities. The basis for this lay in the assumption that sterically hindered *cis* linkages, located in carotenoids at the 11,12 or 11′,12′ and the 7,8 or 7′,8′ double bonds [that is, Pauling's (1939) classes *d* and *e*], force a degree of twisting owing to intramolecular overcrowding, sufficient to break conjugation and in effect divide the conjugated system into segments.

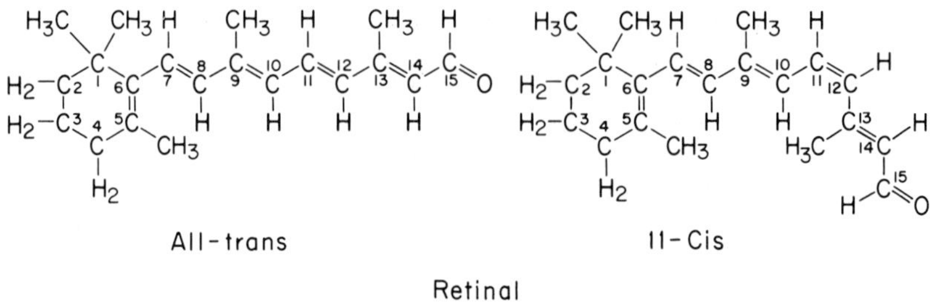

All-trans

11-Cis

Retinal

FIGURE 2.
Structural formulas of all-*trans* and 11-*cis* retinal (formerly retinene). In vitamin A, the terminal aldehyde group is replaced by —CH_2OH; in vitamin A acid, by —COOH; and in retinaldehyde oxime, by —CH=NOH. The 6,7 bond is drawn s-*cis*, because this is its configuration in all-*trans* vitamin A acid (Stam and MacGillavry, 1963); its configuration in the other compounds in this set has not yet been determined.

Resonance theory states that the longer the conjugated system, the greater the increase in resonance energy and hence resonance stabilization. A break in conjugation is therefore expected to lower the stability. Pauling (1949) estimated, for example, that dividing an all-*trans* carotenoid with eleven conjugated double bonds into two segments containing three and eight double bonds, as in the 11-*cis* isomer, would incur a loss of resonance energy of 16.1 kcal per mole. He therefore concluded that though this type of molecule could probably be synthesized, it would tend to isomerize spontaneously to the all-*trans* configuration.

About 1953, the systematic synthesis of polyenes containing sterically hindered *cis* linkages began in the laboratories of Karrer, Oroshnik, and Isler, and since then many such compounds have been perpared (see Eugster et al., 1953; Oroshnik and Mebane, 1954; Isler et al., 1957). It soon became apparent that most of them are quite stable. This may mean only that the hindered *cis* linkages are stabilized by large kinetic barriers to their isomerization, yet it calls into question their purported thermodynamic instability.

No quantitative data exist on the kinetics and thermodynamics of isomerization of sterically hindered *cis* isomers of C_{40} carotenoids. However, preliminary experiments show that 11,11′-*dicis* β-carotene (prepared and crystallized by Isler et al., 1957),* which by analogy with Pauling's (1949) calculations would be expected to be about 28 kcal per mole less stable than the all-*trans* isomer, is so stable that solutions of this isomer in n-heptane have to be warmed to about 85°C before the 11- and 11′-*cis* linkages isomerize, with a half-time of about 10 hours (Hubbard, 1966).

A good deal more is known about the sterically hindered C_{20} carotenoids, 11-*cis* vitamin A and 11-*cis* retinal (see Figure 2), to date the only sterically hindered *cis* carotenoids known to occur in nature. The chromophore of all known visual pigments is 11-*cis* retinal, except for some (mainly freshwater) vertebrates, in which it is in the 3,4-dehydro form, retinal$_2$ (for recent reviews, see Wald, 1960, 1961; Hubbard et al., 1965). Some 11-*cis* vitamin A is stored in the eyes of certain vertebrates (Hubbard and Dowling, 1962). In

*We are indebted to Dr. Otto Isler of Hoffmann-La Roche, Inc., in Basel, for a gift of this compound.

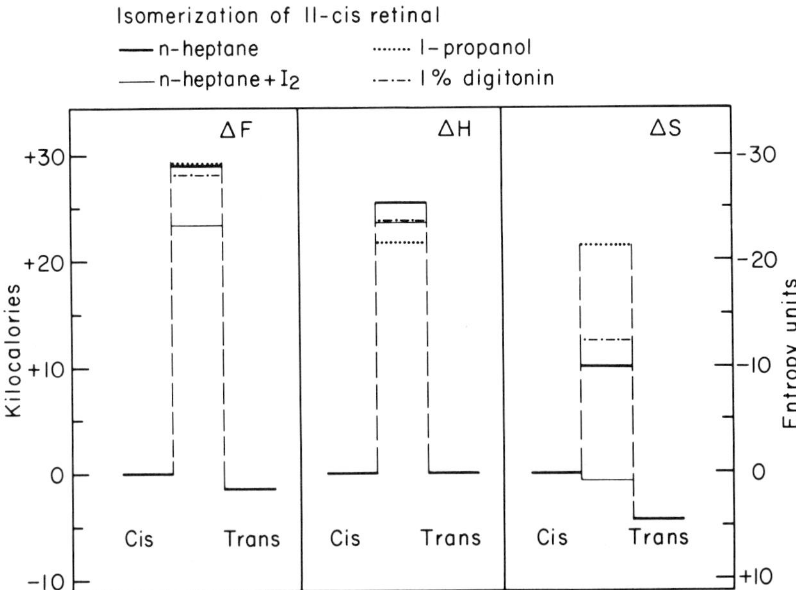

FIGURE 3.
Summary of the relative free energies (ΔF), heats (ΔH), and entropies (ΔS) of 11-*cis* and all-*trans* retinal in *n*-heptane, and of the activation barriers to their interconversion in various solvents. [From Hubbard, 1966.]

some invertebrates, all the vitamin A in the eye is in this form, and may be the only vitamin A in these animals (Wald and Burg, 1956–1957; Wald and Brown, 1956–1957). 11-*cis* retinal is formed enzymatically from all-*trans* retinal by both photic and thermal routes (Hubbard, 1955–1956). The thermodynamics and kinetics of its isomerization to the all-*trans* configuration have been analyzed (Hubbard, 1966), and are summarized in Figure 3.

Pauling's (1949) estimates of the resonance energies for conjugated systems of various lengths predict a difference in free energy (ΔF) of about 8 or 9 kcal per mole between 11-*cis* and all-*trans* retinal. As can be seen from Figure 3, however, the measured difference amounts to only about 1.5 kcal per mole. This discrepancy probably arises from the chief assumption upon which these calculations are based: that the hindered *cis* linkage introduces a complete break in conjugation. It seems appropriate, therefore, to reexamine this assumption in the light of recent observations that bear upon the relationship between steric hindrance and conjugation, derived mainly from X-ray crystallography and absorption spectra.

Little X-ray crystallography has yet been done on carotenoids, and no data are available for their sterically hindered *cis* isomers. In all, four carotenoids or related molecules have been analyzed: all-*trans* 15,15′-dehydro β-carotene (Sly, 1955, 1964; see our Figure 1); all-*trans* vitamin A acid (Stam and MacGillavry, 1963; see our Figure 2); and the two isomers of β-ionylidene crotonic acid shown in Figure 4—the all-*trans* (Eichhorn and MacGillavry, 1959) and the sterically unhindered 9-*cis* isomer (Koch and MacGillavry, 1963).

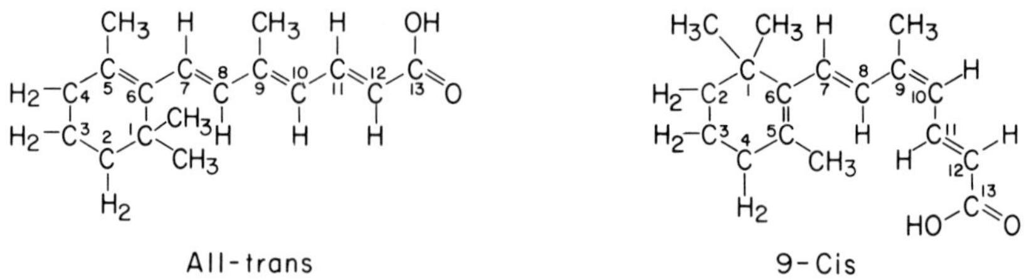

FIGURE 4.
Structural formulas of all-*trans* and 9-*cis* β-ionylidene γ-crotonic acid. X-ray crystallographic analysis has shown that the 6,7 bond is s-*trans* in the all-*trans* isomer (Eichhorn and MacGillavry, 1959) and s-*cis* in the 9-*cis* form (Koch and MacGillavry, 1963).

As might be expected, the side chains of the three all-*trans* molecules are essentially planar. Nonetheless, the successive double and single bonds retain a considerably greater measure of identity than would be predicted by resonance or molecular orbital theory (see Table I). What conjugation exists is difficult to rationalize in terms of molecular structure. For example, in all-*trans* β-ionylidene crotonic acid, the single and double bond lengths tend to approach one another (single bonds shorten, double bonds lengthen) in going from the ring toward the carboxyl group, as though the degree of conjugation increased in that direction. Yet the opposite trend, if any, prevails in all-*trans* vitamin A acid, a molecule that differs from all-*trans* β-ionylidene crotonic acid only in that it possesses one more double bond in the side chain, and that the 5.6-double bond in the ring lies *cis* rather than *trans* to the rest of the conjugated system (Figures 2 and 4).

In 9-*cis* β-ionylidene crotonic acid the analysis, though still incomplete, indicates that "the alternating bond system as a whole is far from plane," with the terminal zigzag from C_9 on "rather flat" but "slightly inclined" to the C_6 to C_9 piece. The β-ionone ring is fairly planar, but is sharply skewed so as to make an angle of about 80° with the side chain (Koch and MacGillavry, 1963). We would predict from theory that these gross departures from planarity should shift the absorption maximum (λ_{max}) of the 9-*cis* isomer to considerably shorter wavelengths than that of the planar all-*trans* form and reduce its specific absorbance. Arens and van Dorp (1947), however, who synthesized both isomers, found that both have λ_{max} 320 mμ, and that at that wavelength the specific absorbance of the 9-*cis* isomer is only about 20 percent lower than that of the all-*trans* molecule [E (1 percent, 1 cm) = 1110 as against 1350].

The ring-to-side chain relationships of the four molecules that have been analyzed are shown in Table 2. These relationships depend upon the *cis-trans* configuration of the 6,7 single bond, which determines whether the 5,6 double bond in the ring lies *cis* or *trans* to the conjugated system of the side chain (Figures 1, 2, and 4). As can be seen from Table 2, the *cis-trans* configuration of the 6,7 single bond is independent of the stereoisomeric configuration of the rest of the side chain. It is s-*trans* in all-*trans* β-ionylidene crotonic acid, but s-*cis* in the other two all-*trans* molecules as well as in 9-*cis* β-ionylidene crotonic acid. Furthermore, the preponderance of the variously twisted (and hence presumably less stable s-*cis* configurations) over the planar s-*trans* form is contrary to theoretical expectations.

TABLE 1.
*Average single and double bond distances in the all-*trans *side chains of conjugated molecules, compared with these bond lengths in molecules with no conjugation and with a prediction from LCAO molecular orbital calculations.*

Compound	C—C distance (Å)	C═C distance (Å)	Reference
no conjugation	1.53–1.54	1.334	Pauling (1960)
β-ionylidene crotonic acid	1.473	1.332	Eichhorn and MacGillavry (1959)
vitamin A acid	1.45	1.36	Stam and MacGillavry (1963)
15,15′-dehydro-β-carotene	1.455	1.345	Sly (1964)
1,8-diphenyl-1,3,5,7-octa-tetraene	1.449	1.350	Dreuth and Wiebenga (1955)
LCAO prediction for same	1.401	1.368	Dreuth and Wiebenga (1955)

TABLE 2.
Geometric relationships between the β-ionone ring and the side chain.

Compound	Configuration about the 6,7 Single Bond	Approximate Angle between Ring and Side Chain	Reference
all-*trans* β-ionylidene crotonic acid	s-*trans*	planar	Eichhorn and MacGillavry (1959)
9-*cis* β-ionylidene crotonic acid	s-*cis*	80°	Koch and MacGillavry (1963)
all-*trans* vitamin A acid	s-*cis*	35°	Stam and MacGillavry (1963)
all-*trans* 15,15′-dehydro-β-carotene	s-*cis*	48.5°	Sly (1964)

Spectroscopic data on β- and γ-carotene suggest that they too have the 6,7 s-*cis* configuration. If we compare the positions of λ_{max} of all-*trans* β-carotene, γ-carotene, and lycopene—all three of which have eleven conjugated double bonds and differ only in that two, one, or none of them are located in β-ionone rings (Figure 1)—we find that each ring opening shifts λ_{max} 10 mμ toward longer wavelengths. (Lengthening the straight conjugated chain by one double bond shifts λ_{max} only slightly more—about 15 to 20 mμ to longer wavelengths.) It is difficult to understand why the 5,6 double bond should contribute so much less to the conjugated system when it is located in a β-ionone ring than when it is part of the open chain, unless the ring were s-*cis* and the 5,6 double bond consequently twisted out of the plane of the side chain.

Let us now return to the sterically hindered 11-*cis* carotenoids. The degree of intramolecular overcrowding in these molecules is similar to that encountered at the 6,7 s-*cis* linkage, for both involve the overlap of a hydrogen with a methyl substituent four carbon atoms away (Figures 2 and 4). Eichhorn and MacGillavry (1959) have taken account of differences in the distribution of bond angles and have calculated that the steric hindrance should be somewhat greater for the 11-*cis* than the 6,7 s-*cis* configuration. We would therefore expect the 11-*cis* isomers to exhibit considerable departures from coplanarity (Table 2).

We have already discussed the relative stabilities of 11-*cis* retinal (Figure 3) and 11,11′-*dicis* β-carotene; the behavior of 11-*cis* vitamin A and 11-*cis* retinaldehyde oxime is qualitatively similar. Furthermore, both 11-*cis* retinal and 11,11′-*dicis* β-carotene readily form well-shaped crystals with sharp melting points. What is more surprising, the absorption maxima of the 11-*cis* isomers of retinal, vitamin A, and retinaldehyde oxime are shifted only 3 to 7 mμ toward shorter wavelengths from those of the all-*trans* isomers, and in this do not differ appreciably from the sterically unhindered 9- and 13-*cis* isomers (see Table 3). Other molecules containing sterically hindered *cis* linkages, among them 11-11′-*dicis* β-carotene, exhibit much greater shifts of λ_{max} toward shorter wavelengths (Oroshnik and Mebane, 1954). The anomaly, however, does not carry over to the molar absorbances, which generally are considerably lower for hindered than unhindered *cis* isomers. In this, 11-*cis* retinal, vitamin A, and retinaldehyde oxime behave like other sterically hindered *cis* isomers (Oroshnik and Mebane, 1954), and are clearly differentiated from the unhindered 9- and 13-*cis* isomers (Table 3). Whatever factors are responsible for the anomalous positions of λ_{max}, therefore, do not similarly affect the molar absorbances.

An interesting point concerning the geometry of the 11-*cis* isomers is raised by the observation that their molar absorbances at λ_{max} increase to the level of those of the unhindered *monocis* forms upon cooling to −190°C (Jurkowitz et al., 1959). (These changes in molar absorbance involve the areas under the absorption curves, not merely the heights of the absorption bands, which are greatly sharpened by cooling.) Cooling to −190°C thus appears to relieve the intramolecular crowding responsible for the steric hindrance, probably by decreasing the van der Waals radii of the overlapping substituent groups (Wald, 1959).

In summary, we should like to stress three points. (1) Departures from planarity represent relative rather than absolute barriers to conjugation, subject also to modification by environmental conditions. (2) We cannot yet predict the extent to which such departures from planarity affect stability, crystallizability, or the positions and heights of absorption spectra; nor need these effects be correlated with one another in particular instances. (3) The difference in free energy between a hindered *cis* and the all-*trans* configuration of a conjugated system may be very small. Thus ΔF between 11-*cis* and all-*trans* retinal is only about 1.5 kcal per mole (Figure 3).

Various degrees of aplanarity are better tolerated by conjugated systems than is commonly supposed, and this realization calls for a more adequate formulation of the quantum-mechanical conditions that govern the interaction of π-orbitals than we yet possess.

TABLE 3.
Absorption maxima (λ_{max}) *and molar absorbances at* λ_{max} (ϵ_{max}) *of the all*-trans, *the sterically unhindered 9- and 13*-monocis, *and the sterically hindered 11*-monocis *isomers of retinal, vitamin A, and retinaldehyde oxime.*

Compound	λ_{max} (mμ)		ϵ_{max} (liter mole^{-1}cm^{-1})	
	Hexane	Ethanol	Hexane	Ethanol
Retinal				
All-*trans*	368	383	48,100	42,900
13-*cis*	363	375	38,800	35,600
9-*cis*	363	375	37,700	36,100
11-*cis*	365	379.5	26,400	24,900
Vitamin A				
All-*trans*	325	325	51,800	52,800
13-*cis*		328		48,300
9-*cis*		323		42,300
11-*cis*	318	319	34,300	34,900
Retinaldehyde oxime				
All-*trans*	355	359	59,600	60,400
13-*cis*	352	356	52,800	54,900
9-*cis*	351.5	354	54,200	56,200
11-*cis*	350	356	35,900	37,100

REFERENCES

Arens, J. F., and D. A. Van Dorp (1947). *Rec. Trav. Chim.* (Pays Bas) **66**, 759.

Dreuth, W., and E. H. Wiebenga (1955). *Acta Cryst.* **8**, 755.

Eichorn, E. L., and C. H. MacGillavry (1959). *Acta Cryst.* **12**, 872.

Eugster, C. H., C. F. Garbers, and P. Karrer (1953). *Helv. Chim. Acta* **36**, 1378.

Garbers, C. F., and P. Karrer (1953). *Helv. Chim. Acta* **36**, 828.

Gillam, A. E., and M. S. El Ridi (1935). *Nature* **136**, 914.

Gillam, A. E., and M. S. El Ridi (1936). *Biochem. J.* **30**, 1735.

Hubbard, R. (1955–1956). *J. Gen. Physiol.* **39**, 935.

Hubbard, R. (1966). *J. Biol. Chem.* **241**, 1814.

Hubbard, R., and J. E. Dowling (1962). *Nature* **193**, 341.

Hubbard, R., D. Bownds, and T. Yoshizawa (1965). *Cold Spring Harbor Symp. Quant. Biol.* **30**, 301.

Isler, O., L. H. Chopard-dit-Jean, M. Montavon, R. Rüegg, and P. Zeller (1957). *Helv. Chim. Acta* **40**, 1256.

Jurkowitz, L., J. N. Loeb, P. K. Brown, and G. Wald (1959). *Nature* **184**, 614.

Karrer, P., R. Schwyzer, and A. Neuwirth (1948). *Helv. Chim. Acta* **31**, 1210.

Koch, B., and C. H. MacGillavry (1963). *Acta Cryst.* **16**, 48.

Oroshnik, W., and A. D. Mebane (1954). *J. Am. Chem. Soc.* **76**, 5719.

Pauling, L. (1939). *Fortschr. Chem. organ. Naturst.* **3**, 203.

Pauling, L. (1949). *Helv. Chim. Acta* **32**, 2241.

Pauling, L. (1960). *The Nature of the Chemical Bond*, 3rd ed., Chapter 7. Ithaca: Cornell Univ. Press.

Sly, W. G. (1955). *Acta Cryst.* **8**, 115.

Sly, W. G. (1964). *Acta Cryst.* **17**, 511.

Stam, C. H., and C. H. MacGillavry (1963). *Acta Cryst.* **16**, 62.

Wald, G. (1959). *Nature* **184**, 620.

Wald, G. (1960), in M. Florkin and H. S. Mason, eds., *Comparative Biochemistry*, vol. 1, chap. 8, p. 311. New York, Academic Press.

Wald, G. (1961), in W. D. McElroy and B. Glass, eds., *Light and Life*, p. 724. Balitimore: Johns Hopkins Press.

Wald, G., and P. K. Brown (1956–1957). *J. Gen. Physiol.* **40**, 627.

Wald, G., and S. P. Burg (1956–1957). *J. Gen. Physiol.* **40**, 609.

Zechmeister, L. (1944). *Chem. Rev.* **34**, 267.

Zechmeister, L. (1962). *Cis-Trans Isomeric Carotenoids, Vitamins A and Arylpolyenes.* New York: Academic Press.

Zechmeister, L., and P. Tuzson (1938). *Biochem. J.* **32**, 1305.

PETER PAULING

William Ramsay and Ralph Forster Laboratories
University College
University of London
London, England

The Structure of Molecules Active in Cholinergic Systems

The simple cation acetylcholine, $(CH_3)_3\,N^+CH_2CH_2O(CO)CH_3$, is the intercellular effector substance in most nervous transmission systems. It is released at the end of a nerve cell, diffuses across a space, and stimulates the next cell, either another nerve cell or an operator cell such as smooth or striated muscle. Of the motor nervous systems, it appears that only the postganglionic autonomic sympathetic system is not mediated by acetylcholine; it is mediated by adrenaline or noradrenaline.

In the past fifty years a good deal has been learned about the properties of various cholinergic nervous transmission systems. Much has been learned by the study of the agonistic and antagonistic effects of various natural and synthetic molecules on such systems. Small doses of L(+)-muscarine, for example, isolated from the mushroom *Amanita muscaria*, mimic the effect of acetylcholine on some tissue—such as the smooth muscle of the gut—whereas large doses inhibit the effect of acetylcholine. These effects are similar to those of acetylcholine itself, which stimulates in small doses and inhibits in large ones. Muscarine, however, is ineffective in other cholinergic systems, as in the junction of nerve and striated muscle. For many years cholinergic transmission systems have been divided into "nicotonic," in which the effect of acetylcholine is mimicked by small amounts of nicotine and inhibited by curare alkaloids, and "muscarinic," in which the effect of acetylcholine is mimicked by muscarine and inhibited by atropine. This differentiation is certainly not specific enough to explain the variations in cholinergic systems that have been found by studying the effects of many other molecules on them.

The available knowledge of the vast range of substances that affect the various cholinergic systems in one way or another has often been used in attempts to specify the stereochemistry of the interaction of these molecules and acetylcholine with the receptors and with acetylcholinesterase. These attempts have been hampered by a lack of information on molecular structure, though it is often argued that the labile molecules that are sometimes involved do not necessarily have the crystalline molecular structure in solution or in vivo. Nonetheless it remains remarkable that though the structure

I am most grateful to Dr. F. G. Canepa for stimulating my interest in this problem.

analysis of crystals by X-ray diffraction analysis (Bragg, 1913) and the nervous transmission effects of acetylcholine (Dale, 1914) were discovered more than fifty years ago, the structures of only two molecules affecting nervous transmission systems have been analyzed.

THE STRUCTURE OF ACETYLCHOLINE

An analysis of the crystal structure of acetylcholine bromide was reported by Sörum in 1959. The spacegroup was reported as $P2_1$ with four molecules per unit cell—that is, with two independent molecules per asymmetric unit. Dunitz (1963) suggested that the unusual systematic absences of the diffraction data observed by Sörum indicate that the crystals used in the investigation were twinned across $(10\bar{2})$, spacegroup $P2_1/c$, with four molecules per unit cell, or one molecule per asymmetric unit. Canepa (1964) has also questioned the analysis. The structure of acetylcholine bromide has recently been reinvestigated (Canepa, Pauling, and Sörum, 1966) with the use of the three-dimensional X-ray diffraction data of Sörum.

Crystals of acetylcholine bromide are monoclinic,

$$a = 7.153 \pm 0.001 \text{ Å} \qquad b = 13.690 \pm 0.003 \text{ Å}$$
$$c = 11.057 \pm 0.002 \text{ Å} \qquad \beta = 109°39' \pm 2',$$

spacegroup $P2_1/c$, four molecules per unit cell, or one molecule per asymmetric unit. The structure has been refined anisotropically by least squares methods: the residual $R = 0.098$ for the 770 observed diffraction data.

The molecular structure and some interatomic distances and angles are shown in Figure 1. The statistical standard deviations of the distances are about 0.06 Å. The $(CH_3)_3N—CH_2—CH_2—$group forms a *trans* extended chain backbone of the molecule. The ester oxygen atom O1 is *gauche* to the nitrogen atom, the C5—O1 bond being approximately coplanar with the C1—N bond. The methyl carbon C1 ester oxygen O1 distance is so short (3.02 Å) that the hydrogen atoms of the methyl group C1 are crowded. Their positions appear fairly fixed and can be observed in glycerylphosphorylcholine (Pascher, personal communication, 1966). One hydrogen atom is in a position to interact with an electron pair of the ester oxygen atom O1. This group of atoms H—C1—N—C4—C5—O1 forms a pseudo six-membered ring in the chair conformation.

The entire ester group,

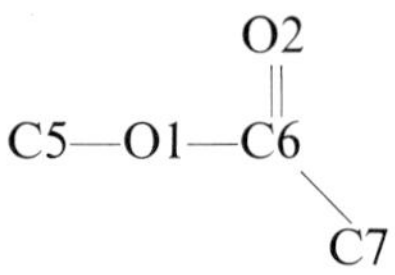

```
          O2
          ||
C5—O1—C6
            \
             C7
```

is coplanar, the carboxyl oxygen O2 being synplanar to C5. The angle between the plane of

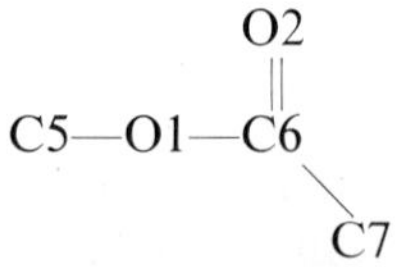

```
          O2
          ||
C5—O1—C6
            \
             C7
```

and the plane of C3—N—C4—C5 is not 90° but about 111°, probably because of the close contact between the ester oxygen atom O1 and the methyl group C1. The nitrogen ester oxygen distance is fairly short (3.29 Å). The planarity

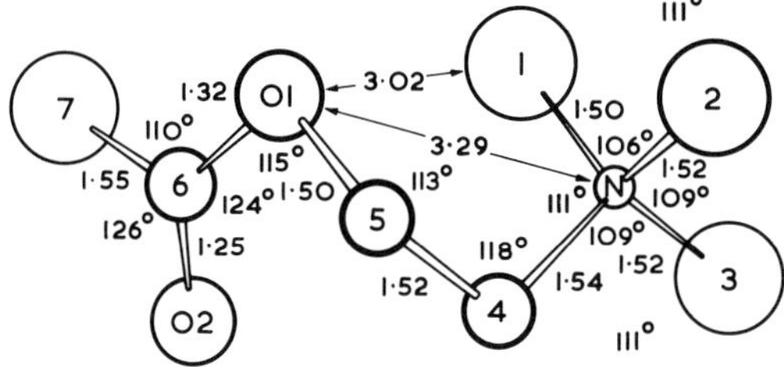

FIGURE 1.
The structure of acetylcholine in crystals of the bromide (Canepa, Pauling, and Sörum, 1966).

of the ester group and the slight (though not statistically significant) shortening of the O1—C6 bond suggest that this bond has partial double-bond character (Pauling, 1960).

THE CORRELATION

The structure of acetylcholine in the solid state can be compared with the known structures of related molecules. Choline chloride (Senko and Templeton, 1960), shown in Figure 2, has a molecular structure very similar to that of acetylcholine bromide, the oxygen atom being *gauche* to the nitrogen atom. The torsion angle O—C5—C4—N is 84°. Bond distances and angles are similar. Choline—O—sulfate (Okaya, 1964), up to the bridging oxygen atom, has a structure similar to the structures of choline and acetylcholine, the bridging oxygen atom again being *gauche* to the nitrogen atom. The S—O—C—C torsion angle, however, appears to be about 180°, compared to the C6—O1—C5—C4 torsion angle of about 60° in acetylcholine. The sulfate group shows fairly unhindered rotation about the S—O1 bond. The nitrogen-bridging oxygen atom distance of 3.06 Å is about 0.2 Å shorter than the observed corresponding distance in acetylcholine.

The structures of glycerylphosphorylcholine and glycerylphosphorylcholine cadmium chloride have been analyzed by Abrahamsson and Pascher (1966) and Sundaralingam and Jensen (1965), respectively. A portion of the molecule of glycerylphosphorylcholine with interatomic distances and angles (which are the mean of those of the two independent molecules in the asymmetric unit) is shown in Figure 3. Up to the first linking oxygen atom, the structure of this portion of the molecule is very similar to that of acetylcholine. The oxygen atom O1 is *gauche* to the nitrogen atom, as in all the other molecules of choline studied. The mean O1–N distance of 3.1375 ± 0.0047 Å is about 0.015 Å less than the corresponding (and much less reliably known) distance of 3.02 ± 0.06 Å in acetylcholine. The close methyl-ester oxygen C1–O1 distances in the various compounds appear identical. The P—O1—C5—C4 torsion angle of about −120° differs from the corresponding angle of about +60° in acetylcholine.* The phosphorus atom thus approximately eclipses a hydrogen atom of C5. One of the nonbonded phosphate oxygen atoms is

*The signs of the torsion angles mentioned refer to one of the two molecules in the asymmetric unit of glycerylphosphorylcholine and to one conformational isomer of acetylcholine. For the other molecule of glycerylphosphorylcholine and the inverted form of acetylcholine, the signs are reversed.

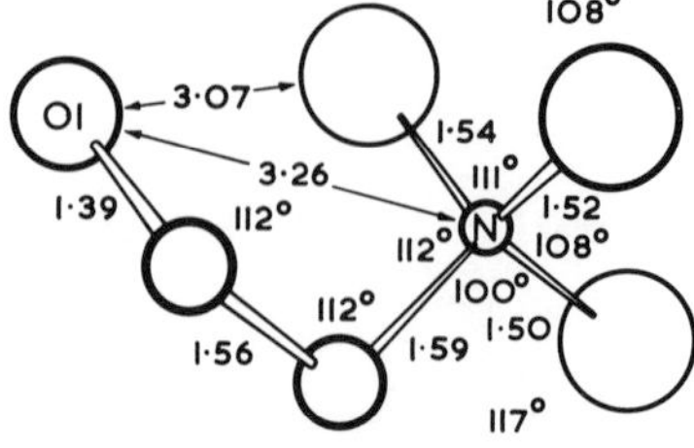

FIGURE 2.
The structure of choline in crystals of the chloride (Senko and Templeton, 1960).

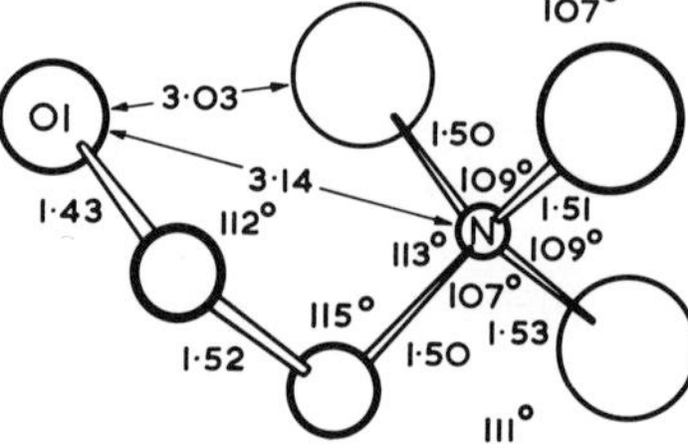

FIGURE 3.
The structure of part of the molecule of glycerylphosphorylcholine (Abrahamsson and Pascher, 1966). Interatomic distances and angles are the mean of those of the two independent molecules in the asymmetric unit.

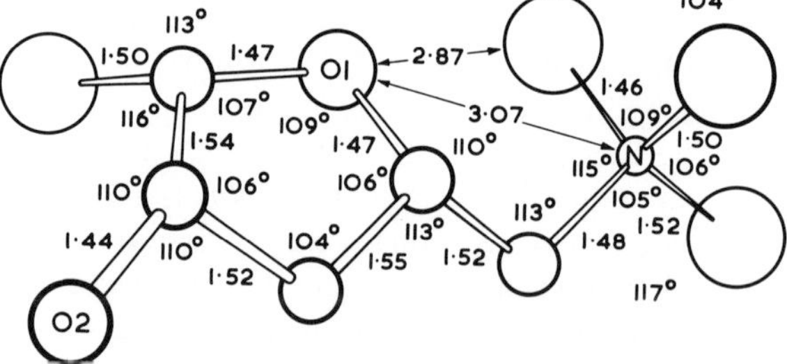

FIGURE 4.
The structure of muscarine in crystals of the iodide (Jellinek, 1957). The configuration shown is that of the active L(+) (2S, 3R, 5S) form.

transplanar to C5, whereas the carboxyl oxygen atom O2 of acetylcholine is synplanar to C5.

The structure of glycerylphosphorylcholine cadmium chloride differs in an interesting way from that of glycerylphosphorylcholine. The choline ends of the molecules are identical, the oxygen atom O1 being *gauche* to the nitrogen atom, C1—N being approximately coplanar with O1—C5. The orientations of the phosphate groups are different, however, the torsion angles O2—P—O1—C5 and P—O1—C5—C4 being about $-60°$ and 180° in glycerylphosphorylcholine cadmium chloride and about 180° and $-120°$ in glycerylphosphorylcholine alone. In the cadmium chloride compound the two free oxygen atoms of the phosphate group coordinate cadmium ions, which are octahedrally coordinated. Evidently coordination of the metal ions causes the conformation of the glycerylphosphorylcholine to change; this change may have an important effect on the biological role of the molecules.

Other than acetylcholine, muscarine iodide is the only molecule active in cholinergic systems whose structure has been determined by X-ray diffraction analysis (Jellinek, 1957). In small doses, muscarine mimics the effect of acetylcholine on certain receptors. The structure of muscarine is shown in Figure 4. The ring oxygen atom O1 is *gauche* to the nitrogen atom as in all known cholines. In muscarine the C1–O1 and N1–O1 distances do not differ from those in acetylcholine. The ring, however, forces the torsion angle C6—O1—C5—C4 to be slightly less than 180°—rather than 60°, as in acetylcholine—and the methyl group C7 is somewhat further away from the onium group. The position of the hydroxyl oxygen atom O2 in muscarine does not in any way correspond to that of the carboxyl oxygen O2 in acetylcholine. The nonplanar ring carbon atom C8 is in the plane of C3—N—C4—C5 and just adds to the *trans* extended chain formed by these atoms. The ring C5—O1—C6—C9 is roughly coplanar with C1—N.

Crystals of acetylcholine are centrosymmetric and molecules thus exist not only in the conformation shown in Figure 1 but also in its mirror image. The conformation shown is the one consistent with the absolute configuration of muscarine as determined chemically.

THE MODEL STRUCTURE

This correlation and comparison allows us to make a start at proposing a model structure for molecules active in cholinergic systems. Since the structures of only two active molecules are known, little can be done to specify the stereochemical differences of the various cholinergic systems as shown by the differing effects of various drugs.

The choline groups $(CH_3)_3N^+CH_2CH_2O$—of these six molecules are essentially identical with the oxygen atom *gauche* to the nitrogen atom, and with one methyl carbon C1 to oxygen O1 distance of about 3.03 Å and a nitrogen to oxygen O1 distance of about 3.14 Å. The group C3—N—C4—C5 is a normal *trans* extended chain. At the (ester) oxygen atom O1 the formulas and structures of the molecules vary. Choline—O—sulfate, glycerylphosphorylcholine cadmium chloride, and muscarine iodide have a torsion angle (S,P,C6)—O1—C5—C4 of approximately 180°. Glycerylphosphorylcholine has a torsion angle P—O1—C5—C4 of about −120° with O2 transplanar to C5; acetylcholine has a torsion angle C6—O1—C5—C4 of +60° with O2 synplanar to C5. The conformation of acetylcholine is not that normally found in primary esters (Mathieson, 1965). The C6—O1—C5—C4 torsion angle is normally 180°, and the plane of the acetyl group bisects the angle between the two hydrogens of —OCH_2— (C5 in acetylcholine). Culvenor and Ham (1966) have argued on the basis of a proton-nuclear magnetic-resonance investigation of acetylcholine chloride in D_2O that the C6—O1—C5—C4 torsion angle in solution is 180°, though their arguments do not appear conclusive. If the conformation of acetylcholine were to change in solution, the correlation of the structure with that of muscarine (which is rigid) would increase. Though a change in this torsion angle might result from the biological activity of acetylcholine, there is at present not enough structural evidence available to enable us to conclude that the active conformation is other than that in the solid state.

PREDICTIONS AND PROBLEMS

We can use the model structure that arises from the correlation of these known structures to predict the structures of other molecules active in cholinergic systems. Beckett, Harper, and Clitherow (1963) and Ellenbroek and

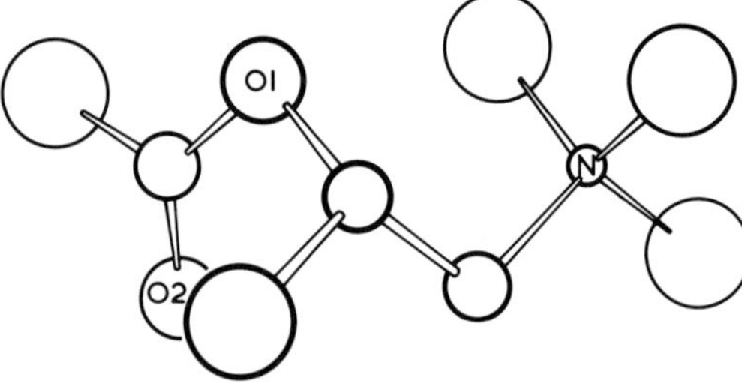

FIGURE 5.
A proposed structure for the active S(+)-acetyl-β-methylcholine. The inactive R form has the inverse structure.

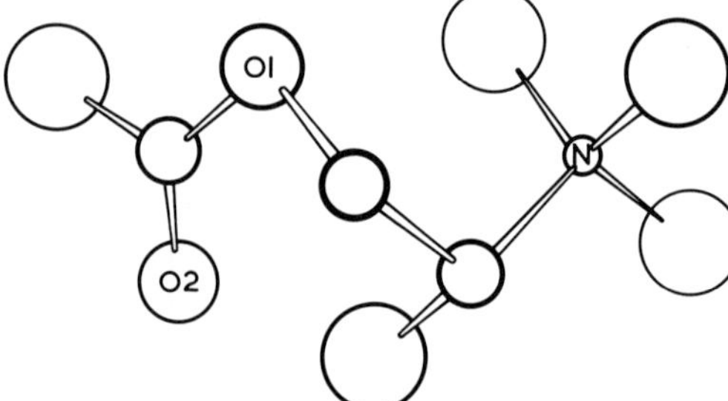

FIGURE 6.
A proposed structure for the reasonably active R(+)-acetyl-α-methylcholine.

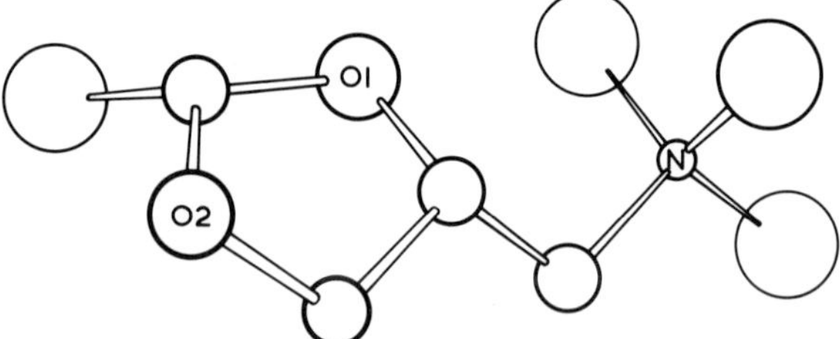

FIGURE 7.
A proposed structure for the highly active muscarinic agent L(+)-*cis*-2R-methyl-4S-trimethylammoniummethyl-1,3-dioxolane.

van Rossum (1960) have synthesized and separated the stereoisomers of acetyl-α and β-methylcholine iodide and have determined their activities as mimics of acetylcholine in various cholinergic systems and as a substrate of acetylcholinesterase. S(+)-acetyl-β-methylcholine is equivalent to acetylcholine in muscarinic systems whereas R(−)-acetyl-β-methylcholine is only 1/200 as effective. A structure of S(+)-acetyl-β-methylcholine consistent with what is known at present of the structure of cholinergic molecules is shown in Figure 5. The extra methyl group on the acetylcholine molecule may occupy a position that exactly corresponds to the position of a ring carbon atom of muscarine that forms part of the extended chain backbone C3—N—C4—C5—C8. In this proposed structure, the torsion angle C6—O1—C5—C4 of 60° with O1 synplanar to C5 is not consistent with the usual orientation of secondary esters determined by Mathieson (1965), in which the torsion angle is −120° with the oxygen atom O2 synplanar to the hydrogen atom. It is of course impossible to satisfy the preferred orientation of both primary and secondary esters with a single structure for acetylcholine and acetyl-β-methylcholine; the structure of acetylcholine and the suggested structure of acetyl-β-methylcholine satisfy neither. The argument stating why the R(−) form of acetyl-β-methylcholine is inactive (given by Canepa, Pauling, and Sörum, 1966), which involves an overcrowded molecule based on the structure of poly-isobutylene (a helix with 16 carbon atoms per turn), is incorrect. The R and S forms of the iodide crystals give identical X-ray diffraction photographs, which indicates that the two forms are mirror images of each other. The S form has the correct conformation to react with receptors; its mirror image, the R form, has not.

The ratio of muscarinic activities of the two stereoisomers of acetyl-α-methylcholine is, in terms of the model structure of cholinergic molecules, a little more difficult to understand than the ratio of activities of acetyl-β-methylcholine. The R(+) isomer is 1/28 as effective as acetylcholine on guinea-pig ileum and the S(−) isomer is 1/232 as effective (Beckett et al., 1963). A structure for R(+)-acetyl-α-methylcholine consistent with the model structure of cholinergic compounds is shown in Figure 6 with the extra methyl group just added to the α-carbon of the conformational isomer of acetylcholine, consistent with the activity of L(+)-muscarine. If this structure is correct, the extra methyl group sticks down on what appears to be the active side of the molecule—the side on which the acetyl group and the short O1–C1 distance lie. One might think that the more active isomer of this molecule would have the extra methyl group on the side of the extended chain backbone that is opposite to the acetyl group in order not to interfere with the interaction of the acetyl and the onium groups with the receptor. The interaction of the methyl group and the acetyl group of acetyl-α-methylcholine may cause the acetyl group to be on the same side of the molecule as the methyl group, as suggested here, in which case the R form would fit the handedness of the receptor and the S form would not—as, indeed, they do. The extra methyl group would interfere with the interaction with the receptor to some extent, a factor of about 28. It is necessary, however, that the crystal structure of this molecule be determined.

Triggle and Belleau (1962) have synthesized and separated all the isomers of 2-methyl-4-trimethyl-ammoniummethyl-1,3-dioxolane, and Belleau and Puranen (1963) and Belleau and Lacasse (1964) have studied their activities on muscarinic receptors and acetylcholinesterase respectively. The most active isomer by a factor of about 100 is the L(+)-*cis* (2R,4S) isomer. This result is entirely consistent with other stereoisomer investigations and with the model structure. A suggested structure for L(+)-*cis*-2R-methyl-4S-trimethyl-ammoniummethyl-1,3-dioxolane is shown in Figure 7. It is seen that it is just L(+)-muscarine with the ring carbon atom (with the hydroxyl group) changed to an oxygen atom. That the active isomer of this 1,3-dioxolane (2R,4S) should be the same as that of muscarine (2S,3R,5S) is the strongest evidence on the absolute configuration of the handedness of cholinergic receptors available at present. It can also be seen that the correlation between this dioxolane and acetylcholine would be almost complete if the C6—O1—C5—C4 torsion angle of acetylcholine in solution were about 180° as suggested by Culvenor and Ham (1966) instead of +60° as it is in crystals of acetylcholine bromide.

A very interesting series of muscarine derivatives has been synthesized by Eugster, and its pharmacological effects have been studied by Waser (1961). Muscarone is muscarine with the 3-hydroxyl changed to a keto oxygen. It has two asymmetric carbon atoms instead of three as in muscarine. Both the L(+) isomer (2S,5S) and the D(−) isomer (2R,5R) of muscarone are more active as mimics than acetylcholine or L(+)-muscarine (2S,3R,5S) on muscarinic receptors. This is the only known example of a molecule in which both stereoisomers are active, though the number of molecules whose stereoisomers have been separated and tested is very small. It is easy to predict a structure for L(+)-muscarone. As shown in Figure 8 it is just L(+)-muscarine with the hydroxyl changed to a keto oxygen. The difficulty comes with D(−)-muscarone. The mirror image of the structure shown in Figure 8 is shown in Figure 9, drawn leaving atoms C1—N and C5—O1 in the same place: that is, reflecting the structure through the plane formed by C1, N, and most of the ring. If the ring is planar, seven of the heavy atoms will occupy the same positions in the two isomers. The backbone, two of the ammonium methyl groups, and the 2-methyl will move. Moving the 2-methyl may not be too important: d,l-allo-muscarone with the methyl and trimethylammoniummethyl groups on opposite sides of the ring is about as effective as the isomers of muscarone, though d,l-2-methyl-muscarone with two methyl groups is only about 1/20 as effective. Moving the α-carbon to the nitrogen may not be important, but moving the two ammonium methyl groups should affect the binding interaction. The real problem is that if the structure shown in Figure 9 is effective, the corresponding isomer D(−)-*cis*-2S-methyl-4R-trimethylammoniummethyl-1,3-dioxolane should be as effective as its 2R,4S isomer instead of 1/100 as effective. Insufficient data are available at present to explain these results.

Much work has been done to clarify the interaction of acetylcholine and acetylcholinesterase during the hydrolysis of the former by the enzyme. Wilson, Bergman, and Nachmansohn (1950) have proposed a mechanism in which a basic group in the esteratic site of the enzyme forms a covalent bond with the carboxyl carbon atom of acetylcholine. Krupka and Laidler (1961) think it unlikely that a basic group forms the covalent bond and suggest that the oxygen atom of the hydroxyl group of a serine residue forms the bond, though they do not propose a complex involving a quadriligant carboxyl carbon atom C6. They have made a detailed proposal of the interaction of the enzyme and substrate, including the relative location of the reacting groups of the enzyme and acetylcholine. Their picture cannot be quite correct (though the general scheme may be) because they have considered a *trans* conformation of acetylcholine, whereas all known cholines have a *gauche* configuration. What appears likely (Triggle, 1965) is that four groups in two sites of the enzyme are involved in the binding of acetylcholine: the anionic site, and an acidic group, a basic group and a serine residue of the esteratic site. The anionic site binds the trimethylammonium group of the substrate and may be a phosphoric acid group (Ariens, 1962). The esteratic site binds and reacts with the acetyl group of the substrate. The basic group may be an imidazole nucleus (Krupka and Laidler, 1961).

If we make a number of unproven but not completely unjustifiable assumptions, it is possible to draw a diagram of an interaction between acetylcholine and the enzyme. The assumptions are (*i*) the structure of acetylcholine is that found in crystals of acetylcholine bromide, (*ii*) the molecule remains in this configuration with a minimum relative movement of atoms, (*iii*) a complex is formed involving a covalent bond to the carboxyl carbon atom, making this atom quadriligant, (*iv*) interactions occur with the oxygen atoms O1 and

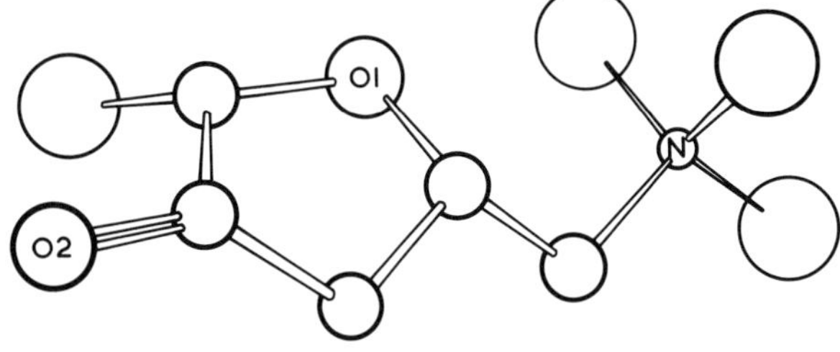

FIGURE 8.
A proposed structure for the highly active muscarinic agent L(+)-muscarone (2S, 5S), based on the structure of L(+)-muscarine.

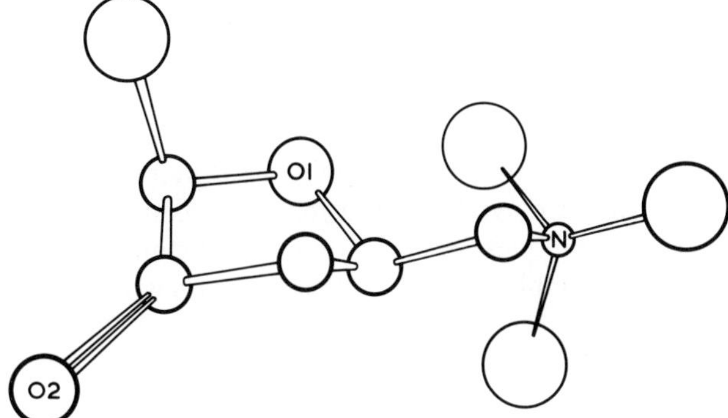

FIGURE 9.
The mirro-image structure of that of Figure 8, or the proposed structure of the highly active muscarinic agent D(−)-muscarone (2R, 5R).

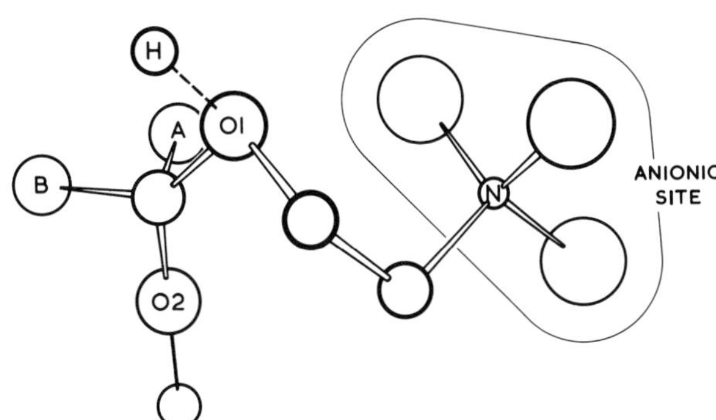

FIGURE 10.
A suggestion for the stereochemical interaction between a molecule of acetylcholine and four sites of the enzyme acetylcholinesterase, based on the structure of acetylcholine in crystals of the bromide. A covalent bond is formed with the carboxyl carbon atom. This bond is either to position A or to position B, and the unused position is the methyl group C7.

O2, holding these atoms in the position of (*i*) and (*ii*). The diagram is shown in Figure 10. The trimethylammonium group is held in place by the anionic site of the enzyme. The ester oxygen atom O1 is held in position by the interaction of an acidic site on the enzyme with an electron pair of the oxygen atom. The interaction is shown on one of the two electron pairs of the oxygen atom because the other pair is involved with a hydrogen atom of methyl group C1, though Dunitz (personal communication, 1966) has argued that even though the region between O1 and C1 is fairly crowded, there is room for a group on the enzyme to form a link to O1 on this side. The carboxyl oxygen atom O2 is held fixed to the enzyme by an unidentified bond. (If the carboxyl carbon atom is quadriligant a bonding electron is available on this oxygen to form a covalent bond.) If our four assumptions are valid, only the methyl group C7 moves. It can move to either position A or position B in

Figure 10. To whichever position it moves, the covalent bond from the enzyme to the carboxyl carbon atom C6 is via the other one. Whether A or B is more likely to be the site of the enzyme interaction is not clear; A is a little more than 4 Å and B a little more than 5 Å from the nitrogen atom.

Much can be learned of the stereochemical interactions involved in nervous transmission by studying the structures of many active molecules, but what would be really worthwhile is to study the structure of a complex formed by acetylcholinesterase or a receptor with a suitable inhibitor. It is not likely that this will be done soon. Acetylcholinesterase is difficult to get, and when obtained is impure and not crystalline (Kremzner and Wilson, 1963, 1964). At least its molecular weight and the number of active sites are decreasing functions of time. About the receptor, Waser (1963) wrote, "I have to admit that we know and can prove only the existence of the cholinergic receptor; we do not know yet its nature, nor do we understand its functioning." The situation is a little better today, but not much. Until these interesting large molecules or groups of molecules are available in a form that can be studied by rigorous three-dimensional techniques, perhaps a good deal can be learned by studying the shapes of the available relevant molecules, as has been done with some success in other fields (for example, Pauling, 1927, 1928, 1929, 1931, 1932, 1939, 1947, 1949, 1965; Pauling and Corey, 1951; Pauling, Corey, and Branson, 1951).

REFERENCES

Abrahamsson, S., and I. Pascher (1966). *Acta Cryst.* **21**, 79.

Ariens, E. J. (1962). In Ciba Foundation Study Group (No. 12) *Curare and Curare-like Agents*, p. 73. Boston: Little, Brown.

Beckett, A. H., N. J. Harper, and J. W. Clitherow (1963). *J. Pharm. Pharmacol.* **15**, 349, 362.

Belleau, B., and G. Lacasse (1964). *J. Med. Chem.* **7**, 768.

Belleau, B., and J. Puranen (1963). *J. Med. Chem.* **6**, 325.

Bragg, W. L. (1913). *Proc. Roy. Soc.* (London) A **89**, 248.

Canepa, F. G. (1964). *Nature* **201**, 184.

Canepa, F. G., P. J. Pauling, and H. Sörum (1966). *Nature* **210**, 907.

Culvenor, C. C. J., and N. S. Ham (1966). *Chem. Communs.* 537.

Dale, H. H. (1914). *J. Pharmacol.* **6**, 147.

Dunitz, J. D. (1963). *Acta Chem. Scand.* **17**, 1471.

Ellenbroek, B. W. J., and J. M. van Rossum (1960). *Arch. Int. Pharmacoydn.* **125**, 216.

Jellinek, F. (1957). *Acta Cryst.* **10**, 277.

Kremzner, L. T., and I. B. Wilson (1963). *J. Biol. Chem.* **238**, 1714.

Kremzner, L. T., and I. B. Wilson (1964). *Biochem.* **3**, 1902.

Krupka, R. M., and K. J. Laidler (1961). *J. Am. Chem. Soc.* **83**, 1445, 1448, 1454, 1458.

Mathieson, A. M. (1965). *Tetrahedron Letters* 4137.

Okaya, Y. (1964). Abstracts ACA Annual Meeting, Bozeman, Montana 26–31 July, p. 54.

Pauling, L. (1927). *J. Am. Chem. Soc.* **49**, 765.

Pauling, L. (1928). *Sommerfeld Festschrift*. Leipzig: S. Hirzel.

Pauling, L. (1929). *J. Am. Chem. Soc.* **51**, 1010.

Pauling, L. (1931). *J. Am. Chem. Soc.* **53**, 1367, 3225.

Pauling, L. (1932). *J. Am. Chem. Soc.* **54**, 988, 3570.

Pauling, L. (1939). *The Nature of the Chemical Bond.* Ithaca: Cornell Univ. Press.

Pauling, L. (1947). *J. Am. Chem. Soc.* **69**, 542.

Pauling, L. (1949). *Proc. Roy. Soc.* (London) A **196**, 343.
Pauling, L. (1960). *The Nature of the Chemical Bond*, 3rd ed., p. 239. Ithaca: Cornell Univ. Press.
Pauling, L. (1965). *Proc. Nat. Acad. Sci. U.S.* **54**, 989.
Pauling, L., and R. B. Corey (1951). *Proc. Nat. Acad. Sci. U.S.* **37**, 235, 241, 251, 256, 261, 272, 282.
Pauling, L., R. B. Corey, and H. R. Branson (1951). *Proc. Nat. Acad. Sci. U.S.* **37**, 205.
Senko, M. E., and D. H. Templeton (1960). *Acta Cryst.* **13**, 281.
Sörum, H. (1959). *Acta Chem. Scand.* **13**, 345.
Sundaralingam, M., and L. H. Jensen (1965). *Science* **150**, 1035.
Triggle, D. J. (1965). *Chemical Aspects of the Autonomic Nervous System*, p. 160 ff. London: Academic Press.
Triggle, D. J., and B. Belleau (1962). *Can. J. Chem.* **40**, 1201.
Waser, P. J. (1961). *Pharm. Rev.* **13**, 465.
Wilson, I. B., F. Bergman, and D. Nachmansohn (1950). *J. Biol. Chem.* **186**, 781.

HANS KUHN
Institute of Physical Chemistry
The University of Marburg
Marburg an der Lahn, Germany

On Possible Ways of Assembling Simple Organized Systems of Molecules

A chemist has no methods for arranging single molecules of different substances in a planned way, even though he needs such methods for synthesizing models in molecular biology (for example, arrangements having the properties of chloroplasts or mitochondria), or for constructing tools with component parts of molecular size (for example, assemblies acting as computers with functional units of molecular size). The synthesis of organized systems of molecules is thus a key problem in an overlapping region of chemistry, molecular biology, and cybernetics, that needs to be attacked from every side.

Let us consider here the following possibility of obtaining, step by step, an assemblage of molecules A, B, C, D, ordered in one direction. Let us deposit a monolayer of substance A on a support, superimpose a monolayer of substance B, then a monolayer of substance C, then one of D; an ordered array of molecules A, B, C, D is thus obtained.

It has been known for a long time that monolayers of a long-chain fatty acid may be spread on a water surface (Rayleigh, 1899) and deposited on a slide (Langmuir, 1917), a procedure that may be repeated many times (Blodgett, 1935) in order to obtain multilayers. Such multilayer systems show a diffusion (Sobotka, 1956) of the fatty acid molecules across the layers, and rearrangements of layers have been observed (Langmuir, 1939). Thus the method at first seems to be inadequate for constructing organized systems of molecules. It has been shown (Bücher et al., 1967), however, that appropriate systems can be obtained with molecules such as I and II, which contain

$C_{18}H_{37}$ $C_{18}H_{37}$
N N
(benzothiazolium, +) C—CH=C (benzothiazoline)
S S
J^- I

$C_{18}H_{37}$ $C_{18}H_{37}$
N N
(benzoxazolium, +) C—CH=CH—CH=C (benzoxazoline)
O O
J^- II

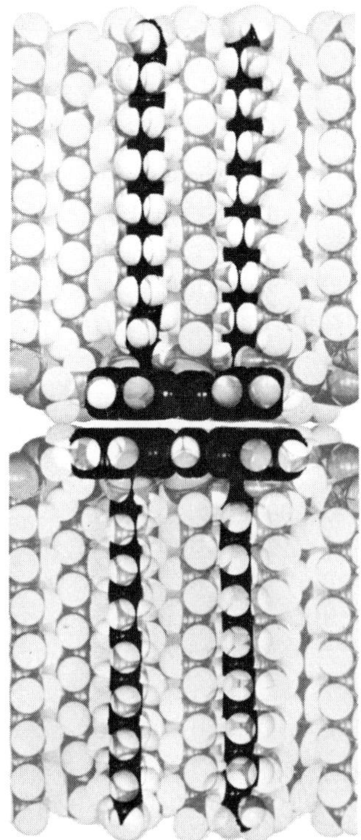

FIGURE 1.

a hydrophilic chromophore and long-chain hydrocarbon substituents. Like long-chain fatty acids, a mixture of these molecules with tripalmitin may be spread on a water surface and the monolayers then deposited on a slide. An example of a simple layer system thus obtained is visualized in Figure 1. It can be shown (Bücher et al., 1967) that the molecular arrangement of such systems will remain stable for months. I shall summarize here very briefly some results and indicate some possibilities of this technique.

In the arrangement shown in Figure 1 the chromophores of the two dyes are kept side by side in direct contact. In this way two arbitrary chromophores may be coupled, and the corresponding changes in the absorption and fluorescence spectra may be studied. The absorption bands of dyes I and II are weakly shifted, the band of I is enhanced, and the band of II is strongly reduced. These spectral changes are quantitatively predicted by the exciton theory when the coupling coefficient is calculated by the refined electron gas model. It is to be expected that exciton or charge-transfer spectra of arbitrary pairs of molecules may be investigated in this way. The neighbors of a molecule may be chosen in a planned manner and the surrounding effects thus studied systematically under controlled conditions. By this rendezvous procedure, chemically reacting groups may be composed and bound specifically, thus giving new possibilities in preparative chemistry.

The dependence of the absorption spectrum on the mixing ratio of dye and tripalmitin has been investigated for both dyes I and II. The results indicate that, during deposition upon the first layer, the dye molecules of the second layer are mobile enough for each molecule to move to a partner in the first layer, thus forming the complex. As soon as a layer is deposited, the molecules are frozen at fixed positions (Bücher et al., 1967).

It should be possible to use this behavior for retrieving information at the molecular level in the following way. If a mixed monolayer of suitable composition is produced on the surface of water in which informational molecules, such as protein molecules, have been dissolved, these molecules will be adsorbed at the monolayer in such a way that the molecules in the monolayer are ordered at the matrix given by each adsorbed molecule (for example, in a mixed monolayer containing stearic acid and stearyl amine we will find free amino groups of a protein bound to stearic acid and carboxyl groups bound to stearyl amine). Let us assume that we deposit this layer on a slide by dipping the slide in the water and then removing the adsorbent molecules (for example, by changing the *p*H). It may be assumed that the matrix of these molecules will be frozen in the pattern of the monolayer. The original conditions in the solvent being restored, these molecules should be adsorbed specifically; that is, the monolayer should act as an antibody.

When using this monolayer technique, we may fix the chromophore of a dye at any given distance from a layer of a substance acting as an acceptor of the excitation energy of the dye, the distance being determined and maintained by long-chain fatty-acid-salt interlayers. The transfer of excitation energy from the sensitizer to the acceptor may be investigated by measuring the fluorescence quenching of the sensitizer or the sensitized fluorescence of the acceptor, or by determining the increase in photochemical stability of the sensitizer or in the sensitized photodecomposition of the acceptor (Barth et al., 1966). By these methods the energy transfer from an excited singlet or triplet state of a dye to an acceptor at a distance of 50 or 100 Å was studied (Bücher et al., 1967). A thin film of a metal or a semiconductor was found to be as suitable an acceptor as a dye monolayer, and compound systems with several acceptors competing with one another were assembled.

The fluorescence measurements of these simple organized systems of molecules may be predicted quantitatively by treating the sensitizing molecule as a classical radiating oscillator and by describing the acceptor as a body that has the property of absorbing energy in an alternating electric field. This approach does not refer to a particular molecular model of the acceptor—such as that given by the well known Perrin-Förster theory (Förster, 1966) of the radiationless resonance mechanism of energy transfer—and it might seem surprising at first that the long-range energy transfer from the sensitizer to the acceptor layer may be explained by this simple approach, since the layers of the dyes and the films of the metals and semiconductors used as acceptors absorb, even at their maximum absorbance, only a few percent of the incident light.

Thus, let us look at an oscillating dipole with amplitude μ_0, and frequency ν. The amplitude F_0 of the electric field at a distance r is (Pauling and Wilson, 1935):

$$F_0 = 4\pi^2\mu_0 \sin\theta / \lambda^2 r \qquad (r \gg \lambda) \qquad (1)$$

$$F_0 = (1 + 3\cos^2\theta)^{1/2}\mu_0/r^3 \qquad (r \ll \lambda) \qquad (2)$$

where θ is the angle between $\vec{r}$ and $\vec{\mu}$. Equation (1) holds for the wave field region ($r \gg \lambda$), where $\lambda = c/\nu$ and c = velocity of light, and Equation (2) holds for the proximity field region ($r \ll \lambda$), in which F_0 agrees with the Coulomb field of μ_0. Let us assume that a sphere of radius r surrounds the oscillator, carrying on its surface a very thin film of an isotropic absorbing substance—perhaps a gold film a few Ångströms thick. If A is the absorbance of the film for light of frequency ν, the average rate of energy absorption of an area f of the film when irradiated with light of intensity J (amplitude of electric field F_0' where $J = c\, F_0'^2/8\pi$) would be

$$\frac{dE'}{dt} = AJf = A\frac{c}{8\pi}F_0'^2 f \tag{3}$$

For the average rate of absorption by our spherical shell about the oscillator, this would give

$$\frac{dE}{dt} = A\frac{c}{8\pi}\int_0^{\pi} F_0^2 2\pi r^2 \sin\theta\, d\theta \tag{4}$$

and, using Equation (1),

$$\frac{dE}{dt} = A\frac{16\pi^4 c\mu_0^2}{3\lambda^4} \qquad (r \gg \lambda) \tag{5}$$

Thus we find that the rate of absorption is here independent of r and is A times the rate of emission of the dipole.

However, by introducing Equation (2) in (4), we find

$$\frac{dE}{dt} = A\frac{c\mu_0^2}{r^4} \qquad (r \ll \lambda) \tag{6}$$

which indicates that in the proximity field region the absorption of the shell is no longer independent of r: according to Equations (5) and (6) the rate of absorption of the shell is increased by a factor $(3/16\pi^4)(\lambda/r)^4$ as we proceed from the case $r \gg \lambda$ to the case $r \ll \lambda$, leaving the thickness of the film (the absorbance A) unchanged. This factor equals 2×10^5 for $r = 50$ Å and $\lambda = 5000$ Å. The rate of absorption is no longer A times the rate of emission, as it was for $r \gg \lambda$, but $2 \times 10^5 A$ times this rate; that is, almost 100 percent of the excitation energy of the oscillator is absorbed by the shell even if A is very small. The result obtained by this simple approach is in agreement with the result obtainable by use of Förster's resonance transfer model.

In the arrangements we have investigated, the energy is transferred from the lowest excited singlet or triplet state of the sensitizer to the acceptor, but the calculation shows that it should be possible, under favorable conditions, to obtain arrangements in which the energy is transferred from a more highly excited state of the sensitizer molecule directly to the acceptor before the molecule reaches its lowest excited state. Since the time necessary for an energy transfer may be calculated, study of such arrangements would give us information about the velocity and the nature of internal conversions.

Layer systems may be built in such a way that the acceptor is at a solid-liquid interface and may initiate a photochemical reaction in the solution while the sensitizer is inside the layer system, and is thus protected against reaction with molecules in the solution. If the sensitizer is in great excess with regard to the acceptor, an acceptor molecule might be doubly excited and then induce reactions of particular interest, such as the decomposition of water. We may also use a mixed monolayer of two or more acceptor dye molecules with different reduction potentials. Properly chosen, these acceptors may induce consecutive photochemical steps in the reaction. The acceptors may be fixed in the top layer by means of long hydrocarbon chains (as with the dyes I and II), or acceptors and reaction catalysts such as enzymes may be adsorbed at the surface of the layer system. We have recently obtained layer arrangements showing energy transfer from chlorophyll to an acceptor (D. Möbius, unpublished results), and it will be of interest to study coupled photochemical processes in such arrangements, particularly electron transport chains similar to those proposed for photosynthesis in the chloroplasts.

Besides energy transfer, electron transfer in monolayers may be studied. A monolayer of a long-chain fatty-acid salt may be deposited on an evaporated metal film and covered by another metal film. The DC conductance of such an arrangement is practically independent of the temperature and at high voltage the current increases exponentially rather than proportionally with voltage; at low voltage conduction is increased by a factor of 10 as we go from the C_{22} (behenic) acid to the C_{20} (arachidic) acid, and again by a factor of 10 as we go further to the C_{18} (stearic) acid (Bücher et al., 1967). All this demonstrates an electron-tunneling mechanism and shows that other conduction mechanisms may be neglected. The conductivity is particularly sensitive to layer imperfections and is a good test of the effectiveness of the technique used to obtain the layer systems.

Because electron transport through biological membranes may be important for the coupling of reactions on the two sides of a membrane, it should be valuable to study the conductance of monolayers of various compositions. A conjugated chain such as a carotinoid stuck in a layer between the fatty-acid chains should strongly reduce the width of the potential barrier through which an electron must tunnel, since the conjugated chain would conduct electrons. An electron would have to tunnel only through the barriers at the ends of the conjugated chain: it would be much more likely to do so than to tunnel through a fatty-acid-salt layer, since the tunneling probability decreases exponentially as the width of a potential barrier is increased. It would be of interest to study systematically the effect of changing the shape of a potential barrier between metal electrodes by investigating arrangements of layers of molecules with unsaturated portions differing in size and position. The energy of the lowest empty states of the conjugated system, as well as the width of the potential barrier, should determine the conductivity.

The probability that an electron will tunnel through a long-chain fatty-acid-salt layer is so small that the layer may stop electron transfer processes. Let us look at an electronically excited dye molecule separated from a trap for the excited electron by an arachidate monolayer. The time required for the electron to tunnel through the fatty-acid-salt barrier is estimated to be about a second. This is long compared to the life time of an excited singlet state. A barrier of two arachidate layers would prevent an electron transfer from an excited triplet state to the trap, although a transfer of the excitation energy to a suitable acceptor separated by several monolayers would still be possible. We may thus distinguish between the different possibilities of a sensitizing process, and by this technique the photosensitization of AgBr (Bücher et al., 1967) was studied.

An electron transfer between monolayers of two different dyes may also be studied by assembling a system consisting of a metal film covered by a fatty-acid-salt layer, a layer of the first dye, and a layer of the second. This arrangement may then be covered by arachidate layers and by a second metal film, or it may be dipped in a solution connected with a reference electrode (K. H. Beck and H. Schreiber, unpublished results; cf. Kuhn, 1967). When using a suitable pair of dyes such as perylene and zinc tetrabenzporphin, a photovoltage is obtained, and with these dyes the action spectrum of the photovoltage corresponds with the absorption spectrum of a perylene layer. The photovoltage is strongly reduced as we interpose a fatty-acid-salt layer between the dye layers, probably because this layer prevents electrons from tunneling between the two dye layers. It would be interesting to study arrangements in which the two dye layers are separated by a layer of molecules with low-lying empty states—for example, a polyene. When interposing such an electron-conducting layer we would expect the photovoltage to increase because of the greater charge separation. It would be of particular interest to study the photovoltage of an arrangement containing a chlorophyll monolayer followed by a monolayer of carotin chains stuck between fatty acid chains, and then by an adsorbed layer of ferredoxin, or of a sandwich of chlorophyll and cytochrome. If the chlorophyll were excited, the arrangements might pump electrons across the membrane. By assembling other species of molecules involved in the photosynthesis of plants we might obtain a possible model for a photosynthetic acting unit. By combining monolayers of molecules that donate, conduct, and trap electrons, we could obtain many different arrangements, with electrical and photoelectrical properties similar to those of assemblies of semiconductors.

The different arrangements described here all have the properties of a switch. Thus, by combining such electrical, optical, or chemical units, computers might be devised. A layer seems to be particularly suitable for processing a large amount of information. It should be technically possible to store a single bit on an area of $(100\ \text{Å})^2$ and thus to store 10^{14} bits (the probable information content of the human brain) on a square decimeter. This information, stored on an adequately perforated gold film, could be quickly retrieved by an electron microscope, but it would be technically important to have an easy method of copying such a giant memory. Let us consider the following possibility. If we bring a monolayer of a dye in contact with the perforated gold film and illuminate the system the excitation energy of the dye will be transferred to the gold film at the points where the dye is less than about 50 Å from the gold. The dye will then be stable at these points, while the dye molecules at the perforations of the gold will be bleached out. The duplicate thus obtained might be removed and developed by one of the techniques used in electron microscopy.

These examples show that the monolayer technique considered here might be a useful tool in future developments in synthetic molecular biology and in molecular engineering.

REFERENCES

Barth, P., K. H. Beck, K. H. Drexhage, H. Kuhn, D. Möbius, D. Molzahn, K. Röllig, F. P. Schäfer, W. Sperling, and M. M. Zwick (1966). 2nd. Int. Farbensymposium 1964. In *Optische Anregung organischer Systeme*. Weinheim: Verlag Chemie. S. 639.

Blodgett, K. B. (1935). *J. Am. Chem. Soc.* **57**, 1007.

Bücher, H., K. H. Drexhage, M. Fleck, H. Kuhn, D. Möbius, F. P. Schäfer, J. Sondermann, W. Sperling, P. Tillman, and J. Wiegand (1967). *Mol. Cryst.* **2**, 199.

Bücher, H., H. Kuhn, B. Mann, D. Möbius, L. von Szentpály, and P. Tillmann (1967). *Photog. Sci. Eng.* **11**, 233.

Förster, Th. (1966). In Riehl and Kallmann, eds., *Internationales Lumineszenz-Symposium über die Physik und Chemie der Szintillatoren*, p. 1. München.

Kuhn, H. (1967). *Naturwiss.* **54**, 429.

Langmuir, I. (1917). *J. Am. Chem. Soc.* **39**, 1848.

Langmuir, I. (1939). *Proc. Roy. Soc.* (London) A **170**, 15.

Pauling, L., and E. B. Wilson (1935). *Introduction to Quantum Mechanics*. New York: McGraw-Hill.

Rayleigh (John William Strutt), Lord (1899). *Phil. Mag.* **48**, 337.

Sobotka, H. (1956). *J. Coll. Sci.* **11**, 435.

J. L. HOARD

Department of Chemistry
Cornell University
Ithaca, New York

Some Aspects of Heme Stereochemistry

The complex physico-chemical mechanisms by which the hemoproteins perform their biological functions have received a good deal of attention, but nothing that approaches detailed elucidation. What is entirely clear is that in every such mechanism a central role is played by one or another of the iron porphyrins. Thus a ferrous protoporphyrin-IX moiety—the heme— is initially and most directly concerned in the reversible oxygenation of hemoglobin or myoglobin. Each heme may utilize its iron atom for the attachment of one molecule of oxygen; the complexing interaction involves a fundamental change in electronic structure that is marked by the transition in spectroscopic state of the iron atom from a (high-spin) quintet with four unpaired electrons to a (low-spin) diamagnetic singlet with no unpaired electrons. It is still uncertain whether the iron atom in the oxygenated species coordinates just one or both of the oxygen atoms of the O_2 group; in structural terms, the choice lies between the standard octahedral configuration that is usual for low-spin iron and a seven-coordinate geometry.

Quantitative stereochemical descriptions of the heme before and after oxygenation are surely prerequisite, if by no means sufficient, to the formulation of the critical initial steps in any plausible mechanism of the hemoprotein function. It might be supposed that this need would be adequately filled by utilizing the powerful three-dimensional techniques of X-ray crystallography for the direct determination of hemoprotein structure; in fact, however, it is highly doubtful whether any such determination can approach the minimum level of accurate definition that is required for detailed theoretical interpretation. The semiquantitative descriptions of hemoprotein structure that come from direct X-ray analysis provide an indispensable background, but it would seem that even the moderately detailed quantitative aspects of heme behavior must be derived—synthesized, in part—from data obtained by independent study of the metalloporphyrins in conjunction with theory already well-tested in its applicability to simpler complexed species.

Indeed, it is the stereochemistry of iron porphyrins as pure compounds and, by extension, as hemes within hemoproteins that constitutes the principal theme of this article. One finds an apparently unorthodox geometry for the coordination groups in high-spin iron porphyrins that is, in fact, quite con-

Studies of porphyrin structure in the Baker Laboratory of Cornell University have been supported by Public Health Service Research Grant GMO9370 from the National Institutes of Health, General Medical Sciences, by National Science Foundation Grant G-23470, and by the Advanced Research Projects Agency under Contract No. SD-68 with Cornell University.

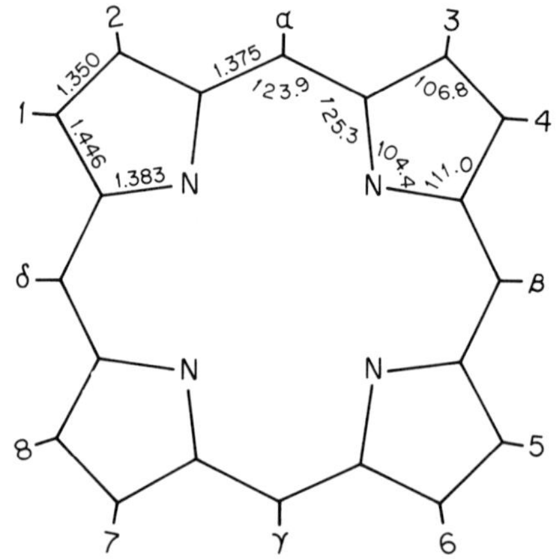

FIGURE 1.
Diagram of the (anionic) porphine skeleton. The bond lengths and angles shown are from the NiDeut molecule.

sistent with established structural principles and approximate bonding theory. One is then led to the confident expectation that striking conformational and, more often than not, configurational alterations in and around the coordination group of the porphyrin must accompany the changing electronic structure that is signalized by transition from the high-spin to the low-spin state. This postulated behavior, as applied to a hemoprotein, does not necessarily—nor even probably—require a substantial movement of the iron atom relative to axes fixed either in the protein molecule or in the crystal. It is in this connection that a careful assessment of the general significance of several recent X-ray studies of the hemoproteins is required; both the virtues and the limitations of the special techniques employed in these studies are deserving of the strongest emphasis.

The two distinctive seven-coordinate configurations that must be considered for the coordination group in the heme of oxymyoglobin or oxyhemoglobin, if symmetrical involvement of both oxygen atoms of the O_2 group be assumed, are each subject to severe stereochemical constraints that are not readily satisfied by any low-spin iron complex. The capped trigonal prism of C_{2v} symmetry, hitherto neglected in theoretical treatments of bonding in the oxygenated heme, lacks much of the stereochemical disability of the pentagonal bipyramid and affords an equally satisfactory basis for bonding considerations.

The discussion in detail of the points made and the problems raised in these introductory paragraphs is preceded, in the following section, by a summary of the more important structural characteristics exhibited by the porphine skeleton in porphyrins.

THE STEREOCHEMISTRY OF THE PORPHINE CORE IN PORPHYRINS

A diagram of the carbon-nitrogen skeleton in the anionic porphine core of metalloporphyrins is given in Figure 1. No classical array of double bonds is displayed; the contributions of resonance forms or, alternatively, of delocalized π bonding to the electronic structure of the skeleton, as evaluated for a typical porphyrin molecule with the aid of structural data, are not readily indicated by any simple combination of classical formulae. Several biologically interesting porphyrins, including protoporphyrin-IX, are for-

TABLE 1.
Bond orders in the porphine skeleton.

Bond type[a]	Length (Å)	MO Bond order	π Bond order	Kekulé-Pauling
$N—C_a$	1.383	1.39	0.2[b]	1/4
$C_a—C_b$	1.446	1.40	0.18	1/4
$C_b—C_b$	1.350	1.92	0.88	3/4
$C_m—C_a$	1.375	1.81	0.67	1/2

[a]Subscripts a and b identify the respective α and β pyrrole carbon atoms; the subscript m identifies methine carbon.

[b]This estimate corresponds to 1.42 Å for the length of a trigonally hybridized C—N σ bond and 1.274 Å for a C=N double bond.

mally derivatives of deuteroporphyrin-IX, the trivial name for 1,3,5,8-tetramethylporphine-6,7-dipropionic acid. Replacement of the two hydrogen atoms that are attached to nitrogen in a neutral porphyrin molecule by a divalent metal atom, or by ClFe(III), CH_3Fe(III), and so on, gives a metalloporphyrin; the central metal atom, to which the four nitrogen atoms are complexed, then assumes a coordination number of four or more as determined by the number of axial ligands.

The bond distances and angles shown in Figure 1 for the porphine skeleton come from the structure determination for triclinic crystals of nickel(II) 2,4-diacetyldeuteroporphyrin-IX dimethyl ester (abbreviated as NiDeut). Although the determination is not accurate enough to exhibit the subtle variations in bond parameters that one expects to be produced by the diverse chemical nature and asymmetric substitution pattern of the side chains, the distances and angles given on the diagram, obtained by averaging in agreement with the fourfold symmetry of C_{4v}, are probably the best available for the porphine skeleton in a metalloporphyrin (Hamor, Caughey, and Hoard, 1965). Essentially equivalent results come from the accurate determination of structure for copper(II) tetraphenylporphine (Fleischer, Miller, and Webb, 1964).

Bond orders within the porphine skeleton of the NiDeut molecule as evaluated by the several approaches that have utilized experimentally determined bond lengths to this end are given in Table 1. (Hamor, Caughey, and Hoard, 1965). The MO bond orders were read from curves constructed by Lofthus (1959) with the aid of molecular orbital theory; orders of 1 and 2 correspond to the respective bond lengths of 1.543 and 1.335 Å for C—C links and to 1.474 and 1.274 Å for C—N links. The π bond orders were calculated by the method of Cruickshank and Sparks (1960); π bond orders of 0 and 1 for C—C links correspond, respectively, to a pure σ bond of 1.48 Å between trigonally hybridized carbon atoms and a standard double bond of 1.335 Å length. Simple enumeration, following Pauling (1960), of the classical Kekulé-type formulae gives the double bond characters in the fifth column.

The observed shortening of the $C_a—C_b$ distance from the pure σ bond (sp^2 hybridization) value of 1.48 Å provides a more sensitive test for delocalization of π bonding within the pyrrole rings than does the close approach of the $C_b—C_b$ bond distance to the double bond value. It is seen, nevertheless, that the significant part of the delocalized π bonding that tends to enforce a strictly planar configuration in the equilibrium state of the porphine skeleton is largely confined to the central sixteen-membered ring (Figure 1) of in-and-out contour. Inasmuch as a planar configuration, in contrast with the case of a benzenoid hydrocarbon, cannot be expected to minimize angular strains in the pattern of σ bonding, one is not sure whether the equilibrium configuration of the skeleton should be planar or modestly ruffled. The usual expectation is that the π bonding will dominate, thus giving a planar configuration, but observations on the isolated molecule are required to prove or disprove the point. Easy deformation of the porphine skeleton normal to the mean plane is to be expected in any case.

That the porphine skeleton is readily deformable by environmental stresses into a ruffled or domed configuration is fully confirmed by several studies of crystalline structure. Thus in tetraphenylporphine (Hamor, Hamor, and Hoard, 1964) and its copper and palladium derivatives (Fleischer, Miller and Webb, 1964), the porphine skeleton is markedly ruffled in agreement with the point group, $\bar{4}$–S_4; methine carbon atoms of α and γ type (Figure 1) lie nearly 0.40 Å above the mean skeletal plane, those of β and δ type the same distance below. Some departure from planarity of the porphine skeleton has been objectively confirmed for every crystalline porphyrin in which the absence of packing disorder permits a definite answer. Even in crystalline porphine (Webb and Fleischer, 1965) there is a quasi-S_4 ruffling of the skeleton that, however trivial it be in a practical sense, is objectively significant in terms of the assigned accuracy with which atomic positions were determined.

THE IRON PORPHYRINS

Stereochemical parameters for the coordination groups in metalloporphyrins as given by X-ray analysis are listed in Table 2; analogous dimensional data for porphine and for the tetraphenylporphine molecule as it exists in tetragonal and in triclinic crystalline forms are listed also. The "radius of the central hole," Ct—N, gives the distance of each porphine nitrogen atom from the center of the grouping; it is identical with the complexing M—N bond distance whenever the metal atom lies in the plane of the nitrogen atoms. For those species, just half of the total number, in which the molecule lacks, or is not required in the crystal to display, a fourfold axis, the values listed for M—N and Ct—N are obtained by averaging in agreement with this symmetry element. Deviations of individual M—N or Ct—N distances from the average for any given porphyrin are trivial, except in the tetraphenylporphine molecule (*tr*-TPP) as it occurs in the triclinic crystalline modification. In the *tr*-TPP molecule, the central pair of hydrogen atoms are attached to a diagonally opposed pair of nitrogen atoms; the separation of these nitrogen atoms is 0.14 larger than that of the other diagonally opposed pair* (Silvers and

*In the *tet*-TPP (Hamor, Hamor, and Hoard, 1964) and porphine (Webb and Fleischer, 1965) molecules in the crystals, by contrast, difference Fourier syntheses of electron density show the hydrogen as four well-defined half-atoms each attached to a nitrogen atom; the frequency of proton jump from one position to another probably is comparable, at least, with that associated with the thermal vibrations of the nitrogen atoms in the plane of the porphine skeleton (Hoard, 1966).

TABLE 2.
Stereochemical parameters for porphyrins and metalloporphyrins (distances in Ångstrøms).

	M—N[a]	Ct.—N[b]	Δ[c]	Axial Ligands
NiEtio[d]	1.957	1.957	—	None
NiDeut[e]	1.960	1.960	—	None
CuTPP[f]	1.981	1.981	—	None
PdTPP[g]	2.009	2.009	—	None
H_2OZnTPP[h]	$\gtrsim$2.05	2.042	$\gtrsim$0.19	H_2O at $\lesssim$ 2.21 Å
ClFeTPP[i]	2.049	2.012	0.383	Cl at 2.19 Å
Chlorohemin[j]	2.062	2.008	0.475	Cl at 2.218 Å
MeOFeMeso[k]	2.073	2.022	0.455	OMe at 1.842 Å
Porphine[l]	—	2.051	—	None
tet-TPP[m]	—	2.054	—	None
tr-TPP[n]	—	2.065	—	None
Fe(III)Mb[o]	~1.9	~1.9	~0.30[p]	N at ~ 1.9, H_2O ~ 2.1 Å

[a]Metal-nitrogen distance.
[b]Radius of central hole: see text.
[c]Out-of-plane displacement of the metal atom.
[d]Nickel(II) etioporphyrin-I (Fleischer, 1963).
[e]Nickel(II) 2,4-diacetyldeuteroporphyrin-IX dimethyl ester (Hamor, Caughey and Hoard, 1965).
[f]Copper(II) tetraphenylporphine (Fleischer, Miller, and Webb, 1964).
[g]Palladium(II) tetraphenylporphine (Fleischer, Miller, and Webb, 1964).
[h]Aquozinc(II) tetraphenylporphine (Glick, Cohen, and Hoard, 1967).
[i]Chloroiron(III) tetraphenylporphine (Hoard, Cohen, and Glick, 1967).
[j](Koenig, 1965).
[k]Methoxyiron(III) mesoporphyrin-IX dimethyl ester (Hoard, Hamor, Hamor, and Caughey, 1965).
[l](Webb and Fleischer, 1965).
[m]Tetraphenylporphine in tetragonal crystals (Hamor, Hamor, and Hoard, 1964).
[n]Tetraphenylporphine in triclinic crystals (Silvers and Tulinsky, 1964).
[o]Ferrimyoglobin (Kendrew *et al.*, 1960).
[p](Watson, 1966).

Tulinsky, 1964). The condensed notations employed for most of the porphyrins are explained in the footnotes to Table 2.

Reserving ferrimyoglobin for later consideration, a number of interesting conclusions emerge from the data in Table 2. Only for a complexing bond length M—N < 2.01 Å does the metal atom lie in plane with the four nitrogen atoms. Ct—N $= 2.015 \pm 0.007$ Å covers the range of experimental values for the radius of the central hole in the three iron porphyrins; this radius (2.015 Å) is close to the mean of the smallest (1.957 Å in NiEtio) and the largest (2.05–2.06 Å in porphine and tetraphenylporphine) values thus far reported for porphyrins, and it probably corresponds to approximate minimization of strain in the porphine skeleton. The out-of-plane displacement (Δ) of the metal atom is substantial in each of the iron porphyrins listed in Table 2; all of these (including ferrimyoglobin) are high-spin Fe^{+++} derivatives.

The key to the understanding of the generally unorthodox stereochemistry of the high-spin iron porphyrins is contained in an earlier hypothesis (Hoard, Hamor, Hamor, and Caughey, 1965) that, in view of more recent experimental and theoretical studies, may be stated as a probably valid principle: A substantial displacement ($>$0.30 Å) of the iron atom from the mean plane of the four porphine nitrogen atoms is a normal structural property of all high-spin iron porphyrins. Based initially upon the premise that the electronic structure of high-spin Fe^{+++} gives this ion too large an effective size to allow its accommodation in the plane of the nitrogen atoms, the principle is expected to apply *a fortiori* to ferrous porphyrins utilizing the still larger high-spin Fe^{++}; it is expected also to apply for any coordination number (4, 5, 6, 7) that may be accessible to the porphyrin. (See also Zerner, Gouterman, and Kobayashi, 1966). A qualitative feeling for the validity of this principle is gained by considering a square-planar model for a high-spin iron porphyrin.

In such a model the $d_{x^2-y^2}$ orbital of the iron ion carries an unpaired electron and has each of its four lobes directed, at extremely close quarters, toward an electron pair on a nitrogen atom. That the dimensional relations with Fe—N ~2.01 Å are a good deal too tight for favorable energy relations is quite in line with earlier experience; thus the Fe—N bond distance in amine-polycarboxylate complexes of high-spin Fe^{+++} ranges from 2.16 to 2.33 Å (Lind, Hamor, Hamor, and Hoard, 1964; Cohen and Hoard, 1966), and the sum of Pauling's ionic radii (for octahedral coordination) is 2.35 Å. Removal of the electron from $d_{x^2-y^2}$ with spin pairing in one of the more stable d orbitals and the formation of dsp^2 bonds (or the molecular orbital equivalent thereof) to give a complex of intermediate spin is an obvious possibility that does not appear to be utilized by either ferrous or ferric porphyrins. The radius of the central hole in phthalocyanine appears to be fully 0.10 Å smaller than that in porphine (Hamor, Caughey, and Hoard, 1965); a high-spin iron phthalocyanine, consequently, would need to take an excessively pyramidal shape. Griffith's (1959, 1961) assignment of chloroiron(III) phthalocyanine and ferrous phthalocyanine to intermediate spin states would seem to be theoretically straightforward and, in the first instance, to be substantiated by his analysis of the electron spin resonance data.*

The square-pyramidal coordination groups of the high-spin MeOFeMeso, chlorohemin, and ClFeTPP molecules have Fe—N bond lengths, lying within the range 2.061 ± 0.012 Å, much longer than the covalent (dsp^2 or d^2sp^3) low-spin value, 1.91 Å (Pauling, 1960). One then asks why the bond lengths, Fe—Cl = 2.19–2.22 Å and Fe—O = 1.84 Å, along the unique axes of these high-spin complexes are as short as, or somewhat shorter than, the values expected for low-spin (octahedrally coordinated) Fe(III) (Pauling, 1960).

From a qualitative consideration of electron correlations in such a square-pyramidal complex, the charge distribution represented by the unpaired electron in the $3d_{z^2}$ orbital of Fe^{+++} must be so strongly polarized in the field of the axial ligand as to be largely concentrated in the axial lobe extending from the Fe^{+++} ion directly away from the ligand. Rephrased in molecular orbital terms, the d_{z^2} orbital of Fe^{+++} can combine with the appropriate σ-type orbital of the axial ligand to give a strongly bonding orbital that accommodates a donor electron pair and a weakly antibonding orbital that houses the unpaired electron; the 4s and 4p_z orbitals of Fe^{+++} can further contribute to the bond-

*Griffith (1959, 1961) erroneously attributes the data for chloroiron(III) phthalocyanine to chlorohemin—a definitely high-spin molecule. The description of these phthalocyanine derivatives as thermal mixtures of high-spin and low-spin forms, in analogy with what appears to be a valid description for an occasional porphyrin (Falk, 1964), is unappealing; the maximum of spin-pairing required for the low-spin state appears to be theoretically unnecessary, and awkward stereochemical relations accompany the high-spin state (Hoard, Hamor, Hamor, and Caughey, 1965).

ing. A recasting that uses, after Pauling, a pair of hybrid valence bond orbitals on the Fe^{+++} ion gives a no less satisfactory description of the bonding. One further notes that closed-shell or nonbonding repulsions between the axial and the basal ligands tend to be relatively small in the pyramidal configuration and, indeed, are nearly trivial in the MeOFeMeso molecule (Hoard, Hamor, Hamor, and Caughey, 1965). Thus it may be concluded that a short axial bond of relatively high stability is to be expected in a high-spin five-coordinate ferric porphyrin provided that, in quantum mechanical terms, the axial ligand species has the appropriately deformable electronic structure that makes it a good donor of an electron pair.

No direct determination of crystalline structure for a pure low-spin iron porphyrin has yet been carried out. (This is simply the practical consequence of the reluctance displayed by the metalloporphyrins, especially those of prime biological interest, to provide single crystals that are suitable for X-ray diffraction analysis). Theory and experiment join in making six the characteristic coordination number for a low-spin iron porphyrin; no alternative need be considered except for the oxygenated heme. Owing to Pauling's theoretical studies and his semi-empirical analyses of the experimental data obtained from a wide variety of the simpler complexed species, one can specify, within a narrow range, the quantitative stereochemical parameters that are confidently expected for the octahedral coordination groups in low-spin iron porphyrins. Inasmuch as the (d^2sp^3) octahedral covalent bond radii (Pauling, 1960) of Fe(II) and Fe(III) differ by only $\sim$0.02 Å, the mean value (1.22 Å) is satisfactory for either; Fe—N and Fe—O bond lengths of 1.92 and 1.88 Å, respectively, are thus predicted. It is pertinent to note that low-spin Fe(II) and Co(III) are isoelectronic, and the Co—N and Co—O bond lengths of, respectively, 1.92 and 1.88–1.90 Å, are observed in the low-spin ethylenediaminetetraacetatocobaltate(III) ion (Weakliem and Hoard, 1959). A near centering, possibly within the accuracy of experimental measurement, of the iron atom among the four porphine nitrogens is expected even when the two axial ligands are chemically and structurally rather different.

Introduction of the obvious modifications appropriate to Fe^{++} into the earlier discussion of electronic structure in the five-coordinate high-spin ferric porphyrins leads to the conclusion that a high-spin ferrous porphyrin also should be stabilized in five-coordination by an (electrically neutral) axial ligand having strong donor properties.* Such a ligand, presumably, is the histidine nitrogen atom that is axially bonded to high-spin Fe^{++} in the square-pyramidal coordination group in deoxymyoglobin (Nobbs, Watson, and Kendrew, 1966). It is probable that the length of the axial Fe—N bond is at or near the low-spin value of $\sim$1.92 Å—a prediction justified by the generally close analogy with the five-coordinate chlorohemin and MeOFeMeso molecules (see above)—and, consequently, that this bond distance undergoes little or no alteration upon oxygenation of the heme. No such approach to invari-

*The absence of the intermediate-spin state in both ferrous and ferric porphyrins—at any rate in those that have at least one axial ligand—can be understood in the following terms. A single axial ligand stabilizes the porphyrin in the high-spin square-pyramidal configuration, and the joint action of two such ligands is required to force spin-pairing. The strong interaction of the iron atom with both axial ligands that is needed for clearing the d_{z^2} orbital of its unpaired electron requires the iron atom to be essentially coplanar with the porphine nitrogens, and this is feasible only if the $d_{x^2-y^2}$ orbital is simultaneously cleared of its unpaired electron.

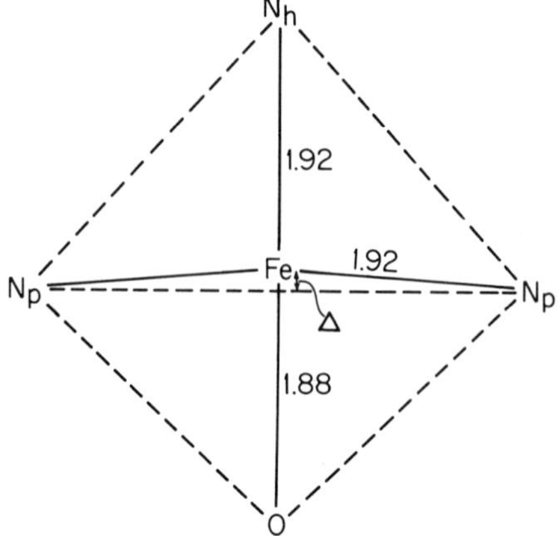

FIGURE 2.
A diagonal section through the proposed octahedral model for the coordination group in the heme of oxymyoglobin.

ance in the length of the complexing bonds that link the iron atom to the four porphine nitrogen atoms is to be expected during oxygenation of the heme; the predicted bond lengths are ≥ 2.06 Å in the high spin deoxymyoglobin, and ~ 1.92 Å in the low-spin oxymyoglobin. For Fe—$N_p \gtrsim 2.06$ Å ($N_p \equiv$ porphine nitrogen; $N_h \equiv$ histidine nitrogen) and Ct—N = 2.01 Å in deoxymyoglobin, an out-of-plane displacement of the high-spin iron atom of $\gtrsim 0.45$ Å is required.

With the aid of Figure 2 and Table 3 one can see that a quantitatively reasonable model for the coordination group in the low-spin oxymyoglobin is obtained if just one atom of the oxygen molecule is coordinated to Fe(II); the possibility of symmetrical attachment of the O_2 molecule with concomitant seven-coordinate geometry requires separate consideration. A diagonal section through the octahedral coordination group in a plane that includes the Fe, N_h, O and two of the four N_p atoms is shown in Figure 2; the N_p—N_p distance in Table 3 refers to nearest neighbors, and is $1/\sqrt{2}$ times the diagonal separation shown in the diagram. The nearest-neighbor nonbonded separations given in Table 3 correspond to Fe—O = 1.88 Å, Fe—N = 1.92 Å, and the listed values (Δ) of the out-of-plane displacement of the iron atom toward the histidine nitrogen atom.

The nonbonded separations (Table 3) are all so short as to imply a very large (> 50 kcal./mole) repulsive energy that is not necessarily minimized for $\Delta = 0$. The N_p—N_p distance, however, has its maximum value for $\Delta = 0$ and should not be appreciably decreased; it is already so small as to produce a considerable strain in the porphine skeleton (see above). $\Delta = 0.05$ Å makes the

TABLE 3.
Computed geometry for octahedral coordination in oxymyoglobin (distances in Ångstrøms).

Δ	N_p—O	N_p—N_h	N_p—N_p
0	2.68	2.72	2.72
0.05	2.65	2.75	2.72
0.12	2.60	2.80	2.72
0.19	2.55	2.84	2.70

difference of the N_p—N_h and N_p—O separations just the 0.10 Å expected from the difference of the van der Waals radii (Pauling, 1960), leaves N_p—N_p unaltered to better than three significant figures, and requires little distortion of the bond angles from the ideal values. N_p—O = 2.65 Å for Δ = 0.05 Å is to be compared with the N—O = 2.66 ± 0.06 Å that defines the observed range in numerous aminepolycarboxylate complexes with coordination numbers of 6, 7, 8, 9, and 10. The effects of further increasing Δ are deleterious in all respects but one, and that no longer of the first importance; the N_p—N_h separation continues to increase. Thus an out-of-plane displacement of ~0.05 Å for the iron atom is not improbable. This conclusion depends primarily on the relative values of the Fe—O and Fe—N bond lengths and of the packing radii of the nitrogen and oxygen atoms. If, as the double bond character in Pauling's (1949, 1964) proposed model might suggest, the Fe—O bond length were shorter than 1.88 Å, Δ would be limited to a still smaller value.

Complexing of the oxygen molecule to the iron atom in the heme of deoxymyoglobin should then be accompanied by a tightening of the Fe—N_p bonds from ~2.06 Å to ~1.92 Å, and a decrease in Δ from ~0.45 Å to $\lesssim$0.05 Å. The shifting of atomic positions is best considered relative to axes fixed in the molecule or in the crystal. Because the iron atom is tightly bonded, before and after oxygenation, to a histidine residue of the protein framework, it is perhaps more likely than not that the coordinates of the iron atom do not change very much during oxygenation. In these terms, the motion that brings the iron and the four porphine nitrogen atoms into near coplanarity must be looked for in the changing coordinates of the nitrogen and other atoms in the flexible skeleton of the heme. This point is directly pertinent to later consideration of the structural results reported from X-ray studies of various myoglobin derivatives.

The foregoing remarks on the probable stereochemical changes that accompany the oxygenation of deoxymyoglobin are clearly intended to apply, with some modification of detail, to such generally analogous processes as the addition of carbon monoxide to deoxymyoglobin and the substitution of cyanide ion for a water molecule in the heme of ferrimyoglobin. One is quite sure that either CO or CN^- attaches through its carbon atom to the iron and, moreover, that the principal axis of the ligand, by contrast with that of O_2, is essentially colinear with the Fe—N_h bond.

The possibility that the two atoms of the O_2 group are symmetrically coordinated to the iron atom in the coordination group of the heme in oxymyoglobin arose from theoretical considerations (Griffith, 1956) in which the stereochemical constraints implicit in a seven-coordinate geometry were not discussed. That such constraints are important, the shortness of the bond within the oxygen molecule notwithstanding, will appear during subsequent discussion.

The maximum effective symmetry of the problem, $C_{2v}-mm2$, allows two different seven-coordinate geometries. Theoretical discussions of the bonding have uniformly implied a pentagonal bipyramidal configuration in which the two oxygen and three nitrogen (N_h, $2N_p$) atoms define the pentagonal girdle around the iron atom (as seen in Figure 3), and two (N_p) nitrogen atoms occupy the apical positions. Noting, however, that in C_{2v} all degeneracy vanishes from the system of orbitals on the iron atom, one sees how to formulate a

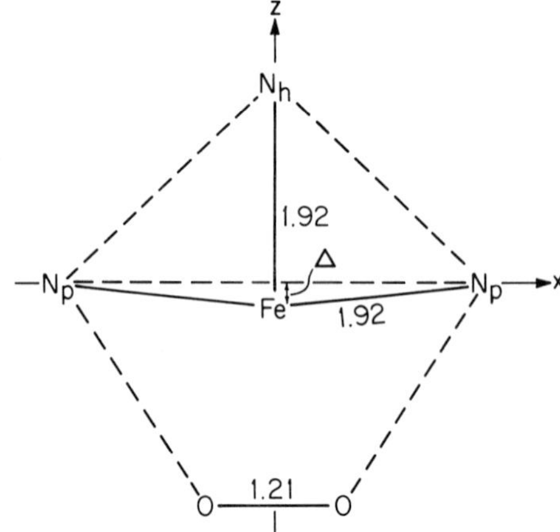

FIGURE 3.
The pentagonal girdle in the proposed pentagonal bypyramidal model for the coordination group in the heme of oxymyoglobin.

bonding scheme that is alternative, but very similar, to that of Griffith with use of the configuration first observed for the TaF_7^{--} and NbF_7^{--} ions (Hoard, 1939). This configuration, derivable by capping a trigonal prism on one of the square faces, affords rather better packing relations than does the pentagonal bipyramid, a modest advantage that may become quite significant in special circumstances. Discussion of the pentagonal bipyramid is simplified by the fact that all of the critical nonbonding repulsions are confined to the plane of the pentagonal girdle of five ligand atoms.

Seven-coordinate *high*-spin Fe^{+++} is observed in two series of aminepolycarboxylate complexes (see Cohen and Hoard, 1966, for complete references). The geometry of the coordination group in the ethylenediaminetetraacetatoaquoferrate(III) ion (Lind, Hamor, Hamor, and Hoard, 1964) is approximately pentagonal bipyramidal with the water molecule, the two nitrogen atoms, and two of the carboxylate oxygen atoms forming a not quite planar girdle around the iron atom. The nonbonded O—O and N—O separations at 2.500 and 2.595 Å, respectively, are essentially the shortest that have been observed in a dozen or more aminepolycarboxylate complexes of various metal atoms; the N—N nonbonded distance is a minimum, 2.72 Å, in the low-spin ethylenediaminetetraacetatocobaltate(III) ion (Weakliem and Hoard, 1959). Recalling that the rapid exponential increase of repulsive energy with decreasing separation becomes the controlling factor at small distances, one concludes that nonbonding separations below 2.60 and 2.70 Å for, respectively, N—O and N—N tend strongly to destabilize configuration.

Calculated nonbonding separations in the pentagonal girdle (Figure 3) of the bipyramidal model for the coordination group in oxymyoglobin are listed in Table 4 for comparison with those in the octahedral model (Table 3). The data of Table 4 are obtained subject to the following constraints. A uniform bond distance, Fe—N_h = Fe—N_p = 1.92 Å is maintained, as in the earlier computations. The bond length, 1.21 Å, in molecular oxygen is employed for the O—O separation in the complex. Using the positioning of the pentagonal girdle in the *x-z* plane shown in Figure 3, the parameter, Δ, is the displacement of the iron atom from the origin (Ct) along negative *z*; this displacement is away from the N_h atom and toward the O_2 group. The required additional parameter can be taken as the distance of the O—O axis (Ax) from the iron

TABLE 4.
Computed geometry of pentagonal bipyramidal coordination in oxymyoglobin (distances in Ångstrøms).

Δ^a	Fe—Ax[b]	Ct[c]—Ax[b]	N_p—O	N_p—N_h	N_p—N_p	Fe—O
0	1.88	1.88	2.29	2.72	2.72	1.98
0.20	1.88	2.08	2.46	2.57	2.70	1.98
0	2.18	2.18	2.55	2.72	2.72	2.27
0.20	1.99	2.19	2.55	2.57	2.70	2.08

NOTE: All computations use Fe—N_h = Fe—N_p = 1.92 Å, O—O = 1.21 Å.
[a]Out-of-plane displacement of iron atom toward the oxygen molecule.
[b]Ax is the axis of the oxygen molecule.
[c]Ct is used for the origin of coordinates.

atom, i.e., the Fe—Ax in Table 4; a sensible alternative choice for this parameter is the nonbonded separation, N_p—O. Calculations are then made by assigning either a maximum value to the Fe—Ax bonding distance or a minimum value to the N_p—O nonbonding separation.

It would seem that the limitation, Fe—Ax $\lesssim$ 1.88 Å, is implicit in Griffith's (1956) theoretical treatment of this complexing bond. Using the coordinate system of Figure 3, the oxygen molecule in its prepared valence state has an empty Π^*2p_z molecular orbital that is to form a π bond with the filled $3d_{xz}$ orbital on the iron atom; Griffith also has a filled $\Pi 2p_z$ molecular orbital on the O_2 group in bonding interaction with the empty $3d^24sp^3$ hybrid orbital on the iron atom that extends along the negative z axis in Figure 3. Both types of bonding, but especially the first, would seem to require for strong interaction an approach distance, Fe—Ax, as small as, or smaller than, the Fe—O bond length, ~1.88 Å, that corresponds to the norm in the standard octahedral case.

It is evident from Table 4 that the simultaneous conditions, Fe—Ax $\lesssim$ 1.88, N_p—O $\gtrsim$ 2.60, and N_p—N_h $\gtrsim$ 2.70 Å, cannot be satisfied or even closely approached for any value of Δ. This striking failure of steric merit relative to the standard octahedral case (Table 3) could be alleviated by allowing one or both of the Fe—N_h and Fe—Ax bonding distances to be substantially longer than, respectively, the 1.92 and 1.88 Å previously assumed. A substantial stretching of the Fe—N_h bond is quite improbable; consequently, the pentagonal bipyramidal model retains plausibility only if the Fe—Ax bonding interaction at a distance $\gtrsim$ 2.10 Å can be strong enough to require the spin-pairings that characterize the oxygenated heme.

The alternative seven-coordinate polyhedron of C_{2v} symmetry, derivable by capping a trigonal prism on one rectangular face (Hoard, 1939), is illustrated in Figure 4. The two orthogonal mirror planes, taken as the x-z and y-z axial planes, intersect in a twofold axis that coincides with the z axis. N_h and Fe atoms lie on the z axis, and the O_2 group lies in the x-z plane. The four N_p atoms define a rectangle (or *quasi*-square for present purposes); the point in which the z axis intersects the plane of this rectangle is taken as the origin (Ct) of coordinates. Δ is the displacement of the iron atom from the origin toward the O_2 group.

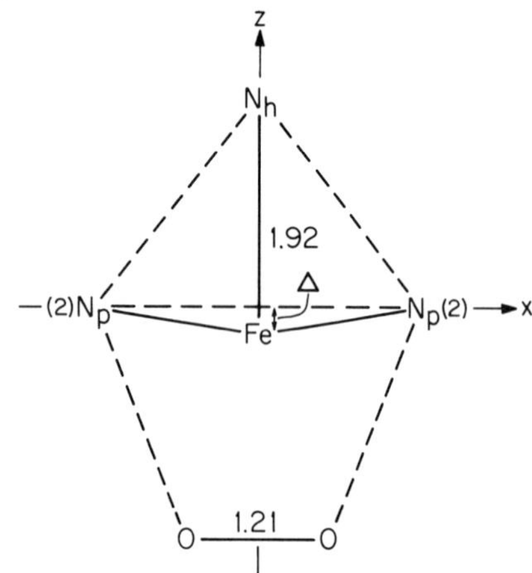

FIGURE 4.
The capped trigonal prism as an alternative model for a seven-coordination group in the heme of oxymyoglobin.

Reference to the computed geometry in Table 5 for the capped trigonal prism shows that this polyhedron is sterically rather poor relative to the standard octahedron (Table 3), but that it comes very much nearer to meeting the simultaneous conditions, Fe—Ax $\lesssim$ 1.88 Å, N_p—O $\gtrsim$ 2.60 Å, and N_p—N_h $\gtrsim$ 2.70 Å, than does the pentagonal bipyramid (Table 4). One then notes that either type of seven-coordinate polyhedron is formally transformed into the other by prescribing a rotation of $\pi/4$ around z for the O_2 group, but that it is convenient to keep the O_2 group in the x-z plane (as in both Figures 3 and 4) and to carry out the $\pi/4$ rotation on the remainder of the complex. It is then seen that Griffith's theoretical description of the complexing in the pentagonal bipyramidal case requires only minimal changes in order that it may apply equally well to the capped trigonal prismatic configuration. The only alteration required in the formal description of the bonding is an interchange—quite permissible in C_{2v}—of the roles assigned to the $3d_{x^2-y^2}$ and $3d_{xy}$ orbitals on the iron atom. Thus in the capped trigonal prismatic case, it is the four lobes of the d_{xy} orbital that are directed toward the N_p ligands, and d_{xy} replaces the usual $d_{x^2-y^2}$ in constructing the d^2sp^3 hybrid valence bond orbitals; also, of course, $d_{x^2-y^2}$ is used (instead of d_{xy}) to house an electron pair in the low-spin complex. Although the energy levels of all orbitals on the iron atom are affected in some degree by this interchange, one's qualitative judgment is that the net effect is not at all unfavorable to the valence state utilized in the capped trigonal prismatic configuration.

TABLE 5.
Computed geometry of capped trigonal prismatic coordination in oxymyoglobin (distances in Ångstrøms).

Δ[a]	Fe—Ax[b]	Ct[c]—Ax[b]	N_p—O	N_p—N_h	N_p—N_p	Fe—O
0	1.88	1.88	2.44	2.72	2.72	1.98
0.11	1.88	1.99	2.53	2.63	2.72	1.98

NOTE: All computations use Fe—N_h = Fe—N_p = 1.92 Å, O—O = 1.21 Å.
[a]Out-of-plane displacement of iron atom toward the oxygen molecule.
[b]Ax is the axis of the oxygen molecule.
[c]Ct is used for the origin of coordinates.

It then appears that the capped trigonal prism is much more probable—or much less improbable—than is the pentagonal bipyramid as a possible coordination polyhedron for the low-spin heme in oxymyoglobin. One other point is to be emphasized. As may be seen from Table 5, passable steric relations in the seven-coordinate polyhedron demand some displacement (Δ) of the iron atom from the mean plane of the porphine nitrogen atoms toward the O_2 group. Thus the shift of the N_p mean plane relative to the iron atom that is expected to accompany oxygenation is larger by $\gtrsim 0.15$ Å if the low-spin product be seven-coordinate than if it be octahedral.

COMPARISONS WITH THE DATA FROM HEMOPROTEIN STRUCTURE ANALYSES

The experimental results given by structure analyses of high-spin iron porphyrins and the predicted stereochemical behavior of the hemes in hemoproteins, as set forth earlier, may now be compared with the structural data reported from X-ray studies of several myoglobin derivatives. The Cambridge Workers have not yet published a definitive account of their direct determination at maximum resolution (1.40 Å) of the ferrimyoglobin structure, nor have they fully delineated the capabilities and the limitations of the specialized form of Fourier difference synthesis that they have utilized in their recent studies of other myoglobin derivatives. (In making his own assessment of the published hemoprotein structural data, this writer has benefited from clarifying discussions with Dr. H. C. Watson (1966); it does not follow that Dr. Watson concurs in the judgments pronounced herein.)

The maximum accuracy with which the parameters of a complex crystalline structure can be determined is fixed by the character of the recordable X-ray data, the dominant factor being, of course, the range of $(\sin\theta)/\lambda$ within which nonvanishing intensities of reflection are given by the crystal. Subject to these limitations, the evaluation of what is, by objective criteria, the best set of structural parameters that are implicit in a given set of X-ray data requires that the later stages of the structure determination be devoted to the cyclic refinement of, simultaneously, the atomic positions, the phases of the reflection amplitudes, anisotropic thermal parameters for each atom, and a general scaling factor. For a noncentrosymmetric structure, in particular, the refinement of the complex phases and, concomitantly, of the atomic positions is wholly dependent upon the simultaneous refinement of a model structure that carries the requisite degree of sophistication.

It has not been feasible to carry out such a refinement—or even a primitive version thereof—on the noncentrosymmetric structure of ferrimyoglobin (Watson, 1966). The data at 1.40 Å resolution are too numerous to be handled all at once in any existing computer, and the difficult problem of selecting appropriate constraints that would permit refinement in stages is unsolved (Diamond, 1965).

It is important to note, however, that the recordable X-ray data for ferrimyoglobin are all comprised within the comparatively narrow range in $(\sin\theta)/\lambda$ that is measurable with the Ti$K\alpha$ wavelength (2.75 Å); the source of this limitation is to be found in the extraordinarily high value, $\sim$10 Å^2, of the averaged thermal parameter of the crystalline structure. Consequently, there is little reason to suppose that ultimate refinement of the structure can give stereochemical parameters that are better than semiquantitative.* Diamond (1966) remarks in this connection that, "It is a difficult matter, even at the highest resolution, to determine atomic coordinates in a protein with an accuracy better than 0.25 Å."

The estimated lenhgts of the complexing Fe—N and Fe—OH_2 bonds in the heme of the high-spin ferrimyoglobin (Table 3) come from a study at only 2.0 Å resolution (Kendrew et al., 1960). The ~1.9 Å value given for all five Fe—N bonds is appropriate only to a low-spin porphyrin; Fe—N = 2.06 ± 0.01 Å is confidently expected for the complexing bonds to porphine nitrogen atoms in any high-spin ferric porphyrin (see above). The ~0.30 Å displacement of the iron atom from the mean plane of the heme—the displacement from the mean plane of the porphine nitrogen atoms may differ significantly (Hoard, Hamor, Hamor, and Caughey, 1965)—comes from the structure at 1.40 Å resolution (Watson, 1966). The effective volume of the reported coordination group approaches that expected for a low-spin complex. One may suggest, consequently, that the unrefined complex phases of the reflection amplitudes, as determined approximately for the real high-spin hemoprotein, are—in this approximation—no less applicable to the parallel, but hypothetical, low-spin material.

Having seen that the accuracy of direct structure analysis of a hemoprotein must be quite low, it is the more surprising to learn that the structural information obtained by applying an approximate version of Fourier difference synthesis to the study of several myoglobin derivatives is rather widely interpreted as demonstrating that the stereochemical parameters of the heme are virtually invariant from one hemoprotein to another, the spin state notwithstanding. Inasmuch as this interpretation is in direct conflict with the stereochemical expectations set forth earlier, it is evidently necessary to appraise the capabilities and the limitations of the approximate form of difference synthesis. This technique, first applied to the azide derivative of ferrimyoglobin (Stryer, Kendrew, and Watson, 1964), takes fullest advantage of the generally close approach to isomorphism that exists between crystalline ferrimyoglobin and each of several other myoglobin derivatives. The unrefined complex phases of ferrimyoglobin at low resolution (2.0 Å in the first application of the method, 2.8 Å in subsequent studies) are carried over without change to the corresponding reflection amplitudes (measured to the same resolution) of the myoglobin derivative under study. The differences between the measured amplitudes of corresponding reflections from ferrimyoglobin

*Ibers' (1961) treatment of peak densities and shapes for the electron density distribution in the carbon atom for various choices of effective wavelength and thermal parameter illustrates the type of quantitative calculation that is pertinent to this matter. His computations, unfortunately, do not extend nearly far enough to include the very high values of either the thermal parameter (10 $Å^2$) or the effective wavelength (2.80 Å) that characterize the 1.40 Å ferrimyoglobin data—much less the effective wavelengths (4.0 and 5.6 Å) employed in applications of a special form of difference synthesis to the myoglobin derivatives (Stryer, Kendrew, and Watson, 1964; Nobbs, Watson, and Kendrew, 1966). Lacking such quantitative calculations, it is nonetheless evident that the combined effects of these factors on a determination of hemoprotein structure are profoundly adverse to definitive resolution in Fourier syntheses and, especially, to the accurate evaluation of atomic positions. These undesirable effects are strongly reinforced by the uncertainties in the complex phases of the Fourier coefficients (see Luzzati, 1953).

and the myoglobin derivative are then used as coefficients in a Fourier synthesis that gives, to some degree of approximation, the point-by-point difference between the electron densities of the two crystals.

The application of this technique to the X-ray data measured to 2.8 Å resolution for deoxymyoglobin and ferrimyoglobin gives a mapping of the excess electron density in the deoxymyoglobin crystal that is briefly characterized as follows: the asymmetric unit of structure, which contains one molecule and constitutes just half the unit cell, shows five peaks and four holes (or negative peaks) having maximum heights or depths $\gtrsim |0.20|$el/Å^3 (Nobbs, Watson, and Kendrew, 1966). According to the interpretation given these results, the significant differences in structure between molecules of deoxymyoglobin and ferrimyoglobin are associated with just the two most prominent features—both holes—in the mapping of the residual electron density. Thus a hole of 0.42 el/Å^3 depth is associated with the loss from each molecule in the ferrimyoglobin structure of the water molecule that is coordinated to the ferric ion, and a hole of 0.68 el/Å^3 depth is associated with the loss of a heme-linked anion (perhaps HSO_4^- or, statistically, $\frac{1}{2}$ SO_4^{--}) that is involved in the balancing of electrical charge during reduction of the ferric ion to the ferrous state.

As a first step toward appraisal of the merits of this form of difference synthesis, let it be assumed that the atomic coordinates in the deoxymyoglobin structure carry over, quite unaltered, to the ferrimyoglobin structure, but that the latter includes also a very simple substructure that specifies the positions in each unit cell of the two heme-linked anions and of the two water molecules that are coordinated to ferric ions; this assumption puts into precise form the conclusions drawn from the experimental study. It is then evident that the assignment of the same set of complex phases to both the deoxymyoglobin and the ferrimyoglobin structures prescribes a quite artificial pattern for the complex phases of the reflection amplitudes associated with the substructure. Each reflection amplitude of the substructure is constrained to accept either the phase angle of the corresponding reflection amplitude from ferrimyoglobin or one differing from this by the angle π. Inasmuch as the analogous study of ferrimyoglobin and its azide derivative led also to the conclusion that significant differences in structure between these hemoproteins are confined to differences between their substructures (Stryer, Kendrew, and Watson, 1964), it follows that assignment of the ferrimyoglobin phases to both hemoprotein structures constrains the complex phases of both substructures into rigidly artificial patterns of the type specified above.

Whereas the substructure in ferrimyoglobin places two water molecules and two heme-linked anions in each cell, the substructure in the azide derivative carries two azide ions instead of the water molecules and the heme-linked anions are eliminated. It is evident that these very simple, but wholly different, distributions of scattering matter must give rise to highly distinctive sets of complex amplitudes that cannot be closely related to one another in respect either to magnitudes or phase angles. Indeed, a quasi-random distribution of the phase angles within the range from 0 to 2π is the normal expectation for any arbitrarily selected substructure. In the absence of any direct correlation between the complex phases of a substructure and those of a deoxymyoglobin crystal, it is appropriate to examine more fully the character of the approximations that enter into the grossly simplified version of Fourier difference synthesis while still allowing it to provide results of qualitative merit.

Let $\mathbf{F}_r = F_r e^{i\alpha_r}$ represent the complex amplitude (or Fourier coefficient) of an arbitrarily chosen reflection (*hkl*) from the ferrimyoglobin crystal, and $\mathbf{F}_v = F_v e^{i\alpha_v}$ represent the complex amplitude of the corresponding reflection from the crystal of the other hemoprotein; the numerical values of the absolute magnitudes, F_r and F_v, and of the phase angles, α_r and α_v, are all determined by the choice of (*hkl*). The corresponding Fourier coefficient that appears in the difference synthesis may be written as $\mathbf{F} = Fe^{i(\alpha_r+\gamma)}$ wherein γ is the difference in phase angle between $\mathbf{F}$ and $\mathbf{F}_r$. Since

$$\mathbf{F} = \mathbf{F}_v - \mathbf{F}_r, \text{ or } Fe^{i\alpha_r}e^{i\gamma} = F_v e^{i\alpha_v} - F_r e^{i\alpha_r}, \qquad (1)$$

it follows that

$$e^{i\alpha_r}Fe^{i\gamma} = e^{i\alpha_r}(F\cos\gamma + iF\sin\gamma)$$
$$= e^{i\alpha_r}[F_v\cos(\alpha_v - \alpha_r) - F_r + iF_v\sin(\alpha_v - \alpha_r)], \qquad (2)$$

or

$$F\cos\gamma = F_v\cos(\alpha_v - \alpha_r) - F_r = (F_v - F_r) - F_v[1 - \cos(\alpha_v - \alpha_r)], \qquad (2a)$$

and

$$F\sin\gamma = F_v\sin(\alpha_v - \alpha_r) \qquad (2b)$$

It is seen that the representative difference vector, $\mathbf{F}$, can be resolved into a component, $F\cos\gamma$, that is either parallel (for $|\gamma| < \pi/2$) or antiparallel (for $|\gamma| > \pi/2$) with the amplitude vector, $\mathbf{F}_r = F_r e^{i\alpha_r}$, of the ferrimyoglobin structure, and a second component, $F\sin\gamma$, that is orthogonal to $\mathbf{F}_r$; the phase angle of the $F\cos\gamma$ component is either α_r or $\alpha_r + \pi$, while that of the $F\sin\gamma$ component is either $\alpha_r + \pi/2$ or $\alpha_r - \pi/2$. Identical relations apply to each of the very numerous complex amplitudes (*hkl*) that enter into the difference synthesis; in view of the quasi-random distribution of the expectation values for γ within the range from 0 to 2π, the expectation values for $|F\cos\gamma|$ and $|F\sin\gamma|$ are equal. Consequently, the complete difference synthesis can be regarded as the superposition of two subsyntheses that are conveniently designed as the **C**-series and the **S**-series accordingly as the Fourier coefficients take the respective forms:

$$e^{i\alpha_r}F\cos\gamma = e^{i\alpha_r}[F_v\cos(\alpha_v - \alpha_r) - F_r] \qquad (\mathbf{C}\text{-series})$$

and

$$e^{i(\alpha_r+\pi/2)}\,F\sin\gamma = e^{i(\alpha_r+\pi/2)}\,F_v\sin(\alpha_v - \alpha_r) \qquad (\mathbf{S}\text{-series})$$

The zonal (*h*0*l*) Fourier coefficients contribute only to the **C**-series, but they are outnumbered by the general (*hkl*) coefficients in the approximate ratio of 10:1 for any specified range of scattering angle.

By taking the difference in phase angles, $\alpha_v - \alpha_r$, between every pair of corresponding reflections from the two hemoprotein crystals, to be identically zero, the **S**-series is made to vanish term by term, and only a drastically simplified form of the **C**-series remains. This procedure yields the version of Fourier difference synthesis that has thus far been applied to the study of hemoprotein structure. Two important points are noted for subsequent discussion. Neither subsynthesis is a quantitatively adequate substitute for the other —indeed, their functions are basically complementary. Furthermore, the qualitatively reliable features of the results given by the primitive version of the **C**-synthesis should provide a practicable basis for proceeding to a higher degree of approximation that involves both subsyntheses in equal measure.

It is convenient for discussion of the first point to make use of such numerical data as are given in the published studies. The $\sim$6,000 pairs of reflections from ferrimyoglobin and its azide derivative that were used in the simplified **C**-synthesis comprised $\sim$65 percent of all those having interplanar spacings $\gtrless$2.0 Å and $\sim$78 percent of those actually observed within the specified range; pairs of reflections having $F_r < 80$ electrons—less, that is, than 40 percent of the averaged value, 200 electrons, assigned both to F_v and F_r—were excluded from the synthesis. The averaged and maximum values of $|F_v - F_r|$ were, respectively, 10 and 97 electrons. By comparison with these data, the 3,212 pairs of reflections from ferrimyoglobin and deoxymyoglobin that were used in the simplified **C**-synthesis seem to have included most or all of the reflections observed for interplanar spacings $\gtrless$2.8 Å—perhaps 90 percent or more of the theoretical yield for this range. Averaged values of F_v and F_r were made equal at 266 electrons; the averaged and maximum values of $|F_v - F_r|$ were, respectively, 24 and 135 electrons.*

Inasmuch as the data in each study include the centrosymmetric (*h*0*l*) amplitudes, the averaged value for $|F_v - F_r|$ presumably is a somewhat high estimate for the averaged values of $|F \cos \gamma|$ and $|F \sin \gamma|$ for the general (*hkl*) coefficients, but is an underestimate for the average F. Rather large fluctuations in F from the most probable value together with the quasi-random distribution of γ in the range from 0 to 2π are characteristic of a simple substructure. There is, by contrast, no implication that the distribution of the phase-angle differences, $\alpha_v - \alpha_r$, can be quasi-random. The simultaneous conditions, $F_v \leq F$ and $F_r \leq F$, that must be satisfied in order to have $|\alpha_v - \alpha_r| \gtrless \pi/2$ will most commonly restrict both F_v and F_r to small fractions of their most probable values. Consequently, the statistical yield of Fourier coefficients having $|\alpha_v - \alpha_r| \gtrless \pi/2$ is relatively small. All terms involving these highly structure-sensitive coefficients were better omitted from the simplified **C**-series, because $F_v - F_r$ is a wholly unrealistic substitute for $F \cos \gamma$. For $|\alpha_v - \alpha_r| > \pi/2$, $F \cos \gamma = -F_r - F_v \cos(\alpha_v - \alpha_r)$; thus the phase angle of the coefficients is always $\alpha_r + \pi$, the sign of $F_v - F_r$ notwithstanding, and the magnitude lies always between F_r and $F_r + F_v$. It is probable that in the ferrimyoglobin azide study all coefficients of this type were excluded from the partial synthesis, but at the cost of rejecting $\sim$1700 pairs of observed reflections having $F_r < 80$ electrons.

Such truncation of the simplified **C**-series also eliminates some part, but by no means all, of the terms in which $|\alpha_v - \alpha_r|$ is sufficiently large to make $F_v - F_r$ a poor approximation to $F \cos \gamma$. Although the large majority of the (*hkl*) coefficients probably have $|\alpha_v - \alpha_r|$ small enough to make the neglected quantities, $F_v[1 - \cos(\alpha_v - \alpha_r)]$, individually small, the resulting errors have always the same sign. Excepting only the (*h*0*l*) coefficients, $|F \cos \gamma|$ is invariably underestimated for $F_v < F_r$, but overestimated for $F_v > F_r$; and for $\cos(\alpha_v - \alpha_r) < F_r/F_v < 1$, $F_v - F_r$ provides neither the correct sign for $F \cos \gamma$ nor a fair approximation to its magnitude. These effects are asymmetrically

*Thus the maximum amplitude of the difference vector is at least 135 electrons, a figure for F that is larger than the substructure discussed earlier for ferrimyoglobin can possibly provide. The total number of electrons carried by $2H_2O$ and $2SO_4^{--}$ is 120; it would be most remarkable if the substructure could give any F larger than $\sim$90 electrons.

cumulative in the simplified **C**-synthesis; all (hkl) coefficients with overlarge magnitudes carry the phase angles, α_r, all with undersized magnitudes the phase angles, $\alpha_r + \pi$.

That the Fourier coefficients in the **S**-series are highly sensitive to the phase-angle differences, $\alpha_v - \alpha_r$, is evident in the relation, $F \sin \gamma = F_v \sin(\alpha_v - \alpha_r)$. The pertinence of the **S**-synthesis to the structural problem of present interest can be explicitly demonstrated by considering first the ferrimyoglobin structure at different stages of refinement. Let α_r and α_v represent corresponding phase angles in, respectively, the structure as presently described and the structure as ultimately refined; let it be assumed also that the errors, $|\alpha_r - \alpha_v|$, are rather small—less than 15° for most reflections. Of course $F_r \equiv F_v$, so that $F \cos \gamma$ reduces to $F_v[\cos(\alpha_v - \alpha_r) - 1]$. For $|\alpha_v - \alpha_r|$ small, sufficiently accurate relations are $F \cos \gamma = \frac{1}{2} F_v(\alpha_v - \alpha_r)^2$ and $F \sin \gamma = F_v(\alpha_v - \alpha_r)$. The amplitude ratio, $|F \sin \gamma|/|F \cos \gamma| = 2/(\alpha_v - \alpha_r)$, is 7.6 for $|\alpha_v - \alpha_r| = 15°$, and 38 for $|\alpha_v - \alpha_r| = 3°$—where an $|F \sin \gamma|$ of 10 electrons calls for an F_v of 191 electrons. In the ferrimyoglobin synthesis, consequently, the orthogonal correction, $e^{i(\alpha_r + \pi/2)} F \sin \gamma$, to each Fourier coefficient is of far greater significance than is the in-phase correction, $-e^{i\alpha_r}|F \cos \gamma|$. In these circumstances, the capacity to refine structure resides mostly in the **S**-series, increasingly so as all $|\alpha_v - \alpha_r|$ decrease toward zero.

The differences in structure between any two of the quasi-isomorphous hemoproteins divide rather cleanly into two classes. The addition or subtraction of substructure constitutes a relatively large structural change that ordinarily makes the dominant contribution to the vectorial difference, $\mathbf{F}_v - \mathbf{F}_r$, between any corresponding pair of complex (hkl) amplitudes from the two hemoproteins, while the accompanying shifts in atomic positions within the heme—that of the iron atom probably not excepted—make generally smaller contributions* to $\alpha_v - \alpha_r$ and $F_v - F_r$. Indeed it is clear that, leaving out substructure, the remainder of the structure in either ferrimyoglobin azide or deoxymyoglobin corresponds to a comparatively small, but distinctive, refinement of the analogous part of the ferrimyoglobin structure. For such refinement the **S**- and the **C**-subsyntheses, both used in cycles of successive approximation, are equally essential.

A quite unacceptable substitute for this refinement program is that offered by the grossly approximate version of the **C**-synthesis, in which experimental

*The general range of these contributions to $F_v - F_r$ can be estimated as follows. In the ferrimyoglobin structure the iron and the porphine nitrogen atoms occur as square-pyramidal FeN_4 groups, two in each cell, with the iron atom ~0.30 Å out from the basal plane. Let the iron atom and the N_4 grouping be translated each ~0.15 Å toward one another to give a planar FeN_4 group—just that expected from the earlier discussion for the low-spin azide derivative—and let it be assumed (in order to maximize the result) that the groups are oriented so that the translations are along the normal to some one of the $(h0l)$ planes having the 2.0 Å minimum spacing. The $|F_v - F_r|$ corresponding to this $(h0l)$ then includes a contribution from the specified conformational change that may be as large as ~4.5 electrons, but has an expectation value of ~3 electrons. The expectation values for similar contributions to the $|F_v - F_r|$ for all other planes are smaller, much smaller for the overwhelming majority of them. The shifts in the positions of other atoms that are involved in the conformational changes must also be taken into account; the probability is small, however, that the maximum contribution to $|F_v - F_r|$ for any (hkl) from conformational alterations—excluding changes in substructure—can reach the ~10 electron level that represents the averaged value of $|F_v - F_r|$ in the ferrimyoglobin azide study. Only an unexpectedly large shift in the coordinates of the iron atom would require a major revision of these estimates.

$F_v - F_r$ data bearing the dominant stamp of the differing substructures are constrained to an asymmetrically simplified pattern that is tailored to requirements set by the complete ferrimyoglobin structure. In view of the deplorably low resolving power of the individual terms in the Fourier syntheses, there is cogent reason to retain every statistical advantage that accrues from the proper use of the very numerous data; the imposition on these data of a strong bias toward the ferrimyoglobin structure, abetted as always by the inadequate resolution, is well calculated to minimize and distort all quantitative aspects of the structural changes, and to obscure altogether the changes that are small.

There are, of course, simple standard procedures for the computation and refinement, in cycles of successive approximation, of both the **S**- and the **C**-series or, perhaps more usefully, of the complete Fourier syntheses for the ferrimyoglobin azide and the deoxymyoglobin structures. The refinement is initiated by using the complex amplitudes calculated for the approximate substructure to get the first approximate set of the $\alpha_v - \alpha_r$, and the continuing process affords opportunities to check on the internal consistency of the paired sets of data and on the interpretations thereof. The obvious first check, one that should be made even when no refinement is contemplated and the first approximation to the **C**-series is to be used for the qualitative detection of substructure, is that of computing the Fourier synthesis of the amplitudes calculated for the approximate substructure; it is clearly desirable to have this evidence for or against the qualitative interpretation of substructure as derived from the simplified **C**-series, and in particular, to see what peak shapes and electron densities actually are obtained when the complex phases are correctly specified.

By such refinement procedures the ferrimyoglobin azide and deoxymyoglobin structures can each be brought to the same level of semiquantitative merit as that of the ferrimyoglobin structure for the same range of the experimental data. The level of quantitative merit of the ferrimyoglobin structure, as it presently stands, is not prepossessing. No objective estimate of the standard deviations in the positions of the atoms in the heme of ferrimyoglobin would put any of these below 0.15 Å, excepting only that of the iron atom; substantially higher values must apply to the ferrimyoglobin azide and deoxymyoglobin structures as determined by way of the oversimplified **C**-syntheses. These bald facts notwithstanding, there has been some inclination on the part of workers in the field of hemoproteins to accept, at face value, the results given by the approximate **C**-syntheses. Consequently, the dangers implicit in such gross oversimplification of the analytical problems are exposed in the rather detailed appraisal of the difference-synthesis technique that has been presented herein. It remains only to emphasize that full acceptance of the results given by the oversimplified **C**-syntheses leads to a heme stereochemistry that is markedly inconsistent with that expected from the study of pure iron porphyrins and other structurally simple complexes.

The formal results of the hemoprotein studies that are pertinent to heme stereochemistry can be summarized in a few words. Apart from the chemically variable ligand that occupies one axial position in the coordination group, the structural parameters of the heme are taken to be essentially invariant from one hemoprotein to another; furthermore, the water molecule and the azide nitrogen atom that occupy the special axial position in, respectively, ferrimyoglobin and its azide derivative are assigned identical positions, ~2.10 Å from the iron atom. The better-defined bond lengths that correspond to this description are now to be compared with the expectation values for these parameters from the earlier discussion.

Five Fe—N bond distances in each hemoprotein are put at ~1.9 Å, a value that (if read ~1.92 Å) is appropriate to all six Fe—N distances in the low-spin ferrimyoglobin azide and to the Fe—N bond involving the histidine nitrogen atom in all of the hemoproteins, but it is quite inappropriate to the four bonds to porphine nitrogen atoms in each of the two high-spin hemoproteins—unless the accurately determined value of this parameter, 2.06 ± 0.01 Å, in three high-spin porphyrins (Table 2) is altogether discounted. The length of the bond to the chemically variable ligand, ~2.10 Å, is not unreasonable for the weak Fe—OH_2 bond that is an addendum to what is effectively a five-coordination group in the high-spin ferrimyoglobin, but it is ~0.20 Å longer than the low-spin Fe—N bond length that, according to the earlier discussion, is concomitant with the strong complexing interaction whereby the low-spin ferrimyoglobin azide is formed. The excessive length of this 2.10 Å bond notwithstanding, the azide nitrogen atom is placed only 2.60 Å from each of the four porphine nitrogen atoms, 0.10 Å below the 2.70 Å set in the earlier discussion as a realistic minimum for N—N nonbonding distances. The indicated N—N separation of 2.66 Å between porphine nitrogen atoms is ~0.06 Å below that expected for a low-spin iron porphyrin and is 0.18 Å below the value observed in each of the three high-spin iron porphyrins of Table 2.

It is probable that the direct analysis of structure for ferrimyoglobin has given fairly accurate positions for the iron atom and the mean plane of the heme in this hemoprotein. It appears, however, that the positions indicated for the porphine nitrogen atoms are in error by ~0.15 Å—more specifically that the Ct—N distance (see Table 2 and the earlier discussion) is too small by $\gtrsim$0.13 Å. The estimate for the out-of-plane displacement of the iron atom, ~0.30 Å, should be reliable. The precise magnitude of this displacement is not critically significant for a high-spin porphyrin; its value, indeed, is extraordinarily sensitive to very minor changes in the Fe—N and Ct—N distances (Table 2). Subject to these reservations, the out-of-plane displacement of the iron atom in the high-spin deoxymyoglobin is expected to be somewhat larger, perhaps 0.15 Å larger, than that in ferrimyoglobin. An increment of 0.15 Å in this displacement, as divided between positional shifts of the iron and the quasi-planar group of four nitrogen atoms, is quite unlikely to be detected in the approximate **C**-synthesis.

It is in the low-spin six-coordination group of the heme in ferrimyoglobin azide that the precise magnitude of the out-of-plane displacement of the iron atom assumes critical importance. With six nitrogen atoms as ligands, according to the earlier discussion, the iron atom is confidently expected to be nearly or exactly centered in the coordination group. The apparent maintenance of the 0.30 Å out-of-plane displacement of the iron atom in ferrimyoglobin azide, as indicated by the approximate **C**-synthesis, is primarily responsible for the excessive length, ~2.10 Å, of the complexing bond to azide nitrogen and for the unacceptably short N—N nonbonding separations, 2.60 Å, between azide nitrogen and porphine nitrogen atoms. The reality of any such displacement that exceeds $\sim$0.05 Å is strongly to be doubted; it is much more probable that the iron and the four porphine nitrogen atoms are shifted from their positions in ferrimyoglobin into a virtually coplanar array in the azide derivative. Furthermore, it is likely that positional shifts of the nitrogen atoms account for at least half—perhaps most—of this conformational alteration in the coordination group, and that the over-all change is quite lost in the oversimplified **C**-synthesis (see footnote, p. 590).

CONCLUDING REMARKS

Magnetic susceptibility and Mossbauer data (Lang and Marshall, 1966) for oxyhemoglobin appear to be compatible with either of two quite different theoretical models for the coordination group. Pauling's (1949, 1964) model requires octahedral coordination, but with angular attachment of the O_2 group to the iron atom; an Fe—O—O bond angle of $\sim$125° is indicated. Griffith's (1956) model and the alternative thereto discussed earlier are, of course, seven-coordinate. The stereochemical evidence favoring the octahedral coordination group is recapitulated as follows.

Of the two possible seven-coordination polyhedra in which the O_2 group is symmetrically attached to the iron atom, the capped trigonal prism appears to be much the more probable. Relative to the standard octahedron, however, the capped trigonal prism carries heavy steric liabilities that are minimized only by displacing the iron atom out-of-plane from the porphine nitrogen atoms toward the axis of the O_2 group. The oxygenation of deoxymyoglobin would then demand movements of the atoms in the central portion of the heme that would seem to be implausibly large and involved, especially if there be little or no shift in the position of the iron atom in the unit cell. Substantially less movement of the heme atoms during oxygenation would be required for octahedrally coordinated low-spin iron; in the equilibrium configuration the iron may lie slightly out-of-plane from the porphine nitrogen atoms away from the O_2 group.

REFERENCES

Cohen, G. H., and J. L. Hoard (1966). *J. Am. Chem. Soc.* **88**, 3228.
Cruickshank, D. W. J., and R. A. Sparks (1960). *Proc. Roy. Soc.* (London) A **258**, 270.
Diamond, Robert (1965). *Acta Cryst.* **19**, 774.
Diamond, Robert (1966). *Acta Cryst.* **21**, 253.
Falk, J. E. (1964). *Porphyrins and Metalloporphyrins.* New York: Elsevier Publishing Co.
Fleischer, E. B. (1963). *J. Am. Chem. Soc.* **85**, 146.
Fleischer, E. B., C. K. Miller, and L. E. Webb (1964). *J. Am. Chem. Soc.* **86**, 2342.
Glick, M. D., G. H. Cohen, and J. L. Hoard (1967). *J. Am. Chem. Soc.* **89**, 1996.
Griffith, J. S. (1956). *Proc. Roy. Soc.* (London) A **235**, 23.
Griffith, J. S. (1959, 1961). *Discussions Faraday Soc.* **26**, 81; *The Theory of Transition-Metal Ions.* pp. 370–373. London: Cambridge University Press.
Hamor, T. A., W. S. Caughey, and J. L. Hoard (1965). *J. Am. Chem. Soc.* **87**, 2305.
Hamor, M. J., T. A. Hamor, and J. L. Hoard (1964). *J. Am. Chem Soc.* **86**, 1938.
Hoard, J. L. (1939). *J. Am. Chem. Soc.* **61**, 1252.
Hoard, J. L. (1966). Stereochemistry of Porphyrins. In Chance, B., R. W. Estabrook, and T. Yonetani, eds. *Hemes and Hemoproteins*, pp. 9–24. New York: Academic Press.
Hoard, J. L., G. H. Cohen, and M. D. Glick (1967). *J. Am. Chem. Soc.* **89**, 1992.

Hoard, J. L., M. J. Hamor, T. A. Hamor, and W. S. Caughey (1965). *J. Am. Chem. Soc.* **87**, 2312.

Ibers, J. A. (1961). *Acta Cryst.* **14**, 538.

Kendrew, J. C., R. E. Dickerson, B. E. Strandberg, R. G. Hart, D. R. Davies, D. C. Phillips, and V. C. Shore (1960). *Nature* **185**, 422.

Koenig, D. F. (1965). *Acta Cryst.* **18**, 663.

Lang, G., and W. Marshall (1966). *J. Mol. Biol.* **18**, 385.

Lind, M. D., M. J. Hamor, T. A. Hamor, and J. L. Hoard (1964). *Inorg. Chem.* **3**, 34.

Loftus, A. (1959). *Mol. Phys.* **2**, 367.

Luzzati, V. (1953). *Acta Cryst.* **6**, 142.

Nobbs, C. L., H. C. Watson, and J. C. Kendrew (1966). *Nature* **209**, 339.

Pauling, L. (1949). The Electronic Structure of Haemoglobin. In *Haemoglobin*, p. 57. London: Butterworths Scientific Publ.

Pauling, L. (1960). *The Nature of the Chemical Bond*, 3rd ed. Ithaca: Cornell Univ. Press.

Pauling, L. (1964). *Nature* **203**, 182.

Silvers, S., and A. Tulinsky (1964). *J. Am. Chem. Soc.* **86**, 927.

Stryer, L., J. C. Kendrew, and H. C. Watson (1964). *J. Mol. Biol.* **8**, 96.

Watson, H. C. (1966). Privately communicated.

Weakliem, H. A., and J. L. Hoard (1959). *J. Am. Chem. Soc.* **81**, 549.

Webb, L. E., and E. B. Fleischer (1965). *J. Am. Chem. Soc.* **87**, 667.

Zerner, M., M. Gouterman, and H. Kobayashi (1966). *Theor. Chim. Acta* **6**, 363.

J. D. DUNITZ
P. STRICKLER
Organic Chemistry Laboratory
Swiss Federal Institute of Technology
Zurich, Switzerland

Preferred Conformation of the Carboxyl Group

The series of crystal structure analyses of amino acids and simple peptides carried out at the California Institute of Technology during the late 1940's and early 1950's provided information about interatomic distances and bond angles, about the conformations of the molecules, and about the dimensions and directional characteristics of the N—H···O hydrogen bonds that determine the packing of the molecules in the crystals. In particular, it led to the recognition that the preferred conformation of the peptide group is the one with the C=O group synplanar with respect to the N—C bond (I):

I

The recognition of this feature, together with the knowledge of the precise dimensions of the peptide group, was of great importance in deriving the stable conformations of polypeptide chains (Pauling and Corey, 1950; Pauling, Corey, and Branson, 1951).

In spite of the recognition of this important conformational preference of the peptide group, little attention has been given to the examination of the ever-increasing body of known crystal structures in the search for preferred conformations in other systems. As Mathieson (1965) has pointed out, if a particular conformational pattern is observed consistently in a range of crystal structures involving different packing arrangements, we may conclude that the intermolecular forces are here of less significance than the intramolecular forces and hence that the same preferred conformation applies in the case of the isolated molecule. Nevertheless, in published descriptions of crystal structures of organic compounds it is still the exception rather than the rule to find much discussion of conformational details. At Zurich, because of our interest in the conformational regularities in molecules of medium-sized cycloalkane and cycloalkene derivatives, we make a routine practice of calculating torsion angles* about all bonds of interest in the molecules we study together with the interatomic distances and bond angles. Perhaps as a result of this, we have recently stumbled on a conformational preference of the carboxyl group, which seems to us sufficiently interesting that we bring it to attention here.

To obtain more information about the deviation of the C—C—C angles in cyclohexane from the ideal tetrahedral angle (Davis and Hassel, 1963; Wohl, 1964; Dunitz and Strickler, 1965), we have made a fairly accurate determination of the crystal structure of cyclohexane-1,4-*trans*-dicarboxylic acid. The X-ray analysis of the crystals presented no difficulties; the structure was solved by direct methods and refined by full-matrix least-squares analysis of some 450 counter-measured F_o values to an R factor of 5.6 percent. The molecular dimensions are shown in Figure 1. Here we draw attention particularly to the rotational conformation of the carboxyl group. To be honest, we had expected—although for no very good reason—to find the equatorial carboxyl group in one of the two symmetrical conformations II or III,

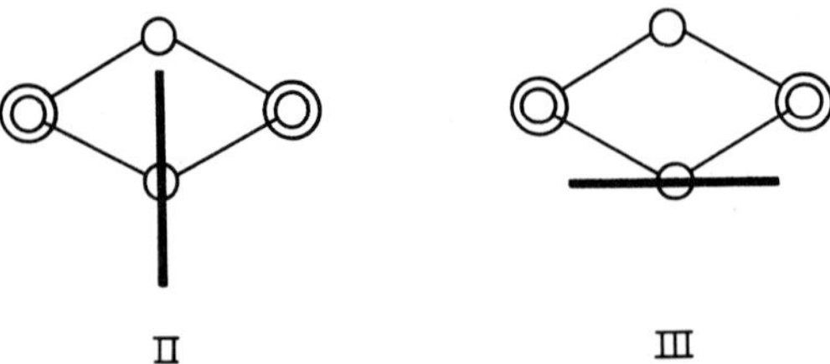

but the analysis shows clearly that it occurs in the unsymmetrical conformation IV, and in such a way that the C═O bond is synplanar with respect to the Cα—Cβ bond, the torsion angle Cβ—Cα—C═O being 5°.

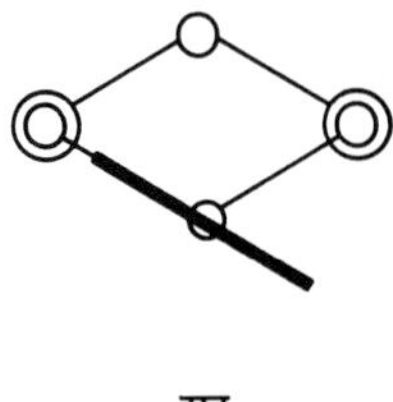

We were sufficiently impressed by this unexpected result to look up the conformations adopted by the carboxyl group in the known crystal structures of other saturated and unsaturated carboxylic acids. In almost every case we found the torsion angle Cβ—Cα—C═O to be close to zero, in agreement with the observed conformation in cyclohexane-1,4-*trans*-dicarboxylic acid. We also found that we were not the first to have noticed this regularity. In a paper on the crystal structures of methyl *m*- and *p*-bromocinnamates, Leiserowitz

*In accordance with a recommendation of Klyne and Prelog (1960), the essential parameter defining the relative positions of A and B in a system A—X—Y—B is called the torsion angle (τ) between A and B across the bond XY. The angle is taken as *positive* when measured *clockwise* from A to B, and *negative* when measured *counterclockwise*. This sign convention is equivalent to taking the sign as that of the vector product $(\overrightarrow{A-X} \times \overrightarrow{X-Y}) \times (\overrightarrow{X-Y} \times \overrightarrow{Y-B})$. Conformations with τ = 0°, 60°, 120°, 180° are called synplanar, synclinal, anticlinal, antiplanar, respectively.

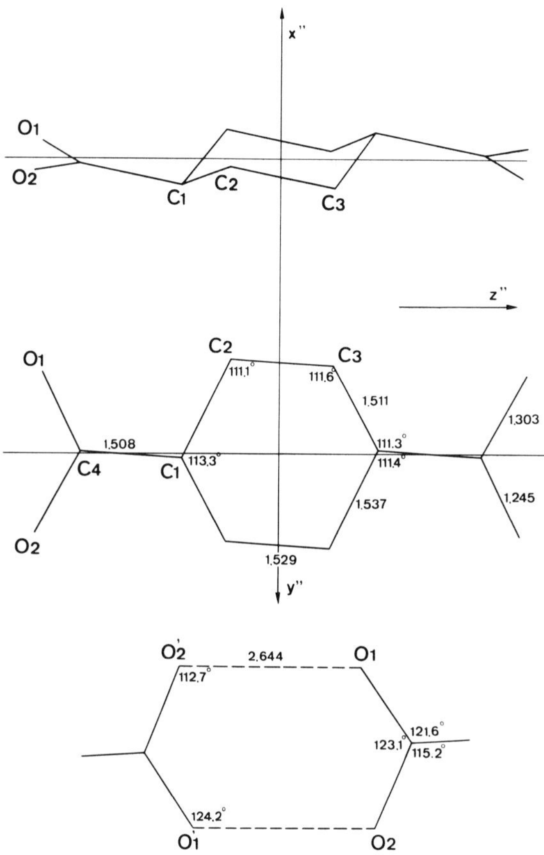

FIGURE 1.
Interatomic distances and bond angles in cyclohexane-1,4-*trans*-dicarboxylic acid.

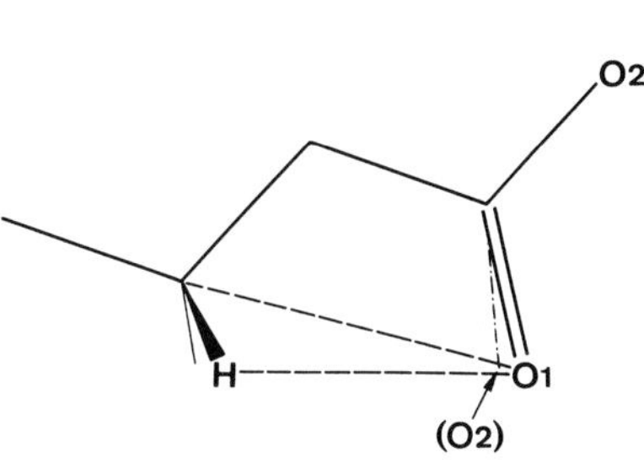

FIGURE 2.
Carboxyl group in normal synplanar and in antiplanar conformation, drawn approximately to scale.

and Schmidt (1965) had already pointed out that the synplanar arrangement Cβ—Cα—C═O seems to be a general feature of the molecular shapes of carboxylic acids, as well as of esters and amides.

Table 1 shows some results, which indicate the general applicability of the rule, at least for the saturated acids. In the case of the unsaturated acids with Cβ—Cα a double bond, exceptions appear to occur; thus, while the conformation of Cβ—Cα—C═O is synplanar in acrylic acid, crotonic acid, and sorbic acid, it is antiplanar (torsion angle ~180°) in α-*trans*-cinnamic acid and in fumaric acid.

Leiserowitz and Schmidt assume that the preferred conformation is determined by the nonbonded interactions between Cβ and its attached hydrogen atoms, on the one hand, and the hydroxyl or carbonyl oxygen atoms, on the other. Mainly because of the difference between the angles Cα—C═O and Cα—C—OH (see Table 1), the nonbonded distances C$\beta\cdots$O and H$\beta\cdots$O are 0.1 to 0.2 Å larger for the usual synplanar than for the alternative antiplanar conformation (Figure 2). A calculation based on the representative parameter values: C═O, 1.23 Å; C—OH, 1.31 Å; Cα—C, 1.50 Å; Cβ—Cα, 1.52 Å; C—H, 1.09 Å; C—C═O, 122°; C—C—OH, 114°; Cβ—Cα—C, 113°; H—C—H, 106°, leads to the nonbonded distances:

		$C_\beta\cdots O$	$H_\beta\cdots O$
synplanar	C_β—C_α—C═O	2.77 Å	2.69 Å
synplanar	C_β—C_α—C—OH	2.63	2.54

TABLE 1.
Geometry of carboxyl groups in saturated and unsaturated acids.

Acid	Bond lengths (Å)		Bond angles (°)		Torsion angle(°)[a]	References
	C=O	C—OH	C—C=O	C—C—OH	C_β—C_α—C=O	
Propionic acid	1.23	1.32	124	114	12	(*a*)
Butyric acid	1.22	1.35	124	113	0	(*b*)
Valeric acid	1.26	1.35	125	117	2	(*c*)
3-indoleacetic acid	1.22	1.30	124	113	12	(*d*)
Succinic acid	1.24	1.31	124	113	0	(*e*)
Adipic acid	1.23	1.29	122	116	8	(*f*)
α-pimelic acid	1.20	1.34	126	114	2	(*g*)
	1.24	1.26	121	118	37	
Suberic acid	1.23	1.31	123	115	5	(*h*)
Cyclohexane 1,4-*trans*-dicarboxylic acid	1.25	1.30	122	115	5	(*i*)
cis-AMCHA·HB[b]	1.25	1.31	126	115	0	(*j*)
cis-AMCHA·HCl	1.21	1.35	125	119	~172	(*k*)
trans-AMCHA·HBr	1.20	1.30	124	112	~33	(*k*)
Acrylic acid	1.26	1.28	122	116	0	(*l*)
Crotonic acid	—	—	121	113	synplanar	(*m*)
Sorbic acid	—	—	122	116	synplanar	(*m*)
α-trans-cinnamic acid	—	—	119	115	antiplanar	(*m*)
Fumaric acid	1.21	1.30	122	114	antiplanar	(*n*)
	1.22	1.29	119	115	antiplanar	
	1.23	1.29	119	117	antiplanar	

[a]Calculated from published atomic positions.
[b]AMCHA = 1-aminomethylcyclohexane-4-carboxylic acid.

REFERENCES: (*a*) Strieter, Templeton, Scheuermann, and Sass (1962). (*b*) Strieter and Templeton (1962). (*c*) Scheuermann and Sass (1962). (*d*) Karle, Britts, and Gum (1964). (*e*) Broadley and Cruickshank (1959). (*f*) Housty and Hospital (1965). (*g*) Kay and Katz (1958). (*h*) Housty and Hospital (1964). (*i*) Dunitz and Strickler (1966). (*j*) Groth and Hassel (1965). (*k*) Kadoya, Hanazaki, and Iitaka (1966). (*l*) Higgs and Sass (1963). (*m*) Leiserowitz and Schmidt (1965). (*n*) Brown (1966).

The C···O distance for synplanar C_β—C_α—C=O is closer to the sum of the van der Waals radii (~3.1 Å) and hence, according to Leiserowitz and Schmidt, this arrangement is the preferred one.

This cannot be the whole story, however. In acetaldehyde and in the acetyl halides, the preferred conformation is also the one with a synplanar arrangement H—C_α—C=O (see Millen, 1962, for a review of restricted rotation about single bonds), and this can hardly be ascribed to a steric effect of the kind mentioned above. Moreover, the steric argument does not explain why twisted conformations of the carboxylic group do not occur more often in the saturated acids, in which the planar conformations ($\tau = 0°$ and 180°) are not stabilized by conjugation. Nor does the steric argument explain why the antiplanar conformation C_β—C_α—C=O is observed in some of the unsaturated acids.

An additional factor of importance in determining the preferred conformation of the carbonyl group in carboxylic acids, aldehydes, esters, and amides becomes apparent if the C=O double bond is regarded as being decomposed into two bent bonds. Pauling (1959) has suggested that the early bent-bond description of double and triple bonds (Baeyer, 1885) has certain advantages over the more fashionable σ, π description. In particular, it accounts satisfactorily for the observed bond angles and rotational barriers in olefinic compounds. If the bent-bond description of the carbonyl double bond is adopted, the synplanar conformations C_β—C_α—C=O (V) and H—C_α—C=O (VI) are seen to correspond to the energetically favorable staggered disposition of bonds about the central C_α—C bond for the saturated compounds. If, however, the C_β—C_α bond is a double bond, then the antiplanar conformation C_β=C_α—C=O (VII), corresponding to the energetically favorable staggered disposition of bonds, should be more stable than the synplanar conformation (VIII).

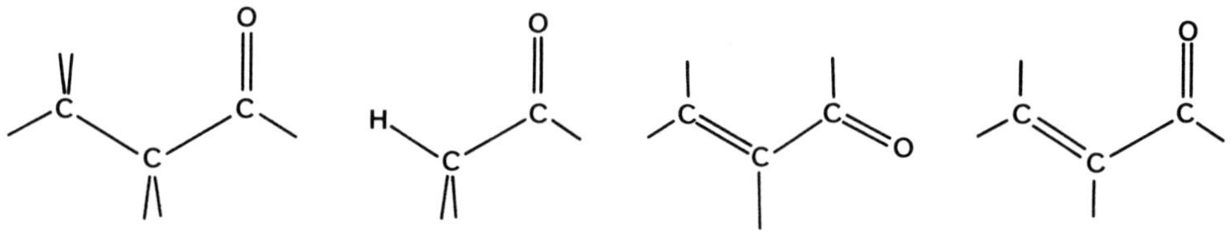

V VI VII VIII

These two factors, nonbonded interactions and the preference for staggered rather than eclipsed dispositions of bonds, are thus both operating in favor of the synplanar conformation C_β—C_α—C=O in saturated systems. If C_β—C_α is a double bond, then one factor favors synplanar, the other antiplanar, so that the preference for synplanar will not be so pronounced.

The synplanar conformation C_β—C_α—C=O is observed in *cis*-1-aminomethylcyclohexane-4-carboxylic acid hydrobromide (Groth and Hassel, 1965), in which the carboxyl group adopts the axial position. (The rotational conformation of the carboxyl group was not discussed by these authors, but calculations based on their published coordinates lead to a torsion angle $\tau(C_\beta—C_\alpha—C{=}O) = 0.3°$.) In contrast, in the corresponding hydrochloride the antiplanar conformation has been found (Kadoya, Hanazaki, and Iitaka, 1966), with C—OH rather than C=O eclipsing C_β—C_α in the ring. It is to be noted that this hydrochloride and hydrobromide do not constitute an isomorphous pair but have completely different crystal structures, the difference in conformation being accompanied by major changes in packing. In the hydrobromide, with the normal synplanar conformation C_β—C_α—C=O, the carboxyl groups appear to be linked in the usual way into centrosymmetric dimers; in the hydrochloride, on the other hand, with its abnormal antiplanar conformation, the packing is characterized by hydrogen bonds of the type C=O···H—$\overset{+}{N}$< and C—OH···$\overset{-}{Cl}$. Evidently, these hydrogen bonds can be formed at the expense of an unfavorable carboxyl group conformation, whereas in the hydrobromide (in which C—OH···Br^- hydrogen bonds would be vanishingly weak) the energy balance favors the normal carbonyl group conformation. Crystals of *trans*-1-aminomethylcyclohexane-4-carboxylic acid hydrobromide (both substituents equatorial) again show a different type of packing, characterized by C—O···H—$\overset{+}{N}$< and C—OH···Br^- hydrogen bonds rather than by centrosymmetric carboxyl group dimers, and the conformation of C_β—C_α—C═O is found to be twisted but closer to synplanar than to antiplanar, with $\tau \sim 30°$ (Kadoya et al., 1966).

In molecules of aromatic carboxylic acids, the carboxyl groups lie in the plane of the aromatic system unless hindered from doing so by the presence of other substituents. This means that, for a given orientation of the aromatic system, there are two equivalent orientations of the carboxylic group (IX) in which the positions of the oxygen atoms are nearly identical:

IX

In a hydrogen-bonded pair of such acids, the carboxyl group of one molecule is correlated with the carboxyl group of the other (X):

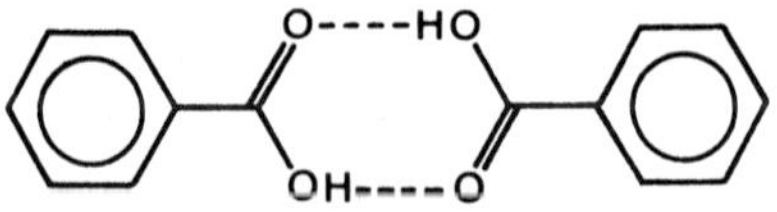

X

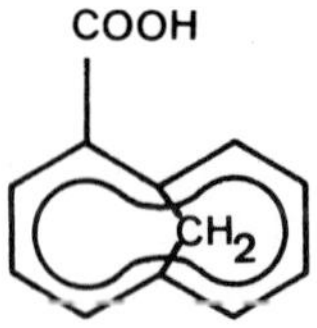

XI

But if these pairs interact only weakly with one another, we would expect the arrangement in any one pair to be largely independent of the arrangement in other pairs. In other words, we anticipate the possibility of disorder with respect to the C═O and C—OH bonds in crystalline aromatic carboxylic acids.

A recent analysis of 1,6-methanocyclodecapentaene-2-carboxylic acid (XI) (Dobler and Dunitz, 1965) indicates that this crystal is probably disordered with respect to the C═O and C—OH bonds. The two C—O bond lengths are equal within experimental error—1.267 Å and 1.257 Å; the oxygen atoms have the largest temperature-factor parameters of any atoms in the molecule; the hydrogen atoms in the hydrogen-bonded pairs are found in a difference map to be exactly midway between the oxygen atoms. These features are most simply accounted for by assuming that the observations apply not to a single molecule but refer to an average over two orientations of the carboxyl group.

The structures of benzoic acid (Sim, Robertson, and Goodwin, 1955) and of 1- and 2-naphthoic acids (Trotter, 1960, 1961) are known only by analyses of two-dimensional data, and in none of them have the atomic positions been determined very precisely. The reported dimensions of the carboxyl groups are shown in Table 2, from which it is clear that the C═O and C—OH bond

TABLE 2.
Geometry of carboxyl group in aromatic acids.

Acid	Bond lengths (Å)		Bond angles		References
	C=O	C—OH	C—C=O	C—C—OH	
1,6-methanocyclodecapentaene-2-carboxylic acid	1.26	1.27	120	118	(*a*)
Benzoic acid	1.24	1.29	122	118	(*b*)
1-naphthoic acid	1.25	1.28	127	122	(*c*)
2-napthtoic acid	1.33	1.37	122	127	(*d*)

REFERENCES: (*a*) Dobler and Dunitz (1965). (*b*) Sim, Robertson, and Goodwin (1955). (*c*) Trotter (1960). (*d*) Trotter (1961).

lengths are much more nearly equal than in the examples of Table 1. In the two naphthoic acids the C—C—OH angles are reported as 122° and 127°, whereas this angle is less than 120° in all other known carboxylic acids. It may well be the case that more accurate, three-dimensional analyses will reveal that these aromatic carboxylic acids are actually disordered with respect to the orientations of the carboxyl groups.

The preferred conformation of carboxylic acids, in which C=O is synplanar to C_α—C_β, is only a special case of a more general preference that applies also in esters and in amides. In the case of esters, the question of the preferred conformation of the ester group about the C—O formally single bond has also been examined on the basis of crystal structure results. It is found (Mathieson and Taylor, 1961) that the preferred conformation is again the one with the C=O bond synplanar with respect to the O—C (ester) bond (XII), quite analogous to the preferred conformation of the peptide group. More recently Mathieson (1965) has shown that for secondary esters the conformation (XIII) is invariably adopted. For primary esters the evidence points to (XIV) as the preferred conformation, with all carbon and oxygen atoms coplanar, reminiscent of the atomic arrangements found in the early studies of β-glycylglycine (Hughes and Moore, 1949) and of acetylglycine (Carpenter and Donohue, 1950) and in subsequent studies of other peptide compounds.

Crystal structure studies cannot be expected to give any information about the energy difference between the two extreme conformations in these compounds, or of the potential energy barrier separating them. In the case of the esters, Tabuchi (1958) has estimated by an acoustic dispersion method, that

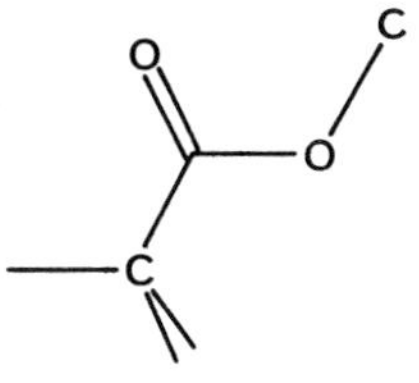

XII 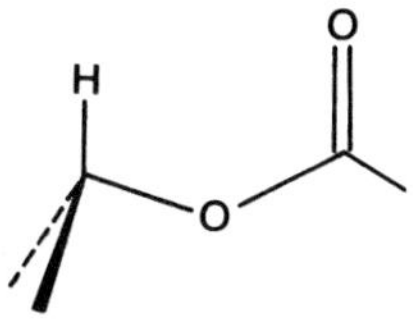 XIII 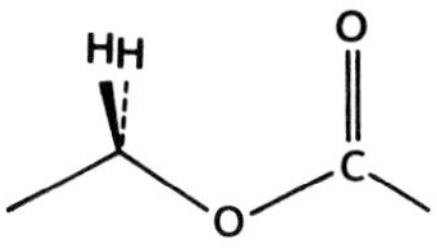 XIV in ethyl formate the two extreme forms differ in energy by 2.5 kcal/mole with a potential barrier of about 3.4 kcal/mol referred to the higher energy state. For methyl formate, Owen and Sheppard (1963), on the basis of an infrared spectral study, estimate the synplanar form to be at least 2.7 kcal/mol more stable than the antiplanar form. As Owen and Sheppard point out, the synplanar conformation of carboxylic esters may be stabilized somewhat by minimization of lone pair–lone pair repulsions between the oxygen atoms, a factor that does not come into play in stabilizing the synplanar $C_\beta—C_\alpha—C=O$ conformation in the acids. We might expect, then, a somewhat smaller energy difference between the synplanar and antiplanar conformations of the acids, but one that should be detectable all the same. Experiments designed to measure this energy difference and the energy barrier between these conformations are under way.

REFERENCES

Baeyer, A. (1885). *Ber. deutsch. chem. Ges.* **18**, 2269.

Broadley, J. S., and D. W. J. Cruickshank (1959). *Proc. Roy. Soc.* (London) A **251**, 441.

Brown, C. J. (1966). *Acta Cryst.* **21**, 1.

Carpenter, G. B., and J. Donohue (1950). *J. Am. Chem. Soc.* **72**, 2315.

Davis, M., and O. Hassel (1963). *Acta Chem. Scand.* **17**, 1181.

Dobler, M., and J. D. Dunitz (1965). *Helv. Chim. Acta* **48**, 1429.

Dunitz, J. D., and P. Strickler (1965). *Helv. Chim. Acta* **48**, 1450.

Dunitz, J. D., and P. Strickler (1966). *Helv. Chim. Acta* **49,** 2505.

Groth, P., and O. Hassel (1965). *Acta Chem. Scand.* **19**, 1709.

Higgs, M. A., and R. L. Sass (1963). *Acta Cryst.* **16**, 657.

Housty, J., and M. Hospital (1964). *Acta Cryst.* **17**, 1388.

Housty, J., and M. Hospital (1965). *Acta Cryst.* **18**, 693.

Hughes, E. W., and W. J. Moore (1949). *J. Am. Chem. Soc.* **71**, 2618.

Kadoya, S., F. Hanazaki, and Y. Iitaka (1966). *Acta Cryst.* **21**, 38.

Karle, I. L., K. Britts, and P. Gum (1964). *Acta Cryst.* **17**, 496.

Kay, M. I., and L. Katz (1958). *Acta Cryst.* **11**, 289.

Klyne, W., and V. Prelog (1960). *Experientia* **16**, 521.

Leiserowitz, L., G. M. J. Schmidt (1965). *Acta Cryst.* **18**, 1058.

Mathieson, A. McL. (1965). *Tetrahedron Letters* no. 46, 4137.

Mathieson, A. McL., and J. C. Taylor (1961). *Tetrahedron Letters* no. 17, 590.

Millen, D. J. (1962). In Klyne, W., and P. B. D. de la Mare, eds., *Progress in Stereochemistry*, vol. 3. Butterworths: London.

Owen, N. L., and N. Sheppard (1963). *Proc. Chem. Soc.*, 264.

Pauling, L. (1959). In *Theoretical Organic Chemistry*, IUPAC Kékulé Symposium. Butterworths: London.

Pauling, L., and R. B. Corey (1950). *J. Am. Chem. Soc.* **72**, 5349.

Pauling, L., R. B. Corey, and H. R. Branson (1951). *Proc. Nat. Acad. Sci. U.S.* **37**, 205.

Scheuermann, R. F., and R. L. Sass (1962). *Acta Cryst.* **15**, 1244.

Sim, G. A., J. M. Robertson, and T. H. Goodwin (1955). *Acta Cryst.* **8**, 157.

Strieter, F. J., and D. H. Templeton (1962). *Acta Cryst.* **15**, 1240.

Strieter, F. J., D. H. Templeton, R. F. Scheuermann, and R. L. Sass (1962). *Acta Cryst.* **15**, 1233.

Tabuchi, D. (1958). *J. Chem. Phys.* **28**, 1014.

Trotter, J. (1960). *Acta Cryst.* **13**, 732.

Trotter, J. (1961). *Acta Cryst.* **14**, 101.

Wohl, R. A. (1964). *Chimia*, 219.

JOHN D. ROBERTS
Division of Chemistry and Chemical Engineering
California Institute of Technology
Pasadena, California

Studies of Conformational Equilibria and Equilibration by Nuclear Magnetic Resonance Spectroscopy

Nuclear magnetic resonance spectroscopy has been extraordinarily fruitful in the study of stereochemical problems, especially those associated with conformational analysis. In reasonably favorable cases, the position of the equilibria between conformational isomers can be determined with a high degree of accuracy, and in many cases it is also possible to make measurements of rates of equilibration of such isomers as a function of temperature. The activation parameters obtained from such measurements are of special theoretical interest because the rate processes involved are not complicated by making and breaking of bonds. This, along with the well-established utility of conformational analysis in natural-product chemistry, has led to much research on conformations and conformational equilibration in a variety of acyclic and cyclic compounds. Many of the principles which are involved are to be illustrated here, using examples drawn from work in the Gates and Crellin Laboratories. There will be no attempt to provide a general review of either conformational analysis or nuclear magnetic resonance spectroscopy. The lack of reference to many studies of other workers of conformational equilibration should not be interpreted as a claim for priority or originality of the ideas and concepts employed here. Eliel (1965a, 1965b) and Feltkamp and Franklin (1965a, 1956b) have provided excellent reviews which give a more balanced perspective of the development of this area of research.

A salient feature of the studies to be described here is the emphasis on the use of the ^{19}F resonance spectra of *gem*-substituted difluorocycloalkanes as an aid to conformational analysis. The hope has been to have the fluorines act as a sort of tracer or label for hydrogen, with advantage to be taken of the greater chemical shift of fluorine, which is ten to fifty times that of hydrogen. Whether fluorine is a reasonable equivalent of hydrogen in conformational problems will be discussed in some detail later. The advantages of fluorine for

Contribution No. 3406 from the Gates and Crellin Laboratories of Chemistry. Much of the work reported herein was supported by the National Science Foundation. This article was prepared in connection with a 1966 Centenary lectureship of the Chemical Society of London and was published also in *Chemistry in Britain*, pp. 529–535 (1966).

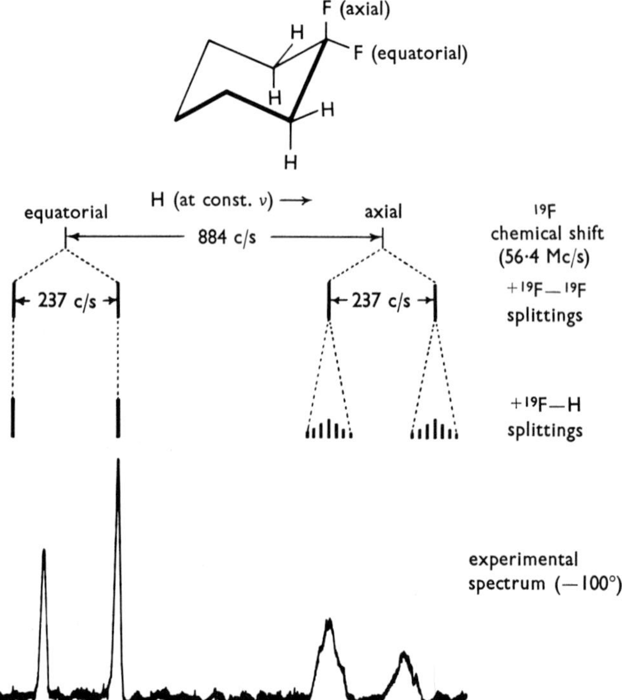

FIGURE 1.
Chemical shift and spin-spin splitting in ^{19}F spectra of 1, 1-difluorocyclohexane at $-100°$.

studies of this kind are well illustrated by the spectrum of 1,1-difluorocyclohexane under conditions where ring inversion is slow (Figure 1) (Spassov, Glazer, Griffith, Nagarajan, and Roberts, 1966; see also Roberts, 1963a, 1963b). The elements of this spectrum, reading down from the top of Figure 1, are first, a chemical-shift difference between the axial and equatorial fluorines of 884 cps at 56.4 Mcps; then a fluorine-fluorine spin-spin splitting of 237 cps; and finally hydrogen-fluorine splittings, assumed here to involve only the four vicinal hydrogens, which are shown as being larger for one of the fluorines than the other. The available theoretical evidence and experimental findings agree that the axial fluorines are more strongly coupled with the adjacent hydrogens. In fact, the difference between these couplings is a convenient diagnostic tool for determining which of a pair of appropriate fluorine resonances on a cyclohexane ring should be taken as arising from axial or equatorial fluorines. It is noteworthy that the *whole* breadth of the chemical-shift difference and couplings between an axial proton and an equatorial proton located on the same carbon is normally substantially less than the breadth of just *one* of the broadened peaks of an axial fluorine.

A very special feature of the experimental spectrum of 1,1-difluorocyclohexane shown in Figure 1 is that it was taken at $-100°$. Between $-100°$ and 30° the spectrum changes dramatically, as shown in the left side of Figure 2 (Spassov et al., 1966). The change in spectrum with increasing temperature occurs, of course, because of an increase in the rate of ring inversion, and on the right side of Figure 2 are shown theoretical spectra which utilize chemical shift and coupling parameters determined at $-100°$ but which

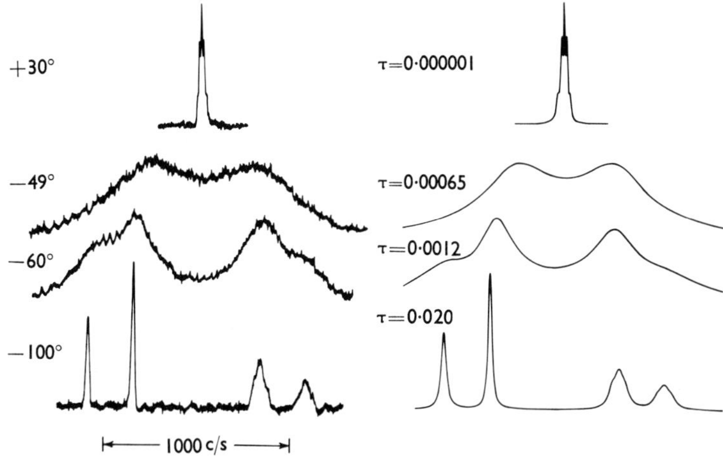

FIGURE 2.
Experimental (*left*) and calculated (*right*) ^{19}F spectra of 1,1-difluorocyclohexane as a function of inversion frequency. Each of the calculated spectra is labeled with the appropriate τ, which is the mean lifetime in seconds before inversion occurs.

were calculated with different values of τ, the mean lifetime of the molecules in seconds, before inversion occurs. Comparison of theoretical and experimental spectra of this sort permits evaluation of τ as a function of temperature and, from this, the activation parameters for the rate processes involved.

It has been common to use peak separations below the so-called "coalescence point" (for example, below $-46°$ for 1,1-difluorocyclohexane, as in Figure 2) to obtain τ values with the aid of the Gutowsky equation,

$$\frac{\delta\nu_\tau}{\delta\nu_\infty} = \left[1 - \frac{1}{2\pi^2\tau^2(\delta\nu_\infty)^2}\right]^{1/2}$$

where τ is the mean lifetime of each state and $\delta\nu_\infty$ is the chemical shift when $\tau = \infty$. The equation is valid for equal populations when spin coupling is absent and there is no overlap of lines at $\tau = \infty$, with $\delta\nu_\infty$ constant.

When τ is to be determined as a function of temperature, this procedure is not valid unless $\Delta\nu_\infty$ is independent of temperature. A much more reliable procedure is to measure the line shape because, as can be seen from Figure 3, the line shape is more sensitive to changes in τ and less sensitive to changes in $\Delta\nu_\infty$ than is the peak separation. In the work to be described here, line shapes were calculated by the method of Alexander.*

Values for the activation energy for inversion of cyclohexane and fluorinated cyclohexanes have been obtained by a number of workers, starting with Jensen and his co-workers (1960); see Table 1. In general, the agreement between different workers is satisfactory, especially considering that quite different procedures were used to evaluate τ. Of particular importance to the

*S. Alexander (1962). The computer programs were developed from Alexander's equations by Dr. J. T. C. Gerig and Mr. J. L. Beauchamp.

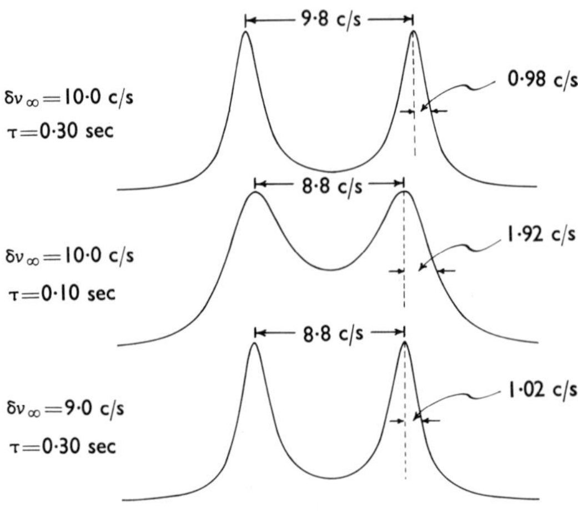

FIGURE 3.
Typical differences produced in theoretical spectra by changes in $\delta\nu_\infty$ and τ. The top spectrum is the standard. The middle spectrum has the same $\delta\nu_\infty$ but a smaller τ. The line separation for this spectrum is 8.8 cps, which is the same as for the bottom spectrum, in which τ is as in the top spectrum but $\delta\nu_\infty$ has been decreased to 9.0 cps. The point is that a decrease in τ produces a much larger change in line width than does a change in $\delta\nu_\infty$.

story here is the close similarity between the activation energies determined for cyclohexane and fluorocyclohexanes. It is evident that substitution of fluorine for hydrogen results in at most a small decrease in the activation energy for inversion. Such a decrease is to be expected because each chair conformation will have axial fluorine-hydrogen interactions which would be expected to make the ground states slightly less stable relative to the transition state for inversion. The same kind of effect on E_a is observed with other substituents, as can be seen from the activation parameters given in Table 2 for several substituted *gem*-fluorocyclohexanes which have been studied by following the temperature dependence of their fluorine spectra (Spassov et al., 1966).

TABLE 1.
Reported activation enthalpies for inversion of cyclohexane and fluorinated cyclohexanes.

Compound	$\Delta H^\ddagger$, kcal.	Investigator
C_6H_{12}	11.5	Jensen et al.
	9.0	Harris and Sheppard (1961)
	11.5 (spin echo)	Meiboom
	10.9	Anet et al.
	10.5	Bovey et al.
	10.3	Harris and Sheppard (1964)
	9.1 (spin echo)	Gutowsky
$C_6H_{11}F$	9.6	Bovey et al.
$C_6H_{11}F_2$	11.6 (CH_2Cl_2)	K. Nagarajan
	10.9 (propene)	S. Spassov
	9.5 (CS_2)	Gutowsky et al.
	9.1 ($CFCl_3$)	Gutowsky et al.
	9.8 (spin echo)	Gutowsky et al.
C_6F_{12}	9.9	Tiers

TABLE 2.
Activation energies for some gem-*fluorocyclohexanes.*

	Solvent	E_a, kcal/mole	A
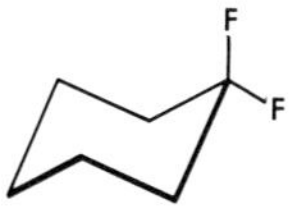 1,1-difluorocyclohexane	propene	10.9 ± 0.5	6.2×10^{13}
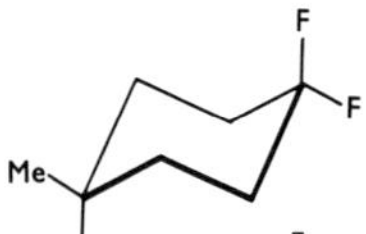 4,4-dimethyl-1,1-difluorocyclohexane (Me, Me)	propene	9.4 ± 0.5	9.1×10^{11}
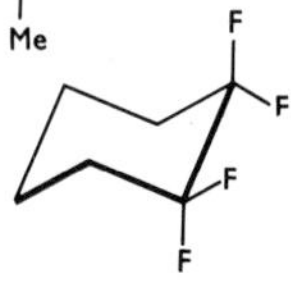 1,1,4,4-tetrafluorocyclohexane	propene	8.0 ± 0.3	1.5×10^{10}

TABLE 3.
Chemical shifts and coupling parameters in ^{19}*F-spectra of alkyl-substituted* gem-*fluorocyclohexanes.*

	30°C		∼ −100°C	
	δ_{F-F}	J_{F-F}	δ_{F-F}	J_{F-F}
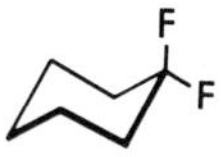 1,1-difluorocyclohexane	0 cps	0 cps	884 cps	237 cps
 4-Me (H)	563	238	610	238
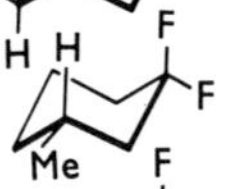 3-Me (H, H)	697	239	713	239
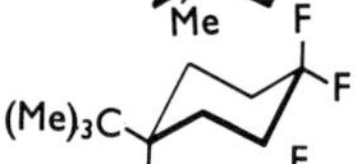 4-(Me)$_3$C (H)	661	236	671	236
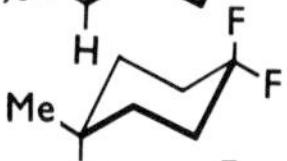 4-Me, 4-Me	0	0	653	235
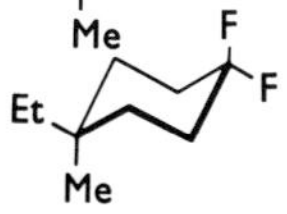 4-Et, 4-Me	154	234	696 (55%) 642 (45%)	235 235

The fluorine-labeling technique is useful for studying conformational equilibria and, in general, the ^{19}F chemical-shift differences observed at room temperature with various 4-substituted-1, 1-difluorocyclohexanes (see Table 3, Spassov, et al., 1966) correspond in a reasonable way to what is already known of the effects of substituents on conformational equilibria (Eliel, 1965a, 1965b; Feltkamp and Franklin, 1965a, 1965b). There are several aspects of the results given in Table 3 that are worthy of comment. In the first place, we note that there is only a very small temperature dependence of the fluorine chemical-shift difference with a 4-*t*-butyl group between 35° and −100°. This is in agreement with the idea that the *t*-butyl group will be very predominantly in the equatorial position. With 4-methyl, the temperature effect on the chemical-shift difference is larger and, if we assume that at −100° the methyl group is solely equatorial, the change in going to room temperature corresponds to an equilibrium mixture containing about 5% of the form with methyl axial. The 3-methyl group is different from 4-methyl in having a much smaller temperature dependence of the fluorine chemical shift, but this is explicable in terms of the expected 3-methyl-fluorine interaction when the methyl group is axial, which would not occur with an axial 4-methyl group.

What may be a very important effect with far-reaching implications is that much smaller chemical-shift differences are observed with the substituted *gem*-fluorocyclohexanes than with 1,1-difluorocyclohexane itself.* The most reasonable explanation is in terms of a distortion of the ring produced by the substituent groups. This explanation is strongly supported by results obtained with *cis*-decalin derivatives, as will be described later.† However, what we do not know yet is the degree of sensitivity of the fluorine chemical shifts to distortion of the ring. The *cis*-decalin results to be discussed below suggest that the sensitivity is likely to be very large; if it is, then these fluorine chemical-shift differences may turn out to be very helpful for diagnosis of subtle conformational effects. What is needed is independent measurement of the postulated ring distortion and a calibration curve of some angular measure of ring distortion versus ^{19}F chemical-shift differences. The absolute values of the chemical shifts may also be significant in this connection; however, this possibility will be discussed elsewhere.

The fluorine-labeling technique is particularly valuable for the study of conformational equilibration in the chair-chair interconversion of *cis*-decalin.‡ With *cis*-decalin itself, the two forms are enantiomers and are therefore energetically equivalent. The situation is quite different for 2,2-difluoro-*cis*-

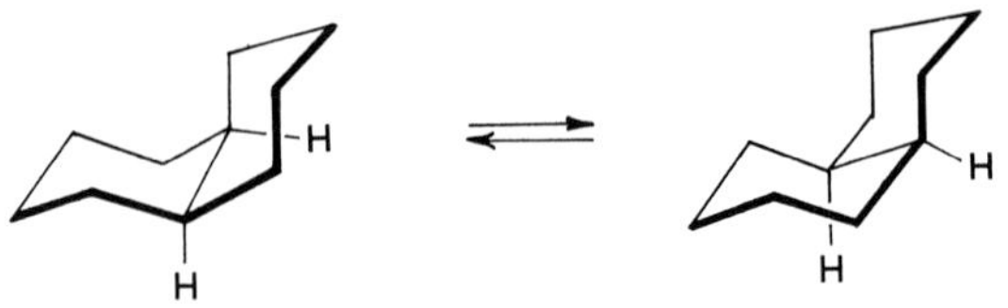

*Similar but much smaller effects have been observed on proton spectra. See H. Booth (1966) and the references cited there.

†Other evidence for distortion of cyclohexane rings by substituents has been discussed by R. A. Wohl (1965) and C. A. Grob and S. W. Tam (1965).

‡A good review of the conformational situation of *cis*-decalin is provided by M. Hanack (1965). Studies of the rather small temperature dependence of the proton spectra of *cis*-decalin and substituted *cis*-decalins have been reported by W. B. Moniz and J. A. Dixon (1961), N. Muller and W. C. Tosch (1962), and F. G. Riddell and M. J. T. Robinson (1965). See also J. T. C. Gerig and J. D. Roberts (1966).

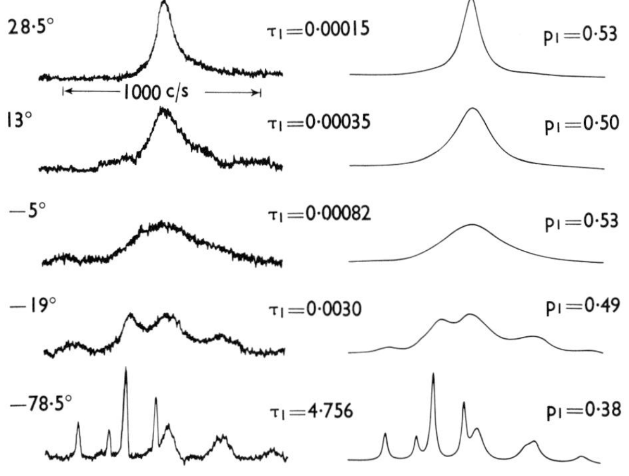

FIGURE 4.
Experimental and calculated changes in line shape for inversion in the ^{19}F spectrum of 9-methyl-2, 2-difluoro-*cis*-decalin. The values given for p_1 correspond to the mole fraction of the conformation III.

decalin where the forms I and II are not expected to be equally probable because of the way the axial fluorine interacts with the hydrogens at 8-position in form II but not in form I. This leads to the expectation that, at low enough temperatures for interconversion to be slow, two separate AB patterns should be observed corresponding to I and II, with the one representing I in greater concentration. When interconversion is fast, an average AB pattern is expected. Figure 4 shows the way the fluorine spectrum of 9-methyl-2,2-difluoro-*cis*-decalin actually changes with temperature. At the lowest temperature, two AB spectra are clearly evident, with axial and equatorial types of resonances. For this compound, the proportions of the forms III and IV are nearly but not exactly equal at −78.5°. The spectra are substantially blurred at the higher temperatures by rapid ring inversion.

I ⇄ II

III ⇄ IV

Calculation of theoretical spectra for this kind of system is much more complex than for cyclohexane inversion, as shown in Figure 2, because four chemical shifts, two coupling constants, four relaxation times, and the equilib-

TABLE 4.
*Equilibrium constants, magnetic resonance shifts and couplings, and activation parameters for interconverting conformers of 2,2-difluoro-*cis*-decalins.*

2,2-di-fluoro-decalin	$K = \frac{I}{II}$, 30°C	δ_{AB}, cps	J_{AB}, cps	E_a kcal/mole	δ'_{AB}, cps	J'_{AB}, cps	E_a' kcal/mole
cis-	2.9	757	233	14.6 ± 0.7	334	239	13.9 ± 0.7
cis-9-methyl-	1.08	544	239	9.1 ± 0.6	397	242	9.2 ± 0.6
cis-10-methyl-	3.4	713	235	10.6 ± 0.6	188	233	10.4 ± 0.6
cis-9-ethyl-	1.00	558	239	9.5 ± 0.5	549	236	9.5 ± 0.5

NOTE: The unprimed values are for the conformation that corresponds to I and the primed values are for the conformation that corresponds to II.

rium constant for I $\rightleftharpoons$ II must be taken into account along with τ. Theoretical curves calculated with the aid of the Gerig program will be seen to reproduce the experimental spectra well. In these calculations, both τ and p, the proportion of isomer I (R = CH_3, R′ = H) was varied until a satisfactory fit was obtained. The data obtained in this way (Gerig and Roberts, 1966) for a number of 2,2-difluoro-*cis*-decalin derivatives are summarized in Table 4.

The activation energy for the conversion of I to II for 2,2-difluoro-*cis*-decalin was found to be 14.6 kcal. Detailed calculation by the Hendrickson-Wiberg procedure (Hendrickson, 1961; Wiberg, 1965) of the energies of various conformations which might be intermediate in this interconversion suggest an activation energy of about 17 kcal, which is reasonably close to the experimental value. The activation energy for inversion is substantially reduced by substitution of alkyl groups at the 9- or 10- positions of *cis*-decalin. This appears to be a consequence of a substantial degree of destabilization of the ground state of the *cis*-decalin ring system by the axial-alkyl against *syn*-axial hydrogen interactions, which occur in a 9- or 10-alkyl substituted *cis*-decalin in either conformation (III or IV and V or VI)—the point being that the alkyl group must be axial to one or the other of the rings in all of these forms. The ascribing of the aforementioned abnormal fluorine chemical shifts to some degree of ring distortion is in general accord with

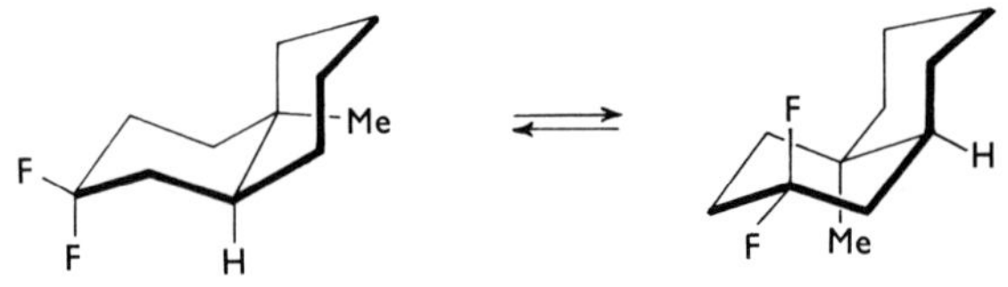

F F H Me $\rightleftharpoons$ F F Me H

V VI

the shifts observed with III–VI. A 10-methyl group, as expected, has a negligible effect on the equilibrium between V and VI and a relatively small effect on the equatorial-axial chemical-shift difference of the fluorines in V.

However, the corresponding chemical shift for VI is profoundly affected, being only 188 cps as compared to 761 cps in I. This is in accord with the idea that the interaction between the axial fluorine and the ring in II produces a distortion leading to the chemical-shift difference decrease of 347 cps, and the additional interaction of the axial-methyl against the *syn*-hydrogens at the 1,3-positions in VI further increases the distortion in the neighborhood of the fluorines and their chemical-shift difference.

Somewhat different effects are observed with 9-methyl substitution (III and IV). In III there is an interaction between the axial methyl and axial fluorine, which increases the equilibrium constant between the forms to 0.95, compared to 0.35 for I $\rightleftharpoons$ II. It is interesting that the interactions of axial methyl versus axial fluorine and of axial ring CH_2 versus axial fluorine are pretty much at a standoff in III and IV so far as energy effects are concerned, and they produce similar but, not identical, chemical-shift differences of 544 and 397 cps.

Substitution of a methyl group at other than the 9- or 10-positions of *cis*-decalin leads to a strong unbalancing of the forms corresponding to I or II, since the methyl will in general be much more favorably located in one form than in the other. Such effects have been observed with the 1-, 3-, and 6-methyl-2,2-difluoro-*cis*-decalins, all of which give spectra corresponding to a single conformation with negligible changes in the equatorial-axial fluorine chemical-shift differences over a 100° or more temperature variation. As expected, the same behavior is observed with 2,2-difluoro-*trans*-decalin which also has only one favorable conformation.

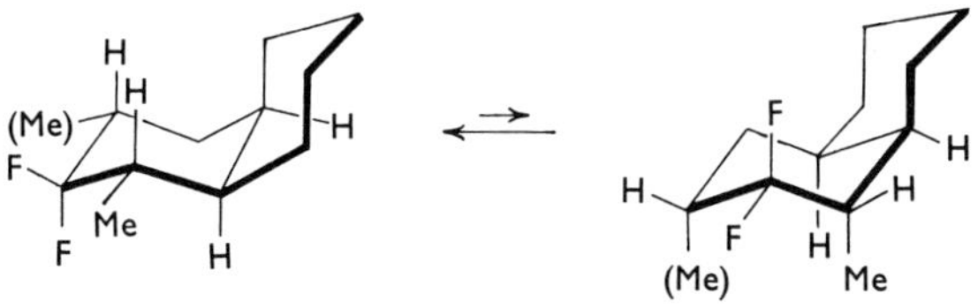

$\Delta\nu_{F-F}(1-Me) = 1131$ c/s

$\Delta\nu_{F-F}(3-Me) = 1040$ c/s

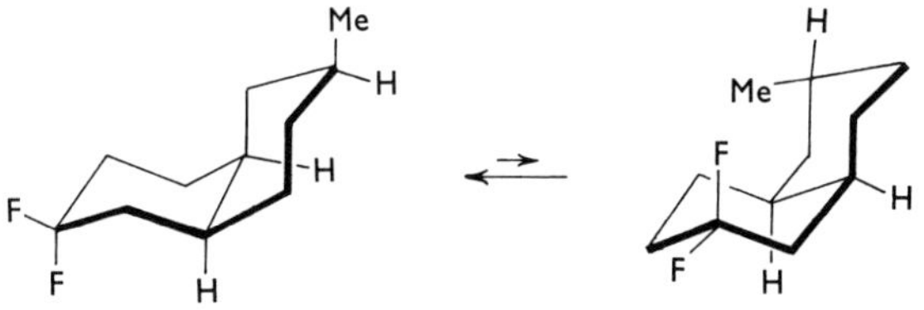

$\Delta\nu_{F-F} = 759$ c/s

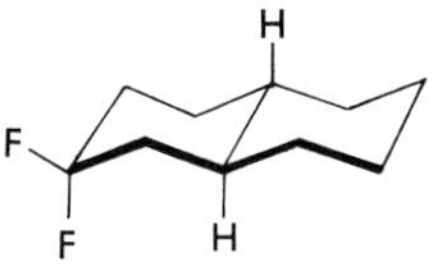

$\Delta\nu_{F-F} = 657$ c/s

VII

The very large axial-fluorine, equatorial-fluorine chemical-shift difference observed with 1- and 3-methyl substitution is especially interesting in that it is substantially larger than the observed shift difference with 1,1-difluorocyclohexane (884 cps). This may be due to a neighboring-group electrical effect or possibly a steric effect of the methyl group, which because of its position in the favorable conformation may act to spread the F—C—F bond angle. The latter notion fits with the 1175 cps shift difference calculated for the stable conformer of 1,1-difluoro-3-phenylcyclobutane (Lambert and Roberts, 1965), wherein the F—C—C bond angle is expected to be larger than the normal value because of the strained ring. Furthermore, the steric interferences which in I-VI lead to decreases in the fluorine chemical-shift differences are of types expected to act as to tend to decrease the F—C—F bond angle.

The general success of the fluorine-labeling technique in dealing with cyclohexane and fused-cyclohexane ring systems has inspired an attack on determining the conformations and rates of conformational equilibration of cycloheptane, which are much less familiar and less well understood than those of cyclohexane. There is little definitive experimental evidence on the favored conformation of the cycloheptane ring and, indeed, the most helpful work that is so far available is the excellent theoretical work of Hendrickson (1961, 1962), wherein a number of interesting a priori predictions were made about the relative stabilities of the various reasonably possible conformations of cycloheptane and substituted cycloheptanes.

A very striking feature of cycloheptane—most clearly evident from molecular models such as the Dreiding or Fieser models, which have free rotation about C—C bond axes—is the general limberness of the ring. There is no conformation like the chair form of cyclohexane, which is rigid so long as no bending of the C—C—C angles from their normal values is allowed. This suggests that conformational equilibration of a type which would interchange the relative positions of *gem* substituents might well occur with substantially greater ease than with cyclohexane. Consequently, it is perhaps not surprising that 1,1-difluorocycloheptane (VII) shows no chemical-shift difference between the fluorines at room temperature or all the way down to −180° (Figure 5).*

*The results reported here on cycloheptane come from unpublished research by E. S. Glazer and R. Knorr.

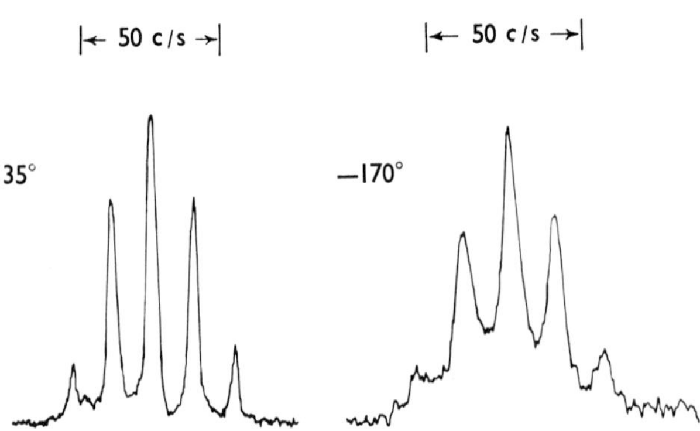

FIGURE 5.
Change in ^{19}F magnetic resonance spectra of 1,1-difluorocycloheptane between 35° and −170°.

The failure to observe a chemical-shift difference between the fluorines of VII at very low temperatures might be taken to indicate that inversion is rapid, but there is another possibility that has to be considered: the fluorines could be in a conformation in which they would be equivalent by symmetry and hence unable to show a chemical-shift difference, no matter how low the temperature. This is particularly important with the twist-chair form, which has an axis of symmetry passing through a carbon atom and the bond opposite (VIII). If the fluorines were located most favorably at the 1 or axis position

2-fold axis

side view

end view

VIII

of VIII, they would be equivalent by symmetry and, to determine whether this is a likely possibility, we may consider Hendrickson's (1962) calculation, in which the interacting energy of a methyl group has been evaluated for each position of the twist-chair conformation (Table 5). It will be noted that, except for the 1-position, there are equatorial-like and axial-like configurations of a substituent group at each of the carbons. The difference in energy between the various equatorial positions is quite small, and all are substantially more favorable than the axial-like positions. If we assume that the interaction energy associated with *gem* substitution is the simple sum of the separate energies for each position, then clearly the most favorable conformation for *gem* substitution will be the one with the groups at the axis position. Two fluorines at this position would be equivalent and would show no chemical-shift difference.

TABLE 5.
Calculated energies resulting from methyl substitution on the twist-chair form of cycloheptane [*Hendrickson* (*1962*)].

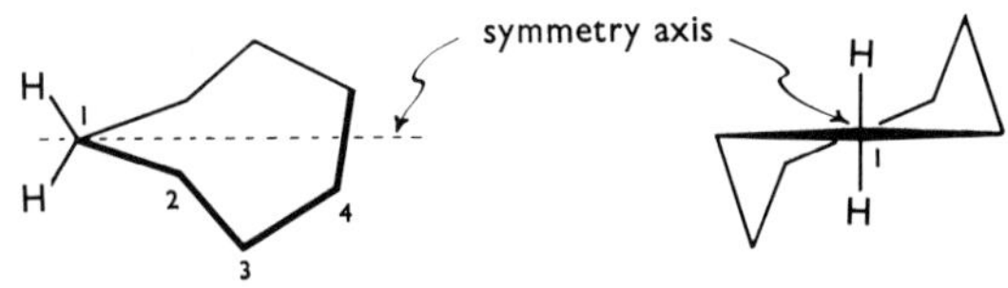

Position	Mono	Gem
1 (axis)	0.03 kcal	0.06
2e	−0.01	
2a	4.27	4.26
3e	0.00	
3a	4.88	4.88
4e	0.01	
4a	1.43	1.44

There are two other modes whereby *gem*-fluorines could be equivalent. One is a more or less conventional inversion whereby the chair (or twist chair) goes to the boat and to a new chair in much the same manner as that postulated for cyclohexane. The other possible mode of inversion is that discussed by Hendrickson, whereby a twist-chair to chair to twist-chair interconversion (pseudorotation, ψ) causes a substituent group to travel around the ring and undergo inversion of configuration whenever the carbon carrying the group goes through the axis position* of the twist-chair form. The sequence of changes for the chair forms is shown here.

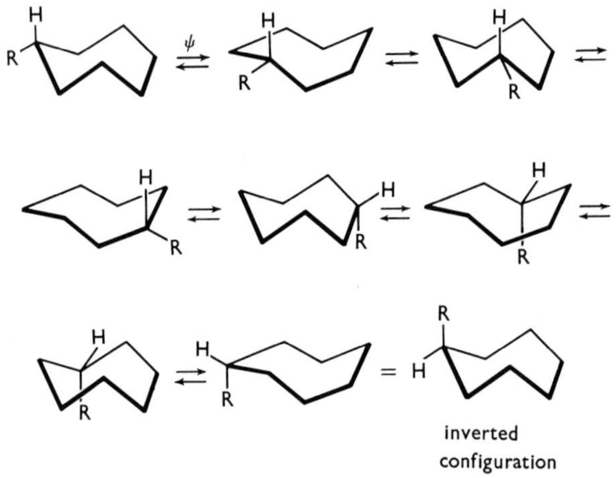

inverted configuration

Pseudorotation is expected to be an especially favorable way of achieving inversion, for Hendrickson's calculations show that the twist-chair form is but 2 kcal more stable than the chair form. Since the chair forms are anticipated to be the least favorable point on the pseudorotation "itinerary" (Hendrickson, 1962), the activation energy for inversion by pseudorotation should not be much more than 2 kcal, even for a *gem*-fluoro compound. The

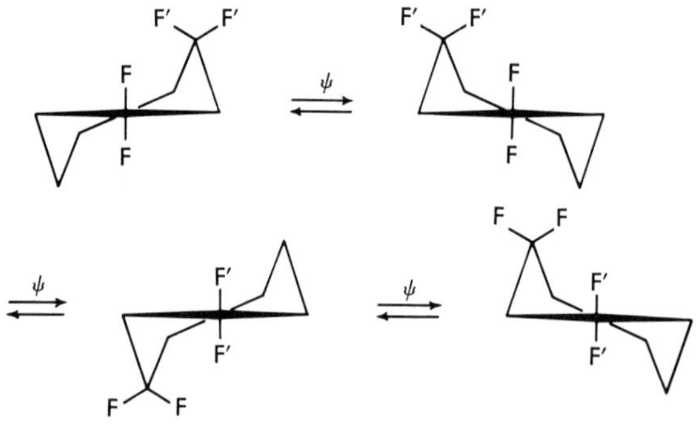

IX

*This position is not really an *axis position* when the ring is asymmetrically substituted because there can be no axis of symmetry for such compounds. Nonetheless, inversion occurs when the substituent(s) pass through the 1-position of VIII.

relatively slight changes in steric interactions that occur in pseudorotation are best seen by actual manipulation of models, preferably of the Fieser-Dreiding type.

To distinguish between the possible modes of achieving equivalent fluorines we have studied some substituted *gem*-fluorocycloheptanes (Glazer and Knorr, unpublished data). Consider 1,1,3,3,-tetrafluorocycloheptane (IX). With this compound we expect at least one pair of fluorines to be nonequivalent unless pseudorotation (or inversion) is very easy, because no more than two fluorines can be on an "axis" position at any one time. The fact is that the fluorines of IX show no magnetic nonequivalence to −180°. We can therefore conclude that pseudorotation and/or ring inversion is fast even at this temperature. Some help in distinguishing between these two possibilities is provided by 1,1-dimethyl-4,4-difluorocycloheptane. With this compound, the methyl groups, being larger than fluorine, are expected to occupy the axis positions, which would make the fluorines take nonequivalent equatorial and axial locations at the 4-position unless pseudorotation or inversion occurs.

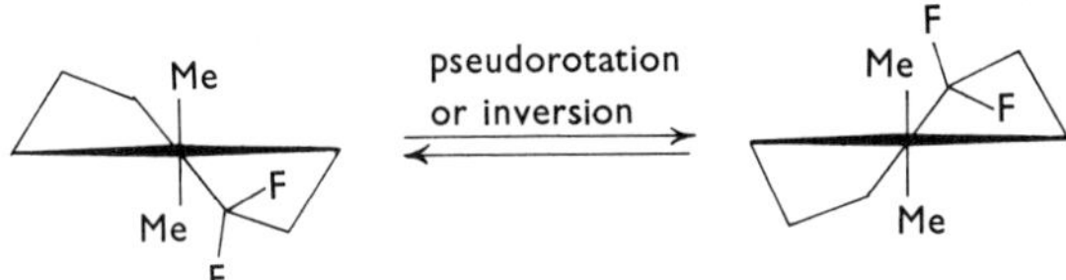

For *gem* methyls, pseudorotation should be far less favorable than for cycloheptane itself, because the chair form with an inside methyl group is expected to have a cross-ring steric repulsion of 9.6 kcal (Hendrickson, 1962).

Interestingly, 1,1-dimethyl-4,4-difluorocycloheptane does show a striking change in its nuclear magnetic resonance spectrum at low temperatures; see Figure 6. At and below −163°, the fluorine chemical-shift difference is 840 cps, and the general appearance of the spectrum indicates the presence of equatorial-like and axial-like fluorines (compare with Figure 1). Preliminary analysis of the rate of change of the line shapes as a function of temperature indicates an E_a value of about 6 kcal for the process that interchanges the

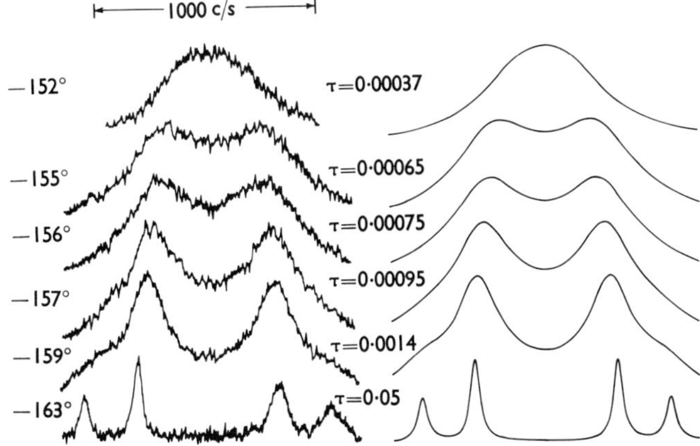

FIGURE 6.
Changes in experimental and calculated ^{19}F spectra of 1,1-difluoro-4,4-dimethylcycloheptane as a function of inversion rate.

fluorines. Since this is so substantially less than the calculated 9.6 kcal for pseudorotation, we conclude that it is likely that the interchange results from ring inversion for this particular system and the exceedingly rapid exchange that occurs in 1,1,3,3-tetrafluorocycloheptane involves pseudorotation.

Although the value of the fluorine-labeling technique for studying conformational equilibria and equilibration by nuclear magnetic resonance spectroscopy still remains to be established by additional point-to-point checks with compounds not containing fluorine, the evidence so far obtained does indicate that the method has considerable advantages and may be of wide applicability.

REFERENCES

Alexander, S. (1962). *J. Chem. Phys.* **37**, 967, 974.

Booth, H. (1966). *Tetrahedron* **22**, 615.

Eliel, E. L. (1965a). *Angew. Chem.* **77**, 784.

Eliel, E. L. (1965b). *Int. Ed.* **4**, 761.

Feltkamp, H., and N. C. Franklin (1965a). *Angew Chem.* **77**, 798.

Feltkamp, H., and N. C. Franklin (1965b). *Angew. Chem., Int. Ed.* **4**, 774.

Gerig, J. T. C., and J. D. Roberts (1966). *J. Am. Chem. Soc.* **88**, 2791.

Grob, C. A., and S. W. Tam (1965). *Helv. Chim. Acta* **48**, 1317.

Hanack, M. (1965). *Conformation Theory*, pp. 56–61 and 180–206. New York: Academic Press.

Hendrickson, J. B. (1961). *J. Am. Chem. Soc.* **83**, 4537.

Hendrickson, J. B. (1962). *J. Am. Chem. Soc.* **84**, 3355.

Jensen, F. R., D. S. Noyce, C. H. Sederholm, and A. J. Berlin (1960). *J. Am. Chem. Soc.* **82**, 1256.

Lambert, J. B., and J. D. Roberts (1965). *J. Am. Chem. Soc.* **87**, 3884.

Moniz, W. B., and J. A. Dixon (1961). *J. Am. Chem. Soc.* **83**, 1671.

Muller, N., and W. C. Tosch (1962). *J. Chem. Phys.* **37**, 1170.

Riddell, F. G., and M. J. T. Robinson (1965). *Chem. Commun.* 227.

Roberts, J. D. (1963a). *Angew. Chem.* **75**, 20.

Roberts, J. D. (1963b). *Angew. Chem., Int. Ed.* **2**, 53.

Spassov, S., E. S. Glazer, D. L. Griffith, K. Nagarajan, and J. D. Roberts. *J. Am. Chem. Soc.* **89**, 88.

Wiberg, K. B. (1965). *J. Am. Chem. Soc.* **87**, 1070.

Wohl, R. A. (1965). *Chimia* **18**, 219.

EDWARD W. HUGHES
Gates and Crellin Laboratories of Chemistry
California Institute of Technology
Pasadena, California

The Past, Present, and Future of Crystal Structure Determination

The diffraction of X-rays by crystals was first observed in 1912 in an experiment suggested by Max von Laue, and by the spring of 1913 the Braggs had shown how, with known crystals, one could engage in X-ray spectroscopy and how, with known X-rays, one could determine the atomic arrangements in crystals. At the time these first crystal structures were being published, Linus Pauling was only a boy of twelve, but as almost all experimentation was halted for a number of years by the first World War, he was a graduate student at Caltech by the time that X-ray crystallography was getting well into its primary development. Instructed in the science and the art of the new technique by R. G. Dickinson, he became a leader in its application to problems of chemical structure. The method remained one of his main experimental approaches in subsequent investigations into many fields: minerals; metals and alloys; coordination compounds; and organic compounds, chiefly those of biological importance. It is appropriate that something general about crystallography should be included in this volume.

I have been asked so often by scientific friends and colleagues who are not trained crystallographers about the reliability of various structure determinations of the past or present or about the feasibility of making a successful attack on some specified structure problem that it seems worthwhile to discuss these matters here. Accordingly, the ensuing discussion is addressed to persons with these kinds of questions, and my colleagues in crystallography will find little that they do not already know in what follows. It will be assumed, however, that readers are familiar with the descriptions of crystals set forth in physics or chemistry texts at the freshman or sophomore levels.

In order to discuss the errors that can be made it is necessary to describe briefly the nature of crystals and some aspects of the processes of X-ray crystallography, and these can perhaps be introduced best by describing and correcting some of the erroneous notions held by some scientists who are not specialists in the field. I have not conducted a poll to ascertain just how prevalent these misconceptions are, but since they appear repeatedly in widely used textbooks in many fields of science and of many levels of sophistication, I can only conclude that a great many students have been contaminated.

Contribution No. 3423 from the Gates and Crellin Laboratories.

An ideal crystal is an array of atoms, molecules, or ions in which there is a relatively small, basic pattern that repeats itself exactly at regular intervals in real three-dimensional space. The precise scheme according to which the basic unit repeats itself in space is called the space-lattice, or simply, the lattice, of the crystal. It is the mathematical framework upon which the real structure, made of atoms, is constructed.

The concept of the lattice is one of the most basic constructs of the crystallographer, and perhaps is the one that is most misunderstood and misused by nonspecialists. Being a mathematical description, it can have no real physical properties. But one sees in texts and papers repeated references to such properties as if they were properties of the lattice. "The electrical conductivity of the copper lattice." "The density of the rock-salt lattice." "The birefringence of the calcite lattice." These represent but a tiny fraction of the common examples. One even sees the large voids that occur in some structures, such as the zeolites, described as "holes in the lattice." *Any* lattice, by itself, is just one big hole marked off into regular cells by the repeat points. In every example these authors are discussing a property of the arrangement of *atoms*. Almost every time an author finds himself about to use the word *lattice* he would be well advised to substitute some other word such as *structure* or *arrangement*.

The commonly used classification of lattices was derived nearly 120 years ago by Bravais and the fourteen types recognized by him are commonly called the Bravais lattices. Seven of these are called simple or primitive lattices; there is only one repeat of the pattern per unit cell. As a matter of fact one can describe *any* crystal in terms of one or another of these simple lattices. However, Bravais recognized that it would be convenient if the lattice used had a certain minimum amount of symmetry in common with the crystal described, and he was able to show that this would require seven additional lattice types, all characterized by the fact that there are either two or four units of pattern per unit of lattice. These are called centered lattices because the extra pattern units occur either at the centers of unit cells or at the centers of faces of the cell. As an example we may note that the rhombohedral lattice (which is always a simple one) is defined by three repeat vectors of equal length diverging from a common origin and enclosing three equal angles. The *unit cell* is described simply by giving the length a of the vectors and the angle α between them. This lattice, by itself, possesses, among other symmetry elements, a single threefold axis of symmetry. It is suitable for discussing structures having a single threefold axis. Bravais pointed out, however, that when α has any one of three certain special values, the lattice, by itself, possesses *four* noncoplanar threefold axes and three fourfold axes and is thus compatible with cubic arrangements. At $\alpha = 60°$ there results the *face-centered* cubic lattice, at $\alpha = 90°$ the usual *simple* cubic lattice, and at $\alpha = 109°28'$ the *body-centered* cubic lattice. These examples are illustrated in Figure 1. Any centered cubic structure of either kind could be described in suitable rhombohedral axes with a simple cell, but any satisfaction achieved by use of a simple cell is usually greatly outweighed by the additional computational difficulties involved. Similar remarks apply to all the other centered lattices introduced by Bravais.

A very common textbook error involves a table showing the restrictions

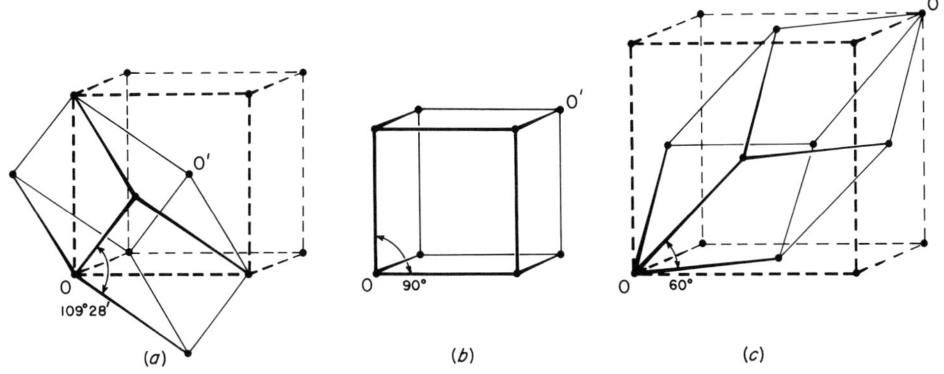

FIGURE 1.
The three Bravais cubic lattices and their relation to a simple rhombohedral lattice. Dashed lines outline the centered cubic cells. The rhombohedral cell (solid lines) has the same cell edge in each drawing. Solid circles are lattice repeat points. (*a*) $\alpha = 109°28'$; body-centered cubic. (*b*) $\alpha = 90°$; simple cubic. (*c*) $\alpha = 60°$; face-centered cubic.

imposed on unit cell dimensions by crystal symmetry. I know of only one text where the table does not imply strongly that the cell dimensions establish the symmetry. Thus they say that for orthorhombic crystals $\alpha = \beta = \gamma = 90°$, which is correct, and then they add $a \neq b \neq c$. This implies that if $a = b$ the crystal cannot be orthorhombic, and this is *not* correct. It is the symmetry of the crystal as a whole that determines the appropriate crystal system. If an orthorhombic crystal wishes to have $a = b$, it may do so *without* becoming tetragonal. But it *must* have $\alpha = \beta = \gamma = 90°$ if it wishes to remain orthorhombic. As a matter of fact, when a structure was first proposed for metallic gallium on the basis of powder patterns, the cell was found to have $\alpha = \beta = \gamma = 90°$; $a = b = 4.51$ KX; $c = 7.51$ KX; and a tetragonal structure was suggested (Jaeger et al., 1927). Later, when single crystals were obtained, they were found to differ considerably from tetragonal symmetry, and the presently accepted orthorhombic structure was worked out (Laves, 1932). But still no difference could be detected between the lengths of the a and b axes. It was only after the development of high-precision lattice-constant techniques that a small difference, about one-and-a-half parts per thousand, could be measured (Bradley, 1935). But by adjusting the temperature and pressure these axes can probably be made equal within any attainable accuracy. The point is that the arrangement of gallium atoms in the cell does not have a fourfold axis of symmetry and a purely accidental *equality* of axial *lengths* does not imply *equivalence* of the two corresponding *directions.*

A common error is to describe the sodium chloride structure as simple cubic. If a simple cubic lattice of proper size and orientation has its origin superimposed on an atom of this structure one does indeed find an atom at all lattice points. But half of them are sodium atoms and half are chlorine atoms, and the important requirement that lattice points shall be *repeat* points is not met. Sodium chloride is face-centered cubic with two nonequivalent atoms in the basic pattern. For example, if one puts the origin of a correctly scaled and oriented face-centered cubic lattice at a point half way between neighboring sodium and chlorine atoms one then finds a similar pair of atoms similarly oriented about *every* lattice point. An even commoner text error is that describing the cesium chloride structure as body-centered cubic. This structure is actually based on a simple cubic lattice with two non-equivalent atoms per repeat unit. On the other hand, α-iron *is* body-centered cubic; the atoms at the cube centers are chemically identical with, and crystallographically equivalent to, those at the cube corners. The difference between the two examples is not a quibble, and is easily demonstrated experimentally. In the X-ray diffraction pattern from α-iron all the observed reflections have triple-order numbers $h\ k\ l$ such that $(h + k + l)$ is even. At the places on the film where a simple cubic lattice would predict reflections with $(h + k + l)$ odd, there are absolutely no spots. For cesium chloride both types of reflections occur, and the film is much richer in spots.

Many texts discuss only the simplest structures and leave the student with the impressions that all atoms in crystals lie at highly symmetrical points in the cell and that all atoms lie in Bragg planes. In fact, some authors discuss at length the density of atoms per unit area in various simple Bragg planes. Such simple structures, although occurring for very many simple substances, are not typical of even moderately complicated substances. Particularly in nonplanar-molecular type crystals (and often even for planar molecules), one rarely finds atoms at symmetrical locations and one rarely finds a Bragg plane that has a single atom lying *exactly* in it. To describe such a structure one must give the dimensions and symmetry of the lattice on which it is based, the symmetry of the structure as a whole, the number of repeat units per cell, the symmetry of the cell contents, and then the x, y, z coordinates (parallel to the cell edges) of every atom that is not related by symmetry to another atom already located. The smallest set of atoms, which need to have their positions specified, is generally called the "asymmetric unit," even where the set consists simply of a single atom. The asymmetric unit may be identical with the repeat unit of the lattice, but is usually much smaller. In molecular crystals the asymmetric unit is not generally a single molecule; it may vary from a fraction of a symmetrical molecule up to some small multiple of such fractions. Thus, for example, in hexamethylene tetramine ($C_6H_{12}N_4$) the asymmetric unit contains one-twenty-fourth of a molecule, in melamine ($C_3N_3(NH_2)_3$) it is just one molecule, in biphenylene ($C_{12}H_8$) it is one-and-one-half molecules and for the simplest form of tri-nitrotoluene ($C_6H_2CH_3(NO_2)_3$) it is two molecules. The asymmetric unit, in spite of its name, may have considerable symmetry: the asymmetric unit for melamine has, within experimental accuracy, a threefold axis perpendicular to a mirror plane which in turn contains three twofold axes. The point is that whatever symmetry this unit has is not being used by the crystal; the symmetry elements of the asymmetric unit do not operate on atoms in neighboring units.

When we turn to consider errors that can and have been made by crystallographers it is not surprising that they have often been mistakes related to the textbook errors described above. In the period when the above notions were not completely familiar to everyone in the field a number of incorrect lattice constants were reported. I am not referring here to lack of accuracy but to gross errors involving, generally, simple integral factors like two, or three, applied to cell edges. Usually in such cases the cell reported was too small in one or more directions; it did not represent a true repeat. During the nineteen-thirties and early forties I happened to examine about a dozen crystals whose cells had been reported by earlier workers after they had abandoned attempts at solving the structure. Four of these had mistakes of the kind described, which might account for the failures to solve the structures. In fairness to most of my predecessors I should say that three of the four errors were all made by one experimenter; in crystallography, as in other branches of science, it is useful to know who does careful work and who does not.

By about 1930 procedures had been worked out which, if conscientiously followed, would make such mistakes impossible for single-crystal work, except in exceptional circumstances. The amplitudes of X-ray waves scattered from an atom depend, among other things, on the number of electrons in the atom. Thus a few heavy atoms may account for most of the scattered intensity, particularly if there are not many light atoms present. The lattice found in such a case may be one that governs the repetition of the heavy atoms considered by themselves only but not that of the light atoms; the true lattice will be compatible with the repetition of all the atoms present, *including* the heavy atoms when their environment of light atoms is considered. What happens experimentally is that the Bragg reflections that would reveal the true larger repeat are so weak that they escape detection. In the early days homemade X-ray tubes were often feeble, and the more powerful "store-bought" outfits were exceedingly expensive and burned out easily. Experimenters were often reluctant, for one reason or another, to make the long heavy exposures necessary to reveal weak reflections, and sometimes were even unwilling to make *enough* pictures to show the true lattice. As noted above, centering can cause a systematic suppression of reflections. Certain symmetry elements do the same sort of thing. In a body-centered case, for example, if an experimenter samples only a portion of the reflections, and happens to collect only some of those for which $(h + k + l)$ is even, and interprets them as if there were no such restrictions, he will arrive at a wrong cell. I am aware of several published errors of this kind.

Such errors are exceedingly rare today, particularly in work published in journals with reputable refereeing systems. As one of the co-editors of "Acta Crystallographica" during a recent seven-year period, I, together with my referees found grounds for questioning only about one manuscript out of each hundred because of doubts about the lattice determination. Errors in recognizing symmetry elements can occur for the same reasons when this recognition depends upon the systematic absence of some reflections. This is sometimes a tricky business, and it may be necessary to make an arbitrary choice at the start and justify it in the end by the excellence of the agreement between observed Bragg intensities and the values calculated from the final structure. Careful workers will usually call attention in their papers to any special circumstances that might have led them to a wrong cell or space-group, and will discuss fully and frankly the probability that this might have happened.

A special caution needs to be added in regard to structures based upon powder diagrams only. Powder patterns are very useful for identifying unknown substances although the method is not very sensitive. And there are a host of other interesting applications for powder work, such as determination of particle size, preferred orientation, thermal expansion coefficients, and many other properties. For simple substances one can also derive very accurate lattice constants by special powder techniques, provided the true approximate lattice can be established by other means. But for the determination of atomic arrangements in even moderately complicated substances, the data obtained are generally quite inadequate. The chief difficulty is lack of resolving power; for the average molecular compound almost every line of the pattern is an approximate superposition of two or more Bragg reflections, and for some of the more symmetrical lattices one finds exact superposition of nonequivalent reflections. Very few of the more complicated diagrams have been even indexed.

The determination of the lattice constants and symmetry are but the start of a structure determination. We observe diffraction *intensities*. For direct calculation of a structure we need to know the *amplitudes*. In general an intensity is the square of the absolute magnitude of the amplitude:

$$I = |F|^2 = F \cdot F^*$$

where F^* is the complex conjugate of the amplitude F. By taking the square root of I, one obtains $|F|$, but the phase angle cannot be determined experimentally by any direct method so far devised. If the crystal possesses a center of symmetry and this center is used as origin, it can be shown that the complex phase angle is either zero or π. That is to say

$$F = \pm \sqrt{I}.$$

But if there is no center the phase angle ϕ is $-\pi \leq \phi \leq \pi$.

The first structures determined were so simple that one could test all possible structures that conformed with the lattice, symmetry, stoichiometry, and crystal density by comparing calculated and observed intensities. This was so despite the fact that at the start only very crude methods were available for calculating intensities; the physical laws governing the scattering process were not known exactly. This trial-and-error method was the standard operating procedure for many years. As more crystal chemistry was learned from the solution of simple structures, it became possible to "guess" the structure of more complicated structures; and as knowledge of the physical laws improved, the intensity-comparison test became more stringent and could establish the correctness of the more complicated structures.

So now one must ask: "Is it possible to have a completely or partially wrong structure that will pass the intensity test?" The question has been discussed chiefly by Patterson (1944). One way of formulating the intensity calculation shows clearly that the intensities of the Bragg reflections depend upon the *interatomic distances* in a unit cell, weighted of course in an appropriate fashion by the atomic numbers of the two atoms associated with each distance. So Patterson asked the question in this form: "Is it possible to have two or more different arrangements of the same set of atoms that will yield the same set of weighted distances?" Patterson called such arrangements "homometric structures" and was forced to admit, somewhat reluctantly, one judges, that they can exist. However, general rules for establishing the circumstances under which they can exist are few and far between, and the best we can do is to surmise that they are very rare for complicated arrays of atoms in three dimensions, and then to be careful. "Being careful" consists of satisfying oneself that a proposed structure meets the intensity test within the known errors in measuring and calculating the intensities; that it complies with all *well-established* facts about crystal chemistry and physics; and that there does not seem to be any other reasonable structure that comes close to meeting all these tests. One valuable check is to calculate the average electron density as a function of position in the cell, using the phase angles calculated from the proposed structure and the observed amplitudes. The resulting calculated density should show at the proposed atom sites maxima that agree in shape with the best current quantum mechanical calculations for atoms, and at other places there should be no false maxima larger than those to be expected from the suspected errors in the intensity measurements. And in particular there should be no negative regions greater than those expected from the errors in the measurement.

There are few indications in the literature of troubles of the kind just discussed. One clear example is the first structure proposed for triphenylene (Klug, 1950). Suspicions were aroused because some of the nonbonded (van der Waals) distances between neighboring molecules were too small as judged by an overwhelming number of previous structures containing similar atoms, chiefly C—H and $C{=}H_2$ groups. Reinvestigation in two different laboratories (Vand and Pepinsky, 1954; Pinnock, Taylor, and Lipson, 1956) showed that the molecular model first proposed was almost right and was correctly oriented, but that its center was displaced a considerable distance (about 1.0 Å) from the correct position. In triphenylene all the carbon atoms lie very close to some of the points of a planar net of equilateral triangles. If neighboring molecules are displaced from their correct positions by displacements parallel to the planar net so that intermolecular vectors are changed by just one edge of a triangle of the net, there will be, after the displacements, duplicates of many of the old distances, although they will be between different atoms. And since all the intramolecular distances remain completely unchanged, the net result is an almost, but not quite, correct set of interatomic distances; that is, almost a homometric set. The new investigators found by trial several positions for the molecular center that yielded almost as good intensity checks as the one judged to be best. The latter yielded van der Waals distances completely in accord with previous experience. From this example one can readily see the sort of situation in which one must be particularly careful. One also sees the justification for testing structures against previous experience. At a number of times in the past authors have claimed to have found structures with abnormally short van der Waals distances and have claimed that these distances were real and were associated with abnormal intermolecular attractions. Usually planar aromatic hydrocarbons were involved, and these abnormal forces were alleged to account for the known tendency of these substances to form complexes with each other, particularly when one of them is substituted with highly polar groups, like the nitro group. In every case where these crystals have been reexamined with great care, and there must be at least four or five such cases, errors have been turned up and structures have been found that are quite normal in every respect and give excellent agreement between calculated and observed intensities, better agreement than that obtained in the original work. The moral is that before rushing into print with peculiar structures one should check all measurements, arguments, and calculations with great care and skepticism, and perhaps seek the criticism of some unbiased outsiders. The following story is perhaps permissible in this book at this point. In 1939, while I was working on the structure of melamine (Hughes, 1941),

```
      NH2
        \
         C—N
        //   \\
       N       C—NH2
        \  ↷  /
         C=N
        /
      NH2
```

I was trying to fit the expected planar ring into the cell so that it would be about perpendicular to the direction, reported in the literature, for the vibration direction in the crystal for the smallest refractive index for light. Each successive Fourier projection synthesis based on these attempts produced what appeared to be a very good projection of the chair form of a saturated ring, presumably of

```
      NH      H
        \\    /
         C—N
        /    \
      HN      C=NH
        \    /
         C—N
        //    \
      NH      H
```

One day Professor Pauling visited my office and looked carefully at all these drawings. On the way out he paused long enough to remark that it was very interesting but that he hoped I would have better judgment than to present *that* structure to the local seminar. Much later, after having abandoned the work for some months (because the structures would not refine properly to *any* kind of ring), a reinspection of the *X-ray data alone* suggested another orientation for the ring. A quick optical experiment showed that the published position for the acute bisectrix of melamine was off by something like 50°. The actual position was almost exactly at the place suggested by the X-ray data, and there was no more trouble in getting a good intensity check. I need hardly add that it yielded a planar aromatic ring as expected.

It may at first appear that I have cited quite a few erroneous structures that have found their way into the literature. Actually these represent only a small fraction of a percent of the total number of structures reported in the past fifty-four years, an overwhelming number of which appear to be reliable. My experience as editor and referee of approximately four-hundred papers has uncovered, in addition to the doubtful lattice determinations already mentioned, just three papers that were rejected because the structure itself seemed in doubt, over and above any mere question regarding the accuracy of the interatomic distances. It is interesting that in two of these cases reinvestigations by the authors verified completely the fears of the referees, and eventually different and very satisfactory structures were published. The third example is too recent to have been resolved.

The matter of accuracy, as differentiated from general correctness, is also of considerable interest. If a structure is a very simple one in which the atoms are all at positions fixed by symmetry, the interatomic distances are fixed fractions of the cell edges and their percent accuracy is the same as that applying to the lattice constants. Suitable examples are the sodium chloride structure, the cesium chloride structure and the diamond structure. With just moderate care and patience, lattice constants can be had to an accuracy of two parts in ten thousand or better. By a special technique Mrs. Lonsdale (1947) has measured the lattice constants of diamonds to about two parts in one-hundred thousand. As she points out, such precision is meaningless unless reported along with the accurate temperature of the experiment and the chemical analyses of the diamonds; traces of impurity affect the lattice constant by detectable amounts. But if some or all of the atoms are not in fixed positions, the accuracy is much less and depends upon the agreement that can be obtained between calculated and observed intensities, considered as a function of the position of the atoms in the cell.

For the first simple structures with variable parameters (that is, with atomic coordinates to be determined from the intensities), it was possible to plot the calculated intensities as functions of the coordinates and to pick from the graphs the values corresponding to the best fit. Estimates of the limit of error on the resulting bond lengths ran in the range ± 0.05 Å. Soon, however, crystallographers became anxious to solve structures more complicated than those. Already in 1915 Sir William Bragg (1915) had shown the relationship between diffraction amplitudes and the amplitudes in the Fourier representation of the electron density in the cell, but it was only late in the twenties that the method began to be used on multiparameter structures. If one had arrived by hook or by crook at a trial structure that yielded calculated intensities in fair to moderate agreement with observed values one assumed that the signs were correct, at least for the larger amplitudes, and a Fourier synthesis using calculated signs and observed amplitudes generally suggested how to improve the model. By successive approximations one approached the situation in which the signs were all determined. Since there are always spots, and particularly outlying spots, too weak to be observed, the Fourier synthesis was "artificially terminated," and this produced small errors in the positions of the peaks representing atoms. Later, corrections were developed for this effect. At first it was thought that the labor of summing such series, calculating electron densities at hundreds of points by adding together a hundred or more trigonometric terms per point, could not be justified unless one had the best possible intensity data from an ionization spectrometer. It was soon demonstrated that intensities estimated visually against a calibrated scale of spots were almost as good (Hughes, 1935a, b); after 1935 and until just recently, most structures have been determined in this way. In 1941 the method of least-squares was adapted to improve the intensity agreement (Hughes, 1941). This required linearizing the intensity equations with regard to a trial structure and so the improved parameters from one refinement could not be the best possible; the process must be repeated until the indicated changes in parameters are smaller than their indicated standard deviations. The labor involved was great and the method was at first used only for determinations in which Fourier projections were unresolved and could not be used. With the advent of high-speed computers this has probably become the most common method of refinement. A single stage of the least-squares refinement that required three days of very tedious work in 1940 (even with using an IBM tabulator as a computer of sorts) would occupy the attention of a modern high-speed computer for about a minute!

The quality of the final result is commonly indicated in two ways. First, a simple index usually called the R-factor, or the agreement index, is computed. There are several formulae in vogue, a common one being (in percent)

$$R = 100\Sigma\,(|F_c - F_o|)/\Sigma(|F_o|)$$

where F_c and F_o are the calculated and observed amplitudes, and the sums are over all the data. This takes no account of the relative reliabilities of the data. The second and more satisfactory procedure is to obtain an estimate of the standard deviations of the observations (this depends upon how the intensities were measured), and then by the regular calculus for the propagation of error, to convert these into standard deviations in atomic coordinates and in bond lengths and angles. In 1945 when people were still usually doing projections with sets of two-dimensional data, typical R values were 15 to 18 percent, and the few standard deviations (*not* limits of error) worked out were something like ± 0.020 Å for a carbon–carbon bond length. Since then there have been great improvements in the scattering factors used in the calculations; they are now based on the most advanced quantum mechanical calculations of electron densities in atoms and are available for the entire periodic table, and for various valence states of some atoms (Cromer and Waber, 1965). Moreover, the increasing speeds and capacities of digital computers have made it feasible to include the small contributions from hydrogen atoms, formerly neglected, and to include the effects of the anisotropic temperature vibrations about the equilibrium point for each atom separately. These refinements have brought R factors for visually estimated data down into the range of 5 to 8 percent, and since most work today is three-dimensional and there is accordingly a larger ratio of measurements to parameters, standard deviations for carbon–carbon bonds have gone as low as ± 0.005 Å. In the past couple of years much attention has been given to the commercial development of fully automated diffractometers that measure the diffracted X-ray intensities with counter techniques, and now the first results are beginning to appear. These instruments not only speed up the data collection considerably and relieve people of the painful chore of estimating visually thousands of spots, but they generally produce more accurate data. It seems likely that R values will be in the 2 to 3 percent range.

The standard deviations quoted throughout the whole period in which such values have been given, are in my opinion too small. Perhaps one should say that they are all right as precision indices but overestimate the real accuracy. I believe this because not even today are we allowing for all the influences known to affect the intensities. There are at least four potentially important effects; only a few tentative trials have been made at allowing for any of them.

1. Our very best scattering factors, mentioned above, are calculated on the assumption that atoms in crystals are spherically symmetric. The effects of bonding and of the electric fields near ions are generally ignored. A few persons have just recently claimed to have improved agreement significantly by making corrections for electrons concentrated into bonds, but so far there is no general agreement of when and how to allow for such effects and they are usually ignored.

2. When the X-ray frequency is close to a natural frequency of the electron cloud of an atom, there is a disturbance in the scattering factor. Both the amplitude and the phase of the scattered rays are changed. This is the "anomalous dispersion" effect. In the past it has often been considered to be important only when the X-ray frequency was exceedingly close to the K absorption edge of an atom. Reasonably good estimates of the effect have recently been published (Cromer, 1965) for nearly all atoms and all commonly used X-ray frequencies, and one sees that, with the possible exception of Mo$K\alpha$ radiation on first-row elements, the effect might always be detectable and that for heavier atoms and longer wavelengths the effect is frequently quite large; not only K edges contribute, but L and M edges also do. Fortunately, the effects are not very sensitive to scattering angle, and it ought to be possible to allow for them in computer programs.

3. Our current methods for allowing for the thermal vibrations of atoms in crystals assume that the atoms vibrate independently of each other, each about its own equilibrium position, and that the vibrations are such as can be represented by a centrosymmetric triaxial ellipsoid. In addition to the three coordinates x, y, z of an atom in a general position one also introduces (and refines in the least-squares process) six parameters giving the orientation in space and the axial lengths of the ellipsoid. For atoms in special positions with symmetry, the number of parameters needed is reduced. In Fourier syntheses the shapes of atoms frequently appear just as if all this were true. But sometimes they do not. One often sees atoms that appear slightly "bent" in one or even two dimensions, something like a very short banana or a very thick curved pancake. The X-rays give only a "time-exposure" picture of the vibrating crystal, and the atoms in question appear to be riding on a rather rigid molecule that not only oscillates in a linear way about its equilibrium position but is also carrying out rotational oscillations about its center of gravity. People have analyzed these temperature effects *after* a structure determination is completed, and have applied some corrections to the apparent positions of such atoms, but seemingly little is known about the direct effect of such coordinated motion on the average scattered intensity, and apparently no effort has been made to allow for it in the actual refinement.

4. When a crystal is in the proper position to reflect the primary X-ray beam into a diffracted beam of given indices this latter beam may be regarded as a new and transitory primary beam. If this new "primary" happens to be oriented properly with regard to the crystal, it can suffer a second diffraction. This possibility was first noted by Renninger (1937) who showed that spots produced by such double reflection must fall at places permitted by the crystal lattice and so cannot produce an error in the lattice determination. But they can fall at places forbidden by the space group and this effect has at times caused doubt about symmetry determinations. In the present discussion we are concerned by the fact that these double reflections can and do cause spurious enhancements of some of the ordinary reflections. If the resulting discrepancy between a calculated and an observed intensity is large (that is, both of the cooperating reflections are strong) many investigators will suspect the cause and make tests to verify and to correct for this effect for that one reflection. But as unit cells increase in size and as beams become more divergent in certain diffractometer applications, the probability of encountering this effect increases and there can be a considerable number of intensity

enhancements produced which are small enough to escape notice but which are large enough and sufficiently numerous to make errors in the interatomic distances at least as big as the accompanying standard deviations, and perhaps bigger. No systematic procedure for detecting or eliminating this effect has been proposed, and it appears probable that many workers are even unaware of its possible importance.

That the routine standard deviations are too optimistic is occasionally demonstrable when dealing with a crystal containing many chemically equivalent but crystallographically nonequivalent interatomic distances. For two isomers of the general formula $C_8F_2(C_6H_5)_6$, the crystals are both triclinic and the molecule quoted is the asymmetric unit (Fritchie, 1963; Beineke, 1966). We need to evaluate 138 positional parameters and 276 vibrational parameters for each crystal. The orientations of the benzene rings about the C_8 nucleus suggest no conjugation and so for each crystal we expect thirty-six independent but equal carbon–carbon benzene-like bonds. About five thousand reflections were measured for each crystal, and the indicated standard deviations came out gratifyingly low: about ± 0.0085 Å and ± 0.0050 Å per bond length for the two crystals. However, the scattering of the phenyl bond lengths about their mean yields standard deviations per bond length of about ± 0.017 Å and ± 0.011 Å, respectively; we believe that these larger values are the ones that count. It seems likely that most published standard deviations need to be multiplied by similar factors before they are used in significance tests. Incidentally the average carbon–carbon benzene bond lengths for the two crystals are 1.381 Å and 1.383 Å, respectively. The overall mean is 1.382 Å and has a standard deviation not greater than ± 0.002 Å, based on internal consistency and not corrected for thermal vibrations.

The remaining topic of interest concerns the methods of obtaining valid trial structures for ultimate refinement. The operation of the trial-and-error method has already been described since it had a bearing on possible mistakes that might be found in published structures.

Two of the more recent methods have been in use for some time, and have received considerable attention recently because of the part they played in the winning of several Nobel prizes. They are the heavy atom and isomorphous replacement methods; they need not be described here. There are some drawbacks to these methods, and they cannot in any general sense be regarded as solutions of the phase problem. In the heavy atom technique, if there is a large enough fraction of the scattering power of the unit cell concentrated in the heavy atom or atoms to make the method work, the light atoms, which we wish to locate, do not contribute heavily to the scattered rays and so the accuracy with which they can be located is reduced. Moreover, if the molecule of interest does not contain a heavy atom it may be undesirable to introduce one by chemical means; the investigator may need to know the structure undisturbed by further chemical action. For example, when Professor Pauling initiated a program of X-ray crystal studies of amino acids and simple peptides in 1937, it was important to discover the configurations of these protein building blocks and the way they packed and hydrogen bonded to one another without causing the disturbance of the introduction of heavy atoms foreign to the proteins. It would have been easy to have turned out any number of structures of these substances complexed with such atoms as copper, nickel, and so on, but the information obtained would have been of little or no use in predicting structures for polypeptides and proteins. It was necessary to grind out the structures without using heavy-atom techniques. The isomorphous replacement method does not suffer from precisely these troubles because generally one of the isomorphous forms has been the heavy-atom-free material and one of the final structures obtained is always for the free material. However, the prospect of finding isomorphous crystals is not always very bright. Fortunately, for proteins, for which the method is so far the only successful one, it seems possible to fasten quite a few different groups containing heavy atoms to a number of different sites on a protein molecule without changing significantly the way in which the molecules pack together in a crystal. It is perhaps worth pointing out, that for simpler substances, potassium and rubidium salts of the same acid are very often isomorphous, and the ammonium salt also is in cases where hydrogen bonding is not important.

If the light atom structure is completely unknown, finding that structure is generally the main concern and one does not worry very much about extreme accuracy of bond lengths or about packing. This is the situation in which the heavy-atom technique is supreme. For example, Mathieson and colleagues have determined the detailed atomic arrangements for several alkaloids with previously unknown structures. The heavy atoms were usually introduced either by forming the hydrobromide or by substituting into the molecule methiodide groups (Fridrichsons and Mathieson, 1965).

Another approach is that of the Patterson-Fourier synthesis first introduced in 1934. It is well enough described in a number of junior level physical chemistry texts. Actually this method is almost an absolute prerequisite for the start of both the heavy-atom and the isomorphous replacement methods; it is used to locate the heavy atoms. The Patterson peaks due to the heavy atoms are usually so few and stand out so well above the light atom peaks that interpretation is relatively easy. But this is not the case for complicated structures in which all the atoms are of about the same scattering power. In recent years a number of ingenious procedures have been developed for unscrambling the Patterson diagrams, and most of these have been or can be adapted to computer operation. As matters stand today it is the method in which it is easiest for the knowledgeable crystal chemist to use his familiarity with earlier structures to guide the interpretation. Every year has seen larger and more complicated structures solved by this method, and one cannot guess where the limit will be. For example, the first of the isomers of $C_8F_2(C_6H_5)_6$, mentioned above (Fritchie, 1963), with 46 atoms per asymmetric unit and 92 per triclinic cell was solved by Patterson diagrams in conjunction with a special interpretational device also developed by Patterson (1949). At the start the molecular structure was uncertain. Trimesic acid, $C_6H_3(CO_2H)_3$, with 6 molecules per asymmetric unit and 720 atoms per monoclinic cell has been solved even more recently, chiefly by fitting a predictable model to Patterson maps (Duchamp, 1965).

It has been implied earlier, particularly in the discussion of the least-squares refinement, that the number of experimentally observed amplitudes exceeds by a considerable factor the number of unknown x, y, z coordinates required to fix the structure. Mathematically, the problem is overdetermined, and this implies that there must be relationships between the amplitudes and phase angles. Banerjee (1933) actually succeeded in determining correctly some signs for a previously reported structure using very special considerations of this kind; but the time was not ripe, and the work was forgotten. In 1947 Harker and colleagues were attempting to solve the structure of crystalline decaborane, $B_{10}H_{14}$. They were having difficulties, partly because of disorder in the crystal but mostly because practically all the previous notions about borohydride structures were wrong and were worse than useless in helping to interpret the X-ray data. In the course of the investigation Harker and Kasper (1948) were finally led to work out their "inequality method," and it was applied successfully to the decaborane problem (Kasper, Lucht, and Harker, 1950). This was a very important result and showed how desirable it was to have, when needed, a method that was completely independent of previous knowledge about structures. Their result is essentially dependent upon the two simple ideas that the structure is made up of atoms and that the electron density throughout the crystal is never negative. These ideas permitted them to deduce inequalities involving amplitudes whose indices are related in specified ways. The relations are such that, if the absolute amplitudes are large enough, one may be forced to give a definite sign to one of them to avoid contradicting the inequality. Many other authors derived more general inequalities and more complicated ones. But it soon became apparent (Hughes, 1949) that as crystals become more complicated, the frequency with which one would find large enough amplitudes to produce sign indications would decrease and that this method, despite its great historical import, would be useful only for relatively simple structures. However, many people began to note that although a given inequality did not quite *require* a given sign for a certain amplitude, nevertheless, more often than not, the amplitude had, in fact, the sign that would be required for a slightly larger amplitude value. This is perhaps why suddenly—from about 1950 to 1953—so many persons began to consider probability relations between amplitudes.

In four different ways (Sayre, 1952; Cochran, 1952; Hauptman and Karle, 1953; Hughes, 1953) it was shown that the triple product $F_{hkl} \cdot F_{h'k'l'} \cdot F_{h'-h,\, k'-k,\, l'-l}$ tends to be positive (for centrosymmetry), and that the larger the absolute size of the product in relation to the largest possible value it might have, the more probable is the positive sign. In work of this kind it has become customary to modify the amplitude, either by dividing it by the total possible scattering power at that scattering angle, or by dividing the intensity by the average intensity at that scattering angle and then taking its square root. This results in the unitary structure factor, $U_{hkl}(|U| \leq 1)$, and the normalized structure factor, E_{hkl}, $(\langle E^2 \rangle = 1)$, respectively. In particular my own result (Hughes, 1953) showed that if all the atoms of the cell are the same (a useful limiting case contrasting with the heavy-atom case) then

$$\langle U_{hkl} \cdot U_{h'-h,\, k'-k,\, l'-l} \rangle = \frac{1}{N} U_{h'k'l'}$$

where N is the number of atoms per cell, and the average is over all available *hkl* values with h', k', l' constant. This result is nearly true even when the atoms are not all exactly the same; for an organic crystal containing only carbon, nitrogen, and oxygen (not counting hydrogen), there is a fictitious value of N such that the equation will be about as accurate as the experimental measurements, except for quite small $U_{h'k'l'}$. One can see by inspection that if $U_{h'k'l'}$ is large, then for any large product in the average, the sign must be, more often than not, the sign of $U_{h'k'l'}$.

Zachariasen (1952) proposed a scheme for solving structures based on these results. First he restricted attention to a set of the larger amplitudes. He then assigned three signs according to certain rules to fix the origin, sought signs available from inequalities, if any, and then used letters, *a*, *b*, *c*, and so on, to represent a few more signs for the larger amplitudes. Finally, on the basis of the probability relation he asserted that

$$s_{h'k'l'} = s\langle s_{hkl} \cdot s_{h'-hk'-kl'-l}\rangle$$

where s_{hkl} stands for the sign of U_{hkl}. Substituting into the average as many products of known sign as he could find, he worked through the data, first finding relations between the letters, and then gradually eliminating them with ± 1. He applied this method to the three-dimensional data for metaboric acid, HBO_2, and was able to solve the structure; he obtained a total of 198 signs for the larger amplitudes, and in a Fourier synthesis these gave a good trial structure. There are three molecules per asymmetric unit, and twelve per monoclinic cell.

Attention was distracted from this method by the nearly simultaneous appearance of more detailed and more complicated statistical procedures (Hauptman and Karle, 1953). Although these looked promising on paper and had some successes with simple structures, their quantitative applications depend upon having data corresponding to a resolved Patterson map and for complicated structures this never happens. The Zachariasen procedure depends only on having a resolved structure, and in three dimensions with CuKα data all structures are resolved, except perhaps for the hydrogen atoms.

More recently Karle and Karle (1966) have gone back to the use of the simple triple-product relation, and have sharpened it somewhat by using actual amplitudes, instead of signs, in the average so that

$$s_{h'k'l'} = s\langle E_{hkl} \cdot E_{h'-h,\ k'-k,\ l'-l}\rangle$$

(one can use an E only if one knows its sign or has given it a letter sign). And they use estimates of the probability that the triple products are positive to decide when the overall sign indication is reasonably certain. They have solved a number of structures, including some for which the molecular structure was previously unknown. Perhaps their most important achievement is the extension of the method to the solution of a noncentric structure (Karle and Karle, 1964). In the centrosymmetric case they have solved a very large structure (Karle and Karle, 1963), cyclo(hexaglycyl) hemihydrate, $(CH_2CONH)_6 \cdot \frac{1}{2}H_2O$. There are 98 atoms in the asymmetric unit and 196 in the triclinic cell. However, there is a strong pseudocell in this structure, and the phase-determining method was actually used only to get the pseudocell structure containing only one-fourth as many atoms. At the moment the largest cell without pseudo-structure to be solved by this Zachariasen procedure seems to be the second of the isomers, $C_8F_2(C_6H_5)_6$, mentioned previously (Beineke, 1966), with 46 atoms per asymmetric unit and 92 per cell. This molecular structure was unknown, at the start, except to the extent that it was known that there were six benzene rings.

From the preceding discussions and the examples quoted, there does not seem to be a *known* limit to the power of crystal diffraction methods, providing that materials are well enough crystallized and adequate computing facilities are available. This is particularly true if it is possible to prepare a suitable isomorphous series of crystals. One must also add the proviso that neutron diffraction is to be considered part of the technique, and must be used to gather the data whenever the scattering powers of atoms of interest make X-ray results ambiguous or impossible.

For the foreseeable future we expect that computers will become faster and will have larger memories and better programming characteristics. The automated diffractometer, newly arrived on the scene, is certain to be improved in speed and perhaps in accuracy. Today new graduate students undertake for their doctoral theses structural determinations that would have been considered impossible ten years ago for a small team of postdoctoral fellows. But the new time-saving facilities are expensive both in initial cost and upkeep, and the graduate students of the future should not tackle bigger and more complicated problems just for the fun of it. More than ever before efforts must be concentrated on problems that may reasonably be expected to lead to results of value and interest to solid-state chemists and physicists, to biologists, to metallurgists and mineralogists, or to scientists in general.

REFERENCES

Banerjee, K. (1933). *Proc. Roy. Soc.* A **141**, 188.
Beineke, T. A. (1966). Ph.D. thesis, California Institute of Technology, Pasadena.
Bradley, A. J. (1935). *Z. Krist.* **91**, 302.
Bragg, W. H. (1915). *Phil. Trans. Roy. Soc. London* **210**, 253.
Cochran, W. (1952). *Acta Cryst.* **5**, 65.
Cromer, D. T. (1965). *Acta Cryst.* **18**, 17.
Cromer, D. T., and J. T. Waber (1965). *Acta Cryst.* **18**, 104.
Duchamp, D. J. (1965). Ph.D. thesis, California Institute of Technology, Pasadena.
Fridrichsons, J., and A. McL. Mathieson (1965). *Acta Cryst.* **18**, 1043. The latest of a series.
Fritchie, C. J., Jr. (1963). Ph.D. thesis, California Institute of Technology, Pasadena.
Harker, D., and J. S. Kasper (1948). *Acta Cryst.* **1**, 70.
Hauptman, H., and J. Karle (1953). *Solution of the Phase Problem, I.* A. C. A. Monograph No. 3. Wilmington, Delaware: The Letter Shop.
Hughes, E. W. (1935a,b). *J. Chem. Phys.* **3**, 1, 650.
Hughes, E. W. (1941). *J. Am. Chem. Soc.* **63**, 1737.
Hughes, E. W. (1949). *Acta Cryst.* **2**, 34.
Hughes, E. W. (1953). *Acta Cryst.* **6**, 871.
Jaeger, F. M., P. Terpstra, and H. G. K. Westenbrink (1927). *Z. Krist.* **66**, 195.
Karle, I. L., and J. Karle (1963). *Acta Cryst.* **16**, 969.
Karle, I. L., and J. Karle (1964). *Acta Cryst.* **17**, 835.
Karle, J., and I. L. Karle (1966). *Acta Cryst.* **21**, 849.
Kasper, J. S., C. M. Lucht, and D. Harker (1950). *Acta Cryst.* **3**, 436.
Klug, A. (1950). *Acta Cryst.* **3**, 165.
Laves, F. (1932). *Naturwiss.* **20**, 472.
Lonsdale, K. (1947). *Phil. Trans. Roy. Soc. London* **240**, 219.
Patterson, A. L. (1944). *Phys. Rev.* **65**, 195.
Patterson, A. L. (1949). *Acta Cryst.* **2**, 339. See also Pepinsky, R., ed. (1952). *Computing Methods and the Phase Problem in X-ray Crystal Analysis*, p. 29. State College, Pennsylvania: Pennsylvania State College.
Pinnock, P. R., C. A. Taylor, and H. Lipson (1956). *Acta Cryst.* **9**, 173.
Renninger, M. (1937). *Z. Krist.* **97**, 107.
Sayre, D. (1952). *Acta Cryst.* **5**, 60.
Vand, V., and R. Pepinsky (1954). *Acta Cryst.* **7**, 595.
Zachariasen, W. H. (1952). *Acta Cryst.* **5**, 68.

JAMES A. IBERS
Department of Chemistry
Northwestern University
Evanston, Illinois

Synthetic Oxygen Carriers

Linus Pauling for many years has been concerned with various aspects of hemoglobin chemistry and structure. In particular, he has proposed models for the structural attachment of molecular oxygen to the heme group in oxyhemoglobin. Because of his interest in this field I have chosen to discuss some of our recent work on the mode of attachment of molecular oxygen to transition metals in adducts of synthetic molecular oxygen carriers.

It seems fair to state that there is no general agreement on the mode of bonding of molecular oxygen to the heme group in oxyhemoglobin. The three configurations that have been considered are:

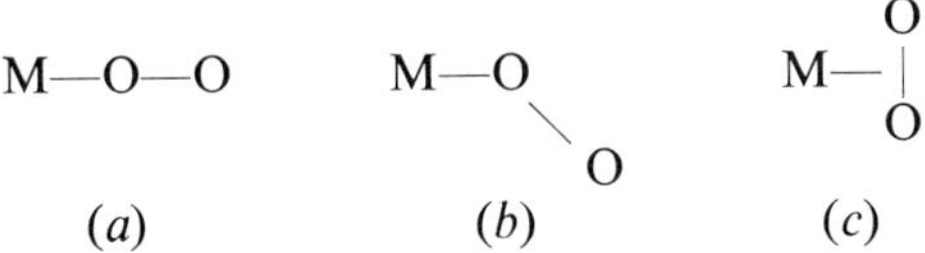

From their interpretations of chemical, spectroscopic, and magnetic data, various workers have favored one or another of these configurations at various times. Pauling and Coryell (1936) originally proposed configuration (*a*); later Pauling (1949) argued in favor of configuration (*b*). Recently Pauling (1964) has reiterated his views that configuration (*b*) is to be favored, and he has suggested that the iron in oxyhemoglobin remains in the ferrous state. Griffith (1956) favored configuration (*c*), in which the oxygen molecule is pi-bonded to the iron in a manner reminiscent of the bonding of the ethylene molecule to platinum in Zeiss's salt. Weiss (1964a, 1964b) suggested that oxyhemoglobin should be described as $Fe^{+++}O_2^-$; that is, with iron in the ferric state. Viale, Maggiora, and Ingraham (1964) have sketched a molecular orbital calculation which they claim supports Weiss's model.

It is conceivable that eventually this configurational problem will be settled by a high-resolution structure determination on oxyhemoglobin or oxymyoglobin. Yet such a single determination will probably not provide much insight into the nature of the bonding. We need to know, in addition to the configuration, the oxygen-oxygen bond length with an accuracy of ± 0.03 Å, and the prospects for this accurate a determination are not good. Table 1 presents information on the O—O bond length as a function of the bond order. The bond length is a sensitive function of the bond order, and this is why an accurate O—O bond-length determination is important. Moreover, a single determination of the oxyhemoglobin structure to high resolution will not lead in a direct way to an understanding of why hemoglobin takes up oxygen reversibly nor to why this oxygen-carrying capacity is easily interfered with.

TABLE 1.
Bond order versus O—O *bond length.*

Species	Bond Order	Bond Length[a]
O_2^{--}	1.0	1.49 Å
O_2^{-}	1.5	1.28 Å
O_2	2.0	1.21 Å
O_2^{+}	2.5	1.12 Å

[a]Abrahams (1956).

It is possible that a study of several derivatives of hemoglobin to high resolution would provide such understanding, but such a study would involve enormous labor.

An approach that is interesting and possibly germane to this problem is the investigation of simpler, synthetic systems in which molecular oxygen is taken up reversibly. In such synthetic systems various chemical changes may be made readily and their effect on oxygen uptake and on molecular structure is readily determined. Martell and Calvin (1952) have summarized the early work on such synthetic systems. Reviews by Vogt, Faigenbaum, and Wiberley (1963) and Connor and Ebsworth (1964) contain additional information. These reviews suggest that the chemistry of most of these synthetic systems is not well defined; in fact, there are still disagreements over whether or not specific complexes take up molecular oxygen reversibly. The best characterized of these systems is probably the bishistidinatocobalt(II) complex, which takes up molecular oxygen in the molar ratio O_2:Co of 1:2. Sano and Tanabe (1963) have isolated the oxygenated complex. Despite considerable effort on our part,

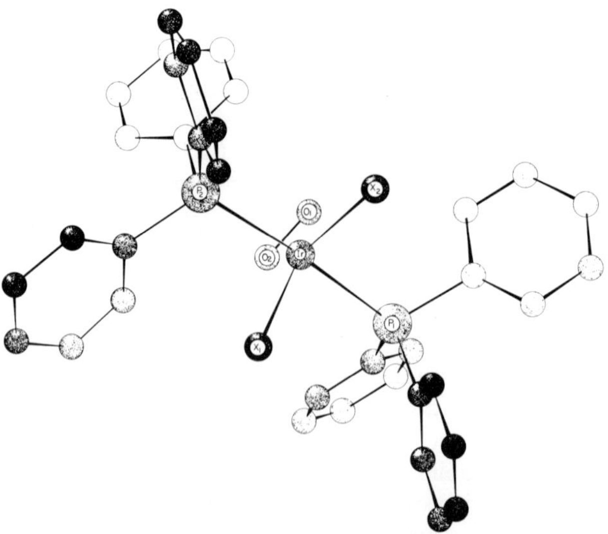

FIGURE 1.
The molecular structure of $IrO_2Cl(CO)(PPh_3)_2$.

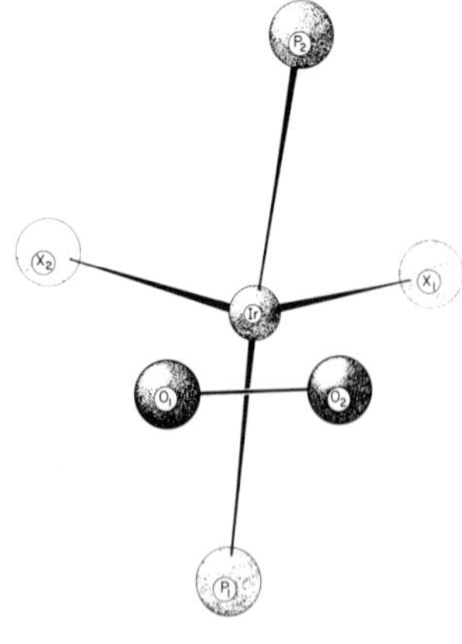

FIGURE 2.
The inner coordination sphere of $IrO_2Cl(CO)(PPh_3)_2$. X denotes the disordered CO and Cl positions.

the crystals we have obtained by their route are an order of magnitude too small to be used in a diffraction study.

Ideally a model synthetic system should be aqueous, as is the bishistidinato-cobalt(II) system. Yet there is no inherent reason why studies of nonaqueous systems should be unprofitable. Accordingly, Vaska's (1963) discovery that the complex $IrCl(CO)(PPh_3)_2$ ($Ph = C_6H_5$) takes up molecular oxygen reversibly in benzene is most important, for the system is very well characterized and the oxygen adduct, $IrO_2Cl(CO)(PPh_3)_2$, can be crystallized as a reasonably stable compound. I will describe our studies of this and related complexes.

THE STRUCTURE OF $IrO_2Cl(CO)(PPh_3)_2$

The structure of $IrO_2Cl(CO)(PPh_3)_2$ has been described in detail (Ibers and La Placa, 1964; La Placa and Ibers, 1965) and only some salient features will be reviewed here.

The molecular structure of the complex is shown in Figure 1; the inner coordination sphere is shown in Figure 2. The configuration found here is clearly of type (*c*). In the figures the disordered positions of Cl and CO are denoted by X. This disorder is not surprising in the solid state, for the packing of the molecules is almost entirely dictated by interactions among the very bulky triphenylphosphine ligands. In fact the volume per triphenylphosphine in this structure is only about 15 percent greater than in triphenylphosphine itself. The salient features of the oxygen attachment are: (1) the two oxygen atoms are equidistant from the Ir atom, the Ir—O distances being 2.04 and 2.09 Å (both ± 0.03 Å); (2) the O—O distance is 1.30 ± 0.03 Å, essentially that found in the O_2^- ion (Table 1). Thus we might describe this compound as $Ir^{++}O_2^-$. Both here and in the formulation of oxyhemoglobin as $Fe^{+++}O_2^-$ a problem arises, since both compounds are diamagnetic. It is probably possible by suitable incantations of molecular orbital theory to explain this diamagnetism.

There is a question of the coordination number of the iridium. The iridium is five-coordinated and has a trigonal bipyramidal configuration if the oxygen molecule is counted as a single ligand, as is the ethylene molecule in Zeiss's salt; the iridium is six-coordinated and has a distorted octahedral configuration if the oxygen atoms are counted separately. Vaska (1966) has reported that $IrCl(CO)(PPh_3)_2$ will also take up SO_2 reversibly in benzene. It occurred to us that the SO_2 adduct would be worth studying, for if the iridium inherently prefers five-coordination then we should find an Ir—S bond; if the iridium prefers six-coordination then the SO_2 group could function as a bidentate ligand with attachment through S and O or possibly through the two O atoms.

THE STRUCTURE OF $IrCl(CO)(SO_2)(PPh_3)_2$

The molecular structure of $IrCl(CO)(SO_2)(PPh_3)_2$ is shown in Figure 3, and the inner coordination sphere in Figure 4 (La Placa and Ibers, 1966). Clearly the configuration around iridium is that of a tetragonal pyramid, with S at the apex. Thus we conclude that in these systems iridium may show an inherent tendency to be five-coordinated. (Two recent reviews on five-coordination (Ibers, 1965; Muetterties and Schunn, 1966) make it clear that five-coordinate

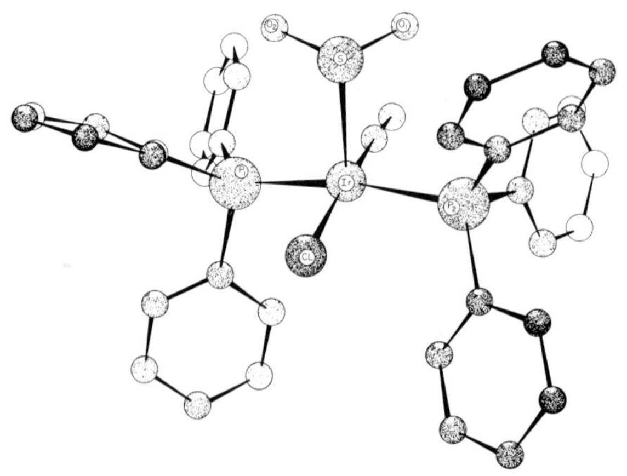

FIGURE 3.
The molecular structure of $IrCl(CO)(SO_2)(PPh_3)_2$.

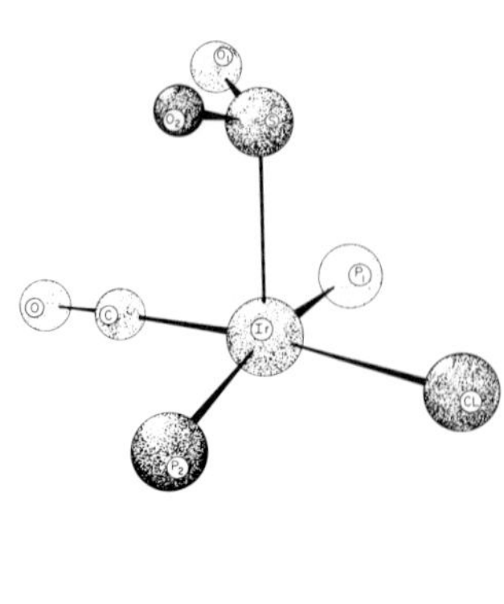

FIGURE 4.
The inner coordination sphere of $IrCl(CO)(SO_2)(PPh_3)_2$.

complexes are far more common than had been supposed by many inorganic chemists.) Several features of the SO_2 structure, in addition to the coordination around the iridium, are of interest. First, the complex is not disordered, possibly because of steric repulsions between the Cl and SO_2 ligands. This repulsion might also account for the nonplanarity of the Ir—SO_2 grouping, as shown in Figure 4. Second, the SO_2 group is very loosely bound to the iridium, the Ir—S bond length being 2.49 Å as compared with a Ru—S bond length of 2.07 Å in $(RuCl(NH_3)_4(SO_2))Cl$ (Vogt, Katz, and Wiberley, 1965). This very long Ir—S bond length is pleasing in view of the reversible nature of the attachment of SO_2 to the parent compound in benzene.

THE STRUCTURE OF $IrO_2I(CO)(PPh_3)_2$

I indicated at the beginning of this article that one of the advantages of synthetic systems is that various chemical changes are readily made and studied. The iodine atom can donate more electrons to iridium than can the chlorine atom. If these additional electrons end up in the Ir—O_2 interaction, then we should find that the oxygen molecule is more strongly bound to iridium in the iodo complex than in the chloro complex. Dr. J. McGinnety carried out the appropriate experiments and found that once the oxygenated iodo complex is formed in benzene solution the parent complex cannot be recovered by reducing the oxygen pressure as it can in the chloro system. In fact unless we carefully de-aerate the benzene we cannot avoid making the oxygenated iodo complex. Thus qualitatively the iodo complex is readily oxygenated irreversibly. Subsequent kinetic studies (Chock and Halpern, 1966) have shown that oxygen uptake is ten times faster for the iodo complex than for the chloro complex. Thus we had an excellent opportunity to examine possible structural implications of reversibility of oxygen uptake. The structure of the oxygenated iodo complex has now been determined (McGinnety, Doedens, and Ibers, 1967).

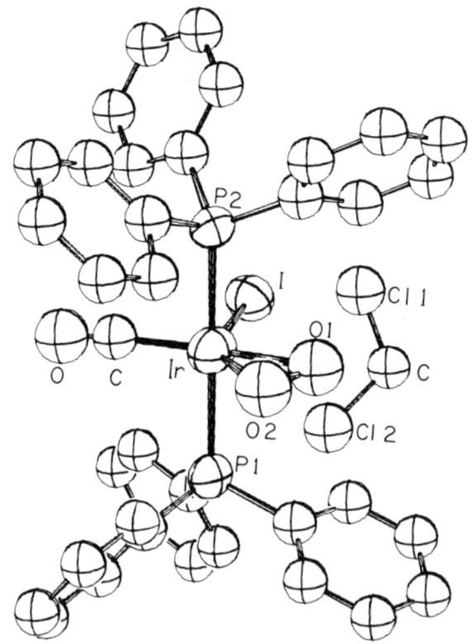

FIGURE 5.
The molecular structure of $IrO_2I(CO)(PPh_3)_2 \cdot CH_2Cl_2$.

The iodo complex crystallizes in the monoclinic system, while the chloro complex crystallizes in the triclinic system. The compounds cannot be isostructural, and we thought initially that, since iodine is larger than chlorine, the iodo complex might not be disordered. In fact, it is and the compound we studied is the 1:1 adduct of $IrO_2I(CO)(PPh_3)_2$ with CH_2Cl_2, the solvent from which the crystals were recrystallized.

The structure of the oxygenated iodo complex is closely similar to the oxygenated chloro complex and is shown in Figure 5. (The disorder of the I and CO positions is not shown.) Again the oxygen atoms are equidistant from the iridium atom, the Ir—O distances being 2.04 and 2.08 Å (both ± 0.02 Å). However, the O—O distance is 1.51 ± 0.03 Å, significantly longer than that in the chloro complex, and essentially the same as that found in the peroxide ion (Table 1). Thus one might formulate this complex as $Ir^{+++}O_2^{--}$.

DISCUSSION

In these studies of oxygen adducts of synthetic oxygen carriers, we have established several features that are of some interest. First, complexes of the type $IrX(CO)Y(PPh_3)_2$ (X = halogen, $Y = O_2$ or SO_2) may be considered five-coordinate complexes of iridium. Second, in both the oxygenated chloro and iodo complexes the Ir—O_2 grouping has configuration (c) and the oxygen molecule may be considered pi-bonded to the iridium. Third, in the chloro complex, which takes up oxygen reversibly in solution, the O—O distance is 1.30 ± 0.03 Å, while in the iodo complex, which takes up oxygen irreversibly in solution, the O—O distance is 1.51 ± 0.03 Å. Insofar as one can extrapolate from the solid state to species in solution it appears that reversibility of oxygen uptake depends upon the molecular oxygen being loosely bound to the metal. If enough electrons are transferred to the oxygen molecule, as in the iodo complex, then oxygen uptake is irreversible, presumably because the metal has been irreversibly oxidized. Thus in these synthetic systems reversibility of oxygen uptake is a sensitive function of the electronic structure and depends upon a delicate balance of ligands and the oxidation potential of the metal. In the analogous rhodium system the chloro complex does not pick up molecular oxygen at all. It seems likely that by a suitable change to more electron-donating ligands, for example, I for Cl or PEt_3 for PPh_3, a rhodium system could be synthesized that would exhibit reversible oxygen uptake.

It is difficult to know how far these results apply to natural oxygen carriers. But it is known that the oxygen carrying properties of such molecules as hemoglobin are easily interfered with, and it is tempting to speculate that the molecular basis for this is the interference with the delicate electronic balance of the system.

Studies are continuing on other synthetic systems, including some involving iridium with bidentate phosphorus ligands.

In conclusion we feel that it would be of great interest to measure the bond-dissociation energies between iridium and oxygen in these systems, and we hope that someone equipped to carry out such measurements can be convinced of their value.

In this work we have benefited from many discussions with our colleagues. In particular Professor F. Basolo has been of immense help to us in discussions of the chemistry of these systems. We would like to thank the National Institutes of Health for their partial support of this work.

POSTSCRIPT

Finally I wish to make a personal comment on the important influence that Linus Pauling has had on structural chemistry. I think the Caltech school, under Linus Pauling, refutes the notion held by some, especially by our European colleagues, that crystallography should be taught as a separate science. There could have been no better place to learn both crystallography *and* chemistry than at Caltech. Students learned to do crystallography well and at the same time learned why they were doing it. The risk I see in schools of crystallography is that students and professors may get so wrapped up in structure solving that the possible chemical implications of such studies are of little import. I feel that the combination of crystallography and chemistry, as developed by Linus Pauling, and as continued by his many associates—Bill Lipscomb and the late Bob Rundle, to cite just two—is the very strength of American structural chemistry and of chemical crystallography in general. Here, as in so many other areas, we are greatly indebted to Linus Pauling.

REFERENCES

Abrahams, S. C. (1956). *Quart. Rev.* **10**, 407.

Chock, P. B., and J. Halpern (1966). *J. Am. Chem. Soc.* **88**, 3511.

Connor, J. A., and E. A. V. Ebsworth (1964). *Adv. Inorg. Radiochem.* **6**, 279.

Griffith, J. S. (1956). *Proc. Roy. Soc.* (London) A **235**, 23.

Ibers, J. A. (1965). *Ann. Rev. Phys. Chem.* **16**, 375.

Ibers, J. A., and S. J. La Placa (1964). *Science* **145**, 920.

La Placa, S. J., and J. A. Ibers (1965). *J. Am. Chem. Soc.* **87**, 2581.

La Placa, S. J., and J. A. Ibers (1966). *Inorg. Chem.* **5**, 405.

Martell, A. E., and M. Calvin (1952). *Chemistry of the Metal Chelate Compounds.* Englewood Cliffs: Prentice-Hall.

McGinnety, J. A., R. J. Doedens, and J. A. Ibers (1967). *Science* **155**, 709.
Muetterties, E. L., and R. A. Schunn (1966). *Quart. Rev.* **20**, 245.
Pauling, L. (1949). In *Hemoglobin* (Sir Joseph Barcroft Memorial Symposium), p. 57. London: Butterworth's Scientific Publ.
Pauling, L. (1964). *Nature* (London) **203**, 182.
Pauling, L., and C. D. Coryell (1936). *Proc. Nat. Acad. Sci. U.S.* **22**, 210.
Sano, Y., and H. Tanabe (1963). *J. Inorg. Nucl. Chem.* **25**, 11.
Vaska, L. (1963). *Science* **140**, 809.
Vaska, L. (1966). *J. Am. Chem. Soc.* **88**, 1333.
Viale, R. O., G. M. Maggiora, and L. L. Ingraham (1964). *Nature* (London) **203**, 183.
Vogt, L. H., H. M. Faigenbaum, and S. E. Wiberley (1963). *Chem. Rev.* **63**, 269.
Vogt, L. H., J. L. Katz, and S. E. Wiberley (1965). *Inorg. Chem.* **4**, 1157.
Weiss, J. J. (1964a). *Nature* (London) **202**, 83.
Weiss, J. J. (1964b). *Nature* (London) **203**, 183.

OTTO BASTIANSEN
Department of Chemistry
University of Oslo
Oslo, Norway

Intramolecular Motion and Conformation Problems in Free Molecules as Studied by Electron Diffraction

The heroic period of gas electron diffraction of the early thirties is above all connected to the Caltech group. Linus Pauling's ingenuity and inspiring enthusiasm greatly influenced both the method itself and the choice of chemical problems that profitably could be studied by the method. A considerable number of eminent scientists have spent shorter or longer periods with the Caltech group. Two of the men at Pauling's side in the earlier days of electron diffraction deserve special honor: Lawrence O. Brockway and Verner Schomaker. The Caltech group has greatly contributed to the creation of the foundation upon which present-day work in the field of electron diffraction and the structure of free molecules are based.

MOLECULAR VIBRATION

In the earlier days the investigation of molecular structure was confined to the study of the geometry of the molecule. For this purpose static or rigid molecular models were used. Though such models undoubtedly played a useful role in the development of structure chemistry, chemists have of course all the time been aware of the fact that the rigid model is nothing but a rather rough approximation. Even for diatomic molecules the molecular vibration leads to root-mean-square deviations from the equilibrium distance of the order of 0.05Å. Approximately the same value is found for bond distances in ordinary molecules, but the vibrational amplitudes for longer internuclear distances may be considerably larger.

To obtain detailed information of the molecular motion it is advantageous to study the compound in the gaseous state. In the condensed phase the intermolecular forces restrict the freedom of intramolecular motion. In some cases lattice forces restrict the intramolecular motion enough to be decisive for the conformational choice of the molecule. For example, derivatives of ethane or cyclohexane often exist in two conformations in the gaseous state, but

Since this article was written, nearly one year ago, the Oslo group has been active in the field of intramolecular motion, and new ideas have been produced. The article, therefore, might have been slightly different if it had been written today.

usually only one conformer is present in the crystal. In biphenyl and in a series of its derivatives, the free molecules exhibit nonplanar conformations —the angle between the two rings being approximately 45°—and the molecules in the crystalline phase are planar.

The intramolecular motion in free molecules has been studied by two different methods: spectroscopy and electron diffraction. The various spectroscopic approaches have the common feature that the observables are frequencies; that is, the spectroscopically observed quantities are closely related to the eigenvalues of the interatomic motion of the molecule. The observables of an electron-diffraction study, on the other hand, are closely related to the eigenfunctions (rather, the squares of the eigenfunctions). This simple fact should be kept in mind when experimental values are interpreted theoretically. Theoretical chemists have to live with the dilemma of quantum mechanics: the general validity of the Schrödinger equation on one side and the impossibility of exact solution of it on the other. The success of the approximations that have to be used is naturally measured by the adjustability of theoretical values to observables. Since the spectroscopists measure values connected with the eigenvalues and electron diffractionists measure values connected with the eigenfunctions, it is obvious that the way of computing and arguing established by spectroscopists cannot always uncritically be adopted by electron diffractionists. In spite of this, the theoretical work based upon the findings of spectroscopists has in fact been the foundation on which the electron-diffraction work on intramolecular motion has been based. Particularly for the interatomic motion characterized by simple molecular vibration, the work of the spectroscopists can be directly adopted for the interpretation of measured electron diffraction data (Herzberg, 1945; Karle and Karle, 1949; Morino, 1950; Wilson, Decius, and Cross, 1955; Cyvin, 1960).

The simplest molecule to be studied is the diatomic molecule. The measured vibrational amplitude obtained from electron-diffraction studies is in good agreement with spectroscopicly deduced values. (The amplitude is measured by the root-mean-square deviation from the equilibrium distance. The notation l or u is usually applied.) So far as the determination of the amplitude goes, the harmonic oscillation approximation is already very good. However, the distance distribution peak observed by electron diffraction exhibits slight asymmetry corresponding to anharmonicity. Though this anharmonicity effect is easily reproducible, it is not well suited for quantitative determinations of anharmonicity constants.

In polyatomic molecules of moderate size the molecular vibration problems are also usually well understood (Cyvin, 1960). The vibrational amplitudes of internuclear distances can for many molecules be determined from the spectra with sufficient accuracy to fulfill the requirements of the electron diffractionists. The vibrational amplitude of the bond distances are particularly well determined. For a bond such as the C—C bond, the amplitude can usually be determined by electron diffraction with an accuracy of a few percent, and with at least this accuracy the amplitude may be deduced from the spectroscopic data. For the longer, nonbonded internuclear distances, the accuracy of the measurement of the vibrational amplitudes is less, and the spectroscopicly computed values are also less reliable. The uncertainty depends upon the size of the molecule, and the length as well as the weight of the internuclear distance. In a molecule like benzene, for example (Bastiansen and Cyvin, 1957), the amplitudes for all the three C—C distances are determined with an accuracy of a few percent or even better. The amplitudes of the C—H distances are less accurate, the error being approximately 5 percent, and for the H···H distances the amplitudes can hardly be quantitatively estimated by electron diffraction. This is because of the relative small contribution of the H···H distances to the total electron diffraction from the entire molecule.

For polyatomic molecules the harmonic oscillation approximation is usually applied and is sufficiently good for vibrational amplitude calculations. However, the peaks of a radial distribution curve as determined by gas-electron diffraction experiments for polyatomic molecules exhibit asymmetry due to anharmonicity. Peaks corresponding to the bond distances have a typical "normal anharmonic" shape; that is, the left side of the peak is steeper than the right side. For larger distances, however, it may be the other way around —the right side of the peak is the steeper one. In a molecule such as carbon suboxide the O···O distance peak shows this effect. The same is the case for the longest C···C distance peak in 1.3-butadiene and in butatriene.

SHRINKAGE EFFECT

The asymmetry of peaks of the radial-distribution curve unfortunately obscures the determination of the interatomic distances. In a diatomic molecule and for a valency bond, the concept of "interatomic distance"—at least within a few thousands of an Ångstrøm—is well defined. The situation is more difficult for the longer interatomic distances. For molecules usually designated as "linear molecules," for example, the long internuclear distances are less than the value obtained by adding the individual bond distances. This effect is often referred to as the "shrinkage effect," and the difference between the expected undistorted distance and that observed is referred to as the "shrinkage." For the longest C···C distance in allene (Almenningen, Bastiansen, and Trætteberg, 1959) and butatriene (Almenningen, Bastiansen, and Trætteberg, 1961) the shrinkage is observed to be 0.006 Å and 0.013Å respectively. The effect may be explained by out-of-linearity vibration. These vibrations should make the longer internuclear distances, on an average, shorter than the corresponding distances in a rigid linear molecule. The effect is theoretically well understood (Morino, 1960; Cyvin and Meisingseth, 1961), and the shrinkage of a series of molecules has been calculated from the vibration spectra. The agreement between such calculated results and those obtained by electron diffraction is in general very good. For carbon suboxide the shrinkage effect is considerably larger than for the other linear molecules studied. The molecule has been submitted to three independent electron-diffraction investigations in this laboratory (Bastiansen and Munthe-Kaas, 1955; Breed, Bastiansen, and Almenningen, 1960). For the O···O distance the shrinkage observed is 0.15 Å. This large effect could not be explained on the earlier known vibration frequencies alone. An unobserved bending frequency (ν_7) was predicted and has later been observed (Leroi, 1965; Smith and Leroi, 1966).

For molecules with linear equilibrium conformation, the shrinkage effect would make the molecule appear bent. It seems difficult, if not impossible, to distinguish between such an apparent unlinearity and a small but real deviation from linearity. This kind of difficulty is not restricted to linear molecules; it may in principle enter into electron-diffraction structure study of any polyatomic molecule. For a precise determination of molecular geometry it is always an advantage (and sometimes a necessity) to use information from all internuclear distances, both the short and the long ones. In such a study all nonbonded distances should in principle be corrected for shrinkage before being used. When applying a routine least-squares calculation on the electron-diffraction intensity curve, this fact should be kept in mind.

INTRAMOLECULAR MOTION DUE TO TORSION

Until now it was tacitly implied that the internal motion in the molecule was restricted to vibration alone. More complicated and also more interesting is the motion that occurs in molecules that offer possibilities of internal rotation.* The simplest example of such a molecule is ethane, though it is itself not a good molecule for studying torsional motion by electron diffraction. Simple ethane derivatives are better suited for that purpose. Molecules of the type $CH_2X—CH_2Y$ offer excellent possibility for electron-diffraction study. Following Pitzer's notions as to the structure of ethane (Kemp and Pitzer, 1936), this kind of ethane derivative should exhibit two conformational possibilities, *trans* and *gauche*, that may coincide in equilibrium. This phenomenon is now well known, and a series of molecules has been studied with the ambition of determining the geometry of the conformers, the energy difference between them, and the total shape of the barriers as a function of the torsional angle. The electron-diffraction method has contributed to the detection of the conformers, to the determination of their structure, and to the determination of the energy difference, but it appears not very well suited for determination of barriers. This is because the two conformers (sometimes only one) prevail to the extent that the intermediates are hardly observable. For most ethane derivatives the determination of the barriers must be left to spectroscopists.

Which of the two possible conformers predominates in the type of ethane derivative used as an example above (CH_2X CH_2Y) seems to vary rather considerably. In the case of ethylene chlorohydrin (X = OH and Y = Cl) the *gauche* form is the only conformer present in a detectable amount (Bastiansen, 1949b). This is apparently because an intramolecular hydrogen bond between the chlorine atom and the hydroxyl group has been formed. The hydrogen bond locks the molecule in the *gauche* form and hampers the torsional motion. The torsion angle is here as usual for *gauche* conformers larger than the ideal 60° angle, corresponding to the *gauche* position of an unsubstituted ethane. (The torsion angle of the *cis* position is here set to zero.) *Gauche* predominance is also observed for ethylene glycol (X = Y = OH), and is likewise explained by a torsion-preventing hydrogen bond.

For the 1,2-dihaloethanes, both conformers are easily detectable by electron

*In most of the literature such motion will also be included in the term "vibration." The semantics adopted here, separating torsion from the rest of the intramolecular motion, may be favorable for electron-diffraction studies.

diffraction. In the case of 1,2-dibromoethane the *trans* conformer predominates, with a stability in favor of *trans* of 1630–1700 cal/mole (Ainsworth and Karle, 1952; Kuratani, Miazawa, and Mizushima, 1952; Almenningen, Bastiansen, Haaland, and Seip, 1965). *Trans* predominance is also found for the corresponding chlorocompound with an energy difference between *gauche* and *trans* of 900–1140 cal/mole. For 1,2-difluoroethane, however, the *gauche* conformation predominates. The electron-diffraction results indicate a *gauche* concentration slightly larger than the 67 percent corresponding to energy equality of the two conformers, and spectroscopic studies (Klæboe and Rud Nielsen, 1960) indicate no energy difference.

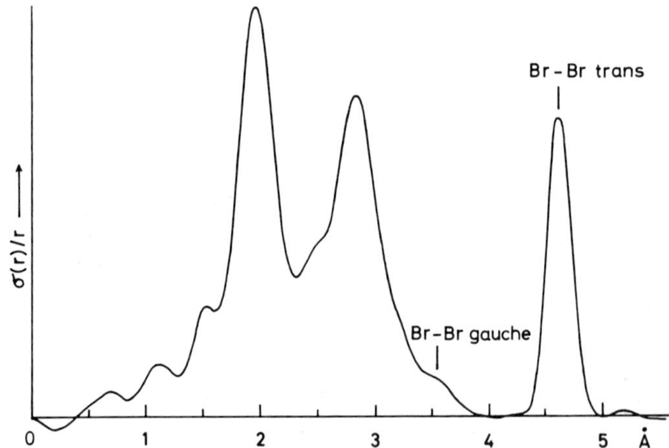

FIGURE 1. Radial-distribution curve of 1,2′-dibromoethane.

As to the possibility of detecting intermediates using electron diffraction study Figure 1 may serve as an illustration. The radial-distribution curve of 1,2-dibromoethane has been chosen. The well defined peak around $r = 4{,}6\text{Å}$ is the Br–Br peak corresponding to the *trans* conformation. The shoulder at approximately 3,5Å is the Br–Br *gauche* contribution. Study of the area of these contributions and comparison with the area under the rest of the peaks of the radial-distribution curve form the basis for the conformation analysis. This analysis should give the relative contribution of the two conformers as well as the total contribution of possible intermediates. Unfortunately the analysis is obscured by lack of experimental information of electron-diffraction intensity data at small diffraction angles, which makes it difficult to determine the zero line in the radial-distribution curve ("envelope"), and the position of the zero line of course influences the area analysis. The sum of the areas under the *trans* and *gauche* Br–Br peak, with the error taken into consideration, is sufficient to explain the total expected Br–Br contribution. Area analysis, therefore, does not seem to be able to yield satisfactory information about intermediates. A critical analysis of the shape of the most pronounced peak, the *trans* Br–Br peak, also seems inadequate for studies of intermediates. The *trans* Br–Br peak is very symmetric. In case of substantial intermediate contribution that should not be the case.*

*In a recent article by J. Karle (1966), a more optimistic view is expressed.

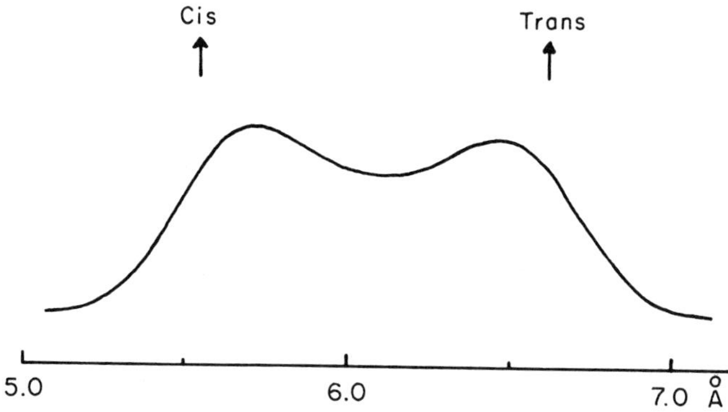

FIGURE 2.
Br-Br contribution to the radial-distribution curve of 1,4-dibromobutyne-2.

The situation is rather similar for other molecules exhibiting equilibrium between two conformers. For example, several cyclohexane derivatives (Bastiansen and Hassel, 1946; Atkinson and Hassel, 1959) exist in two easily detectable conformations, the geometry of which as well as the energy difference between them may be studied with electron-diffraction technique. However, neither the structure of the intermediates nor their relative contribution seems to be easily detectable by electron diffraction, at least at its present state of development.

In contrast to the examples just referred to, the molecules of the 1,4-dihalobutyne-2 type may be mentioned ($CH_2X—C{\equiv}C—CH_2X$). Electron-diffraction studies of the chloro (Kuchitsu, 1957) and the bromo (Almenningen, Bastiansen, and Harshbarger, 1957) compound show that no single rigid conformation predominates. The halogen-halogen distance contribution to the radial distribution curve is a saddle-shaped double peak, as indicated in Figure 2. The inner peak occurs at a position somewhat to the right of the *cis* position and the outer maximum at a position somewhat to the left of the *trans* position. This picture can be brought into accordance with the assumption of free rotation about the axis of the carbon skeleton. A quantitative study of the phenomenon requires a more profound analysis of intramolecular motion. Any treatment of internal motion that includes torsion faces the problem of separability of vibrational and torsional motion. For electron-diffraction studies the problem is often dealt with in the following approximative way: the description of the total intramolecular motion is tried by averaging the effect of vibrating molecules having different torsion angles (Karle and Hauptman, 1949). The averaging is carried out by applying a weight factor given as a function of the torsion angle. The weight factor of course depends upon the degree of restriction of the torsional motion and should in principle be deducible from the torsional potential and the temperature. The ambition of today's electron diffractionists includes the attempt of reversing the procedure—trying to deduce the potentials from electron-diffraction data. In such a case the weight factor would have to be deduced from the data. For the free rotating molecule the weight factor is constant over the entire rotation. For the type of ethane derivatives earlier dealt with, the weight factor is large near the *trans* and *gauche* positions, but elsewhere very small. In this description the internuclear motion of the long distances is expressed as a torsional motion superimposed upon a framework vibration. In principle the framework vibration may vary with the torsion angle, but this variation seems to be unattainable by electron-diffraction measurement. In practice the influence of the vibration is assumed constant during the rotation. But even if this assumption represented a satisfactory approximation, the determination of the framework vibration still presents difficulties. For a molecule such as 1,3-dibromobutyne-2 the simplest approach is to assume a constant Br–Br framework vibration amplitude and then average over the whole torsion-angle range. This procedure is repeated, each time with a different framework vibration. Satisfactory agreement was obtained for this compound with a vibrational *u* value of approximately 0.13Å, together with free rotation. But agreement may also have been reached by introducing a small torsional barrier, together with a torsion-angle dependent framework vibration.

Between the two kinds of molecule—those that can be approximated by free rotation and those that may be described by assuming discrete conformers in equilibrium—there is an interesting group of molecules whose electron-diffraction data cannot be explained unless restricted rotation is assumed. The biscyclopropyl molecule may serve as an example. Raman and infrared studies of the solid state shows that the rule of mutual exclusion is obeyed: the molecule has a center of symmetry in the crystal and must accordingly have the s-*trans* form (Lüttke, de Meijere, Wolff, Ludwig, and Schrötter, 1966). The same result is obtained from X-ray crystallographic studies (Eraker and Rømming, 1967). For the liquid phase spectroscopic studies show that the mutual exclusion rule is not obeyed, which indicates that other rotational isomers are present in the liquid together with the s-*trans* form. Electron-diffraction studies also clearly demonstrate the presence in the gas phase of forms other than the rigid symmetric *trans* form (Bastiansen and de Meijere, 1966). It is easily demonstrated that no simple rigid conformation exists in the gas. It appears that the weight factor earlier described must be nearly constant in a torsion-angle range from approximately 100° to 260°. For the *cis* form (torsion angle equal to 0°) the weight factor must be very small, but a *gauche* contribution with a torsion angle in the range of 38°± 18° has to be included. This picture, derived from electron-diffraction studies, is in excellent agreement with conclusions drawn from simple torsion potential calculations. In Figure 3 the sum of the H—H interactions has been used for calculating the potential as a function of the torsion angle. The formula of Bartell has been used (Bartell, 1960), and already this simple procedure seems to give a satisfying qualitative picture of the restricted rotation. It is a question of semantics how we describe the structure of such a molecule. One alternative would be to describe the molecular structure of biscyclopropyl as a conformational mixture of an infinite number of conformers defined by a continuous change of the torsion angle and assigned a weight factor varying with the angle. Another and certainly more practical alternative would be to describe the molecule as existing in conformational equilibrium involving two conformers, one being the *trans*, the other the *gauche*. The two conformers are described by their average torsion angle (here 180° and 38°, respectively) and by the amplitude of the torsion motion (here 80° and 18°, respectively).

A similar conformational problem is presented by biscyclobutyl (de Meijere,

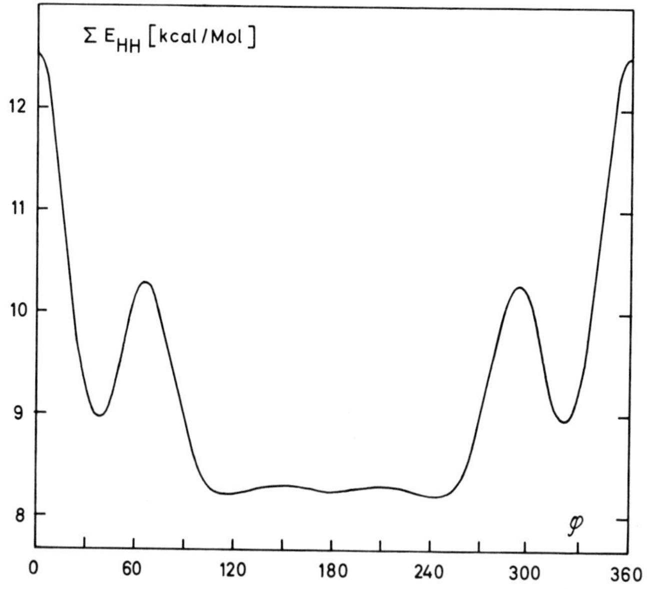

FIGURE 3.
Potential curve of the biscyclopropyl molecule calculated from the H—H interaction terms according to Bartell (1960).

1966), though this molecule is considerably more complicated because of the nonplanarity of the individual four-membered rings. Here four nonrigid conformers are involved: *ee*-s *trans*, *ea*-s *trans*, *ee gauche*, and *ea gauche*. (The notation *e* and *a* from cyclohexane chemistry has here been adopted also for the four-membered ring.) As a whole the torsional freedom is somewhat more restricted than in biscyclopropyl because of the size of the rings.

In the case of aromatic rings, as in the biphenyls, the torsional freedom should also be influenced by the double bond character of the bridge bond. Biphenyl itself and its derivatives are all found to deviate from planarity in the gas phase (Bastiansen, 1949a, 1950; Almenningen and Bastiansen, 1958) in spite of the molecular planarity in the crystal (Dahr, 1932; Toussaint, 1948; Saunder, 1946; Trotter, 1961). The experimentally obtained angle between the two phenyl planes in the gas is about 45°. Theoretical calculations, including both conjugation and nonbonded interaction, suggest a potential curve with a flat minimum around 35–40° (Fischer-Hjalmars, 1962, 1963). For the study of the torsional motion, 3,3′-dibromobiphenyl and 3,5,3′,5′-tetrabromobiphenyl were chosen (Bastiansen and Skancke, 1967). The idea was to study the shape of the radial-distribution peak of the longest Br· · ·Br distance. The width of this peak is so large that it clearly indicates considerable torsional motion. Unfortunately the problem of separating framework vibration and torsional motion obscures the investigation. It is possible to deduce a map correlating framework-vibration *u* values and torsional amplitude. The torsion-motion probability curve as a function of torsion angle was in the calculation arbitrarily assumed to be Gaussian, though there may be good reasons for a somewhat asymmetric distribution. To obtain some evidence as to the framework vibration the molecule 3,5,4′-tribromobiphenyl was also studied. The long Br–Br distance is about as long as the studied Br–Br distance in the other compounds and, being uninfluenced by torsional motion, it was believed to give some idea also about the framework vibration for 3,3′-dibromobiphenyl and for 3,5,3′,5′-tetrabromobiphenyl. Although no real quantitative conclusion can be drawn from this investigation, the work indicates that the torsional amplitude calculated as a root-mean-square deviation is around 15–20°. The nonplanar biphenyl molecule, as it appears in the gas phase, is accordingly far from being a rigid arrangement.

The question of internal torsional motion within cyclic molecules has attracted considerable interest. In addition to the restriction to torsion of the type known from ethane and larger open-chain molecules, the closed cycle itself hampers motion. The smallest aliphatic ring, cyclopropane, must of course be very rigid, but cyclobutane already offers the possibility of some internal motion. The deviation from planarity of the carbon skeleton of four-membered unsaturated rings could be believed to be associated with molecular motion. However, electron-diffraction studies of halogen derivatives of cyclobutane (Almenningen, Bastiansen, and Walløe, 1967) show that the deviation from planarity is the same as for cyclobutane itself and that the unplanar ring appears rather rigid. The five- and seven-membered rings have perhaps not been studied sufficiently in the gas phase, but considerable molecular motion seems to take place, particularly in cyclopentane, and the six-membered ring is again rather rigid. This is probably due to the fact that the H—H distances are favorable for stabilizing the ring. The eight-membered ring has been studied in more detail (Almenningen, Bastiansen, and Jensen, 1966). Electron-diffraction data of gaseous cyclooctane are not compatible with the assumption of any single conformation. To fit the data one has to assume a conformational mixture based on high flexibility of the ring and include asymmetric molecular model species. This result is in agreement with energy studies of the various possible conformations (Hendrickson, 1964). Such studies favor a very mobile conformational mixture at ordinary temperatures.

In the crystalline phase the situation may be different. Here a conformational choice is taken (Dunitz and Mugnoli, 1966; Groth, 1965), though even in the solid the medium-size rings may exhibit considerable internal motion.

What happens in the gaseous phase with rings larger than the eight-membered ring is little known, but most probably flexibility of the same kind should be expected for most of them.

MOLECULAR MOTION IN SANDWICH COMPOUNDS

Sandwich molecules of the ferrocene type exhibit internal motion of a special kind. The following simplified description may give a qualitative idea of the situation. Let us first consider the two organic molecules without the central atom and assume that they are kept parallel with coinciding axes. If the distance between the rings is reduced, contact would first be obtained between π electrons on the periphery of the rings. When contact is established, there is a cavity of low electron density in the central region between the rings. In this cavity the central atom finds its place, where it has the function of keeping the two rings together. The internal motion of the molecule is now depending upon the size of the central atom compared to the size of the cavity. The smallest possible atom that can be used for the formation of a sandwich molecule is beryllium. Electron-diffraction studies of beryllocene (Almenningen, Bastiansen, and Haaland, 1964) has shown that the beryllium atom moves around in the cavity in a double minimum potential barrier. The two minima are situated on the axis of the molecule, 1.48 Å from one ring and 1.98 Å from the other. The beryllium atom spends most of its time in these positions, corresponding to a static picture of nearly one-half beryllium atom in each position. In this molecule the rings are in contact and their relative positions are well fixed, parallel, and staggered. This geometry explains two apparently contradictory observations: the large dipole moment (μ = 2,24 D in cyclohexane solution) (Fischer and Hofmann, 1959), and the fact that the molecule occupies a centrosymmetric site in the crystal (Schneider and Fischer, 1963).

The double minimum potential for the Be atom is easily understood by simple electrostatic consideration. The problem has been treated quantum mechanically (Sundbom, 1966), and the results are compatible with the electron-diffraction findings and the dipole moment value. Since the most pronounced motion (that of the beryllium atom) takes place in the center of the molecule, the motion should not be hampered by neighboring atoms in a lattice.

If the central atom is large compared to the cavity, the molecular motion will be quite different. The two rings may then be considered as moving on a ball. Positions corresponding to contact between the two rings may be favored by attractive London forces between π electrons in the two rings. The choice of a staggered or eclipsed form seems also to depend upon the attractive forces between the rings. In any case the internal motion of sandwich molecules with a large central atom is dominated by motion of the rings. This motion may lead to an average unsymmetric conformation and thus to a dipole moment (Bohn and Haaland, 1965; Haaland, 1965). Such a molecular freedom of motion for sandwich molecules with a large central atom should be severely hampered in the crystalline phase, and we may therefore expect higher molecular symmetry in the solid than in the vapor or liquid.

THE EFFECT OF TEMPERATURE ON INTRAMOLECULAR MOTION

Even in the qualitative description—and certainly in a quantitative treatment of intramolecular motion—the understanding of the temperature-dependence is essential. In principle, the influence of temperature upon intramolecular motion is well understood and can be treated theoretically. However, an experimental study of the effect of temperature by electron diffraction is considerably more difficult. In classical gas electron-diffraction study exclusive attention was paid to the geometry of the molecule, and even in present-day work the effect of intramolecular motion is rarely the main object of the investigation. The experimental conditions are, therefore, usually chosen to minimize the effect of thermal motion. Thermal motion reduces the resolution and the accuracy of the determination of geometrical structure parameters. Most electron-diffraction diagrams are accordingly taken at the lowest possible temperature to give sufficient vapor pressure. The development of the gas electron-diffraction technique has favored long exposure times and low gas densities, partly to minimize thermal motion. Only very few examples of electron-diffraction studies of free molecules at different temperatures have reached the literature (Hedberg and Iwasaki, 1962).

The temperature effect is in most cases rather small and the temperature interval between the lowest and the highest temperatures at which diagrams can be obtained seems often insufficiently large. The upper limit of the temperature is set either by the thermal stability of the molecule or by experimental limitations. Most electron-diffraction laboratories have not developed any special high-temperature technique. Only the Moscow group seems to have a running procedure allowing diagrams to be taken at really high temperatures (Akishin, Vilkov, and Spiridonov, 1955). But even if sufficiently large temperature intervals were obtainable, there are other difficulties to be overcome. The greatest difficulty is perhaps to *measure* the temperature of the vapor at the diffraction point. The temperature of the gas nozzle is not identical with the temperature of the gas. The gas expands into a vacuum and is condensed at the liquid nitrogen trap. The expansion leads to a cooling of the gas, a cooling effect that depends upon the experimental setup, particularly upon the shape of the nozzle opening. But even if this cooling effect were known, we could not consider the vapor at the diffraction point to be in thermodynamic equilibrium. The translation motion of the molecule as a whole is primarily "cooled down" and perhaps also the total molecular rotation. However, the internal motion may not be "cooled down" to the same extent. Further, we may expect the various kinds of internal freedoms of motion to "cool down" differently.

Very little work has been done to attack these difficulties experimentally. One attempt was made in this laboratory to follow the "vibrational temperature" of a diatomic molecule as a function of the nozzle temperature (Bastiansen, 1956, unpublished). Br_2 was chosen, and the idea was to take diagrams at various nozzle temperatures and to calculate the u values. There is a theoretical one-to-one correspondence between u value and temperature, and measurement of the u value should therefore give the "vibrational temperature" of the Br_2 vapor at the diffraction point. Unfortunately the effect of temperature on the u value is small, and the accuracy of u-value determination is not too high. However, the work indicated that the nozzle temperature is a rather good approximation to the vapor "vibrational temperature" around room temperature, but that a considerably lower value than the nozzle temperature is obtained for the "vibrational temperature" when higher nozzle temperatures are used.

For a further exploration of temperature effects, we should look for easily observed molecular motion parameters that are as sensitive to temperature changes as possible. The shrinkage effect in C_3O_2 may be such an effect. The shrinkage of the $O \cdots O$ distance is very well reproducible and should be sufficiently temperature-dependent. In fact, work done in the University of Oslo laboratory indicates that the $O \cdots O$ distance shrinkage varies with the experimental setup. A systematic study of this is on our program.

ACKNOWLEDGMENT

The author wants to thank Førsteamanuensis Per Andersen for having read the manuscript and for valuable comments.

REFERENCES

Ainsworth, J., and J. Karle (1952). *J. Chem. Phys.* **20**, 425.
Akishin, P. A., L. V. Vilkov, and V. P. Spiridonov (1955). *Doklady Akad. Nauk S.S.S.R.* **101**, 77.
Almenningen, A., and O. Bastiansen (1958). *Det Kgl. Norske Videnskabers Selskabs Skrifter.* **No. 4**.
Almenningen, A., O. Bastiansen, and F. Harshbarger (1957). *Acta Chem. Scand.* **11**, 1059.
Almenningen, A., O. Bastiansen, and A. Haaland (1964). *J. Chem. Phys.* **40**, 3434.
Almenningen, A., O. Bastiansen, A. Haaland, and H. M. Seip (1965). *Angew. Chem.* (int. ed.) **4**, 819.
Almenningen, A., O. Bastiansen, and H. Jensen (1966). *Acta Chem. Scand.* In press.
Almenningen, A., O. Bastiansen, and M. Trætteberg (1959). *Acta Chem. Scand.* **13**, 1699.
Almenningen, A., O. Bastiansen, and M. Trætteberg (1961). *Acta Chem. Scand.* **15**, 1557.
Almenningen, A., O. Bastiansen, and L. Walløe (1967). *Universitets-forlaget.* In press.
Atkinson, V., and O. Hassel (1959). *Acta Chem. Scand.* **13**, 1737.
Bartell, L. S. (1960). *J. Chem. Phys.* **32**, 827.
Bastiansen, O. (1949a). *Acta Chem. Scand.* **3**, 408.
Bastiansen, O. (1949b). *Acta Chem. Scand.* **3**, 415.
Bastiansen, O. (1950). *Acta Chem. Scand.* **4**, 926.
Bastiansen, O., and S. J. Cyvin (1957). *Nature.* (London) **180**, 980.
Bastiansen, O., and O. Hassel (1946). *Tidsskr. f. Kjemi, Bergv. Met.* **6**, 96.
Bastiansen, O., and A. de Meijere (1966). *Acta Chem. Scand.* **20**, 516.
Bastiansen, O., and A. Skancke (1967). *Acta Chem. Scand.* **21**, 587.
Bohn, R. K., and A. Haaland (1965). *J. Organometal Chem.* **5**, 470.
Breed, H., O. Bastiansen, and A. Almenningen (1960). *Acta Cryst.* **13**, 1108.
Cyvin, S. J. (1960). *Acta Polytechn. Scand.* Phys. 6.
Cyvin, S. J., and E. Meisingseth (1961). *Acta Chem. Scand.* **15**, 1289, 2021.
Dahr, J. (1932). *Indian J. Phys.* **7**, 43.
Dunitz, J. D., and A. Mugnoli (1966). *Chem. Communs.* 166.
Eraker, J., and C. Rømming (1967). *Acta Chem. Scand.* In press.
Fischer, E. O., and H. P. Hofmann (1959). *Chem. Ber.* **92**, 482.
Fischer-Hjalmars, I. (1962). *Tetrahedron.* **17**, 235.
Fischer-Hjalmars, I. (1963). *Tetrahedron.* **19**, 1805.
Groth, P. (1965). *Acta Chem. Scand.* **19**, 1497.
Hedberg, K., and M. Iwasaki (1962). *J. Chem. Phys.* **36**, 589.
Hendrickson, J. B. (1964). *J. Am. Chem. Soc.* **86**, 4854.
Herzberg, G. (1945). *Infrared and Raman Spectra of Polyatomic Molecules.* New York: Van Nostrand.
Haaland, A. (1965). *Acta Chem. Scand.* **19**, 41.
Karle, I. L., and J. Karle (1949). *J. Chem. Phys.* **17**, 1052.
Karle, J. (1966). *J. Chem. Phys.*, **45**, 4149.
Karle, J., and H. Hauptman (1949). *J. Chem. Phys.* **18**, 875.
Kemp, J. D., and K. S. Pitzer (1936). *J. Chem. Phys.* **4**, 749.
Klæboe, P., and J. Rud Nielsen (1960). *J. Chem. Phys.* **33**, 1764.
Kuchitsu, K. (1957). *Bull. Chem. Soc. Japan.* **30**, 391.
Kuratani, K., T. Miazawa, and S. Mizushima (1952). *J. Chem. Phys.* **21**, 1411.
Leroi, G. E. (1965). *Abstr. 8th European Congress on Molecular Spectroscopy*, Copenhagen, p. 267.
Lüttke, W., A. de Meijere, H. Wolff, H. Ludwig and H. W. Schrötter (1966). *Angew. Chem.* **78**, 141.

de Meijere, A. (1966). *Acta Chem. Scand.* **20**, 1093.
Morino, Y. (1950). *J. Chem. Phys.* **18**, 395.
Morino, Y. (1960). *Acta Cryst.* **13**, 1107.
Munthe-Kaas, T. (1955). Thesis, University of Oslo.
Saunder, D. H. (1946). *Proc. Roy. Soc.* (London) **188**, 31.
Schneider, F., and E. O. Fischer (1963). *Naturwiss.* **50**, 349.
Smith, W. H., and G. E. Leroi (1966). *J. Chem. Phys.*, **45**, 1767, 1784.
Sundbom, M. (1966). *Acta Chem. Scand.* **20**. In press.
Toussaint, J. (1948). *Acta Cryst.* **1**, 43.
Trotter, J. (1961). *Acta Cryst.* **14**, 1135.
Wilson, E. B., J. C. Decius, and P. C. Cross (1955). *Molecular Vibrations*. New York: McGraw-Hill.

S. H. BAUER
KINYA KATADA
KATSUMI KIMURA
Department of Chemistry
Cornell University
Ithaca, New York

The Structures of C_6, B_3N_3, and C_3N_3 Ring Compounds

The attempt to ascertain differences in bond character from measured differences in bond lengths has proved to be a challenging exerçise. The early efforts to establish direct correlations have been replaced during the past decade by arguments in which the measured average interatomic distances between adjacent atoms are related not only to the nature of the atom pair but also to their molecular environment. The challenging aspect of these empirical correlations is the identification of the specific molecular structure factors which influence interatomic distances and bond angles. Those effects centering around carbon-carbon bonds and carbon-hydrogen bonds have been discussed by a number of authors (Stoicheff, 1962; Cumper, 1958; Bartell, 1962). In the study described below, we considered changes in the size and shape of six-membered rings that were due to the substitution of fluorine atoms for hydrogen atoms. We selected four compounds for which the ring skeletons are essentially isoelectronic, all with at least D_{3h} symmetry (or higher) in the inner ring as well as in the outer ring of atoms which serve as the substituents. Their structures have been determined by electron diffraction on the vapors, with the use of the sector-microphotometer technique. The data were analyzed by visual optimization of the agreement between the calculated and experimentally deduced radial distribution curves, and the angular positions of the diffraction intensity curves. This work was completed prior to the development of least squares minimization programs for adjusting observed diffraction intensities and their backgrounds to those calculated. For reference, the structure of benzene is now well established by electron diffraction and high resolution rotational raman spectra (Almenningen et al., 1958; Langseth and Stoicheff, 1956; Kimura and Kubo, 1960); that of s-triazine is known with equal precision from rotational raman spectra (Lancaster and Stoicheff, 1956) and that of borazine is known rather imprecisely from an early electron diffrac-

The present address of Dr. Kinya Katada is Department of Physics, College of General Education, University of Osaka Prefecture, Sakai, Osaka, Japan; that of Dr. Katsumi Kimura is Department of Chemistry, Faculty of Engineering Science, Osaka University, Toyonaka, Osaka, Japan. This work was supported in part by the ONR under Contract No. N onr-401(41) and the Material Science Center at Cornell University, ARPA-SD-68.

tion study (Bauer, 1938). The question posed in this investigation is whether the substitution of fluorine atoms for hydrogen atoms in a symmetric manner onto a conjugated six-membered ring influences the π-electron density in the inner ring of atoms sufficiently to change their interatomic distances. We were led to this question in considering measures of aromaticity. The magnitude of the diamagnetic anisotropy (Lonsdale and Toor, 1959; Watanabe et al., 1960) has been proposed; it shows that borazine has a lesser π orbital delocalization than does benzene. Here we suggest another measure of the extent of delocalization of the π electrons—the degree of retention of structural integrity of the ring in the presence of perturbing substituents. A related effect, but one that is not as straightforward to interpret, is the extent of contraction of C—F or B—F bond lengths due to substitution of F for H on other atoms in the ring.

SOURCES OF SAMPLES AND EXPERIMENTAL DETAILS

The hexafluorobenzene was obtained from the Monsanto Chemical Company. A sample of 1,3,5-trifluorobenzene was kindly given to us by Dr. G. C. Finger of the Illinois State Geological Survey; its index of refraction was $n_D^{25} =$ 1.4710. The B-trifluoroborazine was sent to us by Dr. Kurt Niedenzu, U. S. Army Research Office, Durham University, and the cyanuric trifluoride was provided by Dr. F. S. Fawcett of the E.I. duPont deNemours and Company, Experimental Station. All of these materials were vacuum-distilled before use, and further purified when the infrared or mass spectra indicated that there were minor impurities. Electron diffraction photographs were obtained with the apparatus which had been previously described (Hastings and Bauer, 1950) and subsequently modified. The sample-plate distance was approximately 17 cm. Two sets of photographs were taken for low and high voltages, using two sectors suitably cut for covering the ranges $q = (10\text{–}45)$ and $q =$ (20–120); $q = 40/\lambda \sin(\theta/2)$, in Å^{-1}.

The samples were stored in 12-liter glass bulbs at pressures of 1 to 5 torr. These were attached directly to the nozzle tube. In order to provide for density-intensity calibration, a set of two photographs with different time exposures was taken under otherwise identical condition. The stability of the high voltage was checked frequently with a type K potentiometer. For each set of gas diffraction photographs Au patterns were recorded to establish the scale factor. Kodak lantern slide (medium) plates were used; these were developed in D11 for five minutes. Optical density tracings of the diffraction patterns were obtained with a Leeds and Northrup microdensitometer; during scanning, the plates were oscillated around an axis which passed through the center of the pattern. The optical densities were converted to intensities by means of a procedure described elsewhere (Kimura and Bauer, 1963).

The data were reduced in the following manner. An experimental molecular scattering function was obtained from

$$M_{\text{expt}}(q) = \frac{\pi}{10} Pq[I(q)/B(q) - 1] \tag{1}$$

where P is $\sum_k (Z_k^2 + Z_k)$; $I(q)$ is the observed total intensity; $B(q)$ is the estimated background intensity. In the absence of extraneous scattering, this molecular function may be corrected for nonnuclear scattering and phase-shift factors:

$$M(q) = M_0(q) + N(q) \tag{2}$$

$$\frac{\pi}{10} qPM_0(q) = \sum_{ij}{}'(Z_iZ_j/r_{ij}) \exp(-\pi^2 l_{ij}^2 q^2/200) \sin(\pi q r_{ij}/10) \tag{3}$$

Here $N(q)$ is a "difference" function calculated for an approximate model, l_{ij} the rms amplitude, and r_{ij} the interatomic distance between atoms i and j. In practice, the observed intensities are adjusted according to

$$M_{0\,\mathrm{expt}}(q) = Y^{-1}M_{\mathrm{expt}}(q) - N(q) \tag{4}$$

where Y is the inverse of an index of resolution to allow for the presence of a uniform extraneous background. A radial distribution function is then calculated

$$f(r) = \sum_{q=1,2\cdots}^{q_{\max}} [M_0(q < q_{\min}) + M_{0\,\mathrm{expt}}(q \geq q_{\min})] \exp(-\gamma q^2) \sin(\pi q r/10) \tag{5}$$

$q_{\min}$ and $q_{\max}$ are the extremes of the range of observed scattering angles, and γ is a damping factor set by the condition $\exp(-\gamma q_{\max}^2) = 0.1$. Were the data available for the complete range of q, the model assumed in estimating $N(q)$ entirely correct, and the vibrational motions harmonic, $f(r)$ would consist of a superposition of Gaussian peaks:

$$f(r)_{\mathrm{theor}} = 5 \sum_{ij}{}'(Z_iZ_j/r_{ij})(H_{ij}/\pi)^{1/2} \exp[-H_{ij}(r - r_{ij})^2] \tag{6}$$

with

$$H_{ij} = \frac{1}{2l_{ij}^2 + (400\gamma/\pi^2)} \tag{7}$$

In practice, the shapes of the radial distribution peaks are close to Gaussian, and the areas check to a few percent.

All extended calculations were made with a CDC 1604 digital computer. A structure was obtained after a series of successive refinements using the criteria of a smooth background, $B(q)$, and zero-values for $f(r)$ for distances sufficiently far removed from any that could occur in the molecule.

ANALYSIS OF THE DIFFRACTION DATA

For the hexafluorobenzene diffraction data were available for the range $q = 16$ to 98. The experimental total scattered intensity and the background as deduced from a sequence of successive approximations are given in Figure 1. The refined radial distribution curve calculated with a damping factor $\gamma = 2.1 \times 10^{-4}$ Å^2 is shown in Figure 2. The regions less than 1.0 Å and greater than 5.8 Å are not included; the maximum amplitude of the ripples for the short distances is less than 3 percent of the peak height at 1.4 Å, whereas at the large distances the ripple is less than 1 percent. Resolution of the radial distribution peaks in terms of the contributing atom pairs (syn curve) is entirely compatible with a model of D_{6h} symmetry. From Figure 2 a set of interatomic distances and root mean square amplitudes were obtained. These

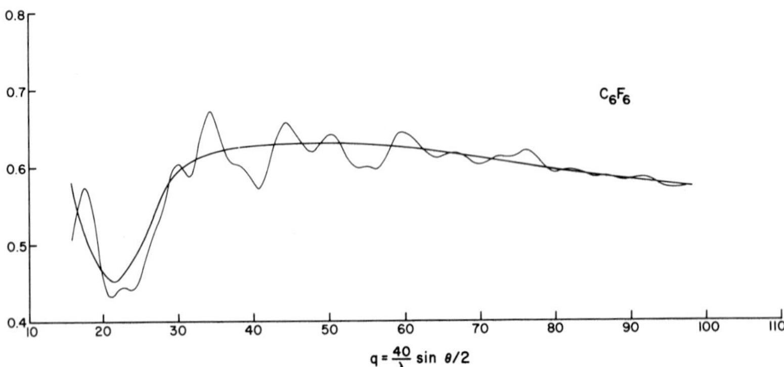

FIGURE 1.
The experimental total scattered intensity for C_6F_6, and the corresponding background function (smooth line).

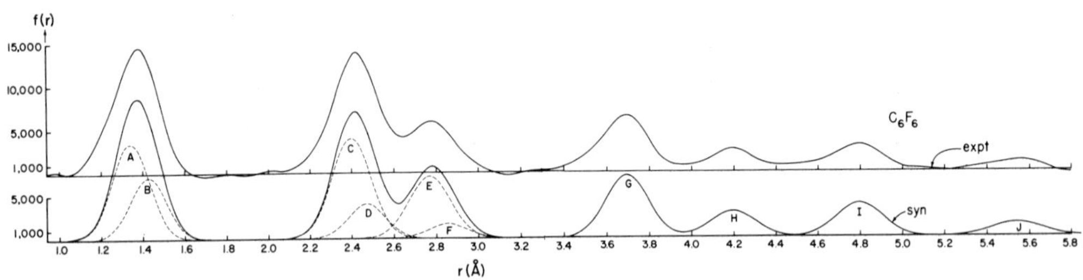

FIGURE 2.
Final radial distribution curve (expt), computed from Equation (5) for C_6F_6. For comparison, the lower curve (syn) was calculated from Equation (6), using the parameters deduced for the "best" model (Z). The resolved Gaussians, identified by letters A–J, are the scattering contributions from the corresponding atoms pairs listed in Table 2. The abscissa readings should be multiplied by 0.9867 to adjust for the apparatus scale factor.

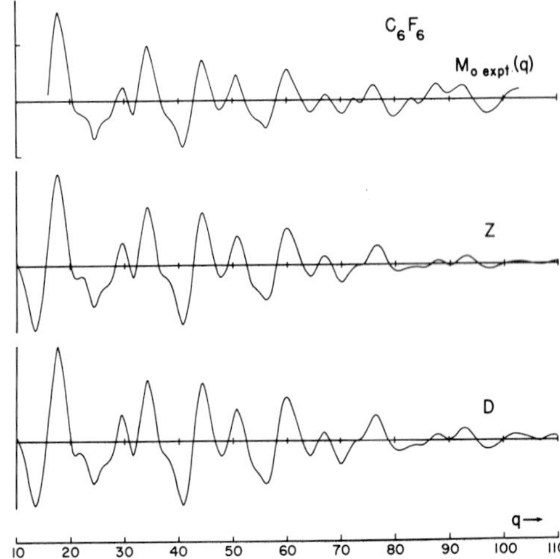

FIGURE 3.
The reduced experimental molecular intensity pattern for C_6F_6, compared with two curves calculated according to Equation (3), for two models:
Z: (C—F)/(C—C) = 0.940; $\sigma_Q = 32 \times 10^{-4}$
D: (C—F)/(C—C) = 0.952; $\sigma_Q = 46 \times 10^{-4}$

were then checked by calculating a series of theoretical intensity curves. Approximately 30 models were tested with ratios of (C—C)/(C—F) = 0.980 to 0.920, in various combinations, with small variations in the root mean square amplitudes, as summarized in Table 1. The field of parameters covered in these calculations is here shown on two scales. The "best" model was selected on the basis of the minimal value for σ_Q, defined by

$$\sigma_Q^2 \equiv \frac{1}{N}\left[\frac{q_c}{q_0} - \left\langle\frac{q_c}{q_0}\right\rangle_{\text{av}}\right]^2 \tag{8}$$

where N is the number of features in the intensity pattern (20 for C_6F_6) for which q values can be precisely measured (q_0's) for comparison with those calculated for the various models.

It appears that model b with (C—F)/(C—C) = 0.940 has the lowest σ_Q. Further slight adjustment of the parameters (model Z) reduced the value of σ_Q to 32×10^{-4}. The theoretical intensity curve calculated for this model by means of Equation (3) is shown in Figure 3, for comparison with $M_{0\ \text{expt}}(q)$ derived from Equation (4). The structure so deduced is close to that reported by Almenningen et al. (1964) for C_6F_6. Their data favor the ratio (C—F)/(C—C) = 0.952; the curve for this model (D) is also included in Figure 3. The final distances for C_6F_6 are summarized in Table 2. The errors were estimated from the overall quality of the data.

The experimental total scattered intensity patterns for 1,3,5-trifluorobenzene are shown in Figure 4, which includes curves for low and high voltage runs. The reduced $M_{0\ \text{expt}}(q)$ is plotted in Figure 5, where it is compared with the theoretical curve for model B. Figure 6 shows the final radial distribution curve

TABLE 1.
Parameter fields for C_6F_6[(C_1—C_2) *held at* 1.420 Å]; *numerical values listed in table are* $\sigma_Q \times 10^4$.

(C_1—F_1)/(C_1—C_2)	0.050 0.055	0.050 0.065	0.050 0.075	← $l_{C_1-C_2}$ ← $l_{C_1-F_1}$
0.980	A	H	O	not acceptable
0.970	B	I	P	
0.960	C	J	Q	
0.950	D	K	R	← indicated by $f(r)$
0.940	E	L	S	
0.930	F	M	T	← not acceptable
0.920	G	N	U	

(C_1—F_1)/(C_1—C_2)	0.050 0.055	0.050 0.060	0.050 0.065	0.050 0.070	0.050 0.075	← $l_{C_1-C_2}$ ← $l_{C_1-F_1}$
0.950	46{D}		48{K}		43{R}	
0.945	44{V}	39{a}	46{X}			
0.940	41{E}	38{b}	40{L}	41{d}	46{S}	
0.935	43{W}	39{c}	44{Y}			
0.930	47{F}		45{M}		51{T}	

TABLE 2.
Interatomic distances in C_6F_6; *symmetry:* D_{6h}.

Distance	r_g(Å)	l_{ij}(Å)	Identified in Figure 2 by:
C_1—C_2	1.408 ± .006	0.055 ± .005	B
C_1—F_1	1.324 ± .006	0.060 ± .006	A
C_1—C_3	2.439	0.060	D
C_1—C_4	2.816	0.065	F
C_1—F_2	2.366	0.068	C
C_1—F_3	3.646	0.075	G
C_1—F_4	4.140	0.083	H
F_1—F_2	2.732	0.075	E
F_1—F_3	4.732	0.085	I
F_1—F_4	5.464	0.095	J

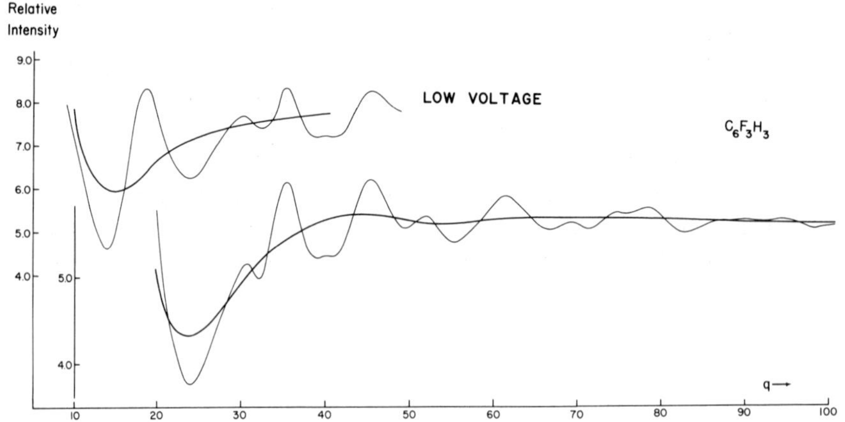

FIGURE 4.
Experimentally determined scattering intensity for sym $C_6F_3H_3$, showing patterns obtained with low and high voltage electrons. The corresponding backgrounds are drawn in.

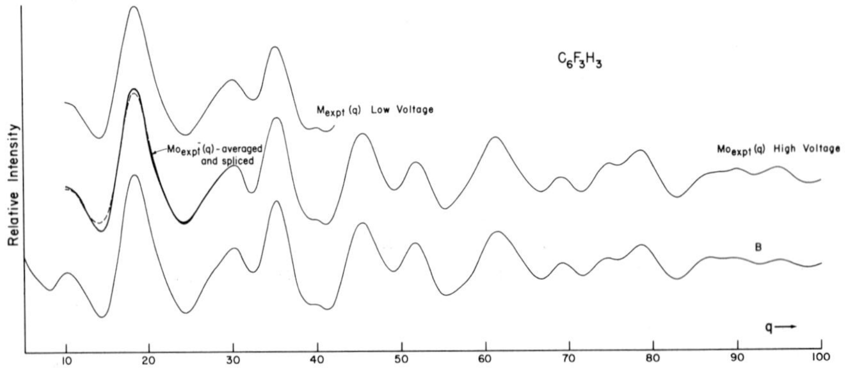

FIGURE 5.
Reduced molecular scattering functions for $C_6F_3H_3$ as deduced from the data in Figure 4. For comparison, note the theoretical intensity curve calculated for model B.

computed with the damping factor $\gamma = 2.3 \times 10^{-4}$ Å². The first peak at 1.37 Å, has been resolved into contributions from the bonded C—C (1.402 Å), the bonded C—F (1.305 Å), and the bonded C—H (1.090 Å), indicated in Figure 6 by Gaussians A, B. The root mean square amplitudes observed were 0.045 Å and 0.051 Å for the C—C and C—F bonds, respectively. The second peak at 2.36 Å is due to scattering by the nonbonded C···F and C···C pairs with slight contributions from the nonbonded C···H and F···H. After subtracting Gaussians for the small contributions involving hydrogen atoms, the remaining radial distribution peak was decomposed into two Gaussians, at 2.388 Å (C) and 2.430 Å (D) for $C_2\cdots F_1$ and $C_1\cdots C_3$. A small shoulder at 2.79 Å (E) is due to $C_1\cdots C_4$. The peak (G) at 4.09 Å was assigned to $C_4\cdots F_1$ and the one at 4.68 Å to $F_1\cdots F_3$.

Theoretical intensity curves were calculated for many models, in which the C—C and C—H distances were kept at 1.40 Å and 1.09 Å, with various C—F distances, ranging from 1.27 to 1.36 Å. It was confirmed that the experimental curve fits best those calculated for C—F between 1.30 and 1.31 Å, particularly with the respect to the features observed at $q = 40$ and 58. A comparison of the observed distances with those calculated for D_{3h} symmetry using the distances derived from the $f(r)$ curve is presented in Table 3. The close agreement between the nonbonded values demonstrates that correct assignments were made and that this molecule does possess D_{3h} symmetry.

The electron diffraction patterns recorded for B-trifluoroborazine are indistinguishable in general appearance from those of sym trifluorobenzene (compare Figures 4 and 7) except that the patterns for $F_3B_3N_3H_3$ extended over a larger q range. The final radial distribution curve is shown in Figure 8 for comparison with the synthetic curve given below it. The contributing Gaussians are labeled to correspond to the atom pairs listed in Table 5 for model Y. There is a slight discrepancy at maximum (M) at 4.22 Å, which must be as-

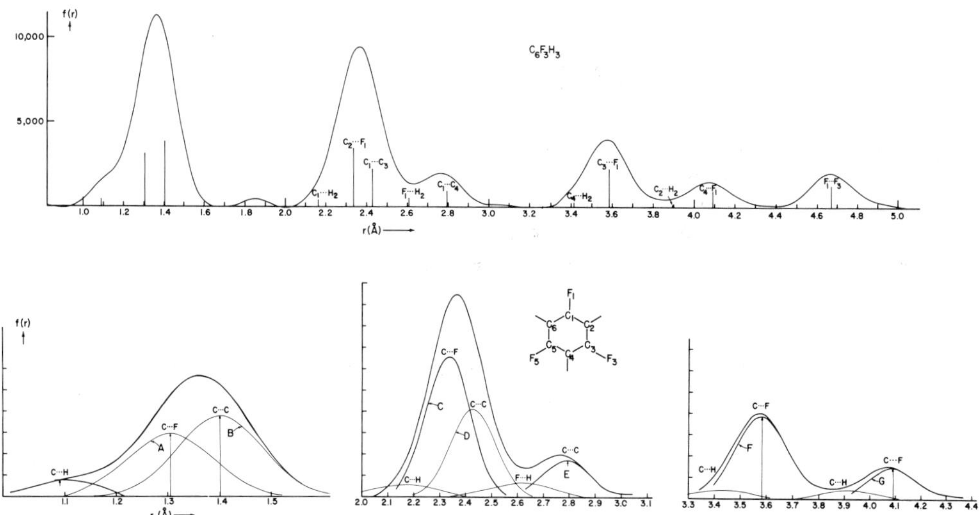

FIGURE 6.
Final radial distribution curve for sym $C_6F_3H_3$. Three peaks are shown resolved into pair contributions on a somewhat enlarged scale.

TABLE 3.
Interatomic distances in $C_6F_3H_3$; *symmetry:* D_{3h}.

Atom pair	r_g(obs), Å	r_g(calc), Å	l_{ij}(Å)
C—H	1.090 ± 0.015	(1.090)	(0.077)[a]
C—F	1.305 ± 0.010	(1.305)	0.051 ± 0.008
C—C	1.402 ± 0.005	(1.402)	0.045 ± 0.005
$C_2 \ldots F_1$	2.338 ± 0.008	2.345	0.053 ± 0.005
$C_1 \ldots C_3$	2.430 ± 0.010	2.428	0.053 ± 0.008
$C_1 \ldots C_4$	2.794 ± 0.015	2.804	0.063 ± 0.01
$C_3 \ldots F_1$	3.585 ± 0.015	3.618	0.070 ± 0.01
$C_4 \ldots F_1$	4.090 ± 0.020	4.109	0.070 ± 0.02
$F_1 \ldots F_3$	4.670 ± 0.020	4.689	0.075 ± 0.02
$C_1 \ldots H_2$		2.165	(0.100)[a]
$C_4 \ldots H_2$		3.415	(0.096)[a]
$C_5 \ldots H_2$		3.894	(0.094)[a]
$F_1 \ldots H_2$		2.607	(0.099)[b]
$F_1 \ldots H_4$		5.199	(0.076)[b]
$H_2 \ldots H_4$		4.300	(0.132)[a]

[a]Values taken from C_6H_6.
[b]Estimated.

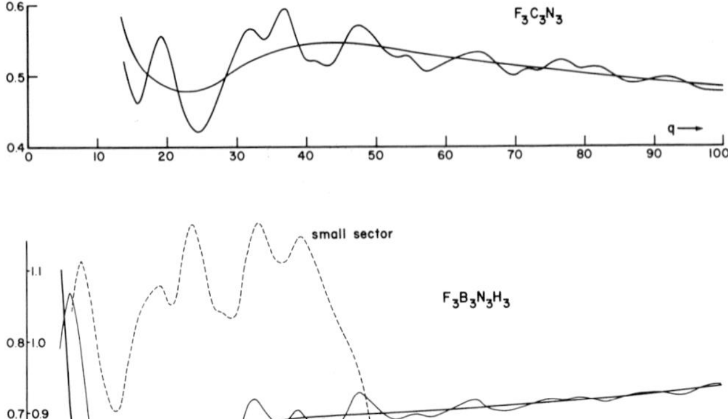

FIGURE 7.
Experimentally determined scattering intensities for $F_3B_3N_3H_3$ (lower curves) and for $F_3C_3N_3$ (upper curve). The tracing marked "small sector" was taken with 25 kv electrons.

signed to the $N_1 \cdots F_2$ contributions. The reason for this is not clear; all these peaks are somewhat distorted by the small nonbonded $B \cdots H$, $N \cdots H$, and $H \cdots H$ contributions.

About 50 intensity curves were calculated for a wide range of models; generally the B—N distance was held at 1.42 Å and the B—F distance was varied from 1.44 to 1.29 Å. Various combinations were tested with ∠BNB ranging from 116° to 122°. All models were assumed to possess D_{3h} symmetry. The N—H distances were assigned 1.01 Å, except in about 10 models for which N—H was set at 1.03 Å. Several intensity curves were obtained for

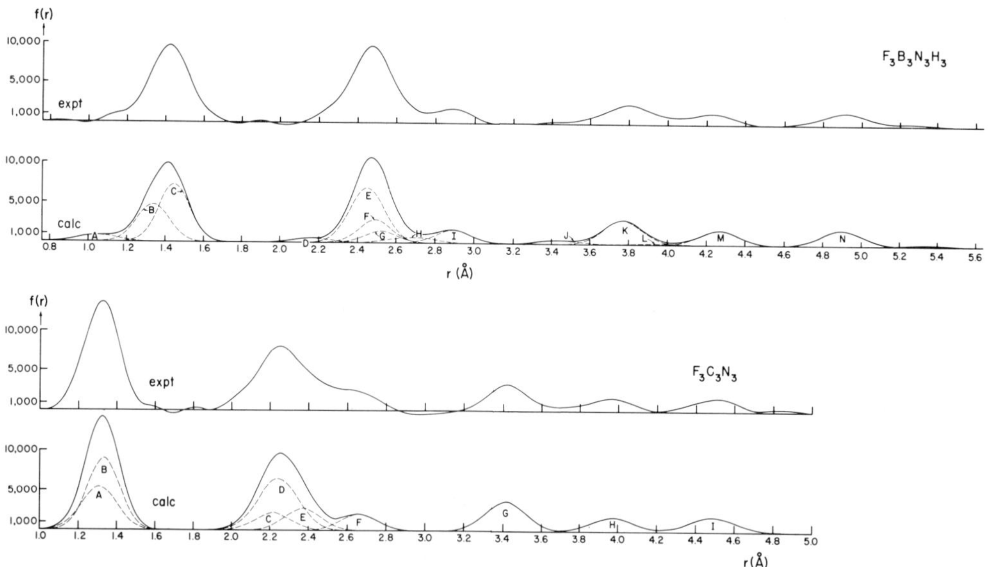

FIGURE 8.
Refined radial distribution curves for $F_3B_3N_3H_3$ and $F_3C_3N_3$. For each pair the upper curve is that deduced from the reduced intensity patterns, and the lower is the synthetic f(r) for the best model. The abscissas should be multiplied by the appropriate scale factors: $F_3B_3N_3H_3 \times 0.990$, $F_3C_3N_3 \times 1.000$.

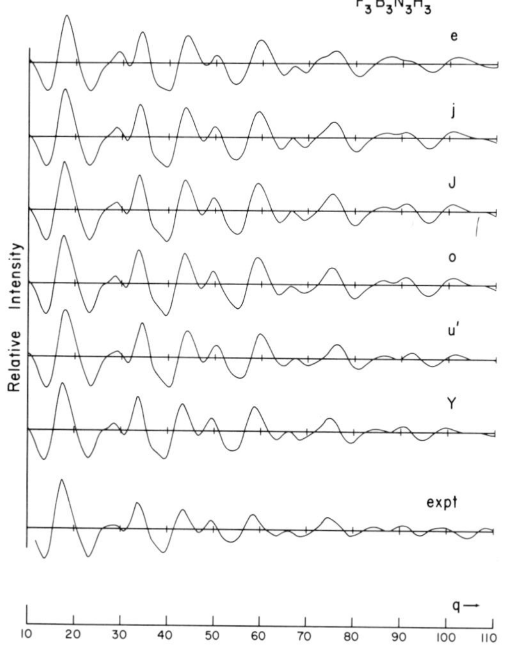

FIGURE 9.
The experimental and calculated molecular scattering functions for $F_3B_3N_3H_3$. All models were assumed planar, D_{3h} symmetry.

	e	j	J	o	u′	Y
N_1—B_1	1.42	1.42	1.42	1.42	1.42	1.42
B_1—F_1	1.40	1.40	1.40	1.40	1.35	1.35
∠BNB	116°	118°	120°	122°	118°	121°

B—N = 1.41 to 1.43 Å. Typical curves are shown in Figure 9, for comparison with the reduced experimental intensity curve. Models e, j, J, and o illustrate the effect of varying ∠BNB from 116° to 122° for a fixed (N—B)/(B—F) ratio 1.42/1.40; models j and u′ show the effect of varying the distance ratio from 1.42/1.40 to 1.42/1.35 at 118°. The shapes of all these patterns are so

TABLE 4.
Portion of parameter field explored for $F_3B_3N_3H_3$ [(B—N) *fixed at* 1.42 Å]; *numerical values listed in the table are* $\sigma_Q \times 10^4$ *for the designated models.*

B—F	∠BNB 118°		120°		122°
1.43	^{63}{g}		^{58}{G}		^{70}{l}
1.41	^{60}{i}		^{48}{I}		^{59}{n}
1.39	^{52}{s}		^{44}{S}		^{59}{s′}
1.37	^{59}{t}		^{41}{T}		^{47}{t′}
1.35	^{63}{u}		^{38}{U}	^{30}{Y}	^{40}{u′}
		CR			
1.33	^{69}{v}		^{48}{V}		^{41}{v′}
1.31	{w}		^{76}{W}		^{56}{w′}
1.29	{x}				{x′}

TABLE 5.
Interatomic distances in $F_3B_3N_3H_3$ (*Model Y*); *symmetry:* D_{3h}; ∠NBN = 119° ± 1°.

Distance	r_g(Å)	l_{ij}(Å)	Identified in Figure 8 by:
$N_1—H_1$	1.04 ± 0.02	0.080 ± .10	A
$B_1—F_1$	1.361 ± 0.010	0.055 ± .005	B
$N_1—B_1$	1.432 ± 0.010	0.050 ± .005	C
$B_1—H_1$	2.143	0.090	D
N_1 F_1	2.424	0.065	E
$N_1—N_2$	2.467	0.055	F
$B_1—B_2$	2.492	0.055	G
$F_1—H_1$	2.647	0.100	H
$N_1—B_2$	2.863	0.065	I
$N_1—H_2$	3.406	0.100	J
$B_1—F_2$	3.732	0.075	K
$B_1—H_3$	3.901	0.110	L
$N_1—F_2$	4.223	0.080	M
$F_1—F_2$	4.849	0.085	N
$F_1—H_3$	5.262	0.120	—

Atom designation

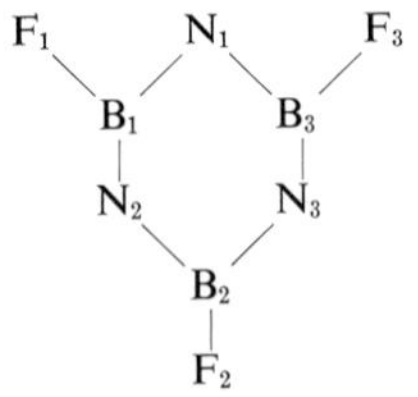

F_1 N_1 F_3
B_1 B_3
N_2 N_3
B_2
F_2

similar that on the basis of qualitative features alone, one cannot select a best model. However, the quantitative criterion for the best model in terms of minimum σ_Q provides the necessary discrimination. Table 4 shows a portion of the parameter field covered in these intensity calculations; the subscripts designate the magnitudes of σ_Q. From this it is evident that model Y is in best agreement with the data. Table 5 summarizes the listing of interatomic distances and root mean square amplitudes for this model.

That B-trifluoroborazine is planar, with D_{3h} symmetry, is clear from the overall agreement between the synthetic radial distribution function for model Y and that deduced from the data. The inner ring of atoms does not quite constitute a regular hexagon, but the departure from 120° bond angles is within the estimated error for this structure determination.

The last compound of this series to be studied was cyanuric trifluoride. The experimental intensity pattern and the corresponding background are shown in the upper curve in Figure 7. The radial distribution curve obtained after a sequence of refinements of the data is given in the lower portion of Figure 8, calculated with a damping factor $\gamma = 2.3 \times 10^{-4}$. The assignment of the peaks is straightforward. The maximum at 1.32 Å is narrow, indicating that the bonded C—N and C—F distances must be nearly equal. The peak at 2.25 Å is due in part to a superposition of the nonbonded C—C and N—N in the ring; however, its shape is determined by the larger contribution from the six nonbonded N—F distances. The shoulder at 2.66 Å is due to the transannular C—N. The remaining peaks at the larger distances are individual atom-pair contributions. The structure is essentially determined by these three outer peaks plus the well-resolved shoulder at 2.66 Å. Note that in Figure 8 (H) occurs at $(A + F)$, and $(F) = 2(B)$. The assignments are unambiguously established by the relative areas.

For this compound, 47 intensity curves covering (C—F)/(C—N) ratios from 1.10 to 0.90, and ∠NCN from 115° to 130°, were calculated. Qualitative comparison of the experimental and calculated intensity curves narrowed the field to (C—F)/C—N) ratios in the interval 1.025 to 0.950, and ∠NCN to the interval 122½° to 130°. The parameter field is shown in Table 6 with σ_Q values listed below the corresponding symbols for the qualitatively acceptable models. Comparison of the theoretical intensity curves for various models with the reduced experimental curve is presented in Figure 10. Model z clearly has the smallest standard deviation for peak positions. The corresponding interatomic distances and mean square amplitudes are listed in Table 7. In this compound

TABLE 6.
Parameter field for $F_3C_3N_3$ [(C—N) *fixed at* 1.330 Å; *subscripts show* $\sigma_Q \times 10^4$].

(C—F)/(C—N)	∠NCN									
	115°		120°		125°	126°	127°	127½°	128°	130°
1.10	A		F		K					P
1.05	B		G	e	L					Q
		a	c	f	i			k		
1.00	C	b	H	g	^{97}{M}	^{83}{r}		^{67}{l}		R
0.990										
					^{85}{q}	^{73}{s}		^{61}{u}		
0.985							^{61}{v}		^{60}{x}	
								^{45}{z}		
0.980							^{60}{w}		^{59}{y}	
0.975			d	h	^{77}{j}	^{69}{t}		^{62}{m}		p
0.950	D		I		N			o		S
0.900	E		J		O					T

as well, our conclusions about the planarity of the structure and its symmetry are strongly supported by the close agreement between the theoretical curves and the reduced molecular intensity pattern.

DISCUSSION OF STRUCTURES AND COMPARISON WITH RELATED COMPOUNDS

The question whether unsymmetrical substitutions of highly electron-attracting or electron-repelling groups onto a benzene ring introduce measurable distortions in the geometry has been asked many times; the answers have been ambiguous. Inspection of Kitaigorodskii's (1957) compendium of crystal structures leads to the general conclusion that since the limits of error for the interatomic distances so determined were $\pm(0.04$ to $0.05)$ Å, there is no evidence that the C_6 ring deviates from D_{6h} symmetry, with (C—C) = 1.40 Å. Even in 1,3,5-triphenylbenzene, there is no indication that the core rings differ from the terminal rings; (C—C) = 1.39 ± 0.03 Å.

During the past decade, as the precision of interatomic distance determinations has increased, indications that departures from D_{6h} symmetry exist became more definite. In $C_{23}H_{22}N_2OS_2$, there are three phenyl groups in nonequivalent positions; Karle and Karle (1965) reported $(C—C)_{arom}$ ranging in magnitude from 1.352 Å to 1.413 Å, and ring bond angles from 117.6° to 123.8°. The cited standard deviations are 0.010 Å in bond lengths and 0.4° to 0.9° in angles. In hexanitrosobenzene (Cady et al., 1966) the sides of the hexagons (uncorrected for thermal motions) range from 1.401 Å to 1.440 Å, with a quoted standard deviation of 0.008 Å; there are corresponding departures of $\pm 3°$ from 120° bond angles.

Gas phase structure determinations of substituted benzenes led to somewhat different conclusions. Table 8 is a summary of available data. In bromobenzene, the microwave data (Rosenthal and Dailey, 1965) were insufficient to

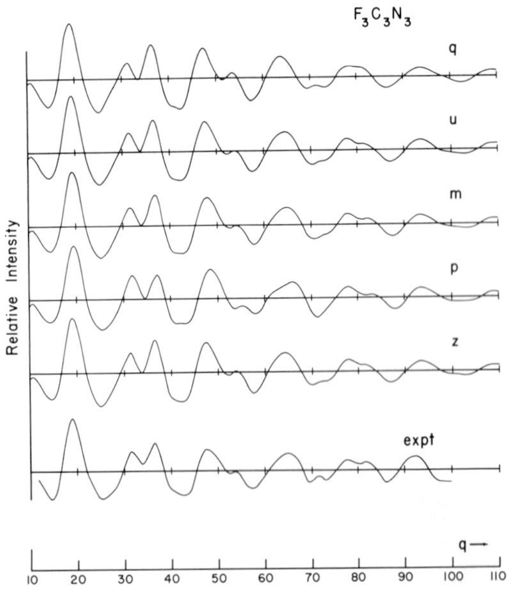

FIGURE 10.
The experimental and calculated molecular scattering functions for $F_3C_3N_3$. All models were assumed planar, D_{3h} symmetry.

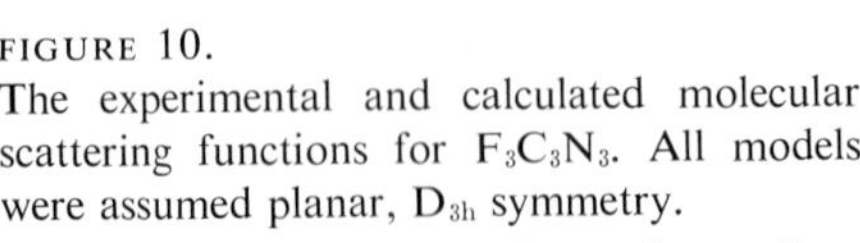

	q	u	m	p	z
$C_1—N_1$	1.330	1.330	1.330	1.330	1.330
$C_1—F_1$	1.313	1.313	1.297	1.297	1.310
∠NCN	125°	$127\frac{1}{2}$°	$127\frac{1}{2}$°	130°	$127\frac{1}{2}$°

TABLE 7.
Interatomic distances in $F_3C_3N_3$ (*Model* z); *symmetry* D_{3h}; $\angle$NCN = 127° ± 1°.

Distance	r_g(Å)	l_{ij}(Å)	Identified in Figure 8 as:
C_1—F_1	1.310 ± 0.008	0.062 ± 0.005	A
C_1—N_1	1.333 ± 0.009	0.050 ± 0.005	B
C_1—C_2	2.217	0.063	C
N_1—N_2	2.371	0.063	E
N_1—F_1	2.244	0.078	D
C_1—N_2	2.660	0.065	F
C_1—F_2	3.414	0.078	G
N_1—F_3	3.970	0.088	H
F_1—F_2	4.485	0.100	I

establish the presence of a distortion. A symmetric ring structure with (C—C) = 1.4020 ± 0.0001 Å and (C—Br) = 1.8674 ± 0.0002 Å is compatible with four sets of moments; however, so is a distorted structure with (C_2—C_3) = 1.375 ± 0.011 Å, (C_4—C_3) = 1.4010 ± 0.0003 Å, and (C—Br) = 1.85 ± 0.03 Å. The available microwave data on fluorobenzene were insufficient for a complete structure determination, but for phenyl cyanide nine isotopic species were measured by Bak et al. (1962), who reported (C_1—C_2) = 1.391 Å; (C_2—C_3) = 1.393 Å; (C_3—C_4) = 1.400 Å; (C_1—C_7) = 1.455 Å and (C_7—N) = 1.159 Å. For the ring dimensions, the error limits ranged from 0.005 to 0.009 Å. They concluded that "no kind of ordinary monosubstitutions, such as by halogens, aliphatic chains, OCH_3, $COCH_3$, CHO, COOH, CF_3, is likely to cause benzene ring distortion exceeding about 0.005 Å for carbon-carbon distances and 2 to 3° for the valence angles." The dimensions of the reference compounds, benzene and perdeuterobenzene, are well established; electron diffraction gave (1.397 ± .003 Å) and rotational raman spectra (1.397_4 ± .001 Å) (Almenningen et al., 1958; Langseth and Stoicheff, 1956; Kimura and Kubo, 1960). For comparison in perfluorobenzene one electron diffraction investigation gave for the C—C distance 1.394 ± .007 Å (Almenningen, 1964) and in this study we obtained 1.408 ± .006 Å. Also, we found in the symmetric trifluorobenzene, (C—C) = 1.402 ± .005 Å. Strand and Cox (1966) reported retention of D_{6h} symmetry in the following rings, with (C—C) = 1.404 ± 0.006 Å in C_6Cl_6; 1.395 ± 0.004 Å in 1,2,4,5-$C_6Cl_4H_2$; 1.401 ± 0.01 Å in C_6Br_6; 1.402 ± 0.005 Å in $C_6Br_4H_2$. It is thus evident that substitution of highly electronegative atoms onto the C_6 ring has no perturbing effect on the ring dimensions in the ground state to within ±0.005 Å. In contrast, the $\sigma^2\pi^2$ bond in ethylene is measurably shortened by successive fluorine substitution (Laurie, 1963), from 1.336_9 ± 0.001_6 Å in C_2H_4 (Bartell et al., 1965) to 1.313 in C_2F_4.

It is interesting to note the closeness in the positions of the first UV absorption bands for C_6H_6, sym $C_6F_3H_3$, and C_6F_6 (Bauer and Aten, 1963). These occur at around 39,000 cm^{-1}. Because of their low intensity, these absorption bands have been assigned to the electronically forbidden ${}^1B_{2u} \longleftarrow {}^1A_{1g}$ but

TABLE 8.
Comparison of distances in related molecules (Assigned limits of error are shown in parentheses, in units of 0.001 Å).

Compound	C—C	C—X	Compound	C—N	C—X	Compound	B—N	B—X
C_6H_6	$1.397_4(1)$	1.084 (5)	$H_3C_3N_3$	$1.338_1(1)$	1.084 (9)	$H_3B_3N_3H_3$	1.437 (5)	1.23 (10)
C_6H_5CN	1.391–1.400 (5–9)	1.455 (5)	H_5C_5N	1.340 (5)	1.080 (5)	$(H_2N)H_2B_3N_3H_3$	1.430 (4)	1.440 (8)
C_6H_5F	1.395 (x)	1.330 (5)				$H_3B_3N_3Me_3$	1.42 (20)	
(s)$C_6F_3H_3$	1.402 (5)	1.305 (10)	$F_3C_3N_3$	1.333 (9)	1.310 (8)	$F_3B_3N_3H_3$	1.426 (6)	1.355 (6)
C_6F_6	1.401 (6)	1.324 (6)						
$C_6Cl_4H_2$	1.395 (4)	1.724 (5)						
C_6Cl_6	1.404 (6)	1.717 (5)	$Cl_3C_3N_3$	1.33 (20)	1.68	$Cl_3B_3N_3H_3$	1.413 (10)	1.760 (15)
$C_6Br_4H_2$	1.402 (5)	1.883 (8)						
C_6Br_6	1.401 (10)	1.879 (7)	$(H_2N)_3C_3N_3$	1.34 (10)		$Me_3B_3N_3H_3$	1.39 (40)	
C_2H_4	$1.336_9(2)$	1.103 (2)	>C=N<	NO ADEQUATE DATA		>B=N<	NO ADEQUATE DATA	
HFC:CH_2	1.337 (2)	1.344 (2)						
(c)HFC:CFH	1.324 (5)	1.335 (5)						
F_2C:CH_2	1.315 (2)	1.323 (2)	$H_2C{=}N$⋮N	1.300	1.319_{NN}	BF_3		1.311 (1)
C_2F_4	{1.313 (35) {1.27 (40)	{1.313 (20) {1.33 (20)	N⋮N	1.280 (10)	$1.132\ (10)_{NN}$	BCl_3		1.742 (4)
C_2Cl_4	$1.32_7(30)$	$1.72_4(30)$	H_2C=, C⋮N	1.165 (5)	$1.424\ (6)_{CC}$			
HC⋮CH	1.207_7	1.058_5	HC⋮N	1.155	1.063			
FC⋮CH	1.198	1.279	FC⋮N	1.159	1.262			
ClC⋮CH	1.204	1.637	ClC⋮N	1.159	1.631			
CF_2		1.301 (1)						
H_2CF_2		1.358 (1)						
CF_4		1.323 (5)						
CCl_4		1.766 (3)						

vibronically allowed transitions. Since the energy separations are about equal, one might argue that, even in the first excited state, symmetric fluorine substitution has little effect on π electron system in the ring. On closer inspection, one finds that whereas the f values for C_6H_6 and $C_6F_3H_3$ (integrated from 37,000 to 44,000 cm^{-1}) are about equal (1.4×10^{-3}), the absorption by C_6F_6 is much stronger with $f = 1.9 \times 10^{-2}$ for the corresponding interval 34,500 – 47,600 cm^{-1}. This is not unexpected in view of the fact that perfluoro substitution introduces a greater vibronic perturbation than does a symmetric trifluoro substitution. By treating quantitatively the inductive and resonance effects associated with the presence of various fluorine substituents, Caldow and Coulson (1966) were able to reproduce in a striking manner the variations in ionization potential with position in substituted fluorobenzenes.

In monofluorobenzene, microwave data suggest that (C—F) = 1.330 ± .005 Å, but no unique model could be established in these studies (Tannenbaum, 1957). Our values for symmetric trifluorobenzene (C—F) = 1.305 ± 0.010 Å and for the perfluoro compound (C—F) = 1.324 ± 0.006 Å. Thus it appears that π electron delocalization in the ring transmits the effect of fluorine substitution, as does the $\sigma^2\pi^2$ bond in ethylene. The corresponding changes in the latter (Laurie, 1963) range from (C—F) = 1.344 Å in CH_2CHF to 1.313 Å in C_2F_4. In saturated compounds, contraction in C—F bond length is observed when substitutions are made on the same carbon atom (The Chemical Society, 1958, 1965). To account for the slightly longer C—F separation in C_6F_6 one must invoke an "ortho effect." This parallels observations on the heats of formation of substituted fluorobenzenes; the heat of formation of o-difluorobenzene is −67.7 kcal/mole, whereas for the meta and para isomers, the heats of formation are −71.4 and −70.7 kcal/mole, respectively (Good and Scott, 1961). Cox et al. (1964) showed that one may account for the nontransferability of average C—F bond energies among the fluoroaromatic compounds on the basis of dipole-dipole repulsions between adjacent C—X bond moments. The interatomic distances reported for the chloro and bromo benzenes (Strand and Cox, 1966) give no indication of an ortho effect: (C—Cl) = 1.717 ± 0.005 Å in C_6Cl_6 and 1.724 ± 0.005 Å in $C_6Cl_4H_2$; (C—Br) = 1.879 ± 0.007 Å in C_6Br_6 and 1.883 ± 0.008 Å in $C_6Br_2H_4$.

For the isoelectronic B_3N_3 ring, our electron diffraction study of $F_3B_3N_3H_3$ and a concurrent X-ray crystal structure analysis of this compound by Parkes and Hughes (1967) lead to essentially the same bond distances; they report (B—N) = 1.420 ± 0.004 Å and (B—F) = 1.350 ± 0.008 Å, compared with our 1.432 ± 0.010 and 1.361 ± 0.010 Å, respectively. They found no distortion from 120° bond angles (∠NBN = 119.9 ± 0.8°); we have an indication of a slight distortion (∠NBN = 119° ± 1°), as given in Table 5, but these differences are not significant. With regard to the effect of substitution on the B_3N_3 ring, the lack of sufficient precision in the available data leaves the conclusion ambiguous. In borazine, (B—N) = 1.44 ± .02 Å (Bauer, 1938);* in $F_3B_3N_3H_3$ an average value is 1.426 ± 0.006 Å; in $H_3B_3N_3Me_3$, 1.42 ± 0.02 Å (Coffin and

*A redetermination of the structure of borazine is in progress (Porter, et al., 1967). Preliminary values for (B—N) = 1.437 Å, (B—H) = 1.23 Å, and (N—H) = 1.02 Å; there are clear indications that ∠NBN > ∠BNB, and that large amplitude out-of-plane vibrations are present. The most recent analysis of its vibrational spectrum was presented by Niedenzu and coworkers (1967).

Bauer, 1955); in $Cl_3B_3N_3H_3$, 1.413 ± 0.010 Å (Coursen and Hoard, 1952); and in a recent X-ray crystal structure analysis of $Me_3B_3N_3H_3$, $(B—N)_{av}$ = 1.39 ± 0.04 Å (Anzenhofer, 1966). As yet an incomplete analysis of current electron diffraction data on B-aminoborazine gives 1.43 ± 0.01 Å (Lee, Porter, and Bauer, 1967). There is then a bare possibility that the B—N distance in the trichloro compound is slightly less than in the others, but the distortability test for aromaticity cannot yet be applied. Trends in the magnitudes of average bond dissociation energies (Smith and Thakur, 1965) are not informative on this point, nor are the positions of the first UV absorption bands for borazine and its substituted derivatives (Coffin and Bauer, 1955; Rector et al., 1949). The electronic structures of various B—N compounds and borazine were recently discussed by Hoffmann (1964), who concluded on the basis of extended Hückel calculations that the borazine ring is relatively ineffective in transmitting electronic effects, and thus is less aromatic than is the benzene ring; Hoffmann's conclusions were in agreement with the diamagnetic anisotropy measurements (Lonsdale and Toor, 1959).

It is interesting to compare the boron-fluorine atom separations in $F_3B_3N_3H_3$ (av. value 1.355 ± .005 Å) and in BF_3 (1.311 ± 0.001 Å; Kuchitzu and Konaka, 1966). The difference is greater than that reported for the pair $Cl_3B_3N_3H_3$ (Coursen and Hoard, 1952) (1.760 ± .015 Å) and BCl_3 (Konaka et al., 1966) (1.742 ± .004 Å). The simplest interpretation starts with the assumption that there is considerably more delocalization of the π electrons in BF_3 than BCl_3. This is consistent with the relative magnitudes of the expansions in the B—F and B—Cl distances as one passes from corresponding tricoordinated to tetracoordinated borons, as well as with the observation that BCl_3 is a stronger Lewis acid than is BF_3 (McCoy and Bauer, 1956). Since the π electrons in B_3N_3 are not as fully delocalized as in benzene, there is more "back-donation" from the B—F bonds than from B—Cl to the ring. This is also consistent with the relative magnitudes of the ring breathing frequencies found in the raman spectra of B-trihaloborazines, as interpreted by Hester and Scaife (1966).

In the C_3N_3 system, the imperturbability by substituents is as clear-cut as it is for benzene. Thus, in s-triazene (C—N) = 1.338 ± .001 Å (Lancaster and Stoicheff, 1956; Wheatley, 1955*); in cyanuric trifluoride, (C—N) = 1.333 ± .009 Å; in the trichloride, 1.33 ± .02 Å and in s-triaminotriazene, 1.34 ± 0.01 Å (Akimoto, 1955). Confirmation of the relative independence of ring dimensions on symmetric substitution is provided by the values reported for the ∠NCN: 127 ± 1°, 127.5 ± 1°, 125°, and 123 ± 3°, respectively; also, confirmation is provided by the consistency of the estimated magnitudes of the force constants for molecular vibrations in $F_3C_3N_3$ and $H_3C_3N_3$ (Long et al., 1962).

The effects of conjugation with aromatic rings on the magnitudes of C—X distances appear in C_3N_3 as they do in C_6 rings. Compare (C—Cl) = 1.68 Å in $Cl_3C_3N_3$ (Akimoto, 1955) with 1.724 ± .005 Å in $C_6Cl_4H_2$ (Strand and Cox, 1966) and with 1.766 ± .003 Å in CCl_4 (The Chemical Society, 1958, 1965). Similarly, (C—F) = 1.310 ± 0.008 Å in $F_3C_3N_3$, 1.305 ± 0.010 Å in $C_6F_3H_3$,

*From an X-ray crystal structure analysis of s-triazene Wheatley found (C—N) = 1.319 Å, with a standard deviation of 0.005 Å. No corrections were made for rigid-body oscillations in the crystal lattice. This may remove the discrepancy as initially observed for gaseous and crystalline benzene.

and $1.323 \pm .005$ Å in CF_4 (The Chemical Society, 1958, 1965). The shortest C—X distances have been observed in X—C≡N and X—C≡CH compounds; for fluorine the distances are 1.262 Å and 1.279 Å, respectively; for chlorine they are 1.631 Å and 1.637 Å, respectively (Tyler and Sheridan, 1963).

Possibly the most intriguing question pertains to the length of the C—N bond distance and the magnitude of the bond angle at the nitrogen atom, when HC is replaced by N in a benzene ring. In pyridine $\angle$CNC = 116° 53′ (Bak, 1958);* in pyrazine it is $115° \pm 1°$ (Wheatley, 1957); in s-triazene $113° \pm 1°$ (Lancaster and Stoicheff, 1956); and in s-tetrazene 116° (Bertinotti et al., 1956). The (C—N) separations remain essentially equal to $1.335 \pm .005$ Å. Theoretical discussions of variations of internal bond angles in the pyridine-s-tetrazene sequence have been presented (de la Vega and Hameka, 1963); the experimentally measured angles are well reproduced in these calculations.

ACKNOWLEDGMENTS

We sincerely thank Dr. G. C. Finger, Dr. K. Niedenzu, and Dr. F. S. Fawcett for giving us samples of their very interesting compounds. Sincere thanks are also due to Professor R. E. Hughes and Professor Roald Hoffmann for stimulating and informative discussions.

*See also Almenningen, A., O. Bastiansen, and L. Hauser (1955), who reported (C—N) ≈ (C—C) = 1.377 Å; (C—H) = 1.078 Å; and Cumper, C. W. N. (1957), who proposed an empirical method for reducing microwave data when these are insufficient for a complete structure determination.

REFERENCES

Akimoto, Y. (1955). *Bull. Chem. Soc. Japan* **28**, 1.

Almenningen, A., O. Bastiansen, and L. Fernholt (1958). *Det. KGL Norske Videnskabers Selskabs Skrifter*, no. 3, p. 3.

Almenningen, A., O. Bastiansen, and L. Hauser (1955). *Acta Chemica Scand.* **9**, 1306.

Almenningen, A., O. Bastiansen, R. Seip, and H. M. Seip (1964). *Acta Chim. Scand.* **18**, 2115.

Anzenhofer, K. (1966). *Mol. Phys.* **11**, 493.

Bak, B., *et al.*, (1958). *J. Mol. Spec.* **2**, 361.

Bak, B., et al., (1962). *J. Chem. Phys.* **37**, 2027.

Bartell, L. S. (1962). *Tetrahedron* **17**, 177.

Bartell, L. S., et al., (1965). *J. Chem. Phys.* **42**, 2683.

Bauer, S. H. (1938). *J. Am. Chem. Soc.* **60**, 524.

Bauer, S. H., and C. F. Aten (1963). *J. Chem. Phys.* **39**, 1253.

Bertinotti, F., J. Giacumello, and A. M. Liquori (1956). *Acta Cryst.* **9**, 510.

Cady, H. H., A. C. Larson, and D. T. Cromer (1966). *Acta Cryst.* **20**, 336.

Caldow, G. L., and C. A. Coulson (1966). *Tetrahedron*, suppl. 7, 128.

The Chemical Society, London (1958). *Tables of Interatomic Distances.* Special Publication No. 11.

The Chemical Society, London (1965). *Tables of Interatomic Distances.* Special Publication No. 18.

Coffin, K. P., and S. H. Bauer (1955). *J. Phys. Chem.* **59**, 193.

Coulson, C. A. (1963). *J. Chem. Soc.*, p. 5893.

Coursen, D. L., and J. L. Hoard (1952). *J. Am. Chem. Soc.* **74**, 1742.

Cox, J. D., H. A. Gundry, and A. J. Head (1964). *Trans. Farad. Soc.* **60**, 653.

Cox, E. G., D. W. J. Cruickshank, and J. A. S. Smith (1955). *Nature* **175**, 766.

Cumper, C. W. N. (1958). *Trans. Faraday Soc.* **54**, 1266.
Cumper, C. W. N. (1958). *Trans. Faraday Soc.* **54**, 1261.
Good, W. D., and D. W. Scott (1961). *Pure and Applied Chem.* **2**, 77.
Hastings, J. M., and S. H. Bauer (1950). *J. Chem. Phys.* **18**, 13.
Hester, R. E., and C. W. J. Scaife (1966). *Spectrochim. Acta* **22**, 455.
Hoffmann, R. (1964). *J. Chem. Phys.* **40**, 2474.
Karle, I. L., and J. Karle (1965). *Acta Cryst.* **19**, 92.
Kimura, K., and S. H. Bauer (1963). *J. Chem. Phys.* **39**, 3172.
Kimura, K., and M. Kubo (1960). *J. Chem. Phys.* **32**, 1776.
Kitaigorodskii, A. I. (1957). *Organic Chemical Crystallography*, Consultant's Bureau Translation, Chapter 5.
Konaka, S., Y. Murata, K. Kuchitzu, and Y. Morino (1966). *Bull. Chem. Soc., Japan* **39**, 1134.
Kuchitzu, K. (1966). *J. Chem. Phys.* **44**, 906.
Kuchitzu, K., and S. Konaka (1966). *J. Chem. Phys.* **45**, 4332.
Langseth, A., and B. P. Stoicheff (1956). *Can. J. Phys.* **34**, 350.
Lancaster, J. E., and B. P. Stoicheff (1956). *Can. J. Phys.* **34**, 1016.
Laurie, V. W., and D. T. Pence (1963). *J. Chem. Phys.* **38**, 2693.
Lee, G., R. F. Porter, and S. H. Bauer (1967). In preparation.
Long, D. A., J. Y. H. Chau, and R. B. Gravenor (1962). *Trans. Faraday Soc.* **58**, 232.
Lonsdale, K., and E. W. Toor (1959). *Acta Cryst.* **12**, 1048; (1959). *Nature* **184**, suppl. 4, 1060.
McCoy, R. E., and S. H. Bauer (1956). *J. Am. Chem. Soc.* **78**, 206.
Niedenzu, K., et al. (1967). *Inorganic Chemistry* **6**, 1453.
Parkes, A. S., and R. E. Hughes (1967). To be published in *Inorganic Chemistry*.
Porter, R., G. Lee, W. Harshbarger, and S. H. Bauer (1967). In preparation.
Rector, C. W., G. W. Schaeffer, and J. R. Platt (1949). *J. Chem. Phys.* **17**, 460.
Rosenthal, E., and B. P. Dailey (1965). *J. Chem. Phys.* **43**, 2093.
Smith, B. C., and L. Thakur (1965). *Nature* **208**, 74.
Stoicheff, B. P. (1962). *Tetrahedron* **17**, 135.
Strand, T. G., and H. L. Cox, Jr. (1966). *J. Chem. Phys.* **44**, 2426.
Tannenbaum, E. (1957). *J. Chem. Phys.* **26**, 134.
Tyler, J. K., and J. Sheridan (1963). *Trans. Faraday Soc.* **59**, 2665.
de la Vega, J. R., and H. F. Hameka (1963). *J. Am. Chem. Soc.* **85**, 3504.
Watanabe, H., K. Ito, and M. Kubo (1960). *J. Am. Chem. Soc.* **82**, 3294.
Wheatley, P. J. (1955). *Acta Cryst.* **8**, 224.
Wheatley, P. J. (1957). *Acta Cryst.* **10**, 182.

HERMAN BRANSON
Department of Physics
Howard University
Washington, D. C.

Negative Ions in the Mass Spectra of the Methylamines

Although the positive ion spectra of organic compounds are a well-known source of information concerning structure (Elliott, 1963), the negative ion spectra have not been much studied, partly because far fewer negative ions are produced from compounds under the usual operating conditions and partly because the negative ion spectra are difficult to interpret (Melton, 1963). In this paper, emphasis will be placed upon an interpretation of some of our results on the negative ions from electron impact on the methylamines: $(CH_3)NH_2$, $(CH_3)_2NH$, and $(CH_3)_3N$. It is an especially apposite interpretation in harmony with one of Linus Pauling's interests, since it leads to a possible new technique for characterizing the hydrogens in certain compounds.

The methylamines have long appealed to me as compounds of primary importance in biophysics. One may contend that somewhere, sometime when the C—N bond was formed, the chemical steps leading to living systems were initiated. The methylamines are among the simplest compounds having this bond. In addition, because they form a series from mono- to tri-, information impossible to derive from a study of any one of these compounds might be obtained from study of the group. Unfortunately they are also many body systems, and attempts at theory have not been especially helpful in the interpretation of such data. The computer enables us to perform certain simplified calculations on the interactions of the impinging electron with the molecule, but the results have been of little assistance. Then too even the Born approximation, which is applicable in an inelastic process when the energy of the bombarding electron is much greater than the energy of the state excited, is not of use to us, because the energies of our electrons are about the same as the states excited, or they differ by, at the most, a factor of 2 or 3.

The measurements were made in an Atlas mass spectrometer, CH_3/IV, in the First Institute for Experimental Physics at the University of Hamburg.* The gases were supplied by the Matheson Company, East Rutherford, N.J. The readings were taken by a modified retarding potential-difference method and by the conventional method (Kraus, 1961). Pressures in the spectrometer tube were 10^{-5} to 10^{-7} mm of Hg, and in the ion source it was from 2 to 4 $\times$

This work has been supported by the NASA project, SC 09-011-(004).

*I am indebted to Professor H. Neuert for the hospitality of his laboratory.

TABLE 1.
The negative ions in the mass spectra of the methylamines.

Amine and Ions	Mass	Appearance potential in ev
Monomethyl amine		
CH_3NH^-	30	4.87
CH_2N^-	28	5.26, 7.72
$CH\ NH^-$		
NH_2^-	16	5.37
NH^-	15	9.37
H^-	1	4.85, 8.27
Dimethyl amine		
$(CH_3)_2N^-$	44	4.75
CH_3NH^-	30	4.86
NH^-	15	7.38
H^-	1	4.66, 7.98
Trimethyl amine		
NH^-	15	8.35
H^-	1	8.16

TABLE 2.
Appearance potentials in ev of the common ions from the four molecules.

Negative ion	Parent molecule			
	NH_3	NH_2CH_3	$NH(CH_3)_2$	$N(CH_3)_3$
NH_2^-	5.36	5.37	—	—
NH^-	9.38	9.37	7.38	8.35
H^-	5.22	4.85	4.66	—
		8.27	7.98	8.16

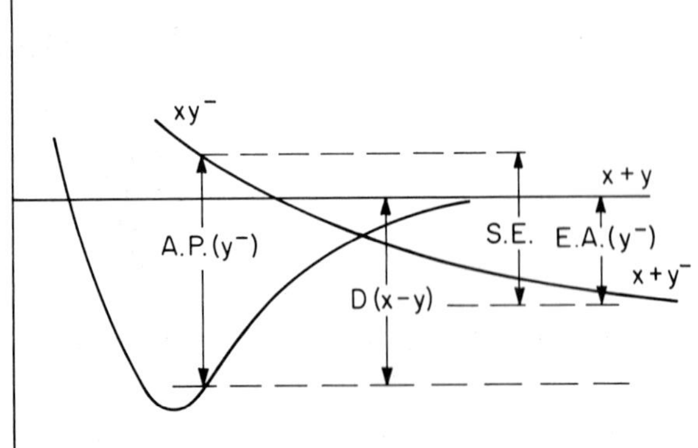

FIGURE 1.
A description of the electron-impact result when the polyatomic molecule can be considered as a two-component system. The XY^- curve intersects the plane of the XY curve where the A.P. (Y^-) vertical line meets it. The remainder of the curve is a projection of the XY^- curve on the plane; thus the XY curve and the XY^- curve do not cross on the right.

10^{-4}. CO was used as the calibrating gas, with 9.66 ev as the appearance potential of the O^-. All the data have standard deviations no greater than ± 0.04 ev. The results are presented in Table 1. In Table 2 are listed the common ions found in all the negative ion spectra.

Where the reaction may be looked upon as

$$XY + e \longrightarrow X + Y^-$$

such as

$$(CH_3)_2NH + e \longrightarrow H + (CH_3)_2N^-$$

or

$$(CH_3)_2NH + e \longrightarrow (CH_3)_2N + H^-,$$

we may schematize it in a Franck-Condon diagram as in Figure 1.

As an example of the quantitative use of the Franck-Condon representation, we consider H^- from $(CH_3)_2N$—H (6):

$$\text{A.P. } (H^-) = D[(CH_3)_2N\text{—}H] - \text{E.A. } (H^-) + \text{S.E.}$$

Here S.E. means "surplus energy," which is either the kinetic energy or the excitational energy of the fragments, if not both. Inserting the numerical values, we have

$$4.66 = 3.69 - 0.72 + \text{S.E.}$$
$$\text{S.E.} = 1.69 \text{ ev}$$

An example leading to a lower limit for the electron affinity for NH^- is

$$(CH_3)_2NH + e \longrightarrow 2\ CH_3 + NH^-$$

Thus

$$\text{A.P. } (NH^-) = D\ [CH_3 - NH - CH_3] - \text{E.A.} + \text{S.E.}$$

whereupon

$$7.38 = 7.90 - \text{E.A.} + \text{S.E.}$$

then

$$\text{E.A. } (NH^-) \geq 0.52 \text{ ev.}$$

The most interesting characteristic of the data for the negative mass spectra is the behavior of the H^- spectra.* From the data in Table 2, we see that the H^- has two maxima in mono- and di- but only one in tri-. We observed that the ratio of the first peak to the second was 1.8 in mono-, 1.6 in di-, and zero in tri-. The interpretation is that the H^- in the first maximum arises from —NH_2 in the mono- and from —NH in the di-. Thus we would expect none from tri-; this is consistent with the findings. This interpretation implies that the H's attached to CH_3 are not released singly without a considerable amount of excitational energy being imparted to the radical. Since the appearance-potential values for the members of the series are fairly close (8.27, 7.98, and 8.16 ev respectively), the energy can be interpreted as being given to the single CH_3 group from which the H^- originates. To assume that this energy is kinetic energy is less reasonable, because of the $2m/M$ factor in the transfer

*A prominent feature of all the negative ion spectra is a large peak at mass 26. This peak is present no matter how carefully the gases are purified and is observed in all experiments with carbon compounds. It is attributed to CN^-. The ease with which this radical is produced may be significant.

of kinetic energy between the electron (m) and the molecule (M), than to assume that it comes from the mechanism that requires that the excitation of the radical be involved. On electron impact an H is ejected and captures the electron; the radical is raised to a higher electronic level (3.5 ev) from which it may fall to a higher vibrational level of the electronic ground state.

The specific mechanism implies that there is an electronic level in CH_2NH_2, CH_3CH_2NH, and $CH_3CH_3CH_2N$ that is about the same distance in energy units ($\sim$3.5 ev) above the ground state for each. Since the common factor is CH_2N, one would expect the electronic level to be associated with it. Irrespective of the details of this model, however, the simplest and most straightforward interpretation of the experimental result is to attribute the H with the lower appearance potential to the NH_2 or NH, and the H with the higher appearance potential to a CH_3.

REFERENCES

Elliott, R. M. ed. (1963). *Advances in Mass Spectrometry*, vol. 2. New York: Macmillan.

Kraus, K. (1961). *Z. Naturforsch.* **16**a, 1378.

Melton, C. E. (1963). In McLafferty, F. W., ed., *Mass Spectrometry of Organic Ions*, pp. 163–202. New York: Academic Press.

Vendeneyev, V. I., L. V. Gurvich, V. N. Kondrat'yev, V. A. Medvedev, and Ye. L. Frankevich (1966). *Bond Energies, Ionization Potentials and Electron Affinities.* New York: St. Martin's Press.

JÜRG WASER
Gates and Crellin Laboratories of Chemistry
California Institute of Technology
Pasadena, California

Pauling's Electroneutrality Principle and the Beginner

Professor Pauling's important contributions to the teaching of chemistry have been, and are, twofold: directly, his basic texts (Pauling, 1948a,* 1950) have enormously stimulated the acceptance of modern ideas in beginning courses; indirectly, his introduction of new concepts and principles has provided insight into the behavior of molecules and has furthered the understanding of the relation between their properties and structure.

As an example of the latter type of contribution I have chosen his electroneutrality principle (Pauling, 1948b, 1960), which has not been given the prominence it deserves in the teaching of beginning chemistry. One of the common problems confronting the beginning student of chemistry is that he has no idea what constitutes a reasonable structural formula for an unfamiliar molecule. To give an extreme example, students confronted by the task of writing a structural formula for the pyrophosphate ion $P_2O_7^{----}$ have been known to write Lewis formulas such as

$$\left[:\ddot{\underset{..}{O}}-\ddot{\underset{..}{O}}-\ddot{\underset{..}{O}}-\ddot{\underset{..}{P}}-\ddot{\underset{..}{P}}-\ddot{\underset{..}{O}}-\ddot{\underset{..}{O}}-\ddot{\underset{..}{O}}-\ddot{\underset{..}{O}}:\right]^{----}$$

which satisfies the octet rule and shows the correct number of electrons, but is, nevertheless, incorrect. Pauling's electroneutrality principle provides a useful guide to correct structural formulas in this and similar situations. Before applying the principle in this way, we shall explore its general nature by considering the distribution of formal charges (Pauling, 1960) in a number of molecules.

STATEMENT OF THE PRINCIPLE, AND SIMPLE APPLICATIONS

According to the electroneutrality principle, *the electronic structure of stable molecules and crystals is such that the electric charge of each atom is close to zero, and varies between at most +1 and −1.* To see in detail how this principle works, a knowledge of electronegativities is required, as well as an understanding of the relationship between the ionic character of a bond and

Contribution No. 3455 from the Gates and Crellin Laboratories. This contribution is an extension of several sections of a projected freshman text.

*A lithoprinted precursor of Professor Pauling's 1948 freshman text was published in 1944.

TABLE 1.
Electronegatives of some elements.

B	2.0	C	2.5	O	3.5
H	2.1	S	2.5	Cl	3.5
P	2.1	N	3.0	F	4.0

TABLE 2.
Relationship between ionic character and electronegativity difference.

Electro-negativity difference	Percent ionic character	Electro-negativity difference	Percent ionic character
0.2	1	1.0	22
0.3	2	1.1	26
0.4	4	1.2	30
0.5	6	1.3	34
0.6	9	1.4	39
0.7	12	1.5	43
0.8	15	1.6	47
0.9	18	1.7	50

the electronegativity difference of the atoms connected by this bond (Pauling, 1960). The information needed is given in Tables 1 and 2 (Pauling, 1960).

The approximate charges that correspond to a dot formula can be found by starting with the formal charges and adding the calculated charges caused by the ionic character of the bonds. To give an illustration, in the hydronium ion

$$\left[\begin{array}{c} \text{H} \\ \quad\diagdown^{(+)} \\ \text{H—}\ddot{\text{O}}\text{:} \\ \diagup \\ \text{H} \end{array}\right]^{+} \tag{1}$$

the electronegativity difference of 1.4 between oxygen and hydrogen corresponds to 39 percent ionic contribution to the bonding, with hydrogen positive and oxygen negative. Since the formal charge of hydrogen in (1) is zero, we find that the ionic contributions leave a charge of $+0.39$ on each atom of hydrogen. The formal charge of oxygen in (1) is $+1$, and the calculated charge from ionic contributions is $+1 - 3 \times 0.39 = -0.17$, because for each of the three hydrogen atoms there is a contribution of -0.39 to the charge on the oxygen atom. As a result, the $+1$ charge of the entire atom is completely transferred to the hydrogen atoms, the final result being

$$\left[\begin{array}{c} \overset{0.39}{\text{H}} \\ \quad\diagdown^{-0.17} \\ {}^{0.39}\text{H—}\ddot{\text{O}}\text{:} \\ \diagup \\ \text{H}^{0.39} \end{array}\right]^{+} \tag{2}$$

It must be stressed that these considerations are not exact and that (2) and a formula in which the oxygen atom has the charge zero and the hydrogen atoms have the charge +0.33 have essentially the same meaning.

As a second example, let us consider nitrous oxide, which is represented as a resonance hybrid between two Lewis formulas

$$\left\{ :N{\equiv}\overset{(+)}{N}{-}\ddot{\underset{\cdot\cdot}{O}}:^{(-)},\quad {}^{(-)}:\ddot{N}{=}\overset{(+)}{N}{=}\ddot{\underset{\cdot\cdot}{O}}: \right\} \qquad (3)$$

(a) (b)

The electronegativity difference of 0.5 between nitrogen and oxygen corresponds to 6 percent ionic character, with nitrogen positive, and oxygen negative. No charge is transferred by ionic contributions between the two atoms of nitrogen, because they have the same electronegativity. The oxygen atom in the Lewis formula (3, a) has, therefore, the charge −1.06, and the central nitrogen atom, the charge +1.06. In the Lewis formula (3, b) a double bond connects the atoms of nitrogen and oxygen, corresponding to a charge transfer of twice 0.06 units, so that charges of −0.12, +1.12, and −1.00 are associated with the oxygen atom, the central nitrogen atom, and the terminal nitrogen atom respectively. If (3, a) contributes about twice as much to the resonance hybrid as does (3, b)—a very rough estimate, based on the less favorable charge distribution in (3, b)—the resulting charges on the atoms are

$$^{-0.33}N{=}\overset{1.08}{N}{=}O^{-0.75} \qquad (4)$$

Dots have been omitted here because the purpose of (4) is to show the estimated charges on the framework of atoms; these charges apply to the final resonance hybrid between (3, a) and (3, b).

Ionic character may change the charges associated with electron-dot formulas considerably if the electronegativity difference of bonded atoms is large, particularly if the atoms are connected by multiple bonds. Substantial modifications of formal charges by the ionic character of bonds must therefore be taken into account when judging the qualifications of dot formulas for resonance. Consider, for example, the resonance formulation of CO. One of the contributing formulas is

$$^{(-)}:C{\equiv}O:^{(+)} \qquad (5)$$

for which the octet rule is satisfied, but in which oxygen has a positive formal charge and carbon a negative one. However, because an electronegativity of 1.0 corresponds to an ionic contribution of 22 percent, the charge on the oxygen atom becomes $1 - 3 \times 0.22 = +0.34$, and that on the carbon atom, −0.34. The charge distribution is, therefore, less extreme than would appear by considering just the formal charges in (5). A second contributing formula is

$$:C{=}\ddot{O}: \qquad (6)$$

in which the carbon atom has only a sextet of electrons. The formal charges are zero, modified to $-2 \times 0.22 = -0.44$ for oxygen, and $+0.44$ for carbon. A final possible formula is

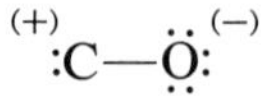

$$\overset{(+)}{:}\text{C}\!-\!\ddot{\underset{..}{\text{O}}}:^{(-)} \qquad (7)$$

in which the carbon atom has only a quartet of electrons. The charge on the oxygen atom is -1.22, that on the carbon atom, $+1.22$. This formula is much less favorable than the other two for three reasons: the carbon atom has only four electrons rather than an complete octet, only a single bond connects the two atoms, and the calculated charges are large. Of the two formulas (5) and (6), the second has a more favorable charge distribution than the first, but in the first both atoms have an octet of electrons, and the formula accordingly shows more bonds. Assuming that (5) and (6) contribute about equally, the charge expected on the oxygen atom would be $(+0.34 - 0.44)/2 = -0.05$, and that on the carbon atom, $+0.05$. This conclusion is supported by the fact that the observed dipole moment of CO is very small.

Another example in which ionic contributions change the formal charges significantly is the metaborate ion, $B_3O_6^{3-}$, for which two formulas can be written,

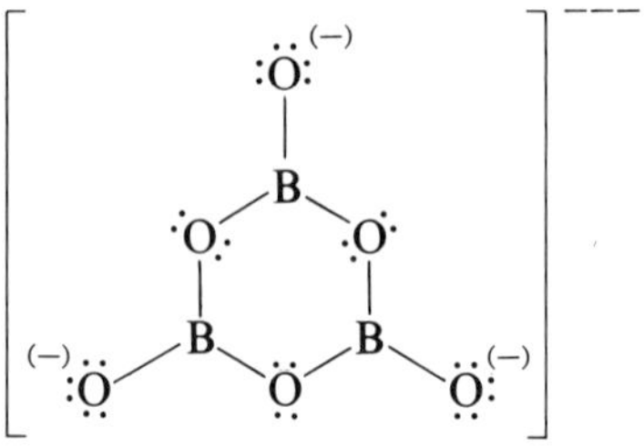

(a) [ring structure: three B and three ring O atoms, each B bearing an exocyclic :Ö:$^{(-)}$ by single bond]$^{---}$ and (b) [ring structure: each B$^{(-)}$ bearing an exocyclic O by double bond]$^{---}$ (8)

In (8, a) the boron atoms have only a sextet of electrons, but three of the oxygen atoms carry a formally negative charge. In (8, b) the boron atoms have an octet of electrons, and three B—O single bonds have been replaced by double bonds, but now the formal negative charges reside on the boron atoms that are less electronegative than the oxygen atoms by 1.5 units. However, this electronegativity difference corresponds to an ionic character of 43 percent, and the charges on (8, b) work out as follows:

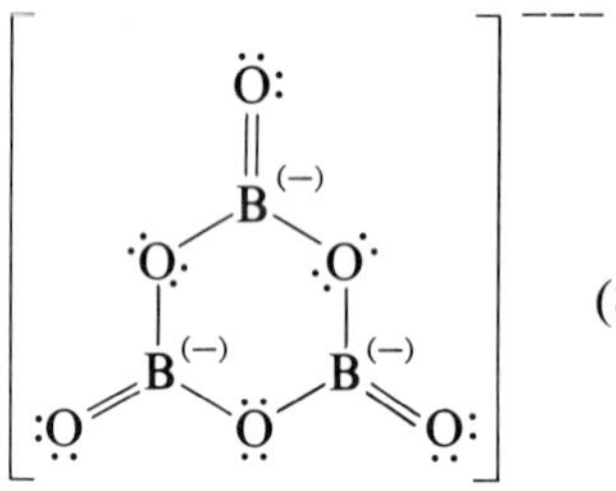

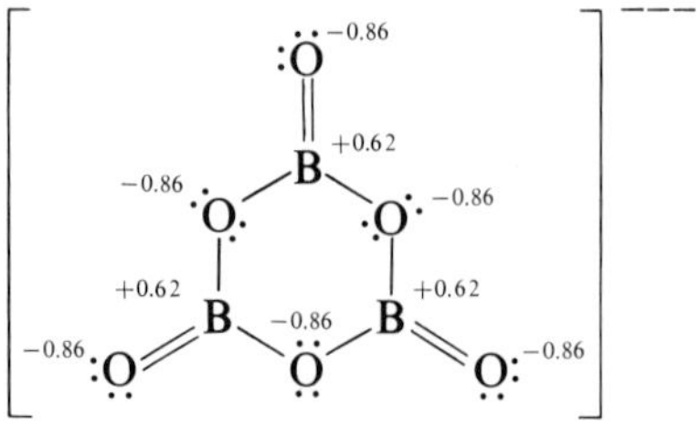

[structure (8, b) with charges: each B $+0.62$, each ring O -0.86, each exocyclic O -0.86]$^{---}$ (9)

This is an excellent distribution of charges. In formula (8, a) the calculated charges are -1.43 for the peripheral oxygen atoms and to $+1.29$ for the boron atoms, not nearly as favorable a distribution. The metaborate ion is thus well described by (9) or (8, b), and (8, a) makes only a minor contribution.

It must be emphasized that the atomic charges calculated in this way are, at best, *crude estimates*, obtained by simple means. Moreover, even the concept of atomic charges in molecules has its limitations. Consider, for example, how such charges might be determined experimentally. One method would be to determine the distribution of the electrons in the molecule from careful X-ray diffraction measurements, and to count the total number of electrons, including fractions, associated with each atom. But here a difficulty arises: As we move from one atom to the next, where shall we stop counting the electrons as belonging to the first atom and start counting them as belonging to the next? Thus, atomic charges are, to some extent, *arbitrary*; nevertheless, this does not detract from their usefulness. Values like those calculated may, for example, be used to make rough estimates of electrical properties of molecules and ions, such as dipole moments.

THE USE OF *d* ORBITALS

Atoms such as Si, P, S, Cl, and their congeners in the long and the extra-long periods may use the *d* orbitals outside the valence shell to satisfy the electroneutrality principle, expanding their valence shells in this way. For example, in the formula

$$\begin{array}{l} \qquad\qquad\quad \text{H} \\ \qquad\qquad\;\, / \\ ^{(+)}\text{:}\ddot{\text{Cl}}\text{—}\ddot{\text{O}}\text{:} \\ \quad\;\, | \\ \;\;\text{:}\ddot{\text{O}}\text{:}^{(-)} \end{array} \tag{10}$$

the chlorine atom and one of the oxygen atoms have formal charges. These charges are modified because the Cl—O bond has 6 percent ionic character, and the O—H bond 39 percent, which leads to the following situation:

$$\begin{array}{l} \qquad\qquad\quad \text{H}^{+0.39} \\ \qquad\qquad\;\, / \\ ^{+1.12}\text{:Cl—}\ddot{\text{O}}\text{:}^{-0.45} \\ \qquad\;\, | \\ \qquad\text{:}\ddot{\text{O}}\text{:}^{-1.06} \end{array} \tag{11}$$

The high charges on the chlorine atom and one of the oxygen atoms may be reduced by the use of a *d* orbital on the chlorine atom to accept a bonding pair from the oxygen atom that is not connected to a hydrogen atom. Because of the 6 percent ionic character of this second bond between chlorine and oxygen, the reduction of the charges amounts to 0.94 units, from $+1.12$ to $+0.18$ for chlorine, and from -1.06 to -0.12 for oxygen:

$$\begin{array}{l} \qquad\qquad\quad \text{H}^{+0.39} \\ \qquad\qquad\;\, / \\ ^{+0.18}\text{:}\ddot{\text{Cl}}\text{—}\ddot{\text{O}}\text{:}^{-0.45} \\ \qquad\;\, \| \\ \qquad\ddot{\text{O}}\text{:}^{-0.12} \end{array} \tag{12}$$

While the charge distribution in this formula is more favorable than in (11), expansion of the valence shell by the use of a $3d$ orbital by the chlorine atom requires energy. It is thus expected that both (11) and (12) contribute to the resonance hybrid. Note that the ionic character of the Cl—O bond does not significantly modify the charge distribution in (12) and thus does not affect the relief afforded to the unfavorable charge distribution in (11) by contributions from (12).

A more extreme case is that of perchloric acid, $HClO_4$, for which the formula showing single bonds between the atom of chlorine and each of the atoms of oxygen leads to a formal charge of +3 on the chlorine atom

```
          (−)
       :Ö:
          \ (+++)
(−) :Ö—Cl—Ö:
          /       \
       :Ö:         H
        (−)
```
(13)

To achieve formal neutrality of chlorine and oxygen atoms, an expansion of the valence shell of the chlorine atom to include *three* $3d$ orbitals is required:

```
      :Ö
         \\
      :Ö═Cl—Ö:
         //     \
      :Ö         H
```
(14)

The resonance hybrid describing $HClO_4$ is expected to include contributions from (13) and (14), as well as from intermediate formulas showing both single- and double-bonded oxygen atoms.

Another example is the structure of sulfuric acid, H_2SO_4, to which the formulas

```
 (−)
    :Ö:           H
       \ (++)    /
(−) :Ö—S—Ö
        /
     :Ö
        \
         H
```
(15)

and

```
      Ö:          H
        \\       /
     :Ö═S—Ö
        /
     :Ö
        \
         H
```
(16)

contribute, as do two others in which one of the oxygen atoms that does not carry a hydrogen atom is attached to the sulfur atom by a single bond, and the other by a double bond.

APPLICATION TO STRUCTURAL QUESTIONS

The rules for judging the stability associated with electron-dot formulas, and in particular the electroneutrality principle, often make it possible to arrive at the correct arrangement of the atoms in an unfamiliar molecule or ion. If several isomers exist, it may be possible to arrange them in order according to stability. In some cases, consideration of the formal charges involved in the different formulas is all that is required, because the refinements obtained by including the effects of ionic contributions do not change the relative stabilities. To illustrate, we shall look at a few general examples, and then formulate rules concerning the structure of oxygen compounds.

We begin with hydrogen peroxide, H_2O_2, where both hydrogen atoms might be linked to the same oxygen atom as in (17, a), or to different ones as in (17, b):

```
H                          H
 \(+)      (−)              \
  :O—Ö:                      :O—Ö:
 /                               \
H                                 H

    (a)                        (b)                    (17)
```

Structure (17, b), in which all atoms have zero formal charge, is the correct one.

In hypochlorous acid, HOCl, the hydrogen atom might be attached to the chlorine atom or to the oxygen atom, leading to the possibilities

```
H                                  H
 \(+)     (−)                     /
  C̈l—Ö:                    :C̈l—O:                    (18)

    (a)                        (b)
```

The second of these formulas is correct. In the first formula a double bond between the atoms of chlorine and oxygen, and thus an expansion of the valence shell of chlorine, would be required to avoid formal charges, while in the correct formula this is not necessary.

Nitrosyl chloride is often given the formula NOCl, but we note that placing either the oxygen atom or the chlorine atom in the middle of the molecule leads to the distributions of formal charges

```
      (+)  (−)                 (++)  (−)
       Ö=N̈:                     C̈l=N̈:
      /                    (−)/                       (19)
  :Cl:                     :O:
```

which are very unfavorable, while a satisfactory formula can be written with the nitrogen atom at the center:

$$\begin{array}{c} :\text{N}{=}\underset{\cdot\cdot}{\text{O}}: \\ / \\ :\underset{\cdot\cdot}{\text{Cl}}: \end{array} \tag{20}$$

We thus expect the sequence of atoms to be ClNO, and experimentally, this is indeed found to be the case.

As an example of possible isomerism, consider the sequence of the atoms in the ion NOC^-. To place the oxygen atom in the middle will not do, because then only unsatisfactory formulas, such as

$$\left[\overset{(-)}{:\ddot{\text{N}}}{=}\overset{(++)}{\text{O}}{=}\overset{(--)}{\ddot{\text{C}}:}\right]^- \quad \text{and} \quad \left[\overset{(--)}{:\ddot{\underset{\cdot\cdot}{\text{N}}}}{-}\overset{(++)}{\text{O}}{\equiv}\overset{(-)}{\text{C}:}\right] \tag{21}$$

can be written. It is an improvement to place the nitrogen atom in the center, leading to the resonance hybrid

$$\left\{\left[\overset{(-)}{:\text{C}}{\equiv}\overset{(+)}{\text{N}}{-}\overset{(-)}{\ddot{\underset{\cdot\cdot}{\text{O}}}:}\right]^-, \quad \left[:\text{C}{=}\underset{\cdot\cdot}{\text{N}}{-}\overset{(-)}{\ddot{\underset{\cdot\cdot}{\text{O}}}:}\right]^-\right\} \tag{22}$$

in which the second formula has only a sextet of electrons associated with the carbon atom. A still more stable structure is one with the carbon atom in the center, with the resonance formulas

$$\left\{:\text{N}{\equiv}\text{C}{-}\overset{(-)}{\ddot{\underset{\cdot\cdot}{\text{O}}}:}, \quad \overset{(-)}{:\ddot{\text{N}}}{=}\text{C}{=}\ddot{\text{O}}:\right\} \tag{23}$$

Consideration of the ionic character of the bonds suggests that the second of the Lewis formulas shown may be about as favorable energetically as the first.

Experimentally, the ions CNO^- and NCO^- are both known to exist, but NOC^- is not. Of the known ions, the cyanate ion, NCO^-, is quite stable; the fulminate ion, CNO^-, is less stable and has explosive properties. The relative stabilities of these ions are therefore consistent with those expected from the formulas that may be written (Pauling and Hendricks, 1926).

It is, further, of interest to ask whether the sequence of atoms in cyanic acid is HNCO or NCOH. The resonance formulation of the first possibility is

$$\left\{\underset{(a)}{\text{H}{-}\overset{(+)}{\text{N}}{\equiv}\overset{(-)}{\text{C}}{-}\ddot{\underset{\cdot\cdot}{\text{O}}}:}, \quad \underset{(b)}{\text{H}{-}\ddot{\text{N}}{=}\text{C}{=}\ddot{\text{O}}:}, \quad \underset{(c)}{\text{H}{-}\overset{(-)}{\ddot{\underset{\cdot\cdot}{\text{N}}}}{-}\text{C}{\equiv}\overset{(+)}{\text{O}}:}\right\} \tag{24}$$

In (24, a), nitrogen has a positive formal charge and oxygen a negative one, the first being a destabilizing and the second a stabilizing factor. In (24, c) the situation is reversed, which is less favorable but not so unfavorable as to rule out this formula. This evaluation is not markedly changed by considering ionic contributions to bonding, and all three formulas are expected to contribute substantially to the resonance hybrid. On the other hand, for

NCOH we may write

$$\left\{ :N{\equiv}C{-}\ddot{\underset{\cdot\cdot}{O}}{-}H \;,\; :\overset{(-)}{\ddot{N}}{=}C{=}\overset{(+)}{\ddot{O}}{-}H \;,\; :\overset{(--)}{\ddot{\underset{\cdot\cdot}{N}}}{-}C{\equiv}\overset{(++)}{O}{-}H \right\} \quad (25)$$

$$\text{(a)} \qquad\qquad \text{(b)} \qquad\qquad \text{(c)}$$

where the contribution of (25, c) is expected to be negligible, while the stability of (25, b) is anticipated to be comparable to that of (24, c). Again, this evaluation is not significantly changed by considering the ionic contributions. Since (24) is favored, cyanic acid is expected to be HNCO, as indeed it is.

THE STRUCTURE OF OXYGEN COMPOUNDS

As was mentioned earlier, beginners frequently write strange formulas for molecules, and they have particular difficulty with oxygen atoms. Such incorrect formulas can be avoided by considering different possible arrangements in the light of the electroneutrality principle. I shall not discuss this in detail, but shall summarize by listing rules for oxygen-containing molecules:

1. Oxygen atoms are not directly linked to form chains. With the exception of ozone, O_3, no more than two oxygen atoms are ever linked directly, and even the linking of two is an unusual and relatively unstable situation that occurs in compounds called "peroxides."
2. In single molecules and ions containing three or more atoms, the central atom is usually not oxygen, but one of the other components. Illustrations are the chlorate ion, ClO_3^-, the cyanate ion, NCO^-, and the fulminate ion, CNO^-. Exceptions to this rule are H_2O, H_3O^+, OF_2, OCl_2, and OBr_2. Another kind of exception is that often there are several "central" atoms to which atoms of oxygen and of other elements are attached. The resulting groups may be bridged by an oxygen atom. An example is the pyrophosphate ion, the formula of which is

$$\left[(:\ddot{O}:)_3P{-}\ddot{O}{-}P(:\ddot{O}:)_3 \right]^{----} \quad (26)$$

rather than the stretched-out monstrosity used earlier as an example of how oxygens do *not* arrange themselves. Two similar characteristic examples from organic chemistry are dimethyl ether, $H_3C{-}O{-}CH_3$, and ethyl acetate, $H_3C{-}CH_2{-}O{-}CO{-}CH_3$.
3. When both hydrogen and oxygen atoms are present, the hydrogen atoms are often attached to the oxygen atoms, and can frequently be detached as H^+ ions with relative ease; that is, the molecule or ion in question represents an acid.

In conclusion, there is, of course, no single concept in structural chemistry that explains all the known facts and applies without exception. The foregoing discussion shows, however, that the electroneutrality principle is important and useful, and that it is simple enough to be taught to the beginner.

REFERENCES

Pauling, L. (1948a). *General Chemistry*. San Francisco: W. H. Freeman and Company; 2nd ed., 1953.

Pauling, L. (1950). *College Chemistry*. San Francisco: W. H. Freeman and Company; 2nd ed., 1955; 3rd ed., 1964.

Pauling, L. (1948b). *J. Chem. Soc.* 1461.

Pauling, L. (1960). *The Nature of the Chemical Bond*, 3rd ed. Ithaca: Cornell Univ. Press.

Pauling, L., and S. B. Hendricks (1926). *J. Am. Chem. Soc.* **48**, 641.

METALS AND MINERALS

STEN SAMSON
Gates and Crellin Laboratories of Chemistry
California Institute of Technology
Pasadena, California

The Structure of Complex Intermetallic Compounds

A short listing of the contents of this paper needs be given first—especially since part of it represents a synopsis of categories of Friauf structures, a classification that may be of some interest or use to the specialist. After a brief historical review of the subject (and, not surprisingly, before I reach the summary and conclusion), the following topics will be treated.

The Friauf Polyhedron and the Icosahedron
Three-Dimensional Close Packing of Friauf Polyhedra: the Two Principal Friauf Phases and the Laves Phase
Layers of Close-Packed Friauf Polyhedra: the μ Phases
Rows of Close-Packed Friauf Polyhedra: the P Phase
Isolated Friauf Polyhedra: the E Phase, αVAl_{10}, AB_{22} Compounds and $Cr_{23}C$
Infinite, Three-Dimensional Frameworks of Friauf Polyhedra
γ $Mg_{17}Al_{12}$, α Manganese, and the χ Phases
ϵ $Mg_{23}Al_{30}$ and the R Phases
$Mg_{32}(Zn,Al)_{49}$ and Related Compounds
The Determination of Cubic Metallic Structures of Extreme Complexity
The Stochastic Method
Structural Principles Associated with Friauf Polyhedra and Icosahedra
The Packing Map and Most Useful Planes
The Crystal Structure of βMg_2Al_3 and $NaCd_2$
On the Nature of the Disorder
The Crystal Structure of Cu_4Cd_3

The state of knowledge about the nature of intermetallic compounds during the early 1920's has been recorded by Jeffries and Archer (1924) in their book *The Science of Metals*: "The constitution of $CuAl_2$ might be represented by the formula

$$\begin{array}{c} Cu \\ \diagup \quad \diagdown \\ Al{=}Al \end{array}$$

. This would indicate that the crystals of this compound are built of molecules as units in the space lattice. It is probable that such is not the case, but that the units of the crystal lattice are the atoms of copper and

Contribution No. 3421 from the Gates and Crellin Laboratories of Chemistry. Acknowledgment is made to the National Science Foundation for its financial support (Grants GP-1701 and GP-4237), which enabled me to carry out a significant number of the investigations described here.

aluminum arranged in definite and repeating patterns, in which the atoms of copper and aluminum are not interchangeable."

At about the same time as this was written, Linus Pauling determined the crystal structure of the first intermetallic compound (1923). He reported in his paper that in crystals of Mg_2Sn eight magnesium atoms are placed around each tin atom at the corners of a cube and that four tin atoms surround each magnesium atom at tetrahedron corners; the structure is the one known as the calcium fluoride arrangement. The first experimental proof that the chemically different atoms arrange themselves in definite and repeating patterns was thus established. (The distinction between interchangeability and noninterchangeability of chemically different atoms has remained a matter of judgment.)

In the same paper (Pauling, 1923) mention was also made of a study of a crystal of $NaCd_2$, which was found to be cubic. The X-ray diffraction photographs were, however, so complicated that it was not then possible to assign indices with certainty to many of the spots. Up to the present this crystal exhibits the largest unit of structure that ever has been observed in intermetallic compounds; the unit cube was later found (Pauling, 1955) to have an edge length slightly over 30Å and to contain about 384 sodium atoms and 768 cadmium atoms.

It was a curious coincidence that this paper, which first described the structure of an intermetallic compound, presented simultaneously one of the most difficult structure problems that lay ahead. The awareness of the existence of this complex compound $NaCd_2$ stimulated a great deal of research on metallic structures, especially among Pauling's co-workers, since it seemed probable that the successful attack on the problem of determining the atomic positions would require knowledge of fundamental structural principles that had to be obtained gradually through structure analyses of less complicated compounds. Research activity on ever more complex structures thus developed; it has become especially intense during the last ten years.

Among the most interesting structures discovered in the course of the earlier investigations were those of $MgCu_2$ and $MgZn_2$, Friauf (1927a, 1927b), who also determined the atomic arrangement in $CuAl_2$. Two atomic groupings observed here, the Friauf polyhedron and the icosahedron, were later found to represent the most important structural elements in intermetallic compounds of extreme complexity—including $NaCd_2$.

THE FRIAUF POLYHEDRON AND THE ICOSAHEDRON

Each of these polyhedra can be derived through relatively simple modifications of the cubic closest-packed arrangement of spheres of equal size (face-centered cube). A configuration of sixteen such spheres is shown in Figure 1, *a*. Removal of a tetrahedron of four contiguous spheres from this aggregate results in the framework of twelve spheres, arranged about the corners of a truncated tetrahedron, as shown in Figure 1, *b*. The central cavity is capable of accommodating a thirteenth sphere with a radius 1.35 times that of the surrounding spheres, as shown in Figure 1, *c*. Out from the center of each of the four hexagons there is an additional sphere of the large kind (35 percent larger in radius), as shown in Figure 1, *d*. The central (large) sphere, accordingly, is surrounded by sixteen spheres, twelve small and four large ones.

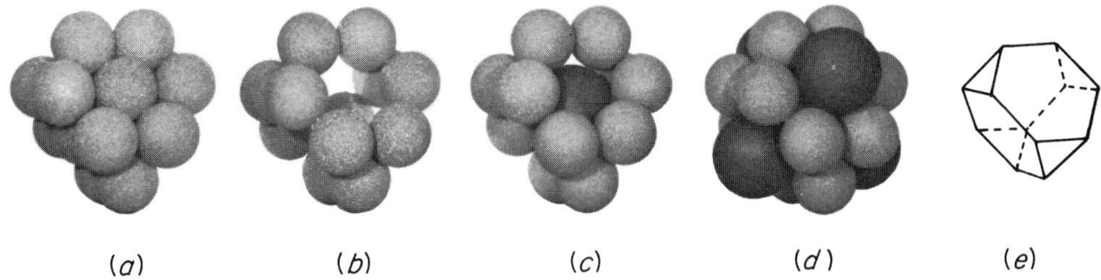

FIGURE 1.
Derivation of the Friauf polyhedron from an aggregate of sixteen spheres of equal size arranged in the cubic closest packing. (*a*) The group of sixteen spheres. (*b*) and (*c*) The truncated tetrahedron. (*d*) The aggregate of seventeen spheres referred to as the Friauf polyhedron. (*e*) A formal representation of the Friauf polyhedron.

This group of seventeen spheres is called a Friauf polyhedron. It consists, as we see, of two integral parts: (1) the truncated (say, positive) tetrahedron bounded by four hexagons and four triangles (twelve vertices); (2) the regular (negative) tetrahedron (four vertices, representing four large atoms).

In the formal representation of the Friauf polyhedron shown in Figure 1, *e*, the spheres out from the centers of the hexagons forming the regular, negative tetrahedron are not indicated, since, in most instances, each such sphere is shared between two adjacent Friauf polyhedra or between a Friauf polyhedron and a different kind of coordination shell.

Figure 2, *a* shows the conventional representation of the cubic closest packing of fourteen spheres arranged at the lattice points of a face-centered cubic lattice. If the origin of the cube is translated one-half of the length of the cube edge along any one of the three axes or one-half the body diagonal, the configuration shown in Figure 2, *b* is brought into view. It is seen (Figure 2, *c*) that the centers of the twelve spheres that surround the central one are located at corners of three mutually perpendicular squares to form the cubo-

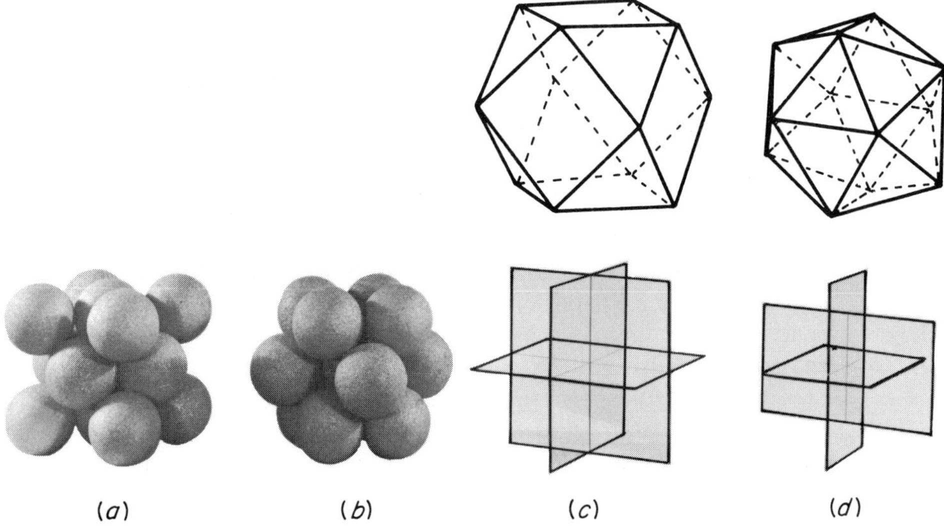

FIGURE 2.
Derivation of the centered icosahedron from a group of thirteen spheres arranged in the cubic closest packing. (*a*) Fourteen spheres arranged about the lattice points of a face-centered cubic lattice. (*b*) The group of thirteen spheres as seen after translation of the origin by $a/2$. (*c*) The cubo-octahedron. (*d*) The icosahedron.

octahedron shown at the top of Figure 2, *c*; it is bounded by eight equilateral triangles and six squares.

The three mutually perpendicular squares in Figure 2, *c* have been replaced by rectangles in Figure 2, *d*; the sides of the rectangles are a and $b = 1.62a$. This results in a polyhedron bounded by twenty equilateral triangles—an icosahedron—shown at the top of Figure 2, *d*. Each corner of this polyhedron is connected with five other corners that form a plane pentagon. Each pentagon has the side a, and there are five longest corner-to-corner distances across the pentagon, each one of length $b = 1.62a$—that is, equal to the long side of the rectangle. There are, accordingly, five different orientations in which the set of three mutually perpendicular rectangles can be fitted into the upper part of Figure 2, *d*.

The icosahedron has 15 twofold axes (30 edges), 10 threefold axes (20 triangles), and 6 fivefold axes (12 vertices), as well as 15 planes of symmetry and other elements of the second kind.

The conversion of the cubo-octahedron, which has 24 nearest-neighbor distances or edges of length a, into an icosahedron of 30 nearest-neighbor distances of the same edge length a results in a shortening of the center-to-vertex distance of 4.9 percent and a corresponding decrease in volume of slightly more than 7 percent. Accordingly, with twelve contiguous spheres of equal size at the vertices, the central sphere of the icosahedron has to be nearly 10 percent smaller in radius.

Of the two coordination shells of ligancy twelve discussed here, surrounding a central sphere that may become 10 percent smaller, the icosahedron is the smaller and hence corresponds to lower energy or higher stability. This is probably the most important feature of the icosahedron.

THREE-DIMENSIONAL CLOSE PACKING OF FRIAUF POLYHEDRA: THE TWO PRINCIPAL FRIAUF PHASES AND THE LAVES PHASE

Figure 3, *a* shows a layer of Friauf polyhedra arranged so as to fill a plane. Each truncated tetrahedron shares three of its four hexagons with three other Friauf polyhedra; the unshared fourth hexagon will be shared by a Friauf polyhedron of the next layer that will be superimposed on this one. It is now seen that each large atom out from the center of a hexagon represents, in turn, the center of an adjacent Friauf polyhedron.

In the close-packed, three-dimensional arrangement of Friauf polyhedra, each truncated tetrahedron shares each of its twelve vertices (small atoms) with six other truncated tetrahedra so as to reduce the average number of small atoms per Friauf polyhedron to two and the number of large atoms to one. The resulting composition of the crystal is AB_2, where A represents a large atom and B a small one.

The three principal structures exhibiting closest packing of Friauf polyhedra are shown in Figures 3, *b*, *c*, and *d*; they differ from each other only in the manner in which the layers are superimposed on one another.

In Figure 3, *b*, each Friauf polyhedron of the second layer is turned 60° with respect to the one of the first layer, but in Figure 3, *c* the second Friauf polyhedron is a mirror image of the first one. Figure 3, *b* represents the structure of $MgCu_2$ (Friauf, 1927a), type *C* 15; it is cubic, with $a = 7.04$ Å, and has 24 atoms per unit cell.

Figure 3, *c* represents the $MgZn_2$ structure (Friauf, 1927b), type *C* 14, which is hexagonal, with $a = 5.18$ Å and $c = 8.52$ Å, and has 12 atoms per unit cell.

Figure 3, *d* represents the structure of the Laves phase $MgNi_2$ (Laves and Witte, 1935), type *C* 36; the first and second layers as well as the third and fourth layers are related to one another by mirror planes, but not the second and third. This arrangement is a combination of the $MgCu_2$ and $MgZn_2$ types (Friauf phases), which results in a hexagonal cell of $a = 4.82$ Å and $c = 15.80$ Å, with 24 atoms per unit cell.

We can think of a large variety of combinations of sequences from the *C* 14 and *C* 15 structures that lead to an ever increasing length of the *c*-axis. Komura (1962) found two modifications of crystals of approximate composition MgCuAl, one with a hexagonal unit cell with $a = 5.14$ Å and $c = 21.05$ Å, the other with a rhombohedral cell, which corresponds to a hexagonal cell with $a = 5.14$ Å and $c = 37.89$ Å. Both structures are combinations of the two Friauf types.

The representations given in Figure 3 show the atomic configurations around the large atoms (the magnesium atoms) only. The coordination polyhedron around each small atom (Cu, Zn, or Ni) is an icosahedron; any centered rectangle of the sides *a* and $b \sim 1.6a$ that can be traced out in Figures 3, *b*, *c*, and *d* belongs to an icosahedron. In fact, each of the three structures has twice as many icosahedra as Friauf polyhedra. The two kinds of polyhedra interpenetrate in such a way that each icosahedron has six magnesium atoms and six small atoms (Cu, Zn, or Ni) at the vertices and one

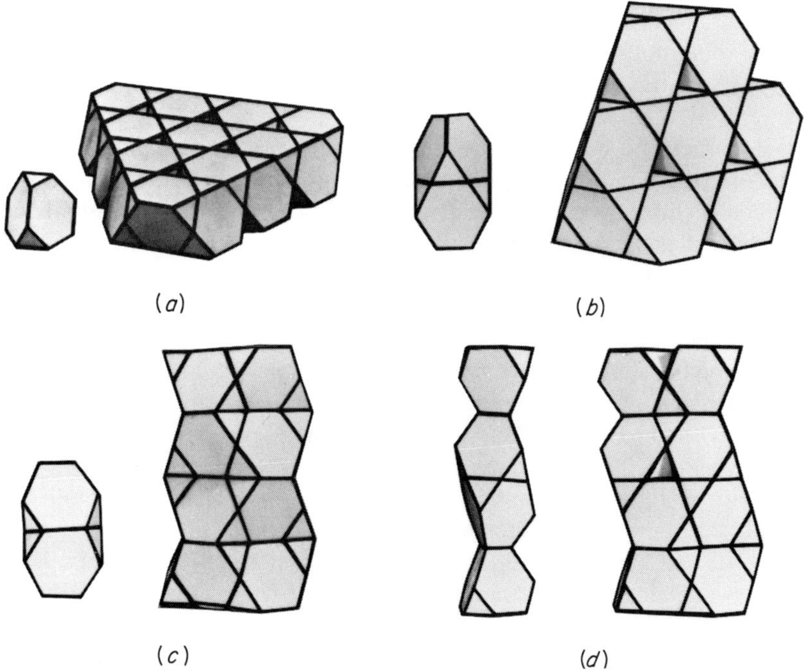

FIGURE 3.
Close packing of Friauf polyhedra. (*a*) A close-packed layer of Friauf polyhedra. (*b*) The $MgCu_2$ structure, type *C* 15. (*c*) The $MgZn_2$ structure, type *C* 14. (*d*) The $MgNi_2$ structure, type *C* 36.

small atom at the center. This distribution of atoms causes the icosahedra to be slightly deformed, and now the *average* size of the vertex atoms is such as to correspond to a central atom that is approximately 10 percent smaller. Deformation provides flexibility with regard to the *effective* size of the atom that is to be accommodated at the center, and in all three structure types the deformations are of slightly different nature. The predicted radius ratios for Mg/Cu, Mg/Zn, and Mg/Ni are 1.28, 1.19, and 1.31, if allowance is made for the proper coordination numbers (Pauling, 1947). Today several hundred intermetallic compounds of these three types of structures are known.

LAYERS OF CLOSE-PACKED FRIAUF POLYHEDRA: THE μ PHASES

Figure 4 represents the structure of a series of intermetallic compounds often referred to as the μ phases. The known representatives are W_6Fe_7, W_6Co_7, Mo_6Fe_7, and Mo_6Co_7. The atoms of the large kind are tungsten and molybdenum and those of the small kind are iron and cobalt; the nominal radius ratios $L16/L12$ (L = ligancy or coordination number) range from 1.13 to 1.14. The structure (type $D8_5$) was discovered by Arnfelt and Westgren (1935). It has a rhombohedral unit cell, which corresponds to a hexagonal cell with a = 4.76 Å and c = 25.83 Å (for W_6Fe_7), containing 39 atoms.

The bottom layer of polyhedra in the model shown to the right in Figure 4 is identical with the layer shown in Figure 3, *a*. Here, one hexagon of each truncated tetrahedron is shared with a hexagonal antiprism of the layer above it (dark). The third layer of polyhedra (also dark) consists, again, of hexagonal antiprisms; and, finally, the fourth layer consists, again, of Friauf polyhedra. Each Friauf polyhedron in this layer is turned 60° relative to that in the first layer. Accordingly, the μ-phase structure can be derived by inserting two contiguous layers of hexagonal antiprisms (dark in Figure 4) between each two contiguous layers of Friauf polyhedra in the $MgCu_2$ structure (Figure 3, *b*).

Each corner of a hexagon that is shared between two contiguous hexagonal antiprisms (that are on top of one another) represents the center of a coordination shell of ligancy 15. This shell may be described as a truncated trigonal prism bounded by three hexagons and eight triangles, four above and four below. Out from the center of each hexagon is a large atom, which often represents the center of another such L 15 shell that shares one of its hexagons with this one. For the sake of brevity, this polyhedron shall be called the μ-phase polyhedron; two of them are shown at the top of Figure 4.

The two layers of hexagonal antiprisms (dark in Figure 4) can, in fact, be replaced with one layer of close-packed μ-phase polyhedra, in which each such polyhedron shares its three hexagons with three others. Accordingly, the dark portion of Figure 4 consists of two kinds of interpenetrating polyhedra, and only one kind can be shown at a time. The center of each μ-phase polyhedron lies out from the center of an equilateral triangle of a Friauf polyhedron, while the center of each hexagonal antiprism lies above a hexagon of a Friauf polyhedron.

This arrangement results in the following sequence of polyhedra in the vertical direction (hexagonal c-axis): (1) Friauf polyhedron; (2) hexagonal antiprism; (3) hexagonal antiprism; (4) Friauf polyhedron; (5) μ-phase poly-

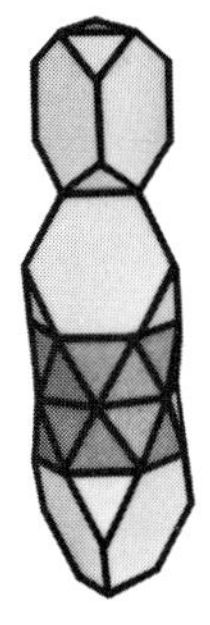
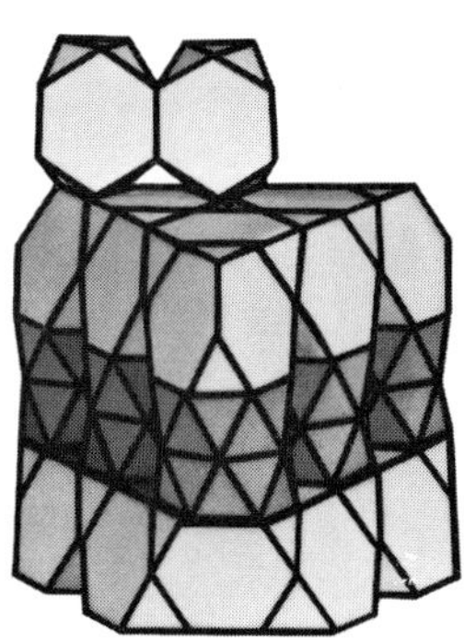

FIGURE 4.
The μ-phase structure, type $D8_5$. Layers of close-packed Friauf polyhedra (light) are separated by layers of hexagonal antiprisms (dark).

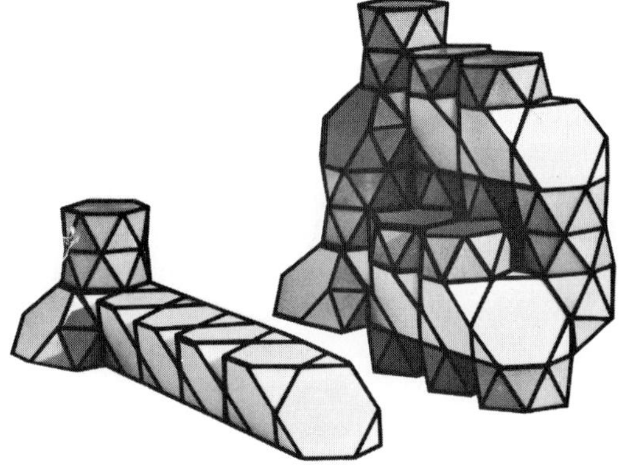

FIGURE 5.
The essential part of the P-phase structure. The Friauf polyhedra form rows parallel with the z-axis (*left*). The rows are connected with each other via μ-phase polyhedra and hexagonal antiprisms (*right*).

hedron; (6) icosahedron; (7) μ-phase polyhedron; (8) Friauf polyhedron. The same sequence starts over again at (8), which is identical to (1); the repeat distance is 25.83 Å.

The icosahedron (6) is not shown in Figure 4. Its center represents the vertex of a truncated tetrahedron of the sixth layer. Each truncated tetrahedron is interpenetrated by twelve icosahedra, and each of these has about half its vertices occupied by small atoms and the other half by large ones, a feature that is also observed in the three structure types described in the preceding section.

ROWS OF CLOSE-PACKED FRIAUF POLYHEDRA: THE P PHASE

If molybdenum is alloyed with nickel and chromium instead of with cobalt, as was the case above, the same kind of polyhedra are formed, but the arrangement of these polyhedra with respect to one another is drastically different. This new type of compound is called the P phase (Rideout et al., 1951). The structure was determined by Shoemaker et al. (1957); it is orthorhombic, with $a = 9.07$ Å, $b = 16.98$ Å, and $c = 4.75$ Å. The unit cell contains 56 atoms, which are distributed among twelve crystallographically different positions, partially in substitutional disorder. Therefore a chemical formula is not given.

The essential part of the structure is shown in Figure 5. It is seen that here the Friauf polyhedra are arranged in rows, connected with one another via μ-phase polyhedra and hexagonal antiprisms. Each truncated tetrahedron shares two of its hexagons with two contiguous truncated tetrahedra; the third hexagon is shared with a μ-phase polyhedron and the fourth with a hexagonal antiprism. This arrangement results in a slight reduction of the number of icosahedra that interpenetrate the Friauf polyhedra; only ten vertices of each truncated tetrahedron represent centers of icosahedra, and each of the remaining two vertices is at the center of a hexagonal antiprism that has two atoms at the extended poles (ligancy 14).

It is of interest to note that in this structure the edge length of the hexagon of the μ-phase polyhedron equals that of the truncated tetrahedron, whereas in the μ-phase itself the edge length is $2/\sqrt{3}$ times that of the truncated tetrahedron, if the hexagon is assumed to be regular. Consequently, the overall size of the μ-phase polyhedron shown in Figure 4 is significantly larger than that shown in Figure 5.

The δ phase (Shoemaker and Shoemaker, 1963) seems to be very closely related to the P phase; it contains the same number and kinds of coordination shells per unit cell although it is pseudotetragonal with $a = b = 9.108$ Å, $c = 8.852$ Å, space group $P2_12_12_1$, 56 atoms per cell. Unfortunately, time did not allow me to build a three-dimensional model to reveal the structural details adequately.

ISOLATED FRIAUF POLYHEDRA: THE E PHASE, αVAl_{10}, AB_{22} COMPOUNDS, AND $Cr_{23}C$

If none of the four hexagons of the truncated tetrahedron is shared with another truncated tetrahedron, the Friauf polyhedron shall be regarded as isolated.

The first metallic structure that was found to exhibit such a feature was that of $Mg_3Cr_2Al_{18}$, the E phase (Erdmann-Jesnitzer, 1940). The structure is cubic, with $a_0 = 14.53$ Å and with 184 atoms per unit cube (Samson, 1958). Each truncated tetrahedron is here connected with four others via four hexagonal prisms that have their prism axes directed toward the corners of a regular tetrahedron, as shown in Figure 6. Out from the center of each approximately square prism face of the hexagonal prism is an aluminum atom that fills part of the cavity; six such atoms form a regular octahedron, which has its center at the center of the cavity. Each chromium atom is shared between two cavities and is at the center of a distorted icosahedron. Hence, the structure incorporates polyhedra of four different sizes: (1) the truncated tetrahedron, which is already known to be appropriate for a central sphere up to 35 percent larger in radius than the spheres at the vertices; (2) the hexagonal prism, which is appropriate for a central sphere that is about 20 percent larger in radius; (3) the icosahedron, which may accommodate a central sphere that is about 10 percent smaller; (4) the pentagonal prisms (around vertices of Friauf polyhedra) with two atoms at the extended poles, corresponding to a radius ratio of nearly unity. At the centers of (1) are magnesium atoms (the largest); at the centers of (2) are probably magnesium and aluminum atoms in substitutional disorder; at the centers of (3) are chromium atoms, which have the smallest metallic radius of the three components; and at the centers of (4) are aluminum atoms, which are of intermediate size.

The intermetallic compound αVAl_{10} (Brown, 1957) is isostructural with $Mg_3Cr_2Al_{18}$ but differs from the latter one inasmuch as it does not contain any atom of the large kind (vanadium atoms are comparable in size to chromium atoms). In fact, the Friauf polyhedra are found to be empty, and the centers of the hexagonal prisms are occupied by aluminum. The formula of this phase may be written $(Al_2, hole)_3V_2Al_{18}$. It seems likely that if in αVAl_{10} some of the aluminum atoms (3-valent) at the centers of the hexagonal prisms were replaced by magnesium (2 valent), the Friauf polyhedra would fill up so as to keep the sum of the valences of the magnesium atoms and the aluminum atoms constant; a complete exchange of these aluminum atoms by magnesium would result in a loss of two metallic valences per Friauf polyhedron, which probably could be made up through addition of magnesium. On the other hand, replacement of part of the magnesium atoms inside the hexagonal prisms by aluminum in the ternary compound ($Mg_3Cr_2Al_{18}$) may bring about partial occupancy of the centers of the Friauf polyhedra; this compound was found, indeed, to be of variable composition with considerable variation in the cell edge (Little et al., 1948). Detailed X-ray studies of single crystals of varying compositions of the two compounds will be necessary to clarify the situation. Such investigations are contemplated.

A third representative type of structure, shown in Figure 6, is $ZrZn_{22}$ (Samson, 1961). Each zirconium atom is at the center of a Friauf polyhedron; the metallic radius of zirconium is almost identical to that of magnesium. An interesting feature here is the large stoichiometric ratio, corresponding to about 4.3 atomic percent zirconium. The few zirconium atoms cause the zinc atoms to arrange themselves in polyhedra of significantly different sizes. Comparison of this compound with $Mg_3Cr_2Al_{18}$ suggests that the zinc atoms are of different sizes. The large zinc atoms take the places of two-thirds of the magnesium atoms inside the hexagonal prisms, the small zinc atoms take the places of the chromium atoms inside the icosahedra, and the zinc atoms of intermediate size take the places of the aluminum atoms (pentagonal prisms and two atoms at the extended poles). We are led to conclude that the zinc atoms differ from one another in the crystallographic sense as well as in the chemical sense. In terms of the valence-bond theory of metals and intermetallic compounds (Pauling, 1949), the $ZrZn_{22}$ phase may be regarded as a quaternary compound of the kind $Zr^{\sim 4}Zn_2^{\sim 3}Zn_2^{\sim 5}Zn_{18}^{\sim 4}$, in which the superscripts indicate the approximate metallic valences.

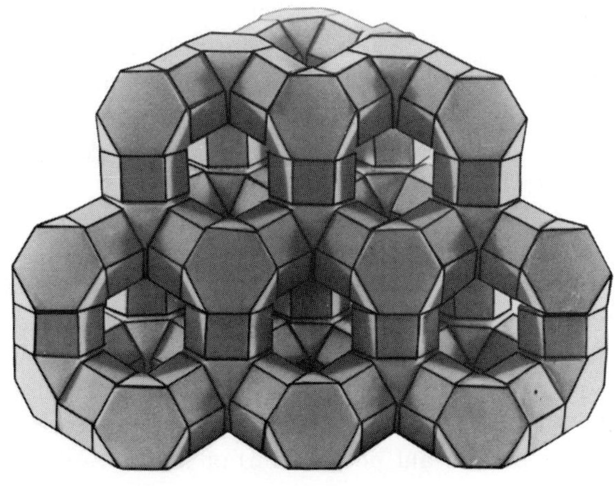

FIGURE 6.
The arrangement of Friauf polyhedra in $Mg_3Cr_2Al_{18}$, αVAl_{10}, $ZrZn_{22}$, and $MeBe_{22}$, in which Me represents molybdenum, rhenium, or tungsten.

A series of compounds $MeBe_{22}$, in which Me represents molybdenum, rhenium, or tungsten, were shown to be isostructural with $ZrZn_{22}$ (Sands et al., 1962).

Figure 7 represents the structure of $Cr_{23}C_6$ (Westgren, 1933), type $D8_4$. It is seen that each truncated tetrahedron is shared between four cubo-octahedra of one kind (large) and four cubo-octahedra of a second kind (small). Each large cubo-octahedron may be regarded as an octahedron truncated by a cube; each small cubo-octahedron represents a cube truncated by an octahedron. This structure has the interesting feature that it does not incorporate any icosahedral coordination shell. True intermetallic compounds having the $A_{23}B_6$ formula, such as $Mn_{23}Th_6$ (Florio et al., 1952) crystallize in the same space group, but the atomic arrangement (type D8a) differs from that of $Cr_{23}C_6$; it is based on icosahedral packing.

The E-phase structure, as well as the $D8_4$ type, is unique inasmuch as none of the Friauf polyhedra is interpenetrated by an icosahedron. It seems that *the isolated Friauf polyhedron represents the only case in which interpenetration with icosahedra does not occur.*

INFINITE, THREE-DIMENSIONAL FRAMEWORKS OF FRIAUF POLYHEDRA

There exist a variety of ways in which the truncated tetrahedra, by sharing hexagons or hexagons and triangles, join together to form infinite, three-dimensional frameworks, and the types of structure that result are very diverse and often very complex. In fact, the most complex metallic structures known to date contain such frameworks, and their determination was made possible largely through studies of the kind presented in this and the preceding sections. Structures of extreme complexity, although belonging to this section, will be discussed later under separate headings, subsequent to the description of the techniques used to apply effectively the information that has been obtained through the studies presented here.

γ $Mg_{17}Al_{12}$, α Manganese, and the χ Phases

The structure of the γ phase in the magnesium-aluminum system, often referred to as the $Mg_{17}Al_{12}$ phase, consists of two different kinds of Friauf polyhedra, which will be called *F1* and *F2*. Each polyhedron of the kind *F1*, the dark one in Figure 8, shares its four hexagons with four polyhedra of the kind *F2* (light), which are arranged about the vertices of a regular tetrahedron; see the left side of Figure 8. These complexes of five Friauf polyhedra (*F1* + 4*F2*) are then arranged about the lattice points of a body-centered cube, as shown on the right side of Figure 8. Each of the four triangles of the dark polyhedron is shared with a triangle of a light polyhedron; accordingly, each complex is connected with four others.

Each of the three atoms out from the center of a hexagon of *F2* (light) that is not shared with *F1* (dark) represents simultaneously a vertex of another truncated tetrahedron of the kind *F2* (light), as shown in Figure 9. This vertex is seen to fill two disparate functions, and it seems possible to assume here either an atom of the large kind (magnesium, which is usually observed out from the center of a hexagon of a Friauf polyhedron) or an atom of the

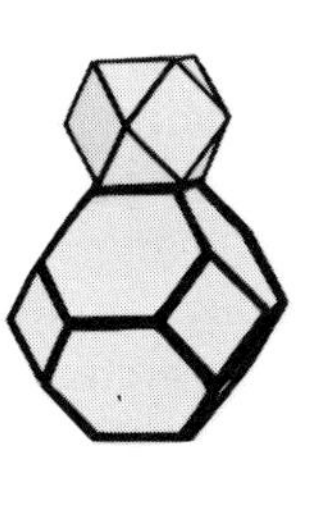
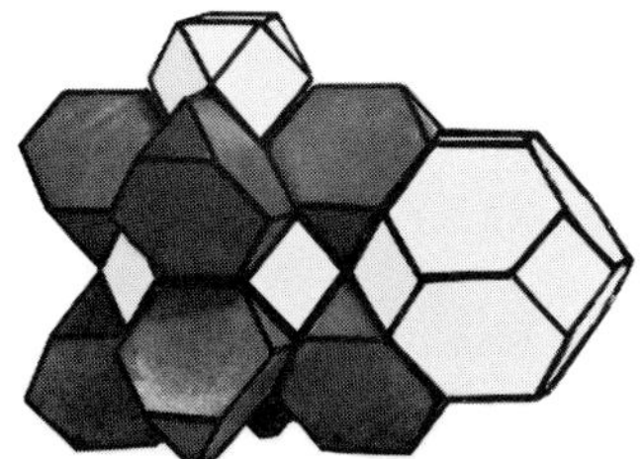

FIGURE 7.
The structure of $Cr_{23}C_6$, type $D8_4$.

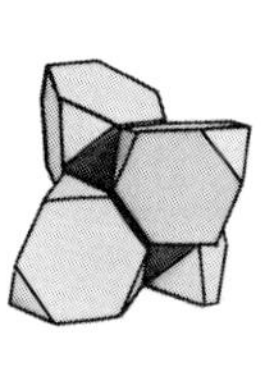
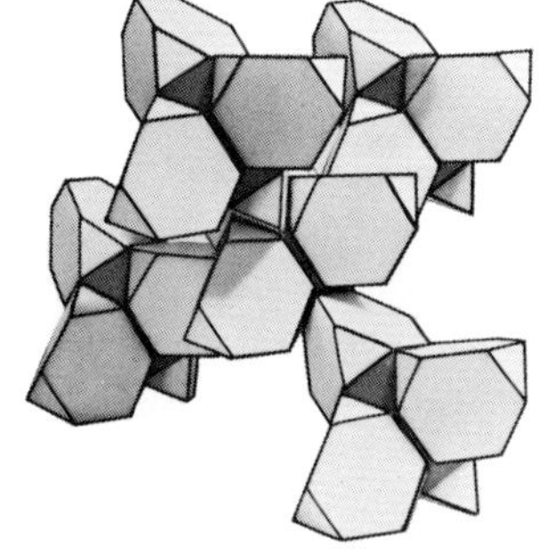

FIGURE 8.
The arrangement of Friauf polyhedra in the γ(MgAl) phase, often called $Mg_{17}Al_{12}$; type *A* 12.

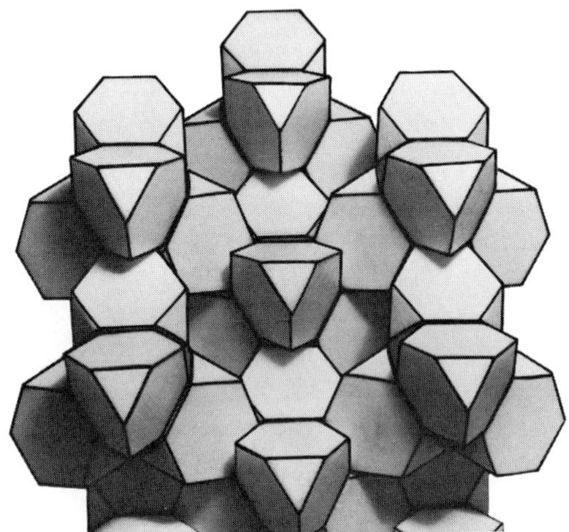

FIGURE 9.
The γ(MgAl) structure with all Friauf polyhedra in place. Here the hybrid vertices (see text) are brought into view.

small kind (aluminum, since it is at the vertex of a truncated tetrahedron). We are, in this case, confronted with a hybrid feature in the structural sense. There are 24 such hybrid vertices in the unit of structure, which is a body-centered cube of edge length 10 Å, containing 58 atoms. The composition $Mg_{17}Al_{12}$ assumed by Laves et al. (1934), who determined the structure, is the one that corresponds to 100 percent occupancy of the bifunctional or hybrid vertices by large atoms (magnesium). The coordination polyhedron around this vertex provides the intermediate ligancy *L* 13, while each aluminum atom is at the center of an icosahedron (*L* 12).

The unit cube of structure contains 10 Friauf polyhedra, 24 coordination polyhedra of ligancy 13, and 24 icosahedra. Each truncated tetrahedron of the dark kind (*F1*) is penetrated by 12 icosahedra and each of the light kind

($F2$) by 9 icosahedra and three L 13 shells. In the idealized ordered case (composition $Mg_{17}Al_{12}$), each icosahedron has three small atoms (aluminum) and as many as nine large atoms (magnesium) at the vertices, an unusually high number as compared to any other structure incorporating Friauf polyhedra and icosahedra.

Riederer (1936) found that this phase is of variable composition. Accurate measurements on single crystals of two drastically different compositions were made by me; the homogeneity range extends at least from $Mg_{13}Al_{16}$, with $a_0 = 10.40$ Å, to $Mg_{17}Al_{12}$, with $a_0 = 10.60$ Å. It is believed that the structural disorder (random occupancy) that must be associated with the variable composition is localized at the hybrid vertices; probably accurate structure analysis of crystals with varying compositions, now in progress, will provide an answer to this question.

In the several types of structures thus far discussed the Friauf polyhedra seem to owe their presence almost entirely to spatial requirements; the nominal radius ratio Mg/Al is 1.14. One exception is the $\alpha(VAl)_{10}$ phase. Another exception is the structure of alpha manganese (Bradley and Thewlis, 1927). Although here the nominal radius ratio is unity, the structure is the same as that of $Mg_{17}Al_{12}$.

Of the 58 manganese atoms in the body-centered cubic cell of edge length 8.91 Å, 10 replace the magnesium atoms inside the Friauf polyhedra, 24 replace the magnesium-aluminum atoms at the hybrid vertices, and 24 replace the aluminum atoms at the centers of the icosahedra. It thus appears that the manganese atoms are of three drastically different sizes; large, intermediate, and small. In terms of the valence-bond theory (Pauling, 1947, 1949), the manganese atoms are, in the chemical sense, of three different kinds: low-valent, normal-valent, and high-valent. Hence, alpha manganese may be regarded as a ternary intermetallic compound.

The χ phases occurring in the systems Mo—Cr—Fe (Kasper, 1954), Re—Cb, Re—Mo, Re—Ta, and Re—W (Greenfield and Beck, 1956) have atomic arrangements identical to those of $Mg_{17}Al_{12}$ and αMn.

ϵ $Mg_{23}Al_{30}$ *and the R Phases*

Annealing of γ MgAl samples rich in aluminum (composition around $Mg_{13}Al_{16}$) at about 350°C results in a conversion* of the above structure into a different one, referred to as the ϵ phase; γ-phase samples rich in magnesium (close to $Mg_{17}Al_{12}$) remain unaltered even after prolonged annealing. The powder X-ray photographs of the ϵ phase are too complex to be indexed with certainty, and single crystals are hard to obtain. Consequently, the nature of this phase remained unknown for about thirty years. Through annealing of a number of samples for about 2700 hours at various temperatures I finally obtained good single crystals. The determination of the structure was completed recently by Samson and Gordon (1967).

The structure consists of an infinite, three-dimensional framework of Friauf polyhedra that are arranged along the edges of a rhombohedral unit cell with $a_0 = 10.33$ Å and $\alpha = 76.4°$, as shown in Figure 10, c. This cell

*The term "transformation" would be misleading, since the γ phase splits up into an eutectoid mixture, which subsequently reacts to form the ϵ phase.

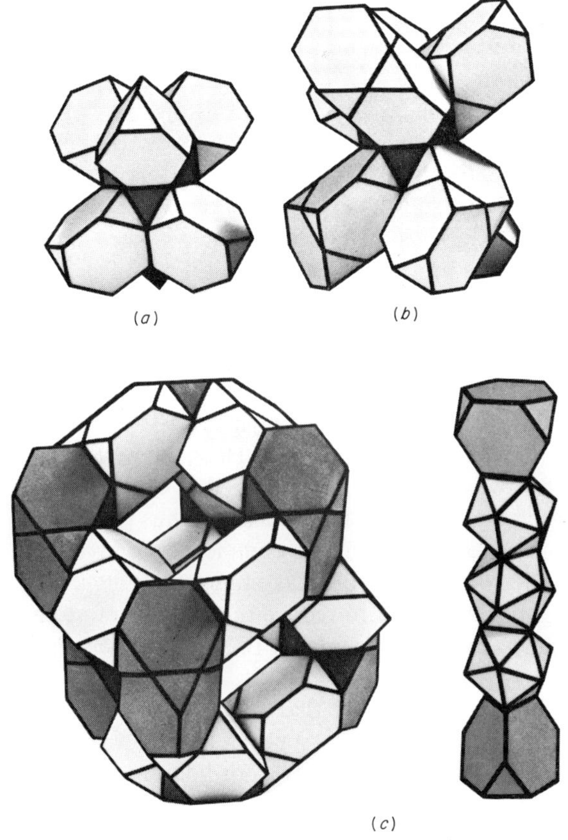

FIGURE 10.
(*a*) and (*b*) The arrangement of Friauf polyhedra in $Mg_{23}Al_{30}$. (*c*) The rhombohedral cell of $Mg_{23}Al_{30}$ and the arrangement of polyhedra along the threefold axis of the rhombohedron in the R phases (*right*).

contains one formula unit of $Mg_{23}Al_{30}$. The hexagonal cell that corresponds to the rhombohedral cell has the size a = 12.83 Å and c = 21.75 Å. The cavity (Figure 10, *c*) is filled by polyhedra other than those of the Friauf type; most of them are icosahedra. The orientation of the Friauf polyhedra (two kinds, dark and light) with respect to one another is shown in Figures 10, *a* and b.

The unit rhombohedron contains 8 Friauf polyhedra, 24 icosahedra, and 21 more-or-less irregular coordination shells, six of which resemble closely the μ-phase polyhedron (*L* 15) discussed previously. Each truncated tetrahedron (integral part of the Friauf polyhedron) of the dark kind (Figure 10) is penetrated by 12 icosahedra, and each truncated tetrahedron of the light kind by 10 icosahedra, one coordination shell of ligancy 14, and one of ligancy 11. Each icosahedron has about half of its vertices occupied by large atoms (magnesium) and the other half by small atoms (aluminum).

The structure was derived simply by fitting into the rhombohedral cell the maximum number of Friauf polyhedra, and there was very limited reasonable latitude for such atomic arrangements to fill the space. The presence of Friauf polyhedra was anticipated on the basis of the phase "conversion" discussed above: $\gamma\ Mg_{\sim 13}\ Al_{\sim 16} \longrightarrow \epsilon\ Mg_{23}Al_{30}$.

At a later date, the $Mg_{23}Al_{30}$ structure was found to be almost isostructural with the R phase in the Mo—Co—Cr system, which is rhombohedral with

$a = 9.01$ Å, $\alpha = 74°28'$ (Komura et al., 1960). There are, however, significant differences in some of the structural parameters, and these result in a drsatic alteration of some of the coordination polyhedra inside the cavity shown in Figure 10, *c*. The arrangement of polyhedra along the threefold axis of the rhombohedron in the R phase is: Friauf polyhedron—icosahedron (1)—icosahedron (2)—icosahedron (3)—Friauf polyhedron; see to the right of Figure 10, *c*. In $Mg_{23}Al_{30}$, the polyhedron (2) is significantly wider and the atoms at the centers of (1) and (3) are in contact with the atom at the center of (2). Thus, in $Mg_{23}Al_{30}$, (1) and (3) are shells of ligancy 13, (2) of ligancy 14, and there are very likely large atoms (magnesium) at the three centers.

$Mg_{32}(Zn,Al)_{49}$ *and Related Compounds*

One of the most interesting arrangements of Friauf polyhedra is observed in the structure of $Mg_{32}(Zn,Al)_{49}$ (Bergman et al., 1957). Here, complexes of 113 atoms each are formed by twenty Friauf polyhedra that are arranged with their centers at the vertices of a pentagonal dodecahedron, thus producing the truncated icosahedron shown in Figure 11. These complexes are arranged about the lattice points of a body-centered cube, as shown in Figure 12; each one shares atoms with eight others so as to reduce the number of atoms per complex to 81. The unit cube accordingly contains 162 atoms; the length of the cube edge is 14.16 Å.

Although it appears from Figures 11 and 12 that the Friauf polyhedra dominate the structure, they are far outnumbered by icosahedra; there are 40 Friauf polyhedra, 98 icosahedra, and 24 irregular coordination shells of ligancy 14 and 15. In fact, when Linus Pauling worked out the structure he anticipated the icosahedron but not the Friauf polyhedron. He recognized that in the less complex structures of $Mg_2Cu_6Al_5$ and Mg_2Zn_{11} [primitive cubic, $a_0 = 8.31$ Å and $a_0 = 8.55$ Å, 39 atoms per cube; Samson (1949a, 1949b)], icosahedral packing is maintained through addition of atoms in successive shells about a central icosahedron in such a manner that they always center the triangles of previous shells. For the details of this structure determination refer to Pauling (1955, 1964).

At the time of the derivation of this structure computer facilities were not readily available; nevertheless, the structure was refined by full-matrix, least-squares calculations with the use of a desk calculator. This was a very tedious process, especially because several positions were occupied randomly by zinc and aluminum, and additional parameters determining the distribution of these atoms had to be included in the matrix; the calculations took more than a year.

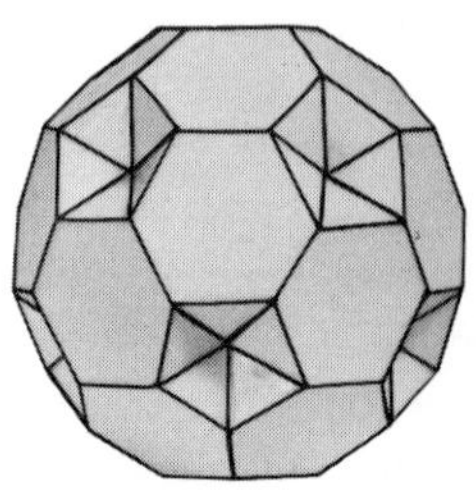
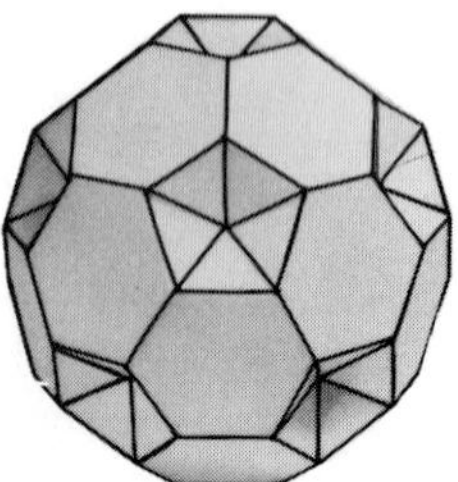

FIGURE 11.
The truncated icosahedron representing a complex of 113 atoms. It is formed by twenty Friauf polyhedra that are arranged with their centers at the vertices of a pentagonal dodecahedron.

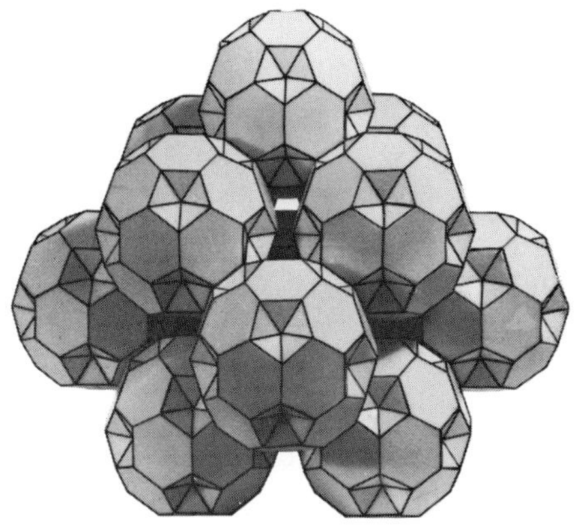

FIGURE 12.
The truncated icosahedra shown in Figure 11, when arranged about the lattice points of a body-centered cube, produce the structure of $Mg_{32}(Zn,Al)_{49}$.

The intermetallic compounds Mg_4CuAl_6 (Laves et al., 1935), Li_3CuAl_5, and $Li_{32}(Zn,Al)_{49}$ (Cherkashin et al., 1964) seem to be isostructural with $Mg_{32}(Zn,Al)_{49}$.

The 20 Friauf polyhedra forming the complexes shown in Figures 11 and 12 are, in the crystallographic sense, of two different kinds: one of them is penetrated by 12 icosahedra, the other by 10 icosahedra and two coordination shells of ligancy 14. All except two of the 98 icosahedra per unit cube have about half their vertices occupied by large atoms (magnesium) and the other half by small atoms (zinc and aluminum). Most of the small atoms occur in substitutional disorder.

THE DETERMINATION OF CUBIC METALLIC STRUCTURES OF EXTREME COMPLEXITY

During the period of research covered in the preceding sections, two additional intermetallic compounds of extreme complexity had become known: β Mg_2Al_3 (Perlitz, 1944, 1946) and Cu_4Cd_3 (Samson, 1965). Each of them has a cubic structure, β Mg_2Al_3 with a cell edge of length $a_0 = 28.239$ Å and with 1168 atoms per smallest cubic unit (Samson, 1965), and Cu_4Cd_3 with $a_0 = 25.871$ Å and with 1124 atoms per smallest cubic unit (Samson, 1967). The more accurate length of the cell edge of $NaCd_2$ was found to be $a_0 = 30.56$ Å (Samson, 1962).

Attempts were made by several investigators to solve the structure of $NaCd_2$ by the stochastic method (Pauling, 1955) and also by the use of three-dimensional Patterson functions. To derive the trial structure by building three-dimensional models seemed a formidable task, and the problem of unraveling many more than half a million Patterson vectors (about 700,000) seemed hopeless. The lack of success during the earlier period of this research, therefore, was not surprising.

The Stochastic Method

Prior to the discovery of the Patterson function, many structure problems were attacked and solved in a manner equal or similar to that described by Bragg and West (1926). This method utilizes the idea to determine all the possible positions of the atoms in the unit cell relative to their symmetry elements through evaluation of the domains of neighbors that must not overlap. It is well known that if a structure is sufficiently simple it can be determined in this way by straightforward, completely logical arguments. If a structure is of considerable complexity, however, the purely geometrical reasoning can only serve as a guide in the search for a reasonable structural motif, and guesses have to be made as to the structural elements (coordination polyhedra and others) as well as to additional features that may be exhibited in the structure. This method of treating structure problems was termed the *stochastic method* by Pauling (1933, 1955).

The *stochastic hypothesis* predicts the existence of certain facts or connections of facts that are immediately verifiable. In the case of crystal structures, verification is produced by the agreement between observed and calculated intensities of X-rays (or neutrons) diffracted from the crystal used. The predictions are usually made with the aid of hints provided by the observed size of the unit cell, the space-group symmetry, the composition of the crystal, and other *known* facts such as atomic-size relationships and fundamental structural principles observed in structures already solved.

Structural Principles Associated with Friauf Polyhedra and Icosahedra

The existence of Friauf polyhedra and icosahedra in the structures discussed presents itself as an important structural principle, but a great deal more can be learned through penetrating studies of these structures. The seemingly most important principles are briefly summarized as follows.

1. Truncated tetrahedra tend to join together to form layers, rows, or infinite, three-dimensional close packing or frameworks of Friauf polyhedra by sharing hexagons in most cases [in one case, hexagons and triangles ($Mg_{17}Al_{12}$)]; the occurrence of isolated Friauf polyhedra is rare.

2. The atoms at the vertices of any polyhedron are, in turn, surrounded by coordination shells that penetrate each other and the central one, and each of these atoms tends to enforce its own coordination requirements. This usually results in distortions of the polyhedra, which sometimes become considerable.

3. In each case, where nonisolated Friauf polyhedra occur, each truncated tetrahedron is penetrated most often by twelve icosahedra and sometimes by nine icosahedra and three coordination shells of ligancy 13 or 14, or ten icosahedra and two other coordination shells.

4. In infinite, three-dimensional frameworks of Friauf polyhedra, each hexagon of a truncated tetrahedron, not shared by another truncated tetrahedron, is usually shared by a hexagonal antiprism that has two atoms at the extended poles (L 14) or by a μ-phase polyhedron (L 15).

The hexagonal antiprisms and the μ-phase polyhedra often terminate groupings of Friauf polyhedra.

5. Truncated tetrahedra have (so far) never been observed to penetrate one another. This is part of the reason why they seem to dominate in most structures, even though they are far outnumbered by other kinds of coordination shells, especially by icosahedra. The Friauf-polyhedra framework, therefore, comprises in most cases the dominant number of crystallographically dif-

ferent positions and represents the most favorable starting point in the search for a trial structure in which it is anticipated.

6. The metrical nature of the icosahedron requires that the average size of the contiguous spheres at the vertices be nearly 1.10 times that of the central sphere. Since structures incorporating Friauf polyhedra (and icosahedra) usually contain atoms differing in radius by significantly more than 10 percent (exceptions, αMn and αVAl_{10}), each icosahedral coordination shell has about half its vertices occupied by large atoms and the other half by small atoms.

7. The distribution of atoms of different sizes at the vertices of the icosahedron causes deformation.

8. It does not seem possible to build a space-filling structure in which each atom is surrounded by an icosahedral coordination shell. In such a case, a central icosahedron must be penetrated by twelve others, and each of these in turn by twelve, and so on, each center atom now being about 10 percent smaller than the atoms at the vertices. In any known structure, the icosahedra are associated with other kinds of coordination shells.

9. Hexagonal antiprisms are more frequently observed in metallic structures than are hexagonal prisms. While the latter give rise to octahedral interstices, the former produce tetrahedral ones, thus reducing the amount of interstitial space.

10. The Friauf polyhedron, the μ-phase polyhedron, and the hexagonal antiprism are related to one another as follows. The Friauf polyhedron can be described as a hexagonal antiprism consisting of a large and a small hexagon; out from the center of the small hexagon is one atom and out from the larger hexagon are three atoms forming an equilateral triangle. The μ-phase polyhedron is obtained through replacement of these three atoms by two atoms. The coordination shells of ligancy 14 are mostly hexagonal antiprisms, each one with two atoms at the extended poles.

11. The icosahedron can be described as a pentagonal antiprism, which has two atoms at the extended poles. It thus provides a condition favorable for the formation of tetrahedral interstices.

The Packing Map and Most Useful Planes

A highly successful method developed in the course of the work on complex metallic structures was my application of packing maps (Samson, 1964). These are two-dimensional graphs that provide at a glance the necessary information regarding the geometrical requirements for packing of coordination polyhedra to fill space. Sections through such polyhedra or through smaller or larger, more-or-less symmetrical atom complexes may be represented with transparent templates; these can be fitted together on the packing map like pieces of a puzzle, and the map then guides the search for a reasonable structural motif much more efficiently than would a three-dimensional model.

The packing map was developed from the idea that, in general, a cubic space-filling structure can be achieved only by utilizing special positions; these are always needed to define the center of any coordination shell described solely or partially by a general position. Hence, if in a cubic crystal

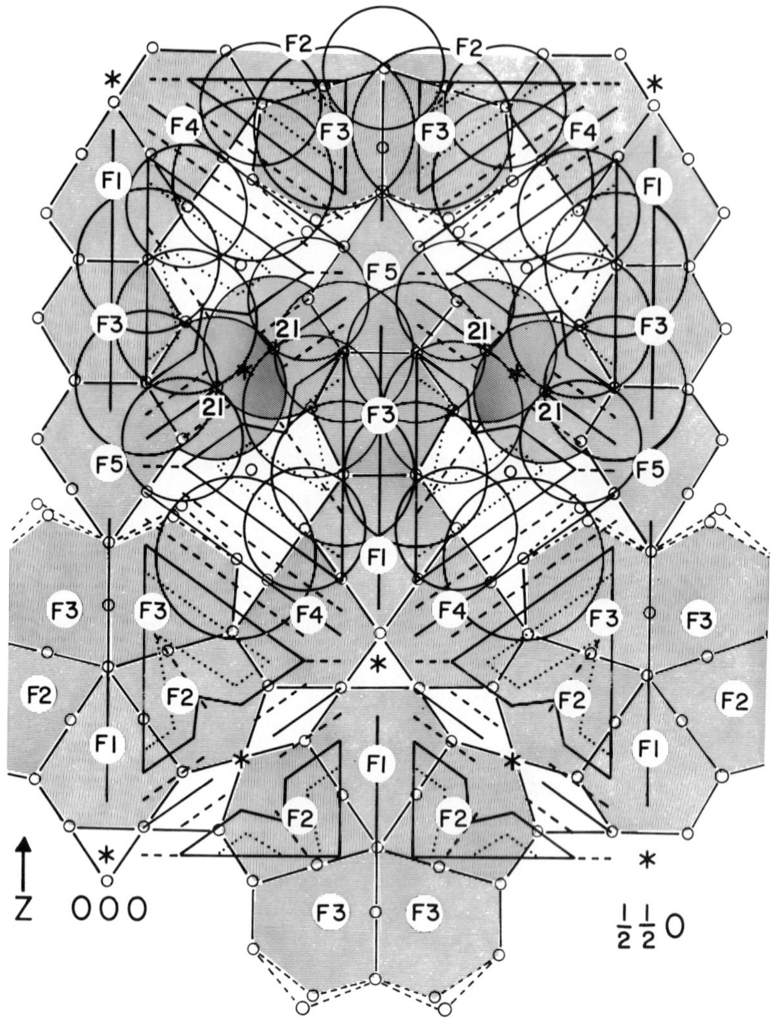

FIGURE 13.
A packing map representing the structure of β Mg_2Al_3. The polygons marked *F1*, *2*, *3*, and so on are sections through Friauf polyhedra.

the configuration of atoms is known around each point that can be defined by a special position, the atomic arrangement of the crystal is completely determined. If, in a cubic space group, every special position places at least one point on one and the same plane—for instance, on a (110) plane—then the problem of solving the complete structure is reduced to the problem of determining the coordination shells around each atom or available site on that plane, which is called *the most useful plane*. The packing map is a means of recognizing the possible configurations of atoms around single atoms on the most useful plane and reduces all the symmetry operations to manageable proportions. (For more details refer to Samson, 1964.)

Each one of the three extremely complex structures referred to above has

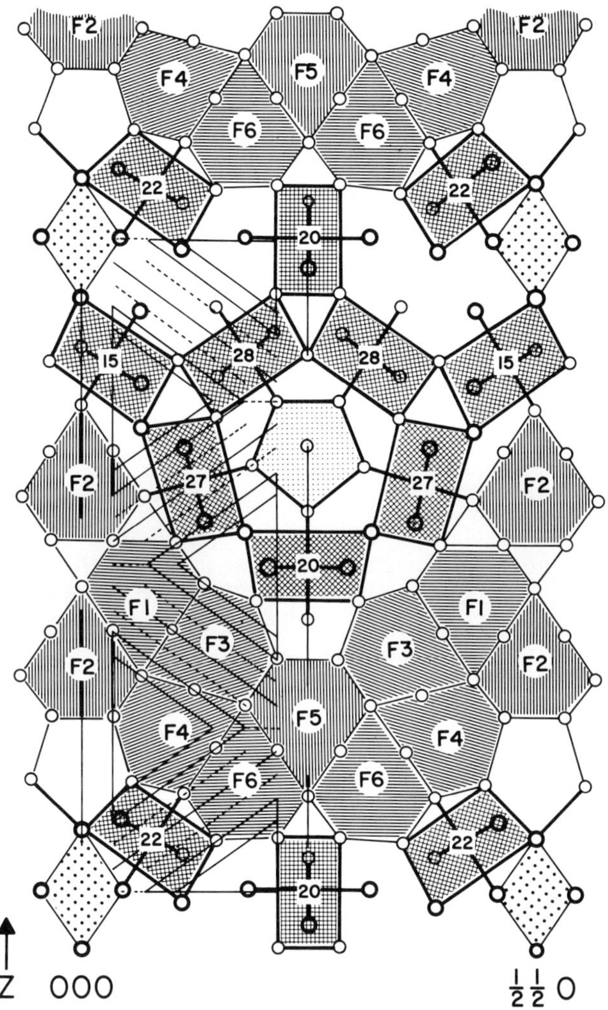

FIGURE 14.
A packing map representing the structure of Cu_4Cd_3. The slightly distorted rectangles marked *15*, *20*, *22*, *27*, and *28* are sectioned icosahedra. The ones marked *20*, *27*, and *28* are arranged about an approximate fivefold axis of symmetry; *F1*, *2*, *3*, ··· are, again, sectioned Friauf polyhedra.

been completely determined and described with the use of a single packing map, applying the structural principles outlined in the preceding section. The derivation of the trial structure of $NaCd_2$ (Samson, 1962) was done in a day, once the packing map was drawn.

Figure 13 shows the packing map of βMg_2Al_3; the templates marked *F1*, *2*, *3*, and so on, represent sections through Friauf polyhedra. These can be recognized after inspection of Figure 1, *e*. Figure 14 shows the packing map of Cu_4Cd_3. Here, we see again the sectioned Friauf polyhedra *F1*, *2*, *3*, . . ., but arranged in a different manner; and, in addition, sectioned icosahedra. These latter are the somewhat distorted rectangles marked *15*, *20*, *22*, *27*, and *28*; compare these sections with Figure 2, *d*. It is seen that the icosahedra

20, *27*, and *28* are arranged about an approximate fivefold axis of symmetry. Each map provides all the atomic positional parameters of the structure, and the well-trained user can read out almost all the structural details.

It seems that structures of certain cubic space groups can be completely determined and described with a single packing map, apparently irrespective of the size of the unit cell and the number of atomic positional parameters involved. For structures other than those of intermetallic compounds, these maps seem to be useful as well. The 2016 framework atoms (silicone, aluminum, and oxygen) of the zeolite mineral Paulingite, which is cubic with $a_0 = 35.093$ Å, were located by fitting a few of the most prominent Patterson vectors into the packing maps (Gordon et al., 1966).

THE CRYSTAL STRUCTURES OF βMg_2Al_3 AND $NaCd_2$

The fundamental structural features are the same in these two compounds. Both structures are partially disordered but presumably in slightly different fashions, as is indicated by the difference between the two stoichiometric ratios. The large atoms are magnesium and sodium, and the small atoms are aluminum and cadmium. The details of the disorder have been worked out for βMg_2Al_3 (Samson, 1965), but, for $NaCd_2$ (Samson, 1962) they are still unknown. The latter compound reacts with oxygen or moisture and gradually decomposes during X-ray examination, making it difficult to obtain X-ray data of a quality that will suffice for the unraveling of the structural details. Therefore, only βMg_2Al_3 is discussed below; the idealized ordered model is described first, and then some details of the disorder are given.

The basic building block consists of five Friauf polyhedra arranged about an approximate fivefold axis of symmetry, as shown in Figure 15. The five polyhedra are, in the crystallographic sense, of three different kinds (*F1*, *F2*, and *F3*). The group of polyhedra lies on a plane of symmetry, and therefore the left half of each figure is a mirror image of the right half. The dihedral angles of a tetrahedron are 70°32′ and hence correspond to nearly one fifth of a complete rotation. The aggregate (Figure 15) consists of 47 atoms and is called the *VF* polyhedron. It is easily recognized on the packing map shown in Figure 13.

FIGURE 15.
The basic building block of the structures of βMg_2Al_3 and $NaCd_2$. Five Friauf polyhedra, arranged about an approximate fivefold axis of symmetry, form the *VF* polyhedron, a formal representation of which is shown to the right.

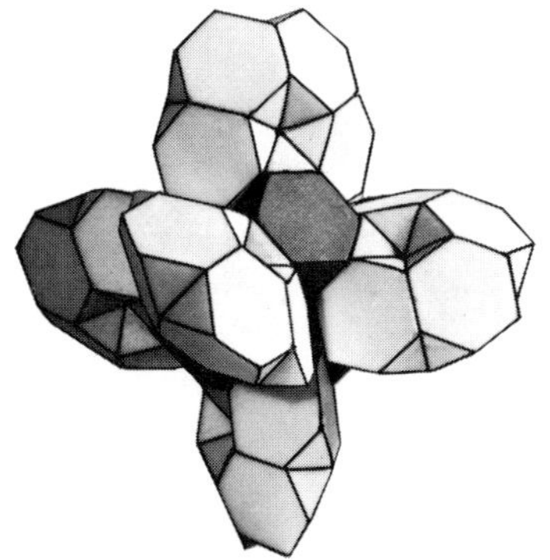

FIGURE 16.
A complex of 234 atoms formed by six *VF* polyhedra that are arranged around the vertices of an octahedron of T_d symmetry. In the disordered model six of the twelve outermost Friauf polyhedra are distorted.

FIGURE 17.
A second complex of 234 atoms inserted into the one shown in Figure 16.

Six *VF* polyhedra are arranged about the vertices of an octahedron in such a way as to produce four additional Friauf polyhedra, *F4*, located at the vertices of a regular tetrahedron and sharing hexagons with polyhedra *F1*. *F4* is dark and only one is seen in Figure 16. The resulting complex consists of 234 atoms and comprises 34 Friauf polyhedra; it has symmetry T_d (Figure 16). The twelve outer Friauf polyhedra are of the type *F3*.

A second such T_d complex can be meshed with the first one, as shown in Figure 17; these two complexes share hexagonal faces of the *F2* polyhedra and are related to one another by a diamond glide. Three more T_d complexes can be added in a similar fashion. Each T_d complex is accordingly connected with four others that are arranged about the vertices of a regular tetrahedron (Figure 18). The atom out from the center of each dark hexagon of an *F4* polyhedron is shared between three truncated tetrahedra, each one belonging to a Friauf polyhedron *F3* of a different T_d complex. This vertex that is shared between these three truncated tetrahedra is bifunctional, like the so-called hybrid vertex in $Mg_{17}Al_{12}$ described earlier in this article.

Continued stacking of T_d complexes leads to an infinite, three-dimensional network (Figure 19) in which each T_d complex of 234 atoms shares atoms with four others, which reduces the average number of atoms per T_d complex to 144. The cubic unit of structure contains eight T_d complexes; they account for 1152 atoms. Eight more atoms (magnesium) have to be added, each of them at the center of a Friauf polyhedron *F5* that is marked out on the packing

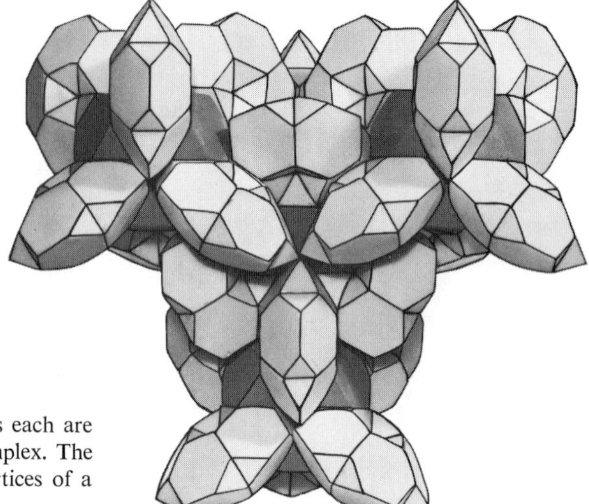

FIGURE 18.
Four complexes of 234 atoms each are arranged about one such complex. The four complexes are at the vertices of a regular tetrahedron.

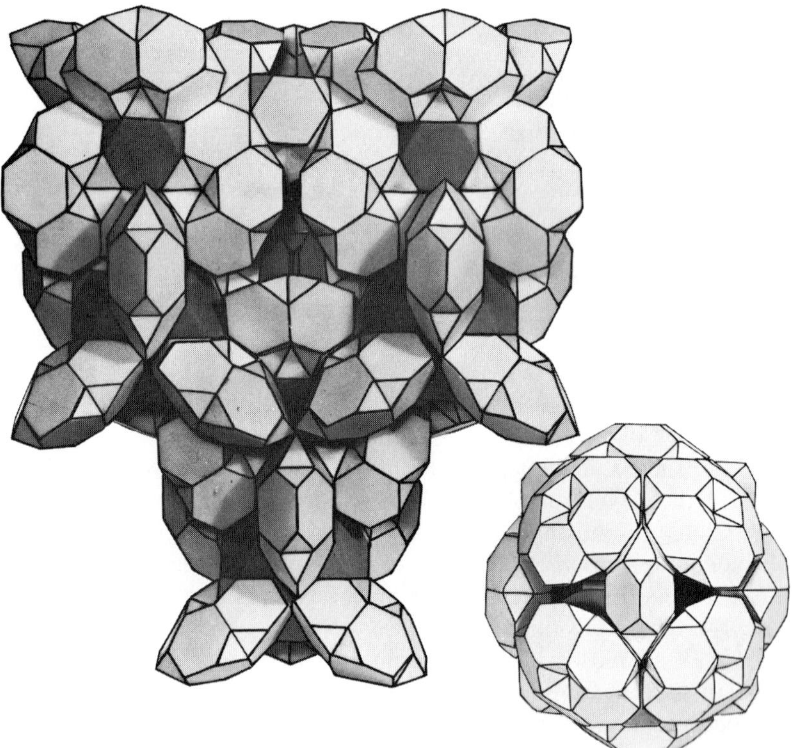

FIGURE 19.
Continued stacking of the T_d complexes of 234 atoms each leads to the configuration shown here. The sphere shown to the right has its center located about two-thirds up the vertical center line of the figure shown to the left. Twelve *VF* polyhedra (4×3) form this sphere, and six additional *VF* polyhedra are arranged about the vertices of a second kind of T_d octahedron. The Friauf polyhedron *F5*, which is shared between these *VF* polyhedra, is at the center of this sphere (*right*).

map (Figure 13) but cannot be brought into view with the opaque models used here. Each *F5* polyhedron shares edges with twelve *F3* polyhedra and lies at the center of the sphere shown to the right of Figure 19. There are eight such spheres in the cubic unit; each one is interpenetrated by four others of the same kind. With the addition of 32 more atoms (aluminum) out from the centers of the triangles of eight such *F5* polyhedra, the entire complement of 1192 atoms in the ordered structure is accounted for.

The disordered model is obtained by replacing every other *F5* polyhedron and the four associated aluminum atoms ($\frac{1}{8} \times 32$) with a centered pentagonal prism that has two atoms at the poles and two atoms out from the centers of two prism faces (complex of 15 atoms). In order that the observed space-group symmetry be retained, it has to be assumed that this 15-atom complex occurs in six orientations and that there is a random interchange of the set of four *F5* polyhedra and the set of four 15-atom complexes in the individual unit cells. The details are given in the original paper (Samson, 1965).

To account for this kind of disorder, the complex of 234 atoms (T_d complex) shown in Figure 16 has now to be modified. Of any two diametrically opposed *VF* polyhedra in this complex, one remains unchanged, while the other contains, instead of two *F3* polyhedra, two modified Friauf polyhedra. It is of interest to note here that the modifications of the *F3* polyhedra involve principally the hybrid vertices described above; see also $Mg_{17}Al_{12}$. In the three-dimensional network of the modified T_d complexes, each second sphere of the kind shown in Figure 19 will then have a 15-atom complex at its center instead of the *F5* polyhedron with the associated four aluminum atoms (21-atom complex). Since there are four spheres of each kind per unit of structure, the total number of atoms is $1192 - 4(21 - 15) = 1168$. A consequence of the occurrence of the 15-atom complexes in six orientations is that certain other atoms in the structure are part of the time displaced (for further details see Samson, 1965).

The atoms of the disordered model occupy 23 different sets of equivalent positions. In an ordered structure the number of different polyhedra is equal to the number of point sets, but here the number of different polyhedra (in the crystallographic sense) is increased to 41.

On the Nature of the Disorder

Although it appears from Figures 16 to 19 that the Friauf polyhedra are dominant in this structure, they are far outnumbered by the icosahedra, as in the structure of $Mg_{32}(Zn,Al)_{49}$.

The disorder observed here seems to result from a tendency toward the formation of the maximum number of icosahedral coordination shells that is compatible with the coordination requirements of the large atoms, the magnesium atoms, most of which are inside the Friauf polyhedra. Eighty-eight of the coordination shells that in the idealized ordered model were more or less irregular have been transformed into icosahedra, 48 of the original number of icosahedra have been lost through partial occupancy, and 8 new icosahedra have been added. Accordingly, there is a gain of 48 icosahedra per unit of structure, and the apparent requirement to have about half the vertices occupied by large atoms (magnesium) and the other half by small atoms (aluminum) is fulfilled through the disorder. While the idealized ordered model contains 280 Friauf polyhedra, 624 icosahedra, and 288 more or less irregular polyhedra, the disordered atomic arrangement corresponds to 252 Friauf polyhedra, 672 icosahedra, and 244 more or less irregular polyhedra of ligancy 10 to 16, of which 48 are modified Friauf polyhedra.

THE CRYSTAL STRUCTURE OF Cu_4Cd_3

Although the unit of structure of this compound contains somewhat fewer atoms than that of β Mg_2Al_3 (and $NaCd_2$), 1124 as compared to 1168 per unit cube, it was considerably more difficult to determine. The reason for this is the lower symmetry and a resulting significant increase in the number of structural parameters. In fact, this structure comprises two different substructures that penetrate one another, and each one is of considerable complexity. Each substructure represents a diamond arrangement, in one case of Friauf polyhedra, in the other case of icosahedra. The structure lacks a center of symmetry, and hence calculations of Fourier syntheses are meaningless unless the trial models of both substructures are nearly correct and are used simultaneously as a basis for the phase-angle determinations. The structure was solved with the use of a packing map (Samson, 1964) exclusively; see also Figure 14.

The diamondlike arrangement of Friauf polyhedra consists of the three types of complexes shown in Figures 20, *a*, *b*, and *c*. The octahedron of T_d symmetry (Figure 20, a) comprises ten Friauf polyhedra, 4 *F1* + 6 *F2*; see packing map, Figure 14. The tetrahedral arrangement shown in Figure 20, *c* consists of the Friauf polyhedra *F5* + 4 *F6* (Figure 14), and the dark polyhedra in Figure 20, *b* are *F3* + 3 *F4*. The cubic unit contains four octahedra (Figure 20, *a*) that are arranged about the points $\frac{1}{2}, \frac{1}{2}, \frac{1}{2}$; $\frac{1}{2}$ 0 0; 0 $\frac{1}{2}$ 0; 0 0 $\frac{1}{2}$ (point set 4b in $F\bar{4}3m$) and four tetrahedra (Figure 20, *c*) that are at the points $\frac{1}{4}, \frac{1}{4}, \frac{1}{4}$, and so on (point set 4c). The layer of the dark Friauf polyhedra serves as a link between the octahedra and the tetrahedra. The infinite, three-dimensional framework of Friauf polyhedra thus formed is shown in Figure 21. It is seen that the tetrahedra, the dark layers, and the octahedra alternate in a zigzag fashion.

The diamondlike arrangement of icosahedra may be described in terms of two kinds of complexes. One of them is shown in Figure 22. It is seen (Figure 22, *a*) that five icosahedra are arranged about an approximate fivefold axis of symmetry, thus enclosing a pentagonal prism. These icosahedra are the ones that have been marked *20*, *27*, and *28* on the packing map (Figure 14). A set of six such fivefold rings is arranged at the vertices of an octahedron of T_d symmetry. All six rings interpenetrate and share icosahedra in such a way that the aggregate (Figure 22, *c*) consists of fourteen icosahedra that enclose six pentagonal prisms of the kind shown in Figure 22, *a*. Figure 22, *b* shows two such fivefold rings interpenetrating at right angles. It is now seen that the pentagonal prism (in Figure 22, *a*) is shared between two icosahedra, one above and the other below the plane of the paper. The two icosahedra have one vertex in common at the center of the pentagonal prism, and each icosahedron center is at an extended pole of that prism. Each additional vertex that is shared between two icosahedra represents the center of a pentagonal

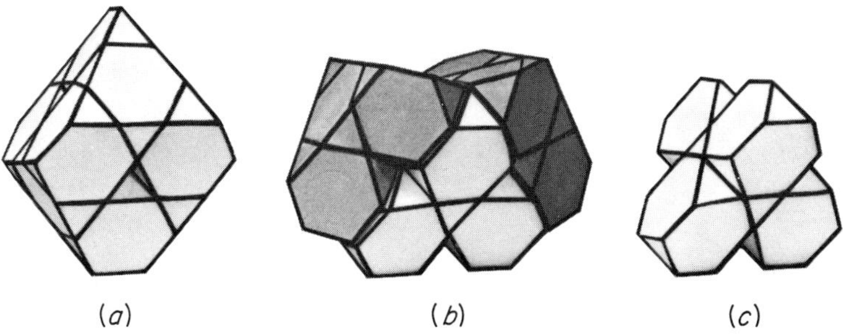

FIGURE 20.
The three types of atom complexes that form the infinite, three-dimensional framework of the Friauf polyhedra shown in Figure 21. (*a*) A complex of ten Friauf polyhedra, 4 *F1* + 6 *F2*, forming an octahedron of T_d symmetry. (*b*) The dark Friauf polyhedra are *F3* + 3 *F4* (see packing map, Figure 14.) (*c*) This tetrahedron consists of *F5* + 4 *F6* (compare with Figure 14).

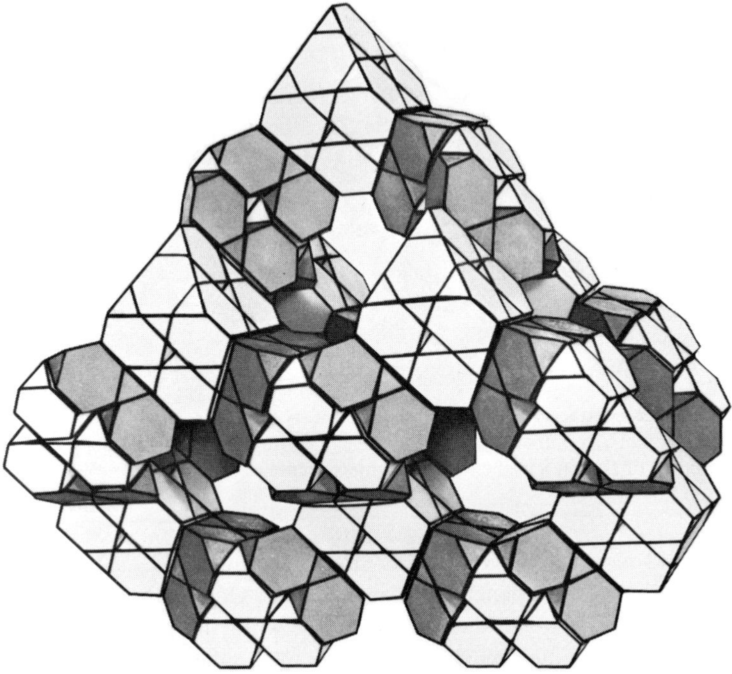

FIGURE 21.
The three atom complexes shown in Figure 20, *a*, *b*, and *c* form the infinite three-dimensional framework of Friauf polyhedra shown here.

prism (which has two atoms at its extended poles), as can be made out on Figure 22, *c*. Accordingly, thirty-six more pentagonal prisms are created. In twelve of these prisms, two prism faces are deformed in such a way that two more atoms are added as ligands (to provide ligancy 14).

The aggregate shown in Figure 22, *c* accordingly represents fourteen icosahedra, thirty pentagonal prisms, each one with two atoms at the poles (ligancy 12), and twelve pentagonal prisms, each of which has two more atoms penetrating two prism faces (ligancy 14).

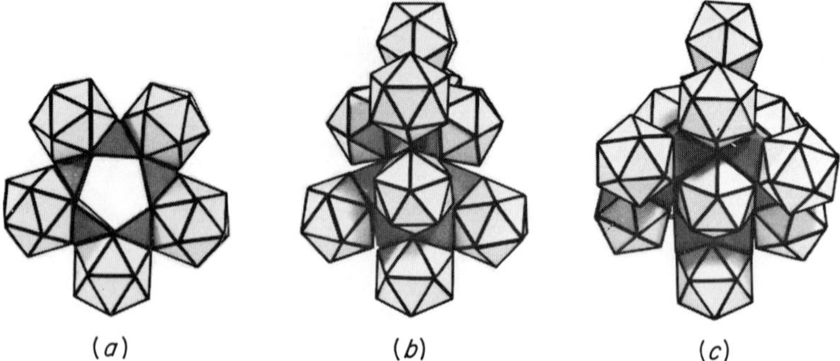

FIGURE 22.
(a) Five icosahedra arranged about an approximate fivefold axis of symmetry. (b) Two such fivefold rings interpenetrate at right angles. (c) Six interpenetrating fivefold rings form a complex of fourteen icosahedra and forty-two centered pentagonal prisms.

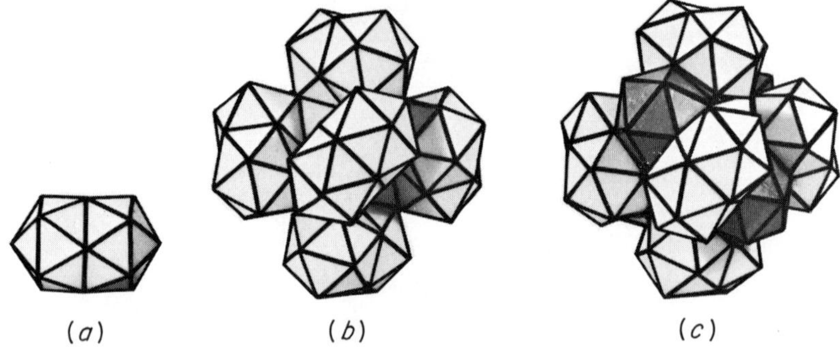

FIGURE 23.
(a) A pair of interpenetrating icosahedra. (b) A set of six such pairs arranged about the vertices of an octahedron of T_d symmetry. (c) Four more icosahedra (the dark ones) have been added to form a complex of sixteen icosahedra and eighteen pentagonal prisms.

The second icosahedral complex is shown in Figure 23, *c*. It consists of a set of six pairs of interpenetrating icosahedra. One such pair is shown in Figure 23, *a*; the two icosahedra are of the kind marked 15 in Figure 14. Each pair has its center at the vertex of an octahedron of T_d symmetry; accordingly, each one of two diametrically opposed pairs has its fivefold axis (long axis) at a right angle to that of the other (Figure 23, *b*). Four more icosahedra have to be inserted into this complex. Their centers are at the vertices of a regular tetrahedron, and each of these icosahedra shares six triangles with three "icosahedron pairs" that surround it, as is shown in Figure 23, *c*. The four icosahedra are of the kind marked 22 in Figure 14 and are the dark ones in Figure 23, *c*. Each vertex that is shared between two icosahedron pairs is, again, the center of a pentagonal prism, which has two atoms at the extended poles and two additional atoms that penetrate two of the prism faces (ligancy 14). Several of the pentagonal prisms can be made out in Figure 23, *b*, especially the one at the upper left.

The aggregate shown in Figure 23, *c* thus represents sixteen icosahedra and eighteen pentagonal prisms.

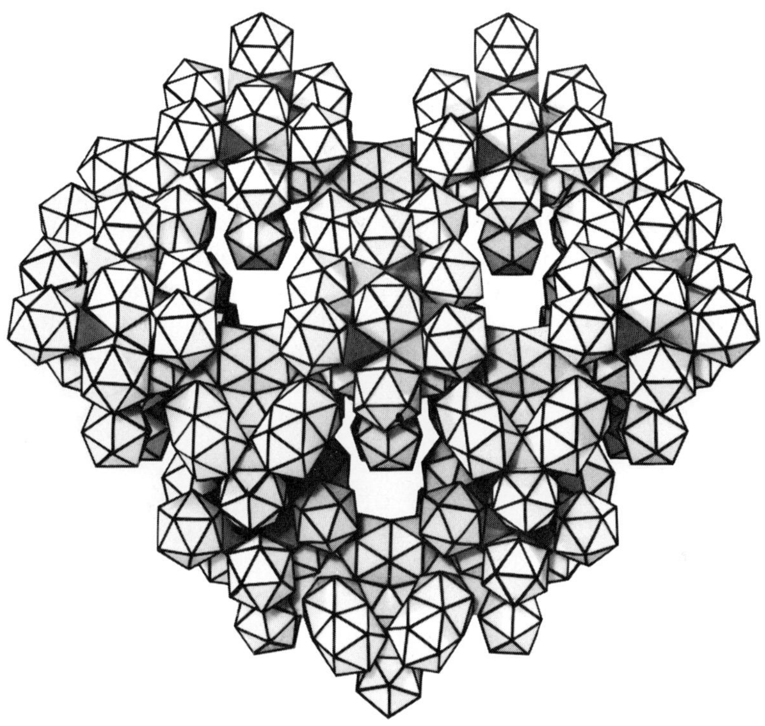

FIGURE 24.
The icosahedral complexes shown in Figures 22 and 23 are connected with one another to form the infinite, three-dimensional framework shown here. This framework fits into the cavities formed by the Friauf polyhedra, as shown in Figure 21.

The cubic unit contains four aggregates of the kind shown in Figure 22, *c* and four of the kind shown in Figure 23, *c*. The former aggregates are arranged about the points $\frac{3}{4}, \frac{3}{4}, \frac{3}{4}$, . . . (point set 4d) and the latter about the points 0, 0, 0, . . . (point set 4a). Both types of aggregates are connected with one another by shared vertices in such a way that twelve more pentagonal prisms (plus two atoms at the poles; ligancy 12) are formed between each two complexes. The two types of aggregates (Figures 22, *c* and 23, *c*) thus form the infinite, three-dimensional framework shown in Figure 24. The dark icosahedra shown in Figure 23, *c* have been omitted in this large model, since they are difficult to insert.

This framework fits into the cavities formed by the Friauf polyhedra, as arranged according to Figure 21. These frameworks share vertices in such a way that additional coordination shells, most of them icosahedra, are produced that interpenetrate the Friauf polyhedra as well as the icosahedra described above.

The unit cube of the structure contains 568 centered icosahedra, 124 Friauf polyhedra, 168 centered pentagonal prisms, each one with two atoms at the extended poles, 120 centered pentagonal prisms in which two more atoms penetrate prism faces such as to produce ligancy 14, and 144 polyhedra of ligancy 15. The latter are hexagonal antiprisms with one atom out from the center of one hexagon and two atoms out from the other hexagon, which is larger (μ-phase polyhedra). Each icosahedron has, again, about half of its vertices occupied by cadmium atoms (large) and the other half by copper atoms (small).

There appears to be a correlation between the two interpenetrating framework structures and the structures of the two phases that are in equilibrium with Cu_4Cd_3. On the more copper-rich side of the phase diagram is $CdCu_2$, with the C 36 type of structure shown in Figure 3, *d*, and on the more cadmium-rich side is the γ Cd_8Cu_5 phase, with the $D8_2$ type of structure (γ brass), which is known to be essentially icosahedral. Cu_4Cd_3 does not form on solidification but appears only after prolonged annealing at 450 to 500°C, apparently through a reaction of a metastable eutectic mixture of $CdCu_2$ and γ Cd_8Cu_5 (Jenkins and Hanson, 1924). Parts of these two structures are retained in the two frameworks.

SUMMARY OF RESULTS

In the course of the discussion of the twelve different types of structure [$MgCu_2$, $MgZn_2$ (Friauf phases), $MgNi_2$ (Laves phase), Mg(Cu,Al), W_6Fe_7 (μ phases), P(Mo—Ni—Cr), $Mg_3Cr_2Al_{18}$ (αVAl_{10}, AB_{22} compounds), γ $Mg_{17}Al_{12}$ (αMn, χ phases), ϵ $Mg_{23}Al_{30}$ (R phase), $Mg_{32}(Zn,Al)_{49}$, β Mg_2Al_3 ($NaCd_2$), Cu_4Cd_3], it has become ever more apparent that the icosahedron and the Friauf polyhedron are to be ascribed significance for the stability of intermetallic compounds. The most important feature of the icosahedron is probably the shortened center-to-vertex distance. The most tangible experimental evidence for the tendency of atoms of unlike atomic radii to arrange themselves in icosahedral configurations has been brought forth in β Mg_2Al_3. Here, the disorder, which provides the stability, brings about a substantial increase in the number of icosahedral coordination shells. This increase occurs in a relatively small region in the structure, just where most of the more or less irregular coordination shells were located before, and the apparent requirement of having about one half of the vertices occupied by large atoms (magnesium) and the other half by small atoms (aluminum) is satisfied. It has thus become obvious that disordered structures, which for a long time have resisted detailed analyses, may provide an extremely important source of information regarding the structural elements that lead to maximum stability.

One of the most important subjects connected with the polyhedral analysis of space-filling structures is the concept of atomic size or metallic radius. We have seen that in the cases of $ZrZn_{22}$ and αVAl_{10} there is hardly any relationship between the atomic sizes observed in the structures of the pure metals and those observed in the structures of their compounds. Another interesting example is the icosahedron observed in $Mg_2Cu_6Al_5$* (Samson, 1949a). In this structure, an aluminum atom is at the center and twelve

*Since this structure does not contain Friauf polyhedra, it is not described here.

copper atoms are at the vertices. This arrangement indicates that the *effective* size of the aluminum atom is 10 percent smaller than that of the copper atoms, whereas in pure aluminum the metallic radius is 12 percent larger than in pure copper. The atomic-size relationships in alpha manganese were also very striking. On the basis of atomic-size relationships, it is also hard to understand why replacement of cobalt by nickel in Mo_6Co_7 does not result in a μ-phase structure, although the atomic radii of Co and Ni are almost identical. We see that "atomic or metallic radius" does not represent a meaningful concept unless it is used in connection with terms that describe the conditions under which it is observed, such as the coordination number, bond number, and valence, or perhaps some other important properties that have, as yet, escaped discovery.

The nominal radius ratio Mg/Al is about 1.14, but the truncated tetrahedron of twelve aluminum atoms is appropriate for the accommodation of a central atom that has a radius about 1.35 times that of the aluminum atoms. Here the apparently large size of the magnesium atoms may be attributed to the fact that the two metallic valences are engaged in sixteen bonds, and thus each represents a weak bond that corresponds to a large interatomic distance. In $Mg_3Cr_2Al_{18}$ the magnesium atom inside each Friauf polyhedron is bonded more strongly to the four surrounding magnesium atoms than to the twelve aluminum atoms; each Mg—Mg bond is 3.15 Å, while each Mg—Al bond is 3.23 Å. In fact, in most of the structures described here, the atom at the center of the Friauf polyhedron appears to be more-or-less tetrahedrally deformed. The degree of deformation depends upon two factors: (1) the nominal radius ratio between the center and the vertex atoms, and (2) the orientation of the Friauf polyhedra with respect to each other in the various frameworks. A more quantitative treatment of this subject is under way and may be published in a separate paper.

An interesting feature of the most complex structures [$Mg_{32}(Zn,Al)_{49}$, $NaCd_2$, β Mg_2Al_3, and Cu_4Cd_3] is the tendency toward the creation of fivefold axes, not only those of the icosahedra but also those of the *VF* polyhedra (Figures 11 and 16) and the icosahedral complexes shown in Figure 22. This feature may be due to repulsions between atoms at nonadjacent corners of polygons (Pauling, private communication). With atoms arranged at a distance a at the corners of a square, the next-nearest-neighbor distance is $1.41a$, while in a pentagonal arrangement it is $1.62\ a$. In a triangular arrangement there are no unbonded neighbors to produce repulsion. This may be a second and perhaps even more important factor than the shortened center-to-vertex distance stabilizing the icosahedral coordination shell with its twenty triangular faces and its six fivefold axes.

It is seen that the polyhedral analysis of structures of intermetallic compounds is useful in arriving at an understanding of the nature of these compounds. Although it does not solve the problem of the metallic bond, it enables us to uncover and to state more clearly a number of important problems; it also provides us with valuable hints as to the nature of the most important geometrical factors that determine stability.

CONCLUSION

In this article, discussion has been confined to those structures that can be described largely with the use of Friauf polyhedra and icosahedra. Most readers will be aware that there exist a considerably larger number of intermetallic compounds that contain icosahedra but no Friauf polyhedra. For instance, the AB_{13} structures ($D2_3$ type) contain icosahedra and snub cubes, the AB_5 compounds with the $D2_d$ structure are combinations of icosahedra and hexagonal prisms, the σ-phase alloys (Bergman and Shoemaker, 1954) contain primarily icosahedra and hexagonal antiprisms, and so forth. A large number of structures contain, in addition to the icosahedra, only more or less irregular coordination shells; the γ-brass alloys (type $D8_2$), the WAl_{12} type alloys (Adam and Rich, 1955), the Mg_2Zn_{11} type alloys (Samson, 1949a, 1949b), and the *A15* types are just a few examples. There also exist many intermetallic compounds in which icosahedra are completely absent; indeed, the first compound to have its structure reported in the literature (Pauling, 1923) is of that nature. In a number of compounds consisting of aluminum and a noble metal, the main building blocks have the cesium chloride configuration (Edshammar, 1965).

The reason for the special interest evidenced here in the combination of icosahedra and Friauf polyhedra is that it leads to the most complex metallic structures as yet known; it also affords an opportunity to demonstrate how we can derive the complex arrangements by starting out with the very simplest one—the cubic closest packing.

REFERENCES

Adam, J., and J. B. Rich (1955). *Acta Cryst.* **8**, 349.

Arnfelt, H., and A. Westgren (1935). *Jernkontor. Ann.* **119**, 185.

Bergman, G., and D. P. Shoemaker (1954). *Acta Cryst.* **7**, 857.

Bergman, G., J. L. P. Waugh, and L. Pauling (1957). *Acta Cryst.* **10**, 254.

Bradley, A. J., and J. Thewlis (1927). *Proc. Roy. Soc.* (London) A **115**, 456.

Bragg, W. L., and J. West (1926). *Proc. Roy. Soc.* (London) A **111**, 691.

Brown, P. J. (1957). *Acta Cryst.* **10**, 133.

Cherkashin, E. E., P. I. Kripyakevich, and G. I. Oleksiv (1964). *Kristallografiya* **8**, 681.

Edshammar, L. E. (1965). *Acta Chem. Scand.* **19**, 871.

Erdmann-Jesnitzer, F. (1940). *Aluminum Archiv.* **29**.

Florio, J. V., R. E. Rundle, and A. I. Snow (1952). *Acta Cryst.* **5**, 449.

Friauf, J. B. (1927a). *J. Am. Chem. Soc.* **49**, 3107.

Friauf, J. B. (1927b). *Phys. Rev.* **29**, 34.

Gordon, E. K., S. Samson, and W. B. Kamb (1966). *Science*, **154**, 1004.

Greenfield, P., and P. A. Beck (1956). *J. Metals*, **8**, 265.

Jeffries, Z., and R. S. Archer (1924). *The Science of Metals*, p. 33. New York: McGraw-Hill.

Jenkins, C. H. M., and D. Hanson (1924). *J. Inst. Metals* **31**, 257.

Kasper, J. S. (1954). *Acta Metallurgica* **2**, 456.

Komura, Y., W. G. Sly, and D. P. Shoemaker (1960). *Acta Cryst.* **13**, 575.

Komura, Y. (1962). *Acta Cryst.* **15**, 770.

Laves, F., K. Löhberg, and P. Rahlfs (1934). *Nachr. Ges. Wiss. Göttingen*, Group 4, **1**, 67.

Laves, F., and H. Witte (1935). *Metallwirt.* **14**, 645.

Laves, F., K. Löhberg, and H. Witte (1935). *Metallwirt.* **14**, 793.
Little, K., H. J. Axon, and W. J. Hume Rothery (1948). *J. Inst. Metals* **75**, 39.
Pauling, L. (1923). *J. Am. Chem. Soc.* **45**, 2777.
Pauling, L. (1933). *Z. Krist.* **84**, 442.
Pauling, L. (1947). *J. Am. Chem. Soc.* **69**, 542.
Pauling, L. (1949). *Proc. Roy. Soc.* (London) A **196**, 343.
Pauling, L. (1955). *Am. Sci.* **43**, 285.
Pauling, L. (1964). *Proc. Nat. Acad. Sci. U.S.* **51**, 977.
Perlitz, H. (1944). *Nature* (London) **154**, 607.
Perlitz, H. (1946). *Chalmers Tekn. Högsk. Handl.* **50**, 1.
Rideout, S., W. D. Manley, E. L. Kamen, B. S. Lement, and P. A. Beck (1951). *J. Metals* **3**, 872.
Riederer, K. (1936). *Z. Metallk.* **28**, 312.
Samson, S. (1949a). *Acta Chem. Scand.* **3**, 809.
Samson, S. (1949b). *Acta Chem. Scand.* **3**, 835.
Samson, S. (1958). *Acta Cryst.* **11**, 851.
Samson, S. (1961). *Acta Cryst.* **14**, 1229.
Samson, S. (1962). *Nature* (London) **195**, 259.
Samson, S. (1964). *Acta Cryst.* **17**, 491.
Samson, S. (1965). *Acta Cryst.* **19**, 401.
Samson, S. (1967). *Acta Cryst.* In press.
Samson, S., and E. K. Gordon (1967). *Acta Cryst.* In press.
Sands, E. D., Q. C. Johnson, A. Zalkin, O. H. Krikovian, and K. L. Kromholtz (1962). *Acta Cryst.* **15**, 832.
Shoemaker, D. P., C. B. Shoemaker, and F. C. Wilson (1957). *Acta Cryst.* **10**, 1.
Shoemaker, C. B., and D. P. Shoemaker (1963). *Acta Cryst.* **16**, 997.
Westgren, A. (1933). *Jernkontor. Ann.* 501.

DAVID P. SHOEMAKER
CLARA BRINK SHOEMAKER
Department of Chemistry and Center for Materials Science and Engineering
Massachusetts Institute of Technology
Cambridge, Massachusetts

Sigma-Phase-Related Transition-Metal Structures with Tetrahedral Interstices

About sixteen years ago at the California Institute of Technology, following the suggestion of Professor Linus Pauling, David P. Shoemaker and Gunnar Bergman undertook to determine the crystal structure of the sigma phase (Shoemaker and Bergman, 1950; Bergman, 1951; Bergman and Shoemaker, 1951, 1954). Previous attempts to determine its structure were frustrated by lack of suitable single crystals (the phase in nearly all systems does not exist in equilibrium with any melt, and is usually obtained through solid-state transformation) and by the extreme complexity of the powder diagram, which defied indexing. The opportunity to undertake this work was provided by Dr. Paul Pietrokowsky and Professor Pol Duwez of the California Institute of Technology, who made available to us powder specimens of σ-FeCr and σ-FeMo that happened to contain single-crystal particles barely large enough (0.05 to 0.1 mm) to permit single-crystal X-ray work to be done.

The crystal structure of the sigma phase (Figure 1, *a*) turned out to be the same in regard to atomic arrangement as the structure found at about the same time for the beta phase of uranium by Tucker (1950), except for space group (Bergman and Shoemaker, 1951); however, that of β-uranium later appeared to be the same as that of the sigma phase (Tucker and Senio, 1953). The sigma-phase structure was considered quite remarkable in that only slightly distorted hexagonal atomic layers are stacked in a tetragonal unit cell, with orientations differing by 90°. Atoms that are superimposed in projection are displaced perpendicularly to positions between the layers, thus "locking" the layers, preventing slip, and presumably accounting in part for the extreme hardness of the sigma phase. Many other features of the structure, and its relations with other structures, were not realized until later. Much the same can be said for the structures of the mu phase (as exemplified by Co_7Mo_6; Arnfelt and Westgren, 1935; Forsyth and da Veiga, 1962) (see Figure 4, *a*) and alpha manganese (Bradley and Thewlis, 1927) (see Figure 1, *b*).

Most of this work was supported by the Army Research Office (Durham). Computations were done in part at the M.I.T. Computation Center. Work on the sigma phase at the California Institute of Technology was sponsored by the Office of Naval Research and the Carbide and Carbon Chemicals Corporation.

It was our good fortune in our subsequent work at the Massachusetts Institute of Technology to establish a close working relationship with Professor Paul A. Beck at the University of Illinois, who was actively engaged in research on phase relationships in ternary transition-metal systems at elevated temperatures (Rideout, Manly, Kamen, Lement, and Beck, 1951). In these ternary systems and others investigated later there appeared several new phases of unknown structure. Professor Beck kindly supplied us with specimens of these phases, and the crystal structures of some of them were determined in our laboratory. These include the P phase of Mo-Ni-Cr (Shoemaker, Shoemaker, and Wilson, 1957) (Figure 2, *a*), the R phase of Mo-Co-Cr (Komura, Sly, and Shoemaker, 1960) (Figure 3, *a*), the delta phase of Mo-Ni (Shoemaker and Shoemaker, 1963) (Figure 3, *b* and *c*), and the D phase of V-Fe-Si (David P. Shoemaker and Clara Brink Shoemaker, unpublished). In addition, we have determined the structure of the M phase of Nb-Ni-Al (Shoemaker and Shoemaker, 1967) (Figure 2, *b*), a specimen of which was kindly given to us by Dr. Bill C. Giessen (Benjamin, Giessen, and Grant, 1966) of this Institute. Like the sigma phase, these phases were formed by solid-state transformations at high temperatures (usually 1200°C) and therefore were found in extremely small crystallites, which made the X-ray diffraction work very difficult; exposures as long as seven days were frequently required for a Weissenberg photograph. The crystal structures of all these phases proved to be closely related to that of the sigma phase.

The nature of the coordination polyhedra (CN12, 14, 15, 16), listed in Table 2 and shown in Figure 5, and the role of these polyhedra as common features in a large class of structures, became clearer to us in our work on the P phase (Shoemaker et al., 1957), the structure of which was the first to be determined by us after that of the sigma phase. In the P phase all four of the polyhedra were found together and without any other kinds of first-

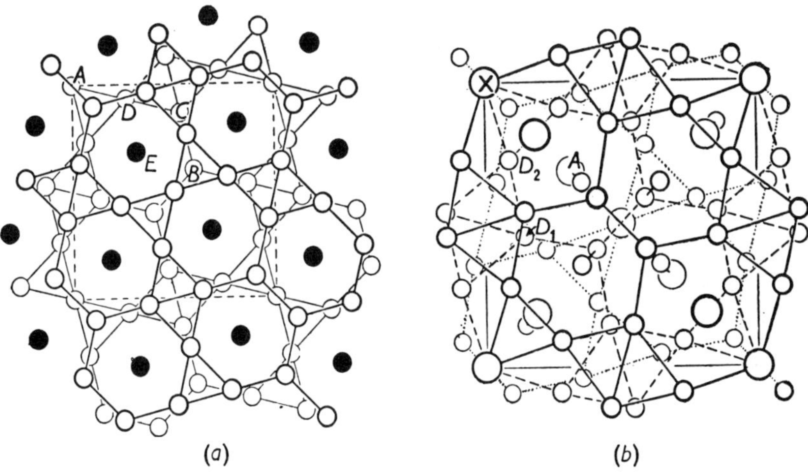

FIGURE 1.
(*a*) Structure of σ phase; Fe-Cr as an example. As also in Figures 2 and 4, open circles represent atoms on planar "main layers" and filled circles represent atoms on "secondary layers" halfway between the main layers. In the σ-phase structure, main layers are half the cell repeat apart. (*b*) Structure of α-Mn or χ phase; layers are rumpled. Note however their resemblance to σ-phase layers. Large circles are CN16 atoms. [Courtesy *Acta Crystallographica.*]

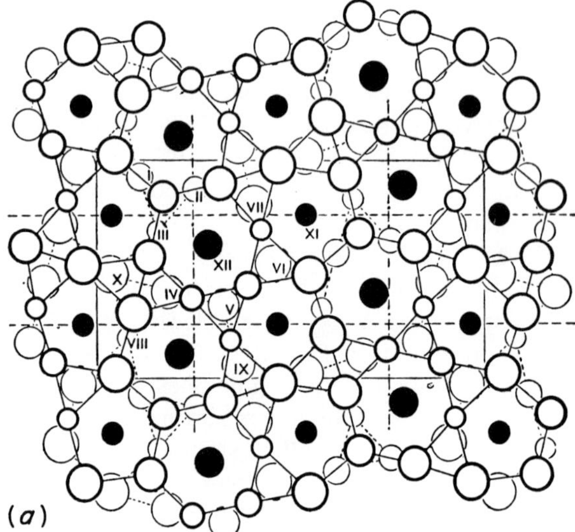

FIGURE 2.
(*a*) Structure of P phase, Mo-Ni-Cr. (*b*) Structure of M phase, Nb-Ni-Al. In both structures the main layers are half the cell repeat apart. Circles increase in size with coordination number. [Courtesy *Acta Crystallographica.*]

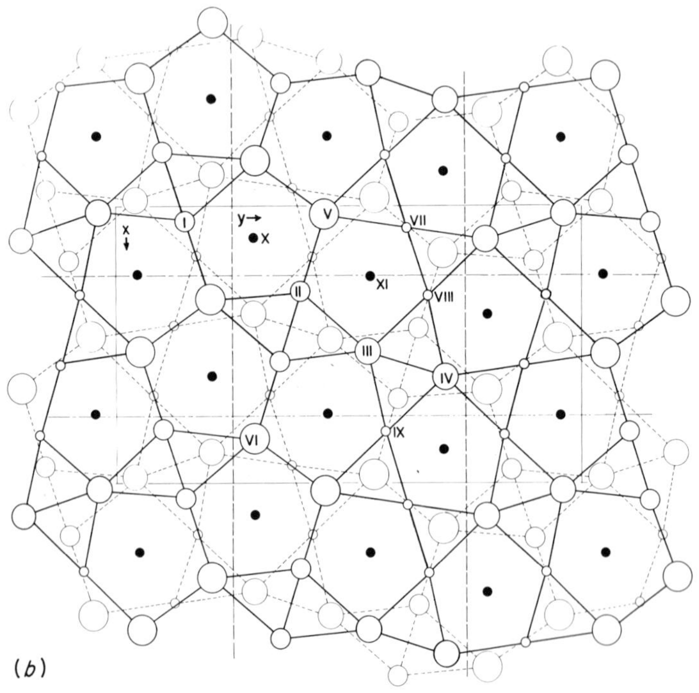

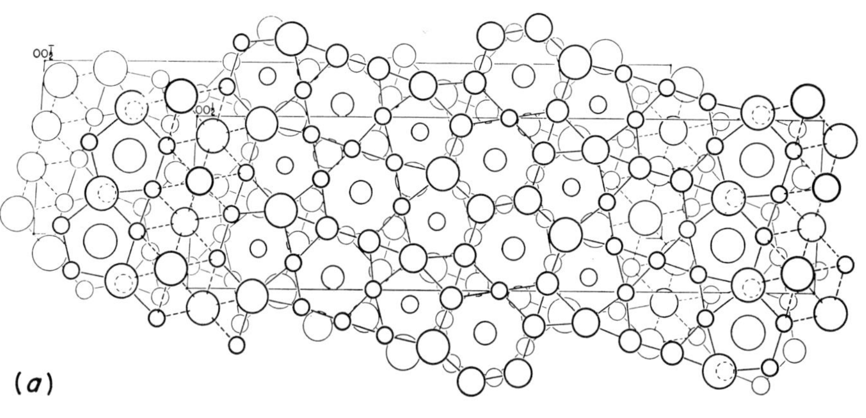

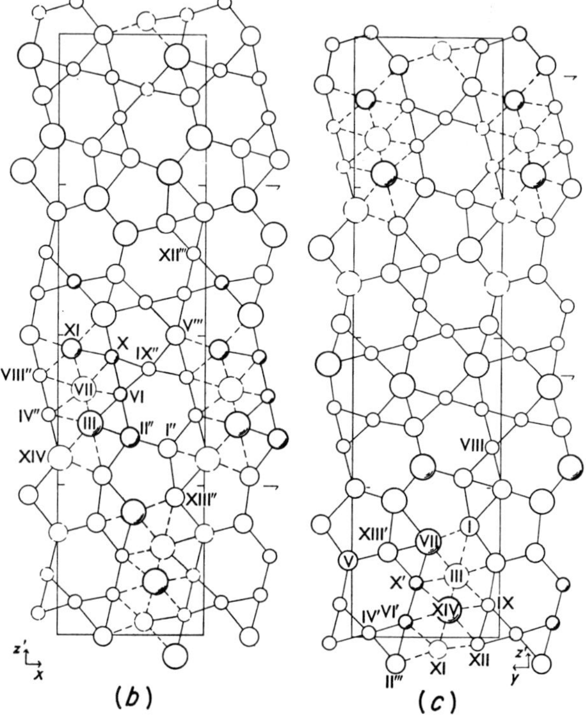

FIGURE 3.
(*a*) Structure of R phase, Mo-Co-Cr, showing two main layers (atoms joined by solid lines) parallel to $(1\bar{3}\bar{5})_{hex}$. Atoms to which no lines are drawn are in secondary layers halfway between main layers. (*b*) Structure of δ phase, Mo-Ni, showing one main layer parallel to $(04\bar{1})$ plane. Atoms in secondary layers are not shown. (*c*) The δ-phase structure, showing one layer parallel to $(40\bar{1})$ plane. Atoms in secondary layers are not shown. In all cases atoms to which only dashed lines are drawn are intermediate between secondary layers and main layers. Atoms not on the mean planes of main layers are projected normally onto them. Circles increase in size with coordination number. [Courtesy *Acta Crystallographica.*]

shell coordination polyhedra for the first time to our knowledge except in the mu phase. The existence and structural significance of these coordination polyhedra, which are the only ones possible with only triangular faces and only fivefold and sixfold vertices, were recognized independently by Kasper (Kasper, 1956; Frank and Kasper, 1958). A crystal structure containing only these coordinations possesses only tetrahedral interstices. Thus, these structures can be described in terms of packings of somewhat distorted regular tetahedra so as to fill all space, so that every tetrahedral edge coincides exactly with an edge of each of four or five additional tetahedra and the longest edge in a given structure is larger than the shortest by no more than a factor of about 4/3. Of the structures so far mentioned, all have only tetrahedral interstices and have only coordination polyhedra of the kinds mentioned—except α-Mn, or chi phase (Kasper, 1954), and the D phase, which contain some 13- and 11-coordinated atoms and distorted octahedral interstices. Other metal structures that also have only tetrahedral interstices and only the polyhedra mentioned include the Friauf-Laves phases (Friauf, 1927a, 1927b; Laves and Löhberg, 1934; Laves and Witte, 1935), the so-called β-tungsten phase (Hartmann, Ebert, and Bretschneider, 1931; Hägg and Schönberg, 1954), Zr_4Al_3 (Wilson, Thomas, and Spooner, 1960) and $Mg_{32}(Al,Zn)_{49}$ (Bergman, Waugh, and Pauling, 1957). The structures of several of these are shown in Figure 4. A large number of additional structures possess coordination polyhedra of this set, particularly the icosahedron (CN12), but polyhedra not in this set predominate; these structures include those of β-manganese and the gamma alloys. Some structures with large cubic unit cells investigated by Sten Samson—namely $NaCd_2$ (1962) and β-Mg_2Al_3 (1965)—possess mainly the regular coordination polyhedra, but in addition about

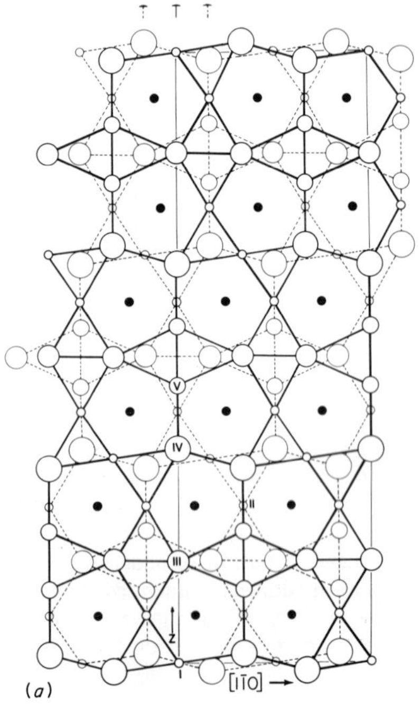

FIGURE 4.
(*a*) Structure of μ phase, Mo_6Co_7, normal projection on (110). (*b*) Structure of Zr_4Al_3, normal projection on (110). (*c*) Structure of Friauf-Laves phase $MgCu_2$ (C15), normal projection on (1$\bar{1}$0). (*d*) Structure of Friauf-Laves phase $MgZn_2$ (C14), normal projection on (110). Circles increase in size with coordination number. [Courtesy *Acta Crystallographica.*]

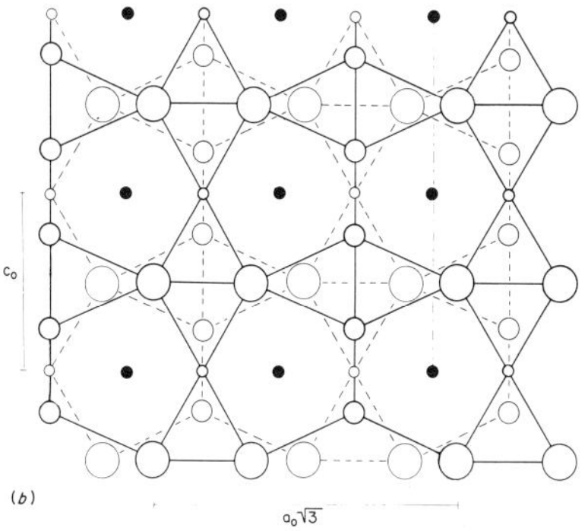

c_0
(b)
$a_0\sqrt{3}$

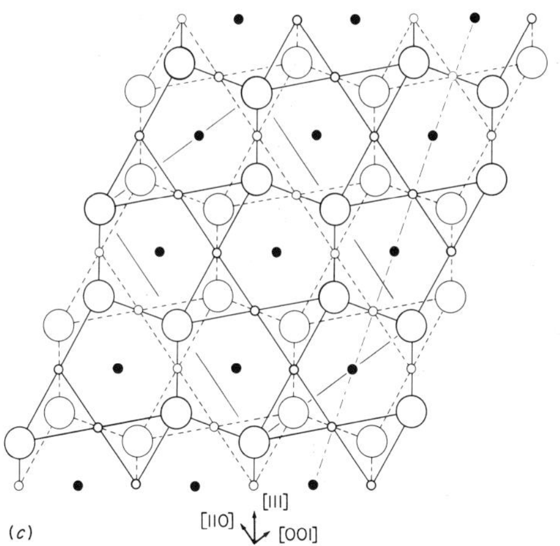

(c)
[111]
[110]
[001]

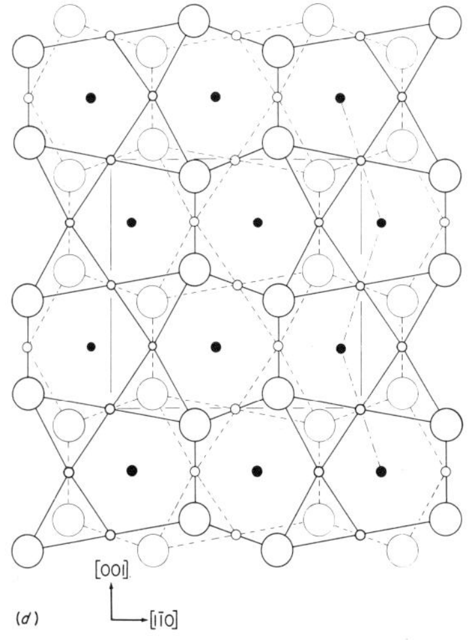

[001]
(d)
[1$\bar{1}$0]

a fifth of the atoms have irregular coordination polyhedra of ligancies 10–16 (Samson, this volume, p. 706). A discussion of polyhedra corresponding to large coordinations, principally triangulated but with some quadrangular faces, has been given by Kripyakevich (1960).

The principles governing the formation of these structures are not yet entirely clear, but some of the broad outlines are apparent. Briefly they are as follows: (1) These structures contains atoms of different sizes and somewhat variable radii, packed so as to give efficient filling of space with only tetrahedral interstices, and thereby giving rise only to the polyhedra mentioned. (2) The interlocking of these polyhedra is subject to geometrical and topological restrictions. (3) The transition metals present divide into two classes: those (B) to the right of the manganese column of the periodic table preferring icosahedral (CN12) sites; those (A) to the left preferring the other sites, manganese itself being ambivalent. We believe that this fact is related to the symmetry properties of the coordination polyhedra. (4) These phases are frequently considered to be "electron compounds," with valence electron concentration (VEC, electrons per atom) that for a given structure is nearly constant regardless of the transition elements present and generally is in the range 6 to 8 electrons per atom. However, much smaller values for the VEC are obtained when nontransition elements such as silicon and aluminum are present, and considerable variability of composition will require wide ranges in the VEC even when only transition elements are present.

That these phases contain atoms of different sizes has been recognized by many workers. The variability of the bonding radius of a given atom to different neighbors was recognized by Pauling (1947) and was ascribed to variability of bond number in accord with his famous equation $R(n) = R(1) - 0.300 \log n$. In our study of the present family of phases we find it necessary to assume at most two values for the radius of a given atom (Shoemaker et al., 1957): a longer radius, r, for bonds to 5-fold vertex atoms—the "minor ligands" of Frank and Kasper (1958)—and (except for CN12) a radius about 0.2 Å shorter, r^*, for bonds to 6-fold vertex atoms—the "major ligands." With the empirical formulas (Shoemaker and Shoemaker, 1964)

$$r_{12} = 0.127 + 0.810(\bar{R}) \quad \text{(in Å)}$$

$$r_{14} = 1.21\, r_{12} \qquad r_{14}^* = 0.99\, r_{12}$$

$$r_{15} = 1.28\, r_{12} \qquad r_{15}^* = 1.11\, r_{12}$$

$$r_{16} = 1.34\, r_{12} \qquad r_{16}^* = 1.22\, r_{12}$$

where $\bar{R}$ is a weighted average CN12 radius for the alloy, calculated with few exceptions from values given by Pauling (1956), distances taken as sums of the radii for the two atoms agree well with those observed, generally to within 0.05 Å. In our work on the M phase (1967), we applied these equations to preliminary adjustment of positional parameters of the trial structure preparatory to refinement with X-ray data. This preliminary least-squares "refinement" without X-ray data to fit interatomic distances calculated with radii obtained with these equations gave parameters for which the conventional agreement index R (subsequently calculated from observed and calculated X-ray structure factors, excluding null observations) was 24 percent. Subsequent least-squares refinement with X-ray data reduced R to 12 per-

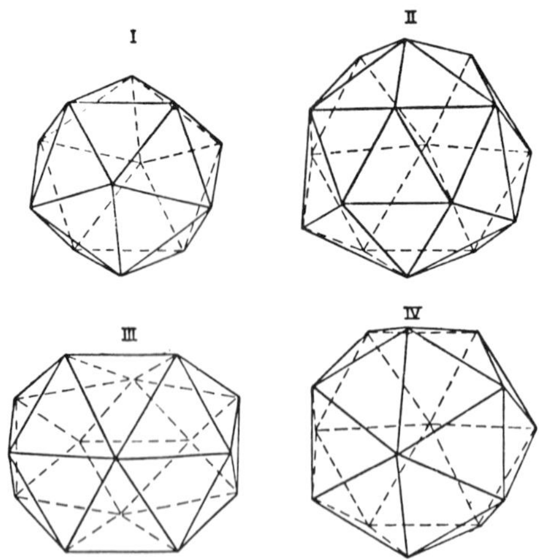

FIGURE 5.
The four triangulated coordination polyhedra with 5-fold and 6-fold vertices: I, CN12 (icosahedron); II, CN16; III, CN15; IV, CN14. [From Bergman, Waugh, and Pauling (1957); courtesy *Acta Crystallographica.*]

cent, with no coordinate shift larger than 0.13 Å and no change in interatomic distance larger than 0.11 Å; the average changes were much smaller. By contrast, first-coordination-shell interatomic distances in a given phase ordinarily span a range of 0.8 or 0.9 Å.

The existence of two separate radii for CN14, 15, and 16, and the increase in radius with coordination probably are primarily consequences of packing geometry, but they quite possibly have important implications for electronic structure. The sites with these larger coordinations preferentially select those elements that have the needed larger atomic radii and also have appropriate symmetry orbitals available for the shorter "major-ligand" bonds, as is discussed further below. It is noteworthy that the sigma phase, which does not contain sites of CN16, is often found in binary systems in which both components belong to the same long period, but transition metal phases having CN16 almost always contain elements from more than one long period. (However, some Laves phases exist with two elements some distance apart in the same long period.) In nearly all ternary phases containing only transition elements, two of the components are "A" atoms and are in different long periods, evidently in order to fill the need for atoms of different sizes to span the coordination range CN14, 15, 16 (Nevitt, 1963). In phases containing silicon or aluminum, however, Si or Al is frequently one of two "B" components occupying icosahedral (CN12) sites.

Geometrical and topological considerations require that at least two different coordinations be present in phases of this type; a structure in which (for example) all atoms are icosahedrally coordinated is apparently impossible. Beyond this, a general treatment of the requirements of geometry and topology that would enable the prediction of all possible structures of this type is not yet in sight. The closest approach is that of Frank and Kasper (1958, 1959), who succeeded in treating those structures that have plane layers in terms of tesselations and of variations upon kagomé tiling. By systematic investigation of stacking and of tesselation faults, these authors have generated a family of hypothetical structures, which happen to include some actual structures already known at the time (those of the Friauf-Laves, σ, P, β-W, and μ phases) and some determined between that time and the present (those of the M

phase and Zr_4Al_3). The structures of the R and δ phases, in which the layers are somewhat warped or buckled, could not be predicted in a systematic fashion on the basis of their criteria. However, the structure of $Mg_{32}(Al,Zn)_{49}$, also with warped layers, was at least meaningfully discussed by Frank and Kasper in terms of the principles they outlined.

The distinct predominance of "A" atoms (to the left of Mn) in CN14, 15, and 16 sites and that of "B" atoms (to the right of Mn) in CN12 sites has been demonstrated by X-ray diffraction work in those cases (Forsyth and da Veiga, 1962; Shoemaker and Shoemaker, 1963), in which the atoms differed significantly in X-ray scattering power. In the case of sigma (Kasper and Waterstrat, 1956), chi (Kasper, 1954), P (Shoemaker, Shoemaker, and Mellor, 1965), and R (Shoemaker et al., 1965) phases, neutron diffraction work gave substantially the same result. As may be seen in Table 1, the percentage of B atoms in some cases deviates considerably from the percentage of CN12 sites, the binary sigma and delta phases being examples. Here some B atoms must occupy non-CN12 sites. Where manganese is present, the compositions are widely variable (Nevitt, 1963), owing presumably to the ability of manganese to act either as an A component or as a B component (as in the α-Mn structure itself).

We believe that the fact that the icosahedral atoms have the more nearly filled *d*-shells is no accident. Indeed, we think that it is connected with the fact that perturbations conforming to the point group of the icosahedron, $Y_h - \bar{5}\,\bar{3}\ (2/m)$, do not affect the degeneracy of the atomic *p* and *d* levels, but with all of the other coordination symmetries at least the *d*-level degeneracy is partially broken; see Table 2 (Shoemaker, 1958; Shoemaker and Shoemaker, 1960). The surroundings of CN12 atoms do not in fact conform to Y_h symmetry with exactidude in the first coordination shell, and deviate widely from this symmetry in outer coordination shells. However, the effective deviations from such symmetry are at any rate small enough to have rather small effect on the r_{12} values (ordinarily less than 0.03 Å), and it seems reasonable to expect that to a first approximation the bonding of CN12 atoms is the same as that which would exist under exact Y_h symmetry. Thus, it seems difficult to justify for icosahedral CN12 atoms any formal bonding scheme involving directional bonds of high *d* character.

It is also noteworthy that when aluminum or silicon combines with transition metals to form structures of this kind, as in the cases of σ-Nb_2Al (Brown and Forsyth, 1961) and the M phase of Nb-Ni-Al (Shoemaker and Shoemaker, 1967), CN12 coordination is strongly predominant for these elements. The same is true in the alloys Zr_4Al_3 (Wilson et al., 1960) and $Mg_{32}(Al,Zn)_{49}$ (Bergman et al., 1957). Thus, it appears that the icosahedral sites show a predominance for atoms with filled or nearly filled, or else empty, *d* shells; atoms with *d* shells close to half filled are mostly found in coordinations of lower symmetry where the *d* orbitals may hybridize with *s* and *p* orbitals to form directed bonds. It is attractive to speculate that with the transition B elements Mn through Ni some degree of electron transfer takes place, so that the CN12 *d* shell is entirely filled or else more nearly filled than would otherwise be the case. Thus, in a formal sense, part of the bonding in these phases may be ionic.

TABLE 1.
Crystal structures of sigma-phase-related phases with tetrahedral interstices.

Phase	Example	Space group	Lattice constants (Å)			Atoms per cell	Sites with CN (%)					
			a_0	b_0	c_0		11	12	13	14	15	16
A15	β-W (W_3O)	O_h^3—Pm3n	5.036			8		25		75		
	Cr_3Si (75, 25)		4.564									
σ	$Cr_{46}Fe_{54}$	D_{4h}^{14}—$P4_2/mnm$	8.800		4.544	30		33		53	13	
Zr_4Al_3	$Zr_{57}Al_{43}$	C_{3h}^1—$P\bar{6}$	5.433		5.390	7		43		28	28	
P	$Mo_{42}Cr_{18}Ni_{40}$	D_{2h}^{16}—Pbnm	9.070	16.983	4.752	56		43		36	14	7
δ	$Mo_{50}Ni_{50}$	D_2^4—$P2_12_12_1$	9.108	9.108	8.852	56		43		36	14	7
R	$Mo_{31}Cr_{18}Co_{51}$	C_{3i}^2—$R\bar{3}$	10.903		19.342	159[a]		51		23	11	15
μ	$Mo_{46}Co_{54}$	D_{3d}^5—$R\bar{3}m$	4.762		25.615	39[b]		54		15	15	15
M	$Nb_{48}Ni_{39}Al_{13}$	D_{2h}^{16}—Pnam	9.303	16.266	4.933	52		54		15	15	15
$Mg_{32}(Al, Zn)_{49}$	$Mg_{40}(Al, Zn)_{60}$	T_h^5—Im3	14.16			162		61		7	7	25
C15	$MgCu_2$(33, 67)	O_h^7—Fd3m	7.080			24		67				33
C14	$MgZn_2$(33, 67)	D_{6h}^4—$P6_3/mmc$	5.16		8.50	12		67				33
C36	$MgNi_2$(33, 67)	D_{6h}^4—$P6_3/mmc$	5.27		13.3	24		67				33
A12	α-Mn	T_d^3—$I\bar{4}3m$	8.912			58		41	41			17
	χ-$Mo_{17}Cr_{21}Fe_{62}$		8.920									
D	$V_{26}Fe_{44}Si_{30}$	D_4^4—$P4_12_12$	8.833		8.646	56	14	29	14	14		29
β-Mg_2Al_3	$Mg_{40}Al_{60}$	O_h^7—Fd3m	28.239			1168[c]		58				22

[a] 53 atoms per primitive rhombohedral cell.
[b] 13 atoms per primitive rhombohedral cell.
[c] 21 percent of the atoms have irregular coordinations with CN10-16.

TABLE 2.
Triangulated coordination polyhedra with 5-fold and 6-fold vertices.

Type	No. of vertices 5-fold	No. of vertices 6-fold	No. of faces	Ideal point symmetry	Ideal sublevel degeneracies *p*	Ideal sublevel degeneracies *d*
CN12[a]	12	0	20	Y_h—$\bar{5}\bar{3}(2/m)$	3	5
CN14	12	2	24	D_{6d}—$\overline{12}\cdot 2\cdot m$	1,2	1,2,2
CN15	12	3	26	D_{3h}—$\bar{6}m2$	1,2	1,2,2
CN16	12	4	28	T_d—$\bar{4}3m$	3	2,3

[a]Regular, or approximately regular, icosahedron.

In the Friauf-Laves phases $MgCu_2$, $MgZn_2$, and $MgNi_2$, the B component (Cu, Zn, and Ni) has a filled or nearly filled *d* shell and may even (in the first two cases) transfer electrons to Mg, which (partly on account of its size) behaves as an A component. In $Mg_{32}(Al,Zn)_{49}$ (Bergman et al., 1957), Mg acts as an A component and Al and Zn are largely interchangeable as CN12 B components. (However, in this structure the Al atoms predominantly occupy the CN12 sites of larger radius.) In Zr_4Al_3 (Wilson et al., 1960), the Zr atoms occupy A-type sites with CN14 and 15, and Al atoms occupy B sites with CN12.

It may be significant that the directions of the "major ligands" in the CN14, 15, and 16 polyhedra are among those most commonly encountered in directed covalent bonding: respectively they are linear digonal, planar trigonal, and tetrahedral, suggesting *sp* or *dp*, sp^2 or dp^2 or d^3, and sp^3 or d^3s hybridization (or various mixtures in each case). Presumably the transition elements make good use of the *d* orbitals available; for Mg in the Friauf-Laves phases and in $Mg_{32}(Al,Zn)_{49}$ the *s* and *p* orbitals must suffice. Beyond this it does not seem worth while to speculate in more detail.

We have been unable to refine the above picture to one that would enable us to predict, on the basis of structure alone, the "valence electron concentrations" of these alloys or the precise composition or composition range of a given phase in a given alloy system. Indeed, there are many instances in which the VEC values differ considerably among alloys having the same structure (for example, β-U and σ-FeCr; μ-W_6Fe_7 and μ-W_6Co_7; $MgZn_2$ and WFe_2), and in which the regions of stability of the phases in the phase diagram are very wide. This suggests that although electronic structure is strongly influential in determining the crystal structure type and in the choice of atomic sites, effects of atomic size are important enough to override fairly large differences in electron content. For these same reasons we do not see much prospect for success of the Brillouin zone treatment as applied to these phases, although it seemed to be very successful in other cases, such as that of the gamma alloys (Jones, 1934; Pauling and Ewing, 1948; Shoemaker and Huang, 1954). Thus the "electron compound" concept is certainly not as clearly applicable as in the case of the gamma alloys.

Our understanding of the electronic structures of these phases can perhaps be helped by further experimental study of those properties of these phases that sensitively depend on electronic structure, notably the locations and magnitudes of atomic magnetic moments. The magnetic moments of iron or cobalt atoms may be quite different in these coordinations than in those found in body-centered or face-centered cubic structures. The small saturation moments and low ferromagnetic Curie temperatures of iron-containing sigma phases (Nevitt and Beck, 1955) may be due to filled or almost filled iron 3*d* shells in CN12, and only small amounts of iron in coordinations of lower symmetry. A neutron diffraction study of sigma phases is in preparation in our laboratory and, although the study must be restricted at least at first to those very few sigma phases that can be prepared in single-crystal form by peripectic transition, we hope that the results will shed light on this problem.

In addition, there remains much to be done in the way of X-ray structure determinations of additional phases of this kind. There are at least a dozen ternary phases prepared by Professor Beck's group, the structures of which remain to be determined and which probably belong to the family of sigma-phase related structures. Hopefully more of the structures predicted by Frank and Kasper (1958, 1959) will be found to exist, and new clues to the principles of atomic structure and electronic structure will be found.

ACKNOWLEDGMENTS

We wish to thank Professor Linus Pauling for motivating us, by his accomplishments in the structural chemistry of metals and by his interest and encouragement, to enter this field of research. We wish also to thank Professor Paul A. Beck, Professor Pol Duwez, Dr. Paul Pietrokowsky, and Dr. Bill C. Giessen for kindly providing us with specimens of the phases on which our work was done.

REFERENCES

Arnfelt, H., and A. Westgren (1935). *Jernkontor. Ann.* **119**, 185.
Benjamin, J. S., B. C. Giessen, and N. J. Grant (1966). *Trans AIME* **236**, 224.
Bergman, B. G. (1951). Ph.D. Dissertation, California Institute of Technology.
Bergman, B. G., and D. P. Shoemaker (1951). *J. Chem. Phys.* **19**, 515.
Bergman, B. G., and D. P. Shoemaker (1954). *Acta Cryst.* **7**, 857.
Bergman, G., J. L. T. Waugh, and L. Pauling (1957). *Acta Cryst.* **10**, 254.
Bradley, A. J., and J. Thewlis (1927). *Proc. Roy. Soc.* A **115**, 456.
Brown, P. J., and J. B. Forsyth (1961). *Acta Cryst.* **14**, 362.
Forsyth, J. B., and L. M. d'Alte da Veiga (1962). *Acta Cryst.* **15**, 543.
Frank, F. C., and J. S. Kasper (1958). *Acta Cryst.* **11**, 184.
Frank, F. C., and J. S. Kasper (1959). *Acta Cryst.* **12**, 483.
Friauf, J. B. (1927a). *J. Am. Chem. Soc.* **49**, 3107.
Friauf, J. B. (1927b). *Phys. Rev.* **29**, 34.
Hägg, G., and N. Schönberg (1954). *Acta Cryst.* **7**, 351.
Hartmann, H., F. Ebert, and O. Bretschneider (1931). *Z. anorg. allgem. Chem.* **198**, 116.
Jones, H. (1934). *Proc. Roy. Soc.* A **144**, 225; **147**, 396.
Kasper, J. S. (1954). *Acta Metallurgica* **2**, 456.
Kasper, J. S. (1956). *Theory of Alloy Phases.* Am. Soc. of Metals Symposium, p. 264.
Kasper, J. S., and R. M. Waterstrat (1956). *Acta Cryst.* **9**, 289.
Komura, Y., W. G. Sly, and D. P. Shoemaker (1960). *Acta Cryst.* **13**, 575.
Kripyakevich, P. I. (1960). *Kristallografiya* **5**, 79.
Laves, F., and K. Löhberg (1934). *Nachr. Akad. Wiss. Göttingen Math. physik. Kl.* IV, new ser. **1**, No. 6, p. 59.
Laves, F., and H. Witte (1935). *Metallwirtschaft* **14**, 645.

Nevitt, M. V. (1963). In Beck, P. A., ed., *Electronic Structure and Alloy Chemistry of the Transition Elements*, pp. 101 ff, esp. p. 118. New York: Interscience.
Nevitt, M. V., and P. A. Beck (1955). *Trans. AIME* **203**, 669.
Pauling, L. (1947). *J. Am. Chem. Soc.* **69**, 542.
Pauling, L. (1956). In *Theory of Alloy Phases.* Am. Soc. of Metals Symposium, Cleveland.
Pauling, L., and F. J. Ewing (1948). *Rev. Mod. Phys.* **20**, 112.
Rideout, S., W. D. Manly, E. L. Kamen, B. S. Lement, and P. A. Beck (1951). *Trans. AIME* **191**, 872.
Samson, Sten (1962). *Nature* (London) **195**, 259.
Samson, Sten (1965). *Acta Cryst.* **19**, 401.
Shoemaker, C. B., and D. P. Shoemaker (1963). *Acta Cryst.* **16**, 997.
Shoemaker, C. B., and D. P. Shoemaker (1964). *Trans. AIME* **230**, 486.
Shoemaker, C. B., and D. P. Shoemaker (1967). *Acta Cryst.* **23**, 231.
Shoemaker, D. P. (1958). Final Report, Office of Ordnance Research Project No. 461, Contract DA-19-020-ORD-1859.
Shoemaker, D. P., and B. G. Bergman (1950). *J. Am. Chem. Soc.* **72**, 5793.
Shoemaker, D. P., and T. C. Huang (1954). *Acta Cryst.* **7**, 249.
Shoemaker, D. P., and C. B. Shoemaker (1950). Abstract 4.4, Fifth International Congress and Symposia of the International Union of Crystallography, Cambridge, England.
Shoemaker, D. P., C. B. Shoemaker, and J. Mellor (1965). *Acta Cryst.* **18**, 37.
Shoemaker, D. P., C. B. Shoemaker, and F. C. Wilson (1957). *Acta Cryst.* **10**, 1.
Tucker, C. W. (1950). *Science* **112**, 448.
Tucker, C. W., and P. Senio (1953). *Acta Cryst.* **6**, 753.
Wilson, C. G., D. K. Thomas, and F. J. Spooner (1960). *Acta Cryst.* **13**, 56.

JOHN L. T. WAUGH
Department of Chemistry
University of Hawaii
Honolulu, Hawaii

Isopolyborates

It has been recognized for many years now that the oxygen chemistry of boron is perhaps more complex than that of silicon. There are common difficulties associated with the classification of complex inorganic compounds. Although generally, among such compounds, stoichiometric composition provides no indication of their structure, it may be plausibly assumed that compounds with the same stoichiometric composition are isostructural. However, BaB_4O_7 and SrB_4O_7 are found to have quite unrelated anionic structural arrangements. Ionic species identifiable in the solid phase do not necessarily have any stable existence in solution, and although it has long been assumed, in order to account for the behavior of boric acid and borate solutions, that two or more polyions coexist in equilibrium with one another in their aqueous solutions, the only ionic species for whose existence in solution there is any convincing evidence is the $[B(OH)_4]^-$ ion. For the borates, there has been the additional difficulty in structure determinations by X-ray crystallographic methods, of precisely locating the positions of the boron atoms; the majority of known borate structures have been reported since 1956. Compared with silicon-oxygen compounds, in almost all of which the silicon atoms are uniformly tetrahedrally coordinated with oxygen only, so that they may be conveniently classified in terms of their Si:O ratio, the boron-oxygen compounds may variously contain boron atoms that are all triangularly coordinated, all tetrahedrally coordinated, or both types may be present in the same compound, coordinated with either oxide ions, hydroxyl ions, or more rarely water molecules. In isopolyborates containing only triangularly coordinated boron atoms, the same principles that determine silicate structures are found to apply, with the important distinction that no cyclical group other than a 6-membered ring has yet been reported in an isopolyborate. There is thus no boron analogue of the amphibole or beryl structures, for example. Monomeric and dimeric anionic units are common, which may polymerize to endless chain, layer, and three-dimensional network type structures; trimeric, tetrameric, and pentameric anionic units are also found in isopolyborate structures.

Recognition of the existence of two different coordination numbers in many isopolyborates and the fact that hydroxyl ions or water molecules may constitute part of the coordination group around a boron atom, has necessarily led to the reformulation of many boron-oxygen compounds, to abandoning any attempt to classify them in terms of their B:O ratio, and makes it particularly inappropriate to attach any significance to the older terminology which refers to ortho-, meta-, and pyroborates. Thus, consistent with the structural information, borax is preferably formulated $[Na_2(OH_2)_8]$ $[B_4O_5(OH)_4]$ rather than $Na_2B_4O_7 \cdot 10H_2O$; the so-called metaborates of sodium, lithium, and zinc are found to contain $[B_3O_6]^{---}$, $[B(OH)_4]^-$, and extended BO_4 network-type anions, respectively; and the three crystallographic modifications of metaboric acid have quite dissimilar structures.

Generally, the distinction between oxide ions, hydroxyl ions, and water molecules in boron-oxygen compounds which also include hydrogen in their composition, has not been made directly on the basis of experimental data; the precise location of hydrogen nucleii in such compounds has actually been accomplished only for boric acid and for metaboric acid-III (Zachariasen, 1954, and Peters and Milberg, 1964, respectively). However, from consideration of the compatibility of observed internuclear distances with the composition, cell dimensions, symmetry, and space group requirements of the crystalline species concerned it is usually possible to infer the location of protons with respect to oxygen nucleii.

Generalizations about the then-known borate structures have been recorded from the points of view of crystallographic data, aqueous phase relationships of the polyborates, conformational similarities of borates and organic compounds, and with special reference to anhydrous borates (Christ, 1960; Edwards and Ross, 1960; Dale, 1961; and Krogh-Moe, 1960, 1962 respectively). In this article, structural information on the synthetic and mineral borates and boric acids, as well as information available from the examination of the behavior and interactions of their aqueous solutions, have been surveyed in an attempt to classify the known isopolyborates and to speculate on the structure of others.

POLYBORATES OF KNOWN CRYSTAL STRUCTURE

About one-third of all of the boron-oxygen compounds for which crystal structural information has been recorded (at least 72 by July 1966) are naturally occurring minerals, many of them hydrated substances. The principal naturally formed borates are calcium, magnesium, and sodium compounds. Of the 45 mineral borates listed in Dana's *System of Mineralogy* (revised by Palache, Berman, and Frondel, 1951), the crystal structure of 18 have now been determined, together with that of 7 other minerals not included in Dana, namely, danburtite, $CaB_2Si_2O_8$, datolite, $CaBSiO_4OH$, tunnelite, $SrB_6O_9(OH) \cdot 3H_2O$, sassolite, the naturally occurring form of boric acid, $B(OH)_3$, reedmargite, $NaBSi_3O_8$, axinite, $H(Fe,Mn)Ca_2Al_2BSiO_{20}$, and tourmaline, $(Na,Ca)(Mg,Fe)_3$ $Al_6B_3Si_6(O,OH,F)_{31}$.

The compounds of known structure are listed below, classified according to whether all of the constituent boron atoms are triangularly coordinated, all are tetrahedrally coordinated, or some of each type are present in the compound. Each of these three groups is subclassified according to the presence of discrete anions, or chain, layer, or infinite network type polyanions. A brief description of the essential features of the structure, the interatomic distances (where available), and the appropriate reference is included for each substance. Attention is focused on the boron-oxygen arrangement; metallic cations are generally present with their usual coordination number, although exceptions are to be noted as in $Na_3B_3O_6$ and $K_3B_3O_6$ [15, 16]*, where both Na^+ and K^+ ions are 7-coordinated, $Zn_4O(BO_2)_6$ where Zn^{++} is 4-coordinated, and CdB_4O_7 where Cd^{++} is 4-coordinated.

Figures 1, 2, and 3 illustrate the significant geometrical features of the known structures. The heavily drawn lines indicate the boron-oxygen bonds, and the predominance of the 6-membered ring structures with 3 boron atoms alternating with 3 oxygen atoms in all of the more complex anionic types is especially emphasized in this manner; the more lightly drawn lines denote the triangular or tetrahedral groups surrounding the boron atoms. The cationic positions have been omitted in all of these illustrations, and they are not all drawn to the same scale.

*Numbers in square brackets, here and in the remainder of this article, refer to compounds in the list that begins on page 736

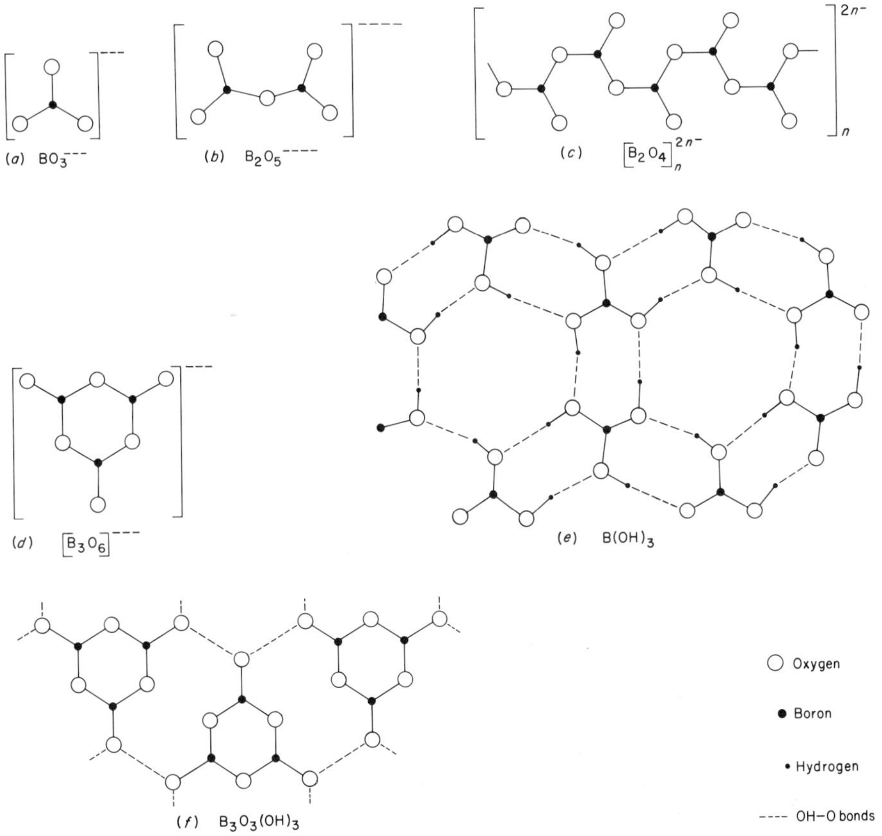

(a) BO_3^{---} (b) $B_2O_5^{----}$ (c) $[B_2O_4]_n^{2n-}$ (d) $[B_3O_6]^{---}$ (e) $B(OH)_3$ (f) $B_3O_3(OH)_3$

○ Oxygen
● Boron
• Hydrogen
---- OH–O bonds

FIGURE 1.
Isopolyborate structures in which all of the constituent boron atoms are triangularly coordinated:
(*a*) discrete monomeric anions, BO_3^{---}, present in compounds 1–12;
(*b*) discrete dimeric anion, $[B_2O_5]^{----}$, found in compounds 13–14;
(*c*) endless chain type anion, $[B_2O_4]_n^{2n-}$, found in compounds 18–19;
(*d*) discrete trimeric anion, $[B_3O_6]^{---}$, present in compounds 15–17;
(*e*) portion of a layer of the $B(OH)_3$ structure;
(*f*) portion of a layer of the α- or metaboric acid-III structure.

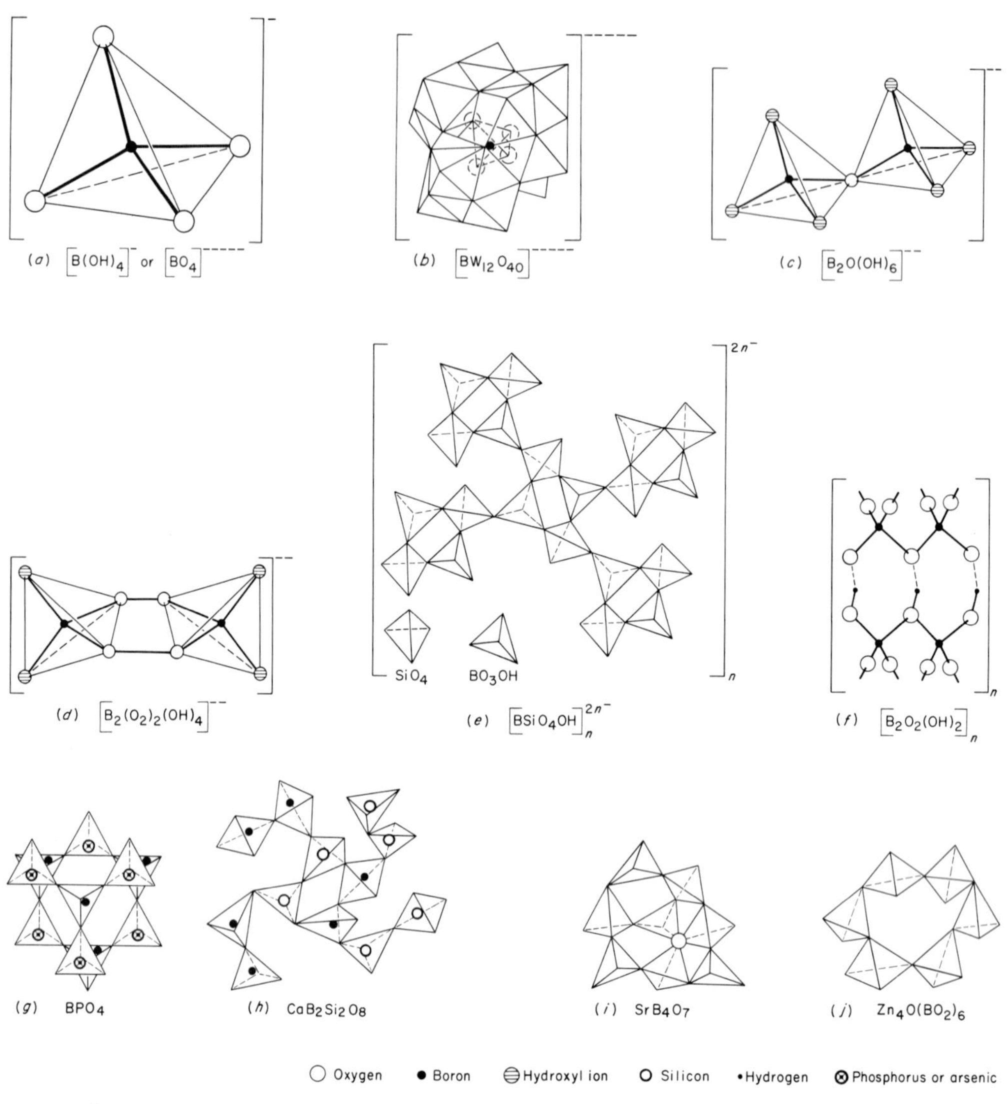

FIGURE 2.
Isopolyborate structures in which all of the constituent boron atoms are tetrahedrally coordinated:
(*a*) discrete monomeric anion, $[BO_4]^{-----}$ or $[B(OH)_4]^-$, found in compounds 29–35;
(*b*) the heteropoly anion, $[BW_{12}O_{40}]^{-----}$, containing a central BO_4 group [36];
(*c*) discrete dimeric anion, $[B_2O(OH)_6]^{--}$, found in mineral pinnoite [37];
(*d*) discrete, dimeric, centro-symmetrical anion, $[B_2(O_2)_2(OH)_4]^{--}$, found in sodium peroxoborate [38];
(*e*) portion of the layer type anion in the mineral datolite, $CaBSiO_4(OH)$ [39];
(*f*) portion of the γ- or metaboric acid-I structure projected on the *ab* plane of the crystallographic axes, showing endless chains of composition, $[B_2O_2(OH)_2]_n$, parallel to the *c*-axis, which in turn are cross linked into a network structure;
(*g*) cristobalite type structure of BPO_4 and $BAsO_4$, showing alternate BO_4 and PO_4 (or AsO_4) tetahedral groups;
(*h*) portion of the danburtite structure, $CaB_2Si_2O_8$, as viewed along the *b* crystallographic axis, indicating that alternate pairs of BO_4 and SiO_4 groups are present in this structure;
(*i*) portion of the three-dimensional network structure found in SrB_4O_7 and PbB_4O_7 viewed along the *c* crystallographic axis showing one of the special position oxygen atoms common to three BO_4 tetrahedra;
(*j*) portion of the network structure of zinc hexaborate, $Zn_4O(BO_2)_6$, showing one group of 6 BO_4 tetrahedra; other 2 oxygen atoms of each tetrahedron are involved in the formation of similar rings of 6 tetrahedra.

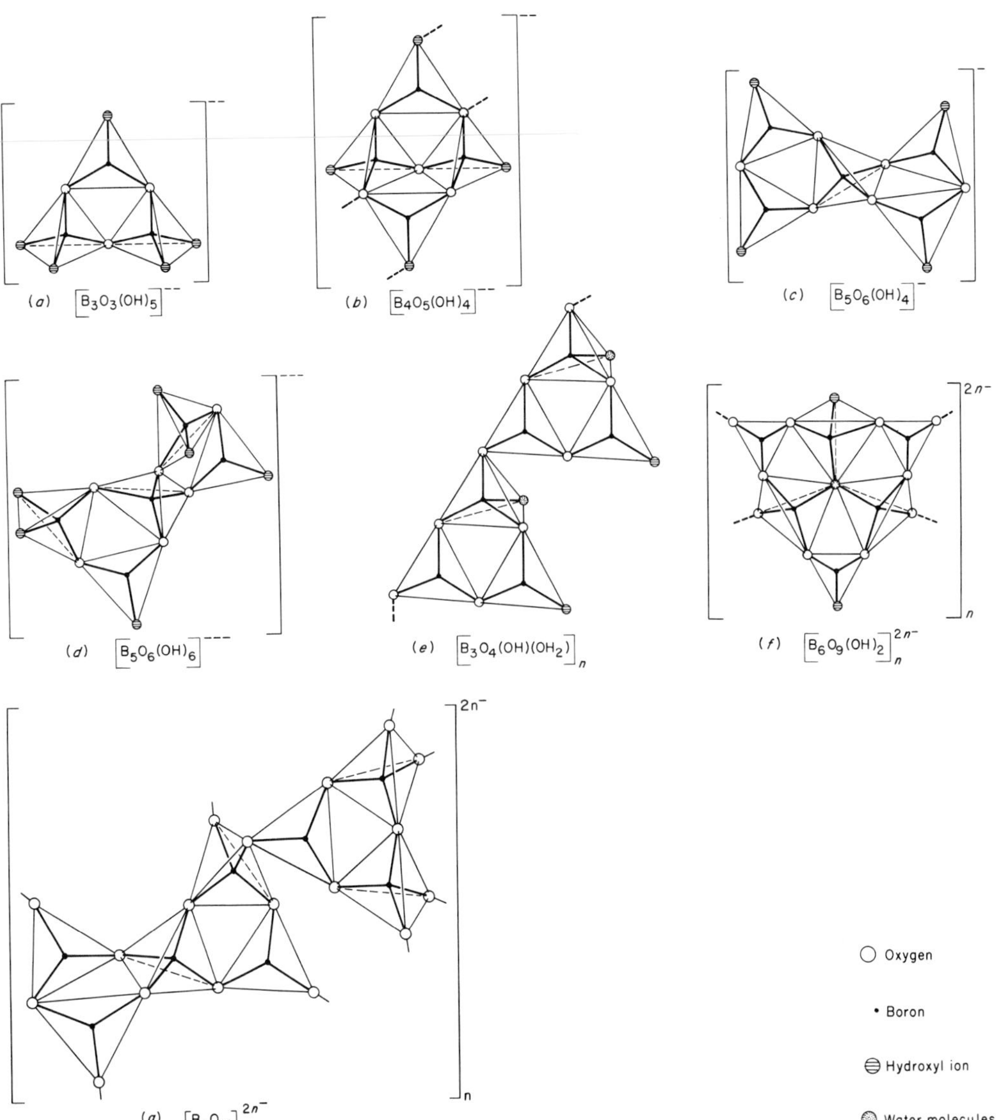

FIGURE 3.
Complex isopolyborate anions containing both triangularly and tetrahedrally coordinated boron atoms:

(*a*) the $[B_3O_3(OH)_5]^{--}$ anion found in the minerals meyerhofferite, inyoite, and the synthetic compound 53 as a discrete anion and condensed into chain type anion, $[B_3O_4(OH)_3]_n^{n-}$, in the mineral colemanite [59] and condensed further as the layer type anion in the synthetic compound 61;

(*b*) the discrete anion, $[B_4O_5(OH)_4]^{--}$, found in borax [54], in potassium tetraborate [55], and cross linked into a three-dimensional network structure in lithium, silver, and cadmium tetraborates [63, 64, 68];

(*c*) the discrete anionic species, $[B_5O_6(OH)_4]^{-}$, found in hydrated potassium pentaborate [56] and cross linked into infinite network in the anhydrous salt [65];

(*d*) the discrete complex anion, $[B_5O_6(OH)_6]^{---}$, found in the mineral ulexite [57], and condensed into chain type anions, $[B_5O_7(OH)_4]_n^{n-}$, in the mineral probertite [60];

(*e*) portion of the extended chain structure found in γ- or metaboric acid-II, showing chains of composition $[B_3O_4(OH)(OH_2)]_n$ projected on (201) crystallographic planes;

(*f*) complex anion found in mineral, tunnelite, as infinite sheets of composition, $[B_6O_9(OH)_2]_n^{n-}$ [62];

(*g*) portion of network structure of anionic group in BaB_4O_7, composed of double ring pentaborate anion as in (*c*) fused with an (*a*) type unit through a common oxygen atom.

All constituent boron atoms triangularly coordinated:

Discrete monomeric anions, BO_3^{---}.

1. $ScBO_3$, scandium borate; isostructural with calcite; no structural parameters available. (Goldschmidt and Hauptmann, 1932.)
2. $InBO_3$, indium borate. (Ibid.)
3. YBO_3, yttrium borate. (Ibid. for one crystallographic modification.)
4. $LaBO_3$, lanthanum borate; isostructural with aragonite. (Ibid.)
5. $GaBO_3$, gallium borate; isostructural with calcite. (Bernal, Struck, and White, 1963.)
6. $CrBO_3$, chromium borate. (Ibid.)
7. $TiBO_3$, titanium borate. (Ibid.)
8. VBO_3, vanadium borate; calcite structure. (Schmid, 1964.)
9. $CaSn(BO_3)_2$, mineral, nordenskjöldine; isostructural with dolomite, $CaMg(CO_3)_2$; rhombohedral, $R\bar{3}$, $Z = 1$, $\rho = 4.20$, power data only. (Zachariasen, 1931; Ehrenberg and Ramdohr, 1934.)
10. $Be_2BO_3(OH)$, mineral, hambergite; orthorhombic, P bca, $Z = 8$, $\rho = 2.359$; array of OH^-, BO_3^{---}, and Be^{++} ions, such that each Be^{++} ion is tetrahedrally coordinated to 3 oxygen atoms of borate ions and one OH^- ion; only known beryllium borate. (Zachariasen, 1931; Zachariasen, Plettinger, and Marezio, 1963.)
11. $Mg_3(BO_3)_2$, mineral, kotoite; orthorhombic, $Z = 2$, $\rho = 3.10$; been synthesized; oxygen atoms packed similarly to those in Mg_2SiO_4; B—O, 1.38 Å. (Berger, 1949.)
12. $Co_3(BO_3)_2$, cobalt borate; isostructural with kotoite above; mean B—O, 1.38 Å. (Ibid.)

Discrete dimeric anions, $B_2O_5^{----}$.

13. $Mg_2B_2O_5$, magnesium diborate or pyroborate; mineral form, $Mg_2B_2O_5 \cdot H_2O$, known as ascherite, szaibelyite, or sussexite, is the most insoluble of the calcium and magnesium borates; angle between planes of BO_3 groups, 22° 19′, B—O, 1.36 Å, central B—O—B angle, 153°. (Takéuchi, 1952.)
14. $Co_2B_2O_5$, cobalt diborate or pyroborate; mean B—O, 1.30 Å, central B—O—B angle, 153°, planes of terminal BO_2 groups at angles of +7° and −7° with respect to central B—O—B unit. (Berger, 1950.)

Discrete trimeric anions, $B_3O_6^{---}$.

15. $Na_3B_3O_6$, sodium metaborate; anion built up of planar cyclic rings composed of 3 BO_3 triangles with common O—O edges, Na^+ ions are 7-coordinated; rhombohedral. $R\bar{3}c$; B—O, 1.382 Å. (Fang, 1938; Marezio, Plettinger, and Zachariasen, 1963b.)
16. $K_3B_3O_6$, potassium metaborate; isostructural with sodium metaborate above, K^+ ions also 7-coordinated; B—O, 1.361 Å. (Zachariasen, 1937a.)
17. BaB_2O_4, barium diborate; high temperature modification has rhombohedral structure, $R\bar{3}c$, $Z = 18$; contains planar $B_3O_6^{---}$ and Ba^{++} ions, one type being 6-fold coordinated (triagonal prismatic) and the

other being 9-fold coordinated; B—O, 1.32 Å, B—O—B angles, 122°, O—B—O internal, 117°, O—B—O, external, 120°. (Mighell, Perloff, and Block, 1966.)

Infinite one-dimensional chain type anions, $[B_2O_4]_n^{2n-}$.

18. CaB_2O_4, calcium diborate or metaborate; anion composed of infinite chains of BO_3 groups, each BO_3 group having 2 oxygen atoms in common with 2 adjacent groups and 1 unshared atom; orthorhombic, P nca, Z = 2; chains bonded together by Ca^{++} ions; boron analogue of pyroxene chain, SiO_3^{--}; B—O, 1.372 Å. (Zachariasen and Ziegler, 1932; Marezio, Plettinger, and Zachariasen, 1963a.)
19. $LiBO_2$, lithium metaborate; endless chains of BO_3 groups held together by Li^+ ions; analagous to cristobalite structure; monoclinic, P 2_1/c, Z = 4; B—O, 1.373 Å. (Zachariasen, 1964.)

Infinite two-dimensional layer type anions of hydrogen-bonded molecules.

20. $B(OH)_3$, boric acid, mineral sassolite; triclinic, P $\bar{1}$, Z = 4; structure very accurately determined, built up of $B(OH)_3$ molecules joined together by O—H···O bonds to form infinite layers of pseudohexagonal symmetry spaced 3.181 Å apart, layers held together by van der Waals forces; B—O, 1.361 Å, O—H, 0.88 Å, O—H···O, 2.720 Å, oxygen bond angle, 114°; unsymmetrical hydrogen bonds. (Zachariasen, 1934; 1954.)
21. $H_3B_3O_6$, metaboric acid-III or α-metaboric acid; structure composed of trimeric $H_3B_3O_6$ molecules bonded together by hydrogen bonds into layers of pseudohexagonal symmetry spaced 3.128 Å apart, layers held together by van der Waals forces; ρ = 1.784; orthorhombic, O—H···, 2.74 Å, some short hydrogen bonds at 2.52 Å. (Tazaki, 1940; Peters and Milberg, 1964.)

Close-packed structures with boron atoms in positions of triangular coordination.

22. $Mg_3(OH, F)_3BO_3$, mineral, fluoborite; close-packed array of O^{--}, (OH^-) or F^-, with boron atoms in positions of triangular coordination and Mg^{++} ions in positions of 6-fold coordination, by 3 O^{--} ions and 3 OH^- (F^-) ions. (Takéuchi, Watanabe, and Ito, 1950a.)
23. $Fe_2^{++}(Mg, Fe^{++})_4B_2O_{10}$, mineral, ludwigite; P bam, ρ = 4.7; has been synthesized; orthorhombic, spinel type structure, with 6-coordinated iron and magnesium ions and triangularly coordinated boron atoms. (Bertaut, 1950.)
24. $Fe_2^{+++}(Fe^{++}, Mg)_4B_2O_{10}$, mineral, paigeite; isostructural with ludwigite. (Ibid.)
25. $(Fe, Mg)_3TiB_2O_{10}$, mineral, warwickite; orthorhombic, P nam, Z = 2, ρ = 3.35; spinel type structure with 6-coordinated Mg^{++} and Fe^{++} ions and triangularly coordinated boron atoms. (Takéuchi, Watanabe, and Ito, 1950b.)
26. $Mg_3Mn_2^{+++}Mn^{++}B_2O_{10}$, mineral, pinakiolite; orthorhombic; P 2_1/m, ρ = 3.88; spinel type structure. (Bertaut, 1950.)

27. $H(Fe,Mn)Ca_2Al_2BSiO_{20}$, mineral, axinite; triclinic, P $\bar{1}$; structural units are Si_4O_{12} and BO_3 groups bonded together by Fe(Mn), Al, and Ca ions; AlO_3OH tetrahedra, and $Fe(Mn)_2O_8(OH)_2$ and Al_2O_{10} double octahedra may be distinguished in structure; (Ito and Takéuchi, 1952.)
28. $(Na,Ca)(Mg,Fe)_3Al_6B_3Si_6(O,OH,F)_{31}$, mineral, tourmaline, of variable composition; structural units are Si_6O_{18} cyclical groups and plane triangular BO_3 groups, bonded together by Na, Ca, Li, Al, Mg, and Fe ions; rhombohedral, R 3; all constituent boron atoms in positions of threefold coordination; (Donnay and Buerger, 1950; Ito and Sadanaga, 1951.)

All constituent boron atoms tetrahedrally coordinated:

Discrete monomeric anions, $[BO_4]^{-----}$.

29. Fe_3BO_6, synthetic iron borate; isostructural with mineral norbergite, $Mg_2SiO_4 \cdot Mg(OH, F)_2$ (Taylor and West, 1929.) If composition written as $Fe_2BO_4 \cdot FeO_2$, compared with norbergite, all boron atoms are tetrahedrally coordinated and all iron atoms are octahedrally coordinated [in norbergite with both O^{--} and $OH^-(F^-)$]; only known borate structure in which discrete BO_4^{-----} ions may be distinguished. (White, Miller, and Nielsen, 1965.)

Discrete monomeric anions, $(B(OH)_4]^-$.

30. $Na_2B(OH)_4Cl$, mineral, teepleite; tetragonal, P 4/nmm, Z = 2, $\rho = 2.076$; has been synthesized; composed of Cl^-, Na^+, and $B(OH)_4^-$ ions; B—O, 1.41 Å. (Fornaseri, 1949; 1950; 1951.)
31. $CuB(OH)_4Cl$, mineral, bandylite; tetragonal, P 4/mmm, Z = 2, $\rho = 2.810$; structure composed of puckered layers of tetrahedral $B(OH)_4^-$ ions such that Cu^{++} ions are coordinated with 4 OH^- ions in square coplanar fashion, the layers held together by long Cu—Cl bonds between Cu^{++} ions of adjacent layers and Cl^- ions situated between them; B—OH^-, 1.42 Å, Cu^{++}—4 OH^-, 1.98 Å, Cu^{++}—2 Cl^-, 2.80 Å. (Collin, 1951.)
32. $NaB(OH)_4 \cdot 2H_2O$, formerly referred to as sodium metaborate, preferably sodium tetrahydroxyborate; structure composed of $B(OH)_4^-$ and Na^+ ions, such that Na^+ ions are octahedrally coordinated with water molecules and hydrogen bonds extend throughout structure linking it into a three-dimensional network; triclinic, P $\bar{1}$, Z = 2; Na^+ ion coordination polyhedra have 2 edges in common; also described as $NaBO_2 \cdot 4H_2O$, or $Na_2O \cdot B_2O_3 \cdot 8H_2O$. (Block and Perloff, 1963.)
33. $LiBO_2 \cdot 8H_2O$, lithium metaborate; trigonal, C 3; structure composed of $B(OH)_4^-$ and $Li(OH_2)_4^+$ ions, hydrogen bonded together. (Zachariasen, 1964.)
34. $Li[B(OH)_4]$ formerly regarded as dihydrate of lithium metaborate but preferably described as lithium tetrahydroxyborate; orthorhombic, P bca, $\rho = 1.825$, z = 8; not a hydrated salt, contains only $B(OH)_4^-$ and Li^+ ions, where each Li^+ is tetrahedrally coordinated with 4 OH^- ions. (Höhne, 1964.)

35. $Sr[B(OH)_4]_2$, strontium tetrahydroxyborate; monoclinic, P $2_1/c$, z = 8, $\rho = 2.58$; not a hydrated salt; contains only $B(OH)_4^-$ and Sr^{++} ions, where the latter are coordinated with 9 OH^- ions. (Kutschabsky, 1965.)

Hetropoly anion containing central boron atom.

36. $H_5BW_{12}O_{40} \cdot 5H_2O$, 12-tungstoboric acid; complex heteropoly anion is built around a central BO_4 group, with each of the four O^{--} ions common to groups to 3 WO_6 octahcdra which in turn have common vertices so that group at each B—O vertex is W_3O_9, thus $[BO_4(W_3O_9)_4]^{5-}$; structure adopted by several 12-heteropoly anions containing central Si, P, or As atoms. (Keggin, 1934.)

Discrete dimeric anions, $[B_2O(OH)_6]^{--}$ and $[B_2(O_2)_2(OH)_4]^{--}$.

37. $Mg[B_2O(OH)_6]$, mineral, pinnoite; previously formulated as $MgB_2O_4 \cdot 3H_2O$; anion composed of 2 tetrahedral groups with a common O^{--} ion, the 3 unshared vertices of the tetrahedra corresponding to OH^- ions; each boron atom at center of tetrahedral group of 3 OH^- ions and 1 O^{--} ion; Mg^{++} ions are octahedrally coordinated. (Paton and MacDonald, 1957.)
38. $Na_2[B_2(O_2)_2(OH)_4] \cdot 6H_2O$, sodium peroxoborate; formerly regarded as sodium perborate, $NaBO_3 \cdot 4H_2O$; contains the centro-symmetrical complex anion,

```
          O—O
         /   \
[(OH)2B       B(OH)2]--. (Hansson, 1961.)
         \   /
          O—O
```

Infinite two-dimensional anions.

39. $CaBSiO_4(OH)$, mineral, datolite; complex sheet type anion composed of layers of tetrahedral groups which are alternately SiO_4 and BO_3OH, each tetrahedron having 3 O^{--} in common with tetrahedra of other kind and 1 free corner, an O^{--}in each SiO_4 group and a OH^- of each BO_3OH group; two-dimensional layers held together by Ca^{++} ions. (Ito and Mori, 1953.)

Infinite three-dimensional network anions, boron atoms coordinated with oxygen only.

40. HBO_2-I, normally stable modification of metaboric acid or γ-metaboric acid; cubic, P $\bar{4}$ 3n, a = 8.886 Å, Z = 24, $\rho = 2.486$, B—O—B angles, 118.6°, 126.7°, B—O, 1.472 Å; most dense of metaboric acids; structure composed of three-dimensional network of BO_4 groups with very strong hydrogen bonding, O—H···O, 2.487 Å; (Zachariasen, 1963a.)
41. BPO_4, boron phosphate; cristobalite structure, with both boron and phosporus atoms tetrahedrally coordinated with oxygen; tetragonal; although composed of 2 most common glass-forming elements shows little tendency to vitrify. (Schulze, 1934.)
42. $BAsO_4$, boron arsenate; isostructural with BPO_4; B—O distance in both structures is 1.49 Å. (Ibid.)

43. $CaB_2Si_2O_8$, mineral, danburtite; orthorhombic, P nma, Z = 4; three-dimensional network structure with both boron and silicon atoms in positions of tetrahedral coordination; B—O, 1.47 Å. (Dunbar and Machatszcheki, 1930; Johansson, 1959.)
44. $Mg_3B_7O_{13}Cl$, mineral, boracite; both high and low temperature modifications have same B—O network structure, differing only in position of Mg^{++} and Cl^- ions; coordination around one set of boron atoms has unusually long B—O distance of 1.78 Å, mean B—O, 1.48 Å; cubic, F $\bar{4}$ 3c, Z = 8, ρ = 2.91–2.97; has been synthesized; strongly pyro- and piezo-electric. (Ito, Morimoto, and Sadanaga, 1951.)
45. B_2O_3, boric oxide; structure consists of two sets of BO_4 tetrahedra forming two types of interconnected spiral chains; hexagonal, C 3, Z = 3, ρ = 2.46, high density modification, ρ = 2.95; similar structure to γ-HBO_2; only obtained crystalline in 1937. (McCulloch, 1937; Berger, 1952; 1953.)
46. SrB_4O_7, strontium tetraborate; orthorhombic, P 2_1 nm, Z = 2; one of most dense of known borate structures; structure composed of chains of BO_4 tetrahedral groups parallel to *a* crystollagraphic direction such that 6-membered rings formed by 3 adjacent BO_4 groups have common edge and chains are joined by nonring oxygen atoms to form layers parallel to *ab* plane, layers connected by special position oxygen atoms, common to 3 BO_4 tetrahedra, to form a 3-dimensional network; B—O, 1.36–1.60 (mean, 1.48 Å); strontium coordination polyhedron not well defined, coordinated with 9 oxygen atoms at 2.52–2.84 Å and 6 at 3.04–3.20 Å. (Krogh-Moe, 1964; Perloff and Block, 1966.)
47. PbB_4O_7, lead tetraborate; isostructural with SrB_4O_7. (Perloff and Block, 1966.)
48. $Zn_4O(BO_2)_6$, zinc hexaborate; structure composed of three-dimensional network of BO_4 tetrahedral groups; zinc atoms also tetrahedrally coordinated, to 3 oxygen atoms bonded to boron atoms and 1 free; cubic, I $\bar{4}$ 3m, Z = 2; also referred to as zinc metaborate. (Smith, Garcia-Blanco, and Rivoir, 1964.)
49. $NaBSi_3O_8$, reedmargite; isostructural with albite, $NaAlSi_3O_8$; all boron atoms tetrahedrally coordinated with oxygen; B—O, 1.46–1.48 Å; (Clark and Appleman, 1960.)
50. $Ca_2BAsO_4(OH)_4$, mineral, cahnite; tetragonal, I $\bar{4}$; zircon type structure with all boron atoms and arsenic atoms tetrahedrally coordinated; B—O, 1.47 Å; structure may alternatively be regarded as similar to that of KH_2AsO_4, with one half of arsenic atoms replaced by boron, so compound probably piezoelectric and pyroelectric; (Prewitt and Buerger, 1961); the compound, $TaBO_4$ (Vaslavskii and Zvinchuk, 1953) and the behierite minerals, (Ta, Nb)BO_4 (Mrose and Rose, 1961) also believed to have the zircon type structures.

Some of boron atoms triangularly and some tetrahedrally coordinated:

Discrete complex anions.

51. $CaB_3O_3(OH)_5 \cdot H_2O$, mineral, meyerhofferite; triclinic, P $\bar{1}$, Z = 2, ρ = 2.12; has been synthesized; formerly described as $Ca_2B_6O_{11} \cdot 7H_2O$;

complex anion composed of 2 tetrahedral and 1 triangular B—O groups, each group with 2 common corners so that an O^{--} is located at common vertex and a OH^- ion at noncommon vertex; coordination group around 2 of boron atoms made up of 2 O^{--} and 2 OH^- ions and around other of 2 O^{--} and 1 OH^- ion; hydrogen atoms not located. (Christ and Clark, 1956.)

52. $CaB_3O_3(OH)_5 \cdot 4H_2O$, mineral, inyoite; monoclinic, P 2_1/a, ρ = 1.875; has been synthesized; complex anion identical to that in meyerhofferite. (Clark, 1959.)
53. $CaB_3O_3(OH)_5 \cdot 2H_2O$, synthetic compound corresponding to higher hydrate of meyerhofferite; contains same complex anion as [51, 52]; P $\bar{1}$, B—O, 1.47 Å (tetrahedral) and 1.38 Å (triangular). (Clark and Christ, 1957.)
54. $Na_2[B_4O_5(OH)_4] \cdot 8H_2O$, mineral borax or tincal; has been synthesized; monoclinic, C 2/c, Z = 4, ρ = 1.715; formerly described as $Na_2B_4O_7 \cdot 10H_2O$; complex anion composed of 2 BO_4 groups with a common corner and 2 BO_3 groups with 2 corners common to 1 vertex of each of the BO_4 groups, so that all unshared vertices correspond to OH^- ions; Na^+ ions octahedrally coordinated with water molecules, each coordination polyhedron having 2 edges in common, forming infinite chains parallel to the *c* direction, which in turn hold isolated complex anions, $[B_4O_5(OH)_4]^{--}$ in between by hydrogen bonds; B—O, 1.48 Å and 1.36 Å for tetrahedral and triangular groups respectively; O—H···, 2.896 Å. (Morimoto, 1956.)
55. $K_2[B_4O_5(OH)_4] \cdot 2H_2O$, potassium tetraborate; orthorhombic, P $2_12_12_1$, Z = 4, ρ = 1.898; contains same complex anion as borax [54]; formerly regarded as $K_2B_4O_7 \cdot 4H_2O$; B—O, 1.368 Å and 1.48 Å (triangular and tetrahedral); O—H···O, 2.65–3.14 Å. (Zachariasen, 1963c.)
56. $K[B_5O_6(OH)_4] \cdot 2H_2O$, potassium pentaborate; orthorhombic, A ba, Z = 4, ρ = 1.898; complex anion contains a central tetrahedral BO_4 group, the boron atom of which is common to 2 6-membered rings which are normal to one another; 4 of the boron atoms in the anion are triangularly coordinated to 2 oxygen atoms and 1 OH^- ion; complex anions are bonded to 4 similar ions by short hydrogen bonds at 2.52 Å and to water molecules by long hydrogen bonds at 2.661–2.920 Å; B—O, 1.360 Å and 1.478 Å. (Zachariasen, 1937b; Zachariasen and Plettinger, 1963.)
57. $NaCaB_5O_6(OH)_6 \cdot 5H_2O$, mineral, ulexite; triclinic, P 1, Z = 2, ρ = 1.955; complex anion composed of 2 6-membered B—O rings, each containing 1 triangularly coordinated boron atom and 2 tetrahedrally coordinated boron atoms, one of which is common to both rings; anion related to pentaborate anion [56] by addition of 2 OH^- ions to 2 opposite BO_3 groups; Ca^{++} and Na^+ ions are octahedrally coordinated; B—O, 1.37 Å and 1.48 Å. (Clark and Appleman, 1964.)

Infinite chain type complex anions.

58. HBO_2-II, β-metaboric acid; monoclinic, P 2_1/a, Z = 2, ρ = 2.044; structure composed of endless chains of composition $[B_3O_4(OH)(OH_2)]_n$

parallel to *b* crystallographic axis; each unit of chain contains 1 boron atom coordinated with 3 oxygen atoms and 1 water molecule, 1 boron atom coordinated with 2 oxygen atoms and 1 OH^- ion, and 1 boron atom coordinated with 3 oxygen atoms; chains are hydrogen-bonded together with O—H···O ranging from 2.707–2.734 Å; the water molecule occurs at the noncommon corner of BO_4 group; B—O, 1.361 Å and 1.472 Å. (Zachariasen, 1963b.)

59. $CaB_3O_4(OH)_3 \cdot H_2O$, mineral colemanite; monoclinic, P 2_1/a, Z = 4, $\rho = 2.423$; complex anion composed of infinite chains of anions of the type found in inyoite and meyerhofferite [52, 51] linked together by a common oxygen atom subsequent to the elimination of a water molecule from 2 adjacent OH^- ions; formerly written as $Ca_2B_6O_{11} \cdot 5H_2O$. (Christ, Clark, and Evans, 1958.)
60. $NaCaB_5O_7(OH)_4 \cdot 3H_2O$, mineral, probertite; complex anion composed of endless chains formed by elimination of water molecules from adjacent OH^- ions in complexes of the type found in ulexite [57]. (Kurbanov, Rumanova, and Belov, 1963.)

Infinite layer type complex anions.

61. $CaB_3O_5(OH)$, synthetic calcium triborate; orthorhombic, P bn 2, Z = 4, $\rho = 2.72$; synthesized from inyoite [52]; anionic unit composed of infinite two-dimensional sheets of composition $[B_3O_5(OH)]_n^{2n-}$ by the cross-linking of the chain type anion found in colemanite [59]; sheets bonded together by Ca—O bonds; B—O, 1.38 Å and 1.48 Å. (Clark, Christ, and Appleman, 1962.)
62. $SrB_6O_9(OH)_2 \cdot 3H_2O$, mineral, tunnelite; monoclinic, P 2_1/a, Z = 4; complex anion composed of infinite sheets, the repeat unit of structure being composed of 3 fused 6-membered rings, each ring containing 2 tetrahedral boron atoms and 1 triangularly coordinated boron atom such that 1 oxygen atom is bonded to 3 boron atoms; adjacent sheets held together by hydrogen bonds (to water molecules) only. (Clark, 1963; Cuthbert, MacFarlane, and Petch, 1965.)

Infinite three-dimensional network complex anions.

63. $Li_2B_4O_7$, lithium tetraborate; tetragonal, I 4. cd; basic unit of structure is complex anion found in borax [54] cross linked into two interlocking three-dimensional networks extending throughout the material; one-half of the boron atoms are thus tetrahedrally coordinated; B—O, 1.38 Å and 1.45 Å. (Krogh-Moe, 1962.)
64. $Ag_2B_8O_{13}$, silver tetraborate; monoclinic, P 2_1/c, z = 4, $\rho = 3.41$; basic structural unit is same sequence of cross linked, interlocking, three-dimensional network anions found in [63]; one-half of silver ions are 7-coordinated with O—O distances between 2.23–3.02 Å, and others are 8-coordinated with O—O distances between 2.39–3.00 Å; B—O, 1.37 Å, 1.47 Å. (Krogh-Moe, 1965.)
65. KB_5O_8, anhydrous potassium pentaborate; complex anionic unit has same skeletal structure as the discrete anion found in the hydrated salt [56] condensed together into two identical, interpenetrating, helices, which are cross linked into a three-dimensional framework;

K^+ ions are octahedrally coordinated; B—O, 1.38 Å, 1.48 Å. (Krogh-Moe, 1960.)

66. CsB_3O_5, caesium triborate; orthorhombic, P $2_12_12_1$; basic structural unit may be regarded as derived from the complex anion in β- or HBO_2-II [58] which is condensed with identical units to form parallel helical chains, cross linked into a three-dimensional network, by each chain being linked to 4 similar chains; B—O, 1.38 Å, 1.48 Å. (Ibid.)
67. RbB_3O_5, rubium triborate; isostructural with the caesium compound [66]. (Ibid.)
68. CdB_4O_7, cadmium tetraborate; orthorhombic, P bca; anionic structure made up of two identical interlocking helical chains composed of the type of units found in the borax structure [54], so that 50 percent of the boron atoms are tetrahedrally coordinated; unusual structure in that the Cd^{++} ions occupy positions of 4-fold coordination rather than octahedral sites. (Ihara and Krogh-Moe, 1966.)
69. BaB_4O_7, barium tetraborate; orthorhombic, P $2_1/c$; this compound has a three-dimensional network structure composed of alternating 6-membered single ring units containing 2 tetrahedrally coordinated boron atoms and 1 trianguarly coordinated boron atom (of the type found in meyerhofferite, [51] and double rings of the pentaborate type [56] containing one tetrahedrally coordinated and four triangularly coordinated boron atoms; single ring units are associated with pentaborate structural units such that an additional spiro-boron atom is formed which is common to both ring systems; each single ring is bonded to double rings only and each double ring is linked directly only with single rings, each type to 5 of the other type to form an infinite three-dimensional network; B—O, 1.368 Å, 1.473 Å. (Block and Perloff, 1965.)

GENERAL STRUCTURAL PRINCIPLES

There are actually only a relatively small number of basic units involved in the structure of the known isopolyborate anions. The polynuclear anions are composed of triangular or tetrahedral groups combined in such a manner that the polyhedral groups have common vertices; in all the known borate structures the common vertix corresponds to an oxygen atom, which is usually bonded to 2 boron atoms. There are, however, three of the structures listed in which oxygen atom forms bonds with 3 boron atoms, in boric oxide [45] (Berger, 1953), the hydrated strontium borate, tunnelite [62] (Clark, 1963; Cuthbert, MacFarlane, and Petch, 1965), and in strontium tetraborate [46] (Perloff and Block, 1966). The number of possible discrete polyborate anions appears to be relatively small and the stable species, found in minerals, all contain at least one tetrahedrally coordinated boron atom. In the hydrated borates the positions in the coordination polyhedra around each boron atom which are not common to 2 polyhedral groups are always occupied by hydroxyl ions, so that many of these compounds are not really hydrates, or at least are not hydrates of such a high order as was formerly believed. In all the known structures the only cyclical group found is one composed of 6 members, 3 boron atoms alternating with 3 oxygen atoms; this 6-membered ring is apparently sufficiently stable to permit quite substantial distortions in some structures. Apart from the cyclic timeric anion, $B_3O_6^{---}$, in which all of the boron atoms are triangularly coordinated, polymerization of simple or complex anions, by the elimination of a water molecule from 2 OH^- groups, gives rise to either dimers or to infinite chains, which may also be cross-linked into infinite sheet and three-dimensional network structures, but intermediate degrees of polymerization are observed in complexes containing tetrahedrally coordinated boron atoms. For all isopolyborate anions, the anionic charge is equivalent to the number of tetrahedrally coordinated boron atoms in the discrete anionic unit. The anhydrous borates containing one or more tetrahedrally coordinated boron atoms in their repeat unit of structure all exist as three-dimensional networks. In anhydrous compounds where the boron atoms are all triangularly coordinated, they occupy specific positions in a close-packed array of oxide, hydroxyl, or fluoride ions. Extensive hydrogen bonding is present in all of the hydrated structures.

The mean observed bond length for triangularly coordinated boron is 1.37 ± 0.02 Å, and for tetrahedrally coordinated boron, 1.48 ± 0.03 Å, compared with the Pauling (1962) B—O single-bond length of 1.43 Å. There is a very considerable range of bond-length values reported, from 1.30 Å for the triangularly coordinated boron atoms in $Co_2B_2O_5$ [14] to 2.14 Å for one of the tetrahedral bonds in B_2O_3. The careful redetermination of the structures of CaB_2O_4 (18), $K[B_5O_6(OH)_4]2H_2O$ [56], $Na_3B_3O_6$ [15], Be_2BO_3OH [10] by Marezio, Plettinger, and Zachariasen (1963a) and of the structure of HBO_2-III [21] by Peters and Milberg (1964), undoubtedly provide the basis for the best values of the B—O bond lengths. One type of BO_4 groups in B_2O_3 is reported to exhibit B—O lengths of 2.14 Å, 1.37 Å, and two at 1.48 Å and a rather similar situation was earlier believed to apply to the mineral, danburtite (Dunbar and Machatscheki, 1930). A redetermination of this structure indicates, however, that all four bond lengths are of the order of 1.47 Å (Johansson, 1959). Tetrahedral bond lengths of 1.78 Å in the mineral, boracite (Ito, Morimoto, and Sadanaga, 1951) and of 1.60 Å in SrB_2O_4 (Perloff and Block, 1966) are reported, although the mean B—O in both is 1.48 Å. In the latter, both the very long and the very short bond lengths are those involved in the 6-membered B—O ring, while the nonring B—O distances fall within the mean ranges given above. Observed O—H···O bond lengths in the hydrated borates range from 2.487 Å in HBO_2-I to 3.14 Å in $K[B_5O_6(OH)_4]2H_2O$. These distances are not remarkable, since it is recognized that O—H···O bond lengths vary considerably from one structure to another.

The four most stable complex polyanions, namely, the species $[B_3O_3(OH)_4]^-$, found in HBO_2-II, $(B_3O_3(OH)_5]^{--}$, found in meyerhofferite-inyoite minerals, $[B_5O_6(OH_4)]^-$, found in hydrated potassium pentaborate, and $[B_4O_5(OH)_4]^{--}$, found in borax, may be regarded as the fundamentally important discrete polyanionic types, from which the more highly condensed complexes are formed either by condensation to give rise to infinite chains, sheets, or three-dimensional networks, or by ring fusion; in the latter event the spiro or tetrahedral boron atom is common to both rings. Actually the discrete complex anion, $[B_3O_3(OH)_4]^-$, is unknown in any borate, although in β- or HBO_2-II, it occurs as an infinite chain with a water molecule occupying one of the tetrahedral positions around the boron atom, and it forms the repeat unit of the structure of the triborates, $[B_3O_5]_n^{n-}$ [66, 67]. This anion may be regarded as being formed either by the fusion of 2 $B(OH)_3$ molecules with the $B(OH)_4^-$ simple ion, or by the neutralization of one of the 3 boron atoms in the trimeric ring contained in α- or HBO_2-III. There is much experimental evidence to indicate that boric acid itself in aqueous solution functions exclusively as an acid in a similar manner to form the $B(OH)_4^-$ species (Ingri, Lagerström, Frydman, and Sillén, 1957; Edwards, Morrison, Ross, and Schultz, 1955). The $[B_3O_3(OH)_5]^{--}$ anion may similarly be produced from the $[B_3O_3(OH)_4]^-$ by the neutralization of a second boron atom. If an additional molecule of $B(OH)_3$ is condensed with $[B_3O_3(OH)_5]^{--}$, there results the species, $[B_4O_5(OH)_4]^{--}$. The pentaborate anion, $[B_5O_6(OH)_4]^-$ is related to the triborate species, $[B_3O_3(OH)_4]^-$, in the same manner that the latter is related to $B(OH)_4^-$; it may be regarded as being formed by condensing another 2 molecules of $B(OH)_3$ with the triborate anion. As pointed out by Dale (1961), the anionic species which would result by the neutralization of all three boron atoms in $B_3O_3(OH)_3$, the species $[B_3O_3(OH)_6]^{---}$, would have a very unfavorable structure, from the steric point of view, due to the mutual repulsion of the 3 axial OH^- groups, and not surprisingly such an ion has not yet been detected. By neutralizing the two triangularly coordinated boron atoms in the tetraborate ion, $[B_4O_5(OH)_4]^{--}$, or by condensing the hypothetical $[B_3O_3(OH)_6]^{---}$ ion with another $B(OH)_4^-$ ion, the end result would be the tetrameric ring anion, $[B_4O_6(OH)_4]^{----}$, which apparently is unstable with respect to the simple $B(OH)_4^-$ species, although this structural arrangement is quite stable for the neutral molecules, P_4O_6, P_4O_{10}, As_4O_6, As_4O_{10}, and $(CH_2)_6N_4$ (hexamethylene tetramine).

ASSOCIATION IN AQUEOUS BORATE SOLUTIONS

The first borate crystal structure to be reported was that of danburtite by Dunbar and Machatscheki (1930). Both before 1930 and since then, as crystal structural data for other polyborates have accumulated, a great amount of effort has been expended on the examination of the aqueous phase relationships of the boric acids and borates and their hydration-dehydration behavior. Cryoscopic, cell potential, electrical conductivity, *p*H, ion-exchange, nuclear magnetic resonance, and spectroscopic methods have all been applied in the attempt to detect the polyborate ions which exist in aqueous solutions and determine their nature. Since the observation by Kolthoff (1926) that the *p*H of aqueous solutions of boric acid decreases with concentration, it has generally been assumed that the formation of condensed polyanions is responsible for this phenomenon. The effect of various polyols on the *p*H of boric acid and borate solutions has been extensively studied, and Dale (1961) has isolated crystalline complexes of several 1:2 and 1:3 diols and of 1:3:5 triols formed with sodium borate. The accumulated chemical evidence for the existence of any specific condensed polyion in aqueous borate solutions is not particularly convincing (Ingri, Lagerström, Frydman, and Sillén, 1957; Sillén, 1959). It would appear that the predominant anionic species in boric acid and borate solutions, at low concentrations, is the $B(OH)_4^-$ ion, irrespective of the solid borate phase from which the solution was made. The achievement, and the alteration, of the equilibrium that presumably exists between any condensed isopolyborates in aqueous solution, must be a process that takes place very rapidly. There is a remarkable similarity between the combination of triangularly coordinated boron with oxygen and olefinic or aromatic carbon atoms on the one hand, and between tetrahedrally coordinated boron with oxygen and saturated carbon atoms on the other (Dale, 1961). The single-bond C—H length of 1.54 Å and the double-bond length of 1.35 Å, compared with B—O of 1.37 Å (triangular coordination with oxygen) and 1.48 Å (tetrahedral coordination), are of such a magnitude that condensation reactions between polyborate anions and suitable polyhydric alcohols would be entirely analagous to the condensation of polyborate anions with one another.

BORATES OF UNKNOWN STRUCTURE

It is presumably significant that the greatest number of borate minerals are calcium- or magnesium-containing compounds, although the major natural deposits are composed of the sodium borates. Remarkably, the compound of composition, $Na_2B_4O_7 \cdot 4H_2O$, was not recognized as one of the stable phases in the sodium borate-water system before the discovery of massive deposits of the fibrous mineral, of this composition, known as kernite or razorite, and a synthesis of this compound at normal pressures was only accomplished in 1961 (Morgan, 1961). The structure of the polyborate anion in kernite and in the related hydrated mineral, tincalconite, $Na_2B_4O_7 \cdot 5H_2O$, from consideration of their aqueous phase relationships, is presumably related to that of the borax anion (Figure 3, b), the latter perhaps containing the same discrete species, $[B_4O_5(OH)_4]^{--}$ and the former the infinite polymer derived from this of composition, $[B_4O_6(OH)_2]_n^{2n-}$. The so-called pyroborates and the hydrated metaborates of the alkali metals, other than lithium, may very well contain dimeric tetrahedral and trimeric ring structural units, respectively. The stable pentaborate anion (Figure 3, c) is perhaps of quite wide occurrence in the structures of the isomeric ammonium pentaborate minerals, larderellite and ammioborite, $(NH_4)_2B_{10}O_{16} \cdot 5H_2O$, in the hydrated sodium pentaborate minerals, sborgite, ezcurrite (Suhr's borate), nasinite, and hafnerite, as well as in the synthetic Auger's borate.

The presence of the anion, $[B_3O_3(OH)_5]^{--}$, in the hydrated calcium minerals meyerhofferite [51], inyoite [52], the synthetic triborate [53], as the discrete species, in colemanite [59] as a linear polymer and in the synthetic triborate [61] as a layer polymer, suggests the presence of the same ionic species in the corresponding hydrated magnesium minerals of analagous composition. A further refinement of the crystal structures of the five calcium triborates, written as $Ca_2B_6O_{11} \cdot xH_2O$, where $x = 1, 5, 7, 9$, and 13 ([61], [59], [51], [53], and [52] above) has been reported (Clark, Appleman, and Christ, 1964). Inderite, $Mg_2B_6O_{11} \cdot 15H_2O$, kurnakovite, $Mg_2B_6O_{11} \cdot 13H_2O$, and inderborite, $CaMgB_6O_{11} \cdot 11H_2O$, occur as massive deposits, while the mineral hydroboracite, $CaMgB_6O_{11} \cdot 5H_2O$, is fibrous, indicating the presence of the meyerhofferite type anion in the first three minerals and the colemanite type linear polymers in the latter. The anion, $[B_3O_3(OH)_5]^{--}$ (Figure 3, a), is presumably stabilized in the solid phase when combined with species of relatively high ionic potential such as calcium and magnesium against the mutual repulsion of the axial OH^- constituents. It has indeed been confirmed that this anion occurs in inderite, $Mg_2B_6O_{11}\cdot 15H_2O$ (Ashirov, Rumanova, and Belov, 1962), but until detailed structural data becomes available for some of the other primary members of the hydrated magnesium borate group of minerals, there would appear to be little possibility of differentiating between the numerous synthetic and naturally occurring compounds in this series (Frondel, Morgan, and Waugh, 1956; Pennington and Petch, 1960; Schaller and Mrose, 1960). The calcium borate minerals, gowerite and nobleite, probably contain polyborate anionic units which are trimeric, and this could also apply to the strontium-containing mineral, veatchite. None of the borate structures reported so far are compatible with the composition of the so-called octaborates of the alkali and alkaline earth metals; paternoite, $MgB_8O_{13}\cdot 4H_2O$, is perhaps the only known mineral representative of this class of compound.

The rare mineral, jeremejevite, $AlBO_3$, which is reported to crystallize in the hexagonal system (Dana, 1951), presumably adopts the calcite structure. Rhodizite, $NaKLi_4Al_4Be_3B_{10}O_{27}$, with reportedly cubic symmetry, has perhaps the spinel type structure.

The variety of structures found for the three metaboric acids [21, 40, 58], for the so-called metaborates [15–19, 32–35, 48], and for the tetraborates [46, 47, 63, 64, 68, 69] invalidates previously formulated generalizations about the borates. There does not appear to be any evidence for the stability of a 6-membered ring containing three tetrahedral boron atoms (Christ, 1960). On the other hand, the structures found for ulexite [57] and tunnelite [62] would be specifically excluded on the basis of the conformational analysis of Dale (1961). The $Zn_4O(BO_2)_6$ [48] structure is closely related to that of the sodalites, with the sodium content of the latter replaced by zinc and the chlorine replaced by free oxygen atoms not involved in coordination with boron.

The formation of three-dimensional network structures in many anhydrous borates, formed by the crosslinking of two identical helical chains [46, 47, 63–69], apparently gives rise to a more efficient space filling arrangement than is possible with the single chain species, where the bond angles of the constituent structural units would permit only very open, low density structures, in the absence of crosslinking and the inclusion of cationic species of relatively large ionic potential.

The influence of the constituent cationic species, of its radius, charge, electronegativity, and polarizability, is conceivably of considerable importance in determining the crystal lattice energy of the borates. The occurrence, however, of only triangularly coordinated boron in minerals of the ludwigite-paigeite type (23, 24), of only tetrahedrally coordinated boron in the synthetic Fe_3BO_6 (isostructural with the silicate mineral, norbergite, [29]), and the quite dissimilar anionic arrangements in the compounds of analagous composition, CdB_4O_7 [68], SrB_4O_7 [46], and BaB_4O_7 [69], does not presently make it appear possible to understand what factors determine the coordination number of boron in the polyborates.

REFERENCES

Ashirov, A., I. M. Rumanova, and N. V. Belov (1962). *Doklady Akad. Nauk S.S.S.R.* **143**, 331.

Berger, S. V. (1949). *Acta Chem. Scand.* **3**, 660.

Berger, S. V. (1950). *Acta Chem. Scand.* **4**, 1054.

Berger, S. V. (1952). *Acta Cryst.* **5**, 359.
Berger, S. V. (1953). *Acta Chem. Scand.* **7**, 611.
Bernal, I., C. W. Struck, and J. G. White (1963). *Acta Cryst.* **16**, 849.
Bertaut, E. F. (1950). *Acta Cryst.* **3**, 473.
Block, S., and A. Perloff (1963). *Acta Cryst.* **16**, 1233.
Block, S., and A. Perloff (1965). *Acta Cryst.* **19**, 297.
Christ, C. L., and J. R. Clark (1956). *Acta Cryst.* **9**, 830.
Christ, C. L., J. R. Clark, and H. T. Evans, Jr. (1958). *Acta Cryst.* **11**, 761.
Christ, C. L. (1960). *Am. Min.* **45**, 334.
Clark, J. R., and C. L. Christ (1957). *Acta Cryst.* **10**, 776.
Clark, J. R. (1959). *Acta Cryst.* **12**, 162.
Clark, J. R., and C. L. Christ (1959). *Z. Krist.* **112**, 213.
Clark, J. R., and D. E. Appleman (1960). *Science* **132**, 1837.
Clark, J. R., C. L. Christ, and D. E. Appleman (1962). *Acta Cryst.* **15**, 207.
Clark, J. R. (1963). *Science* **141**, 1178.
Clark, J. R., and D. E. Appleman (1964). *Science* **145**, 1295.
Clark, J. R., D. E. Appleman, and C. L. Christ (1964). *J. Inorg. Nuclear Chem.* **26**, 73.
Collin, R. L. (1951). *Acta Cryst.* **4**, 204.
Cuthbert, D. J., W. T. MacFarlane, and H. T. Petch (1965). *J. Chem. Phys.* **43**, 173.
Dale, J. (1961). *J. Chem. Soc.*, 922.
Dana, J. D. (1951). *System of Mineralogy*, 7th ed., vol. II. (Rewritten and enlarged by C. Palache, H. Berman, and C. Frondel.) New York, John Wiley & Sons, Inc.
Donnay, G., and M. J. Buerger (1950). *Acta Cryst.* **3**, 379.
Dunbar, C., and F. Machatscheki (1930). *Z. Krist.* **76**, 133.
Edwards, J. O., G. C. Morrison, V. F. Ross, and J. W. Schultz (1955). *J. Am. Chem. Soc.* **77**, 266.
Edwards, J. O., and V. F. Ross (1960). *J. Inorg. Nuclear Chem.* **13**, 329.
Ehrenberg and Ramdohr (1934). *Jahrb. Min.*, Beil-Bd., **69**, 1.
Fang, S.M. (1938). *Z. Krist.* **99**, 1.
Fornaseri, M. (1949). *Period Min.* (Rome) **18**, 103.
Fornaseri, M. (1950). *Period. Min.* (Rome) **19**, 157.
Fornaseri, M. (1951). *Ricerca Sci.* **21**, no. 7.
Frondel, C., V. Morgan, and J. L. T. Waugh (1956). *Am. Min.* **41**, 927.
Goldschmidt, V. M., and H. Hauptmann (1932). *Nachr. Ges. Wiss. Gottingen*, 53.
Hansson, A. (1961). *Acta Chem. Scand.* **15**, 934.
Höhn, E. (1964). *Z. Chem.* **4**, 431.
Ihara, M., and J. Krogh-Moe (1966). *Acta Cryst.* **20**, 132.
Ingri, N., G. Lagerström, M. Frydman, and L. G. Sillén (1957). *Acta Chem. Scand.* **11**, 1034.
Ito, T., N. Morimoto, and R. Sadanaga (1951). *Acta Cryst.* **4**, 310.
Ito, T., and R. Sadanaga (1951). *Acta Cryst.* **4**, 385.
Ito, T., and Y. Takéuchi (1952). *Acta Cryst.* **5**, 202.
Ito, T., and H. Mori (1953). *Acta Cryst.* **6**, 24.
Johansson, G. (1959). *Acta Cryst.* **12**, 522.
Keggin, J. F. (1934). *Proc. Roy. Soc.* (London) A **144**, 75.
Kolthoff, I. M. (1926). *Rec. trav. Chim.* **45**, 501.
Kracek, F. C. (1938). *Am. J. Sci.* (5) A, 143.
Krogh-Moe, J. (1959). *Arkiv. Kemi.* **14**, 439.
Krogh-Moe, J. (1960). *Acta Cryst.* **13**, 889.
Krogh-Moe, J. (1962). *Acta Cryst.* **15**, 190.
Krogh-Moe, J. (1964). *Acta Chem. Scand.* **18**, 2055.
Krogh-Moe, J. (1965). *Acta Cryst.* **18**, 77.
Kutschabsky, L. (1965). *Z. Chem.* **5**, 110.

Kurbanov, H. M., I. M. Rumanova, and N. V. Belov (1963). *Doklady Akad. Nauk. S.S.S.R.* **152**, 1100.
Marezio, M., H. A. Plettinger, and W. H. Zachariasen (1963a). *Acta Cryst.* **16**, 390.
Marezio, M., H. A. Plettinger, and W. H. Zachariasen (1963b). *Acta Cryst.* **16**, 594.
McCulloch, L. (1937). *J. Am. Chem. Soc.* **59**, 2650.
Mighell, A. D., A. Perloff, and S. Block (1966). *Acta Cryst.* **20**, 819.
Morimoto, N. (1956). *J. Min. Soc. Japan* **2**, no. 1, 1.
Morgan, V. (1961). *U. S. P.* **2**, 577, 983.
Mrose, M. E., and H. J. Rose (1961). *Abstr. Paper, Geol. Soc. Amer., Ann. Meeting, Cinn., Ohio*, p. 111.
Pasternak, R. A. (1959). *Acta Cryst.* **12**, 612.
Paton, F., and S. G. G. MacDonald (1957). *Acta Cryst.* **10**, 653.
Pauling, L. (1962). *The Nature of the Chemical Bond*, 3rd ed. Ithaca: Cornell Univ. Press.
Pennington, K. S., and H. E. Petch (1960). *J. Chem. Phys.* **33**, 329.
Perloff, A., and S. Block (1966). *Acta Cryst.* **20**, 819.
Peters, C. A., and M. E. Milberg (1964). *Acta Cryst.* **17**, 229.
Prewitt, C. T., and M. J. Buerger (1961). *Am. Min.* **46**, 1077.
Schaller, W. T., and Mrose (1960). *Am. Min.* **45**, 732.
Schmid, H. (1964). *Acta Cryst.* **17**, 1080.
Schulze, G. E. (1934). *Z. Phys. Chem.* B **24**, 215.
Sillén, L. G. (1959). *Quart. Rev.* **13**, 146.
Smith, P., S. Garcia-Blanco, and L. Rivoir (1964). *Z. Krist.* **119**, 375.
Takéuchi, Y., K. Watanabe, and T. Ito (1950a). *Acta Cryst.* **3**, **98**.
Takéuchi, Y., K. Watanabe, and T. Ito (1950b). *Acta Cryst.* **3**, 208.
Takéuchi, Y. (1952). *Acta Cryst.* **5**, 574.
Taylor, W. H., and J. West (1929). *Z. Krist.* **70**, 961.
Tazaki, H. (1940). *Hiroshima J. Sci.* **10**, 55.
Vaslavskii, A. N., and P. A. Zvinchuk (1953). *Doklady Akad. Nauk S.S.S.R.* **90**, 781.
White, J. G., A. Miller, and R. E. Nielsen (1965). *Acta Cryst.* **19**, 1966.
Zachariasen, W. H. (1931). *Z. Krist.* **76**, 289.
Zachariasen, W. H., and G. E. Ziegler (1932). *Z. Krist.* **83**, 354.
Zachariasen, W. H. (1934). *Z. Krist.* **88**, 150.
Zachariasen, W. H. (1937a). *J. Chem. Phys.* **5**, 919.
Zachariasen, W. H. (1937b). *Z. Krist.* **98**, 266.
Zachariasen, W. H. (1954). *Acta Cryst.* **7**, 305.
Zachariasen, W. H. (1963a). *Acta Cryst.* **16**, 380.
Zachariasen, W. H. (1963b). *Acta Cryst.* **16**, 385.
Zachariasen, W. H. (1963c). *Acta Cryst.* **16**, 975.
Zachariasen, W. H., and H. A. Plettinger (1963). *Acta Cryst.* **16**, 376.
Zachariasen, W. H., H. A. Plettinger, and M. Marezio (1963). *Acta Cryst.* **16**, 1144.
Zachariasen, W. H. (1964). *Acta Cryst.* **17**, 749.

CHEMICAL THEORY

E. BRIGHT WILSON, JR.
Mallinckrodt Chemical Laboratory
Harvard University
Cambridge, Massachusetts

Some Remarks on Quantum Chemistry

On this happy occasion, wherein we have the special privilege of honoring Linus Pauling, the editors have invited us to indulge in speculation and generalization. This presents a dangerous temptation, which can lead to sterile diffuseness, but I shall take the risk in discussing both some broad points of view about quantum chemistry and some more specific areas. In the light of Pauling's tremendous contributions to this field, it seems an appropriate topic for the present volume.

I should like to begin by making some general remarks about theory in chemistry, adopting at first a rather over-strict viewpoint. Chemistry is still very largely an empirical science with a stupendous collection of observed facts, running into the tens of millions at the very least. Relatively little sound theory is available to tie together, account for, and permit predictions from this overpowering mass of data. But there are indeed some very elegant theories that have proven their validity and utility beyond a shadow of a doubt. Foremost among these I would put the atomic theory, the idea of the tetrahedral structure of the carbon atom, and related principles of structural organic chemistry, thermodynamics, statistical mechanics, and, to a lesser extent, quantum mechanics.

Each of these displays certain qualities that, to me at least, are the essential earmarks of a really satisfactory theory. Thus, each has quite definite rules that are widely agreed upon. These rules permit predictions to be made that could not be made by other theories, predictions of results that are not known at the time but that are verifiable later. These predictions are unambiguous, and when tested are found to be correct. Finally, and not least, scientifically trained people have confidence in the validity of these predictions. A theory that is never believed is not very useful, and the more quickly it is understood and accepted, the more useful it is.

Thermodynamics is perhaps the most elegant and esthetically satisfying of these theories and serves well as a model toward which to aspire in other areas. The statistical mechanical calculation of thermodynamic properties of simple molecules from spectroscopic data is also very successful and therefore satisfying.

Quantum mechanics has not yet attained the status of thermodynamics in chemistry. Despite forty years of effort by thousands of investigators, it is still largely an article of faith that the Schrödinger equation is capable of explaining all the facts of chemistry. There are, in fact, a few doubts today, having to do with the importance of relativistic corrections for the heavier elements.

One area in which quantum mechanics does work very convincingly is that of rotational spectra. Microwave spectroscopy provides a great quantity of measurements of pure rotational transitions, most of which are accurate to about one part in 250,000. To account for these, the very simple model of a rigid rotating body with three principal moments of inertia with energy levels and selection rules governed by quantum mechanics is eminently successful, as long as only transitions between states of relatively low angular momentum are considered. Three transitions are required to obtain the values of the moments of inertia; the other transitions are then honest predictions. Table 1 shows how well such predictions work. The entries are representative of a larger set, which was completely unselected except for certain limitations, such as to $J < 3$, limitations specified in advance. For larger J values, the predictions progressively become less accurate (but even for $J = 12$, the agreement is usually better than 0.04 percent) because real molecules are not completely rigid and centrifugal distortion has to be taken into account, which can be done in simple cases.

This and many similar tests of quantum mechanics in the domain of microwave spectroscopy have a considerable philosophical importance for me. As it is still very difficult to make a priori computations of known reliability even for excessively simple molecules, it is hard to be confident that only larger computers are required in order to predict the properties of any molecule. Yet the feeling that the Schrödinger equation just might really contain all of chemistry (if we can make the relativistic corrections!) is strengthened by the phenomenal accuracy with which it can be made to fit the precise observations of microwave and molecular beam spectroscopy, experimental domains that did not exist in 1926. These fits are of course parametric but can be very over-determined.

We should, however, be able to start from the electrons and nuclei, and dispense with experimental data. The classical calculations of James and Coolidge (1933) showed that quantum mechanics did predict accurately a priori the basic properties of the hydrogen molecule. The advent of the computer age might have been expected to permit similar calculations for great numbers of more complicated molecules, but this has not been the case. No single additional molecule has yet been calculated to an accuracy equal to that achieved with a desk computer by James and Coolidge on H_2! This is a measure of the enormous increase in difficulty encountered on adding even two more electrons.

At a somewhat lower level of precision, LiH has been successfully treated (Browne and Matsen, 1964). After than, we must be satisfied with Hartree-Fock calculations; that is, the electrons are treated as noninteracting except for the interaction (including exchange) of each electron with a kind of averaged field due to the others. Great effort and ingenuity have led to the capability of applying the Hartree-Fock method to first row diatomic molecules (Cade, Sales, and Wahl, 1966). Calculations of dipole moments and some other quantities seem to agree well with experiment, but the total energy, though good on a percentage basis, is above the true energy by amounts that are large compared with chemical bond energies. The dissociation energies of some of these molecules can nevertheless be roughly calculated by subtracting the Hartree-Fock atomic energies, thus cancelling out much of the error.

TABLE 1.
Comparison of certain observed and calculated rotational transition frequencies (Rigid Rotor Model).

Transition	Observed ν	Observed ν–Calculated ν
	$C^{12}O^{16}F_2$[a]	
3_{33}–3_{21}	17,525.47	−.58
0_{00}–1_{01}	17,633.95	.00
2_{02}–2_{21}	17,798.14	.27
3_{12}–3_{31}	17,890.51	.07
3_{13}–3_{12}	29,509.27	−.02
3_{02}–3_{22}	29,512.10	.52
	$C^{13}O^{16}F_2$[a]	
3_{22}–3_{21}	17,501.02	−.67
0_{00}–1_{01}	17,627.05	−.03
2_{02}–2_{21}	17,805.26	.12
3_{12}–3_{31}	17,908.80	−.09
	$C^{12}O^{11}F_2$[a]	
3_{22}–3_{21}	14,350.10	.27
0_{00}–1_{01}	16,531.91	.05
	S^{32}=$S^{32}F_2$[b]	
3_{12}–3_{22}	11,324.42	−.43
0_{00}–1_{10}	12,147.40	.13
2_{12}–2_{20}	15,577.32	.33
1_{01}–2_{11}	20,083.62	.05
3_{21}–3_{31}	22,796.30	1.07
3_{22}–3_{30}	23,486.60	−.15
2_{02}–3_{12}	28,532.46	.41
2_{11}–3_{21}	34,312.85	−.27
2_{12}–3_{22}	36,441.63	−.18
	S^{32}=$S^{34}F_2$[b]	
1_{11}–2_{21}	28,304.60	−.54
2_{02}–3_{12}	28,371.20	−1.70
2_{11}–3_{21}	34,109.39	.53
2_{12}–3_{22}	36,205.45	−.17

[a] V. W. Laurie, D. T. Pence, and R. H. Jackson (1962). *J. Chem. Phys.* **37**, 2995.
[b] R. L. Kuczkowski (1964). *J. Am. Chem. Soc.* **86**, 3617.

These very valuable results do, however, emphasize several serious problems. The first is the well-known fact that the dissociation energy is only a minuscule fraction of the total electronic energy of a molecule, so that extraordinary accuracy in the total energy is required in order to get chemically useful numbers. The second difficulty is that almost never have quantum mechanically calculated quantities, whether energies or expectation values, been provided with error limits. This has been possible, though laborious, for a long time for energy (Temple, 1928; Bazley and Fox, 1961; Gay, 1964; Miller, 1965; Wilson, 1965; Löwdin, 1965), and very recently formulas have been obtained for error limits for expectation values (Bazley and Fox, 1966; Jennings and Wilson, 1966). The basis for this is short enough to present here. We wish to compare the true expectation value

$$\langle A \rangle = \int \psi^* A \psi d\tau$$

for some operator A (for example, dipole moment) with its approximate calculated value

$$\langle A \rangle^\circ = \int \phi^* A \phi d\tau$$

where ϕ is a normalized approximation to ψ. Consider

$$\begin{aligned} |\langle A \rangle - \langle A \rangle^\circ| &= |\int \psi^* A \psi d\tau - \int \phi^* A \psi d\tau + \int \phi^* A \psi d\tau - \int \phi^* A \phi d\tau| \\ &= |\int (\psi^* - \phi^*) A \psi d\tau + \int \phi^* A(\psi - \phi) d\tau| \end{aligned}$$

where we have added and subtracted an extra integral. The well-known Schwartz inequality then gives

$$|\langle A \rangle - \langle A \rangle^\circ|^2 \leq \int |\psi - \phi|^2 d\tau [\int |A\psi|^2 d\tau + \int |A\phi|^2 d\tau]$$

The first factor on the right can be estimated from experimental energy data using Eckart's inequality (Eckart, 1930). The bracket on the right can usually be adequately approximated by assuming that the two integrals are approximately equal, or the whole process can be reapplied to $\int |A\psi|^2 d\tau$ itself to give a higher approximation. Finally, the right-hand side can normally be considerably reduced by subtracting a suitable constant from A throughout, a procedure that does not alter the left-hand side.

It is very reasonable to hope that this type of argument will be rapidly extended and made more efficient. The benefits would be considerable. With good error limits, quantum mechanical calculations could, at long last, stand on their own feet as predictors of known accuracy. Up till now, comparison with experiment was the only reliable measure of accuracy, so that true prediction was in a strict sense, impossible.

At the risk of overemphasizing this point, I would expect that the standards demanded of quantum mechanical calculations should and probably will improve and may well soon include the requirement of error limits. Then disagreement with experiment will either mean that the experiments are in error or that there is some fundamental difficulty with the model or with quantum mechanics itself.

To return to a priori calculations of electronic energies, there are recent calculations in which configuration interaction has been added to the Hartree-Fock approximation, bringing the results somewhat closer to exactness (Das and Wahl, 1966). There also are promising attempts to treat the remaining energy (the "correlation energy") empirically so that one can hope by the combination of Hartree-Fock calculations and empirical correlation energies to make energy predictions of chemically interesting accuracy (Allen, Clementi, and Gladney, 1963).

Incidentally, since the calculation of energy to the very high accuracy needed has proven to be very difficult, there has been a tendency to follow Aesop's fox and claim that energy is not really very interesting anyhow. With this I do not agree. Energy seems far the most important quantity in chemistry, since it determines the compounds that exist, their entire thermodynamic properties, their structure, their spectra, and their reactivity. We can indeed measure many of these quantities more easily than they can be calculated but there remain many energies inaccessible by experiment, such as many activation energies, energies of some excited states and of short-lived radicals, and energies of species not yet observed. In attempting to assign mechanisms of reaction, for example, it would be a tremendous advantage to know from theory the energetics of alternate paths.

The formidable obstacle that the many-body problem has proven to be raises the very pertinent question of how far accurate *ab initio* quantum mechanical calculations can ever be pushed. The millionfold increase in computational speed available since the James and Coolidge hydrogen calculation has not moved the frontier very far as yet. Is further progress hopeless? Personally I think not, although I do not clearly see the right route. I suspect that it will require some new ideas. After all, the Hartree-Fock method, configuration interaction, correlated wave functions, and perturbation theory all date from the 1930's or earlier, and it would seem reasonable that, even though the last twenty years have turned up almost nothing in the way of new methods, the next twenty years should do better.

In fact, there has been a tremendous output of methodology from the theoretical physicists in recent decades. This is cast in a mathematical form that is highly unfamiliar and difficult for most chemists to absorb and use. This difficulty causes many to adopt the defensive attitude that probably most of this material is only a way of presenting old methods in fancy and obscure clothing. The present fashions in physics make it unlikely that any important number of younger theoretical physicists will any longer contribute to the direct solution of chemical problems, despite the unlimited range and tremendous challenge they present. Fortunately, formal methods have a great appeal for many younger members of the growing fraternity of chemical physicists, and some of these plus a few hardy physicists are beginning to test the machinery of the many-body theory on atomic and molecular problems. The beginnings look quite hopeful.

What are some other possibilities? Pekeris (1958) revived one in his masterful treatment of the helium atom by the series method—no integrals were evaluated, all the elements of his secular equation were integers or zeros (and most were zeros) and he achieved eight-place accuracy. (Incidentally, he also calculated error limits.) Frost (1964) has been exploring the wider application of this approach.

The avoidance of many-dimensional wave functions by the use of the second-order reduced density matrix or even the ordinary electron density function remains an unattained but enticing objective. The second order reduced density matrix $\Gamma^{(2)}$ is defined in terms of the wave function as

$$\Gamma^{(2)}(1,2;1'2') = \frac{N(N-1)}{2} \int \psi^*(1,2,3\cdots N)\,\psi(1'2',3\cdots N)\,d(3\cdots N)$$

(Husimi, 1940; Löwdin, 1955). Note the integration over the coordinates and spins of all but the first two electrons. This function is sufficient for calculating the energy or the expectation values of other operators that involve no more than two-body forces. Therefore, one can hope to evaluate an approximate $\Gamma^{(2)}$ directly by varying its form so as to minimize the energy. There would thus be no need to obtain the wave function at all! So far, this interesting approach has not been generally realizable because certain mathematical problems have not yet been solved, having to do with ensuring that the Pauli Exclusion Principle is obeyed. Work is in progress on this, and a solution should eventually be found. Some special solutions are already at hand, though unpublished, and they have interesting consequences. While it is not at all certain that the density matrix approach will be any simpler computationally, it is interesting as an alternative scheme that avoids the wave function completely.

Conceivably it will be possible to go even further and reduce everything to the calculation of the first-order density $\rho(xyz)$, that is, the ordinary density of electrons in three-dimensional space. Hohenburg and Kohn (1964) have proven the interesting theorem that the energy, the wave function, and the external electrical field (for example, from the nuclei) are all fixed when the density function $\rho(xyz)$ is specified. Thus the energy must be some universal function of ρ.

The proof is so concise that it is worth repeating here. Assume that the theorem is untrue so that a given electron density function $\rho(xyz)$ could have arisen from two different problems, that is, different external fields described by the potential energies $v(xyz)$ and $v'(xyz)$ for an electron. (These are assumed to differ by more than a constant.) Then since the wave equation must differ, the ground state electronic wave functions ψ and ψ' (assumed nondegenerate) must differ, although both are assumed to lead to the same density $\rho(xyz)$. We can write

$$E = \int \psi^* H \psi\, d\tau, \qquad E' = \int \psi'^* H' \psi'\, d\tau$$

but if we separate out the contribution of the potential energy of interaction of the electrons with the field and evaluate this term with the use of ρ and v, we can write, employing also the variational principle

$$\begin{aligned} E &= \int \Psi^* H \Psi\, d\tau < \int \Psi'^* H \Psi'\, d\tau \\ &= E' + \int \Psi'^* (H - H') \Psi'\, d\tau \end{aligned}$$

or

$$E < E' + \int [v(xyz) - v'(xyz)]\, \rho(xyz) dxdydz$$

since the operators H and H' differ only in the external field terms. By the same reasoning, interchanging primed and unprimed quantities

$$E' < E + \int [v'(xyz) - v(xyz)]\, \rho(xyz) dxdydz$$

Addition leads to the inconsistency

$$E + E' < E + E'$$

thus completing Hohenberg and Kohn's proof that there can be only one external field $v(xyz)$ leading to a specified electron density function $\rho(xyz)$ and hence such a function completely specifies the ground state wave function and energy.* As of now we do not know how to obtain E from ρ and do not have any way of calculating ρ except via the wave function. It remains an interesting problem for the future to find a practical way of obtaining ρ directly.

Despite past and prospective progress in pure theory, I fear that chemistry will continue to be largely empirical. There is no disgrace in empiricism, and no need or advantage in disguising it as pseudotheory. But there is a very great disgrace, or should be, in falsely claiming that a conclusion is supported by theory when it is not, or that it is upheld by experiment when, in fact, it has been especially tailored to fit a very small number of known cases. As chemists, we have no great reason to be proud of our standards in this regard. Reading our journals of fifteen years ago often illustrates not only how rapid true progress has been but also that our editors and referees have often been rather lenient.

It would be a mistake to concentrate all attention, as I have just done, on the discussion of highly precise a priori quantum mechanical calculations. It must be admitted that the prospect of carrying out such calculations for any but the very simplest molecules appears to be far in the future. In the meantime, of course, large numbers of people are working in an attempt to understand the chemical facts and to guide chemical investigations with theories that involve many approximations which are hard to justify mathematically or that have a frankly large empirical content. It is sure that work of this kind will continue and that it will become more and more successful. A theory of chemistry that is limited to extremely small numbers of electrons is useful but, of course, does not scratch the surface of what chemistry really is and cannot even begin to present models of most of the phenomena of chemistry. The necessity to study molecules of interesting size has driven many investigators to overreach themselves by approximations that do not work when finally properly tested or to force results to agree with experiment by the introduction of large numbers of variable parameters, concealed or not. But this regrettable fact should not blind us to the need for such methods or to the fact that more and more careful work is being done and will be done in this area.

There is even a good deal to say for theories that do not have a sound basis in either mathematics or empiricism provided that they are cast in a sufficiently convincing form. Experimentalists need reasons for doing particular experiments, and they very often find interesting unexpected new results when they do experiments motivated by a theory that turns out to be unsound. In this sense many unsound theories have been of very great importance in the history of chemistry, and this situation is sure to continue.

Nor should it be overlooked that most theories are developed in stages and that they must pass through a series of refinements in which the early versions are compared with more and more experimental data and modified to fit the new data. In many situations there has finally emerged from this process a theory that has true predictive value once the modifications and adjustments have been so made that a large enough domain can be correctly encompassed.

*Actually not all functions $\rho(xyz)$ can be derived from an N particle *antisymmetric* wave function, so not all functions ρ are admissible. For the conditions, see Smith (1966).

In short, one can be too pure, and never make any real practical progress. It remains largely a question of personal taste whether one wishes to try for high accuracy and reliability on very simple systems or rougher, more empirical treatments of larger molecules of greater chemical importance. Both areas need people of optimistic dispositions, capable of continuing despite discouraging obstacles, and both need criticism and constant evaluations. Optimists are often not very self-critical, so it is certain that quantum chemistry will continue for some time to be a subject that evokes controversies. This is regrettable, but at least is an indication that there is life and activity and a certain amount of excitement in the field.

REFERENCES

Allen, Clementi, and Gladney (1963). *Rev. Mod. Phys.* **35**, 465.
Bazley, N. W., and D. W. Fox (1961). *Phys. Rev.* **124**, 483.
Bazley, N. W., and D. W. Fox (1966). *J. Math. Phys.* **7**, 413.
Browne, J. C., and F. A. Matsen (1964). *Phys. Rev.* A **135**, 1227.
Cade, P. E., K. D. Sales, and A. C. Wahl (1966). *J. Chem. Phys.* **44**, 1973.
Das, G., and A. C. Wahl (1966). *J. Chem. Phys.* **44**, 87.
Eckart, C. (1930). *Phys. Rev.* **36**, 878.
Frost, A. A. (1964). *J. Chem. Phys.* **41**, 478.
Gay, J. G. (1964). *Phys. Rev.* A **135**, 1220.
Hohenberg, P., and W. Kohn (1964). *Phys. Rev.* B **136**, 864.
Husimi, K. (1940). *Proc. Phys. Math. Soc. Japan* **22**, 264.
James, H. M., and A. S. Coolidge (1933). *J. Chem. Phys.* **1**, 825.
Jennings, P., and E. B. Wilson, Jr. (1966). *J. Chem. Phys.* **45**, 1847.
Löwdin, P.-O. (1955). *Phys. Rev.* **97**, 1474.
Löwdin, P.-O. (1965). *J. Chem. Phys.* **43**, S175.
Miller, W. H. (1965). *J. Chem. Phys.* **42**, 4305.
Pekeris, C. L. (1958). *Phys. Rev.* **112**, 1649.
Smith, D. W. (1966). *Phys. Rev.* **147**, 896.
Temple, G. (1928). *Proc. Roy. Soc.* (London) A **119**, 276.
Wilson, E. B., Jr. (1965). *J. Chem. Phys.* **43**, S172.

MAURICE L. HUGGINS
Stanford Research Institute
Menlo Park, California

Interactions Between Nonbonded Atoms

The concept of spherical atoms of different size for different elements, packed together in condensed systems so that they are in mutual contact, is an old one. After knowledge of the distances between atomic centers in a considerable number of crystals had been accumulated, attempts were made (Bragg, 1920) to deduce the sizes of the atomic spheres of different elements and to show their degree of constancy and additivity. It was soon pointed out (Huggins, 1922) that, in agreement with theoretical expectation, the atomic radii required to give the actual interatomic distances varied with the environments of the atoms and especially with the nature of the bonding between them. Nevertheless, it proved possible to deduce sets of atomic radii for certain standard types of bonding and environment, to show that adding these together gave approximately the correct interatomic distances, and to use departures from strict additivity to study differences in atomic and interatomic properties.

For example, a set of atomic radii, deduced from structures in which each of the two atoms concerned was tetrahedrally bonded by single bonds to four others, with the kernel charges on these two atoms adding up to eight, was found to be approximately additive also in other structures in which the two atoms concerned were connected by single bonds (Huggins, 1926). After many more experimental data on interatomic distances became available, this set of "tetrahedral radii' was slightly modified and extended, and other related sets for covalently bonded atoms in different environments deduced (Pauling and Huggins, 1934).

As data on distances between the centers of pairs of atoms *not* directly bonded together (for example, atoms in different molecules in molecular crystals) were accumulated, it was noted (Hendricks, 1930) that these distances, for two atoms of a given element or elements (two chlorine atoms, for instance), were also roughly the same in different structures. The degree of constancy was not as great as for bonded atoms, and the interatomic distances were much larger than for the same pairs of atoms held together by covalent bonds. Nevertheless, even roughly constant and additive "van der Waals radii" have proved useful in crystal structure analysis and in other fields. A set of such radii derived by Huggins (1932) was later modified and extended by Pauling (1939). Further modifications and extensions have been published by Kitaigorodskii (1961), Klemm and Busmann (1963), and Bondi (1964). These radii were deduced to give, by addition, approximate interatomic distances between two atoms (not held together by covalent bonds nor by significant

Maurice L. Huggins is now retired. Permanent address: 135 Northridge Lane, Woodside, California.

ionic forces) in a condensed system such as a molecular crystal. In such a system, many forces, involving atoms other than those of the pair being considered, affect the equilibrium distance. These other forces, which are primarily attractions of the van der Waals and Coulomb types, usually tend to make the equilibrium distance between the given pair of nonbonded atoms much less than if it were determined only by the attractions and repulsions between the two atoms of the pair. This fact has not been realized by many scientists who have derived and used energy-distance relationships based on the assumption that the minimum for a given pair of atoms must be at a distance equal to the sum of their van der Waals radii.

POTENTIAL ENERGY CURVES

For many problems of current interest, such as those concerned with barriers to rotation about single bonds in organic molecules and with the conformations of macromolecules in their crystals and in solutions, the concept underlying the derivation and use of van der Waals radii involves too crude an approximation. Potential energy curves, relating the interaction energy of two atoms with the distance between their centers, are needed. It is realized that the use of a single such curve for each pair of elements involves the insufficiently tested assumption that the energy at a given distance is practically independent of (1) the orientation of the interatomic centerline relative to the orientations of the bonds by which these atoms are held to other atoms in the structure, and (2) the nature and strengths of these bonds. Nevertheless, those working on this problem have assumed, tentatively, that the assumption of a single energy-distance curve for each type of atom pair will be a reasonably good approximation for their purposes.

The useful applicability of a valid set of such curves (or of the corresponding analytical functions) is beyond question, and many attempts have been made to deduce such curves or functions and to apply them to structural problems. The derivations have often involved very unsound assumptions, yet readers have not usually been apprised of the questionable validity of the results.

To illustrate this point, Figure 1 shows curves corresponding to several functions recently proposed and used for interactions between two nonbonded

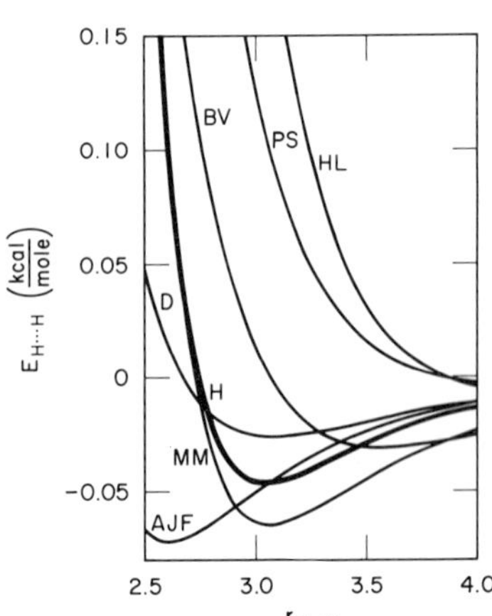

FIGURE 1.
Proposed energy curves for H · · · H interactions.

hydrogen atoms. In another recent publication (Huggins, 1966a) I have shown a similar comparison of other proposed H···H curves. (A considerable number have not been included in these comparisons, because they have been based on the demonstrably incorrect assumption, already referred to, that the equilibrium distance for a pair of atoms equals the sum of the van der Waals radii.) Most of the widely varying functions must be far from the correct one.

Equations for the interaction energy, in kcal/mole, corresponding to the curves shown in Figure 1 (curve labels at left) are the following:

H $$E_{HH} = 4 \cdot 10^5 \exp(-5.4r) - 47r^{-6} - 98r^{-8} - 205r^{-10} \quad (1)$$
(Huggins, this paper)

HL $$E_{HH} = 3716.4 \exp(-3.0708r) - 89.52r^{-6} \quad (2)$$
(Hirschfelder and Linnett, 1950; Mason and Kreevoy, 1955; DeSantis et al., 1963)

PS $$E_{HH} = 1858 \exp(-3.071r) - 44.76r^{-6} \quad (3)$$
(Pritchard and Summer, 1955; Pauncz and Ginsburg, 1960; Opschoor, 1965)

BV $$E_{HH} = 1200 \exp(-2.85r) - 160r^{-6} \quad (4)$$
(Borisova and Volkenstein, 1961)

MM $$E_{HH} = 43473r^{-12} - 105.9r^{-6} \quad (5)$$
(McCullough and McMahon, 1965)

D $$E_{HH} = 1727 \exp(-3.54r) - 49.23r^{-6} \quad (6)$$
(Dows, 1961)

AJF $$E_{HH} = 9950 \exp(-4.54r) - 45.2r^{-6} \quad (7)$$
(Abe, Jernigan, and Flory, 1966)

In addition to the H···H interaction energies, Opschoor (1965) and Abe, Jernigan, and Flory (1966) include a term, $+1.4(1 - \cos 3\varphi)$, for a postulated increase in orbital interaction energy resulting from rotation of a group of atoms at one end of a single bond relative to a group of atoms at the other end.

The functions used by different scientists for C···H and C···C interactions show similar large variations, as is evident from the following examples:

H $$E_{CH} = 25 \cdot 10^5 \exp(-5.4r) - 137r^{-6} - 237r^{-8} - 411r^{-10} \quad (8)$$
(Huggins, this paper)

AJF $$E_{CH} = 8.61 \cdot 10^4 \exp(-4.57r) - 127r^{-6} \quad (9)$$
(Abe, Jernigan, and Flory, 1966)

H $$E_{CC} = 158 \cdot 10^5 \exp(-5.4r) - 418r^{-6} - 574r^{-8} - 787r^{-10} \quad (10)$$
(Huggins, this paper)

AJF $$E_{CC} = 9.086 \cdot 10^5 \exp(-4.59r) - 363r^{-6} \quad (11)$$
(Abe, Jernigan, and Flory, 1966)

B $$E_{CC} = 3.012 \cdot 10^5 r^{-12} - 327.2r^{-6} \quad (12)$$
(Bartell, 1960; DeSantis et al., 1963)

The large differences in the energy-distance functions assumed by different previous investigators would not matter if there were good reason to believe that one function is much more likely to be correct than all the others. However, I have been unable to find such a reason. All of the previous derivations have involved either definite (and important) errors or assumptions of very doubtful validity. It has seemed worthwhile to derive a new relationship—by a procedure used (Huggins, 1937, unpublished) many years ago—taking advantage of the much better data now available.

DERIVATION OF NEW EQUATIONS

It is assumed that the repulsion energy for a given pair of atoms, such as H···H, can be expressed with sufficient accuracy by a simple exponential term, $A \exp(-ar)$, where r is the distance between the atomic centers. Such an exponential form has been successful in dealing with diatomic molecules (for example, Huggins, 1935, 1936), with lattice energies of ionic crystals (Born and Mayer, 1932; Huggins and Mayer, 1933; Huggins, 1937; Huggins and Sakamoto, 1957), and with the dependence of bond lengths on bond energies (Huggins, 1954). It is questionable whether the same constants can properly be used over the whole range of interatomic distances with which we are concerned, and also whether the assumption of independence of orientation (already discussed) is sufficiently valid, but it will tentatively be considered that a more complicated form for the repulsion energy is unwarranted.

For the attraction energy, it seems best, following practically all of the others in the field, to start with the Slater-Kirkwood expression (Slater and Kirkwood, 1931; Pitzer, 1959) for the coefficient of an r^{-6} term, deduced from the atomic polarizabilities, α_H, α_C. The polarizabilities, in turn, can be computed from the atomic refractions for infinite wavelength. The atomic refractions for carbon (0.9831) and hydrogen (0.4063) have been recomputed from National Bureau of Standards data for normal paraffins (Forziati, 1950; Camin, Forziati, and Rossini, 1954; Camin and Rossini, 1955). The effective numbers of electrons, needed for the Slater-Kirkwood equation, were taken as $N_H = 1$ for hydrogen and $N_C = 5.6$ for carbon, the latter being read from a curve published by Scott and Scheraga (1965). This procedure leads to the values of the coefficients of the r^{-6} term shown in equations (1), (8), and (10).

Calculations of lattice energies of ionic crystals (Huggins and Mayer, 1933; Huggins, 1937; Huggins and Sakamoto, 1957) suggest that for our purpose r^{-8} and perhaps even r^{-10} terms might be important. The coefficients (c_8) of the r^{-8} terms for H···H and C···C interactions have been calculated from the r^{-6} coefficients (c_6) and the polarizabilities, using equations derived by Mayer (1933). With our units,

$$c_{8,\mathrm{HH}} = \frac{6c_{6,\mathrm{HH}}^2}{14.3945 \times 10^{12} e^2 \alpha_H N_H} \tag{13}$$

(A similar equation holds for $c_{8,\mathrm{CC}}$.) Here e is the electron charge. The $c_{8,\mathrm{CH}}$ coefficient can be computed from the relationship

$$c_{8,\mathrm{CH}} = \frac{3}{e^2}\left(\frac{c_{6,\mathrm{HH}}}{N_H \alpha_H} + \frac{c_{6,\mathrm{CC}}}{N_C \alpha_C}\right) c_{6,\mathrm{CH}} \tag{14}$$

also derivable from equations given by Mayer (1933; see also Huggins and Sakamoto, 1957).

Coefficients of the r^{10} terms have been obtained from the assumed approximation

$$\frac{c_{10}}{c_8} = \frac{c_8}{c_6} \tag{15}$$

For our purpose, this is believed to be sufficiently accurate.

The coefficients deduced as just outlined are given in equations (1), (8), and (10).

It is difficult or impossible to decide a priori what is the best value to choose for the constant (or constants) in the exponential of the repulsion term. Considering the H···H repulsions, for example, it does not seem reasonable, to me, to assume that the repulsion should vary with the distance in quantitatively the same way for two hydrogens, each strongly bonded to a carbon atom, as for two excited ($^3\Sigma$) atoms, each bonded to nothing. The assumption (Scott and Scheraga, 1965) of a smooth curve dependence on atomic number, passing through points determined for the inert gases, likewise seems questionable, especially for hydrogen atoms that are strongly bonded to carbon or other atoms. Likewise, it does not seem reasonable to use, for nonbonded interactions, expressions for the variation of the repulsion energy with interatomic distance which have been found applicable to diatomic molecules, lattice energies and interionic distances, or bond lengths.

The problem can be somewhat simplified if we assume that the same value of the constant a in the repulsion term can be used for H···H, C···C, and C···H interactions. There is some justification for this in the research on diatomic molecules (Huggins, 1935, 1936), bond lengths (Huggins, 1954), and molecular beam scattering (Amdur and Harkness, 1954; Amdur and Mason, 1955; Scott and Scheraga, 1965).

Another simplification is achieved by making use of the differences in the sums of the "constant energy radii" for HH, CH, and CC interactions (Huggins, 1954). These differences lead to ratios of the A coefficients:

$$\frac{A_{\mathrm{CH}}}{A_{\mathrm{HH}}} = \exp\,[(\overset{*}{r}_{\mathrm{CH}} - \overset{*}{r}_{\mathrm{HH}})a] = \exp\,(0.34a) \tag{16}$$

$$\frac{A_{\mathrm{CC}}}{A_{\mathrm{HH}}} = \exp\,[(\overset{*}{r}_{\mathrm{CC}} - \overset{*}{r}_{\mathrm{HH}})a] = \exp\,(0.68a) \tag{17}$$

For the sublimation energy of polymethylene (linear polyethylene), we can then write

$$-E_{\mathrm{sub}} = A_{\mathrm{HH}}[\Sigma \exp\,(-ar_{\mathrm{HH}}) + \exp\,(0.34a)\,\Sigma \exp\,(-ar_{\mathrm{CH}}) + \exp\,(0.68a)\,\Sigma \exp\,(-ar_{\mathrm{CC}})] - E_{\mathrm{attr}} \tag{18}$$

Billmeyer's (1957) calculation of 1.838 kcal/mole of CH_2 groups for the sublimation energy appears reliable. The relative atomic positions can be assumed to be the same as in crystals of orthorhombic hexatriacontane, $C_{36}H_{74}$, except near the ends of the molecules. Teare's (1959) careful X-ray diffraction study gives these, apparently quite accurately. From his data, the interatomic distances required for the summations in the above equation are readily computed. For the repulsion terms, only the few closest pairs of atoms (in different molecules) contribute. For the attraction terms, the contributions of the atoms close to a given one have been summed directly and, for the more distant atoms, an integration procedure has been followed. This involves the density of atoms of the appropriate type, integration over all distances beyond the radius of a sphere of size depending on the number of atoms summed directly, and correction for the atoms in the same molecule that lie in the integration range. It may be noted that the total contributions of the more distant atom pairs to the attraction energy are of roughly the same magnitude as the contributions of those summed directly. The calculated value of E_{attr} is 3.265 kcal/mole of CH_2 groups.

From Equation (18) we can readily deduce A as a function of a. There are various possible ways of deciding between different pairs of values agreeing with this function. Up to now, only one has been tried. Following Howlett (1957), it has been assumed that one-sixth of the energy barrier for rotation in ethane is due to vibrational energy differences for the staggered and eclipsed states. The remainder has been tentatively assumed to arise entirely from H···H interactions, conforming to Equation (1). We can thus deduce A and a values that are in agreement with both the sublimation energy of polymethamer crystals and the rotational energy barrier for ethane (provided the assumptions just mentioned are valid). The validity of this procedure can perhaps best be tested by determining whether or not the resulting H···H, C···H, and C···C equations will give agreement with other rotational energy barriers, energy differences between *gauche* and *trans* forms for certain molecules, sublimation energies of other substances, and so on. Such tests have not yet been made.

With slight rounding of the figures first deduced,

$$A = 4 \cdot 10^5 \qquad \text{and} \qquad a = 5.4\ \text{Å}^{-1} \tag{19}$$

These values lead to a calculated sublimation energy of 1.87 kcal per mole of CH_2 for polymethylene and to a rotational barrier of 3.03 kcal per mole for ethane. The latter is to be compared with "experimental" values of 2.75 (Kistiakowsky, Lacher, and Stitt, 1939), 2.875 ± 0.125 (Pitzer, 1951), and 3.10 ± 0.45 (Volkenstein, 1963).

DISCUSSION

A possible source of considerable error in the constants of equations (1), (8), and (10) is inaccuracy of the positions of the hydrogen atoms reported by Teare. These positions cannot be accurately determined experimentally by X-ray diffraction methods, and it is doubtful if the assumed 1.07 Å C—H bond length and 107° H—C—H bond angle would be maintained if that required putting the hydrogens in locations giving high intermolecular H···H energies. In other words, it seems likely that the shortest intermolecular H···H distances (2.58 and 2.6 Å) given by Teare are too short, the true distances being somewhat larger as a result of departures from the assumed bond angles. This would of course reduce the calculated repulsion energy in the crystal and so modify the a and A values in the atom-to-atom interaction equations.

Another possible source of error might be the slight twisting of the zigzag chains in polymethylene that I have postulated (Huggins, 1961, 1966a, 1966b). Small amounts of twisting would be expected to have little effect, however, since twisting of one chain in one direction would be largely compensated for by twisting of its neighbors in the opposite direction.

In this connection it may be noted that the H···H function here deduced, like the other functions shown in Figure 1, exhibits an energy minimum that is considerably greater than the H···H distance of 2.54 Å between successive zigs (or successive zags) in a strictly planar zigzag polymethylene chain. In fact, all proposed H···H functions of which I am aware, excepting only those based on the arbitrary (and incorrect) assumption that the minimum in the curve must coincide with the sum of the van der Waals radii, have their minima at distances greater than 2.54 Å. (See, for example, Figure 1; also Figure 10 of Huggins, 1966a). It follows that a slightly twisted (helical) structure should be more stable in crystalline polyethylene than a strictly planar structure. Of course, as the angle of twist (measured from the orientation giving greatest intermolecular stability) increases, the intermolecular repulsion energy must increase. This leads to the concept of a limiting stable length of zigzag chain, with some type of structural irregularity, such as a chain fold, occurring when that limit is reached. This idea I have further developed elsewhere (Huggins, 1966a, 1966b).

SUMMARY

Because of questionable and sometimes demonstrably wrong assumptions in their derivation, most (perhaps all) of the functions that have been proposed for the energy of interaction of two atoms (H···H, C···H, and C···C) not directly bonded together, are of doubtful value. The chief problem is to determine the two constants in the customary exponential repulsion term. In the present approach the sublimation energy of crystalline polymethylene is used to deduce a relationship between these two constants. It is hoped that good values of these constants, individually, can now be determined by seeing what pairs give best agreement with other pertinent experimental data, such as the rotational energy barriers in simple organic molecules. Illustrating the procedure, the two constants in question have been deduced from the previously measured rotational barrier in ethane.

The H···H function deduced here, like those previously proposed by others, has a minimum at a distance greater than that between successive hydrogen atoms on the same side of a planar zigzag chain in a normal paraffin or in crystalline polymethylene. This indicates that a slightly twisted zigzag is more stable than a strictly planar zigzag.

REFERENCES

Abe, A., R. L. Jernigan, and P. J. Flory (1966). *J. Am. Chem. Soc.* **88**, 631.
Amdur, I., and A. L. Harkness (1954). *J. Chem. Phys.* **22**, 664.
Amdur, I., and E. A. Mason (1955). *J. Chem. Phys.* **23**, 415.
Bartell, L. S. (1960). *J. Chem. Phys.* **32**, 827.
Billmeyer, F. W., Jr. (1957). *J. Appl. Phys.* **28**, 1114.
Bondi, A. (1964). *J. Phys. Chem.* **68**, 441.
Borisova, N. P., and M. V. Volkenstein (1961). *Zhur. Strukt. Khim.* **2**, 346; *J. Struct. Chem.* **2**, 324.
Born, M., and J. E. Mayer (1932). *Z. Physik* **75**, 1.
Bragg, W. L. (1920). *Phil. Mag.* [6] **40**, 169.
Camin, D. L., A. F. Forziati, and F. D. Rossini (1954). *J. Phys. Chem.* **58**, 440.

Camin, D. L., and F. D. Rossini (1955). *J. Phys. Chem.* **59**, 1173.
DeSantis, P., E. Giglio, A. M. Liquori, and A. Ripamonti (1963). *J. Polymer Sci.* A **1**, 1383.
Dows, D. A. (1961). *J. Chem. Phys.* **35**, 282.
Forziati, A. F. (1950). *J. Res. Nat. Bur. Stand.* **44**, 373.
Hendricks, S. B. (1930). *Chem. Revs.* **7**, 431.
Hirschfelder, J. O., and J. W. Linnett (1950). *J. Chem. Phys.* **18**, 130.
Howlett, K. E. (1957). *J. Chem. Soc.*, 4353.
Huggins, M. L. (1922). *Phys. Rev.* **19**, 346.
Huggins, M. L. (1926). *Phys. Rev.* **28**, 1086.
Huggins, M. L. (1932). *Chem. Revs.* **10**, 427.
Huggins, M. L. (1935). *J. Chem. Phys.* **3**, 473.
Huggins, M. L. (1936). *J. Chem. Phys.* **4**, 308.
Huggins, M. L. (1937). *J. Chem. Phys.* **5**, 143.
Huggins, M. L. (1954). *J. Am. Chem. Soc.* **75**, 4126.
Huggins, M. L. (1961). *J. Polymer Sci.* **50**, 65.
Huggins, M. L. (1966a). *Makromolekulare Chem.* **92**, 260.
Huggins, M. L. (1966b). Lecture on "Recent Research on Polymer Structure," International Symposium on Macromolecular Chemistry, Tokyo, to be published in *Pure and Applied Chemistry*.
Huggins, M. L., and J. E. Mayer (1933). *J. Chem. Phys.* **1**, 643.
Huggins, M. L., and Y. Sakamoto (1957). *J. Phys. Soc. Japan* **12**, 241.
Kistiakowsky, G. B., J. R. Lacher, and F. Stitt (1939). *J. Chem. Phys.* **7**, 289.
Kitaigorodskii, A. I. (1961). *Organic Chemical Crystallography*. New York: Consultants Bureau.
Klemm, W., and E. Busmann (1963). *Z. anorg. allgem. Chem.* **319**, 297.
Mason, E. A., and M. M. Kreevoy (1955). *J. Am. Chem. Soc.* **77**, 5808.
Mayer, J. E. (1933). *J. Chem. Phys.* **1**, 270.
McCullough, R. L., and P. E. McMahon (1965). *J. Phys. Chem.* **69**, 1747.
Opschoor, A. (1965). *Conformational Analysis of Polyethylene and Isotactic Polypropylene*. Thesis, Technische Hogeschool te Delft.
Pauling, L. (1939). *The Nature of the Chemical Bond*, 1st ed., p. 174. Ithaca: Cornell University Press.
Pauling, L., and M. L. Huggins (1934). *Z. Krist.* A **87**, 205.
Pauncz, R., and D. Ginsburg (1960). *Tetrahedron* **9**, 40.
Pitzer, K. S. (1951). *Disc. Faraday Soc.* **10**, 66.
Pitzer, K. S. (1959). In I. Prigogine, ed., *Advances in Chemical Physics*, vol. 2, p. 59. New York: Interscience.
Pritchard, H. O., and F. H. Summer (1955). *J. Chem. Soc.*, 1041.
Scott, R. A., and H. A. Scheraga (1965). *J. Chem. Phys.* **42**, 2209.
Slater, J. C., and J. G. Kirkwood (1931). *Phys. Rev.* **37**, 682.
Teare, P. W. (1959). *Acta Cryst.* **12**, 294.
Volkenstein, M. V. (1963). *Configurational Statistics of Polymer Chains*. New York: Interscience.

MASSIMO SIMONETTA
Institute of Physical Chemistry
University of Milan
Milan, Italy

Forty Years of Valence Bond Theory

Valence bond theory is a chapter of science to which a definite birthday can be assigned: this was the day—June 30, 1927—on which Heitler and London sent to *Zeitschrift für Physik* their paper on the quantum mechanical treatment of the covalent bond (Heitler and London, 1927). In this paper, among other things, they considered the interaction of two hydrogen atoms. Taking account of the identity of the electrons and of the symmetry of the system, two approximate wave functions were built:

$$\psi_+ = \frac{\varphi_{1sA}(r_1)\varphi_{1sB}(r_2) + \varphi_{1sA}(r_2)\varphi_{1sB}(r_1)}{\sqrt{2 + 2S^2}},$$
$$\psi_- = \frac{\varphi_{1sA}(r_1)\varphi_{1sB}(r_2) - \varphi_{1sA}(r_2)\varphi_{1sB}(r_1)}{\sqrt{2 - 2S^2}} \tag{1}$$

where φ_{1sA} and φ_{1sB} are $1s$ hydrogen orbitals, r_1 and r_2 stand for the space coordinates of the two electrons, and S is the overlap integral. By means of the functions (1) it is possible to calculate the energy $E_\pm$ as a function of the internuclear distance R_{AB}.

The curve for E_+ shows a pronounced minimum, corresponding to attraction of the hydrogen atoms with formation of a stable hydrogen molecule. Proper spin functions must be included to satisfy the Pauli principle so that the wave function for the ground singlet state of H_2 in this approximation is

$$\Psi_+ = \psi_+ \frac{\alpha(\omega_1)\beta(\omega_2) - \alpha(\omega_2)\beta(\omega_1)}{\sqrt{2}} \tag{2}$$

where ω_1 and ω_2 are the spin coordinates for the two electrons. Then ψ_- corresponds to three repulsive states (triplet):

$$\Psi_- = \begin{cases} \psi_- [\alpha(\omega_1)\alpha(\omega_2)] \\ \psi_- \left[\dfrac{\alpha(\omega_1)\beta(\omega_2) + \alpha(\omega_2)\beta(\omega_1)}{\sqrt{2}}\right] \\ \psi_- [\beta(\omega_1)\beta(\omega_2)] \end{cases} \tag{3}$$

Since the energy curve for the function $\varphi_{1sA}(r_1)\ \varphi_{1sB}(r_2)$ (or $\varphi_{1sA}(r_2)\ \varphi_{1sB}(r_1)$) shows that there is only a very weak attraction, it may be concluded that the energy of the covalent bond is mainly due to the interchange of the two electrons between the atomic orbitals. The Heitler and London paper has been called (Pauling and Wilson, 1935) *the greatest single contribution to the clari-*

fication of the chemist's conception of valence, giving a sound quantum mechanical basis to the intuition of Lewis (1916), who had described the formation of a covalent bond as due to the *sharing* of an electron pair between the two bonded atoms.

DEVELOPMENTS AFTER HEITLER-LONDON

The second fundamental paper on valence bond theory is due to Slater (1931), who extended the method to many-electron systems. Suppose that for a molecule with N electrons we can construct a set of n functions $\Psi_k{}^0$, approximate solutions of the Schrödinger equation for the problem. We can try to find better approximations to the true wave functions by making linear combinations of these functions:

$$\Psi_q = \sum_k {}_1^n C_{qk}\Psi_k^0 \tag{4}$$

This can be done by standard techniques—that is, perturbation theory or variation method. We then arrive at the familiar set of n linear equations,

$$\sum_r {}_1^n C_{qr}(H_{kr} - ES_{kr}) = 0, \qquad k = 1, 2, \ldots, n \tag{5}$$

and the secular equation

$$H_{kr} - ES_{kr} = 0 \tag{6}$$

in which

$$H_{kr} = \int \Psi_k{}^{0*}\bar{H}\Psi_r{}^0 d\tau = \langle\Psi_k{}^0|\bar{H}|\Psi_r{}^0\rangle, \qquad S_{kr} = \int \Psi_k{}^{0*}\Psi_r{}^0 d\tau = \langle\Psi_k{}^0|\Psi_r{}^0\rangle$$
$$d\tau = d\tau_1 d\tau_2 \cdots d\tau_N \text{ and } \tau_i = r_i\omega_i.$$

Then $\bar{H}$ is the Hamiltonian operator* with no relativistic or magnetic effects included, and the integration is extended throughout the $4N$-dimensional space. To build the starting functions $\Psi_k{}^0$, we first consider $l \geq N/2$ orbitals $\varphi_i(r)$, where $i = 1, \ldots, l$; φ_i is in fact an atomic orbital or a linear combination of atomic orbitals centered on the same atom (hybrid orbital). Each orbital can be combined with a spin function (α or β) to give two spin orbitals $\sigma_j(r, \omega), j = 1, \ldots, 2l$.

With N different spin orbitals we form products such as $P_t = \sigma_a(r_1,\omega_1)\,\sigma_b(r_2, \omega_2)\cdots\sigma_z(r_N, \omega_N)$. How many such products can be formed depends upon the number of the orbitals we are using. If in all our products we use the same orbitals, and each the same number of times (one or two), the products are different only in the spin part of the product, and there is *spin degeneracy*. Where products differ also in the orbitals, there is *orbital degeneracy*. A particular case of spin degeneracy will now be examined, in which we start with N different orbitals ($i = 1, \ldots, N$) and use all of them once in every product. We can build 2^N different products $P_t(t = 1, \ldots, 2^N)$, since the spin function α or β can be assigned to each orbital.

The products P_t are eigenfunctions of the operator $\bar{S}_z$ (the operator for the z component of the spin angular momentum for the N electrons) with eigenvalues $M(M = N/2, (N/2) - 1, \ldots, -N/2$, in $\hbar$ units, but, in general, they are

*In this paper an overbar indicates an operator, an underline a vector, boldface a matrix.

not eigenfunctions of $\bar{S}^2$ (the operator for the square of the spin angular momentum). Such eigenfunctions can be built as linear combinations of the products P_t in many different ways.

The way suggested by Rumer (1932) will be followed. The eigenvalues of $\bar{S}^2$ are $S(S+1)$, in units of $\hbar^2$, with $S = N/2, (N/2) - 1, \ldots,$ zero (N even), or $\frac{1}{2}$ (N odd). For each value of S the allowed values of M are $S, S-1, \ldots, -S$.

At first only functions with $M = S$ will be considered. The number of different eigenfunctions for a given S and $M = S$ is

$$\chi_S^N = \frac{(2S+1)N!}{\left(\frac{N}{2}+S+1\right)!\left(\frac{N}{2}-S\right)!} \tag{7}$$

and it is equal to the number of different paths by which the point of coordinates N, S on the "branching diagram" shown in Figure 1 can be reached (Kotani, Ohno, and Kayama, 1961). There is a one-to-one correspondence between these paths and the *Rumer diagrams*. These diagrams were originally constructed for singlets only, according to the following rules. Draw a circle and N points on it. Then connect the points two by two with straight lines in all possible ways without any intersection between the connecting segments. Each way gives a Rumer diagram. If $S > 0$, we consider *extended* Rumer diagrams. At an arbitrary site on our circle we mark a cross (pole) and connect $2S = 2M$ points on the circle with the pole, in all possible ways, provided that any two successive points connected to the pole either have no points or have an even number of points on the circle between them. Then, for each case, the remaining $N - 2S$ points are connected two by two in all possible ways without there being any intersection between these segments or with the lines to the pole. The function corresponding to a given Rumer or extended Rumer diagram is obtained by the following; we assign to each point on the circle one of the N orbitals, in an arbitrary sequence, and to all the orbitals connected with the pole we assign an α spin function. Then, starting from one arbitrarily chosen point we traverse the circle, in either a clockwise or counterclockwise direction. To each spinless orbital that we meet we assign a β spin; we assign an α spin to the orbital connected to it. In such a way we get one particular product P. Then we reverse the α and β spins between connected orbitals in all possible ways, changing the sign of the product for each spin

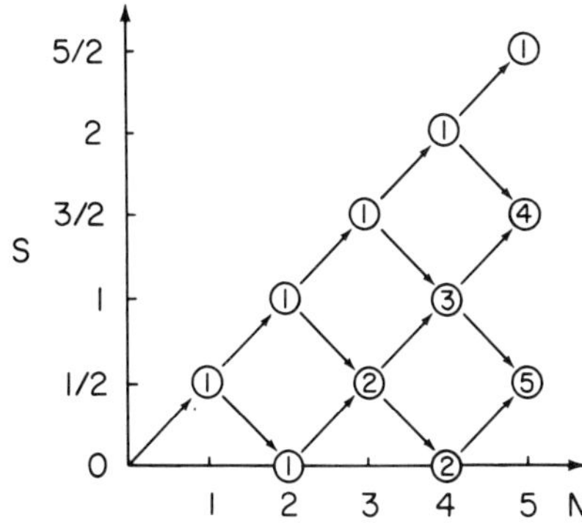

FIGURE 1.
The *branching* diagram.

reversal, and sum all these products. We end with a linear combination of $2^{(N-2S)/2}$ products of spin orbitals:

$$\sum_{\bar{R}} (-1)^r \bar{R}\varphi_a(r_1)\beta(\omega_1)\varphi_b(r_2)\alpha(\omega_2)\cdots\varphi_z(r_N)\alpha(\omega_N) \tag{8}$$

where $\bar{R}$ is the operator for the spin reversal between two connected spin orbitals and r is the number of spin reversals. Functions like (8), however, are not yet antisymmetrized with respect to the exchange of electrons, as they should be. This can be accomplished by considering all the $N!$ possible permutations of the electrons among the spin orbitals and summing all the corresponding functions of type (8), with a plus or minus sign for even or odd permutations. We end with expressions of this kind:

$$\Psi_k^0 = \frac{1}{\sqrt{N!}} \sum_{\bar{P}} (-1)^p \bar{P} \left\{ \frac{1}{2^{(N-2S)/4}} \sum_{\bar{R}} (-1)^r \bar{R}\varphi_a(r_1)\beta(\omega_1)\cdots\varphi_z(r_N)\alpha(\omega_N) \right\} \tag{9}$$

where the constant factors $1/\sqrt{N!}$ and $2^{-(N-2S)/4}$ are included for convenience; if the orbitals φ_i satisfy the orthonormality conditions, the functions (9) are normalized. Functions (9) can also be written in the form

$$\Psi_k^0 = \frac{1}{2^{(N-2S)/4}} \sum_{\bar{R}} (-1)^r \bar{R} \|\varphi_a(r_1)\beta(\omega_1)\varphi_b(r_2)\alpha(\omega_2)\cdots\varphi_z(r_N)\alpha(\omega_N)\| \tag{10}$$

where the notation $\|\cdots\|$ indicates a *Slater determinant*—that is, $1/\sqrt{N!}$ times the determinant built with the spin orbitals $\varphi_a(r_1)\beta(\omega_1)\cdots$. It may be easily shown (McLachan, 1960; Kellogg, 1962; Simonetta, Gianinetti, and Vandoni, 1967) that functions (9) are independent and are eigenfunctions of $\bar{S}^2$; they are usually called *structures*. When the previously mentioned conventions are adopted, they are *canonical covalent structures*. From functions (9), with $M = S$, the functions with $M < S$ are easily obtained by use of the *step-down* operator $\bar{S}_x - i\bar{S}_y$.

We may drop the condition that all the orbitals must be different and allow the use of one, two, or more orbitals twice (coupled with α and β). Then we have singly, doubly, and so on, polar structures, whose number is easily found, as well as the expressions of type (9) and the corresponding Rumer diagrams. We may also drop the condition that $i_{\max} = N$; this does not generate any extra difficulty, except that the number of possible covalent and polar structures increases tremendously. We may also notice that we imposed no restriction on the charge of the core to which the N electron system is connected so that the treatment may be applied to molecules, positive or negative ions, or neutral or charged radicals, in states of any multiplicity.

We can look at Rumer diagrams in a different way. Each dot on the circle represents an orbital that belongs to a given atom in the molecule, and each atom may contribute one or more orbitals. If an orbital is connected with the pole, we assign to it an unpaired electron; then the pairs of different orbitals connected by a segment correspond to bond. If the same orbital appears twice, the corresponding points must be connected; then a lone pair of electrons is assigned to that orbital. This means that each structure represents a particular disposition of bonds in the molecule. The diagrams in which such bonds are shown are also called *structures*. For example, we may consider the H_2O molecule as a four-electron, four-orbital problem. Two orbitals are contributed by the two hydrogen atoms ($1s_A$ and $1s_B$) and two orbitals by the oxygen atom ($2p_x$ and $2p_y$). Let us consider the molecule in the ground state, which is a singlet state. There are two independent covalent structures. Their Rumer diagrams are

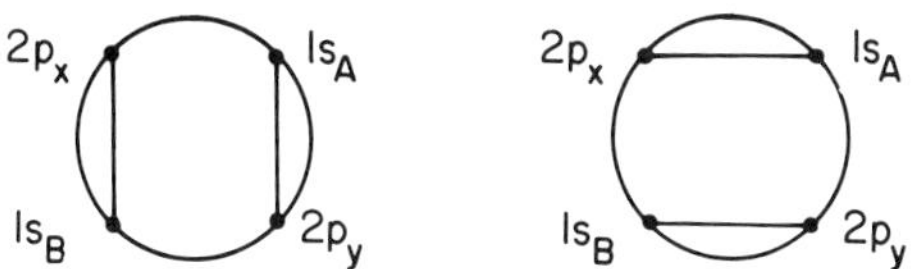

and their corresponding structural formulas are

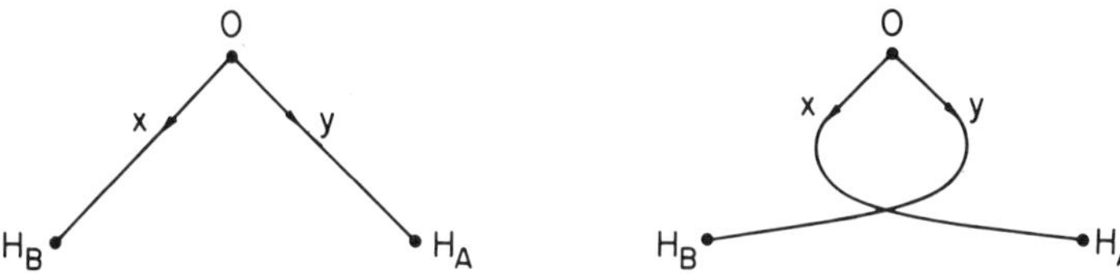

This second way of looking at the structures can be very useful for the chemist. It may in some cases allow us to see immediately that one structure Ψ_k^0 is a much better approximation to the ground-state function than any other; hence we may forget about the variational problem of finding the coefficients C_{qr} and simply use this structure to calculate some properties of the ground state (for example, the energy). We call this the *perfect-pair* approximation.

In many cases the perfect-pair approximation cannot be used if we start with pure orbitals, but the more stable structure becomes evident if we first hybridize the orbitals (Pauling, 1931). The most usual example is CH_4. With $2s$, $2p_x$, $2p_y$, $2p_z$ carbon orbitals, many structures may be equivalent, but if we first hybridize the four carbon orbitals to form sp^3 hybrids, then the perfect-pair approximation becomes applicable, and correct conclusions about the angles between the different carbon-hydrogen bonds can be reached.

The principle to be followed in hybridization is to concentrate the orbitals as much as possible in order to get the maximum overlapping between bonded orbitals. Sometimes (for example, in strained molecules) it may even happen that hybridization has to be carried on in such a way as to produce noncylindrical orbitals (Pauling and Simonetta, 1952).

Hybridization has been very successfully employed in discussing the nature of the chemical bond and the magnetic properties of complex ions of the transition elements (Pauling, 1931, 1948a, 1948b). The tremendous impact of this idea on the development of our understanding of molecular structure is evident from reading *The Nature of the Chemical Bond* (Pauling, 1960).

In many cases, however, it is not possible to couple the orbitals in a unique way, and the perfect-pair approximation is not applicable. The most usual examples are of course benzene and the conjugated systems. Here we have to make the linear combination (4) and solve the variational problem to find the coefficients. We need to calculate the matrix elements H_{kr} and S_{kr}. Since Ψ_k^0 and Ψ_r^0 are linear combinations of Slater determinants, we need in fact to know how to calculate matrix elements over Slater determinants. If the orbitals φ_i are orthonormal, we can use the very simple rules given by Slater (1931). For the energy matrix (and a spinless Hamiltonian) the elements are obtained as linear combinations of integrals of the following types:

Coulomb integrals

$$Q = \langle \varphi_a\varphi_b\varphi_c\cdots\varphi_z|\bar{H}|\varphi_a\varphi_b\varphi_c\cdots\varphi_z\rangle$$

Single-exchange integrals

$$\alpha = \langle \varphi_a\varphi_b\varphi_c\cdots\varphi_z|\bar{H}|\varphi_b\varphi_a\varphi_c\cdots\varphi_z\rangle$$

The Slater determinants turn out to be orthonormal. For sets of nonorthogonal orbitals other integrals are needed (multiple-exchange integrals), and the formulas for matrix elements become more complicated (Lowdin, 1955). In the early applications, especially to aromatic or (in general) conjugated systems, not only was the orthogonality of orbitals assumed, but other simplifications were introduced. First of all (after proper hybridization of the carbon orbitals), for such systems the σ, π approximation was used, greatly reducing the number of electrons to be considered; for example, in benzene the reduction was from forty-two electrons to six π electrons and six $2p$ carbon orbitals. Since the main interest was in the ground state of molecules, only the states of proper multiplicity were considered (singlet states). Also, the single-exchange integrals were assumed to be dependent upon the distance in a sharp way: all equal for bonded orbitals, all zero for nonbonded orbitals. Exchange integrals were not calculated but were assumed to be empirical parameters; Coulomb integrals were also taken as empirical parameters, and usually didn't even need to be estimated.

When only covalent canonical structures were used, the coefficients of the Coulomb and single-exchange integrals in the matrix elements of $\bar{H}$ and $\bar{1}$ could be easily and quickly calculated by means of a graphical method, based on superposition diagrams and permutation diagrams (Pauling, 1933). Superimposing the Rumer diagrams for two structures j and l, a pattern was obtained containing a certain number—say w_{jl}—of closed polygons, called islands. The Coulomb integral coefficient was $2^{w_{jl}-n}$, with $n = N/2$ being the number of bonds in the molecule. The coefficient of the single-exchange integral was $2^{w_{jl}-n}\Sigma f(p)$, where the sum was over all the adjacent pairs of orbitals in the molecule, p was the number of bonds between the two exchanged orbitals in the superposition pattern, and $f(p)$ was a factor with the values $-\frac{1}{2}$ (for $p = 0$), $+1$ (for $p = 1, 3, 5, \ldots$), and -2 (for $p = 2, 4, 6, \ldots$). The method was easily extended to hydrocarbon free radicals (the ground state is usually a doublet) by the use of a phantom orbital (Pauling and Wheland, 1933). Even in this very rough form the valence bond theory has been successfully used to discuss the radical dissociation constants of substituted ethanes. Other properties were also apt to be studied by this treatment. For example, the resonance energy was defined as the difference in energy between the ground state and the energy of the most stable structure, and calculated values of resonance energy for many conjugated hydrocarbons gave satisfactory correlations with empirical values from thermochemical data (Pauling and Sherman, 1933).

Other quantities that can be used to correlate the results of valence bond theory with experiment are (1) the bond order p_{rs} (Pauling, Brockway, and Beach, 1935) for the bond between atoms r and s,

$$p_{rs} = \frac{\sum_k' C_{0k}^2}{\sum_k C_{0k}^2} \tag{11}$$

where the subscript 0 indicates ground-state coefficients, Σ is the sum over all the structures and Σ' the sum over the structures in which adjacent atoms r and s are bonded; (2) the free-valence index (Daudel and Pulmann, 1946), defined as

$$I_r = \frac{\sum_k'' C_{0k}^2}{\sum_k C_{0k}^2} \tag{12}$$

where Σ'' is the sum over all the structures with a bond between r and a nonadjacent atom. The bond order has been empirically correlated with bond lengths and is used to predict or check bond lengths in conjugated hydrocarbons; the free-valence number can be used to predict relative reactivities at different atoms in a molecule or in a series of molecules.

FURTHER DEVELOPMENTS

Valence bond theory, in a qualitative, descriptive form, has also been used to elucidate the metallic state (Pauling, 1948). The basic idea is that the metallic bond is similar in nature to the covalent bond. The characteristic feature of metals is that each atom in a metallic crystal has an extra, usually empty orbital—the metallic orbital—so that unsynchronized resonance of electron-pair bonds from one position to another adjacent position in the crystal is allowed. The external orbitals in a metal may be divided into three groups: valence orbitals, atomic orbitals, and metallic orbitals. Support to this idea was first given by the study of the magnetic properties of the transition metals and their alloys (Pauling, 1938). From the values of the atomic saturation magnetization at absolute zero, μ_B, for iron, cobalt, and nickel, it is possible to calculate the number of orbitals of each type in these metals and predict the dependence of atomic saturation moments from the number of atomic electrons (Figure 2). Predicted values for Fe—Co, Co—Ni, and Ni—Cu alloys

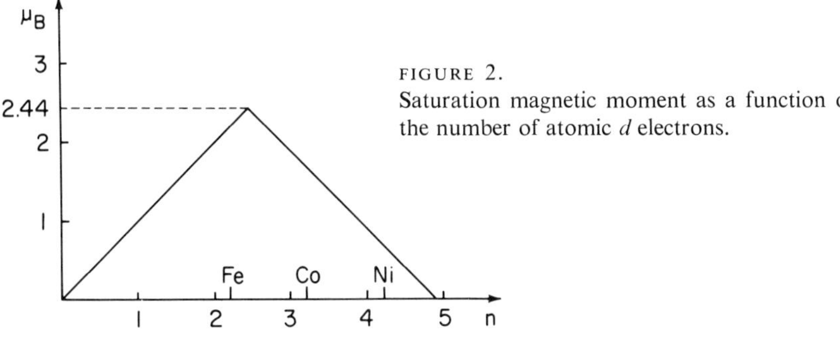

FIGURE 2.
Saturation magnetic moment as a function of the number of atomic d electrons.

fit the experimental curves very nicely. It is also possible to interpret the different behavior of white and gray tin (Pauling, 1949). The number of covalent bonds resonating among the available positions around an atom was used to calculate interatomic distances in metals and their alloys and a consistent table of metallic radii was obtained (Pauling, 1947). Atoms are also divided into three classes: hypoelectronic, hyperelectronic, and buffer atoms, according to their tendency to add or to release electrons. This property can be used to explain the values of observed interatomic distances in intermetallic compounds in which electron transfer occurs (Pauling, 1950). The states of the metallic crystal are described by wave functions, each corresponding to an electron-pair bond resonating along the crystal, and to a certain energy level. The levels can be grouped together and the existence of Brillouin zones recognized (Pauling and Ewing, 1948). A theory of ferromagnetism (Pauling, 1953) is largely based on the same assumptions.

Valence bond theory as formulated by Heitler and London and developed by Slater and Pauling (usually referred to as the HLSP method) has been universally recognized as an extremely useful tool, as it is easily applicable to a wide range of molecules, but it was also subject to some criticisms; the most important perhaps is that the requirement of one electron per orbital overemphasizes electron repulsions. To avoid this deficiency polar structures must be introduced. This problem was considered by Craig (1950) and, independently, by Simonetta and Schomaker (1951). The latter authors introduced juxtaposition diagrams, which allowed them to recognize easily the different kinds of integrals occurring in the matrix elements, and to find their coefficients. The inclusion of polar structures allowed the treatment of heterocyclic molecules, such as piperidine, pyrazine, pyrrol, furan, thiophene (Simonetta, 1952; Mangini and Zauli, 1960), and vinyl chloride (Takekiyo, 1962; Schug, 1965). Consideration of systems involving orbital degeneracy allows the treatment of positive or negative ions or radical ions. New types of integrals appear and their values have to be evaluated by empirical methods, and new prescriptions for calculating their coefficient must be used. This extension has been worked out by Karplus and co-workers (Schug et al., 1961), who also gave a method for evaluating spin and charge density, and by Simonetta and Heilbronner (1964), who discussed the acid-base equilibria between methyl-substituted benzenes and their conjugated acids and the electronic structure of the assumed intermediate in the Birch reduction. Valence bond theory in this approximation has also been used to explain the electronic effects of substituents in aromatic, electrophilic, and nucleophilic substitution reactions. Agreement with experimental data is satisfactory, even in ortho:para ratios (Green, 1954). Detailed analysis of electron populations calculated by use of valence bond wave functions has also been given (Peters, 1962).

Even these more sophisticated treatments, however, are largely empirical. Attempts to develop valence bond theory for many electron systems in the kind of approximation used by Heitler and London for the H_2 molecule, or the assumption of a given geometry and calculation of the integrals, are not numerous. Altman (1951) discussed the σ-π approximation and the singlet spectra of ethylene and derivatives, including only covalent structures, but calculating or evaluating all the integrals. Kopineck (1952) discussed the N_2 molecule, comparing the σ-π model and the "bent-bonds" model; the former gave a more stable ground state. The same result was obtained for the ethylene molecule, treated as a four-electron problem, with inclusion of all multiple-exchange integrals but using covalent structures only (Simonetta, 1959). The main points in the development of an a priori valence bond theory are the inclusion of ionic structures and the problem of nonorthogonality. Many treatments that include ionic structures have already been mentioned. Let us turn to the problem of nonorthogonality, which was presented as fundamental by Lowdin (1950). He suggested that by linear combinations of the initial atomic orbitals a set of orthonormal orbitals should first be constructed. This can easily be obtained by use of the metrical matrix of the one-electron space.

If $a, b, c, \ldots$ are the initial orbitals and $\mathbf{S}$ is the overlap matrix, the new set of orthogonalized orbitals $\bar{a}\bar{b}\bar{c}\cdots$ is given by

$$(\bar{a}\bar{b}\bar{c}\cdots) = (abc\cdots)\mathbf{S}^{-1/2} \tag{13}$$

The matrix $\mathbf{S}^{-1/2}$ can be obtained by the substitution $\mathbf{S} = \mathbf{1} + \mathbf{\Delta}$, where $\mathbf{\Delta}$ has the same nondiagonal elements as $\mathbf{S}$, but all the diagonal elements are equal to zero. Then

$$(\mathbf{1} + \mathbf{\Delta})^{-1/2} = \mathbf{1} - \tfrac{1}{2}\mathbf{S} + \tfrac{3}{8}\mathbf{S}^2 - \tfrac{5}{16}\mathbf{S}^3 + \cdots \tag{14}$$

and $\mathbf{S}^{-1/2}$ can be calculated with the desired accuracy. The aim was then to reformulate valence bond theory, with these orthogonal orbitals as a base. It has been pointed out by Slater (1951), however, that this formulation is rather dangerous, since the Heitler-London treatment of the H_2 molecule with orthogonalized atomic orbitals would lead to a repulsive ground state for this molecule. This is because bonding is due to the overlap of orbitals participating in the bond, with piling up of charge in the region of the bond. If the covalent structure is based on orthogonal orbitals, there is no such overlap and the charge in the region of the bond is decreased. A way out from this difficulty was proposed by McWeeny (1954), who showed that when using orthogonal atomic orbitals polar structures must be admitted in the calculation, since they give an important contribution to all the states. This theory was applied to cyclobutadiene, benzene, and water (McWeeny, 1955; McWeeny and Ohno, 1960).

It appears that in the a priori valence bond theory two ways of carrying out the calculations are possible. (1) Use of nonorthogonal orbitals and inclusion of all multiple exchange integrals and proper normalization constants. Covalent structures may be sufficient to obtain a good wave function for the ground state, although ionic structures may be included to obtain better functions. (2) Use of orthogonal orbitals, which allows all the simplifications in the formulas for matrix elements, as given by Slater. Polar structures must be included, since they give important contributions to wave functions for all the states.

For simple systems, when both treatments can easily be performed, they lead to exactly the same results. For more complex systems, however, the first method seems to be more convenient (Simonetta, Gianinetti, and Vandoni, 1967).* Earlier examples of this treatment can be found in the literature for H_3

*But see Cooper and McWeeny (1966) and Sutcliffe (1966).

(Hirshfelder, Eyring, and Rosen, 1936), allyl radical and allyl cation (Lefkovits, Fain, and Matsen, 1955), and butadiene molecule and cation (Fain and Matsen, 1957). A discussion of nonorthogonality problems has also been given by Simpson (1962).

MOST RECENT APPLICATIONS

In the last decade valence bond theory has found a new field of vast application: the interpretation of magnetic resonance spectra, both nuclear magnetic resonance (N.M.R.) and electron spin resonance (E.S.R.) spectra.

The experimental finding that the splitting of nuclear resonance lines in the N.M.R. spectra of liquids occurs, which is independent of the strength of the external magnetic field, has been interpreted empirically, and it was found that the interaction energy between two nuclei N and N' could be calculated in terms of a Hamiltonian of the form

$$hJ_{NN'}\bar{I}_N \cdot \bar{I}_{N'} \tag{15}$$

where h is Planck's constant and $J_{NN'}$ is a coupling constant between nuclei N and N' with nuclear spins $\bar{I}_N$ and $\bar{I}_{N'}$. When by frequent collisions the molecule undergoes rapid tumbling motions, as in liquids or gases, the direct magnetic interactions between two nuclei in a molecule average to zero. This is not true for the indirect interactions, which are a result of the magnetic interactions between each nucleus and the electrons of the molecule (Ramsey, 1953). Of the many terms that appear in the Hamiltonian for a molecule in a magnetic field, the most important is the so-called contact term (Fermi, 1930):

$$\bar{H}' = \frac{8\beta h}{3} \sum_{k,N} \gamma_N \delta(\underline{r}_{kN}) \bar{S}_k \cdot \bar{I}_N \tag{16}$$

where β is the Bohr magneton, γ_N is the gyromagnetic ratio of nucleus N, $\bar{S}_k$ is the electron spin angular momentum, and $\delta(\underline{r}_{kN})$ is the Dirac delta function, with $\underline{r}_{kN}$ the radius vector from electron k to nucleus N. By second-order perturbation theory the spin-spin interactions between nuclei N and N' involving $\bar{H}'$ can be evaluated and their energy reduced to form (15), in which it is

$$J_{NN'} = \frac{-2}{3h}\left(\frac{8\beta h}{3}\right)^2 \gamma_N \gamma_{N'} \sum_n \frac{1}{E_n - E_0} \langle \Psi_0 | \sum_k \delta(\underline{r}_{kN}) \bar{S}_k | \Psi_n \rangle \langle \Psi_n | \sum_m \delta(\underline{r}_{mN'}) \bar{S}_m | \Psi_0 \rangle \tag{17}$$

Ψ_0 and Ψ_n are the unperturbed wave functions for the ground state, having energy E_0, and the excited states, having energies E_n, and the summation is over all excited states. An approximate form of (17) is generally used:

$$J_{NN'} = \frac{-1}{\Delta E_{\mathrm{Av}}} \frac{2}{3h} \left(\frac{8\beta h}{3}\right)^2 \gamma_N \gamma_{N'} \langle \Psi_0 | \sum_{k,m} \delta(\underline{r}_{kN}) \delta(\underline{r}_{mN'}) \bar{S}_k \cdot \bar{S}_m | \Psi_0 \rangle \tag{18}$$

where ΔE_{Av} is a mean excitation energy, to be estimated empirically. Ψ_0 can be expressed as a linear combination of canonical structures of the form (9). By using the Dirac expression (Dirac, 1929) for the exchange operator,

$$\bar{P}_{km} = \tfrac{1}{2}(1 + 4\bar{S}_k \cdot \bar{S}_m) \tag{19}$$

Karplus and Anderson (1959) obtained the following expression for the coupling constant:

$$J_{NN'} = \frac{1}{4\Delta E_{\text{Av}}} \frac{2}{3h} \left(\frac{8\beta h}{3}\right)^2 \gamma_N \gamma_{N'} \phi_N(0) \phi_{N'}(0) \sum_{j,l} C_{0j} C_{0l} \frac{1}{2^{n-w_{j,l}}} [1 + 2f_{j,l}(P_{NN'})] \tag{20}$$

In this expression $\phi_N(0)$ is the electron density at nucleus N; C_{0j} and C_{0l} are the coefficients of structures j and l in the ground state function, as defined by Equation (4); $w_{j,l}$ is the number of islands in the superposition diagram for structures j and l; and $f_{j,l}(P_{NN'})$ is $2^{n-w_{j,l}}$ times the coefficient of the exchange integral for orbitals centered on nuclei N and N' in the same superposition diagram. The theory has been successfully applied to methane, ethylene, and ethane in different conformations (Karplus, 1959), to the study of dependence of the proton interactions in CH_2 groups from the HCH angle (Gutowsky, Karplus, and Grant, 1959) and from substituents (Barfield and Grant, 1962, 1963), and to the study of long-range proton coupling constants (Barfield, 1964).

Next we consider the E.S.R. spectra of hydrocarbon radicals. The hyperfine structure of E.S.R. spectra has also been interpreted on the basis of the Fermi contact term in the Hamiltonian. For large external magnetic fields only the interaction of the z components of the electron and nuclear spins must be considered and Equation (16) can be expressed in the form

$$H'' = \frac{8\beta h}{3} \sum_{k,N} \gamma_N \bar{S}_{kz} \bar{I}_{Nz} \delta(\underline{r}_{kN}) \tag{21}$$

First-order perturbation theory gives the change in energy due to this term:

$$\Delta E = \frac{8\beta h}{3} \sum_{k,N} \gamma_N \langle \Psi_0 | \delta(\underline{r}_{kN}) \bar{S}_{kz} \bar{I}_{Nz} | \Psi_0 \rangle \tag{22}$$

where Ψ_0 is the unperturbed ground-state wave function. The experimental results for hydrocarbon radicals are usually given in terms of the proton hyperfine splitting constant a_N, which is related to ΔE through the equation

$$\Delta E = h \sum_N a_N I_{Nz} S_z \tag{23}$$

where

$$a_N = \frac{8\pi}{3h} \beta g \beta_N g_N \delta_N \tag{24}$$

and

$$\delta_N = \frac{\langle \Psi_0 | \sum_k \delta(\underline{r}_{kN}) \bar{S}_{kz} | \Psi_0 \rangle}{S_z} \tag{25}$$

β and β_N are Bohr and nuclear magnetons, g and g_N are the g factors for electron and proton, respectively, and S_z is the total z component of the spin angular momentum. It is evident that a Fermi constant interaction occurs only if there is a finite unpaired electron density at the protons. As most of the radicals that have been studied are aromatic radicals, it is not easy to see how this condition may be fulfilled in a π-electron radical, where the umpaired electron belongs to a π orbital with a nodal plane in the plane of the nuclear skeleton, where the protons lie. The answer to this question was given by McConnell (1956b). His proposed mechanism is the transmission of the π-orbital electron spin polarization of the σ-bonding electron system, since the resulting electron spin polarization is s-orbital-like in the region of the protons.

Valence bond theory was used by McConnell and Chesnut (1958) to calculate the electron interaction in a C—H fragment. They obtained a theoretical value for the hfs constant in the C—H fragment of the right order of magnitude. In a conjugated radical the unpaired electron is shared among different carbon atoms; if the odd electron density at carbon atom N is ρ_N, the hfs constant for the proton bonded to that carbon can be assumed to be (McConnell, 1956a)

$$a_N = Q\rho_N \tag{26}$$

where Q is a constant. This relationship is the bridge between experimental and theoretical investigations. Then Q can be calculated in a rather crude approximation or can be deduced from the hfs constant of benzene. A value of 30 gauss gives generally good agreement with experiment, and ρ_N can be calculated from the ground-state wave function.

Brovetto and Ferroni (1957) were perhaps the first to use valence bond theory in the Pauling approximation for calculating ρ_N in an aromatic radical (triphenylmethyl radical); McConnell and Dearman (1958) calculated spin densities by the same method in perynaphtenyl radical. In both cases positive and negative spin densities were obtained, as expected, and the numerical values agreed satisfactorily with experimental ones.

Spin densities for the allyl radical were calculated by the a priori valence bond method, with the theoretical Q value obtained by considering the C—H fragment in the same approximation. The calculated densities are $\rho_a = \rho_c = 0.62$, $\rho_b = -0.13$ (Simonetta, Gianinetti, and Vandoni, 1967), which compare favorably with experimental values: $\rho_a = \rho_c = 0.58$, $\rho_b = -0.16$ (Fessenden and Schuler, 1963).

We may now ask, What are the possibilities of development and application of valence bond theory in the near future? The valence bond theory has not yet been so extensively applied as the molecular orbital theory. The use of ever bigger computers might remove some of the computational difficulties inherent in valence bond calculations and open up some fields of research that up to now have appeared to be out of praticable range. We may mention among others the interpretation of the ultraviolet spectra of aromatic hydrocarbons, the a priori calculation of properties of metals, and an application to the study of energies and geometries of transition states, which should be particularly amenable to interpretation by valence bond theory.

REFERENCES

Altman, S. L. (1951). *Proc. Roy. Soc.* (London) A **210**, 327, 343.
Barfield, M. (1964). *J. Chem. Phys.* **41**, 3825.
Barfield, M., and D. M. Grant (1962). *J. Chem. Phys.* **36**, 2054.
Barfield, M., and D. M. Grant (1963). *J. Am. Chem. Soc.* **85**, 1899.
Brovetto, P., and S. Ferroni (1957). *Nuovo Cimento* **5**, 142.
Cooper, J. L., and R. McWeeny (1966). *J. Chem. Phys.* **45**, 226.

Craig, D. P. (1950). *Proc. Roy. Soc.* (London) **200**, 390.
Daudel, R., and A. Pullman (1946). *J. Phys. Radium*, **7**, 59, 74, 105.
Dirac, P. A. M. (1929). *Proc. Roy. Soc.* (London) A **123**, 714.
Fain, J., and F. A. Matsen (1957). *J. Chem. Phys.* **26**, 376.
Fermi, E. (1930). *Z. Physik* **60**, 320.
Fessenden, R. W., and R. H. Schuler (1963). *J. Chem. Phys.* **39**, 2147.
Green, A. L. (1954). *J. Chem. Soc.* 3538.
Gutowsky, M. S., M. Karplus, and D. M. Grant (1959). *J. Chem. Phys.* **31**, 1278.
Heitler, W., and F. London (1927). *Z. Physik* **44**, 455.
Hirschfelder, J., H. Eyring, and N. Rosen (1936). *J. Chem. Phys.* **4**, 121.
Karplus, M. (1959). *J. Chem. Phys.* **30**, 11.
Karplus, M., and D. H. Anderson (1959). *J. Chem. Phys.* **30**, 6.
Kellogg, R. E. (1962). *J. Chem. Phys.* **37**, 2950.
Kopineck, H. J. (1952). *Z. Naturforsch.* A **7**, 22, 314.
Kotani, M., K. Ohno, and K. Kayama (1961). In *Hand. der Phys.* XXXVII/2, 133. Berlin: Springer.
Lefkovits, H. C., J. Fain, and F. A. Matsen (1955). *J. Chem. Phys.* **23**, 1690.
Lewis, G. N. (1916). *J. Am. Chem. Soc.* **38**, 762.
Lowdin, Per-Olov (1950). *J. Chem. Phys.* **18**, 365.
Lowdin, Per-Olov (1955). *Phys. Rev.* **97**, 1474.
Mangini, A., and C. Zauli (1960). *J.* 2210.
McConnell, H. M. (1956a). *J. Chem. Phys.* **24**, 632.
McConnell, H. M. (1956b). *J. Chem. Phys.* **24**, 764.
McConnell, H. M., and D. B. Chestnut (1958). *J. Chem. Phys.* **28**, 107.
McConnell, H. M., and H. H. Dearman (1958). *J. Chem. Phys.* **28**, 51.
McLachlan, A. D. (1960). *J. Chem. Phys.* **33**, 663.
McWeeny, R. (1954). *Proc. Roy. Soc.* (London) A **223**, 63, 306.
McWeeny, R. (1955). *Proc. Roy. Soc.* (London) A **224**, 288.
McWeeny, R., and K. Ohno (1960). *Proc. Roy. Soc.* (London) A **255**, 367.
Pauling, L. (1931). *J. Am. Chem. Soc.* **53**, 1367.
Pauling, L. (1933). *J. Chem. Phys.* **1**, 280.
Pauling, L. (1938). *Phys. Rev.* **54**, 899.
Pauling, L. (1947). *J. Am. Chem. Soc.* **69**, 542.
Pauling, L. (1948a). *Nature* (London) **161**, 1019.
Pauling, L. (1948b). *J.* 1461.
Pauling, L. (1949). *Proc. Roy. Soc.* (London) A **196**, 343.
Pauling, L. (1950). *Proc. Nat. Acad. Sci. U.S.* **36**, 533.
Pauling, L. (1953). *Proc. Nat. Acad. Sci. U.S.* **39**, 551.
Pauling, L. (1960). *The Nature of the Chemical Bond.* Ithaca: Cornell Univ. Press.
Pauling, L., L. O. Brockway, and J. Y. Beach (1935). *J. Am. Chem. Soc.* **57**, 2705.
Pauling, L., and F. J. Ewing (1948). *Revs. Modern Phys.* **20**, 112.
Pauling, L., and J. Sherman (1933). *J. Chem. Phys.* **1**, 606, 679.
Pauling, L., and M. Simonetta (1952). *J. Chem. Phys.* **20**, 29.
Pauling, L., and G. W. Wheland (1933). *J. Chem. Phys.* **1**, 362.
Pauling, L., and E. B. Wison, Jr. (1935). *Introduction to Quantum Mechanics.* New York: McGraw-Hill.
Peters, D. (1962). *J. Am. Chem. Soc.* **84**, 3812.
Ramsey, N. F. (1953). *Phys. Rev.* **91**, 303.
Rumer, G. (1932). *Göttinger Nach.*, 377.
Schug, J. C. (1965). *J. Chem. Phys.* **42**, 2547.
Schug, J. C., T. H. Brown, and M. Karplus (1961). *J. Chem. Phys.* **35**, 1873.
Simonetta, M. (1952). *J. Chim. Phys.* **49**, 68.
Simonetta, M. (1959). *Gazz. Chim. It.* **89**, 1956.
Simonetta, M., E. Gianinetti, and I. Vandoni (1967). *J. Chem. Phys.* In press.

Simonetta, M., and E. Heilbronner (1964). *Theoret. Chim. Acta* **2**, 228.
Simonetta, M., and V. Schomaker (1951). *J. Chem. Phys.* **19**, 649.
Simpson, W. T. (1962). *Theories of Electrons in Molecules*, Chap. 3. Englewood Cliffs: Prentice-Hall.
Slater, J. C. (1931). *Phys. Rev.* **38**, 1109.
Slater, J. C. (1951). *J. Chem. Phys.* **19**, 220.
Sutcliffe, B. T. (1966). *J. Chem. Phys.* **45**, 235.
Takekiyo, S. (1962). *Bull. Chem. Soc. Japan* **35**, 460.

HARRY B. GRAY
Gates and Crellin Laboratories of Chemistry
California Institute of Technology
Pasadena, California

Electronic Structural Theory for Metal Complexes

The search for an accurate and useful electronic structural theory for molecular transition metal complexes was initiated by Linus Pauling. Pauling's valence bond method (1960) emphasizes a correlation of bond type with ground-state magnetic properties. Because it is undoubtedly familiar to most readers, it will not be reviewed here. In the period 1950–1960, as research in transition metal chemistry increased, Pauling's approach gave way to calculations based on an ionic model. However, the ionic model is no longer fashionable, and the current trend is to make calculations using one of the various molecular orbital methods. In most cases, these molecular orbital calculations give electronic distributions in reasonable agreement with those calculated by Pauling's method. Unfortunately, the valence bond method is not easily adapted to the description of excited states, and thus the interesting electronic spectra of metal complexes are commonly interpreted on a model of molecular orbital energy levels. We shall devote our present discussion to a critical review of the current status of molecular orbital theory as applied to metal complexes. We shall attempt to pinpoint the areas where work is needed and, in addition, make suggestions as to possible molecular orbital interpretations of certain important coordination chemical phenomena.

GENERAL APPROACH

The molecular orbital methods that are in current use necessarily involve a considerable number of severe approximations. *At the outset, we must accept the fact that precise quantitative calculations of molecular orbitals for metal complexes are not within our reach.* Accepting this assertion, we must adopt a sensible approach if we are to provide useful semiquantitative electronic structural results. Such results are highly desirable as vital aids in our continuing quest for an understanding of the various structures and chemical reactivities of different molecules. Many approaches have been tried, and slow progress has been made. Most of the approximate methods of contemporary interest bear some resemblance to the Wolfsberg-Helmholz calculation of tetrahedral complex ions (1952).

Our approach is to try to establish one good model calculation for each im-important class of metal complexes. The model calculation incorporates judiciously chosen experimental data so that the various parameters in the semiempirically determined matrix elements may be fixed. In the better cases, these data include results from single crystal optical spectral and ESR studies. We then assume that the "normalized" parameters can be used to calculate other complexes which are very similar in ligand composition and have the same basic molecular structure as the model complex. For example, we have recently used $Fe(CN)_5NO^{--}$ as a model in a comparative electronic structural study of the many interesting $M(CN)_5NO^{n-}$ complexes (Manoharan and Gray, 1966a, 1966b).

In summary, we emphasize again that the best we can do at this stage of development is to interpret *relative* electronic structural properties. The *absolute* numbers we get from molecular orbital calculations are *not* to be taken seriously in a strictly quantitative sense. Keeping these cautionary remarks firmly in mind, we shall outline the procedure we have used in approximate molecular orbital calculations.

A SEMIEMPIRICAL PROCEDURE FOR CALCULATING MOLECULAR ORBITALS*

The computational framework is taken as the LCAO–MO approximation. The MO's, $\phi_i^{\lambda\alpha}$, are assumed to be linear combinations of atomic symmetry basis orbitals (LCAO's), $X_p^{\lambda\alpha}$, constructed from atomic orbitals centered on each of the nuclei of the molecule and having the symmetry properties of the molecule:

$$\phi_i^{\lambda\alpha} = C_{ip}^{\lambda} X_p^{\lambda\alpha} \tag{1}$$

The superscripts classify the basis atomic orbitals and doubly occupied MO's by their symmetry species (λ, μ, ν, . . .) and subspecies (α, β, Υ, . . .); the latter differentiate rows or columns of a degenerate representation.

The coefficients, C_{ip}^{λ}, and corresponding orbital energies, E_i^{λ}, are found by solving a limited number of secular determinants, one for each symmetry species:

$$|F_{qs}^{\lambda} - E_i^{\lambda} G_{qs}^{\lambda}| = 0 \tag{2}$$

$$G_{qs}^{\lambda} = \int X_q^{\lambda\alpha}(1) X_s^{\lambda\alpha}(1)\, dv(1) \tag{3}$$

The nonvalence or inner-orbital electrons are treated as a nonpolarizable core and are neglected (the "core approximation"). Since the group overlap integral, G_{qs}^{λ}, is calculated exactly, it is necessary initially to choose a set of analytical orbital functions for the basis orbitals; these, the geometry and the bond distance(s), define G_{qs}^{λ}, which can always be algebraically related to simple diatomic overlap integrals.

It now remains to specify the Hamiltonian matrix elements, F_{qs}; diagonal ($q = s$) and off-diagonal ($q \neq s$) elements are evaluated differently:

$$F_{qq}^{\lambda} = -I_q^{\lambda} \tag{4}$$

*See Ballhausen and Gray (1962); Basch, Viste, and Gray (1962); Ballhausen and Gray (1964); Basch (1966).

The diagonal elements are approximated as the negative of the appropriate valence orbital ionization potential (VOIP). The VOIP is defined (Basch et al., 1965) as the free atom or ion average of configuration ionization energy of an electron in orbital q at the computed Mulliken charge and configuration. An iteration procedure is used to calculate the diagonal elements of the central atom, as follows. A charge and configuration are initially assumed and used to compute the diagonal elements. The secular determinants are solved and the calculated charges and configurations compared with the assumed values. This procedure is continued until the assumed and computed values agree to within some predetermined limits.

Extrapolation for partial charge on an atom is accomplished by fitting the experimentally observed integral points to a quadratic function of the charge as, for example, is shown schematically for a metal d orbital in Figure 1. The VOIP for a given metal orbital is considered as a linear combination of three different configurations in order to represent adequately fractional populations. Thus, in Figure 1, interpolation for charge is horizontal and configuration vertical.

The VOIP's for integral points are derived from atomic spectral data, and an extensive calculation and tabulation of VOIP's for the elements Li through Br for many configurations and stages of ionization have been carried out (Basch et al., 1965).

The diagonal elements for the ligand orbitals are treated differently from those of the metal; for the ligands only the configuration s^2p^n is used and *no adjustment* is made for partial charge in the iterating procedure. Further, the ligand p_σ is assumed more stable (by 10,000 cm^{-1}) than p_π, and the latter is set equal to the *neutral atom value*. In a later section we will attempt to justify these two arbitrary procedures.

The off-diagonal elements are computed from the formula,

$$F_{qs}^{\lambda} = -K^{\sigma \text{ or } \pi} \tfrac{1}{2} G_{qs}^{\lambda} (\text{VOIP}_q + \text{VOIP}_s) \tag{5}$$

where $K^{\sigma \text{ or } \pi}$ is an empirical constant—actually two different constants, one for *sigma*- and the other for *pi*-type interactions. This is essentially the innovation suggested by Wolfsberg and Helmholz (1952) and has the advantage over the simpler Hückel calculation of providing an automatic procedure for treating heteroatom molecules.

VOIP_q is equal to I_q^{λ} of Equation (4) for metal orbitals and for ligand symmetry basis orbitals when neglecting ligand-ligand overlap in their normaliza-

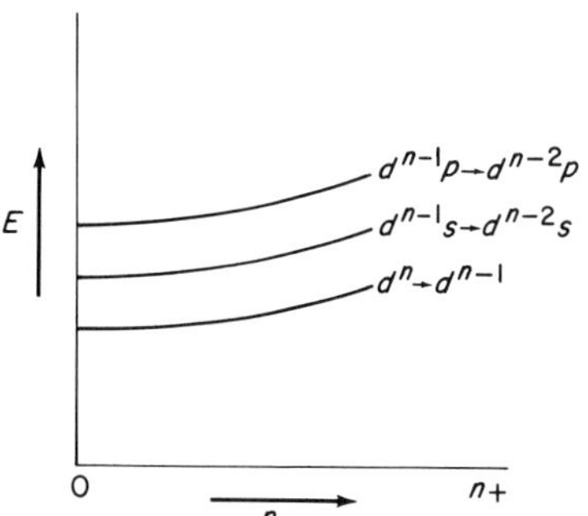

FIGURE 1.
Ionization energy of a d electron as a function of atomic charge and configuration.

tion. For ligand orbitals, including ligand-ligand overlap,

$$I_q^\lambda = \frac{1 + KY_q^\lambda}{1 + Y_q^\lambda}\,\mathrm{VOIP}_q \tag{6}$$

where Y_q^λ appears in the normalization factor $[N(1 + Y_q^\lambda)]^{-1/2}$ for the ligand symmetry orbitals and $N^{-1/2}$ would be the normalization factor in the absence of ligand-ligand overlap. The K in Equation (6) is the same as in Equation (5). These details follow logically from the method as outlined.

The valence orbitals generally used in calculations are the nd, $(n + 1)s$, and $(n + 1)p$ for the metal, and $n's$ and $n'p$ on the ligand. From Equation (5) it would appear that any two atomic orbitals with a finite overlap should be included in the calculation, but it is here that we exercise chemical intuition and choose only those orbitals traditionally considered to be important in chemical bonding. Omitting the nonvalence atomic orbitals may be unduly optimistic but, nevertheless, seems necessary if simplicity is to be preserved. In addition, the method's success may very well be due to a fortuitous cancellation of errors introduced by the particular assumptions made for the core approximation and Equations (4) and (5).

Several modifications of the basic method outlined above have been employed by various investigators. Of particular note are the Pariser-Parr-Pople-type calculations of Oleari and co-workers (1966), the modification of Fenske and co-workers (1966), the procedure of Ros and Schuit (1966), the "angular overlap" method of Schäffer and Jørgensen (1965), and the "Madelung-corrected" calculations of Pearson and Mawby (1967), and Jørgensen, Horner, Hatfield, and Tyree (JHHT, 1967). In discussing the results in certain specific systems, we will attempt to compare the different procedures and evaluate the critical remarks by the sundry investigators in the field.

OCTAHEDRAL COMPLEXES

The SCCC–MO method has been used to calculate the energy levels in a wide selection of octahedral complexes containing the simple ligands F^-, Cl^-, Br^-, O^{--}, and S^{--} (Basch et al., 1966). Some pertinent results are given in Table 1. All the calculations give approximately the same scheme of one-electron energy levels. The diagram for FeF_6^{---} is representative of the results; it is shown in Figure 2. In these calculations, the position of the first ligand field transition was fitted approximately by fixing K^π at 2.10 and varying K^σ as required. It was found that the K^σ values follow a regular pattern with increasing atomic number of the central metal, the relationship being $K^\sigma(n) = (0.027n + 1.546) \pm 0.02$, where $n = 1$ for Ti, $n = 2$ for V, and so on.

For octahedral complexes, the most important levels are $3e_g$ and $2t_{2g}$ for the d-d spectra, and, in addition, the two highest levels of odd parity, the $3t_{1u}$ and t_{2u}, for the ligand-to-metal charge transfer. The t_{1g}, although a strictly ligand π level, as is the t_{2u}, is relatively unimportant, since both $t_{1g} \longrightarrow 2t_{2g}$ and $t_{2g} \longrightarrow 3e_g$ are parity-forbidden transitions. The $3e_g$ and $2t_{2g}$ roughly correspond to the orbitals e_g and t_{2g} of crystal field theory. The $3t_{1u}$ level is mainly composed of ligand π functions.

In calculations on the sixteen octahedral complexes, permutations of levels

occurred within some of the bracketed sets. Only one of these permutations is of importance, that interchanging $3t_{1u}$ and t_{2u}. Since they are both triply-degenerate orbitals of odd parity and of similar composition, it will be difficult to establish the correct order in these cases. However, magneto-optical rotatory dispersion (MORD) studies might decide the issue. For consistency in the present cases, the first $L \longrightarrow M$ charge transfer was calculated from the t_{2u}.

However, with the possible exception of the $t_{2u} - 3t_{1u}$ comparison, these calculations, for reasonable choices of the parameters, yield the expected relative orderings for the level that should be most important in the *d-d* and $L \longrightarrow M$ charge-transfer spectra of octahedral complexes. In particular, we always obtain $3e_g > 2t_{2g}$ for octahedral systems.

As displayed in Table 1, as the metal atomic number increases along the first transition series, the computed charges on the central atom decrease in an analogous series, an example comparison being $VF_6^{----}(+0.93) > MnF_6^{----}(+0.85) > NiF_6^{----}(+0.78)$. Of course, this trend is expected by a freshman chemist and can be directly correlated with the increasing stability of the $3d$, $4s$, and $4p$ valence orbitals of the central metal or, in simpler language, the increasing electronegativities of these elements. An encouraging result is the fact that essentially the same K^σ value accounts for the substantial increase in Δ_0 with metal oxidation number in the hexafluoro complexes. For example, a K^σ in the small range 1.61–1.62 is consistent with $\Delta_0 = 8{,}400\ \text{cm}^{-1}$ for MnF_6^{----} and $\Delta_0 = 21{,}800\ \text{cm}^{-1}$ for MnF_6^{--}. This type of increase in Δ_0 may be correlated primarily with the increasing strength of σ-bonding in the series.

A very clear trend of decreasing Δ_0 values with increasing metal atomic number is evident, one series being TiF_6^{---} ($\Delta_0 = 17{,}500\ \text{cm}^{-1}$) $> VF_6^{---}$ ($\Delta_0 = 15{,}900\ \text{cm}^{-1}$) $> CrF_6^{---}$ ($\Delta_0 = 15{,}200\ \text{cm}^{-1}$) $> FeF_6^{---}$ ($\Delta_0 = 14{,}000\ \text{cm}^{-1}$) $> CoF_6^{---}$ ($\Delta_0 = 13{,}100\ \text{cm}^{-1}$). The calculations allow us to interpret this as a clear *decrease* in the strength of both $L(\pi) \longrightarrow d$ and $L(\sigma) \longrightarrow d$

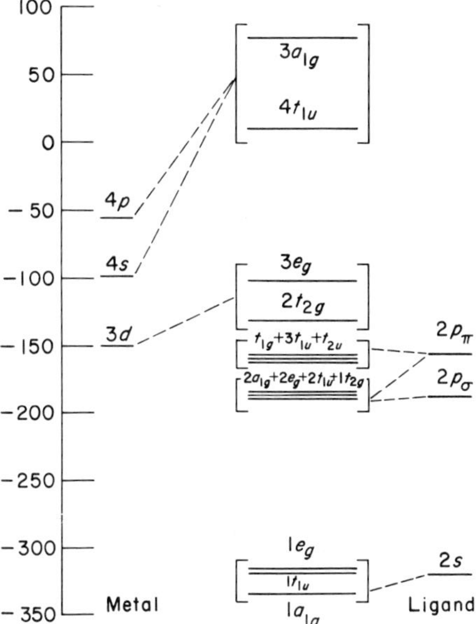

FIGURE 2.
Diagram of molecular orbital energy levels for octahedral complexes containing monatomic ligands. The scheme specifically represents the results of a calculation of FeF_6^{---}. The left-hand energy scale is in $10^3\ \text{cm}^{-1}$ units.

TABLE 1.
Results of SCCC calculations of various octahedral complexes (all energies in 1000 cm^{-1}).

	Complexes							
	TiF_6^{---}	$TiCl_6^{---}$	$TiBr_6^{---}$	VF_6^{---}	VCl_6^{---}	VF_6^{----}	CrF_6^{---}	$CrCl_6^{---}$
Metal charge	+1.12	+0.66	+0.68	+1.02	+0.59	+0.93	+0.97	+0.53
3*d* population	2.81	2.95	3.14	3.77	3.95	3.88	4.76	4.95
4*s* population	0.00	0.15	0.00	0.01	0.10	0.05	0.06	0.16
4*p* population	0.17	0.24	0.18	0.20	0.36	0.14	0.21	0.36
Δ	17.5	13.8	13.0	15.9	13.9	12.0	15.2	13.8
First allowed $L \longrightarrow M$ charge−transfer:								
Calculated orbital energy	59.0	37.0	25.8	48.4	27.7	45.0	38.9	21.8
Observed energy	>50	≳38	<38	>40	≳40	≫40	>37	>37
$K_\sigma(\pm 0.005)$	1.53	1.60	1.51	1.55	1.57	1.59	1.60	1.61

	Complexes							
	$CrBr_6^{---}$	CrO_6^{9-}	MnF_6^{--}	MnF_6^{----}	FeF_6^{---}	CoF_6^{---}	CoO_6^{10-}	NiF_6^{----}
Metal charge	+0.64	+0.75	+1.00	+0.85	+0.87	+0.84	+0.56	+0.78
3*d* population	5.26	4.91	5.65	5.89	6.76	7.72	7.96	8.79
4*s* population	0.10	0.08	0.06	0.10	0.14	0.17	0.22	0.20
4*p* population	0	0.26	0.29	0.16	0.24	0.26	0.26	0.23
Δ								
First allowed $L \longrightarrow M$ charge−transfer:								
Calculated orbital energy	12.8	31.9	36.9	29.6	23.5	16.9	13.6	17.6
Observed energy	<37	≳37	>28	>43	>30		~60	>60
$K_\sigma(\pm 0.005)$	1.58	1.61	1.61	1.62	1.67	1.69	1.71	1.68

SOURCE: Basch et al. (1966).

bonding in going from TiF_6^{---} to FeF_6^{---}, but a sharper decrease in $L(^\sigma) \longrightarrow d$ bonding yields the resultant decreasing trend in Δ_0 values.

The slack in the $L \longrightarrow d$ bonding in going across the transition series is taken up by enhanced $L \longrightarrow 4s$ and $L \longrightarrow 4p$ bonding, as judged from the increasing 4*s* and 4*p* populations. In fact, it is the increasing 4*s* and 4*p* involvement that leads to the net *increase* in covalency that we referred to earlier. In our opinion, the larger potential of 4*s* and 4*p* valence orbitals for bonding in the later transition metals is a very important factor in changing the relative binding characteristics in the direction of the heavier donor atoms, as will be discussed later.

L ⟶ M CHARGE TRANSFER SPECTRA

In the $L \longrightarrow M$ charge transfer spectra reported in Table 1, three distinct trends are evident. One is the order of charge-transfer energies with ligand in a series such as CrX_6^{n-}. The calculated order, of course, follows the ligand VOIP values, giving the prediction S < Br < Cl < O < F. This order is

TABLE 2.
Ligand field parameters and charge distributions for octahedral metal cyanide complexes.

Complex	Metal charge, calculated	Δ_0 (cm^{-1})
$Ti(CN)_6^{---}$	+0.66	22,300
$V(CN)_6^{---}$	+0.56	23,400
$Cr(CN)_6^{---}$	+0.47	26,600
$Mn(CN)_6^{---}$	+0.48	30,000
$Mn(CN)_6^{----}$	+0.46	35,200
$Fe(CN)_6^{---}$	+0.44	35,000
$Fe(CN)_6^{----}$	+0.42	34,000
$Co(CN)_6^{---}$	+0.41	34,900

SOURCE: From Alexander (1966).

borne out by the experimental results where data are available. A second trend is the energy of $L \longrightarrow M$ charge transfer in an isoelectronic series of complexes differing in the central atom oxidation number. In all cases, the calculations give $M(n) > M(n+1)$, an expected result. Experimental data are lacking for first-row octahedral halides, but examples such as $RhCl_6^{---} > PdCl_6^{--}$ confirm the trend (Jørgensen, 1959). A final result is that the orbital energy of $L \longrightarrow M$ charge transfer is calculated to decrease in a series of analogous complexes as the atomic number of the central metal increases across the transition row. Experimental results are sparse for octahedral first-row halides, but in other geometries (such as tetrahedral) this effect is clearly shown.

Molecular orbital calculations have been reported for octahedral complexes containing diatomic ligands such as CO and CN^- (Berthier et al., 1965, and references therein; Alexander, 1966; Beach and Gray, 1964, and unpublished data, 1966). The energy levels calculated for $Fe(CN)_6^{----}$ in a recent study (Alexander, 1966) are shown in Figure 3. With CO and CN^-, three ligand valence levels, the filled π^b and σ^b and the empty π^*, are included in the calculations. The $Fe(CN)_6^{----}$ complex may be used as a model for the several $M(CN)_6^{n-}$ complexes. A summary of results for these complexes is given in Table 2. In striking contrast to the MX_6^{n-} complexes, the Δ_0 values for cyanide complexes show a regular *increase* with an increase in the atomic number of the central metal, and essentially no dependence on oxidation number for the same central metal. For example, $\Delta_0[Cr(CN)_6^{---}] < \Delta_0[Co(CN)_6^{---}]$ and the Δ_0 values of $Fe(CN)_6^{----}$ and $Fe(CN)_6^{---}$ are about the same.

TETRAHEDRAL COMPLEXES

Since the initial work of Wolfsberg and Helmholz on MnO_4^- and CrO_4^{--}, tetrahedral complexes have played a key role in the development of semi-empirical molecular orbital calculations of metal complexes. Although there is still considerable debate as to the assignments of charge-transfer spectra of tetrahedral complexes (Oleari et al., 1966), the principal features of the energy level scheme for these systems is worked out (Basch et al., 1966; Oleari et al.,

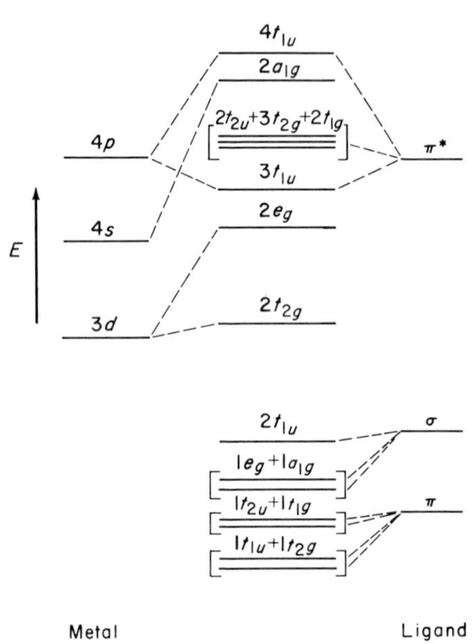

FIGURE 3.
Molecular orbital energy levels for octahedral complexes containing diatomic ligands such as CO and CN^-. The ordering of levels is derived from a calculation of $Fe(CN)_6^{----}$

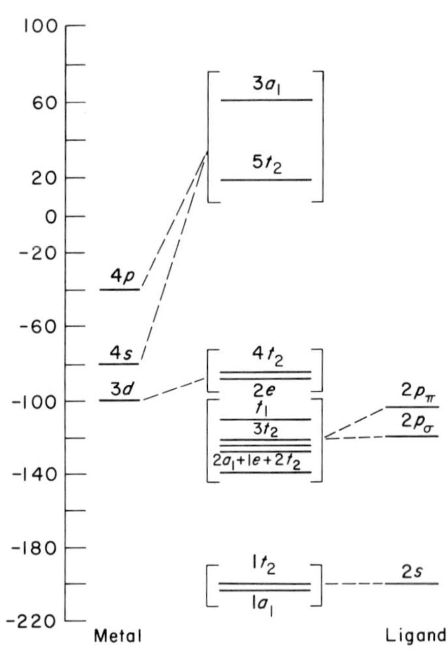

FIGURE 4.
Diagram of molecular orbital energy levels for tetrahedral complexes containing monatomic ligands. The levels were calculated for $FeCl_4^{--}$. The left-hand energy scale is in 10^3 cm^{-1} units.

1966; Ros and Schuit, 1966; R. F. Fenske, private communication). The level scheme shown in Figure 4 is specifically for $FeCl_4^{--}$ but is representative of most tetrahedral complexes containing simple one-atom ligands such as Cl^-, Br^-, and O^{--}. The most important "spectroscopic" levels are labeled $3t_2$, t_1, $2e$, and $4t_2$. The $2e$ and $4t_2$ levels roughly correspond to the d orbitals of crystal field theory, although in this case they are significantly delocalized over the ligands. The t_1 is a strictly ligand π level, whereas $3t_2$ tends to be predominantly composed of ligand π functions.

Table 3 gives a summary of results from calculations and electronic spectral measurements for a number of representative tetrahedral complexes (Basch et al., 1966). Trends in Δ_t and charge transfer energies parallel those found for octahedral halides, and similar arguments can be put forward in explanation. With diatomic ligands there has been some work (Perumareddi et al., 1963, and references therein; A. Viste, unpublished data, 1964; J. Finholt, unpublished data, 1966), with recent calculations devoted to the series $Fe(CO)_4^{--}$, $Co(CO)_4^-$, and $Ni(CO)_4$ (Finholt, unpublished, 1966). These complexes exhibit $M \longrightarrow L$ charge transfer bands with increasing energy along the series $Fe(CO)_4^{--} < Co(CO)_4^- < Ni(CO)_4$, as expected. The π^* level of CO plays a very significant part in the bonding, and the net charges calculated actually show electron density withdrawal by the originally neutral CO ligand. Thus, in the metal carbonyls, the metal is the *donor* atom. Much additional work is needed in tetrahedral systems with diatomic and larger ligands.

TABLE 3.
Results of SCCC calculations of various tetrahedral complexes (all energies in 100 cm^{-1}).

	Complexes							
	$TiCl_4$	$TiBr_4$	VCl_4	VCl_4^-	CrO_4^{--}	MnO_4^-	MnO_4^{--}	MnO_4^{---}
Metal charge	+0.64	+0.52	+0.58	+0.51	$^-$0.67	+0.68	+0.68	+0.67
$3d$ population	2.84	2.93	3.92	3.98	4.45	5.35	5.50	5.63
$4s$ population	0.19	0.15	0.13	0.13	0.29	0.34	0.28	0.24
$4p$ population	0.33	0.41	0.37	0.39	0.58	0.63	0.54	0.47
Δ	8.7	7.6	9.0	5.6	26.0	26.0	19.0	14.8
First allowed $L \rightarrow M$ charge–transfer:								
Calculated orbital energy	40.2	32.0	37.7	37.0	48.2	36.3	34.6	33.7
Observed energy	34.8	29.0	24.2	?	26.8	18.3	22.9	30.8
$K_\sigma(\pm 0.005)$	1.64	1.62	1.60	1.59	1.95	1.96	1.78	1.81

	Complexes							
	$MnCl_4^{--}$	$FeCl_4^-$	$FeCl_4^{--}$	$CoCl_4^{--}$	$CoBr_4^{--}$	CoO_4^{6-}	CoS_4^{6-}	$NiCl_4^{--}$
Metal charge	+0.37	+0.38	+0.33	+0.32	+0.30	+0.54	+0.13	+0.34
$3d$ population	6.03	6.92	7.01	8.01	8.19	7.95	8.06	8.99
$4s$ population	0.24	0.27	0.27	0.28	0.18	0.11	0.26	0.34
$4p$ population	0.36	0.42	0.38	0.38	0.33	0.29	0.45	0.33
Δ	3.6	5.0	4.0	3.7	3.1	3.8	3.2	3.5
First allowed $L \rightarrow M$ charge–transfer:								
Calculated orbital energy	21.5	13.7	17.2	14.9	9.4	20.7	8.5	8.4
Observed energy	30	27.2	45.5	42.5	35.1	?	?	35.8
$K_\sigma(\pm 0.005)$	1.65	1.66	1.68	1.66	1.56	1.71	1.72	1.68

SOURCE: Basch et al. (1966).

DISCUSSION OF OCTAHEDRAL AND TETRAHEDRAL SYSTEMS

Calculations by the SCCC–MO method have been criticized on various grounds. If we accept the fact that for reasonable *relative* results we must "normalize" a model for each set of calculations, the remaining criticism is that the method consistently yields molecular orbitals that have excessive covalent character (Fenske et al., 1966; Cotton, 1966). If the method is as far out of line in the covalent direction as is the crystal field model in the ionic direction, it is the correct target for criticism. Unfortunately, the amount of ionic character in a bond is a very elusive quantity and, indeed, has no meaning except within the framework of some model. We can produce convincing experimental evidence of *some* covalent character in metal complexes from measurements of electron spin resonance (ESR), nuclear magnetic resonance (NMR), nuclear quadrupole resonance (NQR), electronic spectra, magnetic susceptibility, and others, but we *cannot find exactly how much.* The problem stems from the fact that it is impossible to reduce the experimental data to a set of molecular orbitals without introducing *severe* approximations. Thus, the various experiments only serve to confirm the calculated *trends* in the molecular orbital compositions, and in an absolute sense can *at best* only tell if a calculation is "in the ballpark" or not. As an example, it has been demonstrated that molecular orbitals extracted from ESR g values, in a calculation that ignores the low-energy charge-transfer states and the M–L overlap terms, will always be far off in an absolute sense (Kivelson and Lee, 1964; Kon and Sharpless, 1966; Manoharan and Gray, 1966).

Although the criticism concerning the covalent character of molecular orbitals may not always be relevant, the general feeling that improvements are necessary has led to modified methods that have a very useful correction term in the approximation for the diagonal elements. We shall show in a later section that such a correction term is required to calculate the correct ground states in certain diatomic metal oxides. And, unlike covalent character, the orbital symmetry and spin properties of the ground state have more precise definition and can be obtained from experimental data. A calculational method that gives an incorrect ground state within the limits of its parameter framework must be discarded for the system under consideration.

Several investigators have proposed calculations which *explicitly* take into account the fact that, for example, the metal diagonal elements represent the energy of an electron in its own atomic field plus the fields of the ligand atoms (Pearson and Gray, 1963; Jørgensen, 1964; Schäffer and Jørgensen, 1965; Fenske et al., 1966; Oleari et al., 1966; Ros and Schuit, 1966; Basch and Gray, 1967a; Jørgensen et al., in press; Pearson and Mawby, in press). Thus, the charge- and configuration-corrected VOIP's must be further corrected because of two-center terms in the diagonal elements. We shall call this the interatomic-coulomb correction. Actually, it should be pointed out that the SCCC-MO method includes a consideration of the result of an interatomic-coulomb term because all the diagonal elements are *not* taken as the VOIP's for the isolated ions. This is done only for the metal. The ligand, although carrying a negative charge in complexes containing one-atom ligands, is assigned diagonal elements held at *neutral* atom values. This particular move allows the ligand to accept electron density without destabilization of its valence level, and at least partially compensates for the covalent-producing treatment of the metal diagonal elements.

It is interesting to compare in Table 4 a number of calculated charge distributions for octahedral and tetrahedral complexes containing one-atom ligands. Added for enlightenment is the partial ionic character for single bonds which a freshman chemistry student could calculate, using Pauling's formula. It is clear that in the MF_6^{---} complexes the method of Fenske, Caulton, Radtke, and Sweeney (FCRS) (1966) produces more ionic charge distributions than does the SCCC–MO method (Basch et al., 1966). Although the two different calculations on the MF_6^{n-} series give significantly different over-all charge distributions, each is reasonably consistent with fluorine p_σ spin densities calculated from ESR and NMR data for CrF_6^{---} FeF_6^{---} (Shulman and Knox, 1960; Helmholz et al., 1961), FeF_6^{---} (Helmholz, 1959), and NiF_6^{----} (Sugano and Shulman, 1963). One reason for the larger positive charges on the metals in the FCRS series is that the 4s and 4p populations were not assigned to the metal. And the gradual decrease in metal positive charge across the transition series is principally correlated with increasing 4*s* and 4*p* electron density, as calculated from the SCCC–MO model.

The permanganate ion is a good example of the "my-model-fits-the-data-better-than-yours" game. In this case, the SCCC–MO method gives a charge distribution (Ballhausen and Gray, 1964) similar to that obtained from a limited basis set SCF–MO calculation (Basch, unpublished calculations). Also, the Pariser-Parr-Pople SCF–MO calculation of Oleari and co-workers (1966) yields a model with very substantial covalent character. On one side it has been argued (Fenske, 1965) that the ESR results (Schonland, 1959) show only 10 percent ligand involvement in the 3*e* level in MnO_4^{--}, in sharp contrast to the 40–60 percent range obtained for MnO_4^- in the various PPP–SCF–MO and SCCC–MO calculations. On the other side, the SCCC molecular orbitals agree nicely with the MORD data for MnO_4^- recently reported by Stephens and co-workers (Schatz et al., 1966). We offer that, although the MORD results do not establish the correctness of the SCCC–MO calculation of MnO_4^-, the ESR calculation which has been cited is very approximate and is probably substantially off on the ionic side with regard to the 3*e* level.

The complex $TiCl_4$ is another workhorse, calculated net charges on Ti ranging from a low of 0.64 in the SCCC–MO method (Basch et al., 1966) to a high of +2.64 in the JHHT calculations. Here the NQR data (Belford et al., 1956) agree (Cotton, 1966) with the +1.24 calculated by Fenske (private communication). The agreement is so good that it is a reasonable assumption that the Fenske model overcorrects toward the ionic side because, as Pauling noted

TABLE 4.
Comparison of calculated charge distributions in some octahedral and tetrahedral metal complexes.

Complex	Calculated metal positive charge			
	BVG[a]	JHHT[b]	FCRS[c]	Pauling[d]
TiF_6^{---}	1.12	3.00	1.92	1.8
VF_6^{---}	1.02	—	1.94	1.5
VF_6^{----}	0.93	—	—	—
CrF_6^{---}	0.97	—	1.92	1.5
MnF_6^{----}	0.85	—	—	0.8
MnF_6^{--}	1.00	—	—	—
FeF_6^{---}	0.87	2.32	1.96	1.0
CoF_6^{---}	0.84	—	1.91	1.0
NiF_6^{----}	0.78	2.00	—	0.2
$TiCl_4$[e]	0.64	2.64	—	1.5
VCl_4	0.58	2.58	—	1.4

[a]From the Wolfsberg-Helmholz SCCC-MO calculations reported in Basch et al. (1966).
[b]Minimization of total energy by differential ionization method, with Madelung correction; from Jørgensen et al. (in press).
[c]Method of Fenske et al. (1966).
[d]From the ionic character estimated in single bonds from the formula of Pauling (1960, page 98). Inclusion of π-bonding will lower the calculated metal charges.
[e]Other calculations give the following results: +1.12, Fenske (private communication); +2.1, Pearson and Gray (1963); +2.0, Wolfsberg-Helmholz calculation with a Madelung correction, from Jørgensen et al. (in press).

several years ago (1960), the interpretation of quadrupole-coupling constants is not straightforward and gives *more* ionic character to heteronuclear diatomic molecules than obtained from the most direct source, their electric dipole moments. And it is just as probable that the SCCC–MO model is too far over on the covalent side.

In summary, we again suggest that calculational procedures are best judged on how well they reproduce trends in electronic structural properties. In systems in which the interatomic-coulomb terms are not needed to set up a correct ground-state model, it is still an open question as to whether adding various additional "refinements" to the simple method is worth the extra effort. (And we must keep in mind that in calculations of metal complexes containing diatomic and larger ligands, the negative charge on the ligands will be spread out, thus reducing the interatomic-coulomb correction.) Vigorous investigation should decide this issue before long. A possible way to proceed would be to compare the various approximate methods with the best Roothaan SCF–MO calculations available. This involves a retreat to smaller molecules, specifically to diatomic metal oxides.

DIATOMIC METAL OXIDES

A logical way to develop the more approximate and empirical methods is presumably to "calibrate" them with the "better" SCF–MO calculations. One system for which some SCF–MO results are available is the diatomic metal oxides.

The ground states for several metal oxides are given in Table 5. According to crystal field theory, the ordering of the *d*-orbital MO's in transition metal diatomic oxides, where the five degenerate *d* orbitals of the free-metal atom split into three levels of molecular symmetries δ, π, and σ, is predicted to be $\delta < \pi < \sigma$ in order of increasing energy (Berg and Sinanoglu, 1960). However, this scheme is apparently inconsistent with the observed ground states of the oxides of at least the early transition metals. That is, the ground states of ScO, YO, and LaO are all $^2\Sigma$'s (Berg et al., 1965; Kasai and Weltner, 1965),

TABLE 5.
Ground states of diatomic metal oxides.

Metal oxide	Valence configuration of M^{++}	Observed ground state	Reference
ScO	$3d^1$	$^2\Sigma$	Berg et al. (1965); Kasai and Weltner (1965)
YO	$4d^1$	$^2\Sigma$	Ibid.
LaO	$5d^1$	$^2\Sigma$	Ibid.
TiO	$3d^2$	$^3\Delta$	Weltner and McLeod (1965a)
ZrO	$4d^2$	$^1\Sigma$	Ibid.
HFO	$5d^2$	$^1\Sigma$	Ibid.
TaO	$5d^3$	$^2\Delta_r$	Weltner and McLeod (1965b)
CuO	$3d^9$	$^2\Pi$ or Σ^2	T. M. Dunn (private communication)

those of ZrO and HfO, ${}^1\Sigma$'s (Weltner and McLeod, 1965a); TiO has a ${}^3\Delta$ ground state (Weltner and McLeod, 1965a), TaO has a ${}^2\Delta_r$ (Weltner and McLeod, 1965b); CuO has either a ${}^2\pi$ or a ${}^2\Sigma$ ground state (T. M. Dunn, private communication). Assuming ${}^2\pi$ for CuO, the ground states of CuO and TiO are ambiguous with respect to the δ, σ ordering. Otherwise, these observed ground states require $\sigma < \delta < \pi$.

Furthermore, limited basis set SCF–MO calculations on ScO (Carlson et al., 1965) and TiO (Carlson and Moser, 1964; Carlson and Nesbet, 1964) are able to reproduce the $\sigma < \delta < \pi$ order of levels. A population analysis of the SCF wave function for TiO (configuration σ^2, ${}^1\Sigma$) gives, $Ti^{+0.55}$, $3d^{1.25}4s^{1.72}4p^{0.53}$, with the valence σMO predominantly of metal 4*s* character, as reflected in the population analysis (Carlson and Nesbet, 1964). A low-lying σMO of mainly 4*s* character had also been suggested earlier by Jørgensen (1964). Thus, the value of the crystal field theory in assigning ground states in the transition metal diatomics is hampered by the limitation of an exclusively *d* orbital basis set.

The SCCC–MO calculations of metal oxides invariably give the ordering $\delta < \sigma < \pi$ of the ligand field levels (Basch, 1966). This is unsatisfactory for several of the metal oxides, because the wrong ground state would result. These calculations pinpoint probably the most important flaw in the simple MO method, which, as indicated above, is in ignoring two-center contributions to the diagonal elements. And we know that the presence of the oxygen nucleus and electrons will split the σ, π, δ metal orbitals even *before* the covalent bonding is "turned on."

We have suggested (Basch and Gray, 1967a) the following modified approach and tested it against the SCF–MO calculations for TiO. We approximate two-center, off-diagonal elements as in (7):

$$\mathrm{F}_{qasb} = -\mathrm{S}_{qasb}[\mathrm{W}_{qa} + \mathrm{W}_{sb}] + (qa/T/sb) \tag{7}$$

Note that T is defined as the negative of the kinetic energy operator, which is in the form suggested by Ruedenberg (1951). Note further that Equation (7) is not appropriate for one-center, off-diagonal elements.

The diagonal matrix elements are given by

$$\mathrm{F}_{qaqa} = -\mathrm{W}_{qa} - [Q_b + k^\lambda](qa, qa/r_b) \tag{8}$$

where Q_b is the formal charge on atom b and k^λ is an empirically determined constant (λ differentiates σ, π, and δ symmetries).

For extension to polyatomic molecules, it is easily shown that an additional term, $+\sum_{c(\neq a,b)} (qa/R^c/sb)$, appears on the right side of (7). For the diagonal elements, Equation (8) is replaced by

$$\mathrm{F}^\lambda_{qaqa} = -\mathrm{W}^\lambda_{qa} - \sum_{b(\neq a)} [Q_1 + k^\lambda](qa, qa/r_b) \tag{9}$$

It remains to define W_{qa} in (8). If F^a in the expression $(qa/F^a/qa) = -\mathrm{W}_{qa}$ were rigorously the atomic SCF operator appropriate to atom a (for some specified electronic configuration) and qa the appropriate SCF orbital function, then W_{qa} would be exactly the negative of the orbital energy of atomic orbital qa in the free atom or ion for that configuration. Thus, W_{qa} can be taken either from atomic spectral data (Basch et al., 1965) or atomic SCF computations (Clementi, 1965) interpolated for nonintegral charge and configuration.

In Equation (8) the term $-[Q_b + k^\lambda](qa, qa/r_b)$, which represents the interaction of an electron in orbital qa with the electrons and nucleus of atom b, is the interatomic-coulomb term mentioned earlier. It can be reasoned (Basch, 1966) that appropriate values for k^λ in (8) are in the range 0.2 to 0.6. Thus, if $Q_b \sim 0.4$, appearing in a metal orbital diagonal element with a charge of $\sim +0.4$ on the metal, the interatomic-coulomb correction may be close to zero, and thus the (metal) diagonal element equals the charge corrected ionization potential or orbital energy, W_{qa}. In the case of a ligand, in which adjustment for partial charge makes $-W_{qa}$ more positive than the neutral atom value, the interatomic-coulomb correction of $-(0.8 \pm 0.2)(qa, qa/r_b)$ will bring the diagonal element back toward its neutral atom value. Thus, from this development, there is some justification for choosing diagonal elements in the general manner prescribed by the SCCC–MO method; the metal elements are charge-corrected, whereas the ligand elements are for the uncharged species. However, as we have noted, trouble will arise with the SCCC–MO method in the cases in which the metal diagonal elements are split significantly by the interatomic-coulomb term.

Semiempirical MO calculations were undertaken on TiO and CuO to test the usefulness of Equations (7) and (8) as a basis for electronic structure calculations. The computational procedure is similar to that outlined for the SCCC–MO method. The W_q's [Equations (7) and (8)] are identified with the VOIP's given previously, interpolated for partial charge and configuration according to the population analysis. It now remains to choose the k-factors. For TiO (with one electron each in the σ and δ MO's) the values, $k^\sigma = 0.60$, $k^\pi = 0.40$, and $k^\delta = 0.20$, give the ordering $\sigma < \delta < \pi$, with the δ–σ separation at 8,200 cm^{-1}. If, however, the k's are brought closer, with $k^\sigma = 0.50$, $k^\pi = 0.40$, and $k^\delta = 0.30$, the reverse ordering, $\delta < \sigma < \pi$, is obtained. A population analysis of the former result (configuration σ^2 for comparison with the SCF–MO result) shows $Ti^{+0.72}$, $3d^{1.70}\,4s^{1.50}\,4p^{0.08}$; the σMO is predominantly of $4s$ character (73 percent), as predicted by the SCF–MO calculations and expected on the qualitative grounds outlined previously.

In going to CuO, again with $k^\sigma = 0.60$, $k^\pi = 0.40$, and $k^\delta = 0.20$, the ordering of the valence MO levels is found to be $\delta < \sigma < \pi$. The character of the σ MO is roughly divided equally ($\sim\frac{1}{3}$ each) among $3d_\sigma$, $4s_\sigma$, and $2p_\sigma$ (oxygen). Thus, in the second half of the first transition series, we can expect σ MO to move above δ and increase in $3d$ character.

Comparison of the calculated ground-state charge distributions shows an increase in covalent character in going from TiO to CuO, as is expected. The calculated values are Ti(+.67) and Cu(+.60). It is interesting that this modified SCCC–MO calculation again gives molecular orbitals that have slightly less covalent character than those obtained from a limited basis set SCF–MO calculation (Carlson and Nesbet, 1964). Recall that the computed-charge distribution for TiO from SCF–MO is Ti(+0.55), and this for a state slightly more ionic than the ground state.

The ionization potential of TiO (presumably from δ) is probably within

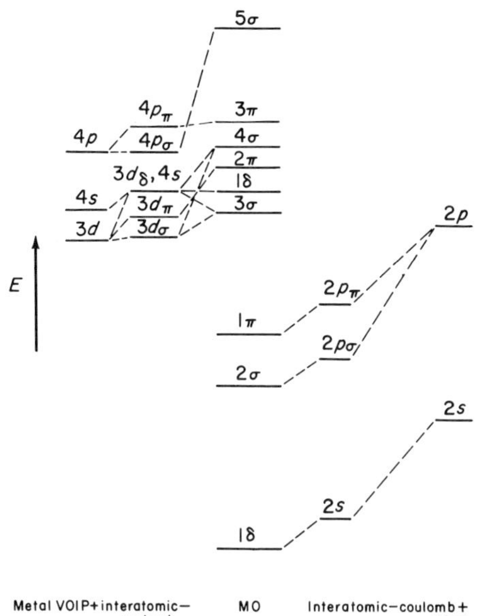

FIGURE 5.
Molecular orbital energy levels for TiO.

± 0.5 eV of the ground state ionization potential of Ti, which is known to be at 55,100 cm^{-1} (Basch and Gray, 1967a). The orbital energy of the δ MO in TiO is calculated here at $-68{,}100$ cm^{-1}; this should roughly correspond to the negative of the first ionization potential of TiO. The agreement is reasonable.

In Figure 5, the molecular orbital energy level diagram for TiO is presented, showing the splitting of the degenerate orbital VOIP's by the interatomic-coulomb energy term. Note that the oxygen p_σ diagonal element is significantly lower (more stable) than the p_π diagonal element. This is a completely general result for the method. It is reasonable that the p_σ orbital on the oxygen should be affected more by the metal nucleus than a corresponding p_π orbital. Therefore, the σ nuclear attraction integral will always be larger than that for the π orbital and the p_σ diagonal element correspondingly more negative than that of the p_π. The same relationship is obtained for the metal diagonal elements. This result, lowering the ligand p_σ level relative to the p_π, has previously been used without apparent justification in the SCCC–MO method to obtain levels in agreement with observed charge-transfer spectra (Basch et al., 1966; Basch and Gray, 1967b).

In summary, the modified method's success in the limited application discussed here is very encouraging since, in fact, the diatomic metal oxides are a difficult case; the simple SCCC–MO method predicts the ordering $\delta < \sigma < \pi$ for all the diatomics. The splitting of the metal d-orbital diagonal elements, in the order $d_\sigma < d_\pi < d_\sigma$, which is neglected in SCCC–MO, is crucial to obtaining the correct MO ordering in the diatomics containing early transition metals. The extremely sensitive position of the empirical parameter in the diagonal elements suggests that it might very well be possible to set a narrow range for the k's that can be transferred from molecule to molecule. This remains to be seen from further applications of the method to polyatomic systems.

SQUARE-PLANAR COMPLEXES

Unlike the situation for octahedral and tetrahedral metal complexes, the ordering of the primarily *d* orbital levels in square planar complexes is not known with certainty. Recent extensive experimental and theoretical studies in $PtCl_4^{--}$ are best interpreted in terms of the ligand field splitting $z^2 < xz, yz < xy \ll x^2 - y^2$ (Basch and Gray, 1967b; Martin et al., 1965, 1966; Mortensen, 1965; Bosnich, 1966). The energy level diagram representing the most recent MO calculation (Basch and Gray, 1967b) of $PtCl_4^{--}$ is shown in Figure 6. Extension has been made to other planar halide complexes (Basch, 1966), and a summary of results is given in Table 6. It should be noted that the K-factor values are not the same as those found for the corresponding first-row metal in octahedral or tetrahedral geometry.

TABLE 6.
Ligand field parameters and K factors for planar halides.

Parameter	$PtCl_4^{--}$	$PtBr_4^{--}$	$PdCl_4^{--}$	$PdBr_4^{--}$	$AuCl_4^{-}$	$AuBr_4^{-}$
K^σ	1.98	2.13	2.04	2.22	1.91	2.07
K^π	1.80	1.70	1.75	1.70	1.70	1.60
δ_1	28,700	26,500	23,600	21,500	20,800	19,700
δ_2	5,700	3,800	3,900	2,700	4,200	2,600
δ_3	7,800	5,800	4,900	3,800	8,600	6,900
charge (metal)	0.24	0.13	0.23	0.09	0.23	0.10

NOTE: Energies in cm^{-1}; the δ values are calculated separations of the appropriate molecular orbitals as defined in Figure 6; from Basch (1966).

FIGURE 6.
Molecular orbital energy levels for square-planar halide complexes such as $PtCl_4^{--}$.

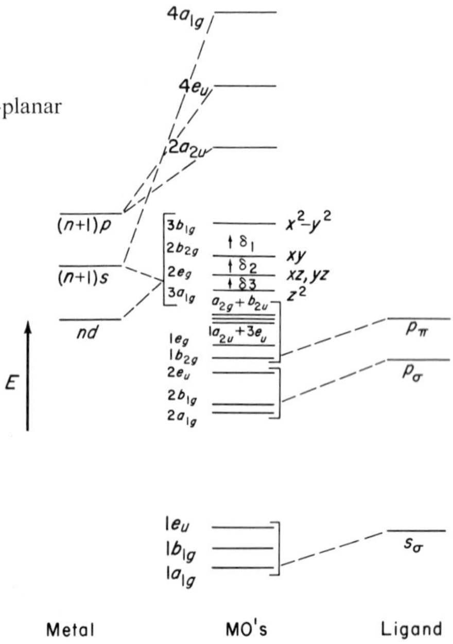

Detailed assignments of the *d-d* and charge-transfer spectra are presented in Table 7 for $PtCl_4^{--}$ and $PdCl_4^{--}$. The *d-d* assignment outlined for $PtCl_4^{--}$ is essentially the alternative A suggested by Martin, Tucker, and Kassman (MTK) (1965).

The observed charge-transfer bands in the square-planar halide ions have been established as being of the $L \longrightarrow M$ type (Gray and Ballhausen, 1963). From Figure 6, the first dipole allowed $L \longrightarrow M$ charge transfer band is ${}^1A_{1g} \longrightarrow {}^1A_{2u}$, ${}^1E_u(1)[b_{2u}, 3e_u \longrightarrow 3b_{1g}]$. Note that the b_{2u} and $3e_u$ levels are calculated to be almost degenerate, which is consistent with the view that both $b_{2u} \longrightarrow 3b_{1g}$ and $3e_u \longrightarrow 3b_{1g}$ transitions give rise to the first $L \longrightarrow M$ band. Introduction of ligand-ligand overlap will not alter this result significantly. A second band is also observed in the Pd(II) complex; this band is assigned ${}^1A_{1g} \longrightarrow {}^1E_u(2)[2e_u \longrightarrow 3b_{1g}]$. In Table 7, the calculated and observed positions of this second band are compared.

On examining the δ values for the planar halides in Table 6, an interesting effect is observed. The calculated δ_1 values actually shrink substantially in progressing from $Pt(II)Cl_4^{--}$ to $Au(III)Cl_4^{-}$, suggesting that the ligand-field Δ_1 value will decrease in the order Pt(II) > Au(III). This apparent inverse Δ-dependency on oxidation number is a trend that appears to be a feature of systems in which the highest-filled ligand field levels have an unusually large component of ligand orbitals. As a result, the first ligand-field band acquires considerable $L \longrightarrow M$ charge-transfer character and thus makes a red shift on an increase in metal oxidation number. Experimental evidence (Baddley et al., 1963; Williams et al., 1966) for the inverse Δ-dependency is set out in Table 8. Both the $M(dien)X^{n+}$ and $M(mnt)_2^{n-}$ systems clearly show the effect. A final observation of interest in the halide series is that the calculated charge distributions in $PtCl_4^{--}$ and $PdCl_4^{--}$ are not significantly different, a result

TABLE 7.
Electronic spectra of $PtCl_4^{--}$ *and* $PdCl_4^{--}$. *Band maxima in 1000* cm^{-1}; *molar extinction coefficients in parentheses.*

Transition	$PtCl_4^{--}$		$PdCl_4^{--}$	
	Observed[a]	Calculated[d]	Observed[e]	Calculated[d]
${}^1A_{1g} \rightarrow {}^3E_g$	17.0(<1)z, 18.0(2)xy, 19.0(<1)z,		Not reported	
${}^1A_{1g} \rightarrow {}^3A_{2g}$	20.9(9)xy, 20.6(10)z		Not reported	
${}^1A_{1g} \rightarrow {}^3B_{1g}$	24.0(7)xy, 24.1(3)z		18.0(19)xy, 17.0(7)z	
${}^1A_{1g} \rightarrow {}^1A_{2g}$	26.3(28)xy	28.7	20.0(67)xy	23.6
${}^1A_{1g} \rightarrow 1E_g$	29.2(37)xy, 29.8(55)z	34.3	22.6(128)xy, 23.0(80)z	27.5
${}^1A_{1g} \rightarrow {}^1B_{1g}$	36.5[b]	42.2	29.5(67)xy	32.4
${}^1A_{1g} \rightarrow {}^1A_{1u}$, ${}^1E_u(1)$	46.0(9580)[c]	Fitted	36.0(12,000)[f]	Fitted
${}^1A_{1g} \rightarrow {}^1E_u(2)$	Not reported	58.0	44.9(30,000)[f]	48.9

[a]Single crystal absorption spectrum of K_2PtCl_4 at 15°K; from Martin et al. (1965).
[b]Reflectance spectrum of K_2PtCl_4; from Day et al. (1965).
[c]Aqueous solution spectrum of K_2PtCl_4; from Gray and Ballhausen (1963).
[d]From Basch and Gray (1967b).
[e]Single crystal absorption spectrum of K_2PdCl_4; from Day et al. (1965).
[f]Aqueous solution spectrum of K_2PdCl_4 with excess KCl; from Gray and Ballhausen (1963).

also obtained from the shifts in the NQR spectra of the chlorides (Ito et al., 1961; Marram et al., 1963).

Some comment is in order regarding the low position of z^2 in the ligand field levels of $PtCl_4^{--}$. As MTK have pointed out (Martin et al., 1965), neither the $xz,yz > z^2$ nor the $xz,yz \approx z^2$ result is compatible with a point-charge crystal field calculation, which gives $x^2 - y^2 > xy > z^2 > xz,yz$ (Fenske et al., 1962). Fortunately, the molecular orbital model is compatible with the low position of z^2 in $PtCl_4^{--}$, possibly because of the very large participation of the $6s$ orbital in the σ-bonding. Since the $6s$ and $5d_{z^2}$ bond with the same ligand combination, the large participation of the $6s$ orbital in the Pt(II)–Cl bonds leaves $5d_{z^2}$ at an unusually stable position.

According to the best available experimental and theoretical work on $Ni(CN)_4^{--}$, the ligand-field pattern shows $z^2 > xz,yz$ from both the *d-d* and charge-transfer spectra (Gray and Ballhausen, 1963; Ballhausen et al., 1965). A very elegant analysis of the electronic spectra of single crystals containing the $Ni(CN)_4^{--}$ anion has been carried out by Ballhausen and co-workers (1965).

The lowest energy charge-transfer bands in all the planar $M(CN)_4^{n-}$ complexes have been assigned as M $\longrightarrow$ CN transitions (Gray and Ballhausen,

TABLE 8.
Position of the first spin-allowed ligand field band in M(dien)X^{n+} *and M*(mnt)$_2^{n-}$ *complexes.*

Complex	Band maximum [$cm^{-1}(\epsilon)$]
Pt(*dien*)Cl^+	37,040 (275)
Au(*dien*)Cl^+	33,110 (815)
Pt(*dien*)Br^+	36,360 (250)
Au(*dien*)Br^+	29,410 (510)
Pt(*dien*)I^+	33,330 (500)
Au(*dien*)I^+	26,110 (1800)
Ni(*mnt*)$_2^{--}$	11,850 (30)
Cu(*mnt*)$_2^{-}$	6,400 (337)
Pt(*mnt*)$_2^{--}$	18,500 (*sh*)
Au(*mnt*)$_2^{-}$	13,400 (44)

SOURCE: Data for *dien* complexes from Baddley et al. (1963); for *mnt* complexes from Williams et al. (1966).

TABLE 9.
Position of the first charge transfer transitions in $M(CN)_4^{n-}$ *complexes.*

Complex	Band maximum [$cm^{-1}(\epsilon)$]
$Ni(CN)_4^{--}$	32,300 (750)
$Pd(CN)_4^{--}$	41,600 (1200)
$Pt(CN)_4^{--}$	35,720 (1590)
$Au(CN)_4^{-}$	$>$50,000

SOURCE: From data compiled in Gray and Ballhausen (1963). The $Au(CN)_4^{-}$ value from Mason and Gray (1967).

1963). A comparison of the positions of the first charge-transfer band in the various complexes is given in Table 9. The energy order Ni(II) < Pd(II) > Pt(II) < Au(III) has been cited as characteristic of $M \longrightarrow L$ charge-transfer bands (Gray and Ballhausen, 1963). There is a nice correlation of $M \longrightarrow L$ charge-transfer energies in the three $M(CN)_4^{--}$ complexes with the observed half-wave potentials for the one-electron oxidation of $M(mnt)_2^{--}$ complexes, because the relative difficulty of oxidation as judged from the $E_{1/2}$ values follows the order Ni(II) < Pd(II) > Pt(II) (Williams et al., 1966). This comparison of energy-level positions from both electronic spectral and polarographic experiments tends to confirm the $M \longrightarrow L$ interpretation of the charge-transfer bands.

In summarizing the results for planar systems, the halides probably represent the case in which z^2 is at its lowest relative position in the ligand field level scheme. We make this suggestion because of the low spectrochemical position of halide ligands, indicating good π-donor and poor $(\sigma \longrightarrow d)$-donor capabilities. With ligands of better π-acceptor or $(\sigma \longrightarrow d)$-donor potential (or both), the z^2 level in many cases should move significantly above the xz,yz orbitals. Simple O- and N-donor ligands would be in this category, as well as CN^- and CO. It appears that in $Ni(CN)_4^{--}$ the ordering $z^2 > xz,yz$ is appropriate.

STABILITIES OF METAL-LIGAND BONDS

A legitimate question to ask of the molecular orbital model is whether anything of use can be extracted in regard to the stabilities and reactivities of metal complexes. Our optimistic view is that certain primitive conclusions may be reached at this stage of development concerning molecular orbital interpretations of *relative* stabilities of metal-ligand bonds and *relative* reactivities of metal complexes. There has been considerable discussion concerning a useful nomenclature for the classification of donor-acceptor bonds (for a summary and references, see Pearson, 1963); it appears well established that there is a definite trend toward enhanced stability of the (heavier-donor-atom-) metal bond in proceeding from the Ti family in the direction of the Zn family in the periodic table. The most recent descriptive nomenclature (Pearson, 1963) designates central metal atoms as "hard" if they show the preference

$$\begin{bmatrix} \text{N} \gg \text{P} > \text{As} \\ \text{O} \gg \text{S} > \text{Se} \\ \text{F} \gg \text{Cl} > \text{Br} > \text{I} \end{bmatrix}$$

in aqueous solutions of the complexes. Metal atoms are "soft" if the reverse binding trend

$$\begin{bmatrix} \text{As} < \text{P} \gg \text{N} \\ \text{Se} \sim \text{S} \gg \text{O} \\ \text{I} \gg \text{Br} > \text{Cl} > \text{F} \end{bmatrix}$$

is followed. Obviously there are many intermediate cases, and the classification is generally good only in a relative sense. For example, Pt(II) is softer than Co(III) because $Pt(NH_3)_3I^+$ is more stable than $Pt(NH_3)_3Cl^+$ in aqueous solution, whereas under similar conditions $Co(NH_3)_5Cl^{++}$ is more stable than $Co(NH_3)_5I^{++}$.

An assortment of explanations of the preference of "soft" metals for the heavier donor atoms has been put forward. A particular favorite is enhanced $d \longrightarrow$ ligand π-bonding in the complexes with a high d electron configuration. There seems to be no unequivocal evidence for this explanation, and we are of the opinion that π-bonding effects, although an important part of the metal-ligand bond, are definitely secondary to the principal σ-bonding interactions. *We propose that the changing stability characteristics of the metal-ligand bond in proceeding across the transition series can be interpreted primarily in terms of two effects. (1) The increasing relative participation of the metal valence s and p orbitals in the over-all bond structure, (2) The relative importance of inter-electronic repulsion terms in the d symmetry molecular orbitals as the d configuration builds up.*

Support for the idea of enhanced $4s$ and $4p$ bonding comes from several sources. First, the energies of the valence s and p orbitals (obtained from atomic spectral data) drop significantly in going across the transition series. For a metal such as Pt, the $6s$ and $6p$ orbitals are sufficiently stable to be involved in tight binding. The calculations discussed above established an increasing role for the s and p orbitals in bonding involving the later transition elements. We also have evidence of a lessened role for the d metal orbitals. Associated with their great stability, the d orbitals are contracted so that their overlap with the ligand valence orbitals is extremely small. (In the later transition elements, the d orbitals are rapidly becoming "nonvalence.") Experimentally, this is reflected in the decreasing Δ_0 values across the series, which itself is an indication of the important decrease in ligand $\longrightarrow$ metal (d) bonding.

The postulated explanation of "hard" and "soft" behavior in terms of relative orbital interactions is shown in Figure 7. Once the ligand valence p orbitals become less stable than the metal d orbitals, the highest-filled d-symmetry levels become more strongly associated with the ligand than the metal, and various types of different properties become evident: the decrease in Δ with increasing oxidation number, as has been found in the Pt(II)–Au(III) series; and the preference for Cl over F, S over O, and P over N.

We have argued for an increasing role for the metal s and p orbitals, but it remains to tie this up to an increased affinity of a metal atom for a heavier donor atom. Calculations on the Pt(II)-L bond indicate increasing overlap of the np valence orbital of the ligand with a metal $6p$ orbital in the order $P > S > Br > Cl > N > O > F$ (Langford and Gray, 1966), from which we can assume $P \gg N$, $S \gg O$, and $Br > Cl > F$, as observed with the "soft"

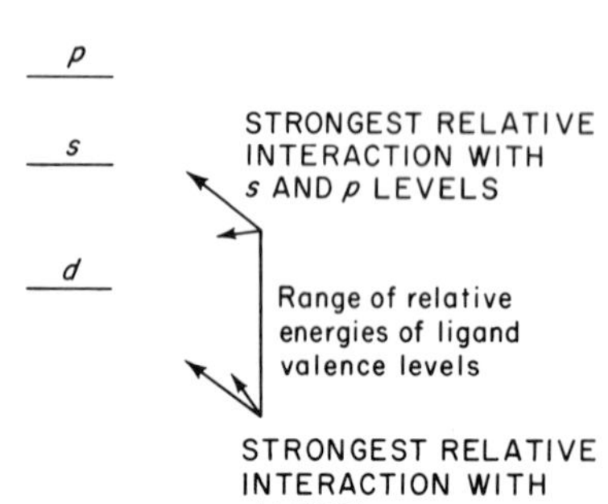

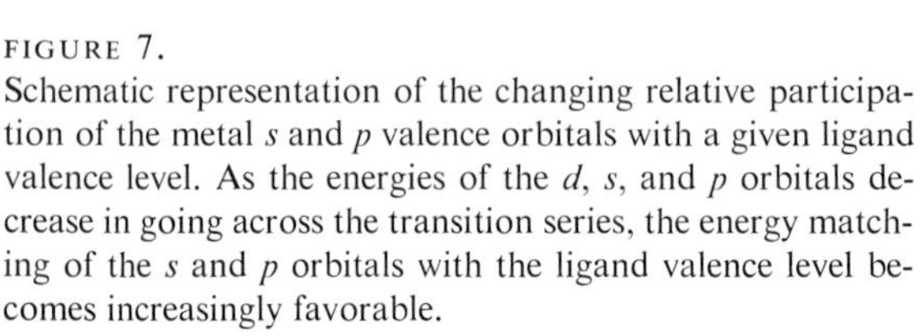
FIGURE 7.
Schematic representation of the changing relative participation of the metal s and p valence orbitals with a given ligand valence level. As the energies of the d, s, and p orbitals decrease in going across the transition series, the energy matching of the s and p orbitals with the ligand valence level becomes increasingly favorable.

metals. One feature of the enhanced valence *s* and *p* bonding is that it nicely accounts for the fact that both Hg(II) and Tl(III) are definitely in the "soft" category.

Three sets of facts are needed in order to predict the effect of interelectronic repulsion on the *M–L* bond stability. (1) As the d^n configuration increases, there are more electrons in *d*-symmetry molecular orbitals and thus a correspondingly larger energy term due to interelectronic repulsion. (2) The interelectronic-repulsion energies in the valence orbitals of the donor atoms, as taken from atomic spectra, decrease in going down a family. For example, the more tightly held electrons in the 2*p* orbitals of oxygen experience larger mutual repulsion than the more loosely held 3*p* electrons in S valence orbitals. Thus, we have $O \gg S > Se$ in interelectronic-repulsion effects. (3) The interelectronic-repulsion effects in valence *d* orbitals decrease in going down a transition metal family. For example, we have $Ni > Pd > Pt$.

Armed with these three items of information, we can construct a role for interelectronic repulsion in the *d*-symmetry molecular orbitals (both σ and π, bonding and antibonding). In the earlier transition metals, with *n* small in d^n, there will be a great advantage in spreading out the bonding electrons to minimize the interelectronic repulsion effects. Donor atoms which particularly benefit are those that have the largest interelectronic repulsion, as, for example, O and F. This is a reasonable interpretation of the unusually great strength of σ- and π-donor bonding in the M–O and M–F complexes of the early transition metals.

In the late transition metals, with high d^n, the gain in stability obtained from spreading out the donor atom valence electrons into *d*-symmetry bonding molecular orbitals (which, of course, include metal orbitals) is offset by the concommitant increase in interelectronic-repulsion effects around the metal. In this situation, the heavily occupied *d* orbitals become less attractive to donor atoms; ligands are favored which can attach more strongly to the outer (and unoccupied) *s* and *p* orbitals and at the same time have the smallest mutual repulsion for the d^n electrons. From the considerations outlined above, and other things being equal, this means that P will become better than N, S will become better than O, and I will become much better than Br. The enhancement of "soft" character in going down a family (Ni, Pd, Pt) can be viewed as a reduction in interelectronic repulsions between the filled metal *d*-symmetry levels and the filled orbitals mainly on the ligands.

In summary, we have presented a model for donor-metal bonding that is consistent with the significant trends in stability characteristics. The model includes the effects of changing valence orbital energies and interelectronic-repulsion energies in proceeding across the transition series. It should be emphasized that these correlations must involve an analogous series of ligands differing *only* in the donor atom. Almost certainly, the comparisons of metal-ligand bonding involving significant changes in the electronic structure of the ligand, as, for example, PF_3 versus PMe_3, will involve important π-bonding contributions.

ACKNOWLEDGMENTS

It is a special pleasure to record my gratitude to Professor C. J. Ballhausen, Dr. Harold Basch, and Dr. Arlen Viste, whose ideas have contributed heavily to much of the work herein presented.

REFERENCES

Alexander, J. J. (1966). Ph.D. thesis, Columbia Univ.

Baddley, W. H., F. Basolo, H. B. Gray, C. Nölting, and A. J. Pöe (1963). *Inorg. Chem.* **2**, 921.

Ballhausen, C. J., and H. B. Gray (1962). *Inorg. Chem.* **1**, 111.

Ballhausen, C. J., and H. B. Gray (1964). *Molecular Orbital Theory*. New York: Benjamin.

Ballhausen, C. J., N. Bjerrum, R. Dingle, K. Eriks, and C. R. Hare (1965). *Inorg. Chem.* **4**, 514.

Basch, H. (1966). Ph.D. thesis. New York. Columbia Univ.

Basch, H., and H. B. Gray (1967a). *Inorg. Chem.* **6**, 639.

Basch, H., and H. B. Gray (1967b). *Inorg. Chem.* **6**, 365.

Basch, H., A. Viste, and H. B. Gray (1965). *Theoret. Chim. Acta.* **3**, 458.

Basch, H., A. Viste, and H. B. Gray (1966). *J. Chem. Phys.* **44**, 10.

Beach, N. A., and H. B. Gray (1964). ACS Meeting Abstracts, Philadelphia.

Belford, R. L., A. E. Martell, and M. Calvin (1956), *J. Inorg. Nucl. Chem.* **2**, 11.

Berg, R. A., and O. Sinanoglu (1960). *J. Chem. Phys.* **32**, 1082.

Berg, R. A., L. Wharton, W. Klemperer, A. Buchler, and J. L. Stauffer (1965). *J. Chem. Phys.* **43**, 2416.

Berthier, G., P. Millie, and A. Veillard (1965). *J. Chem. Phys.* **62**, 8.

Bosnich, B. (1966). *J. Am. Chem. Soc.* **88**, 2606.

Carlson, K. D., and C. Moser (1964). *J. Phys. Chem.* **67**, 2644.

Carlson, K. D., and R. K. Nesbet (1964). *J. Chem. Phys.* **41**, 1051.

Carlson, K. D., E. Ludena, and C. Moser (1965). *J. Chem. Phys.* **43**, 2406.

Clementi, E. (1965). *Tables of Atomic Functions*, Supp. to *IBM J. Res. and Dev.* **9**, 2.

Cotton, F. A. (1966). Dwyer Memorial Lecture, Univ. of New South Wales, Australia.

Day, P., A. F. Orchard, A. J. Thomson, and R. J. P. Williams (1965). *J. Chem. Phys.* **42**, 1973.

Fenske, R. F. (1965). *Inorg. Chem.* **4**, 33.

Fenske, R. F., D. S. Martin, Jr., and K. Ruedenberg (1962). *Inorg. Chem.* **1**, 441.

Fenske, R. F., K. G. Caulton, D. D. Radtke, and C. C. Sweeney (1966). *Inorg. Chem.* **5**, 951, 960.

Gray, H. B., and C. J. Ballhausen (1963). *J. Am. Chem. Soc.* **85**, 260.

Helmholz, L. (1959). *J. Chem. Phys.* **31**, 172.

Ito, K., D. Nakamura, Y. Kurita, K. Ito, and M. Kubo (1961). *J. Am. Chem. Soc.* **83**, 4526.

Jørgensen, C. K. (1964). *Mol. Phys.* **7**, 417.

Jørgensen, C. K. (1959). *Mol. Phys.* **2**, 309.

Jørgensen, C. K. (1964). *Orbitals in Atoms and Molecules*. New York: Academic Press.

Jørgensen, C. K., S. M. Horner, W. E. Hatfield, and S. Y. Tyree, Jr. (1967). *Internat. J. Quantum Chem.* In press.

Kasai, P. H., and W. Weltner (1965). *J. Chem. Phys.* **43**, 2557.

Kivelson, D., and S.-K. Lee (1964). *J. Chem. Phys.* **41**, 1896.

Kon, H., and N. E. Sharpless (1966). *J. Phys. Chem.* **70**, 105.

Langford, C. H., and H. B. Gray (1966). *Ligand Substitution Processes*, p. 27. New York: Benjamin.

Manoharan, P. T., and H. B. Gray (1966a). *J. Am. Chem. Soc.* **87**, 3340.

Manoharan, P. T., and H. B. Gray (1966b). *Inorg. Chem.* **5**, 823.

Marram, E. P., E. J. McNiff, and J. L. Ragle (1963). *J. Phys. Chem.* **67**, 1719.

Martin, D. S., Jr., M. A. Tucker, and A. J. Kassman (1966). *Inorg. Chem.* **4**, 1682; amended in **5**, 1298.

Martin, D. S., Jr., J. G. Foss, M. E. McGarville, M. A. Tucker, and A. J. Kassman (1966). *Inorg. Chem.* **5**, 491.
Mason, W. R., and H. B. Gray (1967). *Inorg. Chem.* In press.
Mortensen, O. S. (1965). *Acta. Chem. Scand.* **19**, 1500.
Oleari, L., G. DeMichelis, and L. DiSipio (1966). *Mol. Phys.* **10**, 111.
Pauling, L. (1960). *The Nature of the Chemical Bond*, 3rd ed., Chapter 5. Ithaca: Cornell Univ. Press.
Pearson, R. G. (1963). *J. Am. Chem. Soc.* **85**, 3533.
Pearson, R. G., and H. B. Gray (1963). *Inorg. Chem.* **2**, 358.
Pearson, R. G., and R. Mawby (1967). *International Review of Halogen Chemistry*. New York: Academic Press. In press.
Perumareddi, J. R., A. D. Liehr, and A. W. Adamson (1963). *J. Am. Chem. Soc.* **85**, 249.
Ros, P., and G. C. A. Schuit (1966). *Theoret. Chim. Acta.* **4**, 1.
Ruedenberg, K. (1951). *J. Chem. Phys.* **19**, 1433.
Schäffer, C. E., and C. K. Jørgensen (1965). *Mol. Phys.* **9**, 401.
Schatz, P. N., A. J. McCaffery, W. Suëtaka, G. N. Henning, A. B. Ritchie, and P. J. Stephens (1966). *J. Chem. Phys.* **45**, 722.
Schonland, D. S. (1959). *Proc. Roy. Soc.* (London) A **254**, 111.
Shulman, R. G., and K. Knox (1960). *Phys. Rev. Letters* **4**, 603; Helmholz, L., A. V. Guzzo, and R. N. Sanders (1961). *J. Chem. Phys.* **35**, 1349.
Sugano, S., and R. G. Shulman (1963). *Phys. Rev.* **130**, 506, 512, 517.
Weltner, W., Jr., and D. McLeod, Jr. (1965a). *J. Phys. Chem.* **69**, 3488.
Weltner, W., Jr., and D. McLeod, Jr. (1965b). *J. Chem. Phys.* **42**, 882.
Williams, R., E. Billig, J. H. Waters, and H. B. Gray (1966). *J. Am. Chem. Soc.* **88**, 43.
Wolfsberg, M., and L. Helmholz (1952). *J. Chem. Phys.* **20**, 837.

J. I. FERNÁNDEZ-ALONSO
J. PALOU
Department of Physical Chemistry
University of Valencia
Valencia, Spain

On the Theoretical Electronic Structure Interpretation of the Pentalene and Heptalene Molecules

During the last few years, a thorough study has been made of the series of bicyclic hydrocarbons that can be drawn with rings of five and seven carbon atoms—not only from the experimental point of view (synthesis, chemical behavior, determination of their different spectra), but also from the theoretical angle (electronic structures, chemical reactivity, electronic spectra).

The first hydrocarbon synthesized from this nonbenzenoid series was azulene (Sherndal, 1915), several decades before theoreticians and experimenters fixed their attention on it in relation to the chemical concept of "aromaticity." The synthesis of fulvalene and heptafulvalene (Doering, 1959), sesquifulvalene (Prinzbach, 1961; Prinzbach and Rosswog, 1961; Prinzbach and Seip, 1961), and heptalene (Dauben and Bertelli, 1961) followed. The synthesis of pentalene has been a challenge in the many attempts made up to the present. About five years ago, two of its closest derivatives were synthesized: pentalene dianion (Katz and Rosenberg, 1961) and hexaphenylpentalene (Le Goff, 1962).

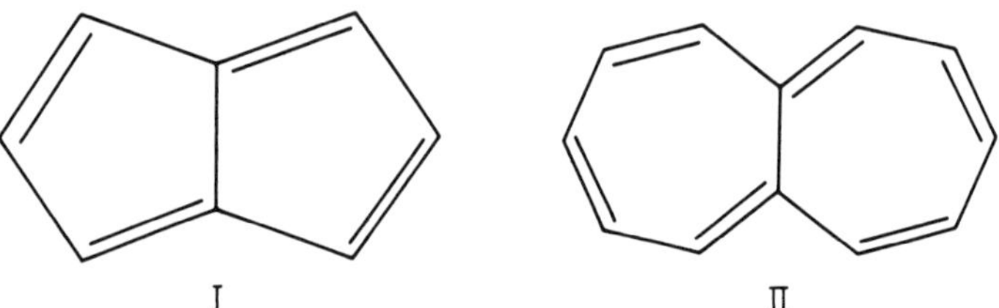

A study of pentalene (I) and heptalene (II) molecules, especially the former, is unusually interesting when considering the contradictions that have arisen in comparing the results of successive experimental and theoretical studies. Recently, these molecules have been studied by means of semiempirical quantum mechanical methods, within the SCF theory in their diverse modalities

Taken in part from the thesis submitted by Juan Palou in partial fulfillment of the requirements for the degree of Doctor of Science. The Department of Physical Chemistry is supported in part by Fomento Investigacion Universidad (Madrid).

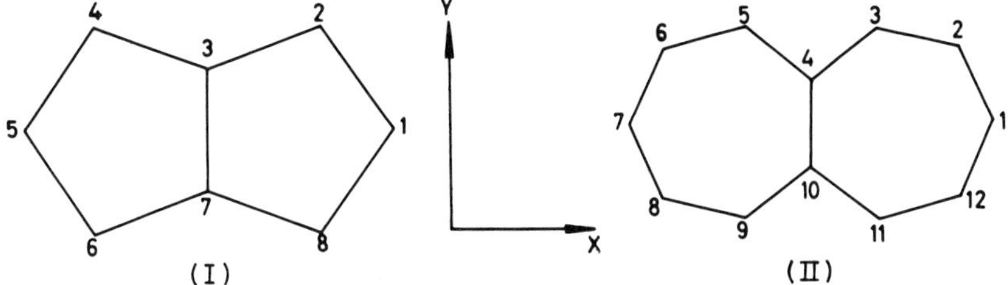

FIGURE 1.
Choice of axes and numbering of the molecules.

(Allinger et al., 1965; Boer-Veenendaal, 1964; Dewar and Gleicher, 1965; Nakajima et al., 1964). Of interest also is their study by means of Roothaan's nonempirical method (1951, 1960).

Our work has been the calculation of the electronic molecular indices for the ground state of pentalene and heptalene molecules for the last method.

METHOD OF CALCULATION AND NUMERICAL RESULTS

The numbering used—as well as the molecular axes chosen—is shown in Figure 1. (It has been assumed that both molecules possess a structure with D_{2h} symmetry.) The structural parameters used are given in Table 1.

The Chirgwin and Coulson (1950) method was employed to calculate the molecular orbitals, using the overlap integrals obtained by means of the expression of Mulliken, Rieke, Orloff, and Orloff (1949) from the STO (atomic orbital of Slater), with an effective charge of 3.25.

Secular determinants were resolved by means of the Frame method (Dwyer, 1960).

Two-center Coulomb integrals $(aa|bb)$ and integrals of penetration $(a{:}ab)$ and $(a{:}bb)$ were calculated from the equation of Scrocco and Salvetti (1953). Three-center $(ab|cc)$, $(ac|bc)$ and four-center $(ab|cd)$ integrals were obtained by means of the Mulliken (1949) approximation, and three-center integrals $(a{:}bc)$ according to Sklar's (1939) approximation.

Table 2 lists the normalized atomic orbitals of symmetry, σ_p, used to simplify the calculation of SCF MO. Table 3 indicates the molecular orbitals of symmetry,

TABLE 1.
Structural parameters used.

Parameters	Pentalene	Heptalene
Bond lengths (Å)[a]	$R_{12} = 1.370$	$R_{12} = 1.396$
	$R_{23} = 1.412$	$R_{23} = 1.383$
	$R_{37} = 1.410$	$R_{24} = 1.420$
		$R_{4\,10} = 1.430$
Bond angles[b]	$\widehat{2\,1\,8} = 108°$	$\widehat{2\,1\,12} = 128°\,42'$

[a]Calculated by Deas' (1955) expression, using the bond orders calculated by Estrellés and Fernández-Alonso (1953) for pentalene and also those for heptalene by Pullman and Pullman (1952).
[b]We reproduced those for regular polygon; the rest were adapted to the bond lengths.

$$\phi_i^\sigma = \sum_p d_{ip}\sigma_p$$

The Mulliken and Parr method (1951) was followed in calculating the integrals of overlap, S_{rs}^σ, of Coulomb, $J_{i,rs}^\sigma$, and of exchange, $K_{i,rs}^\sigma$.

The diagonalization of secular matrices was carried out by the Mayot (1950) method. Iterations were followed through until the point of self-consistency of the elements F_{rs}^σ, with an approximation of 2.10^{-4} a.u. The SCF LCAO-MO are shown in Table 4.

TABLE 2.
Normalized atomic orbitals of symmetry, σ_p.

Molecule	Number of the Orbital (p)	σ_p	Symmetry
Pentalene	1	$0.70691(\chi_1 + \chi_5)$	B_{1u}
	2	$0.48384\|(\chi_2 + \chi_4) + (\chi_6 + \chi_8)\|$	B_{1u}
	3	$0.63492(\chi_3 + \chi_7)$	B_{1u}
	4	$0.70730(\chi_1 - \chi_5)$	B_{2g}
	5	$0.49437\|(\chi_2 - \chi_4) - (\chi_6 - \chi_8)\|$	B_{2g}
	6	$0.81127(\chi_3 - \chi_7)$	B_{3g}
	7	$0.50728\|(\chi_2 + \chi_4) - (\chi_6 + \chi_8)\|$	B_{3g}
	8	$0.51634\|(\chi_2 - \chi_4) + (\chi_6 - \chi_8)\|$	A_u
Heptalene	1	$0.70711(\chi_1 + \chi_7)$	B_{1u}
	2	$0.49300\|(\chi_2 + \chi_6) + (\chi_8 + \chi_{12})\|$	B_{1u}
	3	$0.47962\|(\chi_3 + \chi_5) + (\chi_9 + \chi_{11})\|$	B_{1u}
	4	$0.63695(\chi_4 + \chi_{10})$	B_{1u}
	5	$0.70711(\chi_1 - \chi_7)$	B_{2g}
	6	$0.49300\|(\chi_2 - \chi_6) - (\chi_8 - \chi_{12})\|$	B_{2g}
	7	$0.52134\|(\chi_3 - \chi_5) - (\chi_9 - \chi_{11})\|$	B_{2g}
	8	$0.50731\|(\chi_2 + \chi_6) - (\chi_8 + \chi_{12})\|$	B_{3g}
	9	$0.48148\|(\chi_3 + \chi_5) - (\chi_9 + \chi_{11})\|$	B_{3g}
	10	$0.80708(\chi_4 - \chi_{10})$	B_{3g}
	11	$0.50731\|(\chi_2 - \chi_6) + (\chi_8 - \chi_{12})\|$	A_u
	12	$0.52271\|(\chi_3 - \chi_5) + (\chi_9 - \chi_{11})\|$	A_u

TABLE 3.
Molecular orbitals of symmetry, $\phi_i{}^\sigma$

Molecule	MO	$\phi_i{}^\sigma$	Symmetry
Pentalene	I	$0.3231\sigma_1 + 0.5064\sigma_2 + 0.5451\sigma_3$	B_{1u}
	IV	$0.7065\sigma_1 + 0.2114\sigma_2 - 0.7473\sigma_3$	B_{1u}
	II	$0.5938\sigma_4 + 0.6206\sigma_5$	B_{2g}
	III	$0.4579\sigma_6 + 0.7542\sigma_7$	B_{3g}
Heptalene	I	$0.2254\sigma_1 + 0.3661\sigma_2 + 0.4808\sigma_3 + 0.5041\sigma_4$	B_{1u}
	IV	$0.5523\sigma_1 + 0.4802\sigma_2 - 0.1769\sigma_3 - 0.5310\sigma_4$	B_{1u}
	II	$0.4870\sigma_5 + 0.6038\sigma_6 + 0.3024\sigma_7$	B_{2g}
	VI	$0.6265\sigma_5 - 0.0721\sigma_6 - 0.7988\sigma_7$	B_{2g}
	III	$0.4104\sigma_8 + 0.6763\sigma_9 + 0.3001\sigma_{10}$	B_{3g}
	V	$0.6760\sigma_{11} + 0.5839\sigma_{12}$	A_u

TABLE 4.
SCF LCAO-MO calculated.

Molecule	MO	SCF LCAO-MO	ϵ_i (a.u.)	Symmetry
Pentalene	I	$0.2418(\chi_1 + \chi_5) + 0.2482(\chi_2 + \chi_4 + \chi_6 + \chi_8) + 0.3323(\chi_3 + \chi_7)$	−0.1940	B_{1u}
	II	$0.4365(\chi_1 - \chi_5) + 0.2952(\chi_2 - \chi_4 - \chi_6 + \chi_8)$	−0.0226	B_{2g}
	III	$0.3658(\chi_2 + \chi_4 - \chi_6 - \chi_8) + 0.4050(\chi_3 - \chi_7)$	0.1610	B_{3g}
	IV	$0.4594(\chi_1 + \chi_5) + 0.1255(\chi_2 + \chi_4 + \chi_6 + \chi_8) - 0.5045(\chi_3 + \chi_7)$	0.2505	B_{1u}
	V	$0.5163(\chi_2 - \chi_4 + \chi_6 - \chi_8)$	0.7829	A_u
	VI	$0.6182(\chi_1 - \chi_5) - 0.4390(\chi_2 - \chi_4 - \chi_6 + \chi_8)$	1.2371	B_{2g}
	VII	$0.5501(\chi_1 + \chi_5) - 0.4750(\chi_2 + \chi_4 + \chi_6 + \chi_8) + 0.3088(\chi_3 + \chi_7)$	1.3334	B_{1u}
	VIII	$-0.3910(\chi_2 + \chi_4 - \chi_6 - \chi_8) + 0.7545(\chi_3 - \chi_7)$	1.3506	B_{3g}
Heptalene	I	$0.1487(\chi_1 + \chi_7) + 0.1904(\chi_2 + \chi_6 + \chi_8 + \chi_{12}) + 0.1915(\chi_3 + \chi_5 + \chi_9 + \chi_{11}) + 0.3673(\chi_4 + \chi_{10})$	−0.2254	B_{1u}
	II	$0.2433(\chi_2 + \chi_6 - \chi_8 - \chi_{12}) + 0.2562(\chi_3 + \chi_5 - \chi_9 - \chi_{11}) + 0.3491(\chi_4 - \chi_{10})$	−0.0914	B_{3g}
	III	$0.3217(\chi_1 - \chi_7) + 0.2980(\chi_2 - \chi_6 - \chi_8 + \chi_{12}) + 0.1805(\chi_3 - \chi_5 - \chi_9 + \chi_{11})$	−0.0821	B_{2g}
	IV	$0.4264(\chi_1 + \chi_7) + 0.2304(\chi_2 + \chi_6 + \chi_8 + \chi_{12}) - 0.1275(\chi_3 + \chi_5 + \chi_9 + \chi_{11}) - 0.2758(\chi_4 + \chi_{10})$	0.1008	B_{1u}
	V	$0.3013(\chi_2 - \chi_6 + \chi_8 - \chi_{12}) + 0.3484(\chi_3 - \chi_5 + \chi_9 - \chi_{11})$	0.1335	A_u
	VI	$0.4279(\chi_1 - \chi_7) + 0.0092(\chi_2 - \chi_6 - \chi_8 + \chi_{12}) - 0.4260(\chi_3 - \chi_5 - \chi_9 + \chi_{11})$	0.4117	B_{2g}
	VII	$0.0911(\chi_1 + \chi_7) + 0.0203(\chi_2 + \chi_6 + \chi_8 + \chi_{12}) - 0.4434(\chi_3 + \chi_5 + \chi_9 + \chi_{11}) + 0.4888(\chi_4 + \chi_{10})$	0.9546	B_{1u}
	VIII	$0.4630(\chi_2 + \chi_6 - \chi_8 - \chi_{12}) - 0.2968(\chi_3 + \chi_5 - \chi_9 - \chi_{11}) - 0.1102(\chi_4 - \chi_{10})$	0.9584	B_{3g}
	IX	$0.4297(\chi_2 - \chi_6 + \chi_8 - \chi_{12}) - 0.4135(\chi_3 - \chi_5 + \chi_9 - \chi_{11})$	1.0268	A_u
	X	$0.5969(\chi_1 + \chi_7) - 0.4521(\chi_2 + \chi_6 + \chi_8 + \chi_{12}) + 0.1494(\chi_3 + \chi_5 + \chi_9 + \chi_{11}) + 0.0187(\chi_4 + \chi_{10})$	1.1618	B_{1u}
	XI	$0.0234(\chi_2 + \chi_6 - \chi_8 - \chi_{12}) + 0.3517(\chi_3 + \chi_5 - \chi_9 - \chi_{11}) - 0.7749(\chi_4 - \chi_{10})$	1.2369	B_{3g}
	XII	$0.5312(\chi_1 - \chi_7) - 0.4551(\chi_2 - \chi_6 - \chi_8 + \chi_{12}) + 0.2798(\chi_3 - \chi_5 - \chi_9 + \chi_{11})$	1.2774	B_{2g}

TABLE 5.
Structural indices calculated.

Molecule	Atom	q_r	F_r	Bond	P_{rs}	R(Å)[a]	R(Å)[b]
Pentalene	1	1.142	0.401	1 – 2	0.665	1.391	1.414
	2	0.799	0.561	2 – 3	0.506	1.426	1.428
	3	1.260	0.090	3 – 7	0.630	1.404	1.393
Heptalene	1	1.188	0.474	1 – 2	0.629	1.404	1.394
	2	0.886	0.474	2 – 3	0.629	1.404	1.394
	3	1.069	0.529	3 – 4	0.574	1.414	1.409
	4	0.903	0.229	4 – 10	0.355.	1.453	1.468

[a]Calculated by the expression of Coulson and Golebiewski (1961).
[b]Calculated by the expression of Deas (1955).

Table 5 gathers together the structural indices calculated; q is the π-electronic charge, F the free valence, p_{rs} the bond order, and R the bond length.

The resonance energy, E_R, was calculated from the equation

$$E_R = N^d(E^s - E^d) + E^A_{Nb} + \Big(\sum_{pq} C^s_{pq} - \sum_{uv} C^d_{uv}\Big)N^d$$

following Dewar and Gleicher (1965). In the equation,

E^s = bond energy of a simple bond C—C of length 1.485 Å

E^d = bond energy of a double bond C═C of length 1.334 Å

The energies of the highest occupied molecular orbital (HOMO) and lowest vacant molecular orbital (LVMO) are listed in Table 6.

The π-electron energies for the ground state of these molecules,

$$E_\pi = 2\sum_{i=1}^{n} \epsilon_i - \sum_{ij}^{n} (2J^\sigma_{ij} - K^\sigma_{ij})$$

are shown in Table 6, in which

N^d = number of double bonds

$E^A_{\pi b}$ = binding energy of the calculated structures using SCF methods

C^s_{pq} = compression energy necessary to change the lengths of the simple bonds from their values of equilibrium 1.485 Å to lengths R_{pq}

C^d_{uv} = compression energy necessary to change the lengths of double bonds from their values of equilibrium 1.334 Å to lengths R_{uv}

The nuclear repulsion terms were calculated according to the method indicated by Chung and Dewar (1965).

The values obtained for resonance energies are given in Table 6.

TABLE 6.
Energies of the HOMO and LVMO, π-electron energies and resonance energies (in ev).

Molecules	ϵ_{HOMO}	ϵ_{LVMO}	E_π	E_R (per molecule)
Pentalene	$W_{2p} + 3.41$	$W_{2p} + 10.65$	$8W_{2p} - 194.72$	-7.78
Heptalene	$W_{2p} + 5.60$	$W_{2p} + 12.98$	$12W_{2p} - 354.50$	$+0.30$

TABLE 7.
Singlet states: monoexcited transitions, polarization, and symmetry.

Molecule	Transition (mμ)	Polarization	f (c.g.s.)	Symmetry
Pentalene	1140	(*F*)	Forb.	$A_g \rightarrow B_{1g}$
	238	(*X*)	0.36	$A_g \rightarrow B_{3u}$
	179	(*Y*)	0.23	$A_g \rightarrow B_{2u}$
Heptalene	460	(*F*)	Forb.	$A_g \rightarrow B_{1g}$
	352	(*X*)	0.13	$A_g \rightarrow B_{3u}$
	250	(*Y*)	0.27	$A_g \rightarrow B_{2u}$
	213	(*F*)	Forb.	$A_g \rightarrow B_{1g}$

π-π^* *Electronic transitions*

We used the following expressions to calculate values for energies corresponding to singlet and triplet monoexcited transitions, $E_{V_{i \to a}}$ and $E_{T_{i \to a}}$, respectively:

$$E_{V_{i \to a}} = E_N + \epsilon_a - \epsilon_i - J_{ia} + 2K_{ia}$$

$$E_{T_{i \to a}} = E_N + \epsilon_a - \epsilon_i - J_{ia}$$

The oscillator strengths, f, were calculated by means of the known expression

$$f = 1.085.10^{11} \nu(\text{cm}^{-1})\, Q_{ia}^2(\text{cm}^2)$$

To obtain values of f of the same order of magnitude as the experimental ones, we introduced the correction factor 0.3, which is the ratio f_{obs}/f_{calc} for the lowest allowed singlet transition in the benzene according to the suggestion of Nakajima and Katagiri (1964). In Table 7 we give the values of the nomoexcited transitions and the oscillator strengths, with their polarizations and symmetries.

DISCUSSION

Bond orders. Note, in pentalene, the discordance between our $p_{37} = 0.630$ (Table 5), corresponding to the central bond, and that given by Nakajima and co-workers (1964), 0.528, and, in heptalene, the discordance between our $p_{4\,10} = 0.355$, also corresponding to a central bond, with the value of 0.487 given by Nakajima's group. In other words, our results assign a greater double bond character to the central bond of the pentalene molecule, and a greater simple bond character to the heptalene molecule.

Bond lengths. If the "aromaticity" of a compound is explained by the alternation of the lengths of the peripheral bonds (Table 5), then heptalene is more aromatic than pentalene.

Chemical reactivity. Table 8 shows the predicted probable positions by the attacking electrophilic, nucleophilic, and radical reagents on the pentalene and heptalene molecules, as given by the static method (Fueno, 1961) and the frontier electron theory (Fukui, Yonezawa, and Shingu, 1952; Fukui et al., 1954).

From our own results, we can predict that the pentalene reacts more easily with the electrophilic reagents than does heptalene, and in general, that the

TABLE 8.
Predicted probable positions by different attacking reagents.

Method	Structural Index	Reagent	Position	
			Pentalene	Heptalene
Static	q	Electrophilic	$3 > 1$	$1 > 3$
	q	Nucleophilic	2	$2 > 4$
	F	Radical	$2 > 1$	$3 > 2 = 1$
Frontier electron	f_e	Electrophilic	$3 > 1$	$1 > 3$
	f_n	Nucleophilic	2	$4 > 3$
	f_r	Radical	$2 > 3 > 1$	$3 > 4 = 1$
Experimental[a]	—	Nucleophilic	—	3

[a]Dauben and Bertelli (1961).

same is true with respect to other kinds of reagents. Although there is no complete agreement between the static method and the frontier electron theory, we are able to point out that the greater reactivity predominates in pentalene.

Ionization potential and electron affinity. The theorem of Koopmans (1934) shows, that the $-\epsilon_{\text{HOMO}}$ should be a good approximation to the first ionization potential, and, similarly, we know that the quantitiy $-\epsilon_{\text{LVMO}}$ should be a good approximation to the electron affinity.

Binding energy per π electron. Starting with the binding energies, $E_{\pi b}$, we are able to calculate the so-called binding energy per π electron. From values of $E_{\pi b}$, calculated by the Chung and Dewar method (1965), for both molecules, and keeping in mind that pentalene and heptalene molecules have 8 and 12π electrons respectively, we obtain the following results: pentalene, -1.1425; and heptalene, -0.1750 ev.

From the existing difference between the values of these molecules, we deduced that the heptalene has probably a greater facility than pentalene to form π bonds, and this in turn made us conclude that it is more aromatic.

Resonance energy. As in the preceding case, there is a great difference (Table 6) between the values for heptalene and pentalene, making evident as always the discrepancy between two molecules. According to the criterion of Chung and Dewar (1965), we concluded that pentalene is a less stable molecule, with less tendency to be coplanar than is heptalene.

Electronic spectra. When Dauben and Bertelli (1961) synthesized heptalene, they obtained the ultraviolet-visible absorption spectrum and found the following maxima: 256 and 352 mμ with a long tail through the visible region. They also indicated that these maxima did not coincide with those calculated by the HMO method, because they occur at much shorter wavelengths, but the maxima were in good agreement with those calculated by the Simpson method (1955): 245 mμ, allowed; 387 mμ, allowed.

Allinger and co-workers (1965) have given these values for the spectrum of heptalene: 500 mμ ($f = 0.001$), 325 mμ ($f = 0.13$), 260 mμ ($f = 0.12$). 220 mμ ($f = 0.35$), and 199 mμ ($f = 1.0$); they based these results upon the judgment that the real structure of the heptalene molecule can be interpreted as intermediate between a planar form with alternating bond lengths and another which is not planar.

TABLE 9.
Electronic spectra.

Molecule	Molecular Symmetry	Transition Type		Excitation Energy (ev)			f (c.g.s.)		
				Calc.		Obs.	Calc.		Obs.
		a	b	a	b	c	a	b	c
Pentalene	D_{2h}	$A_{1g} \rightarrow B_{1g}$	$A_{1g} \rightarrow B_{1g}$	0.75	0.84	—	Forb.	Forb.	—
		$A_{1g} \rightarrow B_{3u}$	$A_{1g} \rightarrow B_{3u}$	3.73	5.21	—	0.75 (x)	0.36 (x)	—
		$A_{1g} \rightarrow B_{2u}$	$A_{1g} \rightarrow B_{2u}$	4.99	6.91	—	0.52 (y)	0.23 (y)	—
Heptalene	D_{2h}	$A_{1g} \rightarrow B_{1g}$	$A_{1g} \rightarrow B_{1g}$	0.67	2.57	tail	Forb.	Forb.	Long tail into visible
		$A_{1g} \rightarrow B_{3u}$	$A_{1g} \rightarrow B_{3u}$	3.18	3.52	3.52	1.29 (x)	0.13 (x)	0.15
		$A_{1g} \rightarrow B_{2u}$	$A_{1g} \rightarrow B_{2u}$	4.06	4.96	4.84	0.63 (y)	0.27 (y)	—
		—	$A_{1g} \rightarrow B_{1g}$	—	5.83	—	—	Forb.	Strong absorption below 200 mμ

a. Nakajima, Yaguchi, Kaeriyama, and Namoto (1964).
b. Our results.
c. Dauben and Bertelli (1961)

Our results are summarized in Table 9. As will be noticed, one of our values coincides exactly with one of the maxima found by Dauben and Bertelli (1961), and is almost coincident with the other one. The difference in the latter is very slight. This seems to confirm what Allinger and his collaborators found: it appears from the spectrum that the molecule is near to being planar, a hypothesis which constitutes one of the premises of our research.

However, we should point out that we are presently studying the modifications which would introduce interplay of mono- and bi-excited configurations, at the same time keeping in mind the action of hydrogens, which was neglected in the first part of this study.

REFERENCES

Allinger, A. L., M. A. Muller, L. W. Chow, R. A. Ford, and J. C. Graham (1965). *J. Am. Chem. Soc.* **87**, 3430.

Den Boer-Veenendaal, P. C., D. H. W. Den Boer, and T. H. Goodwin (1964). *Rec. Trav. Chim.* (Pays Bas) **83**, 764.

Chirgwin, B. H., and C. A. Coulson (1950). *Proc. Roy. Soc.* (London) A **201**, 196.

Chung, A. L. H., and M. J. S. Dewar (1965). *J. Chem. Phys.* **42**, 762.

Coulson, C. A., and A. Golebiewski (1961). *Proc. Phys. Soc.* **78**, 1310.

Dauben, H. J., Jr., and D. J. Bertelli (1961). *J. Am. Chem. Soc.* **83**, 4659.

Deas, H. H. (1955). *Phil. Mag.* **46**, 670.

Dewar, M. J. S., and G. J. Gleicher (1965). *J. Am. Chem. Soc.* **87**, 685.

Doering, W. von E. (1959). In *Theoretical Organic Chemistry; Kekulé Symposium*, p. 35. London: Butterworths Scientific Publ.

Dwyer, P. S. (1960). *Linear Computations*, p. 225 New York: Wiley.

Estellés, I., and J. I. Fernández-Alonso (1953). *Anal. Fis. Quim.* (Madrid) B **49**, 267.

Fueno, T. (1961). *Ann. Rev. Phys. Chem.* **18**, 303.

Fukui, K., T. Yonezawa, and H. Shingu (1952). *J. Chem. Phys.* **20**, 722.

Fukui, K., T. Yonezawa, C. Nagata, and H. Shingu (1954). *J. Chem. Phys.* **22**, 1433.

Le Goff, E. (1962). *J. Am. Chem. Soc.* **84**, 3975.

Katz, T. J., and M. Rosenberg (1962). *J. Am. Chem. Soc.* **84**, 865.

Koopmans, T. (1934). *Physica* **1**, 104.

Mayot, M. (1950). *Ann. Astrophys.* **13**, 282.

Mulliken, R. S. (1949). *J. Chem. Phys.* **46**, 691.

Mulliken, R. S., and R. G. Parr (1951). *J. Chem. Phys.* **19**, 1271.

Mulliken, R. S., C. A. Rieke, D. Orloff, and H. Orloff (1949). *J. Chem. Phys.* **17**, 1248.

Nakajima, T., and S. Katagiri (1964). *Mol. Phys.* **7**, 149.

Nakajima, T., Y. Yaguchi, R. Kaeriyama, and Y. Namoto (1964). *Bull. Chem. Soc.* (Japan) **37**, 272.

Prinzbach, H. (1961). *Angew. Chem.* **73**, 169.

Prinzbach, H., and W. Rosswog (1961). *Angew. Chem.* **73**, 543.

Prinzbach, H., and D. Seip (1961). *Angew. Chem.* **73**, 169.

Pullman, B., and A. Pullman (1952). *Les théories électroniques de la Chimie Organique*, p. 571. Paris; Masson.

Roothaan, C. C. J. (1951). *Rev. Mod. Phys.* **23**, 69.

Roothaan, C. C. J. (1960). *Rev. Mod. Phys.* **32**, 179.

Srocco, E., and O. Salvetti (1953). *Ricerca Sci.* **23**, 98.

Sherndal, A. E. (1915). *J. Am. Chem. Soc.* **37**, 567.

Simpson, W. T. (1955). *J. Am. Chem. Soc.* **77**, 6164.

Sklar, A. L. (1939). *J. Chem. Phys.* **7**, 984.

GERHARD BINSCH*
EDGAR HEILBRONNER
Eidgenössische Technische Hochschule
Zürich, Switzerland

Double-Bond Fixation in Linear, Cyclic, and Benzenoid π-Electron Systems

The prediction of the geometry of conjugated hydrocarbons (that is, of their interatomic distances) has always been a major concern of π-electron theories.† It has been known for over thirty years that the carbon-carbon bond lengths are equal in benzene but different in butadiene or naphthalene. The first theoretical explanation of this fact was given by Pauling (1932, 1935, 1937, 1939), who introduced the concept of double-bond character. This was followed by the derivation of bond orders, based on valence-bond (Penney, 1937) or molecular orbital models (Lennard-Jones, 1937; Lennard-Jones and Turkevich, 1937; Coulson, 1938, 1939). There seemed to be no doubt at first that a theory yielding good approximations for the observed bond lengths of small alternant π-electron systems could also be applied to the higher members of the series or to nonalternant systems. Furthermore, it was assumed that, with the exception of open-shell systems with multiplet ground states (susceptible of undergoing pseudo-Jahn-Teller distortions), the symmetry of a closed shell π-electron system would always be that suggested by the superposition with equal weight of its equivalent resonance structures. In particular, it had been predicted by Lennard-Jones (1937) and by Coulson (1938, 1939) that increasing the size of a cyclic conjugated olefin (annulene) should not make the bonds unequal, but increasing the number of links of a linear polyene should eventually equalize the bond lengths in the middle of the chain.

Experimental evidence against such ideas came from different sources. For example, it was noticed that in the electronic spectra of linear polyenes the frequency of the long-wavelength absorption converges to a finite value as the number of atoms in the chain tends to infinity, and not to zero as we should expect. Kuhn (1948, 1949) and Dewar (1952) showed that this observation could be rationalized by postulating that some degree of bond alternation persists even in an infinite polyene. Such a postulate was strongly suggested by the constancy of the single and double bond distances observed in sample

This is the second of a series of projected papers on double-bond fixation. For the first, see Binsch, Heilbronner, and Murrell (1966).

*Present address: Department of Chemistry, University of Notre Dame, Notre Dame, Indiana.

†For an excellent account of the history of this problem and its present state of development, see Salem (1966).

calculations which Kuhn, Huber, and Bär (1958) (Kuhn, 1957, 1959) had carried out for polyenes with 2 to 12 conjugated double bonds. Furthermore such an assumption is supported by the results of structure determinations of long-chain polyenes (carotinoids), which show that alternation of "normal" single-bond and double-bond lengths is conserved even in a β-carotene derivative (Sly, 1964). Another example of the breakdown of simple relationships between bond order and bond length is seen in the nonalternant hydrocarbon heptalene (Dauben and Bertelli, 1961):

I ⇌ II

It does not show an energy minimum for the D_{2h} symmetry suggested by the superposition of I and II, but flips back and forth between two conformations that correspond to one or the other of the Kekulé structures and exhibit strong bond alternation of a C_{2h} type (den Boer-Veenendaal et al., 1962).

There have been several papers in which a theoretical justification for such anomalies has been attempted (Platt, 1956; Labhart, 1957; Ooshika, 1957; Longuet-Higgins and Salem, 1959, 1960; Salem and Longuet-Higgins, 1960; Coulson and Dixon, 1962). The most noteworthy among these papers is that of Labhart (1957), in which it was recognized that the stable configuration of a conjugated molecule cannot be predicted by π-electron theory alone, without making allowance for the effects of σ-bond compression. A similar suggestion has been put forward by Kuhn (1957, 1959). The same author has recently reviewed the history of the problem of bond fixation in linear and cyclic polyenes (Försterling, Huber, and Kuhn, 1967). This idea was further elaborated (Ooshika, 1957; Longuet-Higgins and Salem, 1959, 1960; Salem and Longuet-Higgins, 1960; Coulson and Dixon, 1962), and the resulting treatments have finally been programmed for digital computers, so that both one-electron (Nakajima and Katagiri, 1964) and many-electron schemes (Dewar and Gleicher, 1965) are available for practical applications.

There are three unsatisfactory features in these treatments which have led to a reexamination of the theory of bond fixation (Binsch et al., 1966).

1. A basic postulate common to all these theories is that bond alternation (for example, corresponding to a Kekulé-like structure) is the most favorable distortion of a conjugated molecule. This is clearly a drawback, as, for obvious reasons, such a postulate cannot be extended to open-shell or charged π-electron systems or to odd-membered (nonalternant) cycles such as the ions having $(4n + 2)$ π electrons. The same applies for electronically excited states. But even for cases in which such difficulties do not arise it would be highly desirable that bond alternation emerge from theory as the energetically most favorable distortion rather than being the starting point.

2. The results of the theories used so far depend—as we shall see—very critically on the functional dependence of the resonance integral β and of other two-center integrals on bond length. Sometimes, as in Longuet-Higgins and Salem's treatment (1959, 1960; Salem and Longuet-Higgins, 1960), a single parameter fixes both the choice of $\beta(R)$ and the form of the σ-potential curve uniquely.

3. The effects causing differences in bond lengths of small conjugated molecules are well accounted for by Hückel or SCF bond orders and can therefore be attributed to first-order changes in the molecular energy, whereas the tendency of the π electrons in extended systems to cluster in certain regions of space must arise from changes in the wavefunction and should accordingly be treated on a different footing. Calculations performed as a function of only one bond-alternation parameter, such as those made up to now, obscure this essential difference.

THEORY

Our aim (Binsch et al., 1966) was to develop a general theory of bond fixation that would make a clear distinction between distortions arising from first- and second-order effects. By being able to assess the relative importance of both effects for a variety of molecules, we expected to gain deeper insight into the origin of molecular distortions occurring in π-electron systems. In our first paper (1966) the theory was presented in a rather formal way. We shall restate the important steps of our argument and subsequently compare our basic assumptions and main conclusions with those of previous theories.

To begin with, we assume complete σ-π separation. This means that the total energy of a π-electron system (E_T) can be written as the interaction-free sum of the energy of the σ core (E_σ) and of the π-electron system proper (E_π):

$$E_T = E_\sigma + E_\pi \tag{1}$$

The energy E_σ can be regarded as the sum of the M individual contributions of the σ bonds (M = number of bonds) between bonded centers μ and ν: $V_{\widehat{\mu\nu}} = V(R_{\widehat{\mu\nu}})$. These are functions of the interatomic distances $R_{\widehat{\mu\nu}}$ (Morse functions). For the reduced distance interval of interest (about 1.33 Å to 1.50 Å) we shall approximate these functions by a quadratic potential with a minimum at (for the moment) an unspecified distance R_0, which is the same for all bonds:

$$E_\sigma = \sum_{\mu<\nu} V_{\mu\nu} = \sum_{\widehat{\mu<\nu}} \frac{k}{2}(R_{\mu\nu} - R_0)^2 \tag{2}$$

In (2) the force constant is one appropriate for an $sp^2 - sp^2$ σ bond. [For such a bond we expect (Dewar and Schmeising, 1960) that $k = 6 \times 15^5$ dyn cm^{-1} or, more practically, $k = 864$ kcal mole^{-1} Å^{-2}.]

The π-electron energy is computed according to a Hückel model. The individual π electrons move in orbitals of the type

$$\psi_J = \sum_\mu c_{J\mu}\phi_\mu \tag{3}$$

the energies of which are given by

$$\varepsilon_J = \alpha + 2 \sum_{\widehat{\mu<\nu}} c_{J\mu}c_{J\nu}\beta_{\mu\nu} \tag{4}$$

The resonance integral $\beta_{\mu\nu}$ is an as yet unspecified function of the interatomic distance $R_{\widehat{\mu\nu}}$ of the two $2p$ AO's ϕ_μ and ϕ_ν. The total π-electron energy is taken as the sum over the individual contributions of the single electrons. If b_J electrons have the energy $\mathcal{E}_J (b_J = 0, 1, \text{or } 2)$, then we have

$$E_\pi = \sum_J b_J \mathcal{E}_J = N\alpha + 2 \sum_{\mu<\nu} \beta_{\widehat{\mu\nu}} \sum_J b_J c_{J\mu} c_{J\nu} \tag{5}$$

The mobile bond order between two bonded centers μ and ν is defined as usual,

$$p_{\mu\nu} = \sum_J b_J c_{J\mu} c_{J\nu} \tag{6}$$

so that we have for N π electrons

$$E_\pi = N\alpha + 2 \sum_{\mu<\nu} p_{\mu\nu} \beta_{\mu\nu} \tag{7}$$

Finally we get for the total energy

$$E_T = \sum_{\mu<\nu} \left\{ \frac{k}{2} (R_{\mu\nu} - R_0)^2 + 2 p_{\mu\nu} \beta_{\mu\nu} \right\} + N\alpha \tag{8}$$

Note that the term for $E\sigma$ is the sum of M contributions, each of which depends on one interatomic distance only. There are, of course, interactions in the term for E_π. It is true that $\beta_{\mu\nu}$ depends on $R_{\mu\nu}$ only, but the bond orders at the bond $\mu\nu$ do depend on changes of the $\beta_{\kappa\lambda}$ value for some other bond $\kappa\lambda$.

So long as we do not depart too drastically from the internuclear distance R_0, we shall be allowed to describe the total energy E_T satisfactorily by a Taylor expansion:

$$E_T = E_T(R_0) + \sum_{\mu<\nu} \left(\frac{\partial E_T}{\partial R_{\mu\nu}} \right)_{R_0} \Delta R_{\mu\nu} + \frac{1}{2} \sum_{\mu<\nu} \sum_{\kappa<\lambda} \left(\frac{\partial^2 E_T}{\partial R_{\mu\nu} \partial R_{\kappa\lambda}} \right)_{R_0} \Delta R_{\mu\nu} \Delta R_{\kappa\lambda} + \text{higher terms} \tag{9}$$

The problem is to look for interatomic distances $R_{\mu\nu}$ that make the total energy E_T a minimum or $dE_T = 0$.

We shall do so by a two-step process, chosen in such a way as to bring out the importance of the contributions of first and higher derivatives of E_σ and E_π with respect to the interatomic distances $R_{\mu\nu}$.

First-Order Treatment

If only the first-order terms in (9) are taken into account, then all the M interatomic distances will be independent of each other. Therefore, the total differential dE_T will vanish only if all partial derivatives with respect to $R_{\widehat{\mu\nu}}$ vanish individually:

$$\left(\frac{\partial E_T}{\partial R_{\widehat{\mu\nu}}} \right) = 0 \tag{10}$$

for all bonds $\widehat{\mu\nu}$. Differentiating (8) with respect to all $R_{\mu\nu}$, we get M linearly independent equations, which then yield the following relations

$$R_{\widehat{\mu\nu}}^{(1)} = R_0 - \frac{2\beta'}{k} p_{\widehat{\mu\nu}} \tag{11}$$

for all bonds $\widehat{\mu\nu}$. This is nothing but the usual relationship between bond order and bond distance. By empirically fitting the bond lengths in ethylene, benzene, and graphite, for all of which $p_{\widehat{\mu\nu}}$ is fully given by symmetry, we find $R_0 = 1.50$ Å and $(2\beta'/k) = 0.16$ Å, thus fixing the minimum of our core potential in (2) and at the same time the zero point of our Taylor expansion (9) to about 1.50 Å. Because of (11) the system will suffer first-order bond distortion, resulting in a shift of the equilibrium positions of the nuclei from the common value R_0 in the core to the first-order bond lengths $R_{\mu\nu}^{(1)}$ in the π system. Equation (11) shows that these first-order bond lengths depend on the first derivative of the resonance integral $\beta' = (\partial\beta_{\mu\nu}/\partial R_{\mu\nu})_{R_0} = (d\beta/dR)_{R_0}$. $R_{\mu\nu}^{(1)}$ is reached when the negative derivative of the $V_{\mu\nu}$ curve equals the slope of the E_π curve (which is a straight line in this case). This situation is illustrated in Figure 1. The steeper the slope (that is, the higher the bond order), the further in will be $R_{\mu\nu}^{(1)}$ from R_0. Departure from the new first-order equilibrium bond lengths increases the energy as $(k/2)(R_{\mu\nu} - R_{\mu\nu}^{(1)})^2$, with the same k as in the potential $V_{\mu\nu}$ (Figure 1).

The most important conclusion that we draw from our result is that the first-order distortions are such that the symmetry of our system is not lowered. Indeed, the $p_{\mu\nu}$ reflect the starting symmetry and therefore the same symmetry must be reflected in the $R_{\mu\nu}^{(1)}$ values.

Second-Order Treatment

If we move our reference point from the common value R_0 to $R_{\mu\nu}^{(1)}$ for each bond (Figure 1) and if we develop our Taylor series around $R_{\mu\nu}^{(1)}$, then all first-order energy changes will vanish and, dropping higher-order terms, we obtain for the total energy

$$E_T = E_T(R_{\mu\nu}^{(1)}) + \frac{1}{2}\sum_{\mu<\nu} k(\Delta R_{\mu\nu}^{(1)})^2 + \frac{1}{2}\sum_{\mu<\nu}\sum_{\kappa<\lambda}\left(\frac{\partial^2 E_\pi}{\partial R_{\mu\nu}\partial R_{\kappa\lambda}}\right)_{R^{(1)}} \Delta R_{\mu\nu}^{(1)}\Delta R_{\kappa\lambda}^{(1)} \qquad (12)$$

To decide whether a second-order distortion will lead to a further decrease of the total energy below $E_T(R_{\mu\nu}^{(1)})$, we have to compare the second with the third term in (12) at the geometry of the molecule defined by the first-order bond lengths. The molecule will stabilize itself by second-order bond distortion if the third term in (12) becomes negative and exceeds the second in magnitude (dashed line in Figure 1), and this event will happen if the curvature of $E_\pi(R_{\mu\nu}^{(1)})$ becomes more negative than the critical value $-k$ (solid line in Figure 1). Under those circumstances the gain in π-electron energy is large enough to overcompensate the σ-compression energy, and the first-order geometry of the molecule becomes unstable with respect to second-order distortions. These distortions may be of such a kind as to lower the molecular symmetry, which, as already pointed out, cannot be destroyed by first-order effects. If the curvature of $E_\pi(R_{\mu\nu}^{(1)})$ is smaller in magnitude than k (dotted line in Figure 1), the second-order effects can only be noticed by a flattening of the potential energy curve and could presumably be detected by changes in the force constants of in-plane vibrations of conjugated molecules.

A direct comparison of the crucial terms in (12) is complicated by the fact that the σ part is only associated with pure squares of the displacements,

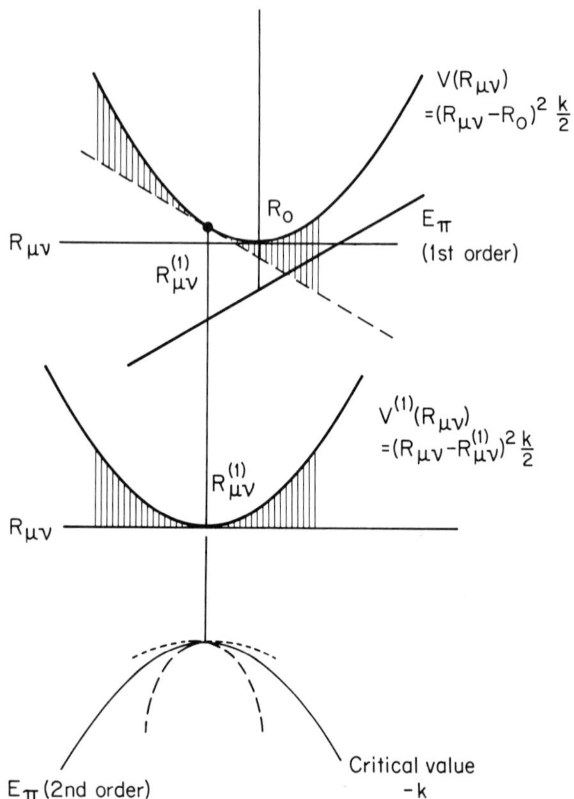

FIGURE 1.
Components E_π and E_σ of the total energy E_T of a π-electron system.

whereas the π part also contains cross-terms. It is therefore convenient to apply a normal-coordinate analysis. If we define a set of displacement coordinates by

$$D_i = \sum_{\mu<\nu} C_{i,\mu\nu}\Delta R_{\mu\nu}^{(1)} \tag{13}$$

we can rewrite (12) as

$$E_T = E_T(R_{\mu\nu}^{(1)}) + \frac{1}{2}\sum_i (k + \Lambda_i)D_i^2 \tag{14}$$

where the Λ_i are the eigenvalues of the matrix

$$\left(\frac{\partial^2 E_\pi}{\partial R_{\mu\nu}\partial R_{\kappa\lambda}}\right)_{R^{(1)}} \tag{15}$$

The largest negative eigenvalue $\Lambda_{\max}$ of this matrix corresponds to the energetically most favorable second-order distortion and its magnitude is a measure of the π-electron energy gained. The distortion is given by the eigenvector belonging to this eigenvalue. All eigenvectors are orthogonal and the corresponding distortions are therefore mutually exclusive, so that only the distortion yielding the largest energy gain has to be considered.

In our first paper (1966) we showed that the matrix elements are given by

$$\left(\frac{\partial^2 E_\pi}{\partial R_{\mu\nu}\partial R_{\kappa\lambda}}\right)_{R^{(1)}} = 2\pi_{\mu\nu,\kappa\lambda}^{(1)}\beta'_{\mu\nu}\beta'_{\kappa\lambda} + 2\delta_{\mu\nu,\kappa\lambda}p_{\mu\nu}^{(1)}\beta''_{\mu\nu} \tag{16}$$

where

$$\beta'_{\mu\nu} = \left(\frac{\partial \beta_{\mu\nu}}{\partial R_{\mu\nu}}\right)_{R^{(1)}}, \qquad \beta''_{\mu\nu} = \left(\frac{\partial^2 \beta_{\mu\nu}}{\partial R^2_{\mu\nu}}\right)_{R^{(1)}} \tag{17}$$

and the $\pi_{\mu\nu,\kappa\lambda}$ are the bond-bond polarizabilities,

$$\pi_{\mu\nu,\kappa\lambda} = \left(\frac{\partial p_{\mu\nu}}{\partial \beta_{\kappa\lambda}}\right) \tag{18}$$

If we neglect the curvature of the $\beta(R)$ curve (that is, $\beta'_{\mu\nu} = \beta'_{\kappa\lambda} = \beta'$ and $\beta'' = 0$), as suggested in our first paper (1966), then second-order bond fixation can be examined by the following procedure.

1. Calculate the bond-bond polarizabilities $\pi_{\mu\nu,\kappa\lambda}$ for the fully symmetrical model of the π-electron system (for example, D_{2h} symmetry for heptalene I, II).

2. Diagonalize the matrix $(\pi_{\mu\nu,\kappa\lambda}) = \boldsymbol{\pi}$ of order $M \times M$

$$||\boldsymbol{\pi} - \lambda \mathbf{I}|| = 0 \tag{19}$$

This will yield M eigenvalues λ_i,

$$\lambda_{\max} > \lambda_2 > \cdots > \lambda_i \cdots > \lambda_M \tag{20}$$

and the corresponding eigenvectors

$$D_{\max} = \begin{bmatrix} \Delta R_{\max,12} \\ \vdots \\ \Delta R_{\max,\mu\nu} \\ \vdots \end{bmatrix}, \qquad \ldots D_i = \begin{bmatrix} \Delta R_{i,12} \\ \vdots \\ \Delta R_{i,\mu\nu} \\ \vdots \end{bmatrix}, \quad \ldots \tag{21}$$

3. Select the largest eigenvalue $\lambda_{\max}$ and compare it with the critical value λ_{crit} (see below). If one finds that $\lambda_{\max}$ exceeds λ_{crit} in magnitude, then the molecule is predicted to suffer a distortion as a consequence of second-order effects. We will refer to this as second-order bond fixation. The point we want to stress here is that second-order bond fixation will in general be of such a kind as to reduce the molecular symmetry. The type of symmetry reduction is given by the irreducible representation $\Gamma^{(\max)}$ to which $D_{\max}$ and $\lambda_{\max}$ belong, relative to the group of the original, fully symmetrical model (for example, irreducible representation B_{1u} for the distortion of heptalene toward one or the other of the Kekulé structures I or II). If $|\lambda_{\max}| < |\lambda_{\text{crit}}|$, the full symmetry is conserved, but second-order effects still influence the curvature of the potential energy.

GENERAL DISCUSSION

In this section we will comment on the validity of some approximations implicit in our model. This will lead to general conclusions regarding the actual significance of the numbers we calculate and furthermore enable us to examine critically the theories of bond fixation that have been proposed by other investigators.

1. The polarizabilities $\pi_{\mu\nu,\kappa\lambda}$ occurring in (16) and (19) should be calculated at the actual first-order bond lengths $R^{(1)}_{\mu\nu}$. For cyclic polyenes, in which the

polarizabilities are fully determined by symmetry, there will be no change and we expect only minor ones in systems of lower symmetry. In fact, it turns out that no significant error is introduced if we use zero-order Hückel values also for linear polyenes and benzenoid hydrocarbons, in which the bond lengths vary over a range of about 1.40 ± 0.05 Å. We have checked this point by computing bond-bond polarizabilities from a Hückel matrix that was self-consistent in first-order bond lengths. The relations (Dewar and Schmeising, 1959)

$$R_{\widehat{\mu\nu}} = 1.504 - 0.166\, p_{\widehat{\mu\nu}} \tag{22}$$

and

$$H_{\widehat{\mu\nu}} = k\, S_{\mu\nu}(R_{\mu\nu}) \tag{23}$$

were used in an iterative procedure to obtain new off-diagonal matrix elements $H_{\widehat{\mu\nu}}$, with $S_{\widehat{\mu\nu}}$ being the $2p\pi - 2p\pi$ overlap integrals between Slater orbitals and k being an empirical constant calibrated on benzene. The λ values thus calculated are referred to as λ^{SCF}.

2. Equation (16) shows that Λ_{crit} depends on the first and second derivatives of the resonance integral β with respect to the bond lengths. In computing λ values the terms in β'' have been neglected; that is, we assume a linear dependence of β on bond length in the reduced distance interval. Remembering that the polarizabilities are expressed in units of β_0^{-1}, the criterion for second-order bond fixation becomes

$$\lambda > \lambda_{\mathrm{crit}} \qquad \text{with} \qquad \lambda_{\mathrm{crit}} = -\frac{k\beta_0}{2(\beta')^2} \tag{24}$$

In this approximation $(\beta')_{R^{(1)}}$ equals $(\beta')_{R_0}$ and can be obtained from $(2\beta'/k) = 0.16$. Table 1 contains a collection of λ_{crit} values for a range of values for β_0 and k. The most reasonable parameter values appear to be $\beta_0 = -20$ kcal and $k = 6 \cdot 10^5$ dyn cm^{-1}, thus giving $\lambda_{\mathrm{crit}} = 1.81$.

TABLE 1.
λ_{crit} *as a function of k and* β_0. (*See formula (24).*)

β_0(kcal)	k (dyn cm^{-1})		
	$5 \cdot 10^5$	$6 \cdot 10^5$	$7 \cdot 10^5$
−15	1.63	1.36	1.16
−20	2.17	1.81	1.55
−25	2.71	2.26	1.94
−30	3.25	2.71	2.32

If the $\beta(R)$ curve deviates slightly from a straight line in the reduced distance interval, then this will not appreciably alter β', but it may nevertheless have a significant influence on the λ values by virtue of the fact that a term in β'' contributes to the diagonal elements in (16). Let us assume that the $p_{\mu\nu}^{(1)}$ in (16) can be approximated by an average value p_{av} ($\approx 2/3$). Our secular problem then becomes

$$||2\pi(\beta')^2 - (\Lambda + 2p_{\mathrm{av}}\beta'')\mathbf{I}|| = 0 \tag{25}$$

and the λ's obtained by diagonalizing the π matrix should now be compared with a revised critical value,

$$\lambda > \lambda_{\text{crit}} - \frac{p_{\text{av}}\beta_0\beta''}{(\beta')^2} \tag{26}$$

A negative β'' value thus favors second-order bond fixation. To investigate how seriously the curvature of $\beta(R)$ will affect our results, we shall derive a "reasonable" value by choosing an exponential behavior for the resonance integral of the form $\beta = Be^{-aR}$, as suggested in the papers by Longuet-Higgins and Salem (1959, 1960) and Salem and Longuet-Higgins (1960). Then (26) simply reads

$$\lambda > \lambda_{\text{crit}} - p_{\text{av}} \tag{27}$$

This shows that the point at which second-order bond distortions set in depends critically on the special functional form of the $\beta(R)$ curve. Whereas β' may be estimated with reasonable accuracy, the curvature of the $\beta(R)$ function must be considered as very uncertain, not only in magnitude but even in sign. Our analysis therefore leads us to the conclusion that the detailed predictions attempted in previous theories must necessarily be very unreliable, and this unfortunate fact is a direct consequence of the very nature of the problem. However, it is nevertheless possible to compare different molecules with respect to their relative second-order localization tendencies and to draw conclusions from such a comparison.

3. The situation becomes even worse when one tries to calculate the actual magnitudes of second-order distortions. From Figure 1 it may be seen that if Λ exceeds Λ_{crit}, our model system will either collapse or fall apart. The fact that the nuclei of the real molecule will of course settle at new second-order equilibrium distances must therefore be due to third-order effects—that is, the anharmonicity of the σ potential and the third and higher derivatives of the resonance integral. Since there is no hope of getting even crude estimates of these quantities, we consider all predictions as to the absolute size of second-order distortions as entirely worthless. We will therefore only give the type of such distortions (that is, the normalized components of the displacement coordinate D), corresponding to the energetically most favorable distortion.

4. Finally, we should emphasize the fundamental difference between our theory and those methods in which bond distances are calculated by some variation of the resonance integral in the framework of a Hückel or SCF model. Dewar and Gleicher (1965) use an iterative scheme of the type already mentioned—Equations (22) and (23). Since bond distances are obtained from bond orders by means of a relationship that is equivalent to Equation (11), the full symmetry with which we start the calculations is maintained during each cycle. As a consequence the procedure can never yield the second-order distortions with which we are concerned in this paper, if the symmetry type of such a distortion is not known beforehand and included in the basis set (3) of the molecular orbitals. For this reason we can use such methods only to predict bond lengths for those molecules in which second-order effects are known to be unimportant. In the next sections we shall attempt to ascertain for which molecules this is expected to be the case.

A different approach has been suggested by Nakajima and Katagiri (1964). In their method the total energy V is calculated as a function of a bond-alternation parameter $k = \beta_s/\beta_d$, the ratio of the resonance integrals β_s of formal single and β_d of formal double bonds in a fictitious Kekulé structure. Again the type of bond fixation (that is, alternation) is preselected by the choice of this structure, and distortion to lower symmetry is postulated to occur if the V versus k curve shows a minimum for $k \neq 1$. [For an application of this procedure to the heptalene problem, see den Boer-Veenendaal et al., 1962.] Since first- and second-order effects cannot be separated in such a treatment, it is extremely difficult to see how the results depend on the various assumptions. This question will be considered in detail in another publication.

LINEAR POLYENES

Bond fixation in even-membered linear chains having C_{2h} symmetry has been discussed previously by Labhart (1957), Ooshika (1957), and Longuet-Higgins and Salem (1960). The results obtained by our method are summarized in Figure 2, where the eigenvalues λ_{max} of the polarizability matrix $\boldsymbol{\pi}$ (19) are given as a function of the number n of $2p\pi$ centers. In the ground state (curve 1 in Figure 2) both first- and second-order distortions belong to the totally symmetric representation A_g of C_{2h} and are such as to favor bond alternation:†

III

From the examples of Table 2 it can be seen that the differences $|p_{\mu-1,\mu} - p_{\mu,\mu+1}|$ of the zero-order Hückel bond orders between successive bonds decrease gradually as the chain length increases and that first-order effects therefore tend to equalize the bond lengths, especially in the middle of a long chain. The

†In the following diagrams a thick solid line (dashed line) stands for a bond that exhibits second-order bond contraction (bond lengthening). A thin solid line stands for a bond showing no second-order bond distortions.

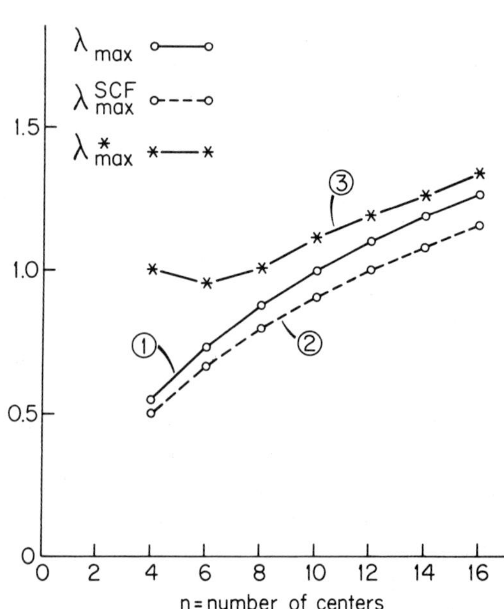

FIGURE 2.
Linear π-electron systems, n = even. Curve 1: λ_{max} for the electronic ground state, according to HMO polarizabilities. Curve 2: λ^{SCF}_{max} for the electronic ground state, according to a calculation which is self-consistent in first-order changes in $R_{\mu\nu}$. Curve 3: $\overset{*}{\lambda}_{max}$ for the $^1B^0$ state.

TABLE 2.
First- and second-order bond fixation in the ground states of butadiene, octatetraene, and hexadecaoctaene.

Butadiene

$p_{\mu\mu+1}$	0.894	0.447	0.894
$p_{\mu\mu+1}^{SCF}$	0.918	0.397	0.918
$\Delta R_{max,\ \mu\mu+1}$	−0.408	0.816	−0.408
$\Delta R_{max,\ \mu\mu+1}^{SCF}$	−0.369	0.853	−0.369

Octatetraene

$p_{\mu\mu+1}$	0.862	0.495	0.758	0.529	0.758	0.495	0.862
$p_{\mu\mu+1}^{SCF}$	0.890	0.445	0.800	0.480	0.800	0.445	0.890
$\Delta R_{max,\ \mu\mu+1}$	−0.239	0.388	−0.403	0.510	−0.403	0.388	−0.239
$\Delta R_{max,\ \mu\mu+1}^{SCF}$	−0.212	0.396	−0.389	0.542	−0.389	0.396	−0.212

Hexadecaoctaene

$p_{\mu\mu+1}$	0.852	0.506	0.735	0.558	0.707	0.574	0.697	0.579	0.697	0.574	0.707	0.558	0.735	0.506	0.852
$p_{\mu\mu+1}^{SCF}$	0.882	0.457	0.777	0.513	0.747	0.532	0.736	0.537	0.736	0.532	0.747	0.513	0.777	0.457	0.882
$\Delta R_{max,\ \mu\mu+1}$	−0.105	0.152	−0.204	0.260	−0.283	0.329	−0.325	0.353	−0.325	0.329	−0.283	0.260	−0.204	0.152	−0.105
$\Delta R_{max,\ \mu\mu+1}^{SCF}$	−0.084	0.139	−0.185	0.260	−0.274	0.345	−0.326	0.375	−0.326	0.345	−0.274	0.260	−0.185	0.139	−0.084

convergence of this process becomes slower if we calculate self-consistent bond orders $p_{\mu\nu}^{SCF}$ by the method described above (see Equations (22) and (23)). This general trend is counteracted by second-order effects as illustrated by the components $\Delta R_{max,\mu\mu+1}$ and $\Delta R_{max,\mu\mu+1}^{SCF}$ of the eigenvectors D_{max} and D_{max}^{SCF} of Equation (21) belonging to the eigenvalues λ_{max} and λ_{max}^{SCF}, respectively. At the point where λ_{max} or λ_{max}^{SCF} become greater than the critical value λ_{crit}, second-order effects take over, and Table 2 shows that the resulting distortions will be most pronounced toward the middle of the chain. Such an event will probably happen somewhere between $n \approx 20$ and $n \approx 40$. We therefore draw the qualitative conclusion that there will always be bond alternation in an even-membered linear polyene, but that the degree of alternation might pass through a flat minimum at a value of n inside the interval $n \approx 20$ to 40.

Electronically excited states (curve 3 in Figure 2) are dealt with in exactly the same fashion as the ground state. The difference consists only in that the computation of the bond orders and of the polarizability matrix (step 1 of our procedure) has to be carried out for a model in which an electron has been promoted from a bonding to an antibonding orbital, yielding bond orders $p_{\mu\nu}^{*}$ and polarizabilities $\pi_{\mu\nu,\kappa\lambda}^{*}$.

In the particular case of a polyene chain, the lowest electronically excited state—$^1B^0$ in the nomenclature of Platt (1956)—is represented by an MO model in which this promotion involves the highest bonding and the lowest antibonding orbital of the π system. As seen from Figure 2, there is a higher tendency toward second-order bond fixation for polyenes in their $^1B^0$ state than in their ground state ($\lambda_{max} < \lambda_{max}^{*}$). The corresponding distortion is no longer totally symmetrical but belongs now to the B_u representation of C_{2h}:

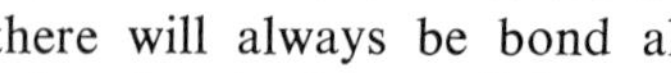

IV

Moreover, the first-order bond fixation, as shown by the bond orders $p_{\mu\mu+1}^{*}$, is much less pronounced. Decapentaene may serve as a typical example:

$\lambda_{max}^{*} = 1.112 \quad (\lambda_{max} = 0.990)$

$p_{\mu\mu+1}^{*}$

0.756 0.593 0.568 0.693 0.522 0.693 0.568 0.593 0.756

$\Delta R_{max,\mu\mu+1}^{*}$

−0.364 0.389 −0.414 0.210 0.000 −0.210 0.414 −0.389 0.364

In the odd-membered linear chains, which exhibit C_{2v} symmetry, the situation is somewhat more complicated. The λ_{max} values are plotted against chain length in Figure 3, for both the ground state and the lowest electronically excited state. Since zero-order molecular orbital theory predicts the same polarizabilities for the cations, radicals, and anions of odd alternant π-electron systems, such as

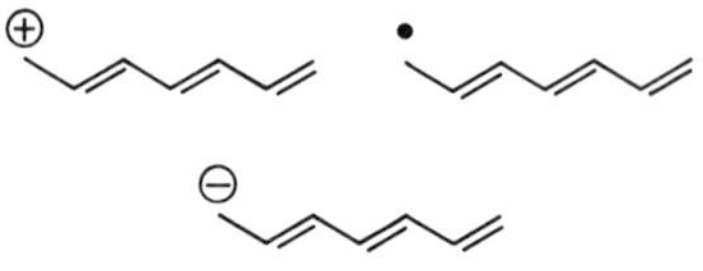

V

the λ_{max} and λ^*_{max} values are the same for charged and uncharged species. Here the first electronically excited state has been identified with a configuration in which an electron is promoted from the highest bonding to the nonbonding molecular orbital or from the nonbonding to the lowest antibonding orbital.

Comparison of Figure 3 with Figure 2 shows that second-order bond fixation is more important in the ground state of odd-membered linear chains than in the even ones, but the reverse is true of their electronically excited states, as shown in the following tabulation.

n	λ_{max}	λ^*_{max}
9	1.188	0.780
10	0.990	1.112
11	1.295	0.882
12	1.092	1.191
13	1.388	0.972
14	1.179	1.264
15	1.469	1.051
16	1.257	1.331
17	1.541	1.122
18	1.326	1.392
19	1.606	1.186
20	1.388	1.449

Therefore, second-order distortions should occur in odd chains about 7 to 11 centers earlier than in their even counterparts when the systems are in their electronic ground states, and at about the same number of centers later when electronically excited ($^1B^0$ state). So far as the ground state is concerned, this result is just the opposite of what Longuet-Higgins and Salem (1959, 1960)

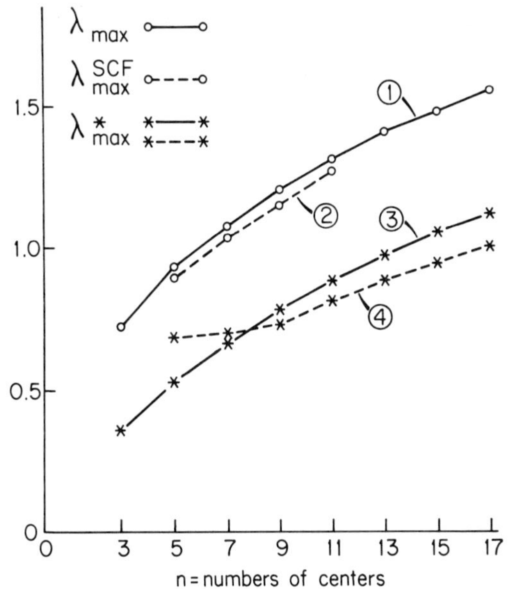

FIGURE 3.
Linear π-electron systems, n = odd. Curve 1 and curve 2 as in Figure 2. Curve 3: λ^*_{max} for $^1B^0$ state, distortion of B_2 type. Curve 4: λ^*_{max} for $^1B^0$ state, distortion of A_1 type.

and Labhart (1957) concluded from their studies. Using the same set of parameters for both series, Longuet-Higgins and Salem calculated that bond alternation sets in at chain lengths of 34 atoms for the even and 55 atoms for the odd polyenes. The discrepancy obviously stems from the a priori assumptions made by these authors concerning the type of distortion to be expected, namely

VI

This is quite at variance with the result we obtain. Whereas in even members we found bond alternations (III) belonging to the A_g representation (which would correspond to A_1-type distortions in C_{2v}), we now get second-order bond fixation belonging to the irreducible representation B_2—that is, corresponding to the structures V.

In the nonatetraenyl radical we have, for instance,

$p_{\mu\mu+1}$:	0.828	0.531	0.682	0.616	0.616	0.682	0.531	0.828
$\Delta R_{\max,\mu\mu+1}$:	−0.227	0.310	−0.339	0.439	−0.439	0.399	−0.310	0.227

In the odd chains with an even number of bonds we have two entirely equivalent situations of second-order bond fixation, differing only in the (arbitrary) sign of the vector $D_{\max}$:

or

VII

Whatever the nature of higher-order terms in the σ potential of the individual bonds, their contribution to the two situations will be precisely the same. In other words, the potential energy curve is of the form (*a*) of Figure 4. The energy barrier between the two symmetrical displacements is probably quite low. Thus second-order effects show up in a "dynamic" distortion of the molecule. This is especially true if the barrier height is of the same order of magni-

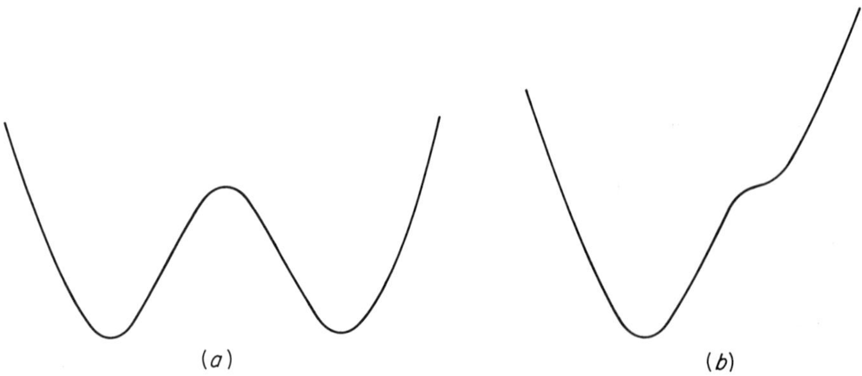

FIGURE 4.
Potentials for dynamic (*a*) and static (*b*) second-order bond fixation.

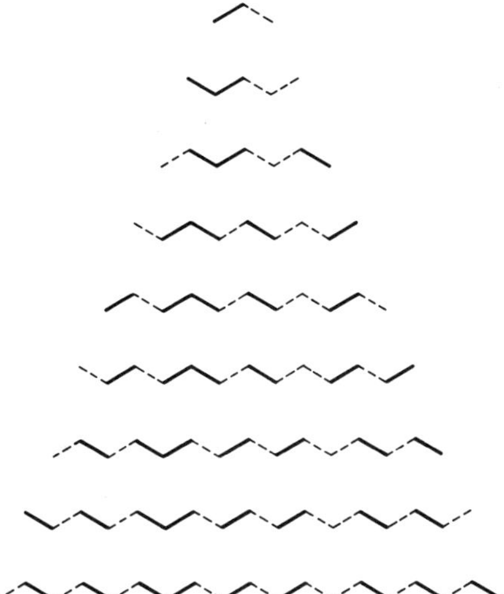

FIGURE 5.
Second-order bond fixation of the B_2 type in odd polyenes.

tude as the zero-point vibrational energy. In the even polyenes, on the other hand, we have an odd number of bonds. Anharmonicity of the σ potential must therefore result in a potential energy curve of the general shape (*b*) of Figure 4. Therefore we may speak of "static" second-order bond fixation in linear polyenes with an even number of atoms. This applies to the ground states. In our model we obtain for the first excited states of even polyenes a second-order bond fixation of the B_u representation, corresponding to curve (*a*) of Figure 4. In the series of odd-atom radicals and ions, second-order distortions in the first excited states transform as B_2. However, there are two types of B_2 distortions yielding similar gains in π-electron energy. Curve 3 of Figure 3 represents the energy gain by a B_u distortion of a bond-alternation type, and curve 4 corresponds to a distortion that is shown schematically in Figure 5. A crossover is observed between chains consisting of 7 and 9 atoms, so that in long polyenes of odd n bond alternation turns out to be again the most favorable second-order bond fixation in the excited state.

CYCLIC POLYENES

By our formalism we can treat the ground state of those cyclic conjugated π-electron systems which, satisfying Hückel's rule, contain $(4r + 2)$ π electrons (r = integer). In all other cases unpaired electrons have to occupy degenerate orbitals and our perturbation expansion will contain infinite terms. Such cyclic π systems will be subject to the well-known Jahn-Teller or pseudo-Jahn-Teller distortions.

The degeneracy of the excited states of cyclic π sytems containing $(4r + 2)$ π electrons is an artifact of the independent-electron theory and this spurious degeneracy could be lifted by explicitly taking electron interaction into account. We shall, however, not deal with this aspect of the problem and therefore limit our discussion to the ground state of such π systems.

The results for uncharged, singly charged, and doubly charged π-electron systems obeying Hückel's rule are collected in Figure 6. In the series of the

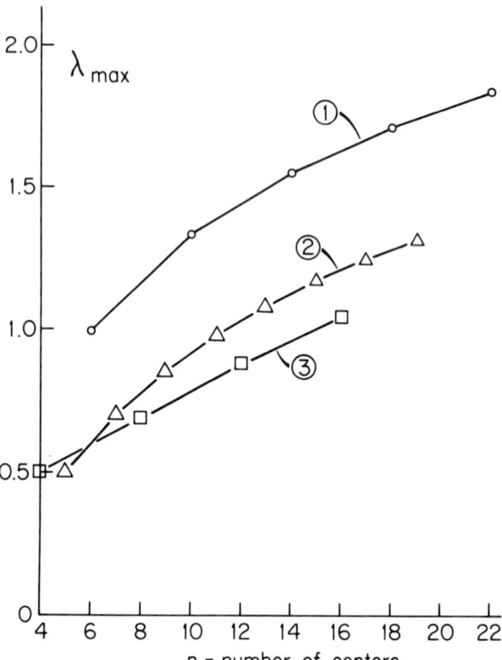

FIGURE 6.
Cyclic π-electron systems satisfying Hückel's rule. Curve 1: neutral systems, $N = n = 4r + 2$. Curve 2: singly charged systems, $n = N - 1 = 4r + 1$ (negative) and $n = N + 1 = 4r + 3$ (positive). Curve 3: doubly charged systems, $n = N \pm 2 = 4r$.

uncharged [n]-annulenes of symmetry D_{nh} with $n = N = 4r + 2$ (that is, n is even), the energy gain resulting from second-order effects increases rapidly with increasing ring size (curve 1 of Figure 6). The λ_{max} values can be obtained in closed form as a function of n (Binsch et al., 1966; Longuet-Higgins and Salem, 1959, 1960; Salem and Longuet-Higgins, 1960).

The most favorable distortion is bond alternation (Binsch et al., 1966), which transforms as the irreducible representation B_{2u} of D_{nh}, and this means that the localized π systems exhibit $D_{(n/2)h}$ symmetry. This is predicted to occur at $n \approx 20$ if $\lambda_{crit} = 1.8$.

Here again second-order effects result in a "dynamic" distortion of the molecule (Figure 4a), symbolized by

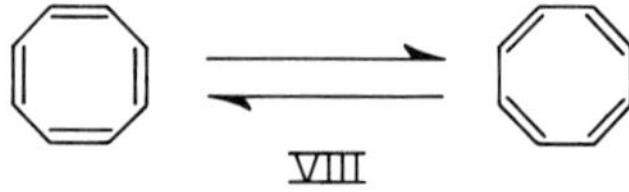

The singly and doubly charged species have a significantly lower tendency to localize their bonds than the uncharged ones. Curve 2 of Figure 6 corresponds to cyclic systems of $n = N - 1 = 4r + 1$ centers (charged negatively) and of $n = N + 1 = 4r + 3$ centers (charged positively), having D_{nh} symmetry with n being odd. In all these cases λ_{max} is doubly degenerate, belonging to the irreducible representation of highest index—for example, E_3 for the tropylium cation ($n = 7$, $N = 6$) or E_4 for the cyclononatetraenide anion ($n = 9$, $N = 10$). The λ_{max} values of cyclic π systems with $n = 4r$ centers, possessing either $N = n + 2$ or $N = n - 2$ π electrons (which gives them either a doubly negative or doubly positive charge), are given in curve 3 of

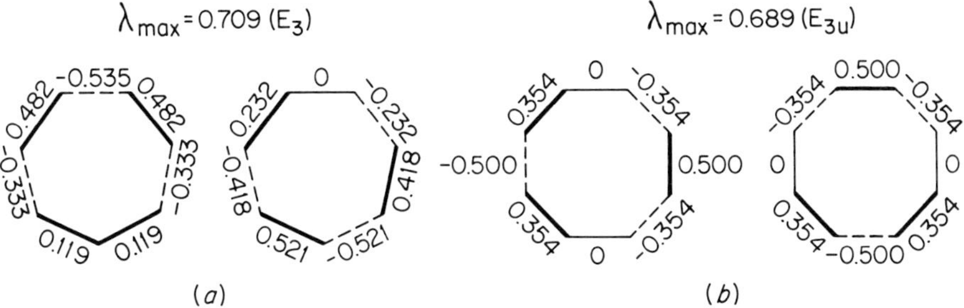

FIGURE 7.
Example for dynamic second-order distortion in cyclic π-electron systems: (*a*) tropylium cation; (*b*) cyclooctatetraene dianion.

Figure 6. In these cases a degenerate nonbonding molecular orbital is either filled ($N = n + 2$) or empty ($N = n - 2$) so that both types of model systems will yield the same polarization matrix $\boldsymbol{\pi}$ (19) and hence the same $\lambda_{\max}$ values. Again the $\lambda_{\max}$ values belong to the degenerate irreducible representation of highest index—for example, E_3 for cyclooctatetraene dianion (or dication). Figure 7 shows two examples of second-order distortions for the tropylium cation (D_{7h}) and cyclooctatetraene dianion (D_{8h}). (Note, however, that $\lambda_{\max} < \lambda_{\text{crit}}$!). It is seen that in these cases—that is, for odd cyclic π systems and for even, doubly charged ones satisfying Hückel's rule—the distortion is again of the dynamic type. However, as any linear combination of the two eigenvectors $D_{\max}$ belonging to the irreducible representation $E_{(n-1)/2}$ (or $E_{(n/2)-1}$) is again an eigenvector, we conclude that the distortions can shift around the ring without any change in the total energy. Therefore such systems must exhibit an extreme tendency toward the accommodation of their bond-fixation pattern under the influence of external perturbations. This, however, will only occur for rather large values of n, perhaps in the region of $n \approx 30$.

BENZENOID HYDROCARBONS

The results concerning second-order double-bond fixation in acenes and phenes, when in their electronic ground state, have already been reported (Binsch et al., 1966). For example, it has been found that for those which have

TABLE 3.
$\lambda_{\max}$ values for acenes (D_{2h}) in their electronic ground and excited ^1L_a state.

Acene	$\lambda_{\max}$	$\overset{*}{\lambda}_{\max}$	$\overset{*}{\lambda}_{\max}$
naphthalene	1.034	0.835	1.549
anthracene	1.062	0.974	1.039
naphthacene	1.083	0.998	0.921
pentacene	1.098	1.021	0.889
hexacene	1.110	1.043	0.884
Irreducible representation	B_{2u}	B_{2u}	B_{1g}

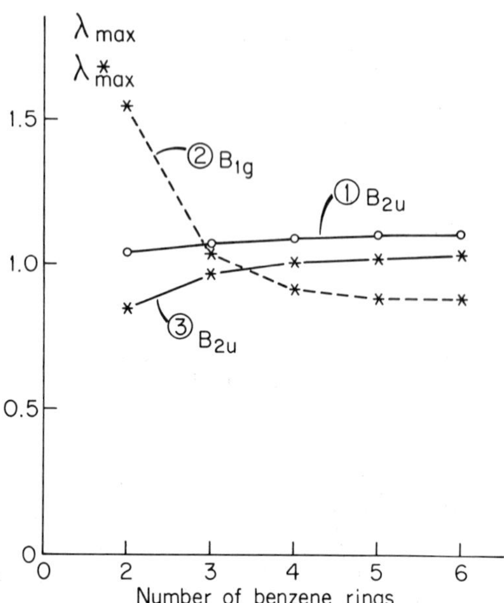

FIGURE 8.
Acenes. Curve 1: λ_{max} as a function of the number of benzene rings. Curves 2 and 3: λ^*_{max} for the ^1L_a state, distortions of the B_{1g} and B_{2u} types, respectively.

D_{2h} symmetry, λ_{max} is always much smaller than λ_{crit}, and furthermore the values are almost constant throughout the series (see Table 3 and Figure 8, curve 1). This suggests that second-order effects are not important in benzenoid hydrocarbons, even if one goes to fairly extended π systems. SCF calculations yield almost identical results (see Binsch et al., 1966). The eigenvectors D_{max}, corresponding to the λ_{max} values given in Table 3, belong to the irreducible representation B_{2u} of D_{2h} (y-axis = long axis of the molecule). The largest components $\Delta R_{max,\mu\nu}$ of D_{max} are those of the bonds in the perimeter of the system. But whereas Longuet-Higgins and Salem assumed that the cross bonds would not be affected by second-order distortions, we find that they are slightly perturbed, so that the molecule—if such distortions become effective—would tend toward a wedge shape:

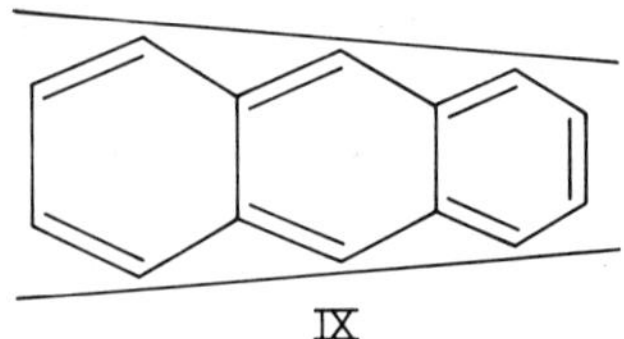

IX

If, in the molecular orbital model of an acene (or other benzenoid hydrocarbon), an electron is promoted from the highest bonding to the lowest antibonding orbital, we obtain a configuration representative of the ^1L_a state of the system (Platt, 1956). The polarization matrix π^* (19) yields the λ^*_{max} values of Table 3, shown in curves 2 and 3 of Figure 8. As can be seen, there is a crossover between distortions of the B_{1g} and B_{2u} type between anthracene and naphthacene, and it is noteworthy that for this reason second-order distortions in the ^1L_a state are more important in naphthalene than in any of the higher acenes. Examination of Figure 8 leads us to the conclusion that, in analogy to our findings for the ground state, we should not expect a lowering of the symmetry from D_{2h} to C_{2v}, even for the higher members of the series.

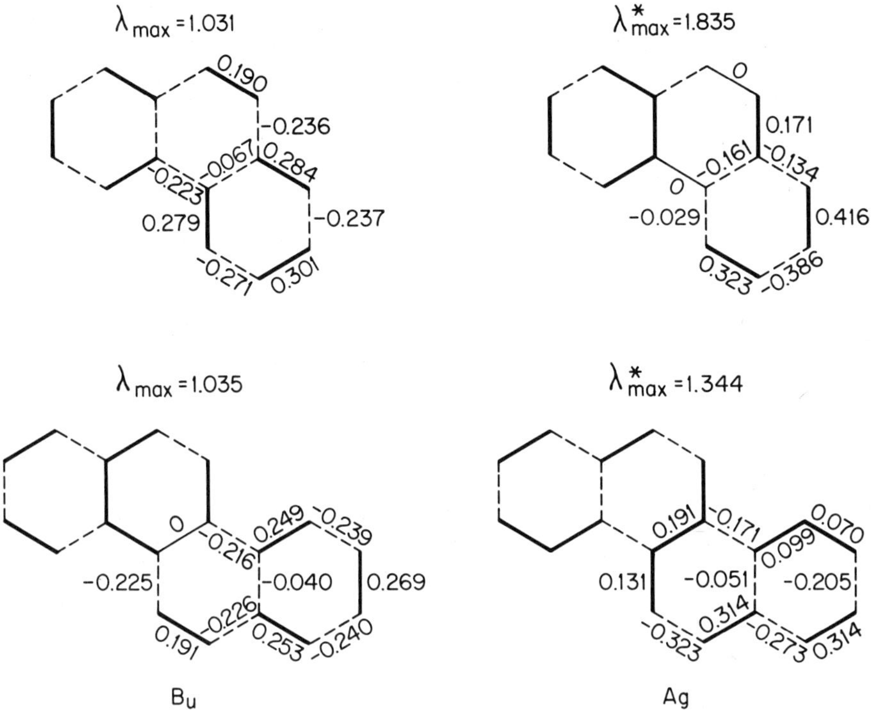

$\lambda_{max} = 1.031$ $\lambda^*_{max} = 1.835$

0.190 -0.236 -0.067 0.284 -0.223 0.279 -0.237 -0.271 0.301

0 0.171 -0.161 -0.134 0 -0.029 0.416 0.323 -0.386

$\lambda_{max} = 1.035$ $\lambda^*_{max} = 1.344$

0 0.249 -0.239 -0.216 -0.225 -0.040 0.269 -0.226 0.191 0.253 -0.240

Bu

0.191 -0.171 0.070 0.099 0.131 -0.051 -0.205 0.314 -0.323 -0.273 0.314

Ag

X

Diagram X summarizes the results obtained for the ground and ^1L_a state of phenanthrene and chrysene. From these one would deduce that phenanthrene may tend to lower its symmetry when electronically excited. However, not too much reliance should be attached to the value 1.8 for λ_{crit}, and we believe that valid predictions can be made only when this value is substantially exceeded.

Modification of these systems by either addition or removal of a π electron leads to the formation of radical anions or radical cations, respectively. In the framework of zero-order molecular orbital theory, the unpaired electron moves in either the highest bonding or the lowest antibonding orbital. The λ_i values and eigenvectors D_i for such systems are obtained in exactly the same manner as before. In view of the pairing properties of the orbitals in alternant π systems, the same results are obtained for the ground state of both radical anions and radical cations. These are given for naphthalene, anthracene and phenanthrene in diagram XI.

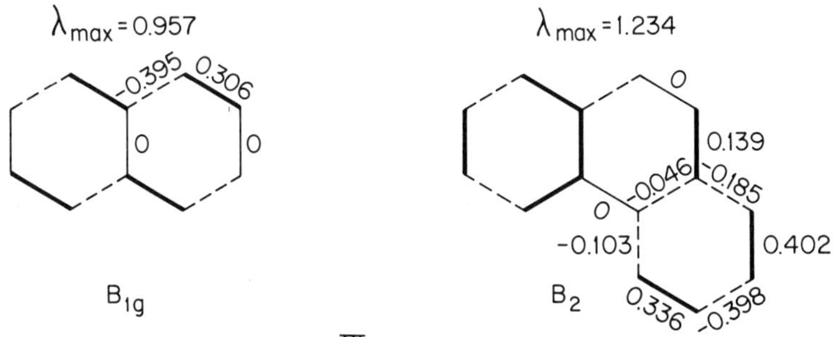

$\lambda_{max} = 0.957$ $\lambda_{max} = 1.234$

-0.395 0.306 0 0

B_{1g}

0 0.139 -0.046 -0.185 0 -0.103 0.402 0.336 -0.398

B_2

XI

No second-order distortions are expected, since all λ_{max} values are much smaller than λ_{crit}. However, as shown in XII, we would expect that the radical cations and anions of naphthalene and phenanthrene will lower their symmetry when electronically excited, the excitation corresponding in both cases to the promotion of an electron from the highest bonding (ψ_b) to the lowest antibonding orbital (ψ_a):

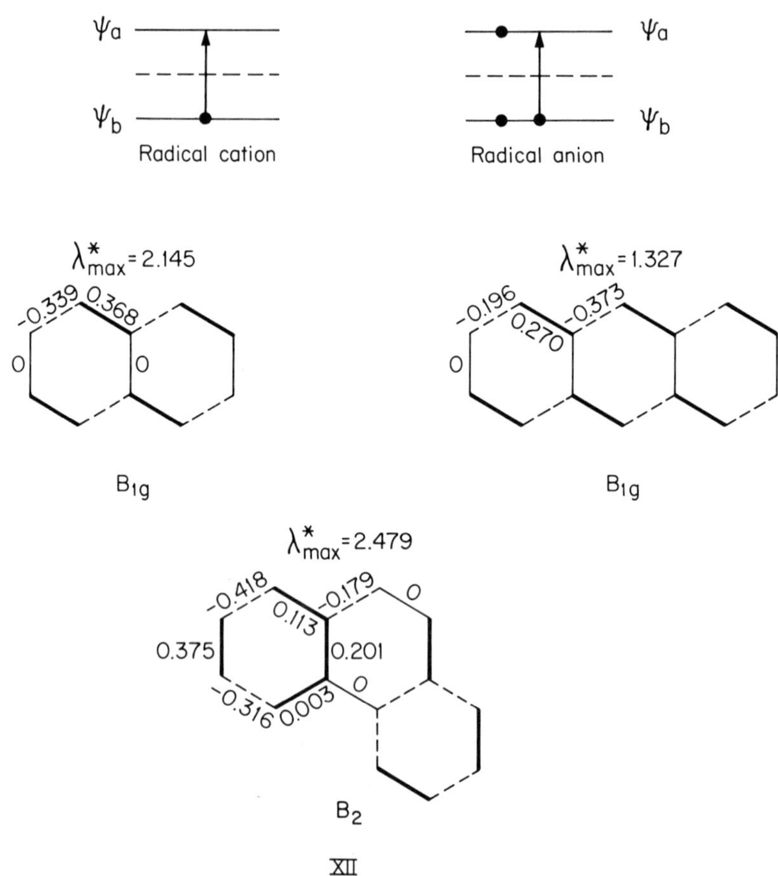

XII

Excitation according to the following schemes

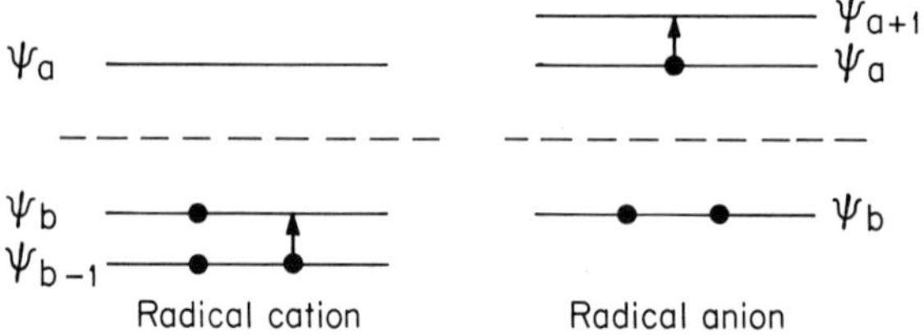

will lead to excited states that show no tendency toward distortion (naphthalene $\lambda^*_{max} = 0.837$, phenanthrene $\lambda^*_{max} = 0.888$). It is evident, however, that a many-electron model will be needed in cases such as these, in which extensive configuration mixing prevails, even in the lowest electronically excited states.

CONCLUSION

So far as their tendency to undergo second-order bond fixation is concerned, conjugated molecules fall into three distinct classes. For linear and monocyclic polyenes the λ_{max} values are found to exceed λ_{crit} only if the π-electron system itself exceeds a critical size. In the ground states of benzenoid hydrocarbons, whatever their size, second-order effects turn out to be unimportant; for these molecules the application of relationships between bond order and bond length will therefore be on safe grounds. Finally, there are certain small nonalternant systems, such as pentalene or heptalene, that show pronounced second-order bond distortions in their ground states. Such systems are sometimes classified as pseudo-aromatic or quasi-aromatic; it appears that the eigenvalues λ_{max} could well serve as a discriminating characteristic. This latter class of compounds will receive detailed treatment in a forthcoming publication.

REFERENCES

Den Boer-Veenendaal, P. C., J. A. Vliegenthart, and D. H. W. den Boer (1962). *Tetrahedron* **18**, 1325.

Binsch, G., E. Heilbronner, and J. N. Murrell (1966). *Mol. Phys.* **11**, 305.

Coulson, C. A. (1938). *Proc. Roy. Soc.* (London) A **164**, 383.

Coulson, C. A. (1939). *Proc. Roy. Soc.* (London) A **169**, 413.

Coulson, C. A., and W. T. Dixon (1962). *Tetrahedron* **17**, 215.

Dauben, H. J., Jr., and D. J. Bertelli (1961). *J. Am. Chem. Soc.* **83**, 4657.

Dewar, M. J. S. (1952). *J. Am. Chem. Soc.*, 3532, 3544.

Dewar, M. J. S., and G. J. Gleicher (1965). *J. Am. Chem. Soc.* **87**, 685, 692.

Dewar, M. J. S., and H. N. Schmeising (1959). *Tetrahedron* **5**, 166.

Dewar, M. J. S., and H. N. Schmeising (1960). *Tetrahedron* **11**, 96.

Försterling, H. D., W. Huber, and H. Kuhn (1967). *Internat. J. Quantum Chem.* **1**, 225.

Kuhn, H. (1948). *J. Chem. Phys.* **16**, 840.

Kuhn, H. (1949). *J. Chem. Phys.* **17**, 1198.

Kuhn, H. (1957). *Angew. Chem.* **69**, 239.

Kuhn, H. (1959). *Angew. Chem.* **71**, 93.

Kuhn, H., W. Huber, and F. Bär (1958). *Calcul des fonctions d'onde moléculaire*, Ed. du Centre National de la Recherche Scientifique, p. 179.

Labhart, H. (1957). *J. Chem. Phys.* **27**, 957, 963.

Lennard-Jones, J. E. (1937). *Proc. Roy. Soc.* (London) A **158**, 280.

Lennard-Jones, J. E., and J. Turkevich (1937). *Proc. Roy. Soc.* (London) A **158**, 297.

Longuet-Higgins, H. C., and L. Salem (1959). *Proc. Roy. Soc.* (London) A **251**, 172.

Longuet-Higgins, H. C., and L. Salem (1960). *Proc. Roy. Soc.* (London) A **257**, 445.

Nakajima, T., and S. Katagiri (1964). *Mol. Phys.* **7**, 149.

Ooshika, Y. (1957). *J. Phys. Soc. Japan* **12**, 1238, 1246.

Pauling, L. (1932). *Proc. Nat. Acad. Sci. U.S.* **18**, 293.

Pauling, L., L. O. Brockway, and J. Y. Beach (1935), *J. Am. Chem. Soc.* **57**, 2705.

Pauling, L., and L. O. Brockway (1937). *J. Am. Chem. Soc.* **59**, 1223.

Pauling, L. (1939). *The Nature of the Chemical Bond.* Ithaca: Cornell Univ. Press.

Penney, W. G. (1937). *Proc Roy. Soc.* (London) A **158**, 306.

Platt, J. R. (1956). In Holländer, A., ed., *Radiation Biology*, vol. 3. New York: McGraw-Hill.

Platt, J. R. (1956). *J. Chem. Phys.* **25**, 80.

Salem, L. (1966). *The Molecular Orbital Theory of Conjugated Systems*. New York: Benjamin.

Salem, L., and H. C. Longuet-Higgins (1960). *Proc Roy. Soc.* (London) A **255**, 435.

Sly, W. (1964). *Acta Cryst.* **17**, 511.

MARTIN KARPLUS
Department of Chemistry
Harvard University
Cambridge, Massachusetts

Structural Implications of Reaction Kinetics

As one of Linus Pauling's graduate students some fifteen years ago, I came away from Cal Tech with an approach to theoretical chemistry that owes much to his view of the subject. During my Cal Tech years, Pauling often stressed that structure must play the primary role in any theoretical approach to chemical problems, an understanding of function following from a knowledge of structure. Also, he pointed out that theory in chemistry, perhaps even more than in other fields of science, must be very closely tied to experiment. These precepts, which have played an important role in Pauling's own research, have helped to guide my work into certain areas of chemistry in which new experimental developments called for theoretical analysis. In analogy to Pauling's use of X-ray crystallography and electron diffraction for determining molecular and crystal structures, I began research by trying to learn about electronic structure through the interpretation of NMR and ESR measurements. Chemical kinetics, which is my present concern, may appear far removed from such structural pursuits. However, since the rate of a chemical reaction is governed by the form of the potential surface for the interacting species, kinetic measurements must contain information about structure. Gas-phase rate constants have for many years been interpreted in terms of the Arrhenius equation to provide an estimate of the interaction range and barrier height involved in the reaction. It is only recently, however, that crossed molecular beam studies of chemical reactions have begun to supply the data required for a more detailed examination of potential energy surfaces. From the beam measurements and their theoretical analysis by classical trajectory methods there is emerging a field of chemistry which may be regarded as an incipient "spectroscopy" of the interactions between reactive species.

For pairs of nonbonded atoms, molecular-beam elastic scattering studies are now yielding considerable information about the interaction potential; that is, by the appropriate classical, semiclassical, or quantum-mechanical analysis of two-body scattering, measurements of total and differential cross sections are being related to the potential parameters of such systems. (For reviews, see Bernstein, 1967; Pauly and Toennies, 1965; Bernstein and Muckerman, 1967.) In most cases, the comparisons are based on an assumed form for the potential (for example, Lennard-Jones 6-12 potential), though some

This paper is based on a lecture given as part of the symposium associated with the presentation of the first Pauling Medal Award to Linus Pauling in Portland, Oregon, December, 1966.

progress has been made in relating certain features of the measured cross sections—rainbow angle in differential cross section (Duren and Schlier, 1965), glory maxima in velocity-dependence of total cross section (Duren et al., 1965), and so on—directly to properties of the potential (slope and curvature at certain points, and so on). Thus, for elastic two-body scattering, one is approaching a situation for which the term "translational spectroscopy" may not be too far-fetched.

In the reactive case, with all of the complications introduced by the presence of three or more atoms and the possibility of a variety of inelastic processes coupled with the rearrangement, our understanding of the connection between the potential energy surface and the measured total and differential reactive cross sections is in a much more primitive state than for the two-body case. Moreover, experimental difficulties have so far prevented cross section measurements over a sufficiently wide range of the initial and final state variables (rotational and vibrational states of molecules, relative velocities, and so on). To illustrate the progress that has been made in spite of these limitations, I shall outline some of the recent developments in our understanding of the $K + CH_3I \longrightarrow KI + CH_3$ reaction. The results demonstrate that by means of the measurements and the interpretive techniques that are now available, it is indeed possible to use kinetic studies for the evaluation of potential energy surfaces. Thus, even in the case of chemical reactions, function is found to serve as the handmaiden of structure.

TRAJECTORY ANALYSIS

In any spectroscopic technique, theory provides the relationship between the experimental data and the significant parameters of the system under investigation. For ordinary rotation-vibration spectroscopy of a stable diatomic molecule, quantum-mechanical calculations based on the rigid-rotor, harmonic-oscillator approximation and its refinements serve as the link by which the potential energy function is extracted from the measured frequencies. In the reaction kinetic studies, the desired interpretation of cross sections is performed by means of classical trajectory calculations. Their use for this problem is predicated on the assumption that classical mechanics is adequate to describe the collision process; that is, given the potential energy of interaction, the behavior of the colliding atoms can be determined by solving the classical equations of motion. Although it is not yet possible to give a rigorous justification of the classical approach to reactive scattering, the form of the potential and the masses of the atoms are such that quantum effects on the measurable properties are expected to be small in the energy range of interest.

The essence of the trajectory method is the determination of whether a reaction does or does not take place during a given collision trajectory by exact integration of the equations of motion with a certain set of initial conditions describing the separated reactants to the final state corresponding to the separated products. Appropriate averaging of the trajectory results over the range of initial conditions dictated by our knowledge of the reactants and the experimental situation leads unequivocally to the desired reaction attributes (cross section, energy and angular momentum distribution, and so on). Thus, for each potential energy surface, whether obtained by theoretical, semitheoretical, or empirical techniques, the trajectory treatment yields unique results for comparison with experiment. However, the complexity of the multiparticle interactions and the limitations in the present measurements are such that the inverse procedure—determination of the potential surface from the cross section data—is far from having a unique solution. Nevertheless, by the introduction of suitable parametrized surfaces, which conform to our knowledge of potentials for the separated reactants and products, considerable progress has been made.

THE REACTION $K + CH_3I \longrightarrow KI + CH_3$

The first investigation of the $K + CH_3I \longrightarrow KI + CH_3$ reaction in crossed thermal beams was made by Herschbach and his coworkers (for reviews, see Herschbach, 1962, 1966). From their measurement of the laboratory distribution of KI product molecules, they were able to conclude (*a*) that the center-of-mass differential reaction cross section (σ_r) is highly anisotropic with the product KI scattered *backward* with respect to the initial direction of the K beam and (*b*) that the heat of reaction (≈ -22 kcal) appears primarily as internal excitation of the products rather than as kinetic energy of relative motion. Also, they estimated that the total reaction cross section (S_r) for this system was on the order of 7 Å^2. In the initial attempt at a trajectory analysis of these conclusions, Blais and Bunker (1962) assumed that the $K + CH_3I$ surface could be represented by an expression of the form

$$\mathcal{V}_B = \mathcal{V}_0(\text{K—I}) + \mathcal{V}_0(\text{I—CH}_3) + \mathcal{A}(\text{I—CH}_3, \text{K—I}) + \mathcal{R}(\text{K—CH}_3) \qquad (1)$$

Here $\mathcal{V}_0$(K—I) and $\mathcal{V}_0$(I—CH_3) are Morse potentials for the K—I and I—C bond, respectively, $\mathcal{A}$(I—CH_3, K—I) is an attenuation factor which reduces the I—C bonding when the K—I distance is small, and $\mathcal{R}$(K—CH_3) is a steep, short-range repulsive interaction between K and CH_3. Although the Morse terms in the potential energy surface are obtained from experiment, the parameters in the attenuation factor were chosen empirically. They were selected to give a reasonable cut-off for the I—CH_3 interaction, to provide a smooth surface with only a small activation barrier, and to insure that the heat of reaction was released during the approach of the K atom rather than during the departure of the CH_3 group. The third of these conditions was introduced because it was anticipated that it would lead to high internal excitation of the products in agreement with experiment (for earlier qualitative discussions of this point, see Glasstone, Laidler, and Eyring, 1949; Polanyi, 1959; Smith, 1959; Bunker, 1962). As to the K—CH_3 repulsion in Equation (1), it is a somewhat arbitrary term for which there is no simple theoretical justification (except at very close range), since it is known that K and CH_3 have an attractive interaction which can lead to bonding. Thus, although surface $\mathcal{V}_B$ has a generally reasonable form, it is essentially empirical and must, therefore, be evaluated by its agreement or disagreement with the molecular beam results. To carry out such a comparison via trajectory calculations, Blais and Bunker (1962), assumed that the $K + CH_3I$ system could be treated as composed of only three particles, with the methyl group reduced to a single atom of mass 15. Furthermore, they simplified the integration of the classical equations by limiting the motion of the three-atom system to a plane. Blais and Bunker showed that the energy partitioning between relative motion and internal excitation was, as expected, in very good agreement with the estimate of Herschbach et al. (1962, 1966). However, because of the restriction of the collisions to a plane, they were unable to determine the total reaction cross section or the angular distribution of products.

To refine the analysis of surface $\mathcal{V}_B$, an extension of the trajectory method from two-dimensional (planar) motion to three-dimensional space was clearly required. This was accomplished without introducing any intrinsic complications or an excessive increase in the computation time for the collision trajectories (Karplus, Porter, and Sharma, 1964). In the three-dimensional treatment (Karplus and Raff, 1964) of potential $\mathcal{V}_B$ [Equation (1)], the energy partitioning was essentially unchanged from that obtained by Blais and Bunker (1962) in their two-dimensional model and so remained in agreement with the experimental estimate. However, the calculated differential reaction cross section disagreed significantly with the measured result; that is, instead of the observed strong backward scattering for the product KI molecules, an almost uniform center-of-mass angular distribution was found. Furthermore, the three-dimensional treatment yields a value of $\sim$400 Å^2 for the total reaction cross section, much larger than the experimental estimate of 7 Å^2. Thus, from the complete classical trajectory analysis, it was obvious that something was wrong with the surface $\mathcal{V}_B$. To determine the nature of the error, it was necessary to examine the details of the surface and of the resulting collision trajectories. From the form of the potential [Equation (1)], it is evident that there is a long-range attraction of K and I due to the Morse term $\mathcal{V}_0$(K—I), independent of the I—CH_3 distance. This runs counter to the concept of chemical saturation, which suggests that, when an I atom is part of stable CH_3I molecule, the KI interaction should be considerably decreased from that between a free K and a free I. Quantitatively, one finds from surface $\mathcal{V}_B$ that the derivative of the potential with respect to the K—I distance [R(K—I)] is on the order of 3.6 kcal/mole-Å for large R(K—I) and for R(I—CH_3) near the equilibrium value for CH_3I. Since the most probable velocity of K relative to CH_3I in the thermal beams used for the experiments corresponds to an energy of only $\sim$1.4 kcal/mole, even K atoms with large impact parameters ($\approx$20. a.u.) would be accelerated sufficiently rapidly toward the CH_3I molecule on surface $\mathcal{V}_B$ that a close approach of K and I could take place and reaction might occur. This effect is shown in Figure 1,*a*, which represents a reactive trajectory by plotting the three internuclear distances along the ordinate as a function of time along the abscissa. Although the impact parameter for the illustrated trajectory is 15 a.u., there is considerable negative curvature in the path of the approaching K atom, corresponding to the attractive acceleration, and a reaction does occur. Since the long-range K—I interaction leads to a reaction probability $P_r(b)$ that is nonzero for impact parameters b significantly larger than those that should contribute, the total reaction cross section S_r,

$$S_r = 2\pi \int_0^\infty P_r(b)\, b db, \tag{2}$$

is considerably greater than the true value. The reactions with large impact

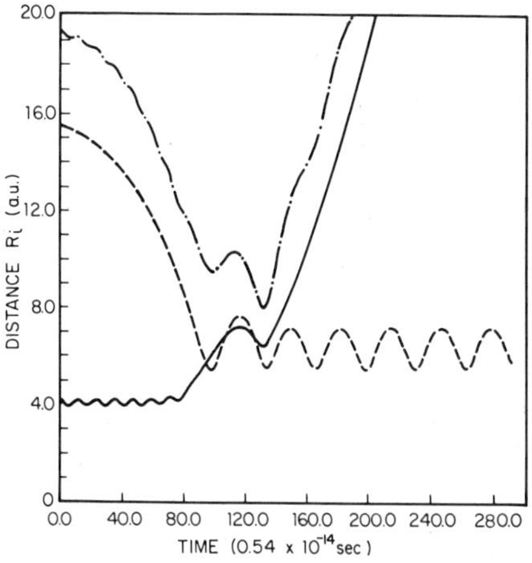

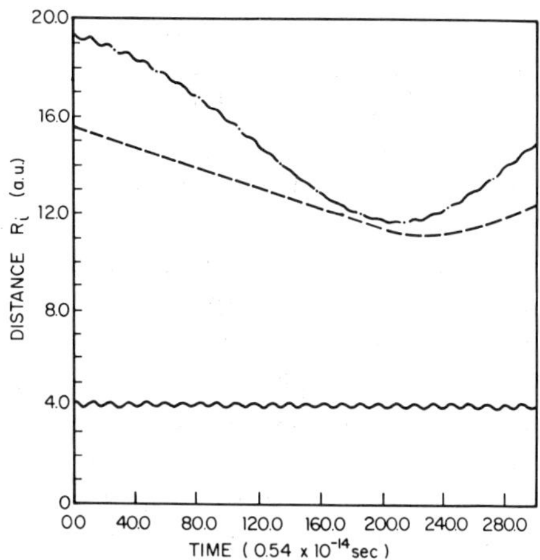

FIGURE 1.
Collision Trajectories on (*a*) Surface $\mathcal{V}_B$ and (*b*) Surface $\mathcal{V}_{MB}$; the initial conditions are the same, with $b = 14.9$ a.u. The ordinate represents the interparticle distances as function of time on the abscissa [(---) = R(K—I), (——) = R(I—CH_3), (·⁻·⁻·) = R(K—CH_3)].

parameter, and high orbital angular momentum also affect the differential cross section because they tend to result in forward scattering, in correspondence with the behavior frequently found in nuclear reactions of the stripping or pick-up type (see, for example, Glendenning, 1963).

The above discussion shows that to bring the trajectory calculations into closer agreement with the experimental data concerning S_r and σ_r, it is necessary to eliminate from $\mathcal{V}_B$ the long-range attraction between K and I for a K atom colliding with a CH_3I molecule. This can be done by adding to $\mathcal{V}_B$ of Equation (1) an attenuation term $\mathcal{A}$(K—I, I—CH_3), analogous to $\mathcal{A}$(I—CH_3, K—I). Studies with such a potential have been made (Raff and Karplus, 1966) and yield results that are very similar to those obtained from an alternative potential, which is somewhat simpler and was the first modification used to replace $\mathcal{V}_B$. The latter potential ($\mathcal{V}_{MB}$) is obtained from $\mathcal{V}_B$ simply by adjusting the Morse term for KI; that is, instead of $\mathcal{V}_0$(K—I),

$$\mathcal{V}_0(\text{K—I}) = D_1\{1 - \exp[-\beta_1(R_1 - R_{01})]\}^2 \tag{3}$$

where D_1, β_1, R_{01} are the standard Morse parameters for KI, $\mathcal{V}_{MB}$ contains $\mathcal{V}_0'$(K—I) which differs from $\mathcal{V}_0$(K—I) only by having β_1 replaced by β_1', the latter being chosen to reduce the long "tail" of the K—I interaction. The effect of this change on large impact parameter collisions is indicated in Figure 1, *b*, which shows a collision trajectory on $\mathcal{V}_{MB}$ with initial conditions identical to those in Figure 1,*a*; as expected, there is now no attractive curvature for large *R*(K—I) and the K atom passes by the CH_3I molecule without reaction. For β_1' equal to 12.59×10^7 cm^{-1}, instead of the Morse potential value of 8.61×10^7 cm^{-1} for an isolated KI molecule, the total cross section is reduced from 400 Å^2 to 20.8 Å^2, in reasonable agreement with the original estimate of 7 Å^2; moreover, recent refinements in the measured value have increased it to 30 Å^2 (Herschbach, 1966). What is of greater significance is that the removal of the long-range KI tail also has produced a drastic alteration in the differential cross section; that is, instead of the uniform center-of-mass distribution found previously, backward scattering now dominates in correspondence with the experimental conclusion. Since the energy partitioning is essentially unchanged in going from $\mathcal{V}_B$ to $\mathcal{V}_{MB}$, all the results calculated with the latter surface appear to be in reasonable agreement with the available data.

In addition to providing a potential energy surface for the $K + CH_3I \longrightarrow KI + CH_3$ reaction that yields reasonable results, the trajectory analysis suggested an important relationship between S_r and σ_r that seems to be of rather general significance. Small values of S_r corresponding to the limitation of reaction to low impact-parameter collisions are shown to be associated with σ_r functions that are highly anisotropic in the backward direction; as S_r increases and larger impact-parameter collisions contribute to reaction, σ_r shifts forward until for large S_r, σ_r becomes uniform (independent of angle). This relationship, which should apply to reactions without activation energy, has recently been confirmed by beam experiments for K reacting with a series of different molecules (for reviews, see Herschbach, 1962, 1966). It has also been found that a further shift of σ_r toward mainly forward scattering occurs for still larger S_r values (Datz and Minturn, 1964; Wilson et al., 1964).*

The above comparisons raise the question of whether the experimental and theoretical results considered so far are sufficient to prove that $\mathcal{V}_{MB}$ is a good approximation to the actual (K, CH_3I) interaction potential. In particular, it is not clear that the K—CH_3 repulsion, which was introduced arbitrarily into $\mathcal{V}_B$ and $\mathcal{V}_{MB}$, is important in determining any of the reaction attributes that have been discussed. To examine this point, we chose to investigate an alternative surface (Raff and Karplus, 1966) which can be written

$$\mathcal{V}_S = \mathcal{V}_0'(\text{K—I}) + \mathcal{V}_0(\text{I—CH}_3) + \mathcal{A}(\text{I—CH}_3, \text{K—I}) + \mathcal{V}_{\text{exp-six}}(\text{K—CH}_3\text{I}) \quad (4)$$

where the first three terms have the same form as in $\mathcal{V}_{MB}$ and the last term provides a long-range ($1/R^6$) attraction and a short-range exponential repulsion between K and the center-of-mass of CH_3I. This type of effective two-body potential has been found useful by Greene, Ross, and their coworkers in the analysis of elastic scattering of potassium by methyl iodide (Airey, Greene, Moursund, and Ross, 1966) and by many other molecules (for a review, see Greene, Moursund, and Ross, 1966). Qualitatively, the essential difference between $\mathcal{V}_{MB}$ and $\mathcal{V}_S$ is that $\mathcal{V}_{MB}$ includes a K—CH_3 repulsion which makes the energy strongly dependent on the CH_3I orientation relative to the incoming K, while $\mathcal{V}_S$ lacks such a repulsion and is much closer to being spherically symmetric than is $\mathcal{V}_{MB}$. Trajectory calculations (Raff and Karplus, 1966) for $\mathcal{V}_S$ showed that the total cross section, the energy partitioning, and the center-of-mass differential cross section are similar to those obtained from $\mathcal{V}_{MB}$, so that both $\mathcal{V}_S$ and $\mathcal{V}_{MB}$ appear to yield results in agreement with the qualitative conclusions of Herschbach et al. (1962, 1966).

Beuhler, Bernstein, and Kramer (1966) and Brooks and Jones (1966) have

*An analysis of this case with particular emphasis on the $K + Br_2 \rightarrow KBr + Br$ reaction has been made (M. Godfrey and M. Karplus, to be published); related work is being done by J. C. Polanyi (private communication).

TABLE 1.
Total cross sections (Å^2) *for oriented and unoriented collisions*[a].

	$\mathcal{V}_{MB}$	$\mathcal{V}_S$[b]
Unoriented	25	13
K ⟶ CH_3I	0	14
K ⟶ ICH_3	35	11

[a]Results based on 500 trajectories with maximum impact parameter of 8 a.u.
[b]The difference among the various $\mathcal{V}_S$ results is not significant.

recently reported a crossed beam study of the reaction of rubidium and potassium atoms with methyl iodide molecules partially oriented by rotational state-selection with an electric six-pole field. They found that in collisions with the I atom of CH_3I directed toward the incoming alkali M (M equal to Rb or K), the reaction cross section for formation of MI is significantly greater than that in collisions with the reverse orientation of CH_3I. To examine the significance of this result for distinguishing between surfaces $\mathcal{V}_{MB}$ and $\mathcal{V}_S$, we have computed trajectories corresponding to collisions of K atoms with oriented CH_3I molecules (Karplus and Godfrey, 1966). The method used is identical to that employed for the previous calculations except that the initial conditions for each collision trajectory are altered so that (*a*) rather than averaging over random initial orientations of CH_3I, only molecules with their axis parallel or antiparallel to the initial K,CH_3I relative velocity vector are included, and (*b*) instead of averaging over a CH_3I rotational state population corresponding to a Boltzmann distribution at the beam temperature, only $J = 0$ molecules are selected. This choice of initial conditions provides a very simple approximate simulation of CH_3I molecules perfectly oriented by an external electric field. It does not prevent reorientation of the CH_3I molecules by its interaction with the incoming K atom. But, this has been shown by trajectory analyses to be unimportant for the system under consideration, in agreement with the observation of Beuhler et al. (1966), that their results are independent of the orienting voltage. Table 1 gives the total cross section values obtained with surface $\mathcal{V}_{MB}$ and $\mathcal{V}_S$ for unoriented CH_3I molecules and for the two initial orientations (K ⟶ CH_3I and K ⟶ ICH_3). There is a sharp distinction between the surfaces: for $\mathcal{V}_{MB}$, S_r is very sensitive to the initial orientation, whereas for $\mathcal{V}_S$ there is no correlation between the initial orientation and the magnitude of S_r. Although refinements in the measurements and calculations are required before quantitative conclusions can be drawn (Karplus and Godfrey, 1966), it is evident that potential $\mathcal{V}_S$ can be eliminated as having too little asymmetry to explain the experimental results and that potential $\mathcal{V}_{MB}$ with its significant (K, CH_3) repulsion has the more appropriate form.

Additional confirmation of the presence of a repulsive (K, CH_3) interaction in the true potential energy surface has been obtained in some recent studies (Airey et al., 1967; Herschbach, 1967) of nonreactive K, CH_3I scattering. By an optical-model analysis of such measurements (Greene et al., 1966) it is possible to estimate the probability of reaction as a function of the impact

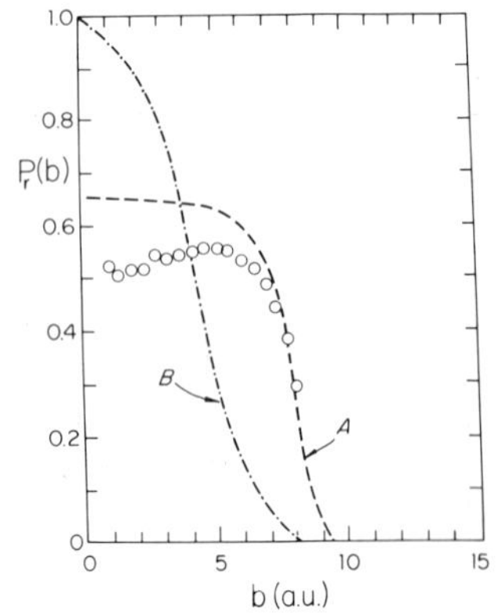

FIGURE 2.
Probability of reaction P_r (*b*) as a function of impact parameter *b*. Curve *A* (---), Surface $\mathcal{V}_{MB}$; Curve *B* (·-·-·), Surface $\mathcal{V}_S$; open circles, experimental results from Herschbach (1967).

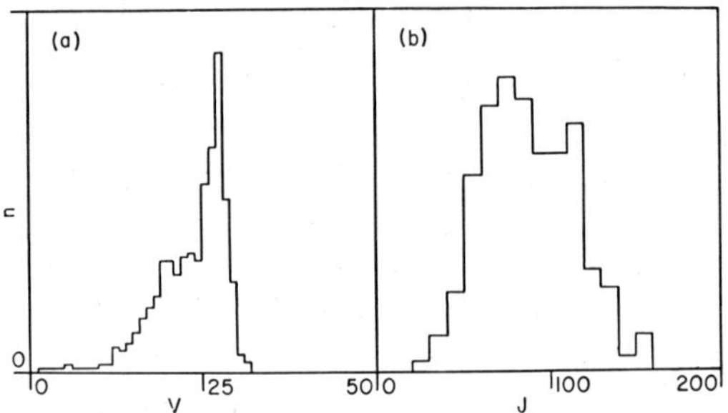

FIGURE 3.
Vibrational and rotational state distribution for product KI molecules on surface $\mathcal{V}_{MB}$: The ordinate represents the number of reactive trajectories with the quantum number on the abscissa. (*a*) Vibrational quantum number *v*, (*b*) rotation quantum number *J*.

parameter. Figure 2 gives the experimental estimates and the results obtained from the trajectory studies of surfaces $\mathcal{V}_{MB}$ and $\mathcal{V}_S$ (Raff and Karplus, 1966). It is clear that the experimental points are in better agreement with the $\mathcal{V}_{MB}$ curve than with that from $\mathcal{V}_S$. Qualitatively, the significant point is that the experimental values and the $\mathcal{V}_{MB}$ curve do not go to unity as the impact-parameter goes to zero, in correspondence with expectation for a potential containing an anisotropic ("steric") repulsion term.

Although the trajectory results from surface $\mathcal{V}_{MB}$ are in accord with all of the experimental data for the K, CH_3I reaction,* additional tests of the surface would be most desirable. One possibility is to examine the final rotational and vibrational state distribution of the newly formed KI molecules. Figure 3 gives the results obtained on surface $\mathcal{V}_{MB}$ by a quasiclassical procedure for the determination of the rotational and vibrational quantum numbers (Raff and Karplus, 1966). Both distributions are non-Boltzmann in character and are indicative of the high degree of internal excitation of KI. The vibrational distribution peaks at $v \cong 27$, and the rotational distribution at $J \cong 90$. A significant fraction of the internal energy appears in molecular rotation (the average rotational energy is 4.4 kcal/mole), as well as in vibration (the average vibrational energy is 16.4 kcal/mole). The results for surface $\mathcal{V}_S$ are significantly different from those for surface $\mathcal{V}_{MB}$ in that the rotational excitation is considerably lower (Raff and Karplus, 1966). As yet there are no experimental data on the partitioning of the internal excitation of KI

*A recent reanalysis (Herschbach, private communication, 1967) indicates that more energy appears in relative translation than is suggested by these comparisons.

between rotation and vibration. However, by the introduction of an inhomogeneous electric field into the molecular beam apparatus (Herschbach, 1962, 1966; Herm and Herschbach, 1965), a measure of the rotational excitation of the dipolar KI product can be obtained.*

Another measurement of interest would be a determination of the velocity dependence of the various reaction attributes. The trajectory calculations predict that the center-of-mass angular distribution depends on the initial velocity of K relative to that of CH_3I. Although no quantitative analysis has been made, the qualitative conclusion is that the strong backward peaking should become less pronounced with increasing relative velocity; that is, since the effective repulsive barrier becomes less important for higher relative velocities, σ_r should be shifted in the forward direction. Also, with increasing relative velocity, the $P_r(b)$ curve should approach nearer to unity for $b = 0$ than in Figure 3 (an indication of this type of behavior is given in Airey et al., 1967), and the total reaction cross section should become less dependent on the CH_3I orientation. There is hope that, by means of the improved technology which is becoming available, the required high velocity measurements can be made. If the experimental data are obtained, detailed calculations for the different potential energy surface could easily be made for a quantitative comparison.

PROGNOSIS

The recent developments in our understanding of the $K + CH_3I \longrightarrow KI + CH_3$ reaction provide but one illustration of the interplay between molecular beam experiments and classical trajectory calculations in the elucidation of the potential energy surfaces for reactive systems. Although much has been learned already about K,CH_3I and other reactions, only the mere beginnings of a "spectroscopy" for reactive surfaces exist at the present time. It is to be hoped that experimental refinements and theoretical advances will soon lead to considerably more quantitative results. Ultimately, it would be desirable to be able to express the measurable reaction attributes as a function of a relatively small number of parameters characterizing suitably chosen potential energy surfaces. Only in this way would the reactive case achieve a position in terms of accuracy and uniqueness that corresponds to that obtaining in the ordinary spectroscopy of stable molecules or metastable molecular fragments. The difficulties inherent in this approach suggest that, concomitant with the analysis of reactive scattering dynamics, strong emphasis must be given to the theoretical evaluation of the potential energy surface itself. Some advances on this quantum-mechanical problem have been made recently for very simple systems—for example, for the H, H_2 potential energy surface (see Edminston and Krauss, 1965; Conroy and Bruner, 1965; Shavitt, Stevens, Minn, and Karplus, 1967), and there is ground for optimism concerning the application of semitheoretical methods to more

*The K, CH_3I rotational analysis experiment is being attempted (D. R. Herschbach, private communication). A recent renalysis of the K, CH_3I experimental data (Entemann and Herschbach, private communication, 1967; Balint and Karplus, 1967) suggests that a somewhat larger fraction of the reaction exothermicity appears in relative translation than is calculated from surface $\mathcal{V}_{MB}$.

complicated systems. The availability of accurate surfaces, when combined with detailed classical trajectory studies (augmented where necessary by quantum-mechanical calculations) would of course make it possible to predict the results of even the most detailed and refined molecular beam experiment. However, the prospect of such a millennium should not inspire fear of obsolescence in the experimentalist, whose atoms and molecules afford him a complete knowledge of both the exact surface and the exact dynamics.

ACKNOWLEDGMENTS

The collaboration of Dr. L. M. Raff and Dr. M. Godfrey on many aspects of the trajectory calculations reviewed here is gratefully acknowledged. Partial support for the work was provided by a contract with the Atomic Energy Commission.

REFERENCES

Airey, J. R., E. F. Greene, G. P. Reck, and J. Ross (1967). *J. Chem. Phys.* **45**, 3295.

Balint, G., and M. Karplus (1967). Unpublished calculations.

Bernstein, R. B. (1967). In M. R. C. McDowell, ed., *Atomic Collision Processes.* Amsterdam: North-Holland Publishing Co.

Bernstein, R. B., and J. T. Muckerman (1967). In I. Prigogine, ed., *Advances in Chemical Physics*, vol. 12. In press.

Beuhler, R. J., Jr., R. B. Bernstein, and K. H. Kramer (1966). *J. Am. Chem. Soc.* **88**, 5331.

Blais, N. C., and D. L. Bunker (1962). *J. Chem. Phys.* **37**, 2713. (See also D. L. Bunker and N. C. Blais, *ibid.*, **41**, 2377 (1964) for a three-dimensional study that is not particularly concerned with K,CH_3I.)

Brooks, P. R., and E. M. Jones (1966). *J. Chem. Phys.* **45**, 3449.

Bunker, D. L. (1962). *Nature* **194**, 1277.

Conroy, H., and B. L. Bruner (1965). *J. Chem. Phys.* **42**, 4047.

Datz, S., and R. E. Minturn (1964). *J. Chem. Phys.* **41**, 1153.

Düren, R., R. Feltgen, W. Gaide, R. Helbing, and H. Pauly (1965). *Phys. Letters* **18,** 282.

Düren, R., and C. Schlier (1965). *Discussions Faraday Soc.* **40**, 56.

Edminston, C., and M. Krauss (1965). *J. Chem. Phys.* **42**, 1119.

Entemann, E. A., and D. R. Herschbach (1967). Private communication.

Glasstone, S., K. J. Laidler, and H. Eyring (1949). *The Theory of Rate Processes.* New York: McGraw-Hill.

Glendenning, N. K. (1963). *Annual Rev. Nuclear Sci.* **13**, 191.

Greene, E. F., A. L. Moursund, and J. Ross (1966). In I. Prigogine, ed., *Advances in Chemical Physics*, vol. 10, p. 135.

Herm, R. R., and D. R. Herschbach (1965). *J. Chem. Phys.* **43**, 2139.

Herschbach, D. R. (1962). *Discussions Faraday Soc.* **33**, 149.

Herschbach, D. R. (1966). In I. Prigogine, ed., *Advances in Chemical Physics*, vol. 10, p. 319.

Herschbach, D. R. (1967). Private Communication.

Karplus, M., and M. Godfrey (1966). *J. Am. Chem. Soc.* **88**, 5332.

Karplus, M., R. N. Porter, and R. D. Sharma (1964). *J. Chem. Phys.* **40**, 2033.

Karplus, M., and L. M. Raff (1964). *J. Chem. Phys.* **41**, 1267.

Pauly, H., and J. P. Toennies (1965). In D. R. Bates and I. Estermann, ed., *Advances in Atomic and Molecular Physics*, vol. I. New York: Academic Press.

Polanyi, J. C. (1959). *J. Chem. Phys.* **31**, 1338.

Raff, L. M., and M. Karplus (1966). *J. Chem. Phys.* **44**, 1212.
Shavitt, I., R. M. Stevens, F. Minn, and M. Karplus (1967). In press.
Smith, F. T. (1959). *J. Chem. Phys.* **31**, 1352.
Wilson, K. R., G. H. Kwei, J. A. Norris, R. R. Herm, J. H. Birely, and D. R. Herschbach (1964). *J. Chem. Phys.* **41**, 1154.

THIS
CALIFORNIA

[Reprint from the Journal of the American Chemical Society, 53, 1367 (1931).]

The Nature of the Chemical Bond. Application of Results Obtained from the Quantum Mechanics and from a Theory of Paramagnetic Susceptibility to the Structure of Molecules

---

By Linus Pauling

the parts s, p_x, p_y, p_z of the eigenfunctions depending on θ and φ, normalized to 4π, are

$$\left.\begin{aligned} s &= 1 \\ p_x &= \sqrt{3}\sin\theta\cos\varphi \\ p_y &= \sqrt{3}\sin\theta\sin\varphi \\ p_z &= \sqrt{3}\cos\theta \end{aligned}\right\} \qquad (2)$$

Absolute values of s and p_x are represented in the xz plane in Figs. 1 and 2. s is spherically symmetrical, with the value 1 in all directions. $|p_x|$ consists of two spheres as shown (the x axis is an infinite symmetry axis), with the maximum value $\sqrt{3}$ along the x axis. $|p_y|$ and $|p_z|$ are similar, with maximum values of $\sqrt{3}$ along the y and z axis, respectively. From Rule 5 we conclude that *p electrons will form stronger bonds than s electrons*, and that *the bonds formed by p electrons in an atom tend to be oriented at right angles to one another.*

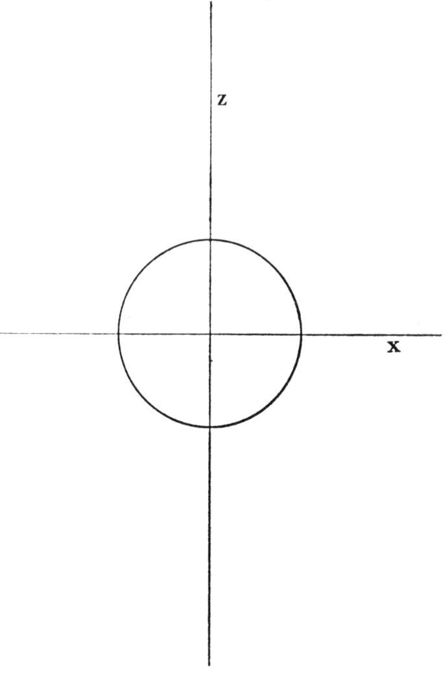

Fig. 1.—Polar graph of 1 in the xz plane, representing an s eigenfunction.

The second conclusion explains several interesting facts. Normal oxygen, in the state $2s^2 2p^4\ ^3P$, contains two unpaired p electrons. When an atom of oxygen combines with two of hydrogen, a water molecule will result in which the angle formed by the three atoms is 90°, or somewhat larger because of interaction of the two hydrogen atoms. It has been long known from their large electric moment that water molecules have a kinked rather than a collinear arrangement of their atoms, and attempts have been made to explain this with rather unsatisfactory calculations based on an ionic structure with strong polarization of the oxygen anion in the field of the protons. The above simple explanation results directly from the reasonable assumption of an electron-pair bond structure and the properties of tesseral harmonics.

It can be predicted that H_2O_2, with the structure $\begin{matrix}:\ddot{O}:\ddot{O}:\\ \ddot{}\,\mathrm{H}\ \ \ddot{}\,\mathrm{H}\end{matrix}$ involving bonds of p electrons, also consists of kinked rather than collinear molecules.

Nitrogen, with the normal state $2s^2 2p^3\ ^4S$, contains three unpaired p

electrons, which can form bonds at about 90° from one another with three hydrogen atoms. The ammonia molecule, with the resulting pyramidal structure, also has a large electric moment.

The crystal skutterudite, $Co_4^{3+}(As_4^{4-})_3$, contains As_4^{4-} groups with a square configuration, corresponding to the structure $\left[\begin{array}{c} :\ddot{\mathrm{As}}:\ddot{\mathrm{As}}: \\ :\ddot{\mathrm{As}}:\ddot{\mathrm{As}}: \end{array}\right]^{4-}$ This complex has bond angles of exactly 90°.

In the above discussion it has been assumed that the type of quantization has not been changed, and that s and p eigenfunctions retain their identity. This is probably true for H_2O and H_2O_2, and perhaps for NH_3 and As_4^{4-} also. A discussion of the effect of change of quantization on bond angles is given in a later section.

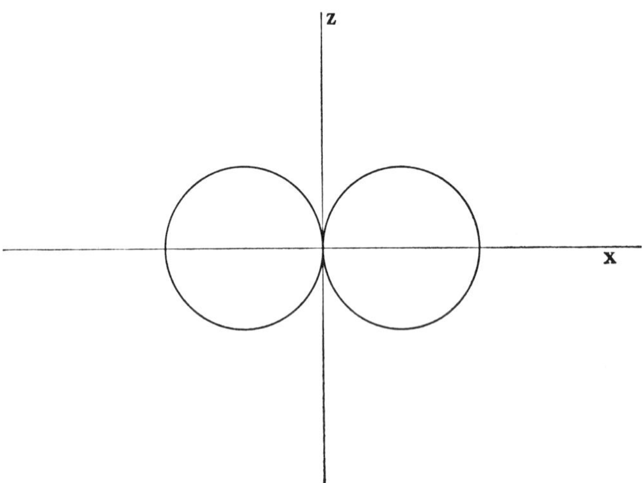

Fig. 2.—Polar graph of $|\sqrt{3}\sin\theta|$ in the xz plane, representing the p_x eigenfunction.

Transition from Electron-Pair to Ionic Bonds. The Hydrogen Bond.—In case that the symmetry character of an electron-pair structure and an ionic structure for a molecule are the same, it may be difficult to decide between the two, for the structure may lie anywhere between these extremes. The zeroth-order eigenfunction for the two bond electrons for a molecule MX (HF, say, or NaCl) with a single electron-pair bond would be

$$\Psi_{MX} = \frac{\psi_M(1)\,\psi_X(2) + \psi_M(2)\,\psi_X(1)}{\sqrt{2 + 2S^2}} \qquad (3)$$

in which $S = \int \psi_M(1)\,\psi_X{}^*(1)\,d\tau_1$. The eigenfunction for a pure ionic state would be

$$\Psi_{M^+X^-} = \psi_X(1)\,\psi_X(2) \qquad (4)$$

In certain cases one of these might approximate the correct eigenfunction closely. In other cases, however, it would be necessary to consider combinations of the two, namely

$$\Psi_+ = a\,\Psi_{MX} + \sqrt{1-a^2}\,\Psi_{M^+X^-}$$

and

$$\Psi_- = \sqrt{1-a^2}\Psi_{MX} - a\,\Psi_{M^+X^-} \quad (5)$$

For a given molecule and a given internuclear separation a would have a definite value, such as to make the energy level for Ψ_+ lie as low as possible. If a happens to be nearly 1 for the equilibrium state of the molecule, it would be convenient to say that the bond is an electron-pair bond; if a is nearly zero, it could be called an ionic bond. This definition is somewhat unsatisfactory in that it does not depend on easily observable quantities. For example, a compound which is ionic by the above definition might dissociate adiabatically into neutral atoms, the value of a changing from nearly zero to unity as the nuclei separate, and it would do this in case the electron affinity of X were less than the ionization potential of M. HF is an example of such a compound. There is evidence, given below, that the normal molecule approximates an ionic compound; yet it would dissociate adiabatically into neutral F and H.[13]

But direct evidence regarding the value of a can sometimes be obtained. The hydrogen bond, discovered by Huggins and by Latimer and Rodebush, has been usually considered as produced by a hydrogen atom with two electron-pair bonds, as in $\left[:\ddot{\underset{..}{F}}:H:\ddot{\underset{..}{F}}:\right]^-$. It was later pointed out[1] that this is not compatible with the quantum mechanical rules, for hydrogen can have only one unpaired 1*s* electron, and outer orbits are so much less stable that strong bonds would not be formed. With an ionic structure, however, we would expect H^+F^- to polymerize and to add on to F^-, to give H_6F_6 and $[F^-H^+F^-]^-$; moreover, the observed coördination number 2 is just that predicted[14] from the radius ratio 0. Hence the observation that hydrogen bonds are formed with fluorine supports an ionic structure for HF. Hydrogen bonds are not formed with chlorine, bromine, and iodine, so that the bonds in HCl, HBr, and HI are to be considered as approaching the electron-pair type.

Hydrogen bonds are formed to some extent by oxygen ($(H_2O)_x$, ice, etc.) and perhaps also in some cases by nitrogen. The electrostatic structure for the hydrogen bond explains the observation that only these atoms of high electron affinity form such bonds, a fact for which no explanation was given by the older conception. It is of interest that there is considerable

[13] There would, however, be a certain probability, dependent on the nature of the eigenfunctions, that actual non-adiabatic dissociation would give ions rather than atoms, and this might be nearly unity, in case the two potential curves come very close to one another at some point. See I. v. Neumann and E. Wigner, *Physik. Z.*, **30**, 467 (1929).

[14] Linus Pauling, THIS JOURNAL, **51**, 1010 (1929).

evidence from crystal structure data for $[OHO]^{=}$ groups. In many crystals containing H and O, including topaz,[15] $Al_2SiO_4(F,OH)_2$; diaspore,[16] $AlHO_2$; goethite,[16] $FeHO_2$; chondrodite,[17] $Mg_5Si_2O_8(F,OH)_2$; etc., the sum of the strengths of the electrostatic bonds from all cations (except hydrogen) to an anion is either 2 or 1, indicating, according to the electrostatic valence rule,[14] the presence of $O^{=}$ and of F^- or $(OH)^-$, respectively. But in some crystals, including[18] KH_2PO_4; staurolite,[19] $H_2FeAl_4Si_2O_{12}$; and lepidocrocite,[16] $FeHO_2$, the sum of bond strengths is 2 or $^3/_2$, the latter value occurring twice for each H; the electrostatic valence rule in these cases supports the assumption of $[O^{=}H^{+}O^{=}]^{=}$ groups, the hydrogen ion contributing a bond of strength $^1/_2$ to each of two oxygen ions.

In other cases, discussed below, the lowest electron-pair-bond structure and the lowest ionic-bond structure do not have the same multiplicity, so that (when the interaction of electron spin and orbital motion is neglected) these two states cannot be combined, and a knowledge of the multiplicity of the normal state of the molecule or complex ion permits a definite statement as to the bond type to be made.

Change in Quantization of Bond Eigenfunctions.—A normal carbon atom, in the state $2s^22p^2\ ^3P$, contains only two unpaired electrons, and can hence form no more than two single bonds or one double bond (as in CO, formed from a normal carbon atom and a normal oxygen atom). But only about 1.6 v. e. of energy is needed to excite a carbon atom to the state $2s2p^3\ ^5S$, with four unpaired electrons, and in this state the atom can form four bonds. We might then describe the formation of a substituted methane CRR′R″R‴ in the following way. The radicals R, R′, and R″, each with an unpaired electron, form electron-pair bonds with the three *p* electrons of the carbon atom, the bond directions making angles of 90° with one another. The fourth radical R‴ then forms a weaker bond with the *s* electron, probably at an angle of 125° with each of the other bonds. This would give an unsymmetrical structure, with non-equivalent bonds, and considerable discussion has been given by various authors to the difference in the carbon bonds due to *s* and *p* electrons. Actually the foregoing treatment is fallacious, for the phenomenon of change in quantization of the bond eigenfunctions, first discussed in the note referred to before,[1] leads simply and directly to the conclusion that *the four bonds formed by a carbon atom are equivalent and are directed toward tetrahedron corners.*

The importance of *s*, *p*, *d*, and *f* eigenfunctions for single atoms and ions

[15] Linus Pauling, *Proc. Nat. Acad. Sci.*, **14,** 603 (1928); N. A. Alston and J. West, *Z. Krist.*, **69,** 149 (1928).

[16] Unpublished investigation in this Laboratory.

[17] W. L. Bragg and J. West, *Proc. Roy. Soc.* (London), **A114,** 450 (1927); W. H. Taylor and J. West, *ibid.*, **A117,** 517 (1928).

[18] J. West, *Z. Krist.*, **74,** 306 (1930).

[19] St. Naray-Szabo, *ibid.*, **71,** 103 (1929).

results from the fact that the interaction of one electron with the nucleus and other electrons can be represented approximately by a non-Coulombian central field, so that the wave equation can be separated in polar coördinates r, θ, and φ, giving rise to eigenfunctions involving tesseral harmonics such as those in Equation 1. The deeper penetration of s electrons within inner shells causes them to be more tightly bound than p electrons with the same total quantum number. If an atom approaches a given atom, forming a bond with it, the interaction between the two can be considered as a perturbation, and the first step in applying the perturbation theory for a degenerate system consists in finding the correct zeroth-order eigenfunctions for the perturbation, one of which is the eigenfunction which will lead to the largest negative perturbation energy. This will be the one with the largest values along the bond direction. The correct zeroth-order eigenfunctions must be certain normalized and mutually orthogonal linear aggregates of the original eigenfunctions. If the perturbation is small, the s eigenfunction cannot be changed, and the only combinations which can be made with the p eigenfunctions are equivalent merely to a rotation of axes. But in case the energy of interaction of the two atoms is greater than the difference in energy of an s electron and a p electron (or, if there are originally two s electrons present, as in a normal carbon atom, of twice this difference), hydrogen-like s and p eigenfunctions must be grouped together to form the original degenerate state, and the interaction of the two atoms together with the deviation of the atomic field from a Coulombian one must be considered as the perturbation, with the former predominating. The correct zeroth-order bond eigenfunctions will then be those orthogonal and normalized linear aggregates of both the s and p eigenfunctions which would give the strongest bonds according to Rule 5.

A rough criterion as to whether the quantization is changed from that in polar coördinates to a type giving stronger bond eigenfunctions is thus that the possible bond energy be greater than the s–p (or, if d eigenfunctions are also involved, s–d or p–d) separation.[20]

This criterion is satisfied for quadrivalent carbon. The energy difference of the states[21] $2s^2 2p^2\ ^3P$ and $2s2p^3\ ^3P$ of carbon is 9.3 v. e., and a similar value of about 200,000 cal. per mole is found for other atoms in the first row of the periodic system. The energy of a single bond is of the order of 100,000 cal. per mole. Hence a carbon atom forming four bonds would certainly have changed quantization, and even when the bond energy must be divided between two atoms, as in a diamond crystal, the criterion is sufficiently well satisfied. The same results hold for quadrivalent

[20] This criterion was expressed in Ref. 1.

[21] States with the same multiplicity should be compared, for increase in multiplicity decreases the term value, the difference between $2s^2 2p^2\ ^3P$ and $2s2p^3\ ^5S$ being only about 1.6 v. e., as mentioned above.

nitrogen, a nitrogen *ion* in the state N^+ $2s2p^3$ ^5S forming four bonds, as in $(NH_4)^+$, $N(CH_3)_4^+$, etc. But for bivalent oxygen there is available only about 200,000 cal. per mole bond energy, and the s–p separation for two s electrons corresponds to about 400,000 cal. per mole, so that it is very probable that the oxygen bond eigenfunctions in H_2O, for example, are p eigenfunctions, as assumed in a previous section. Trivalent nitrogen is a border-line case; the bond energy of about 300,000 cal. per mole is sufficiently close to the s–p energy of 400,000 cal. per mole to permit the eigenfunctions to be changed somewhat, but not to the extent that they are in quadrivalent carbon and nitrogen.

It may be pointed out that the s–p separation for atoms in the same column of the periodic table is nearly constant, about 200,000 cal. per mole for one s electron. The bond energy decreases somewhat with increasing atomic number. Thus the energies of a bond in the compounds H_2O, H_2S, H_2Se, and H_2Te, calculated from thermochemical and band spectral data, are 110,000, 90,000, 73,000, and 60,000 cal. per mole, respectively. Hence we conclude that if quantization in polar coördinates is not broken for a light atom on formation of a compound, it will not be broken for heavier atoms in the same column of the periodic system. The molecules H_2S, H_2Se, and H_2Te must accordingly also have a non-linear structure, with bond angles of 90° or slightly greater.

Let us now determine the zeroth-order eigenfunctions which will form the strongest bonds for the case when the s–p quantization is broken. The dependence on r of s and p hydrogen-like eigenfunctions is not greatly different,[22] and it seems probable that the effect of the non-Coulombian field would decrease the difference for actual atoms. We may accordingly assume that $R_{n0}(r)$ and $R_{n1}(r)$ are effectively the same as far as bond formation is concerned, so that the problem of determining the bond eigenfunctions reduces to a discussion of the θ, φ eigenfunctions of Equation 1. Arbitrary sets of θ, φ eigenfunctions formed from s, p_x, p_y, and p_z are given by the expressions

$$\left.\begin{aligned} \psi_1 &= a_1s + b_1p_x + c_1p_y + d_1p_z \\ \psi_2 &= a_2s + b_2p_x + c_2p_y + d_2p_z \\ \psi_3 &= a_3s + b_3p_x + c_3p_y + d_3p_z \\ \psi_4 &= a_4s + b_4p_x + c_4p_y + d_4p_z \end{aligned}\right\} \qquad (6)$$

in which the coefficients a_1, etc., are restricted only by the orthogonality and normalization requirements

$$\int \psi_i^2 d\tau = 1 \quad \text{or } a_i^2 + b_i^2 + c_i^2 + d_i^2 = 1 \qquad i = 1,2,3,4 \qquad (7a)$$

and

$$\int \psi_i\psi_k d\tau = 0 \quad \text{or } a_ia_k + b_ib_k + c_ic_k + d_id_k = 0 \quad i,k = 1,2,3,4 \quad i \neq k \quad (7b)$$

[22] See the curves given by Linus Pauling, *Proc. Roy. Soc.* (London), **A114,** 181 (1927), or A. Sommerfeld, "Wellenmechanischer Ergänzungsband," p. 88.

From Rule 5 the best bond eigenfunction will be that which has the largest value in the bond direction. This direction can be chosen arbitrarily for a single bond. Taking it along the x axis, it is found that the best single bond eigenfunction is[23]

$$\psi_1 = \frac{1}{2}s + \frac{\sqrt{3}}{2}p_x \quad \text{(8a)}$$

with a maximum value of 2, considerably larger than that 1.732 for a p eigenfunction. A graph of this function in the xz plane is shown in Fig. 3.

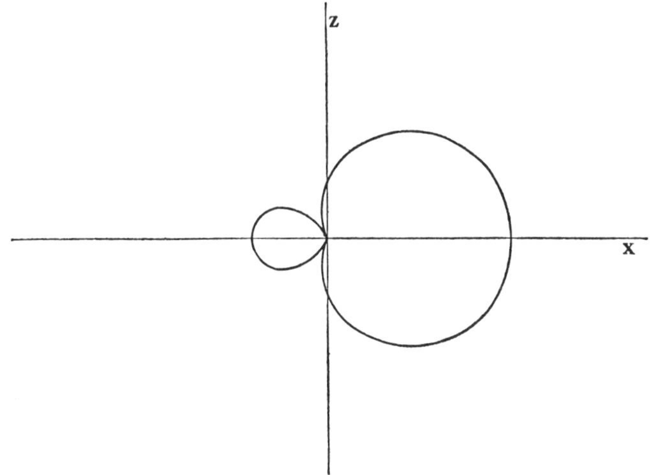

Fig. 3.—Polar graph of $\left| {}^1/_2 + {}^3/_2 \sin\theta \right|$ in the xz plane, representing a tetrahedral eigenfunction, the best bond eigenfunction which can be formed from s and p eigenfunctions.

A second bond can be introduced in the xz plane. The best eigenfunction for this bond is found to be

$$\psi_2 = \frac{1}{2}s - \frac{1}{2\sqrt{3}}p_x + \frac{\sqrt{2}}{\sqrt{3}}p_z \quad \text{(8b)}$$

[23] It is easily shown with the use of the method of undetermined multipliers that the eigenfunction with the maximum value in the direction defined by the polar angles θ_0, φ_0 has as coefficients of the initial eigenfunctions quantities proportional to $\psi_k(\theta_0,\varphi_0)$, and that the maximum value is itself equal to $\{\Sigma_k[\psi_k(\theta_0,\varphi_0)]^2\}^{1/2}$. For let $\psi(\theta,\varphi) = \Sigma_{k=1}^{n} a_k\psi_k(\theta,\varphi)$, with $\Sigma a_k^2 = 1$. We want $\psi(\theta_0,\varphi_0) = \Sigma a_k\psi_k(\theta_0,\varphi_0)$ to be a maximum with respect to variation in the a_k 's. Consider the expression

$$\Lambda = \psi(\theta_0,\varphi_0) - \frac{\lambda}{2}\{\Sigma a_k^2 - 1\} = \Sigma\left\{a_k\psi_k(\theta_0,\varphi_0) - \frac{\lambda}{2}a_k^2\right\} + \frac{\lambda}{2}$$

in which λ is an undetermined multiplier. Then we put

$$\frac{\partial\Lambda}{\partial a_k} = \psi_k(\theta_0,\varphi_0) - \lambda a_k = 0 \text{ or } a_k = \frac{\psi_k(\theta_0,\varphi_0)}{\lambda}, \; k = 1,2\ldots n$$

in which λ has such a value that $\Sigma a_k^2 = 1$; *i. e.*, $\lambda = \{\Sigma[\psi_k(\theta_0,\varphi_0)]^2\}^{1/2}$. $\psi(\theta_0,\varphi_0)$ is itself then equal to $\Sigma[\psi_k(\theta_0,\varphi_0)]^2/\lambda$ or $\{\Sigma[\psi_k(\theta_0,\varphi_0)]^2\}^{1/2}$.

This eigenfunction is equivalent to and orthogonal to ψ_1, and has its maximum value of 2 at $\theta = 19°28'$, $\varphi = 180°$, that is, at an angle of 109°28′ with the first bond, *which is just the angle between the lines drawn from the center to two corners of a regular tetrahedron.* The third and fourth best bond eigenfunctions

$$\psi_3 = \frac{1}{2}s - \frac{1}{2\sqrt{3}}p_x - \frac{1}{\sqrt{6}}p_z + \frac{1}{\sqrt{2}}p_y \tag{8c}$$

and

$$\psi_4 = \frac{1}{2}s - \frac{1}{2\sqrt{3}}p_x - \frac{1}{\sqrt{6}}p_z - \frac{1}{\sqrt{2}}p_y \tag{8d}$$

are also equivalent to the others, and have their maximum values of 2 along the lines toward the other two corners of a regular tetrahedron.

An equivalent set of four tetrahedral eigenfunctions is[24]

$$\left.\begin{aligned} \psi_{111} &= \frac{1}{2}(s + p_x + p_y + p_z) \\ \psi_{1\bar{1}\bar{1}} &= \frac{1}{2}(s + p_x - p_y - p_z) \\ \psi_{\bar{1}1\bar{1}} &= \frac{1}{2}(s - p_x + p_y - p_z) \\ \psi_{\bar{1}\bar{1}1} &= \frac{1}{2}(s - p_x - p_y + p_z) \end{aligned}\right\} \tag{9}$$

These differ from the others only by a rotation of the atom as a whole.

The Tetrahedral Carbon Atom.—We have thus derived the result that *an atom in which only s and p eigenfunctions contribute to bond formation and in which the quantization in polar coördinates is broken can form one, two, three, or four equivalent bonds, which are directed toward the corners of a regular tetrahedron* (Fig. 4). This calculation provides the quantum mechanical justification of the chemist's tetrahedral carbon atom, present in diamond and all aliphatic carbon compounds, and for the tetrahedral quadrivalent nitrogen atom, the tetrahedral phosphorus atom, as in phosphonium compounds, the tetrahedral boron atom in B_2H_6 (involving single-electron bonds), and many other such atoms.

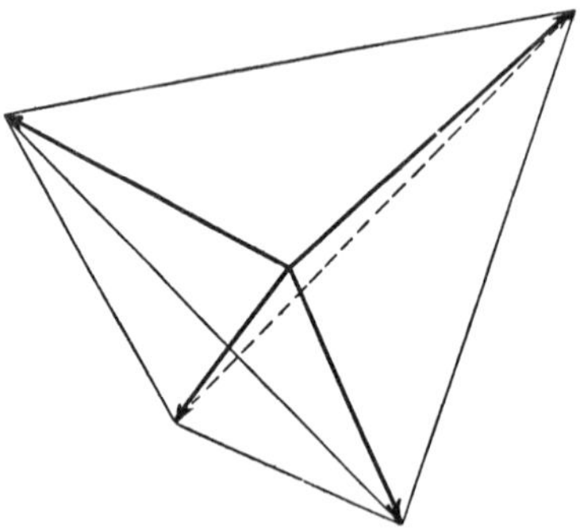

Fig. 4.—Diagram showing relative orientation in space of the directions of the maxima of four tetrahedral eigenfunctions.

Free or Restricted Rotation.—Each of these tetrahedral bond eigen-

[24] It should be borne in mind that the bond eigenfunctions actually are obtained from the expressions given in this paper by substituting for s the complete eigenfunction $\Psi_{n00}(r,\theta,\varphi)$, etc. It is not necessary that the r part of the eigenfunctions be identical; the assumption made in the above treatment is that they do not affect the evaluation of the coefficients in the bond eigenfunctions.

functions is cylindrically symmetrical about its bond direction. Hence the bond energy is independent of orientation about this direction, so that there will be *free rotation about a single bond,* except in so far as rotation is hindered by steric effects, arising from interactions of the substituent atoms or groups.

A double bond behaves differently, however. Let us introduce two substituents in the octants xyz and $\bar{x}\bar{y}z$ of an atom, a carbon atom, say, using the bond eigenfunctions ψ_{111} and $\psi_{\bar{1}\bar{1}1}$. The two eigenfunctions $\psi_{1\bar{1}\bar{1}}$ and $\psi_{\bar{1}1\bar{1}}$ are then left to form a double bond with another such group. Now $\psi_{1\bar{1}\bar{1}}$ and $\psi_{\bar{1}1\bar{1}}$ (or any two eigenfunctions formed from them) are not cylindrically symmetrical about the z axis or any direction, nor are the two eigenfunctions on the other group. Hence the energy of the double bond will depend on the relative orientation of the two tetrahedral carbon atoms, and will be a maximum when the two sets of eigenfunctions show the maximum overlapping. This will occur when the two tetrahedral atoms share an edge (Fig. 5). Thus we derive the result, found long ago by chemists, that there are two stable states for a simple compound involving a double bond, a *cis* and a *trans* state, differing in orientation by 180°. *There is no free rotation about a double bond.*[25]

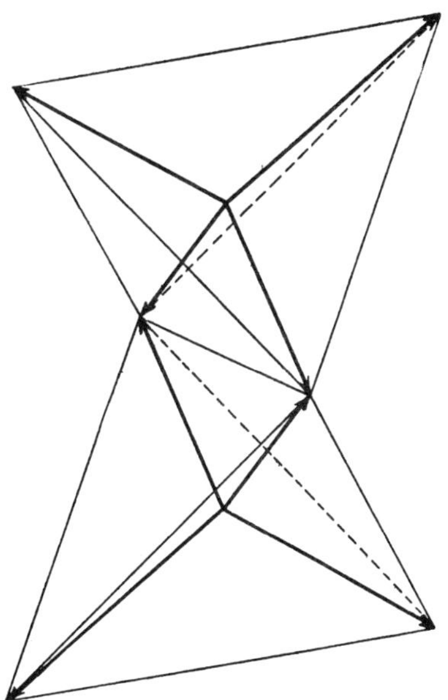

Fig. 5.—Directions of maxima of tetrahedral eigenfunctions in two atoms connected by a double bond.

The three eigenfunctions which would take part in the formation of a triple bond can be made symmetrical about the bond direction, for an atom of the type considered above, with only four eigenfunctions in the outer shell; but since the group attached by the fourth valence lies on the axis of the triple bond, there is no way of verifying the resulting free rotation about the triple bond.

The Angles between Bonds.—The above calculation of tetrahedral angles between bonds when the quantization is changed sets an upper limit on bond angles in doubtful cases, when the criterion is only approximately satisfied. For we can now state that the bond angles in H_2O and NH_3

[25] A discussion of rotation about a double bond on the basis of the quantum mechanics has been published by E. Hückel, *Z. Physik*, **60**, 423 (1930), which is, I feel, neither so straightforward nor so convincing as the above treatment, inasmuch as neither the phenomenon of concentration of the bond eigenfunctions nor that of change in quantization is taken into account.

should lie between 90 and 109°28′, closer to 90° for the first and to 109°28′ for the second compound. The same limits should apply to other atoms with an outer 8-shell (counting both shared and unshared electron pairs). Direct evidence on this point is provided by crystal structure data for non-ionic crystals, given in Table I Every one of the angles given in this table depends on one or more parameters, which have been determined experimentally from observed intensities of x-ray reflections. The probable error in most cases is less than 5°, and in many is only about ±1°. It will be observed that quadrivalent carbon and nitrogen and trivalent nitrogen form bonds at tetrahedral angles, whereas heavier atoms forming only two or three bonds prefer smaller bond angles. The series As, Sb, Bi is particularly interesting. We expect, from an argument given earlier,

TABLE I

ANGLES BETWEEN BONDS, FROM CRYSTAL STRUCTURE DATA[a]

Compound	Atom	Number of bonds	Angles between bonds
$C_6N_4H_{12}$	C	2 C—N, 2 C—H	112° between C—N bonds
$C_6N_4H_{12}$	N	3 N—C	108°
$(NH_2)_2CO$[b]	C	2 single C—N	115° between single bonds
		1 double C═O	
As	As	3	97°
Sb	Sb	3	96°
Bi	Bi	3	94°
Se	Se	2	105°
Te	Te	2	102°
FeS_2[d], MnS_2, CoS_2, NiS_2	S^{++}	1 S—S 3 S—M	103° between S—S and S—M bonds 115° between two M—S bonds
MoS_2[e]	S^+	3 S—Mo	82°
$Co_4(As_4)_3$	As^-	2 As—As	90°
$CaSi_2$	Si	3 Si—Si	103° between Si—Si bonds
HgI_2	I^+	2 Hg—I	103°
GeI_4	Ge	4 Ge—I	109.5°
SnI_4	Sn	4 Sn—I	109.5°
As_4O_6	As	3 As—O	109.5°
	O	2 O—As	109.5°
Sb_4O_6	Sb	3 Sb—O	109.5°
	O	2 O—Sb	109.5°
$NaClO_3$[c]	Cl^{++}	3 Cl—O	109.5°
$KClO_3$	Cl^{++}	3 Cl—O	109.5°
$KBrO_3$	Br^{++}	3 Br—O	109.5°

[a] Data for which no reference is given are from the *Strukturbericht* of P. P. Ewald and C. Hermann. [b] R. W. G. Wyckoff, *Z. Krist.*, **75**, 529 (1930). [c] W. H. Zachariasen, *ibid.*, **71**, 501, 517 (1929). [d] The very small paramagnetic susceptibility of pyrite requires the presence of electron-pair bonds, eliminating an ionic structure $Fe^{++}S_2^{=}$. Angles are calculated for FeS_2, for which the parameters have been most accurately determined. [e] The parameter value (correct value u = 0.371) and interatomic distances for molybdenite are incorrectly given in the *Strukturbericht*.

that the bond eigenfunctions will deviate less and less from pure *p* eigenfunctions in this order, and this evidences itself in a closer approach of the bond angle to 90° in the series. Geometrical effects sometimes affect the bond angles, as in As_4O_6 and Sb_4O_6, where a decrease in the oxygen bond angle would necessarily be accompanied by an increase in that for the other atom, and in molybdenite and pyrite.

Many compounds with tetrahedral structures (diamond, sphalerite, wurzite, carborundum, etc.) are known, in which the four bonds have tetrahedral angles. Tetrahedral atoms in such crystals include C (diamond, SiC), Si, Ge, Sn, Cl^{3+} (in CuCl), Br^{3+}, I^{3+}, O^{++} (in Cu_2O and ZnO), S^{++}, Se^{++}, Te^{++}, N^+ (in AlN), P^+, As^+, Sb^+, Bi^+, $Cu^{\equiv}$, $Zn^{=}$, $Cd^{=}$, $Hg^{=}$, Al^-, Ga^- and In^-.

The Valence of Atoms.—In the last paragraph and in Table I the atoms are represented with electrical charges which are not those usually seen. These charges are obtained by the application of Rule 1, according to which an electron-pair bond is formed by one electron from each of the two atoms (even though as the atoms separate the type of bonding may change in such a way that both electrons go over to one atom). Accordingly in determining the state of ionization of the atoms in a molecule or crystal containing electron-pair bonds each shared electron-pair is to be split between the two atoms. In this way every atom is assigned an electrovalence obtained by the above procedure and a covalence equal to the number of its shared electron-pair bonds.

It is of interest to note that a quantity closely related to the "valence" of the old valence theory is obtained for an atom by taking the algebraic sum of the electrovalence and of the covalence, the latter being given the positive sign for metals and the negative sign for non-metals. For example, oxygen in OH^- is O^- with a covalence of 1, in H_2O it is O with a covalence of 2, in H_3O^+ it is O^+ with a covalence of 3, and in crystalline ZnO it is O^{++} with a covalence of 4; in each case the above rule gives -2 for its valence.

Trigonal Quantization.—We have seen that an atom with *s-p* quantization unchanged will form three equivalent bonds at 90° to one another. If quantization is changed, the three strongest bonds will lie at tetrahedral angles. But increase in the bond angle beyond the tetrahedral angle is not accompanied by a very pronounced decrease in bond strength. Thus three equivalent bond eigenfunctions in a plane, with maxima 120° apart, can be formed

$$\left.\begin{aligned} \psi_1 &= \frac{1}{\sqrt{3}}\,s + \sqrt{\frac{2}{3}}\,p_x \\ \psi_2 &= \frac{1}{\sqrt{3}}\,s - \frac{1}{\sqrt{6}}\,p_x + \frac{1}{\sqrt{2}}\,p_y \\ \psi_3 &= \frac{1}{\sqrt{3}}\,s - \frac{1}{\sqrt{6}}\,p_x - \frac{1}{\sqrt{2}}\,p_y \end{aligned}\right\} \tag{10}$$

and these have a strength of 1.991, only a little less than that 2.000 of tetrahedral bonds (Fig. 6). As a result, we may anticipate that in some cases the bond angles will be larger than 109°28′. The carbonate ion in calcite and the nitrate ion in sodium nitrate are assigned a plane configuration from the results of x-ray investigations. In these ions the oxygen atoms are only 2.25 Å. from one another, so that their characteristic repulsive forces must be large, resisting decrease in the bond angle (the smallest distance observed between oxygen ions in ionic crystals is 2.5 Å.). But repulsion of the oxygen atoms would not be very effective in increasing the bond angle in the neighborhood of 120°, so that we might expect equilibrium to be achieved at a somewhat smaller angle, such as 118°. This would give $CO_3^{=}$ and NO_3^- a

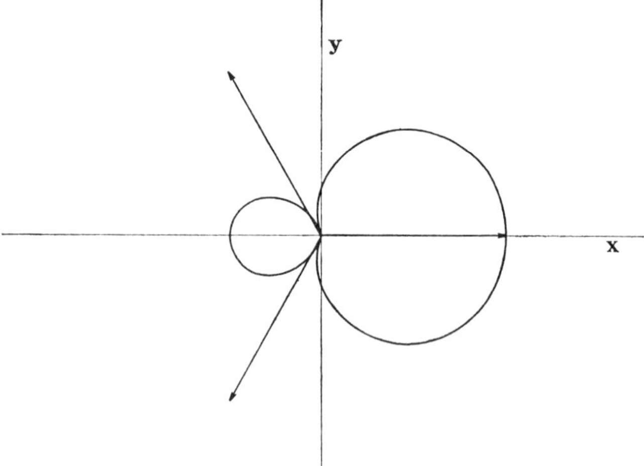

Fig. 6.—Polar graph of $\left|\frac{1}{\sqrt{3}} + \sqrt{2}\cos\varphi\right|$ in the xy plane, representing a trigonal eigenfunction. The maximum directions of the other two equivalent eigenfunctions are also shown.

pyramidal structure, like that of NH_3. There would be two configurations possible for a given orientation of the O_3 plane, one in which the carbon (or nitrogen) atom was a short distance above this plane (taken as horizontal) and one with it below the plane. If there is appreciable interaction between these two, as there will be in case the pyramid is flat, the symmetric and antisymmetric combinations of the two will be the correct eigenfunctions, corresponding to the rapid inversion of the pyramid, with a frequency of the order of magnitude of the vibrational frequency of the complex ion along its symmetry axis. This inversion would introduce an effective symmetry plane normal to the three-fold axis, so that a pyramidal structure with rapid inversion is compatible with the x-ray observations.[26]

[26] Simulation of symmetry by molecules or complex ions in crystals has been discussed by Linus Pauling, *Phys. Rev.*, **36**, 430 (1930).

Thus the x-ray data do not decide between this structure and a truly plane structure. Evidence from another source is at hand, however. A plane $CO_3^=$ or NO_3^- ion should show three characteristic fundamental vibrational frequencies. These have been observed as reflection maxima in the infra-red region. But two of the maxima, at 7μ and 14μ, are double,[27] and this doubling, which is not explicable with a plane configuration, is just that required by a pyramidal structure, the separation of the components giving the frequency of inversion of the pyramid.[28]

In graphite each carbon atom is bound to three others in the same plane; and here the assumption of inversion of a puckered layer is improbable, because of the number of atoms involved. A probable structure is one in which each carbon atom forms two single bonds and one double bond with other atoms. These three bonds should lie in a plane, with angles 109°28′ and 125°16′, which are not far from 120°. Two single bonds and a double bond should be nearly as stable as four single bonds (in diamond), and the stability would be increased by the resonance terms arising from the shift of the double bond from one atom to another. But this problem and the closely related problem of the structure of aromatic nuclei demand a detailed discussion, perhaps along the lines indicated, before they can be considered to be solved.

The Structures of Simple Molecules.—The foregoing considerations throw some light on the structure of very simple molecules in the normal and lower excited states, but they do not permit such a complete and accurate discussion of these questions as for more complicated molecules, because of the difficulty of taking into consideration the effect of several unshared and sometimes unpaired electrons. Often the bond energy is not great enough to destroy s–p quantization, and the interaction between a bond and unshared electrons is more important than between a bond and other shared electrons because of the absence of the effect of concentration of the eigenfunctions.

Let us consider an atom forming a bond with another atom in the direction of the z axis. Then p_z and s form two eigenfunctions designated σ, p_x and p_y two designated π (one with a resultant moment of $+1$ along the z axis, one with -1). If s–p quantization is not broken, the strongest bond will be formed by p_z, and weaker ones by π. If s–p quantization is broken, new eigenfunctions σ_b and σ_o will be formed from s and p_z. In this case the strongest bond is formed by the σ_b eigenfunction, which extends out toward the other atom, weaker ones are formed by $\pi+$ and $\pi-$, and an extremely weak one, if any, by σ_o. We can also predict the stability of

[27] C. Schaefer, F. Matossi and F. Dane, *Z. Physik*, **45**, 493 (1927).

[28] The normal states of these ions are similar to certain excited states of ammonia, which also show doubling. The frequency of inversion of the normal ammonia molecule is negligibly small.

unshared electrons; σ_o, involving s with its greater penetration of the atom core, will be more stable than π.

As examples we may discuss CO, CN, N_2 and NO. CO might be composed of normal or excited atoms, or even of ions. A neutral oxygen atom can form only two bonds. Hence a normal carbon atom, ^3P, which can also form two bonds, is at no disadvantage. We can write the following reaction, using symbols similar to those of Lennard-Jones[29] and Dunkel,[30] whose treatments of the electronic structure of simple molecules have several points of similarity with ours

$$\mathrm{C}\ 2s^22p2p\ {}^3P + \mathrm{O}\ 2s^22p^22p2p\ {}^3P \longrightarrow \mathrm{CO}\ (2\sigma_o{}^2)(2\sigma_o{}^22\pi^2)\{2\sigma_b2\pi + 2\sigma_b2\pi\}{}^1\Sigma$$

$$:\underset{\cdot}{\mathrm{C}}\cdot + \cdot\underset{\cdot}{\ddot{\mathrm{O}}}: \longrightarrow :\mathrm{C}::\ddot{\mathrm{O}}: \qquad \text{Normal state}$$

Here symbols in parentheses represent unshared electrons attached to C and O, respectively, and those in braces represent shared electrons. An excited carbon atom ^5S lies about 1.6 v. e. above the normal state, but can still form only a double bond with oxygen, so that the resultant molecule should be excited. We write

$$\mathrm{C}^*\ 2s2p2p2p\ {}^5S + \mathrm{O}\ 2s^22p^22p2p\ {}^3P \longrightarrow \mathrm{CO}^*\ (2\sigma_o2\pi)(2\sigma_o{}^22\pi^2)\{2\sigma_b2\pi + 2\sigma_b2\pi\}{}^3\Pi \text{ or } {}^1\Pi$$

The resultant states are necessarily Π, for σ_b and one π are used for the bond, leaving on C σ_o and π. These two electrons may or may not pair with one another, giving $^1\Pi$ and $^3\Pi$, respectively. Of these $^3\Pi$ should be the more stable, for the two electrons are attached essentially to one atom, and the rules for atomic spectra should be valid. This is substantiated; the observed excited states $^3\Pi$ and $^1\Pi$ lie at 5.98 and 7.99 v. e., respectively. Another way of considering these three states is the following: to go from $:\mathrm{C}::\ddot{\mathrm{O}}:$ to $\cdot\underset{\cdot}{\mathrm{C}}::\ddot{\mathrm{O}}:$ we lift an electron from the more deeply penetrating σ_o orbit to π; about 6–8 v. e. is needed for this, and the resultant state is either $^3\Pi$ or $^1\Pi$. This viewpoint does not necessitate the discussion of products of dissociation.

CN is closely similar. The normal nitrogen atom, $2s^22p2p2p\ {}^4S$, can form three bonds, and more cannot be formed by an excited neutral atom (with five L electrons), so that there is no reason to expect excitation. But a normal carbon atom can form only a double bond, and an excited carbon atom, only 1.6 v. e. higher, can form a triple bond, which contributes about 3 v. e. more than a double bond to the bond energy. Hence we write

$$\mathrm{C}^*\ 2s2p2p2p\ {}^5S + \mathrm{N}\ 2s^22p2p2p\ {}^4S \longrightarrow \mathrm{CN}\ (2\sigma_o)(2\sigma_o{}^2)\{2\sigma_b2\pi2\pi + 2\sigma_b2\pi2\pi\}{}^2\Sigma$$

$$\cdot\underset{\cdot}{\dot{\mathrm{C}}}\cdot + \cdot\underset{\cdot}{\dot{\mathrm{N}}}: \longrightarrow \cdot\mathrm{C}:::\mathrm{N}: \qquad \text{Normal state}$$

The first excited state of the molecule, $:\mathrm{C}::\underset{\cdot}{\mathrm{N}}:$, is built from normal atoms, and has the term symbol $^2\Pi$. It lies 1.78 v. e. above the normal state.

[29] J. E. Lennard-Jones, *Trans. Faraday Soc.*, **25,** 668 (1929).

[30] M. Dunkel, *Z. physik. Chem.*, **B7,** 81 (1930).

Two normal nitrogen atoms form a normal molecule with a triple bond.

$$2\mathrm{N}\ 2s^22p2p2p\ ^4S \longrightarrow \mathrm{N}_2(2\sigma_o^2)(2\sigma_o^2)\{2\sigma_b2\pi2\pi + 2\sigma_b2\pi2\pi\}\,^1\Sigma$$

$$:\dot{\underset{\cdot}{\mathrm{N}}}\cdot + \cdot\dot{\underset{\cdot}{\mathrm{N}}}: \longrightarrow :\mathrm{N}:::\mathrm{N}: \qquad \text{Normal state}$$

All other states lie much higher.

A normal oxygen atom and a normal nitrogen atom form a normal NO molecule with a double bond.

$$\mathrm{N}\ 2s^22p2p2p\ ^4S + \mathrm{O}\ 2s^22p^22p2p\ ^3P \longrightarrow \mathrm{NO}(2\sigma_o^22\pi)(2\sigma_o^22\pi^2)\{2\sigma_b2\pi + 2\sigma_b2\pi\}\,^2\Pi$$

$$:\dot{\underset{\cdot}{\mathrm{N}}}\cdot + \cdot\ddot{\underset{\cdot}{\mathrm{O}}}: \longrightarrow :\underset{\cdot}{\mathrm{N}}::\underset{\cdot\cdot}{\mathrm{O}}: \qquad \text{Normal state}$$

This treatment sometimes fails for symmetrical molecules. Thus $:\ddot{\mathrm{O}}::\ddot{\mathrm{O}}:$ $^1\Sigma$ would be predicted for the normal state of O_2, whereas the observed normal state, $^3\Sigma$, lies 1.62 v. e. below this. It seems probable that the additional degeneracy arising from the identity of the two atoms gives rise to a new type of bond, the *three-electron bond*, and that in normal O_2 there are one single bond and two three-electron bonds, $:\mathrm{O}\overset{\cdot\cdot\cdot}{\underset{\cdot\cdot\cdot}{:}}\mathrm{O}:$, $^3\Sigma$; a definite decision regarding this question must await a detailed quantum-mechanical treatment. Evidence regarding the oxygen–oxygen single bond is provided by O_4, with the square structure $\begin{matrix}:\ddot{\mathrm{O}}:\ddot{\mathrm{O}}:\\ :\ddot{\mathrm{O}}:\ddot{\mathrm{O}}:\end{matrix}$. The 90° bond angles are expected, since quantization in s and p eigenfunctions is not changed. The equality in energy of O_4 and $2O_2$ leads to an energy of 58,000 cal. per mole per single bond in O_4; the difference between this value and that for a carbon–carbon single bond (100,000 cal.) shows the greater bond-forming power of tetrahedral eigenfunctions over p eigenfunctions. Ozone, which very probably has the symmetrical arrangement $\begin{matrix} & \dot{\mathrm{O}} & \\ \cdot\mathrm{O} & : & \mathrm{O}\cdot \end{matrix}$, has 60° bond angles, and this distortion from the most favorable bond angle of 90° shows up in the bond energy, for the heat of formation of −34,000 cal. per mole leads to 47,000 cal. per mole per single bond, a decrease of 11,000 cal. over the favored O_4 bonds.

For some polyatomic molecules predictions can be made regarding the atomic arrangement from a knowledge of the electronic structure or *vice versa.* Thus $\cdot\mathrm{C}:::\mathrm{N}:$ $^2\Sigma$ can form a bond through the unpaired σ_o electron of carbon, and this bond will extend along the CN axis. Hence the molecules $\mathrm{H}:\mathrm{C}:::\mathrm{N}:$, $:\mathrm{N}:::\mathrm{C}:\mathrm{C}:::\mathrm{N}:$ and $:\ddot{\underset{\cdot\cdot}{\mathrm{Cl}}}:\mathrm{C}:::\mathrm{N}:$ should be linear. This is verified by band spectral data.[31] The isocyanides, RNC, such as H_3CNC, may be given either a triple or a double bond structure: $\mathrm{R}:\mathrm{N}:::\mathrm{C}:$ or $\mathrm{R}:\underset{\cdot\cdot}{\mathrm{N}}::\mathrm{C}:$. The first of these is built of the ions N^+ ^5S and C^- ^4S, which may be an argument in favor of the second structure, built of normal

[31] Private communication from Professor Richard M. Badger of this Laboratory, who has kindly provided me with much information concerning the results of band spectroscopy.

atoms.[32] A decision between the two alternatives could be made by determining the atomic arrangement of an isocyanide, for the triple bond gives a linear molecule, bond angle 180°, and the double bond a kinked molecule, bond angle between 90 and 109°28′.

The molecules and complex ions containing three kernels and sixteen L electrons form an interesting group. Of these CO_2, formed from excited carbon ${}^5S$ and normal oxygen atoms, would have the structure $:\ddot{O}::C::\ddot{O}:$. The two double bonds make the molecule linear, which is verified by both crystal structure and band spectral data. Crystal structure data also show N_2O to be linear, although it is not known whether or not the molecule has oxygen in the middle or at one end, as first suggested by Langmuir[33] and supported by the kernel-repulsion rule.[34] The known linear arrangement eliminates structures built of neutral atoms, $:\dot{N}:O:\dot{N}:$ and $:\dot{N}::N:\ddot{O}:$, for these have bond angles between 90 and 125°. The structures $:N::N::O:$ and $:N::O::N:$, built from N N^+ O^- and N^- O^{++} N^-, respectively, would both be linear, and so compatible with the known arrangement. An *a priori* decision between them is difficult, although previously advanced arguments favor the unsymmetrical structure. Band spectra should soon decide the question.

The trinitride, cyanate, and isocyanate ions, the first two of which are known[35] to be linear, no doubt have identical electronic structures.

$$:\ddot{N}\cdot^{-}\ {}^{3}P + \cdot\dot{N}\cdot^{+}\ {}^{5}S + \cdot\ddot{N}:^{-} \longrightarrow\ :\ddot{N}::N::\ddot{N}:^{-}$$

or

$$N^- + N^+ + N^- \longrightarrow N_3^- \text{ Trinitride ion}$$
$$N^- + C + O \longrightarrow NCO^- \text{ Cyanate ion}$$

The fulminate ion, CNO^-, probably has a structure intermediate between $:\ddot{C}::N::\ddot{O}:^-$ and $:C:::N:\ddot{O}:^-$; for since these two bond types have the same bond angles and term symbols (${}^1\Sigma$), they can form intermediate structures lying anywhere between the two extremes. Which extreme is the more closely approached could be determined from a study of the bond angles in un-ionized fulminate molecules, such as AgCNO or ONCHgCNO, for the first structure would lead to an angle of 125° between the CNO axis and the metal–carbon bond, the second to an angle of 180°.

Bonds Involving *d*-Eigenfunctions.—When d eigenfunctions as well as s and p can take part in bond formation, the number and variety of bonds which can be formed are increased. Thus with an s, a p and a d subgroup as many as nine bonds can be formed by an atom. It is found from a

[32] Thus W. Heitler and G. Rumer, *Nachr. Ges. Wiss. Göttingen, Math. physik. Klasse*, **7**, 277 (1930), in a paper on the quantum mechanics of polyatomic molecules, discuss only the second structure.

[33] I. Langmuir, This Journal, **41**, 1543 (1919).

[34] Linus Pauling and S. B. Hendricks, *ibid.*, **48**, 641 (1926).

[35] S. B. Hendricks and Linus Pauling, This Journal, **47**, 2904 (1925).

consideration of the eigenfunctions that all cannot be equivalent, but six equivalent bonds extending toward the corners of either a regular octahedron or a trigonal pyramid, four extending toward the corners of a tetrahedron or a square, etc., can be formed; and the strength and mutual orientation of the bonds are determined by the number of d eigenfunctions involved in their formation.

There are five d eigenfunctions in a subgroup with $l = 2$ and with given n. They are

$$\left.\begin{aligned} d_z &= \sqrt{5/4}\,(3\cos^2\theta - 1) \\ d_{y+z} &= \sqrt{15}\sin\theta\cos\theta\cos\varphi \\ d_{x+z} &= \sqrt{15}\sin\theta\cos\theta\sin\varphi \\ d_{x+y} &= \sqrt{15/4}\sin^2\theta\sin 2\varphi \\ d_x &= \sqrt{15/4}\sin^2\theta\cos 2\varphi \end{aligned}\right\} \tag{11}$$

or any set of five orthogonal functions formed by linear combination of these. These functions are not well suited to bond formation. d_{y+z}, d_{x+z} and d_{x+y}, which are similarly related to the x, y and z axes, respectively, have the form shown in Fig. 7. Each eigenfunction has maxima in

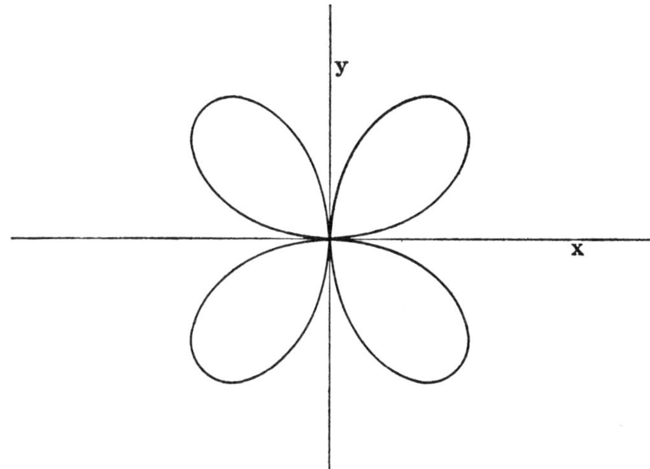

Fig. 7.—Polar graph of $\left|\frac{\sqrt{15}}{2}\sin^2\theta\sin 2\varphi\right|$ in the xy plane, representing the d_{x+} eigenfunction.

four directions. d_x is similar in shape, differing from d_{x+y} only in a rotation of 45° about the z axis. d_z, shown in Fig. 8, has two maxima along the z axis, and a girdle about its waist.

Assuming as before that the dependence on r of the s, p and d eigenfunctions under discussion is not greatly different, the best bond eigenfunctions can be determined by the application of the treatment already applied to s and p alone, with the following results.

The best bond eigenfunction which can be obtained from s, p and d is

$$\frac{1}{3}s + \frac{1}{\sqrt{3}}p_z + \frac{\sqrt{5}}{3}d_z$$

and has a strength of 3. The best two equivalent bond eigenfunctions involving one d eigenfunction

$$\frac{1}{2\sqrt{3}}s + \frac{1}{\sqrt{2}}p_z + \frac{\sqrt{5}}{2\sqrt{3}}d_z \quad \text{and}$$
$$\frac{1}{2\sqrt{3}}s - \frac{1}{\sqrt{2}}p_z + \frac{\sqrt{5}}{2\sqrt{3}}d_z$$

are oppositely directed and have a strength of 2.96.

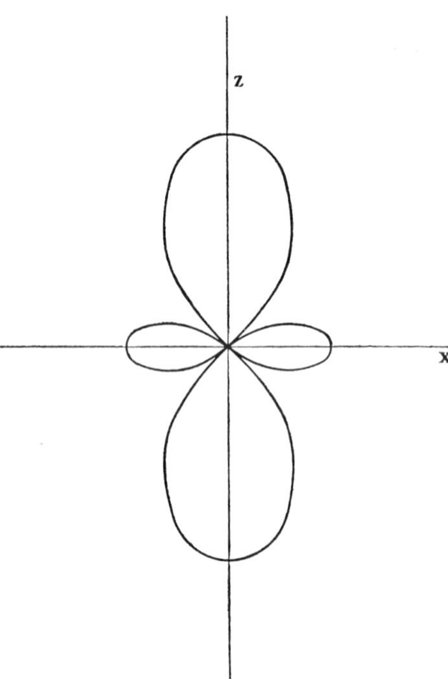

Fig. 8.—Polar graph of $\left|\frac{\sqrt{5}}{2}(3\cos^2\theta - 1)\right|$ in the xz plane, representing the d_z eigenfunction.

The atoms of the transition elements, for which d eigenfunctions need to be considered, are of such a size as usually to have a coördination number of 4 or 6, so that four or six equivalent bond eigenfunctions are here of especial interest. If there is available only one d eigenfunction to be combined with an s and three p eigenfunctions, then no more than five bond eigenfunctions can be formed. One may have the maximum strength 3, in which case the others are weak; or two may be strong and three weak; but *with a single d eigenfunction no more than four strong bonds can be formed, and these lie in a plane.* The fifth bond is necessarily weak. The four equivalent bond eigenfunctions formed from s, p and one d eigenfunction are

$$\left.\begin{aligned} \psi_1 &= \frac{1}{2}s + \frac{1}{2}d_x + \frac{1}{\sqrt{2}}p_x \\ \psi_2 &= \frac{1}{2}s + \frac{1}{2}d_x - \frac{1}{\sqrt{2}}p_x \\ \psi_3 &= \frac{1}{2}s - \frac{1}{2}d_x + \frac{1}{\sqrt{2}}p_y \\ \psi_4 &= \frac{1}{2}s - \frac{1}{2}d_x - \frac{1}{\sqrt{2}}p_y \end{aligned}\right\} \quad (12)$$

One of these is shown in Fig. 9. These all have their maxima in the xy plane, directed toward the corners of a square. The strength of these bond eigenfunctions, 2.694, is much greater than that of the four tetrahedral eigenfunctions formed from s and p alone (2.00). But if three d

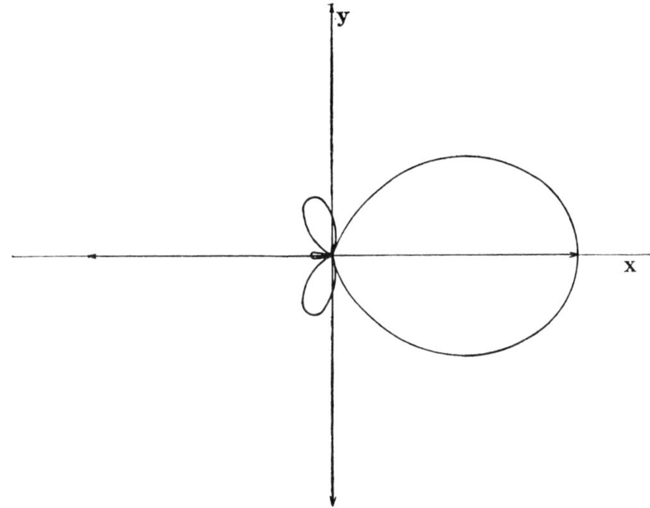

Fig. 9.—Polar graph of $\left| 1/2 + \sqrt{3/2} \cos \varphi + \frac{\sqrt{15}}{4} \cos 2\varphi \right|$ in the xy plane, representing one of the four equivalent dsp^2 bond eigenfunctions. The directions of the maxima of the four are represented by arrows.

eigenfunctions are available, stronger bonds directed toward tetrahedron corners can be formed. The equivalent tetrahedral bond eigenfunctions

$$\psi_{111} = \frac{1}{2} s + \frac{\sqrt{3}}{4\sqrt{2}} (p_x + p_y + p_z) + \frac{\sqrt{5}}{4\sqrt{2}} (d_{y+z} + d_{x+z} + d_{x+y})$$
$$\psi_{1\bar{1}\bar{1}} = \frac{1}{2} s + \frac{\sqrt{3}}{4\sqrt{2}} (p_x - p_y - p_z) + \frac{\sqrt{5}}{4\sqrt{2}} (d_{y+z} - d_{x+z} - d_{x+y})$$
$$\psi_{\bar{1}1\bar{1}} = \frac{1}{2} s + \frac{\sqrt{3}}{4\sqrt{2}} (-p_x + p_y - p_z) + \frac{\sqrt{5}}{4\sqrt{2}} (-d_{y+z} + d_{x+z} - d_{x+y})$$
$$\psi_{\bar{1}\bar{1}1} = \frac{1}{2} s + \frac{\sqrt{3}}{4\sqrt{2}} (-p_x - p_y + p_z) + \frac{\sqrt{5}}{4\sqrt{2}} (-d_{y+z} - d_{x+z} + d_{x+y})$$

have a strength of 2.950, nearly equal to the maximum 3. These leave only two pure d eigenfunctions behind, however, the others being part d and part p. Thus we conclude that if there are three d eigenfunctions available, a transition group element forming four electron-pair bonds will direct them toward tetrahedron corners. Examples of such bonds are provided by $CrO_4^{=}$, $MoO_4^{=}$, etc. Only when one d eigenfunction alone is available will the four bonds lie in a plane. In compounds of bivalent nickel, palladium, and platinum, such as $K_2Ni(CN)_4$, $K_2Pd(CN)_4$, K_2PdCl_4, K_2PtCl_4, etc., there are eight unshared *d* electrons on each metal atom, which occupy four of the five *d* eigenfunctions. Hence the four added atoms or groups lie in a plane at the corners of a square about the metal atom. Such a configuration was assigned to palladous and platinous compounds by Werner because of the existence of apparent *cis* and *trans* compounds, and has been completely substantiated by the x-ray investigation of the chloropalladites and chloroplatinites.[36] The square configuration has not before been attributed to $K_2Ni(CN)_4$; it is supported by the observed isomorphism of the monoclinic crystals $K_2Pd(CN)_4 \cdot H_2O$ and $K_2Ni(CN)_4 \cdot H_2O$, and it will be shown in a following section that it is compatible with the magnetic data.

The non-existence of compounds K_3PtCl_5, etc., is explained by the weak bond-forming power (1.732) of the remaining eigenfunction p_z.

Now if two *d* eigenfunctions are available, six equivalent eigenfunctions

$$\left.\begin{aligned}
\psi_1 &= \frac{1}{\sqrt{6}}s + \frac{1}{\sqrt{2}}p_z + \frac{1}{\sqrt{3}}d_z \\
\psi_2 &= \frac{1}{\sqrt{6}}s - \frac{1}{\sqrt{2}}p_z + \frac{1}{\sqrt{3}}d_z \\
\psi_3 &= \frac{1}{\sqrt{6}}s + \frac{1}{\sqrt{12}}d_z + \frac{1}{2}d_x + \frac{1}{\sqrt{2}}p_x \\
\psi_4 &= \frac{1}{\sqrt{6}}s + \frac{1}{\sqrt{12}}d_z + \frac{1}{2}d_x - \frac{1}{\sqrt{2}}p_x \\
\psi_5 &= \frac{1}{\sqrt{6}}s + \frac{1}{\sqrt{12}}d_z - \frac{1}{2}d_x + \frac{1}{\sqrt{2}}p_y \\
\psi_6 &= \frac{1}{\sqrt{6}}s + \frac{1}{\sqrt{12}}d_z - \frac{1}{2}d_x - \frac{1}{\sqrt{2}}p_y
\end{aligned}\right\} \qquad (13)$$

can be formed. These form strong bonds, of strength 2.923, directed toward the corners of a regular octahedron; and no stronger octahedral bonds can be formed even though more *d* eigenfunctions be available (Figs. 10 and 11). Hence we expect transition group atoms with six or less unshared electrons to form six electron-pair bonds. Examples of such compounds are numerous: $PtCl_6^{=}$, $Fe(CN)_6^{=}$, etc., although the definite assignment of an electron-pair bond structure rather than an ionic structure (as in $FeF_6^{\equiv}$, formed of Fe^{+++} and 6 F^-) can be made only after the discussion of paramagnetic susceptibility.

I have not succeeded in determining whether or not these octahedral eigenfunctions are the strongest six equivalent bond eigenfunctions which can be formed when more than two *d*'s are available. The known structure of molybdenite, MoS_2, suggests that six bonds directed toward the corners of a trigonal prism are stable; but only a small increase in bond strength can possibly be obtained (from 2.923 to not over 3), and the mutual re-

[36] R. G. Dickinson, THIS JOURNAL, **44**, 2404 (1922).

pulsion of the six atoms or groups will in most cases overcome this, if it does exist, and leave the octahedral configuration the stable one.

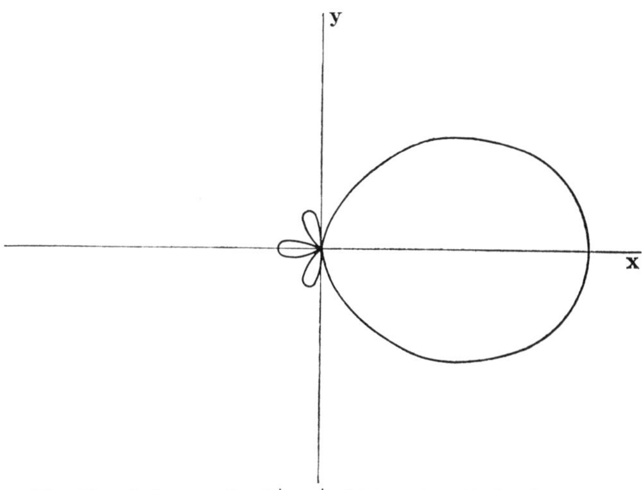

Fig. 10.—Polar graph of $|\psi_3|$ of Equation 13, in the xy plane, representing one of the six equivalent d^2sp^3 bond eigenfunctions (octahedral eigenfunctions).

II. The Magnetic Moments of Molecules and Complex Ions

The theory of the paramagnetic susceptibility of substances has been developed gradually over a long period of years through the efforts of a number of investigators. The theoretical calculation of the magnetic moments of complex molecules and ions has in particular attracted much attention recently, and both theoretical and empirical considerations have been used in developing rules applicable in various cases. The work reported in this paper provides little more than the justification and unification of previously developed rules. This finishing touch is, however, of much significance for the problem of the nature of the chemical bond; for it, in conjunction with the quantum mechanical discussion of the previous sections, permits definite conclusions to be drawn regarding type of bond in many molecules and complex ions from a knowledge of their magnetic moments, and conversely provides the basis for the definite prediction of magnetic moments from a knowledge of the type of bonds and the atomic arrangement.

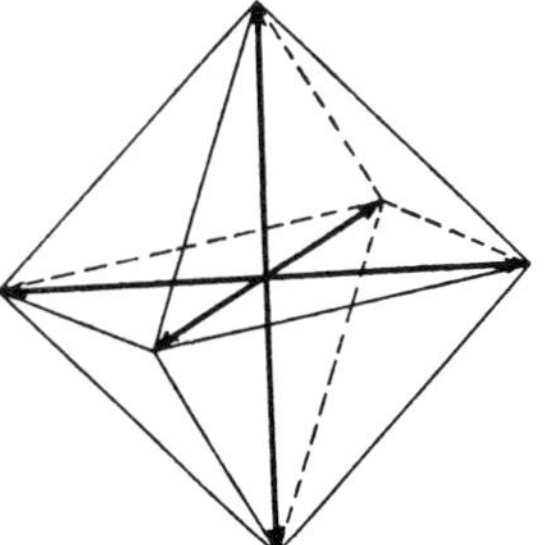

Fig. 11.—Diagram showing relative orientation in space of the directions of the maxima of the octahedral eigenfunctions.

The calculation of the magnetic moments of the rare-earth ions by

Hund[37] in 1926 and of oxygen and nitric oxide by Van Vleck[38] in 1928 were triumphs of the theory of spectra. The magnetic moment of an atom or monatomic ion with Russell–Saunders coupling of the quantum vectors is

$$\mu_J = g\sqrt{J(J+1)}$$

in which g, the Landé splitting factor, is given by

$$g = 1 + \frac{J(J+1) + S(S+1) - L(L+1)}{2J(J+1)}$$

Here L, S, and J are the quantum numbers corresponding to the total orbital angular momentum of the electrons, the total spin angular momentum, and the resultant of these two. Hund predicted values of L, S, and J for the normal states of the rare-earth ions from spectroscopic rules, and calculated μ-values for them which are in generally excellent agreement with the experimental data for both aqueous solutions and solid salts.[39] In case that the interaction between L and S is small, so that the multiplet separation corresponding to various values of J is small compared with kT, Van Vleck's formula[38]

$$\mu_{LS} = \sqrt{4S(S+1) + J(J+1)}$$

is to be used.

But similar calculations for the iron-group ions show marked disagreement with experiment, and many attempts were made to explain the discrepancies. The explanation is simple: *in many condensed systems the perturbing effect of the atoms or molecules surrounding a magnetic atom destroys the contribution of the orbital momentum to the magnetic moment, which is produced entirely by the spin moments of unpaired electrons.*[40]

This conclusion is easily deduced from the consideration of the nature of eigenfunctions giving rise to magnetic moments. In an atom containing unpaired p electrons, say, a component of orbital magnetic moment of $\pm(h/2\pi)\cdot(e/2mc)$ is obtained when an unpaired electron is in a state given by the eigenfunction $p_x \pm i\,p_y$. Now if the perturbing influence of surrounding atoms or molecules is such as to make the perturbation energy for the eigenfunction p_x or p_y or any combination of them other than $p_x \pm i\,p_y$ greater than the field energy, this will be the correct zeroth order eigenfunction, and the atom will show no orbital magnetic moment. In an atom with Russell–Saunders coupling the interaction energy of L and S takes the place of the field energy, so that the criterion to be satisfied in order that the magnetic moment due to L be destroyed is that the perturbation energy due to surrounding atoms and ions be greater than the multiplet separation, which for the iron-group ions is of the order of magnitude of 1 v. e.[41]

---

[37] F. Hund, *Z. Physik*, **33**, 345 (1925).

[38] J. H. Van Vleck, *Phys. Rev.*, **31**, 587 (1928).

[39] The few discrepancies have been accounted for by S. Freed [This Journal, **52**, 2702 (1930)] and J. H. Van Vleck and A. Frank [*Phys. Rev.*, **34**, 1494 (1929), and a paper delivered at the Cleveland meeting of the American Physical Society, December 31, 1930].

[40] This assumption was first made by E. C. Stoner, *Phil. Mag.*, **8**, 250 (1929), in order to account for the observed moments of iron-group ions.

[41] Essentially the same conclusion has been announced by J. H. Van Vleck at the Cleveland meeting of the American Physical Society, December 31, 1930.

If the perturbation function shows cubic symmetry, and in certain other special cases, the first-order perturbation energy is not effective in destroying the orbital magnetic moment, for the eigenfunction $p_x \pm i\, p_y$ leads to the same first-order perturbation terms as p_x or p_y or any other combinations of them. In such cases the higher order perturbation energies are to be compared with the multiplet separation in the above criterion.

In linear molecules only the component of orbital momentum normal to the figure axis is destroyed, that along the figure axis being retained. In non-linear molecules with strong interatomic interactions the concept of orbital angular momentum loses its significance.

The rare-earth ions owe their magnetic moments to an incompleted $4f$ subshell, which lies within an outer shell of $5s$ and $5p$ electrons, and is thus protected from strong perturbations by surrounding atoms. As a consequence the orbital magnetic moment is not destroyed, and the ion is not affected by its environment. But in the iron-group ions and other transition-group ions the incompleted subshell is the outermost one. Hence it is not surprising that the solvent molecules or the surrounding atoms or ions in a complex ion or a crystal interact sufficiently strongly with these atoms or ions to destroy, in whole, or in part, the orbital magnetic moment, leaving the spin moment, with perhaps a small contribution from the orbital moment in border-line cases. We can state with certainty that the formation of electron-pair bonds will destroy the orbital moment.

This greatly simplifies the theory of the magnetic moments of molecules and complex ions. *The magnetic moment of a molecule or complex ion is determined entirely by the number of unpaired electrons, being equal to*

$$\mu_S = 2\sqrt{S(S+1)}$$

in which S is one-half that number. The factor 2 is the g-factor for electron spin.

As a matter of fact, Sommerfeld[42] in 1924, a year before Hund's treatment of the rare-earth ions, noticed that the observed magnetic moments of K^+ and Ca^{++}, Ca^+ (spectroscopic), Ca (spectroscopic), Cr^{3+}, Cr^{++}, Mn^{++}, Fe^{++}, Co^{++}, Ni^{++}, Cu^{++} and Cu^+ are approximately reproduced by the above equation with $S = 0, {}^1/_2, 1, {}^3/_2, 2, {}^5/_2, 2, {}^3/_2, 1, {}^1/_2$ and 0, respectively. But with the development of spectral theory he apparently gave up this simple formula because of lack of a theoretical derivation of it, and it remained for Bose[43] in 1927 to state explicitly the assumption that only S contributes to the moment in these cases, without, however, explaining why L gives no contribution, and for Stoner[40] in 1929 to supply the explanation. The comparison of calculated and observed values is given in Table I. It may be pointed out that S increases to a maximum value of ${}^5/_2$ when the $3d$ subgroup is half filled; Pauli's principle requires that succeeding electrons decrease the spin, so that μ_S is symmetrical about

[42] A. Sommerfeld, "Atombau," 4th ed., p. 639.

[43] D. M. Bose, *Z. Physik*, **43**, 864 (1927).

this point. The agreement with experiment, while much better than for μ_J, is not perfect; ions with more than five $3d$ electrons are found to have moments larger than μ_S, while V^{3+} deviates in the other direction. Bose suggested that perhaps S could in some cases exceed the maximum value allowed by Pauli's principle, but the obviously correct explanation is that the perturbing effect of surrounding atoms is not sufficient completely to destroy the L moment. Hence the observed moment should lie between μ_S and μ_J, which it does in every case.

Since the interaction is not strong enough to destroy the L moment, we conclude that in aqueous solution and in some crystalline salts the atoms[44] Fe^{II}, Co^{III}, Co^{II}, Ni^{II} and Cu^{II} do not form strong electron-pair bonds with H_2O, Cl, or certain other atoms, the bonds instead being ion-dipole or ionic bonds.

The formation of a stable coördination compound involving the four tetrahedral sp^3 eigenfunctions might decrease the L contribution appreciably. It was indeed pointed out by Bose that in the compounds listed in the last column of Table II the observed moments approach more closely the theoretical values μ_S.

The Magnetic Moments of Complexes with Electron-Pair Bonds.—The peculiar magnetic behavior of some complex ions has attracted much attention. $[Fe(CN)_6]^{3-}$ and $[Fe(CN)_6]^{4-}$, for example, have $\mu = 2.0$ and 0.00, respectively, instead of the values 5.9 and 4.9 for Fe^{3+} and Fe^{++} Welo and Baudisch[45] and later Sidgwick and Bose expressed essentially the following rule: the magnetic moment of a complex is the same as that of the atom with the same number of electrons as the central atom of the complex, counting two for each electron-pair bond. Fe^{++} has 24 electrons; adding 12 for the six bonds gives 36, the electron number of krypton, so that the diamagnetism of the ferrocyanide ion is explained. This rule is satisfactory in many cases, but there are also many exceptions. Thus $[Ni(CN)_4]^{=}$ is diamagnetic, although the above rule would make it as paramagnetic as $[Ni(NH_3)_4]^{++}$.

The whole question is clarified when considered in relation to the foregoing quantum mechanical treatment of the electron-pair bond. For the iron-group elements the following rules follow directly from that treatment and from the rules of line spectroscopy.

1. *Bond eigenfunctions for iron-group atoms are formed from the nine eigenfunctions $3d^5$, $4s$ and $4p^3$, as described in preceding sections. One bond eigenfunction is needed for each electron-pair bond.*

2. *The remaining (unshared) electrons are to be introduced into the $3d$ eigenfunctions not involved in bond formation.*

[44] The symbol Fe^{II} is used for bivalent iron, etc., when the type of bond is undetermined.

[45] L. A. Welo and O. Baudisch, *Nature*, **116**, 606 (1925).

TABLE II

MAGNETIC MOMENTS OF IRON-GROUP IONS[a]

Ion	Normal state	μ_J	μ_S	Obs. moment in aqueous soln.	Solid salts, probable coördination number 6		Solid salts, coördination number	
K^+, Ca^{++}, Sc^{3+}, Ti^{4+}	$^1S_0$	0.00	0.00	0.00				
V^{4+}	$^2D_{3/2}$	1.55	1.73	1.7				
V^{3+}	$^3F_2$	1.63	2.83	2.4				
V^{++}, Cr^{3+}	$^4F_{3/2}$	0.78	3.88	3.8–3.9	$Cr_2O_3 \cdot 7H_2O$	3.85		
					$CrCl_3$	3.81		
Cr^{++}, Mn^{3+}	$^5D_0$	0.00	4.90	4.8–4.9				
Mn^{++}, Fe^{3+}	$^6S_{5/2}$	5.91	5.91	5.8	$MnCl_2$	5.75		
					$MnSO_4$	5.87		
					$MnSO_4 \cdot 4H_2O$	5.87		
					$Fe_2(SO_4)_3$	5.86		
					$(NH_4)_2Fe_2(SO_4)_4$	5.86		
Fe^{++}, Co^{3+}	$^5D_4$	6.76	4.90	5.3	$FeCl_2$	5.23	$Fe(N_2H_4)_2Cl_2$	4.87
					$FeCl_2 \cdot 4H_2O$	5.25		
					$FeSO_4$	5.26		
					$FeSO_4 \cdot 7H_2O$	5.25		
					$(NH_4)_2Fe(SO_4)_2 \cdot 6H_2O$	5.25		
Co^{++}	$^4F_{9/2}$	6.68	3.88	5.0–5.2	$CoCl_2$	5.04	$Co(N_2H_4)_2SO_3 \cdot H_2O$	4.31
					$CoSO_4$	5.04–5.25	$Co(N_2H_4)_2(CH_3COO)_2$	4.56
					$CoSO_4.7H_2O$	5.06	$Co(N_2H_4)_2Cl_2$	4.93
					$(NH_4)_2Co(SO_4)_2 \cdot 6H_2O$	5.00		
Ni^{++}	$^3F_4$	5.64	2.83	3.2	$NiCl_2$	3.24–3.42	$Ni(N_2H_4)_2SO_3$	3.20
					$NiSO_4$	3.42	$Ni(N_2H_4)_2(NO_2)_2$	2.80
							$Ni(NH_3)_4SO_4$	2.63
							$Ni(C_2H_4(NH_2)_2)_2(SCN)_2H_2O$	2.63
Cu^{++}	$^2D_{5/2}$	3.56	1.73	1.9–2.0	$CuCl_2$	2.02	$Cu(NH_3)_4(NO_3)_2$	1.82
					$CuSO_4$	2.01	$Cu(NH_3)_4SO_4 \cdot H_2O$	1.81
Cu^+, Zn^{++}	$^1S_0$	0.00	0.00	0.00				

[a] Observed magnetic moments, other than those in the last column, are from "International Critical Tables"

3. *The normal state is the state with the maximum resultant spin S allowed by Pauli's principle.*

These rules apply also to the palladium and platinum groups, the eigenfunctions involved being $4d^5 5s 5p^3$ and $5d^5 6s 6p^3$, respectively.

There are several important types of molecules and complexes to be given separate discussion.

If the bonds are ionic or ion-dipole bonds, the magnetic moments are those of the isolated central ions, given in the first column of moments in Table III. If the complex involves electron-pair bonds formed from sp^3 alone, such as four tetrahedral sp^3 bonds, the magnetic moments are the same, for the five d eigenfunctions are still available for the remaining electrons. The hydrazine and ammonia complexes mentioned above come in this class.

If four strong bonds involving a d eigenfunction are formed (giving a square configuration), only four d eigenfunctions are available for the additional electrons. The magnetic moments are then those given in the second column of the table. Examples of such compounds are $K_2Ni(CN)_4$, $K_2Pd(CN)_4 \cdot H_2O$, K_2PdCl_4, K_2PtCl_4, $K_2Pt(C_2O_4)_2 \cdot 2H_2O$ and $Pt(NH_3)_4SO_4$. With eight unshared d electrons, these should all be diamagnetic. This has been experimentally verified for the first and the last three compounds; data for the others are not available. The square configuration has been experimentally verified for the chloropalladites and chloroplatinites, as mentioned before. It can be predicted that in the $[Pt(C_2O_4)_2 \cdot 2H_2O]^{=}$ complex the two oxalate groups lie in a plane, each attached to the platinum atom by two electron-pair bonds of the type dsp^2. The two water molecules, if attached to the complex, are held by ion-dipole bonds.

In complexes in which the central atom forms a coördinated octahedron of six atoms or groups, the bonds may be any of several types. If they are all ionic or ion-dipole bonds, the moments are those in the first column. If four electron-pair bonds are formed, these must be dsp^2 and lie in a plane (sp^3 gives tetrahedral bonds); the $[Pt(C_2O_4)_2 \cdot 2H_2O]^{=}$ ion is of this type, assuming that the water molecules are part of the complex. The moments are then those of the second column. If six electron-pair bonds are formed, only three d eigenfunctions are left for the additional electrons, giving the magnetic moments of the third column. It is seen that in atoms with three or fewer unshared electrons magnetic data provide no information as to bond type with coördination number six, but that in other cases a definite statement can be made as to the type of bond when magnetic data are available. The observed magnetic moments are collected in Table IV. From them we deduce that trivalent and bivalent manganese, chromium, iron, and cobalt form six strong electron-pair bonds with cyanide groups, and in some cases with other groups, including NH_3, Cl and NO_2.[46] Tri-

[46] An electron-pair bond with a water molecule may perhaps be formed when induced by other strong bond-forming groups in the complex.

TABLE III

PREDICTED MAGNETIC MOMENTS OF COMPLEXES CONTAINING TRANSITION ELEMENTS

			For ion or $4sp^3$ bonds	For 4 dsp^2 bonds	For 6 d^2sp^3 bonds	For 8 d^4sp^3 bonds
K^{I} Ca^{II} Sc^{III} Ti^{IV}, etc.	Rb^{I} Sr^{II} Y^{III} Zr^{IV} Nb^{V} Mo^{VI}	Cs^{I} Ba^{II}—Hf^{IV} Ta^{V} W^{VI}	0.00	0.00	0.00	0.00
V^{IV}	Nb^{IV} Mo^{V}	W^{V}	1.73	1.73	1.73	1.73
V^{III} Cr^{IV}	Mo^{IV} Ru^{VI}	W^{IV} Os^{VI}	2.83	2.83	2.83	0.00
V^{II} Cr^{III} Mn^{IV}	Mo^{III}		3.88	3.88	3.88	
Cr^{II} Mn^{III} Fe^{IV}	Mo^{II} Ru^{IV}	Os^{IV}	4.90	4.90	2.83	
Mn^{II} Fe^{III} Co^{IV}	Ru^{III}	Os^{III} Ir^{IV}	5.91	3.88	1.73	
Fe^{II} Co^{III}	Ru^{II} Rh^{III} Pd^{IV}	Ir^{III} Pt^{IV}	4.90	2.83	0.00	
Co^{II} Ni^{III}	Rh^{II}		3.88	1.73		
Ni^{II}	Rh^{I} Pd^{II} Ag^{III}	Pt^{II} Au^{III}	2.83	0.00		
Cu^{II}			1.73			
Cu^{I} Zn^{II} Ca^{III} Ge^{IV}, etc.	Ag^{I} Cd^{II} In^{III} Sn^{IV} Sb^{V} Te^{VI}	Au^{I} Hg^{II} Tl^{III} Pb^{IV} Bi^{V} Po^{VI}	0.00			

valent iron apparently does not form electron-pair bonds with fluorine (in $[FeF_5 \cdot H_2O]^-$); although investigation of $(NH_4)_3FeF_6$ is to be desired in order to be sure of this conclusion. Ir^{III} and Pt^{IV} form six electron-pair bonds with Cl, NO_2 or NH_3.

TABLE IV

OBSERVED MAGNETIC MOMENTS OF COMPLEXES CONTAINING TRANSITION ELEMENTS[a]

	μ		μ
$K_3[Mn(CN)_6]$	3.01	$[Co(NH_3)_6]Cl_3$	0.00
$K_4[Cr(CN)_6]$	3.3	$[Co(NH_3)_5Cl]Cl_2$	.00
$K_3[Fe(CN)_6]$	2.0	$[Co(NH_3)_4Cl_2]Cl$	.00
$K_4[Mn(CN)_6]$	2.0	$[Co(NH_3)_3(NO_2)_3]$	.00
$K_4[Fe(CN)_6] \cdot 3H_2O$	0.00	$[Co(NH_3)_5H_2O]_2(C_2O_4)_3$	.00
$Na_3[Fe(CN)_5NH_3]$	.00		
$K_3[Co(CN)_6]$	.00	$K_2Ni(CN)_4$	0.00
$(NH_4)_2[FeF_5 \cdot H_2O]$	5.97	$K_2Ni(CN)_4 \cdot H_2O$	.00
$K_4[Mo(CN)_8]$	0.00	K_2PtCl_4	.00
$K_4[W(CN)_8] \cdot 2H_2O$	.00	$K_2Pt(C_2O_4)_2 \cdot 2H_2O$	.00
$Na_3[IrCl_2(NO_2)_4]$	.00	$Pt(NH_3)_4SO_4$	.00
$[Ir(NH_3)_5NO_2]Cl_2$	.00		
$[Ir(NH_3)_4(NO_2)_2]Cl$	.00	$Na_2[Fe(CN)_5NO] \cdot 2H_2O$	.00
$[Ir(NH_3)_3(NO_2)_3]$	.00	$[Ru(NH_3)_4 \cdot NO \cdot H_2O]Cl_3$	.00
$K_2[PtCl_6]$	.00	$[Ru(NH_3)_4 \cdot NO \cdot Cl]Br_2$	.00
$[Pt(NH_3)_6]Cl_4$	.00	$[Co(NH_3)_5NO]Cl_2$	2.81
$[Pt(NH_3)_5Cl]Cl_3$	.00		
$[Pt(NH_3)_4Cl_2]Cl_2$	.00	$Ni(CO)_4$	0.00
$[Pt(NH_3)_3Cl_3]Cl$	.00	$Fe(CO)_5$	.00
$[Pt(NH_3)_2Cl_4]$	.00	$Cr(CO)_6$	.00

[a] Values quoted are from "International Critical Tables" or from W. Biltz, *Z. anorg. Chem.*, **170**, 161 (1928), and D. M. Bose, *Z. Physik*, **65**, 677 (1930). I am indebted to Mr. P. D. Brass for collecting from the literature some of the data in this table.

The moments of complexes containing NO offer a puzzling problem. The diamagnetism of compounds of iron and ruthenium suggests that Fe^{IV} and Ru^{IV} form a double bond with NO, making seven bonds in all, which woud lead to $\mu = 0$. But this structure cannot be applied to $[Co(NH_3)_5NO]Cl_2$, which has a moment corresponding to a triplet state. Further study of such complexes is needed.

The observed diamagnetism of the ions $[Mo(CN)_8]^{4-}$ and $[W(CN)_8]^{4-}$ shows that the central atom forms eight electron-pair bonds, involving the eigenfunctions d^4sp^3 (fourth column of Table III).

The metal carbonyls $Ni(CO)_4$, $Fe(CO)_5$, and $Cr(CO)_6$ are observed to be diamagnetic. This follows from the theoretical discussion if it is assumed that an electron-pair bond is formed with each carbonyl; for the nine eigenfunctions available ($3d^54s4p^3$) are completely filled by the n bonds and $2(9-n)$ additional electrons attached to the metal atom ($n = 4, 5, 6$). The theory also explains the observed composition of these unusual sub-

GUSTAV ALBRECHT
Division of Chemistry and Chemical Engineering
California Institute of Technology
Pasadena, California

Scientific Publications of Linus Pauling

The number assigned each paper indicates its place in the chronological sequence of all the scientific papers, not including the books.

In addition to the publications listed here, Dr. Pauling has written about 100 papers on science and world affairs, including "Science and Peace" (Nobel Prize Lecture for 1962), published in *The Nobel Prizes for 1963* and elsewhere.

I. Papers on the Structure of Crystals

1. The Crystal Structure of Molybdenite. *J. Am. Chem. Soc.* **45**, 1466 (1923). [Roscoe G. Dickinson and L. P.]
2. The Crystal Structure of Magnesium Stannide. *J. Am. Chem. Soc.* **45**, 2777 (1923).
3. The Crystal Structure of Uranyl Nitrate Hexahydrate. *J. Am. Chem. Soc.* **46**, 1615 (1924). [L. P. and Roscoe G. Dickinson]
4. The Crystal Structures of Ammonium Fluoferrate, Fluoaluminate and Oxyfluomolybdate. *J. Am. Chem. Soc.* **46**, 2738 (1924).
5. The Crystal Structures of Hematite and Corundum. *J. Am. Chem. Soc.* **47**, 781 (1925) [L. P. and Sterling B. Hendricks]
6. The Crystal Structure of Barite. *J. Am. Chem. Soc.* **47**, 1026 (1925). [L. P. and Paul H. Emmett]
7. The Crystal Structures of Cesium Tri-Iodide and Cesium Dibromo-Iodide. *J. Am. Chem. Soc.* **47**, 1561 (1925). [Richard M. Bozorth and L. P.]
10. A New Crystal for Wave-Length Measurements of Soft X-Rays. *Proc. Nat. Acad. Sci. U.S.* **11**, 445 (1925). [L. P. and Albert Björkeson]
12. The Crystal Structures of Sodium and Potassium Trinitrides and Potassium Cyanate and the Nature of the Trinitride Group. *J. Am. Chem. Soc.* **47**, 2904 (1925). [Sterling B. Hendricks and L. P.]
17. Über die Kristallstruktur der kubischen Tellursäure. *Z. Krist.* **63**, 502 (1926). [L. Merle Kirkpatrick and L. P.]
18. Die Struktureinheit und Raumgruppensymmetrie von β-Aluminiumoxyd. *Z. Krist.* **64**, 303 (1926). [S. B. Hendricks and L. P.]
19. An X-Ray Study of the Alloys of Lead and Thallium. *J. Am. Chem. Soc.* **49**, 666 (1927). [Edwin McMillan and L. P.]
22. The Sizes of Ions and the Structure of Ionic Crystals. *J. Am. Chem. Soc.* **49**, 765 (1927).
26. The Sizes of Ions and Their Influence on the Properties of Salt-like Compounds. *Z. Krist.* **67**, 377 (1928).

27. The Influence of Relative Ionic Sizes on the Properties of Salt-like Compounds. *J. Am. Chem. Soc.* **50**, 1036 (1928).
28. The Crystal Structure of Brookite. *Z. Krist.* **68**, 239 (1928). [L. P. and J. H. Sturdivant]
30. The Crystal Structure of Potassium Chloroplatinate. *Z. Krist.* **68**, 223 (1928). [F. J. Ewing and L. P.]
31. The Crystal Structure of Topaz. *Proc. Nat. Acad. Sci. U.S.* **14**, 603 (1928).
32. The Coordination Theory of the Structure of Ionic Crystals. In *Festschrift zum 60. Geburtstage Arnold Sommerfelds.* Leipzig: Verlag Hirzel, 1928, pp. 11–17.
33. Note on the Pressure Transitions of the Rubidium Halides. *Z. Krist.* **69**, 35 (1928).
34. The Principles Determining the Structure of Complex Ionic Crystals. *J. Am. Chem. Soc.* **51**, 1010 (1929).
35. Note on the Paper of A. Schröder: Beiträge zur Kenntnis des Feinbaues des Brookits usw. *Z. Krist.* **69**, 557 (1929). [J. H. Sturdivant and L. P.]
36. The Crystal Structure of the *A*-Modification of the Rare Earth Sesquioxides. *Z. Krist.* **69**, 415 (1929).
38. On the Crystal Structure of the Chlorides of Certain Bivalent Elements. *Proc. Nat. Acad. Sci. U.S.* **15**, 709 (1929).
39. The Molecular Structure of the Tungstosilicates and Related Compounds. *J. Am. Chem Soc.* **51**, 2868 (1929).
41. On the Crystal Structure of Nickel Chlorostannate Hexahydrate. *Z. Krist.* **72**, 482 (1930).
42. The Crystal Structure of Pseudobrookite. *Z. Krist.* **73**, 97 (1930).
43. The Structure of Some Sodium and Calcium Aluminosilicates. *Proc. Nat. Acad. Sci. U.S.* **16**, 453 (1930).
44. The Structure of Sodalite and Helvite. *Z. Krist.* **74**, 213 (1930).
45. Note on the Lattice Constant of Ammonium Hexafluoaluminate. *Z. Krist.* **74**, 104 (1930).
46. The Structure of the Micas and Related Minerals. *Proc. Nat. Acad. Sci. U.S.* **16**, 123 (1930).
48. Über die Kristallstruktur des Rubidimazids. *Z. physik. Chem.* B **8**, 326 (1930).
49. The Crystal Structure of Bixbyite and the *C*-Modification of the Sesquioxides. *Z. Krist.* **75**, 128 (1930). [L. P. and M. D. Shappell]
50. The Structure of the Chlorites. *Proc. Nat. Acad. Sci. U.S.* **16**, 578 (1930).
51. The Crystal Structure of Cadmium Chloride. *Z. Krist.* **74**, 546 (1930). [L. P. and J. L. Hoard]
52. The Determination of Crystal Structure by X-Rays. *Annual Survey of Am. Chem.* **5**, 118 (1931).
57. Objections to a Proof of Molecular Asymmetry of Optically Active Phenylaminoacetic Acid. *J. Am. Chem. Soc.* **53**, 3820 (1931).
58. The Crystal Structure of Magnesium Platinocyanide Heptahydrate. *Phys. Rev.* **39**, 537 (1932). [Richard M. Bozorth and L. P.]
59. The Packing of Spheres. *The Chemical Bulletin* (Chicago Section Am. Chem. Soc.) **19**, 35 (1932).
60. The Crystal Structure of Chalcopyrite, $CuFeS_2$. *Z. Krist.* **82**, 188 (1932). [L. P. and L. O. Brockway]

62. Determination of Crystal Structure by X-Rays. *Annual Survey of Am. Chem.* **6**, 116 (1932).
67. The Crystal Structure of Sulvanite, Cu_3VS_4. *Z. Krist.* **84**, 204 (1933). [L. P. and Ralph Hultgren]
68. Note on the Crystal Structure of Rubidium Nitrate. *Z. Krist.* **84**, 213 (1933). [L. P. and J. Sherman]
69. The Crystal Structure of Zunyite, $Al_{13}Si_5O_{20}(OH, F)_{18}Cl$. *Z. Krist.* **84**, 442 (1933).
74. The Crystal Structure of Ammonium Hydrogen Fluoride, NH_4HF_2. *Z. Krist.* **85**, 380 (1933).
80. Covalent Radii of Atoms and Interatomic Distances in Crystals Containing Electron-Pair Bonds. *Z. Krist.* **87**, 205 (1934). [L. P. and M. L. Huggins]
81. The Crystal Structure of Enargite, Cu_3AsS_4. *Z. Krist.* **88**, 48 (1934). [L. P. and Sidney Weinbaum]
82. The Crystal Structure of Binnite, $(Cu,Fe)_{12}As_4S_{13}$, and the Chemical Composition and Structure of Minerals of the Tetrahedrite Group. *Z. Krist.* **88**, 54 (1934). [L. P. and E. W. Neuman]
84. The Structure of the Carboxyl Group. II. The Crystal Structure of Basic Beryllium Acetate. *Proc. Nat. Acad. Sci. U.S.* **20**, 340 (1934). [L. P. and J. Sherman]
85. The Structure of Calcium Boride, CaB_6. *Z. Krist.* **87**, 181 (1934). [L. P. and Sidney Weinbaum]
88. The Unit of Structure of Telluric Acid, $Te(OH)_6$. *Z. Krist.* **91**, 367 (1935).
93. The Crystal Structure of Swedenborgite, $NaBe_4SbO_7$. *Am. Mineralogist* **20**, 492 (1935). [L. P., H. P. Klug, and A. N. Winchell]
94. The Structure and Entropy of Ice and of Other Crystals with Some Randomness of Atomic Arrangement. *J. Am. Chem. Soc.* **57**, 2680 (1935).
98. Atomic Scattering Factors. In *Internationale Tabellen zur Bestimmung von Kristallstrukturen*, vol. 2. Berlin: Gebrüder Borntraeger, 1935, pp. 568–575.
99. Ionic and Atomic Radii. In *Internationale Tabellen zur Bestimmung von Kristallstrukturen*, vol. 2. Berlin: Gebrüder Borntraeger, 1935, pp. 610–617.
104. The Crystal Structure of Metaldehyde. *J. Am. Chem. Soc.* **58**, 1274 (1936). [L. P. and D. C. Carpenter]
107. The Crystal Structure of Aluminum Metaphosphate, $Al(PO_3)_3$. *Z. Krist.* **96**, 481 (1937). [L. P. and J. Sherman]
108. The X-Ray Analysis of Crystals. *Current Science.* (Special Number on "Laue Diagrams"), p. 20, January 1937.
117. The Crystal Structure of Cesium Aurous Auric Chloride, $Cs_2AuAuCl_6$, and Cesium Argentous Auric Chloride, $Cs_2AgAuCl_6$. *J. Am. Chem. Soc.* **60**, 1846 (1938). [Norman Elliott and L. P.]
118. The Structure of Ammonium Heptafluozirconate and Potassium Heptafluozirconate and the Configuration of the Heptafluozirconate Group. *J. Am. Chem. Soc.* **60**, 2702 (1938). [G. C. Hampson and L. P.]
120. The Crystal Structure of Ammonium Cadmium Chloride, NH_4CdCl_3. *J. Am. Chem. Soc.* **60**, 2886 (1938). [Henri Brasseur and L. P.]
132. The Crystal Structures of the Tetragonal Monoxides of Lead, Tin, Palla-

dium, and Platinum. *J. Am. Chem. Soc.* **63**, 1392 (1941). [Walter J. Moore, Jr., and L. P.]

190. The Problem of the Graphite Structure. *Am. Mineralogist* **35**, 125 (1950). [J. S. Lukesh and L. P.]
237. The Structure of Chlorine Hydrate. *Proc. Nat. Acad. Sci. U.S.* **38**, 112 (1952). [L. P. and Richard E. Marsh]
248. The Crystal Structure of β-Selenium. *Acta Cryst.* **6**, 71 (1953). [Richard E. Marsh, L. P., and James D. McCullough]
264. On the Structure of the Heteropoly Anion in Ammonium 9-Molybdmanganate, $(NH_4)_6MnMo_9O_{32} \cdot 8H_2O$. *Acta Cryst.* **7**, 438 (1954). [John L. T. Waugh, David P. Shoemaker, and L. P.]
326. Problems of Inorganic Structures. In P. P. Ewald, ed., *Fifty Years of X-Ray Diffraction*, dedicated to Max von Laue, Utrecht: N. V. A. Oosthoek's Uitgeversmaatschappij, 1962, pp. 136–146.
327. Early Work on X-Ray Diffraction in the California Institute of Technology. In P. P. Ewald, ed., *Fifty Years of X-Ray Diffraction*, dedicated to Max von Laue, Utrecht: N. V. A. Oosthoek's Uitgeversmaatschappij, 1962, pp. 623–628.
362. The Structure and Properties of Graphite and Boron Nitride. *Proc. Nat. Acad. Sci. U.S.* **56**, 1646 (1966).

II. Papers on Quantum Mechanics

14. The Quantum Theory of the Dielectric Constant of Hydrogen Chloride and Similar Gases. *Proc. Nat. Acad. Sci. U.S.* **12**, 32 (1926).
16. The Quantum Theory of the Dielectric Constant of Hydrogen Chloride and Similar Gases. *Phys. Rev.* **27**, 568 (1926).
20. The Electron Affinity of Hydrogen and the Second Ionization Potential of Lithium. *Phys. Rev.* **29**, 285 (1927).
21. The Influence of a Magnetic Field on the Dielectric Constant of a Diatomic Dipole Gas. *Phys. Rev.* **29**, 145 (1927).
23. The Theoretical Prediction of the Physical Properties of Many-Electron Atoms and Ions. Mole Refraction, Diamagnetic Susceptibility, and Extension in Space. *Proc. Roy. Soc.* (London) A **114**, 181 (1927).
24. Die Abschirmungskonstanten der relativistischen oder magnetischen Röntgenstrahlendubletts. *Z. Physik* **40**, 344 (1926).
29. The Application of the Quantum Mechanics to the Structure of the Hydrogen Molecule and Hydrogen Molecule-Ion and to Related Problems. *Chem. Rev.* **5**, 173 (1928).
37. The Momentum Distribution in Hydrogen-like Atoms. *Phys. Rev.* **34**, 109 (1929). [Boris Podolsky and L. P.]
40. Photo-Ionization in Liquids and Crystals and the Dependence of the Frequency of X-Ray Absorption Edges on Chemical Constitution. *Phys. Rev.* **34**, 954 (1929).
47. The Rotational Motion of Molecules in Crystals. *Phys. Rev.* **36**, 430 (1930).
61. Screening Constants for Many-electron Atoms. The Calculation and Interpretation of X-Ray Term Values, and the Calculation of Atomic Scattering Factors. *Z. Krist.* **81**, 1 (1932). [L. P. and J. Sherman]

70. The Normal State of the Helium Molecule-Ions He_2^+ and He_2^{++}. *J. Chem. Phys.* **1**, 56 (1933).
75. The Calculation of Matrix Elements for Lewis Electronic Structures of Molecules. *J. Chem. Phys.* **1**, 280 (1933).
86. A Wave-mechanical Treatment of the Mills-Nixon Effect. *Trans. Faraday Soc.* **31**, 939 (1935). [L. E. Sutton and L. P.]
91. The van der Waals Interaction of Hydrogen Atoms. *Phys. Rev.* **47**, 686 (1935). [L. P. and J. Y. Beach]
92. A Quantum Mechanical Discussion of Orientation of Substituents in Aromatic Molecules. *J. Am. Chem. Soc.* **57**, 2086 (1935). [G. W. Wheland and L. P.]
101. Quantum Mechanics and the Third Law of Thermodynamics. *J. Chem. Phys.* **4**, 393 (1936). [L. P. and E. D. Eastman]
106. The Diamagnetic Anisotropy of Aromatic Molecules. *J. Chem. Phys.* **4**, 673 (1936).
128. A Theory of the Color of Dyes. *Proc. Nat. Acad. Sci. U.S.* **25**, 577 (1939).
197. The Electronic Structure of Excited States of Simple Molecules. *Z. Naturforschung* A **3**, 438 (1948).
219. Quantum Theory and Chemistry. *Science* **113**, 92 (1951).
315. Quantum Theory and Chemistry. In W. Frank, ed., *Max Planck Festschrift*. Berlin: Deutscher Verlag der Wissenschaften, 1959, pp. 385–388.

III. Papers on the Nature of the Chemical Bond

13. The Prediction of the Relative Stabilities of Isosteric Isomeric Ions and Molecules. *J. Am. Chem. Soc.* **48**, 641 (1926). [L. P. and Sterling B. Hendricks]
15. The Dynamic Model of the Chemical Bond and Its Application to the Structure of Benzene. *J. Am. Chem. Soc.* **48**, 1132 (1926).
25. The Shared-Electron Chemical Bond. *Proc. Nat. Acad. Sci. U.S.* **14**, 359 (1928).
53. Quantum Mechanics and the Chemical Bond. *Phys. Rev.* **37**, 1185 (1931).
54. The Nature of the Chemical Bond. Application of Results Obtained from the Quantum Mechanics and from a Theory of Paramagnetic Susceptibility to the Structure of Molecules. *J. Am. Chem. Soc.* **53**, 1367 (1931).
55. The Nature of the Chemical Bond. II. The One-Electron Bond and the Three-Electron Bond. *J. Am. Chem. Soc.* **53**, 3225 (1931).
56. The Nature of the Chemical Bond. III. The Transition from One Extreme Bond Type to Another. *J. Am. Chem. Soc.* **54**, 988 (1932).
63. Interatomic Distances in Covalent Molecules and Resonance between Two or More Lewis Electronic Structures. *Proc. Nat. Acad. Sci. U.S.* **18**, 293 (1932).
64. The Additivity of the Energies of Normal Covalent Bonds. *Proc. Nat. Acad. Sci. U.S.* **18**, 414 (1932). [L. P. and Don M. Yost]
65. The Electronic Structure of the Normal Nitrous Oxide Molecule. *Proc. Nat. Acad. Sci. U.S.* **18**, 498 (1932).
66. The Nature of the Chemical Bond. IV. The Energy of Single Bonds and the Relative Electronegativity of Atoms. *J. Am. Chem. Soc.* **54**, 3570 (1932).

76. The Resonance of Molecules among Several Electronic Structures. *The Nucleus* (Northeastern Section Am. Chem. Soc.) **9**, 183 (1932).
77. The Nature of the Chemical Bond. V. The Quantum-Mechanical Calculation of the Resonance Energy of Benzene and Naphthalene and the Hydrocarbon Free Radicals. *J. Chem. Phys.* **1**, 362 (1933); Errata, *J. Chem. Phys.* **2**, 482 (1934). [L. P. and G. W. Wheland]
78. The Nature of the Chemical Bond. VI. The Calculation from Thermochemical Data of the Energy of Resonance of Molecules among Several Electronic Structures. *J. Chem. Phys.* **1**, 606 (1933). [L. P. and J. Sherman]
79. The Nature of the Chemical Bond. VII. The Calculation of Resonance Energy in Conjugated Systems. *J. Chem. Phys.* **1**, 679 (1933). [L. P. and J. Sherman]
89. Remarks on the Theory of Aromatic Free Radicals. *J. Chem. Phys.* **3**, 315 (1935). [L. P. and G. W. Wheland]
97. The Dependence of Interatomic Distance on Single Bond-Double Bond Resonance. *J. Am. Chem. Soc.* **57**, 2705 (1935). [L. P., L. O. Brockway, and J. Y. Beach]
113. A Quantitative Discussion of Bond Orbitals. *J. Am. Chem. Soc.* **59**, 1450 (1937). [L. P. and J. Sherman]
114. The Structure of Cyameluric Acid, Hydromelonic Acid, and Related Substances. *Proc. Nat. Acad. Sci. U.S.* **23**, 615 (1937). [L. P. and J. H. Sturdivant]
116. The Significance of Resonance to the Nature of the Chemical Bond and the Structure of Molecules. In Henry Gilman, ed., *Organic Chemistry. An Advanced Treatise*, vol. 2. New York: Wiley, 1938, pp. 1850–1890.
126. Recent Work on the Configuration and Electronic Structure of Molecules; with some Applications to Natural Products. *Fortschr. Chem. organ. Naturstoffe* **3**, 203 (1939).
170. Modern Structural Chemistry. (Acceptance Talk for the Willard Gibbs Medal, awarded 14 June 1946 by the Chicago Section of the American Chemical Society.) *Chem. Eng. News* **24**, 1788 (1946).
173. Unsolved Problems of Structural Chemistry. (Acceptance Address for the Theodore William Richards Medal for 1947, awarded by the Northeastern Section of the American Chemical Society) *Chem. Eng. News* **25**, 2970 (1947).
176. The Valences of Transition Elements. In *Contribution à l'Etude de la Structure Moléculaire* (Victor Henri Memorial Volume). Liege: Desoer, 1948, pp. 1–14.
196. The Modern Theory of Valency. (Liversidge Lecture, Chemical Society of London, 3 June 1948.) *J. Chem. Soc.* **1948**, 1461.
199. The Valence-State Energy of the Bivalent Oxygen Atom. *Proc. Nat. Acad. Sci. U.S.* **35**, 229 (1949).
200. On the Stability of the S_8 Molecule and the Structure of Fibrous Sulfur. *Proc. Nat. Acad. Sci. U.S.* **35**, 495 (1949).
201. The Dissociation Energy of Carbon Monoxide and the Heat of Sublimation of Graphite. *Proc. Nat. Acad. Sci. U.S.* **35**, 359 (1949). [L. P. and William F. Sheehan, Jr.]
217. Bond Orbitals and Bond Energy in Elementary Phosphorus. *J. Chem. Phys.* **20**, 29 (1952). [L. P. and Massimo Simonetto]

236. Interatomic Distances and Bond Character in the Oxygen Acids and Related Substances. *J. Phys. Chem.* **56**, 361 (1952).
241. Resonance in the Hydrogen Molecule. *J. Chem. Phys.* **20**, 1041 (1952)
250. The Structural Chemistry of Molybdenum. In D. H. Killeffer and Arthur Linz, *Molybdenum Compounds, Their Chemistry and Technology.* New York: Interscience Publishers, 1952, pp. 95–109.
266. The Dependence of Bond Energy on Bond Length. (Debye 70th Birthday Symposium, Cornell University, Ithaca, New York, March 1954.) *J. Phys. Chem.* **58**, 662 (1954).
274. The Energy Change in Organic Rearrangements and the Electronegativity Scale. In *Biochemistry of Nitrogen.* (A collection of papers dedicated to Artturi Ilmari Virtanen.) Helsinki: Suomalaisen Tiedeakatemian, 1955, pp. 428–432.
277. Modern Structural Chemistry. (Nobel Lecture, 11 December 1954.) In M. G. Liljestrand, ed., *Les Prix Nobel en 1954.* Stockholm: Kungl. Boktryckeriet P. A. Norstedt and Söner, 1955, pp. 91–99. *Angew. Chem.* **67**, 241 (1955). *Science* **123**, 255 (1956).
287. The Nature of the Theory of Resonance. In Sir Alexander Todd, ed., *Perspectives in Organic Chemistry.* (Dedicated to Sir Robert Robinson.) New York: Interscience, 1956, pp. 1–8.
300. The Nature of Bond Orbitals and the Origin of Potential Barriers to Internal Rotation in Molecules. *Proc. Nat. Acad. Sci. U.S.* **44**, 211 (1958).
302. The Structure of Water. In D. Hadzi and H. W. Thompson, *Hydrogen Bonding.* (Papers presented at the Symposium on Hydrogen Bonding held at Ljubljana, 29 July to 3 August 1957.) New York: Pergamon Press, 1959, pp. 1–6. Also read at Royal Society meeting on "Physics of Water and Ice," November 1957.
309. Kekulé and the Chemical Bond. In *Theoretical Organic Chemistry.* (Papers presented to the Kekulé Symposium organized by the Chemical Society, London, September 1958. Sponsored by International Union of Pure and Applied Chemistry, Section of Organic Chemistry.) London: Butterworths Scientific Publ., 1959, pp. 1–8.
313. The Discussion of Tetragonal Boron by the Resonating-Valence-Bond Theory of Electron-Deficient Substances. *Z. Krist.* **112**, 472 (1959). [L. P. and Barclay Kamb]
323. The Carbon-carbon Triple Bond and the Nitrogen-nitrogen Triple Bond. *Tetrahedron* **17**, 229 (1962).
330. Valence Bond Theory in Coordination Chemistry. *J. Chem. Educ.* **39**, 461 (1962).
332. The Theory of Resonance in Chemistry. *J. Mendeleev All-Union Chemical Society* 7, 462 (1962).
336. The Electroneutrality Principle and the Structure of Molecules, *Ann. física y quimica*, Madrid B **60**, 87 (1964).
337. The Structure of Methylene and Methyl. In *Molecular Orbitals in Chemistry, Physics, and Biology.* New York: Academic Press, 1964, pp. 207–213.
339. Methylene and the Carbenes (El Metileno y los Carbenos). *Afinidad* **20**, 393–396, November–December 1963 [L. P. and Gustav Albrecht]
340. The Architecture of Molecules. *Proc. Nat. Acad. Sci. U.S.* **51**, 977 (1964).

346. The Energy of Transargononic Bonds. In *The Law of Mass Action, a Centenary Volume 1864–1964.* Det Norske Videnkaps-Akademi i Oslo. Oslo: Universitet forlaget. 1964, pp. 151–158.
349. The Nature of the Chemical Bonds in Sulvanite, Cu_3VS_4. *Tschermaks mineral. u. petro. Mitteilungen* **10**, 379 (1965).

IV. Papers on the Determination of the Structure of Molecules by the Diffraction of Electrons

71. The Determination of the Structures of the Hexafluorides of Sulfur, Selenium and Tellurium by the Electron Diffraction Method. *Proc. Nat. Acad. Sci. U.S.* **19**, 68 (1933). [L. O. Brockway and L. P.]
73. The Electron-Diffraction Investigation of the Structure of Molecules of Methyl Azide and Carbon Suboxide. *Proc. Nat. Acad. Sci. U.S.* **19**, 860 (1933). [L. O. Brockway and L. P.]
83. The Structure of the Carboxyl Group. I. The Investigation of Formic Acid by the Diffraction of Electrons. *Proc. Nat. Acad. Sci. U.S.* **20**, 336 (1934). [L. P. and L. O. Brockway]
87. A Study of the Methods of Interpretation of Electron-Diffraction Photographs of Gas Molecules, with Results for Benzene and Carbon Tetrachloride. *J. Chem. Phys.* **2**, 867 (1934). [L. P. and L. O. Brockway]
95. The Radial Distribution Method of Interpretation of Electron Diffraction Photographs of Gas Molecules. *J. Am. Chem. Soc.* **57**, 2684 (1935). [L. P. and L. O. Brockway]
96. The Electron Diffraction Investigation of Phosgene, the Six Chloroethylenes, Thiophosgene, α-Methylhydroxylamine and Nitromethane. *J. Am. Chem. Soc.* **57**, 2693 (1935). [L. O. Brockway, J. Y. Beach, and L. P.]
109. The Structure of the Pentaborane B_5H_9. *J. Am. Chem. Soc.* **58**, 2403 (1936). [S. H. Bauer and L. P.]
110. The Adjacent Charge Rule and the Structure of Methyl Azide, Methyl Nitrate, and Fluorine Nitrate. *J. Am. Chem. Soc.* **59**, 13 (1937). [L. P. and L. O. Brockway]
112. Carbon-Carbon Bond Distances. The Electron Diffraction Investigation of Ethane, Propane, Isobutane, Neopentane, Cyclopropane, Cyclopentane, Cyclohexane, Allene, Ethylene, Isobutene, Tetramethylethylene, Mesitylene, and Hexamethylbenzene. Revised Values of Covalent Radii. *J. Am. Chem. Soc.* **59**, 1223 (1937). [L. P. and L. O. Brockway]
115. The Electron Diffraction Study of Digermane and Trigermane. *J. Am. Chem. Soc.* **60**, 1605 (1938). [L. P., A. W. Laubengayer, and J. L. Hoard]
123. The Electron Diffraction Investigation of Methylacetylene, Dimethylacetylene, Methyl Cyanide, Diacetylene, and Cyanogen. *J. Am. Chem. Soc.* **61**, 927 (1939). [L. P., H. D. Springall, and K. J. Palmer]
124. The Electron Diffraction Investigation of the Structure of Benzene, Pyridine, Pyrazine, Butadiene-1,3,Cyclopentadiene, Furan, Pyrrole, and Thiophene. *J. Am. Chem. Soc.* **61**, 1769 (1939). [V. Schomaker and L. P.]
134. The Alkyls of the Third Group Elements. II. The Electron Diffraction Study of Indium Trimethyl. *J. Am. Chem. Soc.* **63**, 480 (1941). [L. P. and A. W. Laubengayer]
135. The Electron-diffraction Method of Determining the Structure of Gas Molecules. *J. Chem. Educ.* **18**, 458 (1941). [Robert Spurr and L. P.]

139. The Electron Diffraction Investigation of Propargyl Chloride, Bromide, and Iodide. *J. Am. Chem. Soc.* **64**, 1753 (1942). [L. P., Walter Gordy, and John H. Saylor]
145. The Molecular Structure of Methyl Isocyanide. *J. Am. Chem. Soc.* **64**, 2952 (1942). [Walter Gordy and L. P.]
207. The Determination of the Structures of Complex Molecules and Ions from X-Ray Diffraction by their Solutions: The Structures of the Groups $PtBr_6^{--}$, $PtCl_6^{--}$, $Nb_6Cl_{12}^{++}$, $Ta_6Br_{12}^{++}$, and $Ta_6Cl_{12}^{++}$. *J. Am. Chem. Soc.* **72**, 5477 (1950). [Philip A. Vaughan, J. H. Sturdivant, and L. P.]

V. Papers on Electronic Theory and the Structure of Metals and Intermetallic Compounds

122. The Nature of the Interatomic Forces in Metals. *Phys. Rev.* **54**, 899 (1938).
168. The Nature of the Bonds in Metals and Intermetallic Compounds. (Talk given before Section 1 of 11th International Congress of Pure and Applied Chemistry. London, 17–24 July 1947.) *Proc. Intern. Congr. Pure and Applied Chem.* (London) **11**, 249 (1947).
171. Atomic Radii and Interatomic Distances in Metals. *J. Am. Chem. Soc.* **69**, 542 (1947).
178. The Structure of Uranium Hydride. *J. Am. Chem. Soc.* **70**, 1660 (1948). [L. P. and Fred J. Ewing]
181. The Ratio of Valence Electrons to Atoms in Metals and Intermetallic Compounds. *Rev. Modern Phys.* **20**, 112 (1948). [L. P. and Fred J. Ewing]
184. The Nature of the Bonds in the Iron Silicide FeSi and Related Crystals. *Acta Cryst.* **1**, 212 (1948). [L. P. and A. M. Soldate]
185. The Resonating Valence-Bond Theory of Metals. *Physica* **15**, 23 (1949).
191. The Metallic State. *Nature* **161**, 1019 (1948).
192. A Resonating Valence-Bond Theory of Metals and Intermetallic Compounds. *Proc. Roy. Soc.* (London) A **196**, 343 (1949).
193. La Valence des Métaux et la Structure des Composés Intermétalliques. *J. chim. Phys.* **46**, 276 (1949).
206. Compressibilities, Force Constants, and Interatomic Distances of the Elements in the Solid State. *J. Chem. Phys.* **18**, 747 (1950). [J. Waser and L. P.]
210. Interatomic Distances in Co_2Al_9. *Acta Cryst.* **4**, 138 (1951).
212. Electron Transfer in Intermetallic Compounds. *Proc. Nat. Acad. Sci. U.S.* **36**, 533 (1950).
218. Discussion of Paper by Roland Kiessling on the Borides of Some Transition Elements. *J. Electrochem. Soc.* **98**, 518 (1951).
229. The Structure of Alloys of Lead and Thallium. *Acta Cryst.* **5**, 39 (1952). [You-Chi Tang and L. P.]
239. Interatomic Distances and Atomic Valences in $NaZn_{13}$. *Acta Cryst.* **5**, 637 (1952). [David P. Shoemaker, Richard E. Marsh, Fred J. Ewing, and L.P.]
245. Crystal Structure of the Intermetallic Compound $Mg_{32}(Al, Zn)_{49}$ and Related Phases. *Nature* **169**, 1057 (1952). [Gunnar Bergman, John L. T. Waugh, and L. P.]
246. The Atomic Arrangement and Bonds of the Gold-Silver Ditellurides. *Acta Cryst.* **5**, 375 (1952). [George Tunell and L. P.]
262. A Theory of Ferromagnetism. *Proc. Nat. Acad. Sci. U.S.* **39**, 551 (1953).

280. The Electronic Structure of Metals and Alloys. In *Theory of Alloy Phases.* Cleveland: American Society for Metals, 1956, pp. 220–242.
284. On the Valence and Atomic Size of Silicon, Germanium, Arsenic, Antimony, and Bismuth in Alloys. *Acta Cryst.* **9**, 127 (1956). [L. P. and Peter Pauling]
291. A Set of Effective Metallic Radii for Use in Compounds with the β-Wolfram Structure. *Acta Cryst.* **10**, 374 (1957).
296. The Use of Atomic Radii in the Discussion of Interatomic Distances and Lattice Constants of Crystals. *Acta Cryst.* **10**, 685 (1957).
297. The Crystal Structure of the Metallic Phase $Mg_{32}(Al, Zn)_{49}$. *Acta Cryst.* **10**, 254 (1957). [Gunnar Bergman, John L. T. Waugh, and L. P.]
321. Nature of the Metallic Orbital. *Nature* **189**, 656 (1961).
322. The Nature of the Metallic Orbital and the Structure of Metals. (Centenary Souvenir Volume of the Indian Chemical Society.) *J. Indian Chem. Soc.* **38**, 435 (1961).
363. Electron Transfer and Atomic Magnetic Moments in the Ordered Intermetallic Compound $AlFe_3$. In Per-Olov Löwdin, ed., *Quantum Theory of Atoms, Molecules, and the Solid State, A Tribute to John C. Slater.* New York: Academic Press, 1966, p. 303.
371. The Dependence of Bond Lengths in Intermediate Compounds on the Hybrid Character of the Bond Orbitals. *Acta Cryst.*, December (1967).

VI. Papers on the Structure and Properties of Hemoglobin and Other Proteins

90. The Oxygen Equilibrium of Hemoglobin and Its Structural Interpretation. *Proc. Nat. Acad. Sci. U.S.* **21**, 186 (1935).
102. The Magnetic Properties and Structure of the Hemochromogens and Related Substances. *Proc. Nat. Acad. Sci. U.S.* **22**, 159 (1936). [L. P. and Charles D. Coryell]
103. The Magnetic Properties and Structure of Hemoglobin, Oxyhemoglobin and Carbonmonoxyhemoglobin. *Proc. Nat. Acad. Sci. U.S.* **22**, 210 (1936). [L. P. and Charles D. Coryell]
105. On the Structure of Native, Denatured, and Coagulated Proteins. *Proc. Nat. Acad. Sci. U.S.* **22**, 439 (1936). [A. E. Mirsky and L. P.]
111. The Magnetic Properties and Structure of Ferrihemoglobin (Methemoglobin) and Some of its Compounds. *J. Am. Chem. Soc.* **59**, 633 (1937). [Charles D. Coryell, Fred Stitt, and L. P.]
119. The Magnetic Properties of Intermediates in the Reactions of Hemoglobin. *J. Phys. Chem.* **43**, 825 (1939). [Charles D. Coryell, L. P., and Richard W. Dodson]
125. The Structure of Proteins. *J. Am. Chem. Soc.* **61**, 1860 (1939). [L. P. and Carl Niemann]
127. The Magnetic Properties of the Compounds Ethylisocyanide-Ferrohemoglobin and Imidazole-Ferrihemoglobin. *Proc. Nat. Acad. Sci. U.S.* **25**, 517 (1939). [Charles D. Russell and L. P.]
129. A Structural Interpretation of the Acidity of Groups Associated with the Hemes of Hemoglobin and Hemoglobin Derivatives. *J. Biol. Chem.* **132**, 769 (1940). [Charles D. Coryell and L. P.]
157. The Adsorption of Water by Proteins. *J. Am. Chem. Soc.* **67**, 555 (1945).

183. The Interpretation of Some Chemical Properties of Hemoglobin in Terms of Its Molecular Structure. *Stanford Med. Bull.* **6**, 215 (1948).
198. The Electronic Structure of Haemoglobin. In *Haemoglobin.* London: Butterworths Scientific Publ. 1949, pp. 57–65.
211. Two Hydrogen-Bonded Spiral Configurations of the Polypeptide Chain. *J. Am. Chem. Soc.* **72**, 5349 (1950). [L. P. and Robert B. Corey]
221. The Structure of Proteins: Two Hydrogen-Bonded Helical Configurations of the Polypeptide Chain. *Proc. Nat. Acad. Sci. U.S.* **37**, 205 (1951). [L. P., Robert B. Corey, and H. R. Branson]
222. Atomic Coordinates and Structure Factors for Two Helical Configurations of Polypeptide Chains. *Proc. Nat. Acad. Sci. U.S.* **37**, 235 (1951). [L. P. and Robert B. Corey]
223. The Structure of Synthetic Polypeptides. *Proc. Nat. Acad. Sci. U.S.* **37**, 241 (1951). [L. P. and Robert B. Corey]
224. The Pleated Sheet, A New Layer Configuration of Polypeptide Chains. *Proc. Nat. Acad. Sci. U.S.* **37**, 251 (1951). [L. P. and Robert B. Corey]
225. The Structure of Feather Rachis Keratin. *Proc. Nat. Acad. Sci. U.S.* **37**, 256 (1951). [L. P. and Robert B. Corey]
226. The Structure of Hair, Muscle, and Related Proteins. *Proc. Nat. Acad. Sci. U.S.* **37**, 261 (1951). [L. P. and Robert B. Corey]
227. The Structure of Fibrous Proteins of the Collagen-Gelatin Group. *Proc. Nat. Acad. Sci. U.S.* **37**, 272 (1951). [L. P. and Robert B. Corey]
228. The Polypeptide-Chain Configuration in Hemoglobin and Other Globular Proteins. *Proc. Nat. Acad. Sci. U.S.* **37**, 282 (1951). [L. P. and Robert B. Corey]
230. The Structure of Proteins. (Phi Lambda Upsilon Lecture, Second Annual Series.) Columbus: Ohio State University, February 1951.
231. The Combining Power of Hemoglobin for Alkyl Isocyanides, and the Nature of the Heme-Heme Interactions in Hemoglobin. *Science* **114**, 629 (1951). [Robert C. C. St. George and L. P.]
232. Configuration of Polypeptide Chains. *Nature* **168**, 550 (1951). [L. P. and Robert B. Corey]
233. The Configuration of Polypeptide Chains in Proteins. *Record Chem. Progress* **12**, 155 (1951).
234. Configurations of Polypeptide Chains with Favored Orientations around Single Bonds: Two New Pleated Sheets. *Proc. Nat. Acad. Sci. U.S.* **37**, 729 (1951). [L. P. and Robert B. Corey]
235. Configurations of Polypeptide Chains with Equivalent Cis Amide Groups. *Proc. Nat. Acad. Sci. U.S.* **38**, 86 (1952). [L. P. and Robert B. Corey]
240. The Lotmar-Picken X-Ray Diagram of Dried Muscle. *Nature* **169**, 494 (1952). [L. P. and Robert B. Corey]
242. Structure of the Synthetic Polypeptide Poly-gamma-methyl-L-glutamate. *Nature* **169**, 920 (1952). [Harry L. Yakel, Jr., L. P., and Robert B. Corey]
243. Fundamental Dimensions of Polypeptide Chains. *Proc. Roy. Soc.* (London) B **141**, 10 (1953). [R. B. Corey and L. P.]
244. Stable Configurations of Polypeptide Chains. *Proc. Roy. Soc.* (London) B **141**, 21 (1953). [L. P. and R. B. Corey]
247. The Configuration of Polypeptide Chains in Proteins. (A Report for the Ninth Solvay Congress, University of Brussels, 6–14 April 1953.) In

R. Stoops, ed., *Les Proteines. Rapports et Discussions.* Bruxelles: Secretaires du Conseil sous les Auspices du Comite Scientifique de l'Institut, 76–78 Coudenberg, 1953, pp. 63–99.

253. Molecular Models of Amino Acids, Peptides, and Proteins. *Rev. Sci. Instr.* **24**, 621 (1953). [Robert B. Corey and L. P.]

254. The Planarity of the Amide Group in Polypeptides. *J. Am. Chem. Soc.* **74**, 3964 (1952). [L. P. and Robert B. Corey]

255. Compound Helical Configurations of Polypeptide Chains: Structure of Proteins of the α-Keratin Type. *Nature* **171**, 59 (1953). [L. P. and Robert B. Corey]

256. Protein Interactions. Aggregation of Globular Proteins. *Discussions Faraday Soc.* **13**, 170 (1953).

260. Two Pleated-Sheet Configurations of Polypeptide Chains Involving Both Cis and Trans Amide Groups. *Proc. Nat. Acad. Sci. U.S.* **39**, 247 (1953). [L. P. and Robert B. Corey]

261. Two Rippled-sheet Configurations of Polypeptide Chains, and a Note about the Pleated Sheet. *Proc. Nat. Acad. Sci. U.S.* **39**, 253 (1953). [L. P. and Robert B. Corey]

265. The Configuration of Polypeptide Chains in Proteins. *Fortschr. Chem. organ. Naturst.* **11**, 180 (1954). [L. P. and Robert B. Corey]

267. The Stochastic Method and the Structure of Proteins. In *13th International Congress of Pure and Applied Chemistry: Plenary Lectures*, Stockholm, 1954, pp. 37–52. *American Scientist* **43**, 285 (1955).

268. The Structure of Protein Molecules. *Scientific American* **191**, 51 (1954). [L. P., Robert B. Corey, and Roger Hayward]

269. An Investigation of the Structure of Silk Fibroin. *Biochim. Biophys. Acta* **16**, 1 (1955). [Richard E. Marsh, Robert B. Corey, and L. P.]

272. The Crystal Structure of Silk Fibroin. *Acta Cryst.* **8**, 62 (1955). [Richard E. Marsh, Robert B. Corey, and L. P.]

275. The Structure of Tussah Silk Fibroin (with a note on the structure of β-poly-L-alanine). *Acta Cryst.* **8**, 710 (1955). *Proceedings of International Wool Textile Research Conference*, Australia, B, 176 (1955). [Robert B. Corey and L. P.]

276. The Configuration of Polypeptide Chains in Proteins. *Rend. ist. lombardo sci.*, Pt. I **89**, 10 (1955). *Proceedings of International Wool Textile Research Conference*, Australia, B, 249 (1955). [Robert B. Corey and L. P.]

278. Calculated Form Factors for the 18-Residue 5-Turn α-Helix. *Acta Cryst.* **8**, 853 (1955). [L. P., Robert B. Corey, Harry L. Yakel, Jr., and Richard E. Marsh]

281. The Combining Power of Myoglobin for Alkyl Isocyanides and the Structure of the Myoglobin Molecule. *Proc. Nat. Acad. Sci. U.S.* **42**, 51 (1956). [Allen Lein and L. P.]

289. The Probability of Errors in the Process of Synthesis of Protein Molecules. In *Arbeiten aus den Gebiet der Naturstoffchemie.* (Festschrift, Prof. Dr. Arthur Stoll. Zum Siebzigsten Geburtstag, 8 January 1957.) Basel: Birkhauser, 1957, pp. 597–602.

293. The Configuration of Polypeptide Chains in Proteins. In George Ross Robertson, ed., *Modern Chemistry for the Engineer and Scientist.* New York: McGraw-Hill, 1957, pp. 422–434.

298. Factors Affecting the Structure of Hemoglobins and Other Proteins. In

Albert Neuberger, ed., *Symposium on Protein Structure.* (Presented at the International Union of Pure and Applied Chemistry, Paris Meeting, 25–29 July 1957.) London: Methuen; New York: Wiley, 1958, pp. 17–22.

299. The Structure of Proteins. In Edward Hutchings, Jr., ed., *Frontiers in Science, A. Survey.* New York: Basic Books, 1958, pp. 28–36. [Robert B. Corey and L. P.]

303. Review of C. H. Bamford, A. Elliott, and W. E. Hanby, *Synthetic Polypeptides: Preparation, Structure, and Properties.* New York: Academic Press, 1956. *Arch. Biochem. Biophys.* **72**, 250 (1957).

305. The Configuration of Polypeptide Chains in Proteins. In G. Stainsby, ed., *Recent Advances in Gelatin and Glue Research.* (Proceedings of a Conference sponsored by The British Gelatine and Glue Research Association held at the University of Cambridge, 1–5 July 1957.) London: Symposium Publications Division, Pergamon Press, 1958, pp. 11–13.

320. A Comparison of Animal Hemoglobins by Tryptic Peptide Pattern Analysis. *Proc. Nat. Acad. Sci. U.S.* **46**, 1349 (1960). [Emile Zuckerkandl, Richard T. Jones, and L. P.]

325. Thalassemia and the Abnormal Human Hemoglobins. *Nature* **191**, 398 (1961). [Harvey A. Itano and L. P.]

333. Chemical Paleogenetics: Molecular "Restoration Studies" of Extinct Forms of Life. *Acta Chem. Scand.* vol. 17, suppl. no. 1, pp. S9–S16 (1963).

341. Nature of the Iron-Oxygen Bond in Oxyhaemoglobin. *Nature* **203**, 182 (1964).

347. Les documents moleculaires de l'evolution. *Atomes* **20**, 339–343 (1965). [Emil Zuckerkandl and L. P.]

348. Molecules as Documents of Evolutionary History. *J. Theoret. Biol.* **8**, 357 (1965). [Emil Zuckerkandl and L. P.]

VII. Papers on the Structure of Antibodies and the Nature of Serological Reactions

130. A Theory of the Structure and Process of Formation of Antibodies. *J. Am. Chem. Soc.* **62**, 2643 (1940).

133. Serological Reactions with Simple Substances Containing Two or More Haptenic Groups. *Proc. Nat. Acad. Sci. U.S.* **27**, 125 (1941). [L. P., Dan H. Campbell, and David Pressman]

137. Complement Fixation with Simple Substances Containing Two or More Haptenic Groups. *Proc. Nat. Acad. Sci. U.S.* **28**, 77 (1942). [David Pressman, Dan H. Campbell, and L. P.]

138. The Agglutination of Intact Azo-Erythrocytes by Antisera Homologous to the Attached Groups. *J. Immunology* **44**, 101 (1942). [David Pressman, Dan H. Campbell, and L. P.]

140. The Serological Properties of Simple Substances. I. Precipitation Reactions between Antibodies and Substances Containing Two or More Haptenic Groups. *J. Am. Chem. Soc.* **64**, 2994 (1942). [L. P., David Pressman, Dan H. Campbell, Carol Ikeda, and Miyoshi Ikawa]

141. The Serological Properties of Simple Substances. II. The Effects of Changed Conditions and of Added Haptens on Precipitation Reactions

of Polyhaptenic Simple Substances. *J. Am. Chem. Soc.* **64**, 3003 (1942). [L. P., David Pressman, Dan H. Campbell, and Carol Ikeda]

142. The Serological Properties of Simple Substances. III. The Composition of Precipitates of Antibodies and Polyhaptenic Simple Substances; the Valence of Antibodies. *J. Am. Chem. Soc.* **64**, 3010 (1942).
143. The Production of Antibodies *in vitro*. *Science* **95**, 440 (1942). [L. P. and Dan H. Campbell]
144. The Manufacture of Antibodies *in vitro*. *J. Exp. Med.* **76**, 211 (1942). [L. P. and Dan H. Campbell]
146. The Serological Properties of Simple Substances. IV. Hapten Inhibition of Precipitation of Antibodies and Polyhaptenic Simple Substances. *J. Am. Chem. Soc.* **64**, 3015 (1942). [David Pressman, David H. Brown, and L. P.]
147. The Serological Properties of Simple Substances. V. The Precipitation of Polyhaptenic Simple Substances and Antiserum Homologous to the *p*-(*p*-Azophenylazo)-phenylarsonic Acid Group and its Inhibition by Haptens. *J. Am. Chem. Soc.* **65**, 728 (1943). [David Pressman, John T. Maynard, Allan L. Grossberg, and L. P.]
148. The Nature of the Forces between Antigen and Antobody and of the Precipitation Reaction. *Physiol. Rev.* **23**, 203 (1943). [L. P., Dan H. Campbell, and David Pressman]
150. An Experimental Test of the Framework Theory of Antigen-Antibody Precipitation. *Science* **98**, 263 (1943). [L. P., David Pressman, and Dan H. Campbell]
151. The Serological Properties of Simple Substances. VI. The Precipitation of a Mixture of Two Specific Antisera by a Dihaptenic Substance Containing the Two Corresponding Haptenic Groups; Evidence for the Framework Theory of Serological Precipitation. *J. Am. Chem. Soc.* **66**, 330 (1944). [L. P., David Pressman, and Dan H. Campbell]
152. A Note on the Serological Activity of Denatured Antibodies. *Science* **99**, 198 (1944). [George G. Wright and L. P.]
153. The Serological Properties of Simple Substances. VII. A Quantitative Theory of the Inhibition by Haptens of the Precipitation of Heterogeneous Antisera with Antigens, and Comparison with Experimental Results for Polyhaptenic Simple Substances and for Azoproteins. *J. Am. Chem. Soc.* **66**, 784 (1944). [L. P., David Pressman, and Allan L. Grossberg]
154. The Serological Properties of Simple Substances. VIII. The Reactions of Antiserum Homologous to the *p*-Azobenzoic Acid Group. *J. Am. Chem. Soc.* **66**, 1731 (1944). [David Pressman, Stanley M. Swingle, Allan L. Grossberg, and L. P.]
156. The Serological Properties of Simple Substances. IX. Hapten Inhibition of Precipitation of Antisera Homologous to the *o*-, *m*-, and *p*-Azophenylarsonic Acid Groups. *J. Am. Chem. Soc.* **67**, 1003 (1945). [L. P. and David Pressman]
158. The Serological Properties of Simple Substances. X. A Hapten Inhibition Experiment Substantiating the Intrinsic Molecular Asymmetry of Antibodies. *J. Am. Chem. Soc.* **67**, 1219 (1945). [David Pressman, John H. Bryden, and L. P.]
159. The Serological Properties of Simple Substances. XI. The Reactions of Antisera Homologous to Various Azophenylarsonic Acid Groups and

the *p*-Azophenylmethylarsonic Acid Group with Some Heterologous Haptens. *J. Am. Chem. Soc.* **67**, 1602 (1945). [David Pressman, Arthur B. Pardee, and L. P.]

160. Molecular Structure and Intermolecular Forces. In Karl Landsteiner, *The Specificity of Serological Reactions* (rev. ed.). Cambridge: Harvard University Press, 1945, pp. 275–293.
163. The Serological Properties of Simple Substances. XII. The Reactions of Antiserum Homologous to the *p*-Azophenyltrimethylammonium Group. *J. Am. Chem. Soc.* **68**, 250 (1946). [David Pressman, Allan L. Grossberg, Leland H. Pence, and L. P.]
167. Analogies between Antibodies and Simpler Chemical Substances. (The first Harrison Howe Lecture before the Rochester Section of the American Chemical Society, 4 February 1946.) *Chem. Eng. News* **24**, 1064 (1946).
174. The Serological Properties of Simple Substances. XIII. The Reactions of Antiserum Homologous to the *p*-Azosuccinanilate Ion Group. *J. Am. Chem. Soc.* **70**, 1352 (1948). [David Pressman, John H. Bryden, and L. P.]
177. The Serological Properties of Simple Substances. XIV. The Reaction of Simple Antigens with Purified Antibody. *J. Am. Chem. Soc.* **71**, 143 (1949). [Arthur B. Pardee and L. P.]
182. Antibodies and Specific Biological Forces. *Endeavor* **7**, 43 (1948).
189. La Structure des Anticorps et la Nature des Réactions Sérologiques (The Structure of Antibodies and the Nature of Serological Reactions). *Bull. Soc. Chim. Biol.* **30**, 247 (1948).
194. The Serological Properties of Simple Substances. XV. The Reactions of Antiserum Homologous to the 4-Azophthalate Ion. *J. Am. Chem. Soc.* **71**, 2893 (1949). [David Pressman and L. P.]

VIII. Papers on Chemistry in Relation to Biology and Medicine

131. The Nature of the Intermolecular Forces Operative in Biological Processes. *Science* **92**, 77 (1940). [L. P. and Max Delbrück]
169. Molecular Architecture and Biological Reactions. *Chem. Eng. News* **24**, 1375 (1946).
172. Molecular Architecture and Medical Progress. (Radio talk sponsored by U.S. Rubber Co., 13 October 1946.) Published by U.S. Rubber Co., 1946.
175. Molecular Structure and Biological Specificity. (Presidential Address at Section 2, 11th International Congress of Pure and Applied Chemistry, London, 17–24 July 1947.) *Chemistry and Industry* (Supplement), 1948, pp. 1–4.
187. A Rapid Diagnostic Test for Sickle Cell Anemia. *Blood* **4**, 66 (1949). [Harvey A. Itano and L. P.]
188. The Nature of Forces between Large Molecules of Biological Interest. (Friday Evening Discourse, The Royal Institution of Great Britain, London, on 27 February 1948.) *Nature* **161**, 707 (1948).
195. Molecular Architecture and the Processes of Life. (21st Sir Jesse Boot Foundation Lecture, 28 May 1948.) Published by Sir Jesse Boot Foundation, Nottingham, England, 1948.
202. Sickle Cell Anemia, A Molecular Disease. *Science* **110**, 543 (1949). [L. P., Harvey A. Itano, S. J. Singer, and Ibert C. Wells]

205. Structural Chemistry in Relation to Biology and Medicine. (Second Bicentennial Science Lecture of the City College Chemistry Alumni Association, New York, 7 December 1949.) *Baskerville Chemical Journal* **1**, 4 (1950). *Medical Arts and Sciences* (Sci. J. of the Coll. of Med. Evangelists) **4**, 84 (1950).
214. The Preparation and Properties of a Modified Gelatin (Oxypolygelatin) as an Oncotic Substitute for Serum Albumin. *Texas Reports on Biology and Medicine* **9**, 235 (1951). [Dan H. Campbell, J. B. Koepfli, L. P., Norman Abrahamsen, Walter Dandliker, George A. Feigen, Frank Lanni, and Arthur LeRosen]
220. Sickle Cell Anemia Hemoglobin. *Science* **111**, 459 (1950). [L. P., Harvey A. Itano, Ibert C. Wells, Walter A. Schroeder, Lois M. Kay, S. J. Singer, and R. B. Corey]
238. On a Phospho-tri-anhydride Formula for the Nucleic Acids. *J. Am. Chem. Soc.* **74**, 1111 (1952). [L. P. and Verner Schomaker]
249. Structure of the Nucleic Acids. *Nature* **171**, 346 (1953). [L. P. and Robert B. Corey]
251. On a Phospho-tri-anhydride Formula for the Nucleic Acids. *J. Am. Chem. Soc.* **74**, 3712 (1952). [L. P. and Verner Schomaker]
257. The Hemoglobin Molecule in Health and Disease. *Proc. Am. Phil. Soc.* **96**, 556 (1952).
259. A Proposed Structure for the Nucleic Acids. *Proc. Nat. Acad. Sci. U.S.* **39**, 84 (1953). [L. P. and Robert B. Corey]
271. Abnormality of Hemoglobin Molecules in Hereditary Hemolytic Anemias. (Harvey Lecture, 29 April 1954.) In *The Harvey Lectures*, 1953–1954, series 49. New York: Academic Press, 1955, pp. 216–241.
273. The Duplication of Molecules. In Dorothea Rudnick, ed., *Aspects of Synthesis and Order in Growth.* (The Thirteenth Symposium of the Society for the Study of Development and Growth.) Princeton: Princeton Univ. Press, 1954, pp. 3–13.
283. The Future of Enzyme Research. (The Fourth Edsel B. Ford Lecture.) In Oliver H. Gaebler, ed., *Enzymes: Units of Biological Structure and Function.* New York: Academic Press, 1956, pp. 177–182. *Henry Ford Hospital Med. Bull.* **4**, 1 (1956).
285. Specific Hydrogen-Bond Formation between Pyrimidines and Purines in Deoxyribonucleic Acids. In Linderstrom-Lang Festschrift. *Arch. Biochem. Biophys.* **65**, 164 (1956). [L. P. and Robert B. Corey]
286. The N-Terminal Amino Acid Residues of Normal Adult Human Hemoglobin: A Quantitative Study of Certain Aspects of Sanger's DNP-Method. *J. Am. Chem. Soc.* **79**, 609 (1957). [Herbert S. Rhinesmith, W. A. Schroder, and L. P.]
290. The Molecular Basis of Genetics. *Am. J. Psychiatry* **113**, 492 (1956).
292. Abnormal Hemoglobin Molecules in Relation to Disease. *Svensk Kemisk Tidskrift* **69**, 509 (1957). [Harvey A. Itano and L. P.]
294. A Quantitative Study of the Hydrolysis of Human Dinitrophenyl (DNP) globin: The Number and Kind of Polypeptide Chains in Normal Adult Human Hemoglobin. *J. Am. Chem. Soc.* **79**, 4682 (1957). [Herbert S. Rhinesmith, W. A. Schroder, and L. P.]
295. The Nature of the Forces Operating in the Process of the Duplication of

Molecules in Living Organisms. In F. Clark and R. L. M. Synge, eds., *The Origin of Life on the Earth.* (Proceedings of the First International Symposium on the Origin of Life on the Earth, Moscow, 19–24 August 1957.) New York: Pergamon Press, 1959, pp. 215–223.

301. Summary and Discussion. In Linus Pauling and Harvey A. Itano, eds., *Molecular Structure and Biological Specificity.* (A Symposium sponsored by the Office of Naval Research and arranged by the American Institute of the Biological Sciences, held in Washington, D.C., 28–29 October 1955.) American Institute of Biological Sciences, Washington, D.C., Publication No. 2, 1957, pp. 186–195.
306. The Relation between Longevity and Obesity in Human Beings. *Proc. Nat. Acad. Sci. U.S.* **44**, 619 (1958).
307. Emoglobine Anormali in Rapporto alle Malattie. *Rendi. ist. super. sanità* (Roma) **21**, 30 (1958).
308. Genetic and Somatic Effects of Carbon 14. *Science* **128**, 1183 (1958).
311. The Effects of Strontium 90 on Mice. *Proc. Nat. Acad. Sci. U.S.* **45**, 54 (1959). [Barclay Kamb and L. P.]
312. Molecular Structure and Disease. In Henry Knowles Beecher, ed., *Disease and the Advancement of Basic Science.* (1958 Lowell Institute Lectures at Harvard University.) Cambridge: Harvard Univ. Press, 1960, pp. 1–7.
314. Molecular Disease. *Am. J. Orthopsychiat.* **29**, 684 (1959).
316. Review of P. W. Bridgman, *The Ways Things Are.* Cambridge: Harvard Univ. Press, 1959. *Perspectives in Biology and Medicine*, vol. 3, no. 1, Autumn 1959.
318. Molecular Structure in Relation to Biology and Medicine. In G. E. W. Wolstenholme, Cecilia M. O'Connor, and Maeve O'Connor, eds., *Significant Trends in Medical Research.* Proceedings of the Tenth Anniversary Symposium of the Ciba Foundation. Boston: Little, Brown, 1959, pp. 3–10.
319. Observations on Aging and Death. *Eng. and Sci.* **23**, 9 (1960).
324. A Molecular Theory of General Anesthesia. *Science* **134**, 15 (1961).
328. Molecular Disease, Evolution, and Genic Heterogeneity. In *Horizons in Biochemistry.* (Szent-Gyorgyi Dedicatory volume.) New York: Academic Press, 1962, pp. 189–225. [Emile Zuckerkandl and L. P.]
329. Une Théorie Moléculaire de l'Anesthésie Générale. *J. Chim. Phys.* **59** (1962).
331. Our Hope for the Future. In Morris Fishbein, ed., *Birth Defects.* Philadelphia: J. B. Lippincott, 1963, pp. 164–170.
335. The Molecular Basis of Genetic Defects. In *Congenital Defects.* (Presented at the First Inter-American Conference on *Congenital Defects,* Los Angeles, Calif., 22–24 January 1962.) Philadelphia: J. B. Lippincott, 1963, pp. 15–21.
338. The Hydrate Microcrystal Theory of General Anesthesia. (First Baxter Lecture of the International Anesthesia Research Society.) *Anesthesia and Analgesia* **43**, 1 (1964).
343. Die Hydrat-Mikrokristall-Theorie der Narkose. *Der Anaesthesist*, **13**, 245 (1964).
344. The Possibilities for Further Progress in Medicine through Research. *Univ. Toronto Med. J.* **42**, 7 (1964).

345. Molecular Disease and Evolution. (Rudolf Virchow Medical Society Lecture, included in the Karger Gazette, Basel, Switzerland, 1963.) *Bull. N. Y. Acad. Med.* **40**, 334 (1964).
350. Anesthesia of Artemia Larvae: Method for Quantitative Study. *Science* **149**, 1255 (1965). [Arthur B. Robinson, Kenneth F. Manly, Michael P. Anthony, John F. Catchpool, and L.P.]
351. Chemical and Molecular Philogeny. In *Problems of Evolutional and Technical Biochemistry* (dedicated to A. I. Oparin). (In Russian.) Moscow: Nauka, 1964.
353. Albert Schweitzer, Médico y Humanitario. *Folia Humanistica* **3**, 963 (1965). [Frank Catchpool and L. P.]
354. Evolutionary Divergence and Convergence in Proteins. In Vernon Bryson and Henry J. Vogel, eds., *Evolving Genes and Proteins.* New York: Academic Press, 1965, pp. 97–166. [Emile Zuckerkandl and L. P.]
364. Die Beziehungen zwischen Molekülstruktur und medizinischen Problemen. *Naturwissenschaftliche Rundschau* **19**, 217 (1966).
365. Biological Treatment of Mental Illness: Academic Address. In Max Rinkel, ed., *Biological Treatment of Mental Illness.* New York: L. C. Page, 1966, pp. 30–37.
366. Biochemical Aspects of Schizophrenia: Discussion of Plasma Protein Factors and Serum Factors. In Max Rinkel, ed., *Biological Treatment of Mental Illness.* New York: L. C. Page, 1966, pp. 424–425.

IX. Papers on the Structure of Atomic Nuclei

342. Helion. *Nature* **201**, 61 (1964).
355. The Structural Significance of the Principal Quantum Number of Nucleonic Orbital Wave Functions. *Phys. Rev. Letters* **15**, 499 (1965).
356. The Structural Basis of Neutron and Proton Magic Numbers in Atomic Nuclei. *Nature* **208**, 174 (1965).
357. The Close-packed-spheron Model of Atomic Nuclei and its Relation to the Shell Model. *Proc. Nat. Acad. Sci. U.S.* **54**, 989 (1965).
358. Structural Basis of the Onset of Nuclear Deformation at Neutron Number 90. *Phys. Rev. Letters* **15**, 868 (1965).
360. The Close-Packed-Spheron Theory and Nuclear Fission. *Science* **150**, 297 (1965).
367. The Close-packed-spheron Theory of Nuclear Structure and the Neutron Excess for Stable Nuclei. *Revue Roumaine de Physique* **11**, 825 (1966).
369. Baryon Resonances as Rotational States. *Proc. Nat. Acad. Sci. U.S.* **56**, 1676 (1966).
372. Orthomolecular Methods in Medicine. In A. Oparin, ed., *Essays on Basic and Applied Biochemistry.* (Sissakian Memorial Volume.) Akad. Nauk. Moscow, 1967.
373. Orthomolecular Somatic and Psychiatric Medicine. *Internat. J. Vital Sub. and Nutr.*, November (1967).
374. Magnetic-moment Evidence for the Polyspheron Structure of the Lighter Atomic Nuclei. *Proc. Nat. Acad. Sci. U.S.*, December (1967).
375. Geometric Factors in Nuclear Structure. In *Marie Sklodawska-Curie Centenary Volume.* Warsaw, Acad. Sci. (1967).

X. Papers in Other Fields of Science

8. The Inter-Ionic Attraction Theory of Ionized Solutes. IV. The Influence of Variation of Dielectric Constant on the Limiting Law for Small Concentrations. *J. Am. Chem. Soc.* **47**, 2129 (1925). [P. Debye and L. P.]
9. The Entropy of Supercooled Liquids at the Absolute Zero. *J. Am. Chem. Soc.* **47**, 2148 (1925). [L. P. and Richard C. Tolman]
11. The Dielectric Constant and Molecular Weight of Bromine Vapor. *Phys. Rev.* **27**, 181 (1926).
72. The Formulas of Antimonic Acid and the Antimonates. *J. Am. Chem. Soc.* **55**, 1895 (1933). Errata *J. Am. Chem. Soc.* **55**, 3052 (1933).
100. Note on the Interpretation of the Infra-red Absorption of Organic Compounds Containing Hydroxyl and Imino Groups. *J. Am. Chem. Soc.* **58**, 94 (1936).
121. The Future of the Crellin Laboratory. *Science* **87**, 563 (1938).
136. Prolycopene, a Naturally Occurring Stereoisomer of Lycopene. *Proc. Nat. Acad. Sci. U.S.* **27**, 468 (1941). [L. Zechmeister, A. L. LeRosen, F. W. Went, and L. P.]
149. Spectral Characteristics and Configuration of Some Stereoisomeric Carotenoids Including Prolycopene and Pro-γ-carotene. *J. Am. Chem. Soc.* **65**, 1940 (1943). [L. Zechmeister, A. L. LeRosen, W. A. Schroeder, A. Polgár, and L. P.]
155. The Light Absorption and Fluorescence of Triarylmethyl Free Radicals. *J. Am. Chem. Soc.* **66**, 1985 (1944).
161. The Use of Punched Cards in Molecular Structure Determinations. I. Crystal Structure Calculations. *J. Chem. Phys.* **14**, 648 (1946). [P. A. Shaffer, Jr., Verner Schomaker, and L. P.]
162. The Use of Punched Cards in Molecular Structure Determinations. II. Electron Diffraction Calculations. *J. Chem. Phys.* **14**, 659 (1946). [P. A. Shaffer, Jr., Verner Schomaker, and L. P.]
164. Roscoe Gilkey Dickinson, 1894–1945. *Science* **102**, 216 (1945).
165. An Instrument for Determining the Partial Pressure of Oxygen in a Gas. *Science* **103**, 338 (1946). [L. P., Reuben E. Wood, and James H. Sturdivant]
166. An Instrument for Determining the Partial Pressure of Oxygen in a Gas. *J. Am. Chem. Soc.* **68**, 795 (1946). [L. P., Reuben E. Wood, and James H. Sturdivant]
179. Chemical Achievement and Hope for the Future. *Am. Scientist* **36**, 51 (1948).
180. Silliman Lecture. (Sigma Xi National Lectureships, 1947 and 1948. Sixth Series.) In George A. Baitsell, ed., *Science in Progress.* New Haven: Yale University Press, 1949, pp. 100–121.
186. La Structure des Complexes et l'Influence de cette Structure sur les Réactions d'Echange. *Colloques Internationaux du Centre National de la Recherche Scientifique* **5**, 142 (1948).
203. Zur *cis-trans*-Isomerisierung von Carotinoiden. *Helv. Chim. Acta* **32**, 2241 (1949).
204. Chemistry and the World of Today. *Chem. Eng. News* **27**, 2775 (1949).

208. The Place of Chemistry in the Integration of the Sciences. *Main Currents in Modern Thought* **7**, 108 (1950).
209. Chemistry. *Scientific American* **183**, 32 (1950).
213. Academic Research as a Career. *Chem. Eng. News* **28**, 3970 (1950).
215. The Significance of Chemistry to Man in the Modern World. *Eng. and Sci.* **14**, 10 (1951).
216. It Pays to Understand Science. *Science Digest* **29**, 52 (1951). Appeared as The Significance of Chemistry, in Edward Hutchings, Jr., ed., *Frontiers in Science, A Survey*. New York: Basic Books, 1958, pp. 278–88.
252. Status of the Values of the Fundamental Constants for Physical Chemistry as of July 1, 1951. *J. Am. Chem. Soc.* **74**, 2699 (1952). [Frederick D. Rossini, Frank T. Gucker, Jr., Herrick L. Johnston, L. P., and George W. Vinal]
258. Use of Propositions in Examinations for the Doctor's Degree. *Science* **116**, 667 (1952).
263. The Strengths of the Oxygen Acids. *School Science and Mathematics* **53**, 429 (1953).
270. Hugo Theorell. *Science* **122**, 1222 (1955).
279. Why is Hydrofluoric Acid a Weak Acid? *J. Chem. Ed.* **33**, 16 (1956).
282. A Simple Theoretical Treatment of Alkali Halide Gas Molecules. (Paper read at the Silver Jubilee Session of the National Academy of Sciences of India, 27 December 1955, Univ. of Lucknow.) *Proc. Nat. Acad. Sci. India* A **25**, 1 (1956).
288. Amedeo Avogadro nel centenario della morte. (Talk given at La solenne celebrazione di Avogadro in Campidoglio.) *La Chimica e l'Industria* **38**, 678 (1956). *Rendiconti Accademia Nazionale dei Quaranta* **6** e **7**, 175 (1955–1956).
304. Amedeo Avogadro. *Science* **124**, 708 (1956).
310. Quelques precisions sur la nature et les propriétés radioactives résultant des explosions atomiques depuis 1945. *Compt. rend. acad. sci.* **249**, 982 (1959). [L. P., Shoichi Sakata, Sin-Itiro Tomonaga, Jean-Pierre Vigier, and Hideki Yukawa]
317. Arthur Amos Noyes—1866–1936. (A Biographical Memoir). In *Biographical Memoirs*, vol 31. New York: Columbia University Press (For the National Academy of Sciences of the United States), 1958, pp. 322–346.
334. The Genesis of Ideas. In *Proceedings of the Third World Congress of Psychiatry*, 1963.
352. Fifty Years of Physical Chemistry in the California Institute of Technology. *Ann. Rev. Phys. Chem.* **16**, 1 (1965).
359. Foreword to *Behavioral Science and Human Survival.* (Edited by Milton Schwebel for the American Orthopsychiatric Association and the World Federation for Mental Health.) Palo Alto, California: Science and Behavior Books, pp. v and vi, 1965.
361. Foreword to Bodo Manstein's *Atomare Gefahr und Bevölkerungsschutz.* Stuttgart: J. Fink Verlag, 1965.
368. The Social Responsibilities of Scientists and Science. *The Science Teacher*, **33**, No. 5, (1966).
370. Molecular Chemistry. *Industrial Research* January 1967, p. 74.

XI. Books

The Structure of Line Spectra. New York: McGraw-Hill, 1930. [L. P. and S. Goudsmit.]

Introduction to Quantum Mechanics, with Applications to Chemistry. New York: McGraw-Hill, 1935. [L. P. and E. Bright Wilson, Jr.]

The Nature of the Chemical Bond, and the Structure of Molecules and Crystals. Ithaca: Cornell Univ. Press, 1939; 2nd ed. 1940; 3rd ed., 1960.

General Chemistry. San Francisco: W. H. Freeman and Company, 1947; 2nd ed., 1953.

College Chemistry. San Francisco: W. H. Freeman and Company, 1950; 2nd ed., 1955; 3rd ed., 1964.

No More War! Dodd, Mead and Co., New York, 1958; paperback edition, Liberty Prometheus Book Club, New York, 1959; revised paperback edition, Dodd, Mead and Co., New York, 1962.

The Architecture of Molecules. San Francisco: W. H. Freeman and Company, 1964. [L. P. and Roger Hayward]

Neue Moral und Internationales Recht. Berlin: Union Verlag, 1965.

The Chemical Bond. Ithaca: Cornell Univ. Press, 1967.